Foundations of Fisheries Science

Foundations of Fisheries Science

Edited by

Greg G. Sass
Northern Unit Fisheries Research Team Leader
Wisconsin Department of Natural Resources
Escanaba Lake Research Station
3110 Trout Lake Station Drive, Boulder Junction, Wisconsin 54512, USA

Micheal S. Allen
Professor, Fisheries and Aquatic Sciences
University of Florida
7922 NW 71st Street, PO Box 110600, Gainesville, Florida 32653, USA

Section Edited by

Robert Arlinghaus
Professor, Humboldt-Universität zu Berlin and
Leibniz-Institute of Freshwater Ecology and Inland Fisheries
Department of Biology and Ecology of Fishes
Müggelseedamm 310, 12587 Berlin, Germany

James F. Kitchell
A.D. Hasler Professor (Emeritus), Center for Limnology
University of Wisconsin-Madison
680 North Park Street, Madison, Wisconsin 53706, USA

Kai Lorenzen
Professor, Fisheries and Aquatic Sciences
University of Florida
7922 NW 71st Street, P.O. Box 110600, Gainesville, Florida 32653, USA

Daniel E. Schindler
Professor, Aquatic and Fishery Science/Department of Biology
Harriett Bullitt Chair in Conservation
University of Washington
Box 355020, Seattle, Washington 98195, USA

Carl J. Walters
Professor, Fisheries Centre
University of British Columbia
2202 Main Mall, Vancouver, British Columbia V6T 1Z4, Canada

AMERICAN FISHERIES SOCIETY
BETHESDA, MARYLAND
2014

A suggested citation format for this book follows.

Sass, G. G., and M. S. Allen, editors. 2014. Foundations of Fisheries Science. American Fisheries Society, Bethesda, Maryland.

© Copyright 2014 by the American Fisheries Society

All rights reserved. Photocopying for internal or personal use, or for the internal or personal use of specific clients, is permitted by AFS provided that the appropriate fee is paid directly to Copyright Clearance Center (CCC), 222 Rosewood Drive, Danvers, Massachusetts 01923, USA; phone 978-750-8400. Request authorization to make multiple copies for classroom use from CCC. These permissions do not extend to electronic distribution or long-term storage of articles or to copying for resale, promotion, advertising, general distribution, or creation of new collective works. For such uses, permission or license must be obtained from AFS.

Printed in the United States of America on acid-free paper.

Library of Congress Control Number 2014936128
ISBN 978-1-934874-37-0

American Fisheries Society Web site address: *www.fisheries.org*

American Fisheries Society
5410 Grosvenor Lane, Suite 100
Bethesda, Maryland 20814
USA

Table of Contents

Editor Biographies..xi
Preface..xv

Introduction..1
GREG G. SASS AND MICHEAL S. ALLEN

1   Managing Fish Stocks
CARL J. WALTERS
    1.1   Synthesis..5
    1.2   Reprinted Articles..10

        On the Question of the Biological Basis of Fisheries......................11
            Th. I. Baranoff

        An Epitaph for the Concept of Maximum Sustained Yield............73
            P. A. Larkin

        Maximum Reproductive Rate of Fish at Low Population Sizes....85
            Ransom A. Myers, Keith G. Bowen, and Nicholas J. Barrowman

        Canada's Recreational Fisheries: the Invisible Collapse?.........101
            John R. Post, Michael Sullivan, Sean Cox, Nigel P. Lester,
            Carl J. Walters, Eric A. Parkinson, Andrew J. Paul,
            Leyland Jackson, and Brian J. Shuter

        Stock and Recruitment...113
            W. E. Ricker

        Big Effects from Small Causes: Two Examples from Fish
           Population Dynamics...179
            W. E. Ricker

        Compensatory Density Dependence in Fish Populations:
           Importance, Controversy, Understanding and Prognosis.....187
            Kenneth A. Rose, James H. Cowan, Jr.,
            Kirk O. Winemiller, Ransom A. Myers, and Ray Hilborn

        Some Aspects of the Dynamics of Populations Important to
           the Management of Commercial Marine Fishes....................223
            Milner B. Schaefer

        Adaptive Control of Fishing Systems...255
            Carl J. Walters and Ray Hilborn

Rebuilding Global Fisheries..271
  Boris Worm, Ray Hilborn, Julia K. Baum, Trevor A. Branch,
  Jeremy S. Collie, Christopher Costello, Michael J. Fogarty,
  Elizabeth A. Fulton, Jeffrey A. Hutchings, Simon Jennings,
  Olaf P. Jensen, Heike K. Lotze, Pamela M. Mace,
  Tim R. McClanahan, Cóilín Minto, Stephen R. Palumbi,
  Ana M. Parma, Daniel Ricard, Andrew A. Rosenberg,
  Reg Watson, and Dirk Zeller

1.3　Honorable Mention Full Citations and Abstracts..279

2　Managing People
ROBERT ARLINGHAUS

2.1　Synthesis..281
2.2　Reprinted Articles..292

The Lobster Fiefs: Economic and Ecological Effects of
  Territoriality in the Maine Lobster Industry...293
  James M. Acheson

Leisure Value Systems and Recreational Specialization:
  the Case of Trout Fishermen..319
  Hobson Bryan

The Struggle to Govern the Commons...333
  Thomas Dietz, Elinor Ostrom, and Paul C. Stern

The Economic Theory of a Common-Property Resource:
  the Fishery..339
  H. Scott Gordon

A Model of Regulated Open Access Resource Use....................................359
  Frances R. Homans and James E. Wilen

Fisheries Co-Management: Delegating Government
  Responsibility to Fishermen's Organizations..381
  Svein Jentoft

The Fishery: the Objectives of Sole Ownership..399
  Anthony Scott

Economic Impacts of Marine Reserves: the Importance of
  Spatial Behavior..409
  Martin D. Smith and James E. Wilen

2.3　Honorable Mention Full Citations and Abstracts..433

3 Managing Fish Habitat
Daniel E. Schindler

    3.1   Synthesis..445

    3.2   Reprinted Articles...450

        Habitat Structural Complexity and the Interaction between Bluegills and their Prey..451
            Larry B. Crowder and William E. Cooper

        Magnification of Secondary Production by Kelp Detritus in Coastal Marine Ecosystems..463
            D. O. Duggins, C. A. Simenstad, and J. A. Estes

        When can Marine Reserves Improve Fisheries Management?.................467
            Ray Hilborn, Kevin Stokes, Jean-Jacques Maguire, Tony Smith, Louis W. Botsford, Marc Mangel, José Orensanz, Ana Parma, Jake Rice, Johann Bell, Kevern L. Cochrane, Serge Garcia, Stephen J. Hall, G. P. Kirkwood, Keith Sainsbury, Gunnar Stefansson, and Carl Walters

        The Flood Pulse Concept in River-Floodplain Systems..........................477
            Wolfgang J. Junk, Peter B. Bayley, and Richard E. Sparks

        A Pacific Interdecadal Climate Oscillation with Impacts on Salmon Production..495
            Nathan J. Mantua, Steven R. Hare, Yuan Zhang, John M. Wallace, and Robert C. Francis

        The Natural Flow Regime...507
            N. LeRoy Poff, J. David Allan, Mark B. Bain, James R. Karr, Karen L. Prestegaard, Brian D. Richter, Richard E. Sparks, and Julie C. Stromberg

        The River Continuum Concept..523
            Robin L. Vannote, G. Wayne Minshall, Kenneth W. Cummins, James R. Sedell, and Colbert E. Cushing

        An Experimental Test of the Effects of Predation Risk on Habitat Use in Fish..531
            Earl E. Werner, James F. Gilliam, Donald J. Hall, and Gary G. Mittelbach

    3.3   Honorable Mention Full Citations and Abstracts..541

4 Managing Fish Communities and Ecosystems
JAMES F. KITCHELL

    4.1 Synthesis..........543
    4.2 Reprinted Articles..........549

        Cascading Trophic Interactions and Lake Productvity..........551
            Stephen R. Carpenter, James F. Kitchell, and James R. Hodgson

        Interactions between Yellow Perch Abundance, Walleye Predation, and Survival of Alternate Prey in Oneida Lake, New York..........557
            John L. Forney

        Fishing Down Marine Food Webs..........567
            Daniel Pauly, Villy Christensen, Johanne Dalsgaard, Rainer Froese, and Francisco Torres, Jr.

        Patterns in Species Composition and Richness in Fish Assemblages in Northern Wisconsin Lakes..........571
            William M. Tonn and John J. Magnuson

        Cultivation/depensation Effects on Juvenile Survival and Recruitment: Implications for the Theory of Fishing..........589
            Carl Walters and James F. Kitchell

        Possible Ecosystem Impacts of Applying MSY Policies from Single-species Assessment..........601
            Carl J. Walters, Villy Christensen, Steven J. Martell, and James F. Kitchell

        Competition and Habitat Shift in Two Sunfishes (Centrarchidae)..........613
            Earl E. Werner and Donald J. Hall

        Patterns of Life-history Diversification in North American Fishes: Implications for Population Regulation..........621
            Kirk O. Winemiller and Kenneth A. Rose

    4.3 Honorable Mention Full Citations and Abstracts..........645

5 Managing Fisheries Enhancements
KAI LORENZEN

    5.1 Synthesis..........649
    5.2 Reprinted Articles..........657

        A Responsible Approach to Marine Stock Enhancement..........659
            H. Lee Blankenship and Kenneth M. Leber

Stocking Strategies..669
　I. G. Cowx

A Review of the Hatchery Programs for Pink Salmon in
　Prince William Sound and Kodiak Island, Alaska..................685
　Ray Hilborn and Doug Eggers

Testing the Importance of Fish Stocking as a Determinant
　of the Demand for Fishing Licenses and Fishing Effort in
　Colorado..703
　John Loomis and Peter Fix

Population Dynamics and Potential of Fisheries Stock Enhancement:
　Practical Theory for Assessment and Policy Analysis..........719
　Kai Lorenzen

Understanding how the Hatchery Environment Represses or
　Promotes the Development of Behavioral Survival Skills......739
　Bori L. Olla, Michael W. Davis, and Clifford H. Ryer

Genetic Differences in Growth and Survival of Juvenile
　Hatchery and Wild Steelhead Trout, *Salmo gairdneri*..........759
　R. R. Reisenbichler and J. D. McIntyre

Factors Influencing Survival and Growth of Stocked Walleye
　(*Stizostedion vitreum*) in a Centrarchid-dominated
　Impoundment..765
　Victor J. Santucci, Jr. and David H. Wahl

Experiments with Various Rates of Stocking Bluegills,
　Lepomis macrochirus Rafinesque, and Largemouth Bass,
　Micropterus salmoides (Lacepede), in Ponds......................777
　H. S. Swingle

5.3　Honorable Mention Full Citations and Abstracts..........................791

6　Summary
GREG G. SASS AND MICHEAL S. ALLEN
6.1　What Have We Learned?..795

Editor Biographies

Dr. Greg G. Sass

Greg Sass is a Northern Unit Fisheries Research Team Leader and the Director of the Escanaba Lake Research Station with the Wisconsin Department of Natural Resources. Sass earned his B.S. with honors in Biology from the University of South Florida in 1999. He earned his M.S. (2001) and Ph.D. (2004) in Zoology from the Center for Limnology, University of Wisconsin-Madison (Jim Kitchell, major advisor). Sass was a Research Associate at the Center for Limnology with Jim Kitchell and Steve Carpenter during 2004-2006. He is the former Director of the Illinois River Biological Station with the Illinois Natural History Survey, University of Illinois at Urbana-Champaign (2006–2011). Sass currently holds Adjunct Professor appointments at the University of Illinois at Urbana-Champaign (Associate), Eastern Illinois University, Western Illinois University, and the University of Wisconsin-Stevens Point. He maintains an Honorary Fellowship at the University of Wisconsin-Madison, Center for Limnology. He has advised six M.S. students and one Ph.D. student. He served as vice president and president of the Mississippi River Research Consortium in 2009 and 2010, respectively. Sass has published over 30 peer reviewed journal articles, along with several book chapters. His primary interests include ecosystem-based fisheries management, invasive species ecology, bioenergetics, restoration ecology, fish predator-prey interactions, whole-lake experiments, and fish habitat ecology.

Dr. Micheal S. Allen

Mike Allen is a Professor of Fisheries at the University of Florida. He obtained his B.S. from Texas A&M University in fisheries ecology, his M.S. from Auburn University in fisheries, and his Ph.D. at Mississippi State University in fisheries and statistics. His research evaluates problems in recreational fisheries and fish ecology. He uses a combination of field collections and computer modeling to draw inferences about management actions (e.g., regulations, habitat manipulations, stock enhancement) that can improve fisheries and fish communities. He has served as the advisor for twenty-two M.S. students and six Ph.D. students. Allen has over 100 peer reviewed journal articles and book chapters, and he served as the President of the Southern Division, American Fisheries Society in 2013. He has done extensive fisheries work in the United States and Australia. Dr. Allen received the Award of Excellence from the Fisheries Management Section of the American Fisheries Society in 2011.

Dr. Robert Arlinghaus

Robert Arlinghaus is a Professor for Integrative Fisheries Management at Humboldt-Universität zu Berlin and is a fisheries scientist and group leader at the Leibniz-Institute of Freshwater Ecology and Inland Fisheries in Berlin. Since the completion of his doctoral degree in Human Dimension of Fisheries at Humboldt-Universität zu Berlin in 2004, Robert has worked at the interface of natural and social science, focusing on recreational fisheries. His main interests involve studying fisheries from a social-ecological research perspective, and he strives to understand how recreational anglers interact with fisheries resources. He has authored over 130 peer-reviewed journal articles and has produced over 250 publications, including books, edited books, and popular scientific monographs on recreational fisheries. In 2011–2012, Arlinghaus led the development of the United Nations guidelines for sustainable recreational fisheries on a global scale. He is an Associate Editor of the *North American Journal of Fisheries Management*, and serves on the Editorial Boards of *Human Dimensions of Wildlife* and the *Journal of Outdoor Recreation and Tourism*. Arlinghaus has been the recipient of various awards, including the Award of Excellence from the Fisheries Management Section of the American Fisheries Society (2008), the Medal for Young Scientists of The Fisheries Society of the British Isles (2012), and is a member of the German Dream Team of Young Scientists (2006). For his outreach activities, Robert received the Bscher Media Award in 2004.

Dr. James F. Kitchell

Jim Kitchell received his B.S. from Ball State Teachers College (1964), his Ph.D. from the University of Colorado (1970), and completed a post-doc with the International Biological Program. In 1974, he joined the Department of Zoology faculty of the University of Wisconsin-Madison. He served as Director of the Center for Limnology (2000–2009) and is currently the A. D. Hasler Professor (Emeritus). In 2003, the American Fisheries Society selected him for the Award of Excellence in Career Achievements. In 2010, the American Society of Limnology and Oceanography presented him with the Redfield Lifetime Achievement Award. In 2011, the Great Lakes Fishery Commission presented him with the Christie/Loftus Award. His peer-reviewed publications total a bit more than 200, include serving as editor of *Food Web Management: a Case Study of Lake Mendota*, and as co-editor of *The Trophic Cascade in Lakes*. His research projects include food webs in the lakes of Wisconsin, the Laurentian Great Lakes, Africa's Lake Victoria, the Central Pacific, and most recently, climate change in Lake Superior. He retired in August 2010 but continues to profit from interactions with highly motivated, smart colleagues such as those represented by the contents of this volume. He also goes fishing when he can.

Dr. Kai Lorenzen

Dr. Kai Lorenzen is a Professor of Integrative Fisheries Science at the University of Florida. He received his M.S. from Kiel University, Germany and his Ph. D. from the University of London. Lorenzen started his career as a fisheries development consultant, working mostly in Asia. He joined the faculty of Imperial College London in 1997 and moved to his current position in 2010.

Dr. Lorenzen conducts interdisciplinary, problem-oriented fisheries research that integrates quantitative ecology with human dimensions and often engages closely with management initiatives. A particular focus of his research has been the development of methods for assessing and managing fisheries enhancement and restoration programs involving hatcheries. In the course of this research, he also developed empirical generalizations about size and density-dependent processes in fish populations that have found wide application in fisheries science. Lorenzen has also conducted research on the governance of small-scale and recreational fisheries, conservation of fisheries in agricultural landscapes, and the design of aquaculture systems.

Lorenzen's research has been published widely in peer-reviewed journals, book chapters, manuals, and a software package (*EnhanceFish*). He serves on the editorial boards of *Reviews in Fisheries Science* and *Open Fish Science* and on the Scientific Committee of the Gulf of Mexico Fisheries Management Council. Dr. Lorenzen was the 2007–2008 Mote Eminent Scholar at Florida State University and the Mote Marine Laboratory.

Dr. Daniel E. Schindler

Daniel Schindler is the Harriet Bullitt Endowed Chair of Conservation in the School of Aquatic and Fishery Sciences at the University of Washington. Most of his research focuses on understanding the functioning of watersheds that support Pacific salmon in western Alaska, and the dynamics of fisheries that operate in these ecosystems. He is a principal investigator of the UW-Alaska Salmon Program that has studied salmon ecosystems in Alaska since the 1940s, and he spends several months of the year in the field in the Bristol Bay region. Schindler has been a recipient of the Distinguished Research Award from the UW College of Ocean and Fishery Sciences, and of the Carl R. Sullivan Fishery Conservation Award that was awarded to the UW-Alaska Salmon Program from the American Fisheries Society in 2012. He has provided service to a wide variety of governmental and non-governmental organizations, and serves on the editorial boards of the journals *Ecology* and *Ecosystems*. He earned a B.S. with Honours from the University of British Columbia (1990), and a M.S. (1992) and Ph.D. (1995) from the University of Wisconsin-Madison. He joined the faculty at the University of Washington in 1997 and teaches undergraduate and graduate students in Limnology, Aquatic Sciences, and Ecology.

Dr. Carl J. Walters

Dr. Carl Walters is currently a Professor of Zoology and Fisheries at the University of British Columbia, Vancouver, Canada. Walters received his B.S. from Humboldt State College, and his M.S. and Ph.D. from Colorado State University. He has worked at the University of British Columbia since 1969.

Dr. Walters is a specialist in fisheries stock assessment, adaptive management, and ecosystem modeling. He uses mathematical modeling and computer simulation techniques to better understand the dynamics of exploited marine ecosystems and to find more effective methods to manage them in the face of natural variability and high uncertainty. He advocates cooperative arrangements between governments and fishing industries to provide improved information for stock assessment and management via methods such as industry-based surveys. His main research is on the theory of harvesting in natural resource management, with a primary interest in the basic problem of how to behave adaptively in the face of extreme uncertainty. He is one of the main developers of the ecosystem simulation program known as EcoSim, which is being used to test ideas about organization of trophic interactions in marine systems, and the implications of these interactions for sustainable harvesting theory.

He has written over 190 peer-reviewed journal articles and three books, including *Adaptive Management of Renewable Resources* (MacMillan Publishing Company), *Quantitative Fisheries Stock Assessment and Management* (with Ray Hilborn, Chapman-Hall Publishing Company), and *Fisheries Ecology and Management* (with Steve Martell, Princeton University Press). He also serves on the Editorial Boards of a number of journals, including the *Canadian Journal of Fisheries and Aquatic Sciences*, *Conservation Ecology*, *Ecosystems*, *The Open Fish Science Journal*, and *Marine and Coastal Fisheries*.

Dr. Walters is a Fellow of the Royal Society of Canada (1998) and a Pew Fellow in Marine Conservation (2001). He was also the 2001–2002 Mote Eminent Scholar at Florida State University and the Mote Marine Laboratory. He has received the Volvo Environment Prize, American Fisheries Society Award of Excellence, Timothy Parsons Medal, and the Murray A. Newman Award.

Preface

The critical social, economic, and ecological implications of fisheries have made this science an increasingly popular, complex, and multi-disciplinary field. Fish provide protein to much of the world's population (about three billion people rely upon fish for >20% of their animal protein; FAO 2012), have the highest biodiversity of any vertebrate (over 32,700 species and counting, www.fishbase.org), can be strong indicators of aquatic ecosystem health, and provide highly valuable subsistence, commercial, recreational, and aesthetic opportunities for humans. Ecosystem services provided by fishes will continue to be challenged as the world's population grows and aquatic and marine environments are altered by anthropogenic perturbations. As such, the challenge of managing fisheries in the future will require well-trained and multi-disciplinary students and professionals with the capability of understanding quantitative aspects of fisheries, human dimensions and socio-economic perspectives, and a strong background in aquatic and basic ecology. As students and fisheries professionals, we should always consider the roots of our field and the research that helped shape the discipline of fisheries science.

With this critical need of understanding seminal works from various disciplines within fisheries science in mind, in January 2010, we were fishing on a Florida lake in the Ocala National Forest trying to fool trophy Largemouth Bass *Micropterus salmoides floridanus* into eating a live Golden Shiner *Notemigonus crysoleucas* when discussion of this book began. Very simply, we discussed whether marine and freshwater fisheries researchers were reading the classic fisheries works. Not knowing the answer, we reasoned that a book compiling seminal articles from all areas of fisheries science, similar to *Foundations of Ecology* (Real and Brown 1991), would be beneficial for our discipline.

In today's world, much of our communication is rapid and brief. A great amount of focus is placed upon recent articles that garner media attention and are published in high impact factor journals. Clearly, it would be unwise as a discipline to ignore the foundational studies that served as the basis for what fisheries science is today. Using a retrospective view of foundational articles, we may be better able to address fisheries challenges today and in the future.

The purpose of *Foundations of Fisheries Science* is to identify and bridge gaps in the field of fisheries science to help students and professionals appreciate the seminal works in this discipline. Recognition of such foundational articles will be essential for tackling challenging and emerging issues in the management of fish stocks, populations, and communities. With the help of fisheries students and professionals from around the world, we present 43 reprinted articles and 30 honorable mention full citations and abstracts that have helped to mold the discipline of fisheries science. We and our five section editors hope that you find the syntheses and articles included in *Foundations of Fisheries Science* helpful for advancing this discipline.

Greg G. Sass
Micheal S. Allen
April 2014

Introduction

GREG G. SASS AND MICHEAL S. ALLEN

1.1. SYNTHESIS

A fishery is a system that includes the target organisms (fish, decapods, shellfish), the community of species on which the target organism depends, the habitat in which they exist, and the humans who exploit or affect the target organism within the ecosystem. Management of a fishery system requires an interdisciplinary skill set that includes quantitative methods to assess fish stocks, understanding human behavior and economics, and applied aspects of ecology. An elegant depiction of a fishery system was created by Nielsen (1993), and we adopted a similar conceptual framework to define the five sections used in *Foundations of Fisheries Science*; managing fish stocks, managing people, managing fish habitat, managing fish communities and ecosystems, and managing fisheries enhancements.

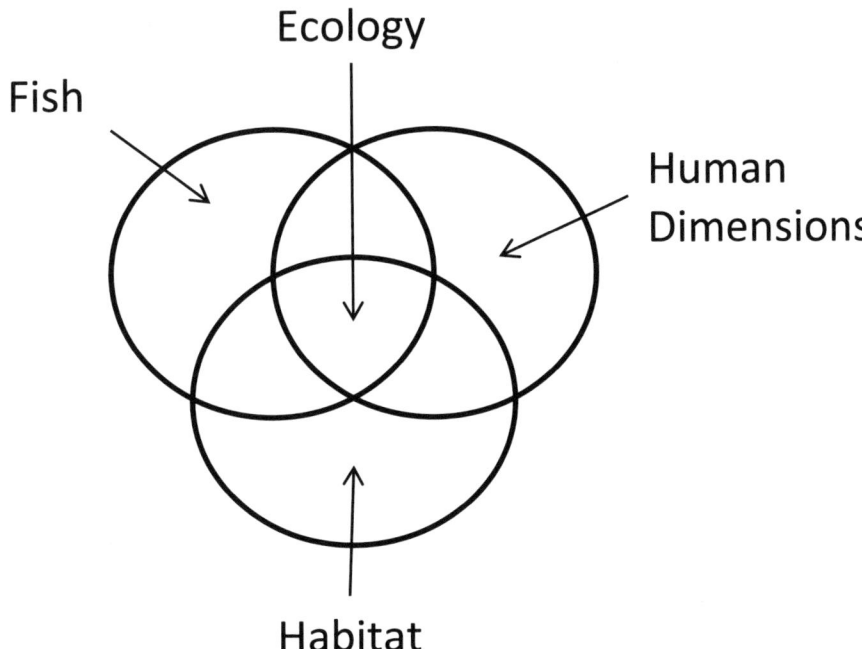

Book Objectives

The objective of *Foundations of Fisheries Science* is to highlight the classic and critical works associated with fisheries management. The book is organized into five sections (1. Managing Fish Stocks, 2. Managing People, 3. Managing Fish Habitat, 4. Managing Fish Communities and Ecosystems, and 5. Managing Fisheries Enhancements), which represent the critical components of fisheries (fish, humans, habitat) and the most common management approaches (regulations, stocking, habitat protection/restoration). *Foundations of Fisheries Science* is solely focused on articles with direct relevance to fisheries management. *Foundations of Fisheries Science* can be used as a reference or text book to lead undergraduate and graduate courses and discussions. Our goal for *Foundations of Fisheries Science* was to provide a compilation of the most influential articles in order to better understand how fisheries science has progressed as a discipline and to identify challenges in the future.

Managing Fish Stocks

Fishes have been exploited since the beginning of human civilization. Overfishing and the challenge of sustaining fish stocks have become critical over the past century as technology has advanced and the human population has expanded. Currently, about two-thirds of marine fisheries in developed countries are below their target biomass (Worm et al. 2009), and there is reason to believe that the situation is worse in developing countries (Costello et al. 2012) and freshwater fisheries (Allan et al. 2005). Use of regulations to manage fisheries has created new insights. Quantitative methods have become powerful tools used to predict sustainable yields and prevent overfishing. This section of the book highlights the critical works developed for stock assessment and to predict and evaluate the effects of regulations (quota, size limits) to provide benefits to humans and allow sustainable harvest.

Managing People

Because fishes are exploited for food and other products, human behavior influences fisheries sustainability. Understanding human behavior and the economic aspects of fisheries have become a critical consideration in fisheries management. This section of the book highlights the critical works developed to assess human behavior, motivation, satisfaction, beliefs, and the socio-economic considerations that are required for effective fisheries management. This section also includes sub-disciplines of human dimensions and economics, which span a broad spectrum of investigations that are critical to fisheries science.

Managing Fish Habitat

All fish stocks require certain environmental conditions to complete their life cycle in order to sustain their populations. Anthropogenic effects on fish stocks are not restricted to harvest and include degradation or loss of appropriate physical and chemical attributes of fish habitat. These influences can occur within systems and across landscapes, and often include perturbations at the land/water interface. Recent fisheries management policies emphasize the importance of habitat in sustaining fish stocks. This section of the book highlights the critical works developed to understand essential fish habitat and to evaluate habitat manipulations as a management tool.

Managing Fish Communities and Ecosystems

The sustainability of a fish stock is integrally linked to aquatic and terrestrial communities and the trophic dynamics of the system. Perturbations to aquatic and terrestrial communities can change food webs and directly influence fishery attributes (e.g., yield, species composition). Single species management actions may also change community dynamics. Thus, managers have recognized that fisheries cause food web alterations, and that deliberate food web manipulations can be used as a management tool. This section of the book highlights the critical works developed to understand how management of fish communities and ecosystems are important for the overall sustainability of fish stocks.

Managing Fisheries Enhancements

Artificial propagation and stocking are long-standing management tools that have been used to create fisheries, offset the effects of harvest, and to manipulate ecosystem properties (e.g., invasive species introductions). The efficacy of stocking to enhance or sustain fisheries has been debated for decades, and the ecological and genetic implications of the practice (e.g., outbreeding depression) have been emphasized as problems. This section of the book highlights the critical works developed to examine the positive, null, and/or negative effects of artificial propagation and stocking in the management of fisheries.

Article Selection

We used a hierarchical approach to select the seminal articles included within *Foundations of Fisheries Science*. The first and primary decision for article selection was to solicit nominations by popular vote from the American Fisheries Society membership and other fisheries societies around the world. Using this approach, we remained objective in the article selection process and aimed to broadly represent the diversity of people (e.g., students, consultants, professionals, managers, scientists) actively involved in fisheries management and their opinions about the most influential articles within the discipline. We purposely chose a survey for article selection to remain as objective and unbiased as possible in the process. Still, a level of subjective selection was required by the editors and section editors in the case of equally-ranked articles, space limitations within the book, maintaining a balance of topics, and/or whether enough articles were received for each section of the book within our survey. Thus, we used the discretion of the editors and section editors when necessary to reach a final decision on article inclusion within *Foundations of Fisheries Science*.

Section 1

Managing Fish Stocks

CARL J. WALTERS

1.1. SYNTHESIS

The articles in this section are mainly about the regulation of harvesting, a critical component of management for sustainable fisheries. Harvest regulation requires specification of quantitative limits or targets, and any such quantification is necessarily based on some sort of mathematical model. Hence, most of the articles in this section are either model-based or about the risks of using models. It is important to keep in mind when reading articles that involve modeling that the critical issue is not whether a given model is right or wrong (they are all wrong; the complexity of nature ensures that), but whether the model is likely to give a useful result when comparing quantitative policy options (e.g., different allowable catches). Any such comparison of options or choices logically requires making a prediction about how each choice will/would perform, i.e., the choice is going to involve some model whether that model is clearly articulated (used with eyes open about its assumptions) or not. Thus, there is a long history of modeling for making predictions needed in harvest management that is done not because anyone believes the models, but because there really is no choice but to try.

Baranov (1918) proposed the first model that could be used for harvest management prediction. His simple "catch equation" predicts that in a seasonal fishery, increasing or cumulating fishing effort will lead to an upper limit on catch set by the biomass or number of fish in the stock (or by age of fish) at the start of the season. The form of the equation was derived by explicitly thinking about how a unit of effort deployed later in the season will encounter fewer fish than a unit of effort occurring earlier, both because there are fewer fish present later and because the later effort units will tend to "sweep" areas that have already been fished. In fact, we still use the Baranov (1918) equation for prediction of catch in seasonal fisheries. But, in its original formulation, the equation did not account for cumulative changes in biomass or number of fish over multiple seasons, due to processes of growth, recruitment, and natural mortality. Much of the thinking after Baranov (1918) has been centered on how to predict such cumulative changes.

By the mid-1950s, two major approaches had been developed for using historical data to predict effects of fishing and sustainable yields. One examined aggregate effects on biomass using the idea of "surplus production" meaning net gain in biomass each year that could be translated either into biomass growth or yield. Following earlier research by Graham (1935) and Gulland (1955), Schaefer (1954) developed methods for estimating surplus production rates over time using data on catch and changes in relative abundance. Schaefer's (1954) approach avoided dangerous assumptions about stock size staying near equilibrium with fishing effects; equilibrium calculations can grossly overestimate surplus production. Dynamic calculations of surplus production are still used today, mainly to evaluate possible non-stationarity (persistent change in biological parameters) for stocks where detailed data on changes in recruitment have not been collected. The second approach explicitly represented how population age-size structure varied with size-selective fishing mortality, using predictions of recruitment, body growth, and natural mortality. Such "dynamic pool" models were first articulated clearly by Beverton and Holt (1957), along with one of the most popular models for predicting how mean recruitment rate is likely to change as spawning stock sizes are reduced through fishing. Beverton and Holt (1957) were the first to clearly distinguish between two kinds of overfishing; "growth overfishing" where fish are harvested at smaller sizes than would maximize production and "recruitment overfishing" where spawning stock size is reduced enough to severely impair recruitment.

Ricker (1954) developed an alternative model for predicting changes in recruitment with spawning stock size. His work emphasized that recruitment is the product of two components; egg production and survival from egg to the (arbitrary) age at recruitment. Ricker (1954) pointed out that just having high egg production (in very fecund fish species) does not imply that recruitment will remain high as spawning stock and egg production are reduced. Rather, maintenance of high recruitment rates also depends on there being strong "compensatory" improvements in early juvenile survival rates. Using his stock-recruitment model, Ricker (1963) warned about a potentially serious practical problem in managing highly productive stocks (like Sockeye Salmon *Oncorhynchus nerka*). Namely, that the exploitation rate (proportion of recruits harvested each year), which results in maximum sustained yield (MSY), can be very close to the rate that would cause biological extinction. Thus, he warned that small errors in the assessment of MSY exploitation rates from historical data could lead to rapid stock collapse.

By the mid-1970s, Larkin (1977) warned that the use of emerging population models and assessment methods were leading to what he considered to be a dogmatic belief in the assumption that the main aim of harvest management should be to achieve MSY. His warnings are echoed today in various prescriptions for ecosystem-based fisheries management. Larkin (1977) had three main concerns. First, as noted by Ricker (1963), errors in estimation of MSY can easily lead to overfishing implying the need for precautionary adjustments in harvest rates and targets. Second, focus on sustainable yield of target stocks can result in longer-term deterioration in biological stock structure and biodiversity, ultimately leading to non-sustainability should the stocks initially targeted for management become unproductive for various reasons (e.g., climate change). Third, there are situations where maximization of biological yield should not be an objective in the first place (e.g., recreational fisheries) or would require fishing efforts greater than would be best for maximizing total economic profits from fishing.

Walters and Hilborn (1976) were also concerned about the adequacy of historical data for estimating key population dynamics relationships, particularly the parameters of stock-recruitment relationships. They pointed out that management decisions that affect stock size have what are called "dual effects of control" (i.e., those decisions effect immediate fishery value and also the legacy information) in terms of stock sizes available to future scientists and managers for improving parameter estimates. Walters and Hilborn (1976) argued for what we now call "actively adaptive" management, where policy choices are treated as deliberate management experiments aimed at ensuring sustainability and also providing better information. They used optimization methods to show that the best experimental policy choice is often a "probing experiment" that pushes stock sizes into ranges where historical experience is lacking, but where there is a possible opportunity for increasing fishery value. Their recommendations have not been widely adopted in fisheries. There have been a few "probing experiments" where harvest rates were deliberately reduced (on salmon populations *Oncorhynchus* spp.) to test whether increased spawning abundance would produce greater recruitment, but in general, fisheries scientists and managers today do not account for the value of information associated with alternative policy choices.

Studies of growth and natural mortality rates proliferated in the wake of the theory developed by Beverton and Holt (1957), allowing Pauly (1980) and Hoenig (1983) to carry out key meta-analyses of data from a large number of studies. Natural mortality rates (M) are very difficult to measure, and these authors showed that M is reasonably predictable from information on growth. Fishes that have high metabolic rates (von Bertalanffy K) and low maximum body sizes also suffer higher natural mortality rates, especially in warmer water. To this day, most stock assessments do not pretend to estimate the natural mortality rate, rather basing stock reconstructions and predictions on the values predicted from these early meta-analyses.

Myers et al. (1999) is arguably one of the most important contributions to fisheries sustainability. Myers and his colleagues assembled a large, time-series database of spawning biomass and recruitment estimates, and fit each of these data sets to stock-recruitment models. The "Myers legacy database" continues to grow and is widely used by researchers seeking empirical relationships involving recruitment and production patterns (Ricard et al. 2012). The key finding in Myers et al. (1999) is that almost all fish populations studied have exhibited compensatory improvement in juvenile survival rate as spawning abundance has been reduced through fishing, but there are definite limits to this improvement. Thus, Myers et al. (1999) has allowed us to definitively reject the "millions of eggs hypothesis" that recruitment is independent of spawning stock size, and almost all major stock assessment models and harvest policy analyses used today include a stock-recruitment relationship so as to recognize the risk of recruitment overfishing. In fact, most stock assessment models now use a "stock synthesis" approach for reconstructing historical recruitment changes using information on catches, size-age composition, and trend in relative abundance. Using this approach, recruitment for each year is estimated as a stock-recruit predicted mean value plus or minus an "anomaly" for the year; methods for fitting such integrated assessment models trace back to the research of Fournier and Archibald (2002).

Rose et al. (2001) pointed out that compensatory improvement in survival and growth rates when populations are reduced through fishing is the ecological basis for sustainable fisheries. In a broad review of historical work on the topic, they noted that there has been much controversy about the existence and strength ("compensatory limits") of compensation, and that most evidence of it comes from long-term population studies, rather than carefully controlled manipulative experiments. They suggested that understanding of the mechanisms that cause compensation would give more confidence in models that assume it in assessments of sustainable harvest rates. They also reviewed a variety of approaches for predicting compensatory limits ranging from analysis of historical data to careful examination of processes including predation and feeding ecology. They pointed out that knowledge of basic life history strategies (e.g., early vs. late maturity, slow vs. fast growth) provided at least some weak ability to predict the strength of compensation. They reviewed a variety of case examples of field and modeling studies that give insights about specific mechanisms that cause compensation, ranging from changes in growth rates and fecundity to simple improvements in juvenile survival rates when competition for various resources is reduced.

Post et al. (2002) pointed out that, while there has been much research on the limits to sustainable harvesting for commercial fisheries, little attention has been paid to the issue of overfishing in recreational fisheries. There has been a tradition in recreational fisheries management of assuming that reductions in stock size will lead to reductions in fishing effort (as anglers give up when catch rates are poor) that will prevent severe overfishing. They argued, from a set of Canadian case examples, that this assumption can be incorrect. Unregulated recreational effort, particularly in highly accessible situations, can lead to severe overfishing and to increased risk of driving populations low enough to exhibit collapse toward low equilibrium levels (or extinction) from which natural recovery may not occur. Mechanisms involved in such collapses range from strong density-dependence in catchability (so that angler success rates can remain high even when fish abundance is very low) to changes in predator-prey interactions that result in depensatory decreases in reproductive success when abundance of preferred target stocks is reduced by fishing.

Worm et al. (2009) provided a major synthesis of information about status and trends in a large number of fish stocks for which there have been reasonably credible stock assessments. The study was developed in response to a set of controversial papers predicting global fishery collapse largely on the basis of examination of trends in global fish catch statistics. Instead of uniformly supporting the findings of those catch-based analyses, Worm et al. (2009) described a greatly mixed situation where some stocks were still collapsing, while others were stable or recovering through deliberate management aimed at ending overfishing. There has been much debate about the study because most of the stock assessments included in the analysis are from pelagic fisheries and fisheries in developed nations where there has been substantial investment in data collection and regulatory systems. However, there is at least some evidence from catch trends that the situation is more dire for developing nations, where management is dubious, but where there is stronger dependence on fisheries as critical food sources.

There is a rapidly growing conservation literature that echoes and expands on Larkin's (1977) warning that fisheries assessments and predictions are often based on relatively optimistic models that do not adequately account for factors such as genetic selection for changes in growth rate and importance of the contribution of large females to population fecundity. Two widely cited examples of this literature have been accorded "Honorable Mention" status here. Conover and Munch (2002) reviewed evidence of selective effects of fishing on growth, and provide an experimental demonstration that fishing selected for reduced growth at least in cases where the age at sexual maturity is fixed. Jackson et al. (2001) argued that coastal fisheries have been causing profound changes in abundances, particularly of large predators, and that these changes have in turn resulted in severe changes in food web structure and species interaction patterns that make it difficult to interpret short-term research data and may involve irreversible changes in ecosystem structure and function.

References

Beverton, R. J. H., and S. J. Holt. 1957. On the dynamics of exploited fish populations. Fishery Investigations Series 2: Sea Fisheries 19:1–533.

Graham, M. 1935. Modern theory of exploiting a fishery, and application to North Sea trawling. J. Cons. Int. Explor. Mer. 10(3):264–274.

Gulland, J. A. 1955. On the estimation of growth and mortality in commercial fish populations. Fishery Invest. London (2) 18(9):1–46.

Ricard, D., C. Minto, O. P. Jensen, and J. K. Baum. 2012. Examining the status of commercially exploited marine species with the RAM legacy stock assessment database. Fish and Fisheries 13:380–398.

1.2 REPRINTED ARTICLES

Baranov, F. I. 1918. Kvoprosu o biologicheskikh osnovaniyakh rybnogo khozyaistva. [On the question of the biological basis of fisheries.] Izvestia Nauchnyi Issledovatelskii Icktiologischeskii Institut 1(1):81–128.

Larkin, P. A. 1977. An epitaph for the concept of maximum sustained yield. Transactions of the American Fisheries Society 106(1):1–11.

Myers, R. A., K. G. Bowen, and N. J. Barrowman. 1999. Maximum reproductive rate of fish at low population sizes. Canadian Journal of Fisheries and Aquatic Sciences 56:2404–2419.

Post, J. R., M. Sullivan, S. Cox, N. P. Lester, C. J. Walters, E. A. Parkinson, A. J. Paul, L. Jackson, and B. J. Shuter. 2002. Canada's recreational fisheries: the invisible collapse? Fisheries 27(1):6–17.

Ricker, W. E. 1954. Stock and recruitment. Journal of the Fisheries Research Board of Canada 11(5):559–623.

Ricker, W. E. 1963. Big effects from small causes: two examples from fish population dynamics. Journal of the Fisheries Research Board of Canada 20:257–264.

Rose, K. A., J. H. Cowan, Jr., K. O. Winemiller, R. A. Myers, and R. Hilborn. 2001. Compensatory density dependence in fish populations: importance, controversy, understanding and prognosis. Fish and Fisheries 2:293–327.

Schaefer, M. B. 1954. Some aspects of the dynamics of populations important to the management of commercial marine fisheries. Bulletin of the Inter-American Tropical Tuna Commission 1(2):27–56.

Walters, C. J., and R. Hilborn. 1976. Adaptive control of fishing systems. Journal of the Fisheries Research Board of Canada 33:145–159.

Worm, B., R. Hilborn, J. K. Baum, T. A. Branch, J. S. Collie, C. Costello, M. J. Fogarty, E. A. Fulton, J. A. Hutchings, S. Jennings, O. P. Jensen, H. K. Lotze, P. M. Mace, T. R. McClanahan, C. Minto, S. R. Palumbi, A. M. Parma, D. Ricard, A. A. Rosenberg, R. Watson, and D. Zeller. 2009. Rebuilding global fisheries. Science 325(5940):578–585.

On the Question of the Biological Basis of Fisheries

By

Th. I. Baranoff

> An hypothesis, refuted by new facts, dies an honorable death. If it has done nothing more than evoke the facts which refute it, it has earned the right to become a monument esteemed forever.
>
> Henle

With 12 figures in the text

Original title: K voprosu o biologicheskiĭ osnovaniĭakh rybnovo khoziāĭstva.

Original publication: Nauchnyĭ issledovatelskiĭ iktiologisheskiĭ Institut, Izvestiīa, 1(1): 81-128. Izvestiīa otdela rybovodstva i nauchnopromyslovykh issledovanii, T. I, 1. (Institute for Scientific Ichthyological Investigations, Proceedings, 1(1): 81-128. Reports from the Division of Fish Management and Scientific Study of the Fishing Industry, Vol. I, 1.). Moscow, 1918.

Foreword

The present work, undertaken as an attempt to elucidate theoretically some questions of fish bionomics, does not have as its goal the creation of a definitive theory about these questions. It undoubtedly is in need of further verification, and so it is devoted mainly to a more modest task: that of presenting material for working hypotheses, in the light of which it may be possible to continue, systematically and successfully, the analysis of the problems put forward.

We will not dwell in the present work upon the general question of the utilization of the organic production of a body of water, or of the conditions for equilibrium among the different groups of its inhabitants. A solution of this question would provide a firm basis for all bio-economic calculations, but it requires knowledge of a whole series of quantitative correlations which are as yet completely unstudied.

Moscow
February, 1916

Th. Baranoff

CONTENTS

	Pages	
	Original	Here

A formal theory of the life of fishes . . . . 84 . . . 4

    The curve of mortality. The curve of population. The curve of catch. The equation of these curves. Data for the plaice. Annual mortality. Mean weight of the fish caught. Determination of the coefficients of natural mortality and of exploitation. The joint effect of natural mortality and exploitation.

The trawl fishery and the theory of utilization of fish stocks . . . . 95 . . . 16

    The geometric intensity of fishing. The efficiency of fishing gear. The real elemental intensity of fishing. The total real intensity of fishing. Empirical data.

    The general course of the curve of population in relation to the state of the fishery. The period of expansion of the fishery. The period of equilibrium of the fishery, and the effect of its intensity. Overfishing. Dependence of the weight of the catch on the intensity of the fishery. Size limits for the fish. Application of the method to fisheries for anadromous fishes.

The stability of fisheries and their periodic fluctuations . . 109 . . . 29

Appendix I. The true curve of population and a comparison with the hypothetical one . . . . 113 . . . 34

Appendix II. The error of the method of random sampling . . . 116 . . . 37

List of works cited . . . . 128 . . . 53

A Formal Theory of the Life of Fishes.

Let us imagine the ideal case of an isolated body of water, in which the fishing is carried on with some steady intensity during the time under consideration. Suppose further, that in this body of water no epidemics occur, and no sharp fluctuations in hydrological factors, or any similar occurrence, such as would cause fortuitous changes in the composition of the fish population. Let us look at the elements of the "curve of population" for one of the commercial species of fish.

Let there originally hatch from the eggs A fry; as these fry grow their number will gradually diminish from various causes, including, eventually, removal by fishing; so representing on the abcissal axis the length of the fish, and on the ordinate axis their number, we will get a "curve of mortality" (figure 1) representing the gradual decrease in the number of the fish of one brood as they increase in size. In this we postulate the same rate of growth for all the fish, and the absence of periodicity in their growth (retardation of growth in winter).

Suppose now that the spawning of the fish takes place continuously, and that new broods of fry (of the same number A) are continuously produced in the body of water. Then at any given moment the population of the reservoir will consist of the groups (broods) of fish of successively greater ages, while the length of the fish, from one group to another, will change continuously; and the number of fish in a group, with the increase of their length, will continuously decrease. Having determined, at any given moment, the number of fish of different lengths in our body of water, combining them into groups according to their length, and then constructing the corresponding <u>curve of population</u>, we obviously will get a curve identical with the curve of mortality (figure 1), for both in the former and in the latter case, the fish of a given length b belong to one brood, the original number of which was equal to A, and has decreased to a, according to the operation of a law which is the same for all broods. And so /page 85/ as things occur under our postulates, the fish population of the body of water is in a state of equilibrium, so that its composition remains always the same, and the mortality curve coincides with the population curve.

Let us assume further that the fishing is done with equipment of the type of a seine or trawl, and that the fish of commercial size are distributed in the body of water sufficiently uniformly, and therefore the length-frequency curve of fish in the catch (curve of catch) will reflect the distribution of fish of commercial size in the body of water.[1]

---

[1] In order that the curve of the catch should coincide with the curve of the population, it is necessary that either (1) the fish of different age-classes be evenly distributed over the whole area of the body of water; in this case the intensity of the fishing in the different parts of the water can be unequal; or else (2) if the fish of different age-classes are distributed in the body of water unevenly, then it is necessary that the intensity of fishing be uniform over the whole area of the body of water.

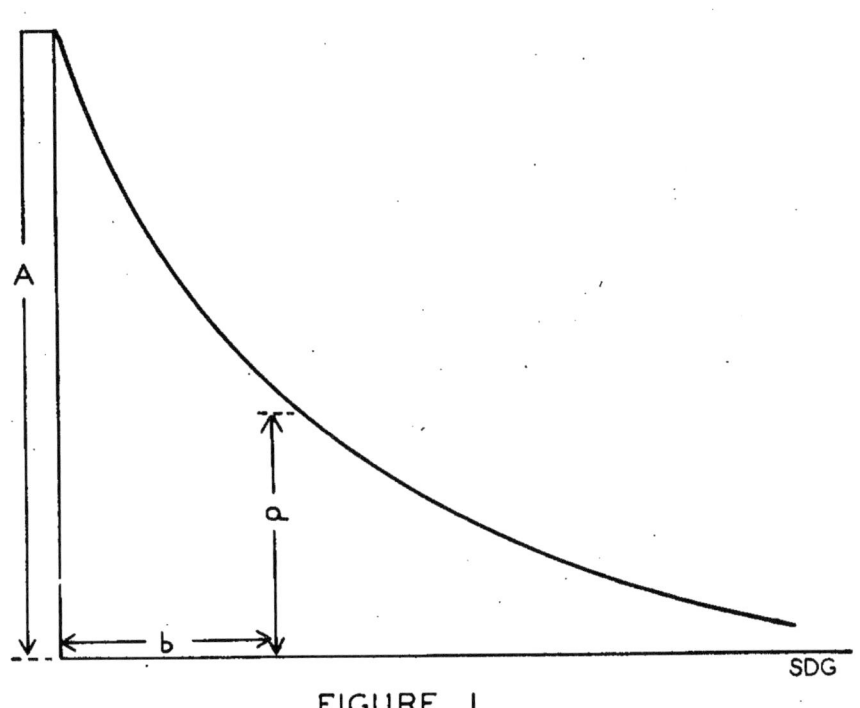

FIGURE 1

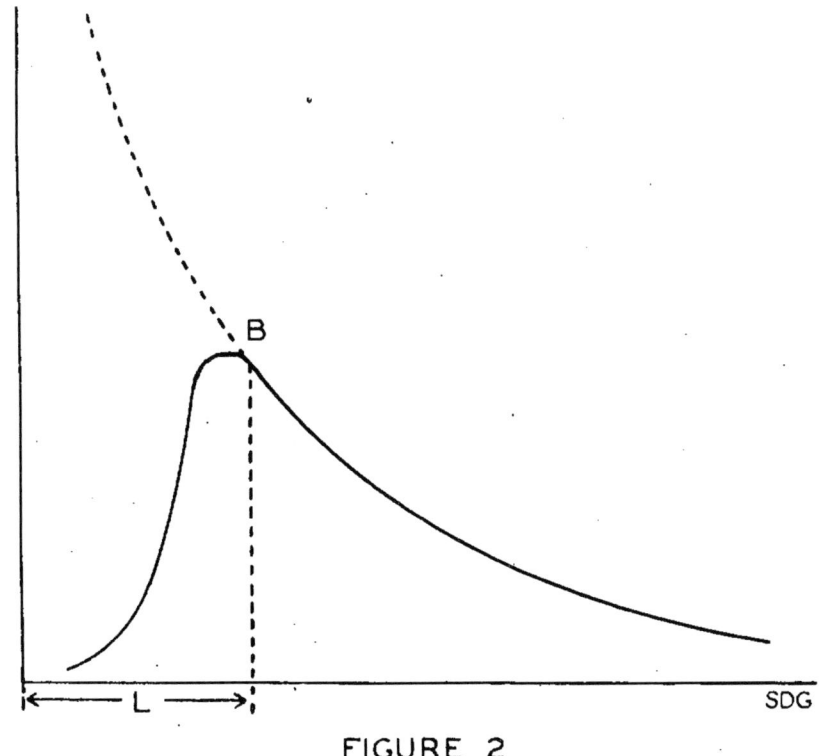

FIGURE 2

Let us construct this curve, putting, as before, the length of the fish on the abcissa and their number on the ordinate. The resulting curve, naturally, can be separated into two principal parts. The small fish caught by the fishing gear often slip through the meshes, and in the larger numbers, the smaller is their size; thus the left part of the curve (curve of selection) is not a good reproduction of the true population curve. The right part of the curve, starting from a certain point B representing the length of the fish, L, at which they can no longer slip through the gear, coincides with the population curve (its ordinates are proportional to the corresponding ordinates of the population curve). Therefore the analysis of the catches affords an opportunity to follow the mortality curve of fish of commercial size.

In general, during an examination of mortality curves two factors appear: it is suggested that mortality results from two principal groups of causes;

(1) from external and accidental causes of death, having nothing to do with the physical condition, and consequently also with the age, of each single individual; and

(2) from causes depending exclusively on age. (1)[2]

---

[2] See the list of works cited.

In this way the well-known formula of Gompertz-Makeham is obtained. In our case, when the larger part of the industry depends on the capture of comparatively young fish, as a rule not more than 4-5 years of age, that is, far from approaching /page 86/ their natural age limit, we can disregard the second group of causes. And equally, analysis of the material available, though it is very limited, fails to yield any hint that such complication of the formulae is necessary. As long as we deal with that part of the catch which is indicated by the right part of the curve of figure 2, then what is involved in the decrease of the fish from exploitation, does not depend on the age (size) of the fish.

Thus, denoting by the letter n the abundance of any group of fish (of one age), and time by the letter t, we come to the conclusion that the decrease dn in the number of fish of this group is a small space of time dt is proportional to the abundance of that group and therefore

$$\frac{dn}{dt} = -k_1 n,$$

where k_1 is a coefficient, the same for all groups. Let us integrate this expression. We get

$$\frac{dn}{n} = -k_1 \, dt, \text{ or}$$

1) .... $\log n = -k_1 t + \log C$, or

2) .... $n = C e^{-k_1 t}$

Where C is a derivative constant, introduced in the integration.

Formula 2 is the theoretical expression of the right part of the curve of catch (as developed from our assumptions). As is evident from formula 1), this curve will be portrayed by a straight line, if the ordinate axis represents the logarithm of n, and the abcissal axis the age of the corresponding groups of fish. Such a figure gives a convenient means for determining how well a real curve corresponds to our theoretical law. Here we may take advantage of the following circumstance.

As a whole series of investigations shows, the average growth of fish of ages from 2-3 to 5-6 years is proportional to their age. For example figure 3, constructed by Thompson (2) on the basis of an analysis of the catches of cod from the Firths of Forth and Moray (on the ordinate axis of this figure is represented the length of the cod in centimeters, and on the abcissa is shown age of the fish; vertical /page 87/ lines separate the age groups, and the divisions marked between them correspond to the months of February, May, August and November) shows the extreme regularity of the growth of the cod up to a length of 80 centimeters, during which time, in this example, the cod grows 18-19 centimeters per year.[1] The growth of the plaice too can be represented

---

[1] From Helland-Hansen's (3) data cod (in another habitat) grow 9-10 centimeters per year.

---

in the same way, accurately enough. Heincke, in his capital work on the plaice, unfortunately not yet completed, shows that:

plaice of 25 cm. length have an age of 3-1/2 years
"   " 30 "   "   "   " " " 4   "
"   " 35 "   "   "   " " " 5   "
"   " 40 "   "   "   " " " 6   "
"   " 45 "   "   "   " " " 7   "
"   " 50 "   "   "   " " " 9   "
"   " 55 "   "   "   " " " 11   "

so that at ages from 4 to 7 years they grow 5 centimeters per year.

Supposing therefore,

$$t = rl,$$

where l represents the length of the fish, and r some coefficient, we reduce equations (1) and (2) to the form

1') ...$\log n = \log C - kl$, and

2') ... $n = C e^{-kl}$,

where coefficient $k = rk_1$ we will call the <u>coefficient of decrease</u>.

Formula 2 is the theoretical expression of the right part of the curve of catch (as developed from our assumptions). As is evident from formula 1), this curve will be portrayed by a straight line, if the ordinate axis represents the logarithm of n, and the abcissal axis the age of the corresponding groups of fish. Such a figure gives a convenient means for determining how well a real curve corresponds to our theoretical law. Here we may take advantage of the following circumstance.

As a whole series of investigations shows, the average growth of fish of ages from 2-3 to 5-6 years is proportional to their age. For example figure 3, constructed by Thompson (2) on the basis of an analysis of the catches of cod from the Firths of Forth and Moray (on the ordinate axis of this figure is represented the length of the cod in centimeters, and on the abcissa is shown age of the fish; vertical /page 87/ lines separate the age groups, and the divisions marked between them correspond to the months of February, May, August and November) shows the extreme regularity of the growth of the cod up to a length of 80 centimeters, during which time, in this example, the cod grows 18-19 centimeters per year.[1] The growth of the plaice too can be represented

---

[1] From Helland-Hansen's (3) data cod (in another habitat) grow 9-10 centimeters per year.

---

in the same way, accurately enough. Heincke, in his capital work on the plaice, unfortunately not yet completed, shows that:

plaice of 25 cm. length have an age of 3-1/2 years
"   " 30 "   "   "   " "   " 4   "
"   " 35 "   "   "   " "   " 5   "
"   " 40 "   "   "   " "   " 6   "
"   " 45 "   "   "   " "   " 7   "
"   " 50 "   "   "   " "   " 9   "
"   " 55 "   "   "   " "   " 11.   "

so that at ages from 4 to 7 years they grow 5 centimeters per year.

Supposing therefore,

$$t = rl,$$

where l represents the length of the fish, and r some coefficient, we reduce equations (1) and (2) to the form

1') ...$\log n = \log C - kl$, and

2') ... $n = C e^{-kl}$,

where coefficient $k = rk_1$ we will call the <u>coefficient of decrease.</u>

From this it follows, that for the construction of a curve of catch it is possible to plot on the abcissal axis either the age of the fish, or the length corresponding to it, and if the logarithm of the number of the fish n is plotted on the ordinate axis, then the curve of catch is represented by a straight line. Indeed, this property of curves of catch of plaice from the southern part of the North Sea was noticed by Edser (5), and was mentioned in the above work of Heincke's. This observation indicates that for plaice the assumptions are in large measure justified, which we made in the course of our theoretical consideration of the question; and since the plaice fishery has been studied with special attention on the part of investigators and is the one which is best understood, in what follows we too will have in mind this fishery principally.

In Heincke's work there is presented detailed information concerning the composition of the catches of plaice unloaded at English ports. These data are adjusted by Heincke in such a manner that they represent the average composition of the whole catch (this adjustment, as the author himself points out, is based on a number of conventions) and are grouped into tables supplementing the work. Using these data /page 88/ figure 4 has been constructed. On the abcissa is represented the length of the fish in centimeters (from 15 centimeters to 65 centimeters) and beside the perpendiculars, at 5 centimeter intervals, is noted the corresponding age of the fish according to Heincke. On the ordinate axis is shown the number of fish, given at the left on a logarithmic scale. The continuous irregular line in the figure represents the composition of the catch of the year 1905-1906 (table X of Heincke's work) on the logarithmic (left) scale; by the large circles the summary data for 1905-1908 are represented (table XIII of Heincke's work). The dotted curve with small circles similarly shows the composition of the 1905-1906 catch on the ordinary (right) scale. So, taking for example the vertical line corresponding to a length of 45 centimeters, we see that it cuts the dotted curve at a point which corresponds on the right scale to about 2000; and likewise the point of intersection of this ordinate with the continuous curve gives, on the left scale, approximately 2000 also.

As we see, the general course of the right branch of the continuous curve really differs little from the straight line indicated by the dotted line.

Turning to formula 2') and substituting in it $l = 0$, we obtain:

$$n_0 = C,$$

/page 89/ that is, the coefficient C is equal to the number of fish in the group at the first moment of their life (this is, general speaking, a fictitious quantity, for the mortality of the fry during the first period of their life is probably different from the mortality of the older fish).

For the determination of the quantities k and C, let us examine two groups of fish. Let the first group consist of n_1 fish of length $\underline{l}_1$ and the second group consists of n_2 fish of length $\underline{l}_2$. Inserting these values in equation 1'), we get:

$$\log n_1 = -k\underline{l}_1 + \log C, \text{ and}$$

$$\log n_2 = -k\underline{l}_2 + \log C;$$

from which,

3') $$k = \frac{\log n_1 - \log n_2}{\underline{l}_1 - \underline{l}_2}$$

or, if ordinary logarithms to the base 10 are being used:

3) $$k = \frac{\log n_1 - \log n_2}{0.434(\underline{l}_2 - \underline{l}_1)}$$

Having determined, in this manner, the value of k, we also find the value of C from the formula:

$$C = n_1 e^{-k\underline{l}_1} = n_2 e^{-k\underline{l}_2}$$

For the determination of the values k and $\underline{l}$ from empirical data it is necessary to bear in mind that in the system of coordinates of figure 4 these quantities determine a straight line, and their determination is equivalent to fitting a straight line through the points $(n_1, \underline{l}_1)$ and $(n_2, \underline{l}_2)$, wherefore it is necessary to select these points in such a way that the straight line running through them corresponds as closely as possible to the curve of catch[1]).

---

[1])It is obvious that there is no advantage, in the case at hand, in resorting to the method of least squares.

To do this it is necessary to take even the first point on the second half of the curve, where the influence of selection is not felt; in choosing the second point it is necessary to remember that the right end of the curve is unreliable, for two reasons: 1) the rate of growth among the old fish slows down, therefore the relationship between growth and age--which was posulated as the basis for constructing of the diagram-- is violated; this, however, is somewhat masked by the diffuseness of the corresponding age groups (see appendix 1), and 2) the abundance of these groups is comparatively small, wherefore among them accidental variations are especially noticeable. Therefore the second point should be taken some distance from the end of the curve.

If therefore the quantities k and C are determined on the basis solely of the numerical data, without constructing a curve, then it is desirable to determine them from 2-3 pairs of values for each case.

[page 90] Applying formula 3) to our case, we get:

First point ..... for $l_1 = 30$ cm., ... $n_1 = 43800$; log $n_1 = 4.64$

Second point .... for $l_2 = 60$ cm., ... $n_2 = 53$; log $n_2 = 1.72$

Hence: $k = \dfrac{4.64 - 1.72}{0.434(60-30)} = \dfrac{2.92}{0.434 \times 30} = 0.22$

Turning to the catch of the years 1905-1908, we see that its composition (indicated in figure 4 by the large circles) up to the length of 40 centimeters agrees well with the curve for 1905-1906, but that in the interval from 40 to 50 centimeters the slope of the curve sharply decreases, and from 50 centimeters on it again goes approximately parallel to the curve for 1905-06 (traces of similar anomalies are seen also even in the curve for 1905-06). For an explanation of the nature of this break it is necessary to bear in mind that, as is evident from formula 1'), the coefficient k determines the slope of the straight line (into which the curve of catch is transformed, when represented on the system of coordinates of figure 4) but coefficient C determines its first ordinate. Thus the parallel shift of the location of the line portraying the distribution of the fish 50-60 centimeters long in the catches of 1905-1908, shows that the general coefficients of decrease of the fish of length 30-40 centimeters and fish of length 50-60 centimeters are the same, but that within the limits of the region 40-50 centimeters some anomaly has appeared. I think the most natural hypothesis will be that, at the origin of the whole curve of the coefficient of decrease, the relative abundance of the groups of fish less than 40 centimeters and more than 50 centimeters long is evaluated inaccurately by the commercial statistics (it may be because Heincke's tables are based only on English data; in which connection it may be observed that the English own the greater part of the steam trawlers, which fish the Dogger Bank and the more northern part of the North Sea, where the average length of the plaice is 40 centimeters and greater).

Therefore there is complete justification for adopting the value $k = 0.22$, as really representing the law of change in the mortality curve of the plaice; if it is desired to estimate the coefficient k so that it represents, as well as possible, the empirical curve as it stands, we should adopt a value of k somewhat smaller, about $k = 0.17$.

In later computations we will adopt an average value for the coefficient k, the rounded figure $k = 0.20$.

Having determined the magnitude of n_1 and n_2 corresponding to the difference $l_2 - l_1$, which is the annual growth of the fish, we may find the corresponding annual decrease of the fish, given by the expression:

$$\varphi = \dfrac{n_1 - n_2}{n_1}$$

We get
$$\varphi = \frac{n_1 - n_2}{n_1} = \frac{Ce^{-kl_1} - Ce^{-kl_2}}{Ce^{-kl_1}}$$

or
$$\varphi = 1 - \frac{Ce^{-kl_2}}{Ce^{-kl_1}} = 1 - e^{-k(l_2-l_1)} = 1 - 10^{-0.43k(l_2-l_1)}$$

/page 91/ The results of the calculation by this formula for the plaice ($l_2 - l_1 = 5$ centimeters) are shown in figure 5, in which the ordinate axis marks the value of coefficient k, and the abcissal axis the value of the coefficient φ.

In this figure we see that the value of coefficient $k = 0.20$ corresponds to $\varphi = 0.63$, and the values $k = 0.17$ and $k = 0.22$ correspond to $\varphi = 0.57$ and $\varphi = 0.67$ respectively.

Let us now make clear the relationship between the composition of a fish population (Aggregate A) and the composition of that part of it (Aggregate B), which perishes from the fishery and other causes. Let us represent by the letter N the "abundance" of some group in division A, and the "abundance" of the corresponding group of aggregate B by the letter n. Then the distribution of frequency in aggregate A is shown by the formula

$$5) \quad N = N_0 e^{-kl},$$

and the distribution of frequency in aggregate B is:

$$n = n_0 e^{-kl}.$$

The decrease in abundance dN of fish of length l during an elemental interval of time corresponding to an increase in length of the fish of magnitude dl, is given by the formula

$$dN = -kN_0 e^{-kl} dl$$

and must be equal the number of fish of aggregate B, whose length lies between the limits l and $l + dl$; that is, it must be equal:

$$n \cdot dl = n_0 e^{-kl} dl,$$

from which: $kN_0 e^{-kl} dl = n_0 e^{-kl} dl$;

and therefore in general, for any given length l, we have:

$$6) \quad \left| n \right|_l = k \left| N \right|_l$$

Consequently it is evident that, on our postulates, the curves of composition of the aggregates A and B are "similar" to one another, that is, the relation between the numbers of corresponding groups among them is constant and equal to the coefficient k. Therefore, since this is the case (and only in this case) we can, from the composition of aggregate B, learn about the composition of aggregate A.

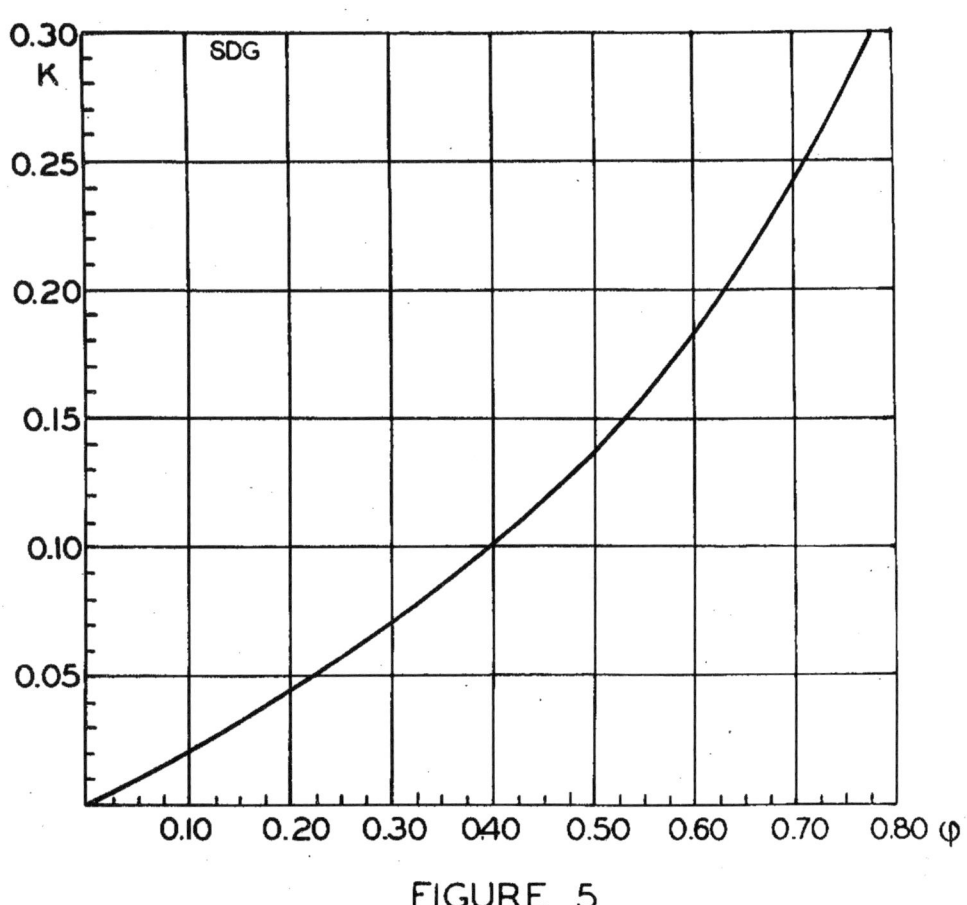

FIGURE 5

From inspection of figure 2 it was noticed above that the curve of catch is divided into two parts, of which the right part constitutes the subject of our discussion. An especial importance, in what follows, attaches to that length (L in figure 2) at which the fish first completely enter the curve of catch; we will call this length L the **minimal length**, and postulate that fish whose length is less than this minimum do not enter into the composition of the catch and are not taken by the fishery; but that, starting at this length, fish are completely represented in the catch.

/page 92/ Under these conditions the number R of fish of **commercial size** (that is, fish whose length is greater than L) in aggregate A is given by the integral:

$$7) \quad R = \int_L^\infty N_0 e^{-kl} \, dl = \frac{N_0}{k} e^{-kL}.$$

In the same way the number r of fish of commercial size of aggregate B is given by the formula:

$$r = \frac{n_0}{k} e^{-kL},$$

or, from formula 6)

$$7') \quad r = N_0 e^{-kL},$$

where $N_0 e^{-kL}$ is the number of fish of minimal length in aggregate A.

In this way we come to an extremely important conclusion, which, if we have in mind the situation over a finite period of time, is at once apparent, and can be generalized as follows: if a fishery is in a condition of equilibrium, then the number of fish of commercial size which die annually equals the number which annually grow up to the minimal size, quite independently of the nature of the curve of population.

In the formulae considered, expressing the relationship between the aggregates A and B, the coefficient of decrease k shows only the decrease in abundance of fish in relation to their growth; neither the rate of growth of the fish, nor the rate of their decrease with time, enter, as such, into these formulae. However, in the deduction of formula 2'), we assumed that the length of the fish was proportional to their age, and thus the choice of a unit of length defines also the unit for measuring time. In particular, we assumed that an interval of time of 12 months represents an increase of 5 centimeters in length of the fish, and therefore our chosen unit of length -- 1 centimeter -- represents an interval of time of 73 days (that is, about 2-1/2 months). This unit of length comes into the determination of the magnitude of k (formula 3) and what follows; therefore the above numerical values of the magnitude of k and the results obtained from them, concerning the composition of aggregates A and B, correspond to an interval of time of about 2-1/2 months.

If the decrease of the fish is in great part the result of fishing, then the thesis stated above leads to the conclusion that: the annual number of fish caught out is equal (for a constant fishery) to the number annually recruited at the minimal length, and consequently does not depend on the intensity of the fishery or the organization of the industry. However, the intensity of fishing, as it affects the form of the curve of population and coefficient k, has an effect upon the distribution, in the catch, of fish of different lengths, and therefore has an effect upon the weight of the catch. For a determination of the weight of catch we will make use of the following circumstances:

Let us postulate that the individual fish are geometrically similar. Then their volume will be proportional to the cube of their length, and, if their specific gravity remains the same, their weight will also be proportional to the cube of their length. This hypothesis is very satisfactorily realized by the empirical data, and Heincke decides that the weight of plaice in grams is given by the formula:

$$P = 0.01\, \underline{l}^3$$

where $\underline{l}$ is in centimeters.

/page 93/ And so, given that the weight of the fish is given by the formula

$$p = w\underline{l}^3,$$

where w is some coefficient, we obtain the whole weight of the fish population (of commercial size) of aggregate A:

$$P = \int wN_0 e^{-k\underline{l}} \underline{l}^3 d\underline{l} = wN_0 \int e^{-k\underline{l}} \underline{l}^3 d\underline{l} =$$

8) . . . . $\dfrac{wL^3 N_0 e^{-kL}}{k} \cdot \left(1 + \dfrac{3}{kL} + \dfrac{6}{(kL)^2} + \dfrac{6}{(kL)^3}\right),$

or, in view of formula 7),

8') . . . . $P = RwL^3 \left(1 + \dfrac{3}{kL} + \dfrac{6}{(kL)^2} + \dfrac{6}{(kL)^3}\right)$

Here R is the total number of fish of commercial size in Aggregate A; wL^3 is the weight of a fish whose length equals L; and $\left(1 + \dfrac{3}{kL} + \dfrac{6}{(kL)^2} + \dfrac{6}{(kL)^3}\right) = q$ is some coefficient.

As it is easy to see, according to formula 8') the product $wL^3 q$ is the mean weight of the fish of aggregate A (and aggregate B).

Values of coefficient q, calculated from its dependence on the magnitude of kL, are shown in figure 6. On the abcissal axis of this figure are given the values of the quantity kL, and on the ordinate, values of the coefficient q.

/page 94/ In estimating the present condition of a fishery there inevitably arises the question of what was the condition of the stocks of fish at the time when the fishery was not of great importance, and the

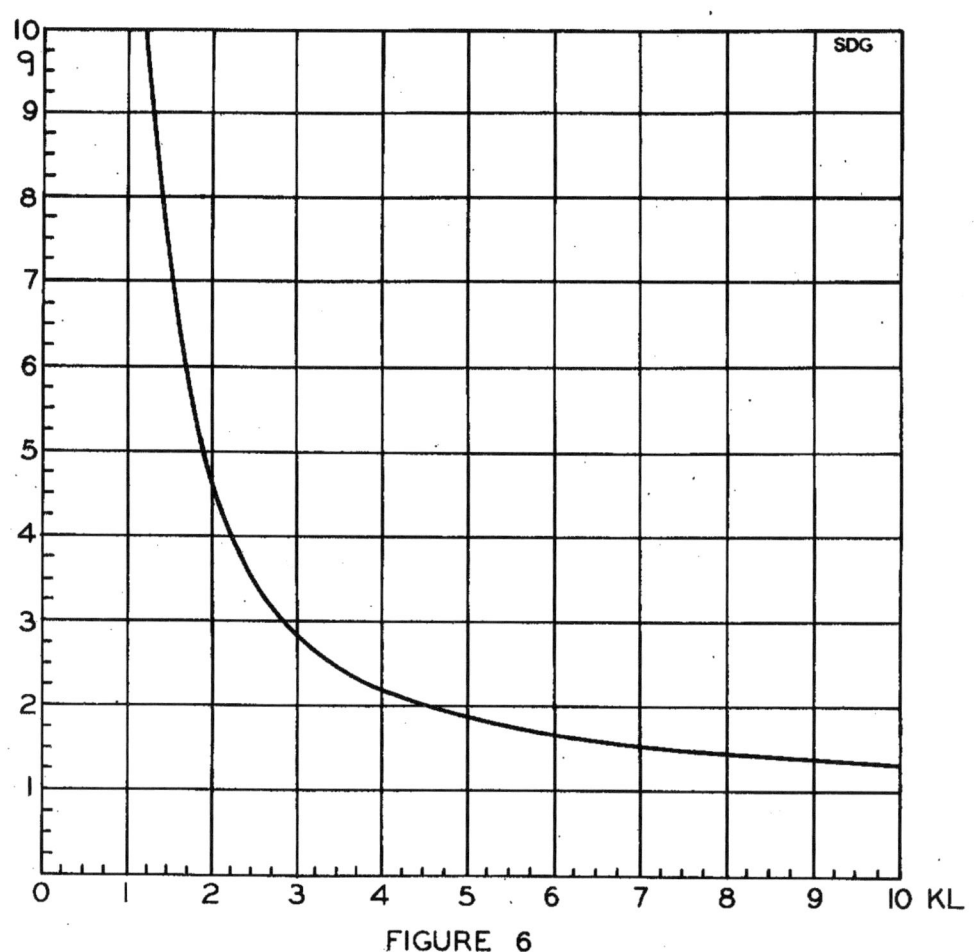
FIGURE 6

decrease of the fish population took place solely from the action of natural causes. The solution of this question is the more difficult, since the study of fisheries was begun only quite recently, and we do not have, in as much detail as we should like, information concerning more remote times. Under such circumstances it may be permissible to use formula (8'), thus providing an opportunity of computing the coefficient k, if the weight of fish of the minimum commercial size and the average weight of the fish in the catch be known.[1]

---

[1] In the practical application of this method, however, there is this consideration, that in determining the mean weight of the fish in the catch, there is also taken into account that part of the curve of catch, in which is reflected the influence of selection caused by the fishing gear.

For plaice, it is extremely convenient to apply the method outlined to the fishery for this species in the Kattegat, for there there exists a legal minimum size for plaice, and this size can be considered as our minimum size L. Some evidence concerning this industry is given in a paper by Johansen (6). He shows that the average weight of all plaice in the catch in recent times equals 320-340 grams, and the minimum weight corresponding to the established minimal size of plaice of 26 centimeters equals 180 grams. Further, according to this evidence, the average weight of plaice in the Skagerrak in the 1880's, when the fishery for plaice was very little developed (catches were made only in nets set near the shore), equalled 1250 grams.[2]

---

[2] This fishery began to develop considerably beginning with the year 1880, when the bottom seine (snurrevaad) came into use for the capture of plaice and the catch began to be taken from the whole Kattegat; and especially in the 1890's, when hand windlasses were replaced by steam or motor-driven ones; and then, toward the end of the 1890's the boats themselves were equipped with auxiliary motors.

From these data we get:

Coefficient $q = \frac{340}{180} = 1.9$,

which by figure 6 corresponds to a value of $kL = 5.0$, or, for $L = 26$ centimeters, $k = 0.19$.

As we see, the coefficient of decrease k, derived by an essentially different method and from the analysis of different material, proves to be amazingly close to the value which we estimated earlier (see page 9).

Given that in the 1880's the minimal size of the plaice was 26 centimeters, with an average weight of the plaice caught at 1250 grams we get:

$q = \frac{1250}{180} = 6.9$,

which corresponds to a value of $kL = 1.5$, and a value of $k = 0.058$; or, the annual decrease of the fish (by figure 5):

$\varphi = 0.25$.

In the absence of more detailed information, we may use this value of k as applying to the upper limit of mortality of the fish at the time when the industry was only feebly developed; therefore in future we will designate as the coefficient of natural mortality of the fish, the rounded value $k_0 = 0.06$.

Turning back to the conclusions stated as the basis for the deduction of formulae 1) and 1'), we will analyze the case of the joint action of natural death /page 95/ and of fishing. Suppose that natural mortality is characterized by coefficient k_0, and an element of decrease of the fish is expressed by:

$$-k_0 n d l,$$

while the fishery is characterized by coefficient k_2 and an element of decrease of the fish is expressed by:

$$-k_2 n d l.$$

Then, as we know from differential calculus, the common action of natural death and fishing is an element of the decrease of the fish, which will be expressed by:

$$-(k_0 n d l + k_2 n d l) = -(k_0 + k_2) n d l.$$

and can be characterized by the common coefficient:

$$k = k_0 + k_2,$$

to which 1') and the other formulae of this work are applicable. But these formulae cannot be applied to coefficients k_0 and k_2 separately, once we suppose that both these factors act concurrently. Therefore, in order to separate, from the common annual decrease, which part is due to natural mortality of the fish, and which part is due to exploitation, it is impossible to make use of figure 5 directly. For the solution of this question we notice that the common decrease for each small interval of time is distributed between the decrease from natural death and decrease from fishing in proportion to the coefficients k_0 and k_2. Since the annual decrease is the sum of these elements, it is obvious that in it also exactly the same proportionality obtains. Therefore, designating by the letter φ the common annual decrease of the fish, as determined from coefficient k using figure 5, we have:

$$\text{annual decrease from natural death} = \frac{\varphi k_0}{k_2 + k_0}$$

$$\text{annual decrease from fishing} = \frac{\varphi k_2}{k_2 + k_0}$$

In our case, for $k = 0.20$, $k_0 = 0.06$ and $k_2 = k - k_0 = 0.14$, we have:

Total annual decrease (from figure 5) 0.63

" " " from natural death $\frac{0.63 \times 0.06}{0.20} = 0.19$

" " " from catch $\frac{0.63 \times 0.14}{0.20} = 0.44$

Before proceeding to make further deductions from the theory described above, it will be useful to compare the coefficient of exploitation which we have found for plaice, and which, generally speaking, is quite large, with the conclusions to which a consideration of information existing in the literature will lead us, in respect to the organization of the fishery for these fish in the North Sea. In doing this we will first of all dwell on some theoretical considerations.

The Trawl Fishery and the Theory of Utilization of Fish Stocks

Let us consider an ideal example of a trawl fishery.

Assume that on some area, S, fish are distributed evenly on the bottom and do not move from place to place; and that the trawl takes all the fish which occur on the area, s, which is fished by it in the course of a haul.

[page 96] Assume further, that the area s is only a small part of area S, so that each haul is made afresh, not covering an earlier spot. Then the ratio of the area covered during the course of any interval of time, to the whole area S, can be taken as a measure of the intensity of the fishing. This value, which involves only the spatial relationship, we will call the geometric intensity of the fishing. In reality however the trawl catches only part of the fish which occur on the area of its haul; the ratio of the size of the catch to the whole quantity of fish occurring on the area of the haul we will call the fishing efficiency of of the trawl, assuming it to be a more or less constant quantity. The product of the geometrical intensity of fishing, times the fishing efficiency, defines, it is easy to see, the real elemental intensity of fishing -- the elemental, because in determining geometrical intensity we proceeded from the postulate that all hauls take place under uniform conditions and that the catches of all of them are the same. Such independence of the hauls could occur in the above-mentioned artificial situation; but, under ordinary conditions - only in the event that the intensity of fishing is elementally small, so that the catch does not appreciably affect the general abundance of the fish.

Let us postulate then that the total quantity of fish in the body of water equals R, and that the real intensity of fishing for a small interval of time, for example, for one day, equals the value p and remains constant. Then, after the lapse of the first day the catch in the body of water consists of the fish . . . . $R(1-p) = R_1$
after the second day . . . . $R_1(1-p) = R(1-p)^2 = R_2$
. . . .
after the nth day . . . . $R_{n-1}(1-p) = R(1-p)^n$

Consequently after n days the fishermen will have caught in all:

$$R - R(1-p)^n \text{ fish,}$$

and the total real intensity of fishing after an interval of n days fishing is given by the formula:

$$9)\dots\dots \frac{R - R(1-p)^n}{R} = 1 - (1-p)^n = np - \frac{n(n-1)p^2}{1.2} + \frac{n(n-1)(n-2)p^3}{1.2.3} - \dots$$

Consequently only for very small values of p and not very large values of n could the general intensity approximate the value np, disregarding the second and later terms.

In practice we are obliged usually to work from a value of intensity which corresponds to some comparatively large interval of time, and such a method is the only valid one, for the intensity of fishing during a given interval of time can vary greatly simply from the state of the weather, hence only the average value of intensity for a relatively large interval of time can give an idea of the degree of utilization of fish resources.

So let us assume that the geometric intensity of fishing for some interval of time T days has the value U (that is, during T days' fishing an area equal to US is covered) and that the fishing efficiency of the gear equals 1. /page 97/ Then the elemental real intensity of fishing (for 1 day) will equal:

$$\frac{U}{T},$$

and if the fishing efficiency of the gear does not equal 1, but f, then

$$\frac{f\ U}{T} = \frac{U_1}{T},$$

where $U_1 = f U$. The final intensity of fishing for the whole interval of time T is given by formula 9):

$$1 - (1 - \frac{U_1}{T})^T$$

For the calculation of this expression, we will decrease the magnitude of the interval of time, approaching the elemental. Then the numerical value of T will increase, and in the end we have:

$$10)\ldots\ldots \lim \left| 1 - (1 - \frac{U_1}{T})^T \right|_{T = \infty} = 1 - e^{-U_1}$$

The results of calculations from this formula are shown in figure 7, in which the ordinate axis gives values of the expressions $1 - e^{-U_1}$, and the abcissal axis has values of U_1 (on an ordinary scale).

Thus if in each individual haul only an insignificant part of the body of water is covered, but in total the whole body of water is covered once a year (that is, the geometric intensity of fishing per year equals unity), then if in addition the fishing efficiency of the gear is unity, the real intensity of fishing per year will be equal to 0.63, as is shown by the diagram.

This result refutes the argument of Heincke (7), who supposed that if the geometric intensity of fishing (for a year's lapse of time) equals unity, then the real intensity is equal to the fishing efficiency of the gear in use.

The assumptions involved in the derivation of formula 9) do not contradict our earlier postulate that the fish population exists in a condition of equilibrium, and that therefore the abundance of fish in the body of water is numerically constant. /page 98/ As a matter of fact, although, owing to the growth of the young, the general abundance of the

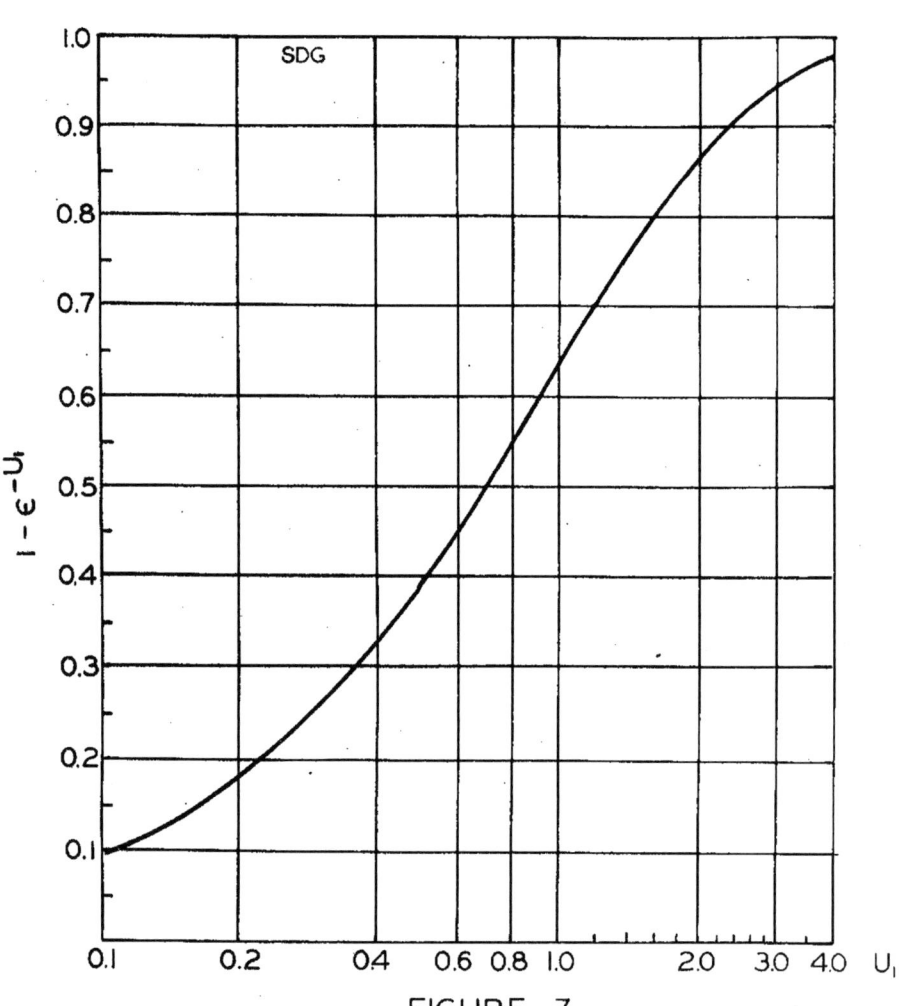

FIGURE 7

fish does remain constant in spite of the fishing toll, nevertheless, for determining the intensity of fishing it is natural to select some definite group of fish and to watch its gradual decline, to determine its abundance at the end of the period under observation, and to compare it with the abundance which characterized the group at the start of the period. In this way we determine the magnitude of the intensity of fishing, which is the same for all groups of fish subject to the fishery, during the course of the whole period under consideration. Then too, in this argument we have disregarded the decrease in the number of fish from natural causes (those not caused by fishing); in a detailed accounting the appropriate correction would have to be added to the value of the elemental intensity.

The most detailed information concerning the trawl fishery in the North Sea is furnished by the English statistics, which are analyzed in a publication of Masterman's (8), among others. In this paper is shown the total duration M of all trawling cruises for different parts of the North Sea (counting in the time consumed in getting to and from the grounds), and also the average duration m of cruises in the same parts of the North Sea. Thus it is possible to calculate the average number of voyages in a given part of the sea, and, dividing the area of the part by the total number of voyages to ascertain how many square miles of the given region are visited in the course of one year's voyages.

Most frequently visited is the deep part of the English channel, where (not counting Belgian and other trawlers) a beam-trawl fishery by English sailing trawlers is concentrated. Here for each square mile there are four voyages per year (with duration of the voyage 6 days). In the North Sea proper no section is more often visited than the Dogger Bank, where there is one voyage to 2 square miles (per year), with the duration of the voyage about 8 days; the average for the whole North Sea is 1 voyage per 6 square miles.

Assuming that the distance between the boards of an otter trawl (while it is fishing) is 20 meters and that the rate of trawling is 3 knots, we compute that per hour the trawl covers an area of about 0.034 square miles. During the course of a week's cruise approximately 3-4 days[1] are devoted to actual fishing (deducting the time for getting to

---

[1] Fulton (20) shows that ordinarily the duration of a haul (in the trawl fishery) is 5 hours, there are 4 hauls in a day and in the course of a week's voyage about 5-1/2 days are devoted to fishing operations, i.e. 100 hours or 4 days of continuous fishing. However, observations included in the work cited, applying to the trawlers, show that such continuous fishing is possible only under especially favorable conditions, when it is not necessary to waste time moving in search of better fishing grounds, and so on. On the other hand, the statistical data under consideration include information on the catch of only the English trawlers, and among them, only those which, at the end of a short cruise, land their catch themselves, i.e. they do not take into account the so-called fishing fleet.

According to the data of Lee (21) the steam trawlers in the course of a cruise of 5 days spend 16 - 17 hours a day in fishing (i.e. about

3-1/2 full days out of the whole cruise), and in the course of a cruise of 10 days, 13 - 14 hours per day (i.e. about 5-1/2 days per cruise).

The sailing trawlers, according to this author's information, have an average cruise duration of 7 days and spend 11 - 12 hours per day in fishing (i.e. about 3-1/2 days per cruise).

the fishing grounds and back, etc.) so that during one cruise an area of about 2.5 - 3.0 square miles is fished. The area covered by a beam trawl (considering its slower speed of trawling etc.) will be three times less.

[page 99] In this way we can calculate that the yearly geometric intensity of fishing on the Dogger Bank is not less than 1-1/2; in the whole North Sea, about 0.5; in the English Channel, about 3 - 4.

Supposing then that the fishing efficiency of the trawl is unity[1])

---

[1]) Heincke (7), on the basis of his experiments, takes the coefficient of fishing efficiency of a trawl to be 0.25, but I do not feel justified in making use of this figure.

---

we get provisional values for the average intensity of the fishery, from figure 7:

```
For the whole North Sea . . . . . . . . . 0.4
 "    "   Dogger Bank . . . . . . . . . . 0.75
 "    "   English Channel . . . . . . . . 0.95
```

It is extremely interesting to compare these data with the results of certain experiments.

In order to determine the course of currents near the bottom, bottles were released by Mr. Bidder (9) in the region where the sailing trawlers fish (in the English Channel), weighted in such a manner that they moved at a distance of 2 feet above the bottom. These bottles were encountered subsequently in the course of the trawl fishing, and in a year 56-57% were picked up, on the average (from 2 series). However in the course of time the bottles became overgrown, sank closer to the bottom, or might become stuck in places inaccessible to the trawls. During the very first month after release of the bottles, the percentage of returns was significantly greater and corresponded to a catch of 20-25% of the bottles in the course of 1-1/2 months. This gives, by formula 9):

$$[1 - (1 - 0.2)^8] \text{ to } [1 - (1 - 0.25)^8] = 0.8 \text{ to } 0.9,$$

that is, nearly the same as the figure given above for the general intensity of fishing in the English Channel.

In 1904, extensive experiments were carried out by the English with the marking of plaice, from which it was found that during a year's time there were fished out over 40% of the marked plaice liberated on the Dogger Bank, and about 30% from the southern part of the North Sea; and from a number of marked plaice liberated in the Kattegat (1904-05), in the course of a year there were recaptured from 40% (Swedish experiments) to 60% (Danish experiments).

The fishery for plaice in the Kattegat is carried on by means of bottom seines of special construction (the trawl fishery here is relatively little developed), and direct comparison of the organization of this industry with the trawl fishery of the North Sea is impossible. However, in the case at hand it is possible to obtain a basis of comparison by comparing the average size of the annual catch per unit area in the North Sea and in the Kattegat. Thus, Heincke gives the magnitude of the whole catch of plaice as 47,300,000 kilograms for the year 1908; the whole area of the shallower part of the North Sea, in which the principal fishery is concentrated, is equal to about 100,000 square miles, and hence the average catch is 500 kilograms of plaice per square mile per year (i.e. about 0.1 pood per desyatina). From Johansen's (11) data, the fishery for plaice in the Kattegat and Skagerrak yields about 600 kilograms per square mile per year, and therefore we must consider that the intensity of the fishery for this fish is not greatly different in the North Sea and in the Kattegat.

Comparing the above data with the results which we obtained in analyzing the curve of population, we must grant that these results /page 100/, and hence also the theory itself, seem to be very reasonable. In passing we may come to the conclusion that the trawl fishery (at least for plaice) is apparently considerably more intense than Heincke believed.

As was demonstrated above, in the case where the fish population of a body of water - it does not matter whether a fishery exists in this water or not - exists in a condition of equilibrium (for which it is necessary and sufficient that, on the one hand, the number of young produced yearly always remain the same, and on the other hand, that the decrease among the grown broods be not subject to abrupt fluctuations), then the curve of distribution of the fish population corresponds with the curve of mortality.[1]

---

[1] The conclusions just stated become much simpler and clearer if the age composition of the fish population is known. Let us assume that we do in fact know the abundance $n_1, n_2, n_3, n_4 \ldots$ of the separate age groups of the population. The expressions:

$$\frac{n_2}{n_1}; \frac{n_3}{n_2}; \frac{n_4}{n_3} \ldots$$

are related to the relative yearly decrease of the fish (page 9) for a given age group. If these expressions are equal to each other:

$$\frac{n_2}{n_1} = \frac{n_3}{n_2} = \frac{n_4}{n_3} = 1 - \omega$$

then it means that in the example at hand the rate of decrease of the fish is constant and the fish population is in a condition of equilibrium.

In that event the abundance of the separate age groups constitutes an indefinite decreasing geometric series with $1 - \varphi$ as the common ratio:

$$n_1; \quad n_2 = n_1(1-\varphi); \quad n_3 = n_1(1-\varphi)^2; \quad n_4 = n_1(1-\varphi)^3 \ldots$$

and the total number of fish N is the sum of the terms of this series:

$$N = n_1 + n_1(1-\varphi) + n_1(1-\varphi)^2 + n_1(1-\varphi)^3 \ldots = \frac{n_1}{1-(1-\varphi)} = \frac{n_1}{\varphi},$$

from which $\varphi = \frac{n_1}{N}$ -- an extremely convenient formula for calculating the annual mortality of the fish.

From this it follows that, even in the absence of fishing, the ordinates of the curve of population will decrease in proportion to the age (length) of the corresponding groups of fish (and, for example, the number of 4 year old fish will always be smaller than the number of 3 year old's) in which case the general character of this curve, as indicated in appendix I, is not violated by anomalies which could arise in consequence of sharp changes in the rate of growth of the older fish. It is necessary to emphasize this condition, which is really quite obvious, for in the literature there can be found the inaccurate idea that in the absence of fishing there would occur an excessive accumulation of fish of very large size.[1]

---

[1] See, for example, the figure on page 79 of Johansen's (6) work.

The sudden appearance of a fishery of any considerable intensity is accompanied by an increase in the "mortality" of the fish of such a nature that there is not any change in the general character of the curve of population, but only an increase in the coefficients k and φ. Here, however, it is necessary to distinguish two cases:

1) the first years of the existence of the fishery, when the curve of distribution is affected by the change caused by the sudden change in the coefficient k,

2) the condition of equilibrium which follows, when all groups of the fish population achieve the composition which corresponds to the new regime.

It is not difficult to see what character this change in the curve of population will have in the first years of the fishery, and by what path it will approach the curve corresponding to the settled fishery.

/page 101/ In a long-established fishery, for example affecting the fish from the time they are two[1] years old, fish of the different age

---

[1] /From what follows it seems obvious that the author meant to say <u>one year old</u>./

---

groups are exposed to the influence of the fishery during the whole of their future life. Namely:

Fish that are 3 years of age – during their 2nd (and 3rd) year of life
" " " 4 " " " " " 2nd, 3rd (and 4th) years of life.
" " " 5 " " " " " 2nd, 3rd, 4th (and 5th) yrs. of life.

and so on.

If however we have to do with a new fishery which, for example, is in only the third year of its existence, then we have only the groups of three- and four-year-old's completely subject to the influence of the fishery, corresponding to the condition of the preceding scheme. Namely:

3 year old's for the duration of their 2nd (and 3rd) year of life
4 " " " " " " " 2nd, 3rd (and 4th) year of life
5 " " " " " " " 3rd, 4th (and 5th) year of life
6 " " " " " " " 4th, 5th (and 6th) year of life

Thus the older groups will bear the characteristics of the period of transition; and in the future state of the fishery, gradually, one after another, will be replaced by groups corresponding to the new regime. This process of successive change in the curve of population is illustrated in figure 8. On it, the horizontal axis represents the year-classes of fish; the /page 102/ vertical axis is the relative abundance of the separate groups, with 100% representing the number of fish of 2 years of age. Assume that the annual coefficient of natural mortality $\varphi = 0.2$, and that all at once a fishery is established, raising the total annual coefficient of decrease to 0.5. Curve ABC is the curve of population before the establishment of the fishery. Curve AB_1C_1 is the curve of population at the expiration of one year of the fishery; here the abundance of three-year old's has already reached its definitive value and will not change during the later existence of the fishery. Curve $AB_1B_2C_2$ is the curve of population at the end of the second year of existence of the fishery; here the abundance of the 4-year-old's has reached its final value. The significance of the later curves is similar. Thus it is evident how, gradually, the curve of population changes from ABC to $AB_1B_2B_3$....., corresponding to the new value of the coefficient φ.

The curves corresponding to different values of coefficient of annual decrease (for a long-established fishery) are shown in figure 9 for the plaice, taking 30 centimeters as the initial length of the fish, and 5 centimeters as the annual increase. On the horizontal axis is the length of the fish in centimeters, on the vertical is the abundance of each group as a percentage, the abundance of fish of 30 centimeters length being taken as 100 percent. The values of the coefficient φ to which the different curves correspond, are shown on each curve.

From all this account it follows, as already shown above, that for a steady regime the stability of the curve of population does not depend on the coefficient φ, that is, does not depend on the intensity of fishing. In the case of a disturbance /page 103/ of the established regime

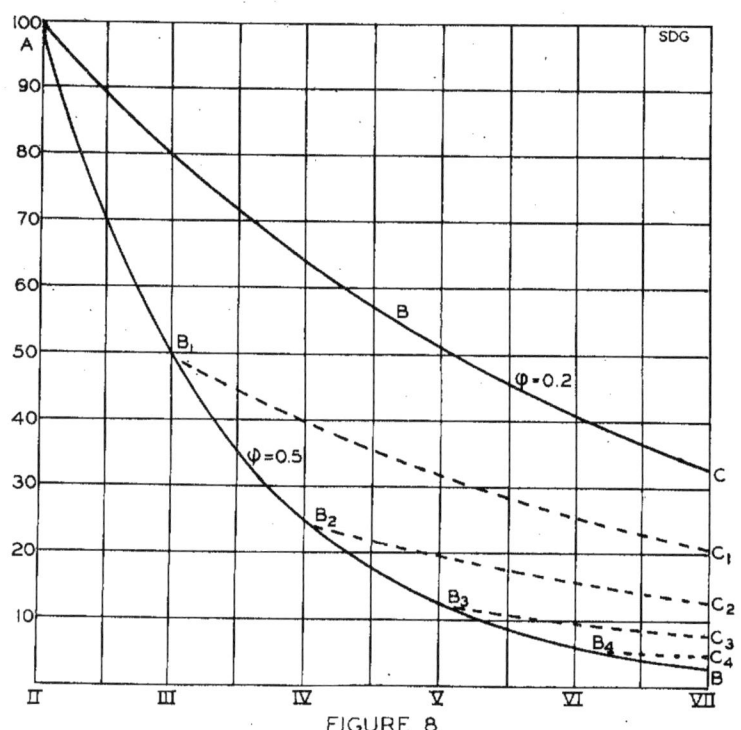

FIGURE 8

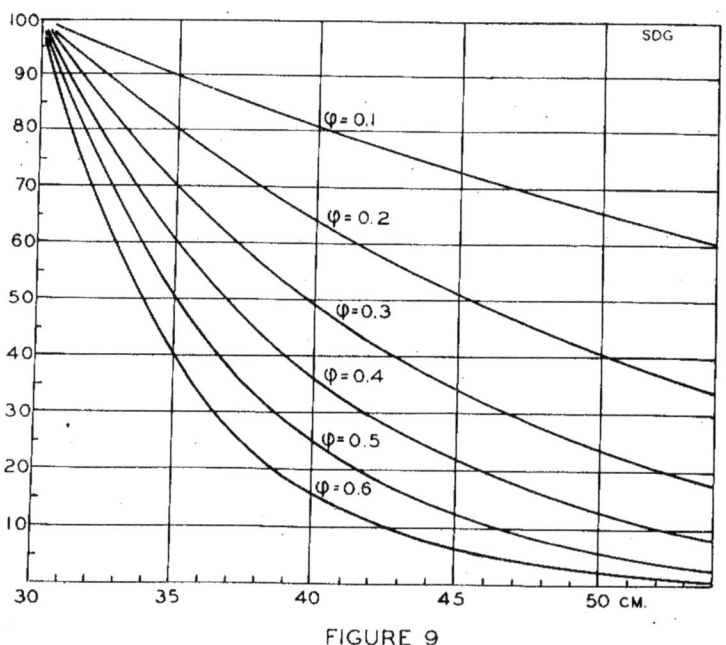

FIGURE 9

(i.e. increase or decrease in the intensity of fishing) the established distribution of the fish of different lengths in the water changes, and gradually comes to a new steady distribution corresponding to the new regime, in which, for each age group of fish, equilibrium is achieved, roughly speaking, after a number of years equal to the age of the group in question.

Thus we come to the question of what significance should be attached to the term overfishing, and to what extent the extremely gloomy associations are justified, which are sometimes connected with this term. Generally the concept "overfishing", in a situation, naturally, where there has been little study of the circumstances which characterize a fishery, is notable for its great vagueness. "Overfishing", says Heincke, "can be considered in different ways: a steady decrease in the annual catch of a given species with a constant or increasing fishing intensity; a steady decrease in absolute abundance of large, old individuals of a given species; a steady increase in the relative abundance of small young fish, compared to the number of larger older ones." Indeed by its very derivation, the word overfishing suggests that intensity of fishing has exceeded some limit beyond which there follows a sharp decrease in the catch of the fishery, "exhaustion of the fish stock", accompanied, as is evident from the words of Heincke just quoted, by a steady uninterrupted decrease in the annual catch, and so on. However, as is shown by the theory developed above, such a conception is profoundly erroneous.

Generally, in an investigation of the development of a fishery, it is necessary to distinguish clearly a transition period arising as a result of a sudden change in intensity of the fishery, and a period of stabilization of the fishery.

The transition period, represented schematically in figure 8, is characterized by a steady decrease in the catch. In fact, the average quantity of fish of commercial size is proportional to an area in the figure; in our example (figure 8) this quantity, in the absence of fishing, corresponds to the area included between the axes and curve ABC, and for the duration of the transition period evidently it is represented by the area included between the axes and $AB_1B_2\ldots\ldots$, according to the intensity of the fishery. Since by hypothesis this intensity remains constant from the time the fishery is started, then throughout the entire time some fixed part of the total available number of fish of commercial size is being fished out, and therefore the size of the catch must, in the course of the transition period, decrease to the size corresponding to the new regime.

Such a decrease in catch, however, could scarcely be described by the term overfishing, and of course it would be a mistake to base economic calculations on the size of the catches which can be made during the time of this transition period. Of course, a fishery of severe intensity does not often spring into being in a body of water where a fishing industry has previously been completely lacking. However in the history of the plaice fishery it is possible to distinguish two such cases. Thus, in this connection we can cite the plaice fishery of Norway, which arising at the end of the 1880's, in 2 or 3 years reached its maximum development and as quickly fell off (12); and the plaice fishery on the Kanin /page 104/ Banks (13), which suffered the same fate. Now the rapid decline of these fisheries can be explained in accordance with the above theory in this way: that the important mass of the catch consists of fish belonging to 2 or 3 of the youngest age groups (above the minimum commercial size), and therefore the fishery, after 2 or 3 years, had practically reached the state of equilibrium corresponding to its intensity, in which state it had already lost its original attractive superiority in yield as compared to other fisheries which had earlier been subjected to steady exploitation. Passing then to an examination of the fishery in its steady state, we see that if fishing is continued with a steady intensity the position taken by Heincke is wholly erroneous. Whatever may be the intensity of the fishery - very small or very great - within a comparatively short time an equilibrium is established, expressed by one of the curves of figure 9, and there will not take place any steady decrease in the catch or any change in the curve of population. The positions of equilibrium (as regards the size of the catch and the form of the curve of population) corresponding to different intensities of fishing are different, but the change, in the case of a gradual alteration in intensity of fishing, also occurs gradually, without any jumps. And so in the case before us there is no reason to make use of the term "overfishing". Finally, if the increased fishing disturbs the equilibrium of the fishery so that, owing to the removal of the spawners, the quantity of young produced annually in the body of water is found to be inconstant and diminishes from year to year (actually we have excluded such a contingency from our consideration, since among sea fisheries this situation, as is shown in more detail later, scarcely occurs, and among anadromous fishes it is too much complicated by accessory factors, so that it would be necessary to understand each concrete case separately) then, notwithstanding the opinion of Heincke, the abundance of large individuals, in comparison with the abundance of smaller ones, must increase. This directly follows from the discussion set forth on page 22, if we turn our attention to the fact that a decrease in young first of all is reflected in the youngest group, and only gradually, year after year, is expressed among the older groups; so that if finally the decrease of young stops and equilibrium is established (with the general abundance of the fish population significantly less than before), then the character of the curve of population and hence also the relative abundance of different age groups, will again be determined by the coefficient k.

Considering the case of a long-established fishery, it is interesting to compare the sizes of the catches corresponding to fishing of different intensities.

The total weight of fish of commercial size in a body of water, being a function of the area included between the curve of population and the coordinate axes, will decrease in proportion to any increase in the intensity of the fishery, but the ratio of size of the catch to the total number of fish will be increased; with the result that the weight of the catch will be altered and for some particular intensity of fishing will reach a maximum. In order to determine this it is possible to make use of the formula (8') developed earlier, which gives the total weight of a fish population:

$$P = RwL^3\left(1 + \frac{3}{kL} + \frac{6}{(kL)^2} + \frac{6}{(kL)^3}\right) = R w L^3 q.$$

/page 105/ Substituting in this formula, in place of the quantity R, the number r of fish which die each year from the fishery and from natural causes, we get the quantity P_1 – the yearly decrease of the fish population. As has already been shown on page 11, r does not depend on the intensity of the fishing; similarly the quantity wL^3 too is invariable. Thus the size of the annual decrease depends only on the coefficient q, values of which are to be found in figure 6 presented above, and are distributed between the decrease from natural death and decrease from fishing in proportion to k_0 and k_2, as shown on page 14.

Thus the weight of the catch which interests us is given by the formula:

$$8'') \quad P_2 = \frac{P_1 k_2}{k_0+k_2} = rwL^3 \cdot \frac{k_2 \cdot q}{k_0+k_2}$$

and is proportional to the quantity $\frac{k_2 q}{k_0+k_2}$; the results of the calculation of a series of determinations of this quantity are shown in figure 10, in which 30 centimeters is taken as the minimal length, L. On the ordinate axis are given the values of $\frac{k_2 q}{k_0+k_2}$; on the abcissa are given the values of the coefficient k_2. The curves constructed correspond to different values of the coefficient of natural mortality k_0, as marked on the respective curves. Thus each curve shows how, for a given value of natural mortality, the weight of the catch changes with a change in the intensity of the fishery. For clarity there are marked on the curves the values[1] Φ_0 of the yearly natural mortality rate, and the corresponding value of k_0 taken from figure 5; /page 106/ and vertical dotted lines

---
[1] I.e. in the absence of fishing.

---

are superimposed, which give the values Φ_2 of the magnitude of the fishing mortality per year for the corresponding values of k_2. These values are indicated conditionally, for a provisional orientation, for, as was emphasized on page 14, in the case of the combined action of natural mortality and exploitation figure 5 gives only approximate values for these quantities, and as for defining them for each pair of values of k_0 and k_2 separately, that obviously is not permitted by the construction of the figure. Relative to the value of Φ_2 it must be observed that, so to speak, the "mechanical intensity" of the fishery (as represented by, for example, the number of units of gear) is proportional to the size of coefficient k_2, and it is quite appropriate to cite the value Φ_2, which is that value of the annual intensity of exploitation which would correspond to a given "mechanical intensity" of the industry in the absence of natural mortality of the fish.

From examination of these curves it follows that for a certain intensity of fishing (which varies with changes in the magnitude of the natural mortality) the weight of the catch reaches a maximum figure; and this maximum, which is very sharply expressed in the case of small coefficients of natural mortality, is, for large values of this coefficient, very much less noticeable. For example, if the natural mortality per year

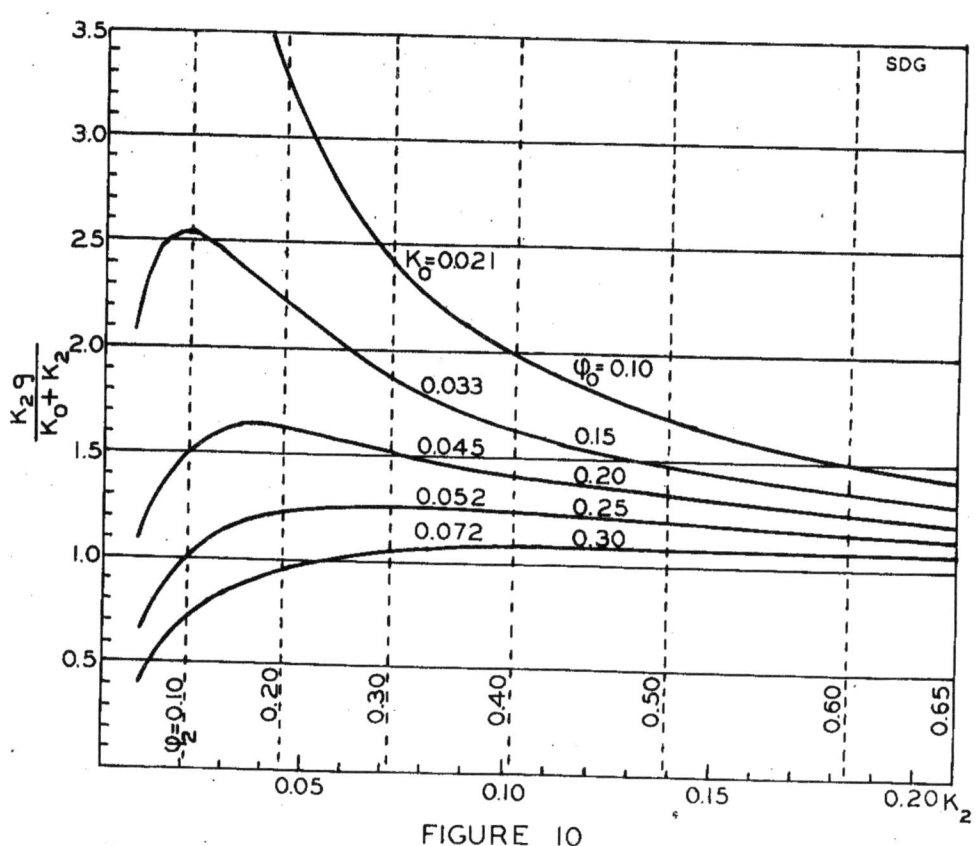

FIGURE 10

is 15%, for the largest possible value of the catch (2.55 on the vertical scale of figure 10) we would have caught out per year a little less than 10% of the fish population; for a fishing mortality rate of 25% we would have a catch (2.03) approximately 20% less; and for a fishing mortality rate of 50% it is 40% less (1.49). Thus under these conditions (little natural mortality among the fish) we can consider as efficient that organization of the fishery in which about 10% of the fish population is caught out yearly, and further increase in intensity of fishing can be called overfishing. The situation is different for large natural mortalities of the fish. For example, if natural mortality reaches 20% ($\varphi_0 = 0.20$), then the catch will attain its largest value at 15-20% fishing mortality (1.62), and if the fishing mortality increases to 50% the decrease by weight is only about 17% (1.34). The situation is even more favorable when the natural mortality of the fish is assigned the value $\varphi_0 = 25-30\%$. Then, as is evident from the figure, the yield achieves its greatest magnitude when the catch is equal to 30-40% of the fish population, and with greater intensity of fishing[1] practically does not diminish. However in this situation too it would be economic to remain at

---

[1] An increase of φ_2 from 0.30 to 0.65 corresponds to an increase in k_2 (mechanical intensity) from 0.07 to 0.21, i.e. by three times.

---

a rather small intensity of fishing, as then the average weight of the fish caught (as determined by coefficient k_2) would be greater, and accordingly the industry would obtain a more valuable product.

From examination of figure 10 we can see a noteworthy fact, that if the natural mortality of the fish reaches 25%, then with continuous increase in the intensity of fishing the total weight of the catch will remain at one and the same level; and as a matter of fact such a condition is observed in the plaice fishery. A whole succession of investigators, convinced of the existence of overfishing in general, and aspiring to find evidence for its existence, have not /page 107/ been able to discover any decrease in the total weight of the catch of plaice in the North Sea. This circumstance has also been responsible, probably, for so much obscurity in understanding of the term overfishing. And as for the plaice fishery of the Kattegat, there exists definite information (6) that in spite of the rapidly-increased intensity of the fishery the size of the catch has remained just about the same (4-5 million kilograms) from 1886 to 1904.[1]

---

[1] Often in investigations of fishing industries, when it has become evident that notwithstanding a general complaint about the decrease in catch, the total quantity of fish caught, as shown by statistical data, has not decreased, the smaller catches of individual fishermen are explained by the fact that, because of an increase in their number, fewer fish than formerly fall to the share of each fisherman. Such a - it would seem - natural explanation stands in need, however, of a great many reservations without which it could lead to a completely inaccurate conception of the subject.

Such a distribution of the catches would occur only in the event that there were caught out of a body of water the whole of the fish living in it (for example, by lowering a pond); then the total size of the catch would not depend on the number of fishermen, and the average catch of a fisherman would be inversely proportional to their number. A fishery however catches out only a part of the fish living in a body of water (this part being the larger, the larger is the number of fishermen taking part in the fishing), and the size of the catch of the individual fishermen depends only on the abundance of the fish in the body of water, the effectiveness of the gear used, and so on, and does not depend directly on the number of fishermen (the number of fishermen has an effect only indirectly, in so far as it causes a decrease in the abundance of fish in the body of water), and actually there is no distribution of the catches.

And only in the important special case when the total weight of the catch remains steady while fishing intensity changes, do things happen as though a catch were being distributed among a larger or a smaller number of fishermen.

Thus we obtain still another independent confirmation that the size of the coefficient of natural mortality which we determined earlier for plaice is not far from the truth. If that is indeed the case, then it is impossible not to agree with Kyle (12) when he states that "the practical solution of the question of overfishing narrows itself down to the determination of whether, under the circumstances, an intensively exploited area would support a larger (or smaller) number of fishing units than it has at the present moment, in doing which it is necessary to take into consideration the value of the fish and the average earnings of a fishing unit."

One of the most frequently proposed measures for improving the fishing industry is a size limit for the fish, and it is of interest to determine its theoretical importance. A size limit on fish established by law, if it is strictly observed, so that fish not yet of the established length are not taken, while the fish exceeding this length are taken with the complete intensity, corresponds, as was mentioned above, to the characteristics which the quantity L in formula (8'') must possess:

$$P_2 = r w L^3 \frac{k_2 q}{k_0 + k_2}$$

For the solution of our problem - the determination of the change in weight of catch in relation to changes in the value of L, for a given value of natural mortality k_0 and intensity of fishing k_2 - it is necessary to bear in mind that with an increase in the length L the size of r decreases (because of natural mortality of the fish). Making use, therefore, of formula (7') deduced earlier, we substitute in formula (8'') for the value r:

$$r = N_0 e^{-k_0 L}$$

/page 108/ Then formula (8'') takes the final form:

$$P_2 = \frac{w k_2 N_0}{k_0 + k_2} L^3 q e^{-k_0 L},$$

where the first part:

$$\frac{w\, k_2\, N_0}{k_0 + k_2}$$

is a constant quantity, and a change in the weight of the catch with change in length L will be proportional to the multiplier:

$$v = q\, L^3\, e^{-k_0 L},$$

depending solely on L.

The results of calculations using this formula are shown in figure 11, in which the values $k_0 = 0.06$ and $k_2 = 0.14$ are used, which were earlier determined for plaice. On the abcissal axis is plotted the length L in centimeters, on the ordinate the corresponding value of the multiplier v.

This figure shows that with an increase in the minimum length of plaice to 45 centimeters, the total weight of the catch increases, but not to such a significant degree that it would bring about a radical change in the condition of the fishery. Thus, for an increase of the minimal length from 25 centimeters to 30 centimeters, the weight of the catch increases by 14%, and for an increase of the minimal length from 30 to 35 centimeters, by 9%; and this is assuming perfect compliance with the legal size for the fish.

In concluding the theoretical part of our work, notice that the method which we have made use of for the determination of ideal catches in a fishery for marine fish, can be extended also to the investigation of a fishery for anadromous fish. In fact the essential peculiarity of a fishery for anadromous fish consists in this, that the fish which serve as the object of the fishery live in the sea under ordinary conditions, where they are not vulnerable to fishing; and only in the course of very brief periods, generally speaking, (their movement to rivers for spawning), are they exposed to exploitation, under conditions favorable to the prosecution of extraordinarily intensive fishing. Bearing in mind that the fish grows only during the time of its life in the sea, and that at the time of its journey for spawning it does not feed and does not increase in size, it is easy, knowing the coefficient of natural mortality of the fish in the sea k_0 and the intensity of exploitation during one migration, to construct directly a curve of mortality of each age-group of fish. Beginning the calculation of time from the moment of the return migration of the fish into the sea we discover that in the course of /page 109/ a year the curve will run parallel to one of the curves of figure 8; at the end of the year it falls with a sudden drop (for there return to the sea and pass over into the next age group only φ% of the whole number of fish which went up to the river), but afterward it again follows the original law, and so on. Hence it is not difficult to be assured that in this case also, if the size of the annual production of young, and also the magnitude of k_0 and φ, do not vary, then the resulting curve coincides in every respect with the curve of distribution at the end of the year, that is, at the time of fishing; and as not more than 5-6 age groups have any significance in the fishery, the summing and all the

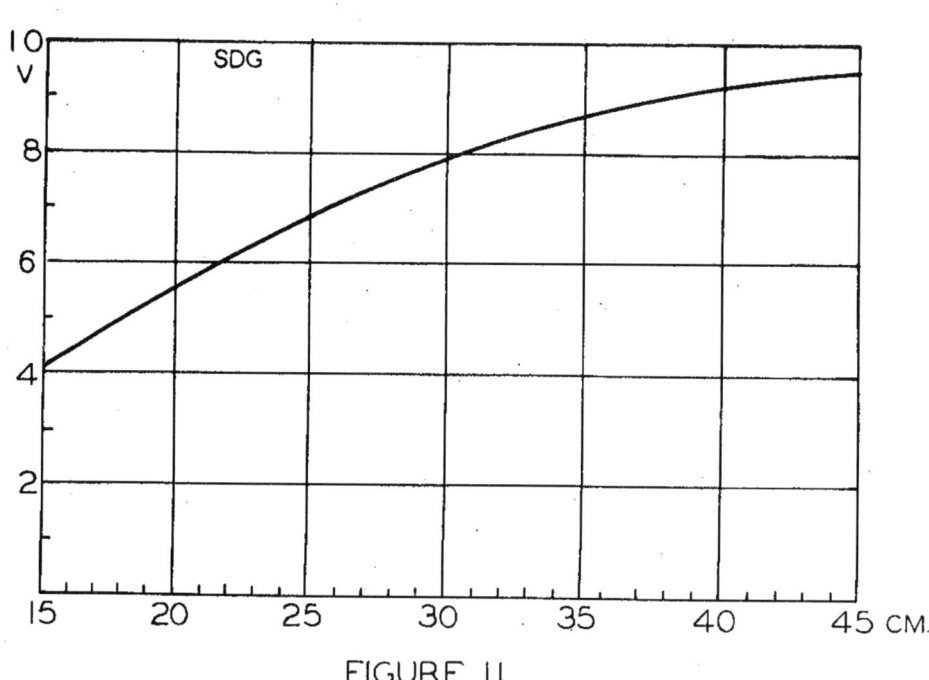

FIGURE 11

further operations can be carried out without having recourse to the formulae directly; in which manner it will be possible, without difficulty, to make all the peculiarities, by which the fishery under consideration is distinguished, susceptible to analysis. However, in the study of fisheries involving migratory fishes great interest attaches precisely to the individual divergence of a fishery from an ideal scheme, for the ideal scheme set forth does not differ essentially from a fishery for marine fish,[1] and for that our conclusions above will

---
[1] For, the peculiarity of a fishery for an anadromous species consists only in the fact that natural mortality and catch, instead of acting concurrently as in marine fisheries, act consecutively.

---

provide just accounts in every respect.

The Stability of Fisheries and Their Periodic Fluctuations.

"The study of the particular means by which fertility on the one hand, and counteracting forces on the other, maintain the different kinds of fishes in a state of equilibrium, constitutes one of the most outstanding problems of ichthyology. It is remarkable, for example, that Solea vulgaris, whose fertility exceeds the fertility of plaice more than two times, and Rhombus maximus, whose fertility exceeds the fertility of plaice by more than 10 times, constitute a minor part of catches, by comparison with the plaice.

Among all commercial fishes, the herring is the least fertile; and nevertheless the catch of herring alone surpasses the total catch of all other species."

Kyle

The total quantity of fish of a given species may be divided into two parts: (a) fish whose length is less than L, and (b) fish whose length is greater than L. We have described the influence exerted by the fishery on this second part, and mentioned in passing that ordinary conceptions about overfishing are not correct, and that stability in a fishery depends not on its intensity, but on how constant is the number of young recruits which every year pass from part a to part b. Thus it remains to be discovered in what manner the composition of part (a) is determined and in what manner it would be affected by this or the other organization of the fishery. In this connexion it is necessary to notice that the subject under consideration is very little studied; for the fishery, which gives us an important mass of information relative to part (b), cannot give information relative to part (a), /page 110/ and detailed study of such questions by means of special investigation is as yet not possible.

In a survey of the conditions regulating the numbers of young fish (section a) in the course of which it is unfortunately necessary to be satisfied with only general reasoning, we can distinguish two factors affecting this abundance: (1) the quantity of eggs deposited, and (2) mortality among the eggs and the fry. We can, on adequate grounds, postulate that the fishery does not have an important influence on the mortality of eggs and fry directly; therefore the influence of the fishery is limited solely to the decrease in the magnitude of the annual deposition of eggs which it undoubtedly causes (as a result of the removal of spawners).

Independently of the well known data concerning the extraordinary fertility of cod and plaice, the actual data of fishery investigations show that in the period of egg laying there is observed an extraordinary abundance of pelagic eggs and young, when the number of larvae attains to 500,000,000 per square kilometer /Helland-Hansen (3)/. This circumstance shows both that the abundance of young in the early stages of their development must be extremely great, and that it would be extremely important to investigate the laws governing the quantity of young which survive. Here again two possibilities are conceivable:

1) It is possible that under the influence of the interaction of many factors there survives to grow to a later stage some definite percentage of all eggs laid (independently of their number). In this event the abundance of young would be proportional to the number of eggs laid, and consequently the removal of the adult fish would be reflected in it.

2) It is possible that the total abundance of surviving young is dependent on some constant factor (for example, the total area of shallow banks suited for the existence of young, etc.), and therefore the absolute abundance of young could not exceed some maximum, however great might be the number of eggs deposited. In that case (if the number of eggs laid exceeded a certain minimum) the abundance of surviving young which annually pass from part a to part b would not be affected by the intensity of the fishery.

As far as the plaice fishery is concerned, at the present time there is evidence to favor the second hypothesis. A whole series of investigators, admitting the intensive removal of adult plaice, remark that the shallow banks are overpopulated with young plaice, suffering, apparently from a scarcity of food. An experiment of W. Garstang, done in the year 1904, will serve as a direct confirmation of the condition described. He transplanted young plaice, caught in proximity to shore and provided with a mark, to the Dogger Bank, and discovered that fish which near shore grow not more than 5-7 centimeters a year, on the Dogger Bank grew in 7 months an average of 12-13 centimeters. So too Peterson (14) reports that the young of plaice which near shore grow 4-5 centimeters a year at most, grow 8-11 centimeters a year in the Skagerrak, where the number of plaice is markedly less.

Hence the data of fishery biology are not contradicted by the findings of the theory set forth and, in its turn, the above-mentioned agreement /page 111/ of the theoretical conclusions with the observed course of evolution of the industry constitutes still another confirmation of the correctness of the assumptions followed. However, it is necessary once again to emphasize that the theory outlined above is not all-inclusive.

The conditions of the fishing industry, in general, are extremely diverse, and in many instances diverge widely from what is described above, as is evident merely from the well-known fact that some fisheries are subject to so-called "periodic" fluctuations.

A work of Hjort's (15) is devoted to the question of such fluctuations in fisheries; in particular, a study of the herring and cod fisheries of Norway (for the most part the former). The author reports in this work that the Norwegian fishery for cod and herring is based on the catch of comparatively large fish (fat-herring -- 3-5 years old; large-herring -- 6-10 years); and makes the extremely important observation that in the course of a series of years the great bulk of the catch, from 50% to 70%, consists of fish of one dominant age group, whereas adjacent age groups are represented in the catch very weakly; in recent years such a dominant group, both for herring and for cod, was hatched from the spawning of 1904. In that year, in the course of ichthyological investigations, an unusual abundance of young was observed. Later, in 1908, there appeared in the catches a mass of young fish (which caused a decrease in the average size of the fish in the catch), a direct determination of the age of which showed that they belonged to the spawning of 1904; and year after year, as the fish of this group grew, the mean size of the fish in the catch increased: this condition was traced up to the year 1914. From this the conclusion follows, that years with an abundance of fish appear as a consequence of conditions which have proved unusually favorable for spawning of the fish in one of the preceding years, and such an abundance continues (gradually diminishing) for a succession of several years until little by little this abnormally abundant group of fish is fished out and dies off. Then there intervenes a period of decline, when the fishery will be held at a low level, corresponding to the average biological production.

Turning to the causes which might result in an extraordinarily abundant production of fry, the author reports that it is impossible to associate this circumstance with especially profuse egg production; on the contrary, from the quantity of cod eggs secured at the time of the Lofoten fishery, the year 1904 appeared to be one of the very worst. Thus, apparently, a year with a small spawning can produce a great quantity of fry. And hence it is possible to conclude that the cause of the poor production of young in ordinary years lies not in a scarcity of eggs, but in a frightful destruction of fry. The cause is seen by the author in the fact that fry in large numbers require a terrific stock of food, and can die from lack of it, especially in the early spring months. In general, plankton organisms develop periodically, and if such a "pulse" of plankton corresponds with the period when the young begin to be hard up for food, then a greater than average quantity of young will survive and make possible a /large/ fish year. Another very important factor in the life of the young are ocean currents, which are capable of carrying away the fry (the pelagic stages of the development of which last some months) to places /page 112/ not suited to their existence. Changes in the course of currents can therefore be accompanied by fluctuations in the numbers of surviving fry.

As we see, the author presents a very well-balanced picture, which sheds light on a difficult and extremely important question: unfortunately this work leaves room for certain doubts. For example, there is given in it a series of curves, showing the age distribution of the fish in the catches of successive years, and these curves show surprisingly distinctly the predominant influence of the fish hatched in 1904. However the author himself shows (l.c., p. 14 and 18) that "very rarely in one and the same catch are there found herring of all sizes, therefore the fish of different sizes form separate schools, moving separately from each other"; also it is mentioned that for cod, too, "frequently the fish caught will vary greatly from day to day, a fact which the fishermen explain - and probably they are not far from the truth - by the existence of different schools"; and it remains unclear by what means he succeeded in compensating for the errors caused by these occurrences and obtaining a table of representative material.

Another source of error is the circumstance that, as follows from the statistical data furnished by the author, the catch of herring in Norway has continuously increased (with unimportant fluctuations) in the period from 1904 to 1913. However if the size of the catch depends on a group of fish of the 1904 year-class, then, obviously, as a consequence of the dying out and fishing out of this group, the size of the catches ought to have gradually and very noticeably declined.

Nonetheless, regardless of whether Hjort's theory corresponds to the true state of affairs in the case at hand[1], it is necessary to emphasize

---

[1] The observations of the next few years should decide this.

---

the great importance of the thesis put forward by him:

If the size of the catch over a series of years depends on the unusually large survival of fry in one of the preceding years, then, by means of systematic study, the composition of these large catches could be determined.

Comparing this situation with the results of our study of the plaice fishery, we come to the following conclusion:

For the elucidation of the condition of a fishery it is necessary and it is sufficient to make a study of the "biological statistics" (D. Damas' term), namely a study of the statistics of the catches and their qualitative composition (i.e. ideally, a knowledge of the age composition of the catch); it is necessary however to conduct the investigation in such a manner that the material obtained really gives a true picture of the whole catch. Such reliability (representativeness) of statistical material is of course attainable (see appendix II); and if it is not usually actually attained, that is only for this reason: that the energy of investigators is being dissipated on very diverse problems, instead of being applied to the one which leads directly to the goal. Very important also is the fact that it is not necessary to consume an indefinite number of years in such an investigation: if a fishery is in a state of equilibrium, then actually it is sufficient to study one year's catch (if however the industry is subject to "periodic fluctuations", then, naturally, it is necessary to follow these fluctuations over the course of a series of years). Such a systematic study /page 113/ of "biological statistics" gives a means - and the only one - for establishing control over the condition of a fishery, and, it would seem, such control should be one of the principal problems of existing ichthyological (fisheries) laboratories.

As concerns the second question touched on above: the question of the adequacy of egg deposition and survival of young, it is precisely here, it would seem, that there is a practical path leading to the solution of the problem. If, as we can assume, the total abundance of young is regulated by the conditions of their nourishment, then the study of the laws of their growth will give the answer to the question of whether there is an overpopulation of young in the body of water, or whether, on the contrary, there is insufficient utilization of its food production. In either event such a path will lead to a solution of the problem more quickly than will a study of the productivity of the body of water in itself. It is only necessary that the study of the growth of the young should be sufficiently comprehensive and systematic, and should really provide material for a judgment concerning the conditions of growth of the young in the given body of water.

Appendix I

The True Curve of Population and a Comparison with the Hypothetical One

The assumption that egg deposition takes place continuously, and that all fish of one age have a uniform length, naturally does not correspond to reality. However, as a consequence of the fact that different individual fish grow at different rates, the bounds of the different year-classes little by little spread out, the groups begin to overlap each other and they merge finally into a continuous curve of population. But the curve made up in this manner, though in superficial features it may be like the hypothetical curve constructed on the assumptions above-mentioned, is profoundly different from the latter in its internal structure. Therefore it is very interesting to examine the law of superposition of separate age groups and, having constructed the corresponding curve of population, to compare it with the theoretical one.

On page 42 of Heincke's (4) work, there is shown a figure from which it follows that the distribution of fish of different lengths, belonging to one age group, is very symmetrical; therefore it can, as a good approximation, be represented by a normal curve of error. Now an appropriate calculation shows that:

The mean length of plaice of the 1st age group (i.e., in their second year), on the basis of measurements of 135 individuals, is equal to 14 centimeters, while the mean square deviation in length is 18 millimeters.

The mean length of plaice of the second age group, on the basis of measurements of 695 individuals, equals 20 centimeters, with a mean square deviation of 22 millimeters.

The mean length of plaice of the 3rd group, on the basis of the measurement of 157 individuals, equals 25.5 centimeters, with a mean square deviation of 24 millimeters.

As no similar information about the growth of plaice of commercial size can be found in Heincke (and generally in the literature), we must be satisfied with the assumption that in their later growth also the distribution of fish of different lengths, belonging to one age group, approximately /page 114/ corresponds to a normal curve of error, of which the mean square deviation does not greatly differ among the different groups.

Thus constructing a common curve of error for each separate class gives a result defined by the well known formula:

$$n = \frac{N}{r\sqrt{2\pi}} e^{-1/2(x/r)^2} = N_1 e^{-1/2(x/r)^2}$$

where: n is the number of fish of a given group whose length differs by x centimeters from the mean length of the group,

N is the total number of fish in the group,

Managing Fish Stocks

r is the mean square deviation in length of the fish of the group from the mean, in centimeters.

Let us suppose that the general coefficient of yearly decrease of fish equals $1 - q$, that r is constant size for all groups, and that in the course of a year the fish grow $\underline{l}$ centimeters, on the average.

Then, taking some group of fish as a starting point, we have the result:

Number of the group	0	1	II	III	IV	....
Mean length of the fish	L	L + $\underline{l}$	L + 2$\underline{l}$	l + 3$\underline{l}$	l + 4$\underline{l}$	....
Number of fish	N_1	$q N_1$	$q^2 N_1$	$q^3 N_1$	$q^4 N_1$	....

Therefore the total number of fish, of some length $L + x$ will be equal to:

$$N_1 \underline{l}^{-1/2}\left(\tfrac{x}{r}\right)^2 + qN_1 \underline{l}^{-1/2}\left(\tfrac{x-\underline{l}}{r}\right)^2 + q^2 N_1 \underline{l}^{-1/2}\left(\tfrac{x-2\underline{l}}{r}\right)^2$$

$$+ q^3 N_1 \underline{l}^{-1/2}\left(\tfrac{x-3\underline{l}}{r}\right)^2 + \ldots\ldots + q^n N_1 \underline{l}^{-1/2}\left(\tfrac{x-n\underline{l}}{r}\right)^2 ;$$

and looking at the total number of fish whose length is $L + k\underline{l}$ (i.e. corresponding to the mean length of fish of the kth group), we get

$$N_k = \ldots\ldots q^{k-2} N_1 \underline{l}^{-1/2}\left(\tfrac{2\underline{l}}{r}\right)^2 + q^{k-1} N_1 \underline{l}^{-1/2}\left(\tfrac{\underline{l}}{r}\right)^2 + q^k N_1$$

$$+ q^{k+1} N_1 \underline{l}^{-1/2}\left(\tfrac{-\underline{l}}{2}\right)^2 + q^{k+2} \underline{l}^{-1/2}\left(\tfrac{-2\underline{l}}{r}\right)^2 + \ldots\ldots$$

9) .... or, $N_k = q^k N_1 \Big\{ \ldots\ldots \tfrac{1}{q^2} \underline{l}^{-1/2}\left(\tfrac{-2\underline{l}}{r}\right)^2 + \tfrac{1}{q} \underline{l}^{-1/2}\left(\tfrac{\underline{l}}{r}\right)^2 + 1$

$$+ q \underline{l}^{-1/2}\left(-\tfrac{\underline{l}}{r}\right)^2 + q^2 \underline{l}^{-1/2}\left(\tfrac{-2\underline{l}}{r}\right)^2 + \ldots \Big\}$$

where $q^k N_1 = q^k \dfrac{N}{r\sqrt{2\pi}}$ is proportional to the number of fish of the kth group, and the expression in brackets does not depend on the number of groups.

Thus if the annual increase in length of the fish is a fixed quantity, and the mean square error is also fixed, then the number of fish in the whole curve of population, whose length is equal to the mean length of the fish of a given age group, is proportional to the number of fish in that group; consequently such a curve of population will coincide with the curve of mortality and to the theory associated with it as set forth in the present work.

Taking into consideration that in the case of the plaice r (for the younger age groups) is less than $\underline{l}$, and therefore, in calculations using formula 9) it is sufficient to take 2 groups on each side of the /page 115/

group under consideration, we can see that if the magnitude of r and $\underline{l}$ is not fixed, but gradually changes from one group to the next, then the whole common course of the true curve of population will be close to the path of the theoretical curve (for a detailed study of such a situation, naturally, it is necessary to know the numerical values of r and $\underline{l}$ and to analyze the curve, breaking it into parts). If the rate of growth of the fish makes a sudden change[1]), then this will be in-

---

[1]) According to Heincke's data cited above on page 8, plaice from 4 to 7 years grow 5 centimeters a year, from 7 to 11 years - 2-1/2 centimeters, and from 11 to 22 years - 1 centimeter; but such jumps are probably the result of imperfections of the empirical data.

---

dicated as a break in the curve of population, delimiting two parts of the curve of distribution, corresponding to different intervals of growth.

In the accompanying figure 12 are shown 2 curves of population calculated theoretically. The curves shown by dots are the distribution of fish of different lengths by separate age groups (the number of the corresponding age group is given in Roman numerals) in which it is assumed that the abundance of a group decreases by half each year ($\varphi = 0.5$) and that the yearly interval of growth (equal to the distance between ordinates) is, after seven years of age, only half of what it is up to the seventh year.

The general curve of population, obtained as the resultant of summing the dotted curves, is shown by the line A B, and, on a logarithmic scale, /page 116/ by the line $A_1 B_1$. As we can see, in agreement with the earlier formula 3), the decrease in the yearly interval $\underline{l}$ is accompanied by an increase in the slope of the curve; the intermediate part, separating the right and left parts of the curve, has an extent corresponding in all to two age groups.

Curve C D is analogous to curve A B, but is constructed on the assumption that the magnitude of the yearly mortality rate is 0.2 ($\varphi = 0.2$). In it the transition zone is somewhat more noticeable (at first glance it looks like a hump; in order to make the meaning of this hump clearer, the right and left parts of the curve are produced as dotted lines).

Figure 12 shows that:

1) The superposition of curves of distribution corresponding to separate age groups gives as a result an extremely regular curve, in which, as is very clearly evident on the right and left parts of curve $A_1 B_1$ (plotted on a logarithmic scale), this curve completely coincides with the theoretical curve; however with small values of φ corresponding to the natural mortality of the plaice (curve C D), there is a certain waviness in the curve.

2) A sharp change in rate of growth (i.e. yearly length interval) results in a certain "accumulation" of fish in the transition zone, but this hump even in most extreme cases does not have any special significance, for to the right of the hump the slope of the curve is increased.

Managing Fish Stocks

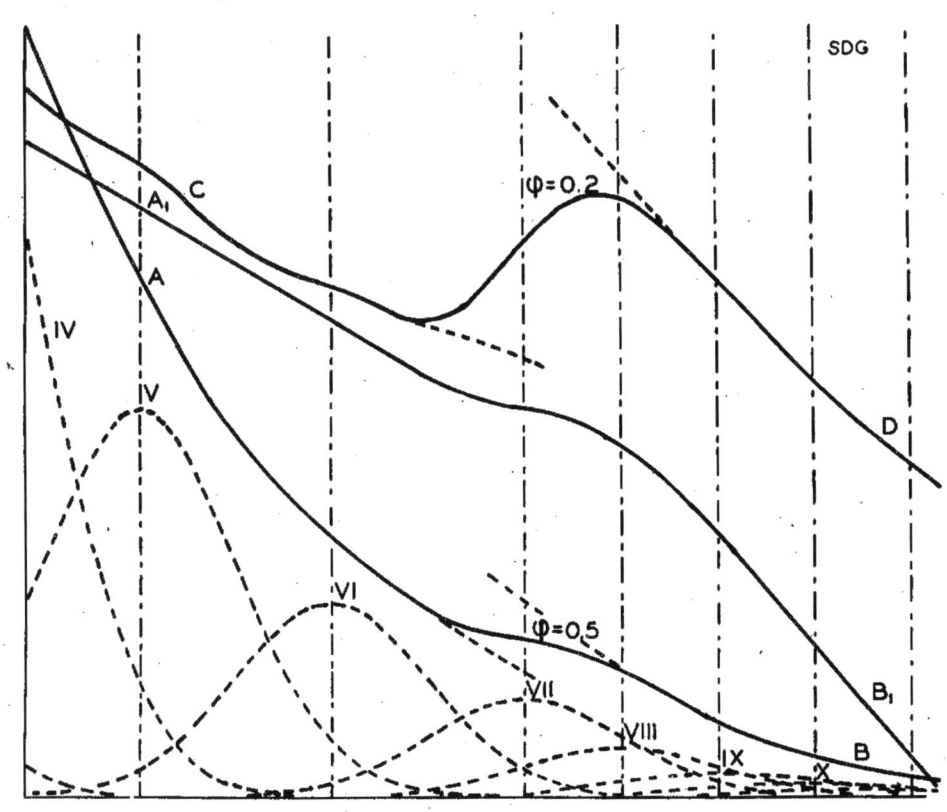

FIGURE 12

We may observe that the anomaly in the course of the curves of figure 4, which was emphasized in its place, might thus be explained by the existence of a sharp decrease in rate of growth of plaice during the 7th-8th years of their life, but of course the empirical material available is not sufficient for a decision on this question.

Appendix II

The Error of the Method of Random Sampling

The principal, inevitable, source of errors in fishery investigations consists in the fact that the only means for studying a catch of any considerable importance is the analysis of a comparatively small part of it taken at random.[1]) Having constructed, for example, the

---

[1]) If the catch is so small that it can all be subjected to analysis, it is in any event only a sample taken from the body of water.

---

curve of distribution of fish of different sizes in the sample analyzed, and having ascribed this same composition to the whole catch, we introduce unavoidable error into all deductions from there on, for the composition of the sample differs somewhat, and it may be considerably, from the composition of the catch. Therefore a theoretical investigation of the question of the size of the possible difference between the composition of a sample taken at random from a catch, and the composition of the whole catch, is a prime problem of the methodology of fishery investigations.

Suppose that we are examining a catch of N individual fish grouped by length, let us say at intervals of 0.5 centimeter, into k groups, in which the number of fish in any group, F, we will call the size of the group. We have:

Number of the group	I	II	III	IV	....	m	....	k
Size of the group	F_1	F_2	F_3	F_4	....	F_m	....	F_k

/page 117/ Suppose further, that we took from this catch at random N individuals and separated them into the same k groups at 0.5 centimeter intervals. We will designate the size of the sample groups by the letter f:

Number of the group	I	II	III	IV	....	m	....	k
Size of the group in the sample	f_1	f_2	f_3	f_4	....	f_m	....	f_k

Comparing the composition of the sample with the composition of the catch, we will say that the sample <u>represents</u> the catch, if the size of any given group in the sample f_m is to the size of the corresponding catch group F_m, as the number of fish in the sample n is to the number of fish in the catch N; i.e.

if $\dfrac{f_m}{F_m} = \dfrac{n}{N}$ is a constant quantity.

Thus in each separate instance it is possible to determine what number of fish must be in a given group of the sample, if this sample can be represented by the formula

$$f_m = \frac{n}{N} F,$$

and to define the difference between the size of a given group and the representative value as:

$$d_m = f_m - f_m'.$$

This difference can also serve as a measure of the discrepancy of a given group in the sample when compared with the representative value; we will call this the deviation of the group.

It is also not hard to see that with moderate to large magnitudes of N and n the number of different possible samples is very great, and therefore there is an extremely small probability that a sample taken at random will exactly correspond to the representative one; however it can be shown also that there is not any great probability that such a sample will deviate significantly from it.

To obtain the probability that a sample taken at random deviates from the representative one by not more than some definite limit, we can proceed from the conclusions of Pearson, published in his memoir "On Deviations from the Probable in a Correlated System of Variables" (16). Pearson compares the theoretical distribution of frequencies in some system of variables with the observed distribution and defines the probability P of the event that, under the action of accidental causes, there may arise a distribution of frequencies deviating from the theoretical by more than the observed. It is shown that the unknown probability P is related in a rather intricate manner to the expression u:

$$10)....u = \frac{(f_1 - f_1')^2}{f_1} + \frac{(f_2 - f_2')^2}{f_2} + \ldots + \frac{(f_n - f_m')^2}{f_m} \ldots + \frac{(f_k - f_k')^2}{f_k};$$

We will not, however, undertake to examine this relationship, for it is possible to avoid the calculations involved in obtaining the magnitude of P, by using the tables calculated by Elderton (18). Using Elderton's data, the following table was prepared which defines the size of P in relation to k and u:

Table I

Number of groups, k		5	10	15	20	30
Probability P	0.5	u = 3	u = 8	u = 13	u = 18	u = 28
	0.1	u = 7	u = 14	u = 21	u = 27	u = 39
	0.01	u = 13	u = 21	u = 29	u = 37	u = 50

From this table it follows that if, for example, our sample is divided into

15 groups[1] and calculations by formula 10) show that its deviation from

---
[1] /The end of a line of the text is missing; "5 groups" has been added to the "1" already there./

---

the representative sample corresponds to a magnitude of u = 21, then if a second sample were taken from this very same material /page 118/ we would in only one case out of 10 get a sample which deviated from the representative one by more than did the first. Conversely, if, as always happens in practice, we do not know the composition which the representative sample would have, but, by means of an analysis of the sample taken, we wish to infer its deviation from the composition of the catch, then Table I makes it possible to get an idea of the magnitude of such a deviation. That is, if, for example, the sample is divided into 20 groups, then it is possible to affirm that if we take from the observed catch 100 samples one after another, then 99 of them will deviate from the representative one less than would correspond to a value of u = 37, and therefore error corresponding to a value of u = 37 can for practical purposes be considered the limit of error of our sample.

Let us assume that the fish in the catch being investigated are distributed in size according to the normal law of error (Gauss's curve); then it is possible to determine for the representative sample the number f_0 of fish in the middle group, that is the largest in size; which number depends on the number of fish in the sample, n, and the number of groups, k, into which the sample[1] is divided. We get:

---
[1] /The text says "group", but "sample" is evidently intended./

Table II

Number of fish in sample, n	5000	1000	500	200	100	50
Size of f_0 for k = 10	1500	300	140	50	23	9
" " " " k = 30	480	85	40	14	7	3

From formula (10) it is evident that the degree of discrepancy of the sample, which depends on the total number of individuals in excess or in defect (compared to the representative sample) in all groups of the sample, does not completely determine the probability P; the latter depends also on the distribution of the individuals in excess and in defect among the groups, and the same value of P can be obtained by different arrangements of these individuals. Assuming, for the purpose of general orientation, that all the deviations are concentrated in the middle group of the sample, which is the largest in number, we obtain:

$$u = \frac{(f_0 - f_0')^2}{f_0}$$

or, 11)......$d_0 = f_0 - f_0' = \sqrt{uf_0}$

from which, for given values of n and k which make up the probability P, it is possible to calculate the number of individuals corresponding to P, namely $d_o = f_o - f_o'$. The results of such calculations are shown in Table III for the cases of 10 and 30 groups, and for probabilities P equal to 0.5, 0.1 and 0.01.

Table III

n	5000		1000		500		200		100		50	
k	10	30	10	30	10	30	10	30	10	30	10	30
f_o	1500	480	300	85	140	40	50	14	23	7	9	3
P = 0.5	110	116	49	49	33	33	20	20	14	14	8	9
P = 0.10	145	137	65	57	44	39	26	23	18	17	11	11
P = 0.01	177	155	79	65	54	45	32	26	22	19	14	12

The data of this table are fictitious, strictly speaking, for we are proceeding from the false assumption that all deviation is concentrated into /page 119/ one group, whereas it is clear that a decrease in the number of individuals in one group (in comparison with the representative sample) must cause a corresponding increase in others; they are set forth in order to illustrate a property of the magnitude d_o which is very important in a practical way, and which comes into all further calculations:

Since u varies almost proportionally to k (Table 1), and f_o varies inversely with k (Table II), the magnitude $d_o = \sqrt{u f_o}$, for a given n and P, for practical purposes (within limits of accuracy necessary in such questions) does not change with changes in the size of k; therefore Table III also provides a point of departure for estimating possible errors of samples divided into a different number of groups.[1]

---
[1] /That is, values of k other than 10 and 30.7

Turning now to the determination of this quantity P, which we must get as a basis for further developments, we will discuss first of all what constitutes the meaning of a determination of the probability of this or the other error in a sample. In practice the question of the error of a sample arises in most cases under these conditions: we have two samples taken from separate catches, and finding that these samples differ from each other in composition, we ascribe the same difference to the composition of the catches concerned. The question is, what is the assurance that the difference between the two samples really results from different catch compositions and is not simply a matter of chance? Particularly when such a conclusion is to serve as evidence from which consequences of a practical character will result later, it should certainly be required that the evidence be completely trustworthy. However the degree of reliability of a judgment is measured by the probability of its being in error, and theory shows that in such cases what is absolutely certain is only the fact that the error cannot be infinitely great, and the assurance measured by the size of P is connected with the size of the possible error (depending on the size of d_o), from which we are guaranteeing ourselves. Thus if we are not satisfied with the assurance which given values of P and d_o afford us, and wish to increase the size of P, then at the same time we widen the permissible bounds of error (depending on d_o); doing this we increase the diffuseness of our judgment, and in this way may lose a considerable part of its practical value. Therefore in such cases we are not obliged to be excessively exacting, and a magnitude of P = 0.01 (to which the last line of Table III pertains, showing that there are 99 chances out of 100 that the error does not exceed the limit shown in the table) can be considered perfectly sufficient; the more so, as in the improbable event that the error exceeds the limit indicated, the probability of its exceeding it by any considerable amount is completely insignificant. In order more clearly to appreciate the degree of assurance provided by a magnitude of P = 0.01 we may observe that, comparing some deviation with the possible error in order to evaluate its size, we ascribe significance to the deviation only in the event that it considerably exceeds the size of the possible error; if it does not, our judgment automatically becomes considerably less /page 120/ categorical, and consequently some vagueness in the size of the limit of the possible error does not make any difference to the character of a judgment. If, however, it is a case of comparing two samples, then we are obliged to consider the possible errors of them both, in doing which, following a well-known theorem of the theory of probability, the probability of the error of both samples simultaneously reaching their limits is equal to the product of their probabilities, i.e. = 0.01 x 0.01 = 0.0001, which is absolutely insignificant[1].

---

[1] Therefore in comparing samples there is some foundation for proceeding from a magnitude of P = 0.1, supposing that they are separated by $2d_o$ (where d_o is taken from the second last line of Table III).

And so the magnitude P = 0.01 gives a perfectly adequate guaranty. If, however, it were a question of being satisfied with a smaller magnitude of P and in that way narrowing the limit of error postulated, then as is apparent from Table III, a decrease of P to 0.1 and less does not give a large decrease in the size of d_o. Therefore, we consider it perfectly suitable to take the value P = 0.01 as the one from which we must proceed in our further discussion.

Returning to our earlier argument, suppose that all the deviations in a given sample are distributed nearly equally in s central groups, whose abundance is uniform and close to the abundance of the largest group, and can be taken as equal to f_o.

Then, by formula 10), the quantity u is equal to:

$$u = \frac{s(f_o - f'_o)^2}{f_o}$$

and, in order that this quantity shall equal that which was under consideration in calculating Table III, it is necessary to equate to this magnitude the value of u obtained by formula (11):

$$u = \frac{s(f_o - f'_o)^2}{f_o} = \frac{d_o^2}{f_o},$$

or

$$f_o - f'_o = \frac{d_o}{\sqrt{s}};$$

and the total deviation of the sample is equal to the sum of the deviations of the separate groups:

$$s(f_o - f'_o) = d_o \sqrt{s}$$

Thus if the deviation of a sample consists of the deviations of several of its groups, then the size of the deviation permitted in each separate group is less than is shown in Table III (but the total deviation of the sample is greater than what is given in Table III). Therefore the deviation in an individual group reaches its maximum size in the case where the number of these deviating groups is a minimum, and that number is two. Since the size of the sample n is a fixed quantity in each individual case, this last is possible only if in one group the number of individuals f'_n is greater than f_n by v, and in another group f'_m is v less than f_m. Then by formula 10) we have, as before:

$$\frac{v^2}{f_n} + \frac{v^2}{f_m} = \frac{d_o^2}{f_o},$$

or

$$v^2 \left(\frac{1}{f_n} + \frac{1}{f_m} \right) = \frac{d_o^2}{f_o}.$$

It is easy to see that v attains its maximum value when the expression $\frac{1}{f_n} + \frac{1}{f_m}$ reaches its minimum size, or, when f_n and f_m have the least possible value, $f_n = f_m = f_o$. If, however, some definite group of frequency f_n interests us, then the minimum possible value of v for it will be $f_m = f_o$. Then we get:

$$v^2 \left(\frac{1}{f_n} + \frac{1}{f_o} \right) = \frac{d_o^2}{f_o}$$

or, 12)............$v = d_o \sqrt{\dfrac{f_n}{f_o + f_n}} = \dfrac{d_o}{\sqrt{1 + \dfrac{f_o}{f_n}}}$

Substituting in this formula $f_n = f_o$, we get the minimum possible deviation of one group v:

$$v = \dfrac{d_o}{\sqrt{2}},$$

that corresponds to the total deviation of the sample

$$2v = d_o \sqrt{2}$$

If now the frequency of group f_n is less than f_o, then v will be less than this size and can be calculated by using formula *).

As for a maximum size for the total deviation of the sample, it is not hard to be assured that this will occur if:

*) .... $\dfrac{f_1 - f_1'}{f_1} = \dfrac{f_2 - f_2'}{f_2} = \dfrac{f_3 - f_3'}{f_3} = \dfrac{f_m - f_m'}{f_4} \ldots =$

$$\dfrac{f_n - f_n'}{f_n} = \ldots = \dfrac{f_k - f_k'}{f_k} = \psi.$$

In fact, an increase in the total deviation can occur only in such a way that in one group, in which f_m' is greater than f_m, the number of specimens f_m' increases still more and becomes equal to say $f_m' + 1$, and in some other group in which $f_n' < f_n$, the number f_n' becomes less by precisely the same amount (for the fixed number n in the sample necessitates this), i.e. $f_n' - 1$. Then the terms corresponding to these groups in formula 10):

$$u = \sum \dfrac{(f_i - f_i')^2}{f_i} = \dfrac{(f_1 - f_1')^2}{f_1} + \dfrac{(f_2 - f_2')^2}{f_2} + \ldots$$

$$+ \dfrac{(f_m - f_m')^2}{f_m} + \ldots + \dfrac{(f_n - f_n')^2}{f_n} + \ldots$$

which are equal to:

$$\dfrac{(f_m - f_m')^2}{f_m} + \dfrac{(f_n - f_n')^2}{f_n} = \dfrac{(\psi f_m)^2}{f_m} + \dfrac{(\psi f_n)^2}{f_n} = \psi^2 (f_m + f_n),$$

turn into:

$$\frac{(\psi f_m + 1)^2}{f_m} \frac{(\psi f_n + 1)^2}{f_n} = \psi^2(f_m + f_n) + 4\psi + \frac{1}{f_m} + \frac{1}{f_n} ;$$

and therefore the magnitude of u grows by $4\psi + \frac{1}{f_m} + \frac{1}{f_n}$.

/page 122/ Thus an increase in the deviation of a sample beyond what is defined by expression *) leads in itself to an increase in the quantity u, and therefore this expression[1] shows the maximum size of

---
[1] In doing such calculations, in practice, it is of course necessary to substitute for f_o and f_n the inexact values f_o' and f_n'.

---

the total deviation which can occur for a given value of u.

Bringing the values of the deviations $f_i - f_i'$ defined by means of expression *) into formula 10) we get:

$$u = \sum \frac{(f_i - f_i')^2}{f_i} = \psi^2(f_1 + f_2 + f_3 + \ldots f_k) = \psi^2 n;$$

and, comparing with the value of u found earlier,

$$u = \frac{d_o^2}{f_o} = \psi^2 n$$

we obtain finally:

13) . . . . $\psi = \frac{d_o}{\sqrt{n f_o}}$,

corresponding to the deviation of the most numerous group:

$$\psi f_o = d_o \sqrt{\frac{f_o}{n}}$$

and the total deviation of the sample:

$$\psi n = d_o \sqrt{\frac{n}{f_o}} ,$$

where the quantities under the square-root sign depend[1] only on the

---
[1] For the given law of distribution.

---

number of groups k into which the sample is divided, and are shown in Table IV (for the normal frequency distribution).

Table IV

Number of groups, k	5	10	20	30
Value of $\sqrt{\dfrac{f_o}{n}}$	0.8	0.5	0.4	0.3
Value of $\sqrt{\dfrac{n}{f_o}}$	1.3	1.8	2.5	3.3

The results obtained up to this point make it possible to decide in each separate case the question of the maximum error of a sample, and this independently of what law the distribution of the fish in the catch and in the sample conforms to:

In practice:

From the data of the last line of Table I we determine the size of u, which depends on the number of groups k into which our sample is divided.

Next we obtain the size of d_o from formula 11), substituting in place of f_o the number of individuals in the most numerous group of our sample.

Then we can find the limit of error of any group in the sample which is of interest to us, substituting the frequency of this group in place of f_n in formula 12):

$$v = \frac{d_o}{\sqrt{1 + \dfrac{f_o}{f_n}}}$$

which gives for the deviation of the most numerous group in the sample a limiting value:

$$v_m = \frac{d_o}{\sqrt{2}}$$

/page 123/ Formula 12) defines the magnitude u, using the assumption that all deviation is divided between 2 groups. If however the sample is divided into a considerable number of groups, such an hypothesis is obviously improbable, and we will obtain a more plausible value by supposing that the error is distributed among all the groups proportionately to their frequency. In this event the relative error ψ is defined by formula 13)

$$\psi = \frac{d_o}{\sqrt{n f_o}} ,$$

which provides us with the magnitude of the deviation of any group in the sample (using our earlier symbols)

$$v' = d_0 \sqrt{\frac{f_n^2}{n\, f_0}},$$

and the deviation of the most numerous group:

$$v'_m = d_0 \sqrt{\frac{f_0}{n}}$$

If it be postulated that the distribution of the frequencies in the different groups of the sample follows the normal curve of distribution (the exact form of the curve makes no difference, and can even be asymmetrical), then it is possible to express the above results in an even more convenient form, which permits the deduction of important conclusions of a general character.

In this case, knowing the size of the sample n, we at once obtain from Table III the magnitude of d_0 and, from the formulae just given, the magnitude of v, v_m, v' and v'_m, among which the values of v_m and v'_m and the total deviation of the sample ψ_n can be calculated beforehand from the size of the sample n and the number of groups k into which it is divided. The results of such calculations are shown in Table V, in which, for convenience of comparison, the size of the deviation is expressed as a percentage of the whole size of the sample n.

Table V

Number of fish in the sample		5000	1000	500	200	100	50
Maximum deviation u_m		3%	6%	7%	12%	16%	20%
k = 5	v'_m	3	6	7	12	16	20
	ψ_n	4	10	13	22	32	40
k = 10	v'_m	2	4	5	9	12	14
	ψ_n	6	14	18	31	45	54
k = 20	v'_m	1.5	3	4	7	10	12
	ψ_n	9	20	25	44	62	74
k = 30	v'_m	1	2	3	5	7	10
	ψ_n	10	26	33	55	77	(110)

Perusing this table we see that unfortunately the method of random samples, which cannot be avoided, gives a less-than-satisfactory result, especially if the sample consists of a small number of individuals. Inevitably there arises the question whether the data presented in the table are not too much /page 124/ exaggerated, and present a limit of possible error which in practice it is not necessary to take into account. Table III gives the answer to this question. From it, it follows that for an increase of P from 0.01 to 0.5, the size of d_0 decreases to 0.6 of its original value if the sample is divided into 10 groups, and to 0.7 if it is divided into 30 groups. However the probability P = 0.5 indicates complete indefiniteness of judgment. It shows only that if we are considering a number of samples, then probably in half of them the deviation is less than 0.6-0.7 of what is given by Table V, and in the other half it is greater than this. Consequently even if it is possible to postulate that in some particular example the deviation is a little less than that given by Table V, it would be an obvious mistake to make any kind of deductions without taking into account a possibility of error greater than 0.5-0.6 of what is given by the table. A consideration of some empirical data also leads to exactly the same conclusion.

The writer having some time ago become interested in the question of the error of the method of random samples, tried to determine it by an experimental method. He took several samples from a seine catch, one after the other, approximately 100 fish in each, and, comparing them among themselves, and also with the distribution of the catch by species as determined from the fishery, tried to find the limit of possible error. However, the inconsistency of the results (which leads to the hypothesis of irregularity of distribution of the fish in the haul, and so on) compelled him to give up this work. As an example of the material with which we had to deal, we may present the three following samples, which were taken at one time:

Length of the fish (roach)	17 cm	18	19	20	21	22	23	24	25	26	27
Sample I	1	5	8	3	13	17	30	12	7	4	-
Sample II	5	17	13	5	13	12	15	9	9	1	1
Sample III	4	7	15	3	11	15	25	13	5	2	-

Comparing the deviations of the separate groups in different samples, and determining the sum of these deviations, we get: I deviates from II by 52; II from III by 40; and I from III by 26. Comparing these data with Table V, which gives in our case (k = 10, n = 100) a total deviation of ψ_n = 45, we see that these data conform to the forecast of the table (see also the footnote on page 41).

As a further confirmation and a useful illustration of the method we can cite a paragraph of Einar Lea's (19). In comparing random samples and considering the distribution in them of 4 age groups, he finds a sample consisting of 200 individuals to be completely satisfactory, at least if the problem is to determine which of the age groups is the most numerous. We will follow the path indicated on page 45.

/page 125/ If a sample of 200 individuals is divided into 4 groups and one of the groups is conspicuous by reason of its abundance, then (since the mean number of individuals belonging to one group equals $\frac{200}{4}$ = 50) probably there will be about 100 individuals in it, and not more than 50 in any of the others. Let us consider f_o = 100, f_n = 50.

Table I does not supply data for $k = 4$; however it is obvious from it that u must be less than 13 and much nearer to 10. Let us take $u = 10$. Assuming further that all individuals in excess are concentrated in the group f_o, and all those in defect are in group f_n, we get the magnitude of d_o:

$$d_o = \sqrt{u f_o} = \sqrt{1000} = 32,$$

and the magnitude of v:

$$v = \frac{d_o}{\sqrt{1 + \frac{f_o}{f_n}}} = \frac{32}{\sqrt{1 + \frac{100}{50}}} = 19$$

Therefore we can expect that in the very worst event the sample taken will turn out to have, in place of $f_o = 100$ and $f_n = 50$, $f'_o = 100 - 19 = 81$ and $f'_n = 50 + 19 = 69$; and the first group is still the most abundant.

If however we were to take, under these circumstances, a sample of only 100 individuals in all ($f_o = 50$, $f_n = 25$), then, repeating the same calculation, we would find $v = 13$, i.e. we might get $f'_o = 37$ and $f'_n = 38$, instead of $f_o = 50$ and $f_n = 25$, and be lead to an erroneous result; therefore the second sample is not large enough.

In the theory set forth the size of the whole catch from which the sample is taken makes no difference at all, and the limit of deviation of a sample from the representative one, and so on, does not depend on how large a fraction of the catch the given sample is - whether it is an appreciable or an insignificant one.

In a situation where we are concerned with a determination of the average size of a given characteristic, the fact has long since been established that the result achieved by means of an analysis of a sample group does not depend on its relation to the abundance of the whole general aggregation; and this fact is widely used in statistics (for example in investigations of games of chance). However this is a view that always seems paradoxical[1] and contrary to sound judgment; therefore it is useful

---

[1] The more so, as authors sometimes really offend our good sense, for example by affirming that "the accuracy of the result when 500 events are selected will be the same, whether these 500 are selected from 5200 or from 520 events" (22).

---

to dwell a little on this question in connection with the question of the application of the method to fishery investigations.

In essence the method of random samples is a corollary of the theory of repeated trials. That is, we have to do with some aggregate consisting of N elements, distributed in k groups, and the number of elements in the first group is F_1, in the second, F_2, and so on. Assume that the elements are distributed uniformly in the aggregate, so that, taking one element at random from the aggregate, we have a probability $\frac{F_1}{N}$ that this element will belong to the first group, and so on. Let us take from this aggregate /page 126/ a number of elements, assuming that all the while the probabilities $\frac{F_1}{N}$, $\frac{F_2}{N}$ ... remain fixed. The question is, what will be the composition of a sample taken in this manner. In this statement of the problem the composition of the sample is determined only by the values of the probabilities $\frac{F_1}{N}$, $\frac{F_2}{N}$ ..., and the size of N has no significance; and therefore the relation of the size of the sample to the size of the catch has no significance either.

Consequently, in practice, in order to have the right to make use of conclusions drawn from the above theory, it is necessary that the probabilities $\frac{F_1}{N}$, $\frac{F_2}{N}$ ... should remain constant during the whole time of the taking of the sample, to accomplish which it is necessary above all that fish of different sizes be uniformly distributed in the catch. However, this last condition is necessary in any event, independently of the theory put forward, if we wish to obtain a result which is deserving of credence[1].

---

[1] In fact, let us assume that we have made a catch of fish and from its result we draw conclusions concerning the composition of the population of the given body of water; such an assumption will be permissible if the physico-geographical conditions in all parts of the given body of water are uniform, and the fish are distributed uniformly through it. If however in the different parts of the body of water there are found non-uniform conditions, then it is necessary to divide the body of water into parts which are alike within themselves and investigate each of them separately (in the course of which it will be necessary to work with uniform material), and this is the only correct procedure; otherwise neither an increase in the number of fish studied, nor any other kind of procedure whatever, will make it possible to place confidence in the correctness of the deductions made, if there is reason to fear an unpredictable lack of uniformity in the distribution of the fish. The same sort of considerations are pertinent also when taking samples from a catch: it is necessary to arrange the work in such a way that the sample is taken from homogeneous material; to accomplish which it is necessary either to mix up the catch (for example the fish in a haul) or to separate the parts having different origins and to take samples separately from them. However, the example cited above shows that an apparent absence of uniformity in the composition of a catch can result simply from an insufficiently large sample.

However, let us assume that we are dealing with a catch of uniform origin. Then, as was shown, the probability p of the event that, in taking the sample, a fish of some length L is included in it, is equal to the ratio of the number of individuals in the corresponding group in the catch to the size of the whole catch:

$$p = \frac{F}{N}$$

If the sample meets the requirement which we set up for a representative one, i.e. that in each of its groups the following proportionality should hold:

$$\frac{f}{n} = \frac{F}{N} = p,$$

then after taking the sample the remaining part of the catch will have its former relative composition:

$$\frac{F - f}{N - n} = \frac{F}{N} = p.$$

Consequently, the probability p during the whole time the sample is taken will remain fixed, and in that event the ratio of the size of the sample to the size of the catch will not have any importance.

The situation is different, if the sample taken differs in composition from the representative one; then the part of the catch which remains will have a composition different /page 127/ from what it originally had, in which event it is not hard to see that a change in the probability p during the time of taking the sample will tend to compensate for the deviation of the sample from the representative one. That is, assume that in taking the first half of the sample there did not occur in it even one individual belonging to the group characterized by the probability p. Then in the catch there remain $N - 0.5n$ individuals and of these F belong to the group under consideration. Therefore, in taking the second half of the sample the probability p is changed to the value:

$$p_1 = \frac{F}{N - 0.5n},$$

or putting $N = tn$, and remembering that $\frac{f}{n} = p$,

$$p_1 = p \frac{t}{t - 0.5} > p,$$

which, for a ratio of catch to sample $t = 10$, gives a value $p_1 = 1.05\, p$; for $t = 5$, $p_1 = 1.11 p$ and for $t = 2$, $p_1 = 1.34\, p$.

Hence, if in taking the first half of a sample, fish of a given size are completely missed, we may expect in the second half of the sample $0.5\, n\, p_1$ such fish.

However, in reality, as is obvious from Table V, the deviation of a sample, and consequently of the first half of it too, cannot exceed a certain limit. Taking the greatest possible deviation (for the whole sample) as 0.1 f, which means the limiting deviation for the first half of the sample is 0.5 f, we arrive at the formula:

$$p_1 = \frac{t - 0.45}{t - 0.50} p,$$

which gives for $t = 5$ a value $p_1 = 1.01\ p$, and for $t = 2$ a value $p_1 = 1.03\ p$.

Thus the influence of the ratio of the size of the sample to the size of the catch expresses itself only in the event that the sample constitutes not less than a fifth part of the catch; but on the other hand if the sample is equal to half of the whole catch, this influence has no practical significance.

The conclusions from this section are as follows:

(1) Random samples give only an approximate picture of the composition of a catch; therefore conclusions based on such samples should not be overvalued.

(2) Samples consisting of less than 200 fish are not satisfactory.

(3) It is desirable that a sample should consist of not less than 1000 fish. With such a sample, if it is divided into 10 groups, the possible error of each group can be estimated as about 15% of the size of the group itself ($\psi = 0.14$), or the possible error of the group having the largest number of fish is about 4% - 5% of the whole sample.

(4) Division of a sample into an excessive number of small parts is undesirable; for samples of 500 - 1000 fish the material should not be divided into more than 10 - 15 groups.

LITERATURE CITED

1. D. Grave. Matematika strakhovovo dela. 1912.

2. D'Arcy Wentworth Thompson. Second Report on the Distribution of the Cod and other Round Fishes. (Conseil permanent international pour l'exploration de la mer. Rapports et procès-verbaux. Vol. XIII).

3. B. Helland-Hansen. Statistical Research into the Biology of the Haddock and Cod in the North Sea. (Rapports et procès-verbaux. Vo. X).

4. Fr. Heincke. The Plaice Fishery and Protective Regulations. Rapp. etc. Vol. XVII A).

5. T. Edser. Note on the Number of Plaice at each Length. (Journ. of the Royal Statistical Society. Vol. LXXI. 1908).

6. Johansen. Ueber die Schollenfischerei im Kattegat und die Mittel, sie zu heben. (Rapp. etc. Vol. V).

7. Fr. Heincke. Investigations on the plaice. Preliminary brief summary of the most important points of the report. (Rapp. etc. Vol. XVI).

8. A. T. Masterman. Report on the later stages of the Pleuronectidae. (Rapp. etc. Vol. XII).

9. Geo. B. Bidder. Rapport sommaire sur les experiments avec "bottom trailers". (Rapp. etc. Vol. IV).

   " Résultats principaux des expériments avec flotteurs de fond. (Rapp. etc. Vol. VI).

10. Rapports etc. Vol. VII, Année C.

11. Johansen. Vierter Bericht uber die Pleuronectiden in der Ostsee. (Rapp. etc. Vol. XVI).

12. H. M. Kyle. Summary of the available Fisheries statistics and their value for the solution of the Problems of Overfishing. (Rapp. etc. Vol. III).

13. N. Knipovich. Tralovyĭ promysel v Barentsovom more. (Vestnik Rybopromyshlennosti, 1914).

14. Rapp. etc. Vol. II. Proces-verbaux des réunions de commissions spéciales. Commission B.

15. I. Hjort. Fluctuations in the great fisheries of Northern Europe. (Rapp. etc. Vol. XX).

16. K. Pearson. On the Criterion that a Given System of Deviations from the Probable in the Case of a Correlated System of Variables is such that it can be reasonably supposed to have arisen from Random Sampling. (Philosophical Magazine, Vol. 50, 1900).

    The contents of this memoir are given in § 31 and § 32 of Slutskiĭ's work (17).

17. E. Slutskiĭ. Theoriĭa korreliatsii i elementy ucheniĭa o krivykh raspredeleniĭa. Kiev, 1912.

18. A. Leontovich. Elementarnoe posobie k primeniĭu metodov Gauss'a i Pearson'a pri otsenke oshibok v statistikei biologii. Kiev. 1909. Ch. I tabl. VII.

19. Einar Lea. A Review of the question as to requisite number and size of representative samples. (Rapp. etc. Vol. XXI).

20. Dr. T. Wemyss Fulton. North Sea Investigations. (Report of the Fishery Board for Scotland for the Year 1901. Part III).

21. Rosa M. Lee. Report on the Grimsby Steam Trawlers Records. (Mar. Biol. Assoc. Internat. Investigations. III Report (Southern Area) 1911).

22. A. A. Kaufman Teoriĭa i metody statistiki. 1912.

An Epitaph for the Concept of Maximum Sustained Yield[1]

P. A. LARKIN

*Institute of Animal Resource Ecology, University of British Columbia
Vancouver, British Columbia V6T 1W5*

About 30 years ago, when I was a graduate student, the idea of managing fisheries for maximum sustained yield was just beginning to really catch on. Of course, the ideas had already been around for quite a while. Baranov (1918) was the first to combine information on growth and abundance to develop a catch equation, and Russell (1931) and Graham (1935) brought the dynamic pool model to the forefront, but they were working from a base of natural history and fishery biology that had been growing for several decades.

By the late 1930s, in North America, the conservation movement was in full cry and fisheries, like other resources, were being illuminated in the glow of the Gospel of Efficiency (Hays 1969). In dozens of states and provinces, fish and game regulations were proliferated, commercial fisheries were increasingly documented, and there was a growing awareness of the necessary scientific base for management. Thompson and Bell (1934) came to the conclusion that too much fishing effort was at the heart of the halibut problem; Hile (1936) produced his classic on the cisco in Wisconsin; and the first steps were being taken to restore the Fraser River sockeye from the effects of overfishing and the Hell's Gate blockage.

The ten years following World War II were the golden age for the concept of maximum sustained yield. Ricker (1948) produced his famous "green book," the first version of his handbook (Ricker 1958); Fry (1947) developed the virtual population idea; and Schaefer (1954) proposed his method for estimating surplus production under nonequilibrium conditions. The literature crackled with new information and new ideas. The solidification of the concept of MSY, its application to fisheries here, there, and everywhere, was just under way. World fisheries catch was a mere 20 million tons, and there were signs in lots of places of irreligious practices such as harvesting more or less than should be harvested. In a mood of excitement about opportunities, coupled with determination to do it properly, the FAO emerged as a major actor in the international fisheries scene.

It was in consequence of this flowering of activity that the graduate students of those days had a missionary zeal about them, and as more than one wit has said, "They had a fine vocabulary of stained glass language." Briefly, the dogma was this: any species each year produces a harvestable surplus, and if you take that much, and no more, you can go on getting it forever and ever (Amen). You only need to have as much effort as is necessary to catch this magic amount, so to use more is wasteful of effort; to use less is wasteful of food. Basically, it was a puritanical philosophy in which the supreme powers were pretty harsh on people who enjoyed themselves rather than doing precisely the Right Thing. Armed with scientific knowledge about the number of fishermen and technological advances, the

---

[1] Keynote address to the American Fisheries Society Annual Meetings, Dearborn, Michigan, September 19–24, 1976.

manager could use regulations to prevent the catch from exceeding the maximum, even if it meant telling fishermen they could only use bare hooks from sailboats on alternate Tuesdays between 6 and 7 p.m. The various laws of supply and demand, marginal revenue, alternative options, and psychological dissatisfaction, were mostly misty mumblings of the social sciences. It was generally assumed that the fishermen would look after themselves. Moreover, it was assumed that the animals were well aware of what was being organized for them as their role in the scheme of things. Organisms were allowed to breed with those of their own species, or interact with individuals of other species, but not in ways that might upset the maximum sustained yield.

As I am sure you realize, I am considerably dramatizing the way it was; but, when speaking in retrospect, one is usually to be allowed that privilege. Certainly, it is to be understood that the people who generated these ideas were appropriately modest and were well aware of the dangers of oversimplification. Their protégés were perhaps no less critical, but in selling the idea to administrators it was essential to make the main argument forcefully. And this they did, with clear conscience, for they all knew that the main idea was correct and it was only necessary to do a bit more research, to get a bit more experience, and then the basic theme could be appropriately fine tuned to perfection.

Like all religious movements, the doctrine of MSY had effects on other doctrines, and the most notable was the impact on traditional limnology. For almost 100 years, working from a European base, limnologists had been developing holistic schemes of trophic status in which fish were part of a complex community for which the rate of harvest was best expressed in pounds per acre or kilos per hectare. I vividly recall being proselytized by Bill Kennedy, a disciple of the new doctrine of population dynamics, about the futility of the old-fashioned limnological approaches of my Master's degree supervisor, Don Rawson, just as I am sure that most others of my year class can remember similar arguments about the limnology versus fisheries approaches. The Langlois-Van Oosten debate about Lake Erie was typical (Van Oosten 1948; Langlois 1954). The believers in MSY had little patience for the systematics of zooplankton or the subtleties of lake classification. The fish, they argued, were the integrators of their environment and the object of our crass interest. "Study the Fish" was the motto.

In addition to their disrespect for traditionalists, the proponents of MSY were highly intolerant of heretical views. Most of you may never have heard of Harden Taylor, who reviewed the fisheries of Maryland (1951) and concluded that the inexorable laws of economics could curtail the rates of harvesting long before any species of fish was faced with extinction. His message was that the fish could recover from whatever we were likely to do to them and, with dollars being the real yield, what was so special about MSY? I vividly recall the frigid silence with which he was greeted whenever he got up to speak.

The emphasis on population dynamics gained increasingly in strength, and throughout the forties and fifties both the theory and practice of maximum sustained yield became widespread. The basic idea was enshrined in national policy documents, incorporated in international treaties, and, in effect, became synonymous in most people's minds with sound management. Most fishery managers and politicians engaged in a steady dialogue of explaining why they had to compromise a bit on MSY for "social reasons" but, in so doing, they usually sounded apologetic. They *knew* they were sinning.

Statisticians, of course, had a heyday, because the estimation of population parameters inevitably involved sampling, and woe betide the budding young fishery manager who could not master the mysteries of regression and analysis of variance! (Just as it should be, I might say, for there's nothing more dangerous than a man who doesn't appreciate the limitations of his data, unless it's a mathematician who hasn't any data.)

In short, the mid-fifties were a fine time to be a fisheries biologist because you could be so single-minded about your job. The object was to get out there and get the harvest of the maximum sustained yield, and there was a healthy bag of theoretical and statistical tools

LARKIN—AN EPITAPH FOR MAXIMUM SUSTAINED YIELD

to draw on. Or at least that's the way it seemed to an impressionable young guy like me.

The crowning achievement of the whole movement was the magnificent work of Beverton and Holt (1957). Their book did four important things: (1) it brought everything together (which in itself was important); (2) it produced a theory of fishing which illustrated that for each specified rate of fishing there is an age of entry corresponding to *a* maximum sustained yield, and that there are therefore as many maxima as there are rates of fishing, all provided, of course, that recruitment is constant; (3) it provided a stock-recruitment relation if recruitment wasn't constant, a relation which could be coupled with the simple theory to give a self-regenerating model of an exploited population; and (4) it anticipated a large number of refinements to the model system, by speculating on such things as spatial variation in the values of parameters, movements of fish within the exploited area, and the relationships among food consumption, the availability of food, and the density and growth of the fish population. Much of this was far ahead of its time when it was published and, indeed, some of it is today still ahead of the time. It is no wonder that Benny Schaefer remarked to Ray Beverton long ago that he "liked his book of flute music."

Since that time much has been done in preparing variations on the basic themes. For example, Cushing (1973) put one of the finishing touches on the whole picture by his distinction between "growth overfishing" (catching them younger than is consistent with MSY at a given level of effort), and "recruitment overfishing" (catching more than will be replaced).

Today, many more people have assimilated the MSY paradigms, or at least the elementary ones, and using such primers as Gulland's (1969) handbook, are daily grinding through the rituals of estimating F, k, and l_∞ and wishing they could get M in some other way than by subtraction. As a matter of fact, many of them are using computer programs for all their calculations, so they are saved the numbing hours of arithmetic that paralyzed the older generation.

Unfortunately, most of them don't see the buried phrase in Gulland's manual: "... it is very doubtful if the attainment of the maximum sustained yield from any one stock of fish should be the objective of management except in exceptional circumstances."

In many ways, it is a pity that now, just when the concept of maximum sustained yield has reached a worldwide distribution and is on the verge of worldwide application, it must be abandoned. But that's the way it goes with the things we believe.

THE BIOLOGICAL COMPLICATIONS

No one can deny that hypothetical animal populations can produce hypothetical maximum sustained yields, but the same cannot be said of real animal populations that are really being harvested. For most species the critical age for harvesting is close to first age of maturity, reflecting the common biological characteristic of animals: that, as maturity approaches, growth in weight is rapid and natural mortality is low. It is thus inevitable that for most kinds of fishing gear, as fishing intensity increases to levels close to the MSY that can be sustained by recruitment, spawning populations will be predominantly made up of fish that are young and first time spawners. In consequence of this and perhaps other qualitative changes in the spawning population, the quality of eggs deposited may be reduced. This has been documented for a number of species of fishes (Nikolsky 1965; Bagenal 1973), and is probably a widespread effect of harvesting. Moreover, with the reduction in the number of spawning age classes, a failure in egg or larval survival for any reason is potentially far more catastrophic in its effect on long-term abundance. Clupeid fisheries are prime examples. Thus, MSY involves greater elements of potential instability than are characteristic of unexploited stocks.

The obvious ways out of these problems are: (1) to obtain information on pre-recruit abundance that can be used as an early warning signal that effort should be reduced; and/or (2) NOT to go for MSY, but for something less that involves a lesser element of risk and that is an optimum in a narrow biological sense (Doubleday 1976); and/or (3) considering much more sophisticated techniques for optimization and adaptive control in fisheries management (Walters 1975; Walters and Hilborn 1976).

The only appropriate response for the manager who is committed to MSY is to devise a system for quick curtailment of effort when there is a recruitment failure. If this system works, prayers for recovery are likely to be more successful. Without quick reduction of effort, stock recovery is likely to be influenced by the mysterious phenomena of depensatory mortality, which are probably related to the effects of predation (perhaps including fishing) at low prey densities (Neave 1954; Larkin 1973; Holling 1973; and an important paper by Clark 1974). Once depressed to certain levels, populations either become extinct, or persist at a low level where they await some happy coincidence of favorable effects before exploding to a higher equilibrium abundance. Catastrophe theory is, of course, interesting (e.g., Jones and Walters 1976), but it is cold comfort for a manager who doesn't know how long he will have to wait until he can again have his MSY.

Another general concern is the likelihood that in the range over which a population or stock of a species occurs, there will be genetic variability with local subpopulations or substocks adapted to the local environment they occupy. We need to know about each of these subpopulations if we are going to harvest each of them in an appropriate way. This is especially so in the circumstances that, except in general terms, the people who harvest the fish do not initially consult the regulators in detail about what kind of gear is going to be used, or where and when they are going to use it.

Pacific salmon are a prime example being, in the aggregate, a group of subpopulations with different capacities for supporting harvests. With the fishery only imperfectly regulated, and with the added problem that some stocks are fished jointly, it is small wonder that we now have an odd assortment of salmon substocks, a lower annual production of salmon, and concern for the future of what we have left, which is now less than one-half of what we had a century ago. (There is a substantial literature. A good starting point is the group of papers edited by Simon and Larkin 1972.)

Loftus (1976) has recently presented a large body of evidence to suggest that this phenomenon of removal of less productive components of natural populations is probably much more widespread than has generally been realized. In Pacific salmon, of course, since the spawning areas are discrete and conspicuous, it is rather obvious when a substock disappears. When the substocks are lake trout or whitefish, the losses may not be as apparent, but may nevertheless be just as real. In fact, the phenomenon is probably a very general one and the recent paper by Wellington (1976) discusses it in relation to insects! Moreover, as Ricker (1973) pointed out, in a period when a fishery is getting started, because the stock is larger than it would be at MSY, a given level of effort in an expanding fishery catches more than a similar level of effort after stabilization. Combined with the elimination of more vulnerable substocks, the illusion of a larger than actual MSY is exaggerated. If there is such a thing as an MSY, then, it must be the yield that the residue of a population can continue to support when its less productive components have been reduced below their individual MSYs. Putting it another way, it may be necessary to compromise MSY in order to preserve genetic variability.

For the purpose of the present discussion, it is to be stressed that in virtually all fisheries that have been prosecuted in the world today, fisheries scientists have not controlled to a high degree of refinement, the technique, amount, and distribution of fishing effort. It is therefore inevitable, in my view, that fishing has eliminated some substocks, and this applies to herring, or cod, or ocean perch, as much as to salmon or lake trout. Indeed, I would argue that it is best to assume that it is true of all species until it is demonstrated to be otherwise.

To recapitulate, for even a single species population it does not seem likely that an MSY based on the analysis of the historic statistics of a fishery is really attainable on a sustained basis. If there is an MSY, it is a yield associated with a high risk of recruitment failure in a population in which the less productive substocks have been depressed or eliminated.

LARKIN—AN EPITAPH FOR MAXIMUM SUSTAINED YIELD

It is also to be underlined that this same process applies to mixtures of species that are caught in the same gear. Many of the world's fisheries are based on catching more than one species in the same gear at the same time. For these fisheries, *species* of lower productivity are progressively eliminated or pushed close to extinction as the fishery harvests the more productive species to the level of their supposed MSY. The ultimate effect of using gear that harvests many species must be to reduce a community to whatever can persist when the most productive species is/are harvested to MSY rates. The saga of the Great Lakes is a sufficient reminder (Regier et al. 1969; Smith 1968). For mixed fisheries, then, if there is such a thing as MSY, it must be that harvest that can be sustained when the less productive *species* have been eliminated or reduced below their MSYs.

It is a relatively easy exercise in algebra to combine a bunch of yield equations to sort out what mesh size will give the maximum sustained *aggregate* yield for any given level of effort, but to imagine techniques of fishing that would get the MSY for each species is mind-boggling. It would be necessary to regulate, from the outset of a fishery, where, when, and how much of what kind of gear was to be used, and using that gear in some way that harvested each substock of species in proportion to its capacity to sustain a yield. To accomplish this objective would almost certainly require research and management expenditures that were greater than the value of the resources to be harvested.

Moreover, it would still assume that species were ecologically separate, feeding neither on the same foods, nor on each other, which is, of course, not so. Since Volterra (1931) looked at the theory of relations between competing species and between predators and their prey, an abundant literature has documented that a lot of things are theoretically possible, but that field observations to confirm what actually happens are few and far between. The recent paper by Lett and Koehler (1976) is probably a landmark in that it provides enough evidence to convince the most skeptical that mackerel and herring are party to a highly complex interrelationship. If this is typical, and there is no reason to suppose it isn't, then it is truly impossible to imagine the scientific effort that would be required to manage a community of fishes species by species, each for an MSY in the context of its associations with other species.

As an aside, but an important aside, it is also useful to remind ourselves of the realities of contemporary statistics on fisheries. While it is true that the statistics of the world's fisheries are better now than they have ever been, it is also true that they are still incomplete and riddled with guesses, inadvertent errors, omissions, and even, perhaps, some perjuries. They are generally, as a statistician would say, more precise than accurate, and that's saying something when you bear in mind that the imprecision of fisheries statistics is notorious. Management from this sort of factual basis requires a certain flair.

In short, just considering fish population dynamics, there is precious little prospect of achieving MSY either for one species or for any number of species in the aggregate. A large recent literature on modelling abundantly demonstrates that a wide variety of unexpected consequences can flow from what seem to be simple management strategies. With the benefit of simulation techniques we can see just how difficult it is even to manage systems that are simplified versions of nature. In another 20 years, the understanding of community dynamics may have proceeded to the point that we could be rather cute at manipulating species compositions while preserving the stability and qualitative integrity of aquatic communities. But we are a long distance from that goal now, and to the extent we can see it, it seems improbable that the perfect strategy would be to take MSY from each species.

Meanwhile, the limnologists who were shunted aside by management biologists 20 to 30 years ago, have been plugging away at their studies of whole aquatic ecosystems, and by a rather direct route have converged on much the same conclusion. Since Hrbacek et al. (1961) and Brooks and Dodson (1965) demonstrated that fish influence the species composition of the zooplankton community, a substantial series of papers has confirmed that the species assemblage at each trophic level is profoundly influenced by the predation imposed from the level above. To any modern day limnologist, the impact of fishing on fish communities is, in general terms, much as would be expected from consideration of the effect of adding fish to a community of zooplankters, or even, perhaps, from adding zooplankters to a community of phytoplankters. On this basis, there is little doubt that in many parts of the world the species assemblages of fishes that we observe today must be profoundly different in their composition and interrelationships from the assemblages of a century ago, and so are the communities of organisms on which the fish feed. From this perspective, to speak of an MSY for any one of the fish species in effect argues that somehow or other the interrelations among the species won't have any effect. While this could be true in the short-term, it is difficult to imagine in the long-term. From the viewpoint of fish communities, the S in MSY, for any species, can't possibly mean more than 50 to 100 years. It certainly isn't forever and ever.

There is one redeeming feature. Leaving aside such considerations as pollution and climatic change, it seems likely that the total production of aquatic systems is more or less constant, albeit distributed among different species. To the degree that we could be completely flexible about what we eat, it is conceivable that for a particular community of species, we could speak of the maximum sustained yield of organisms above a specified size; it being understood that, as the size went down, the MSY would go up. Biologically, that's about the only concept of maximum sustained yield that can stand up in the light of contemporary evidence. It might even suggest that perhaps the preferable technique of harvesting is to take the same proportion of everything above a certain size.

In summary, from a biological point of view the concept of MSY is simply not sufficient. Nevertheless, it should be stressed that it provides a valuable rough index of production potential. As a first rough cut at management policy for major commercial species, MSY is probably acceptable. But once the level of MSY is attained, it should be expected that it may not be sustained.

THE ECONOMIC IMPLICATIONS

Once Michael Graham (1935) had pointed out that the same equilibrium catch could be taken at two different levels of effort, the way was open for economic analysis of commercial fisheries. Scott Gordon (1954) made the first thorough study, and by 1965, Christy and Scott had produced "The Common Wealth in Ocean Fisheries." It was apparent that what happened in fisheries made less than economic sense. In the first place, the economists told us, the real yield from fisheries is not fish, but dollars. While I wouldn't want to try to eat a can of dollars in tomato sauce, it's easy to see what they were getting at—if you owned all the rights to fish in the sea, and you wanted to make money, you wouldn't necessarily want to take the MSY. You'd take the amount of fish that would make you the most profit.

More technically, depending on the relation between yield and effort, and depending on how much the price goes up as the supply goes down, there is a level of harvesting associated with maximum sustained economic revenue. Since the same catch can be taken at two levels of effort, it is obviously more economically rewarding to fish at the lower level of effort. From the perspective of the individual fisherman, this is a fine prospect because he makes a good living.

In the real world, what happens is that more fishermen are attracted into the business, the individual fishermen try harder, and in a very short time the paradise of maximum economic revenue is lost. Left to its own devices, one might suppose that this system would come to its economic senses, ultimately reverting to some equilibrium that would probably *not* be the biological MSY (Clark 1971), but which at least looked healthy to an economist. Unfortunately, fishermen vote; and once a person has become a fisherman, he can almost be counted on to vote against anyone who doesn't help him continue to be a fisherman and ensure him a decent standard of living. From such simple human responses there may flow a long mane of hairy subsidies which directly or indirectly sustain an economic monstrosity. There are almost as many examples as there are fisheries in the world.

Thus, to an economist, the concept of biolog-

LARKIN—AN EPITAPH FOR MAXIMUM SUSTAINED YIELD

ical maximum sustained yield has an entirely different meaning—it isn't a holy duty, but an indicator of biological pressure, and only one of many factors influencing the smooth running of economic systems. An economist may be more than somewhat irritated when there is an insistence on achieving MSY. He has his own holy duties to perform.

The best way of reconciling the MSY and economic religions has been held to be the limitation of entry into a commercial fishery; if there is a continued regulation of the *number* of fishing units and their fishing power, then at least MSY can be taken inexpensively. But, inasmuch as commercial fisheries have not been so regulated and are not characterized by regulation of entry, it is true today that commercial fisheries generally are close to or have gone beyond their biological MSYs to lower levels, and that commercial fisheries generally are not a source of great joy to economists.

Bearing in mind also that some fisheries are international, and that different countries have different economies, a stage is set in which differing economic monstrosities combine to generate the biggest monstrosities of them all—the world's international fisheries.[2]

As an aside, it is to be noted that, when it comes to international negotiations, MSY is something of a mixed blessing. On the one hand it may be presented as an appeal for rational long-term resource use, conceivably the sort of case the fish might make if they were present at the bargaining table. On the other hand, if you plead for MSY for protection, you may be stuck with achieving it as an obligation. Inasmuch as your own national economic disorders are not necessarily going to be cured by taking MSY, you can end up being nicely hoisted in your own petard.

Turning to recreational fisheries, it has long been evident that MSY is not the best economic strategy. With the object being to maximize recreation, it has proven difficult for economists to pinpoint just what the economic values are, but it is nevertheless clear that MSY has rather little relevance. At one extreme anglers may take the MSY ten times over from a suburban small pond which is routinely stocked from a hatchery for the benefit of old-age pensioners and their grandchildren. At another extreme anglers may be required to use only the less efficient lures and gear, and to wait their turn to enjoy the recreational benefits, even though their combined efforts won't take a fraction of the MSY. Wrapped in questions of aesthetics, ethics, distribution of catch, and the various mystiques of angling, it is little wonder that MSY has rather little meaning in recreational fisheries.

And when commercial and recreational fisheries collide, there is potentially no limit to the confusion surrounding economic discussions. In the last analysis, the comparison is between two equations, one of which, the recreational one, contains a variable called X which assumes values of zero to infinity, depending on who you ask. In this debate, MSY is a useful anchor for the commercial interests, but a dead weight to the sports fishermen.

To summarize, for economists MSY is interesting, perhaps, but irrelevant except as a potential constraint.

OPTIMUM YIELD

It was with these kinds of undercurrents that about 10 years ago many people began to have misgivings about MSY, and about maximum economic return, and started to speak of maximizing other things. Just as fish serve economic ends, economics serve social ends, and therefore the objective should be to get a maximum sustained yield of social benefits. In consequence, in recent years economists have been busy trying to put dollar signs on all sorts of social activities and, in some instances, may have even deluded themselves into thinking they have succeeded. But, as you and

---

[2] The International Commission for Northwest Atlantic Fisheries has recently made some decisions that will prove to be of major historic interest. Total allowable catches have been reduced below levels of MSY. The development of the double quota system in ICNAF has been spoken of as a "blunt instrument that has the effect of bringing about a reduction in fishing effort" (remarks by Donald L. McKernan *in* Mundt 1975), which suggests that in some international circumstances you can control entry by controlling effort by using quotas. While this may be true in the ICNAF area, it is a doubtful proposition *within* a country such as Canada, and it remains to be seen whether Canada will limit entry in her east coast fisheries and, if so, try to do it by a system of quotas. It doesn't seem likely it would work.

I know, humans are sufficiently perverse that the only way to judge whether *they* perceive that the social benefits exceed the social costs is to listen to what they say and see how they vote.

From all this sugary murk there crystallized, like fudge, the concept of optimum yield, in which optimum is whatever you wish to call it. In his superlative summary of this Society's Symposium on Optimum Sustainable Yield, Philip Roedel (1975) defined *optimum yield* as

> a deliberate melding of biological, economic, social, and political values designed to produce the maximum benefit to society from a given stock of fish;

and *optimum sustained yield* as a subset of *optimum yield* defined as

> a deliberate melding of biological, economic, social, and political values designed to produce the maximum benefit to society from stocks that are sought for human use, taking into account the effect of harvesting on dependent or associated species.

I do not know what these definitions mean. First, optimum seems to come about from "deliberate" melding, rather than from inadvertent melding. It is somewhat akin to the idea of being a virgin by intent. To say the least, the concept is potentially subject to abuse, and would almost certainly be used primarily as a way of justifying a political course of action. Indeed, it brings clearly to mind the very practices of a generation ago which were the target of the missionaries for MSY.

Second, the two definitions together imply that what you do to a single stock is called "optimum," whereas what you do to a community of species is called "optimum sustained," the idea apparently being that for a single species you may wish to take more than the level you could sustain. Unfortunately, though, the definition of *optimum sustained yield* doesn't say anything about sustaining anything.

Inasmuch as these definitions are virtually meaningless, it is fortunate that Roedel spelled out how they would "likely work out in the real world," so that we can see how really meaningless they are. Without going into each of his ten points in detail, suffice it to say that sometimes optimum yield will be almost zero; other times it will be MSY except when it is more; still other times it will be maximum net economic yield; and for some species it will be all they can stand without becoming extinct.

Rather evidently, as a summarizer and editor of the Symposium, Roedel was struggling in one of the first concerted efforts to find an alternative to MSY, and the result, predictably perhaps, was an eclectic mishmash that was all things to all people. Nevertheless, his summation provided some bases for some concepts for the future.

First, the optimum yield concept recognizes the fact that, because species are interrelated and jointly fished, it is difficult, if not impossible, to contrive for MSY for each. For trawl fisheries, especially in the tropical seas, this is the only realistic attitude (Marr 1976).

Second, it has at last been recognized that there is no obligation to harvest a species just because it is there. After all, if you think about it, there is a good crop of robins to be harvested, and a potential yield from cats and dogs, if protein is the only consideration. The point is made dramatically by considering sport fisheries in which the object is to maximize recreation, and in which the elitist would argue that the maximum yield of benefits comes from the least efficient gear used with the greatest skill to produce the smallest catch at the greatest personal satisfaction. Taking underwater photographs of fish could be even better, for the less consumptive the use of the resource, the more who can enjoy it.

Third, in recognizing the need for joint consideration of biological, economic, social, and political factors, Roedel's definition used the word "deliberate." To me, "deliberate" means that someone will not only *deliberate*, but in so doing will *document* the reasons for the decisions made. If there is one sure criticism to be made of what we have done in the past, it is that we have compromised on MSY and have not objectively documented why we did so. It is crucial for future development of the concept of optimum yield that there be a rigorous attempt to record why particular decisions were taken.

In my view, the major stumbling block in all concepts of optimum sustained yield as discussed at the Law of the Sea Conference

LARKIN—AN EPITAPH FOR MAXIMUM SUSTAINED YIELD

and elsewhere, is that they have yet to provide an operational basis for making decisions. The chances that your optimum is my optimum are nearly zero. This difficulty flows from the fact that natural systems are sufficiently diverse and complex that there is no single, simple recipe for harvesting that can be applied universally. When there is added in the complexity and variety of social, economic, and political systems, the number of potential recipes is just too enormous to be easily summarized by simple dogma.

Perhaps the best we can hope for is a general statement of principles with accompanying guidelines that should be applied in the hope of ensuring that we will trend in the best direction. This seems to be the intent of the draft United States "Fishery Conservation and Management Act of 1976," as outlined in the Report of the Committee of Conference on H.R. 200. Inasmuch as this document advocates "optimum yield," the definition of optimum yield is crucial, and it is the amount of fish:

(A) which will provide the greatest overall benefit to the Nation, with particular reference to food production and recreational opportunities; and
(B) which is prescribed as such on the basis of the maximum sustainable yield from such fishery, as modified by any economic, social, or ecological factor.

In short, it's a recipe for achieving heaven or hell, and what is achieved will depend on how the definition is variously interpreted.

OPTIONS FOR THE FUTURE

The foregoing has demonstrated, I hope, that MSY is not attainable for single species and must be compromised: (1) to reduce the risk of catastrophic decline and reduction of genetic variability; and (2) to accommodate the interactions among the species of organisms that comprise aquatic communities. Moreover, MSY is not necessarily desirable from an economic point of view, and is certainly not so in the circumstances of unlimited entry. We are therefore struggling with rubber-edged concepts such as optimum yield and wondering about ways of managing in the future.

Basically, there are two extreme paths that might be followed, and each presumes an underlying political philosophy. If one starts from a purely technocentric model for human society, then it is quite clear what to do. You measure the various biological risks and set rates of harvesting by species, area, season, type of gear, and so on, bearing in mind what it costs to get the information you need and the risk you take of having incomplete information. You then set the number of fishermen and their fishing power, and place the rest of the fishermen in other activities that are seen as gainful for the state. This approach is technically complicated, but socially simple, and would probably appeal to people who like order.

The alternative extreme path is to intervene as little as possible, only provided that the fish should be protected from total extermination by advanced technologies. This path is also clearly marked. You set permissible catches at moderately safe levels of biological risk and then let the economic and social problems resolve themselves within the biological restraints. Specifically, you do NOT subsidize fishermen or the construction of their vessels; you do NOT provide any incentive for people to stay in the fishing business, NOR do you discourage them from staying in if that's what they want to do. In short, you put your trust in what economists would call natural market forces, and you hope that politicians will live up to their reputations for not keeping their promises. This approach is technically relatively simple and socially chaotic, and appeals to people who prize individual initiative.

In between these two extremes there is a wide spectrum of alternatives that are variously labelled as "middle-of-the-road" philosophies. They are characterized by various mixes of orderliness and initiative, by national policies that are sufficiently vague and/or complicated as to allow quite contradictory actions in different places at the same time, or at different times in the same place, and that in essence preserve future options by maximizing flexibility and confusion. The current Canadian approach is typical (Anonymous 1976), for it says (in only 302 words) that the goals are to maximize food production, preserve ecological balance, allocate access optimally, provide for economic viability and growth, optimize distribution and minimize instability in returns, ensure prior recognition of economic and social impact of technological change, minimize dependence on paternalistic industry and government, and protect national security and sovereignity—it being kept in mind that there is no priority implied in the order things are listed; that there are interactions in the objectives; and that trade-offs and compromise will be necessary. These goals are striking in implying that there is no single optimum policy, for as we all know, one cannot optimize for two things at the same time, let alone a dozen. They are humorous because they so accurately reflect the real difficulties of managing human affairs.

After having had so much fun in commenting on what others have done, I regret that I don't have an inspired personal vision for the future. My personal preference is for a technocentric approach, with the fish first, the economics second, and the social problems a distant third—something we must resolve, and quickly, with sympathy and good sense. I believe our first obligation is to our grandchildren, that we should be quite stern about abusing resources, and almost equally stern about being inefficient economically, if only to save on energy resources. I have this bias because I belong to a particular year class—for which I can't take credit or blame. Representatives of the more recent year classes, particularly Carl Walters and Henry Regier, have contributed much to my reeducation and ensuing middle-aged ambivalence.

FAREWELL TO MSY

Whatever lies ahead in the development of new concepts for harvesting the resources of the world's fresh waters and oceans, it is certain that the concept of maximum sustained yield will alone not be sufficient. The concept has served an important service. It arrived just in time to curb many fisheries problems. To appreciate what MSY has done, we need only ask what the world's fisheries would have looked like today if the concept had not been developed and advocated with such fervor. The fish, I'm sure, would shudder to think of it. Like the hero of a western movie, MSY rode in off the range, caught the villains at their work, and established order of a sort. But it's now time for MSY to ride off into the sunset. The world today is too complex for the rough justice of a guy on a horse with a six-shooter. We urgently need the same kind of morality, but we also need much more sophistication.

Accordingly, I tender the following epitaph:

M. S. Y.
1930s–1970s

Here lies the concept, MSY.
It advocated yields too high,
And didn't spell out how to slice the pie.
We bury it with the best of wishes,
Especially on behalf of fishes.
We don't know yet what will take its place,
But hope it's as good for the human race.

R. I. P.

LITERATURE CITED

ANONYMOUS. 1976. Policy for Canada's Commercial Fisheries. Fisheries and Marine Service, Ottawa. May 1976.

BAGENAL, T. B. 1973. Fish fecundity and its relations with stock and recruitment. Rapp. Cons. Expl. Mer 164.

BARANOV, F. I. 1918. [On the question of the biological basis of fisheries.] Nauch. Issled. Ikhtiol. Inst. Izu. 1(1): 81–128.

BEVERTON, R. J. H., AND S. J. HOLT. 1957. On the dynamics of exploited fish populations. U.K. Min. Agr. and Fish. Fish. Invest. (Ser. 2) 19. 533 pp.

BROOKS, J. L., AND S. I. DODSON. 1965. Predation, body size and composition of plankton. Science 150: 28–35.

CLARK, C. W. 1974. Possible effects of schooling on the dynamics of exploited fish populations. J. Cons. Cons. Int. Explor. Mer 36(1): 7–14.

———. 1971. Economically optimal policies for the utilization of biologically renewable resources. Math. Biosci. 12: 245–260.

CHRISTY, FRANCIS T., JR., AND ANTHONY SCOTT. 1965. The Common Wealth in Ocean Fisheries. John Hopkins, Baltimore, Md. 281 pp.

CUSHING, D. H. 1973. Dependence of recruitment on parent stock. J. Fish. Res. Board Can. 30(12), Part 2: 1965–1976.

DOUBLEDAY, W. G. 1976. Environmental fluctuations and fisheries management. Int. Comm. Northwest Atl. Fish. Sel. Pap. 1. 1: 141–150.

FRY, F. E. J. 1947. Statistics of a lake trout fishery. Biometrics 5: 27–67.

GORDON, H. SCOTT. 1954. The economic theory of a common property resource: the fishery. Jour. Political Economy 62: 124–142.

GRAHAM, M. 1935. Modern theory of exploiting a fishery, and application to North Sea trawling. J. Cons. Cons. Int. Expl. Mer 10: 264–274.

GULLAND, J. A. 1969. Manual of methods for fish stock assessment. Part 1. Fish Population Analysis. F.A.O. Manuals in Fisheries Science 4. 154 pp.

HAYS, SAMUEL P. 1969. Conservation and the Gospel of Efficiency. Atheneum, College Edition. 297 pp. (Originally published by Harvard University Press, 1959.)

HILE, RALPH. 1936. Age and growth of the cisco, *Leucichthys artedi* (Le Sueur), in the lakes of the northeastern highlands, Wisconsin. U.S. Bur. Fish. Bull. 48: 211–317.

HOLLING, C. S. 1973. Resilience and stability in ecological systems. Annu. Rev. Ecol. Syst. 4: 1–23.

HRBACEK, J., M. DVORAKOVA, V. KORINEK, AND L. E. PROCHAZKOVA. 1961. Demonstration of the effect of the fish stock on the species composition of zooplankton and the intensity of metabolism of the whole plankton association. Verh. Int. Verein. Limnol. 14: 192–195.

JONES, DIXON D., AND CARL J. WALTERS. 1976. Catastrophe theory and fisheries regulation. Working Paper 9, Institute of Resource Ecology, Univ. B.C., Vancouver, Canada. 11 pp.

LANGLOIS, THOMAS H. 1954. The western end of Lake Erie and its ecology. J. W. Edwards, Ann Arbor. 479 pp.

LARKIN, P. A. 1973. Some observations on models of stock and recruitment relationships for fishes. Rapp. Cons. Expl. Mer 164: 316–324.

LETT, P. F., AND A. C. KOHLER. 1976. Recruitment: a problem of multi-species interaction and environmental perturbations, with special reference to Gulf of St. Lawrence Atlantic herring (*Clupea harengus harengus*). J. Fish. Res. Board Can. 33(6): 1353–1371.

LOFTUS, K. H. 1976. Science for Canada's fisheries rehabilitation needs. J. Fish. Res. Board Can. 33(8): 1822–1857.

MARR, JOHN C. 1976. Fishery and resource management in southeast Asia. Resources for the Future PISFA Paper 7. 62 pp.

MUNDT, J. CARL. 1975. Limited entry into the commercial fisheries. Univ. Wash., Inst. Mar. Stud. Pub. Ser. IMS-UW-75-1.

NEAVE, F. 1954. Principles affecting the size of pink and chum salmon populations in British Columbia. J. Fish. Res. Board Can. 9(9): 450–491.

NIKOLSKY, G. U. 1965. [Theory of fish population dynamics as the biological background for national exploitation and management of fishery resources.] Nauka Press, Moscow. 382 pp. English Edition, Oliver & Boyd, Edinburgh, 1969. 323 pp.

REGIER, HENRY A., VERNON C. APPLEGATE, AND RICHARD H. RYDER. 1969. The ecology and management of the walleye in western Lake Erie. Great Lakes Fish. Comm. Tech. Rept. 15. 101 pp.

RICKER, W. E. 1948. Methods of estimating vital statistics of fish populations. Indiana Univ. Publ. Sci. Ser. 15. 101 pp.

———. 1958. Handbook of computations for biological statistics of fish populations. Bull. 119, Fish. Res. Board Can. Ottawa. 300 pp.

———. 1973. Two mechanisms that make it impossible to maintain peak period yields from stocks of Pacific salmon and other fishes. J. Fish. Res. Board Can. 30(9): 1275–1286.

ROEDEL, PHILIP M. 1975. A summary and critique of the Symposium on Optimum Yield. Pages 79–89 *in* Optimum sustainable yield as a concept in fisheries management. Am. Fish. Soc. Spec. Publ. 9.

RUSSELL, F. S. 1931. Some theoretical considerations on the "overfishing" problem. J. Cons. Cons. Int. Expl. Mer 6: 3–27.

SCHAEFER, M. B. 1954. Some aspects of the dynamics of populations important to the management of the commercial marine fisheries. Bull. Inter-Amer. Trop. Tuna Comm. 1(2): 27–56.

SIMON, RAYMOND C., AND P. A. LARKIN (eds.). 1972. The Stock Concept in Pacific Salmon. H. R. MacMillan Lectures in Fisheries, Univ. B.C., Vancouver, Canada. 231 pp.

SMITH, STANFORD H. 1968. Species succession and fishery exploitation in the Great Lakes. J. Fish. Res. Board Can. 25(4): 667–693.

TAYLOR, H. F. 1951. Survey of marine fisheries of North Carolina. Univ. North Carolina Press, Chapel Hill.

THOMPSON, W. F., AND F. H. BELL. 1934. Biological statistics of the Pacific halibut fishery. (2). Effect of changes in intensity upon total yield and yield per unit of gear. Rept. Int. Fish. Comm. 8. 49 pp.

VAN OOSTEN, JOHN. 1948. Turbidity as a factor in the decline of Great Lakes fishes with special reference to Lake Erie. Trans. Am. Fish. Soc. 75: 281–322.

VOLTERRA, VITO. 1931. Variations and fluctuations of the number of individuals in animal species living together. Pages 409–448 *in* Animal Ecology (Royal N. Chapman), McGraw-Hill.

WALTERS, C. J. 1975. Optimal harvest strategies for salmon in relation to environmental variability and uncertain production parametres. J. Fish. Res. Board Can. 32: 1777–1784.

———, AND R. HILBORN. 1976. Adaptive control of fishing systems. J. Fish. Res. Board Can. 33: 145–159.

WELLINGTON, W. G. 1976. Returning the insect to insect ecology: some consequences for pest management. *In* Symposium on Population Quality, XV International Congress of Entomology, Washington, D.C. August 20, 1976.

Maximum reproductive rate of fish at low population sizes

Ransom A. Myers, Keith G. Bowen, and Nicholas J. Barrowman

Abstract: We examine a database of over 700 spawner–recruitment series to search for parameters that are constant, or nearly so, at the level of a species or above. We find that the number of spawners produced per spawner each year at low populations, i.e., the maximum annual reproductive rate, is relatively constant within species and that there is relatively little variation among species. This quantity can be interpreted as a standardized slope at the origin of a spawner–recruitment function. We employ variance components models that assume that the log of the standardized slope at the origin is a normal random variable. This approach allows improved estimates of spawner–recruitment parameters, estimation of empirical prior distributions for Bayesian analysis, estimation of the biological limits of fishing, calculation of the maximum sustainable yield, and impact assessment of dams and pollution.

Résumé : Nous étudions une base de données comptant plus de 700 séries géniteur-recrutement à la recherche de paramètres qui sont constants, ou presque, au niveau de l'espèce ou à un niveau hiérarchique supérieur. Nous avons trouvé que le nombre de géniteurs produits par géniteur chaque année dans des populations peu abondantes, c'est-à-dire le taux de reproduction annuel maximal, est relativement constant dans une espèce, et qu'il y a peu de variation d'une espèce à l'autre. Cette valeur peut être interprétée comme une pente normalisée à l'origine d'une fonction géniteur-recrutement. Nous avons recours à des modèles de composantes de la variance selon lesquels le log de la pente normalisée à l'origine est une variable aléatoire normale. Cette approche permet d'obtenir de meilleures estimations des paramètres du rapport géniteur-recrutement, une estimation des données empiriques avant les répartitions pour l'analyse bayésienne, une estimation des limites biologiques de la pêche, un calcul de la production maximale équilibrée, et une évaluation des impacts des barrages et de la pollution.

[Traduit par la Rédaction]

Introduction

Perhaps the most fundamental parameter in population biology is the reproductive rate at low population size. We will analyze this parameter in terms of the maximum reproductive rate, which we define as the average rate at which replacement spawners are produced per spawner at low abundance in the absence of anthropogenic mortality (after a time delay for the age at maturity). The maximum reproductive rate is central to the following: the population growth rate r (Cole 1954; Pimm 1991; Myers et al. 1997b), limits to overfishing (Mace 1994; Cook et al. 1997; Myers and Mertz 1998), estimation of the dynamic behaviour of the population, i.e., whether the population has oscillatory or chaotic behaviour, extinction models and population viability analysis (Lande et al. 1997), establishment of biological reference points for management, e.g., most of the commonly used reference points for recruitment overfishing require estimates of the maximum lifetime reproductive rate (Myers et al. 1994), and estimation of the long-term consequence of mortality caused by pollution, dams, or entrainment by powerplants (Barnthouse et al. 1988).

The purposes of this paper are (*i*) to provide a comprehensive analysis of the maximum reproductive rate in terms of a relatively simple statistical model, (*ii*) to attempt to determine under what conditions this parameter is invariant, i.e., constant for a species or group of species, and (*iii*) to provide empirical Bayesian priors for the estimates (Hilborn and Liermann 1998; Millar and Meyer 2000). We use the extensive database of stock and recruitment data compiled in Myers et al. (1995) and Myers and Barrowman (1996).

Formulation

Estimating reproductive rate

Semelparous species, whose members conveniently die after reproduction, immensely simplify the lives of students of their population biology. For example, in many insects and in pink salmon (*Oncorhynchus gorbuscha*), one generation follows the next in easy units, e.g., the number of spawning females. The relationship between the numbers in year t, N_t, and the numbers in year t plus the age at maturity, a_{mat}, is typically given in the form

(1) $N_{t+a_{mat}} = \alpha N_t e^{-f(N_t)}$

where the density-dependent mortality, $f(N_t)$, is a non-negative function such that $f(N_t) \to 0$ as $N_t \to 0$.

Received June 1, 1999. Accepted September 24, 1999.
J15156

R.A. Myers.[1] Killam Memorial Chair in Ocean Studies, Department of Biology, Dalhousie University, Halifax, NS B3H 4J1, Canada.
K.G. Bowen and N.J. Barrowman. Department of Mathematics and Statistics, Dalhousie University, Halifax, NS B3H 4J1, Canada.

[1] Author to whom all correspondence should be addressed. email: ransom.myers@dal.ca

Myers et al.

The dynamics of iteroparous species are more complicated. Typically, the number of recruits belonging to year-class t, R_t, is a function of the egg production or a proxy such as weight of spawners at time t, S_t, as in the form

(2) $\quad R_t = \alpha S_t e^{-f(S_t)}$

where $f(S_t)$ is the density-dependent mortality as before.

The Ricker model has the form

(3) $\quad R_t = \alpha S_t e^{-\beta S_t}$

where α is the slope at the origin (measured perhaps in recruits per kilogram of spawners). Density-dependent mortality is assumed to be the product of β and the spawner biomass. Dividing by S_t and taking logarithms gives

(4) $\quad \log \dfrac{R_t}{S_t} = \log \alpha - \beta S_t$

i.e., a linear model for log survival.

For the forthcoming calculations, the slope at the origin, α, must be standardized. First consider

$$\hat{\alpha} = \alpha \cdot SPR_{F=0}$$

where $SPR_{F=0}$ is the spawning biomass resulting from each recruit (perhaps in units of kilograms of spawners per recruit) in the limit of no fishing mortality ($F = 0$). This quantity, $\hat{\alpha}$, represents the number of spawners produced by each spawner over its lifetime at very low spawner abundance, i.e., assuming absolutely no density dependence. The quantity $\tilde{\alpha}$ required for our calculations is the number of spawners produced by each spawner per year (after a lag of a years, where a is the age at maturity). If adult survival (the proportion surviving each year, which in the absence of fishing is e^{-M}) is p_s, then $\hat{\alpha} = \sum_{i=0}^{\infty} p_s^i \tilde{\alpha}$, or summing the geometric series:

(5) $\quad \tilde{\alpha} = \hat{\alpha}(1 - p_s) = \alpha \cdot SPR_{F=0}(1 - p_s)$.

This quantity $\tilde{\alpha}$ is the maximum annual reproductive rate and will be the main focus of this study.

A word of warning is needed in the interpretation of the maximum annual reproductive rate. The above formulation is for the deterministic case. However, if stochastic variations in survival are included, then the quantity $\tilde{\alpha}$ would be interpreted as the maximum of the average annual reproductive rate. In other words, the reproductive rate may be higher or lower for any given year.

We also provide estimates of "steepness", denoted by z and first defined by Mace and Doonan (1988), because this is the parameter actually used in many assessments (Hilborn and Walters 1992). The steepness parameter z for the Beverton–Holt model is defined to be the proportion of recruitment, relative to the recruitment at the equilibrium with no fishing, when the spawner abundance or biomass is reduced to 20% of the virgin level. This is related to the maximum lifetime reproductive rate $\hat{\alpha}$ by

$$z = \dfrac{\hat{\alpha}}{4 + \hat{\alpha}}$$

where $0.2 < z < 1$.

Note that at the limit of small population size, the Ricker and Beverton–Holt models coincide, i.e., the slope at the origin, α, is the same. In this context, z can be estimated from either model; however, it can only be applied directly to the dynamics of the Beverton–Holt model. Our estimate of steepness should be viewed as conservative (see Appendix 2).

To summarize, we have introduced notations for the three most common ways that the maximum reproductive rate is used in the analysis of fish population dynamics. First, we have defined $\hat{\alpha}$ to be the maximum lifetime reproductive rate (the adjective "lifetime" would not usually be used but is important here). This quantity is used in many calculations to determine maximum sustainable yield (MSY) and the limits of fishing mortality. The second is the quantity of steepness, z, which is a simple transformation of $\hat{\alpha}$. The third quantity, $\tilde{\alpha}$, is used in calculations where an annual recruitment rate is needed, e.g., for estimating the maximum population growth rate (Myers et al. 1997b).

The Ricker model provides a reasonable model for estimating the slope at the origin

The simplest form of density-dependent mortality is linear, i.e., $f(S) = \beta S$, in eq. 1. We will show that under reasonable conditions, this is perhaps the best first approximation. A simple generalization of the Ricker model is

(6) $\quad f(S) = \beta S^\gamma$

where γ controls the degree of nonlinearity in the functional form of density dependence (Bellows 1981). For most of the data sets that we examine, there are not sufficient data to estimate γ; however, our purpose is only to ensure that estimates of α are robust to our assumptions about γ. We will examine data for Atlantic cod (*Gadus morhua*) because there are excellent data for these populations and all have been reduced to low levels, which will enhance our ability to estimate α. We held γ fixed at values of 0.5, 0.75, 1, 1.25, and 1.5 (Figs. 1 and 2) and estimated $\tilde{\alpha}$ and β.

The functional fits are displayed in terms of survival $(\log(R/S))$ versus S, where R has been multiplied by $SPR_{F=0}(1 - p_s)$.

If $\gamma < 1$, then survival is a convex function of spawner biomass, and the limit of survival is infinity as $S \to 0$. Thus, this model is unrealistic for this case. Furthermore, an examination of the survival versus spawner curve reveals that it does not become appreciably convex until below the lowest observed spawner abundance (Fig. 1). For $\gamma > 1$, survival is a concave function, and the derivative of survival as $S \to 0$ will always be zero.

In practice, the Ricker model is a reasonably cautious estimate of the limit for management purposes. If $\gamma < 1$ is assumed, then a greater α is estimated, while the assumption of $\gamma > 1$ results in only a slight decrease in the estimate of α (Figs. 1 and 2). If we examine the four cod populations with the largest range in observed spawner biomass, the estimate of the slope at the origin appears reasonable in all cases for the Ricker model, while the estimate for $\gamma = 0.5$ is inflated commensurately with the gap between the origin and the lowest observation of spawner abundance.

© 1999 NRC Canada

Fig. 1. Survival, $\log(R/S)$, versus spawner abundance for six cod stocks. The modeled density-dependent mortality of the form $f(S) = \beta S^\gamma$ is shown for $\gamma = 1.5$ (dashed line), $\gamma = 1$ (Ricker case, dotted line), and $\gamma = 0.5$ (solid line). We have standardized recruitment by multiplying by $\text{SPR}_{F=0}(1 - p_s)$, which allows survival to be interpreted as the annual replacement of spawners per spawner. Thus, the extrapolation of the fitted curves to zero spawner abundance provides an estimate of $\log \tilde\alpha$, i.e., the logarithm of the maximum annual reproductive rate.

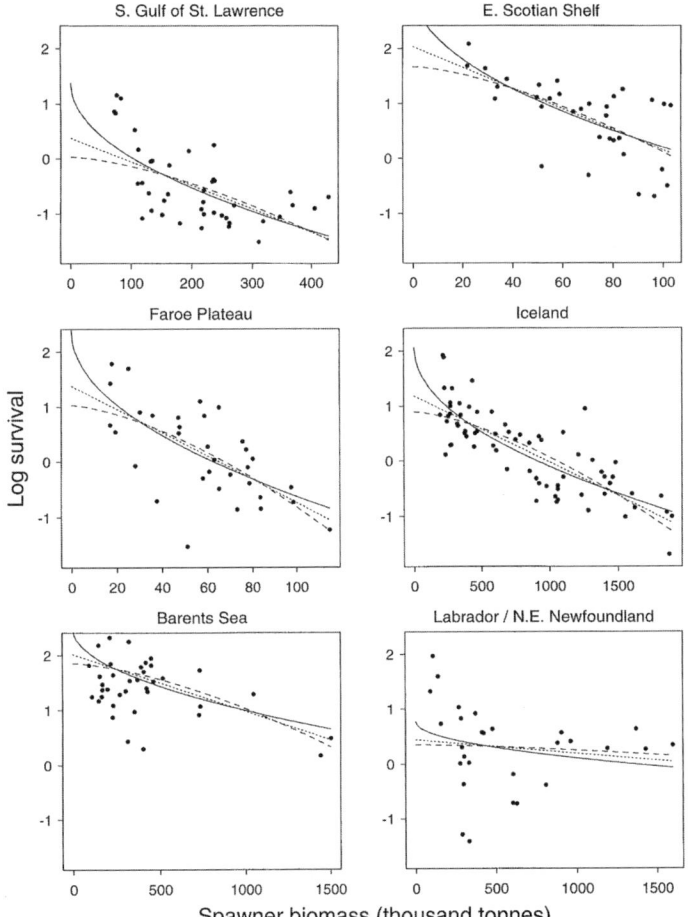

We also considered another common three-parameter model, the "Shepherd function," i.e.:

(7) $$R = \frac{\alpha S}{1 + (S/K)^\delta}.$$

This model was first proposed by Maynard Smith and Slatkin (1973) and was discussed by Bellows (1981). The parameter K has dimensions of biomass and may be interpreted as the "threshold biomass" for the model. For values of biomass S greater than the threshold K, density-dependent effects dominate. The parameter δ may be called the "degree of compensation" of the model, since it controls the degree to which the (density-independent) numerator is compensated for by the (density-dependent) denominator. If $\delta = 1$, then the Beverton–Holt model is recovered. However, for $\delta < 1$, survival is infinity as $S \to 0$; again, in this case the model cannot be considered as a reliable method for extrapolation to low population sizes. For $\delta > 1$, the derivative of survival as $S \to 0$ will always be zero. However, even in the

© 1999 NRC Canada

Fig. 2. Box plots of the logarithm of the scaled slope at the origin, log $\tilde{\alpha}$, for the 20 major cod stocks in the North Atlantic as a function of the form of density-dependent mortality $f(S) = \beta S^\gamma$. For each box plot, the median is marked with a white line and the gray area shows the 95% confidence interval for the median location. When $\gamma = 1$, the Ricker model is recovered.

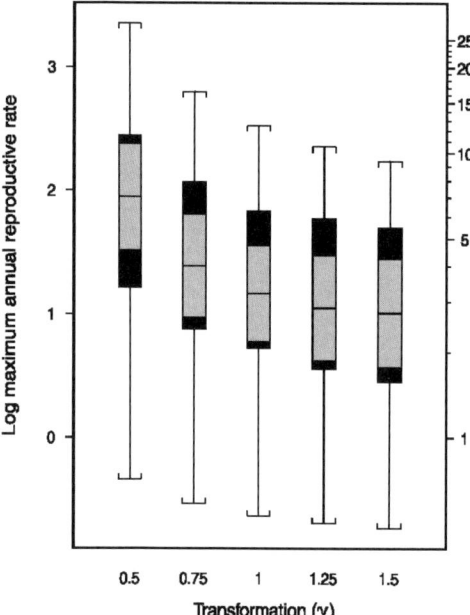

Beverton–Holt case ($\delta = 1$), many estimates of the slope at the origin will be infinity. That is, if $K \to 0$, then $\alpha \to \infty$ is a perfectly feasible solution.

The Deriso–Schnute model (Hilborn and Walters 1992), an alternative three-parameter model, has the Ricker and the Beverton–Holt as special cases. However, it suffers from the same problems that we described above: survival is not constrained to be finite except when the model is a Ricker model, or it has the derivative of survival as $S \to 0$ constrained to be zero.

Any estimation of the slope at the origin is necessarily an extrapolation, since there cannot be observations arbitrarily close to zero spawner abundance. The simplest extrapolation is a linear one (in the relationship between log survival and spawner abundance), while alternative assumptions will often produce unreasonable estimates.

One situation in which a Ricker model would not give a reliable estimate would be if mortality increased at low spawner abundances, known as depensation or the Allee effect. Myers et al. (1995) carried out a metaanalysis and could find no convincing evidence that depensation occurred for exploited fish populations. However, Liermann and Hilborn (1997), using a Bayesian approach, demonstrated that the data were consistent with moderate levels of depensation for several taxa. We conclude that the estimate of the α from the Ricker model will usually provide a reasonable estimate for both ecological and management needs, e.g., when F_τ (sometimes called $F_{\text{extinction}}$), the smallest fishing mortality associated with extinction, is needed.

In this section, we have argued that the Ricker model is often a reasonable model for the estimation of the $\tilde{\alpha}$ (some alternative approaches are discussed below). For the cod populations in the North Atlantic, we have seen that the estimates are only slightly modified if survival is a concave function of spawner biomass. The alternative assumption, that log survival is a convex function, which usually results in the assumption that survival greatly increases at low spawner biomass (Fig. 1), is not strongly supported by the data and may be very dangerous for management decisions in extrapolations to low abundance.

Estimation method

Mixed effects models

Our contention is that focusing on one population at a time can be misleading. In this section, we shall demonstrate how this can be avoided by incorporating the estimation of the Ricker model into a standard linear mixed model. Parameter estimation is easy using widely available software, e.g. SAS or S-PLUS.

We will change the notation slightly to put the results in the standard notation of variance components and mixed models. We consider p populations, subscripted by i, for each of which we want to estimate the parameters of a Ricker model (eq. 4) of the form

(8) $\qquad \log \dfrac{R_{i,t}}{S_{i,t}} = \log \tilde{\alpha}_i + \beta_i S_{i,t} + \varepsilon_{it}$

where $R_{i,t}$ is recruitment to year-class t in population i, $S_{i,t}$ is spawner abundance in year t in population i, $\tilde{\alpha}_i$ and β_i are the Ricker model parameters for population i, and ε_{it} is estimation error, assumed normal. We assume that $\log \tilde{\alpha}_i$ is a normal random variable and define $\mu + a_i \equiv \log \tilde{\alpha}_i$, where μ is the mean of the log-transformed maximum annual reproductive rates and a_i is the random effect for population i. (Note that we will repeat the above calculations using the lifetime reproductive rate instead of the annual reproductive rate.)

We consider the log survival, $\log(R/S)$, of a year-class from a given population as an element of a vector $\mathbf{y}$. If there are n_i observations for population i, then the first n_1 elements of the vector $\mathbf{y}$ will be the n_1 log survivals for the first population, followed by the n_2 log survivals for the second population, and so on.

We consider the fixed effects of the model first. The parameters that we estimate are the overall mean, μ, and p regression parameters, β_i. We consider the spawner abundances, $S_{i,t}$, as known and estimate the density-dependent regression parameter β_i for each population. The standard mixed model notation for the vector of fixed effects parameters is β. The unknown vector β consists of the overall mean μ and the $p\beta_i$s. The vector β is related to $\mathbf{y}$ by the known model matrix $\mathbf{X}$, whose elements are 0, 1, and $S_{i,t}$; the form of this matrix is given below.

For the vector of random effects composed of the a_i, we shall use the standard mixed model notation $\mathbf{u}$. The vector $\mathbf{u}$

is related to **y** by a known model matrix **Z** whose form is given below.

In standard mixed model notation, we have

(9) $\quad \mathbf{y} = \mathbf{X}\boldsymbol{\beta} + \mathbf{Z}\mathbf{u} + \boldsymbol{\varepsilon}.$

Here, ε is an unknown random error vector. For example, consider the simple case of two populations, each of which is observed for 3 years with the first year denoted by 1. The above equation can then be written as

(10) $\quad \mathbf{y} = \begin{bmatrix} y_{11} \\ y_{12} \\ y_{13} \\ y_{21} \\ y_{22} \\ y_{23} \end{bmatrix} = \begin{bmatrix} 1 & S_{11} & \cdot \\ 1 & S_{12} & \cdot \\ 1 & S_{13} & \cdot \\ 1 & \cdot & S_{21} \\ 1 & \cdot & S_{22} \\ 1 & \cdot & S_{23} \end{bmatrix} \begin{bmatrix} \mu \\ \beta_1 \\ \beta_2 \end{bmatrix} + \begin{bmatrix} 1 & \cdot \\ 1 & \cdot \\ 1 & \cdot \\ \cdot & 1 \\ \cdot & 1 \\ \cdot & 1 \end{bmatrix} \begin{bmatrix} a_1 \\ a_2 \end{bmatrix} + \begin{bmatrix} \varepsilon_{11} \\ \varepsilon_{12} \\ \varepsilon_{13} \\ \varepsilon_{21} \\ \varepsilon_{22} \\ \varepsilon_{23} \end{bmatrix}$

where $y_{i,t} = \log(R_{i,t}/S_{i,t})$. The generalization provided by the mixed model enables one not only to model the mean of **y** (as in the standard linear model), but to model the variance of **y** as well. We assume that **u** and ε are uncorrelated and have multivariate normal distributions with expectations **0** and variances **D** and **R**, respectively. The variance of **y** is thus

(11) $\quad \mathbf{V} = \mathbf{ZDZ}' + \mathbf{R}.$

One can model the variance of the data, **y**, by specifying the structure of **D** and **R**. We assume that $\mathbf{D} = \sigma_a^2 \mathbf{I}$ (where **I** is the identity matrix), i.e., that the variability among populations of $\log \tilde{\alpha}_i$ is normally distributed with variance σ_a^2. In the simplest case, one might assume that the error variance is the same for all populations, i.e., $\mathbf{R} = \sigma^2 \mathbf{I}$. (Note that when $\mathbf{R} = \sigma^2 \mathbf{I}$ and $\mathbf{Z} = 0$, the mixed model reduces to the standard linear model.) However, we estimate a separate estimation error variance, σ_i^2, for each population. We also test whether the residuals are autocorrelated. If they are, we can estimate a separate autocorrelation parameter, ρ_i, for each population. This results in a block diagonal structure for **R**, with blocks

(12) $\quad \sigma_i^2 \begin{bmatrix} 1 & \rho_i & \rho_i^2 & \cdots \\ \rho_i & 1 & \rho_i & \cdots \\ \rho_i^2 & \rho_i & 1 & \cdots \\ \vdots & \vdots & \vdots & \ddots \end{bmatrix}.$

Estimation of variance components

Now that we have transformed the problem into this form, estimation is trivial because high-quality software exists for this problem (Appendix 1). The likelihood function for the data vector $\mathbf{y} \sim \mathcal{N}_N(\mathbf{X}\boldsymbol{\beta}, \mathbf{V})$ is

(13) $\quad L = L(\boldsymbol{\beta}, \mathbf{V}|\mathbf{y}) = \dfrac{e^{-\frac{1}{2}(\mathbf{y}-\mathbf{X}\boldsymbol{\beta})'\mathbf{V}^{-1}(\mathbf{y}-\mathbf{X}\boldsymbol{\beta})}}{(2\pi)^{\frac{N}{2}}|\mathbf{V}|^{\frac{1}{2}}}$

where N is the number of fixed effects estimated, i.e., $N = 1 + p$. There are two common approaches to the estimation of variance components based on this function: maximum likelihood (ML) and restricted maximum likelihood (REML) (Searle et al. 1992). REML differs from ML for this model in that it takes into account the degrees of freedom used for estimating the fixed effects, whereas ML does not. Furthermore, in the case of balanced data, REML solutions are identical to ANOVA estimators, which have known optimality properties. For these reasons, we will use REML but will consider ML to check sensitivity. Denote the resulting estimates of **D** and **R** by $\hat{\mathbf{D}}$ and $\hat{\mathbf{R}}$, respectively.

It is possible for the estimate of the variance among populations, $\hat{\sigma}_a^2$, to be zero. This often occurs when only a few populations are available for analysis and should not be interpreted as implying that there is no variability among populations in the maximum annual reproductive rate.

Estimation of individual population parameters

The use of mixed models allows us to obtain improved estimates of parameters for any one population. In general, we wish not only to estimate the fixed model parameters, but also to predict the random variables for each population. In our case, we wish to estimate the density-dependent parameter β and predict the slope at the origin for each population, which is assumed to be a random variable. The terminological distinction between estimation of fixed effects and prediction of random effects is awkward and unnecessary (Robinson 1991); we will "estimate" both fixed and random effects, with the understanding that for random effects, we are in fact obtaining estimates of their realized values.

The best linear unbiased estimators (BLUEs) $\tilde{\boldsymbol{\beta}}$ of the fixed effects $\boldsymbol{\beta}$ and the best linear unbiased predictors (BLUPs) $\tilde{\mathbf{u}}$ of the random effects **u** may be obtained from the mixed model equations

(14) $\quad \begin{bmatrix} \mathbf{X}'\mathbf{R}^{-1}\mathbf{X} & \mathbf{X}'\mathbf{R}^{-1}\mathbf{Z} \\ \mathbf{Z}'\mathbf{R}^{-1}\mathbf{X} & \mathbf{Z}'\mathbf{R}^{-1}\mathbf{Z} + \mathbf{D}^{-1} \end{bmatrix} \begin{bmatrix} \tilde{\boldsymbol{\beta}} \\ \tilde{\mathbf{u}} \end{bmatrix} = \begin{bmatrix} \mathbf{X}'\mathbf{R}^{-1}\mathbf{y} \\ \mathbf{Z}'\mathbf{R}^{-1}\mathbf{y} \end{bmatrix}.$

Without the $\mathbf{D}^{-1}$ in the lower right-hand submatrix of the matrix on the left, eq. 14 would be the ML equations for the model treated as if **u** represented fixed effects, rather than random effects. Although the above equation has been discussed in terms of classical methods, the same result is arrived at using a formal Bayes analysis of incorporating prior information into the analysis of data (Searle et al. 1992).

Since we do not know the variance–covariance matrices **D** and **R**, we substitute $\hat{\mathbf{D}}$ and $\hat{\mathbf{R}}$ into eq. 14 to obtain empirical BLUEs and BLUPs.

The variance–covariance matrix for $[\tilde{\boldsymbol{\beta}} \, \tilde{\mathbf{u}}]'$ is

(15) $\quad \mathbf{C} = \begin{bmatrix} \mathbf{X}'\mathbf{R}^{-1}\mathbf{X} & \mathbf{X}'\mathbf{R}^{-1}\mathbf{Z} \\ \mathbf{Z}'\mathbf{R}^{-1}\mathbf{X} & \mathbf{Z}'\mathbf{R}^{-1}\mathbf{Z} + \mathbf{D}^{-1} \end{bmatrix}^{-}$

where the superscript minus on the above matrix represents a generalized inverse. An approximate variance–covariance matrix $\hat{\mathbf{C}}$ may be obtained by substituting $\hat{\mathbf{D}}$ and $\hat{\mathbf{R}}$ into eq. 15. The approximate standard error for any linear combination **L** of the vector $[\tilde{\boldsymbol{\beta}} \, \tilde{\mathbf{u}}]'$ may be obtained from

(16) $\sqrt{L'\hat{C}L}$.

Note that these standard errors will tend to be underestimates of the standard errors of the empirical BLUPs and BLUEs (Searle et al. 1992).

Our estimation methods above provide estimates of μ for each species and (the realized values of) the a_i. As long as the log-transformed values are considered, the interpretation is simple; however, the interpretation of the values on an untransformed scale is more complex. For a species, the median $\tilde{\alpha}$ is $\exp(\mu)$, and the expectation is $\exp(\mu + 0.5\sigma_a^2)$, where the expectation and median are taken with respect to the distribution of the random effects. This estimate is complicated by the estimation error of the μ and the σ_a^2, which we will ignore here. To keep things simple, we will discuss our results in terms of the log-transformed values and the medians of $\tilde{\alpha}$ for a species, except where noted.

Data sources and treatment

The data that we used are estimates obtained from assessments compiled by Myers et al. (1995). The database is available from the first author. For marine populations, population numbers and fishing mortality were usually estimated using sequential population analysis (SPA) of commercial and (or) recreational catch at age data for most marine populations. SPA techniques include virtual population analysis (VPA), cohort analysis, and related methods that reconstruct population size from catch at age data (see Hilborn and Walters (1992, chaps. 10 and 11) for a description of the methods used to reconstruct the population history). Briefly, the catch at age is combined with estimates from research surveys and (or) commercial catch rates to estimate the numbers at age in the final year and to reconstruct previous numbers at age under the assumption that catch at age is known without error and that natural mortality at age is known and constant.

For salmon stocks, spawner abundance is the estimate of the number of fish reaching the spawning grounds, and recruitment is estimated by combining catch and the number of upstream migrants.

SPA techniques were used for the freshwater species except for brook trout (*Salvelinus fontinalis*). The brook trout populations were from introduced populations in California mountain lakes (DeGisi 1994); these populations were estimated using research gill nets and ML depletion estimation.

Time series of less than 10 paired spawner–recruit observations are not included in this analysis. The $SPR_{F=0}$ was calculated using estimates of natural mortality, weight at age, and maturity at age. Maturity and weight at age were usually estimated from research surveys carried out for each population.

A major source of uncertainty in the SPA estimates of recruitment and spawning stock biomass (SSB) is that they usually assume that catches are known without error. This is particularly important when estimates of discarding and misreporting are not included in the catch at age data used in the SPA. These errors are clearly important for some periods of time for some of the cod stocks (Myers et al. 1997*a*), and these errors will affect our estimates of the number of replacements that each spawner can produce at low population densities ($\tilde{\alpha}$).

The data for this analysis are available at R.A. Myers' web site (http://fish.dal.ca/welcome.html).

Results

We first used ML for a standard Ricker model to obtain single-population estimates of $\log \tilde{\alpha}$ for each population (Fig. 3). Then, for each species in our database, we applied the mixed model to the data from all of the populations belonging to that species, obtaining estimates and predictions as follows. We used REML to estimate σ_a^2, the true variability among populations in the log-transformed maximum annual reproductive rate, and for each population, σ_i^2, the estimation error variance. These estimated variance components were then used to obtain the empirical BLUE of the mean log-transformed maximum annual reproductive rate, μ, for the species and the empirical BLUP of $\log \tilde{\alpha}$ for each population (Table 1). For completeness, we have given the mixed model estimates for the mean and variability at the family level; however, they should be used with great caution because they may not be representative of any given species. More detailed results (which include stock-level estimates) are available at R.A. Myers' web site.

Note that there is less variance among the BLUP estimates than among the single-population estimates (Fig. 4).

The estimates for populations with large estimation error variances (e.g., due to relatively few data points) and that are far from the mean for the species, e.g. Gulf of Maine cod (MLE:$\log \tilde{\alpha} = 2.85$, BLUP:$\log \tilde{\alpha} = 1.84$), are pulled towards the mean more than those for populations with lower estimation error variance and that are close to the species mean, e.g. Iceland cod (MLE:$\log \tilde{\alpha} = 1.19$, BLUP: $\log \tilde{\alpha} = 1.19$).

As expected, the estimate of the true variability in the maximum annual reproductive rate is much less than the sample variability because individual estimates contain estimation error. For example, for pink salmon, if $\tilde{\alpha}$ is estimated separately for each stock, then there is an order of magnitude range of the estimates. However, if $\tilde{\alpha}$ is assumed to be a random variable, then the mixed model estimates suggest that the true range is very small, with all the true values being very close to 3 (Fig. 3). Cod show a similar picture. The median number of replacement spawners per spawner per year for cod at low abundance is between 3 and 4, resulting in a maximum net reproductive rate (if there is no fishing mortality) of between 15 and 20. The maximum annual reproductive rate for Atlantic herring (*Clupea harengus*) appears to be slightly less, and for hakes of the genus *Merluccius*, e.g., silver hake (*Merluccius bilinearis*) and Pacific hake (*Merluccius productus*), it is around 1. Some anadromous species, e.g., sockeye salmon (*Oncorhynchus nerka*), appear to have a maximum annual reproductive rate of around 4 or 5, while others, e.g., pink salmon, have a much lower rate.

The most remarkable aspect of the results is the relative constancy of the estimates of the maximum annual reproductive rate. For species for which we have more than one population in our analysis, the median of the estimated maximum reproductive rate is almost always between 1 and 7 (Fig. 5*a*).

For the species with multiple populations, only Pacific ocean perch (*Sebastes alutus*) and silver hake have an esti-

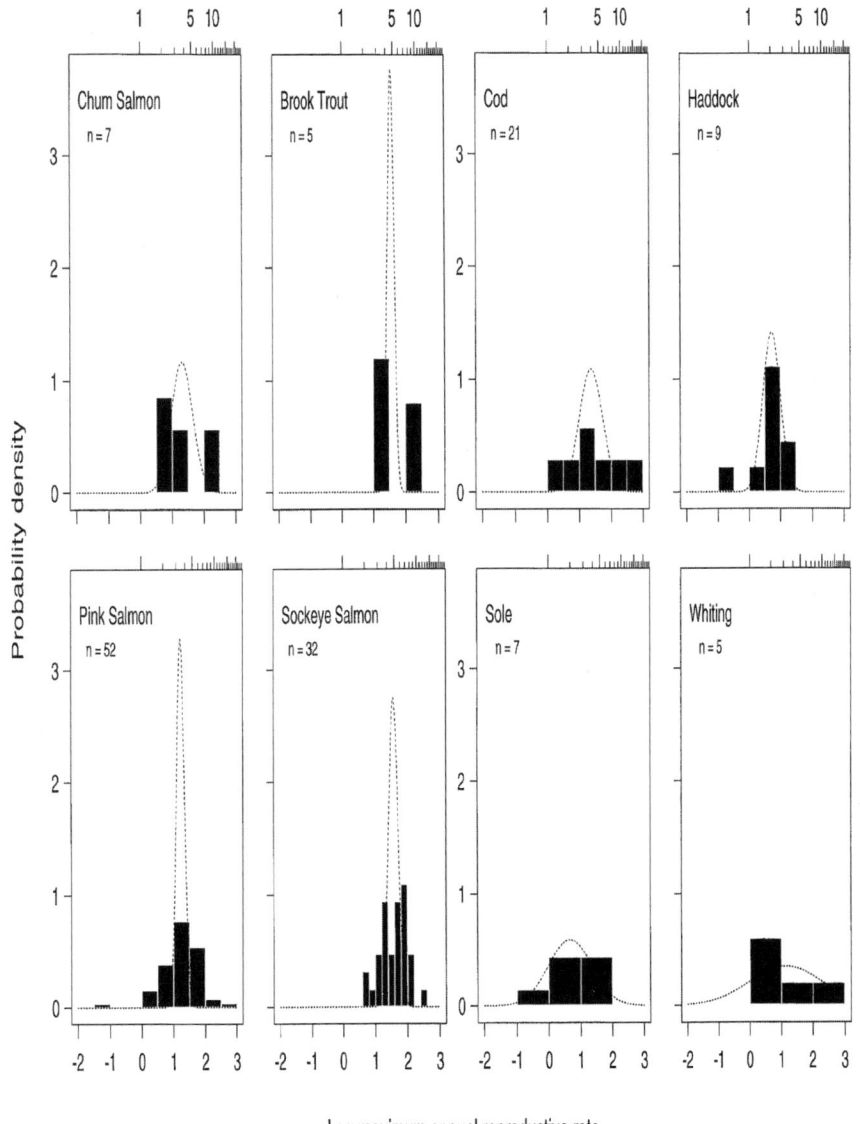

Fig. 3. Histograms by species of the individual ML estimates of the log of the maximum annual reproductive rate, log $\tilde{\alpha}$, compared with probability densities based on REML estimates of the true variability in log $\tilde{\alpha}$ from our mixed model analysis (dotted line). Note that the top axis of each plot shows the untransformed annual reproductive rate. The number of populations (n) for each histogram is also given.

mated maximum annual reproductive rate of less than 1. The low estimate for Pacific ocean perch results in a very low estimate for the expected maximum lifetime reproductive rate, i.e., it is about 3 (Table 1; Fig. 6a). Relatively low estimates of the maximum lifetime annual reproductive rate and steepness were estimated for the other *Sebastes* species (Table 1). We do not know whether these low estimates are real, e.g., are somehow related to their low natural mortality and oviviparous reproduction, or an artifact. The age-based assessments of the *Sebastes* species are unusually uncertain because of aging difficulties. It is also possible that the environmental conditions in recent years, when the low estimates of spawning biomass and recruitment were made, have been unusual and have resulted in lower than average estimates. In any case, it is crucial to determine if the assessments are correct and exploit these species more cautiously than other species.

The estimates of the maximum annual reproductive rate for species for which we have only one population are much more variable than for the species with many populations (Figs. 5b and 6b). The greater variability in these estimates is at least partially caused by estimation error. However, several species have maximum reproductive rates that suggest that they cannot sustain intense fisheries. In some cases, this is certainly true. The southern bluefin tuna (*Thunnus maccoyii*) in the Southern Ocean has been greatly reduced by overfishing. In other cases, there may be serious problems with the assessments.

Despite the large variation in the individual estimates, our general conclusion about the relative constancy of the maximum annual reproductive rate stands; the estimates are usually around 3. There are exceptions for individual stocks, but these usually have large standard errors.

Note that herring has a smaller maximum reproductive rate than many species. The lower mean is due to a few stocks in the northern North Atlantic that have been reduced to very low levels (the Iceland stocks, the Norway stock (often called the "Arcto-Norwegian" stock), and the Georges Bank stocks).

We also considered a model that allowed the residuals to be autocorrelated (see Appendix 1 for the computer code used in the estimation). This approach is probably preferable if autocorrelation is substantial in the model residuals, but may pull the individual estimates too far towards the population mean.

We repeated the above analysis for the lifetime maximum reproductive rate and display the results in terms of the expected lifetime maximum reproductive rate and the steepness. Among taxonomic groups, the lifetime reproductive rate appears to be more variable than the annual rate; however, for species with similar natural mortalities after reproduction, the results are again relatively constant.

The results for the steepness are displayed as the median, the 20th percentiles, and the 80th percentiles (Table 1). These can be be used to approximate priors for Bayesian analyses that commonly use steepness (Punt and Hilborn 1997).

It is useful to compare the median for a species of the maximum annual reproductive rate (the uncorrected value) with the expectation, where the median and expectation are taken with respect to the distribution of the random effects (Fig. 7). The corrected estimates are higher by a factor of $\exp(0.5\sigma_a^2)$. This effect is usually small except for species where the estimate of the variability among populations is unusually large, e.g., for blueback herring (*Alosa aestivalis*).

A generalization

The unexpected generalization that comes from our analysis is that the annual reproductive rate within a species often shows relatively little variation and that the variation in annual reproductive rate among species is surprisingly small, usually ranging from 1 to 7 for species for which we have several populations represented in our analysis.

This is a broad generalization that may have great implications for the management and conservation of fish populations. Although the generalization appears to be firmly established for many well-studied species, these are primarily temperate-zone species.

Possible exceptions

In this section, we will discuss several populations that appear to have anomalously high or low annual reproductive rates. It is unclear whether these rates are real or due to limitations in the assessments.

The blueback herring appears to have a relatively high maximum annual reproductive rate. This relatively high number is consistent with the fast growth rate experienced by this species and other *Alosa* species when they recolonize former habitat (Crecco and Gibson 1990). However, it is possible that these high rates of population growth are caused by movement of fish upstream over obstructions and not population growth per se. This hypothesis needs to be evaluated.

Chinook salmon (*Oncorhynchus tshawytscha*) also appears to have relatively high maximum annual reproductive rates. These appear to be real and are probably conservative. The values that are in the figures for chinook salmon are from the northern limit of the range. The values for the Columbia River appear to be much higher, but it is impossible to estimate the "natural" rates because of dam-induced mortality.

The ayu (*Plecoglossus altivelis*, Plecoglossidae, Salmoniformes) from Lake Biwa, Japan, is the only univoltine species in the database, and it appears to have a very high annual reproductive rate (Table 1) (Suzuki and Kitahara 1996). The analysis appears to be sound and is backed up by fishery-independent survey data, but it is possible that the application of VPA for this species may have led to biases.

Among the lowest estimates of the maximum reproductive rates are those for several species on the west coast of North America: Pacific ocean perch, sablefish (*Anoplopoma fimbria*), and chilipepper rockfish (*Sebastes goodei*). These stocks all are assessed in a similar manner. The assessments on these stocks do not have reliable fishery-independent estimates of abundance, do not have a large amount of aging data, and often assume that the population is at the unfished equilibrium at the beginning of the fishery. It is critical for the management of these stocks to determine whether their actual maximum reproductive rate is as low as it appears to be, or if the assessments are reliable.

These cases represent anomalies, which may represent fundamental inconsistencies with our broad generalization about the reproductive rate, or may well be explained by

Table 1. Mixed model estimates at the species and family levels and corresponding estimates of the percentiles of z (the steepness parameter).

Species	n	$\widehat{\log \bar{\alpha}}$	SE	$\hat{\sigma}_a^2$	$\hat{\bar{\alpha}}$	z_{20}	z_{med}	z_{80}
Aulopiformes								
Synodontidae	1	**0.31**	**0.07**		2		**0.34**	
Bombay duck (*Harpodon nehereus*)	1	0.31	0.07		2		0.34	
Clupeiformes								
Clupeidae	34	**1.06**	**0.19**	**1.16**	**17.1**	**0.49**	**0.71**	**0.86**
Anadromous alewife (*Alosa pseudoharengus*)	4	1.29	0.09	0	5.7		0.59	
Anadromous American shad (*Alosa sapidissima*)	1	1.65	0.3		18.5		0.82	
Atlantic menhaden (*Brevoortia tyrannus*)	1	2.2	0.12		24.8		0.86	
Blueback herring (*Alosa aestivalis*)	3	2.6	0.55	0.81	31.9	0.71	0.84	0.92
Gulf menhaden (*Brevoortia patronus*)	1	1.25	0.16		5.3		0.57	
Atlantic herring (*Clupea harengus*)	18	0.73	0.28	1.31	22.1	0.52	0.74	0.88
Pacific sardine (*Sardinops sagax*)	2	0.66	0.89	1.56	12.7	0.34	0.59	0.81
Spanish sardine (*Sardina pilchardus*)	1	−0.56	0.75		2.1		0.34	
Sprat (*Sprattus sprattus*)	3	0.87	0.55	0.71	10.7	0.48	0.65	0.79
Engraulidae	4	**1.28**	**0.57**	**1.14**	**11.5**	**0.4**	**0.62**	**0.8**
Anchovy (*Engraulis encrasicolus*)	2	0.7	0.13	0	3.6		0.47	
Gold-spotted grenadier anchovy (*Coilia dussumieri*)	1	2.73	0.19		17.6		0.81	
Northern anchovy (*Engraulis mordax*)	1	0.33	0.41		3.1		0.43	
Gadiformes								
Gadidae	49	**1.01**	**0.12**	**0.51**	**19.6**	**0.67**	**0.79**	**0.87**
Blue whiting (*Micromesistius poutassou*)	2	0.59	0.33	0	10		0.71	
Atlantic cod (*Gadus morhua*)	21	1.37	0.15	0.37	26	0.76	0.84	0.9
Haddock (*Melanogrammus aeglefinus*)	9	0.72	0.21	0.28	13	0.64	0.74	0.82
Hake (*Merluccius hubbsi*)	1	1.18	0.45		18		0.82	
Pacific hake (*Merluccius productus*)	1	−0.95	0.83		1.9		0.32	
Pollock or saithe (*Pollachius virens*)	5	1.16	0.14	0.05	18	0.78	0.81	0.84
Silver hake (*Merluccius bilinearis*)	3	−0.18	0.29	0.16	2.7	0.31	0.39	0.47
Walleye pollock (*Theragra chalcogramma*)	2	0.28	0.24	0.01	5	0.53	0.55	0.58
Whiting (*Merlangius merlangus*)	5	1.14	0.51	1.16	30.8	0.64	0.81	0.91
Lophiiformes								
Lophiidae	1	**−0.07**	**0.32**		6.7		**0.64**	
Black anglerfish (*Lophius budegassa*)	1	−0.07	0.32		6.7		0.63	
Perciformes								
Carangidae	3	**0.27**	**0.21**	**0**	**4**		**0.5**	
Horse mackerel (*Trachurus trachurus*)	2	0.52	0.8	0	12.1		0.75	
Mediterranean horse mackerel (*Trachurus mediterraneus*)	1	0.25	0.22		3.5		0.47	
Lutjanidae	1	**1.9**	**0.9**		47.8		**0.95**	
Red snapper (*Lutjanus campechanus*)	1	1.9	0.9		47.8		0.92	
Percichthyidae	1	**0.95**	**0.16**		18.6		**0.82**	
Striped bass (*Morone saxatilis*)	1	0.95	0.16		18.6		0.82	
Percidae	2	**0.91**	**0.57**	**0.28**	**9.5**	**0.57**	**0.67**	**0.76**
Walleye (*Stizostedion vitreum*)	2	0.91	0.57	0.28	9.5	0.57	0.67	0.76
Scianidae	1	**1.88**	**0.28**		26.1		**0.87**	
White croaker (*Argyrosomus argentatus*)	1	1.88	0.28		26.1		0.87	
Scombridae	8	**0.34**	**0.39**	**1.12**	**7.5**	**0.3**	**0.52**	**0.72**
Atlantic bluefin tuna (*Thunnus thynnus*)	1	−0.4	0.23		5.2		0.56	
Bigeye tuna (*Thunnus obesus*)	2	0.73	0.08	0	5.3		0.57	
Chub mackerel (*Scomber japonicus*)	1	−0.05	0.33		2.4		0.38	
Atlantic mackerel (*Scomber scombrus*)	2	1.11	0.91	1.29	31.8	0.62	0.81	0.92
Southern bluefin tuna (*Thunnus maccoyii*)	1	−1.5	0.09		2.9		0.42	
Yellowfin tuna (*Thunnus albacares*)	1	1.43	0.21		9.3		0.7	
Sparidae	3	**2.48**	**0.41**	**0**	**65.9**		**0.95**	
New Zealand snapper (*Pagrus auratus*)	2	1.34	1.31	0	65.6		0.94	
Scup (*Stenotomus chrysops*)	1	2.6	0.38		74.6		0.95	

Myers et al.

Table 1 (concluded).

Species	n	$\widehat{\log \tilde{\alpha}}$	SE	$\hat{\sigma}_a^2$	$\hat{\tilde{\alpha}}$	z_{20}	z_{med}	z_{80}
Xiphiidae	1	1.7	0.05		30.1		0.88	
Swordfish (*Xiphias gladius*)	1	1.7	0.05		30.1		0.88	
Pleuronectiformes								
Pleuronectidae	14	**0.79**	**0.18**	**0.34**	**18.8**	**0.71**	**0.8**	**0.87**
European flounder (*Platichthys flesus*)	1	−0.03	0.42		5.3		0.57	
Greenland halibut (*Reinhardtius hippoglossoides*)	3	0.75	0.68	1.32	29.3	0.59	0.79	0.91
Plaice (*Pleuronectes platessa*)	8	0.92	0.17	0.08	25.1	0.83	0.86	0.88
Yellowtail flounder (*Pleuronectes ferrugineus*)	2	0.79	0.34	0.14	13	0.69	0.75	0.81
Soleidae	7	**0.66**	**0.35**	**0.68**	**28.7**	**0.72**	**0.84**	**0.91**
Sole (*Solea vulgaris*)	7	0.66	0.35	0.68	28.7	0.72	0.84	0.91
Salmoniformes								
Esocidae	2	**0.51**	**0.19**	**0.03**	**6.1**	**0.57**	**0.6**	**0.64**
Northern pike (*Esox lucius*)	2	0.51	0.19	0.03	6.1	0.57	0.6	0.64
Plecoglossidae	1	**4.73**	**0.16**		**123.5**		**0.97**	
Ayu (*Plecoglossus altivelis*)	1	4.73	0.16		123.5		0.97	
Salmonidae	106	**1.43**	**0.05**	**0.18**	**25.1**	**0.8**	**0.85**	**0.89**
Atlantic salmon (*Salmo salar*)	3	1.46	0.25	0.16	5.1	0.46	0.54	0.62
Chinook salmon (*Oncorhynchus tshawytscha*)	6	1.99	0.13	0	7.3		0.65	
Chum salmon (*Oncorhynchus keta*)	7	1.31	0.24	0.34	4.4	0.36	0.48	0.6
Freshwater brook trout (*Salvelinus fontinalis*)	5	1.55	0.24	0.11	27.4	0.83	0.87	0.89
Lake trout (*Salvelinus namaycush*)	1	0.92	0.08		24.1		0.86	
Pink salmon (*Oncorhynchus gorbuscha*)	52	1.22	0.07	0.12	3.6	0.39	0.46	0.53
Sockeye salmon (*Oncorhynchus nerka*)	32	1.57	0.08	0.15	5.2	0.47	0.55	0.62
Scorpaeniformes								
Anoplopomatidae	1	**−2.35**	**0.47**		**1.4**		**0.28**	
Sablefish (*Anoplopoma fimbria*)	1	−2.35	0.47		1.4		0.26	
Hexagrammidae	1	**1.13**	**0.49**		**12**		**0.77**	
Atka mackerel (*Pleurogrammus monopterygius*)	1	1.13	0.49		12		0.75	
Scorpaenidae	5	**−1.57**	**0.24**	**0.17**	**2.8**	**0.31**	**0.39**	**0.48**
Chilipepper (*Sebastes goodei*)	1	−0.85	0.57		2.1		0.35	
Pacific ocean perch (*Sebastes alutus*)	3	−1.93	0.18	0	3		0.43	
Deepwater redfish (*Sebastes mentella*)	1	−1.08	0.18		3.6		0.47	

Note: Listed are the empirical BLUE of the mean value of the log-transformed maximum annual reproductive rate ($\widehat{\log \tilde{\alpha}}$), its standard error, the estimated variance among populations ($\hat{\sigma}_a^2$) (where possible), the estimated expected maximum lifetime reproductive rate for a species, where the expectation is taken over the distribution of the random effects ($\hat{\tilde{\alpha}}$), the 20th percentile of z (z_{20}) (where possible), the median of z (z_{med}), and the 80th percentile of z (z_{80}) (where possible). The mixed model estimates are given at the species and family levels, but the family-level estimates (shown in boldface) should be used with caution.

other factors, e.g., assessment problems or the possibility that the data for these species come from only a relatively short time period, which may not be representative of the average reproductive rate.

Limitations and alternative approaches

Our approach to estimating the standardized slope at the origin of spawner–recruit functions is based on well-studied statistical methods and has intuitive appeal and appears to be a promising method for categorizing species in terms of their vulnerability to overfishing. However, researchers should be aware of its limitations and of alternative approaches.

The first limitation is the functional form assumed for density-dependent mortality. The Ricker model and the non-linear Ricker model (eq. 6) are not appropriate for some species. This is a serious limitation of the methods described here. For example, we did not consider coho salmon (*Oncorhynchus kisutch*) in this analysis because the shape of the spawner–recruitment curve was clearly asymptotic, simi-

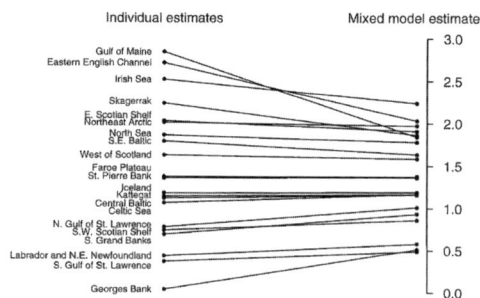

Fig. 4. Comparison of the maximum annual reproductive rate, log $\tilde{\alpha}$, obtained from individual regressions on each cod population in the North Atlantic with the empirical BLUPs obtained from a mixed model analysis. Notice that the mixed model estimates have lower variance than the individual estimates.

© 1999 NRC Canada

Fig. 5. Estimates of the log of the maximum annual reproductive rate for (*a*) species with multiple populations in the database, where the error bars represent the estimated standard deviation of the log of the maximum annual reproductive rate (this estimate is sometimes zero if only two or three populations are used in the analysis), and (*b*) species with only one population in the database, where the error bars represent the standard error of the estimate.

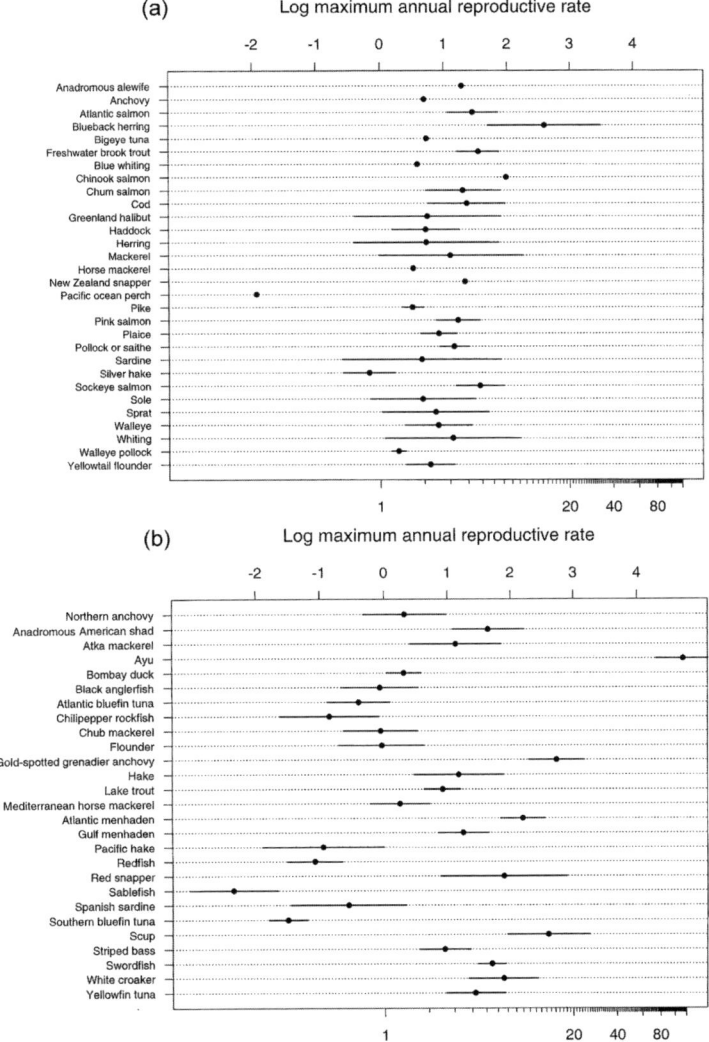

lar to a Beverton–Holt function. For such functions, nonlinear mixed models are required and will be considered in a future paper using the methods of Lindstrom and Bates (1990).

The second limitation is that we have assumed that the distribution of $\tilde{\alpha}$ is approximately lognormal. This appears to be a reasonable approximation in most cases considered here, but violations of the assumption may cause biases (Verbeke and Lesaffre 1996).

A third limitation is the assumption that the recruitment distribution for given spawner abundance is lognormal. This

Fig. 6. Estimates of the log of the maximum lifetime reproductive rate for (*a*) species with multiple populations in the database, where the error bars represent the estimated standard deviation of the log of the maximum lifetime reproductive rate, and (*b*) species with only one population in the database, where the error bars represent the standard error of the estimate.

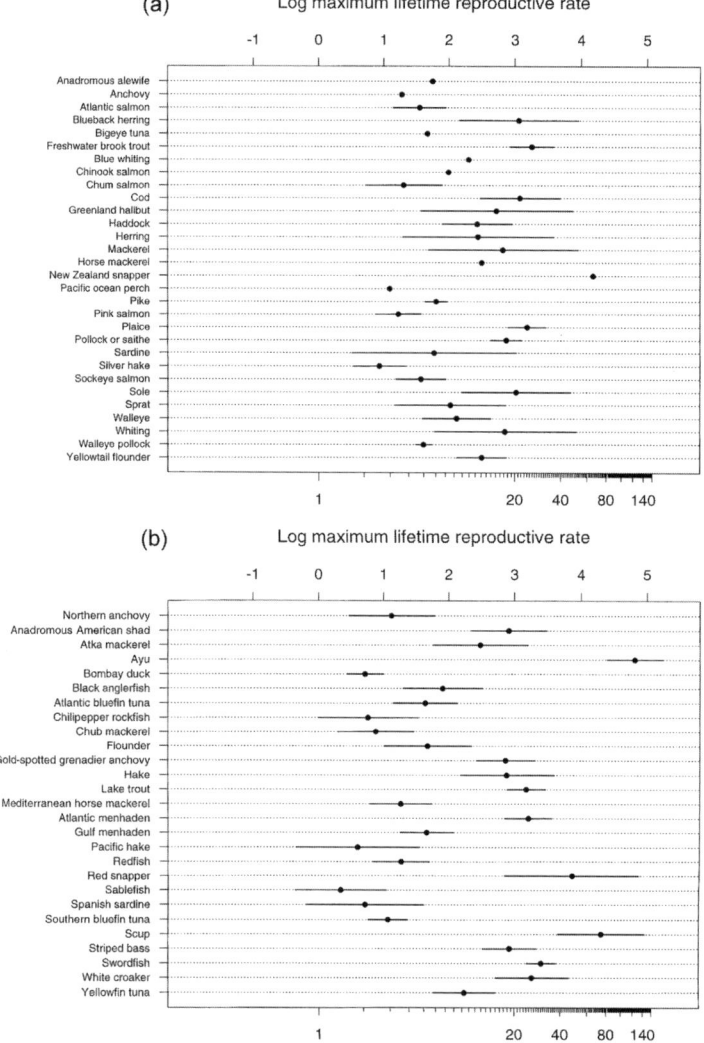

is by far the most common assumption used in fitting spawner–recruitment models (Hilborn and Walters 1992); however, it may not always be the most appropriate assumption. The gamma distribution appears to give more reasonable fits to some stock–recruitment data (Myers et al. 1995; R.A. Myers, K.G. Bowen, and I.A. Zouros, unpublished data).

A fourth limitation is the assumption that all populations within a taxon are comparable, i.e., the maximum reproductive rate for populations within a species (or higher taxon) is described by a lognormal distribution. It is possible that this parameter may vary in a systematic way among populations, e.g., populations in colder conditions may have a lower max-

Fig. 7. Comparison of the uncorrected (simple exponential transform) and corrected (lognormal) values for the maximum annual reproductive rate. Notice that without the lognormal correction, the maximum annual reproductive rate will be underestimated.

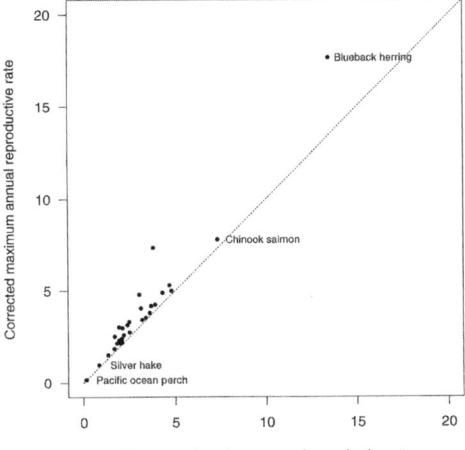

imum reproductive rate. Such hypotheses can be investigated by letting the maximum reproductive rate be a random variable whose mean is a function of a covariate such as temperature or latitude.

If any of the above four assumptions appears to be violated seriously, then an alternative approach is needed. Perhaps the most convenient alternative framework for this type of model is either a Bayes or empirical Bayes hierarchical approach (Efron 1996). Punt and Hilborn (1997) have recently reviewed these approaches in fisheries management. McAllister (1994) implemented an empirical Bayes approach to estimating a parameter functionally related to the slope at the origin, viz. the steepness parameter, using an earlier version of the database used here.

The ML estimators that we have used to estimate the underlying distribution of annual reproductive rates may result in estimates that are less "heavy tailed" than they should be (Searle et al. 1992; Efron 1996).

It should be remembered that this analysis does not circumvent known biases, e.g., estimation error in spawner abundance and time series bias in the treatment of spawner–recruitment relationships (for a review, see Hilborn and Walters 1992).

Discussion

The analysis presented in this paper suggests a new and unsuspected finding: the maximum annual reproductive rate for any of the species examined is typically between 1 and 7. This number may be less for some species and more for others, but the relative constancy of the annual reproductive rate is an unanticipated, and very important, finding. This analysis is consistent with our preliminary analysis (Myers et al. 1996).

The common belief that there is no relationship between spawner biomass and recruitment is founded on the notion that the maximum reproductive rate for fish is essentially infinite, a belief based on the observation that fecundity of fish is often large and cursory examination of spawner–recruitment plots that often show no strong reduction of recruitment at low spawner abundances over the range of the observations. This erroneous belief is caused by the lack of attention paid to the information content of different data sets (Myers and Barrowman 1996; Myers 1997).

Hypotheses

This broad generalization demands an explanation. First, consider the lower limit of the annual reproductive rate at low abundance. This represents the "average" value that should occur at low abundance. Clearly, if this value is much less than 1, then the population may very well go extinct because the value would probably be below 1 for considerable lengths of time because of variation in the environment.

Why, then, would the annual reproductive rate be bounded at the upper end? A reasonable, but speculative, answer is that a very high value of the reproductive rate would imply an excess of resources that are not exploited. In this case, other competitors would be expected to evolve to exploit these resources.

Reducing uncertainty

The uncertainty of the biological processes underlying the population dynamics of exploited species can be greatly reduced by combining data from many studies. The relative constancy of the maximum reproductive rate allows for simple, broad conclusions to be reached on the management of fish stocks. That the maximum reproductive rate is typically around 1–7 replacement spawners per spawner per year is a powerful tool for the management of fish stocks. It allows the maximum exploitation rate to be estimated quickly (Mace 1994; Myers and Mertz 1998) and the recovery rates of exploited fish populations to be calculated (Myers et al. 1997b).

Many of the crucial parameters needed for fisheries management can be estimated using the maximum reproductive rate analyzed here together with simple approximations (Myers et al. 1997b; Myers and Mertz 1998). For example, the steepness parameter that we compiled is now commonly used in assessments, and our analysis provides reasonable ML estimates that can be used in empirical Bayes assessment procedures. All that is required to use these approximations are data on natural mortality, age at maturity, and the maximum reproductive rate. These approximate formulas will require testing and verification, but this approach should allow progress to be made on critical issues. Thus, even if the maximum reproductive rate is not known for a species, the estimates compiled in this paper allow it to be approximated, or our estimates can be used in forming priors for a Bayesian analysis.

Acknowledgments

We wish to thank the Killam Foundation, the Canadian Foundation for Innovation, and the Natural Sciences and Engineering Research Council of Canada for financial support. We thank Stacey Fowlow for programming assistance. We thank the hundreds of assessment biologists whose hard work made this meta-analysis possible. Special thanks to Pamela Mace whose input was key to developing these ideas. This paper is dedicated to the memory of my (R.A.M.) great friend, Gordon Mertz.

References

Barnthouse, L.W., Klauda, R.J., Vaughan, D.S., and Kendall, R.L. 1988. Science, law, and the Hudson River power plants: a case study in envioronmental assessment. Am. Fish. Soc. Monogr. 4.

Bellows, T.S. 1981. The descriptive properties of some models for density dependence. J. Anim. Ecol. 50: 139–156.

Cole, L.C. 1954. The population consequences of life history phenomena. Q. Rev. Biol. 29: 103–137.

Cook, R.M., Sinclair, A., and Stefansson, G. 1997. Potential collapse of North Sea cod stocks. Nature (Lond.), 358: 521–522.

Crecco, V.A., and Gibson, M. 1990. Stock assessment of river herring from selected Atlantic coast rivers. Spec. Rep. 19. Atlantic States Marine Fisheries Commission, Washington, D.C.

DeGisi, J.S. 1994. Year class strength and catchability of mountain lake brook trout. Master's thesis, University of British Columbia, Vancouver, B.C.

Efron, B. 1996. Empirical Bayes methods for combining likelihoods. J. Am. Stat. Assoc. 91: 538–565.

Hilborn, R., and Liermann, M. 1998. Standing on the shoulders of giants: learning from experience in fisheries. Rev. Fish Biol. Fish. 8: 1–11.

Hilborn, R., and Walters, C.J. 1992. Quantitative fisheries stock assessment: choice, dynamics and uncertainty. Chapman and Hall, New York.

Lande, R., Engen, S., and Saether, B.E. 1997. Optimal harvesting, exonomic discounting and extinction risk in fluctuating populations. Nature (Lond.), 372: 88–90.

Liermann, M., and Hilborn, R. 1997. Depensation in fish stocks: a hierarchic Bayesian metaanalysis. Can. J. Fish. Aquat. Sci. 54: 1976–1985.

Lindstrom, M.J., and Bates, D.M. 1990. Nonlinear mixed effects models for repeated measures data. Biometrics, 46: 673–687.

Mace, P.M. 1994. Relationships between common biological reference points used as threshold and targets of fisheries management strategies. Can. J. Fish. Aquat. Sci. 51: 110–122.

Mace, P.M., and Doonan, I.J. 1988. A generalized bioeconomic simulation model for fish population dynamics. N.Z. Fish. Assess. Res. Doc. 88/4.

Maynard Smith, J., and Slatkin, M. 1973. The stability of predator–prey systems. Ecology, 54: 384–384.

McAllister, M.K., Pikitch, E.K., Punt, A.E., and Hilborn, R. 1994. A Bayesian approach to stock assessment and harvest decisions using the sampling/importance resampling algorithm. Can. J. Fish. Aquat. Sci. 51: 2673–2688.

Millar, R.B., and Meyer, R. 2000. Nonlinear state space modelling of fisheries biomass dynamics by using Metropolis–Hastings within Gibbs sampling. Appl. Stat. In press.

Myers, R.A. 1997. Comment and reanalysis: paradigms for recruitment studies. Can. J. Fish. Aquat. Sci. 54: 978–981.

Myers, R.A., and Barrowman, N.J. 1996. Is fish recruitment related to spawner abundance? Fish. Bull. U.S. 94: 707–724.

Myers, R.A., and Mertz, G. 1998. The limits of exploitation: a precautionary approach. Ecol. Appl. 8(Suppl. 1): s165–s169.

Myers, R.A., Rosenberg, A.A., Mace, P.M., Barrowman, N.J., and Restrepo, V.R. 1994. In search of thresholds for recruitment overfishing. ICES J. Mar. Sci. 51: 191–205.

Myers, R.A., Bridson, J., and Barrowman, N.J. 1995. Summary of worldwide stock and recruitment data. Can. Tech. Rep. Fish. Aquat. Sci. No. 2024.

Myers, R.A., Mertz, G., and Barrowman, N.J. 1996. Invariants of spawner–recruitment relationships for marine, anadromous, and freshwater species. ICES C.M. 1996/D:11.

Myers, R.A., Hutchings, J.A., and Barrowman, N.J. 1997a. Why do fish stocks collapse? The example of cod in eastern Canada. Ecol. Appl. 7: 91–106.

Myers, R.A., Mertz, G., and Fowlow, S. 1997b. Maximum population growth rates and recovery times of Atlantic cod, *Gadus morhua*. Fish. Bull. U.S. 95: 762–772.

Pimm, S.L. 1991. The balance of nature. University of Chicago Press, Chicago, Ill.

Punt, A., and Hilborn, R. 1997. Fisheries stock assessment and decision analysis: the Bayesian approach. Rev. Fish Biol. Fish. 7: 35–65.

Robinson, G.K. 1991. That BLUP is a good thing: the estimation of random effects. Stat. Sci. 6: 15–51. With comments and a rejoinder by the author.

Searle, S.R., Casella, G., and McCulloch, C.E. 1992. Variance components. John Wiley & Sons, New York.

Suzuki, N., and Kitahara, T. 1996. Relation of recruitment to the number of caught juveniles in the ayu population of Lake Biwa. Fish. Sci. 62: 15–20.

Verbeke, G., and Lesaffre, E. 1996. A linear mixed-effects model with heterogeneity in the random-effects population. J. Am. Stat. Assoc. 91: 217–221.

Appendix 1. Estimation in SAS

This appendix demonstrates how to fit the proposed model to data for a single species. In the SAS data step, a data set is created with three variables per observation: the name of the stock (i.e., population), `stock`, the number or biomass of spawners, `s`, and the survival, `surv`, respectively. The survival is $\log(R/S)$, where recruitment, R, has been multiplied by $SPR_{F=0}(1-p_s)$, so we will obtain estimates of $\tilde{\alpha}$ in the appropriate units.

The SAS code for fitting the model with autocorrelated recruitment is

```
proc mixed method=reml;
  class stock;
  model surv= s*stock/solution;
  random int /subject=stock;
  repeated /subject=stock group=stock type=AR(1);
```

This model assumes autocorrelated errors and fits a separate first-order autocorrelation parameter and error variance for each stock. The method of estimation is REML.

Appendix 2. Robustness simulations

In order to test the robustness of the estimates from our model, we used simulations based on real data. The approach was to mimic the data for the 20 cod populations in the North Atlantic as closely as possible. In all cases, we used the observed spawner abundances, the estimated density-dependent parameters ($\hat{\beta}_i$), the estimated residual variance for each population, and the estimated mean (1.37) and variance (0.37) of the log $\tilde{\alpha}$ parameters. For each simulation, we randomly generated values for log $\tilde{\alpha}$. Then, using the observed spawner abundances, the $\hat{\beta}_i$s and the residual error variance that we produced simulated recruitment values that would closely match the actual data. We generated 1000 "realizations" of the 20 populations, randomly generating new log $\tilde{\alpha}$ values and then simulating recruitment in each case.

Fig. A1. Histograms of estimates of log α and σ_a^2 from simulations described in Appendix 2. The vertical dotted lines indicate the true estimates obtained using the data on 20 cod populations. Results from three different simulations are shown: (*a*) data simulated to match the model exactly, i.e., residuals and slope at the origin simulated from a lognormal distribution and expected recruitment given by a Ricker model, (*b*) residuals and slope at the origin simulated from a gamma distribution and expected recruitment given by a Ricker model, and (*c*) residuals and slope at the origin simulated from a lognormal distribution and expected recruitment given by a Beverton–Holt model.

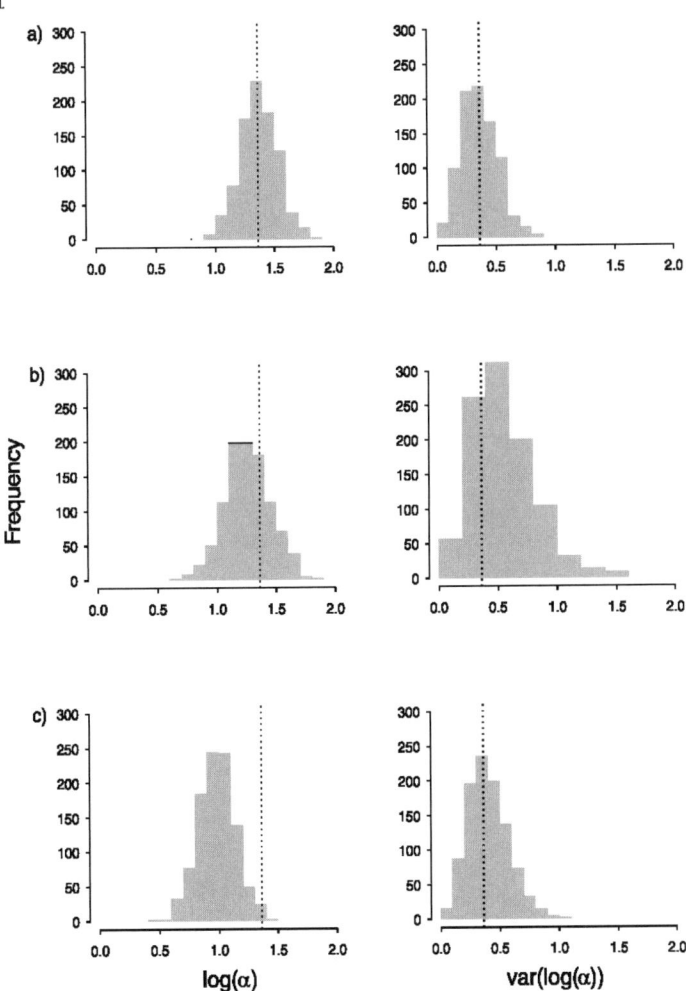

We conducted three simulations. In each case, the model that we fit was a Ricker with lognormal distributions for both the error and the slope at the origin. In the first simulation, we generated data that matched this model exactly, i.e., a parametric bootstrap. In the second simulation, we generated data that was from a Ricker but had gamma (instead of lognormal) distributed residuals and slope at the origin. This was done to test the robustness of our inferences to misspecification of the probability distributions. In the last simulation, we generated data from a Beverton–Holt model with lognormal errors and slope at the origin. This was done to test the robustness of our inferences to misspecification of the spawner–recruitment model. From the histograms of the estimated log $\tilde{\alpha}$ values, we can see that when our model was matched exactly by the data, the estimates were unbiased (Fig. A1a). When the distributional assumptions were violated, the value for log $\tilde{\alpha}$ was underestimated, while the variance was overestimated (Fig. A1b). When the form of the model was incorrect, log $\tilde{\alpha}$ was underestimated and the variance was relatively unbiased (Fig. A1c).

Canada's Recreational Fisheries:
The Invisible Collapse?

ABSTRACT

Fishing for recreation is a popular activity in many parts of the world and this activity has led to the development of a sector of substantial social and economic value worldwide. The maintenance of this sector depends on the ability of aquatic ecosystems to provide fishery harvest. We are currently witnessing the collapse of many commercial marine fisheries due to over-exploitation. Recreational fisheries are typically viewed as different from commercial fisheries in that they are self-sustaining and not controlled by the social and economic forces of the open market that have driven many commercial fisheries to collapse. Here we reject the view that recreational and commercial fisheries are inherently different and demonstrate several mechanisms that can lead to the collapse of recreational fisheries. Data from four high profile Canadian recreational fisheries show dramatic declines over the last several decades yet these declines have gone largely unnoticed by fishery scientists, managers, and the public. Empirical evidence demonstrates that the predatory behavior of anglers reduces angling quality to levels proportional to distance from population centers. In addition, the behavior of many fish species and the anglers who pursue them, the common management responses to depleted populations, and the ecological responses of disrupted food webs all lead to potential instability in this predator-prey interaction. To prevent widespread collapse of recreational fisheries, fishery scientists and managers must recognize the impact of these processes of collapse and incorporate them into strategies and models of sustainable harvest.

Introduction

By the end of the 20th century, many of the world's largest commercial fisheries have collapsed (Roughgarden and Smith 1996; Cook et al. 1997; Lauck et al. 1998). Reasons vary, but typically include economic or social incentives that drive fishers to over-exploit fish stocks (Steele 1996; Myers et al. 1997; Hutchings et al. 1997; Masood 1997). Furthermore, technologies have increased the efficiency of modern fishing such that we can search for, find, and exploit fish populations even when their abundance is low. Combined with incentives to maximize profits and employment, this technological ability leads to depensatory mortality (i.e., increases in per capita fishing mortality as populations decline in abundance) that acts to drive fish populations into commercial, if not biological, extinction.

In contrast, without the economic incentives of fishers to over-exploit, or social systems that demand employment, recreational anglers might be expected to abandon fishing opportunities that do not satisfy their expectation of quality angling and instead choose other recreational opportunities as fisheries decline (Johnson and Carpenter 1994; Hansen et al. 2000). Quantifying such a numerical response for recreational fisheries remains elusive because it appears to depend on complex angler behaviors and decisions and not a tightly coupled interaction as in typical predator-prey interactions (Johnson and Carpenter 1994; Smith 1999). Despite these uncertainties, such a numerical response should lead to compensation in the fish

---

John R. Post, Michael Sullivan, Sean Cox, Nigel P. Lester, Carl J. Walters, Eric A. Parkinson, Andrew J. Paul, Leyland Jackson, and Brian J. Shuter

Post is a professor in the Division of Ecology, Department of Biological Sciences, University of Calgary, Alberta and can be contacted at jrpost@ucalgary.ca, 403/220-6937. Sullivan is a biologist at the Alberta Fish and Wildlife Service, Edmonton. Cox is a post-doctoral fellow at the Center for Limnology, University of Wisconsin, Madison. Lester is a biologist at the Ontario Ministry of Natural Resources, Peterborough. Walters is a professor in the Fisheries Centre, University of British Columbia, Vancouver. Parkinson is a biologist at the B.C. Ministry of Fisheries, Fisheries Centre, University of British Columbia, Vancouver. Paul is a post-doctoral fellow and Jackson is a biologist at the Division of Ecology, Department of Biological Sciences, University of Calgary, Alberta. Shuter is a biologist at the Ontario Ministry of Natural Resources, Peterborough.

population (i.e., reductions in per capita mortality as populations decline in abundance) and stability in the angler-fish population interaction. If recreational fisheries behave in this self-regulating fashion, the dramatic collapses observed in many of the world's commercial fisheries should not occur in recreational fisheries.

Consistent with that expectation, the leading North American fisheries journals, *Canadian Journal of Fisheries and Aquatic Sciences*, *Transactions of the American Fisheries Society*, and *North American Journal of Fisheries Management*, published only 13 papers that refer to declines or collapses of recreational fisheries out of a total of 4,904 papers published in the 1990s. Yet, when you ask recreational fishers with several decades of experience for their assessment of the quality of fisheries resources, they invariably say "It ain't as good as it used to be!" Critical examination of the scientific and popular literature does reveal some evidence of declines in some recreational fisheries, although much of the evidence is anecdotal (Pearse 1988; Ryerson and Sullivan 1998; Schindler 1998; Brewin 1999; Sullivan 1999; Walters and Cox 1999; Cook et al. 2001; Radomski 2001; Schindler 2001). In addition to the uncertainty in the value of anecdotal information, and in contrast to commercial fisheries where a collapse simply means the elimination of the economic incentives to fish, a definition of collapse in recreational fisheries remains elusive. How low does the fish population abundance and angler catch-per-unit-effort have to be driven by overfishing until we call it a collapse? The Committee on the Status of Endangered Wildlife in Canada (COSEWIC) and the International Union for the Conservation of Nature (IUCN) risk categories and criteria of a 50% and 80% decline in 10 years, or 3 generations, which are used to define endangered or critically endangered respectively, provide us with some guidelines (Hutchings 2001). But these guidelines are not really sufficient because it is likely that it is the total magnitude of the decline from pristine conditions to the present, not the short-term rate of decline, that should be used to define a collapse. Certainly a decline in fish abundance or angler catch rates by 80% from some earlier benchmark, equivalent to the IUCN critically endangered criteria, must be acceptable as a conservative definition of a recreational fishery collapse in this era of precautionary management.

In this article we examine data for several high profile Canadian recreational fisheries and conclude that at least some recreational fisheries fit this definition of collapse. We present empirical evidence of several mechanisms inherent in recreational fisheries that are depensatory and should lead to declining fish populations and ultimately to collapse. We conclude that recreational fisheries are not necessarily self-regulating, and can operate in a manner similar to commercial fisheries and will likely suffer the same fate unless there is rapid and substantial intervention.

Evidence of Collapse

The quantitative data required for a definitive assessment of the overall state of Canada's recreational fisheries do not exist. A qualitative survey of Canadian freshwater fisheries, based largely on expert opinion, identified substantial areas showing general declines in a number of recreational fish species (Pearse 1988). These areas of decline tend to be adjacent to urban areas and the Canada-United States southern border. An assessment of trends in abundance of 14 recreational fish species in the 4 continental-scale drainage basins of Canada identify declines due to overfishing and habitat deterioration as common in salmonids, percids, and esocids (Table 1). A number of these fisheries are maintained, at least partly, by stocking (Table 1). In addition, a more quantitative assessment of temporal trends in several high profile fisheries, including 2 salmonid species, 1 percid species, and 1 esocid species (Figure 1) demonstrates that at least some Canadian recreational fisheries are collapsing. Because these 3 families make up approximately 70% of the total harvest of recreational fishes in Canada (DFO 1998), we believe these cases are representative of many recreational fisheries.

The rainbow trout (*Oncorhynchus mykiss*) fishery in south-central British Columbia includes approximately 800 trout populations (Figure 1).

Table 1. Trends in the abundance of recreational fish species by continental scale drainage basins in Canada (adapted from Pearse 1988). Symbols: declining due to overfishing ↓; declining due to habitat deterioration ⇣; stable →; increasing ↑; maintained at least partially by stocking **S**. The trends depicted in this table were extracted from tables in Pearse (1988) for all species but bull trout which was described by Post and Johnston (2001).

	Drainage Basin			
Species	Pacific	Arctic	Hudson Bay	Atlantic
Salmonids				
Rainbow trout	↓ S		↓ S	
Steelhead trout	↑ S			
Pacific salmon	↓ S			
Bull trout	↓ ↓	→	↓ ↓	
Arctic grayling	→	↓		
Lake trout	↓	↓	↓ S	↓ ↓ S
Brook trout			S	↓ ↓
Arctic char		→	↓	
Atlantic salmon				↓ ↓
Whitefish		↓	↓	↑ S
Percids				
Yellow perch			↓	→
Walleye	↓ ↓		↓ ↓ S	↓ ↓ S
Esocids				
Northern pike		↓	↓	↓ ↓
Centrarchids				
Bass				→

fisheries management **feature**

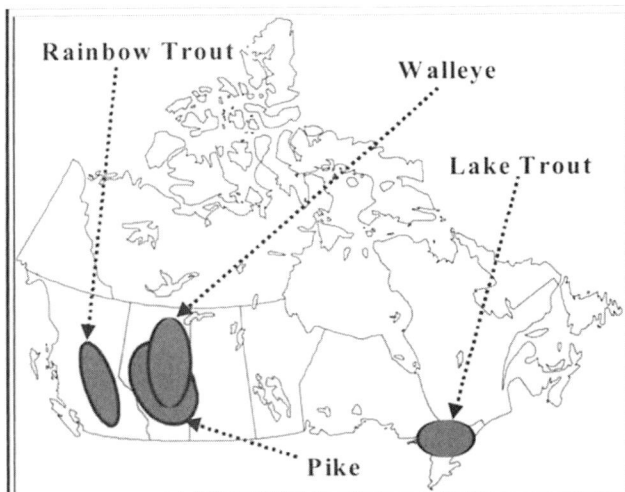

Figure 1. Canadian recreational fisheries for which there is evidence of collapse include the rainbow trout of south central British Columbia, walleye and pike of Alberta, and lake trout of southeastern Ontario.

Two populations for which time series are available show that over the last 3-4 decades, substantial increases in total angler effort are coupled with >6-fold reductions in catch rates. For example, Puntzi Lake angling effort has more than doubled from the 1960s to 1980s while catch-per-unit-effort (CPUE) has declined from 5.6 to 0.25 fish/h. Carp Lake has also seen a doubling of angler effort from the 1970s to 1980s coupled with a decline in CPUE from 2.8 to 0.5 fish/h (Eric Parkinson, B.C. Fisheries, unpublished data). Although there may be a habitat deterioration component, high angling effort is almost certainly a large factor leading to the decline.

In Alberta, 21 of 27 walleye (*Stizostedion vitreum*) populations for which we have data have collapsed because of overfishing (Sullivan in press a) (Figure 1). In Wolfe Lake, Alberta, for example, 2,000 anglers per year enjoyed a CPUE of 0.25 fish/h in the early 1980s but by the 1990s, 10,000 anglers/year experienced a catch rate of only 0.02 fish/h.

Pike (*Esox lucius*) populations in Alberta also show strong evidence of over-exploitation leading to collapse; catch rates in the 1990s were only 15% of what they were two decades earlier in 9 pike populations for which we have data (Michael Sullivan, Alberta Fish and Wildlife Service, unpublished data). For example, an angler at Kehiwin Lake would have fished approximately 2.5 hours to catch a pike in 1969 whereas in 1995 it would have taken an average of 25 hours. Associated reductions in average age, size, number of age-classes in the catch, and failed year-classes all indicate severe overfishing of Alberta pike populations.

In the Ontario lake trout (*Salvelinus namycush*) fishery of the southeast region (Figure 1), nearest the largest human population centers of the province, 60% of formerly naturally sustained lake trout populations are now maintained through stocking (Evans and Wilcox 1991). This is in sharp contrast to <1 % of lake trout lakes that are stocked in northwest Ontario, far from large urban centers.

None of these fisheries are considered collapsed by their respective management agencies, in spite of the large decline in fishing quality that has occurred. Yet these data lead one to question the hypothesis that recreational angling cannot collapse fisheries. It may be no coincidence that two of the examples come from the province of Alberta. This should not be interpreted as an indictment of Albertans or of their fishery management, but is probably a result of geography and human distribution and density. Of the four inland provinces in Canada, Alberta has ~375 licensed recreational anglers per lake, whereas Saskatchewan, Manitoba, and Ontario have <3 licensed anglers per lake (Sullivan in press a). If angling pressure is a factor responsible for Alberta's collapsed fisheries, the state of Alberta's fisheries may be a precursor of what to expect in other jurisdictions as human populations continue to grow and the pressure on fish populations increase.

Why Invisible?

Why are such collapses largely invisible in the scientific literature, public perception, and management action? Why are disgruntled anglers, outfitters, bait dealers, and fishing lodge owners not up in arms? It cannot be because recreational fisheries have a trivial economic impact at home and abroad. The recreational fishing "industry" in Canada has been evaluated at CAN$ 4.4–7 billion annually (DFO 1998). Non-Canadians contributed

26% of the expenditures directly attributed to recreational fishing in Canada, through the participation of 750,000 American, 10,000 British and European, and 1,600 other anglers (DFO 1998). These non-Canadian anglers spent 5.3 million angler-days taking advantage of the perceived unlimited recreational fishing opportunities in Canada. Direct expenditures by anglers on recreational fishing in 1995 exceeded the cumulative value of the landings of all Canadian commercial fisheries in that year by 1.4 times (DFO 1998, 2000).

We propose several reasons for the invisibility of collapses. First, recreational fisheries in Canada number in the hundreds of thousands and tend to be small and diffuse across the landscape. Each one impacts only a relatively small number of anglers and localized economies and therefore receives little regional or national exposure as it collapses. Second, fish populations display considerable spatial and temporal variability and it is difficult for individual anglers or management agencies to develop an accurate picture of processes occurring at scales longer and larger than their own experience. Declines in populations of long-lived species can be slow, and poor intergenerational memory may lead to declining angler expectations as fish populations decline (coined the "shifting baseline" syndrome by Pauly 1995). Photographs of anglers and their catch from the first half of the 20th century provide a sobering perspective of the decline of recreational fisheries over the century (see photo). Third, the declining quality of fisheries can be temporarily masked by fish stocking, fertilization of water bodies, or other management initiatives intended to artificially maintain angling opportunities in the face of increasing demand coupled with dwindling natural stocks. And finally, managers of recreational fisheries have literally thousands of discrete populations to manage with woefully inadequate resources (in proportion to the economic and social value of the resource) to do the assessments necessary to inventory and characterize the status of the recreational fisheries within their jurisdiction.

Fisheries as Predator-Prey Systems

Recreational fisheries can be considered as a series of resources distributed spatially across a landscape and subject to exploitation by human predators with substantial capacity for mobility and communication. Therefore, the system is analogous to the ideas of functional and numerical responses from predator-prey theory (Hilborn and Walters 1992; Smith 1999). Human predator behavior interacts with fish populations at two spatial scales. At the larger regional scale, anglers respond numerically by allocating effort in response to quantitative measures of fishing quality such as fish density (Figure 2a). This numerical response is stronger if the travel time to a prey patch is short. For example, rainbow trout lakes close to the main population center in British Columbia, the greater Vancouver area, attract approximately 2.5 times more anglers-per-unit fish density than do more distant lakes (Figure 2a). This effort response, which depends on resource quality, has a direct consequence on fish populations through exploitation rate (Cox 2000) and therefore mortality and population growth. As a consequence, measures of the functional response, such as angler catch rate, are low in patches characterized by short travel time and high in patches characterized by longer travel time (Figure 2b). These processes result in a reduction in fish abundance as an inverse function of travel distance. Further, homogenization of angling quality within strata of similar travel time may occur based on low variance in catch rate within travel distance strata (Figure 2b). Nevertheless, there is no evidence that the angler effort response at this regional scale would lead to collapse of recreational fisheries. In fact, the numerical response relationships would suggest that as fish abundance declines, angler effort should dissipate in a compensatory manner, allowing stabilization of the predator-prey dynamics, albeit in a low quality state. It is this general observation that leads to the supposition that recreational fisheries tend towards self-regulation (Johnson and Carpenter 1994; Hansen et al. 2000).

At a smaller spatial scale, that of individual fish populations, angler behavior interacts with fish populations in a manner that results in density-dependent catchability. Catchability is the proportion of the fish stock removed per-unit-effort. For fish species that aggregate, catchability should increase as fish population abundance declines if fishers are capable of successfully locating aggregations and exploiting them. Aggregation can be behavioral or habitat-mediated. Behavioral aggregation (shoaling or schooling) is common in pelagic fishes whereas many other species aggregate in association with preferred and spatially limited habitat. A reduction in overall abundance would result in a reduced number of aggregations or an overall range contraction as fish abandon less preferred habitat; however, the density of fish within localized aggregations would remain unchanged or even increase as abundance decreases (Rose and Kulka 1999). This interaction between anglers and fish populations can be predicted by ideal free distribution theory, which assumes the predators have knowledge of prey aggregation behavior (Rosenzweig 1991). Information on fish location and effective communication among anglers provides this knowledge (Smith 1999). Aggregation behavior is common in freshwater fish species targeted by recreational fishers and should result in inverse density-dependent catchability. Data from

12 Ontario lake trout populations that range approximately 25-fold in density, provides strong evidence of inverse density-dependent catchability (Shuter et al. 1998) (Figure 3). This same process, based on the spatial behavior of individual fishers and fish, has led to the dramatic collapse of many commercial fisheries (Rose and Kulka 1999) and likely functions no differently in recreational fisheries.

The occurrence of density-dependent catchability is key to our understanding of the dynamics of recreational fisheries. First, it can lead to population collapse since the proportion of the fish population caught per-unit-effort increases as densities decline causing an increased rate of decline as populations are fished down. Second, angler catch-per-unit-effort, a commonly used measure of population abundance, will initially be invariant as populations decline. Ontario lake trout populations show density-invariant CPUE over broad ranges of density (Figure 3). Therefore, invariant CPUE is not evidence of population stability in fisheries with density-dependent catchability. Density-invariant CPUE has the pathological effect of lulling managers into the belief that everything is fine until the system is well down

Changes in the catches of species of fish at one family's fishing lodge at a large (20,000 ha) northern Saskatchewan lake illustrate the hidden decline in recreational fisheries. The young girl in this 1942 photograph is second author M. Sullivan's mother, balancing a stringer of large walleye. At that time, the highly-desired lake trout fishery of her grandfather's time had collapsed through overfishing, and the family's guiding business had shifted to the large, abundant, and popular walleye. During the mid-1970s, lake trout were extirpated, walleye of the size and abundance in this photograph were unheard-of, and guiding was exclusively for large, but least-desired, northern pike. In recent years (1990s), pike remain abundant but small, a clear sign of growth overfishing. In spite of these spectacular collapses, sport fishermen have continued to travel to the lake (sustaining the guiding business), but with expectations that mirror the decline and change in species availability throughout lakes in this popular Canadian fishing area. In this manner, the collapse of highly-desired prey items such as trout and walleye is masked by the human predators' shift to other less-preferred species such as pike. If the pike fishery collapses (as has happened at lakes in nearby Alberta), the guiding business will also then collapse, because no sportfish species remain to be exploited. This will appear to be a "sudden" economic loss, but in reality it has been in progress for many decades and is only clearly understood when considering the changes spanning five generations of this family. This long-term and critical perspective is seldom available to fisheries managers.

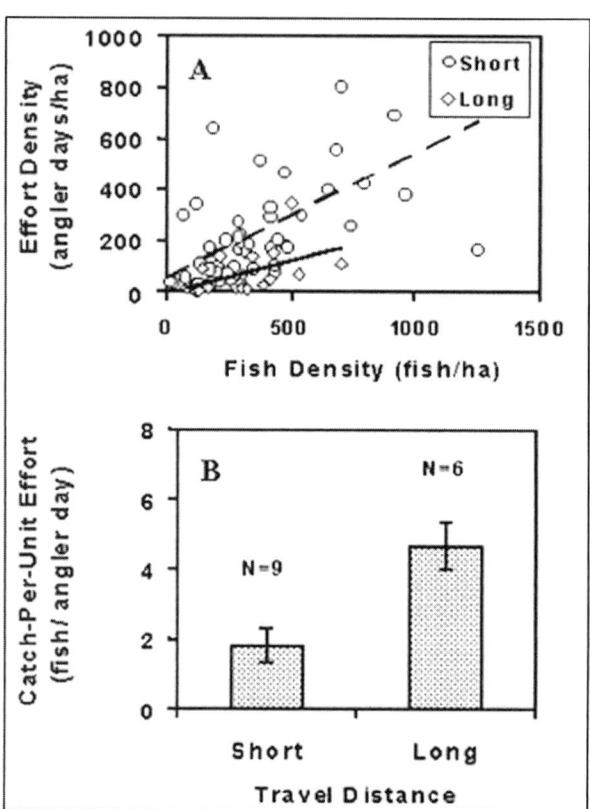

Figure 2. (a) Numerical response of anglers (angler days/ha) to fish density (fish/ha) of rainbow trout in two regions of south central British Columbia that differ in distance from Vancouver (52 lakes in the Kamloops and Okanagan regions with a short travel distance of 3–5 hours indicated by circles and dashed line and 20 lakes in the Williams Lake region with a long travel distance of 5–8 hours indicated by diamonds and solid line). The two numerical responses have the same slopes (F-ratio 0.776, df=1, p=0.382) but different intercepts (F-ratio 9.033, df=1, p=0.004). (b) Catch-per-unit-effort (fish/angler/day) in short and long travel distance regions differed significantly (F=47.88, df=1, p<0.001). Fish density was estimated from mean annual stocking rate of rainbow trout; angling effort density was estimated from aerial boat counts; catch-per-unit-effort was estimated from a creel census on a subset of 15 lake-years (Cox 2000).

the slippery slope towards collapse. The reductions in CPUE over two to three decades for the several fisheries presented above are likely indicative of dramatic population collapses, not merely subtle changes in population abundance.

Management Actions and Human Behavior

A common management response to declines in angling catch rates is the imposition of regulations to reduce harvest (Radomski et al. 2001). Regulations limiting fish size for harvest or maximum per-angler daily harvest limits provide only a blunt instrument to control total harvest because they only limit the harvest by individual anglers and not the total number of anglers using the resource. Imposition of minimum-size limits for harvest is a common regulation intended to reduce total harvest by protecting juvenile fish. Unfortunately, these regulations are not always successful. In Alberta walleye fisheries, noncompliance with regulations increased with declining catch rates (Figure 4) leading to greater per capita mortality at lower fish abundance (Sullivan in press b). If general, this human behavior will also contribute to depensatory mortality as fish populations decline.

Another common management response to declines in recreational fisheries is artificial propagation in hatcheries and stocking into natural waters (Hilborn 1992; Radomski et al. 2001). Indeed, the development of the huge hatchery infrastructure in North America in the second half of the last century may itself be credible evidence of the decline of native stocks (Pearse 1988; Hilborn 1992) and also one reason for the apparent invisibility of collapses. We stock a diversity of native and non-native species in waters containing native species despite a large literature on the negative ecological and genetic impacts of these stocking programs (Hilborn 1992). Ontario lake trout are prized recreational fish, yet 60% of formerly viable natural lake trout populations in southeastern Ontario are now maintained partially or exclusively by hatchery propagation (Evans and Wilcox 1991). Stocking on top of depleted wild stocks of lake trout leads to the loss of wild

fisheries management **feature**

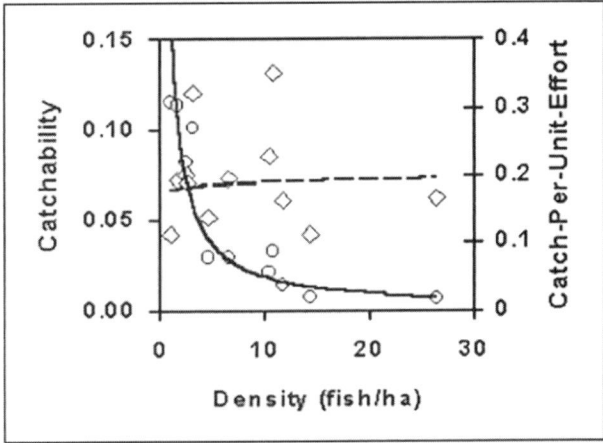

Figure 3. Catch-per-unit-effort (fish/day) did not vary significantly with fish density (fish/ha) for 12 Ontario lake trout fisheries (F=0.072, df=11, p=0.794) (diamonds and dashed line). Catchability (proportion caught/effort) varied significantly as a negative exponential with fish density (F=44.52, df=1, p<0.001) (circles and solid line) (data from Shuter et al. 1998).

stocks because: (1) the number of artificially produced recruits can easily exceed the number of natural recruits at low natural stock densities, particularly in small lakes, (2) high angling exploitation rates differentially reduces the reproductive potential of the wild stocks, and (3) juvenile hatchery-produced lake trout will cannibalize the smaller-bodied naturally produced juveniles (Evans and Wilcox 1991). Stocking is capable of maintaining exploitation rates well above that which is sustainable by wild stocks, thereby compounding the angling effort imposed on natural stocks. From the standpoint of maintaining natural gene pools, stocking depleted populations is a management response that also acts in a depensatory manner and can hasten the collapse of native stocks.

Altered Food Web Structure

Recreational fish species are often imbedded in complex food webs. The large-bodied freshwater fish species that are the primary targets of many recreational fisheries are successful, in part, due to "cultivation effects," where they crop down forage species that are competitors and/or predators on the juveniles of their own species (Walters and Kitchell 2001). When the abundance of adults of a recreational species is depressed due to exploitation, compensatory increases in the forage species may limit the ability of recreational species to rebound due to suppression of juveniles through predation and competition. We have evidence from food webs in which walleye populations have been reduced in abundance by exploitation that there has been a predatory release on small-bodied fishes (of the family Cyprinidae) (Figure 5a) and that these small-bodied fishes both eat and compete with larval and juvenile walleye. Since fisheries tend to reduce the abundance of the larger and more fecund individuals in a population, total population fecundity declines more quickly than numerical abundance (Figure 5a). As a consequence of this decline in population fecundity, there is a substantially higher potential for predation by small-bodied fishes on eggs or juveniles of the targeted species (Figure 5b). If these depensatory food web processes are as common in aquatic systems as has been recently suggested (Walters and

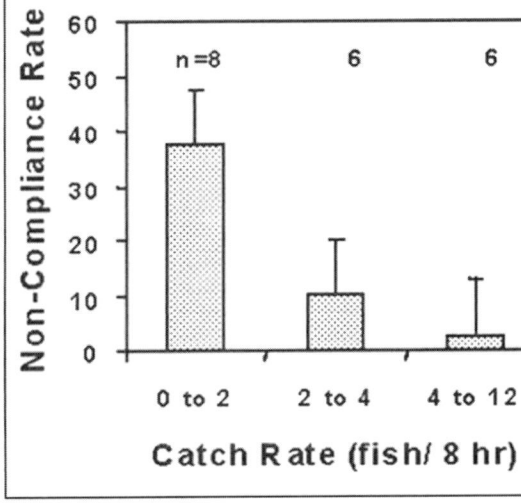

Figure 4. Non-compliance rate (percent of sub-legal sized fish caught that are not returned) as a function of angler catch rate (fish/8 hr) from Alberta walleye fisheries. The sample size in lake-years and standard errors are indicated for each catch rate class. Angler non-compliance with size-limits was determined for 20 walleye fisheries in Alberta that varied widely in walleye abundance, between 1992 and 1998, by contrasting size ratios of illegal-legal size ratios in a test fishery with ratios reported by anglers during creel surveys (data from Sullivan in press b).

Kitchell 2001), they produce another depensatory mechanism that will exacerbate rates of collapse and would likely foil attempts to rehabilitate collapsed fish stocks through simple reductions in harvest.

Uncertainties, Dynamics and the Future

Dynamics of angler effort alone should not lead to collapse of recreational fisheries but instead force the quality of fisheries to lower levels that are inversely proportional to anglers' travel distance. However, the combination of an angler effort response and several processes capable of producing depensatory fish mortality should lead to a dynamic predator-prey system for which instability and collapse are likely. The presence of depensatory dynamics in commercial fisheries has been assessed rigorously with the benefit of hundreds of data time series, although the occurrence, strength, and mechanisms of depensation are under debate (Myers et al. 1995; Liermann and Hilborn 1997; Myers et al. 1999; Shelton and Healey 1999; Frank and Brickman 2000). What is clear, is that most overexploited commercial stocks showed only limited or no recovery within three generations (Hutchings 2000). It is unclear to what degree depensatory mechanisms are responsible for this lack of recovery, but depensation may be more prevalent than formerly thought (Liermann and Hilborn 1997). We don't yet know how prevalent depensatory dynamics are in recreational fisheries because of the paucity of time series from these systems. But, by demonstrating here the potential of four mechanisms of depensation in recreational fisheries, we suggest that depensatory dynamics are likely a general feature of these fisheries.

Figure 5. (a) The relationship between an index of Cyprinid abundance and an index of walleye abundance ($r^2=0.90$, $p=0.034$) and an index of walleye population fecundity versus an index of walleye abundance ($r^2=0.98$, $p<0.004$) in 4 Alberta walleye lakes that have been differentially overexploited by angling. (b) The relationship between an index of the number of Cyprinid predators per walleye egg and an index of walleye abundance ($r^2=0.97$, $p<0.011$). Catches of Cyprinids in beach seine hauls near the mouths of walleye spawning rivers in the spring provided estimates of relative abundance among lakes. Laboratory feeding trials demonstrated that Cyprinids would consume larval walleye. Walleye population fecundity was estimated from direct counts of eggs of females of a range of sizes, size structure of mature female walleye and an index of walleye abundance across lakes (unpublished data from M. Sullivan, Alberta Fish and Wildlife Service, Edmonton, Alberta).

Many models of fishery dynamics used by managers assume logistic population growth (Hilborn and Walters 1992):

$$\frac{dN_t}{N_t dt} = r\left(1 - \frac{N_t}{K_U}\right)$$

This representation of population growth assumes linear negative density-dependence with the maximum per capita rate of population growth (r) as the population size (N) approaches zero with a carrying capacity (K_U) at which per capita growth rate is zero. Population densities above K_U are characterized by a negative per capita growth rate that has the effect of returning N to K_U. This implies a stable unfished equilibrium at the environmental carrying capacity and compensation in birth and death rates that result in negative density-dependence in the per capita growth rate (Figure 6a). This form represents pure compensation with the highest positive per capita growth at the lowest population size, zero per capita growth at the carrying capacity and negative per capita growth at population sizes exceeding carrying capacity. To this model of compensation we add a negative exponential depensation term, similar in form to that which we observed for density-dependent catchability and human cheating behavior:

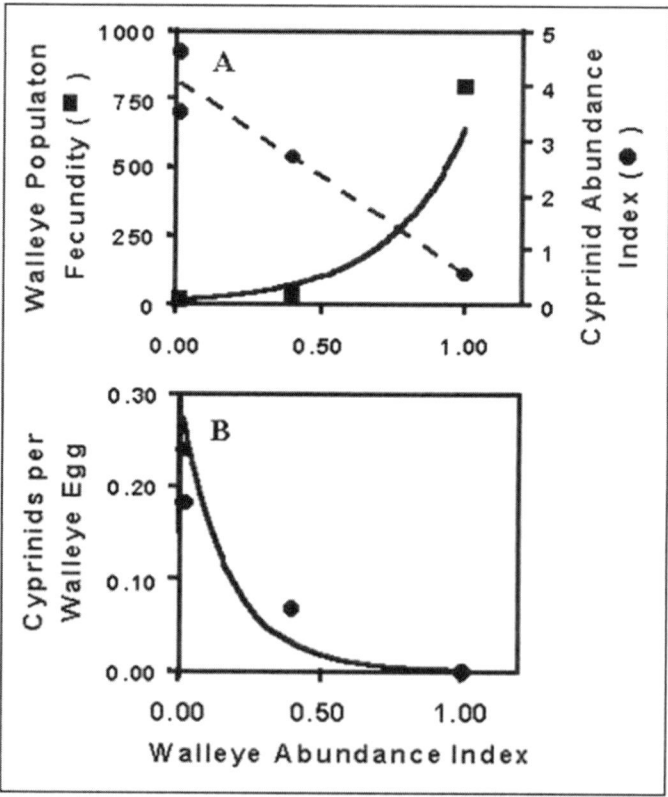

fisheries management
feature

Figure 6. (A) Predicted per capita population growth rates as a function of population size for populations with logistic growth (dashed line) and logistic growth plus exponential depensatory mortality (solid line). The arrows indicate the predicted direction of population change across ranges of population size. Both models predict an upper stable equilibrium (upper carrying capacity K_U). The depensatory model also has a lower unstable equilibrium (K_L). (B) The combined compensatory and depensatory model has three phases noted by net compensatory growth, depensatory growth, and critical depensatory negative growth. (C) The occurrence of multiple additive depensatory mechanisms increases the population size at the lower unstable equilibrium (from weak to strong depensation indicated by the arrow). (D) The population size that maximizes production increases from logistic population growth (dashed line) to increasing strengths of depensatory mechanisms (solid lines) and the maximum production decreases. The black arrow indicates declines in production as the strength of depensatory processes increases and the open arrow indicates the resultant change in the population size that maximizes production.

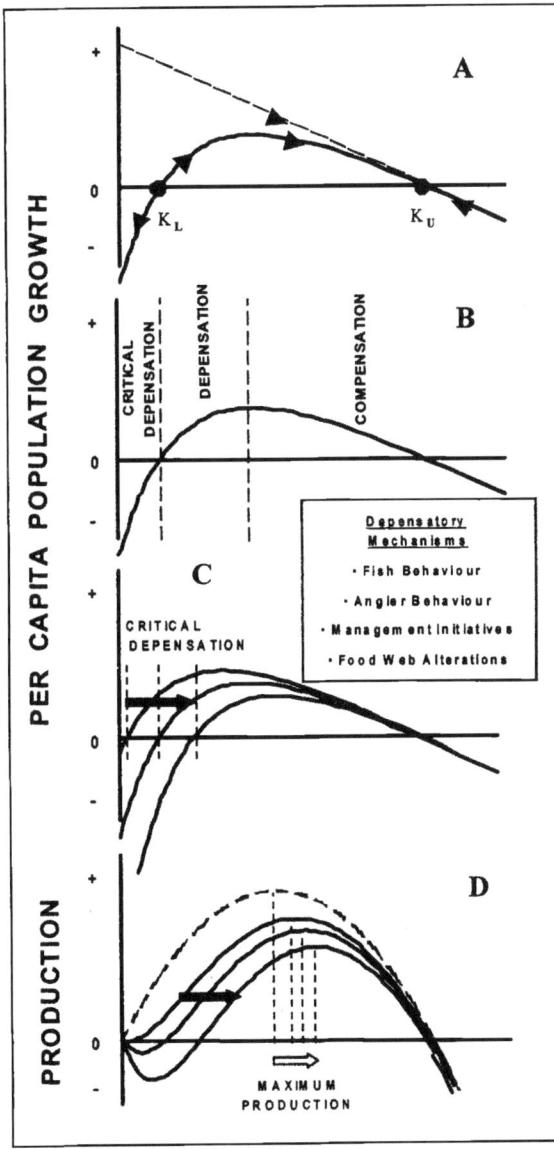

$$\frac{dN_t}{N_t dt} = r\left(1 - \frac{N_t}{K_U}\right) - \alpha e^{\beta N_t}$$

where α represents the mortality rate as $N \rightarrow 0$ and β is the rate at which mortality declines as N increases. The presence of depensatory processes alters the simple dynamics observed in their absence and has important implications for how we manage recreational fisheries. An upper unfished stable equilibrium still exists with depensation (K_U), but in addition, there is also a lower and unstable equilibrium (K_L), above which reductions in fishing will allow recovery (Figure 6a). If exploitation reduces population size below this lower equilibrium, the population will progress to extinction because the strength of depensation exceeds the compensatory abilities of the population. These alternate caricatures of dynamics thus appear qualitatively similar at intermediate-to-high population densities, where the net effect is compensation (i.e., per capita population growth declines with increasing population abundance). Above the lower unstable equilibrium, though, is a region of depensation where declines in density are exacerbated by reduced per capita population growth and increasing sensitivity to overexploitation (Figure 6b). Critical depensation occurs below the lower unstable equilibrium, where population growth becomes negative and populations are predicted to decline to extinction (Figure 6b).

The depensatory mechanisms that we characterize, involving fish behavior, angler behavior, management initiatives, and food web alterations should all be capable, when coupled with sufficient angler effort, of reducing fish populations to the lower unstable equilibrium and therefore to collapse. The occurrence of multiple depensatory mechanisms likely would be additive, increasing population sizes at which this critical depensation is reached and population collapse becomes inevitable (Figure 6c).

The presence of depensatory processes therefore has three important consequences for managers of recreational fisheries. As fisheries decline in response to exploitation, depensation will accelerate the decline towards collapse. A second, and more insidious implication, is that with depensation, there is no guarantee that if populations are reduced below this lower unstable equilibrium, the effects of overfishing can be reversed. Third, if depensatory processes occur in harvested systems, not only will production be lower than predicted in purely compensatory systems, but the population size required to support maximum production will increase with the strength of depensation (Figure 6d). Therefore, management models not incorporating depensatory processes will underestimate population size at maximum sustainable production, overestimate sustainable exploitation, and expose fisheries to risk of collapse.

To ignore the potential for depensation in recreational fisheries would be inconsistent with precautionary approaches to management and conservation. Identification of the threshold that delimits critical depensation provides a minimum population size target for conservation. The maximum net population growth rate that results from the difference between the intrinsic rate of population growth and depensatory mortality provides a higher minimum threshold for precautionary management target population size. Research aimed at determining the extent and strength of the depensatory processes that we identified in specific recreational fisheries is necessary to identify these population size thresholds for conservation and precautionary management.

All recreational fish species, and populations of species, are not likely to be equally vulnerable to depensatory processes and collapse. Species with age-dependent survival and fecundity schedules that lead to high net reproductive rate and short generation times have a higher capacity for compensation (Winemiller and Rose 1992), reducing the range of population sizes susceptible to critical depensation and collapse. Species with early maturity and high juvenile survivorship, such as many centrarchids, are less likely to be driven to collapse through depensatory processes. Latitudinal variation in life history traits within species, such as age-dependent growth and survival and age-at-maturity (Shuter and Post 1990), imply decreasing compensatory ability with latitude and therefore increasing susceptibility to depensation and collapse. As an example, age-at-maturity of walleye in North America varies markedly with latitude and climate (Baccante and Colby 1996). Southern populations (>5,000 growing degree days) mature at 2–3 years of age and northern populations (1,000 growing degree days) at >8 years of age. This broad variation in life history traits across latitude leads to decreasing compensatory abilities, greater impacts of depensatory processes, and to increased risk of collapse as latitude increases.

What is the prognosis for Canada's recreational fisheries? We argue that if the collapses that we observe in some high profile recreational fisheries are real and general, and remain largely invisible, that many recreational fisheries are headed in the same direction as are the world's commercial fisheries. Unfortunately, the quantitative data required to assess the general state of recreational fisheries are not available. The National Recreational Fisheries Survey of Canada (DFO 1998) supplies data describing the state of the resource from the perspective of the angler, but not of the fish populations. We have seen that data on angler catch rates are not necessarily reflective of fish abundance until populations are near collapse. This lag between observed angler catch rates and fishery declines ensures management inaction until it is too late. We therefore require: (1) fishery independent assessments of the status of fish populations, and (2) changes in the management of recreational fishing that increase the visibility of fish population declines to agency biologists, the public, and politicians. We must also recognize, quantify, and incorporate depensatory processes, where and when they exist, into dynamic management models to identify thresholds of population abundance that are necessary to sustain fish populations and the social and economic value that they provide. Only then can fisheries management, and society as a whole, hope to respond in a timely fashion to avoid the collapses and costly mistakes that have characterized the science and management of many of the world's commercial fisheries.

Acknowledgements

We thank Ed McCauley, Lars Rudstam, and Paul Radomski for their reviews of an earlier draft of the manuscript. This work was supported by a Strategic Research Grant from the Natural Science and Engineering Research Council of Canada (CJW, JRP, EAP).

References

Baccante, D. A., and P. J. Colby. 1996. Harvest, density and reproductive characteristics of North American walleye populations. Annales Zoologici Fennici 33:601-615.

Brewin, M. K. 1999. Relationships between recreational angling and native salmonids in Alberta. Fisheries Centre Research Reports 7:51-57.

Cook, M. F., T. J. Goeman, P. J. Radomski, J. A. Younk, and P. C. Jacobson. 2001. Creel limits in Minnesota: a proposal for change. Fisheries 26(5):19-26.

Cook, R. M., A. Sinclair, and G. Stefansson. 1997. Potential collapse of North Sea cod stocks. Nature 385:521-522.

Cox, S. 2000. Angling quality, effort response, and exploitation in recreational fisheries: field and modelling studies on British Columbia rainbow trout (*Oncorhynchus mykiss*) lakes. Ph.D. thesis, University of British Columbia, Vancouver.

DFO (Canada Department of Fisheries and Oceans). 1998. 1995 Survey of recreational fishing in Canada. www.nrc.dfo/communic/statistics/recfish95/content2.htm. Ottawa, Ontario.

_____. 2000. Summary of Canadian commercial catches and values. www.nrc.dfo/communic/statistics/landings/smry9295.htm. Ottawa, Ontario.

Evans, D. O., and C. C. Wilcox. 1991. Loss of exploited, indigenous populations of lake trout, *Salvelinus namaycush*, by stocking of non-native stocks. Canadian Journal of Fisheries and Aquatic Sciences 48(S1):134-147.

Frank, K. T., and D. Brickman. 2000. Allee effects and compensatory population dynamics within a stock. Canadian Journal of Fisheries and Aquatic Sciences 57:513-517.

Hansen, M. J., T. D. Beard, Jr., and S. W. Hewett. 2000. Catch rates and catchability of walleyes in angling and spearing fisheries in northern Wisconsin lakes. North American Journal of Fisheries Management 20:109-118.

Hilborn, R. 1992. Hatcheries and the future of salmon in the northwest. Fisheries 17(1):5-8.

Hilborn, R., and C. J. Walters. 1992. Quantitative fisheries stock assessment. Chapman and Hall, London.

Hutchings, J. A. 2000. Collapse and recovery of marine fishes. Nature 406:882-885.

_____. 2001. Conservation biology of marine fishes: perceptions and caveats regarding assignment of extinction risk. Canadian Journal of Fisheries and Aquatic Sciences 58:108-121.

Hutchings, J. A., C. J. Walters, and R. L. Haedrich. 1997. Is scientific inquiry incompatible with government information control? Canadian Journal of Fisheries and Aquatic Sciences 54:1198-1210.

Johnson, B. M., and S. R. Carpenter. 1994. Functional and numerical responses: a framework for fish-angler interactions? Ecological Applications 4:808-821.

Lauck, T., C. W. Clark, M. Mangel, and G. R. Munro. 1998. Implementing the precautionary principle in fisheries management through marine reserves. Ecological Applications 8:S72-S78.

Liermann, M., and R. Hilborn. 1997. Depensation in fish stocks: a hierarchic Bayesian meta-analysis. Canadian Journal of Fisheries and Aquatic Sciences 54:1976-1984.

Masood, E. 1997. Fisheries science: all at sea when it comes to politics? Nature 386:105-106.

Myers, R. A., N. J. Barrowman, J. A. Hutchings, and A. A. Rosenberg. 1995. Population dynamics of exploited fish stocks at low population levels. Science 269:1106-1108.

Myers, R. A., K. G. Bowen, and N. J. Barrowman. 1999. Maximum reproductive rate of fish at low population sizes. Canadian Journal of Fisheries and Aquatic Sciences 56:2404-2419.

Myers, R. A., J. A. Hutchings, and N. J. Barrowman. 1997. Why do fish stocks collapse? The example of cod in Atlantic Canada. Ecological Applications 7:91-106.

Pauly, D. 1995. Anecdotes and the shifting baseline syndrome of fisheries. Trends in Ecology and Evolution 10:430.

Pearse, P. 1988. Rising to the challenge. Canadian Wildlife Federation. Vancouver, British Columbia.

Post, J. R., and F. D. Johnston. 2001. The status of bull trout (*Salvelinus confluentus*) in Alberta. Wildlife status report. Alberta Environment, Fisheries and Wildlife Management Division and Alberta Conservation Association,

Edmonton.

Radomski, P. J., G. C. Grant, P. C. Jacobson, and M. F. Cook. 2001. Visions for recreational fishing regulations. Fisheries 26(5):7-18.

Rose, G. A., and D. W. Kulka. 1999. Hyperaggregation of fish and fisheries: how catch-per-unit-effort increased as the northern cod (*Gadus morhua*) declined. Canadian Journal of Fisheries and Aquatic Sciences 56(S1):118-127.

Rosenzweig, M. L. 1991. Habitat selection and population interactions: the search for mechanisms. American Naturalist 137:S5-28.

Roughgarden, J. and F. Smith. 1996. Why fisheries collapse and what to do about it. Proceedings of the National Academy of Science 93:5078-5083.

Ryerson, D., and M. Sullivan. 1998. Where have all the northern pike gone? Federation Alberta Naturalists 28:12-13.

Schindler, D. W. 1998. Sustaining aquatic ecosystems in boreal regions. Conservation Ecology 2:18.

_____. 2001. The cumulative effects of climate warming and other human stresses on Canadian freshwaters in the new millennium. Canadian Journal of Fisheries and Aquatic Sciences 58:18-29.

Shelton, P. A., and B. P. Healey. 1999. Should depensation be dismissed as a possible explanation for the lack of recovery of the northern cod (*Gadus morhua*) stock? Canadian Journal of Fisheries and Aquatic Sciences 56:1521-1524.

Shuter, B. J., M. L. Jones, R. M. Korver, and N. P. Lester. 1998. A general life history based model for regional management of fish stocks: the inland lake trout (*Salvelinus namaycush*) fisheries of Ontario. Canadian Journal of Fisheries and Aquatic Sciences 55:2161-2177.

Shuter, B. J., and J. R. Post. 1990. Climate, population viability and the zoogeography of temperate fishes. Transactions of the American Fisheries Society 119:314-336.

Smith, B. D. 1999. A probabilistic analysis of decision-making about trip duration by Strait of Georgia sport anglers. Canadian Journal of Fisheries and Aquatic Sciences 56:960-972.

Steele, J. H. 1996. Regime shifts in fisheries management. Fisheries Research 25:19-23.

Sullivan, M. 1999. Using social and biological reference points in managing sport fisheries. Fisheries Centre Research Reports 7:164-165.

_____. in press a. The collapse and controversial recovery of walleye fisheries in Alberta. North American Journal of Fisheries Management.

_____. in press b. Illegal sport harvest of walleye protected by size limits in Alberta. North American Journal of Fisheries Management.

Walters, C., and S. Cox. 1999. Maintaining quality in recreational fisheries: how success breeds failure in management of open access sport fisheries. Fisheries Centre Research Reports 7:22-29.

Walters, C. J., and J. F. Kitchell. 2001. Cultivation-depensation effects on juvenile survival and recruitment: a serious flaw in the theory of fishing? Canadian Journal of Fisheries and Aquatic Sciences 58:39-50.

Winemiller, K. O., and K. A. Rose. 1992. Patterns of life-history diversification in North American fishes: implications for population regulation. Canadian Journal of Fisheries and Aquatic Sciences 49:2196-2218.

Stock and Recruitment[1]

By W. E. Ricker
Pacific Biological Station, Nanaimo, B.C.

ABSTRACT

Plotting net reproduction (reproductive potential of the *adults* obtained) against the density of stock which produced them, for a number of fish and invertebrate populations, gives a domed curve whose apex lies above the line representing replacement reproduction. At stock densities beyond the apex, reproduction declines either gradually or abruptly. This decline gives a population a tendency to oscillate in numbers; however, the oscillations are damped, not permanent, unless reproduction decreases quite rapidly *and* there is not too much mixing of generations in the breeding population. Removal of part of the adult stock reduces the amplitude of oscillations that may be in progress and, up to a point, *increases* reproduction.

CONTENTS

INTRODUCTION	
General	560
Theory of population regulation	560
Age incidence of compensatory mortality	561
Kinds of population control mechanisms	562
TYPES OF REPRODUCTION CURVES	563
REPRODUCTION IN THE ABSENCE OF DENSITY-INDEPENDENT VARIABILITY	567
VARIATIONS IN REPRODUCTION PRODUCED BY FACTORS INDEPENDENT OF DENSITY	572
COMBINATIONS OF COMPENSATORY AND NON-COMPENSATORY MORTALITY	579
EFFECTS OF REMOVAL OF MATURE STOCK	580
EXAMPLES OF REPRODUCTION CURVES	586
Pacific herring (*Clupea pallasi*)	587
Pink salmon (*Oncorhynchus gorbuscha*)	588
Coho salmon (*Oncorhynchus kisutch*)	590
Sockeye salmon (*Oncorhynchus nerka*)	590
Haddock (*Melanogrammus aeglifinus*)	593
Fruit fly (*Drosophila melanogaster*)	595
Water-flea (*Daphnia magna*)	596
Starfish (*Asterias forbesi*)	601
Sinuous reproduction curves	602
Discussion	604
OTHER REPRODUCTION SITUATIONS	606
Larger immature fish taken by the fishery	606
Exploitation during the compensatory phase	607
Compensation by immature members of the population	608
TOWARD A THEORY OF RECRUITMENT	609
Theory of predation	609

[1]Received for publication June 1, 1953; as revised, April 30, 1954. Portions of this paper appeared in the *Journal of Wildlife Management* for January, 1954, as a contribution to a symposium on cycles in animal populations.

Cannibalism	610
Predation by other organisms	613
Other compensatory agents	617
Fitting the curve $z = we^{1-w}$	618
SUMMARY	619
ACKNOWLEDGMENTS	621
REFERENCES	621

INTRODUCTION

GENERAL

There exists today a considerable body of knowledge which goes by the name of "the theory of fishing" or "the modern theory of fishing"—the work of a succession of the most distinguished fishery biologists of our time. It is concerned mainly with predicting what catch can be obtained from a given number of young fish recruited to a fishery, if their initial size and the growth and natural mortality rates prevailing are known. That is, methods have been developed for computing the effects of different rates of exploitation, of changes in rate of exploitation from year to year, of different minimum size limits, etc., upon the yield obtained. Not only that, but much progress has been made in developing methods of determining the actual magnitudes of the population statistics required to make these calculations.

Valuable as the above contributions have been, they comprise only half of the biological information needed to assess the effects of fishing and an optimum level of exploitation. Fishing changes the absolute and the relative abundance of mature fish in a stock, and the effect of this upon the number of recruits in future years has often been considered only in the most general manner. The points of view encountered usually range from an assumption of direct proportion between size of adult stock and number of recruits, to the proposition that number of recruits is, for practical purposes, independent of the size of the adult stock. The possibility of a *decrease* in recruitment at higher stock densities has less often been considered.

The scarcity of information on this subject is quite explicable, since it usually requires many years of continuous observation to establish a relation between size of stock and the number of recruits which it produces. However, it has become an urgent problem to have a scientific description of the regulation of abundance of fish stocks, in order to complete the basis for predicting optimum levels of exploitation. This paper attempts to summarize some of the theoretical and factual information available, both from fish populations and from other animals, and to provide a stimulus to studies which will eventually put the subject on a solid foundation.

THEORY OF POPULATION REGULATION

Basic in any stock-recruitment relationship is the fact that a fish population, even when not fished, is limited in size; that is, it is held at some more or less fluctuating level by natural controls. Ideas concerning the nature of such controls were first clarified and systematized by the Australian entomologist Nicholson

(1933). He showed that, while the level of abundance attained by an animal can be affected by any element of the physical or biological environment, the immediate mechanism of control must always involve competition, using that word in a broad sense to include any factor of mortality whose effectiveness increases with stock density[2]. The term *density-dependent* mortality was used for the same concept by Smith (1935). More strictly, density-dependent causes of mortality should include both those which become more effective as density increases and those which become less so. The former are the ones which provide control of population size; and they have been called *concurrent* (Solomon, 1949), *compensatory* (Neave, 1953) or *negative* (Haldane, 1953). The opposed terms are *inverse, depensatory* and *positive*, all referring to density-dependent factors which become *less* effective as density increases.

No sharp line can be drawn between the kinds of mortality which are compensatory and those which are not, although Nicholson, Smith and others have felt that as a rule biological factors tend to predominate among the former, and physical agents among the latter, for insects at least. Among fishes, extremes of water temperature, drought and floods, are physical agents which may often cause mortality whose effectiveness is independent of stock density; whereas deaths from such biological causes as disease, parasitism, malnutrition and predation will usually become relatively more frequent as stock density increases. Yet exceptions to the rule above are sufficiently numerous to make the rule itself of doubtful applicability to fishes. For example, the biological factor of predation may have a uniform effectiveness over a considerable range of prey abundance, or at times may even become more effective at lower prey densities; and most if not all physical causes of mortality are compensatory when stock becomes dense enough that some of its members are forced to live in exposed or unsuitable environments. In addition, it is of course often difficult to ascribe a death to any single cause.

There is no necessary relation between the relative magnitudes of the causes of mortality existing at a given time, as measured by the fraction of the stock which each kills, and their relative contribution to compensation. An important and deadly agent of mortality may be strongly density-dependent, or weakly so, or not at all; and different agents may have their maximum compensatory effect over quite different ranges of density.

AGE INCIDENCE OF COMPENSATORY MORTALITY

Density-dependent causes of mortality could affect the abundance of either the existing adult stock or the young which it produces. In this paper we will

[2]This almost axiomatic proposition is implied in the writing of various earlier authors back as far as Malthus, but Nicholson was the first to formulate it explicitly and to emphasize its importance: Haldane (1953) calls his inspiration "a blinding glimpse of the obvious". The theorem and its diverse consequences were elaborated mathematically by Nicholson and Bailey (1935). Subsequent writers have developed it with varying emphasis, but nothing very substantial seems to have been added. Solomon (1949) gives a useful review of this literature, and Varley's (1947) quantitative assessment of various agents controlling the abundance of a trypetid fly population is outstanding.

consider mainly the effects of density dependence in the mortality which strikes the younger members of a population—among fishes, the eggs, larvae, fry and fingerlings. That is, the relative abundance of a brood will be considered to be determined by the time the first of its female members begin to mature: subsequent mortality is assumed to be non-compensatory.

This distinction between immature and mature stages of the life history, and the restriction of compensatory mortality to the former, is the principal difference between the thesis of this paper and that of earlier treatments of effects of density upon reproduction (e.g., Hutchinson, 1948; Haldane, 1953; Fujita and Utida, 1953; and many others). These have usually taken the Verhulst logistic equation as a point of departure:

$$\frac{dN_t}{dt} = \frac{bN_t(K - N_t)}{K}$$

(N_t is abundance at time t; K is equilibrium abundance; b is the instantaneous rate of increase of the population at densities approaching zero.) This expression implies a continuous tendency for the population to adjust itself toward the equilibrium size K, whether it is currently less than or greater than K. The adjustment would evidently have to involve compensatory mortality among the adult stock when its density is greater than K. Though such mortality is not impossible, there is in fish populations, at least, little indication of it; and as a matter of fact this aspect of the logistic equation has never been applied to any concrete biological situation.

Our assumption that *no* compensatory mortality occurs among the mature stock is unlikely to be strictly true of any species, and may not be even approximately true of some. Nevertheless it has seemed worth while to follow out the consequences of making this distinction between mature and immature, with fishes particularly in mind. The opportunities for compensatory effects are so much greater during the small, vulnerable early stages of a fish's life, that restriction of compensation to those stages seems likely to have wide applicability as a useful approximation.

To further simplify the initial approach to the problem, fishing is considered to attack only mature individuals, so that the *recruitment* produced by a given density of mature stock means both the number of commercial-sized fish, and the number of maturing fish, which result from its reproductive activity. The term *reproduction* will be used in a similar but slightly more general sense, to mean the number of young surviving to any specified age after compensation is practically complete; it will *not* mean the initial number of eggs or newborn young.

The above conditions and definitions will be assumed to apply unless exception is made specifically.

KINDS OF POPULATION CONTROL MECHANISMS

Compensatory types of mortality can be of various kinds. Some of the more likely possibilities are as follows:

1. Prevention of breeding by some members of large populations because

all breeding sites are occupied. Note that territorial behaviour may restrict the number of sites to a number less than what is physically possible.

2. Limitation of *good* breeding areas, so that with denser populations more eggs and young are exposed to extremes of environmental conditions, or to predators.

3. Competition for living space among larvae or fry, so that some individuals must live in exposed situations. This too is often aggravated by territoriality—that is, the preemption of a certain amount of space by an individual, sometimes more than is needed to supply necessary food.

4. Death from starvation or indirectly from debility due to insufficient food, among the younger stages of large broods, because of severe competition for food.

5. Greater losses from predation among large broods because of slower growth caused by greater competition for food. It can be taken as a general rule that the smaller an animal is, the more vulnerable it is to predators, and hence any slowing up of growth makes for greater predation losses. Since abundant year-classes of fishes have often been found to consist of smaller-than-average individuals (Hile, 1936), this may well be a very common compensatory mechanism among fishes (cf. Ricker and Foerster, 1948; Johnson and Hasler, 1954).

6. Cannibalism: destruction of eggs or young by older individuals of the same species. This can operate in the same manner as predation by other species, but it has the additional feature that when eggs or fry are abundant the adults which produced them tend to be abundant also, so that percentage destruction of the (initially) denser broods of young automatically goes up—provided the predation situation approaches the type in which kills are made at a constant fraction of random encounters (cf. page 609, below).

7. Larger broods may be more affected by macroscopic parasites or microorganisms, because of more frequent opportunity for the parasites to find hosts and complete their life cycle.

8. In limited aquatic environments there may be a "conditioning" of the medium by accumulation of waste materials that have a depressing effect upon reproduction, increasingly as population size increases.

Not all of the above compensatory effects need exist, or be important, in any given population, or in the same population every year. To a considerable extent they are likely to be complementary, so that if, for example, exceptionally favourable conditions permitted a good hatch of even a large spawning of eggs, a reduced growth rate of the fry would permit increased predation and so reduce survival in that way.

TYPES OF REPRODUCTION CURVES

Whatever the various kinds of compensatory and non-compensatory mortality acting on a brood may be, the average resultant of their action, over the existing range of environmental conditions, is represented by the average size of maturing brood, or recruitment, which each stock density produces. A graph of this relationship between an existing stock, and the future stock which the existing

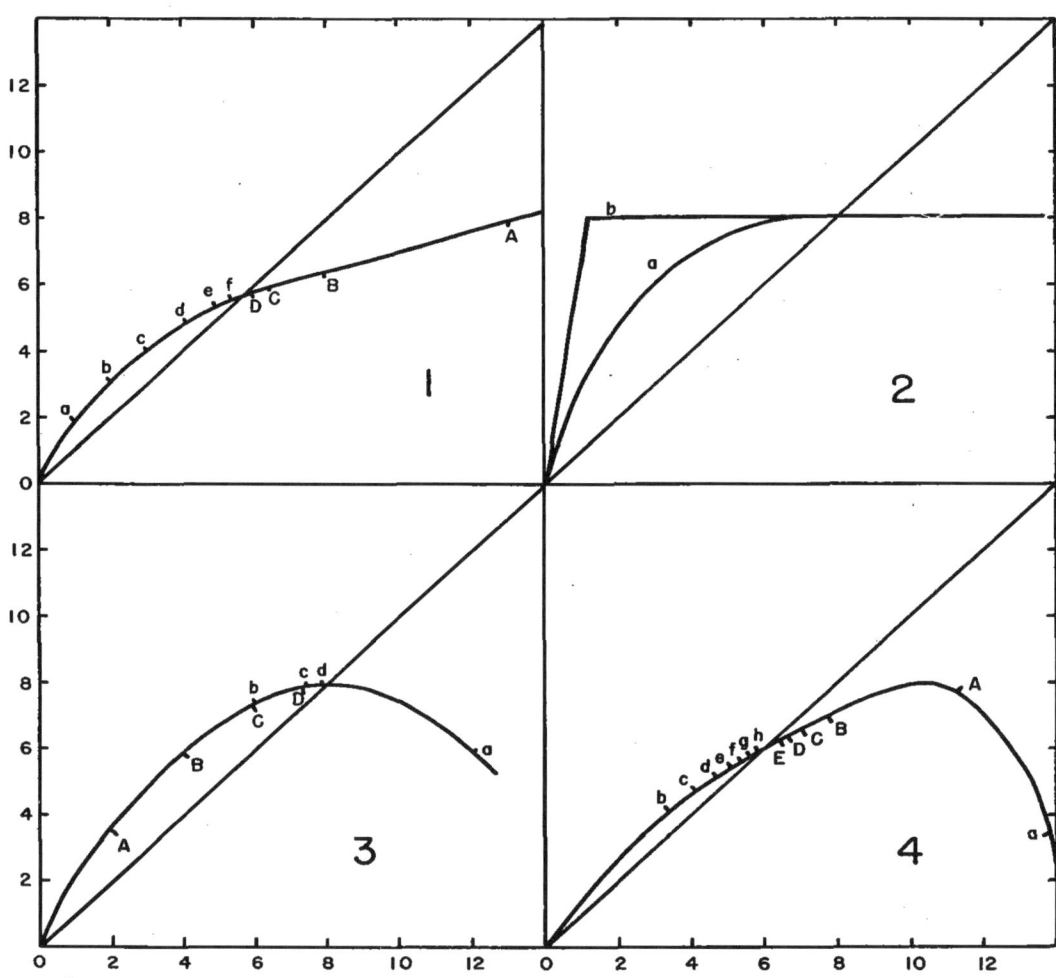

Figures 1–4. Stock-reproduction relationships characterized by a stable equilibrium. *Abscissa*— number of eggs produced by parent stock in a given year; *ordinate*—number of eggs produced by the progeny of that year.

stock produces, will be called a "reproduction curve". It is most convenient to label the axes in terms of the *eggs* in present and future generations, respectively[3]. The abscissa represents the mature eggs produced by the current year's stock. The ordinate represents the total of mature eggs produced by the progeny re-

[3]The argument is developed here in terms of populations of oviparous fishes which spawn once a year, but it can readily be modified to apply to other kinds of animals; an example for a viviparous animal is given on page 596. Among mammals, choice of the most suitable census age for plotting on reproduction curves may require care. If newborn young are used, effects of stock density upon frequency of conception and uterine mortality may be overlooked. If number of mature females is used, it should preferably be adjusted to take care of age variation in litter size or frequency, as the age structure of the population changes.

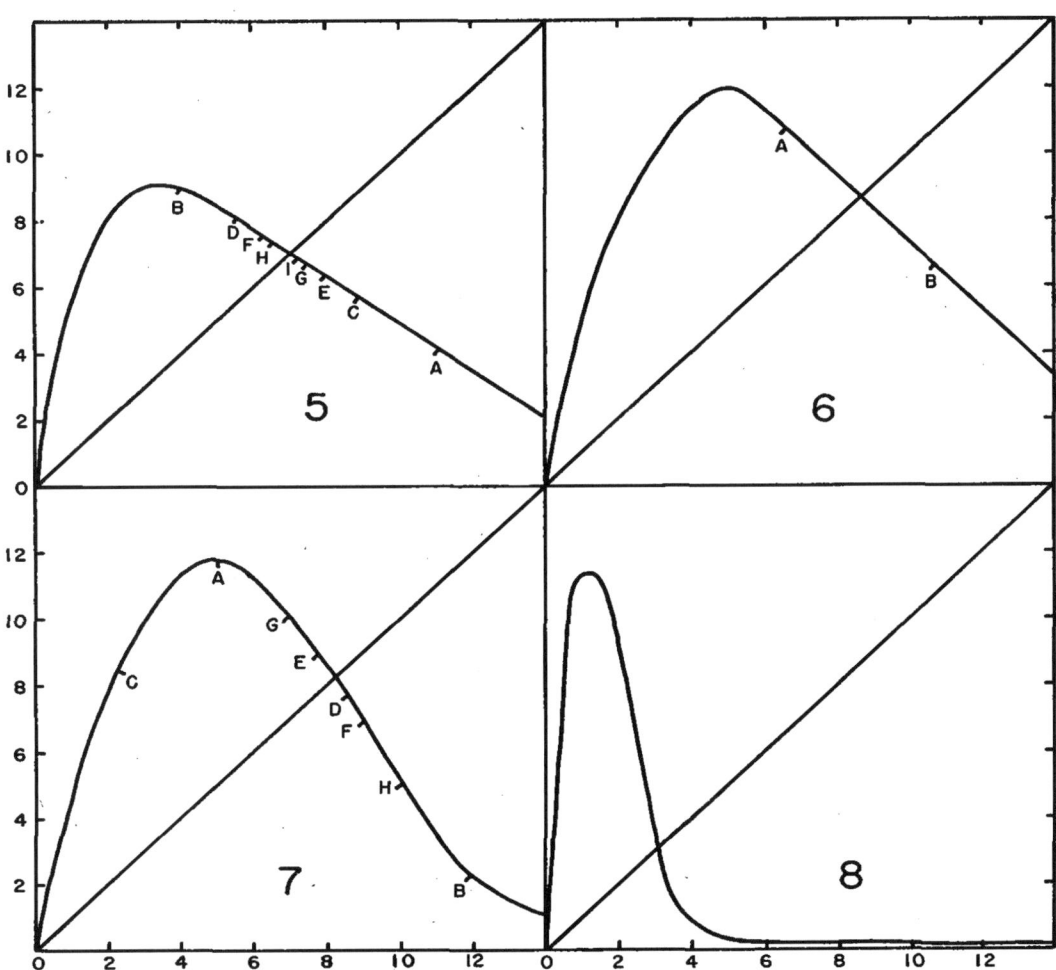

FIGURES 5–8. Stock-reproduction relationships in which there is an oscillating equilibrium, or (in 5) an oscillating approach to stable equilibrium. Axes as in Figures 1–4.

sulting from the current year's reproduction (obtained by summing over such period of time as the current year's hatch is a component of future years' stocks). In a state of nature, or with a stable fishery, the average size of parental and filial egg production, defined as above, tends to be equal over any long period of years, although striking changes may occur between individual years, or generations.

Figures 1–8 show a number of possible types of reproduction curve. In each of them the straight diagonal constitutes a useful boundary of reference which will be called the "45-degree line". Any curve lying wholly above this line describes a stock which is increasing without limit, hence such a curve cannot exist in practice. Similarly a curve below the 45-degree line describes a stock that will decrease to zero in a few generations[4]. The 45-degree line itself would describe a stock in which density dependence is absent, the filial generation tending always to be equal to the parental, except as factors independent of density deflect it. Such a stock would have no mechanism for the regulation of its numbers: if only density-independent causes of mortality exist, the stock can vary without limit, and must eventually by chance decrease to zero. Thus the first qualification of a reproduction curve is that it must cut the 45-degree line at least once—usually only once—and must end below and to the right of it.

Figures 1–8 indicate some of the types of recruitment curve that might exist in actual populations. All are characterized by having a region above the 45-degree line in which reproduction is more than adequate to replace the existing stock, and a region below the 45-degree line in which reproduction is inadequate to replace existing stock.

In the population of Figure 1, rate of reproduction (ratio of filial to parental eggs) decreases continuously as size of stock increases, although the slope of the curve becomes stabilized soon after the 45-degree line is crossed, and the actual number of young produced continues to rise indefinitely. Beyond the 45-degree line, however, this number is inadequate to fully replenish the stock.

In the population of Figure 2a, rate of reproduction also declines continuously; the actual number of recruits produced reaches an asymptotic level and thereafter does not change. Curve 2b differs from 2a in that it rises more steeply and it is initially straight; i.e., at low stock densities rate of reproduction is large, and constant. The range of densities over which there is constant recruitment is quite broad in b, simulating the condition loosely described by the statement "there are always as many recruits as the grounds can support".

Figures 3–8 show reproduction curves in which numbers of recruits begin to decrease after stock reaches some large magnitude. Such curves have an ascending left limb, a dome and a descending right limb. The descending right limb provides of course a more severe control of stock size at the higher densities. The differences in position of the dome and in the slope of the two limbs are of importance in determining changes in stock abundance, as described below.

All these curves are meant to represent the net effect of the sum total of density-dependent mortality factors acting upon the population. The reproduction of any actual year is affected also by density-independent factors, so that the actual number of young produced will deviate from the number indicated by the curve. In a later section the effect of such density-independent factors will be

---

[4]This statement is likely to be literally true only if the reproduction curve is plotted and fitted on logarithmic axes. On arithmetic axes a curve lying wholly below the 45-degree line *could* describe a stable population (having more than one age in the breeding stock) provided there were moderate to large random deviation from the average relationship. This point is discussed further on page 578, where it is evident that it would be advantageous to fit reproduction curves on logarithmic axes so that deviations above and below the curve would be in better balance. However such plotting has the serious disadvantage that statements of parent-progeny relationships in terms of the slopes of the two limbs of the curve become too complex to be practical. The best way out of this difficulty would be to fit the curve on a logarithmic plot, then transform it to arithmetic axes for interpretation.

examined. The first task will be to consider how a population behaves, under the strict control of a reproduction curve, when it is given some initial randomly chosen position.

This subject, and others later, will be considered for two situations separately. The first is where the currently hatched brood will constitute the whole of the breeding stock of a subsequent generation; that is, there is no mixing of ages in the spawning stock. This situation is fairly common among insects; for example, many mayflies, stoneflies, caddis flies, etc., have a single brood per year, usually with a one-year life cycle. Among vertebrates this condition is exceptional, but it exists in some fishes. There the length of life is commonly more than one year, so that two or more separate populations or "lines" exist concurrently.

The other situation to be considered is of course that where two or more broods contribute to the spawning stock at any given moment. Among vertebrates at least, this is the usual state of affairs.

REPRODUCTION IN THE ABSENCE OF DENSITY–INDEPENDENT VARIABILITY

SINGLE-AGE SPAWNING STOCKS

In all of the relationships shown in Figures 1–8, the stock is in equilibrium at the density at which the reproduction curve cuts the 45-degree line—that is, the stock is then producing enough progeny, and only enough, to replace its current numbers. In Figure 1, an initial deflection of abundance to either side of the equilibrium point is compensated by a gradual, asymptotic return to equilibrium, for example by paths A–D or a–f. In Figure 2a things work the same way to the left of the equilibrium point; to the right of it any deflection, no matter how great, is returned to equilibrium in a single generation. In Figure 2b deviations to the left, as far as 1.5 units of stock, are also returned immediately to the equilibrium level.

In Figure 3 the equilibrium point is at the top of the dome of the reproduction curve. The curve resembles 1 as regards deflections to the left of the equilibrium point; but a displacement to the right is followed by an immediate return across the 45-degree line to the ascending limb, after which it "climbs" this limb to the equilibrium point (a–d). The curve of Figure 4 is similar, but the dome lies to the right of the 45-degree line; A–E and a–h are possible paths to the equilibrium point.

In Figure 5, a stock deflected from equilibrium to any position along the descending limb will oscillate back and forth about the equilibrium point with decreasing amplitude, for example by the route A–I. If the deflection is great enough, the stock may have to climb the left limb before it is swung over to the right limb and begins the oscillating phase.

In Figure 5 the right limb has a downward (negative) slope numerically less than -1. In Figure 6 this slope is exactly -1 after the dome is passed. Here any moderate deflection along the straight part of the right limb results in a swing back across the 45-degree line of exactly the same magnitude, so that the deflection tends to be perpetuated indefinitely; for example, A and B are two such conjugate points. The intersection of the 45-degree line is a point of "indifferent" equilibrium; it is itself stable, but there is no inherent tendency for the stock to return to that level.

Finally, when the slope of the right limb of a reproduction curve lies between -1 and $-\infty$, equilibrium at the 45-degree line is not merely indifferent, it is unstable. That is, any deflection from equilibrium, no matter how small, initiates a series of oscillations along the right limb whose amplitude increases until the dome of the curve is reached or surpassed. The latter event usually sends the stock back to the right limb and the cycle begins again. No matter where they begin, all such cycles eventually reach the dome of the curve, and a stable oscillation series is established for which the dome is a convenient starting point. The cycle in Figure 7 is A, B, C, D, E, F, G, H, A, etc. (cf. Fig. 11D, below); the number of stages in this cycle depends upon the exact shape of the curve, chiefly upon whether or not one stage lands close to the 45-degree line. Figure 8 represents a more extreme situation, in which substantial reproduction is obtained over only a narrow range of stock densities considerably below the equilibrium level, and the stock would be subject to violent oscillations.

MULTIPLE-AGE SPAWNING STOCKS

When a spawning population consists of two or more age-groups, the young produced in a given year contribute to the stock of more than one future year, and the results of a deflection from equilibrium abundance are much modified. A fairly plausible example is where each brood contributes to the spawning stocks of four future years, in the ratio 2:3:3:2, and first spawning occurs 4 years after a brood was produced (Table I). Figure 9 shows the result of an initial deflection of such a stock to an abundance of 12, on some of the reproduction curves of Figures 1–8. The course of events for the most part reflects what was learned in the single-spawning situation. Those based on Figures 1–6 all end up at the stable equilibrium level; this being reached by direct approach for 1, 2 and 3, with one hesitation for 4, and by a series of damped oscillations for 5 and 6. From Figures 7 and 8 series of undamped oscillations are obtained, that is, permanent cycles of abundance.

It may seem surprising that curve 6 too does not generate permanent oscillations, but the mixing of year-classes gradually brings the stock to a steady level, in the absence of any tendency toward divergence.

The permanent oscillations of the type produced by curves 7 and 8 will repay more extended discussion. Figure 10 depicts series of oscillations, based on Figure 7, but with each brood contributing to the spawning stock for only two years. Time of first maturity is successively delayed one, two and more years, so that the average contribution to reproduction is made 1.5, 2.5, 3.5, etc., years after the brood in question existed as mature eggs. In every case stable cyclical fluctuations exist, just as in Figure 9F. Their *period* is always double the mean interval from egg to egg, that is, 3, 5, 7, etc., years.

The *amplitude* of the cycle varies. If the fish spawn first in the year after their appearance, amplitude is very small, but it quickly increases if maturity

TABLE I. An illustration of how to obtain the definitive distribution of population abundance for given conditions and a given reproduction curve, in this case Figure 7. Four years are assumed to elapse between a brood's existence as fertilized eggs and its first year of contribution to reproduction, while its total egg production is divided among the 4 successive years of its mature existence in the ratio 2 : 3 : 3 : 2. An arbitrary stock density (measured in terms of its egg production) is taken as a starting point, in this case 2.0 units (column 5). The reproduction corresponding to this abscissal value is read off from the ordinate of Figure 7, namely 7.9 egg units (column 6). These eggs are divided among the 4 years in which they are actually laid, beginning 4 years after the year in which the fish that carry them were hatched. For example, the 10.2 units of "recruitment" in line 5 are divided among line 9 (2.0 units), line 10 (3.1 units), line 11 (3.1 units), and line 12 (2.0 units). Breeding potential for each year (column 5) is obtained by horizontal addition of the contributions of the 4 age-groups in columns 1–4. In this example the definitive 11-year period of oscillation is achieved in the first peak-to-peak interval, while the average definitive amplitude, approximately 3.6 to 11.3, is apparent in the second peak-to-peak interval. Adjustment to the conditions imposed by the reproduction curve is not always as rapid as this.

Eggs produced at successive ages						Eggs produced at successive ages					
IV	V	VI	VII	Stock	Recruitment	IV	V	VI	VII	Stock	Recruitment
0.4	0.6	0.6	0.4	2.0	7.9	0.6	1.3	2.0	2.0	5.9	11.4
0.4	0.6	0.6	0.4	2.0	7.9	0.6	0.8	1.3	1.4	4.1	11.5
0.4	0.6	0.6	0.4	2.0	7.9	1.0	0.9	0.8	0.8	3.5	10.7
0.4	0.6	0.6	0.4	2.0	7.9	1.7	1.4	0.9	0.6	4.6	11.8
1.6	0.6	0.6	0.4	3.2	10.2	2.3	2.6	1.4	0.6	6.9	10.3
1.6	2.4	0.6	0.4	5.0	11.9	2.3	3.4	2.6	1.0	9.3	6.4
1.6	2.4	2.4	0.4	6.8	10.4	2.1	3.4	3.4	1.7	10.6	4.2
1.6	2.4	2.4	1.6	8.0	8.7	2.4	3.2	3.4	2.3	11.3	3.1
2.0	2.4	2.4	1.6	8.4	8.0	2.1	3.5	3.2	2.3	11.1	3.5
2.4	3.1	2.4	1.6	9.5	6.1	1.3	3.1	3.5	2.1	10.0	5.3
2.1	3.6	3.1	1.6	10.4	4.6	0.8	1.9	3.1	2.4	8.2	8.3
1.7	3.1	3.6	2.0	10.4	4.6	0.6	1.3	1.9	2.1	5.9	11.4
1.6	2.6	3.1	2.4	9.7	5.8	0.7	0.9	1.3	1.3	4.2	11.6
1.2	2.4	2.6	2.1	8.3	8.2	1.1	1.0	0.9	0.8	3.8	11.1
0.9	1.8	2.4	1.7	6.8	10.4	1.7	1.6	1.0	0.6	4.9	11.9
0.9	1.4	1.8	1.6	5.7	11.5	2.1	2.5	1.6	0.7	6.9	10.3
1.2	1.4	1.2	1.2	5.2	11.9	2.3	3.1	2.5	1.1	9.0	7.0
1.6	1.7	1.4	0.9	5.6	11.7	2.2	3.5	3.1	1.7	10.5	4.4
2.1	2.5	1.7	0.9	7.2	9.9	2.4	3.3	3.5	2.1	11.3	3.1
2.3	3.1	2.5	1.2	9.1	6.8	2.1	3.6	3.3	2.3	11.3	3.1
2.4	3.5	3.1	1.6	10.6	4.2	1.4	3.1	3.6	2.2	10.3	4.8
2.3	3.6	3.5	2.1	11.5	2.8	0.9	2.1	3.1	2.4	8.5	7.8
2.0	3.5	3.6	2.3	11.4	3.0	0.6	1.3	2.1	2.1	6.1	11.2
1.4	3.0	3.5	2.4	10.3	4.8	0.6	0.9	1.3	1.4	4.2	11.6
0.8	2.0	3.0	2.3	8.1	8.5	1.0	0.9	0.9	0.9	3.7	11.0

is delayed. More generally, appreciable amplitude in cycles of this type depends upon a preponderance of the reproduction of a brood occurring after one or more spawning periods have elapsed from the time of its birth.

With the longer intervals of lag between hatching and first spawning the cycles of Figure 10 become less regular; minor peaks appear, and the pattern is duplicated more closely at 2-cycle intervals, a feature which can be detected also in Figure 9F. However, these tendencies are much less apparent when more than two ages occur in the spawning stock, and only the dominant peaks would be detectable with any combination of average age of spawners and average age at first maturity that is apt to occur in nature.

Endless examples of reproduction-curve cycles can be constructed. Any desired combination of period and amplitude can be obtained, in more than one way, by selecting appropriate combinations of reproduction curve and age distribution of breeding. The general characteristics of such cycles can be summarized as follows:

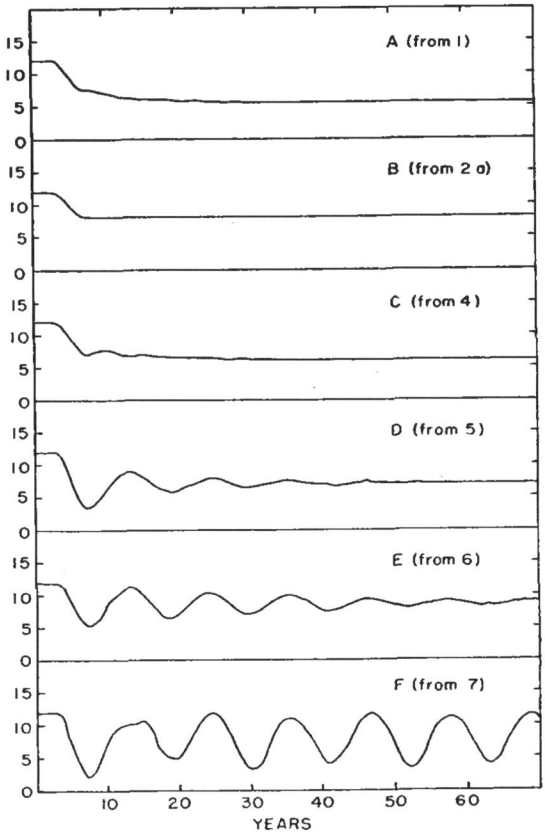

FIGURE 9. Change in abundance of the stocks of Figures 1, 2a and 4-7, following an initial sustained deflection to an abundance of 12 units, when the spawning stock is composed of 4 year-classes and first spawning is in the fourth year after hatching. *Abscissa*—years (generations); *ordinate*—relative abundance (egg production) of the mature stock.

1. Cycles occur when the outer part of the reproduction curve slopes downward, provided this slope begins at some point above the 45-degree line.

2. Cycles are damped and eventually disappear when the slope of the outer limb of the reproduction curve lies between 0 and −1. They are permanent when the slope is numerically somewhat more than −1, the exact critical limit depending upon the amount of mixing of generations in the spawning stock and the interval to first spawning.

3. Period of oscillation is determined by the mean length of time from parental egg to filial eggs, being twice that interval or close to it. It is independent of the exact shape of the reproduction curve, and also independent of the number of generations in the spawning stock, provided there is more than one.

4. Amplitude of oscillation depends partly on the exact shape of the reproduction curve.

5. Amplitude of oscillation tends to decrease with increase in the number of generations comprising the spawning stock.

6. Amplitude of oscillation increases rapidly with increase in number of generations between parental egg and the first production of filial eggs, up to a

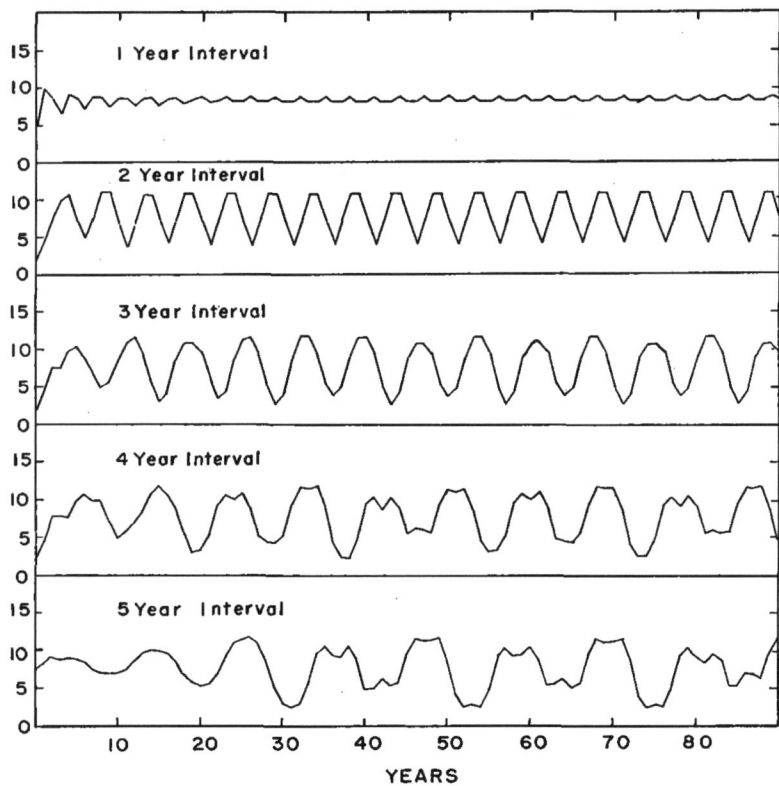

FIGURE 10. Population oscillations determined by the reproduction curve of Figure 7, when there are two ages in the spawning stock and spawning first occurs after 1, 2, 3, 4 and 5 years, respectively, from deposition of the parental eggs. Axes as in Figure 9.

limit imposed by the shape of the reproduction curve. When reproduction by a brood begins strongly in the generation following its birth, the oscillations are so weak that they could not be recognized in practice.

VARIATIONS IN REPRODUCTION PRODUCED BY FACTORS INDEPENDENT OF DENSITY

General

A comparison of density-dependent and density-independent reproduction is desirable in order to find possible means of distinguishing the two by their effects on population abundance, particularly since it has been suggested that some of the apparently periodic variations in animal numbers may reflect random variability alone (Hutchinson, 1948; Palmgren, 1949; Cole, 1951, 1954).

As an introduction, it is known that quantitative events selected completely at random tend to have an average peak-to-peak interval of exactly 3, provided they are classified finely enough that like values do not occur in adjacent positions. Cole (1951) demonstrates this mathematically and found approximately this period in a selection from Tippett's table of random numbers.

Cole uses a sequence of Tippett's numbers as a model with which to compare cycles of animal abundance in nature. Such a model, however, is not appropriate for our present purpose, because it must be interpreted as a reflection of random variation in the capacity of the environment to sustain the animal in question, rather than random variation in number of mature progeny produced per female. This is true for the following reason. When using Tippett's table or any similar assemblage, the number of numbers available is finite. The smallest number can as easily be followed by the largest as by any other; the largest cannot be followed by anything larger. The biological characteristics of a corresponding population are that it must have a very large potential rate of increase (since the greatest possible abundance can follow directly on the least), and it must have severe compensation at the higher stock densities (since they are close to a ceiling of abundance which cannot be exceeded). Few if any populations could meet both these conditions in the course of a single reproductive period, so that Cole's model, to have verisimilitude, must be applied only to situations where a census is taken at intervals of several generations. Thus it is not appropriate, for example, to fish populations when censused yearly, since these usually spawn only once a year.

In any event, a simple series of random numbers is not a suitable model of random variation *in success of reproduction*. What then *is* a suitable model? Since extremes of environmental conditions are less common than conditions approaching normal, our first postulate will be that a normal frequency distribution provides a fairly realistic picture of the relative frequencies of the resultants of the various independent factors making for success or failure in reproduction; from which resultants a random selection is to be made. Table 8.6 of Snedecor (1946) was used for this purpose, selection of t-values being made corresponding to the figure in the ".00" column closest to each of a sequence of selections from Tippett's random table. The series obtained begins as follows: 1.8, 0.5, 0.2, 0.1, 0.2, 0.0,

1.0, 0.1, etc. Tippett's table was again used to divide these into items representing reproduction above and below the replacement level, respectively; this gave the series below, decreases being followed by d.

$$1.8d,\ 0.5,\ 0.2,\ 0.1,\ 0.2d,\ 0.0,\ 1.0d,\ 0.1,\ \text{etc.}$$

The next question is to decide how these are to be applied to an existing reproductive potential. An illustration will be of assistance. Suppose that a breeding population matures 10 billion (10^9) eggs; and let us define "average" environmental conditions here as those which will permit the maintenance level of reproduction, that is, 10 billion future mature eggs produced by the progeny of the current year's sexual activity. Variations in environmental conditions add to or delete from this average production. The most severe conditions imaginable can reduce reproduction to zero. The most favourable conditions possible will permit the survival of all the eggs and thus, assuming equality of the sexes, produce future reproductive units to the number of 5 billion times the total expectation of egg production of a female fish just maturing—which expectation might easily be 2,000,000 in the case of the cod, for example. The mean and the extreme limits of reproduction would thus be:

	Mature eggs	*No. of fish (at first maturity)*
Lower limit:	0	0
Average:	10,000,000,000	10,000
Upper limit:	10,000,000,000,000,000	10,000,000,000

This shows that in a fairly typical instance the stock produced by average environmental conditions is located very asymmetrically with respect to the two extreme limits, on this arithmetic scale. On a logarithmic scale the average production becomes much more nearly central, if a minimum of 1 fish is assumed:

	Log No. of fish
Lower limit:	0
Average:	4
Upper limit:	10

From this and other considerations it appears likely that our symmetrically distributed environmental variations will tend to act in relative rather than absolute fashion; so that, for example, if a given negative deviation produces a decrease in reproduction to half of the average level, the same positive deviation will increase reproduction to twice the average level. In other words, the figures representing environmentally caused deviations from the reproductive norm must be multiplicative rather than additive. To put them into this form, unity is added to each, so that deviations indicating below-average conditions (the d-items above) become divisors, those indicating above-average conditions become multipliers, and zero deviation is multiplication or division by unity. The result is shown in Table II. The absolute magnitude of the items shown is of course arbitrary, and it can be varied at will by multiplying the original random series (before unity was added) by any desired integer or fraction.

TABLE II. Series of numbers selected randomly from a population having the positive half of a normal frequency distribution with a standard deviation of unity, each number augmented by unity, and the series divided randomly into multipliers and divisors (the latter indicated by the suffix "d").

2.8d	1.1d	3.0d	2.2	1.9d	1.3d
1.5	1.1d	1.9	1.1d	1.0	1.6d
1.2	1.9d	1.4	1.0	2.0d	1.2
1.1	2.0d	1.1	1.4	1.1d	2.2d
1.2d	2.4	2.4d	1.7d	2.8	1.3d
1.0	1.5d	1.7d	1.7d	2.7	2.3d
2.0d	2.2	1.3	1.6	2.8d	1.2
1.1	1.3d	2.7d	1.8	3.5d	1.3
2.3d	1.2	1.5d	1.4d	1.2d	3.0
1.8d	1.9	1.9	1.0	1.2d	1.3
1.1d	2.0d	1.9d	1.8	1.7	2.2d
1.7	2.1d	2.1d	1.0	1.9d	3.2d
1.8d	1.5	1.8d	2.2	2.6	1.4d
2.5	1.3d	1.8	1.0	2.4d	1.5
2.0	1.6d	2.6d	1.5d	2.5	1.9d
2.0d	1.5d	2.3	2.2d	2.0d	1.1d
2.0	1.0d	1.7d	1.6d	2.2d	2.6
2.2d	1.9d	1.1	1.9	2.1d	2.5
2.2	1.9d	2.8	1.7	1.9	2.1d
1.3	1.3	1.3	2.6	1.8d	2.6d
1.4	2.1d	1.1d	2.4d	2.8	2.1d
1.2d	2.1	1.6	1.5	1.3	1.8
2.0d	1.2	1.2d	2.0d	1.7d	2.6d
1.7d	1.8	1.0d	1.7d	1.6	1.3
1.1d	2.0	1.7	1.0	1.7d	2.5d
1.5	1.1d	1.4	1.2d	1.8	2.7d
2.2	2.3	1.1	1.5	1.0	1.1
1.6	1.2	2.1d	2.0	2.0d	1.3
1.4d	2.0d	2.1	1.3d	2.0	1.7d
2.0d	1.6d	1.9d	1.8d	1.6	1.1d
1.3d	2.4d	1.2d	2.3	1.0d	2.0
3.4	1.1d	1.1d	1.6	1.3d	1.6
1.3	1.7d	1.6	1.3d	3.7d	3.3d
3.2d	1.9d	1.5d	1.7d	1.3	2.3d
1.3	2.7d	1.3	1.3d	1.0	1.5d
2.2d	1.5	2.2d	1.7	1.3	1.6d

SINGLE-AGE SPAWNERS

Since the figures of Table II constitute a model of variation in success of reproduction under the influence of density-independent factors alone, a picture of corresponding population fluctuations is obtained by multiplying some initial stock density by each item in succession. This is done in the lower line of

Figure 11A; the upper line is from a different random series. The lines fluctuate a good deal and at times diverge considerably from the original abundance: at the end of the 70 generations shown in Figure 11 the Table II line has changed by a factor of 260; while if the same is continued through the whole 216 generations of the Table the net change is 5.5×10^{-7}, representing a relative decrease

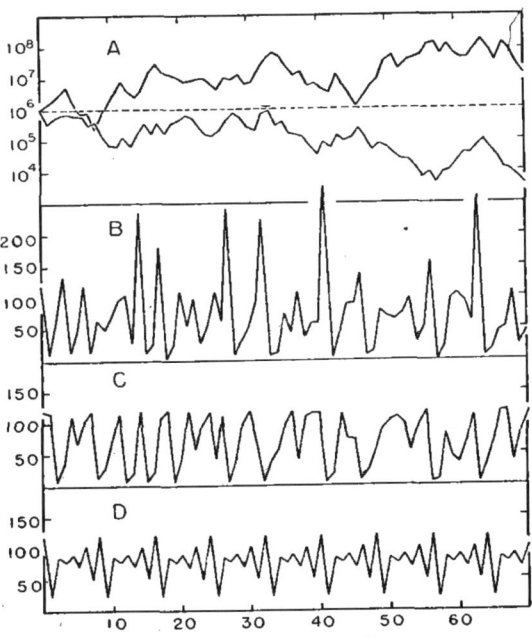

FIGURE 11. Fluctuations in ideal populations of single-age spawners. *Abscissa*—generations; *ordinate*—number of units of stock.

A. Two series of changes in a population from an initial density of 1,000,000, when it varies under the influence of density-independent factors only. The lower line corresponds to the first 70 entries of Table II, multiplied in sequence.

B. Changes in a population from an initial abundance of 119, as determined by the random factors of Table II and the reproduction curve of Figure 7, when the action of the latter precedes the former.

C. As in B, but with the random factors acting before the compensatory mortality shown by the reproduction curve.

D. Population cycle generated by Figure 7 alone.

from 55,000,000 to 1. This large change occurs in spite of the fact that Table II is constructed in such a way that, if it were continued over any really long period, the population should be above its initial level about as often and as much as it is below it[5]. The net change results partly from an excess of d-items in Table II, and partly from the fact that the d's happen to be larger numbers, on the whole.

[5]A series of 1,000 "generations" constructed in a different but analogous fashion by Hutchinson and Deevey (1949, fig. 4) shows similar major long-period trends.

By chance the change indicated by Table II is considerably greater than would be expected to occur often in a series of only 216 generations. However it is easy to show that as the number of generations increases, the most probable divergence of population size from the original level also increases continuously. When the number of generations approaches anything appropriate to a geological time scale the likely change is very great indeed. For example, after 10,000 generations the standard deviation of the number of increases (or decreases), from the most probable number 5,000, is $\sqrt{10,000 \times \frac{1}{2} \times \frac{1}{2}} = 50$. This means that there is about 1 chance in 3 that the excess of increases or decreases will be 50 or more at that time, while the chances of the excess having been 50 or more at some time *during* the 10,000 generations are much greater. With an average change factor of 1.67 from one generation to the next (as in Table II), this means that the population is fairly certain to have been increased or decreased by a factor of about 1.67^{50}, or 140,000,000,000, somewhere along the line. Without wasting thought on the impossibility of a population of any sizable organism *increasing* this much, such a *decrease* would obviously bring even the most numerous fish stock, for example, to extinction. Thus no population of single-age spawners can survive if its reproduction is completely at the mercy of density-independent influences in the environment. In the same way a gambler will in the long run lose everything if his opponent has equal skill and unlimited resources. (In practice, of course, the effects of random environmental factors would usually vary as between different parts of the animal's range, and local extinction could be followed by recolonization from adjacent areas.)

The average interval between peaks of Table II should be close to the 3 which is characteristic of a random series; in actuality it was 3.09 when ties were randomly divided into higher and lower values. However this is not what determines the period of a series such as Figure 11A. In the latter a peak occurs whenever a multiplier is followed by a divisor, and the average interval between peaks is nearly 4 years, again adjusting for ties. The theoretical or long-term average period can be determined as follows: since a peak or trough occurs whenever there is a change from multiplier to divisor or divisor to multiplier, the expectation that the first such change will occur between adjacent cells is one-half, the expectation that it will occur between a cell and the next but one is one-fourth, that it will occur at the third cell is one-eighth, and so on. The average of these intervals 1, 2, 3, etc., weighted as to frequency, is exactly 2, which represents the average peak-to-trough interval. Since peaks and troughs are equally common, the average peak-to-peak or trough-to-trough interval is therefore 4.

However it must be emphasized that this 4 represents the average interval between *all* peaks, regardless of size. Casual inspection of Figure 11A might give a different impression. The eye tends to ignore the smaller humps, and to impose a certain regularity among the rest by magnifying those of intermediate size when they happen to fall into a sequence with large ones, and diminishing them when they do not. In this manner the lower line could "suggest" a cycle of 11 or 12 years, with 4 peaks and 4 troughs actually showing.

MULTIPLE-AGE SPAWNERS

To obtain a model of the consequences of random fluctuations upon reproduction of multiple-age stocks, the random values of Table II have been applied to a "population" constructed on the same basis as Figure 9—that in the spawning stock there are 4 ages, that each year-class produces eggs in the ratio 2:3:3:2 in the 4 years of its reproductive activity, and that each fish spawns for the first time 4 years after it was itself a fertilized egg. The resulting curve is shown in Figure 12A (it starts from an age distribution characteristic of the reproduction-curve cycle of 12D).

The most distinctive characteristic of the line of Figure 12A is that it tends to rise. This increase occurs in spite of the fact that divisors happen to be in excess in Table II, and if the series is continued a little farther much higher values are encountered. The fact is that, if continued, the population will increase without limit. The reason is that the contributions of the several year-classes to a given year's spawning must be added arithmetically, while expectation of

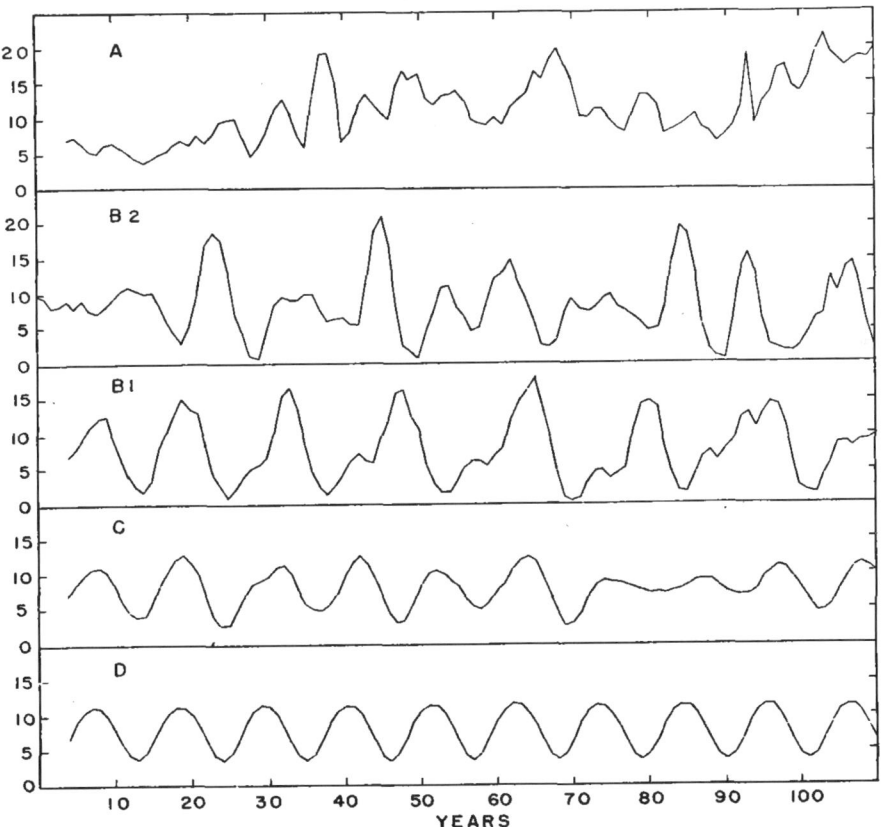

FIGURE 12. Fluctuations in a population of multiple-age spawners, as determined by the random series of Table I (curve A), by the reproduction curve of Figure 7 (curve D), and by combinations of these (curves B and C). Axes as in Figure 9.

survival of an egg is proportional to random environmental effects having a symmetrical distribution so constituted that any increase is greater, absolutely though not relatively, than the corresponding decrease. It would be difficult, of course, to prove that environmental effects upon reproduction are actually distributed in just this manner. On the other hand it was shown above that they are very unlikely to be merely additive; and any intermediate condition will result in a slower but still unlimited upward trend to a graph such as Figure 12A.

It thus appears that any likely random sequence of environmental changes, even one which for single-age spawners causes a catastrophic decline, can be made to produce instead a large increase in abundance of stock by the simple device of dividing the spawning of each year-class among several calendar years. Even though wholly non-compensatory reproduction cannot occur in nature, this circumstance may have considerable advantage for any population in which random variability is large and/or the left limb of the reproduction curve not very steep (e.g., as in Figs. 1, 2a, 3 or 4). Indeed it is easy to show that in the presence of random variability a multiple-age spawning stock can maintain itself even when its reproduction curve lies wholly below the 45-degree line—though of course it cannot be *too far* below. Thus there is a sound ecological basis for the customary occurrence, in nature, of reproductive assemblages consisting of more than one age-group.

As regards period, we find again that the peaks and troughs of Figure 12A have a certain apparent regularity in spite of their random origin. Discounting the first 15 years as being still somewhat under the influence of the reproduction-curve cycle, and counting all peaks no matter how small, there are 17 peak-to-peak intervals in the figure, of which no less than 10 are of 5 years' duration—the average interval is 5.2 and the range 2 to 7. The fact that the average interval is longer than the 3 which is characteristic of a simple random series (and longer than in the series for single-age spawners) is explained by the method used to obtain it. This is quite similar to taking a "running average" of four items at a time, and Cole (1951, fig. 1) has shown that applying such a procedure to a random series increases the mean peak-to-peak interval. "Artificial" trends of really long period are not as apparent in Figure 12A as in 11A, but a longer series might have confirmed them.

In summary, the characteristics of population changes which would result from density-independent factors acting alone (if that were possible) are as follows:

1. A population of single-age spawners would vary widely above and below its initial abundance; eventually, after a few thousand generations at most, it would either become extinct or, more likely, be fragmented temporarily into small independent units.

2. Populations of multiple-age spawners would increase in abundance indefinitely, though not without ups and downs.

3. The average peak-to-peak period for lines of single-age spawners would be 4.

4. The average peak-to-peak period for multiple-age spawners would be

more than 3, and would be the greater, the greater was the number of ages in the spawning stock; in the example used it was about 5 years.

5. For single-age spawners, and probably for multiple-age spawners as well, "cycles" having longer periods than these would tend to be apparent in graphs of abundance, because of conscious or unconscious mental suppression of small peaks and troughs, and regularization of large ones.

COMBINATIONS OF COMPENSATORY AND NON-COMPENSATORY MORTALITY

In natural populations non-compensatory mortality is superimposed upon the reproduction expected from density-dependent factors, and the curve of population fluctuation is the resultant of both. The random effect can be introduced either before or after the amount of reproduction indicated by the curve is written down, according as the random factors are thought to act before, or after, the compensatory ones. The latter procedure gives greater influence to the random element, and may correspond better to events in nature, though the two types of mortality may of course act concurrently, and to a considerable extent probably do.

SINGLE-AGE SPAWNERS

Combinations of Table II or similar series can be made with the various kinds of reproduction curve shown in Figures 1–8. It seems unnecessary to reproduce examples of most of these. With a flat-topped curve such as Figure 2b, and the scale of random variability shown in Table II, the resultant reproduction is practically always equal to the product of the equilibrium abundance and the multiplier or divisor for the year in question. Only an improbably large deviation could shift the population over to the ascending part of the curve for a year. Relatively slight modifications of this situation are obtained when broad-domed curves having a stable equilibrium point are used (e.g., Figs. 3 or 4); the principal difference being that, on the whole, reproduction is somewhat less.

Fluctuations produced by a combination of the reproduction curve of Figure 7 with the random series of Table II have some of the characteristics of either component (Fig. 11). The regular series A–H of Figure 7, repeating at 8-year intervals, is replaced by an irregular one. The average peak-to-peak period is about 3.5—slightly less than that of the random series (which was close to 4) and greater than the 2 of the reproduction-curve sequence. The average amplitude of the "cycles" (ratio of peak to preceding trough) is greater than what either component series exhibits, but there are of course no extreme trends in abundance such as resulted from random causes alone. There is a pronounced tendency for each peak to be followed immediately by a trough, which is a peculiarity also of the random-curve series.

The above description applies whether the random influence operates before the reproduction-curve (Fig. 11C) or after it (Fig. 11B). The two kinds of series are quite similar, but the latter of course has the greater amplitude of changes.

Multiple-Age Spawners

Combination of compensatory and non-compensatory reproduction in multiple-age stocks will be considered only for the case where the former precedes the latter, and will be illustrated by means of a population in which contributions to reproduction are spread over four years in the ratio 2:3:3:2, as in Figure 9.

In general, the kind of fluctuation which results from any combination of compensatory and non-compensatory mortality factors depends upon the relative magnitudes of the two components. The action of the same series of random factors is shown in Figures 12B and 12C at two levels of intensity which are in the ratio of 5:1; Table II is used for B, and values one-fifth as large for C, i.e., $1.36d$, 1.10, 1.04, etc. In each case they are combined with the reproduction indicated by Figure 7. The same initial age distribution is used for all series of Figure 12.

By itself, Figure 7 yields the steady oscillation of Figure 12D. At the lower intensity of random effects (line C) the population cycle determined by the reproduction curve is not too seriously altered: there is variation in amplitude, and the peaks move out of phase by as much as a fourth at times, yet the prevailing periodicity can be determined fairly accurately from even a short series. At the higher intensity of random effects, however, much greater disturbance is evident; it is followed through in two lines of Figure 12, beginning with B1 and continuing in B2. As regards *period*, the 11-year cycle is first increased to 15–17 years, then reduced to 7 or 8 (counting major peaks only). Though the average over a long period may thus tend to 11 years, peaks and troughs move out of and back into phase with the reproduction-curve cycle and also with the random "cycle", which are its determiners. More serious is the fact that the smaller peaks and troughs of what appears to be the main series are in some cases impossible to distinguish from the "artifacts" resulting from random fluctuation: hence the average 11-year period above could not be discovered in practice, even with a very long record. As regards *amplitude*, the maximum ratio of peak to adjacent trough is much greater than is found in either the reproduction-curve cycle or the random series.

When the random element is given still greater relative importance, the reproduction-curve element becomes unidentifiable as such. However the latter continues to make an important contribution to the resulting population changes. It makes peaks and troughs much less numerous than they would otherwise be, and it provides a control of limits of abundance—that is, the progressive upward tendency of the random curve is effectively curbed.

The relative importance of random and compensatory factors in determining population abundance also depends somewhat upon the number of ages represented in the spawning stock. When this is large, random factors are less effective in disturbing reproduction-curve cycles.

EFFECTS OF REMOVAL OF MATURE STOCK

Single-Age Spawning Stocks

Many natural populations have a part of their members removed by man, either because they are useful to him or, in the case of harmful species, because he hopes to reduce their abundance. The case where only *mature* stock is taken by the fishery is considered in examples to follow. Many fisheries take a portion of the immature stock also, but this appears not to add any new principle to what is considered below, provided the region of compensatory mortality is not invaded.

The effect upon the various population parameters of removing a constant percentage of the mature stock, before reproduction, is shown for four types of single-age-group populations in Figures 13–16. In Figure 14, for example, a spawning stock OQ produces filial spawners equal to PQ. However, after the fishery has removed 40 per cent of the mature stock the number remaining is equal to OQ, which again produces PQ, and so on. Thus P is the equilibrium position for 40 per cent

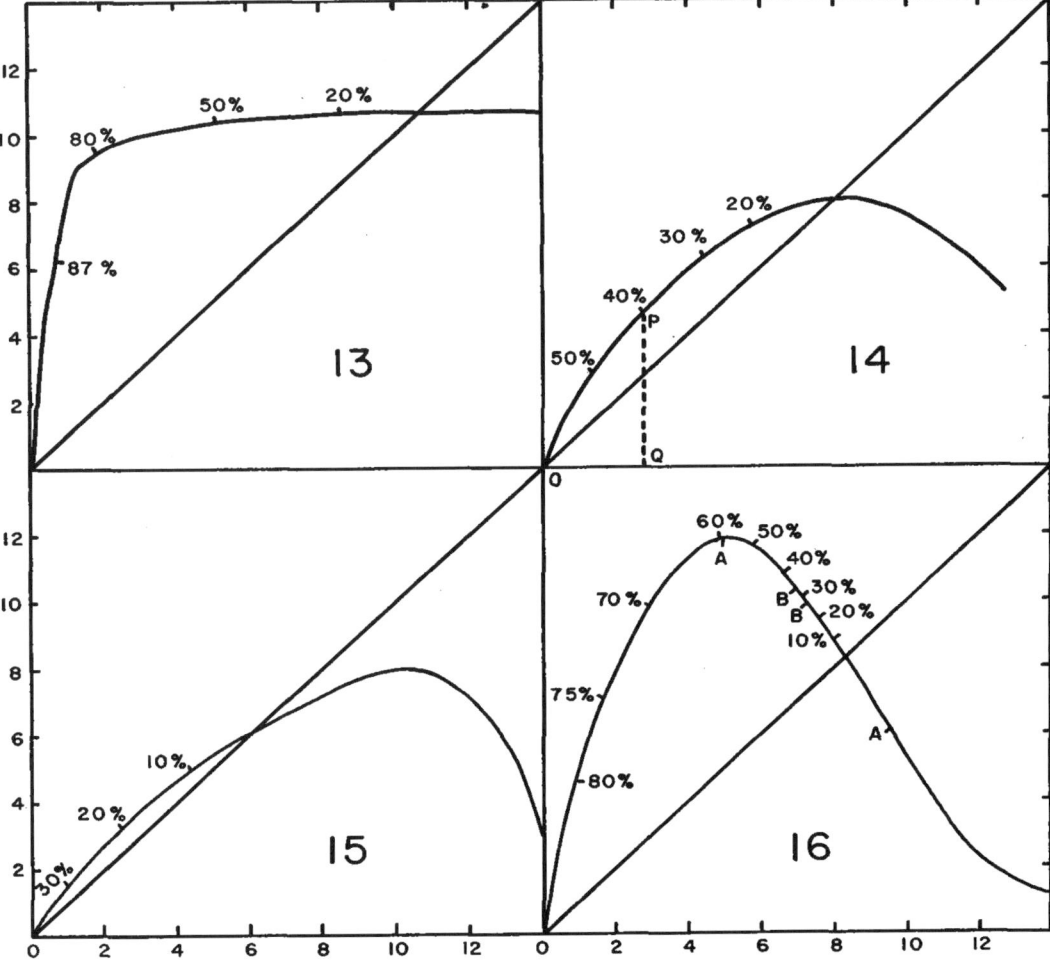

FIGURES 13–16. Equilibrium densities of stock for the various rates of exploitation indicated by percentages on the reproduction curve. Stock density before fishing is the ordinate value corresponding to the point on the curve opposite the percentage exploitation in question. Axes as in Figures 1–4.

exploitation. The fraction of the mature stock which must be removed prior to spawning, in order to maintain equilibrium at any point on the reproduction curve, is equal to the complement of the reciprocal of the slope of a line joining that point to the origin.

The general effect of exploitation is to move the point of equilibrium abundance to the left on the reproduction curve. In Figures 13–15 this means that under exploitation the equilibrium population is always smaller than it is under natural conditions. In Figure 16 (which is the same as 7), removal of part of the spawning stock at first results in a *larger* equilibrium level of stock, and the same is true of Figures 5, 6 and 8. When rate of exploitation becomes large enough, however, the stock is again reduced. Equilibrium points for various rates of exploitation are indicated on Figures 13–16.

Oscillating equilibria are much changed by exploitation. Their amplitude is reduced, and the number of positions through which they swing is decreased, or oscillation may be eliminated entirely. However, stable oscillations persist when exploitation is light to moderate. In Figure 16, for example, with 10 per cent removal the stock traces the cycle 11.9, 4.0, 10.9, 5.5, 11.9, etc.; with 20 per cent removal it alternates between the levels of 6.1 and 11.9 units (A–A); with 30 per cent removal the oscillation is reduced to between 9.9 and 10.3 (B–B), so that it would not be distinguishable in practice; and with 60 per cent removal it disappears completely[6].

The point of maximum sustained yield can easily be computed from graphs like Figures 13–16. For example, in Figure 16 it is at 65 per cent exploitation; in Figure 13 it is close to 80 per cent; in 15 only about 18 per cent. In general, the greater the area between the reproduction curve and the 45-degree line above the latter, the greater is the optimum rate of exploitation.

Beyond the level of maximum yield lies a level of maximum permissible exploitation above which the stock is progressively reduced to zero. This limit is a function of the maximum angle made by any line joining a point on the reproduction curve to the origin; and as noted above, the maximum rate of removal which can be sustained is equal to the complement of the cotangent of that angle.

If a man's interest in a single-age population, of a noxious insect for example, is to reduce it to as low a level as possible by increasingly intensive destruction, he can be sure of making *direct* progress toward the desired goal only if the reproduction curve is one of the types shown in Figures 1 and 2. If it is like Figures 3 or 4, and stock happened to be on the outer limb, some moderate rate of removal would dampen a crash which otherwise was imminent. With curves of types 5–8 a paradoxical situation develops. Moderate destruction of adults will in general tend to stabilize the population at or about some magnitude *greater* than its primitive average abundance. In Figure 16 maximum abundance is con-

[6]When equilibrium is unstable the *average* size of the populations of the cycle tends to be less than the equilibrium level. In Figure 16 the average is 7.7 for no exploitation, 7.8 for 10 per cent, and 9.0 for 20 per cent, as compared with the equilibrium values of 8.2, 8.6 and 9.4. For 30 per cent or greater exploitation, equilibrium stock and average stock are the same.

sistently achieved when as much as 60 per cent of the spawning stock is destroyed each year. To reduce the population below the primitive average, more than 73 per cent must be destroyed. Thus although sufficiently intensive effort will be successful, moderate destruction of the mature population is worse than no action at all[7]. Furthermore, if an intensive campaign of continuing control of a variable species is decided upon, it is most efficient to begin it at a time that the population is at a low point of its cycle, i.e., when the pest may be doing no particular damage.

MULTIPLE-AGE SPAWNING STOCKS

The effect of a fishery upon stocks containing more than one age-group of spawners parallels what has just been found for single-age-group stocks. If the stock is one for which a stable equilibrium exists (Figs. 1–5), adding a fishery does not disturb the stability, but the abundance of the population is changed (in 2b the change begins only after the inflection point is reached). With the curves of Figures 1–4 the change is in the direction of a decrease in abundance, but with Figure 5 exploitation at a moderate rate *increases* the adult stock, until the dome of the reproduction curve is reached.

Stocks which perform regular oscillations in the absence of a fishery (Figs. 7 and 8) are changed in three respects when exploitation begins: (1) their average equilibrium abundance is at first increased, but later decreases again if exploitation becomes sufficiently intensive; (2) the amplitude of oscillation decreases; and (3) the period of oscillation tends to decrease slightly.

1. The first-named effect can be estimated rather easily, since the average abundance of multiple-age stocks proves to be practically the same as the equilibrium abundance. The latter has been calculated for Figure 16 using various rates of exploitation, as shown on that Figure and in Table III. The maximum abundance is at the dome of the curve, and maximum catch is obtained slightly to the left of it, at 65 per cent exploitation.

2. The advent of fishing not only affects the total abundance of a stock, but also gives it a younger age composition, because survival rate from year to year is reduced. In the example of Table IV, based upon Figure 16, the fish are assumed to mature first at age III. Fishing takes place just prior to spawning and rate of fishing is the same for all ages. Natural mortality occurs between successive fishing-and-spawning seasons, and is 20 per cent from age III to age IV, 20 per cent from IV to V, 30 per cent from V to VI, and 100 per cent after the spawning at age VI. The average weight of a fish at time of fishing-and-spawning is 2 units at age III, 4.28 units at IV, 7.14 at V and 9.52 at VI; and egg production is proportional to weight. Under these conditions the *equilibrium* distribution of the contributions of the several age-groups to egg production, at different rates of exploitation, is shown in Table IV.

---

[7] A similar conclusion is reached in Nicholson and Bailey's (1935) detailed analysis of host-parasite interaction in insects. Varley (1947) concluded that a trypetid fly population was maintained at about 10 times the density it would otherwise have had, because of the presence of non-compensatory causes of mortality which killed host and parasite equally.

TABLE III. Average or equilibrium abundance of stock, by weight, in a stock having more than one age at maturity, and the catches at various rates of exploitation; from Figure 16.

Rate of exploitation	Stock before fishing	Catch
%		
0	8.2	0
10	8.8	0.9
20	9.5	1.9
30	10.1	3.0
40	10.8	4.3
50	11.5	5.8
60	11.9	7.1
65	11.4	7.4
70	9.7	6.8
75	7.2	5.4
80	5.0	4.0

TABLE IV. Relative weights and contributions to reproduction of the age-groups in the stock described on page 583.

Rate of exploitation	Fraction of eggs contributed by age			
	III	IV	V	VI
%				
0	0.200	0.300	0.300	0.200
10	0.233	0.314	0.283	0.170
20	0.273	0.327	0.261	0.139
30	0.320	0.336	0.235	0.109
40	0.377	0.339	0.204	0.081
50	0.446	0.333	0.165	0.055
60	0.526	0.315	0.127	0.032
70	0.619	0.278	0.083	0.020
80	0.737	0.221	0.042	0.000

The first line of Table IV was arranged to have the same distribution of contributions to spawning as used for Figure 9F. In Figure 17 are shown cycles obtained from the 10 per cent and 20 per cent lines of Table IV, starting from the equilibrium distribution characteristic of no fishery[8]. At 10 per cent exploitation the oscillation of the population is maintained, though at reduced amplitude and about a higher mean level. At 20 per cent exploitation oscillation gradually decreases to an inconsiderable amplitude, and at higher rates of exploitation it

[8]This procedure is the only practicable one, but it ignores the fact that the age distributions of Table IV would themselves be completely realized only after a period of 4 years from the beginning of fishing. Consequently the equilibrium situations in Figure 17 would actually be approached more gradually than the figure indicates.

disappears completely. However even at 20 to 40 per cent exploitation the oscillation persists through several cycles.

3. The length of a cycle was found earlier to be twice the mean length of a generation (interval from egg to egg). In the first line of Table IV this is 4½ years, and the cycle is 9 years. With the higher exploitations of Table IV a shift of age distribution toward younger fish occurs, and we accordingly expect

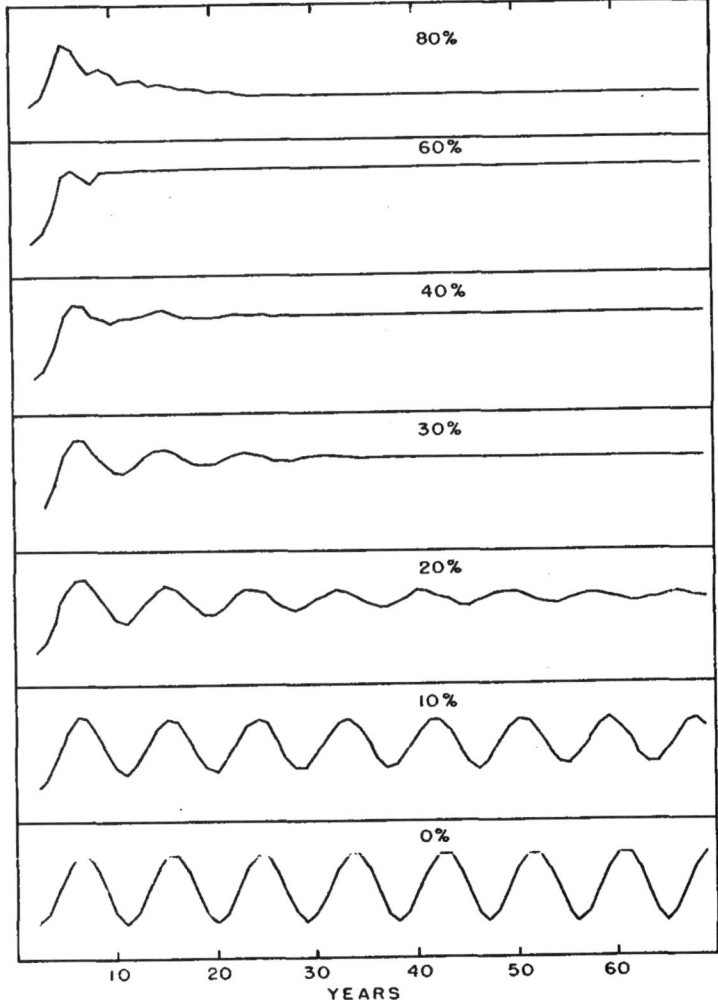

FIGURE 17. Reactions of an hypothetical population to exploitation. The population is one regulated by the reproduction curve of Figure 7, with ages III, IV, V and VI contributing to egg production in the ratio 2:3:3:2 when there is no exploitation. The stable cycle characteristic of no exploitation is shown by the 0% line. The other lines show the effect of annual removal of the indicated percentage of the mature stock from a population which initially had the age distribution characteristic of the start of the ascending phase of the no-exploitation cycle. (See also footnote 8.)

a shorter cycle to appear. In this example, however, the cycles disappear before much decrease in period can take place. At 10 per cent exploitation average age of contribution to reproduction is reduced to only 4.39, and at 20 per cent it is 4.27. The corresponding predicted cycle lengths are 8.8 and 8.5 years, and the peaks shown in Figure 17 do in fact become progressively a little closer together. In a somewhat more realistic example there would have been more than four age-groups in the stock under conditions of light exploitation or none, and in that event the reduction in period of oscillation with increased fishing would be greater. However it is doubtful whether in nature the reduction would ever be great enough to be identifiable with certainty, before increasing exploitation removed oscillation of this type entirely.

When the reproduction curve is dome-shaped, *control* of an undesirable population of multiple-age spawners presents some of the same difficulties as described earlier for single-age stocks. However the initial favourable reduction in abundance will tend to last longer—i.e., until the increased broods of young can grow to the size at which they cause damage. Only when this first phase is over, and the fishing effort has to cope with the larger broods which come from the reduced spawning populations, will it be apparent whether or not control can be permanently effective. Few recorded attempts to reduce nuisance fishes have lasted beyond the initial stage of removal of old stock. Foerster and Ricker (1941) have described the rather easy success achieved in this preliminary phase of control of squawfish (*Ptychocheilus*) in a small lake. The experiment was subsequently continued until one or more large broods of young began to grow into the damaging size range, but it did not last long enough to see if these could be kept sufficiently reduced by appropriate effort. In Lesser Slave Lake, Alberta, unlimited fishing for ciscoes began in 1941 with the hope of thus reducing the lake's population of the tapeworm *Triaenophorus*, and as a result the spawning population has been changed gradually from one 5–7 years old to one now (1951) mostly 2 years old. However recruitment and rate of growth have increased sufficiently to maintain a high level of catch and a fairly large population mass, so that control is not yet effective for the purpose intended (Miller, 1950, 1952).

When non-compensatory variability is added to the effects of exploitation and the reproduction curve, the kind of result obtained differs little from what was discovered from Figure 11. The amount of disturbance introduced depends of course upon the relative magnitude of the random factors.

EXAMPLES OF REPRODUCTION CURVES

A wider recognition of the different possible kinds of reproduction curves, and their different properties, will be of value mainly as a guide to current and future research. Such curves should be useful not only as a possible clue to the nature of observed fluctuations, but still more in estimating expected average recruitment when stock is reduced (or increased!) by fishing it, at different intensities. Knowledge of the reproduction curve will thus complement information on rate of growth and natural mortality, and permit more accurate prediction and regulation of fish catches.

Some examples are given below of reproduction curves, for fish and a few other animals, which can be plotted from available data. In several instances information is available concerning only the first part of the life history, but since that is where most compensatory mortality is expected, the data are included. A disadvantage of such curves, however, is that the "45-degree line" can usually be located only approximately, if at all.

PACIFIC HERRING (*Clupea pallasi*)

The herring of the west coast of Vancouver Island are a natural unit which has been given special study. Recently Tester (1948) examined the relation between the intensity of spawning in the brood year, and an estimate of resulting year-class strength based on catch per unit effort in the various years that each brood contributes to the fishery. Tester had two indices of spawning; using "Index A", which was available for five years longer than "Index B", he computed an inverse correlation of −0.30 between spawn deposited and resulting year-class strength for the period 1931–43, though this figure does not differ significantly from zero.

Mr. J. C. Stevenson has very kindly added data for more recent years, and these are plotted in Figure 18. In this figure the more elaborate "Index B" is used, but the five years 1931–34 are included (open circles) using values from the regression of B on A for the years of overlap. The negative correlation −0.16, for the years 1931–49, differs from zero even less than the one computed by Tester, principally because a series of abundant year-classes from 1943 to 1947 has increased the range of variability. However it is still true that the largest

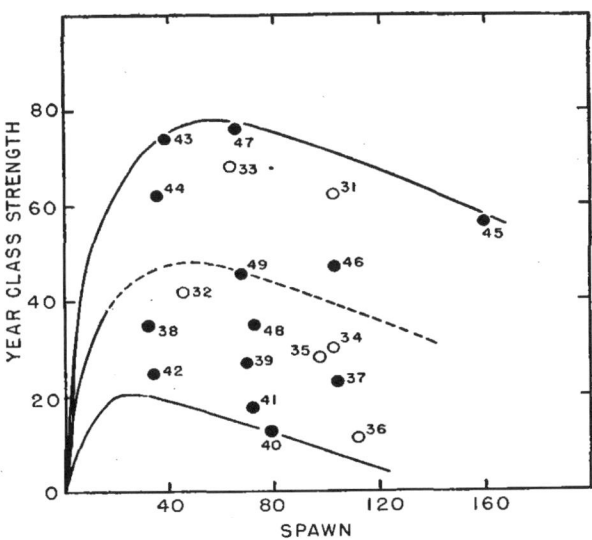

FIGURE 18. Relative strength of year-classes of a Pacific herring stock (ordinate) plotted against relative amount of spawn deposited in the brood year (abscissa). The two solid lines include the observed limits of variation, while the dotted one is an approximate middle value. Data from Tester (1948) and Stevenson (MS).

broods have been produced by smaller-than-average spawnings, whereas the smallest broods have been produced by larger-than-average spawnings.

Dr. Tester and others have also considered the possibility that trends in herring recruitment may exist corresponding to some as yet unidentified rhythm of the ocean environment; and the 1943–47 stanza of good year-classes is in fact fairly well set off from earlier (1934–42) and later (1948–49) stanzas of lower production (Fig. 18). If there is really any systematic effect here, one may notice that *within* each of the two longer stanzas a negative correlation between spawning and recruitment is indicated.

PINK SALMON (*Oncorhynchus gorbuscha*)

Among pink salmon there are two completely separate reproductive lines[9], so that is is very easy to identify the progeny of any year's spawning when it matures two years later. From Pritchard's (1948a, b) summary of the reproduction of the stock which spawns in McClinton Creek, Graham Island, British Columbia,

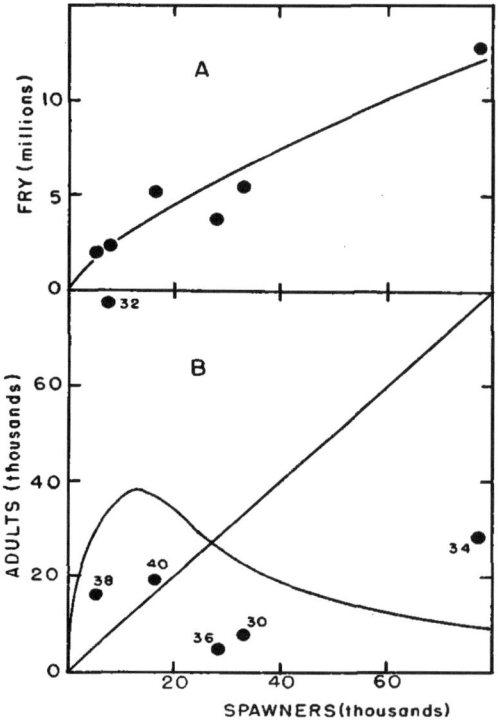

FIGURE 19. A—Fry produced in millions; and B—returning adults in thousands; plotted against the number of pink salmon spawners in the brood years indicated, at McClinton Creek, B.C. Data from Pritchard (1948a, b).

---

[9]Separate or nearly separate breeding stocks of salmon are commonly called "cycles", "years" or even "races" in western North America. The term "lines" was proposed by Huntsman (1931), and is used here because it avoids confusion with other kinds of cycles, etc. (cf. Clemens, 1952).

the reproduction curve of Figure 19B is plotted. The progeny shown does not include the commercial catch, which is a considerable but unknown quantity, and hence all points would be moved upward on a graph that represented total adult production of the spawnings indicated. No very clear idea of the shape of the reproduction curve can be obtained, but it seems to slope downward from 15–20 thousand spawners.

Dr. Pritchard also had records of fry production, and these are plotted in Figure 19A. The fact that *percentage* fry production decreases as population increases indicates that compensatory mortality occurs in the freshwater part of the life history. That compensation occurs also during saltwater life is suggested by the change from a generally upward slope in Figure 19A to a generally downward slope in 19B. Obviously too there is much random variation in mortality at both stages, but particularly in the ocean. The ocean phase includes the effect of exploitation by the fishery, which is certainly variable and might tend to be

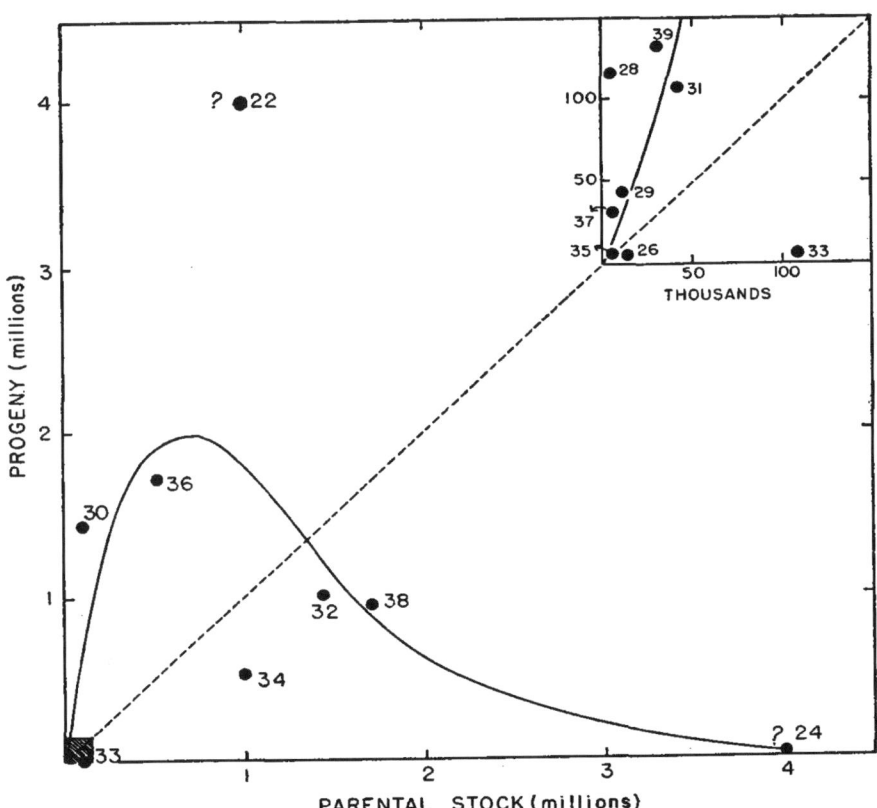

FIGURE 20. Reproduction curve for pink salmon of the Karluk River, Alaska, based upon fence counts except in 1922 and 1924 (see text). The shaded square in lower left is shown on a larger scale in the upper right corner. Commercial catches are not included in the progeny. (From unpublished data of the United States Fish and Wildlife Service, Pacific Salmon Investigations.)

compensatory, because more seiners are attracted to the bay off McClinton Creek when fish are numerous there.

A reproduction curve having a very wide range of parent-stock densities is that for pink salmon of the Karluk River, Alaska. This is shown in Figure 20, plotted from data supplied by Messrs. C. E. Atkinson and C. J. Burner of the U. S. Fish and Wildlife Service. "Progeny" figures do not include the commercial catch; adding the latter could not alter the general form of the curve, but would raise it farther above the 45-degree line. The 1926–40 data are from complete or nearly-complete fence counts. The 1924 escapement was estimated as "over 4,000,000" (Barnaby, 1944, p. 257). Unfortunately there is no numerical estimate of the 1922 run; however, since it is called "large" on the same page as the 1924 run is called "tremendous" (Barnaby, p. 249), it might be assigned a figure of about one million, and on that basis a point for 1922 is plotted in order to better indicate the vertical range of reproduction.

Coho Salmon (*Oncorhynchus kisutch*)

The experiment at Minter Creek, near Tacoma, Washington, has provided information on the relation of spawning stock to smolt output (Smoker, 1954). Production of year-old smolts is related to number of female spawners in Figure

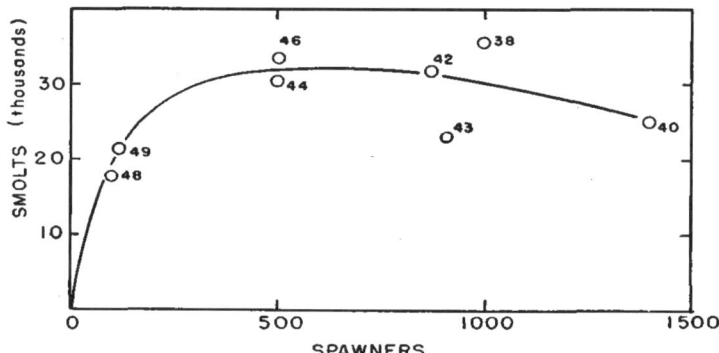

Figure 21. Smolts produced by adult female coho salmon spawners of the brood years indicated, at Minter Creek, Washington. Data from Smoker (1954).

21, based on the same data as the upper half of Smoker's figure 29. The productive capacity of the stream appears to be 25–35 thousand smolts, depending mainly on volume of flow (Smoker's fig. 29, lower half), and if adjustment is made for the latter, the already-flat outer limb of Figure 21 becomes even flatter, with the deviations of the points from the line considerably reduced.

Estimates of total adult production from these smolts are not possible; there is no record of catch, and even the number of returning spawners is available for only four years.

Sockeye Salmon (*Oncorhynchus nerka*)

The sockeye of Karluk Lake, Alaska, mature at 3 to 8 years of age, the mean being a little over 5 years. Barnaby (1944, fig. 3) has plotted a graph of spawning stock against adult progeny produced; the latter was obtained by summing the contributions of the 1921–29 year-classes to subsequent catches and breeding stocks. This graph is practically equivalent to a reproduction curve based upon eggs, the only difference being that possible variation in the number of eggs produced by females of different ages is not taken into account.

Barnaby separates "spring" and "fall" fish in his plotting. Either in this form, or when the data for each year are combined (Fig. 22), there is much variability in production, but a peak apparently occurred when the total spawning stock (spring plus fall) was between 1 and 1½ million fish. Unfortunately, both before and after this period the sockeye stock of the lake decreased, so the 1921–29 results cannot apply either to the earlier period when *catches* of 2–3.5 million fish were customary, or to the modern period when the total stock has fallen below a million. The causes of this decrease are still under study.

An opportunity to document the *ascending* phase of a sockeye reproduction curve is afforded by the "late Shuswap" or Adams River race on the Fraser River. A blockade in 1913 reduced this and other Upper Fraser races of sockeye to a level that was only a small fraction of their natural abundance. These fish mature preponderantly at 4 years of age, so that an adequate reproduction curve

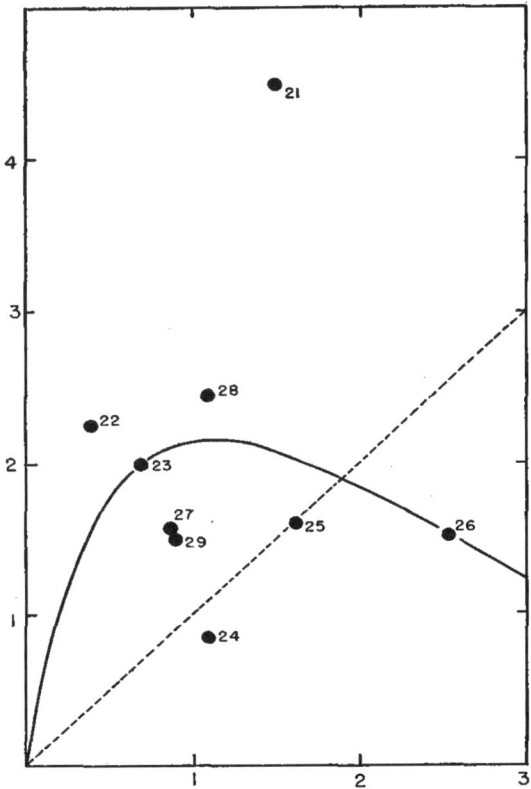

FIGURE 22. Number of mature sockeye (catch plus escapement) produced by the year-classes of 1921–29, at Karluk Lake, Alaska. Data from Barnaby (1944).

can be plotted using estimates of spawners and catches every fourth year (Fig. 23). Sizes of the spawning population are based on estimates of Fishery Inspectors and, in recent years, on figures given in Annual Reports of the International Pacific Salmon Fisheries Commission; the catch is taken as the total Fraser commercial catch each year, less 1,200,000 fish representing the approximate production of other races in this series of years. The curve has a much steeper ascending limb than the Karluk curve; there is a suggestion of a maximum at 1–1.5 million spawners, and possibly a start on the downward path beyond that point.

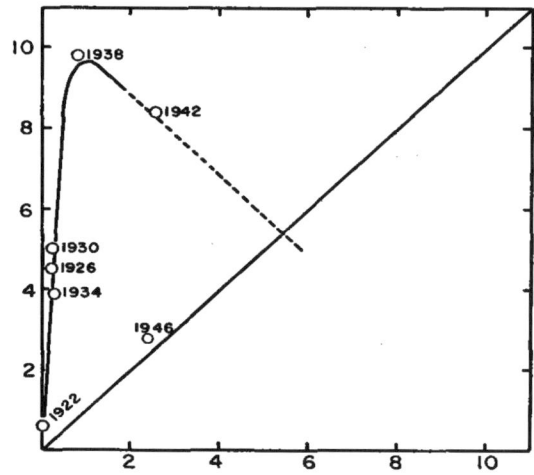

FIGURE 23. Reproduction curve for the late Shuswap race of sockeye salmon, based upon estimated numbers of spawners and an approximate division of the commercial pack. *Abscissa*—estimated number of spawners in the years indicated; *ordinate*—resulting catch plus escapement. Both scales are in millions of fish. The year of spawning is shown beside each point. From data of the B.C. Department of Fisheries and the International Pacific Salmon Fisheries Commission.

The steepness of the left limb of the Shuswap curve is worth noting, being of the same order as that of Figure 8. If the right limb has any considerable slope downward, it would have a tendency to mould chance fluctuations into swings back and forth across the 45-degree line. In this connexion one may note that the original dominant line of sockeye on the Fraser, of which the late Shuswap race was a component, is supposed to have been considerably less abundant in 1909 than in either 1905 or 1913 (Babcock, 1914). This might represent two swings across an equilibrium position, under the then conditions of a fairly low percentage exploitation of this line by the fishery (compare A–A of Fig. 16). Under present conditions the fishery normally takes enough to keep the spawning stock near the dome of the curve, and special measures are taken to ensure this much escapement when below-normal reproduction has occurred, as from the 1946 spawning (Royal, 1953).

HADDOCK (*Melanogrammus aeglifinus*)

Herrington's (1941, etc.) analysis of the George's Bank population is particularly valuable because he was able to obtain from market reports an estimate of the smallest commercial size (scrod—mostly age III) independently of and in addition to the larger sizes. This yields an estimate of recruitment from year to year, which in turn is a reflection of success of reproduction. When size of spawning stock in the years 1912 to 1929 is plotted against scrod produced (solid circles, Fig. 24), a steep outer limb of a recruitment curve is apparent. The 45-degree line has been located very approximately by considering that a pound of scrod during those years had an expectation of egg production, throughout its life, equivalent to the actual egg production in one year of 10 pounds of mature haddock; and by assuming that scrod were half as vulnerable as were adult fish

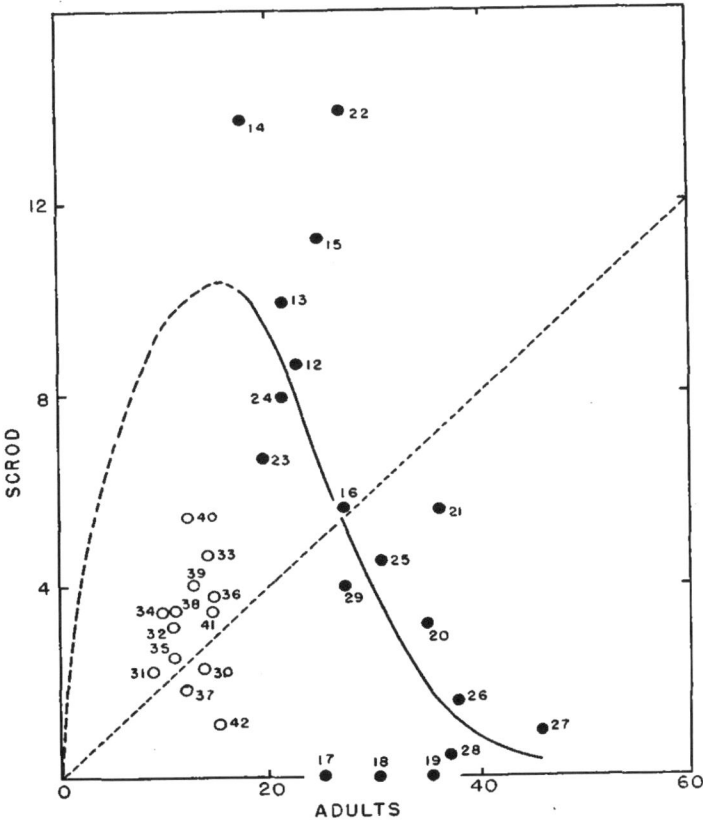

FIGURE 24. Abundance of adult haddock on George's Bank during spawning season (February–April), related to abundance of scrod (age III, mostly) at the same season three years later. Both scales are in terms of thousands of pounds per day caught by trawlers of a certain size class, the figures for adults being smoothed. Black circles represent the years 1912–29, open circles 1930–43. The reproduction curve is drawn for the earlier period only. The diagonal is an *approximate* 45-degree line for that period (see text). Data from Herrington (1948).

to the trawlers whose catches supply these data. Figure 24 is the same as Herrington's figure 11 of 1948, except that we have used catches of the February-April period only, for both scrod and adults. Herrington was of the opinion that the critical time for competition between adults and young, during those years, was when the young were about a year old (his fig. 12); however the two graphs are much the same and do not provide a basis for decision on this point. Either one supports Herrington's interpretation of the fluctuations in adult stock over the period in question, which consisted of two complete oscillations (Fig. 25). The trough-to-trough periods were 9 and 8 years, and trough:peak amplitude was about 1:2. The observed period implies a median age of 4 or 5 years for the production of eggs by the fish of each brood, and this seems consistent with what is known of the size- and probable age-structure of the haddock population in those years.

Thus Herrington has described the only example yet known in which oscillation of a multiple-spawning-age fish population might plausibly be ascribed to the simple effect of a steep reproduction curve. His account of the probable cause of the inverse stock-recruitment relationship emphasizes competition for food between adults and young. Intra-brood competition at an earlier age seems another possible factor, but the feeding habits of haddock are said to preclude significant cannibalism.

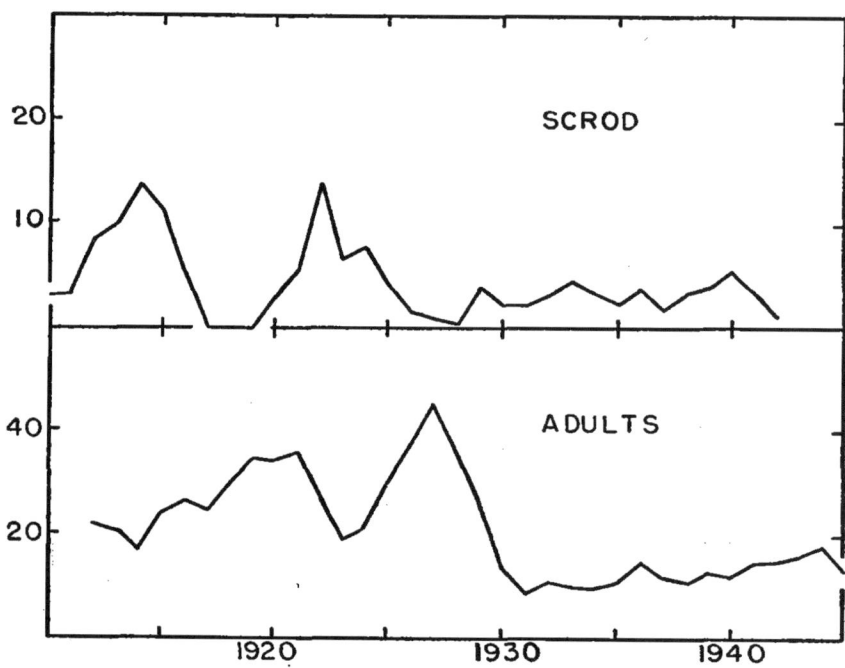

FIGURE 25. Catch per unit effort of adult haddock, related to that of the scrod (age-III fish) produced by them. The years shown on the abscissa are those in which the adult catch was made; the scrod on the same ordinate were taken three years later. From the same data as Figure 24.

Points for the years 1930–43 are plotted as open circles in Figure 24. About 1930 the haddock fishery changed radically within a few years; total effort increased, and otter trawls to a large extent superseded baited-hook methods so that smaller fish became relatively more intensively exploited. Although adjustments for these effects were made as well as possible by Herrington, quantitative comparisons between the period before and after 1930 appear uncertain. In particular, the relative level of recruitment indicated after 1930 seems too low to have sustained the continuing fairly high level of catch—that is, the 45-degree line indicated on Figure 24 probably does not apply to these later years.

What is of most interest for us is that the increase in rate of exploitation about 1930 was sufficient to eliminate oscillation from the population (Fig. 25). It was discovered in Figure 17 above that, in a stock subjected to increasingly heavy fishing, reproduction-curve oscillation may disappear while rate of exploitation is still rather low; and further, that the transition from marked oscillation to unrecognizably small oscillation occurs rather abruptly. Possibly accidental, but also in accord with expectation in a fishery of increasing intensity, is the fact that the second cycle in Figure 25 is shorter than the first (*two* years shorter in Herrington's representation).

FRUIT FLY (*Drosophila melanogaster*)

An optimum population density for reproduction was found by R. Pearl and S. Parker in a series of experiments made during the early 1920's. The reproduction curve of Figure 26A is synthesized from graphs summarizing their work, given by Lotka (1925).

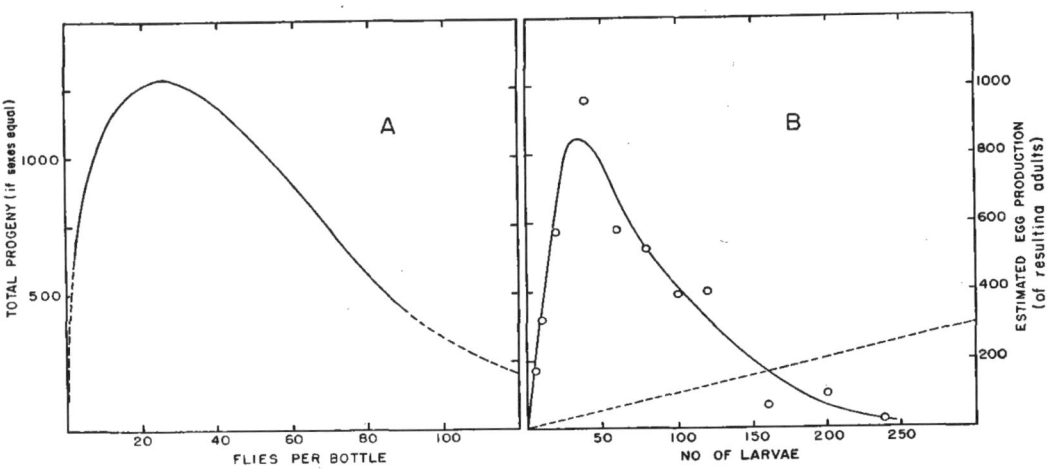

FIGURE 26. Reproduction curves for *Drosophila melanogaster*. A—From data of Pearl and Parker, as presented by Lotka (1925). The ordinate represents half the product of the density in question, times the progeny produced per female per day, times the mean length of life. The range over which observations were actually made is shown as a solid line. B—From data of Chiang and Hodson (1950, table 9 and fig. 13). The abscissa is the number of 24-hour larvae used to start a culture, and the ordinate is the potential egg deposition of the resulting adults. The dotted line is an approximation to the 45-degree line, which ignores possible mortality in the egg to 24-hour larva stages.

Recently Chiang and Hodson (1950) have investigated *Drosophila* reproduction in more detail. Figure 26B shows their graph of "potential fecundity" against initial larval density. Some of the factors involved in decreasing the reproduction with increasing density were greater larval mortality, increased failure to pupate, and smaller size and lesser fecundity of adults produced. The flattening of the curve near the right end was due partly to a secondary *increase* in pupal and adult size at the highest densities.

Azuki Bean Weevil (*Callosobruchus chinensis*)

Extensive studies of this species have been made by S. Utida, but the only account I have been able to consult is that of Fujita and Utida (1953). There is no overlapping of generations. A reproduction curve is shown in Figure 27, obtained under experimental conditions. The right limb trends downward with a slope numerically less than unity, which would make the population subject to damped oscillations. Such oscillations were actually observed by Utida (Fujita and Utida, p. 494).

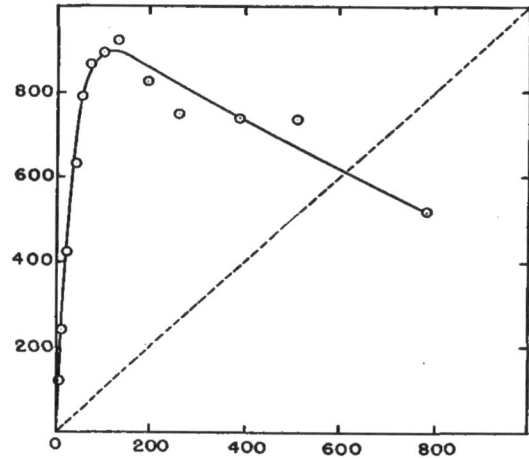

Figure 27. Reproduction curve for the azuki bean weevil under experimental conditions. *Abscissa*—size of parental generation; *ordinate*—number of mature progeny. Replotted from figure 4 of Fujita and Utida (1953).

Water-Flea (*Daphnia magna*)

Pratt (1943) performed experiments on the effect of stock density upon the production of young and upon longevity, in *Daphnia* grown in 50 cc. of culture medium with excess food. These can be used to construct reproduction curves, shown in Figure 28.

The ordinates in Figure 28 are equivalent to the total number of young produced in an average lifetime at the temperature and density in question (Pratt, p. 135). The identity of these values with those required for a reproduction curve can be shown as follows. Consider the life of a *Daphnia* to be divided into a series of "generations", corresponding to the years of an ordinary fish's life. These could be made of any length, but for convenience they are taken as equal to the approximate time from hatching to first maturity (6 days at 25°C., 14 days at 18°C.). Then consider a 10-*Daphnia* population at 25°. It produces young which live 26 days on the average (Pratt's fig. 6), or 4.3 "generations". During the first of these the *Daphnia* are immature, so that their mature life is 3.3 "generations". The total young produced by an individual in an average lifetime is, however, 23.1 (Pratt, p. 135), or 7.0 per generation. Consequently a population of 10 mature individuals produces $10 \times 7.0 = 70$ young per "generation", and these survive through 3.3 generations of maturity on the average, so that the total reproductive potential produced by the parent generation is $70 \times 3.3 = 231$ units. These "units" are numerically the same as the total number of young which would be produced throughout the life of the 10 parent individuals living at constant density. However the "units" are to be regarded not as young actually produced, but as a measure of the generation's potential contribution to the mature population of several future generations, in exactly the same way as eggs in mature females were used for fishes. In actuality, if stock density were to change by the time these 231 "units" were in the mature population, the actual production of young by the progeny of the original generation would be greater or less than 231.

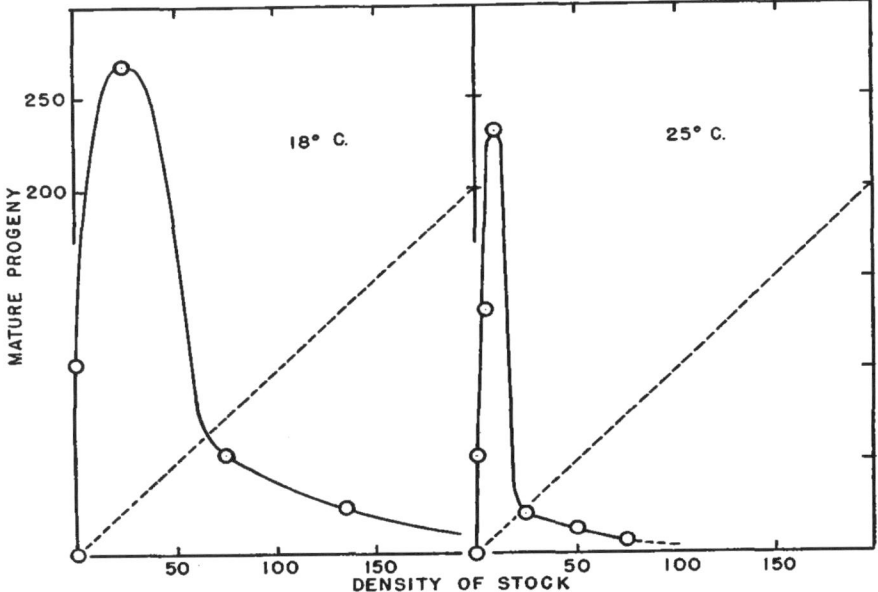

FIGURE 28. Relation between density of adult *Daphnia magna* and their expectation of progeny, at 18° and 25°C. Computed from data of Pratt (1943): 18° curve from his figs. 7 and 8; 25° curve from his figs. 5 and 7.

The reproduction curves of Figure 28 are of a steeper type than any of those discovered among fishes, particularly the 25° curve. Both curves are in fact likely to be even steeper than they are drawn, since the highest point of observed production of young is not necessarily the maximum. This steepness has been believed to be associated with the small volume of habitat and consequent likelihood of waste accumulation—although the medium was changed every second day.

Pratt (1943, figs. 1, 2) also followed the growth of *Daphnia* populations over a period of 6–7 months under the same culture conditions as outlined above. Starting in each case from 2 parthenogenetic females not more than a day old,

TABLE V. Simplified calculation of ideal population changes of *Daphnia magna* at 25°C. under the conditions of Pratt's experiments, by 6-day periods. Average mortality is considered as zero for the juvenile 6 days, one-fourth from the first to the second 6 days of maturity, one-third from the second to third, one-half from the third to fourth, and 100 per cent after the fourth. The number of "reproductive units" produced by a given stock in 6 days is defined as the product of the number of mature animals produced times the average number of 6-day periods each lives. Column 6 shows the total units produced by the stock of column 5, values being taken from Figure 29B. These units are divided among the appropriate periods to give the stock present, as shown in the first four columns. The population starts as two newly mature adults, one of which survives for two 6-day periods, the other for four. Total adult stock thereafter is the horizontal sum of the units in the first four columns. The total stock (column 7) is the adult stock (column 5) plus the number of juveniles, the latter being the same as the initial number of matures, as indicated in column 1 one 6-day period later.

1	2	3	4	5	6	7	1	2	3	4	5	6	7
Fraction mature during successive 6-day intervals				Adult stock	Reproductive units produced	Total stock	Fraction mature during successive 6-day intervals				Adult stock	Reproductive units produced	Total stock
0.4	0.3	0.2	0.1				0.4	0.3	0.2	0.1			
..	..	..	..	2	65	2	6	2	1	0	9	230	101
..	..	..	..	2	65	28	92	4	1	0	97	4	189
26	..	..	..	27	20	53	92	69	3	0	164	4	166
26	20	..	..	47	15	55	2	69	46	2	119	4	121
8	20	13	..	41	15	47	2	1	46	23	72	6	74
6	6	13	6	31	20	37	2	1	1	23	27	20	30
6	4	4	6	20	40	28	3	1	1	0	5	160	13
8	4	3	2	17	90	33	8	2	1	0	11	230	75
16	6	3	2	27	20	63	64	6	1	0	71	6	163
36	12	4	2	54	10	62	92	48	4	0	144	4	147
8	27	8	2	45	15	49	3	69	32	2	106	4	108
4	6	18	4	32	15	38	2	2	46	16	66	6	68
6	3	4	9	22	25	28	2	1	1	23	27	20	30
6	4	2	2	14	200	24	3	1	1	0	5	160	13
10	4	3	1	18	65	98	8	2	1	0	11	230	75
80	8	3	2	93	4	119	64	6	1	0	71	6	163
26	60	5	2	93	4	95	92	48	4	0	144	4	147
2	20	40	2	64	6	66	3	69	32	2	106	4	108
2	1	13	20	36	15	39	2	2	46	16	66	6	68
3	1	1	6	11	230	17	2	1	1	23	27	20	30

the 18° cultures rose to a peak of about 200 animals usually after about 70 days, decreased to about 120 animals, then showed a slight increase. At 25° the cultures grew more rapidly, reached a peak in 20 days, and then as rapidly declined; sometimes they died out, but in two instances the oscillation was repeated four times before the experiment terminated. Figure 29 shows one example at each temperature.

The steep reproduction curves of Figure 28 immediately suggest the possibility of oscillations, and they have been used to obtain the computed population curves (B and D) of Figure 29. These start from 2 newborn animals, like Pratt's figures 1 and 2.

Construction of the curves of Figure 29 required certain assumptions and simplifications. The life of a *Daphnia* was divided into "generations" as before. The total number of "reproductive-units" indicated by the ordinate of Figure 28 was divided among successive generation-long periods of life as follows: for 25°C., in the ratio 0:4:3:2:1; for 18°C., in the ratio 0:22:20:18:14:11:9:5:1. These ratios correspond approximately to the average survival indicated by Pratt's figures 5 and 8. The calculation then proceeded on the same basis as in Table I, the sequence for 25° being shown in Table V. Because Pratt's censuses include immature individuals, the mature stock of column 5 is increased by the number of immatures in the culture (considered equal to the first generation of matures, in column 1 one line below) to give the total stock of column 7.

Obviously this computation differs from actuality in the assumption of an age distribution of mortality that is independent of density (cf. Pratt's figs. 5 and 8). The difference is most serious at 25°, and if a more complex calculation were made to take account of it, the initial effect would be to make the troughs deeper because of the shorter lives of denser stocks.

The predicted course of events in Figure 29 has considerable resemblance to what was actually observed in respect to period, and to a less extent in amplitude. (Pratt's observed curves also vary a lot among themselves, as would be expected when the stock starts out from, and at 25° is periodically reduced to, a very few individuals.) Any judgment concerning the significance of this resemblance would have to be of the subjective type which Smith (1952) rightly deplores, but the comparison has the merit that the factual bases of the two curves of each pair are from different sources—there are no "fitted constants".

At 18° the computed cycles are soon damped to an inconsiderable amplitude, but the suggestion of a second low peak in the observed series (Fig. 29A) would probably have been confirmed if the cultures had been maintained a month or so longer. At 25° the computed oscillation reached a stable equilibrium pattern having a large amplitude by the time of the fifth peak, and the two surviving observed series were still oscillating strongly when they were discontinued. The striking difference between 18° and 25°, in both the calculated and the observed curves, is ascribable partly to the steeper reproduction curve at 25°, partly to the fact that a larger number of "generations" are mixed in the 18° mature population.

However there is another factor which enters into these *Daphnia* fluctuations. At the higher temperature the young produced under crowded conditions are permanently less fertile than those from uncrowded cultures[10]. This phenomenon

[10]This was demonstrated by segregating newly mature specimens from crowded cultures, keeping them in isolation, and comparing their production with that of specimens reared singly, but the data cannot be applied quantitatively to these examples. Pratt did not perform the complementary experiment of crowding mature specimens which had been reared singly. However, the very sharp restriction of reproduction which occurs in such event is sufficiently indicated by the small number of progeny produced by the first young born in a culture. These live their adult (but not their juvenile) days under crowded conditions (cf. Pratt's fig. 3).

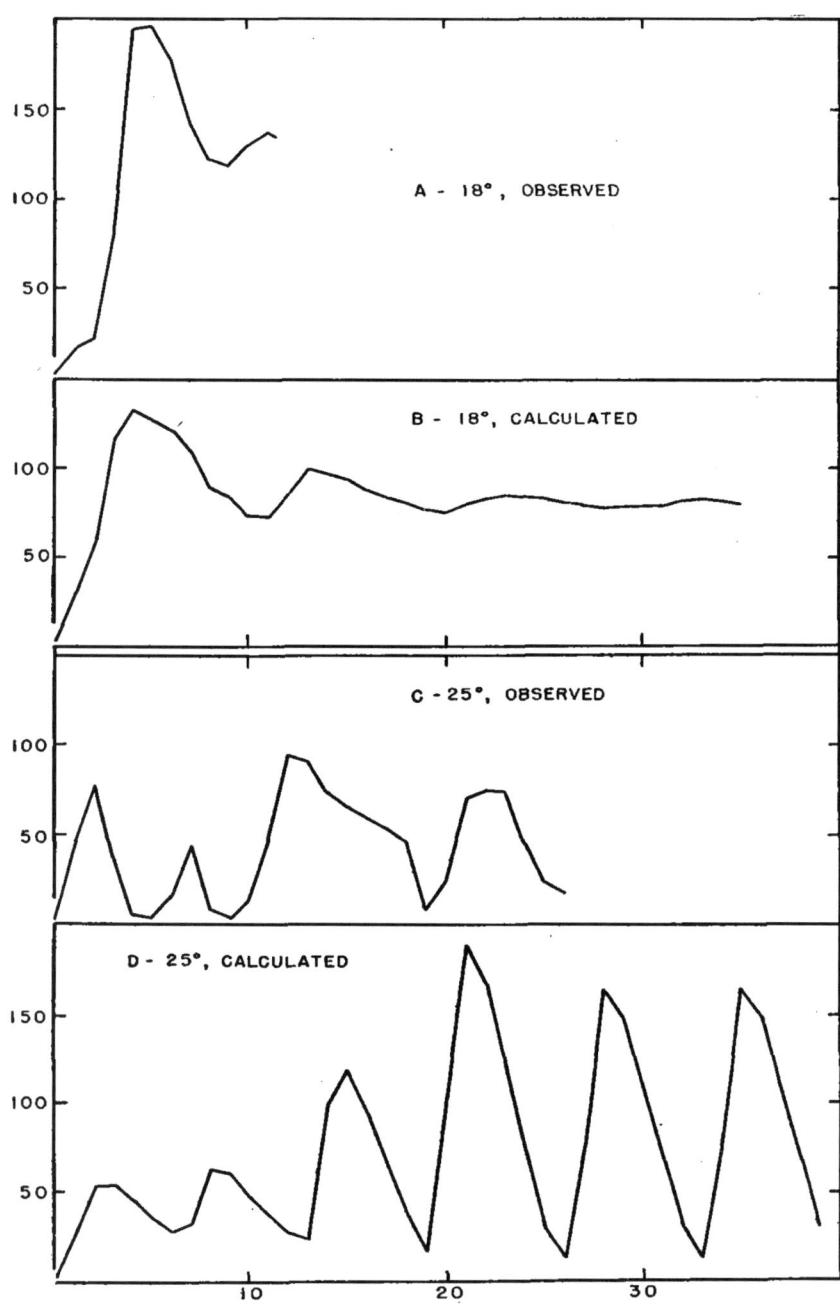

FIGURE 29. Observed and calculated fluctuations in *Daphnia* populations, from data of Pratt (1943). Curves A and C are directly from Pratt's observations, grouped by 14-day intervals at 18°C. and 6-day intervals at 25°C. Curves B and D are calculated from the reproduction curves of Figure 28. *Ordinate*—number of *Daphnia*; *abscissa*—successive 6-day or 14-day intervals.

has been considered the sole basis for the oscillations observed at 25°, both by Pratt and by Hutchinson and Deevey (1949). The present analysis makes it seem more likely that this effect is not an essential feature of the system, and that *actual* density is the more important depressor of reproduction.

STARFISH (*Asterias forbesi*)

No reproduction curve is available for this starfish or any other. However Burkenroad (1946) describes a series of oscillations of its abundance along the New England coast, which have had an average period of 14 years over about 90 years. These seem sufficiently remarkable to justify some speculation concerning whether they could be ascribed to a steep reproduction curve like Figure 7 or 8. If so, the implication would be that the starfish make their median contribution to reproduction 7 years after the zygote stage, and also that they make a relatively small contribution during the first few years after they are hatched. During the peaks of its cycle A. *forbesi* apparently exists in the massive numbers which would make a steep outer limb seem plausible, having in mind that starfish readily eat their own young and that reproduction is partly suppressed among individuals of crowded and poorly fed populations (cf. Vevers, 1949).

The more doubtful aspects are whether starfish live as long as a 14-year cycle would imply, and whether there is sufficient time lag before reproduction. In respect to the latter, Galtsoff and Loosanoff (1939) found the most abundant size group of A. *forbesi* at Wood's Hole in April to have a mode at only 3 cm. (between tips of arms); these would scarcely mature the same year, and even the next group, with a mode at 7 cm., could not be very fecund. The maximum size for this species is about 20 cm., which suggests a life span of several years at least, since rate of linear growth presumably falls off rapidly at the larger sizes, particularly when the stock is dense. Concerning their year-to-year mortality rate little is known except that adult starfish seem to have few natural enemies, so that for the most part they should live out a normal physiological life-span, whose length would of course probably vary a great deal from individual to individual. While they live, the number of eggs produced by each presumably increases more or less in proportion to body volume, so the number produced by a given year-class probably reaches a maximum only after it has been in existence for several years.

On the whole it seems possible, if perhaps not too probable, that the median contribution of a year-class to reproduction could be made 7 years after it was hatched, which is what a 14-year reproduction-curve cycle implies. Because so many ages would contribute to spawning, the curve would have to be very steep —probably even steeper than Figure 8—in order to produce oscillations of the amplitude described by Burkenroad, i.e., having a peak to trough ratio of about 20 to 1. There would likely be minor complications resulting from differences in growth rate between times of abundance and times of scarcity, since growth of starfish is known to vary sharply in response to availability of food. Systematic sampling for size and (if possible) for age, over a period of one or more cycles, should provide the key to the situation. Galtsoff and Loosanoff's samples were taken in 1936-37 at a low point in the cycle, and contained large numbers of young, as expected. At a peak of abundance large individuals should predominate, and young of the year should be very scarce toward the end of the year, or sooner.

SINUOUS REPRODUCTION CURVES

Dr. Pritchard's data on pink salmon reproduction at one site have already been summarized, as well as the adult counts made on the Karluk River. More recently, Neave's (1952, 1953) review of pink salmon reproduction in a number of localities has led him to suggest a rather complex reproduction situation. The species is peculiar in that there are a few areas where one of the two "lines" is completely absent in a series of streams. In other areas one line exists at a level much below the other, and the two lines will have maintained approximately the same relative position for decades, though not necessarily for the whole period of record. In still other areas there is no consistent difference between the lines in respect to abundance. Furthermore there have been cases where one line has dropped suddenly from a high to a low level of abundance and has remained at that level for a considerable period; in one or two instances such a reduced stock has bounced back up to its former level.

Dr. Neave feels that density-dependent mortality in this species must occur mainly during its freshwater life—from the time the adults appear in spawning streams in August-October to the time the fry leave in April or May. Further, there are factors causing both compensatory and "depensatory" mortality at this stage, the latter being mortality which kills a relatively larger fraction of the fish present when their numbers are small than when they are large. Experiments have shown that the activity of stream predators (trout, etc.) upon pink fry is depensatory —the fry apparently being so vulnerable that the predators eat all they can, then leave the rest alone.

This hypothesis accounting for two fairly discrete and fairly stable levels of population in pink salmon implies the existence of anomalous reproduction curves; two possible types are shown in Figures 30 and 31. (The part to the right of the dome does not enter into the argument.) In Figure 31, at some intermediate level of stock, depensatory mortality is at a maximum relative to compensatory mortality, producing the dip in the left limb of the reproduction curve. For example, a population which adhered closely to the curve of Figure 30 could be in equilibrium with a fishery taking 50 per cent of the adult fish at the two different levels of abundance indicated, but between those levels equilibrium rate of exploitation falls to as low as 20 per cent. Since 50 per cent is the maximum permissible limit of exploitation at the higher level of stock, any slightly greater exploitation will send the stock tumbling to the lower level. There it will stay unless exploitation is sharply restricted or temporarily discontinued—merely reducing it to 50 per cent again will not restore the higher equilibrium position.

More generally, when random variation in reproductive success is taken into account, the existence of a fishery makes it easier for a population to slip down

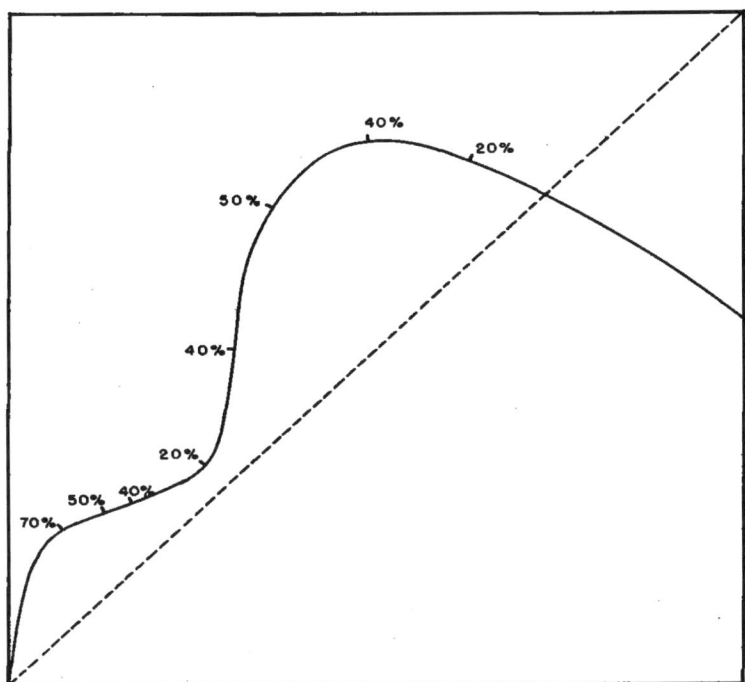

FIGURE 30. Hypothetical reproduction curve, with equilibrium positions for several rates of exploitation. See text.

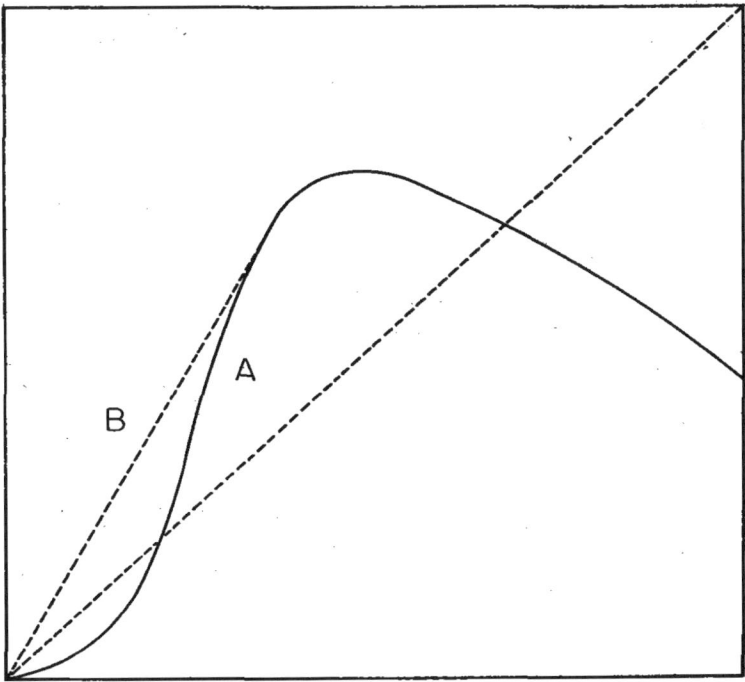

FIGURE 31. Hypothetical reproduction curves. See text.

from the higher to the lower abundance level in a bad year, and harder for it to get back up in a good one. Of course, if the reproduction curve falls off to the right of the 45-degree line, as is actually suggested in Figure 30, the *first* effect of a new fishery would be to increase population abundance.

Figure 31A differs from 30 in that there is no lower level of strong resilience to exploitation. It might represent pink salmon reproduction in areas where one of the two lines ("cycles") does not exist at all. Accidental reduction of the population past a certain low level in that case initiates a progressive decrease and results in eventual extinction. A similar situation is implied in a hypothetical net-change curve of Haldane (1953, fig. 3), and for several animals there is some experimental evidence of reduced effectiveness of reproduction at very low densities (cf. Allee, 1931; Hutchinson and Deevey, 1949).

Dr. Neave's ingenious hypotheses, illustrated by the above curves, are supported by several lines of evidence, but have not as yet been firmly established for any pink salmon stock. McClinton Creek, described earlier, is a stream which has no pinks at all in odd-numbered years, hence might be expected to have a reproduction curve like Figure 31A. The actual curve, however, is of a simple type as far as it is known (Fig. 19). If the left-hand portion of this curve falls below the 45-degree line, it must do so at densities considerably less than 5,000 female spawners—that is, in the extreme lower left corner of Figure 19B. At Karluk there is some indication of an inflexion of the curve at low stock densities (Fig. 20, upper right), but it is not clear that it is sufficient to play the role described above.

A difficulty with the curves of Figures 30 and 31A is that their initial phases are quite different, and hence it seems improbable that both should be applicable to the same species in the same general region. They might be reconciled if the beginning phase of 31A, in varying degrees of development, were commonly found tacked onto the initial part of the Figure 30 curve, producing a doubly sinuous ascending limb for the curve as a whole. If so, hypotheses as to the nature of the compensatory and depensatory mortality would have to be correspondingly more complex. A simpler but less satisfactory compromise type would have merely a straight left limb, as in Figure 31B; this would account for any observed lack of resilience in a population after a certain level of exploitation had been reached, since one density would be as stable as another once the equilibrium point had been shifted to the ascending limb of the curve.

Discussion

Doubtless other investigators will be able to add to the stock of reproduction curves from data now available, and we may expect more in the future. Considering the curves for fishes, cited above, it is unfortunate that in most cases only a section of the curve is available.

A steep ascending limb, as in Figure 2b or 8, is implied whenever a fishery consistently takes some large fraction of a stock before it has a chance to reproduce, as notably among various salmon stocks. Actual points on this ascending limb are available in a few instances where a population has risen from or declined to a very low ebb; the Shuswap sockeye are the best example.

No curve yet discovered suggests that the maximum of recruitment is at or to the right of the 45-degree line (as in Figs. 1, 3, 4), but on several of the curves this line cannot be located.

Concerning the right limb, an indefinite horizontal extension like that of Figure 2 has often been suggested for sea fishes on *a priori* grounds (e.g., Baranov, 1918; Kesteven, 1947). Tester (1948) found this hypothesis consistent with the Vancouver Island west coast herring data, and this is still true (Fig. 18). The same can be said of some of the other curves examined (Figs. 19B, 20, 22). However, without any exception, the lines which best fit these outer limbs slant downward at least slightly, and the weight of their combined evidence

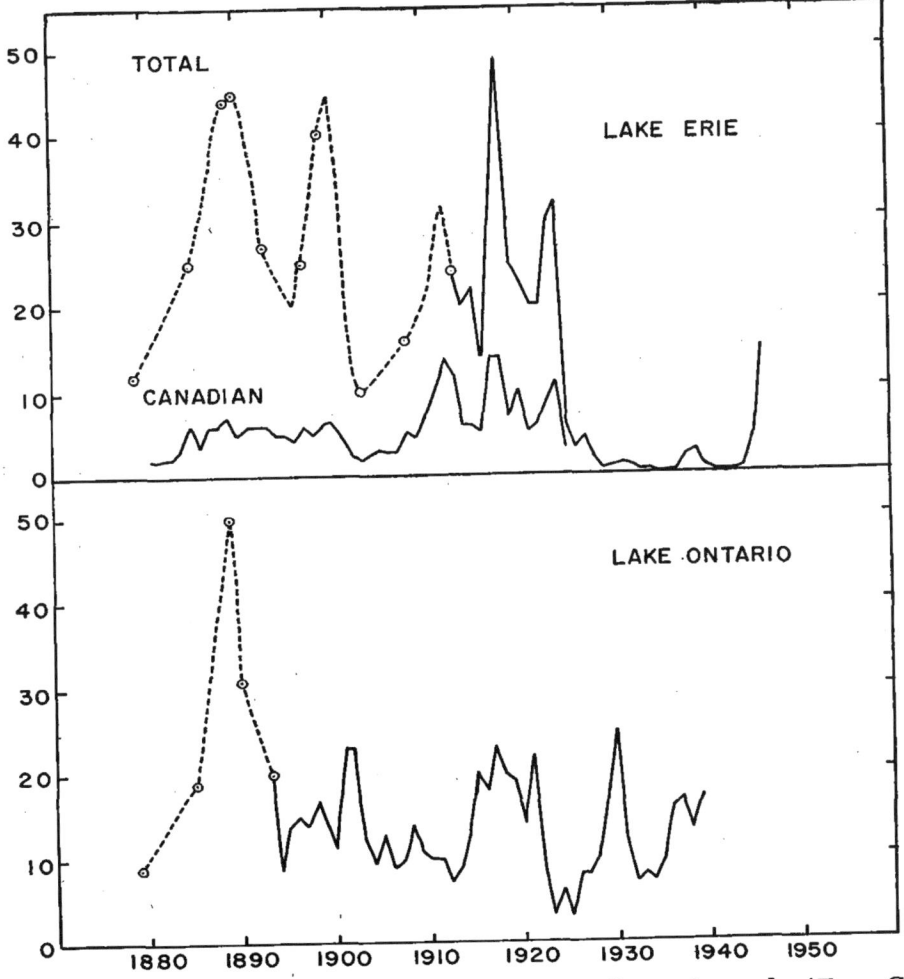

FIGURE 32. Cisco catch from Lakes Erie and Ontario, in millions of pounds. (From Gallagher et al., 1943; and Scott, 1951.)

indicates that negative slopes of at least moderate magnitude are real and are the common situation. On the other hand, an approximation to the flat curve of Figure 2 might well be expected in a species like the coho, which is aggressively territorial and whose stream habitat is limited.

The best-documented example of a really steep descending limb in a natural fish population is provided by the 1912-29 New England haddock (Fig. 24), where the curve is approximately like Figure 7. Even steeper curves characterize the invertebrate populations (Figs. 26-28).

Left to themselves, steep reproduction curves produce such beautifully regular oscillations in population abundance that it is rather disappointing to discover how easily this regularity is disturbed by non-compensatory mortality (Fig. 11B). This means that we should rarely expect to be able to identify the contribution of a reproduction curve to any given series of observations of animal abundance, even though such contribution may be a major factor in many examples of marked fluctuation. And it is a fact that most of the well-known examples of fluctuating abundance cannot be a *simple* resultant of a steep reproduction curve. The 9- or 10-year cycle of grouse and hares, for example, could not be of the simple type shown in Figure 10 because there is not the necessary lag in reproduction—the young are mature in the next reproductive period after they are born; in any event it is doubtful if average longevity in these species is sufficient to put a brood's mean time of contribution to reproduction at 4.5 or 5 years. Certain major changes in fish populations—for example, cod off Greenland, herring off Norway, or mackerel off eastern North America—have occurred on too great a scale and with too long a period to have been governed by any single reproduction curve. Fish population changes having shorter period, like those of the ciscoes of Lake Ontario (Fig. 32), might be strongly influenced by their reproduction curve, but the peaks and troughs are too irregular, and basic population information is insufficient, to permit even a good guess. If the cisco rate of exploitation is moderately great, as seems probable, then direct reproduction-curve periodicity is not likely to be recognizable (cf. Fig. 17).

We are perhaps lucky to have, in Herrington's haddock, even one example of an oscillation in which the effect of the reproduction curve can be identified with fair likelihood.

OTHER REPRODUCTION SITUATIONS

Up to this point all examples and interpretations have been predicated on the postulates that it is only animals of mature size whose abundance affects compensatory mortality; and that where there is human exploitation of the stock, it takes only mature individuals. What is the effect of relaxing each of these restrictions?

Larger Immature Fish Taken by the Fishery

If a fishery takes fish of smaller than mature sizes, it makes little difference to the theory of population control outlined above *as long as the sizes subject to compensatory mortality are not touched*. Details of the equilibrium level of population at different intensities of exploitation are of course affected, but not the general course of events. In actuality it has fairly often been found that spawning populations can be reduced to a very low level while the catch is maintained at a fairly high level by immature fish, as for example the cod in the North Sea. In this and similar instances data for plotting a reproduction curve do not seem to be available, but the general situation strongly suggests that the right limb of the curve must slope downward and that the left limb must be very steep. Until reproduction curves are available and future recruitment can be predicted, proposals for increasing catch by reducing the fishing intensity in the North Sea and elsewhere will lack that final symmetry which would make them wholly convincing.

EXPLOITATION DURING THE COMPENSATORY PHASE

Exploitation that takes fish at an age when natural mortality is still compensatory means, for practical purposes, a fishery for young during the first year or two of their life—the earlier the better. The removal of such young is at least partly balanced by increased survival and/or growth of the remainder; in fact, the effects of removals at this stage are equivalent to reduction of the spawning stock which produced the brood in question. If the reproduction curve for the population is of any of the types 3–8, such reduction will at first *increase* net production of recruits, which will produce more eggs and permit a larger catch of young in future years. This ascending spiral of abundance may continue until the level of stock is reached which produces maximum recruits.

There can of course be no general rule indicating a single optimum time for exploitation for all fisheries. In any particular instance, maximum yield may be obtained by taking young, or adults, or both; and in balancing the alternatives, consideration should be given to the relative value per pound of the two sizes and the ease with which they are caught. But it is clear that any *general* prejudice against exploiting young fish is unsound. Each case should be considered on its merits. The situation most favourable for juvenile exploitation is evidently that where the fish reach a *relatively* large size before compensatory mortality ceases to be important, so that the catch taken during this period can be large in total bulk.

Relatively few fisheries now exist in the temperate parts of the world which attack really young individuals, but this can be ascribed mainly to the (usually) lesser value per pound of small fish as compared with large ones, and the (usually) greater difficulty of catching these young. In Japan a number of species are eaten as larvae—for example the sand lance (*Ammodytes*). There and elsewhere the exploitation of various clupeids in post-larval, but still early, life is fairly common.

One such fishery on this continent is that for "Quoddy" herring in southern New Brunswick (Huntsman, 1952, 1953). A large primitive stock of mature herring in this region apparently fluctuated in abundance during the middle years of the last century, but no details of period or amplitude are available. About 1880 mature herring declined rather abruptly and they have not reappeared in comparable numbers since. About the same time the weir fishery for "sardines" (age 0 to age II herring, mainly) developed enormously and has yielded large and fairly steady catches ever since—catches which bulk two or three times as great as did the early fishery for mature individuals.

If we postulate that the weirs take the young partly during the time that compensatory mortality is operative, and partly after compensation is no longer important, a formal explanation of this situation can be made. The sardines taken young, while the mechanism of compensation is still in operation, represent a diversion to human use of stock which would otherwise be lost to predators or other natural causes. The sardines taken at larger sizes, after compensation is largely finished, plus any larger stock taken, reduce the potential breeding population to a point far below the numbers it would otherwise attain (and did actually attain in years gone by). This reduced spawning stock produces many more young fish than a large one would, the implication being that in decreasing in abundance it has climbed the right limb of a dome-shaped reproduction curve such as Figure 8. In this way the present distribution of sizes and numbers could be maintained.

Compensation by Immature Members of the Population

In the restricted habitat of ponds and small lakes cannibalism is one of the most likely methods of population regulation for abundant predacious species, e.g., bass (*Micropterus*) or crappies (*Pomoxis*). Indeed, the first stages of all centrarchids are so tiny that even a normally insect-eating species like the bluegill (*Lepomis macrochirus*) may eat eggs or fry of its own species during their very early life. Immature individuals tend to be most active in this, since they normally frequent the shallow water where most spawning occurs, and gather around nests of their own or other species. Adults are less frequently in contact with young, and the males actually guard eggs and young as long as they are in the nest, or even longer; hence adults are less effective in reducing the oncoming generation. As an example, two similar ponds at the Tri-lakes Hatchery, Indiana, were stocked in spring with similar weights (13–15 kg.) of bluegills: in Pond 2 they consisted almost wholly of mature fish, in Pond 4 about half (by weight) were small fish of the previous year's brood. By November the young of the year produced by Pond 2 numbered 19,150 and weighed 36.5 kg., whereas Pond 4 produced only 230 fingerlings weighing 0.18 kg.

To adequately illustrate situations of this sort it would be necessary to construct a reproduction diagram showing the three-way relationship between quantity of mature stock, quantity of immature stock, and resulting recruitment. Present information does not permit this, but the general effect of participation of immatures in compensation must be to spread the reproduction-depressing influence of any successful brood over a longer period of time, and to make it almost impossible to have the "lag" effect which produces population oscillation. In such a situation, however, cycles of fish *size*, and to some extent of poundage, might occur if individuals of a successful brood nearly all died off during the same calendar year (cf. Thompson, 1941, p. 209).

TOWARD A THEORY OF RECRUITMENT

The justification for using any reproduction curve must in the long run come from observation. However it would be most useful, if it were possible, to formulate some general theory of reproduction which might lead to a standard type of reproduction curve applicable in a majority of situations.

Theory of Predation

An approach to this goal can be made by way of a consideration of predation upon the young of a species, whether by other animals or by older individuals of its own species. The theory of predation was briefly considered by the writer in a recent paper (1952), and the following quotation will provide a basis for the further argument here:

> It is convenient to distinguish three types of numerical relationship between predators and a species of prey which they attack.
>
> A. Predators of any given abundance take a fixed number of the prey species during the time they are in contact, enough to satiate them. The surplus prey escapes.
>
> B. Predators at any given abundance take a fixed fraction of prey species present, as though there were captures at random encounters.
>
> C. Predators take all the individuals of the prey species that are present, in excess of a certain minimum number. This minimum may be determined in different ways: (1) There may be only a limited number of secure habitable places in the environment, so that some prey are forced to live in exposed situations where capture is inevitable. The number of such secure niches may be partly governed by territorial behaviour of the prey. (2) The maximum "safe" density of prey may be the one at which predators no longer find it sufficiently rewarding to forage for them, and move to other feeding grounds.

The three situations above tend to intergrade, of course, but it is useful to keep their differences in mind.

SITUATION A

This is likely to occur when a prey species is temporarily massed in unusual numbers, for example, adult herring in spawning schools, or newly-emerged fry of pink and chum salmon going downstream. The main characteristic of such situations is that the number of prey eaten depends on the abundance of predators, but not on the abundance of prey. Hence such situations cannot last long, and the predators cannot make the prey in question their principal yearly food; otherwise they would almost surely increase in abundance and the situation would change to type B or type C. If a type A situation persisted for long, it would come to an abrupt end with the extermination of the prey.

SITUATION B

Here the number of the prey species eaten is proportional to the abundance of predators and to the abundance of prey. Unlike A, this type of predation can easily occur over long portions of the year; and the prey species may comprise the larger portion of the predators' annual ration. This situation was observed at Cultus Lake, British Columbia, for predation of squawfish, char, coho and trout upon fingerling sockeye, over a wide range of abundance of the latter (Foerster and Ricker, 1941).

SITUATION C

The classical example of this situation was described by Errington for bob-white in Iowa, where a given range would winter safely a fixed number of birds, practically independently of the number which were present in autumn, the surplus being taken by predators. Studies conducted from the Atlantic Biological Station of the Fisheries Research Board of Canada,

St. Andrews, N.B., suggest that in some rivers predation upon Atlantic salmon parr tends toward type C, because a marked increase in the number of young fish planted was followed by only a relatively small increase in number of surviving smolts (Elson, 1950). In type C situions the predators must tend to have ample alternative foods, and they are often mobile so that they can leave an area where the food supply has been "cleaned up" for the season.

Consider a prey species which is subject to type B predation over some part of its life history. Assume first that predation causes the whole of the mortality that the prey is subject to during that period. Then the prey species decreases in abundance according to the well-known exponential formula

$$N/N_0 = e^{-it}, \qquad (1)$$

where: N_0 is initial abundance, N is abundance at time t, $e = 2.718\ldots$, and i is a statistic representing the fraction of the prey which would be eaten in a unit of time if its abundance were held constant for that long; i is often called the *instantaneous* mortality rate.

Under the conditions postulated, instantaneous mortality rate is directly proportional to the abundance of predators. This can be illustrated by considering a situation where predators attack a prey population of 1,000,000 individuals under type B conditions, and they inflict losses corresponding to $i = 0.8$, where the unit of time, t, is the whole season that predator and prey are in contact. Equation (1) indicates that in, for example, 1/1000 of the season, the predators will eat $i/1000 = 0.0008$ of the prey present, or 800 fish. During the next thousandth of the season, the predators eat $i/1000$ of the surviving prey, or $999,200 \times i/1000 = 799$ fish. The following interval they eat $998,401 \times i/1000 = 799$; then $997,602 \times i/1000 = 798$. This continues until all the thousand time-intervals have elapsed, at the end of which there are:

$$1,000,000(1 - 0.0008)^{1000} = 472,400$$

survivors, or 47.2 per cent. What happens if the number of predators, is doubled? In that event, during the first thousandth of the season twice as many fish will be eaten, i.e., 1,600, leaving 998,400. In the next thousandth the fraction eaten is likewise double, namely 0.0016; multiplied by the number of survivors this gives 1,597; and so on. At the end of the season $1,000,000(1 - 0.0016)^{1000}$ survive, which is 201,900, or 20.2 per cent.

Thus doubling the number of predators doubles the instantaneous mortality rate (which is true generally), but it increases actual mortality by only $27.0/52.8 = 51$ per cent (which is true of only this particular example).

The general relation between predator abundance and actual mortality, or survival, is most conveniently expressed as:

$$\frac{p_2}{p_1} = \frac{\log s_2}{\log s_1}, \qquad (2)$$

where p_2 and p_1 represent two levels of predator abundance, and s_2 and s_1 are the corresponding survival rates for the prey. This expression can easily be derived from (1), since p is proportional to i, and $s = N/N_0$ when $t = 1$.

CANNIBALISM

Of all the methods of population regulation listed earlier, cannibalism is the one in which the abundance of the control agent is most closely and inseparably allied to that of the population controlled. That is, an increase in mature stock not only increases the number of eggs laid or young born in a given reproductive season, but it also decreases the rate of survival of those young. What is the combined effect of these two opposed influences?

Let the size of a stock be measured by the number of eggs it lays, and let this size be proportional to its *instantaneous* efficiency in cannibalism (the fraction of young eaten by the adults in a short interval of time). Consider two situations characterized by different sizes of stock and let the ratio of the second to the first be w, corresponding to the p_2/p_1 of equation (2) of the quotation above. Finally, let all other sources of mortality, whether they occur before, after or during the period of the cannibalism, be density-independent, their total effect being to reduce survival rate of eggs and young to the fraction k of what it would otherwise be. The number of eggs laid, in situation 1, is E_1; in situation 2 it is therefore $E_2 = wE_1$. The reproduction (absolute number of recruits produced) in situation 1 is:

$$R_1 = E_1 \, ks_1; \tag{3}$$

and in situation 2 it is:

$$R_2 = E_2 ks_2 = wE_1 ks_2. \tag{4}$$

From equation (2) of the quotation above,

$$w = \frac{\log s_2}{\log s_1}; \quad \text{hence } s_2 = s_1^w. \tag{5}$$

From (4) and (5), the reproduction in situation (2) becomes:

$$R_2 = wE_1 ks_1^w. \tag{6}$$

Expressed as a fraction of the reproduction in situation 1, this is:

$$ws_1^{w-1}. \tag{7}$$

Plotting numerical values of (7) against w, for various values of s_1, yields a family of curves each of which has an origin at zero, a dome, and an extended right limb which approaches the abscissa asymptotically. No matter what survival rate is chosen for s_1, each curve describes the whole range of possible survival values as w is varied; hence only one curve of the family is needed. The convenient one to choose is that for which (7) is a maximum when $w = 1$. To locate this maximum, (7) is differentiated with respect to w and equated to zero:

$$ws_1^{w-1} \log_e s_1 + s_1^{w-1} = 0; \tag{8}$$

whence,

$$-\log_e s_1 = 1/w; \quad \text{or, } s_1 = e^{-1/w} \tag{9}$$

The value of s which makes (7) a maximum is $e^{-1/w}$, and if this is taken as the initial value of s (that for which $w = 1$), s becomes equal to $1/e$ or 0.3679. Substituting $1/e$ for s_1 in (7), we thus finally obtain the expression

$$we^{1-w}. \tag{10}$$

This shows the actual level of reproduction as a fraction of the maximum, when w represents the ratio of the actual density of mature stock to the density which gives maximum reproduction. Values of (10) are plotted as curve B of Figure 33.

From the equations and the curve the following conclusions are evident:

1. Since the curve approximates to the abscissa as w is increased, then *if the breeding stock is made sufficiently large, cannibalism reduces reproduction practically to zero*, in spite of the greatly augmented egg deposition. In theory this is true no matter how small the original rate of cannibalism; in practice, a type B predation situation between adults and young could not be maintained over *too* wide a range of stock densities.

2. Since, as w approaches 0, e^{1-w} approaches e, it follows that at minimal densities the survival rate from cannibalism is e times the survival characteristic

of maximum reproduction. In other words, the instantaneous mortality rate at maximum reproduction is greater by unity than it is at vanishingly small stock densities (since $-\log_e (1/e) = 1$; cf. expression (1) of the quotation above).

Curve B of Figure 33 can be changed into a reproduction curve by changing the scale of each axis to represent actual numbers of parents and progeny measured in comparable units—for example the egg production unit described in an earlier section. Equation (10) gives no information concerning the steepness of this actual reproduction curve, which will depend chiefly on the magnitude of k, the survival rate from factors other than cannibalism. On Figure 33 a number of possible curves have been drawn, all based on equation (10) but using different ratios of abscissal to ordinate scales. Referring each to the 45-degree line indicated, shapes reminiscent of several of the arbitrarily drawn types of Figures 3–8 can be identified, as well as most of the curves observed in actual populations. Curve E, of course, could not describe a stable population because it lies well below the 45-degree line, while even D would be rather precarious.

In addition to cannibalism, the argument of this section can be extended to the analogous situation where strife between mature and young animals occurs and results in frequent killing of the latter, but without necessarily any eating of one by the other; this has been described among muskrats, for example (Errington, 1954). In so far as these contacts are of type B—with the older animals killing a fixed fraction of the young they encounter—they would have the same population status as described above.

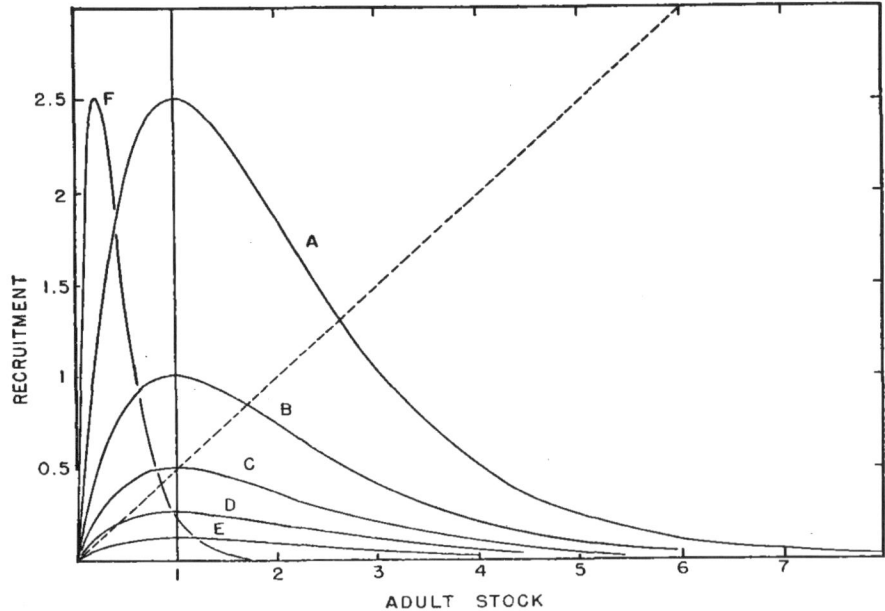

FIGURE 33. Graphs of we^{1-w} ("recruitment") plotted against w ("adult stock"). Curve F corresponds to the axes as labelled; the other curves are obtained by varying the ordinate and (for F) abscissal scales.

It is not easy to assess the probable importance, in nature, of cannibalism and allied effects. In most of the fish populations whose reproduction curves were examined earlier there is little or no evidence of cannibalism or of other direct injury to young by their parents, and among *Oncorhynchus* this is impossible because the parents die. However, in assessing the possible role of cannibalism (or any other agent of predation) in population regulation, it is not possible to weight it directly on the basis of the number of units which it kills. The latter will depend mainly upon the stage at which it is active. For example, suppose that cannibalism by mature trout fell upon their fingerling progeny following an average density-independent egg mortality of 50 per cent and fry mortality of 90 per cent (95 per cent in all); and that the adult trout ate 63 per cent of the 5 per cent remaining, or 3 per cent of the original brood. In that event cannibalism would seem to be quite insignificant either from the point of view of the percentage of the total young which it killed, or from the point of view of the frequency of occurrence of small trout in stomachs of larger trout. In spite of this, under the conditions just postulated cannibalism would be the sole mechanism regulating the abundance of the population.

PREDATION BY OTHER ORGANISMS

Even granting that lethal contact between adults and young is the most direct and least fallible population-regulating mechanism, and that it is quite apt to exist undetected, the writer's present opinion is that it will probably not prove to be important in more than a minority of populations. In its absence, other compensatory agents must take over the role of regulators which determine the shape of the reproduction curve.

TYPE B SITUATIONS. A predator can qualify as an agent of population control only if it can increase in density or effectiveness as the abundance of prey increases, and vice versa. One way this can occur is by migration of additional predator units to the region of predation when prey abundance is large. For example, greater-than-average abundance of herring spawn might attract birds from a wider area than usual. It is possible (not necessarily probable) that, after gathering together, such aggregations would remain in the vicinity for a few days longer than the supply of eggs justified it, with the result that an initially superior spawning would yield a less-than-average number of eggs hatched.

A predator might also actually increase its total abundance with that of the prey in question. Consider, for example, a plankton predator or assemblage of predators which feeds on the pelagic eggs and young of mackerel. Suppose that in years when mackerel eggs approach zero abundance these predators (which have additional foods) are numerous enough to consume the fraction $(1 - s_1)$ of the eggs. Suppose also that for each unit increase in egg numbers the average predator population increases by one-tenth of its "basic" number. (This average would be taken over the whole time that predators and mackerel are in contact because, just before the eggs are available, initial predator abundance is assumed to be the same in all years.) During the later stages of the predator-prey contact the abundance of predators would have "overshot" that of the prey, with the result again that mackerel broods initially larger than average would end up smaller than average. From expression (2) or (7) it is easy to calculate that relative reproduction is equal to

$$Zs_1^{1+Z/10}, \qquad (11)$$

where Z is the density of mackerel eggs immediately after spawning in a given year, in the unit mentioned above, and the predator density corresponding to no eggs is taken as unity. Values of (11) are calculated in Table VI, which shows the relative number of mackerel surviving the pelagic stage, for a series of initial prey densities and for two values of s_1. Sette (1943) estimated total survival of

TABLE VI. Survival rate and absolute number of survivors when average predator abundance varies with initial prey density in the manner indicated by columns 1 and 2. Pairs of survival rates and survivor numbers are shown for initial survival rates of 0.1 and 0.0001. The last two rows show the computation of the maximum number of survivors for each situation, which occurs when survival is 0.3679 (1/e) of its initial value.

1 Prey abundance	2 Predator abundance	3 Survival rate of prey	4 Relative number of survivors	5 Survival rate of prey	6 Relative number of survivors
			$(1) \times (3) \times 10^3$		$(1) \times (5) \times 10^7$
0	1.00	0.1000	0	0.0001000	0
0.2	1.02	0.0955	19	0.0000832	166
0.4	1.04	0.0912	37	0.0000692	277
0.6	1.06	0.0871	52	0.0000575	345
0.8	1.08	0.0832	67	0.0000479	383
1.0	1.10	0.0794	79	0.0000398	398
1.2	1.12	0.0759	91	0.0000331	397
1.4	1.14	0.0724	101	0.0000275	385
1.6	1.16	0.0692	111	0.0000229	367
1.8	1.18	0.0661	119	0.0000191	343
2.0	1.20	0.0631	126	0.0000158	317
2.5	1.25	0.0562	141	0.00001000	250
3.0	1.30	0.0501	150	0.00000631	189
4	1.4	0.0398	159	0.00000251	100
5	1.5	0.0316	158	0.000001000	50
6	1.6	0.0251	151	0.000000398	24
8	1.8	0.0158	127	0.000000063	5
10	2.0	0.0100	100	0.000000010	1
15	2.5	0.0032	47	...	...
20	3.0	0.0010	20	...	...
30	4.0	0.0001	3	...	...
1.086	1.1086	...	...	0.00003679	399
4.343	1.4343	0.03679	160	...	...

Atlantic mackerel eggs and pelagic larvae, in 1932, to lie between 0.000001 and 0.00001, so that the $s_1 = 0.0001$ column of Table VI would not be unrealistic.

The distributions of columns 4 and 6 of Table VI are of course the same as those of Figure 33. Expression (11), like (7), is a maximum when the rate of survival from the predator in question is $1/e$ of what obtains when egg abundance is close to zero (cf. Table VI). Thus we again conclude that *the instantaneous mortality rate at maximum reproduction is greater by unity than the rate characteristic of a very small population density.* For example, if $s_1 = 0.01$, the instantaneous mortality rate at minimum stock density, from the action of the controlling predator, is equal to $-\log_e 0.01 = 4.61$. At the density of maximum reproduction, mortality from this predation becomes 5.61. Mortality from other factors ($= -\log_e k$) must be added to this 5.61 to give the total average egg-to-egg instantaneous mortality rate at maximum reproduction. Note however that at the *replacement* density of stock the compensatory mortality is greater than 1, its exact value depending on the steepness of the reproduction curve (relative to the 45-degree line).

The form of expression (11) indicates that the magnitude of the arbitrary factor 10, relating initial egg density to mean predator density, does not affect the general shape of the recruitment curve, though it of course affects its steepness: the larger this factor, the broader is the range of initial egg densities which afford substantial reproduction. Furthermore the rule italicized above holds even if the relation between prey and predator is quite irregular.

Similarly, the magnitude of s_1 affects the shape of the reproduction curve only in that the smaller s_1 is, the greater is the percentage change in reproduction produced by a given change in predator abundance (Table VI).

SITUATIONS OF TYPES A AND C. So far the theory presupposes type B predation situations. Type A situations need not be discussed in detail: for reasons given earlier, they tend to be restricted in space or time, and they lack the qualifications of a population control mechanism. Their effect on reproduction curves would be to introduce irregularities such as are shown in Figures 30 and 31.

Type C situations however *can* regulate a population, and in a pure form they produce reproduction curves like Figure 2b. The horizontal part of the curve corresponds to the limit of surviving young which the habitat will sustain. Type C situations may grade into type B, in which case it would in practice be difficult to distinguish a broad-domed type B curve from the mixed type.

It is also possible for type B and C situations to follow each other, and this is examined in Figure 34. If the type B situation comes first, its typical reproduction curve is truncated by the limit of reproduction imposed later by Type C (curve A of Fig. 34). If the type C situation comes first, it will at a certain point limit the brood to a density which may be less than, equal to, or more than what gives maximum reproduction in the subsequent type B situation; these three possibilities are illustrated by curves B–D of Figure 34. Note that while the type C situation can *reduce* the difference in instantaneous mortality rate between minimal and maximal reproduction, it cannot *increase* this difference.

The various curves of Figure 34 should be looked for in nature when the biology of the animal in question points in their direction. Among the examples given earlier, a limit of environmental capacity is strongly suspected for coho salmon; actually Figure 21 could readily be fitted with a truncated curve like 34A, but there is only one point for the outer decreasing phase.

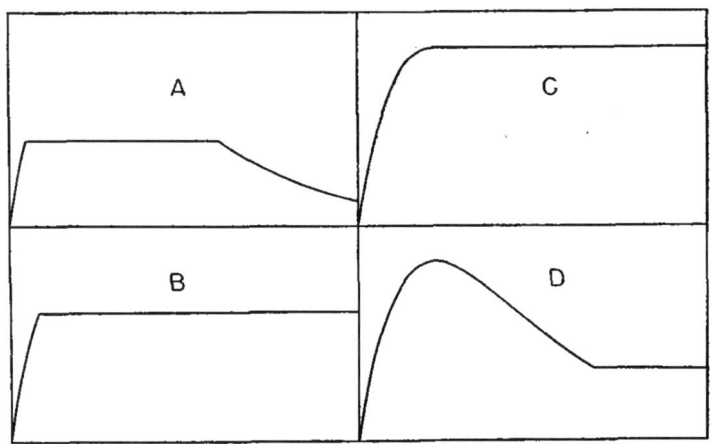

FIGURE 34. Reproduction curves when predation situations of types B and C occur in succession. *Curve A*—when type B precedes type C; *curves B–D*—when type C precedes type B. The type B reproduction curve is the same for all four situations.

SIZE OF PREDATOR AND SIZE OF PREY. It has sometimes been said that large predators (e.g., foxes) cannot control or regulate the abundance of a smaller prey species (e.g., mice), because the rate of reproduction of the prey is so much greater than that of the predator. This argument would be valid only if the predation situation were of type A, with the foxes always able to kill as many mice as interested them—a situation which could scarcely last indefinitely. However if the fox-mouse predation situation were of type B, calculations from (7) show that any increase at all in the number of foxes reduces the survival rate of mice more than proportionately, provided the foxes initially are the cause of an instantaneous mortality rate of 1 or more in the mouse broods; and the greater the increase of the foxes, the less this initial mortality need be to accomplish the result stated. To put it another way, the effectiveness of foxes in reducing the survival of mice will increase more rapidly than does the actual number of foxes, whenever they are already reducing mouse survival to 37 per cent or less of what it would be in the absence of foxes. This effect gives the foxes an advantage which tends to counterbalance the greater rate of reproduction of the mice.

More generally, there is evidently no reason why a large predator, or assemblage of predators, might not regulate the abundance of a smaller and more prolific prey species, provided their abundance can change with that of the prey to *some* extent. Indeed, since a given change in predator abundance has a greater effect upon survival, the smaller the average survival rate of the prey (and hence the greater its fecundity), it could be argued that great fecundity should make a species more—not less—easily controlled by small changes in predator abundance. Against this must be laid the fact that great fecundity makes the prey more likely to occasionally "slip out from under" by becoming so numerous that the type B predation situation cannot be maintained.

OTHER COMPENSATORY AGENTS

We have seen that both cannibalism and "ordinary" predation lead to the Figure 33 type of reproduction curve under type B conditions. For other compensatory agents only a qualitative examination can be attempted here, but they must be mentioned briefly.

The special type of predation situation in which insect "parasites" eat other insects regularly leads to low host reproduction at high host densities, as shown in Nicholson and Bailey's (1935) detailed analysis.

"True" parasites (those which do not regularly and consistently destroy their hosts) and disease organisms can also exhibit non-linear effectiveness which can seriously reduce reproduction at high stock densities—though their activity in this respect must usually be very irregular, if the human situation is a guide.

In the *Daphnia* cultures described earlier no predation or cannibalism existed, nor apparently did disease. The most plausible suggestion which has been made is that some metabolic product acted as a depressor of reproduction. If, for example, a metabolic waste killed a fixed fraction of *Daphnia* embryos per unit concentration and per unit time throughout prenatal life, it would have exactly the same effect upon survival as random cannibalism, and lead to the reproduction curve of Figure 33. The points of Figure 28 can be fitted by expression (10) with good apparent agreement.

Again, consider the compensatory situation where a brood, initially more numerous than average, grows more slowly than average because of competition for limited food; and because of this smaller individual size it is more vulnerable to predation than a faster growing brood at corresponding ages. It is fairly clear that here too there *could* be a lag effect whereby the brood would lose by increased predation more than the numerical advantage which it originally had; though it would be difficult to show that this should (or should not) occur as a general rule.

Finally, consider what happens when spawning facilities are limiting, as for some salmon or trout populations, for example. Natural selection will probably see to it that the population uses the best spawning facilities first. At increasing densities there is both a spread to less suitable gravel in other parts of the stream and a crowding and superimposition of eggs on the favourable grounds. With extremely heavy seeding, fungus may cover and kill almost the whole of the eggs in even the best redds; something of the sort probably decimated the Karluk pinks in the disastrous year 1924 (Barnaby, 1944). The sum of the above effects would usually add up to a domed reproduction curve with an extended right limb. Though an *exact* fit to expression (10) would not be likely, the latter might serve as a picture of the normal expectation.

FITTING THE CURVE $z = we^{1-w}$

A first approximation to a fit of expression (10) to any body of data can be obtained very quickly, because the distribution is completely determined by the position of the maximum and the latter can be selected by eye. The work consists merely of dividing each observed estimate of reproduction (Z) by the estimated maximum Z, and each observed parental stock density W by the density which produced the estimated maximum W; this gives w and z, respectively. From (10),

$$\log_e \hat{z} = \log_e w + (1 - w). \tag{12}$$

A sample calculation is shown in the last line of Table VII.

To find the *best*-fitting curve of this type it would be necessary to calculate the residuals (differences between observed and calculated values) for a series of trial positions of the maximum point. The latter can be varied in two dimensions, so the process might be fairly protracted. The accepted criterion of best fit would be that for which the sum of the squares of the residuals is least, but an easier

TABLE VII. Computation of expected reproduction ($\hat{Z}$) and residuals (R) for a set of stock-reproduction (W–Z) observations, using $z = we^{1-w}$. W = 13, $Z_{max.}$ = 38, is the trial position of the dome of the curve. The last line is a computation of $\hat{Z}$ for an arbitrary W. (See text for further details.)

1	2	3	4	5	6	7	8	9	10	11	12
W	Z	z	w	$\log_e z$	$\log_e w$	Unity	R	R^2	$\log_e \hat{z}$	$\hat{z}$	$\hat{Z}$
5	18	0.47	0.38	−0.75	−0.95	1.00	−0.42	0.18	−0.33	0.72	27
8	78	2.05	0.61	+0.72	−0.49	1.00	+0.82	0.67	−0.10	0.90	34
16	20	0.53	1.23	−0.64	+0.21	1.00	−0.62	0.38	−0.02	0.98	37
28	5	0.13	2.15	−2.02	+0.77	1.00	−1.64	2.69	−0.38	0.68	26
33	8	0.21	2.54	−1.56	+0.93	1.00	−0.95	0.90	−0.61	0.54	21
78	28	0.74	6.00	−0.30	+1.79	1.00	+2.91	8.47	−3.21	0.04	1.5
Sum	..	..	..	..	..	..	+0.10	13.29	..	..	..
50	..	..	3.85	..	+1.35	1.00	..	..	−1.50	0.22	8

criterion—that the sum of the residuals should approximate to zero—would be useful in the earlier stages. Table VII shows a computation of residuals (R) for the pink salmon data of Figure 19. The dome of the curve is estimated to be at a parental abundance of W = 13 and filial abundance of $Z_{max.}$ = 38. Dividing the observed values of columns 1 and 2 by these values gives z and w in columns 3 and 4. The residual (R) is the observed value $\log_e z$ less the estimated $\log_e \hat{z}$ from (12), which is equivalent to

$$R = \log_e z + w - \log_e w - 1. \tag{13}$$

The four right-hand terms of (13) are in the order of columns 4–7 of Table VII; adding columns 4 and 5 and subtracting 6 and 7 gives R in column 8, and R^2 is shown in 9. The sum of the R's is close to zero, showing that the trial position of the dome of the curve cannot be too far from the best one; however since more than half of the R^2 total is contributed by the last observation, a somewhat better fit could probably be obtained by moving the dome a little to the right or upward.

Expected reproduction values, $\hat{Z}$, are calculated in the last three columns of Table VII. Column 10 is equal to 8 less 5; or it can be obtained directly by adding columns 6 and 7 and subtracting 4 (cf. formula 12). The latter method is used to calculate $\hat{Z}$ for arbitrary values of W, as shown in the last line of the Table.

Base-10 logarithms can be used in Table VII if columns 4 and 7 are multiplied by 0.4343, but the procedure shown is more convenient when a comprehensive table of natural logarithms is at hand.

Fitted curves have not been used on the observed reproduction figures above (Figs. 18–28). Apart from the work involved, the theory of expression (10) needs further examination, and uncritical use of it now might conceal variant or alternative situations, such as that suggested for Figure 21. All the curves of Figures 18–28 were drawn freehand before expression (10) had been developed, and it is interesting that most of these observations suggested a concave and tapering right limb. Even the original hypothetical curves of Figures 7 and 8 were drawn this way, because of a feeling that reproduction couldn't really decline right to zero as spawners became increasingly numerous.

Returning to the question introduced at the start of this division of the paper, no universally applicable theory of reproduction has been discovered, or is likely to be. However several possible reasons are apparent why recruitment can decline, and usually will decline, at higher stock densities. Also, from reasonable assumptions a simple mathematical expression has been developed which could be used to represent most of the observed reproduction data over the range of densities which they cover.

SUMMARY

1. The general theory of reproduction indicates that density-dependent causes of mortality set a limit to the size which a population achieves. The "reproduction curve" for a fish species is defined as a graph of the average number of eggs produced by a filial generation against the number produced by its parental spawning assemblage, under the existing frequency distribution of environmental conditions for survival (Figs. 1–8).

2. The level of adult stock at which a *maximum number* of recruits (on the average for existing variations in environmental conditions) is obtained is not necessarily the level at which the *replacement number* is obtained (on the average). Maximum recruitment may occur either at the replacement level of adult stock (Figs. 2, 3), or at a higher level (Figs. 1, 4), or at a lower one (Figs. 5–8).

3. When, with increasing stock density, the maximum of recruits exceeds and precedes the replacement number, a population tends to oscillate in abundance. These oscillations are stable if the reproduction curve crosses the 45-degree line (representing the replacement level of reproduction) with a slope between -1 and $-\infty$ (Figs. 7, 8); they are damped if the curve crosses at any numerically lesser slope (Fig. 5).

4. Under the conditions of Figures 7 and 8 a population of single-age spawners has an irregular but permanent cycle of abundance, such as shown in Figure 11D for example, if enivronmental conditions are stable.

5. Under the same conditions a population of multiple-age spawners tends to have a more regular cycle, whose peak-to-peak period is close to twice the median length of time from oviposition by the parent generation to oviposition by all its progeny (Figs. 9F, 10).

6. For such populations to have cycles of appreciable amplitude, it is necessary that reproduction be absent or light during the first year or two of the animal's life (assuming that reproduction occurs once a year).

7. Random fluctuation in reproductive success, *by itself*, leads to very large changes in abundance and eventual extinction or fragmentation, for populations of single-age spawners (Fig. 11A), and to unlimited increase for multiple-age spawners (Fig. 12A).

8. Combinations of random fluctuation in reproduction with reproduction curves produce populations which neither increase indefinitely nor decline to extinction. The cyclical changes produced by steep reproduction curves are maintained when random fluctuation is small or moderate, but they become variable in period and eventually unrecognizable as random effects become more important.

9. Under the combined influence of a steep reproduction curve and variable non-compensatory mortality, populations fluctuate more widely than when controlled by either of these factors alone.

10. When the dome of a population's reproduction curve lies above the 45-degree line, the result of light or moderate exploitation is to *increase* the abundance of the stock in subsequent generations; more intensive exploitation will decrease it (Fig. 16).

11. Another result of exploitation is to reduce the amplitude and complexity of any reproduction-curve oscillations that may be in progress; sufficiently intensive exploitation eliminates such oscillation entirely (Fig. 17).

12. Among multiple-age spawners exploitation tends also to reduce the *period* of oscillation somewhat, because the spawning stock gradually becomes younger.

13. Reproduction curves, or approximations to them, are plotted for fish populations and four invertebrate populations in Figures 18–28. No example was discovered where the maximum of recruitment is at or to the right of the 45-degree line (as in Figs. 1, 3, 4). The left limb was usually steep. The most probable position for the right limb was always sloping downward, sometimes only slightly, sometimes quite steeply; the most typical situation has a slope about as in Figure 5.

14. Cyclic population changes that are apparently the direct result of a steep reproduction curve are those of Herrington's haddock (1912–29) and the *Daphnia* cultures of Pratt. Most other well-known cycles seem not to be of a *simple* reproduction-curve type, but the shape of this curve must profoundly influence the course of population abundance in any animal.

15. More complex reproduction curves have been suggested by indirect evidence (Figs. 30, 31), and a depression of percentage reproduction at extremely low densities of stock may be fairly common.

16. An asymmetrical dome-shaped reproduction curve can be developed from simple assumptions involving random cannibalism or compensatory preda-

tion: if w represents adult stock density as a fraction or multiple of the density which provides maximum reproduction, we^{1-w} represents the actual reproduction at density w, as a fraction of the maximum reproduction. In this event the survival rate at maximum reproduction is always $1/e$ or 37 per cent of what it is at densities near zero; or in other words, at maximum reproduction the instantaneous mortality rate from compensatory activity is 1.

17. Curves computed from the above formula, shown in Figure 33, include most of the types actually observed. However if "environmental capacity" sets an upper limit to number of survivors at some stage of the pre-adult life history, this "typical" curve could either be levelled off at any point of its course, or else truncated (Fig. 34).

ACKNOWLEDGMENTS

A number of individuals have read the manuscript, and numerous changes, deletions and additions have resulted from their suggestions. Those who cooperated in this manner include Y. M. M. Bishop, L. M. Dickie, J. R. Dymond, R. E. Foerster, S. D. Gerking, J. L. Hart, D. J. Milne, F. Neave, A. L. Pritchard, W. B. Scott, W. M. Sprules, J. C. Stevenson, F. H. C. Taylor, F. C. Withler and D. E. Wohlschlag. I am indebted to Messrs. C. E. Atkinson, C. J. Burner and J. C. Stevenson for permission to peruse, and in some cases use, unpublished or incompletely published data. Mrs. D. Gailus has had the difficult task of maintaining the legibility and accuracy of the manuscript through a long series of revisions.

REFERENCES

ALLEE, W. C. 1931. Animal aggregations. Univ. Chicago Press, 431 pp.
BABCOCK, J. P. 1914. Annual Report of the British Columbia Commissioner of Fisheries for 1913, Victoria, B.C.
BARANOV, F. I. 1918. [On the question of the biological basis of fisheries.] *N.-i. Ikhtiologicheskii Inst., Izvestiia*, 1(1): 81–288.
BARNABY, J. T. 1944. Fluctuations in abundance of red salmon, *Oncorhynchus nerka* (Walbaum), of the Karluk River, Alaska. *U. S. Fish and Wildlife Serv., Fish Bull.*, 50: 237–295.
BURKENROAD, M. D. 1946. Fluctuations in abundance of marine animals. *Science*, 103(2684): 684–686.
CHIANG, H. C., AND A. C. HODSON. 1950. An analytical study of population growth in *Drosophila melanogaster*. *Ecological Monogr.*, 20: 173–206.
CLEMENS, W. A. 1952. On the cyclic abundance of animal populations. *Canadian Field-Naturalist*, 66: 121–123.
COLE, LaMONT C. 1951. Population cycles and random oscillations. *J. Wildlife Management*, 15: 233–252.
   1954. Some features of random population cycles. *Ibid.*, 18:2–24.
ELSON, P. F. 1950. Increasing salmon stocks by control of mergansers and kingfishers. *Fish. Res. Bd. Canada, Atlantic Prog. Repts.*, No. 51, pp. 12–15.
ERRINGTON, P. L. 1954. On the hazards of overemphasizing numerical fluctuations in studies of "cyclic" phenomena in muskrat populations. *J. Wildlife Management*, 18: 66–90.
FOERSTER, R. E., AND W. E. RICKER. 1941. The effect of reduction of predaceous fish on survival of young sockeye salmon at Cultus Lake. *J. Fish. Res. Bd. Canada*, 5: 315–336.

Fujita, H., and S. Utida. 1953. The effect of population density on the growth of an animal population. *Ecology*, 34: 488–498.

Gallagher, H. R., A. G. Huntsman, J. Van Oosten and D. J. Taylor. 1944. Report of the International Board of Enquiry for the Great Lakes Fisheries, pp. 1–24, Govt. Printing Office, Washington, D.C.

Galtsoff, P. S., and V. L. Loosanoff. 1939. Natural history and method of controlling the starfish (*Asterias forbesi* Desor). *Bull. U. S. Bur. Fish.*, 49(31): 75–132.

Haldane, J. B. S. 1953. Animal populations and their regulation. *New Biology*, 15: 9–24. Penguin Books, London.

Herrington, W. C. 1941. A crisis in the haddock fishery. *U. S. Fish and Wildlife Serv., Fish. Circ.*, No. 4, 14 pp.

———— 1944. Factors controlling population size. *Trans. 9th North Am. Wildlife Conf.*, pp. 250–263.

———— 1947. The role of intraspecific competition and other factors in determining the population level of a marine species. *Ecol. Monogr.*, 17: 317–323.

———— 1948. Limiting factors for fish populations. Some theories and an example. *Bull. Bingham Oceanogr. Coll.*, 9(4): 229–279.

Hile, R. 1936. Age and growth of the cisco, *Leucichthys artedi* LeSueur, in the lakes of the northeastern highlands, Wisconsin. *Bull. U. S. Bur. Fish.*, 48: 211–317.

Huntsman, A. G. 1952. How Passamaquoddy produces sardines. *Fundy Fisherman*, 24(24): 5. St. John, N.B.

———— 1953. Movements and decline of large Quoddy herring. *J. Fish. Res. Bd. Canada*, 10: 1–50.

Hutchinson, G. E. 1948. Circular causal systems in ecology. *Annals New York Acad. Sci.*, 50: 221–246.

Hutchinson, G. E., and E. S. Deevey, Jr. 1949. Ecological studies on populations. *Survey of Biol. Progress*, 1:325–359. Academic Press, New York.

Johnson, W. E., and A. D. Hasler. 1954. Rainbow trout production in dystrophic lakes. *J. Wildlife Management*, 18: 113–134.

Kesteven, G. L. 1947. Population studies in fisheries biology. *Nature*, 159: 10–13.

Lotka, A. J. 1925. Elements of physical biology. Williams and Wilkins, Baltimore, 460 pp.

Miller, R. B. 1950. Observations on mortality rates in fished and unfished cisco populations. *Trans. Am. Fish. Soc. for 1949*, 79: 180–186.

———— 1952. The role of research in fisheries management in the Prairie Provinces. *Canadian Fish Culturist*, No. 12, pp. 13–19.

Neave, F. 1952. "Even-year" and "odd-year" pink salmon populations. *Trans. Royal Soc. Canada* (V), Ser. 3, 46: 55–70.

———— 1953. Principles affecting the size of pink and chum salmon populations in British Columbia. *J. Fish. Res. Bd. Canada*, 9: 450–491.

Nicholson, A. J. 1933. The balance of animal populations. *J. Animal Ecol.*, 2: 132–178.

Nicholson, A. J., and V. A. Bailey. 1935. The balance of animal populations. Part I. *Proc. Zool. Soc. London* for 1935, pp. 551–598.

Palmgren, P. 1949. Some remarks on the short-term fluctuations in the numbers of northern birds and mammals. *Oikos*, 1: 114–121.

Pratt, D. M. 1943. Analysis of population development in *Daphnia* at different temperatures. *Biol. Bull.*, 85: 116–140.

Pritchard, A. L. 1948a. Efficiency of natural propagation of the pink salmon (*Oncorhynchus gorbuscha*) in McClinton Creek, Masset Inlet, B.C. *J. Fish. Res. Bd. Canada*, 7: 224–236.

———— 1948b. A discussion of the mortality in pink salmon (*Oncorhynchus gorbuscha*) during their period of marine life. *Trans. Royal Soc. Canada* (V), Ser. 3, 42: 125–133.

Ricker, W. E. 1952. Numerical relations between abundance of predators and survival of prey. *Canadian Fish Culturist*, No. 13, pp. 5–9.

1954. Effects of compensatory mortality upon population abundance. *J. Wildlife Management*, 18: 45–51.

RICKER, W. E., AND R. E. FOERSTER. 1948. Computation of fish production. *Bull. Bingham Oceanogr. Coll.*, 11(4): 173–211.

ROYAL, L. A. 1953. The effects of regulatory selectivity on the productivity of Fraser River sockeye. *Canadian Fish Culturist*, No. 14, pp. 1–12.

SCOTT, W. B. 1951. Fluctuations in abundance of the Lake Erie cisco (*Leucichthys artedii*) population. *Contr. Royal Ontario Mus. Zool.*, No. 32, 41 pp.

SMITH, F. E. 1952. Experimental method in population dynamics: a critique. *Ecology*, 33: 441–450.

SMITH, H. S. 1935. The role of biotic factors in the determination of population densities. *J. Econ. Entomol.*, 28: 873–898.

SMOKER, W. A. 1954. A preliminary review of salmon fishing trends on Inner Puget Sound. *Washington Dept. Fish., Res. Bull.*, No. 2, 55 pp.

SNEDECOR, G. W. 1946. Statistical methods. 4th ed., Iowa State College Press, Ames, Iowa.

SOLOMON, M. E. 1949. The natural control of animal populations. *J. Animal Ecol.*, 18: 1–35.

TESTER, A. L. 1948. The efficacy of catch limitations regulating the British Columbia herring fishery. *Trans. Royal Soc. Canada* (V), Ser. 3, 42: 135–163.

THOMPSON, D. H. 1941. The fish production of inland streams and lakes. *In* "A Symposium on Hydrobiology", pp. 206–217. Univ. Wisconsin Press, Madison.

TIPPETT, L. H. C. 1927. Random sampling numbers. *Tracts for Computers*, No. 15.

VARLEY, G. C. 1947. The natural control of population balance in the knapweed gall-fly (*Urophora jaceana*). *J. Animal Ecol.*, 16: 139–187.

VEVERS, H. G. 1949. The biology of *Asterias rubens* L.: Growth and reproduction. *J. Marine Biol. Assn. U. K.*, 28: 165–187.

Big Effects from Small Causes:
Two Examples from Fish Population Dynamics[1]

By W. E. Ricker

Fisheries Research Board of Canada
Biological Station, Nanaimo, B.C.

ABSTRACT

1. A population which, before exploitation, includes upwards of 12–15 age-groups in appreciable quantities, is very sensitive to fishing. For example, even as little as 5% catch per annum eventually causes a major reduction in the relative weight of older fish in the stock. With unchanging recruitment the absolute size of the total stock also declines markedly, or even catastrophically, under moderate exploitation—so much so that it seems likely that recruitment must usually increase when the stock is first thinned, partly compensating for the removals.

2. A population that produces its greatest sustained yield at a high rate of exploitation—say 75% or more at average recruitment—is then close to a point where recruitment will fall off rapidly if utilization becomes only slightly more intensive. This fact, and the aggravating effect of environmentally-caused variations in recruitment, suggest that any really close approach to the point of optimum yield will usually be too dangerous to be practical.

INTRODUCTION

IN 1928 Professor J. R. Dymond initiated the writer into the cult of those who apply quantitative methods in the broad science of ichthyology. In the loft of Lou Joyce's fish house at Port Credit, Ontario, he set me the task of measuring and computing body proportions of some scores of fourhorn sculpins, then called *Triglopsis thompsoni*. This was a congenial occupation, partly because I was keen to put to practical use the slide-rule know-how recently acquired in Professor John Satterly's first-year physics course—well remembered by several generations of University of Toronto students.

This early experience with fish morphometry led eventually and by circuitous routes to the realm of fish population dynamics. Exploratory excursions with slide-rule and calculator, made in the latter region over the past 10 years or so, have turned up numerous odd bits of information in addition to some that did contribute to the current objective. Most of these unassorted conclusions are not of much general interest. A few, however, seem likely to shed light on events or problems in the known history of fish populations, or in their future prospects.

Two relationships of the latter type are presented herewith. The general background of the models and computations used, as well as the symbols employed, are as given in several earlier papers [2], [3] and [4].

---

[1]Received for publication October 29, 1962.

EFFECT OF DIFFERENT RATES OF FISHING ON POPULATION STRUCTURE

In the year of the armistice, 1918, A. G. Huntsman and F. I. Baranov independently called attention to the *molozhenie* or juvenescence of a fish population that occurs when a fishery is introduced. Since then several additional models have been published to illustrate this phenomenon (references in [3]), and it has often been observed in nature. However, no one seems to have made a systematic examination of the quantitative effects of different levels of fishing on a population. Thus it is logical to compute and compare, even if somewhat belatedly, the effects of a graded series of different rates of fishing on the weight of a fish stock.

Weight rather than numbers has been chosen for illustration here, because the weight of a stock is of more direct commercial interest, and it is also more directly related to reproductive potential. Changes in numerical abundance of a stock resulting from increased fishing are similar to but less extreme than the changes in weight.

In the population model examined, recruitment at age 1 is held at a constant level. The growth rates and natural mortality rates used (Table I) are essentially those quoted by Chatwin [1] for a population of lingcod (*Ophiodon elongatus*). The exact figures of course make little difference here; this range of natural mortality rates is of the right order for a number of other important stocks of marine fishes—for example, cod and redfish off the Atlantic coast of Canada—except that the increase in mortality rate among the oldest fish is probably too abrupt.

TABLE I. Growth rates, natural mortality rates, and average weight of the fish of the stock of Fig. 1. Fish weights are shown for the start of each year indicated. Based on the estimates for lingcod in tables 7 and 8 of Chatwin (1958); values beyond age 16 are extrapolated.

Year of life	Weight (kg)	Instantaneous rates		Year of life	Weight (kg)	Instantaneous rates	
		Growth	Mortality			Growth	Mortality
1–2	0.17	1.56	0.35	11–12	8.3	0.09	0.30
2–3	0.81	0.66	0.30	12–13	9.1	0.08	0.35
3–4	1.56	0.40	0.25	13–14	9.8	0.07	0.40
4–5	2.33	0.29	0.20	14–15	10.6	0.06	0.50
5–6	3.12	0.24	0.20	15–16	11.2	0.05	0.60
6–7	3.96	0.20	0.20	16–17	11.8	0.04	0.80
7–8	4.94	0.16	0.20	17–18	12.3	0.03	1.10
8–9	5.8	0.14	0.20	18–19	12.6	0.02	1.50
9–10	6.7	0.12	0.20	19–20	12.9	0.02	2.20
10–11	7.5	0.10	0.25	20–21	13.2	...	...

Figure 1 shows the age structure of the unexploited population in terms of weight, and also its equilibrium structure at several rates of fishing. The latter figures also represent the age structure of the catch quite well, and will do so exactly if fishing occurs during a short period of time each year, before natural mortality or recruitment begin to operate.

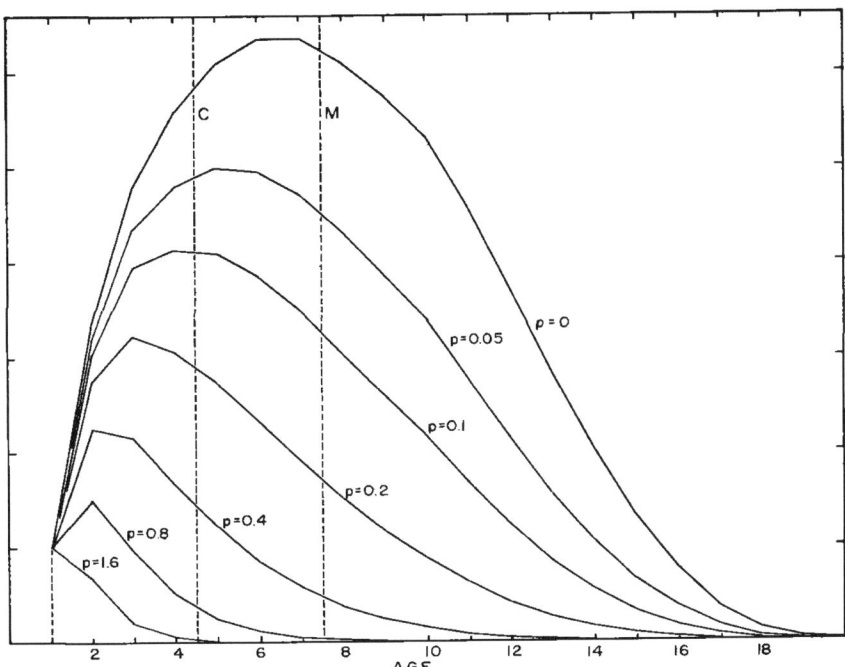

FIG. 1. Weight of the fish present at successive ages, for different rates of fishing (p), in terms of a unit weight of recruits at age 1, for the population of Table I. With constant recruitment, the equilibrium weight of the stock is proportional to the area under each shell. C: average minimum age limit of usable fish; M: average age of first maturity.

What strikes the eye in Fig. 1 is the tremendous effect that even a little bit of fishing has, on the total bulk of the stock, and (still more) on its distribution by age. A 10% per annum fishery, for example, reduces the weight of the stock to little more than half of the original; while 33% utilization, which is not usually considered excessive, reduces its bulk to 16% of the original (Table II).

If the smaller fish are not marketable, yet are unavoidably caught (and are discarded dead), the reduction in *usable* stock is even greater. Though it doesn't actually apply to lingcod, a cull limit between ages 4 and 5 is indicated by line C in Fig. 1, and its effect is examined in column 4 of Table II. The size of the usable stock is reduced well past 50% by a 10% fishery, and it is down to 7% when exploitation reaches 33%.

Finally, the weight of the *mature* stock is of interest, since this represents the reproduction potential. In lingcod first maturity for females is mostly at age 7 or 8 (Fig. 1, line M), and column 5 of Table I shows the weight of all fish older than 7. In this case a 10% fishery reduces the spawners to 38% of their original bulk, while 33% annual removal gets them down below 3%. At 55% utilization, which is by no means impossibly large, the bulk of the spawners is reduced to $\frac{1}{850}$ of the original.

TABLE II. Relative weight of stock for each of the equilibrium populations of Fig. 1, and for two age ranges within those populations.

1	2	3	4	5	6
		Relative weight of stock			"Usable" catch per unit effort
Rate of fishing p	Rate of exploitation u	Total (All ages)	"Usable" (Ages 5–20)	Mature (Ages 8–20)	
0.	0.	100.	100.	100.	100.[a]
0.05	4.9%	72.7	67.6	61.7	66.0
0.10	9.5%	54.5	46.6	38.6	44.4
0.20	18.1%	33.3	23.4	15.6	21.2
0.40	33.0%	16.1	6.95	2.80	5.73
0.80	55.1%	6.68	0.883	0.117	0.608
1.60	79.8%	2.91	0.025	0.0003	0.012

[a]This applies to a vanishingly small rate of fishing.

It is obvious that these effects result from the cumulative action of individually-small percentage removals, operating throughout many years of the life of these fish: it is compound interest in reverse. If a fish species is naturally shorter-lived than the one used here (i.e., has larger natural mortality rates somewhere within the range of ages shown in Table I), then the effects of fishing on stock density are less spectacular. Contrariwise, there are a number of species with considerably smaller natural mortality rates than those used here—to judge from the very old specimens that were formerly rather common, and are still caught occasionally, among halibut, *Hippoglossoides*, etc. For these species the effects of a small rate of utilization will be even more startling than those shown in Fig. 1.

DECLINE IN CATCH PER UNIT OF FISHING EFFORT

With recruitment still maintained at a steady level, column 6 of Table II shows approximately the decrease in usable catch, per unit of fishing effort, that occurs as fishing effort is increased.[2] The decline in fishing success is large even at low fishing intensities: a mere 5% per annum fishery eventually reduces catch per hour by 34%, while the 33% fishery reduces fishing success to less than 6% of the original.

How can a moderately intensive fishery survive, in the face of so great a decline in revenue? The answer is that most of them could not survive, but in practice catch per hour does not usually fall so low. In some cases, though not in all, the rate of growth of the individual fish increases somewhat, partly

---

[2]Column 6 of Table II is computed on the basis that effort is proportional to rate of fishing, p, where $p = -\log_e(1-m)$. The "conditional" rate of fishing, m, is the same as the rate of utilization or exploitation, u, in the situation where all the catch is made early in the statistical year, before natural mortality operates. This is the u shown in column 2 of Table II, so that column 6 of that Table is equal to column 4 multiplied by column 2 and divided by column 1. If however natural and fishing mortality operate concurrently to some extent, the true rates of exploitation would be progressively somewhat less than those shown in column 2, and the catch per unit effort would decline somewhat more rapidly than the series of figures in column 6. With complete overlap of times of fishing and natural mortality, at $p = 0.8$ the catch per unit effort is about two-thirds of what is shown in Table II.

compensating for reduced abundance. What is most likely to have happened generally, however, is that the (absolute) annual recruitment has increased as a result of the major thinning of the stock that occurs at quite an early stage of exploitation. This phenomenon has not been documented by direct estimates of recruitment made as rate of fishing increases, but I believe this is because for most fisheries the increase in recruitment is largely completed during the early stages of their development, when few or no biological observations were being made on the stock. By the time the rate of exploitation has become no more than 9–10% per year, the stock of Fig. 1 has already been reduced to about half its original bulk, and the spawners to much less than half (Table I). By this time a population will usually have reached a plateau on its graph of recruitment against stock size, or be near the top of a dome. Hence observations made from that time onward will not detect any negative slope in the relation of recruitment to parental stock. Nevertheless the general quantitative aspects of the situation often suggest strongly (for naturally long-lived fish) that there must have been a substantial increase in average annual recruitment during the time of the initial thinning of the stock, while the fishery was still young.

EFFECT OF HEAVY FISHING ON RECRUITMENT TO A POPULATION

In the absence of positive information, estimates of the role of stock abundance in governing recruitment to a population have usually varied between the assumption of direct proportionality and that of effective independence. It is recognized that neither extreme can be applicable over the whole range of possible stock sizes: unlimited direct proportionality would increase a population without limit, while complete independence implies that recruits can be obtained from no spawners at all. In recent years observations on the actual or relative magnitude of recruitments have become available, and there are now a number of oceanic fish stocks for which recruitment has not been demonstrated to have any trend upward or downward over quite a wide range of spawning stock sizes. As a result there is a tendency to use independence as the norm in discussing the dynamics of exploited marine fishes. Even if an early phase of negative slope occurs, as was postulated in the previous section, it is no longer accessible to observation and it is unlikely to reappear as long as fishing continues at a moderate, or even rather low, level of activity.

No one, of course, goes to the extreme of obtaining progeny from no parents. This means that there must be an initial rapid rise in recruitment at stock levels close to zero.

Thus a common picture of the average relationship between stock and recruits among sea fishes is of the type shown by the solid line in Fig. 2. Figure 3 is the curve of equilibrium catches taken from this average recruitment, as rate of fishing is increased. The long slowly-ascending portion shows the increase in catch as rate of fishing is increased from zero to its best value. The steep right-hand portion is the rapid decline that follows when exploitation exceeds this level.

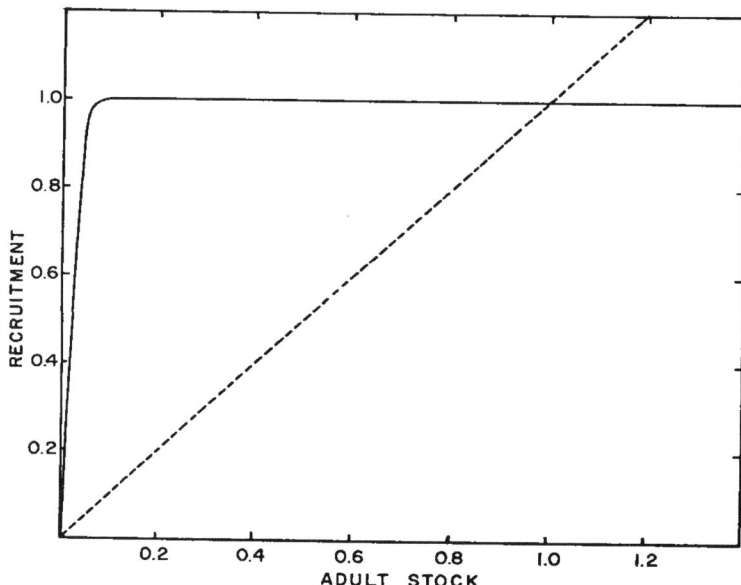

Fig. 2. A recruitment curve of the type where recruitment is independent of spawning stock density over most of the range of the latter. In this example, R = 1 between W = 0.1 and W = 1; R/W = exp[3.153(1−W)] between W = 0 and W = 0.05, which makes a slightly curved ascending line; the short transition interval between W = 0.05 and W = 0.1 is sketched in freehand.

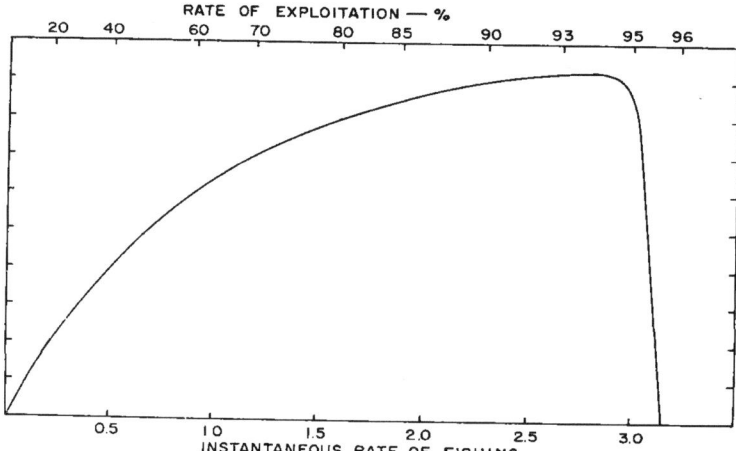

Fig. 3. Relative size of catches obtained from the population of Fig. 2, as fishing effort is increased (assuming all fishing is at one age, or that growth just balances natural mortality in later years). The rate of exploitation (u — top scale) is the fraction of the total recruitment (Fig. 2) taken by the fishery at equilibrium. The rate of fishing (p — lower scale) in most fisheries would be approximately proportional to the fishing effort necessary to take the catch indicated.

RICKER: FISH POPULATION DYNAMICS

The rapidity of this decline, and dangers associated with it, seem to have escaped special comment. In the example used, an increase from 94 to 96% utilization spans the change from maximum yield to no yield at all. Assuming fishing effort is proportional to rate of fishing, this means an increase in effort from 2.8 to 3.2, or by only 14%. If the ascending limb of the reproduction curve of Fig. 2 be made straight, the descending limb of Fig. 3 becomes completely vertical; in which event only a knife-edge separates full utilization from complete collapse of the fishery[3].

There are buffering factors of course. Figure 3 represents *equilibrium* conditions, which (on the ascending limb of the Fig. 2 curve) are reached asymptotically rather than immediately, following any increase in fishing mortality to a new steady level. Also, when more than one age of the fish contributes importantly to the fishable stock (which can be consistent with a high rate of exploitation if recruitment to the fishable stock occurs during more than one year of life), a collapse of reproduction in just one season would not be immediately fatal. Nevertheless, if ever the rate of exploitation exceeds the optimum for a short series of years, even if only slightly, the decline in stock and catch will evidently be swift and drastic.

This conclusion is not restricted to stocks characterized by flat-topped reproduction curves. It obviously must apply to any population where best yield is obtained from a small spawning stock: that is, where optimum rate of exploitation is large—say 75% or more. Indeed for any population "obtaining maximum yield will always be a bit tricky—with a more drastic penalty for too much fishing than for too little"[3].

If reproduction of a population adhered closely to average conditions, precise regulation of the fishery on these stocks would naturally be much easier. In practice, the recruitments that determine a curve such as that of Fig. 2 may easily vary by a distance of up to say 0.9 on either side of it. This means that no single optimum rate of utilization will be applicable every year. But there is a difference in the room for manoeuvre on either side of the norm. In years of better than average stocks, rate of utilization need be increased only slightly over the best level for average recruitment, since the latter rate is already quite close to 100% (it is 94% in Fig. 3). In years of poor stocks, on the other hand, rate of utilization will have to be reduced more substantially in order to maintain a spawning stock close to the optimum. For example, if recruitment "accidentally" fell to 0.1 in Fig. 3, the fishery should not take more than 50% if the stock is to stay out of the region below 0.05 where reproduction becomes disastrously low.

The argument is sometimes heard, concerning stocks whose average reproduction is of the type described by Fig. 2, that because environmental variability causes such great variation in recruitment, there is no need to worry about the size of the spawning stock. This will be true as long as the stock

---

[3]On page 244 of [2] there is a yield diagram for a flat-topped reproduction curve with a less abrupt increasing phase: the broken curve plus the B curve to the right of it represents the catch. In this case the optimum rate of fishing was 1.58, corresponding to 79% catch each year.

remains well away from the "breaking point" of reproduction, at the left end of the recruitment curve. But for maximum utilization the fishery for such a population must be adjusted with the greatest care to the size of the stock present each year. If this is not possible, utilization at considerably less than the maximum rate is the only way to ensure the continued existence of a good stock and a productive fishery.

REFERENCES

[1] CHATWIN, B. M. 1958. Mortality rates and estimates of theoretical yield in relation to minimum commercial size of lingcod (*Ophiodon elongatus*) from the Strait of Georgia. *J. Fish. Res. Bd. Canada*, **15**(5): 831–849.

[2] RICKER, W. E. 1954. Stock and recruitment. *J. Fish. Res. Bd. Canada*, **11**(5): 559–623.

[3] 1958a. Handbook of computations for biological statistics of fish populations. *Bull. Fish. Res. Bd. Canada*, No. 119, 300 pp.

[4] 1958b. Maximum sustained yields from fluctuating environments and mixed stocks. *J. Fish. Res. Bd. Canada*, **15**(5): 991–1006.

Compensatory density dependence in fish populations: importance, controversy, understanding and prognosis

Kenneth A Rose[1], James H Cowan Jr[1], Kirk O Winemiller[2], Ransom A Myers[3] & Ray Hilborn[4]

[1]Coastal Fisheries Institute and Department of Oceanography and Coastal Sciences, Wetland Resources Building, Louisiana State University, Baton Rouge, LA 70803, USA; [2]Department of Wildlife and Fisheries Sciences, Texas A and M University, College Station, TX 77843–2258, USA; [3]Department of Biology, Dalhousie University, Halifax, Nova Scotia, Canada B3H 4J1; [4]School of Aquatic and Fishery Sciences, Box 355020, University of Washington, Seattle, WA 98195, USA

Abstract

Density-dependent processes such as growth, survival, reproduction and movement are compensatory if their rates change in response to variation in population density (or numbers) such that they result in a slowed population growth rate at high densities and promote a numerical increase of the population at low densities. Compensatory density dependence is important to fisheries management because it operates to offset the losses of individuals. While the concept of compensation is straightforward, it remains one of the most controversial issues in population dynamics. The difficulties arise when going from general concepts to specific populations. Compensation is usually quantified using some combination of spawner–recruit analysis, long-term field monitoring or manipulative studies, and computer modelling. Problems arise because there are limitations to each of these approaches, and these limitations generally originate from the high uncertainty associated with field measurements. We offer a hierarchical approach to predicting and understanding compensation that ranges from the very general, using basic life-history theory, to the highly site-specific, using detailed population models. We analyse a spawner–recruit database to test the predictions about compensation and compensatory reserve that derive from a three-endpoint life-history framework designed for fish. We then summarise field examples of density dependence in specific processes. Selected long-term field monitoring studies, manipulative studies and computer modelling examples are then highlighted that illustrate how density-dependent processes led to compensatory responses at the population level. Some theoretical and empirical advances that offer hope for progress in the future on the compensation issue are discussed. We advocate an approach to compensation that involves process-level understanding of the underlying mechanisms, life-history theory, careful analysis of field data, and matrix and individual-based modelling. There will always be debate if the quantification of compensation does not include some degree of understanding of the underlying mechanisms.

Keywords compensation, density dependence, fish populations, fisheries management, life-history theory, sustainable harvest

Correspondence:
Kenneth A Rose,
Coastal Fisheries
Institute and
Department of
Oceanography and
Coastal Sciences,
Wetland Resources
Building, Louisiana
State University, Baton
Rouge, LA 70803, USA
Fax: +1 225-578-6346
E-mail:
karose@lsu.edu

Received 3 Jul 2001
Accepted 6 Aug 2001

Introduction	**294**
Compensation and compensatory reserve	**295**
Compensation, fisheries management and controversy	**295**
Statement of the problem	**296**
Why direct measurement of compensation can be difficult	**297**
Life-history theory	**299**
Strategies	299
Expectations from life-history theory	299
Comparison of life-history expectations with data	300
Understanding compensation	**301**
Density-dependent processes	302
Empirical field studies	304
Japanese anchovy	305
Population cycles of vendace	305
Controlled exploitation of lake whitefish	305
Orange roughy	306
Walleye recovery in western Lake Erie	306
Plaice in the North Sea	306
Haddock recruitment in the North Atlantic	306
Migratory brown trout	307
Computer modelling examples	307
Yellow perch in Oneida Lake	307
Food webs in Lake Mendota	309
Bay anchovy in Chesapeake Bay	311
Cause for optimism: recent technological and theoretical advances	**312**
Synthesis and prognosis	**315**
Acknowledgements	**316**
References	**316**

Introduction

Processes such as growth, survival, reproduction and movement are density dependent if their rates change as a function of the density or number of individuals in a population. Density-dependent processes are said to be compensatory if their rates change in response to variation in population density such that they result in a slowed population growth rate at high densities and promote a numerical increase of the population at low densities. Density-dependent processes are depensatory if they change with population density such that they slow the rate of population growth at low densities.

Density dependence is a fundamental concept in the study of fish population dynamics. Depensatory density dependence is a positive feedback on population size, and therefore tends to destabilise populations. Depensatory density dependence is especially important for depleted populations and for endangered species because it acts to accelerate further population decline and can delay recovery (e.g. Myers *et al.* 1995b; Shelton and Healey 1999). Compensatory density dependence is a negative feedback on population size, and therefore acts to stabilise populations. Compensatory density dependence is important to management because it operates to offset the losses of individuals, which can occur from natural fluctuations in environmental conditions or from anthropogenic activities such as power-plant operations and fishing. Lowered population density temporarily results in increased survival or reproduction of the remaining individuals which favours an increase in population size.

We will focus in this paper on compensatory density dependence. Myers et al. (1995b) analysed spawner–recruit relationships for a variety of marine fish species, and found little evidence of depensatory density dependence at low population sizes. However, Liermann and Hilborn (1997) reexamined the data and Shelton and Healey (1999) evaluated the analytical method, and both urged caution before dismissing the possibility of depensation. Liermann and Hilborn (2001) recently reviewed depensatory density dependence in fish and other taxa. We use the term density dependence in this paper to mean compensatory density dependence.

Compensation and compensatory reserve

Two terms used throughout this paper are: compensation and compensatory reserve. Compensation refers to the net effect of the density-dependent processes that cause negative feedback on population size. Strong compensation means that the processes are highly responsive to density changes. Strong compensation does not necessarily imply high population stability (i.e. population fluctuations within relatively narrow numerical bounds). Specific processes may be tightly coupled to variation in density at particular life stages, but the variation from environmental fluctuations affecting other processes or life stages may dominate overall population variability. Similarly, populations with weak compensation can exhibit low interannual variation if they inhabit relatively stable environments. The magnitude of compensation can be inferred from how tightly coupled the responses of specific processes are to changes in density.

Christensen and Goodyear (1988) defined compensatory reserve as the "excess reproductive capacity under 'ideal' conditions for individual reproduction (i.e. when population size is so low that compensatory mortality does not operate)." Compensatory reserve is the capacity of the population to offset variation in mortality, but is not necessarily related to the stability (interannual variation) of population size over time. Populations with high compensatory reserve can also exhibit large interannual fluctuations in abundance; the degree of fluctuation depends upon the net effects of environmental variation and compensation in multiple life stages. Compensatory reserve can be estimated via the maximum reproductive rate, which is the slope of the spawner–recruit relationship near the origin, where compensation effects would be small (Myers et al. 1999).

Compensation, fisheries management and controversy

The theory of compensatory density dependence underlies the management of fish populations. Two situations that permit sustainable harvest can be envisioned: a population with a net positive growth rate and a quasi-stable population. While sustainable harvest is theoretically possible without compensatory density dependence, it is ecologically unrealistic. In theory, a population that has a positive rate of increase can be harvested until the population growth rate is reduced to almost zero, and the population would persist. While such a population does not require compensation for sustainable harvest, in the absence of harvesting this population would eventually increase unbounded.

The second situation of sustainable harvest from a quasi-stable population requires compensation. Compensatory density dependence permits populations to persist under conditions of an increase in mortality, and is the basis of the concepts of surplus production and sustainable harvest (Schaefer 1954; Beverton and Holt 1957; Ricker 1975; Gulland 1977; Sissenwine 1984; Fogarty et al. 1991). Population stability, which can include bounded fluctuations (Turchin 1995), implies that, averaged over a long enough time period, reproduction is balanced by mortality. Despite the wide fluctuations in abundance typical of many fish populations, many populations have persisted (some with harvesting) within some defined upper and lower abundances for many generations, implying at least a modest degree of long-term stability. Without compensatory responses, any increase in mortality owing to harvesting in an already stable population would eventually result in population decline because, in the long term, mortality would exceed reproduction. Thus, compensatory density dependence must exist for naturally stable populations to persist under harvesting. The basis for surplus production and sustainable harvest is that populations have the ability, at some densities, to increase in numbers at a rate greater than that required for replacement (Goodyear 1993). However, inclusion of compensation in fishery management analyses does not automatically result in predictions of increased yield or higher sustained fishing rates (e.g. Helser and Brodziak 1998).

An illustration of the general magnitude of compensatory reserve in fish populations is that it is considered risk-averse and long-term sustainable to remove 60–70% of a virgin stock's reproductive potential (Goodyear 1993; Mace and Sissenwine 1993; Mace 1994). One of the principal sources of uncertainty and risk in fisheries management is determining the actual magnitude of the compensatory reserve in a specific population (Fogarty *et al.* 1992). The debates over compensation and compensatory reserve are rarely ever resolved, and often act to delay the initiation of needed management actions.

While the concept of compensatory density dependence is straightforward, it remains one of most controversial issues in population dynamics. As stated by Fletcher and Deriso (1988), "Why is it [density-dependent compensation] so easy to imagine but so hard to find?" In the aftermath of a long court case concerning the permitting of Hudson River power plants, Barnthouse *et al.* (1988) stated "The most controversial of the issues raised at these hearings was the potential importance of density-dependent regulatory mechanisms in offsetting direct mortality caused by power plants." The debate over the management of red snapper (*Lutjanus campechanus*) in the Gulf of Mexico has been contentious (e.g. Macaluso 1999). This has been partially fuelled by uncertainty about the appropriate assumptions concerning the magnitude of compensatory reserve in red snapper stock assessment. In lieu of better information, Schirripa and Legault (1999) assumed a range of compensation levels, which resulted in a five-fold variation in estimates of maximum sustainable yield (MSY). Everyone agrees that intraspecific competition for one or more resources ultimately will limit the size of a fish population (i.e. populations can not increase unbounded), and that, at least intuitively, survival and reproduction can be negatively related to density. The difficulties arise when extrapolating from density dependence, identified in a particular process or life stage, to its effect on long-term population size, and when going from very general statements about compensation and compensatory reserve to cases involving specific populations.

Extrapolation of density dependence in a particular life stage to the population level requires understanding of how density-independent and density-dependent factors interact to affect all life stages in the life cycle. Fish exhibit complex life-history strategies that make the comprehensive study of their full life cycle difficult. Quantifying the magnitude of compensation for a particular population requires years to decades of study to uncover the nuances of site-specific dynamics. The universal shortage of adequate data for specific populations results in alternative interpretations of available data, and has lead to acrimonious debate. Use of population dynamics and other models, in lieu of adequate data, only changes the tone of the debate – from inadequate data to uncertain models. The root cause of the debate remains our inability to adequately quantify and understand compensation for a particular population. In the absence of experimental controls, we cannot determine how much of population fluctuation is owing to environmental factors versus density-dependent responses. This is further confounded by environmental fluctuations not simply acting as density-independent factors, but also potentially causing density-dependent responses. Environmental variation can alter the quality and quantity of habitat, which can in turn cause changes in population density and thereby trigger density-dependent growth, survival, reproduction and movement.

Marine fisheries worldwide are in a dismal state (Garcia and Newton 1997; NMFS 1998), with many populations experiencing overfishing and harvesting rates that may not be sustainable (Botsford *et al.* 1997; Pauly *et al.* 1998; NRC 1999). The situation for many freshwater fish species is similarly pessimistic (Warren and Burr 1994), with the need for widespread use of management practices to limit harvest and extensive use of stocking to augment populations (e.g. Schramm *et al.* 1995; Fenton *et al.* 1996). Proper and effective management and restoration of fishery resources require progress on the issues surrounding the quantification and understanding of compensatory processes.

Statement of the problem

In our view, the problem is how to determine the magnitude of compensation and compensatory reserve accurately, and not how to incorporate known compensation and compensatory reserve into population analyses. The mathematics exist for including density-dependent processes in population dynamics models. For example, compensatory spawner–recruit relationships for egg to age-1 survival are often used in age-structured matrix models (Saila *et al.* 1991; Hilborn and Walters 1992); compensatory density dependence has also been included in other types of populations models (Jensen 1993; Marschall and Crowder 1996; Nisbet *et al.* 1996; Rose *et al.* 1996, 1999).

While the mechanics for including compensation in population and stock assessment models exist, limited data and lack of process-level understanding for a particular population often leads to dispute. The default assumption of no compensation is protective of the population (e.g. Ginzburg *et al.* 1990; Fogarty *et al.* 1992), but is so overly conservative that it becomes impractical and economically inefficient, given so many competing demands on fishery resources. The other extreme assumption of recruitment which is independent of spawning population size corresponds to effectively infinite compensation (Hilborn and Walters 1992), and puts the fish population at risk.

This review is organised as follows. We begin with a discussion of why compensation is so difficult to measure in the field. We suggest that life-history theory provides a framework for understanding and predicting the magnitude of compensation and compensatory reserve in populations. We use a three-endpoint life-history framework, developed specifically for fishes, to make first-order predictions of the magnitude of compensation and compensatory reserve. We then analyse an extensive spawner–recruit database to test the predictions about compensation from the life-history framework. This is followed by a description of how compensatory responses can arise from changes in the processes of mortality, growth, reproduction and movement. Selected empirical and computer modelling examples are then presented which illustrate how compensation operates at the process level and why detailed understanding of compensation can be elusive. Examples were chosen to span a wide variety of both freshwater and marine species. Theoretical and empirical advances that offer hope for progress in the future on the compensation issue are next discussed. Finally, we conclude with a synthesis and some suggestions for addressing the compensation problem.

Why direct measurement of compensation can be difficult

Compensation is usually quantified using a combination of spawner–recruit analysis, long-term field monitoring or manipulative studies, and computer modelling. A relationship between spawners and recruits that shows a less than proportional increase in recruitment with increasing spawning implies compensatory density dependence (Cushing 1975; Fogarty *et al.* 1992). Long-term field monitoring and manipulative studies allow for direct determination of population stability and quantification of density dependence in mortality, reproduction and other processes. Field measurements showing positive population growth rates at low densities imply that compensation exists. Ignoring how one arrives at a specific model formulation, realistic models of population dynamics can be developed that include documented compensatory mechanisms.

Problems arise because there are limitations to each of these approaches, and these limitations generally arise from the high uncertainty associated with field measurements (e.g. Bradford 1992). Uncertainty in field measurements is due primarily to lack of experimental controls and the often high measurement error associated with field data. The merits and drawbacks to spawner–recruit data have been debated for decades. Some of the drawbacks include the generally high variability of the data (Walters and Ludwig 1981; Goodyear and Christensen 1984), use of improper proxy variables for egg production (Goodyear and Christensen 1984; Rothschild and Fogarty 1989), the often poor fit of the assumed deterministic spawner–recruit function (Koslow 1992) and weaknesses in the statistical fitting methods (Christensen and Goodyear 1988). The lack of convincing relationships between spawners and recruits led some to assume no relationship, which implies extremely high compensation (Fogarty *et al.* 1992; Hilborn and Walters 1992), and others to propose flexible functional forms (Getz and Swartzman 1981; Mackinson *et al.* 1999). Because spawner–recruit data are annual, many years of monitoring are required before meaningful interpretation is possible. Also, while properly analysed spawner–recruit data provide direct evidence of the magnitude of compensation and compensatory reserve, spawner–recruit data often do not include definitive information on the specific processes that underlie the compensatory responses.

Inferring density dependence from long-term field and manipulative studies is convincing, although rarely adequate for definitive conclusions for most specific populations. For example, despite long-term data on Hudson River striped bass (*Morone saxatilis*), Pace *et al.* (1993) cautioned that the apparent density-dependent response could also be attributed to sampling limitations. Monitoring *in situ* is difficult because fish populations typically exhibit wide interannual variation in their numbers (Sissenwine 1984; Rothschild 1986; Fogarty *et al.* 1991). Often, much of this variation in abundance results from variability in hydrographic and climatic factors that affect individuals prior to recruitment (Shepherd et al. 1984; Laevastu 1993). Furthermore, these environmental factors vary simultaneously in nature, often exhibiting interactive (nonadditive) effects on recruitment (Rose and Summers 1992). When variation in abundances and in processes is large, and is strongly influenced by environmental variables, monitoring can often be unrepresentative of long-term conditions. Furthermore, detection of population stability or density-dependent components in these processes amongst environmental variation becomes very difficult (Bailey and Houde 1989; Bromley 1989; Hixon 1998). Monitoring is also complicated by many species being relatively long-lived with life stages that inhabit different habitats. Furthermore, the magnitude and process that shows density dependence can vary over time and space in a population (Goodyear 1980; Shima 1999), and changes in growth or mortality in one life stage can be offset by changes in subsequent life stages (e.g. Bertram et al. 1993). Thus, use of monitoring to document population stability or to detect density dependence in processes requires long-term and extensive data, which is impractical for many populations.

Manipulative approaches (laboratory experiments, mesocosms and field experiments) attempt to address some of the problems inherent with *in situ* monitoring. By addressing the issue of the lack of experimental controls in field data, experimental manipulation enables observed responses to be attributed to known causes. However, because manipulative studies are usually short-term and under contrived conditions, extrapolation to the broader temporal and spatial scales that operate in nature can be problematic. Also, detection of density dependence in a process for a particular life stage is necessary, but not sufficient, to conclude that the process regulates population size. The realism of manipulative approaches has been much discussed and debated (de Lafontaine 1987a,b; Fausch 1988).

Using models in place of good data simply shifts the focus from inadequate data to uncertain models. Population models have been at the centre of many ecological conflicts (e.g. Barnthouse et al. 1984; Swartzman 1996). The controversies associated with the use of models usually centre on what constitutes model validation (Rykiel 1996), alternative model formulations leading to different conclusions (e.g. Barnthouse et al. 1984), the inability to rigorously test the basic assumptions or the predictive power of a model (e.g. Mathur et al. 1985; Lindenmayer et al. 2000), and the high uncertainty often associated with model forecasts (Ludwig et al. 1993; Botsford et al. 1997). There is no generally accepted theory of population dynamics that would allow clear and unequivocal specification of the structure of a model (i.e. what processes and the correct biological, temporal and spatial scales) in specific situations (Getz 1998). Further, in many studies, the available data were deemed inadequate to address the questions of interest (thus models were needed); yet, the same data are used to develop and evaluate the model.

Measuring compensation and compensatory reserve is further complicated by compensatory responses being potentially site-specific. The site-specificity of compensatory responses can make the use of information from other populations or species problematic. The processes and life stages that respond to changes in abundance can vary among different populations of the same species (Rose and Cowan 2000). Differences in the composition and arrangement of the food web can also affect the compensatory responses of a population (McDermot 1998).

Measuring compensation and compensatory reserve is also hindered by the general lack of sufficient data at the extremes of population densities. Compensatory responses are strongest under conditions of high population densities, while compensatory reserve is estimated from spawner–recruit data at low population densities. Some have suggested that compensatory density dependence operates most strongly only at the extremes of densities (Strong 1986; Murray 1994). Unfortunately, the data available for many fish populations either do not adequately span the full range of densities, or if they do, the data usually cover decades and are therefore confounded with long-term environmental changes. All of these factors make it difficult to obtain accurate and precise long-term data sufficient to definitely quantify and understand compensation and compensatory reserve for specific populations.

We do not believe, however, that it is necessary or even possible to know everything to resolve the compensation quagmire. Further, we believe that with the proper care, spawner–recruit data, field monitoring, manipulative studies and modelling are useful approaches for quantifying and understanding compensation and compensatory reserve. Indeed, we use these approaches extensively later in this paper. We suggest that a hierarchical approach be used to study compensation. The detail to which compensation needs to be understood depends on the specifics of the population of interest and the questions being asked. Effective resource management is possible, whereby the risk taken by managers is commensurate with the level of understanding about the magnitude of compensation. Rough estimates of compensatory reserve, without detailed understanding of the underlying processes, may be adequate for a healthy population supporting a well-managed stable fishery, or when extensive data are available for other similar populations. At the other extreme, detailed understanding of compensation at the process level may be required to target management actions for the rebuilding of an overexploited, declining population. Toward this end, we offer in this paper a hierarchical approach to predicting and understanding compensation that ranges from the very general, using basic life-history theory, to the highly site-specific, using detailed population models. As one proceeds from the general to the specific in our hierarchy, the level of generality of predictions decreases while (hopefully) the level of accuracy of predictions increases.

Life-history theory

Strategies

Winemiller and Rose (1992) analysed data on 16 life-history parameters for 216 North American freshwater and marine species, and found that variation among life histories of these species could be described in terms of three general strategies (Fig. 1). These three strategies are defined as: (i) opportunistic – small, rapidly maturing, short-lived fishes; (ii) periodic – larger, highly fecund fishes with long life spans; and (iii) equilibrium – fishes of intermediate size that often exhibit parental care and produce relatively few, large offspring. For convenience, we use the three strategies here, keeping in mind that they represent endpoints; species can fall anywhere on the surface shown in Fig. 1.

The opportunistic strategy is associated with early maturation, frequent reproduction over an extended spawning season, rapid larval growth and high adult mortality. Examples of opportunistic life-history strategists include many anchovies (family Engraulidae), silversides (family Atherinidae) and killifishes (family Cyprinodontidae).

Equilibrium strategists tend to be intermediate-sized fish that inhabit relatively stable environments, produce small numbers of large eggs, and provide high parental investment in their young. Examples of equilibrium life-history strategists are tropical cichlids, sculpins (family Cottidae), many gobies (family Gobiidae) and some salmonids. Equilibrium life-history strategists are similar to the traditional K-strategy of adaptation to life in resource-limited and density-dependent environments (Pianka 1970), except for their smaller body size.

The periodic life-history strategy identifies fishes that delay maturation to attain a size sufficient for production of a large clutch and to improve adult survival during periods of suboptimal environmental conditions. Red snapper, striped bass and American shad (*Alosa sapidissima*) typify the periodic life-history strategy. Winemiller and Rose (1992) view the periodic strategy as the perennial tactic for spreading reproductive effort over many years (or over a large area), so that high larval or juvenile survivorship during one year (or in one spatial zone) offsets the many bad years (or zones).

Expectations from life-history theory

We formulate expectations for compensation and compensatory reserve based on the three endpoint life-history strategies. We expect compensation to be strongest in equilibrium strategists, while we expect compensatory reserve to be highest in periodic strategists. We use information derived from spawner–recruit relationships to evaluate these expectations. Compensatory reserve can be estimated from the slope of the spawner–recruit relationship (Myers *et al*. 1999). Use of fecundity alone to infer compensatory reserve is overly simplistic because compensatory reserve also includes the survival from egg to

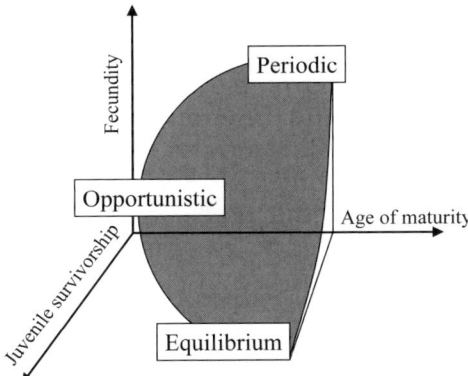

Figure 1 Three endpoint life-history strategies for fish derived by Winemiller and Rose (1992). The endpoint strategies are: opportunistic, equilibrium and periodic. Species can fall anywhere on the surface shown here.

recruitment. Estimating the magnitude of compensation from data is problematic, as it is difficult to obtain consistent measures of the magnitude of compensation that can be compared across many species. Instead, we use an indirect measure based on the magnitude of interannual variation in recruitment derived from an assumed spawner–recruit relationship. We recognise that interannual variation in recruitment reflects both the general magnitude of compensation in the population (ability to adjust to new densities) and fluctuations in recruitment owing to environmental variation. Thus, we state our expectations in terms of the magnitude of interannual variation, rather than in terms of the magnitude of compensation.

Equilibrium strategists should have the lowest interannual variation because they inhabit relatively stable environments and because they generally exhibit the strongest compensation. Equilibrium strategists invest heavily in individual offspring (by trading off against fecundity) to enhance early survivorship. Their population abundance should tend to track a long-term mean value near the carrying capacity of the environment. Equilibrium strategists have relatively low fecundity, which, despite high early life stage survivorship, should result in low compensatory reserve. Interannual variation in recruitment should be higher in periodic and opportunistic strategists. Periodic strategists are typically broadcast spawners that rely on wide fluctuations in year-class strengths, with the occasional exceptional year carrying the population through poor years. Because they inhabit highly variable environments and seldom approach environmental carrying capacity, opportunistic strategists should also show high interannual variation. We note that Mertz and Myers 1996) had difficulty relating fecundity to recruitment variability in marine fishes, although more recent analyses have yielded stronger relationships (Rickman *et al.* 2000). The longevity and high fecundity of periodic strategists should more than offset their low early survivorship, resulting in periodic strategists having the highest compensatory reserve. Opportunistic strategists should have low-to-intermediate compensatory reserve, as they are small and short-lived with relatively low fecundity and low early survivorship.

Comparison of life-history expectations with data

Myers and colleagues assembled and analysed several hundred spawner–recruit data sets (see Myers *et al.* 1995a; Myers and Barrowman 1996; Myers *et al.* 1999). Below we further analyse the Myers' spawner–recruit database by examining measures of compensatory reserve and interannual variation in recruitment, grouped by life-history strategies. The spawner–recruit relationships of a total of 249 populations consisting of 57 species were analysed. A Ricker spawner–recruit relationship was fitted to each population using linear regression (Myers *et al.* 1999); recruitment and spawners were specified in the same units (thousand tonnes for all species, except numbers for salmonids).

We used the fitted spawner–recruit relationships to estimate the compensatory reserve and magnitude of interannual variation for each of the 249 populations. The estimate of the slope at the origin of the fitted spawner–recruit curve (i.e. the maximum rate of recruits per spawner) was converted from an annual rate to a lifetime rate by dividing by one minus the natural adult mortality rate (Myers *et al.* 1999), and then used to compute a steepness value (Hilborn and Walters 1992). The slope at the origin corresponds to the maximum rate of recruitment per spawner because spawners are at very low values where compensatory density dependence would least affect recruitment. The steepness parameter is the fraction of the maximum recruitment expected when the stock is at 20% of its unfished state. The magnitude of interannual variation in recruitment for each population (which we term sigma) was computed as the standard deviation of the natural log-transformed residuals between predicted and observed recruitment values (Mertz and Myers 1996). Use of sigma to infer interannual variation presumes that the spawner–recruit model is a reasonable representation of the true relationship between spawners and recruits, and that the deviations between predicted and observed recruitment are indicative of differences in population dynamics among species (i.e. process error) and are not overly influenced or biased by measurement error.

Each species in the spawner–recruit database was also assigned to a life-history strategy, based upon their egg diameter (mm) and fecundity (total annual egg production per female). Egg diameter was used as a proxy for the degree of parental investment in individual offspring. Species with fecundity ≤ 25 000 eggs and egg diameter > 2 mm were assigned to the equilibrium strategist group. Species with fecundity ≤ 25 000 eggs and egg diameter ≤ 2 mm were definned as opportunistic strategists. Species with fecundity > 25 000 eggs and egg diameter < 2 mm were deemed periodic strategists. There were three exceptions: three

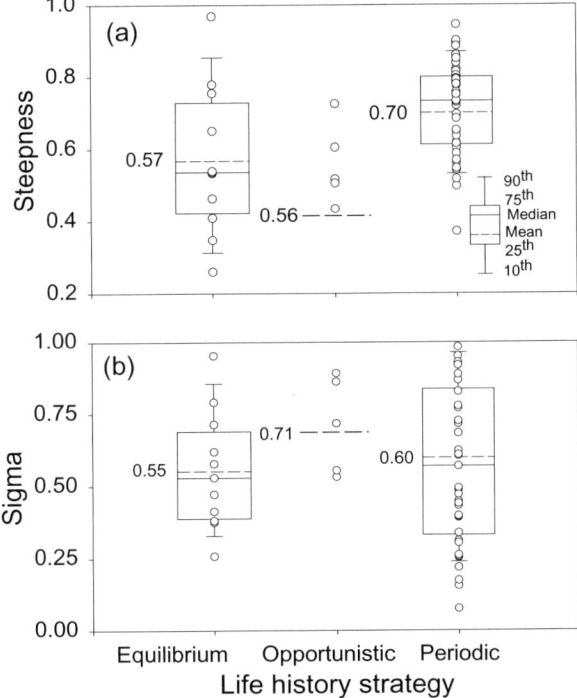

Figure 2 Steepness estimates (a) and sigma values (b), averaged over populations for each species, by life-history strategy. Steepness estimates were determined from the slope at the origin of Ricker spawner–recruit curves fitted to observed data for each population, and are interpreted as indicating the compensatory reserve in a population. Sigma values are the standard deviation of natural log-transformed residuals (observed recruitment minus predicted from the spawner–recruit curve), and are interpreted as reflecting the combined effects of environmental variation and the degree of compensation.

species with egg diameters > 2 mm were considered periodic strategists because of their high fecundity [pike (*Esox lucius*): 2.75 mm and 32 000 eggs; American shad (*Alosa sapidissima*): 3.0 mm and 125 000 eggs; Greenland halibut (*Reinhardtius hippoglossoides*): 4.25 mm and 165 000 eggs]. These rules resulted in 11 species (110 populations; 106 Salmonidae) being equilibrium strategists, 5 species (8 populations) being opportunistic strategists and 41 species (131 populations) being periodic strategists.

Steepness (of recruitment curves) and sigma values were generally consistent with our expectations from life-history theory. Single steepness and sigma values were obtained for each species by averaging over populations to reduce the effects of unevenness in the number of populations per species (Fig. 2). The overall average steepness value was highest for periodic species (0.70) and lower for equilibrium species (0.57) and opportunistic species (0.56). Overall average sigma values were lowest for equilibrium species (0.55), intermediate for periodic species (0.60) and highest for opportunistic species (0.71).

Our initial analysis of the Myers group spawner–recruit database generally supports the expectations from life-history theory. Our analysis is encouraging but admittedly superficial, as we did not perform statistical tests of group differences and we ignored any confounding effects of measurement error. Further analysis should focus on reducing the high variation in steepness and sigma values within each strategy by allowing species to have intermediate strategies, adding more equilibrium and opportunistic species to the database, and including a formal statistical analysis of the results.

Understanding compensation

Population regulation via compensatory responses must ultimately occur through density-dependent changes in rates of mortality or in reproductive success. Mortality and reproduction directly affect the numbers of individuals. Other responses, such as density-dependent changes in growth rate or movement are indirect, and must lead to changes in mortality rate or reproductive success if they are to affect population size. If growth or movement do not ultimately affect numbers of individuals, then changes in these processes would simply result in larger or smaller individuals (for growth) or the same number individuals, but in different locations (for movement).

There are multiple pathways by which mortality, reproduction, growth and movement can interact, leading to compensatory responses at the population level (Goodyear 1980; Jude et al. 1987; Saila et al. 1987). Compensation is achieved by density-dependent factors that may appear weak relative to the environmental influences, but act cumulatively over life stages and generations to stabilise the population (Hixon 1998). Below we describe several of the more common pathways that have been documented. We use reported studies of particular processes, results of long-term field and manipulative studies, and computer modelling examples to illustrate compensatory responses. The studies of particular processes provide evidence of density-dependent mortality, growth, reproduction and movement. The selected long-term empirical studies and model analyses demonstrate how density-dependent processes in life stages can lead to compensatory responses at the population level. The field studies involve real fish populations, but at the cost of the underlying processes causing the compensation to be weakly documented. The modelling examples provide explicit understanding of how compensation resulted from density-dependent processes, but in a virtual world.

Density-dependent processes

Density-dependent mortality can arise from density-dependent responses by predators, or via density-dependent growth, reproduction and movement affecting survival. While density-dependent mortality can lead directly to compensation, detecting density-dependent mortality from field data is difficult. For large-number fish populations (e.g. most marine commercial stocks), estimates of mortality rates and fish densities have large measurement errors (Beverton and Holt 1957; Ricker 1975; Houde 1987). Nevertheless, Myers and Cadigan (1993) explicitly accounted for measurement errors and isolated a density-dependent component in mortality rates of juvenile-stage demersal marine species. Crecco and Savoy (1987) used otolith data to infer that survival of American shad cohorts (grouped into 5-day hatching periods) was inversely related to initial cohort size. Density-dependent mortality in young-of-the-year (YOY) life stages has also been detected for other species (Crozier and Kennedy 1995; Bailey et al. 1996; Michaletz 1997; Planes et al. 1998).

Predation as the cause of density-dependent mortality results from the numerical or functional responses of the predators to prey density (Murdoch and Oaten 1975; Hassell 1978; Bailey 1994; Rose et al. 1999). Numerical responses involve increases in the number of predators when prey densities are high. Functional responses involve changes in the consumption rates of predators such that mortality rate of prey increases with increasing prey density. Recently, Hixon (1998) suggested that an aggregative response, whereby the local distribution of predators shifts in response to local prey density, is reasonable for density-dependent predation of reef fish (e.g. see Schneider 1989). Predation has been implicated for density-dependent mortality in a variety of juvenile fishes (Murdoch 1969; Murdoch and Oaten 1975; Peterman and Gatto 1978; Lockwood 1980; van der Veer 1986; Forrester and Steele 2000; Anderson 2001). Forrester (1995) and Steele (1997) manipulated both juvenile and adult goby densities over realistic ranges, and in both cases, recruit density was not linearly related to adult density, indicating density-dependent survival. Because growth rates of individuals were similar across all densities, the authors implicated predation as the likely cause of the density-dependent mortality. In a more recent study, Forrester and Steele (2000) showed that the intensity and cause of density-dependent mortality differed among three closely related goby species. Reductions in mortality rates due to density-dependent responses by predators are also possible if predators target other forage species owing to reduced profitability when densities are low (Werner and Hall 1974; Townsend and Hughes 1981; Werner and Mittlebach 1981; Hartman and Brandt 1995). Shima (2001) showed that per capita mortality rates of the reef-dwelling six-bar wrasse (*Thalassoma hardwicke*) were dependent on both the density and number of individuals.

Cannibalism, which can result in a strong density-dependent feedback, has been documented for a variety of fish species (Smith and Reay 1991). For example, adults eating their young is prevalent in walleye (Forney 1977) and northern pike (Treasurer et al. 1992). Egg cannibalism has been shown to be a significant cause of density-dependent predation mortality in some fishes via positive relationships between daily egg mortality rates and spawning biomass densities (Hunter and Kimbrell 1980; Smith et al. 1989). For example, simultaneous samples of eggs and adults for estimates of daily egg production indicate that cannibalism can amount to 20–30% of total daily egg mortality in northern anchovy (*Engraulis mordax*) and Peruvian anchovy (*E. ringens*) (MacCall 1981; Alheit 1987).

Density-dependent growth is not simply food limitation of growth rate. Rather, density-dependent growth refers to a situation where the feeding rate of an individual is reduced by the presence of other members of the same population, cohort, or year-class, i.e. intraspecific competition for food increases with increasing density of individuals (Heath 1992). In a recent review, Cowan *et al.* (2000) concluded that density-dependent regulation of cohort growth and biomass via feedbacks derived from reductions in prey resources is more likely to occur during the late-larval to juvenile stage than in the high-density early larval stage. Examples also exist which indicate slower individual growth rates in large cohorts or year classes (e.g. Bannister 1978; Rauck and Zijlstra 1978; Craig and Kipling 1983; Mills and Forney 1983; Rijnsdorp and van Leeuwen 1992). Adults also can exhibit density-dependent growth. Rieman and Myers (1992) found density-dependent growth in adult kokanee (*Oncorhynchus nerka*), but not in yearlings.

Density-dependent growth can result in density-dependent mortality; slower growth leads to prolonged stage duration (Houde 1989; Hovenkamp 1992), and mortality often decreases with body size (Cushing 1975; Dahlberg 1979; Anderson 1988; Sogard 1997). Larger individuals at the end of their first growing season often experience decreased probability of overwinter mortality (Post and Evans 1989; Johnson and Evans 1990), and increased survival to recruitment (e.g. Marshall and Frank 1999b). Growth rate can also affect mortality because size is important in determining the susceptibility of prey to predators and can influence competitive interactions (Miller *et al.* 1988; Bailey and Houde 1989; Fuiman and Magurran 1994; Bystrom and Garcia-Berthou 1999; Lundvall *et al.* 1999). The behaviour of individuals depends on their balancing of energy gain with predation risk; more time spent foraging generally translates into higher predation risk. Selection by individuals for spatial and temporal restriction of foraging activity to reduce predation risk can lead to density-dependent growth and mortality (Walters and Korman 1999; Walters 2000).

Density-dependent reproduction includes changes in fecundity (egg production, eggs per gram of female), maturation, spawning frequency, egg quality and spawning location. Changes in reproduction can arise directly from how individuals allocate their energy (Kjorsvik *et al.* 1990; Henderson *et al.* 1996; Van Winkle *et al.* 1997), and indirectly via density-dependent changes in growth and movement (e.g. Peterman and Bradford 1987). Within morphological upper limits (Roff 1982), the weight of mature ovaries in most fish species is an exponential allometric function of body weight (MacKinnon 1972; Delahunty and deVlaming 1980; Erickson *et al.* 1985; Sibly and Calow 1986). This suggests that fecundity is highly size-dependent (Rijnsdorp 1990, 1994; Koslow *et al.* 1995).

Examples of changes in fecundity that could provide a compensatory mechanism serving to stabilise fish populations have been reported (Nikolsky 1969; Nikolsky *et al.* 1973; Rothschild *et al.* 1989). However, several authors have expressed doubts that density-dependent changes in adult growth affecting fecundity can be large enough alone to stabilise recruitment (Craig and Kipling 1983; Koslow 1992; Koslow *et al.* 1995; Trippel 1995). Interannual changes in fecundity of northern anchovy (Lasker 1985) and Atlantic herring (*Clupea harengus*) (Almatar and Bailey 1989) were unrelated to changes in adult abundance; however, variation in fecundity consistent with density dependence has been reported for other species (Nikolsky 1969; Bagenal 1973; Nikolsky *et al.* 1973; Rothschild 1986; Rothschild and Fogarty 1989).

Density-dependent growth can affect fecundity via growth effects on lifetime egg production owing to either earlier maturation (Muth and Wolfert 1986; Peterman and Bradford 1987; Funakoshi 1992) or delayed maturation (Bowen *et al.* 1991; DeLeo and Gatto 1996), depending on how survival and fecundity vary with age. Trippel (1995) reviewed data from north-west Atlantic groundfish populations for 1959–92 and reported that age and size at maturity decreased by 15–55% in cod and by as much as 30% in haddock (*Melanogrammus aeglefinus*). He presumed the observed shift to a younger age and smaller size at maturity to be a compensatory response to reduced population size to maintain or achieve maximal reproductive output. Perrow *et al.* (1990) attributed 2-year cycles in roach (*Rutilus rutilus*) to density-dependent growth of YOY resulting in their reduced fecundity in the next year.

Serial or batch spawners may also respond to changes in density-dependent growth (actually per capita ration) by changing the frequency of spawning and the number of spawns. Both the interval between successive spawnings (Wootton 1977) and the numbers of spawnings during a season (Townsend and Wootton 1984) have been related to ration. Experimental results with Japanese anchovy (*Engraulis japonicus*) confirmed the dependence of batch-fecundity and interspawning interval on ration, with a lag of 11–21 days between increased ration and increased fecundity (Tsuruta and Hirose 1989). Based upon studies in Tampa Bay, Florida, Peebles et al. (1996) inferred that bay anchovy may be an "income breeder" (following Stearns 1993) that spawns soon after energy for egg production becomes available, implying that observed seasonal and spatial patterns in egg production could largely be explained by variation in metabolic rate and adult ration.

Density-dependent food availability (per capita ration) during the spawning season may also influence the size and quality of eggs spawned (reviewed by Kjorsvik et al. 1990; Bromage 1995). Egg size has been shown to increase with spawner body size for many species (e.g. Buckley 1967; Rogers and Westin 1981; Lobon-Cervia et al. 1986; Hislop 1988; Kjesbu 1989; Zastrow et al. 1989; Monteleone and Houde 1990; Buckley et al. 1991b; Kjesbu et al. 1992). Number of eggs spawned, and their condition and hatchability, have been related to the condition of female Japanese sardine (Sardinops melanostictus) spawners, which in turn, is influenced by the quantity and quality of their food (Lasker and Theilacker 1962; Takeuchi et al. 1981; Watanabe et al. 1984a,b,c; Morimoto 1996). In contrast, increasing egg number, rather than egg quality, in response to low densities was exhibited by Atlantic herring (Bradford and Stephenson 1992) and Japanese anchovy (Funakoshi 1992). Larger eggs leading to increased survival has been documented for some species (Rosenberg and Haugen 1982; Miller et al. 1988; Buckley et al. 1991a), but not for others (Chambers et al. 1989; Kjorsvik et al. 1990; Pepin and Myers 1991; Rijnsdorp and Vingerhoed 1994).

Density-dependent movement affecting mortality and reproductive success is well documented in fishes. Whenever these vital rates directly (or indirectly via growth) depend upon habitat that is both limiting in quantity and variable in quality, the movement of individuals to suboptimal habitats when densities are high can result in compensatory density dependence. For example, competition for feeding and spawning sites in many salmonids in rivers and streams involves fine-scale movement of individuals related to their body size (Caron and Beaugrand 1988; Grant et al. 1989; Hughes 1998; Reinhardt 1999); crowded conditions force individuals to occupy lesser quality habitats, where they presumably experience higher mortality or slower growth (Fausch 1984; Grant and Kramer 1990; Hughes 1992; Elliot 1994). Selection of pool, run, or riffle habitats by juvenile Atlantic salmon (Salmo salar) in experimental river enclosures depended on density (But l et al. 1999). Superimposition of nests whereby, under high densities of spawners, late spawners destroy earlier spawned nests has been documented for some stream and riverine dwelling salmonids (Beall and Marty 1987; Maunder 1997). Large-scale movements related to density has been observed in the contraction and expansion of the distribution of populations that track fluctuations in their abundance (Swain and Sinclair 1994; McConnaughey 1995); presumably such movements force individuals to occupy inferior habitats when crowded (MacCall 1990). Density-dependent mortality of postsettlement coral reef fishes is related to the availability of, and movement between, optimal and suboptimal habitats (Hixon 1998).

There are also less-well-studied processes, such as sex change, parasitism and disease, that can lead to compensatory density dependence. Sex change in protogynous species, such as the gag grouper (*Myctoperca microlepis*), depends on population density (Lutnesky 1994; Collins et al. 1998). Sex change being both density dependent and socially mediated has led to speculation that species in the grouper-snapper complex in the Gulf of Mexico may be sperm limited (Coleman et al. 1996) and have diminished genetic variation (Chapman et al. 1999) owing to a long history of overfishing that targets males (Koenig et al. 1996). Parasitism and disease can be direct sources of density-dependent mortality and fecundity in fishes (Ivlev 1961; Moller 1990; Adlard and Lester 1994), and their prevalence is often facilitated by poor nutritional condition that could arise from density-dependent growth. For example, in a study of flounder (*Platichthys flesus*) in the Elbe estuary, Moller (1990) found that local differences in condition factor and disease prevalence seemed to be negatively correlated with local food supply. In contrast, Poulin (1995) concluded that larger (well-fed) hosts consume greater quantities of food and are therefore exposed to a wider range of parasite-infective stages.

Empirical field studies

Empirical evidence for compensatory responses can be found in long-term and manipulative studies of particular fish populations. However, while there are many examples of density dependence in particular processes and life stages (see above; Jude et al. 1987; Saila et al. 1987; Myers 1995; Cowan et al. 2000), there are fewer examples where the evidence is sufficient to infer how the density-dependent process in a life stage actually influenced population dynamics. To be useful for understanding how compensation works, population-level studies of density dependence must be long-term (multigenerational), and should involve multiple life stages and multiple processes (Hixon 1998). While not meant to be an exhaustive review, we describe below several examples that include species from marine, estuarine and freshwater environments. These examples include a variety of data analysis methods, the use of experimental manipulation, and they are field-orientated and focus on single populations. These studies are included because they provided insights about the underlying causes of the observed compensatory response.

Japanese anchovy
In response to long-term declines (1970–84) in stock levels and the numbers of large adults, Japanese anchovy (*Engraulis japonicus*) underwent a change in 'mode of life' whereby the spawning potential of small anchovy (7–12 cm in length) increased dramatically (Funakoshi 1992). Nutritional condition and the proportion of small females that matured and spawned increased. Spawning frequency changed from once every 4–5 days for large females (12–16 cm in length) to every 2–3 days for small anchovy. Batch fecundity increased by 1.2- to 2.2-fold, although egg size decreased. In combination, these changes in reproductive potential of smaller fish resulted in almost constant population egg production over the period of decline (Funakoshi 1992). Experimental results have confirmed that batch-fecundity and the intervals between spawnings in the Japanese anchovy depend on ration (Tsuruta and Hirose 1989).

Population cycles of vendace
The long-term study of vendace (*Coregonus albula*) in Lake Pyhajarvi in south-west Finland showed a 2-year cycle in year-class strength (Helminen *et al.* 1993; Helminen and Sarvala 1994). Vendace enter the fishery in their first autumn (defined as recruitment) and dominate the catch during the next winter. Individuals spawn the following year, with larval development in the summer. Abundances in the first autumn were estimated for 1971–90. A 2-year cycle was documented using autocorrelation analysis, which showed alternating signs with increasing lags (Helminen *et al.* 1993), and using multiple regression analysis with lagged abundances (Helminen and Sarvala 1994). The multiple regression model related autumn abundance to the following explanatory variables: autumn abundance lagged one year, summer water temperature lagged 2 years (index of abundance of age-2 perch, a major predator on YOY vendace), and a temperature-derived estimate of the duration of the larval stage of vendace. All three regression coefficients were negative and the model explained 77% of the interannual variation in autumn abundance. Additional analyses using bioenergetics modelling showed that vendace can affect the densities of their zooplankton prey (Helminen *et al.* 1990).

The cause of the 2-year cycles was asymmetric competition in which YOY out competed age-1 and older individuals for zooplankton prey (Hamrin and Persson 1986). Thus, a high autumn YOY abundance in one year would lead to lowered zooplankton prey for age-1 and older individuals in that year. However, in the previous year these age-1 individuals had experienced low abundances as YOY, and thus had faster growth rates, making them relatively large at the end of their first growing season. Larger size requires a larger ration, which further lowered their competitive abilities in the next year when compared with relatively small, but numerous, YOY. These age-1 and older individuals therefore had less surplus energy to devote to reproduction and produced fewer eggs, leading to low YOY abundance the next year. Similar 2-year cycles have been documented for vendace in other systems, and for other species with a similar life-history strategy (Hamrin and Persson 1986; Perrow *et al.* 1990).

Controlled exploitation of lake whitefish
Healey (1978, 1980) controlled the annual exploitation rates of lake whitefish (*Coregonus clupeaformis*) in four similar lakes at levels of 0, 10, 20 and 30%. Gill nets were used to sample each lake during various combinations of spring, summer and autumn from 1971 to 1978. Captured fish were aged using scales. Length-at-age (growth), fecundity and year-class strength (recruitment) were compared before and after exploitation was begun for each lake. Length-at-age data showed that growth rates of adults increased in relation to the level of exploitation, such that: (i) little or no change in growth rate of adults occurred in the unexploited lake; (ii) a small but temporary increase in growth rate (5 mm over ages) occurred in the lightly exploited lake; (iii) growth rate increased (12 mm over ages) and was maintained in the moderately exploited lake; and (iv) growth rate was still increasing at the end of the study for the heavily exploited lake. Relationships between fecundity and exploitation and between recruitment and exploitation were suggestive, but less definitive. Individual fecundity increased in response to exploitation, although the increases were not related to the magnitude of exploitation. Year-to-year patterns

between lakes indicated that the most heavily exploited lake had a higher frequency of strong recruitments, especially just after exploitation was begun. Responses of recruitment to exploitation in the other lakes was less consistent. Removal of adults via fishing tended to stimulate good (but variable) recruitment of younger fish. The author proposed that adult suppression of young could be owing to relegation to marginal habitats, or inhibition of normal exploratory and foraging behaviour.

Orange roughy
Koslow et al. (1995) showed that individual fecundity of orange roughy (*Hoplostethus atlanticus*) females increased by 20% after the stock was fished down to around 50% of its virgin biomass. The authors suggested that the compensatory increase in individual fecundity was attributable to increased per capita food availability. Increased fecundity, combined with an apparent 17% increase in the proportion of females spawning, limited the expected decline in population egg production during 1987–92 to 15%, despite the 50% reduction in spawning biomass.

Walleye recovery in western Lake Erie
Muth and Wolfert (1986) documented changes in walleye (*Stizostedion vitreum*) growth and maturity during the recovery of the stock in western Lake Erie. The population declined during the 1960s, and a catch-quota system on recreational fishing was implemented in 1976. The estimated standing stock of yearling and older walleye increased from 14.6 million in 1976 to 44.7 million in 1983. Associated with this increase in abundance were downward annual trends in mean lengths and condition factors of YOY in the autumn, and in age-1 and age-2 fish sampled from the trap-net fishery. Percent-mature-by-age also declined with increasing abundance. The percent mature of age-2 females dropped from 90% in 1976 to 45% in 1977, plateaued until the 1980s, then dropped from 31% in 1981 to 7% in 1983. Nearly all age-1 males were mature (99%) in 1975, 78% were mature during 1978–81, followed by a rapid drop to 45% in 1982 and to 32% in 1983. The authors suggested that the piscivorous YOY walleye exceeded the carrying capacity of their forage fish prey in years of high walleye abundance. Thus, forage fish biomass declined in parallel with increasing walleye abundance. With less food per capita, walleye were unable to maintain growth, which ultimately resulted in delayed maturation and lowered reproductive potential.

Plaice in the North Sea
Beverton and Iles (1992) examined how density-dependent mortality during the first 16 months of life can explain the relatively low variation in recruitment observed in North Sea plaice (*Pleuronectes platessus*). They used regression analysis, and carefully estimated the density-dependent component of mortality for three time periods between settlement (June) and October 1 of the second year. While North Sea data were used to estimate the density-dependent components for the earliest and latest time periods, data both from within and outside of the North Sea were used for the middle period. They then determined that the dampening effect of the density-dependent mortalities was sufficient to reduce a 200-fold variation in abundance at settlement to 4-fold variation on October 1 of the second year. The dramatic reduction generally agreed with the decrease in variation observed in various surveys performed during the 16 months after settlement. They further concluded that all three periods of density-dependent mortality were important, but that the strong density dependence estimated for the earliest period from Wadden Sea data may be higher than is typical for other nursery areas. The explanations for the density-dependent mortality are only well understood for the earliest period, during which the functional response of shrimp predators in the Wadden Sea have been implicated as the causative agent (van der Veer 1986; van der Veer and Bergman 1987).

Haddock recruitment in the North Atlantic
Marshall and Frank (1999a,b) presented empirical evidence for compensatory control of haddock (*Melanogrammus aeglefinus*) recruitment in the southwest Scotian Shelf in the North Atlantic. They analysed summer surveys from 1970 to 1995 and autumn surveys from 1963 to 1986 and showed that: (i) mean length at age-1 was negatively related to adult (age-4 and older) abundance; (ii) differences in mean length at age-1 persisted through the adult stage; and (iii) recruitment was positively related to the mean length of age-4 adults (an indicator of spawning stock condition). The negative feedback on recruitment operated as follows: high adult abundance led to short mean length at age-1, which led, 3 years later, to short mean length at age-4, which led to low recruitment the next year, which led to low adult abundance 3 years later. Likewise, low adult abundance in a year resulted in a tendency towards high adult abundance 8 years later. Marshall and Frank (1999a) caution that this compensatory

feedback occurs only during periods when both density-dependent growth of age-1 and recruitment dependent on mean length of age-4 adults were operating. They further suggested that this feedback could led to cyclic population behaviour, which has been observed in a variety of stocks, and hypothesised that competition for food between juveniles and adults was the cause of density-dependent age-1 growth.

Migratory brown trout
Elliott (1994) synthesised 25 years of study of anadromous brown trout (*Salmo trutta trutta*) in Black Brows Beck, a stream in the English Lake District. Abundances at successive life stages from eggs to spawning for each year-class were estimated for 1966–1990. Analysis of mortality rates between successive life stages (K-factor analysis) using spawner–recruit-type functions showed that there was a critical period from emergence to 30–70 days postemergence during which mortality was density-dependent. After this critical period, mortality was much lower and density-independent. More detailed studies of some year classes indicated that the likely cause of density-dependent mortality during the critical period was the effects of locating and defending feeding territories. Under high egg abundances, smaller young trout experienced reduced food because they could not locate good territories, and the larger young trout also suffered higher mortality because they incurred the greatest costs of territory defense. Locating and defending territories provided a simple explanation for Eliott's observation that both the smaller and larger individuals were lost under high abundances (i.e. variability of mean lengths at the end of the critical period decreased with increasing abundance).

Computer modelling examples

We present the results of three simulation models to illustrate how compensation works at the process level, and to illustrate several features of compensation that make understanding compensation from field measurements so difficult. The three models are: yellow perch (*Perca flavescens*)-walleye in Oneida Lake, New York; bay anchovy in the Chesapeake Bay, Maryland; and fish food web models in Lake Mendota, Wisconsin. The three examples are from our own work. They were selected because they illustrate important points about compensation and were convenient. There are numerous other examples we could have used, including examples that use modelling approaches other than the individual-based approach. The yellow perch-walleye example illustrates how compensation arises from complex relationships among multiple processes operating in multiple life stages. The bay anchovy example demonstrates how small, probably undetectable, density-dependent changes in multiple processes can combine to result in significant compensatory responses. The food web analysis demonstrates how density-dependent relationships in a species can depend on the specifics of the situation.

The models are all site-specific and individual-based. The advantage of site-specific models is that they provide a method for tracking the complex pathways of compensation under relatively realistic (albeit virtual world) conditions. These models were developed for specific locations, which were selected because of the extensive long-term data available for those populations. Thus, model predictions were extensively compared with observed data to ensure model realism. The reader is referred to the original publications cited under each example for the details of model calibration and validation. The individual-based approach used has several advantages for modelling fish population dynamics and predicting compensatory responses (DeAngelis *et al.* 1993, 1994). One such advantage is that density dependence can be an emergent property of the model, resulting from the summed effects over individuals, rather than having to be specified *a priori* in relationships (e.g. spawner–recruit curves) used to construct the model.

Yellow perch in Oneida Lake
Rose *et al.* (1999) used an extensive database to develop, calibrate and corroborate an individual-based model of yellow perch–walleye dynamics for Oneida Lake, New York. The model begins with spawning of individual females of yellow perch and walleye, and simulates growth and mortality of both species progeny as they develop through successive life stages (egg, yolk-sac larva, feeding larva, YOY juvenile, yearling and adult). YOY juvenile and yearling yellow perch are the dominant prey for adult walleye, and walleye predation is the major source of mortality of young yellow perch. Density-dependent growth, survival and reproduction emerge from the size-based interactions represented in the model. Predicted annual adult abundances of yellow perch and walleye under baseline conditions show weak, but consistent, prey–predator cycling (Fig. 3).

We use three simulations to illustrate how growth, survival and reproduction processes can lead to

Compensation in fish populations *K A Rose et al.*

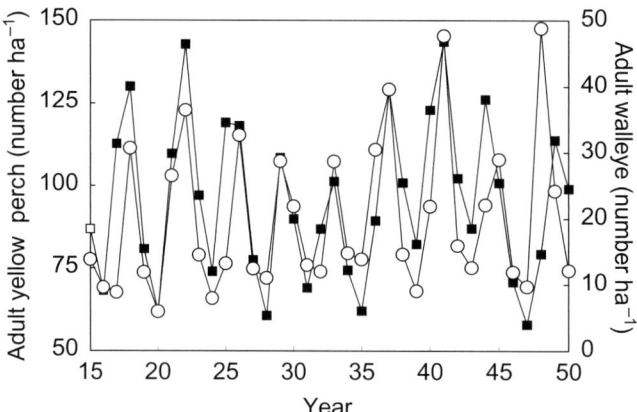

Figure 3 Simulated annual adult densities of yellow perch (■) and walleye (○) for years 15–50 of the baseline simulation of Oneida Lake (modified from Rose *et al.* 1999).

Table 1 Simulated responses of yellow perch to decreased and increased egg mortality rate based on an individual-based model of yellow perch–walleye in Oneida Lake (from Rose *et al.* 1999)

Variable	Units	Decreased egg mortality	Baseline	Increased egg mortality
Increasing egg mortality	Percentage	71	90	96
Lower number of first feeders	No. ha^{-1}	178 325	85 036	40 894
Lower 18 mm abundance	No. ha^{-1}	60 594	30 329	15 325
Increased 18 mm to recruitment survival	Percentage	0.41	0.85	1.56
Higher (slightly) recruitment	No. ha^{-1}	205	210	221
Longer recruitment mean length	mm	141.8	144.5	151.8
Faster adult growth	mm year^{-1} (age-4)	18.5	20.9	22.0
Longer adult mean lengths	mm (age-4)	205	215	221
Younger maturation	Percentage mature (age-4)	24	41	54
Higher (slightly) fecundity	Eggs per spawner	27 885	30 313	30 989
Similar adult abundance	No. ha^{-1}	103	106	112
More spawners	No. ha^{-1}	29.1	35.4	44.2
Higher egg production	$\times 10^6$ eggs ha^{-1}	0.80	1.11	1.35

The value of each variable is the average computed over the last 35 years of the 50-year simulation.

compensation in yellow perch. The baseline yellow perch egg mortality rate of 0.11 day^{-1} was increased to 0.15 day^{-1} and decreased to 0.06 day^{-1}. Predicted average values of abundances, mean lengths, growth rates, survival rates, fecundity and percent mature at age for yellow perch were analysed for density-dependent responses. To match how Oneida Lake data were reported, Rose *et al.* (1999) used several conventions in reporting model predictions: abundance at 18 mm roughly corresponds to metamorphosis (i.e. number of juveniles), recruitment is abundance at age-2, and age-4 responses are considered representative of adults. All simulations were for 50 years duration; average values of all variables were computed over the last 35 years of each simulation.

Yellow perch exhibited density-dependent responses to increasing egg mortality (Table 1). Increasing egg mortality led to the expected lower total number of yellow perch first feeders, and lower yellow perch abundance at 18 mm. The survival rate of yellow perch from 18 mm to recruitment increased with increasing egg mortality, resulting in slightly increased recruitment with increasing egg mortality. Yellow perch survival from 18 mm to recruitment increased owing to decreased walleye predation pressure; average adult walleye abundances in the three simulations decreased with increasing yellow perch egg mortality (20.2 per hectare (ha)$^{-1}$ to 19.9 ha^{-1} to 16.1 ha^{-1}). Walleye adult abundance decreased because there was less food available to the walleye. Fewer 18 mm yellow perch lead to faster growth rates of YOY yellow perch (density-dependent growth) and longer mean recruit lengths. Longer YOY yellow perch were less vulnerable to predation by walleye because yellow perch predation risk decreased with length. Fewer and longer YOY yellow perch resulted in less food for walleye, which resulted in increased cannibalism of YOY walleye by adults, and shorter walleye adults that produced fewer eggs; both of which caused walleye adult abundances to decrease with increasing yellow perch egg mortality. Adult yellow perch growth rates also increased with increasing egg mortality which lead to longer yellow perch adults at age, and resulted in a younger age-of-maturation and slightly higher fecundity of adult yellow perch. With adult yellow perch abundances near or slightly above baseline levels, younger age-of-maturation resulted in more spawners with increasing egg mortality. More spawners and higher fecundity resulted in increased egg production to compensate for increased egg mortality.

Thus, despite a two-fold variation in the egg mortality rate between the high and low mortality simulations, predicted average adult abundances of yellow perch were similar (103 ha^{-1} and 112 ha^{-1}).

These results illustrate how understanding compensation requires knowledge of how multiple processes interact over the full life cycle, and how difficult it would be to try to monitor these process-level responses in nature. Recall that the values reported in Table 1 span all yellow perch life stages and are averages over 35 years, with the only difference among the three simulations being the egg mortality rate of the yellow perch. Sufficient monitoring of the variables listed in Table 1 would require an enormous amount of sampling over several decades.

Food webs in Lake Mendota
We configured six different food webs from a general, multispecies individual-based model (Fig. 4; McDermot

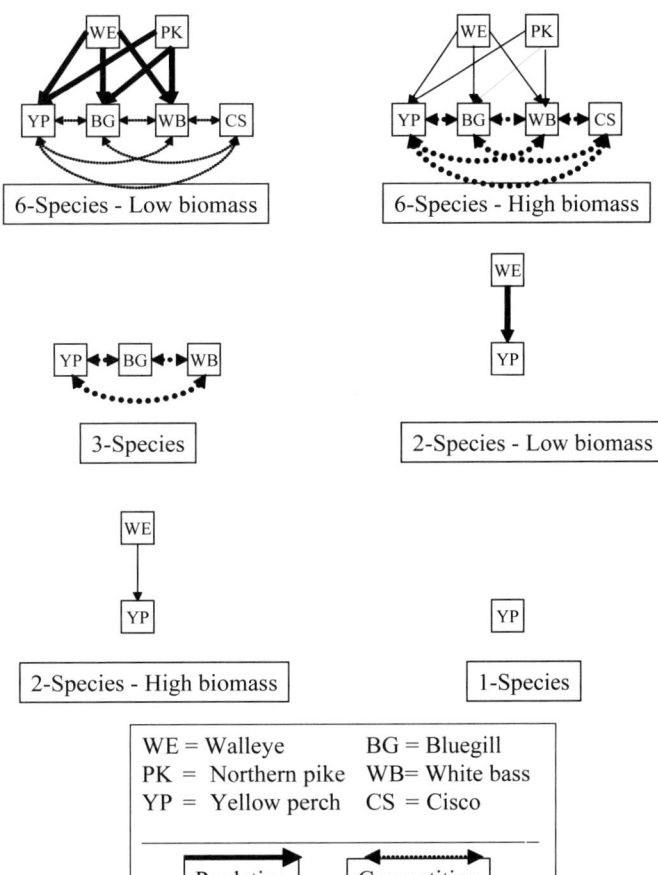

Figure 4 The six food webs used to compare density-dependent survival of juvenile yellow perch. The low- and high-biomass food webs differ in their average yellow perch biomass and their degree of coupling between piscivores (walleye and pike) and their prey. Differences in the feeding-related parameters of walleye and pike resulted in competition among the yellow perch and other planktivores being more important in the high-biomass versions, while piscivore predation was more important in controlling the planktivores in the low-biomass versions. Predator–prey interactions are shown as solid lines; competitive interactions are shown as dashed lines. Heavier lines indicate greater intensity of the interactions.

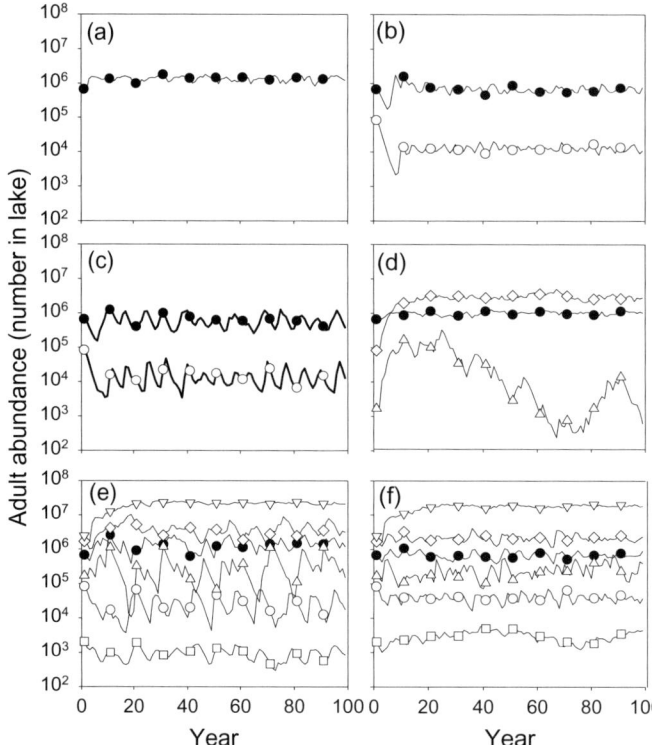

Figure 5 Simulated annual adult abundances of the species in each of the six food webs. (a) 1-species; (b) high-biomass, 2-species; (c) low-biomass, 2-species; (d) 3-species; (e) high-biomass, 6-species; and (f) low-biomass, 6-species. Yellow perch (●); walleye (○); bluegill (△); northern pike (□); white bass (◇); and cisco (▽). Symbols are shown every 10 years (Modified from McDermot 1998).

1998). McDermot and Rose (2000) present a detailed description of the model and analysis of the two of six food webs. All six of the food webs included yellow perch; two versions of the 6-species and 2-species food webs (denoted high and low biomass) were configured such that they differed in their average yellow perch biomass and degree of predator-prey coupling. Differences in the feeding-related parameters of walleye and pike resulted in competition among the yellow perch and other planktivores being more important in the high-biomass versions, while piscivore predation was more important in controlling the planktivores in the low-biomass versions. All six of the food webs shared the same environmental conditions, which were based on data from Lake Mendota, Wisconsin.

The six different food webs were standardised to permit cross-food web comparisons. Selected parameters were calibrated for each version until: (i) all species in the food web persisted for 100 years; (ii) mean lengths of adults (age-5) for all species were biologically reasonable; and (iii) average yellow perch adult biomass was similar to the value from the low-biomass 6-species food web. The 6-food webs used here are similar but not identical to those reported in McDermot (1998); some of the Northern pike parameter values were adjusted. Annual adult abundances for each of the food webs used in this paper are shown in Fig. 5. We examined the relationship between yearling survival and the number of entering yearlings for yellow perch in each of the 100-year simulations of the 6-food webs.

The strength of density-dependent mortality in the yearling stage varied among the different food webs (Fig. 6). Strong negative relationships ($r^2 = 0.34$ and 0.49) were observed in the 1-species (Fig. 6a) and the high biomass, 6-species (Fig. 6e) food webs. A statistically significant, but very weak, negative relationship ($r^2 = 0.04$) was observed in the high biomass, 2-species food web (Fig. 6b), and yearling survival was unrelated to the number of entering yearlings for the 3-species food web (Fig. 6d). Yearling survival actually increased with increasing numbers of entering yearlings for the low biomass 2-species (Fig. 6c) and the low biomass 6-species (Fig. 6f) food webs.

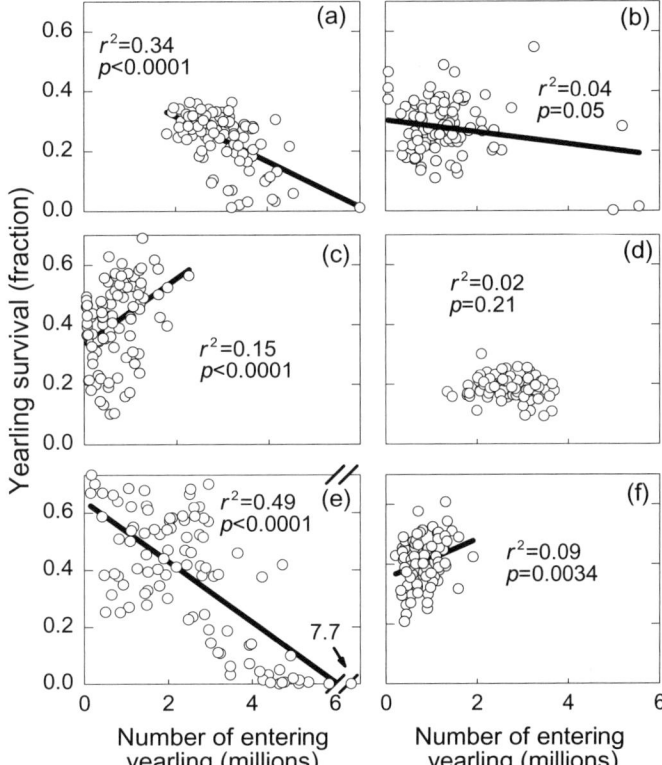

Figure 6 Simulated relationship between annual values of yearling yellow perch stage survival and the numbers of yearlings entering. Values are from the 100-year simulations shown in Fig. 6. Regression lines, and associated r^2 and P-values, are shown. (a) 1-species; (b) high biomass, 2-species; (c) low biomass, 2-species; (d) 3-species; (e) high biomass, 6-species; (f) low biomass, 6-species.

With a little imagination, these 6-food webs can be viewed as yellow perch in different systems or as yellow perch in a single system that is changing over time. Detecting density-dependent yearling mortality would then depend on the population studied, or on quantifying density dependence from a mixture of the relationships observed in various food webs.

Bay anchovy in Chesapeake Bay

Bay anchovy has been the subject of much research (Houde and Zastrow 1991) and has been shown to be important in the Chesapeake Bay as forage for striped bass, weakfish (*Cynoscion regalis*) and bluefish (*Pomatomus saltatrix*) (Hartman and Brandt 1995), and to the overall energy cycling of the bay (Baird and Ulanowicz 1989). The available data on bioenergetics, growth, survival and reproduction were sufficient to develop and verify an individual-based model of the bay anchovy population in the mesohaline region of Chesapeake Bay (Rose *et al*. 1999).

Cowan *et al*. (1999) used the model to examine the maximum compensatory response of bay anchovy, measured in terms of the largest amount of additional larval mortality that could be offset by compensatory increases in survival or reproduction in other life stages. First, a baseline simulation was performed in which the average fecundity, size-at-age, and population abundance closely matched values observed for the Chesapeake Bay. The maximum compensatory response of the population was then simulated by assuming that fecundity, egg mortality, and juvenile and adult mortality would all respond to lowered anchovy densities. The magnitude of the changes in each of these parameters was derived from an extensive review of the literature on bay anchovy population dynamics (Cowan *et al*. 1999). It was assumed that decreased density of bay anchovy would result in (i) an increase in reproduction by 3 batches per year (from 60 under baseline conditions to 63); (ii) a 10% increase in the number of eggs produced per gram of spawning anchovy; (iii) a decrease in egg mortality rate from 0.2 to 0.1 day^{-1}; and (iv) 10% decreases in the daily mortality rates of juvenile and adult bay anchovy.

A total of eight 50-year simulations were performed under maximum compensation conditions,

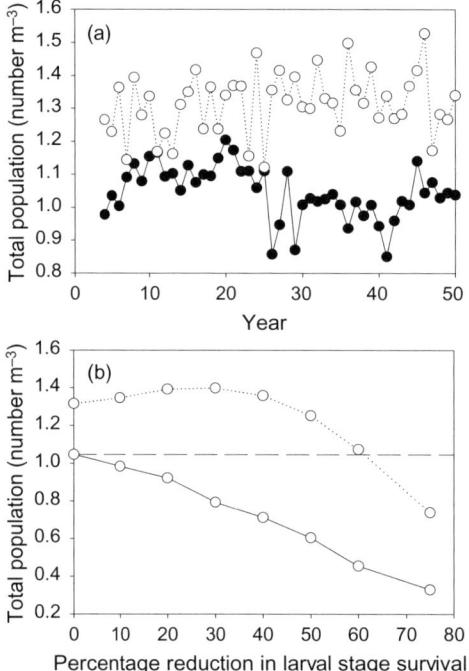

Figure 7 (a) Predicted annual total population density of bay anchovy for the baseline (●) and maximum compensation (○) simulations. (b) Average total population density (computed over the 50 years) for simulations with 0, 10, 20, 30, 40, 50, 60 and 75% reductions in larval stage survival. Maximum compensation conditions result from the imposition on baseline conditions of 3 more spawning batches per year, a 10% increase in individual fecundity, egg mortality rate being decreased from 0.2 to 0.1 day^{-1}, and a 10% decrease in the daily mortality rates of juvenile and adult bay anchovy. Note the different vertical scales on panels (a) and (b). (Modified from Cowan et al. 1999).

and average population abundance was computed for each simulation. The eight simulations corresponded to larval stage survival being reduced by 0% (baseline), 10, 20, 30, 40, 50, 60 and 75%; a 75% reduction meant that larval survival was 25% of the baseline value. Predicted average population abundances were compared with average abundance predicted under baseline conditions. The purpose of the simulations was to see how much larval survival had to be reduced before the compensatory reserve in the population was exceeded, and before predicted population abundances would drop below the baseline value.

The assumed compensatory increases in survival and reproduction increased the simulated density of the bay anchovy by 30% compared with the baseline simulation. Predicted population density under maximum compensation were higher than under baseline conditions (0.87 m^{-3} vs. 0.67 m^{-3}; see Fig. 7a). Higher densities were predicted because of the increased fecundity and lowered mortality rates assumed under maximum compensation conditions. Under the assumed changes in processes, baseline or higher population densities were predicted for up to 60% reductions in larval survival (Fig. 7b). Only at the highest (75%) reduction in larval survival was the compensatory reserve from lowered mortality and increased reproduction exceeded, resulting in a still stable population but at an average density 25% lower than baseline. Whether the changes in processes assumed under maximum compensation would occur in a particular situation is not known, but the changes imposed were relatively small compared with the variation observed in these processes *in situ* (see Cowan et al. 1999). Thus, relatively small changes in fecundity and survival of bay anchovy, which would probably be undetectable in field monitoring amongst the variation from density-independent sources, combined to produce substantial compensation.

Cause for optimism: recent technological and theoretical advances

There have been several technological and theoretical advances in the past decade that make us optimistic that further progress on the compensation issue is possible. These advances are: otolith ageing, new methods for marking of individuals, physiological measures of fish health, microchemical analysis of skeletal material, genetic and molecular approaches to stock identification, recognition of the importance of spatial heterogeneity in population dynamics, new acoustic and video measurement methods, new statistical estimation and time series methods, incorporation of stochasticity into population- and stock-assessment models, tailoring of classical life-history theory to fishes, assembly of large databases, and individual-based modelling.

Some of these advances are directly related to better measurement and modelling of compensation. Other advances do not involve compensation directly, but help by improving our knowledge of the causes of population dynamics and variation. Some of these advances have already had impact on population dynamics, while others are still in their infancy and their full impacts have yet to been realised. Below we briefly discuss these advances.

The developments of otolith ageing, new methods for marking of individuals, physiological measures of fish health, and microchemical analysis of skeletal material have enabled improved estimation of mortality, growth, movement rates and stock structure. Otolith ageing has added a new layer of resolution (i.e. cohorts within a year) in our ability to tease apart year-class strength. By separating cohorts within a year-class, otolith analysis has enabled much greater understanding of density dependence of growth and mortality (Crecco et al. 1983; Crecco and Savoy 1987), and of how survivorship of individuals during the critical first year of life relates to temporal and spatial variation in spawning and environmental variables (Rice et al. 1987; Zastrow et al. 1991; Limburg et al. 1999). The traditional mark-recapture approach to estimating mortality and movement has been greatly expanded by the use of chemicals for mass marking (Ennevor 1994; Secor et al. 1995), tags that permit information on individual fish (Bryan and Ney 1994; Achord et al. 1996), and ultrasonic telemetry that allow continuous tracking of marked individuals (Szedlmayer 1997). The rapidly growing field of biological indicators (see Adams 1990; Holdway et al. 1995) has increased our ability to detect the presence of sublethal effects and environmental stressors affecting individuals in a population. Microchemical analysis of skeletal material can yield vital information on broad-scale movement patterns and stock structure (Campana et al. 1994; Thorrold et al. 1998; Campana 1999), and on pollution exposure (Coutant 1990). These new advances have the potential to greatly improve our ability to estimate mortality rates and habitat usage, and to identify possible environmental causes for population variability not apparent from traditional monitoring.

Developments in the field of genetics and molecular biology have greatly increased our ability to identify fish populations or stocks (Carvalho and Hauser 1994; Lincoln 1994; Ferguson et al. 1995). Density dependence on a local scale may or may not translate into population compensation, depending on the degree of mixing with individuals from other locations (Stepien 1995; Policansky and Magnuson 1998). Being able to define the spatial extent of the unit of the population seems fundamental to understanding population dynamics, but is a continuing problem in many fish population studies. Continued work in the genetics area, as well in microchemical analysis of otoliths (Campana 1999), may ultimately resolve one of the major issues hindering our understanding of fish population dynamics.

The importance of spatial heterogeneity to understanding density dependence and population dynamics is receiving increasing attention (Kareiva 1990a; Stewart-Oaten and Murdoch 1990; Dunning et al. 1995). Space has been called the final frontier in ecological theory (Kareiva 1990b). Explicit consideration of spatial variability is possible from developments in measurement and analysis methods. The general availability of global positioning systems has permitted much greater accuracy in determining where samples are collected than previously possible. Geographic information systems has provided a means for storage and interpretation of spatially explicit data (Keleher and Rahel 1996; Johnson and Gage 1997; Wiley et al. 1997; Clark et al. 2001). Remote sensing provides simultaneous measurement of environmental variables over km scales (Cole and McGlade 1998; Polovina et al. 1999). Recognising that fish population dynamics are influenced by large-scale oceanic patterns and decadal climate patterns such as El Nino-Southern Oscillation (Sinclair et al. 1985; Lluch-Belda et al. 1989, 1992; Lenarz et al. 1995; Gargett 1997) and the North Atlantic Oscillation (Hoffmann and Powell 1998), has aided in explaining variation and trends that may appear random on the local level. There is increasing interest in Lagrangian-type models, where individual fish are imbedded into spatially explicit hydrodynamics models (Walters et al. 1992; Hinckley et al. 1996; Werner et al. 1996; Heath and Gallego 1997).

Video and acoustic technologies are two examples of new measurement methods that have recently become available. Video and acoustic methods have been used to augment and calibrate traditional sampling gear (Luo and Brandt 1993; Banneheka et al. 1995; Greene et al. 1998), and to permit sampling in situations where traditional gear cannot (Parker et al. 1994). They have also been used to simultaneously monitor biological and environmental variables on a meter scale over large areas (Brandt and Mason 1994), and for recording of detailed behavioural interactions between individuals (Fuiman and Batty 1994; Barnett and Pankhurst 1996).

New statistical methods are available that are well suited for analysing fish population dynamics data. Statistical analysis in ecology has in general been moving from hypothesis testing and linear models to multiple hypothesis evaluation (e.g. likelihoods) and nonlinear models (Maurer 1998). New methods, such as generalised additive models (Swartzman et al. 1995; Daskalov 1999), nonlinear time series (Dixon et al. 1999), neural networks (Lek et al. 1996; Thorrold et al. 1998), fuzzy mathematics (Saila 1992; Mackinson et al. 1999), geostatistical methods (Fletcher and Summer 1999) and methods that explicitly account for sampling and measurement error (Mertz and Myers 1995), are being applied to fisheries-related datasets. Statistical techniques are now available for valid detection of density dependence in time series (den Boer and Reddingius 1989; Dennis and Taper 1994; Dennis et al. 1998; Bjornstad et al. 1999), but not without controversy (see Wolda et al. 1994). Recent advances also allow for much greater flexibility in time series model formulation (Dixon et al. 1999; Sanderson et al. 1999). A promising trend is the focus on the interaction between environmental stochasticity and density dependence, and how they combine to control the long-term dynamics of populations (Fogarty et al. 1991; Higgins et al. 1997; Hixon and Carr 1997). Bayesian approaches using maximum likelihood methods are being used to estimate the many unknown parameters in population dynamics and stock assessment models (Hilborn et al. 1994; McAllister and Kirkwood 1998; Schirripa and LeGault 1999). Synthetic analyses involving diverse studies can now be rigorously analysed statistically using meta-analysis methods (Hilborn and Liermann 1998; Myers and Mertz 1998).

Explicit treatment of stochasticity in population models has increased their realism. Population modelling has moved from deterministic models of simple equilibrium to nonequilibrium approaches that explicitly include stochasticity and uncertainty (Turchin 1995; Uchamnski and Grimm 1996). The definition of a regulated population has been expanded from simple statements about equilibrium densities to more encompassing definitions appropriate for highly stochastic populations, such as a bounded variance of population densities and a long-term stationary probability distribution of population densities (Turchin 1995). The influence of stochastic modelling can be seen in the increasing use of uncertainty and risk in stock assessment (Hilborn and Walters 1992) and in fisheries management (Rosenberg and Restrepo 1994; Francis and Shotton 1997). Embracing the stochasticity that is characteristic of almost all fish populations, rather than using models that attempt to average the variability away and produce precise but inaccurate predictions, will increase model realism and credibility.

Finally, three areas of advance emphasised in this paper are life-history theory, large databases and individual-based modelling. Winemiller and Rose (1992) reconfigured the traditional 'r and K' life-history framework to a 3-endpoint framework specifically designed for fishes. We encourage continued efforts in relating fish population responses to stress within a life-history framework (e.g. Garrod and Knights 1979; Schaaf et al. 1987; Armstrong and Shelton 1990); others have since expanded on the Winemiller and Rose framework (McCann and Shuter 1997). Significant progress is occurring on the assembly of large databases (e.g. Myers et al. 1995a; Froese and Pauly 2000) and synthetic analyses (e.g. Miller et al. 1988; Li et al. 1996; Shuter et al. 1998). Individual-based modelling offers a promising approach for modelling population and community dynamics (Huston et al. 1988; DeAngelis et al. 1994; Judson 1994), and has features that should help to quantify compensatory responses of fish populations. Representing local interactions in space, size–based interactions, episodic effects, movement and stochasticity, all of which are important to realistic simulation of fish population dynamics and compensation, is relatively easy in individual-based models (DeAngelis and Rose 1992; DeAngelis et al. 1994; Tyler and Rose 1994). Additional critical review (sensu Grimm 1999) is needed to further the usefulness of individual-based modelling. Intuitively, if one can realistically represent how individuals grow, survive, reproduce and move, then population-level phenomenon such as compensation can be obtained by simply summing over all of the individuals in the model. There are also exciting advances occurring within the individual-based approach, including the use of state-dependent methods to allow individuals to respond dynamically to changing internal and external conditions (e.g. Huse and Giske 1998; Railsback 2001). In this paper, we used a large database to test some of the predictions from life-history theory, and used individual-based models to illustrate several features of compensation.

We have briefly mentioned some of the technical and theoretical advances that make us optimistic that significant progress on the compensation issue is possible in the near future. Our list is not meant to be comprehensive, but rather to illustrate that we are at a point where diverse advances are providing an opportunity for a major leap in our understanding of compensation. The growing emphasis on synthetic and comparative analyses, coupled with advances in measurement, statistical, and population modelling methods, is encouraging and critical for progress in understanding compensation and in the effective management of fish populations.

Synthesis and prognosis

Compensation must be included in management analyses that involve long-term predictions of fish populations. The fact that some resource will eventually limit the size of a fish population is not disputed. We also know how to incorporate density dependence into population dynamics and stock assessment models. Difficulties and controversy arise when the magnitude of compensation and compensatory reserve must be specified for a particular population. The details of compensation, such as the mix of magnitude, life stages, and processes, that exhibit density dependence can greatly influence the resulting population response. The consequences of over- or under-estimating compensation and compensatory reserve are serious; these include risk of population decline, unnecessary spending of monetary resources, and unneeded restrictions on fishers and energy generation.

Analysis of a particular population becomes difficult because model predictions are very sensitive to how compensation is represented, direct measurement of compensation and compensatory reserve in the field require many years of data, and the mechanisms underlying compensation can be highly site-specific. The sensitivity of model predictions to assumptions about compensation is practically legendary (e.g. Barnthouse et al. 1984). Simply measuring the details of compensation in the field is very difficult. Long-term studies that span multiple generations and involve multiple processes and all life stages are needed. As shown with the empirical examples of compensation presented in this paper, even with adequate long-term data, tracking density dependence through the full life cycle in the midst of large variation owing to environmental variables often results in conjecture as to the underlying causes of the compensatory response. The yellow perch-walleye and bay anchovy modelling examples illustrated how extensive monitoring must be to detect density-dependent processes and to track compensatory responses through the life cycle. Finally, we used a food web model to demonstrate that compensation can be site specific, and, thus, that using information from other species and locations can be problematic.

Turchin (1995), in a review chapter on population dynamics in general, asks "Is there no end to the density dependence debate?" He then emphatically answers that "the fundamental issues of the population regulation debate have been resolved ... all recent empirical analyses agree that the frequency of detecting density dependence increases with the length of the data series. Thus, most field populations are regulated, and previous failures to show this were owing to inadequate data sets...." Unfortunately, this statement resembles similar earlier proclamations. For example, Royama began a 1977 paper with "A basic concept in many theories to explain the persistence of animal populations is the notion of density-dependent regulation, which is now widely accepted in spite of much controversial literature during the last half century." We suspect that as long as there is lack of definitive proof of how compensation is operating in a specific population, the debates over the long-term consequences of harvesting, power plant effects and other stressors will continue.

We advocate an approach to compensation that involves process-level understanding of compensation, life-history theory, synthetic analyses, and matrix projection and individual-based models. There will always be debate if the quantification of compensation, from any source, does not include some degree of understanding of the underlying mechanisms. Quantifying compensation without adequate understanding is analogous to correlation without supporting information on cause and effect. Estimates of compensation, like correlative relationships, will hold for a while, but will eventually fail. Understanding compensation and compensatory responses is the key to making progress on this important issue.

We also advocate viewing compensation and compensatory reserve in a life-history framework, and endorse the trend towards synthetic analysis of large databases. Life history theory provides the general framework so that one can quickly appreciate (albeit qualitatively) the scope of compensation and compensatory reserve generally associated with a given configuration of attributes. Life history theory also permits interpopulation and interspecific comparisons of the scope of compensatory responses and compensatory reserve. We hope our initial attempt at a synthetic analysis of steepness parameters by life-history strategies using the Myers group spawner-recruit database will stimulate further similar investigations. One area ripe for extending the life-history framework using synthetic analysis is the inclusion into the framework of the plasticity of life-history traits (e.g. Belk 1995; Rodd and Reznick 1997). The Winemiller and Rose framework does not incorporate differences in the degree of flexibility in life-history characteristics among species and strategies.

This flexibility or phenotypic plasticity is probably important for understanding compensatory responses. Numerous other extensions to the life-history framework and potential synthetic analyses can also be envisioned.

When modelling is required for prediction, we suggest using aggregate models for screening and using individual-based models for detailed analyses. Aggregate models, such as age- and stage-structured matrix projection models (Caswell 2000), provide a screening tool for determining populations at risk. Matrix projection models use fecundity and survival data that are generally available for many species, and can applied to populations using site-specific data. Individual-based models should be developed for select species, either the species shown to be potentially adversely impacted by the screening analyses or for which understanding how compensation works is important. Individual-based modelling provides a powerful way to model compensation that avoids the pitfalls of more aggregate approaches by allowing individuals in the model to respond to changing conditions. Matrix projection-type models are relatively easy to apply and analyse but require compensation to be specified *a priori*; individual-based models are well-suited for detailed modelling of compensation at the process level, but are data hungry. A major role of individual-based modelling is to understand compensation and thereby bound the magnitude of compensation that is assumed *a priori* in aggregate models.

We are optimistic, although a bit worried. Focus on how compensation is working in specific populations can lead to constructive debate and better understanding of population regulation. Also, we described some recent advances that have greatly improved our ability to quantify and understand compensatory density dependence in fish populations; these advances are continuing to be refined and new advances are on the horizon. Yet, we still suffer from a lack of a unified theoretical framework for understanding and modelling population dynamics (Getz 1998). Greater attention to understanding compensation and to synthetic analyses and life-history theory, coupled with these technical advances and efforts towards advancing ecological theory of population dynamics, should enable progress on the important issue of compensation. Now is the time for some soul searching and for asking why important fish populations are seemingly overfished worldwide. Demands on fish stocks will only increase in the future.

Acknowledgements

Financial support for the preparation of this paper was provided by the Electric Power Research Institute. Some of the ideas in this paper originated during discussions among the authors during their involvement with permit preparation activities for a Public Service Enterprise Group (PSEG) facility. The comments of Jamie Gibson on an earlier draft greatly improved the paper.

References

Achord, S., Matthews, G.M., Johnson, O.W. and Marsh, D.M. (1996) Use of passive integrated transponder (PIT) tags to monitor migration timing of Snake River chinook salmon smolts. *North American Journal of Fisheries Management* **16**, 302–313.

Adams, S.M. (1990) Status and use of biological indicators for evaluating the effects of stress on fish. *American Fisheries Society Symposium* **8**, 1–8.

Adlard, R.D. and Lester, R.J.G. (1994) Dynamics of the interaction between the parasitic isopod, *Anilocra pomacentri*, and the coral reef fish, *Chromis nitida. Parasitology* **109**, 311–324.

Alheit, J. (1987) Egg cannibalism versus egg predation: their significance in anchovies. *South African Journal of Marine Science* **5**, 467–470.

Almatar, S.M. and Bailey, R.S. (1989) Variation in the fecundity and egg weight of herring (*Clupea harengus* L.). Part 1. Studies in the Firth of Clyde and northern North Sea. *Journal du Conseil International pour l'Exploration de la Mer* **45**, 113–124.

Anderson, J.T. (1988) A review of size dependent survival during pre-recruit stages of fishes in relation to recruitment. *Journal of Northwest Atlantic Fisheries Science* **8**, 55–66.

Anderson, T.W. (2001) Predator responses, prey refuges, and density-dependent mortality of a marine fish. *Ecology* **82**, 245–257.

Armstrong, M.J. and Shelton, P.A. (1990) Clupeoid life-history styles in variable environments. *Environmental Biology of Fishes* **28**, 77–85.

Bagenal, T.B. (1973) Fish fecundity and its relations with stock and recruitment. *Rapports et Proces-Verbaux des Reunions, Conseil International pour l'Exploration de la Mer* **164**, 186–198.

Bailey, K.M. (1994) Predation on juvenile flatfish and recruitment variability. *Netherlands Journal of Sea Research* **32**, 175–189.

Bailey, K.M., Brodeur, R.D. and Hollowed, A.B. (1996) Cohort survival patterns of walleye pollack, *Theragra chalcogramma*, in Shelikof Strait, Alaska: a critical factor analysis. *Fisheries Oceanography* **5** (Suppl. 1), 179–188.

Bailey, K.M. and Houde, E.D. (1989) Predation on eggs and larvae of marine fishes and the recruitment problem. *Advances in Marine Biology* **25**, 1–83.

Baird, D. and Ulanowicz, R.E. (1989) The seasonal dynamics of the Chesapeake Bay ecosystem. *Ecological Monographs* **59**, 329–364.

Banneheka, S.G., Routledge. R.D., Guthrie, I.C. and Woodey, J.C. (1995) Estimation of in-river fish passage using a combination of transect and stationary hydroacoustic sampling. *Canadian Journal of Fisheries and Aquatic Sciences* **52**, 335–343.

Bannister, R.C.A. (1978) Changes in plaice stocks and plaice fisheries in the North Sea. *Rapports et Proces-Verbaus des Reunions, Conseil International pour l'Exploration de la Mer* **172**, 86–101.

Barnett, C.W. and Pankhurst, N.W. (1996) Effect of density on the reproductive behaviour of the territorial male demoiselle *Chromis dispilus* (Pisces: Pomocentridae). *Environmental Biology of Fishes* **46**, 343–349.

Barnthouse, L.W., Boreman, J., Christensen, S.W., Goodyear, C.P., Van Winkle, W. and Vaughan, D.S. (1984) Population biology in the courtroom: the Hudson River controversy. *Bioscience* **34**, 14–19.

Barnthouse, L.W., Klauda, R.J. and Vaughan, D.S. (1988) Introduction to the Monograph. *American Fisheries Society Monograph* **4**, 1–8.

Beall, E. and Marty, C. (1987) Optimization of the natural reproduction of Atlantic salmon in a spawning channel: effect of female density. In: *La Restauration Des Rivieres a Saumons (Restoration of Salmon Rivers)* (Proceedings of a Conference Held in Bergerac, France, May, 1987). M. Thibault and R. Billard, eds. Institut Natl. de la Recherche Agronomique, Jouy-en-Josas, France, pp. 231–238.

Belk, M.C. (1995) Variation in growth and age at maturity in bluegill sunfish: genetic or environmental effects? *Journal of Fish Biology* **47**, 237–247.

Bertram, D.F., Chambers, R.C. and Leggett, W.C. (1993) Negative correlations between larval and juvenile growth rates in winter flounder: implications of compensatory growth for variation in size-at-age. *Marine Ecology Progress Series* **96**, 209–215.

Beverton, R.J.H. and Holt, S.J. (1957) On the Dynamics of Exploited Fish Populations. *United Kingdom Ministry of Agriculture, Food, and Fisheries Investigations Series 2* **No. 19**, 533 pp.

Beverton, R.J.H. and Iles, T.C. (1992) Mortality rates of 0-group plaice (*Platessa platessa* L.), dab (*Limanda limanda* L.) and turbot (*Scophthalmus maximus* L.) in European Waters 3. Density-dependence of mortality rates and some demographic implications. *Netherlands Journal of Sea Research* **29**, 61–79.

Bjornstad, O.N., Fromentin, J.-M., Stenseth, N.C. and Gjosaeter, J. (1999) A new test for density-dependent survival: the case of coastal cod populations. *Ecology* **80**, 1278–1288.

den Boer, P.J. and Reddingius, J. (1989) On the stabilization of animal numbers. Problems of testing: 2. Confrontation with data from the field. *Oecologia* **79**, 143–149.

Botsford, L.W., Castilla, J.C. and Peterson, C.H. (1997) The management of fisheries and marine ecosystems. *Science* **277**, 509–515.

Bowen, S.H., D'Angelo, D.J., Arnold, S.H., Keniry, M.J. and Albrecht, R.J. (1991) Density-dependent maturation, growth, and female dominance in Lake Superior lake herring (*Coregonus artedii*). *Canadian Journal of Fisheries and Aquatic Sciences* **48**, 569–576.

Bradford, M.J. (1992) Precision of recruitment predictions from early life stages of marine fishes. *Fishery Bulletin* **90**, 439–453.

Bradford, R.G. and Stephenson, R.L. (1992) Egg weight, fecundity, and gonad weight variability among northwest Atlantic herring (*Clupea harengus*) populations. *Canadian Journal of Fisheries and Aquatic Sciences* **49**, 2045–2054.

Brandt, S.B. and Mason, D.M. (1994) Landscape approaches for assessing spatial patterns in fish foraging and growth. In: *Theory and Application of Fish Feeding Ecology* (eds D.J. Stouder, K.L. Fresh and R.J. Feller). University of South Carolina Press, Columbia, pp. 211–240.

Bromage, N. (1995) Broodstock management and seed quality – general considerations. In: *Broodstock Management and Egg and Larval Quality* (eds N.R. Bromage and R.J. Roberts). Blackwell Science, Ltd, Oxford, pp. 1–14.

Bromley, P.J. (1989) Evidence for density-dependent growth in North Sea gadoids. *Journal of Fish Biology* **35** (Suppl. A), 117–123.

Bryan, R.D. and Ney, J.J. (1994) Visible impact tag retention by and effects on condition of a stream population of brook trout. *North American Journal of Fisheries Management* **14**, 216–219.

Buckley, R.V. (1967) Fecundity of steelhead trout (*Salmo gairdneri*) from Alsea River, Oregon. *Journal of the Fisheries Research Board of Canada* **24**, 917–926.

Buckley, L.J., Smigielski, A.S., Halavik, T.A., Caldarone, E.M., Burns, B.R. and Laurence, G.C. (1991a) Winter flounder *Pseudopleuronectes americanus* reproductive success. I. Among location variability in size and survival of larvae reared in the laboratory. *Marine Ecology Progress Series* **74**, 117–124.

Buckley, L.J., Smigielski, A.S., Halavik, T.A., Caldarone, E.M., Burns, B.R. and Laurence, G.C. (1991b) Winter flounder *Pseudopleuronectes americanus* reproductive success. II. Effects of spawning time and female size on composition and viability of eggs and larvae. *Marine Ecology Progress Series* **74**, 125–135.

Butl, T.P., Riley, S.C., Haedrich, R.L., Gibson, R.J. and Heggenes, J. (1999) Density-dependent habitat selection by juvenile Atlantic salmon (*Salmo salar*) in experimental riverine habitats. *Canadian Journal of Fisheries and Aquatic Sciences* **56**, 1298–1306.

Bystrom, P. and Garcia-Berthou, E. (1999) Density dependent growth and size specific competitive interactions in young fish. *Oikos* **86**, 217–232.

Campana, S.E. (1999) Chemistry and composition of fish otoliths: Pathways, mechanisms and applications. *Marine Ecology Progress Series* **188**, 263–297.

Campana, S.E., Fowler, A.J. and Jones, C.M. (1994) Otolith elemental fingerprinting for stock identification of Atlantic cod (*Gadus morhua*) using laser ablation ICPMS. *Canadian Journal of Fisheries and Aquatic Sciences* **51**, 1942–1950.

Caron, J. and Beaugrand, J.P. (1988) Social and spatial structure in brook chars (*Salvelinus fontinalis*) under competition for food and shelter/shade. *Behavioral Processes* **16**, 173–191.

Carvalho, G.R. and Hauser, L. (1994) Molecular genetics and the stock concept in fisheries. *Reviews in Fish Biology and Fisheries* **4**, 326–350.

Caswell, H. (2000) *Matrix Population Models: Construction, Analysis, and Interpretation*, 2nd edn. Sinauer Associates, Sunderland, Massachusetts.

Chambers, R.C., Leggett, W.C. and Brown, J.A. (1989) Egg size, female effects, and the correlations between early life history traits of capelin, *Mallotus villosus*: an appraisal at the individual level. *Fishery Bulletin* **87**, 515–523.

Chapman, R.W., Sedberry, G.R., Koenig, C.C. and Eleby, B.M. (1999) Stock identification of Gag, *Mycteroperca microlepis*, along the southeast coast of the United States. *Marine Biotechnology* **1**, 137–146.

Christensen, S.W. and Goodyear, C.P. (1988) Testing the validity of stock-recruitment curve fits. *American Fisheries Society of Monograph* **4**, 219–231.

Clark, M.E., Rose, K.A., Levine, D.A. and Hargrove, W.W. (2001) Predicting climate change effects on Appalachian trout: combining GIS and individual-based modeling. *Ecological Applications* **11**, 161–178.

Cole, J. and McGlade, J. (1998) Clupeoid population variability, the environment and satellite imagery in coastal upwelling systems. *Reviews in Fish Biology and Fisheries* **8**, 445–471.

Coleman, F.C., Koenig, C.C. and Collins, L.A. (1996) Reproductive styles of shallow-water groupers (Pisces: Serranidae) in the eastern Gulf of Mexico and the consequences of fishing spawning aggregations. *Environmental Biology of Fishes* **47**, 129–141.

Collins, L.A., Johnson, A.G., Koenig, C.C. and Baker, M.S. (1998) Reproductive patterns, sex ratio, and fecundity in gag, *Mycteroperca microlepis* (Serranidae), a protogynous grouper from the northeastern Gulf of Mexico. *Fishery Bulletin* **96**, 415–427.

Coutant, C.C. (1990) Microchemical analysis of fish hard parts for reconstructing habitat: practice and promise. *American Fisheries Society Symposium* **7**, 574–580.

Cowan, J.H. Jr, Rose, K.A. and DeVries, D.R. (2000) Is density-dependent growth in young-of-the-year fishes a question of critical weight? *Reviews in Fish Biology and Fisheries* **10**, 61–89.

Cowan, J.H. Jr, Rose, K.A., Houde, E.D., Wang, S.-B. and Young, J. (1999) Effects of increased larval mortality on bay anchovy population dynamics in mesohaline Chesapeake Bay: evidence of compensatory reserve. *Marine Ecology Progress Series* **185**, 133–146.

Craig, J.F. and Kipling, C. (1983) Reproductive effort versus the environment; case histories of Windemere perch, *Perca fluviatilis* L. and pike, *Esox lucius* L. *Journal of Fish Biology* **22**, 713–727.

Crecco, V. and Savoy, T. (1987) Effects of climatic and density-dependent factors on intra-annual mortality of larval American shad. *American Fisheries Society Symposium* **2**, 69–81.

Crecco, V., Savoy, T. and Gunn, L. (1983) Daily mortality rates of larval and juvenile American shad (*Alosa sapidissima*) in the Connecticut River with changes in year-class strength. *Canadian Journal of Fisheries and Aquatic Sciences* **40**, 1719–1728.

Crozier, W.W. and Kennedy, G.L.A. (1995) The relationship between a summer fry (0+) abundance index, derived from semi-quantitative electrofishing, and egg deposition of Atlantic salmon, in the River Bush, Northern Ireland. *Journal of Fish Biology* **47**, 1055–1062.

Cushing, D.H. (1975) *Marine Ecology and Fisheries*. Cambridge University Press, Cambridge, UK.

Dahlberg, M.D. (1979) A review of survival rates of fish eggs and larvae in relation to impact assessments. *Marine Fisheries Review*, **March**, 1–12.

Daskalov, G. (1999) Relating fish recruitment to stock biomass and physical environment in the Black Sea using generalized additive models. *Fisheries Research* **41**, 1–23.

DeAngelis, D.L. and Rose, K.A. (1992) Which individual-based approach is most appropriate for a given problem? In: *Individual-Based Approaches in Ecology: Populations, Communities, and Ecosystems* (eds D.L. DeAngelis and L.J. Gross). Routledge, Chapman, and Hall, New York, pp. 67–87.

DeAngelis, D.L., Rose, K.A., Crowder, L., Marschall, E. and Lika, D. (1993) Fish cohort dynamics: application of complementary modeling approaches. *American Naturalist* **142**, 604–622.

DeAngelis, D.L., Rose, K.A. and Huston, M.A. (1994) Individual-oriented approaches to modeling populations and communities. In: *Frontiers in Mathematical Biology, Lecture Notes in Biomathematics*, Vol. 100 (ed. S.A. Levin). Springer-Verlag, New York, pp. 390–410.

Delahunty, G. and deVlaming, V.L. (1980) Seasonal relationships of ovary weight, liver weight and fat stores with body weight in the gold fish (*Carassius auratus*). *Journal of Fish Biology* **16**, 5–13.

DeLeo, G.A. and Gatto, M. (1996) Trends in vital rates of European eel: evidence for density dependence? *Ecological Applications* **6**, 1281–1294.

Dennis, B., Kemp, W.P. and Taper, M.L. (1998) Joint density dependence. *Ecology* **79**, 426–411.

Dennis, B. and Taper, M.L. (1994) Density dependence in time series observations of natural populations: estimation and testing. *Ecological Monographs* **64**, 205–224.

Dixon, P.A., Milicich, M.J. and Sugihara, G. (1999) Episodic fluctuations in larval supply. *Science* **283**, 1528–1530.

Dunning, J.B., Stewart, D.J., Danielson, B.J., Noon, B.R., Root, T.L., Lamberson, R.H. and Stevens, E.E. (1995) Spatially explicit population models: current forms and future uses. *Ecological Applications* **5**, 3–11.

Elliott, J.M. (1994) *Quantitative Ecology and the Brown Trout*. Oxford University Press, Oxford.

Ennevor, B.C. (1994) Mass marking coho salmon, *Oncorhynchus kisutch*, fry with lanthanum and cerium. *Fishery Bulletin* **92**, 471–473.

Erickson, D.L., Hightower, J.E. and Grossman, G.D. (1985) The relative gonadal index: an alternative index for quantification of reproductive condition. *Comparative Biochemistry and Physiology* **A 81**, 117–120.

Fausch, K.D. (1984) Profitable stream positions for salmonids: relating specific growth rate to net energy gain. *Canadian Journal of Zoology* **62**, 441–451.

Fausch, K.D. (1988) Tests of competition between native and introduced salmonids in streams: what have we learned? *Canadian Journal of Fisheries and Aquatic Sciences* **45**, 2238–2246.

Fenton, R., Mathias, J.A. and Moodie, G.E.E. (1996) Recent and future demand for walleye in North America. *Fisheries (Bethesda)* **21**, 6–12.

Ferguson, A., Taggart, J.B., Prodohl, P.A., McMeel, O., Thompson, C., Stone, C., McGinnity, P. and Hynes, R.A. (1995) The application of molecular markers to the study and conservation of fish populations, with special reference to Salmo. *Journal of Fish Biology* **47** (Suppl. A), 103–126.

Fletcher, R.I. and Deriso, R.B. (1988) Fishing in dangerous waters: remarks on a controversial appeal to spawner-recruit theory for long-term impact assessment. *American Fisheries Society Monograph* **4**, 232–244.

Fletcher, W.J. and Summer, N.R. (1999) Spatial distribution of sardine (*Sardinops sagax*) eggs and larvae: an application of geostatistics and resampling to survey data. *Canadian Journal of Fisheries and Aquatic Sciences* **56**, 907–914.

Fogarty, M.J., Rosenberg, A.A. and Sissenwine, M.P. (1992) Fisheries risk assessment: a case study of Georges Bank haddock. *Environmental Science and Technology* **26**, 440–447.

Fogarty, M.J., Sissenwine, M.P. and Cohen, E.B. (1991) Recruitment variability and the dynamics of exploited marine populations. *Trends in Ecology and Evolution* **6**, 241–246.

Forney, J.J. (1977) Evidence of inter- and intraspecific competition as factors regulating walleye (*Stizostedion vitreum vitreum*) biomass in Oneida Lake, New York. *Journal of the Fisheries Research Board of Canada* **34**, 1812–1820.

Forrester, G.E. (1995) Strong density-dependent survival and recruitment regulate the abundance of a coral reef fish. *Oecologia* **103**, 275–282.

Forrester, G.E. and Steele, M.A. (2000) Variation in the presence and cause of density-dependent mortality in three species of reef fishes. *Ecology* **81**, 2416–2427.

Francis, R.I.C.C. and Shotton, R. (1997) 'Risk' in fisheries management: a review. *Canadian Journal of Fisheries and Aquatic Sciences* **54**, 1699–1715.

Froese, R. and Pauly, D. (2000) *FishBase 2000: Concepts, Design, and Data Sources*. ICLARM, Laguna, Philippines, 344 pp.

Fuiman, L.A. and Batty, R.S. (1994) Susceptibility of Atlantic herring and plaice larvae to predation by juvenile cod and herring at two constant temperatures. *Journal of Fish Biology* **44**, 23–34.

Fuiman, L.A. and Magurran, A.E. (1994) Development of predator defences in fishes. *Review in Fish Biology and Fisheries* **4**, 145–183.

Funakoshi, S. (1992) Relationship between stock levels and the population structure of the Japanese anchovy. *Marine Behaviour and Physiology* **21**, 1–84.

Garcia, S.M. and Newton, C. (1997) Current situation, trends, and prospects in world capture fisheries. In: *Global Trends: Fisheries Management* (eds E.K. Pikitch, D.D. Huppert and M.P. Sissenwine), Symposium 20. American Fisheries Society, Bethesda, Maryland, pp. 3–27.

Gargett, A.E. (1997) The optimal stability 'window': a mechanism underlying decadal fluctuations in North Pacific salmon stocks? *Fisheries Oceanography* **6**, 109–117.

Garrod, D.J. and Knights, B.J. (1979) Fish stocks: their life-history characteristics and response to exploitation. *Symposium of the Zoological Society of London* **44**, 361–382.

Getz, W.M. (1998) Thinking of biology: an introspection on the art of modeling in population ecology. *Bioscience* **48**, 540–552.

Getz, W.M. and Swartzman, G.L. (1981) A probability transition matrix model for yield estimation in fisheries with highly variable recruitment. *Canadian Journal of Fisheries and Aquatic Sciences* **38**, 847–855.

Ginzburg, L.R., Ferson, S. and Akcakaya, H.R. (1990) Reconstructibility of density dependence and the conservative assessment of extinction risks. *Conservation Biology* **4**, 63–71.

Goodyear, C.P. (1980) Compensation in fish populations. In: *Biological Monitoring of Fish* (eds C.H. Hocutt and J.R. Stauffer). Lexington Books, Lexington, Massachusetts, pp. 253–280.

Goodyear, C.P. (1993) Spawning stock biomass per recruit in fisheries management: foundation and current use. In: *Risk Descriptive Evaluation and Biological Reference Points for Fisheries Management* (eds S.J. Smith, J.J. Hunt and D. Rivard), Canadian Special Publication in Fisheries and Aquatic Sciences No. 120, National Research Council of Canada, Ottawa, pp. 67–81.

Goodyear, C.P. and Christensen, S.W. (1984) On the ability to detect the influence of spawning stock on recruitment. *North American Journal of Fisheries Management* **4**, 186–193.

Grant, J.W.A. and Kramer, D.L. (1990) Territory size as a predictor of the upper limit to population density of juvenile salmonids in streams. *Canadian Journal of Fisheries and Aquatic Sciences* **47**, 1724–1737.

Grant, J.W.A., Noakes, D.L.G. and Jonas, K.M. (1989) Spatial distribution of defence and foraging in young-of-the-year brook charr, *Salvelinus fontinalis*. *Journal of Animal Ecology* **58**, 773–784.

Greene, C.H., Wiebe, P.H., Pershing, A.J. *et al.* (1998) Assessing the distribution and abundance of zooplankton: a comparison of acoustic and net-sampling methods with D-BAD MOCNESS. *Deep-Sea Research*, **45** (Part II), 1219–1237.

Grimm, V. (1999) Ten years of individual-based modelling in ecology: what have we learned and what could we learn in the future? *Ecological Modelling* **115**, 129–148.

Gulland, J.A. (1977) *The Management of Marine Fisheries*. University of Washington Press, Seattle, Washington.

Hamrin, S.F. and Persson, L. (1986) Asymmetrical competition between age classes as a factor causing population oscillations in a obligate planktivorous fish species. *Oikos* **47**, 223–232.

Hartman, K.J. and Brandt, S.B. (1995) Predatory demand and impact of striped bass, bluefish, and weakfish in the Chesapeake Bay: applications of bioenergetics models. *Canadian Journal of Fisheries and Aquatic Sciences* **52**, 1667–1687.

Hassell, M.P. (1978) *Arthropod Predator-Prey Systems*. Princeton University Press, Princeton, New Jersey.

Healey, M.C. (1978) Fecundity changes in exploited populations of lake whitefish (*Coregonus clupeaformis*) and lake trout (*Salvelinus namaycush*). *Journal of the Fisheries Research Board of Canada* **35**, 945–950.

Healey, M.C. (1980) Growth and recruitment in experimentally exploited lake whitefish (*Coregonus clupeaformis*) populations. *Canadian Journal of Fisheries and Aquatic Sciences* **37**, 255–267.

Heath, M.R. (1992) Field investigations of the early life stages of marine fish. *Advances in Marine Biology* **28**, 1–174.

Heath, M. and Gallego, A. (1997) From the biology of the individual to the dynamics of the population: bridging the gap in fish early life studies. *Journal of Fish Biology* **51** (Suppl. A), 1–29.

Helminen, H., Auvinen, H., Hirvonen, A., Sarvala, J. and Toivonen, J. (1993) Year-class fluctuations of vendace (*Coregonus albula*) in Lake Pyhajarvi, southwest Finland, during 1971–90. *Canadian Journal of Fisheries and Aquatic Sciences* **50**, 925–931.

Helminen, H. and Sarvala, J. (1994) Population regulation of vendace (*Coregonus albula*) in Lake Pyhajarvi, southwest Finland. *Journal of Fish Biology* **45**, 387–400.

Helminen, H., Sarvala, J. and Hirvonen, A. (1990) Growth and food consumption of vendace (*Coregonus albula* L.) in Lake Pyhajarvi, SW Finland: a bioenergetics modeling analysis. *Hydrobiologia* **200/201**, 511–522.

Helser, T.E. and Brodziak, J.K.T. (1998) Impacts of density-dependent growth and maturation on assessment advice to rebuild depleted U.S. silver hake (*Merluccius bilinearis*) stocks. *Canadian Journal of Fisheries and Aquatic Sciences* **55**, 882–892.

Henderson, B.A., Wong, J.L. and Nepszy, S.J. (1996) Reproduction of walleye in Lake Erie: allocation of energy. *Canadian Journal of Fisheries and Aquatic Sciences* **53**, 127–133.

Higgins, K., Hastings, A., Sarvela, J.N. and Botsford, L.W. (1997) Stochastic dynamics and deterministic skeletons: population behavior of dungeness crab. *Science* **276**, 1431–1435.

Hilborn, R. and Liermann, M. (1998) Standing on the shoulders of giants: learning from experience in fisheries. *Reviews in Fish Biology and Fisheries* **8**, 273–283.

Hilborn, R., Pikitch, E.K. and McAllister, M.K. (1994) A Bayesian estimation and decision analysis for an age-structured model using biomass survey data. *Fisheries Research* **19**, 17–30.

Hilborn, R. and Walters, C.J. (1992) *Quantitative Fisheries Stock Assessment: Choice, Dynamics, and Uncertainty*. Chapman and Hall, New York.

Hinckley, S., Hermann, A.J. and Megrey, B.A. (1996) Development of a spatially explicit, individual-based model of marine fish early life history. *Marine Ecology Progress Series* **139**, 47–68.

Hislop, J.R.G. (1988) The influence of maternal length and age on the size and weight of the eggs and the relative fecundity of the haddock, *Melanogrammus aeglefinus*, in British waters. *Journal of Fish Biology* **32**, 923–930.

Hixon, M.A. (1998) Population dynamics of coral-reef fishes: Controversial concepts and hypotheses. *Australian Journal of Ecology* **23**, 192–201.

Hixon, M.A. and Carr, M.H. (1997) Synergistic predation, density dependence, and population regulation in marine fish. *Science* **277**, 949–949.

Hoffmann, E.E. and Powell, T.M. (1998) Environmental variability effects on marine fisheries: four case histories. *Ecological Applications* **8** (Suppl.), S23–S32.

Holdway, D.A., Brennan, S.E. and Ahokas, J.T. (1995) Short review of selected fish biomarkers of xenobiotic exposure with an example using fish hepatic mixed-function oxidase. *Australian Journal of Ecology* **20**, 34–44.

Houde, E.D. (1987) Fish early life dynamics and recruitment variability. *American Fisheries Society Symposium* **2**, 17–29.

Houde, E.D. (1989) Comparative growth, mortality, and energetics of marine fish larvae: temperature and implied latitudinal effects. *Fishery Bulletin* **87**, 471–495.

Houde, E.D. and Zastrow, C.E. (1991) Bay anchovy, *Anchoa mitchilli*. In: *Habitat Requirements for Chesapeake Bay Living Resources* (eds S.L. Funderburk, J.A. Mihursky, S.J. Jordan and D. Riley), 2nd edn. Living Resources Subcommittee, Chesapeake Bay Program, Annapolis, Maryland, pp. 8.1–8.14.

Hovenkamp, F. (1992) Growth-dependent mortality of larval plaice *Pleuronectes platessa* in the North Sea. *Marine Ecology Progress Series* **82**, 95–101.

Hughes, N.F. (1992) Selection of positions by drift feeding salmoninds in dominance hierarchies: model and test for

Arctic grayling (*Thymallus arcticus*) in subarctic mountain streams, interior Alaska. *Canadian Journal of Fisheries and Aquatic Sciences* **47**, 2039–2048.

Hughes, N.F. (1998) A model of habitat selection by drift-feeding stream salmonids at different scales. *Ecology* **79**, 281–294.

Hunter, J.R. and Kimbrell, C.A. (1980) Egg cannibalism in the northern anchovy, *Engraulis mordax*. *Fishery Bulletin* **78**, 811–816.

Huse, G. and Giske, J. (1998) Ecology in Mare Pentium: an individual-based spatio-temporal model for fish with adapted behaviour. *Fisheries Research* **37**, 163–178.

Huston, M., DeAngelis, D. and Post, W. (1988) New computer models unify ecological theory. *Bioscience* **38**, 682–691.

Ivlev, V.S. (1961) *Experimental Ecology of the Feeding of Fishes*. Yale University Press, New Haven, Connecticut.

Jensen, A.L. (1993) Dynamics of fish populations with different compensatory processes when subjected to random survival of eggs and larvae. *Ecological Modelling* **68**, 249–256.

Johnson, T.B. and Evans, D.O. (1990) Size-dependent winter mortality of young-of-the-year white perch: climate warming and invasion of the Laurentian Great Lakes. *Transactions of the American Fisheries Society* **119**, 301–313.

Johnson, L.B. and Gage, S.H. (1997) Landscape approaches to the analysis of aquatic ecosystems. *Freshwater Biology* **37**, 113–132.

Jude, D.J., Mansfield, P.J., Schneeberger, P.J. and Wojcik, J.A. (1987) *Compensatory Mechanisms in Fish Populations, Vol. 2: Compensation in Fish Populations Subject to Catastrophic Impact*. Electric Power Research Institute Report EA-5200, Electric Power Research Institute, Palo Alto, California.

Judson, O.P. (1994) The rise of the individual-based model in ecology. *Trends in Ecology and Evolution* **9**, 9–14.

Kareiva, P. (1990a) Population dynamics in spatially complex environments: theory and data. *Philosophical Transactions of the Royal Society of London, Series B* **330**, 175–190.

Kareiva, P. (1990b) Space: the final frontier for ecological theory. *Ecology* **75**, 1.

Keleher, C.J. and Rahel, F.J. (1996) Thermal limits to salmonid distributions in the Rocky Mountain region and potential habitat loss due to global warming: a geographic information system (GIS) approach. *Transactions of the American Fisheries Society* **125**, 1–13.

Kjesbu, O.S. (1989) The spawning activity of cod *Gadus morhua* L. *Journal of Fish Biology* **34**, 195–206.

Kjesbu, O.S., Kryvi, H., Sundby, S. and Solemdal, P. (1992) Buoyancy variations in eggs of Atlantic cod (*Gadus morhua* L.) in relation to chorion thickness and egg size: theory and observations. *Journal of Fish Biology* **41**, 581–599.

Kjorsvik, E., Mangor-Jensen, A. and Holmefjord, I. (1990) Egg quality in fishes. *Advances in Marine Biology* **26**, 71–113.

Koenig, C.C., Coleman, F.C., Collins, L.A., Sadovy, Y. and Colin, P.L. (1996) Reproduction in gag (*Mycteroperca microlepis*) in the eastern Gulf of Mexico and the consequences of fishing spawning aggregations. In: *Biology, Fisheries and Culture of Tropical Groupers and Snappers* (ICLARM Conference Proceedings 48, Laguna, Philippines. International Workshop on Tropical Snappers and Groupers. 26–29 October 1993, Campeche, Mexico). F. Arreguin-Sanchez, J.L. Munro, M.C. Balgos and D. Pauly, eds. ICLARM, Manila, Philippines, pp. 307–323.

Koslow, J.A. (1992) Fecundity and the stock-recruitment relationship. *Canadian Journal of Fisheries and Aquatic Sciences* **49**, 210–217.

Koslow, J.A., Bell, J., Virtue, P. and Smith, D.C. (1995) Fecundity and its variability in orange roughy: effects of population density, condition, egg size, and senescence. *Journal of Fish Biology* **47**, 1063–1080.

Laevastu, T. (1993) *Marine Climate, Weather and Fisheries*. Halsted Press, New York.

de Lafontaine and Leggett, W.C. (1987a) Evaluation of situ enclosures for larval fish studies. *Canadian Journal of Fisheries and Aquatic Sciences* **44**, 54–65.

de Lafontaine and Leggett, W.C. (1987b) Effect of container size on estimates of mortality and predation rates in experiments with macrozooplankton and larval fish. *Canadian Journal of Fisheries and Aquatic Sciences* **44**, 1534–1543.

Lasker, R.E. (1985) What limits clupeoid production? *Canadian Journal of Fisheries and Aquatic Sciences* **42** (Suppl. 1), 31–38.

Lasker, R.E. and Theilacker, G.H. (1962) The fatty acid composition of the lipids of some Pacific sardine tissues in relation to ovarian maturation and diet. *Journal of Liposome Research* **3**, 60–64.

Lek, S., Delacoste, M., Baran, P., Dimopoulos, I., Lauga, J. and Aulagnier, S. (1996) Application of neural networks to modelling nonlinear relationships in ecology. *Ecological Modelling* **90**, 39–52.

Lenarz, W.H., Schwing, F.B., Ventresca, D.A., Chavez, F. and Graham, W.M. (1995) Explorations of El Nino events and associated biological population dynamics off central California. *Reports of California Cooperative Oceanic Fisheries Investigations* **36**, 106–119.

Li, J., Cohen, Y., Schupp, D.H. and Adelman, I.R. (1996) Effects of walleye stocking on year-class strength. *North American Journal of Fisheries Management* **16**, 840–850.

Liermann, M. and Hilborn, R. (1997) Depensation in fish stocks: a hierarchic Bayesian meta-analysis. *Canadian Journal of Fisheries and Aquatic Sciences* **54**, 1976–1984.

Liermann, M. and Hilborn, R. (2001) Depensation: evidence, models and implications. *Fish and Fisheries* **2**, 33–58.

Limburg, K.E., Pace, M.L. and Arend, K.K. (1999) Growth, mortality, and recruitment of larval *Morone* spp. in relation to food availability and temperature in the Hudson River. *Fishery Bulletin* **97**, 80–91.

Lincoln, R. (1994) Molecular genetics applications in fisheries: Snake oil or restorative? *Reviews in Fish Biology and Fisheries* **4**, 389–392.

Lindenmayer, D.B., Lacy, R.C. and Pope, M.L. (2000) Testing a simulation model for population viability analysis. *Ecological Applications* **10**, 580–597.

Lluch-Belda, D., Crawford, R.J.M., Kawasaki, T., MacCall, A.D., Parrish, R.H., Schwartzlose, R.A. and Smith, P.E. (1989) World-wide fluctuations of sardine and anchovy stocks: the regime problem. *South African Journal of Marine Science* **8**, 195–205.

Lluch-Belda, D., Schwartzlose, R.A., Serra, R., Parrish, R., Kawasaki, T., Hedgcock, D. and Crawford, R.J.M. (1992) Sardine and anchovy regime fluctuations of abundance in four regions of the world oceans: a workshop report. *Fisheries Oceanography* **1**, 339–347.

Lobon-Cervia, J., Montanes, C. and de Sosta, A. (1986) Reproductive ecology and growth of a population of brown trout (*Salmo trutta* L.) in an aquifer-fed stream of Old Castile (Spain). *Hydrobiologia* **135**, 81–94.

Lockwood, S.J. (1980) Density-dependent mortality in 0-group plaice (*Pleuronectes platessa* L.) populations. *Journal du Conseil International pour l'Exploration de la Mer* **39**, 148–153.

Ludwig, D., Hilborn, R. and Walters, C. (1993) Uncertainty, resource exploitation, and conservation: lessons from history. *Science* **260**, 17 and 36.

Lundvall, D., Svanback, R., Persson, L. and Bystrom, P. (1999) Size-dependent predation in piscivores: interactions between predator foraging and prey avoidance abilities. *Canadian Journal of Fisheries and Aquatic Sciences* **56**, 1285–1292.

Luo, J. and Brandt, S.B. (1993) Bay anchovy *Anchoa mitchilli* production and consumption in mid-Chesapeake Bay based on a bioenergetics model and acoustic measures of fish abundance. *Marine Ecology Progress Series* **98**, 223–236.

Lutnesky, M.M.F. (1994) Density-dependent protogynous sex change in territorial-haremic fishes: models and evidence. *Behavioral Ecology* **5**, 375–383.

Macaluso, J. (1999) Red snapper season offered. *The Advocate*, November 14, 1999, Baton Rouge, LA.

MacCall, A.D. (1981) The consequences of cannibalism in the stock-recruitment relationship of planktivorous pelagic fishes such as *Engraulis*. *IOC Workshop Report* **28**, 201–220.

MacCall, A.D. (1990) *Dynamic Geography of Marine Fish Populations*. University of Washington Press, Seattle, Washington, 153 pp.

Mace, P.M. (1994) Relationships between common biological reference points used as thresholds and targets of fisheries management strategies. *Canadian Journal of Fisheries and Aquatic Sciences* **51**, 110–122.

Mace, P.M. and Sissenwine, M.P. (1993) How much spawning per recruit is enough? *Risk Evaluation and Biological Reference Points for Fisheries Management* (eds S.J. Smith, J.J. Hunt and D. Rivard). *Canadian Special Publication in Fisheries and Aquatic Sciences* No. 120, National Research Council of Canada, Ottawa, 101–118.

MacKinnon, J.C. (1972) Summer storage of energy and its use for winter metabolism and gonad maturation in American plaice (*Hippoglossoides platessoides*). *Journal of the Fisheries Research Board of Canada* **29**, 1749–1759.

Mackinson, S., Vasconcellos, M. and Newlands, N. (1999) A new approach to the analysis of stock-recruitment relationships: 'model-free estimation' using fuzzy logic. *Canadian Journal of Fisheries and Aquatic Sciences* **56**, 686–699.

Marschall, E.A. and Crowder, L.B. (1996) Assessing population responses to multiple anthropogenic effects: a case study with brook trout. *Ecological Applications* **6**, 152–167.

Marshall, C.T. and Frank, K.T. (1999a) The effect of interannual variation in growth and condition on haddock recruitment. *Canadian Journal of Fisheries and Aquatic Sciences* **56**, 347–355.

Marshall, C.T. and Frank, K.T. (1999b) Implications of density-dependent juvenile growth for compensatory recruitment regulation of haddock. *Canadian Journal of Fisheries and Aquatic Sciences* **56**, 356–363.

Mathur, D., Bason, W.H., Purdy, E.J. and Silver, C.A. (1985) A critique of the instream flow incremental methodology. *Canadian Journal of Fisheries and Aquatic Sciences* **42**, 825–831.

Maunder, M.N. (1997) Investigation of density dependence in salmon spawner-egg relationships using queueing theory. *Ecological Modelling* **104**, 189–197.

Maurer, B.A. (1998) Ecological science and statistical paradigms: at the threshold. *Science* **279**, 502–503.

McAllister, M.K. and Kirkwood, G.P. (1998) Using Bayesian decision analysis to help achieve a precautionary approach for managing developing fisheries. *Canadian Journal of Fisheries and Aquatic Sciences* **55**, 2642–2661.

McCann, K. and Shuter, B. (1997) Bioenergetics of life history strategies and the comparative allometry of reproduction. *Canadian Journal of Fisheries and Aquatic Sciences* **54**, 1289–1298.

McConnaughey, R.A. (1995) Changes in geographic dispersion of eastern Bering Sea flatfish associated with changes in population size. *Proceedings of the International Symposium on North Pacific Flatfish, Alaska* (International Symposium on North Pacific Flatfish. Anchorage, AK, 26–28 October 1994). Sea Grant College Report AK-SG-95-04, Fairbanks, Alaska. pp. 385–405.

McDermot, D. (1998) An Individual-based Modeling Study of Lake Fish Communities in Lake Mendota Wisconsin: the Importance of Considering Food Web Interactions. PhD Thesis, University of Tennessee, Tennessee, 149 pp.

McDermot, D. and Rose, K.A. (2000) Individual-based modeling of lake fish communities: application to piscivore stocking in Lake Mendota. *Ecological Modelling* **125**, 67–102.

Mertz, G. and Myers, R.A. (1995) Estimating the predictability of recruitment. *Fishery Bulletin* **93**, 657–665.

Mertz, G. and Myers, R.A. (1996) Influence of fecundity on recruitment variability of marine fish. *Canadian Journal of Fisheries and Aquatic Sciences* **53**, 1618–1625.

Michaletz, P.H. (1997) Factors affecting abundance, growth, and survival of age-0 gizzard shad. *Transactions of the American Fisheries Society* **126**, 84–100.

Miller, T.J., Crowder, L.B., Rice, J.A. and Marschall, E.A. (1988) Larval size and recruitment mechanisms in fishes: toward a conceptual framework. *Canadian Journal of Fisheries and Aquatic Sciences* **45**, 1657–1670.

Mills, E.L. and Forney, J.L. (1983) Impact on Daphnia pulex of predation by young yellow perch in Oneida Lake, New York. *Transactions of the American Fisheries Society* **112**, 154–161.

Moller, H. (1990) Association between diseases of flounder (*Platichthys flesus*) and environmental conditions in the Elbe estuary, FRG. *Journal du Conseil International pour l'Exploration de la Mer* **46**, 187–199.

Monteleone, D.M. and Houde, E.D. (1990) Influence of maternal size on survival and growth of striped bass *Morone saxatilis* Walbaum eggs and larvae. *Journal of Experimental Marine Biology and Ecology* **140**, 1–11.

Morimoto, H. (1996) Effects of maternal nutritional conditions on number, size and lipid content of hydrated eggs in the Japanese sardine from Tosa Bay, southwestern Japan. In: *Survival Strategies in Early Life Stages of Marine Resources* (Proceedings of an International Workshop, Yokohama, Japan, 11–14, October, 1994) Y. Watanabe, Y. Yamashita and Y. Oozeki, eds. A.A. Balkema Publishers, Rotterdam, Netherlands, pp. 3–12.

Murdoch, W.W. (1969) Switching in general predators: experiments on predator specificity and stability of prey populations. *Ecological Monographs* **39**, 335–354.

Murdoch, W.W. and Oaten, A. (1975) Predation and population stability. *Advances in Ecological Research* **9**, 1–131.

Murray, B.G. (1994) On density dependence. *Oikos* **69**, 520–523.

Muth, K.M. and Wolfert, D.R. (1986) Changes in growth and maturity of walleye associated with stock rehabilitation in western Lake Erie, 1964–83. *North American Journal of Fisheries Management* **6**, 168–175.

Myers, R.A. (1995) Recruitment of marine fish: the relative roles of density-dependent and density-independent mortality in the egg, larval, and juvenile stages. *Marine Ecology Progress Series* **128**, 305–310.

Myers, R.A. and Barrowman, N.J. (1996) Is fish recruitment related to spawner abundance? *Fishery Bulletin* **94**, 707–724.

Myers, R.A., Barrowman, N.J., Hutchings, J.A. and Rosenberg, A.A. (1995b) Population dynamics of exploited fish stocks at low population levels. *Science* **269**, 1106–1108.

Myers, R.A., Bowen, K.G. and Barrowman, N.J. (1999) Maximum reproductive rate of fish at low population sizes. *Canadian Journal of Fisheries and Aquatic Sciences* **56**, 2404–2419.

Myers, R.A., Bridson, J. and Barrowman, N.J. (1995a) Summary of worldwide spawner and recruitment data. *Canadian Technical Report of Fisheries and Aquatic Sciences* No. 2024, 331 pp.

Myers, R.A. and Cadigan, N.G. (1993) Density-dependent juvenile mortality in marine demersal fish. *Canadian Journal of Fisheries and Aquatic Science* **50**, 1576–1590.

Myers, R.A. and Mertz, G. (1998) Reducing uncertainty in the biological basis of fisheries management by meta-analysis of data from many populations: a synthesis. *Fisheries Research* **37**, 51–60.

Nikolsky, G. (1969) *Theory of Fish Population Dynamics as the Background for Rational Exploitation and Management of Fishery Resources*. Oliver and Boyd, Edinburgh, 323 pp.

Nikolsky, G., Bogdanov, A. and Lapin, Y. (1973) On fecundity as a regulatory mechanism in fish population dynamics. *Rapports et Proces-Verbaux des Reunions Conseil International pour l'Exploration de la Mer* **164**, 174–177.

Nisbet, R.M., Murdoch, W.W. and Stewart-Oaten, A. (1996) Consequences for adult fish stocks of human-induced mortality on immatures. In: *Detecting Ecological Impacts: Concepts and Applications in Coastal Habitats* (eds R.J. Schmitt and C.W. Osenberg). Academic Press, San Diego, California, pp. 257–277.

NMFS (National Marine Fisheries Service) (1998) *Report to Congress: Status of Fisheries of the United States*. National Marine Fisheries Service, Washington, DC, 94 pp.

NRC (National Research Council) (1999) *Sustaining Marine Fisheries*. National Academy Press, Washington, DC, 164 pp.

Pace, M.L., Baines, S.B., Cyr, H. and Downing, J.A. (1993) Relationships among early life stages of *Morone americana* and *Morone saxatilis* from long-term monitoring of the Hudson River estuary. *Canadian Journal of Fisheries and Aquatic Sciences* **50**, 1976–1985.

Parker, R.O., Chester, A.J. and Nelson, R.S. (1994) A video transect method for estimating reef fish abundance, composition, and habitat utilization at Gray's Reef National Marine Sanctuary, Georgia. *Fishery Bulletin* **92**, 787–799.

Pauly, D., Christensen, V., Dalsgaard, J., Froese, R. and Torres, F. (1998) Fishing down marine food webs. *Science* **279**, 860–863.

Peebles, E.B., Hall, J.R. and Tolley, S.G. (1996) Egg production by the bay anchovy *Anchoa mitchilli* in relation to adult and larval prey fields. *Marine Ecology Progress Series* **131**, 61–73.

Pepin, P. and Myers, R.A. (1991) Significance of egg and larval size to recruitment variability of temperate marine fish. *Canadian Journal of Fisheries and Aquatic Sciences* **48**, 1820–1828.

Perrow, M.R., Peirson, G. and Townsend, C.R. (1990) The dynamics of a population of roach (*Rutilus rutilus* L.) in a shallow lake: is there a 2-year cycle in recruitment? *Hydrobiologia* **193**, 67–73.

Peterman, R.M. and Bradford, M.J. (1987) Density-dependent growth of age 1 English sole (*Parophrys vetulus*) in Oregon and Washington coastal waters. *Canadian Journal of Fisheries and Aquatic Sciences* **44**, 48–53.

Peterman, R.M. and Gatto, M. (1978) Estimation of functional responses of predators on juvenile salmon. *Journal of the Fisheries Research Board of Canada* **35**, 797–808.

Pianka, E.R. (1970) On r- and k-selection. *American Naturalist* **104**, 592–597.

Planes, S., Jouvenel, J.-Y. and Lenfant, P. (1998) Density dependence in post-recruitment processes of juvenile sparids in the littoral of the Mediterranean Sea. *Oikos* **83**, 293–300.

Policansky, D. and Magnuson, J.J. (1998) Genetics, metapopulations, and ecosystem management of fisheries. *Ecological Applications* **8** (Suppl.), S119–S123.

Polovina, J.J., Kleiber, P. and Kobayashi, D.R. (1999) Application of TOPEX-POSEIDON satellite altimetry to simulate transport dynamics of larvae of spiny lobster, *Panulirus marginatus*, in the northwestern Hawaiian Islands. *Fishery Bulletin* **97**, 132–143.

Post, J.R. and Evans, D.O. (1989) Size-dependent overwinter mortality of young-of-the-year yellow perch (*Perca flavescens*): laboratory, in situ enclosure, and field experiments. *Canadian Journal of Fisheries and Aquatic Sciences* **46**, 1958–1968.

Poulin, R. (1995) Phylogeny, ecology, and the richness of parasite communities in vertebrates. *Ecological Monographs* **65**, 283–302.

Railsback, S.F. (2001) Concepts from complex adaptive systems as a framework for individual-based modelling. *Ecological Modelling* **139**, 47–62.

Rauck, G. and Zijlstra, J.J. (1978) On the nursery aspects of the Wadden Sea for some commercial fish species and possible long term changes. *Rapports et Proces-Verbaux des Reunions, Conseil International pour l'Exploration de la Mer* **172**, 266–275.

Reinhardt, U.G. (1999) Predation risk breaks size-dependent dominance in juvenile coho salmon (*Oncorhynchus kisutch*) and provides growth opportunities for risk-prone individuals. *Canadian Journal of Fisheries and Aquatic Sciences* **56**, 1206–1212.

Rice, J.A., Crowder, L.B. and Holey, M.E. (1987) Exploration of mechanisms regulating larval survival in Lake Michigan bloater: a recruitment analysis based on characteristics of individual larvae. *Transactions of the American Fisheries Society* **116**, 703–718.

Ricker, W.E. (1975) Computation and interpretation of biological statistics of fish populations. *Bulletin of the Fisheries Research Board of Canada* **191**, 382 pp.

Rickman, S.J., Dulvy, N.K., Jennings, S. and Reynolds, J.D. (2000) Recruitment variation related to fecundity in marine fishes. *Canadian Journal of Fisheries and Aquatic Sciences* **57**, 116–124.

Rieman, B.E. and Myers, D.L. (1992) Influence of fish density and relative productivity on growth of kokanee in ten oligotrophic lakes and reservoirs in Idaho. *Transactions of the American Fisheries Society* **121**, 178–191.

Rijnsdorp, A.D. (1990) The mechanism of energy allocation over reproduction and somatic growth in female North Sea plaice, *Pleuronectes platessa* L. *Netherlands Journal of Sea Research* **25**, 279–290.

Rijnsdorp, A.D. (1994) Population-regulating processes during the adult phase in flatfish. *Netherlands Journal of Sea Research* **32**, 207–223.

Rijnsdorp, A.D. and van Leeuwen, P.I. (1992) Density-dependent and independent changes in somatic growth of female North Sea plaice *Pleuronectes platessa* between 1930 and 1985 as revealed by back-calculation of otoliths. *Marine Ecology Progress Series* **88**, 19–32.

Rijnsdorp, A.D. and Vingerhoed, B. (1994) The ecological significance of geographical and seasonal differences in egg size in sole *Solea solea* (L.). *Netherlands Journal of Sea Research* **32**, 255–270.

Rodd, F.H. and Reznick, D.N. (1997) Variation in the demography of guppy populations: the importance of predation and life histories. *Ecology* **78**, 405–418.

Roff, D.A. (1982) Reproductive strategies in flatfish: a first synthesis. *Canadian Journal of Fisheries and Aquatic Sciences* **39**, 1686–1698.

Rogers, B.A. and Westin, D.T. (1981) Laboratory studies of the effects of temperature and delayed initial feeding on development of striped bass larvae. *Transactions of the American Fisheries Society* **110**, 100–110.

Rose, K.A. and Cowan, J.H. (2000) Predicting fish population dynamics: compensation and the importance of site-specific considerations. *Environmental Science and Policy* **3**, S433–S443.

Rose, K.A., Rutherford, E.S., McDermott, D., Forney, J.L. and Mills, E.L. (1999) An individual-based model of walleye and yellow perch in Oneida Lake, New York. *Ecological Monographs* **69**, 127–154.

Rose, K.A. and Summers, J.K. (1992) Relationships among long-term fish abundances, hydrographic variables, and gross pollution indicators in northeastern US estuaries. *Fisheries Oceanography* **1**, 281–293.

Rose, K.A., Tyler, J.A., Chambers, R.C., MacPhee, G. and Danila, D.J. (1996) Simulating winter flounder population dynamics using coupled individual-based young-of-the-year and age-structured adult models. *Canadian Journal of Fisheries and Aquatic Sciences* **53**, 1071–1091.

Rosenberg, A.A. and Haugen, A.S. (1982) Individual growth and size-selective mortality of larval turbot (*Scopthalmus maximus*) reared in enclosures. *Marine Biology* **72**, 73–77.

Rosenberg, A.A. and Restrepo, V.R. (1994) Uncertainty and risk evaluation in stock assessment advice for U.S. marine fisheries. *Canadian Journal of Fisheries and Aquatic Sciences* **51**, 2715–2720.

Rothschild, B.J. (1986) *Dynamics of Marine Fish Populations*. Harvard University Press, Cambridge, Massachusetts, 277 pp.

Rothschild, B.J. and Fogarty, M.J. (1989) Spawning-stock biomass: a source of error in recruitment/stock relationships and management advice. *Journal du Conseil International pour l'Exploration de la Mer* **45**, 131–135.

Rothschild, B.J., Osborn, T.R., Dickey, T.D. and Farmer, D.M. (1989) The physical basis for recruitment variability in fish populations. *Journal du Conseil International pour l'Exploration de la Mer* **45**, 136–145.

Royama, T. (1977) Population persistence and density dependence. *Ecological Monographs* **47**, 1–35.

Rykiel, E.J. (1996) Testing ecological models: the meaning of validation. *Ecological Modelling* **90**, 229–244.

Saila, S.B. (1992) Application of fuzzy graph theory to successional analysis of a multispecies trawl fishery. *Transactions of the American Fisheries Society* **121**, 211–233.

Saila, S.B., Chen, X., Erizini, K. and Martin, B. (1987) Compensatory mechanisms in fish populations, Vol. 1: Critical evaluation of case histories of fish populations experiencing chronic exploitation or impact. *Electric Power Research Institute Report EA-5200*, Electric Power Research Institute, Palo Alto, California.

Saila, S., Martin, B., Ferson, S., Ginzburg, L. and Millstein, J. (1991) Demographic Modeling of Selected Fish Species with RAMAS. *Electric Power Research Institute Report EN-7178*. Electric Power. Research Institute, Palo Alto, California.

Sanderson, B.L., Hrabik, T.R., Magnuson, J.J. and Post, D.M. (1999) Cyclic dynamics of a yellow perch (*Perca flavescens*) population in an oligotrophic lake: evidence for the role of intraspecific interactions. *Canadian Journal of Fisheries and Aquatic Sciences* **56**, 1534–1542.

Schaaf, W.E., Peters, D.S., Vaughan, D.S., Coston-Clements, L. and Krouse, C.W. (1987) Fish population responses to chronic and acute pollution: the influence of life history strategies. *Estuaries* **10**, 267–275.

Schaefer, M.B. (1954) Some aspects of the dynamics of populations important to the management of the commercial marine fisheries. *Bulletin of the Inter-American Tropical Tuna Commission* **1**, 27–56.

Schirripa, M.J. and Legault. C.M. (1999) *Status of the Red Snapper in U.S. Waters of the Gulf of Mexico: Updated Through 1998*. Sustainable Fisheries Division Contribution SFD-99/00–75, Southeast Fisheries Science Center, National Marine Fisheries Service, Miami, Florida, 86 pp.

Schneider, D.C. (1989) Identifying the spatial scale of density–dependent interaction of predators with schooling fish in the southern Labrador Current. *Journal of Fish Biology* **35** (Suppl. A), 109–115.

Schramm, H.L., McKeown, P.E. and Green, D.M. (1995) Managing black bass in Northern waters: summary of the workshop. *North American Journal of Fisheries Management* **15**, 671–679.

Secor, D.H., Houde, E.D. and Montelone, D.M. (1995) A mark-release experiment on larval striped bass *Morone saxatilis* in a Chesapeake Bay tributary. *ICES Journal of Marine Science* **52**, 87–101.

Shelton, P.A. and Healey, B.P. (1999) Should depensation be dismissed as a possible explanation for the lack of recovery of the northern cod (*Gadus morhua*) stock? *Canadian Journal of Fisheries and Aquatic Sciences* **56**, 1521–1524.

Shepherd, J.G., Pope, J.G. and Cousens, R.D. (1984) Variations in fish stocks and hypotheses concerning their links with climate. *Rapports et Proces-Verbaux des Reunions, Conseil International pour l'Exploration de la Mer* **185**, 255–267.

Shima, J.S. (1999) Variability in relative importance of determinants of reef fish recruitment. *Ecology Letters* **2**, 304–310.

Shima, J.S. (2001) Regulation of local populations of a coral reef fish via joint effects of density- and number-dependent mortality. *Oecologia* **126**, 58–65.

Shuter, B.J., Jones, M.L., Korver, R.M. and Lester, N.P. (1998) A general, life history based model for regional management of fish stocks: the inland lake trout (*Salvelinus namaycush*) fisheries of Ontario. *Canadian Journal of Fisheries and Aquatic Sciences* **55**, 2161–2177.

Sibly, R. and Calow, P. (1986) Why breeding earlier is always worthwhile. *Journal of Theoretical Biology* **123**, 311–319.

Sinclair, M., Tremblay, M.J. and Bernal, P. (1985) El Nino events and variability in a Pacific mackerel (*Scomber japonicus*) survival index: Support for Hjort's second hypothesis. *Canadian Journal of Fisheries and Aquatic Sciences* **42**, 602–608.

Sissenwine, M.P. (1984) Why do fish populations vary? *Exploitation of Marine Communities* (ed. R.M. May). Springer-Verlag, New York, pp. 59–94.

Smith, C. and Reay, P. (1991) Cannibalism in teleost fish. *Reviews in Fish Biology and Fisheries* **1**, 41–64.

Smith, P.E., Santander, H. and Alheit, J. (1989) Comparison of the mortality rates of Pacific sardine, *Sardinops sagax*, and Peruvian anchovy, *Engraulis ringens*, eggs off Peru. *Fishery Bulletin* **87**, 497–508.

Sogard, S.M. (1997) Size-selective mortality in the juvenile stage of teleost fishes: a review. *Bulletin of Marine Science* **60**, 1129–1157.

Stearns, S. (1993) *The Evolution of Life Histories*. Oxford University Press, New York, 249 pp.

Steele, M.A. (1997) Population regulation by post-settlement mortality in two temperate reef fishes. *Oecologia* **112**, 64–74.

Stepien, C.A. (1995) Population genetic divergence and geographic patterns from DNA sequences: examples from marine and freshwater fishes. *American Fisheries Society Symposium* **17**, 263–287.

Stewart-Oaten, A. and Murdoch, W.M. (1990) Temporal consequences of spatial density dependence. *Journal of Animal Ecology* **59**, 1027–1045.

Strong, D.R. (1986) Density-vague population change. *Trends in Ecology and Evolution* **1**, 39–42.

Swain, D.P. and Sinclair, A.F. (1994) Fish distribution and catchability: what is the appropriate measure of distribution? *Canadian Journal of Fisheries and Aquatic Sciences* **51**, 1046–1054.

Swartzman, G. (1996) Resource modeling moves into the courtroom. *Ecological Modelling* **92**, 277–288.

Swartzman, G., Silverman, E. and Williamson, N. (1995) Relating trends in walleye pollack (*Theragra chalcogramma*) abundance in the Bering Sea to environmental factors. *Canadian Journal of Fisheries and Aquatic Sciences* **52**, 369–380.

Szedlmayer, S.T. (1997) Ultrasonic telemetry of red snapper, *Lutjanus campechanus*, at artificial reef sites in the northeast Gulf of Mexico. *Copeia* **4** (December), 846–850.

Takeuchi, M., Idhii, S. and Ogiso, T. (1981) Effect of dietary vitamin E on growth, vitamin E distribution and mortalities of the fertilized eggs and fry in ayu *Plecoglossus altivelis*. *Bulletin of Tokai Regional Fisheries Research Laboratory* **104**, 111–122.

Thorrold, S.R., Jones, C.M., Swart, P.K. and Targett, T.E. (1998) Accurate classification of juvenile weakfish *Cynoscion regalis* to estuarine nursery areas based on chemical signatures in otoliths. *Marine Ecology Progress Series* **173**, 253–265.

Townsend, C.R. and Hughes, R.N. (1981) Maximizing net energy returns from foraging. In: *Physiological Ecology: an Evolutionary Approach to Resource Use* (eds C.R. Townsend and P. Calow). Blackwell Science, Ltd, Oxford, pp. 86–108.

Townsend, T.J. and Wootton, R.J. (1984) Effects of food supply on the reproduction of the convict cichlid, *Cichlasoma nigrofasciatum*. *Journal of Fish Biology* **24**, 91–104.

Treasurer, J.W., Owen, R. and Bowers, E. (1992) The population dynamics of pike, *Esox lucius*, and perch, *Perca fluviatilis*, in a simple predator-prey system. *Environmental Biology of Fishes* **34**, 65–78.

Trippel, E.A. (1995) Age at maturity as a stress indicator in fisheries. *Bioscience* **45**, 759–771.

Tsuruta, Y. and Hirose, K. (1989) Internal regulation of reproduction in the Japanese anchovy (*Engraulis japonica*) as related to population fluctuation. *Canadian Special Publications of Fisheries and Aquatic Sciences* **108**, 111–119.

Turchin, P. (1995) Population regulation: old arguments and a new synthesis. In: *Population Dynamics: New Approaches and Synthesis* (eds N. Cappuccino and P.W. Price). Academic Press, San Diego, California, pp. 19–40.

Tyler, J.A. and Rose, K.A. (1994) Individual variability and spatial heterogeneity in fish population models. *Reviews in Fish Biology and Fisheries* **4**, 91–123.

Uchamnski, J. and Grimm, V. (1996) Individual-based modelling in ecology: what makes the difference? *Trends in Ecology and Evolution* **11**, 437–441.

Van Winkle, W., Holcomb, B.D., Jager, H.I., Tyler, J.A., Whitaker, S. and Shuter, B.J. (1997) Regulation of energy aquisition and allocation to respiration, growth, and reproduction: conceptual framework and example using rainbow trout. In: *Early Life History and Recruitment in Fish Populations* (eds R.C. Chambers and E.A. Trippel). Chapman and Hall, New York, pp. 126–178.

van der Veer, H.W. (1986) Immigration, settlement, and density-dependent mortality of a larval and early postlarval 0-group plaice (*Pleuronectes platessa*) population in the western Wadden Sea. *Marine Ecology Progress Series* **29**, 223–236.

van der Veer, H.W. and Bergman, J.N. (1987) Predation by crustaceans on a newly settle 0-group plaice *Pleuronectes platessa* population in the western Wadden Sea. *Marine Ecology Progress Series* **35**, 203–215.

Walters, C. (2000) Natural selection for predation avoidance tactics: implications for marine population and community dynamics. *Marine Ecology Progress Series* **208**, 309–313.

Walters, C.J., Hannah, C.G. and Thomson, K. (1992) A microcomputer program for simulating effects of physical transport on fish larvae. *Fisheries Oceanography* **1**, 11–19.

Walters, C. and Korman, J. (1999) Linking recruitment to trophic factors: revisting the Beverton-Holt recruitment model from a life history and multispecies perspective. *Reviews in Fish Biology and Fisheries* **9**, 187–202.

Walters, C.J. and Ludwig, D. (1981) Effects of measurement errors on the assessment of stock-recruitment relationships. *Canadian Journal of Fisheries and Aquatic Sciences* **38**, 704–710.

Warren, M.L. and Burr, B.M. (1994) Status of freshwater fishes of the United States: overview of an imperiled fauna. *Fisheries* **19**, 6–18.

Watanabe, T., Itoh, A., Murakami, A., Tsukashima, Y., Kitajima, C. and Fujita, S. (1984c) Effect of nutritional quality of diets given to broodstock on the verge of spawning on reproduction of Red Sea bream. *Bulletin of the Japanese Society of Scientific Fisheries* **50**, 1023–1028.

Watanabe, T., Ohhashi, S., Itoh, A., Kitajima, C. and Fujita, S. (1984a) Effect of nutritional composition of diets on chemical components of red sea bream broodstock and eggs produced. *Bulletin of the Japanese Society of Scientific Fisheries* **50**, 503–515.

Watanabe, T., Takeuchi, T., Saito, M. and Nishimura, K. (1984b) Effect of low protein-high calory or essential fatty deficiency diet on reproduction in rainbow trout. *Bulletin of the Japanese Society of Scientific Fisheries* **50**, 1207–1215.

Werner, E.E. and Hall, D.J. (1974) Optimal foraging and the size selection of prey by the bluegill sunfish (*Lepomis macrochirus*). *Ecology* **55**, 1042–1052.

Werner, E.E. and Mittlebach, G.G. (1981) Optimal foraging: field tests of diet choice and habitat switching. *American Zoologist* **21**, 813–822.

Werner, F.E., Perry, R.I., Lough, R.G. and Naimie, C.E. (1996) Trophodynamic and advective influences on Georges Bank larval cod and haddock. *Deep-Sea Research II* **43**, 1793–1822.

Wiley, M.J., Kohler, S.L. and Seelbach, P.W. (1997) Reconciling landscape and local views of aquatic communities:

lessons from Michigan trout streams. *Freshwater Biology* **37**, 133–148.

Winemiller, K.O. and Rose, K.A. (1992) Patterns of life-history diversification in North American fishes: implications for population regulation. *Canadian Journal of Fisheries and Aquatic Sciences* **49**, 2196–2218.

Wolda, H., Dennis, B. and Taper, M.L. (1994) Density dependence tests, and largely futile comments: answers to Holyoak and Lawton (1993) and Hanski, Woiwod and Perry (1993). *Oecologia* **98**, 229–234.

Wootton, R.J. (1977) Effect of food limitation during the breeding season on the size, body components and egg production of female sticklebacks (*Gasterosteus aculeatus* L.). *Journal of Animal Ecology* **46**, 823–834.

Zastrow, C.E., Houde, E.D. and Morin, L.G. (1991) Spawning, fecundity, hatch-date frequency and young-of-the-year growth of bay anchovy *Anchoa mitchilli* in mid-Chesapeake Bay. *Marine Ecology Progress Series* **73**, 161–171.

Zastrow, C.E., Houde, E.D. and Saunders, E.H. (1989) Quality of striped bass (*Morone saxatilis*) eggs in relation to river source and female weight. *Rapports et Proces-Verbaux des Reunions, Conseil International pour l'Exploration de la Mer* **191**, 34–42.

INTER-AMERICAN TROPICAL TUNA COMMISSION
COMISION INTERAMERICANA DEL ATUN TROPICAL

Bulletin — Boletin

Vol. I, No. 2

SOME ASPECTS OF THE DYNAMICS OF POPULATIONS IMPORTANT TO THE MANAGEMENT OF THE COMMERCIAL MARINE FISHERIES

by

MILNER B. SCHAEFER

La Jolla, California

1954

Contents

	Page
Introduction	27
The law of population growth in populations which tend to stability	28
Effects of fishing	30
Catch per unit of effort	31
Maximum equilibrium catch	31
Determination of the status of the fish population and estimation of equilibrium yields	32
An application to the halibut fishery of the North Pacific	33
The nature of the growth of the amount of fishing	37
Stabilization of an unregulated fishery	38
The course of development of an unregulated fishery and the manner of approach to stable equilibrium	40
Examples from the commercial fisheries	48
Pacific halibut	49
California sardine (Pacific pilchard)	51

Introduction

A population of oceanic fish under exploitation by a fishery may be influenced by a great number of elements in the complex ecological system of which it forms a part. Of these, however, only one, predation by man, is capable of being controlled or modified to any significant degree by man's actions. Any management or control of the fishery, to the extent this may be possible at all, must, therefore, be effected through control of the activities of the fishermen. It seems important to elucidate some of the basic principles of the effect of fishing on a fish population and, conversely, the effect of the fish population on the amount of fishing, in order to understand in what circumstances and in what manner such control of the activities of the fishermen can influence the fish population and the yield obtained therefrom.

Management of a fishery has as its purpose the modification or limitation of the activities of the fishermen in order to realize a change in the fish population, or the catch, or both, which in some manner is preferable to that which would obtain if the fishermen were allowed to operate without these modifications or limitations. What may be "preferable" involves in the general case a great many economic and sociological matters difficult or impossible to treat objectively, and not susceptible to quantitative reasoning. We must, therefore, confine our attention to a less general case, but one most often met in practice, where the purpose of management is to obtain a larger average total catch per unit of time than would be obtained without management. An important special case of this is management directed toward obtaining the maximum average total catch per unit of time, which is often referred to, somewhat ambiguously, as the "optimum catch."

The Inter-American Tropical Tuna Commission has the task, specified by the Convention under which it is organized, of gathering and interpreting factual information to facilitate maintaining the populations of the tropical tunas and of the tuna-bait fishes at levels which will permit maximum sustained catches year after year. Information respecting these populations is not at the present time adequate for this purpose. An analysis of the fundamental relationships between population size, intensity of fishing, and catch is a valuable, if not indispensable, basis upon which to plan the efficient collection and interpretation of the information required to accomplish the purposes of the Commission.

The staff of the Commission has directed a large share of its attention since its inception to the collection and compilation of reliable data respecting the total catch and catch per unit of fishing effort of each tuna species over the period of growth and development of the fishery in the Eastern Tropical Pacific. This task is nearing completion. The next step in the investigation is to employ these data together with such ancillary vital statistics as may be required and may be obtainable, to the estimation of the level of maximum sustained yield of each tuna stock and the determination of the present condition of the fishery with relation thereto. This step requires the employment of a suitable mathematical model describing the effect of fishing on the tuna stocks. Models which have been applied in the past to other fisheries are not satisfactory for this purpose. It has, therefore, proven necessary to undertake the investigations reported in this paper directed toward the development of a suitable model, and of methods of its application to fisheries data, which can be applied to the data of the tuna fishery. These studies, although of a theoretical nature, are of the most direct practical importance to the objectives of the Commission, since they are fundamental to the interpretation of the catch data and related information being collected by the staff.

It is well known that in dealing with oceanic fisheries we have to do with very complex ecological systems and that, therefore, the effects of the amount of fishing on the size of the fish population and on the catch is difficult to estimate. Some recent and current controversies bear adequate witness to this. The very complexity of these systems tends, however, to divert attention from consideration of the fundamental laws of population growth which make it possible for a species to survive increases in predator populations, and, by the same token, make possible that extensive predation by man which is commercial fishing.

In this investigation it will be attempted to indicate the manner in which the fundamental laws of population growth operate in the case of a commercial fishery, and so, perhaps, clarify some of the important considerations basic to the management of the oceanic fisheries. These will be shown by means of mathematical models. Certain parts of these models or very similar ones have been employed in predator-prey investigations of other organisms (Gause 1934, Lotka 1925), and there have been limited attempts to apply somewhat similar techniques to the fisheries, as will be noted subsequently. There is rather good reason to believe that the models sufficiently describe reality to be useful in furthering our understanding.

In pursuing the investigation we wish to elucidate the dynamics of a population of oceanic fish not related to environmental variations, that is the dynamics of the "mean" population under average environmental conditions. We shall, therefore, consider the situation in which all the factors of the environment are constant except predation by man, i.e. the amount of fishing. In application, the effect of variation due to environmental changes is treated as a random variable, independent of size of population.

The law of population growth in populations which tend to stability

Populations of organisms living in a constant environment with a limited food supply may be of one of two kinds. In one kind of population, exemplified in particular by some insects, different stages of which are in

DYNAMICS OF POPULATIONS

competition with each other for the means of life, the number of adults fluctuates periodically and continuously (Nicholson 1949, 1950). In the other type, the population tends to stability, so that for a particular set of environmental conditions the population has at each size a definite potential rate of increase which is dependent only on the existing size of the population. A great many populations from the yeasts and protozoa to man have ben shown to be of this sort. Most, at least, of the populations of fishes are believed to be of this kind.*

The general law of population growth for such a population, P, may be expressed as

$$\frac{dP}{dt} = f(P) \quad \quad (1)$$

where $f(P)$ is continuous, positive and single valued between $P = 0$ and $P = L$, the maximum population which the living space and food supply can support, and zero at these limiting values of P. We will call $f(P)$ the *natural rate of increase*.

A particular function which has been shown to fit experimental data as well as data from populations in nature for a good many organisms is the Verhulst-Pearl logistic

$$\frac{dP}{dt} = k_1 P(L - P) \quad \quad (2)$$

where k_1 is a constant

In this case, of course, $f(P)$ is a parabola with its axis along $P = L/2$. It is shown graphically in Figure 1. Integrating, we may obtain P as a function of t, which is a sigmoid curve with an upper asymptote at $P = L$ and an inflection point at the value of P for which $\frac{dP}{dt}$ is a maximum, i.e. at $P = L/2$.

This law has been employed to describe the growth of a considerable variety of organisms, for example yeasts (Gause 1934, p. 78, Pearl 1925, p. 9), protozoa (Gause 1934, p. 36, p. 93 et. seq.), fruitflies (Pearl 1925, p. 11), and humans (Pearl 1925).

Büchman (1938) has considered the general dynamics of commercial fish populations based on this relationship, as has Graham (1939). Graham (1935) employed this growth law in an analysis of the effect of World War I on the abundance and landings of demersal fishes from the North Sea, and Baerends (1947) has made similar analyses.

It is possible that for fish populations the special case of (1) represented by the logistic (2) is not, in general, an exact representation of the

---

*One notable exception may be some species of Pacific Salmon, which tend to periodic fluctuations characteristic of the species. These may be found to be the result of the direct or indirect competition of different year classes for the means of subsistence. This has been little investigated, however.

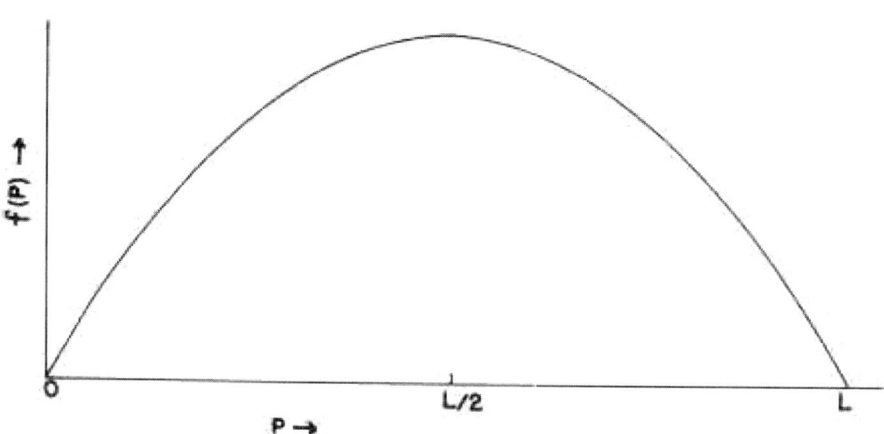

Figure 1. Natural rate of increase of a population which grows according to the Verhulst-Pearl logistic.

population growth law. In particular, the relationship (2) is a parabola, symmetrical with respect to its axis, from which it follows that the maximum natural rate of increase occurs at a value of P half way between zero and the maximum population value, L. There is reason to believe that in at least some populations of fishes, the curve is actually somewhat asymmetrical, with the maximum value of $\frac{dP}{dt}$ at a value of P less than $L/2$. Experimental data have also shown this is sometimes the case for other organisms, for example the yeast data of Gause (1934, p. 68).

Effects of Fishing

A fishery, that is removal of fish from the stock by man, has the effect of subtracting from that increase in stock which would occur at the existing level of population if no fishing were taking place. In other words, the rate of change in the stock will be less than the natural rate of increase by an amount equal to the rate of catching of the fish. That is

$$\frac{dP}{dt} = f(P) - P\,\phi(F) \qquad (3)$$

where $P\,\phi(F)$ is the *rate of catching*, depending on the size of the fish stock and some positive single valued function of the number of units of fishing effort, F.

It is obvious from (3) that whenever the rate of catching is less than the natural rate of increase, the population will increase. Conversely, when the rate of catching exceeds the natural rate of increase, the population shrinks in size. When the rate of catching is exactly equal to the natural rate of increase, $\frac{dP}{dt} = 0$, the population remains unchanged, and the fishery is said to be in equilibrium for that level of population and fishing effort. The annual catch made under such a condition of equilibrium has been

DYNAMICS OF POPULATIONS

called the stabilized catch, the equilibrium yield, and other things. We shall call it the *equilibrium catch*.

From the equation (3) and the general form of $f(P)$ certain conclusions of importance in fishery research and management may be immediately perceived:

(1) As the fishery increases in intensity (as F increases), the stock P, decreases. This decrease in population is a necessary consequence of increasing fishing intensity, and, thus, is an inevitable result of the development of a fishery.

(2) The stock, and the corresponding equilibrium catch, can be held constant, by regulating the amount of fishing, at any value less than $P = L$. Stability or instability of the population and catch over a given period of time has, therefore, no necessary relationship to the level of abundance, but merely reflects whether the rate of catching is changing or is constant.

Catch per unit of effort

Let it be assumed that the fishery operates on the stock in such a manner that one unit of fishing effort produces the same relative effect on the stock, that is it catches the same percentage of the stock, regardless of the time or place it is applied. Then

$$\phi(F) = k_2 F \quad \text{where } k_2 \text{ is a constant}$$
$$\text{and } P \phi(F) = k_2 P F \qquad (4)$$

Under these circumstances, the rate of catching per unit of fishing effort is

$$\frac{k_2 P F}{F} = k_2 P \qquad (5)$$

and is, thus, proportional to the stock. The average catch per unit of effort during a given period of time will be proportional to the average size of the fish stock encountered by the fishery during the period. The average catch per unit of effort per year, or some other short time period, has been extensively used by fishery scientists to measure changes in the size of fish populations.

Maximum equilibrium catch

As has been shown above, when the rate of catching is just equal to the natural rate of increase, the stock will remain unchanged and the catch obtained will, of course, be stabilized also. The size or sizes of stock at which the equilibrium catch may be maximized are levels of maximum equilibrium catch. In general, it is supposed that a fish population has a growth law at least similar to (2) in that there is but a single maximum. In this case, there is but a single size of population at which the equilibrium catch may be maximized. This size of population has been referred to as the optimum stock and the corresponding rate of catching as the optimum catch. I prefer the expression *maximum equilibrium catch* as being more descriptive of exactly what is meant.

Determination of the status of the fish population and estimation of equilibrium yields

In the practical consideration of management of a fishery we are interested in finding out whether the fish population has been driven below the point at which maximum equilibrium catch may be obtained. If so, curtailment of the intensity of fishing will result in increased average catches. It is also of interest to estimate, if possible, the maximum equilibrium catch and the size of population at which it may be obtained.

The investigation of these matters involves, essentially, estimating the equilibrium catch at various levels of population. This may be accomplished by application of equations (3) and (4) where the assumptions underlying these equations are sufficiently nearly realized.

From (3) and (4) we have

$$\frac{dP}{dt} = f(P) - k_2 PF \quad (6)$$

Integrating over the year, we obtain

$$\int_{P_0}^{P_1} dP = \int_0^1 f(P) dt - \int_0^1 k_2 PF dt \quad (7)$$

where $P = P_0$ at $t = t_0$
and $P = P_1$ at $t = t_1$
from which

$$P_1 - P_0 = \triangle P = \overline{f(P)} - k_2 F_t \overline{P} \quad (8)$$

Where $\overline{f(P)}$ is the annual natural rate of increase and, hence, the annual equilibrium catch corresponding to the mean stock $\overline{P}$ encountered by the fishery during the year*. F_t is the total fishing intensity for the year, $F_t = \int_0^1 F dt$. $k_2 F_t \overline{P}$ is, of course, the total catch during the year.

The average catch per unit of effort is

$$U = \frac{k_2 F_t \overline{P}}{F_t} = k_2 \overline{P} \quad (9)$$

If we have adequate statistical records of the fishery we know the amount of effort, the catch, and the catch per unit of effort year-by-year. If we can evaluate k_2 in (9) we shall be able to compute $\overline{P}$ for each year from the catch statistical data. From values of $\overline{P}$ we may estimate P_1 and P_0

---

*$\overline{P}$ is the average of P taken with respect to the units of effort applied during the year.

That is, $\overline{P} = \dfrac{\int_0^1 PF dt}{\int_0^1 F dt}$

DYNAMICS OF POPULATIONS

approximately by interpolating between values of $\overline{P}$ for successive years. Given $P_1 - P_0$ and the catch, we can estimate $\overline{f(P)}$, the annual equilibrium catch corresponding to $\overline{P}$ during each year of the series.

One estimate of k_2 is provided by data from tagging experiment, since $F_t k_2$ is simply the instantaneous rate of fishing mortality, that is $f = 1 - e^{-k_2 F_t}$, where f is the annual fishing mortality rate, which may be determined from the recovery rates of marked fish. Other means also exist, of course, for estimating k_2.

An application to the Halibut fishery of the North Pacific

The manner in which this procedure may be applied can be illustrated by the example of the fishery for Pacific Halibut, using for our example the population of Area 2 (the region south of Cape Spencer). Statistics of catch and catch per unit are given by Thompson and Bell (1934) and by Thompson (1950). Revised, and presumably more accurate, values have been furnished recently by Bell to Dr. R. VanCleve (MS) from which I have taken the values employed here; see the first three columns of Table 1.

Tagging experiments (Thompson and Harrington, 1930) conducted in Area 2 indicated an annual fishing mortality rate of approximately 40% in 1926. Subsequently Thompson and Bell (1934) found that 47% was perhaps more realistic. Using 47% as the annual fishing mortality rate in 1926, we have

$e^{-F_t k_2} = 0.53$, and $F_t = 494,078$ skates (Thompson 1950, table 2)

Then $F_t k_2 = 0.635$

and, $1/k_2 = \frac{494}{635} \times 10^6 = 778 \times 10^3$*

Multiplying the values of U for each year by $1/k_2$ we obtain estimates of $\overline{P}$ (Table 1, column 4). Interpolating between successive values of $\overline{P}$, we obtain estimates of the stock at the beginning of each year (column 5). Differences of the values for successive years indicate the increase or decrease of the stock which resulted from the catch taken during the year ($\triangle P$, column 6). In accordance with (8) we add $\triangle P$ to the annual catch to obtain $\overline{f(P)}$, the annual equilibrium catch corresponding to $\overline{P}$ (column 7).

We now have estimates of the stock and equilibrium catch obtainable from that stock for the series of years from 1916 to 1946. Plotting $\overline{f(P)}$

---

*By considering changes in catch per unit of effort and total catch over the period 1926 to 1933, during which period the stock fell and then returned again to the original level, Thompson (1950) arrived at a value for $1/k_2$ of 335×10^3. This corresponds to a fishing mortality rate of about 77% in 1926, which seems unreasonably high from the tagging results, age composition data, and other information respecting this fishery. An indication of why his analysis gives this result will be given later (p. 37).

against $\overline{P}$ we would expect the points to fall on a curve (a parabola, if $f(P)$ is the logistic) in the absence of other influences. Actually, due to unknown effects of variable environmental factors, measurement errors, and other unaccounted-for sources of variation, the points will tend to scatter about an average curve. By observing the trend of the plot of $\overline{f(P)}$ against $\overline{P}$, we may ascertain, however, how the equilibrium catch for this population varies, on the average, with the size of the population. This has been done in Figure 2, where the small, solid points represent the annual values from Table 1. The centers of the crosses represent the mean values calculated for each 10 units of U.

It is quite obvious that the equilibrium catch increases, on the average, up to a catch per unit of effort of about 80 pounds per skate, at least, corresponding to a mean population of some 62,000,000 pounds. Data beyond this population level are not available (the single point for 1916 at 114 lbs. per skate is not deemed adequate for extending the relationship). Certainly it appears that, contrary to the contention of Burkenroad (1951, 1953), this halibut population was driven below its point of maximum equilibrium catch, and the curtailment of fishing had a beneficial effect on the subsequent catches.

It is unfortunate that reliable data are not available for earlier years when the population was, presumably, larger, which would enable us to estimate equilibrium catches for higher population values and so find out where the maximum occurs. It appears that it might be desirable, if possible, in order to find this out, to curtail fishing to permit higher levels of population to be reached.

This example points out clearly the desirability of obtaining adequate statistical data on a fishery during its early stages so that the maximum equilibrium catch may be estimated, approaching it from those population levels which are too high to give the maximum equilibrium catch. It is practically difficult, once the maximum has been passed, to drive the stock back up past the point of maximum return for purposes of investigation, since the immediate economic welfare of the industry must always be considered in practical regulations.

In the analysis of the halibut data thus far, we have not specified the form of $f(P)$ beyond the general restrictions on (1). As a matter of illustrating methodology, it is of interest to see what results are obtained if we specify that the curve be the logistic (page 29), so that

$$\overline{f(P)} = k_1 \overline{P} (L - \overline{P}) \text{ or,}$$

since $\overline{P} = \dfrac{U}{k_2}$,

$$\overline{f(P)} = \dfrac{k_1}{k_2^2} U (L_u - U),$$

where $L_u = k_2 L$

Fitting a curve of this form to the mean values (crosses) of Figure 2, (with

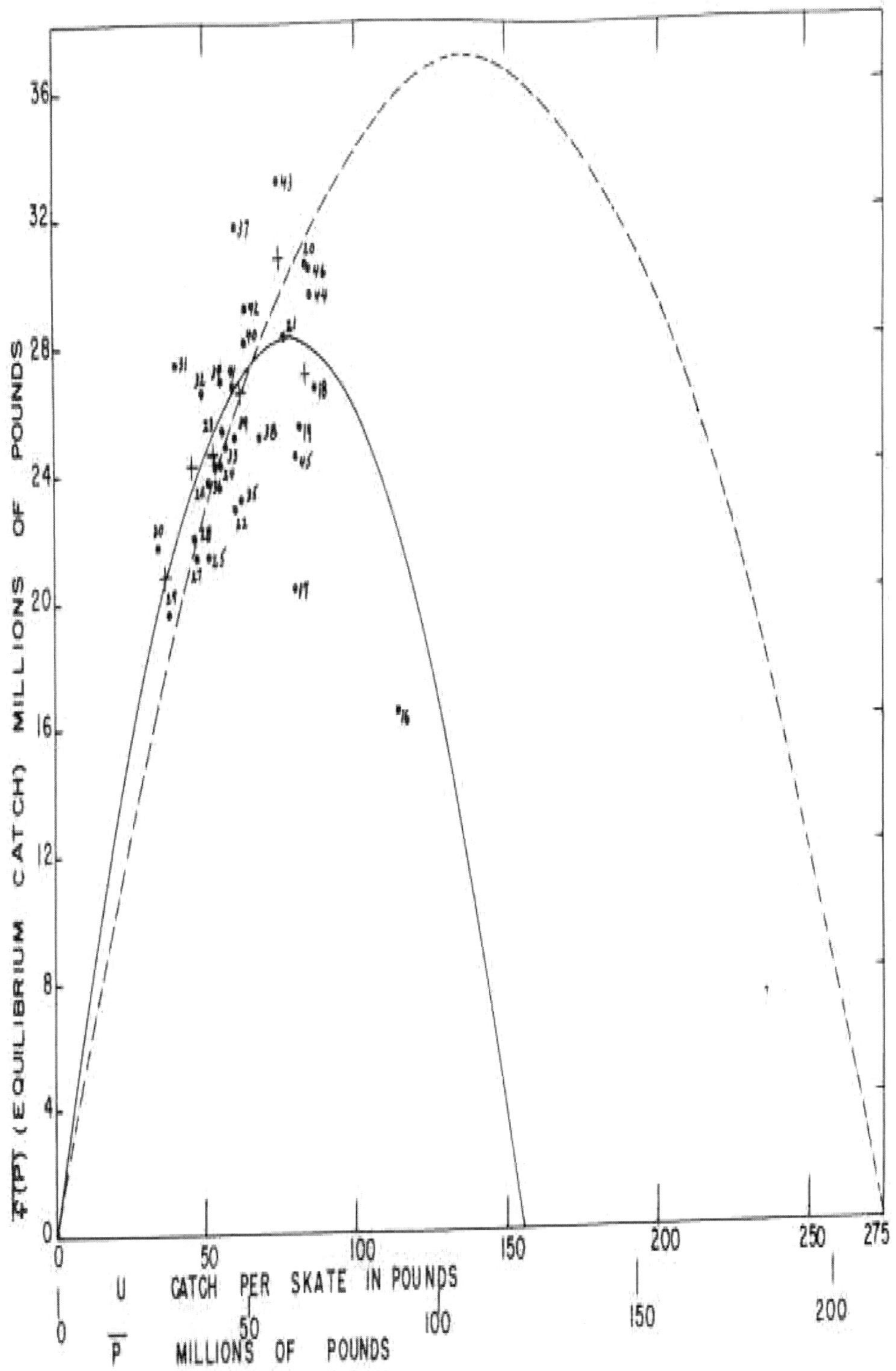

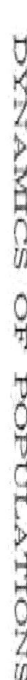

Figure 2. Relationship between mean population and estimated equilibrium catch, Pacific halibut south of Cape Spencer.

DYNAMICS OF POPULATIONS 35

TABLE 1. Estimation of Equilibrium Catches for the Population of Pacific Halibut of the Region South of Cape Spencer (I. F. C. Area 2)

Year	Catch in 1000's of pounds (k_1F_tP)	Catch per unit of effort in pounds per skate $(U = k_2P)$	$\overline{P}$	P_0	$\triangle P$	$\overline{f(P)}$
1915	44,023	117.5	91,415			
1916	30,278	114.1	88,770	90,092	−14,082	16,196
1917	30,803	81.3	63,251	76,010	−10,542	20,261
1918	26,270	87.0	67,686	65,468	+ 195	26,465
1919	26,602	81.8	63,640	65,663	− 1,323	25,279
1920	32,358	83.6	65,041	64,340	− 2,100	30,258
1921	36,572	76.4	59,439	62,240	− 8,364	28,208
1922	30,482	62.1	48,314	53,876	− 7,663	22,819
1923	28,008	56.7	44,113	46,213	− 2,645	25,363
1924	26,155	55.3	43,023	43,568	− 2,101	24,054
1925	22,637	51.3	39,911	41,467	− 1,360	21,277
1926	24,711	51.7	40,223	40,107	− 974	23,737
1927	22,934	48.9	38,044	39,133	− 1,722	21,212
1928	25,416	47.3	36,799	37,411	− 3,530	21,886
1929	24,565	39.8	30,964	33,881	− 4,900	19,665
1930	21,387	34.7	26,997	28,981	+ 272	21,659
1931	21,627	40.5	31,509	29,253	+ 5,718	27,345
1932	21,988	49.4	38,433	34,971	+ 4,279	26,267
1933	22,530	51.5	40,067	39,250	+ 2,217	24,747
1934	22,638	55.1	42,868	41,467	+ 4,240	26,878
1935	22,817	62.4	48,547	45,707	− 311	22,506
1936	24,911	54.3	41,245	45,396	− 778	24,133
1937	26,024	60.4	46,991	44,618	+ 5,640	31,664
1938	24,975	68.8	53,526	50,258	+ 39	25,014
1939	27,354	60.5	47,069	50,297	− 2,372	24,982
1940	27,615	62.7	48,781	47,925	+ 233	27,848
1941	26,007	61.1	47,536	48,158	+ 622	26,629
1942	24,321	64.3	50,025	48,780	+ 4,707	29,028
1943	25,311	73.2	56,950	53,487	+ 7,819	33,130
1944	26,517	84.4	65,633	61,306	+ 2,840	29,357
1945	24,378	80.5	62,629	64,146	+ 39	24,417
1946	29,678	84.5	65,741	64,185	+ 2,100	31,778
1947	28,652	85.9	66,830	66,285		

$\overline{P}$ and $\overline{f(P)}$ in thousands of pounds, U in pounds per skate) under the criterion of least squares, we obtain

$$\frac{k_1}{k_2} = 4.64 \qquad L_\mu = 156.1$$

This curve is plotted as the solid line in Figure 2*. It may be seen that it has a maximum value of 28.25 million pounds for the equilibrium catch at $k_2\overline{P} = 78.05$ pounds per skate.

This curve depends, of course, only on the points to which it is fitted, and may be rather different beyond those points from the curve which would be obtained if we had some values of $\overline{f(P)}$ for higher population levels. The calculated maximum population, corresponding to 156.1 pounds

*See footnote, Page 37.

DYNAMICS OF POPULATIONS

per skate, is much less than is shown by the available data of catch per skate for the early years of the fishery. From the few data available it is indicated by the International Fisheries Commission (Thompson and Bell 1934, table 1) that in the early 1900's the catch per skate was as high as 270 or 280 pounds. This is not, however, necessarily inconsistent with our results, since in the early days of the fishery the vessels may have been operating on local concentrations of halibut more abundant than the average for the entire area fished in later years. Thompson (1950, p. 2) states of the records on which these values are based: "It is my opinion, from personal experience, that such records showed a higher catch per set than the present comprehensive methods of collecting would have shown."

On the other hand, if we assume that the data from the 1900's are representative of the population in an almost unfished condition, so that the maximum population which the area will support corresponds to a catch per skate of, say, 275 pounds, we may fit a logistic to the available points, as before, but with the further restriction that $L_\infty = 275$. This results in a value of $k_1/k_2^2 = 1.95$. This curve is plotted as a broken line in Figure 2. It will be seen that now the estimated maximum equilibrium catch is 36.9 million pounds at a population corresponding to 137.5 pounds per skate. This result is not entirely unreasonable in the light of the total catches of 50 to 60 million pounds per year which were actually obtained by the fishery at its peak of production. (Thompson and Bell, table 1).

It is, it seems, not possible from the data to estimate precisely the population level giving the maximum equilibrium yield. We can, however, state with some certainty that it is at least as high as about 62 million pounds, corresponding to a catch per skate in the neighborhood of 80 pounds, and that at lower values the stock is overfished. This limited conclusion is, however, of very great interest in view of current controversy over the effect of regulation on the halibut stocks.

The nature of the growth of the amount of fishing

The intensity of fishing also may be expected to increase or decrease according to some regular law in response to economic factors. In general, as in any business, new investment of capital and effort will be attracted

---

*It may now be indicated why Thompson's method of determining $1/k_2$ gives a value higher than that from the tagging data. He assumed that the equilibrium catch for the years 1926 to 1933 was a constant. Actually the equilibrium catch was not constant over this period. The deviations of actual catches for this series of years from the equilibrium catches estimated from the logistic with the constants indicated are, on the average, greater than the deviations from the average of $\overline{f(P)}$ over the same period of years. As may be observed from Thompson's formulae on p. 20 of his paper, this will result in a higher value of his "K", which is the same as our $1/k_2$.

to come into the fishery as long as the expected return is equal to or greater than that from alternative enterprises in which the investment might be made. Put in another form, we may state this according to the theory of the "marginal" factor, according to which the cost of the last unit of fishing effort applied will, in general, be equal to the return from that unit.

Under the economic system in effect in most parts of the world, in which the above type of law holds true, as the fishery proves profitable, vessels and fishermen are attracted to it, increasing the rate of catching. This, of course, results in a decrease in the population of fish, lowering the return to each unit of fishing effort, and making the fishery less attractive to new investment. Ultimately, as the fishery grows, that level of fish population will be reached at which the return per unit of effort is so low that the cost of the next unit will be greater than the return from it. If the population falls below this level, vessels will tend to leave the fishery. This may be formulated

$$\frac{dF}{dt} = \psi(F, P - b) \quad (10)$$

where ψ is positive when $P > b$ and negative when $P < b$, F being, as before, the number of units of fishing effort, and b the critical level of fish population at which further investment in fishing becomes unprofitable.

To arrive at a particular function to describe the change of the intensity of fishing with the size of the population, we may consider that the incentive for new investment is proportional to the return to be expected, in which case there will be a linear relation between the percentage rate of change of fishing intensity and the difference between the level of fish population and its economically critical level, b. This function will, then, be

$$\frac{dF}{dt} = k_2 F(P - b) \quad (11)$$

where k_2 is a constant.

It may be noted that this is the law of growth of predator populations which has beeen arrived at in various predator-prey studies, for example Lotka (1925, p. 88), Volterra and d'Ancona (1935).

Stabilization of an unregulated fishery

Equations (3) and (10), taken simultaneously, describe the mutual interaction of a population of fish, the growth law of which is specified in (1) and a "population" of fishermen, the growth law of which is specified by equation (10). A very general model of a fishery is, then, given by the simultaneous equations

$$\left. \begin{array}{l} \frac{dP}{dt} = f(P) - P \phi(F) \\ \frac{dF}{dt} = \psi(F, P - b) \end{array} \right\} \quad (12)$$

An important special case is a population of fish the growth law of which is the logistic, being fished under economic circumstances such that

DYNAMICS OF POPULATIONS

the intensity of fishing has the growth law (11), and where the rate of catching is proportional to the number of units of fishing effort. In this case, the interaction of the fish and fishermen is described by the simultaneous equations

$$\left. \begin{array}{l} \dfrac{dP}{dt} = k_r P(L - P) - k_z PF \\ \dfrac{dF}{dt} = k_s F(P - b) \end{array} \right\} \quad \quad (13)$$

For the reasons which have been given in previous discussion, it is believed that this pair of equations is sufficiently descriptive of the actual laws under which a commercial fishery operates to be of utility in analysis of its dynamics, and will be employed in investigation of the nature of the development of fisheries. Certain important results may be obtained, however, from consideration of the more general pair of equations (12).

As has been pointed out previously, by the first equation of (12) the population of fish and the corresponding equilibrium catch may be stabilized at any level by regulating the amount of fishing, since $\dfrac{dP}{dt} = 0$ whenever $f(P) = P\phi(F)$. The change in fishing intensity in an unregulated fishery will be zero, however, only at $P = b$, so that the system can only be in equilibrium naturally at $P = b$. If the system is such that it will come to a stable equilibrium at all, it will, under no regulation, reach stability of itself at the economically critical population level $P = b$.

This has implications of importance to fishery management:

(1) If b is above the value of P at which $f(P)$ is maximum, the intensity of fishing will cease to increase at a level of fish population greater than that at which the maximum equilibrium catch might be obtained. In this case regulation of the fishery cannot increase the average yield of the fishery.

(2) If b is below the value of P at which $f(P)$ is maximum, it will be possible to increase the equilibrium catch by curtailing the amount of fishing and, if sufficient information is available, to establish that rate of fishing which will result in the fish population which will give maximum equilibrium catch.

It is to be noted that the system defined by the simultaneous equations (13), which seem realistic for describing existing commercial marine fisheries, is such that it will come to stability of itself with $P = b$ and $F = \dfrac{k_r}{k_z}(L - b)$. It will be shown later (p. 41) that this is a point of stable equilibrium. The *manner* in which it arrives at stability will also be discussed subsequently.

SCHAEFER

The course of development of an unregulated fishery and the manner of approach to stable equilibrium

It is of considerable importance to our understanding of the fisheries to investigate the manner in which the population of fish and the amount of fishing react with each other in the course of development of the fishery. We may take as our mathematical model the pair of differential equations (13) and investigate the nature of the solutions. The initial conditions, when the fishery starts, are that P is equal to L and F is small.

There does not seem to be any formal solution of these equations giving P and F as functions of time. It is possible, however, to obtain approximate solutions by means of numerical procedures. First, however, it will be profitable to investigate some of the general properties of the solutions.

By dividing the first equation of (13) by the second, we may obtain an equation in P and F.

$$\frac{dP}{dF} = \frac{k_1 P(L - P) - k_2 PF}{k_3 F(P - b)} \quad \quad (14)$$

This equation does not have, so far as I can ascertain, a formal solution. The general nature of the solution may be investigated, however. It may be seen that there is a line of horizontal tangents, $\frac{dP}{dF} = 0$, when

$$k_1(L - P) = k_2 F, \text{ or } F = \frac{k_1}{k_2}(L - P), \quad P \neq b$$

there is a line of vertical tangents, $dF/dP = 0$

when $\quad P = b, \quad F \neq \frac{k_1}{k_2}(L - b)$

At $F = \frac{k_1}{k_2}(L - b)$, $P = b$ there is a singular point.

Furthermore, it may be seen that

when $P > b$ and $F > \frac{k_1}{k_2}(L - P)$ $\quad\quad\quad \frac{dP}{dF}$ is negative

when $P > b$ and $F < \frac{k_1}{k_2}(L - P)$ $\quad\quad\quad \frac{dP}{dF}$ is positive

when $P < b$ and $F > \frac{k_1}{k_2}(L - P)$ $\quad\quad\quad \frac{dP}{dF}$ is positive

when $P < b$ and $F < \frac{k_1}{k_2}(L - P)$ $\quad\quad\quad \frac{dP}{dF}$ is negative

From this it may be seen that (14) might represent either a family of closed curves or spirals about the singular point. The corresponding solutions of (13) for P and F as functions of t are in the first case an undamped oscillatory function, and in the second a damped oscillatory function, approaching $P = b$ and $F = \frac{k_1}{k_2}(L - b)$ in the limit. In the latter case, the singular point of (14) is a point of stable equilibrium, in the former it is not.

We may examine the behaviour of the solution of (14) in the vicinity of the singular point to determine which it may be.

DYNAMICS OF POPULATIONS

Lotka (1923) has investigated a more general pair of differential equations, which include ours as a special case. He has shown that if a certain descriminant is less than, equal to, or greater than zero, the solution is a spiral winding inward toward the singular point, a family of closed curves, or a spiral winding outward. (The last solution is, of course, impossible in our case from the physical conditions). The value of Lotka's discriminant "R" of our equations is

$$R = \frac{-\frac{1}{2}(k_2 b)^{\frac{1}{2}}\left[\frac{k_1 k_2}{k_2}(L - b)\right]^{-\frac{1}{2}}}{\frac{k_1}{k_2}(L - P)}$$

Since $(L - b)$ and $(L - P)$ are always positive in our case, R is always less than zero.

Therefore, according to Lotka's analysis, the solution would be a spiral winding inward toward the singular point. A diagram of this solution is shown in Figure 3.

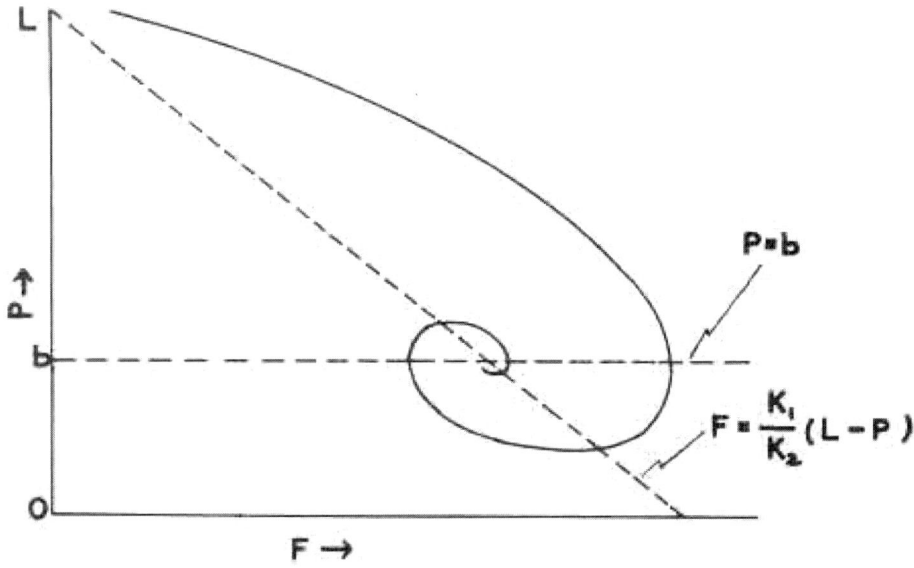

Figure 3. Solution of equation (14), according to Lotka's analysis.

It appears from this that in general P, the fish population, taken as a function of time, follows some damped oscillatory function, fluctuating above and below $P = b$, but the amplitude of the fluctuations getting smaller and smaller. Similarly F, the amount of fishing effort, follows a similar function, approaching $F = \frac{k_1}{k_2}(L - b)$ in the limit. The singular point is a point of stable equilibrium. It is the point which F and P tend to approach, and is the only point where the unregulated fishery will stay in natural equilibrium.

We may note in passing that the line of horizontal tangents $F = \frac{k_1}{k_2}(L - P)$ is the locus of the values of F and P corresponding to the equilibrium condition, that is if we hold F constant at any given value, by regulation, the corresponding value of P lies on this line, when the catch and the natural rate of increase are in equilibrium.

The general nature of the solutions of (13) may also be investigated more directly. If we differentiate the first equation of the pair and substitute from the second, we obtain a differential equation of the second degree in P alone:

$$\frac{d^2P}{dt^2} - \frac{1}{P}\left(\frac{dP}{dt}\right)^2 + \left[k_1 P - k_3(P - b)\right]\frac{dP}{dt} + k_1 k_3 P(P - b)(L - P) = 0 \quad (15)$$

This equation cannot, so far as I can see, be solved formally. However, we may investigate the solutions in the neighborhood of the singular point of (14). Taking a new origin at b, by taking $P = N + b$, we obtain

$$\frac{d^2N}{dt^2} = \frac{1}{N+b}\left(\frac{dN}{dt}\right)^2 - (k_1 - k_3) N \frac{dN}{dt} - k_1 b \frac{dN}{dt}$$
$$- k_1 k_3 \left[(L - b) bN + (L - 2b) N^2 - N^3\right] \quad (16)$$

In the vicinity of the origin, i.e. for very small values of N, we may neglect all terms of the second and higher degree, provided also that $\frac{dN}{dt}$ is small in the vicinity of the singular point of (14). We obtain:

$$\frac{d^2N}{dt^2} \approx - k_1 b \frac{dN}{dt} - k_1 k_3 b(L - b)N \quad (17)$$

Or

$$\frac{d^2N}{dt^2} + k_1 b \frac{dN}{dt} + k_1 k_3 b(L - b)N \approx 0 \quad (18)$$

which is a homogeneous linear equation with constant coefficients. The roots of the characteristic equation are

$$\frac{- k_1 b \pm \sqrt{k_1^2 b^2 - 4k_1 k_3 (L - b)b}}{2} \quad (19)$$

The form of the solution will depend on whether the roots are real or complex, i.e. whether the term under the radical is positive or negative. If the roots are complex, the solution will be of the form

$$N = e^{\frac{-k_1 bt}{2}} \left(C_1 \cos\frac{a}{2}t + C_2 \sin\frac{a}{2}t \right) \quad (20)$$

where $a = \sqrt{4k_1 k_3 b(L-b) - k_1^2 b^2}$ and C_1, C_2 are constants of integration. This solution is, of course, a damped harmonic. As $t \to \infty$, $N \to 0$, approaching the limit by oscillating above and below $N = 0$. This solution is the same kind obtained by Lotka's analysis.

DYNAMICS OF POPULATIONS

On the other hand, if the roots are real, the solution will be of the form

$$N = C_1 e^{-\left(\frac{k_1 b}{2} + \frac{a}{2}\right)t} + C_2 e^{-\left(\frac{k_1 b}{2} - \frac{a}{2}\right)t} \quad \quad (21)$$

where $a = \sqrt{k_1^2 b^2 - 4k_1 k_2 b(L-b)}$

Here, as $t \to \infty$, $N \to 0$, so that the origin is a point of stability, but in this case it is approached from one side only, the equilibrium condition being approached smoothly without oscillations.

The solution will be oscillatory or not depending on whether

$$k_1^2 b^2 \lesseqgtr 4k_1 k_2 b(L-b)$$

or

$$k_1 \lesseqgtr 4k_2 \frac{(L-b)}{b} \quad \quad (22)$$

In the case of real roots, where the point of stable equilibrium is approached from one side only, P is always greater than b, and, correspondingly, F is always greater than $\frac{k_1}{k_2}(L-P)$. Thus, the resulting relationship between F and P, which is a solution of (14), is, in this case, not a spiral, but is a curve remaining always on the positive sides of the lines of horizontal and vertical tangents, and terminating in the limit in the singular point.

These considerations tell us something about the general nature of the solutions and their behavior near the point of stable equilibrium. In order to find out in more detail the changes in the fish population, amount of fishing, and catch, recourse may be taken to approximation methods for solving the equations (13). We have employed the method due to Lord Kelvin, as described by Willers (1948, p. 394 et. seq.) to obtain a graphical solution to the equation (15) and the corresponding equation for F as a function of t. Solutions have been computed for two examples, one of which has complex roots and one of which has real roots for equation (18).

In the first example we have taken

$$k_1 = k_2 \quad \quad b = 0.3L$$
initial conditions, $P = L \quad \quad F = 0.1$

The resulting solution, showing F and P as functions of t, is traced out in Figure 4. It may be seen that the fish population and the intensity of fishing approach the condition of stable equilibrium as a series of damped oscillations. The catch, which is proportional to the product of F and P, also oscillates about its point of stable equilibrium, as may be seen from the graph of catch in the same figure. It is of interest to note that on the first swing the catch rises far above its ultimate position of stable equilibrium, and also above the level of maximum equilibrium catch, which also

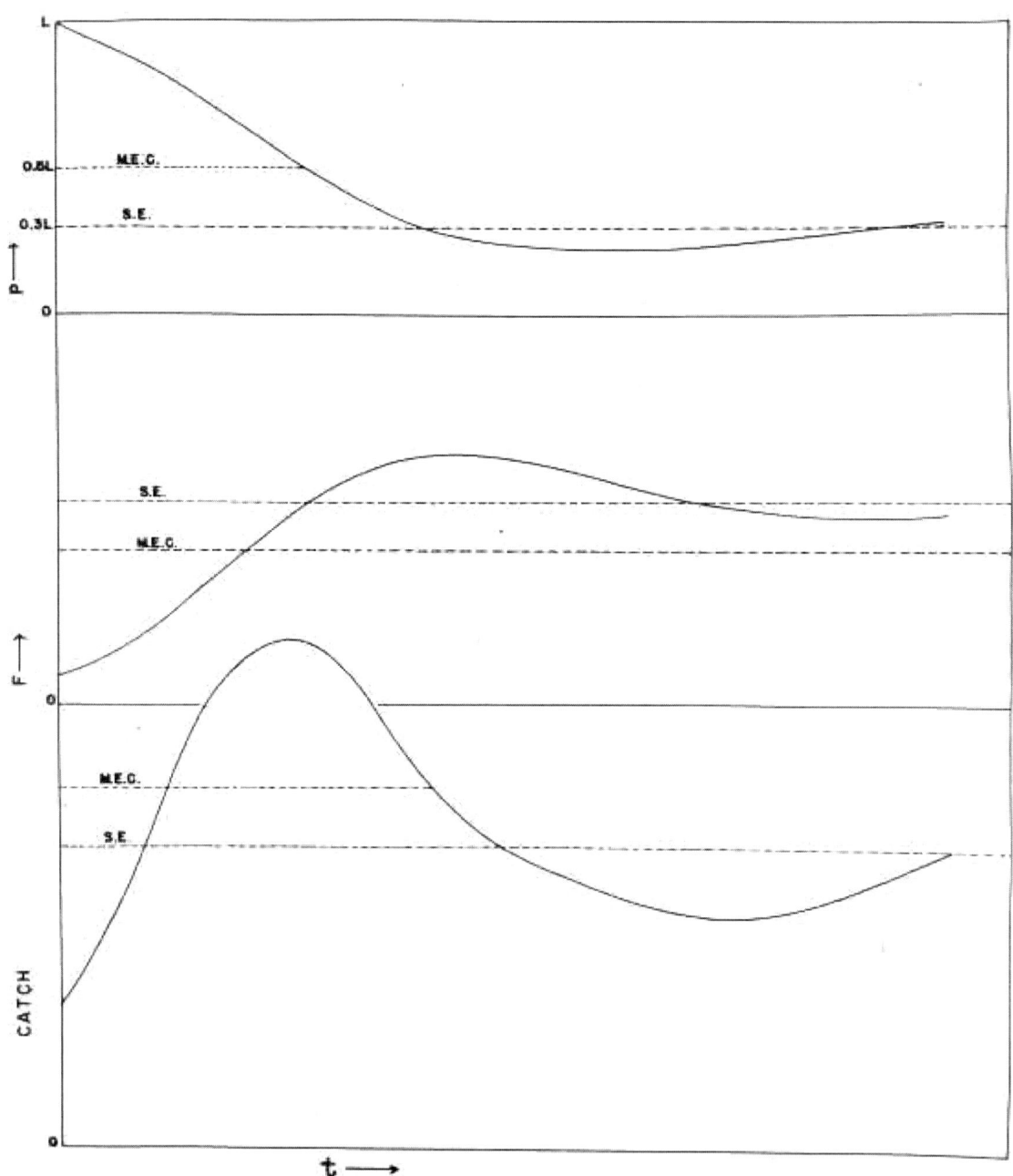

Figure 4. Solution of equations (13) for $k_1 = k_2$, $b = 0.3L$. (S.E. indicates level of stable equilibrium. M.E.C. indicates level of maximum equilibrium catch).

DYNAMICS OF POPULATIONS

is indicated in the figure. The relationship between F and P is plotted in Figure 5 for the values of the two variables which have been computed for this example. It may be seen that, as was deduced, it forms a spiral winding inward toward the singular point of stable equilibrium.

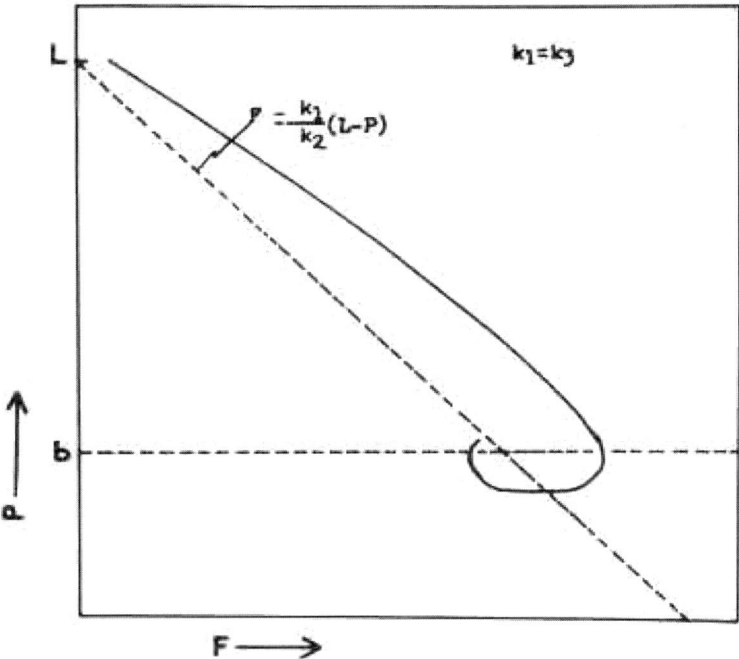

Figure 5. Solution of equations (13) for $k_1 = k_3$, $b = 0.3L$.

For the second example, for which the roots of equation (18) are real, we have taken
$$k_1 = 15k_3 \qquad b = 0.3L$$
initial conditions, $P = L \qquad F = 0.1$

The resulting solution, traced out by the method of approximation cited, showing F and P as functions of t, is graphed in Figure 6. In this case, as we expected, F and P approach the condition of stable equilibrium asymptotically from one side only. The curve of catch, however, rises well above the final stabilization level (and also somewhat above the level of maximum equilibrium catch) then approaches it asymptotically from above.

Finally, for this second example, the relationship between F and P is plotted in Figure 7, showing that its form corresponds to what we expected from the general considerations.

It is of some interest to note that Volterra and D'Ancona (1935, pp. 44-45) stated as a theorem for a system of equations which includes ours as a special case, that these two types of solutions would be found.

DYNAMICS OF POPULATIONS

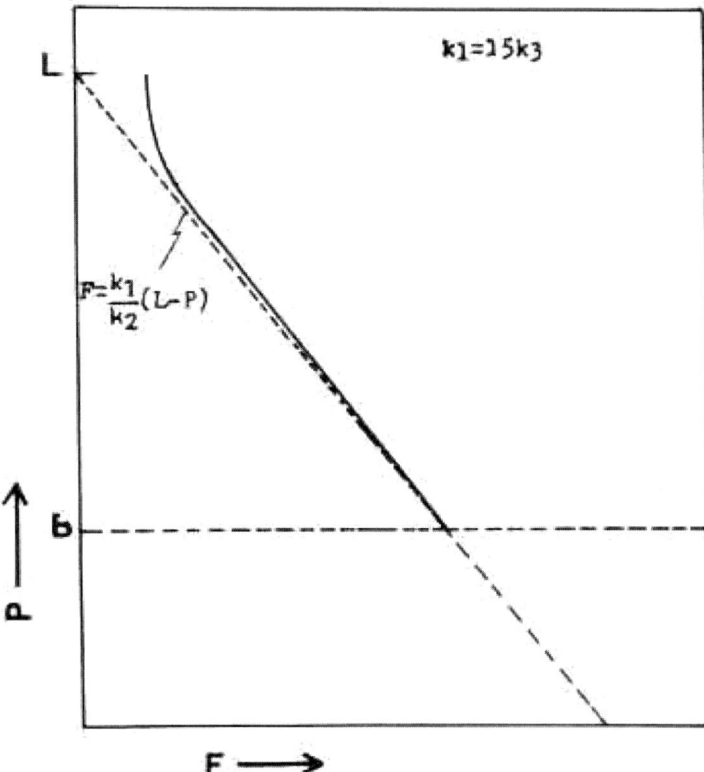

Figure 7. Solution of equations (13) for $k_1 = 15k_3$, $b = 0.3L$.

Two conclusions of importance to fisheries management follow from this analysis, if the mathematical model (13) fairly represents a commercial fishery:

1) Large scale fluctuations in fish population and catch can arise as a result of the interaction of the forces of growth of the fish population and growth of the intensity of fishing, with all other conditions constant.

2) During the development of a fishery, it is to be expected that in the course of reducing the stock of fish from its virgin condition, the catch will rise for a short time well above the level at which it will reach natural stable equilibrium, and also well above the maximum equilibrium catch. The task which conservationists have sometimes set themselves of restoring a fishery to the highest historical levels of production is, in this event, unobtainable on a permanent basis.

SCHAEFER

Examples from the commercial fisheries

It is of interest to see how well our model may be applied to the actual data of some commercial fisheries. This may be best accomplished, perhaps, by presenting the data of some well-documented fisheries in the form of Figure 5 (or 7) by plotting the observed intensity of fishing against the corresponding observed population values. It is convenient to plot U rather than $\overline{P}$, since $U = k_r\overline{P}$, and U is the datum which is obtained directly from the statistical records of the fishery. This form has the further advantage that the product of the ordinate and abscissa for any point along the line of values of $\frac{dU}{dt} = 0$ $\left[\text{i.e. the line } F_t = \frac{k_1}{k_2'}(L_u - U)\text{ which is equivalent to } F_t = \frac{k_1}{k_2}(L - \overline{P})\right]$, and which we will call the line of equilibrium conditions, is the equilibrium catch.

There have been noted previously certain properties, under our theory, of this sort of diagram which should be remembered here. When $\frac{dU}{dt}$ is negative, that is when the population is declining, the catch being greater than the equilibrium catch, the values of F_t, U will fall to the right of the line of equilibrium conditions. Conversely, when $\frac{dU}{dt}$ is positive, that is when the population is increasing due to the catch being less than the equilibrium catch, the values of F_t, U will fall to the left of the line of equilibrium conditions. This property, as well as the equation of the line of equilibrium conditions depends *only* on the first equation of (13) and is completely independent of the second equation. So long as the first equation of (13) correctly describes the natural rate of increase and the catch, the line of equilibrium conditions is determined, and the properties mentioned hold regardless of the way in which F_t varies in relation to P and t. The points F_t, U will fall to the right or left of the line depending on whether the catch is greater or less than the equilibrium catch for the particular value of P.

If, *in addition*, the second equation of (13) is true, the successive values of F_t, U will form a curve corresponding to one of the joint solutions of the pair of equations, as indicated above. It should be noted here that we have assumed that the economically critical level is constant. This may be expected to be true only over a relatively limited time, since it will be influenced by technical developments as well as by the general business cycle.

We may expect the data of commercial fisheries to correspond to the properties of the model if the assumptions underlying the model are fairly well fulfilled by the fisheries. The most basic assumption was that the growth of the fish population is a function of the size of the population, and is not, therefore, subject to important variations due to other causes. If there exist other causes influencing the growth of the population of fish,

DYNAMICS OF POPULATIONS

such as variations in the environment, which give rise to variations in the population growth and which are large in comparison to the changes due to population size alone, we shall expect our plot of F_t, U to exhibit a quite different pattern. In this case we should have large changes in population size in either direction quite independent of antecedent changes in the amount of fishing. As a result, the increases or decreases in population will have little orderly sequence, except as they may be related to cyclic phenomena, and will bear little relation to the location of any line of average equilibrium conditions.

Pacific Halibut

We may first consider the fishery for Pacific halibut on the Southern Grounds, which was the subject of an earlier example (p. 33). Thompson (1950, table 2) gives values of F_t and U for this fishery from 1916 through 1947. Until 1931 the fishery operated essentially without regulation of the catch or intensity of fishing, so that the population of halibut and the intensity of fishing were free to interact according to natural and economic laws. After 1931, the fishery was regulated by placing quota limits on the catch in order to build up the halibut population.

If one examines Figure 2 of Thompson and Bell (1934), in which is depicted the historical record of this fishery up to the time of regulation, it will be observed how similar the curve of landings is to our theoretical curve in Figure 4, as well as the general similarity of the curves for fishing intensity and fish population, which however, are available only for the period after the peak of the catch had been reached.

Values of F_t, U for successive years are plotted from Thompson's (1950) table 2 in our Figure 8. There has also been drawn the line of equilibrium conditions, using the values of the constants estimated by our Previous analysis (p. 36) $\frac{k_1}{k_2^2} = 4.64$, $L_u = 156.1$.

It may be seen that the picture is not inconsistent with the theory. From 1916 until 1930, the stock was falling and the plotted points remain generally to the right of the line of equilibrium conditions. It may be that by 1930, the stock had fallen below an economically critical level and the fishing intensity had commenced, in consequence, to decrease; however, regulation of the fishery commenced in 1931 so we cannot tell much about this. From 1932 to 1947, during which the stock was being built up, the plotted points remain to the left of the line of equilibrium conditions, approaching it closely in the last years of the series when the fishery was becoming stabilized under regulation. This corresponds to just what would be expected from the theoretical considerations discussed above.

We have also plotted on this diagram, as a light broken line, the line of equilibrium conditions corresponding to the fish population logistic with upper asymptote at $L_u = 275$, which, it will be remembered, was obtained

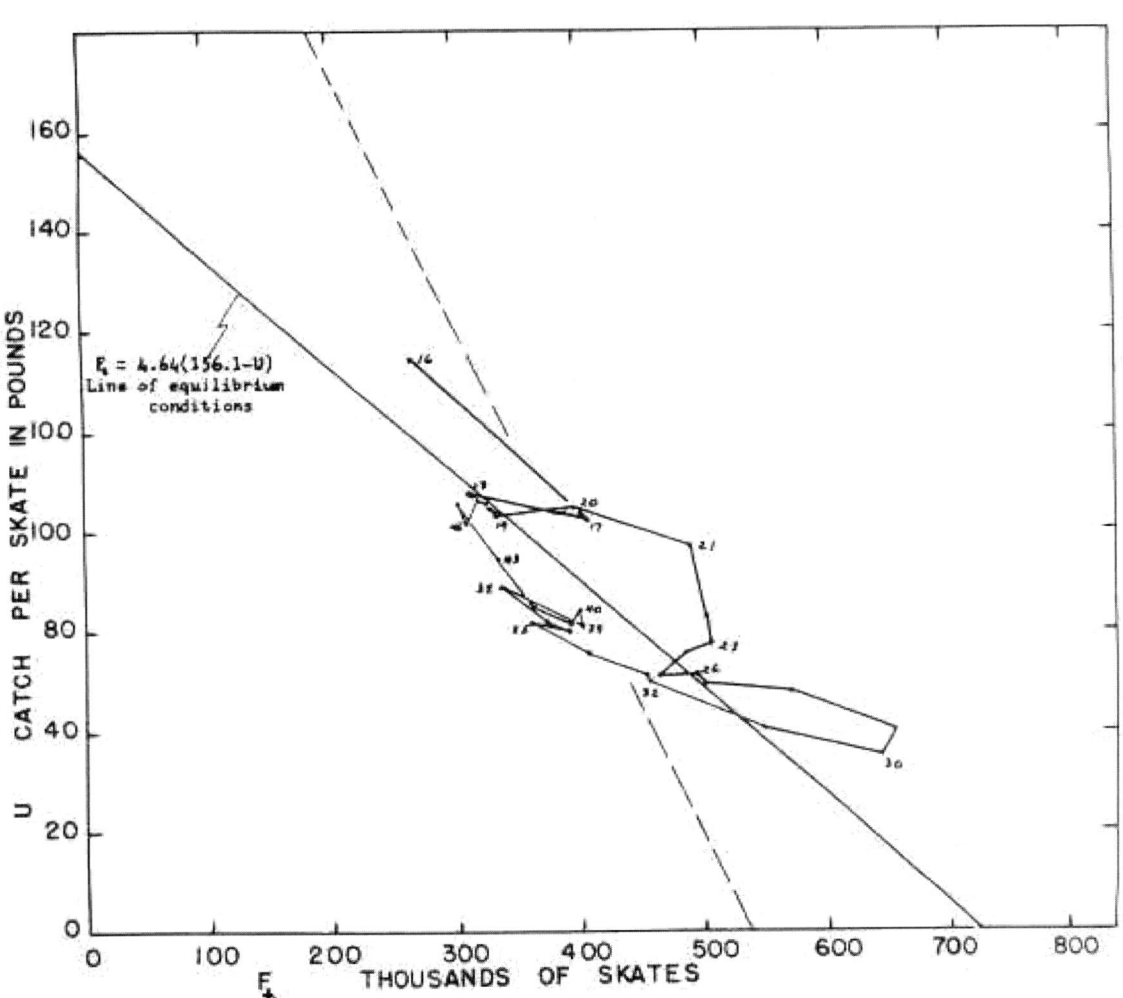

Figure 8. Relationship between intensity of fishing and mean population, Pacific halibut south of Cape Spencer.

DYNAMICS OF POPULATIONS

from the data of the fishery if we specified that the curve should go through this value for $\overline{P} = L$. It may be seen that, employing this line, the plotted points do not form a very reasonable pattern in the light of the theory. It seems therefore most probable that the fish population growth curve with $L_s = 156.1$ corresponds more nearly to reality. If this line of equilibrium conditions is valid for this stock of halibut, the maximum equilibrium catch is 28.2 million pounds with a stabilized fishing intensity of 362 thousand skates. Due to lack of data at higher levels of population this estimate may not be quite correct, however, as pointed out previously.

California Sardine (Pacific Pilchard)

The fishery for the California Sardine, or Pacific Pilchard, has been the subject of much study, and of some notable differences of opinion, during the course of its growth and subsequent decline. It may be instructive to see what sort of results are obtained from considering the population statistics of this fishery in the light of the theory herein developed. In view of the widely held opinion that the major changes in the sardine population have been due to variation in environmental conditions, this is, of course, a bold attempt. In our treatment, variation due to environment is treated as a random variable, independent of P.

Statistics of total catch along the Pacific Coast are available since the early days of the fishery (Schaefer, Sette and Marr 1951, Clark 1952), but data on catch-per-unit-of-effort are available only since the 1932-33 season. The period covered by data on abundance commences, as may be seen from Figure 1 of Schaefer, Sette and Marr, after the fishery was well along in its development, but yet considerably prior to the peak of total catch. In this respect, the data are available for a period commencing when the fishery is "younger" than the initial point of the series for Pacific halibut.

Figures of total catch along the Pacific Coast were taken from Table 1 of Clark (1952). Figures of abundance (catch per California boat month linked to the base year 1941) were taken from Table 3 of Clark and Daugherty (1952). The total catch divided by the catch per boat month gives the apparent intensity of fishing in terms of 1941 California boat-months.

In order to translate values of abundance, U, into values of $\overline{P}$ we need to evaluate the constant $1/k_2$ in (9). To do this, we have referred to the results of California tagging experiments from 1936 to 1943, from which it appears that the average annual fishing mortality rate during that period was about 43 percent (Clark and Jansen 1945) for the California fishery. Using the average of the intensity of fishing in California, as given by Clark and Daugherty (1952), during the same period we have obtained $1/k_2 = 2.197 \times 10^5$ (for U in tons per boat month and $\overline{P}$ in tons). Proceeding as before (p. 33) we then estimated the equilibrium catch for each season from 1934-35 to 1949-50. Fitting a curve of the form $\overline{f(P)} =$

$\frac{k_1}{k_2^2} U(L_\infty - U)$ to the resulting data under a criterion of least squares (the curve was fitted to the points for each season and not to the means of several as in the halibut), Figure 9, we obtained

$$\overline{f(P)} = 1.253\ U\ (1385 - U)$$

where U is in tons per boat month and $\overline{f(P)}$ is in thousands of tons.

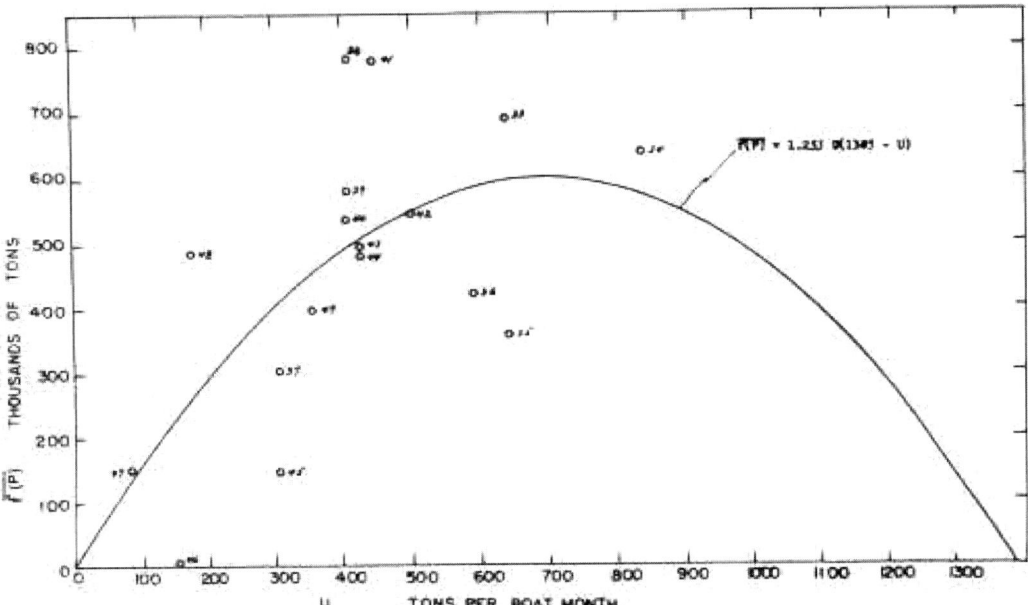

Figure 9. Relationship between mean population and estimated equilibrium catch, California sardine.

In Figure 10 we have plotted successive values of F_t, U for each season, the points being labeled by the first year of the pair for a season. We have also drawn the estimated line of equilibrium conditions

$$F_t = 1.253\ (1385 - U)$$

It may be perceived that the main features of the changes in population over this series of years are in accordance with what we would expect if they were determined by the amount of fishing, but that there are some notable aberrancies attributable to other causes between 1938 and 1942.

During the first two years of this series, the population is increasing, the catches being less than the equilibrium catch. This is doubtless a result of the curtailment of landings, and probably therefore of fishing effort, during the immediately preceding years, as a result of the economic depression (see Figures 1 and 6 of Schaefer, Sette and Marr). In other words, there appears to have been temporarily an economically critical level some-

DYNAMICS OF POPULATIONS

where near 600 tons per boat month. From 1934 to 1937, the population was declining, the decline being associated with catches above the equilibrium catches for the corresponding population sizes. From 1938 to 1942, the pattern is somewhat confused. During this period two things transpired which may account for much of this: (1) There were unusually good year classes entering the fishery from the spawnings of 1938 and 1939 (Clark 1952, Clark and Daugherty, 1952), so that the population growth was greater than would be expected from the average relation of population size to growth of population. (2) The fishery was affected by some restrictions and disturbances at the outbreak of the war. From 1942 to 1947, the population is again rather steadily declining, with the catch above the equilibrium catch. From 1947 to 1949, the catch is below the equilibrium catch and the population is increasing. In 1950 the population again shows some decline, which is associated with a catch above the equilibrium value. From this it would appear that there may be an economically critical level near 150 tons per boat month, and if left to stabilize by itself the fishery would tend to fluctuate about the corresponding population size, at which the equilibrium catch is, on the average, about 232,000 tons per year.

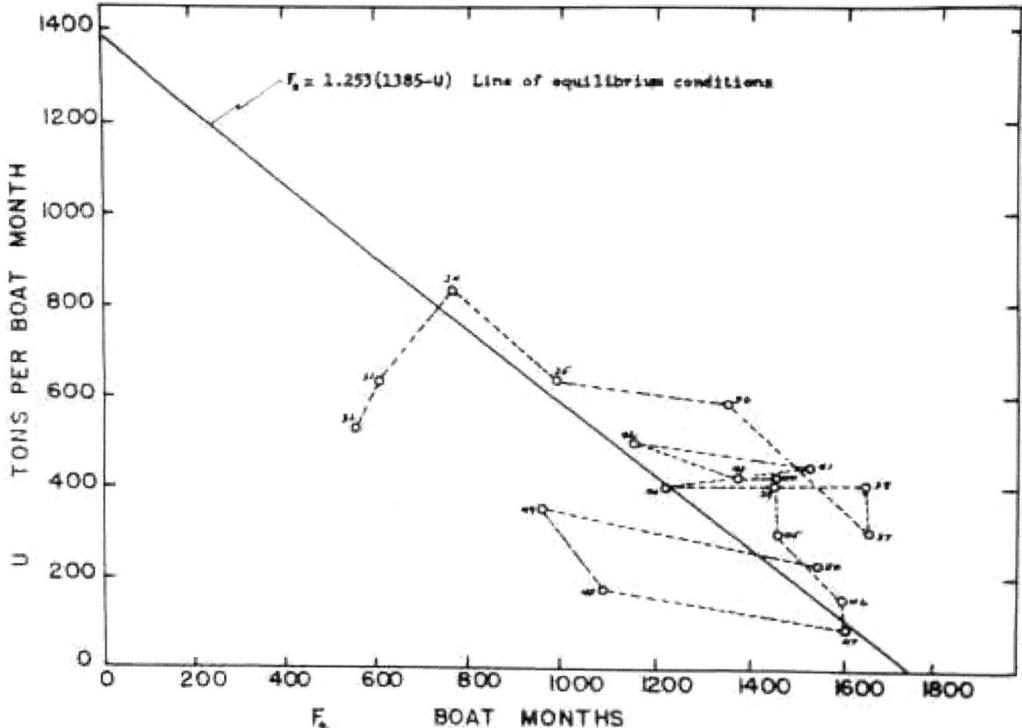

Figure 10. Relationship between intensity of fishing and mean population, California sardine.

SCHAEFER

The pattern of changes in the sardine population, then, over the period for which data are available, appears to be in general consistent with the hypothesis that one of the major causes of the changes has been the associated changes in the intensity of fishing. If the estimated line of equilibrium conditions is correct for this population, the average maximum equilibrium catch will be about 601 thousand tons at a population level corresponding to 692 tons per boat month (a mean population of about 1,520,000 tons), with a stabilized fishing intensity of 868 boat months per year.

The foregoing examples illustrate how the theory developed here may be employed to make estimates concerning the condition of a commercial marine fishery. The examples employed, although having perhaps as complete information as any available for this purpose, leave something to be desired. In particular, in both of these examples, very little or no data are available concerning intensity of fishing and abundance for the early period of development of the fishery, well before the maximum catches are reached. A great deal of precision would be added to the estimate if such information were available.

We may emphasize, therefore, the desirability of obtaining detailed information on the total catch and catch-per-unit-of-effort from as early in the development of a commercial fishery as may be possible. Measurements of fishing mortality rates at more than one level of population would also be desirable, since they would make possible verification of the adequacy of the form of equation (13a) for describing the changes in population under the joint influences of growth and fishing.

In order to apply the theory developed here to the tropical tuna fishery, it will be necessary to compile statistics of catch, abundance and intensity of fishing over a considerable series of years, beginning as early in the history of the fishery as possible. This task is well under way. It will also be necessary to obtain some estimate of the rate of fishing mortality, or to devise some other means of estimating the constant k_2. Estimation of fishing mortality from tagging promises to be a difficult problem for the tunas. Exploration of other means of obtaining the relationship between U and P appears, therefore, to constitute an important line of investigation.

Literature Cited

Baerends, G. P.

1947 De rationeele exploitatie van den zeevischstand, in het bijzonder van den vischstand van de Noordzee.

    Min. U. Landb., Visscherij en Voldselvoorz Verslagen en Mededeelingen van de Afdieling Visscherijen No. 36, 80 pp.

    (Eng. translation in U. S. Fish and Wildlife Service Special Scientific Report—Fisheries, No. 13)

DYNAMICS OF POPULATIONS

Büchmann, A.
- 1938 Ueber den Höchstertrag der Fischerei und die Gesetze organischen Wachstums.
  Ber. der Deutschen Wissench. Komm.f. Meersef. N.F., Vol. 9, pp. 16-48

Burkenroad, M.
- 1951 Some principles of marine fishery biology.
  Publ. Inst. Mar. Sci., Vol. II, No. 1, pp. 177-212
- 1953 Theory and practice of marine fishery management.
  Jour. du Conseil, Vol. 18, No. 3, pp. 300-310

Clark, F. N.
- 1952 Review of the California Sardine Fishery.
  Cal. Fish and Game, Vol. 38, No. 3, pp. 367-380

Clark, F. N. and A. E. Daugherty
- 1952 Average lunar month catch by California sardine fishermen, 1949-50 and 1950-51.
  Cal. Fish & Game, Vol. 38, No. 1, pp. 85-97

Clark, F. N. and J. F. Janssen
- 1945 Movements and abundance of sardine as measured by tag returns.
  Cal. Fish & Game, Bull. No. 61, pp. 7-42

Gause, G. F.
- 1934 The struggle for existence.
  Baltimore, Williams & Wilkins Co., 163 pp.

Graham, M.
- 1935 Modern theory of exploiting a fishery, and applications to North Sea trawling.
  Jour. du Cons., Vol. 10, No. 3, pp. 263-274
- 1939 The sigmoid curve and the overfishing problem.
  Cons. Perm. Int. Expl. Mer., Rapp. et Proc. Verb., Vol. 110, pp. 15-20

Lotka, A. J.
- 1923 Contribution to quantitative parasitology.
  Jour. Wash. Acad. Sci., Vol. 13, No. 8, pp. 152-158
- 1925 Elements of physical biology.
  Baltimore, Williams & Wilkins Co., 460 pp.

Nicholson, A. J.
- 1947 Fluctuation of animal populations.
  Austr. and N. Z. Assoc. Adv. Sci., Perth Meeting, August 1947 Section D. Zoology, pp. 134-147
- 1950 Competition for food among *Lucilia caprina* larvae.
  8th Int. Congress of Entomology, Proceedings, Stockholm 1948, pp. 277-281

Pearl, Raymond
   1925   The biology of population growth.
          New York, Alfred A. Knopf, 260 pp.

Schaefer, M. B., O. E. Sette, and J. C. Marr
   1951   Growth of the Pacific Coast pilchard fishery to 1942.
          U. S. Fish & Wildlife Svc., Research Rept. 29, 31 pp.

Thompson, W. F.
   1950   The effect of fishing on stocks of halibut in the Pacific.
          Publ. Fish Res. Inst., U. of Wash., March 10, 1950, 60 pp.

Thompson, W. F. and F. H. Bell
   1934   Biological statistics of the Pacific halibut fishery.
          Report Int. Fish. Comm. No. 8, 49 pp.

Thompson, W. F. and W. C. Herrington
   1930   Life history of the Pacific halibut
          (1) Marking experiments.
          Rept. Int. Fish. Comm. No. 2, 137 pp.

Volterra, Vito
   1926   Variazioni e fluttuazioni del numero d'individui in specie animali conviventi.
          Mem. R. Acad. Naz. dei Lincei. Ser. VI, Vol. 2.

Volterra, Vito and D'Ancona, Umberto
   1935   Les associations biologiques au point de vue mathematique.
          Actualites Scientifiques et Industrielles 243, 96 pp.

Willers, F. A.
   1948   Practical analysis, graphical and numerical methods.
          (Translated by R. T. Beyer)
          New York, Dover Publications, 1948, 422 pp.

Adaptive Control of Fishing Systems

CARL J. WALTERS AND RAY HILBORN

Institute of Animal Resource Ecology, University of British Columbia, Vancouver, B.C. V6T 1W5

WALTERS, C. J., AND R. HILBORN. 1976. Adaptive control of fishing systems. J. Fish. Res. Board Can. 33: 145–159.

This paper discusses some formal techniques for deciding how harvesting policies should be modified in the face of uncertainty. Parameter estimation and dynamic optimization methods are combined for the Ricker stock-recruitment model to show how exploitation rates should be manipulated to give more information about the model parameters; in general, harvesting rates should be lower than would be predicted by the best fitting recruitment curve unless this curve predicts that the stock is very productive. A decision procedure is developed for comparing alternative stock-recruitment models; when applied to the Fraser River sockeye salmon (*Oncorhynchus nerka*), the procedure indicates that an experimental increase in escapements would be quite worthwhile. It appears that there is considerable promise for extending these methods and procedures to cases where the stock size is unknown and where fishing effort is poorly controlled.

WALTERS, C. J., AND R. HILBORN. 1976. Adaptive control of fishing systems. J. Fish. Res. Board Can. 33: 145–159.

Cet article analyse certaines méthodes formelles permettant de décider comment modifier les politiques de récolte face à l'incertitude. Nous combinons, dans le modèle de Ricker de recrutement des stocks, les méthodes d'estimation des paramètres et d'optimalisation dynamique, et démontrons comment les taux d'exploitation devraient être manipulés pour en arriver à une meilleure connaissance des paramètres du modèle; en général, les taux de récolte devraient être inférieurs à ceux que la meilleure courbe de recrutement pourrait prédire, à moins que cette courbe ne prédise que le stock est très productif. Nous élaborons un protocole de décision permettant de comparer des modèles alternatifs de recrutement des stocks; appliqué au saumon nerka (*Oncorhynchus nerka*) du fleuve Fraser, ce protocole indique qu'il vaudrait la peine d'augmenter expérimentalement la survie des saumons. L'application de ces méthodes et protocoles à des cas où la grandeur des stocks n'est pas connue et où l'effort de pêche est mal contrôlé semble très prometteuse.

Received July 26, 1975
Accepted October 20, 1975

Reçu le 26 juillet 1975
Accepté le 20 octobre 1975

A variety of dynamic models have been used recently to help establish fishery regulations and catch quotas, and increasingly elaborate monitoring systems have been developed to provide sound statistical estimates of model parameters. Most models are used only to predict optimum equilibrium harvest rates, though there have been a few attempts recently to develop harvest strategy curves or "control laws" that specify optimum harvest rates for nonequilibrium situations (Allen 1973; Walters, 1975). Given a time series from which model parameter estimates have been derived, it has often been assumed that the best management strategy is to act as though these estimates were actually correct; that is, we insert the estimates into the model (or into several alternative models), generate some yield curve or isopleth diagram that reveals an apparently optimum harvest policy, and then we recommend that this policy be followed. Little attention has been paid to the problem that by following the apparently optimum policy, the fishery might be brought to an equilibrium that is neither truly optimal nor productive of the kind of data necessary to determine the true optimum. Luckily, most fisheries have gone through a period of more or less uncontrolled development to the point of obvious overexploitation, allowing us to interpolate an optimum regime from data on a wide range of stock sizes. But then another problem arises; when a fishery has been held near some equilibrium for a long period of time, how much confidence can we have in the older data from the nonequilibrium period? Environmental carrying capacities may have changed, selection by the fishery may have produced new genetic types capable of responses, or the old data may be simply unreliable. Implicit in many research programs is the assumption that detailed biological studies on populations near equilibrium will allow a priori determination of optimum harvest-

Printed in Canada (J3920)
Imprimé au Canada (J3920)

ing policies, thus making it unnecessary to introduce trial and error changes or large-scale experiments in harvesting rates; this assumption appears to be naive and unjustified at present.

This paper addresses the question of how harvesting decisions should be modified to take account of statistical uncertainty. In seeking a formal framework for dealing with this question, we have been drawn to the literature on control system theory, where the problem is addressed under the heading of "adaptive" or "dual" control (Larson 1968). To simplify the discussion, we will concentrate primarily on situations where the stock-recruitment relation (rather than growth and natural mortality) is the critical determinant of protential yield. The analysis is divided into two parts: we first look at the case when a simple model, the Ricker stock-recruitment curve, is assumed to be the correct functional form and only the model parameters are uncertain; we then examine more general cases where the form or shape of the stock-recruitment function is uncertain, stock sizes are not directly measurable, and fishing effort is poorly controlled.

Adaptive Control with the Ricker Model

As indicated by several authors with many examples in Parrish (1973), the simple model developed by Ricker (1954) has seen wide use in the analysis of stock-recruitment relations:

$$R_t = S_{t-1} \exp\{\alpha - \beta S_{t-1} + \nu_t\} \quad (1)$$

where R_t = recruits (adults) at end of generation t,
S_{t-1} = spawners at the start of generation t,
α = a stock production parameter,
β = equilibrium stock parameter (equilibrium stock in absence of fishing is equal to α/β),
ν_t = a random environmental factor, normally distributed with mean 0.0 and variance σ^2.

For the discussion that follows, it is indeed critical that ν_t, the noise factor, be normally distributed; there is good empirical evidence for this assumption for a sockeye salmon (*Oncorhynchus nerka*) population on the Skeena River (Allen 1973), and a theoretical justification can be constructed by noting that e^{ν_t} can be viewed as a random survival factor resulting from several independent and multiplicative environmental factors operating in series (thus ν_t represents a sum of several random factors and should be normally distributed by the central limit theorem).

Let us also assume that the management objective is simply to maximize the sum of discounted catches over time:

$$\max \sum_{t=0}^{\infty} C_t e^{-\delta t} \quad (2)$$

where C_t = catch = $R_t - S_t$,
δ = a discount rate.

The discount rate is critical in adaptive control problems, since without it ($\delta = 0$) we would put all management emphasis on getting better information for the far future, no matter what the cost in terms of lost yields in the near future. It is known (Allen 1973; Walters 1975) that for the objective in equation (2) when α and β are not uncertain, the optimum management policy is to allow a fixed escapement each year:

$$\text{choose } C_t = \begin{cases} R_t - \tilde{S} & \text{if } R_t > \tilde{S} \\ 0 & \text{if } R_t \leq \tilde{S} \end{cases} \quad (3)$$

where $\tilde{S}$ is the optimum escapement, computed from α and β (Ricker 1973).

Ordinarily we would recommend that management actions be based on estimates of $\tilde{S}$ computed from regression estimates $\hat{\alpha}_t$ and $\hat{\beta}_t$. The Ricker model can be rewritten as (after Dahlberg 1973):

$$\ln\left(\frac{R_t}{S_{t-1}}\right) = \alpha - \beta S_{t-1} + \nu_t \quad (4)$$

which is a linear regression form ($y = \alpha + \beta x$) with

$$y = \ln\left(\frac{R_t}{S_{t-1}}\right) \quad \text{and} \quad x = -S_{t-1}.$$

We would quite likely ignore some very useful information that comes from regression analysis, namely the parameter covariance matrix

$$\hat{P}_t = \begin{bmatrix} \hat{\sigma}_\alpha^2 & \hat{\sigma}_{\alpha\beta} \\ \hat{\sigma}_{\alpha\beta} & \hat{\sigma}_\beta^2 \end{bmatrix} \quad (5)$$

that measures our uncertainty about the parameter estimates given data up to time t. Further, under the assumption that the ν_t are normally distributed and independent of one another, it can be shown that $\hat{\alpha}_t$, $\hat{\beta}_t$, and the elements of $\hat{P}_t$ constitute a set of "sufficient statistics," meaning that there is no other function or manipulation of the data that can give us more information about the underlying true Ricker parameters.

The objective of adaptive control analysis in this case is to show how the choice of escapement S_t should be related to $\hat{P}$ as well as to $\hat{\alpha}$ and $\hat{\beta}$. The analysis can be formulated as a problem in stochastic dynamic optimization (Walters 1975); given the system state at any time as measured by $\{R_t, \hat{\alpha}_t, \hat{\beta}_t, \hat{P}_t\}$, what choice of C_t will give the best expected combination of present return and future returns, recognizing that a variety of possible future states may occur due to random events? In order to solve problems of this kind, we must be able to formulate a model to specify how each state variable R, α, etc.) will change in relation to the variety of stochastic outcomes that may occur between any times t and $t+1$, and we must be able to assign probabilities to each of these stochastic outcomes. Future recruitment states (R_{t+1}) can be predicted with the Ricker model, but we require analogous predictive formulae for the statistical parameters; the next section shows how these formulae can be derived from a special form of regression analysis.

RECURSIVE OR ADAPTIVE PARAMETER ESTIMATION

Suppose we begin at some time $t = 0$ with no data but with some prior estimates $\hat{\alpha}_o$ and $\hat{\beta}_o$. We might wish to assign no confidence to these estimates, which is equivalent to saying that we recognize σ_α^2 and σ_β^2 to be very large, or

$$\hat{P}_o = \begin{bmatrix} L & 0 \\ 0 & L \end{bmatrix} \quad (6)$$

where L is some large number (e.g. 10^6). In Bayesian statistical terms, we are in effect assigning a "diffuse prior" distribution for α and β (Raiffa and Schlaifer 1961). With starting conditions like this, it can be shown that ordinary regression analyses can be written in a special, "recursive" format (Young 1974). The general format is presented here, since it may be of interest outside the adaptive control context.

In general, linear regression equations are written as:

$$y_i = \sum_{j=1}^{m} a_j x_{ij} + e_i$$

where Y_i are dependent observations, x_{ij} are independent variables, and e_i are error terms. This form can be written more compactly in vector notation as:

$$Y_i = a^T x_i + e_i \quad (7)$$

where a and x_i represent the vectors ($a_1 \; a_2 \ldots a_m$) and ($x_{i1} \; x_{i2} \ldots x_{im}$). Using this notation, it turns out that common regression formulae can be written in recursive form as

$$\hat{a}_n = \hat{a}_{n-1} - \frac{\hat{P}_{n-1} x_n}{\sigma^2 + x_n^T \hat{P}_{n-1} x_n} (\hat{a}^T_{n-1} x_n - y_n) \quad (8a)$$

$$\hat{P}_n = \hat{P}_{n-1} - \frac{\hat{P}_{n-1} x_n x_n^T \hat{P}_{n-1}}{\sigma^2 + x_n^T \hat{P}_{n-1} x_n} \quad (8b)$$

where $\hat{a}_n$ and $\hat{P}_n$ refer to the parameter and parameter error covariance estimators after the nth data point is acquired, and σ^2 is the regression error variance. These formulae allow new data points to be added to a regression analysis without tedious computations involving matrix inversion. Estimation for the Ricker model can be written in the recursive form with

$$y_n = \ln\left(\frac{R_t}{S_{t-1}}\right)$$
$$x_n = \begin{bmatrix} 1 \\ -S_{t-1} \end{bmatrix} \quad (9)$$
$$\hat{a}_n = \begin{bmatrix} \hat{\alpha}_t \\ \hat{\beta}_t \end{bmatrix}$$

and similar transformations can be developed for a variety of other fisheries models.

Equations (8a) and (8b) are critical for the adaptive control formulation developed in the next section. Note that the change in parameter uncertainty from any observation or time step to the next, as measured by $\hat{P}_n - \hat{P}_{n-1}$, depends only on σ^2, $\hat{P}_{n-1}$, and the choice of x_n, i.e. on the choice of S_{t-1} for the Ricker model. On the other hand, changes in the parameter estimates as measured by $\hat{a}_n + \hat{a}_{n-1}$ depend on:

1) the level of uncertainty as measured by $\hat{P}_{n-1}$ and σ^2;
2) the choice of x_n;
3) the a priori prediction error, $D_n = (a^T_{n-1} x_n - Y_n)$. The priori prediction error D_n is the difference between the observed Y_n and its predicted value using the latest x_n data but the older or prior parameter estimates, $\hat{a}_{n-1}$. This prediction error, which is the only uncontrolled or stochastic input into the $\hat{a}$ and $\hat{P}$ changes for any time step, can be rewritten as two error components:

$$D_n = (a^T x_n - y_n) + (\hat{a}^T_{n-1} - a^T) x_n \quad (11)$$

The first component is the deviation of Y_n from the true model, while the second component represents deviation of the parameter estimates from the true value. If the regression errors v_t are normally distributed, both of these error components are normally distributed; thus D_n should have a normal distribution with mean 0.0 and variance

$$\sigma^2_{D_n} = \sigma^2 + x_n^T \hat{P}_{n-1} x_n \quad (11)$$

Thus with the data available up to any time step, we can compute probabilities for different values of D_n, and thus for having different parameter estimates at the next time step. In statistical decision theory this is known as preposterior analysis (Raiffa 1968). For the Ricker model, D_n is interpreted as

$$D_n = \hat{\alpha}_{t-1} - \hat{\beta}_{t-1} S_{t-1} - \ln\left(\frac{R_t}{S_{t-1}}\right) \quad (12)$$

Having chosen a value for D_n, with its associated probability, we can predict R_t by solving equation (12) as

$$\hat{R}_t = S_{t-1} \exp\{\hat{\alpha}_{t-1} - \hat{\beta}_{t-1} S_{t-1} - D_n\}$$
$$(\text{given } D_n) \quad (13)$$

This is the original Ricker model, but with an error component that reflects not only the noise v_t but also the uncertainty about α and β.

The adaptive regression equations (8) can be modified to "forget" older data. There are two simple types of modifications: 1) exponential past weighting of data, based on the assumption that all data become progressively less reliable as they become older; and 2) parameter variance incrementation, based on the more specific assumption that the parameters really do vary in some random or unspecifiable systematic way over time. For exponential past weighting, we define a discount factor V_d that represents the value of any observation relative to the next one that is obtained; for example, if we want to assume that an observation at time $t-1$ is worth 90% as much as an observation at time t, then $V_d = 0.9$. Using this discount factor, equations (8) are simply modified by changing the denominator terms $\sigma^2 + x_n^T \hat{P}_{n-1} x_n$ to

$$V_d \sigma^2 + x_n^T \hat{P}_{n-1} x_n \quad (14)$$

When there is reason to believe that one or more parameters are changing over time, the estimation is modified by introducing a parameter variation matrix Q, where the elements of the matrix are chosen to reflect the expected rate of change in the parameters. For example, if we believe that the Ricker β parameter may change by around 10% per year from an average value around 10^{-6}, while the α parameter is stable, we could set

$$Q = \begin{bmatrix} 0 & 0 \\ 0 & 10^{-14} \end{bmatrix} \quad (15)$$

In statistical terms, the elements of Q are interpreted as variances on a random walk process; thus a 10% change from a base of 10^{-6} represents a standard deviation of 10^{-7}, or a variance of $(10^{-7})^2$. The Q matrix is introduced into equations (8) simply by replacing every $\hat{P}_{t-1}$ with $\hat{P}_{t-1}^*$ where

$$\hat{P}_{t-1}^* = \hat{P}_{t-1} + Q \quad (16)$$

The choice of V_d or Q is not particularly critical; the major effect in both cases is to prevent $\hat{P}_t$ from going to zero over time, so that new observations can continue to effect changes in $\hat{\alpha}$.

ADAPTIVE DECISION STRUCTURE AND OPTIMIZATION

The problem of adaptive control and optimization for the Ricker model can be visualized in terms of the decision tree shown in Fig. 1. At any point in time, the manager is faced with a recruitment R_t, a summary of past data in terms of $\hat{\alpha}_t$ and $\hat{\beta}_t$, and uncertainties in terms of σ^2 and the elements of $\hat{P}_t$. He must then choose a harvest C_t; here there are many possible choices, but the optimization problem can be approximated by looking at a reduced, discrete set of possi-

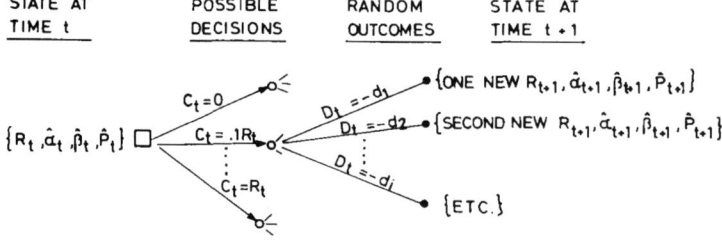

FIG. 1. Decision possibilities and random outcomes for each time step in adaptive control analysis for Ricker model. Symbols in text.

bilities (e.g. $C_t = 0$, $C_t = 0.1R_t$, $C_t = 0.2R_t$, etc.).

Given any choice of C_t, there are many possible random outcomes, but these can be summarized in terms of discrete deviations D_t from the regression predictors of $y = \ln(R_{t+1}/S_t)$. The reasoning is as follows:

1) Given C_t, S_t is calculated as $R_t - C_t$;
2) $\hat{\alpha}$, $\hat{\beta}_t$, and S_t are used to make a regression prediction $\hat{y}_{t+1}$ (equations [7] and [9]);
3) Probabilities for different outcomes $y_{t+1} = \hat{y}_{t+1} + D_t$ are computed from the probability distribution for D_t, which is normal with mean zero and variance given by equation (11);
4) Each outcome y_{t+1} is inserted, along with C_t, into the recursive regression equations (8) to obtain new estimates $\hat{\alpha}_{t+1}$, $\hat{\beta}_{+1}$, $\hat{P}_{t+1}$;
5 Since $y_{t+1} = \ln(R_{t+1}/S_t)$, R_{t+1} for each outcome y_{t+1} is given as $S_t e y_{t+1}$ (equivalent to equation (13)).

If we know the total future value for being in any next state $\{R_{t+1}, \hat{\alpha}_{t+1}, \hat{\beta}_{t+1}, \hat{P}_{t\,+1}\}$, we can calculate the expected value for making any C_t decision. This expected value is the sum of products of probabilities of next states times the values associated with these next states, plus the value of C_t itself. The difficulty is that we cannot immediately assign a value for each next state, since that state is itself a starting point for another decision tree like Fig. 1. If we look ahead a few time steps, the number of branching possibilities becomes essentially infinite. Luckily there is a partial way out of this problem using the "backward recursion" procedure of dynamic programming. A simplified discussion of this procedure is given in Walters (1975); the basic idea is that we begin the optimization calculations at some time point far enough in the future so that the discounted values after that point can be neglected. We then move backwards toward the present, evaluating decisions at each time step in terms of future values that have just been computed for the next time steps forward.

Unfortunately even dynamic programming involves formidable computation problems. If at each time step we examine only 10 discrete values for each of the six state variables (R_t, $\hat{\alpha}_t$, $\hat{\beta}_t$, σ_α^2, σ_β^2, $\sigma_{\alpha\beta}$), 10 catch levels, and 10 values of D_t, we must compute about 10^8 solutions for equations (8) and (13). The problem can be reduced

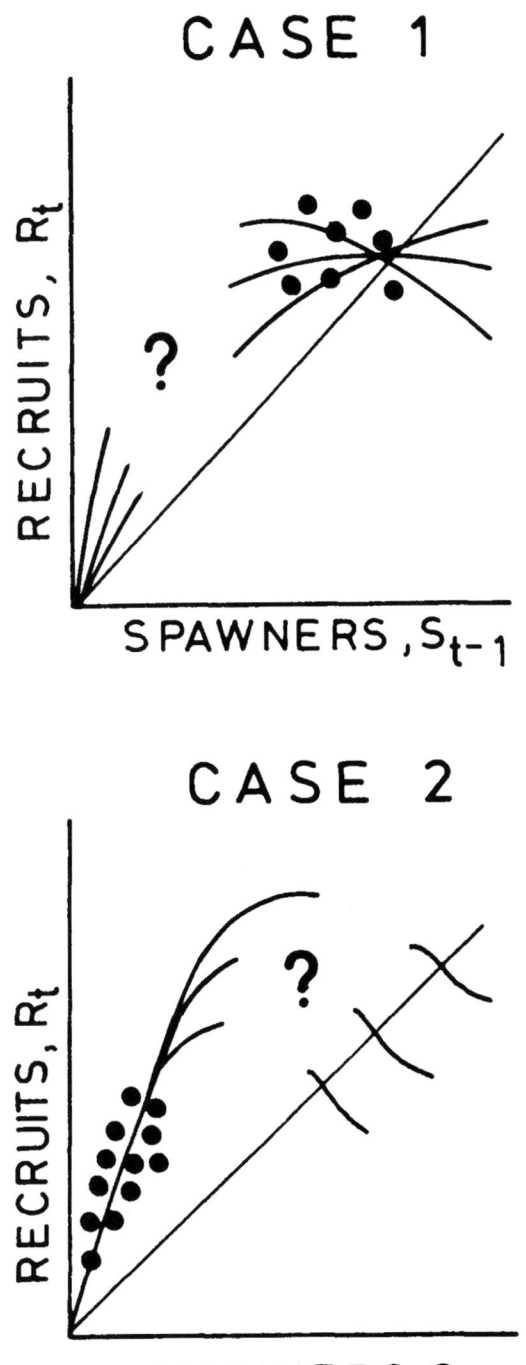

FIG. 2. Management situations that permit simplified adaptive optimization. Case 1: developing fishery, β known and α uncertain; Case 2: older fishery with unreliable data on natural stocks, α known and β unknown.

somewhat by using special computation procedures (Larson 1968), but there is a clear need for entirely different ways of looking at the problem. On the other hand, the optimization need only be carried out for a few representative values of the environmental variance σ^2 and the discount rate δ in order to get a complete adaptive picture for the Ricker model; given σ^2 and δ, the optimization automatically arrives at best harvest rates for all stock size–parameter value–parameter uncertainty combinations, in the form of a multidimensional "control law" (Allen 1973 and Walters 1975 have referred to one dimensional versions of this control law as "strategy curves").

A further point is worth noting about the size of adaptive optimization problems. Suppose instead of the Ricker model that we wish to analyze some model with three parameters (say, α_1, α_2, and α_3). Even if we can put this model into the linear regression form with normally distributed errors, the number of state variables for the dynamic optimization is 10 (R_t, $\hat{\alpha}_1$, $\hat{\alpha}_2$, $\hat{\alpha}_3$, $\sigma_{\alpha_1}^2$, $\sigma_{\alpha_2}^2$, $\sigma_{\alpha_3}^2$, $\sigma_{\alpha_1\alpha_2}$, $\sigma_{\alpha_1\alpha_3}$, $\sigma_{\alpha_2\alpha_3}$). This is too large a problem for even the best modern computers to handle.

Solutions for Special Cases

Instead of carrying out the tedious and expensive computations for the full adaptive optmization, we elected to examine two special cases that appear to be of management interest and that should reveal the general flavor of the full solution. These cases are shown in Fig. 2, and reflect two extreme situations:

Case 1 — The fishery is just beginning, and the stock is near its natural equilibrium. In this case β can be treated as known, and the main uncertainty is about α.

Case 2 — The fishery has been holding spawning stocks at low levels for many years. In this case α is well known and the main uncertainty is about β. Many Pacific salmon fisheries seem to fit this case; environmental carrying capacities may have changed considerably in recent years. In either case the size of the dynamic optimization problem is reduced considerably by treating one parameter as known. In Case 1, the stock and recruitment data can be expressed in stock units relative to the natural equilibrium, the Ricker model can be written simply as

$$R_{t+1} = S_t e^{\alpha(1-S_t)+v_t},$$

the system state vector for optimization becomes $\{R_t, \hat{\alpha}_t, \sigma_\alpha^2\}$, and the variables in the adaptive regression equations become:

$$y_t = \ln\left(\frac{R_t}{S_{t-1}}\right)$$
$$x_t = (1 - S_{t-1}) \quad (17)$$
$$\hat{P}_t = \hat{\sigma}_\alpha^2$$

In case 2, the Ricker model is assumed to maintain its usual form, the optimization state vector becomes $\{R_t, \hat{\beta}_t, \sigma_\beta^2\}$, and the adaptive regression variables become

$$y_t = \ln\left(\frac{R_t}{S_{t-1}}\right) - \tilde{\alpha}$$
$$x_t = -S_{t-1} \quad (18)$$
$$\hat{P}_t = \hat{\sigma}_\beta^2$$

where $\tilde{\alpha}$ is the reasonably certain estimate of α.

Several dynamic programming solutions for the simplified cases were carried out on a PDP 11/45 computer system. Each solution required about 5 hr of computer time, which is not excessive considering the wide range of stock–parameters–uncertainty combinations that must be evaluated. By trial and error we discovered that it was necessary to use 10 discrete levels for each variable (R_t, $\hat{\alpha}_t$ or $\hat{\beta}_t$, $\hat{\sigma}_\alpha^2$ or $\hat{\sigma}_\beta^2$), and to move backwards in time around 20 steps (generations); finer state intervals and more time steps did not change the solutions.

Representative results for case 1 (α uncertain) are shown in Fig. 3. Each isopleth diagram shows optimal harvest rates for a cross section through the $R_t - \hat{\alpha}_t$ plane at one uncertainty (σ_α^2) level. The most striking feature of these results is that optimal harvest rates are nearly independent of $\hat{\alpha}$ for large σ_α^2. What we expected to see was some indication that spawning populations should be reduced (high exploitation) when α is uncertain; by equations (8) and (17), we would expect the greatest reduction in uncertainty by conducting such an "experiment." As the cross section for high uncertainty ($\sigma_\alpha^2 = 4.0$) in Fig. 3 shows, experiments involving high exploitation rates are optimal only if $\hat{\alpha}_t$ is also large; indeed it appears when α is low and the stock size is large. On the other hand, the optimization also takes into account the possibility that low spawning stocks will reveal that α is small and thus that a period of recovery without harvest will be necessary. Examining the low harvest rate isoclines in Fig. 3, it is apparent that the optimal harvest rate for low stock sizes are quite insensitive to α, no matter what the uncertainty about α. We should hardly expect the optimal harvest strategy to depend

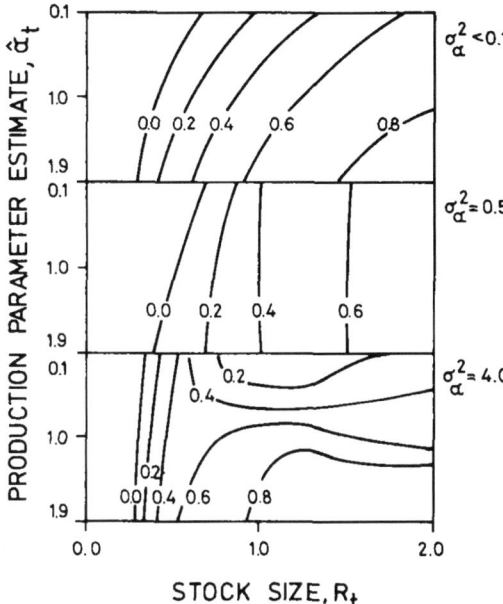

Fig. 3. Optimal exploitation rates for various stock sizes, production rate estimates ($\hat{\alpha}_t$), and uncertainties about α (σ_α^2) assuming the Ricker model form is correct and equilibrium stock is known (case 1 of Fig. 2). These results were obtained with environmental variance $\sigma^2 = 0.5$ and discount rate $\delta = 4\%$ per generation.

greatly on σ_α^2 if this strategy is nearly independent of α in the first place.

The results for case 2 (β uncertain) indicate a similar pattern; the optimum harvest strategy is quite insensitive to β when σ_β^2 is high (Fig. 4). Examining equations (8) and (18), we would expect high spawning stocks to produce the greatest reduction in uncertainty about β; yet the optimization balances the value of low exploitation (high S_t) experiments against the loss in immediate yields that such experiments would entail. Low harvest experiments are called for only when there is intermediate uncertainty about β.

Selection Among Alternative Models

The analysis in the preceding section took two sources of uncertainty into account: random environmental variation and uncertainty about production parameters. This section explores a third type of problem: uncertainty about the basic functional form of the stock-recruitment relation. As an example, consider the data in Fig. 5 on "off-cycle" runs of sockeye salmon in the Fraser River. Several subpopulations of sockeye in this river system exhibit cyclic dominance (Ward and Larkin 1964), with very large "cycle" runs every 4 yr (1962, 1966, ...) that apparently follow a different stock-recruitment relation from the off-cycle runs. Escapement levels in the off-cycle years have apparently been chosen under the assumption that high spawning populations may result in lowered recruitment due to overutilization (space, oxygen, etc.) of spawning areas; Fig. 5 does not support this assumption, at least when the whole river system is treated as a single population unit. Also, the overspawning phenomenon should have resulted in damping or destruction of higher cycle years during the early development of the fishery, and there is no evidence of this (Ward and Larkin); the off-cycle years sustained annual catches of around 4 million fish until the destructive Hell's Gate Slides of 1911. The best fitting Ricker curve for the data (curve η_1 in Fig. 5; $\alpha = 1.9$, $\beta = 0.44$) does predict that production would decline for spawning stock above 2 million, but it seems equally reasonable to assume that the correct relation is a saturating curve of the "Beverton Holt" type (Ricker 1973):

$$R_t = \frac{e^{v_t}}{\beta + \alpha/S_{t-1}} \quad (19)$$

where $\alpha = 1/$(maximum recruits per spawner),

$\beta = 1/$(maximum recruits ever possible),

e^{v_t} = random environmental survival factor as in equation (1). A visual fit to this relation is shown in Fig. 5 as curve η_2; the parameters ($\alpha = 0.1237$, $\beta = 0.1025$) were chosen so as to 1) closely match the Ricker curve through the available data, and 2) predict an equilibrium stock (8.5 million) that seems reasonable considering early catch records. Whatever the fitting procedure and even allowing for some decrease in production for high spawning stocks (dotted lines off curve n_2 in Fig. 5), significant improvements in yield could be obtained if the η_2 curve is correct. The question is, should an experiment (reduced harvests for one or more years) be conducted to test this possibility?

In principle this question could be addressed with the optimization approach introduced in the previous section. The stock-recruitment relation can be written as

$$R_t = \theta_1 f_1(S_{t-1}) + \theta_2 f_2(S_{t-1}) + \ldots \quad (20)$$

where θ_i represent model selection parameters that take only the values 0 or 1 and are constrained as $\Sigma \theta_i = 1$ (so all the θ_i but one must equal 0), and

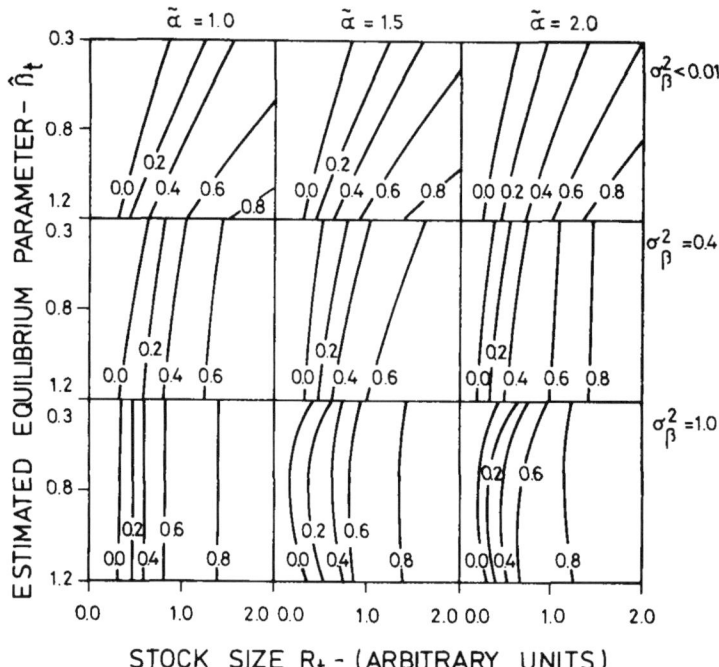

Fig. 4. Optimal exploitation rates for various stock sizes, equilibrium population sizes ($1\hat{\beta}_t$), and uncertainties about $\hat{\beta}$ (σ_β^2), assuming Ricker form is correct and α is known (case 2, Fig. 2). Environmental variance and discount rate as in Fig. 3.

the f_i are alternative models such as the Ricker (equation 1) and the Beverton-Holt (equation 19). Wood (1974), Smallwood (1968), and others have shown that it is possible to calculate the probability that each $\theta_i = 1$ (model i is correct), given that the true model is among the alternatives represented. These probabilities along with parameter estimates and measures of uncertainty for each alternative model can be formed into an extended vector of state variables. Unfortunately the number of variables involved makes dynamic programming optimization impractical, so some drastic simplifications and approximations are necessary in order to trace the most likely statistical outcomes and most promising decision possibilities.

Since full adaptive control analysis is not feasible, the remainder of this section attempts to develop a simplified procedure for designing and evaluating experimental harvesting regimes intended to discriminate between alternative production models. The procedure is modified from a general approach suggested by Bard (1974), and involves five basic steps.

1) Identification of a series of models, or alternative "states of nature" $\eta_1, \eta_2, \ldots \eta_m$ that are to be compared. Alternative η_1 might be the Ricker model, η_2 might be the Beverton-Holt model, η_3 might be a simple free hand curve extrapolating from existing data, and so forth;

2) Assignment of prior or judgemental probabilities $p^*(\eta_1), p^*(\eta_2) \ldots$ to each of the alternative states of nature. These probabilities might be derived through some statistical procedure, or they may represent simple intuition. Reflecting on the Fraser River data, we might for example assign p^* (Ricker model) $= 0.7$, p^* (Beverton-Holt model) $= 0.3$; though both models fit the data about equally well, this probability assignment would give some weight to the common arguments about overspawning;

3) Identification of a series of alternative harvesting experiments $S_1, S_2, \ldots, S_n$, each of which would be reasonably certain to discriminate between the alternative models but would require different lengths of time to complete. For the Fraser River example, some reasonable alternatives are:

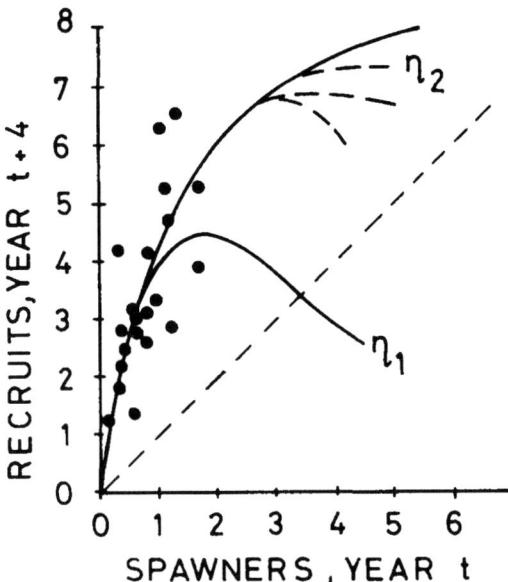

FIG. 5. Alternative stock-recruitment models for Fraser River sockeye salmon (*Oncorhynchus nerka*), off-cycle years. Data shown is for 1939–73, omitting every fourth or cycle year beginning in 1942. η_1, least squares fit to Ricker model; η_2, visual fit to Beverton-Holt model. Graph axis in millions of fish.

S_1 — Continue the present escapement policy (S_t around 1.0 million/yr) forever. That is, do not experiment and hope that luck will eventually provide the necessary data;

S_2 — Allow escapements intermediate between the optima for the alternative models; considering all environmental variability, this experiment would probably not give definite results for at least 20 yr;

S_3 — Allow the optimum escapement ($S_t = 2.0$ million) for the Beverton-Holt model; at this escapement level, any tendency for overspawning should be apparent within 5 yr while it is unlikely that environmental circumstances would combine for that long to give consistently low recruitments if the saturating model were true.

In general, the experiments should reflect tradeoff between small harvest manipulations that require a long time to give definitive results vs. large harvest manipulations that give results quickly. The length of experiment required at any escapement or stock level can be assessed by examining expected variability around the alternative stock-recruitment models at that escapement level.

4) Calculation of expected long-term returns for each combination of experiment and state of nature. The elements of the following table must be evaluated:

		Experiment		
		S_1	S_2	S_n
True state	η_1	V_{11}	V_{12} ..... V_{1n}	
	η_2	V_{21}	V_{22} .	
	η_m	V_{ml} .... V_{mn}		

(21)

Here V_{ij} represents the expected total value of all future harvests given that experimental strategy S_j is applied and the true state of nature is η_i. Suppose that strategy S_j involves allowing a certain escapement $\tilde{E}_j$ for τ_j years. If τ_j is chosen carefully, we should be reasonably certain of detecting that η_i is the true state of nature after the τ_j years, and we should be unwilling to accept η_i is the true state until the τ_j years have elapsed (any experiment not meeting these qualifications should not be included in the first place). Thus V_{ij} can be calculated as the expected value of two discounted sums: 1) the sum of catches during the τ_j experiment years, given that escapement $\tilde{E}_j$ is applied and model η_i is correct; plus 2) the sum of catches after the τ_j'th year, given that the optimum escapement for model η_i is followed thereafter. The first component reflects the short-term impacts of the experiment, while the second component reflects long-term benefits. The simplest way to calculate V_{ij} is to do a whole series of simulation trials, each using model η_i but a different sequence of random environmental inputs. For each trial we calculate

$$V'_{ij} = \sum_{t=1}^{\tau_j} (C_t|\tilde{E}_j)e^{-\delta t} + \sum_{t=\tau_j}^{T} (C_t|\hat{E}_i)e^{-\delta t}$$

where $(C_t|\tilde{E}_j)$ is defined as the catch in the year t using experimental escapement $\tilde{E}_j$, and $(C_t|\hat{E}_i)$ is defined as the catch in year t given the optimum escapement $\hat{E}_i$ for model η_i (δ is the discount rate). It should not be necessary to perform more than about 20 trials, of length $T \approx 50$ generations for reasonable discount rates. V_{ij} is found as the average of V'_{ij} across these trials. To develop the entire strategy-experiment table, it is thus necessary to do about $m \times n \times 20$ simulation trials, but this is a trivial computing exercise.

5) Selection of the experiment with maximum expected benefits. Each column of the strategy-experiment table (21) gives the returns to be expected from one experiment for each possible state of nature. The overall value for the experiment is simply the sum of these returns weighted by the prior probabilities for the η_i:

expected value of experiment $j = \sum_{i=1}^{m} p^*(\eta_i)V_{ij}$ (22)

The best experiment is that one which has the maximum expected value. The key point about this selection procedure is that it takes into account all the possible states of nature in evaluating each proposed experiment.

The five steps outlined above lend themselves well to a gaming situation, in which the resource manager is asked to devise alternative recruitment models, assess their probabilities, and evaluate alternative experimental schemes. The most critical point in the analysis is in the identification of appropriate durations for alternative experiments. If the manager is overly pessimistic (e.g. assumes that some experiment will require τ_j years to be certain which model is correct when in fact fewer years are required), perfectly good experiments may appear very bad in relation to those states of nature for which the experimental escapement $\tilde{E}_j$ is far from optimal. On the other hand, the manager may be overconfident, and may suggest a short experiment which in reality would simply result in loss of yield with no improvement in understanding about the system. In a gaming situation, the best τ_j for any proposed $\tilde{E}_j$ can be evaluated quite quickly by facing the manager with several stochastic simulations from each possible true model while noting how long it takes for him to be sure which model is being used in each simulation. More precisely, the analyst sets up a series of trials. For each trial the manager chooses an escapement $\tilde{E}_j$, and the analyst secretly chooses a model η_i. A stochastic simulation with η_i is then initiated and carried forward in time until the manager positively identifies the results as coming from η_i. The number of simulation steps required for the various trials can then be plotted in relation to the choices of $\tilde{E}_j$. This plot will reflect the manager's subjective definition of "positive" results, analogous to his choice of confidence limit probabilities (e.g. 90% vs. 95%) in ordinary statistical problems. The degree of random variation introduced in each trial should reflect uncertainty about model parameters as well as expected environmental variation, by using the variance relation in equation (11) or its subjective equivalent.

To test the procedure, we carried out a gaming analysis on the Fraser River problem with one of the authors acting as manager and the other as analyst. Two alternative models were considered:

η_1 = Ricker curve from Fig. 5;
η_2 = Beverton-Holt curve from Fig. 5.

By examining the data and following the trial procedure for τ_j outlined in the previous paragraph, we arrived at the following set of experiments:

S_1 — allow an escapement of 1.0 million forever (do not experiment)
S_2 — allow an escapement of 1.5 million for 15 yr
S_3 — allow an escapement of 2.0 million for 5 yr
S_4 — allow an escapement of 3.0 million for 3 yr

Stimulation trials to evaluate the V_{ij} were then performed, assuming a discount rate of 4% per generation (i.e. 1%/yr for Fraser sockeye), to give the following discounted sums of catches, in millions of fish:

		Experiment			
		S_1	S_2	S_3	S_4
True	η_1	77.2	77.8	75.4	71.7
state	η_2	92.4	108.2	110.9	110.6

When only two or three alternative states of nature are to be compared as in the Fraser example, the analysis can be presented in an elegant form that simplifies the problem of assigning subjective probabilities to the alternative models. Suppose we make a graph whose abscissa is V_{1j} (expected value of experiment j given that η_1 is true) and ordinate is V_{2j} (expected value of experiment j given that η_2 is true). Then each experiment can be plotted as a point on this graph (Fig. 6). Points that are close to the ordinate represent experiments or policies that

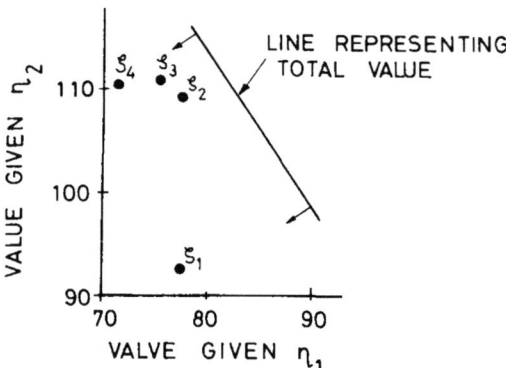

FIG. 6. Values of alternative Fraser River harvesting experiments if the alternative models η_1 and η_2 in Fig. 5 are correct. Graph units are discounted (4%/generation) sums of catches, in millions of fish. S_1–S_4 denote escapement experiments defined in text. Total value line is given by Value = $p^*(\eta_1$ [value given η_1]) + $p^*(\eta_2$ [value given η_2]).

are good if η_2 is true but poor if η_1 is true, while points near the abscissa represent policies that are good if η_1 is true but poor if η_2 is true. If we designate those experiments that can distinguish between the η_i as "effective experiments," then the graphical representation allows us to sort out a smaller subset of "efficient experiments" which are best for at least some values of $p^*(\eta_i)$. In Fig. 6, experiments S_2 and S_3 are efficient, while experiments S_1 and S_4 are not since the others have better expected values for all assignments of $p^*(\eta_i)$. We are trying to find that experiment S_j which maximizes the "objective function" $y = p^*(\eta_1)V_{1j} + (1 - p^*(\eta_1))V_{2j}$. This objective function can be represented as a series of straight lines on Fig. 6, with higher lines representing bigger y values and greater slopes representing larger values of $p^*(\eta_1)$. To find the optimum experiment, we move the lines downward (choose lower y values) until it first touches a "feasible point" representing some S_j. The key point is that we can identify ranges of $p^*(\eta_1)$ for which any efficient S_j is optimal, simply by changing $p^*(\eta_1)$ so as to steepen or flatten the objective function line. For the Fraser example, this process results in the following:

		Range of p^* (Ricker) for which S_j is best
Experiment	S_1	none
	S_2	0.4–1.0
	S_3	0.0–0.4
	S_4	none

Thus the manager does not need to precisely specify his judgement about $p^*(\eta_1)$ as a single number.

The Fraser River test results suggest the following conclusions:
1) The present management policy for off-cycle years is not optimal for either of the models recomended in Fig. 5; either some alternative (and perhaps not clearly specified) model is being used, or the real management objectives are quite unrelated to maximization of discounted long-term catches; 2) A modest experiment involving increased escapements (1.5 million) should cause no serious problems if the Ricker model is correct, and may result in considerably higher yields in the long run; 3) A more drastic experimental policy involving escapements of 2.0 million for several years would be a better gamble if there is considerable confidence that the true stock-recruit relation is like the curve η_2 in Fig. 5. To test the effect of discounting rate on these conclusions, we reevaluated Table 1 for $\delta = 1\%$, 10%, 20%, and 30% per generation. For δ greater than 20%, the tests suggested that the modest experiment is best unless p^*(Ricker) is less than 0.2. For $\delta = 1\%$, the drastic experiment becomes the best alternative unless p^*(Ricker) is greater than 0.6.

Extensions and Generalizations

The previous sections have dealt primarily with uncertainty about stock-recruitment relations. Implicit in the discussion have been two major assumptions: that stock size is directly measurable without error, and that fishing effort is fully controllable to conform with biological recommendations. These assumptions are often not justified, so this section attempts to show how the concepts and methods introduced previously could be extended to include these additional sources of uncertainty.

Schaeffer Production Model

The idea of using logistic population growth assumptions as a basis for production modelling was first made popular by Schaeffer (1954), in his studies on Pacific tunas. In its simplest form, the "Schaeffer Model" can be written as

$$N_{t+1} = N_t + \alpha N_t(1 - \beta N_t) - C_t \quad (23)$$

where N_t = stock size, usually in biomass units,
α, β = production parameters with similar definitions as in the Ricker Model,
C_t = total catch.

Noting that equation (23) is usually not directly usable since N_t is not observable for most populations, Schaeffer and others have assumed a simple "observation model" to accompany the dynamic model:

$$N_t = \frac{q_t}{c} = \frac{C_t/E_t}{c} \quad (24)$$

where q_t = catch per unit effort,
E_t = some effort measure having units (boats) $\times$ (time fishing per boat),
c = catchability coefficient.

Substituting the observation model (24) into (23), we get an expression containing only observable quantities and parameters:

$$\frac{q_{t+1}}{c} = \frac{q_t}{c} + \alpha \frac{q_t}{c}\left(1 - \beta \frac{q_t}{c}\right) - C_t$$

which can be simplified to give

$$q_{t+1} - q_t = \alpha q_t - \frac{\alpha\beta}{c}q_t^2 - cC_t \quad (25)$$

For parameter estimation and adaptive control analysis, this version of the Schaeffer Model can be cast into the recursive regression format, with

$$y_t = q_t - q_{t-1} \quad (\text{or } y_t = q_t)$$

$$x_t = \begin{bmatrix} q_{t-1} \\ q^2_{t-1} \\ C_{t-1} \end{bmatrix} \quad (26)$$

$$\hat{a}_t = \begin{bmatrix} \hat{\alpha}_t \\ \widehat{\alpha\beta/c_t} \\ \hat{c}_t \end{bmatrix} \quad \text{or } \hat{a}_t = \begin{bmatrix} \widehat{(1+\alpha)_t} \\ \widehat{\alpha\beta/c_t} \\ \hat{c}_t \end{bmatrix}$$

Though there is no reason in this case to expect nice statistical properties such as normally distributed errors, the regression format at least provides a unified framework for evaluating parameter uncertainties. Also, it automatically provides a simple state estimator, $N_t = \hat{q}_t/\hat{c}_t$. More complicated, statistically nonlinear versions of equation (25) can be devised by using more realistic observation models than equation (24), and recursive nonlinear estimation techniques are beginning to appear in the literature under the general heading of "extended Kalman filters" (Young 1974).

The Schaeffer Model gives a remarkably good fit to historical data for many large fisheries, as shown in Fig. 7. In fitting these data, we used the linear regression scheme in equations (25) to obtain estimates of α, β, and c. In each case the data had already been corrected for changing vessel efficiency (changing c), so it was unnecessary to introduce discounting of old data or a parameter variation matrix Q (see Section II) into the regression equations. It seems clear that further adaptive control work for the Schaeffer Model is justified, and we intend to develop more complete analyses in a future paper.

INCOMPLETE CONTROL OF FISHING EFFORT

The fisheries literature abounds with biological models and equilibrium yield analysis, yet almost no attention has been paid to the dynamics of the predator–prey system that results from incomplete control of economic investment. Fishery fleets have basic "reproductive" (investment) and "mortality" (disinvestment) relations that in principle make them like any predator population (Smith 1968; Gatto et al. 1975). In the absence of investment control, many fishing fleets have developed to the point where pressure for short-term economic and social welfare benefits has made it virtually impossible to implement biologically sound long-term policies; the current state of the International Whaling Commission is a good example.

To pursue the predator–prey analogy, it may be useful to think of management controls directed at fishing effort as generating a "reachable region" of stock size and investment combinations (Fig. 8) around the "nominal trajectory" of development that would occur without management. Investment control may occur in the form of subsidies to increase the rate of investment, or taxes and direct regulations to reduce it. If the fishery fleet can operate economically at very low stock sizes, and if only small control decreases in effort are possible each year, it may be impossible to move the fishery to any state where maximum sustained yield is possible; this problem may become especially serious if the first incremental controls are not applied until the fishery is well developed. One is reminded of the adage about ounces of prevention and pounds of cure.

Limitations on control changes from one time step to the next may be represented in dynamic optimization by including the control level (effort, harvest rate, etc.) as an additional state variable. For example, if the system state without effort limitation were represented as (R_t, $\hat{\alpha}_t$, σ_α^2), then the state with limitation would be (U_{t-1}, R_t, $\hat{\alpha}_t$, σ_α^2) where U_{t-1} is the exploitation rate from the previous time step. Then instead of looking at all possible harvest rates at each time step for each (R_t, $\hat{\alpha}_t$, σ_α^2) combination, the optimization would only examine harvest rates U_t over an interval:

$$U_{t-1} - k_1 \leqslant U_t \leqslant U_{t-1} + k_2$$

where k_1 = maximum permissible annual decrease in exploitation rate,
k^2 = maximum permissible annual increase in exploitation rate.

Similar constraints can be applied in generating experimental harvest regimes for the analyses in section III.

A series of interesting issues arise concerning the selection of appropriate values for the control limits k_1 and k_2. The maximum rate of increase in harvest, k_2, will depend on private and public willingness to invest in the fishery and on the availability of fishing gear to be transferred from alternative fisheries. The maximum rate of decrease, k_1, will depend on the regulatory power vested in the management agency, the willingness of the agency to accept responsibility for immediate economic and social hardships, and the expected profitability of the fishery. Fisheries agencies are beginning to face these political and

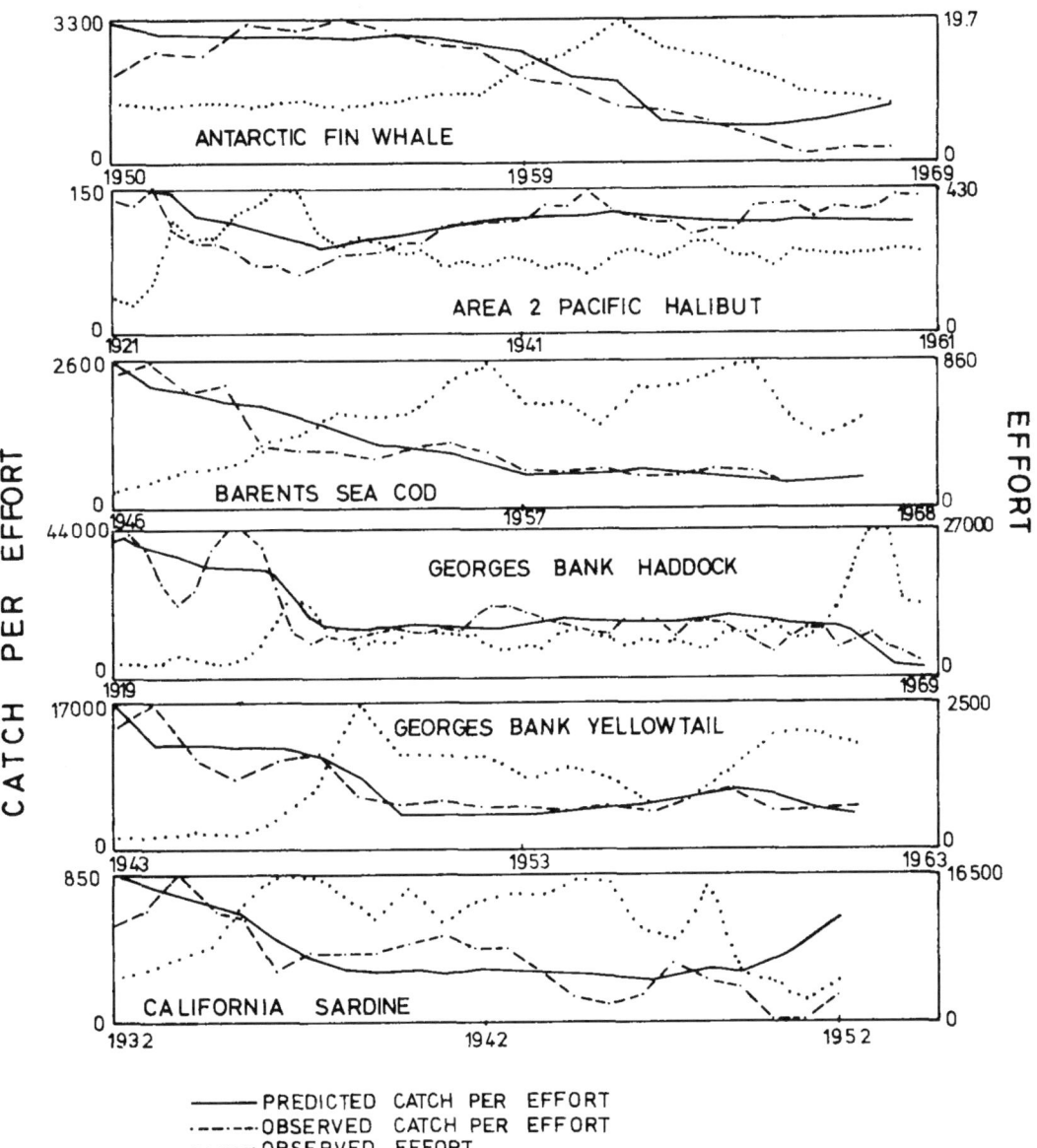

FIG. 7. Observed changes in some fisheries compared to simulated trends predicted by the Schaeffer model (equations 13, 24) using parameter estimator from regression analysis. The entire data set for each case was used to obtain parameter estimates; simulated value for the previous year (rather than from previous year's data). Data were obtained from following sources: fin whales, International Whaling Commission (1969); halibut, Southward (1968); cod, Garrod (1969); haddock, Grosslein and Hennemuth (1973); yellowtail, Lux (1969); sardine, Marr (1959).

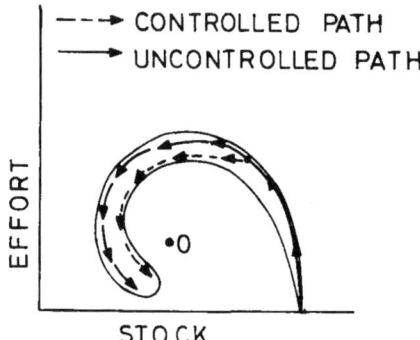

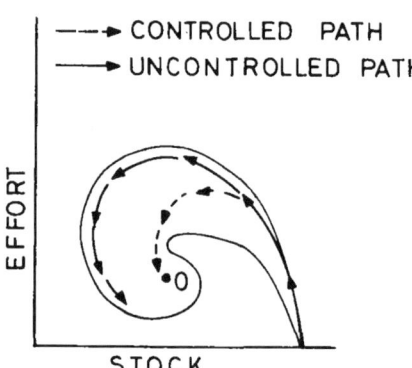

Fig. 8. Imperfect control of fishing effort may result in a restricted "reachable region" of stock size and effort combinations as a fishery develops. The point "O" denotes optimum equilibrium state.

economic issues, and optimization formulations may provide a useful focus for formal debate even if no quantitative solutions are attempted.

Summary

This paper has attempted to move from formal optimization analysis of trivial models to broader approaches for experimental management. The formal analysis was conducted in hopes of discovering simple principles that might be applied in more complex and realistic fisheries situations where uncertainty is a key factor. The principles that emerged are intuitively reasonable: 1) when production parameters are uncertain, lower harvest rates should be used than would be supposed if only the available parameter estimates were considered; 2) when the equilibrium stock size is uncertain, but production rates are well understood, it is worthwhile only under certain conditions to lower the harvest rate in order to obtain better estimates of the equilibrium size. When the general form of the production relation in uncertain and when management control is limited, formal optimization becomes impossible but a gaming procedure may be helpful in trying to devise and evaluate alternative management strategies. The gaming procedure involves defining a series of possible models, selection of a series of effective experiments, and calculation of the optimum experiment under subjective prior probabilities for each model. This technique is an immediately useable solution for complex fisheries problems. Future work will continue to try to overcome the computational obstacles encountered in formal optimization of complex models.

Acknowledgments

Research supported by International Institute for Applied Systems Analysis, Laxenburg, Austria.

ALLEN, K. R. 1973. The influence of random fluctuations in the stock-recruitment relationship on the economic return from salmon fisheries. Cons. Int. Explor. Mer Rapp. 164: 351–359.

BARD, YONATHAN, 1974. Nonlinear parameter estimation. Academic Press, Inc., New York N.Y. 341 p.

DAHLBERG, M. L. 1973. Stock-and-recruitment relationships and optimum escapements of sockeye salmon stocks of the Chignik Lakes, Alaska. Cons. Int. Explor. Mer Rapp. 164: 98–105.

GARROD, D. J. 1969. Empirical assessments of catch/effort relationships in North Atlantic cod stocks. Int. Comm. Northwest Atl. Fish. Res. Bull. 6: 26–34.

GATTO, M., S. RINALDI, AND C. WALTERS. 1975. A predator–prey model for discrete-time commercial fisheries. Int. Inst. Appl. Syst. Anal., Laxenburg, Austria, Res. Rep. Ser. 75-5: 38 p.

GROSSLEIN, M. D., AND R. C. HENNEMUTH. 1973. Spawning stock and other factors related to recruitment of haddock on Georges Bank. Cons. Int. Explor. Mer Rapp. 164: 77–88.

INTERNATIONAL WHALING COMMISSION. 1969. Annual report for 1967–68.

LARSON, R. E. 1968. State increment dynamic programming. American Elsevier Publishing Co., New York, N.Y. 300 p.

LUX, F. E. 1969. Landings per unit of effort, age composition, and total mortality of yellowtail flounder, *Limanda ferruginea* (Storer), off New England. Int. Comm. Northwest Atl. Fish. Res. Bull. 6: 35–43.

MARR, J. C. 1959. The causes of major variations in the catch of the Pacific sardine, *Sardinops caerulea* (Girard). Proc. World Science Meeting on the Biology of Sardines and Related Species, FAO, Rome. p. 667–791.

PARRISH, B. B. [ed.]. 1973. Fish Stocks and recruitment. Cons. Int. Explor. Mer Rapp. 164: 372 p.

RAIFFA, H. 1968. Decision analysis: introductory lectures on choices under uncertainty. Addison-Wesley, Reading, Mass. 309 p.

RAIFFA, H., AND R. SCHLAIFER. 1961. Applied statistical decision theory. MIT Press, Cambridge, Mass. 356 p.

RICKER, W. E. 1954. Stock and recruitment. J. Fish. Res. Board Can. 11: 559–623.

1973. Critical statistics from two reproduction curves. Cons. Int. Explor. Mer Rapp. 164: 333–340.

SCHAEFFER, M. B. 1954. Some aspects of the dynamics of populations important to the management of the commercial marine fisheries. Bull. Int. Trop. Tuna Comm. 1: 27–56.

SMALLWOOD, R. 1968. A decision analysis of model selection. IEEE Trans. on Systems Science and Cybernetics, Vol. SSC-4 No. 4.

SMITH, V. L. 1968. Economics of production from natural resources. Am. Econ. Rev. 58: 409–431.

SOUTHWARD, G. M. 1968. A simulation of management strategies in the Pacific halibut fishery. Int. Pac. Halibut Comm. Rep. 47.

WALTERS, C. J. 1975. Optimal harvest strategies for salmon in relation to environmental variability and uncertainty about production parameters. J. Fish. Res. Board Can. 32: 1777–1784.

WARD, F. J., AND P. A. LARKIN. 1964. Cyclic dominance in Adams River sockeye salmon. Int. Pac. Salmon Fish. Comm., Prog. Rep. 11: 116 p.

WOOD, E. F. 1974. A Bayesian approach to analyzing uncertainty among stochastic models. Int. Inst. Appl. Syst. Anal. Laxenburg, Austria, Res. Rep. 74-16: 19 p.

YOUNG, P. 1974. Recursive approaches to time series analysis. Inst. Math. Appl. 10: 209–224.

RESEARCH ARTICLES

Rebuilding Global Fisheries

Boris Worm,[1]* Ray Hilborn,[2]* Julia K. Baum,[3] Trevor A. Branch,[2] Jeremy S. Collie,[4] Christopher Costello,[5] Michael J. Fogarty,[6] Elizabeth A. Fulton,[7] Jeffrey A. Hutchings,[1] Simon Jennings,[8,9] Olaf P. Jensen,[2] Heike K. Lotze,[1] Pamela M. Mace,[10] Tim R. McClanahan,[11] Cóilín Minto,[1] Stephen R. Palumbi,[12] Ana M. Parma,[13] Daniel Ricard,[1] Andrew A. Rosenberg,[14] Reg Watson,[15] Dirk Zeller[15]

After a long history of overexploitation, increasing efforts to restore marine ecosystems and rebuild fisheries are under way. Here, we analyze current trends from a fisheries and conservation perspective. In 5 of 10 well-studied ecosystems, the average exploitation rate has recently declined and is now at or below the rate predicted to achieve maximum sustainable yield for seven systems. Yet 63% of assessed fish stocks worldwide still require rebuilding, and even lower exploitation rates are needed to reverse the collapse of vulnerable species. Combined fisheries and conservation objectives can be achieved by merging diverse management actions, including catch restrictions, gear modification, and closed areas, depending on local context. Impacts of international fleets and the lack of alternatives to fishing complicate prospects for rebuilding fisheries in many poorer regions, highlighting the need for a global perspective on rebuilding marine resources.

Overfishing has long been recognized as a leading environmental and socioeconomic problem in the marine realm and has reduced biodiversity and modified ecosystem functioning (1–3). Yet, current trends as well as future prospects for global fisheries remain controversial (3–5). Similarly, the solutions that hold promise for restoring marine fisheries and the ecosystems in which they are embedded are hotly debated (4–6). Such controversies date back more than a hundred years to the famous remarks of Thomas Huxley on the inexhaustible nature of sea fisheries (7) and various replies documenting their ongoing exhaustion. Although management authorities have since set goals for sustainable use, progress toward curbing overfishing has been hindered by an unwillingness or inability to bear the short-term social and economic costs of reducing fishing (8). However, recent commitments to adopting an ecosystem approach to fisheries may further influence progress because they have led to a reevaluation of management targets for fisheries and the role of managers in meeting broader conservation objectives for the marine environment (9).

In light of this debate, we strive here to join previously diverging perspectives and to provide an integrated assessment of the status, trends, and solutions in marine fisheries. We explore the prospects for rebuilding depleted marine fish populations (stocks) and for restoring the ecosystems of which they are part. In an attempt to unify our understanding of the global fisheries situation, we compiled and analyzed all available data types, namely global catch data (Fig. 1A), scientific stock assessments, and research trawl surveys (Fig. 1B), as well as data on small-scale fisheries (10). We further used published ecosystem models (Fig. 1B) to evaluate the effects of exploitation on marine communities. Available data sources are organized hierarchically like a Russian doll: Stock assessments provide the finest resolution but represent only a subset of species included in research surveys, which in turn represent only a small subset of species caught globally. These sources need to be interpreted further in light of historical fisheries before data collection and illegal or unreported fisheries operating today (11). We focus on two leading questions: (i) how do changes in exploitation rates impact fish populations, communities, and yields, and (ii) which solutions have proven successful in rebuilding exploited marine ecosystems?

Models. A range of models is available to analyze the effects of changes in exploitation rate on fish populations, communities, and ecosystems. Exploitation rate (u_t) is defined as the proportion of biomass that is removed per year, i.e., $u_t = C_t/B_t$ where C is the catch (or yield) and B is the available biomass in year t. Single-species models are often used to determine the exploitation rate u_{MSY} that provides the maximum sustainable yield (MSY) for a particular stock. Fishing for MSY results in a stock biomass, B_{MSY}, that is substantially (typically 50 to 75%) lower than the unfished biomass (B_0). It has been a traditional fisheries objective to achieve single-species MSY, and most management regimes have been built around this framework. Recently this focus has expanded toward assessing the effects of exploitation on communities and ecosystems (9).

Multispecies models can be used to predict the effects of exploitation on species composition, size structure, biomass, and other ecosystem properties. They range from simpler community models to more-complex ecosystem models (12). Figure 2 displays equilibrium solutions from a size-based community model, which assumes that fishing pressure is spread across species according to their size and that a subset of species remains unfished (13). Results of more-complex ecosystem models across 31 ecosystems and a range of different fishing scenarios were remarkably similar (fig. S1 and table S1). With increasing exploitation rate, total fish catch is predicted to increase toward the multispecies maximum sustainable yield (MMSY) and decrease thereafter. In this example, the corresponding exploitation rate that gives maximum yield u_{MMSY} is ~0.45, and total community biomass B_{MMSY} equilibrates at ~35% of unfished biomass (Fig. 2). Overfishing occurs when u exceeds u_{MMSY}, whereas rebuilding requires reducing exploitation below u_{MMSY}. An increasing exploitation rate causes a monotonic decline in total biomass and average body size, and an increasing proportion of species is predicted to collapse (Fig. 2). We used 10% of unfished biomass as a definition for collapse. At such low abundance, recruitment may be severely limited, and species may cease to play a substantial ecological role. This model suggests that a wide range of exploitation rates (0.25 < u < 0.6) yield ≥90% of maximum catch but with very different ecosystem consequences: whereas at $u = 0.6$ almost half of the species are predicted to collapse, reducing exploitation rates to $u = 0.25$ is predicted to rebuild total biomass, increase average body size, and strongly reduce species collapses with little loss in long-term yield (Fig. 2). In addition to reconciling fishery and conservation objectives, setting exploitation rate below u_{MMSY} reduces the cost of fishing and increases profit margins over the long term (14). This simple model does not incorporate fishing selectivity; however, in practice the proportion of collapsed species could be reduced further by increasing selectivity through improved gear technology (15), by closing areas frequented by vulnerable species, or through offering incentives to improve targeting practices (16). Such strategies allow for protection of vulnerable or collapsed species, while allowing for more intense exploitation of others.

---

[1]Biology Department, Dalhousie University, Halifax, NS B3H 4J1, Canada. [2]School of Aquatic and Fishery Sciences, University of Washington, Seattle, WA 98195–5020, USA. [3]Scripps Institution of Oceanography, University of California-San Diego, La Jolla, CA 92093–0202, USA. [4]Graduate School of Oceanography, University of Rhode Island, Narragansett, RI 02882, USA. [5]Donald Bren School of Environmental Science and Management, University of California, Santa Barbara, CA 93106–5131, USA. [6]National Marine Fisheries Service, National Oceanic and Atmospheric Administration, Woods Hole, MA 02543, USA. [7]Commonwealth Scientific and Industrial Research Organisation (CSIRO) Marine and Atmospheric Research, General Post Office Box 1538, Hobart, TAS 7001, Australia. [8]Centre for Environment, Fisheries and Aquaculture Science, Lowestoft NR33 OHT, UK. [9]School of Environmental Sciences, University of East Anglia, Norwich NR4 7TJ, UK. [10]Ministry of Fisheries, Post Office Box 1020, Wellington, New Zealand. [11]Wildlife Conservation Society Marine Programs, Post Office Box 99470, Mombasa, Kenya. [12]Hopkins Marine Station, Stanford University, Pacific Grove, CA 93950, USA. [13]Centro Nacional Patagónico, 9120 Puerto Madryn, Argentina. [14]Institute for the Study of Earth, Oceans, and Space, University of New Hampshire, Durham, NH 03824–3525, USA. [15]Fisheries Centre, University of British Columbia, Vancouver, BC V6T 1Z4, Canada.

*To whom correspondence should be addressed. E-mail: bworm@dal.ca (B.W.); rayh@u.washington.edu (R.H.)

These results suggest that there is a range of exploitation rates that achieve high yields and maintain most species. To test whether current fisheries fall within this range, we evaluated trends in 10 large marine ecosystems for which both ecosystem models and stock assessments were available (10). Figure 3A shows exploitation rate and biomass trajectories derived from 4 to 20 assessed fish or invertebrate stocks per ecosystem. These stocks typically represent most of the catch, and we assumed that trends in their exploitation rates represent the community as a whole. Ecosystem models were used to calculate u_{MMSY} (light blue bars) and the exploitation rate at which less than 10% of the fished species are predicted to be collapsed ($u_{conserve}$, dark blue bars). Across the 10 examined ecosystems, MMSY was predicted at multispecies exploitation rates of $u_{MMSY} = 0.05$ to 0.28 (mean of 0.16), whereas avoiding 10% collapse rates required much lower exploitation rates of $u_{conserve} = 0.02$ to 0.05 (mean of 0.04).

Up to the 1990s, assessed species in 6 of the 10 ecosystems had exploitation rates substantially higher than those predicted to produce MMSY (Fig. 3A). Only the eastern Bering Sea has been consistently managed below that threshold. Since the 1990s, Iceland, Newfoundland-Labrador, the Northeast U.S. Shelf, the Southeast Australian Shelf, and California Current ecosystems have shown substantial declines in fishing pressure such that they are now at or below the modeled u_{MMSY}. However, only in the California Current and in New Zealand are current exploitation rates predicted to achieve a conservation target of less than 10% of stocks collapsed (Fig. 3A). Declining exploitation rates have contributed to the rebuilding of some depleted stocks, whereas others remain at low abundance. Averaged across all assessed species, biomass is still well below B_{MSY} in most regions. However, biomass has recently been increasing above the long-term average in Iceland, the Northeast U.S. Shelf, and the California Current, while remaining relatively stable or decreasing elsewhere (Fig. 3A).

Scientific stock assessments. Stock assessments quantify the population status (abundance, length, and age structure) of targeted fish or invertebrate stocks. We explored the status of 166 stocks worldwide for which we were able to obtain estimates of current biomass and exploitation rate (Fig. 3B). For about two-thirds of the examined stocks (63%), biomass (B) has dropped below the traditional single-species management target of MSY, that is, $B < B_{MSY}$. About half of those stocks (28% of total) have exploitation rates that would allow for rebuilding to B_{MSY}, that is, $u < u_{MSY}$, whereas overfishing continues in the remainder ($u > u_{MSY}$ in 35% of all stocks). Another 37% of assessed stocks have either not fallen below B_{MSY} or have recovered from previous depletion; most stocks in this category (77%) are in the Pacific. The weight of the evidence, as shown by the kernel density plot in Fig. 3B, indicates that most assessed stocks have

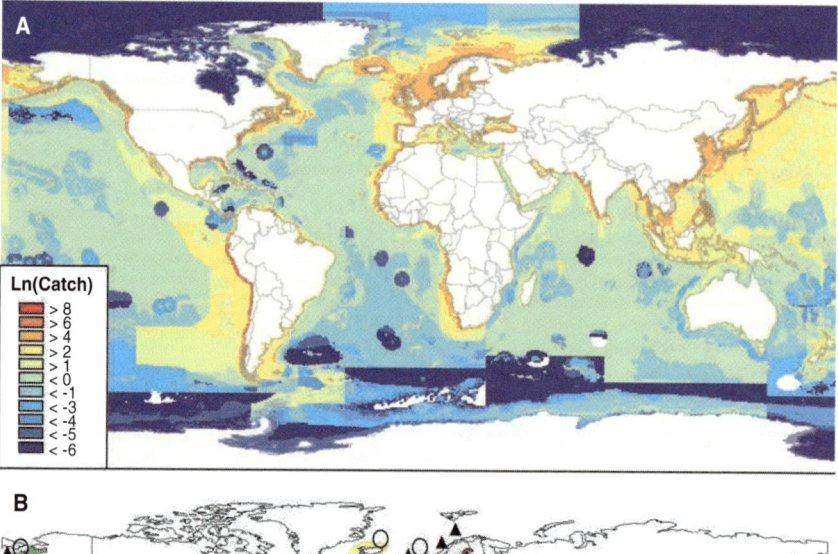

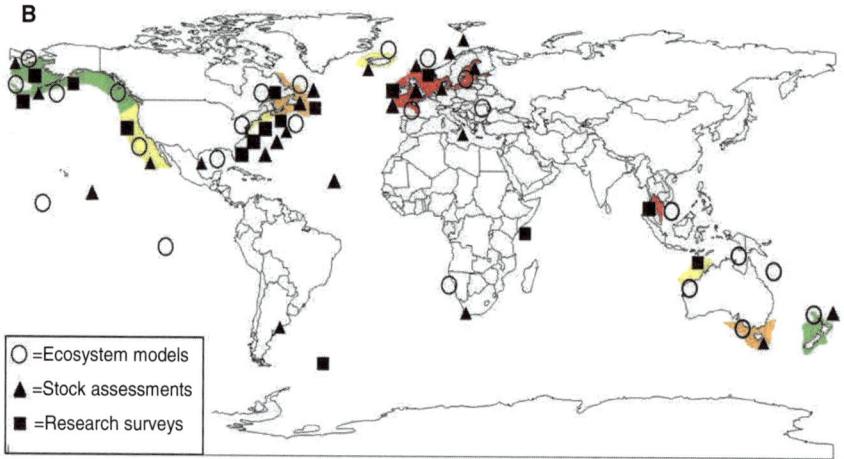

Fig. 1. Data sources used to evaluate global fisheries. (**A**) Global catch data; colors refer to the natural logarithm of the average reported catch (metric ton km^{-2} year^{-1}) from 1950 to 2004. (**B**) Other data: Stock assessments quantify the status of exploited populations; research trawl surveys are used to estimate fish community trends; ecosystem models are used to assess responses to fishing. Ecosystems that were analyzed in some detail are highlighted in green (not overfished), yellow (low exploitation rate, biomass rebuilding from overfishing), orange (low to moderate exploitation rate, not yet rebuilding), or red (high exploitation rate).

RESEARCH ARTICLES

fallen below the biomass that supports maximum yield ($B < B_{MSY}$) but have the potential to recover, where low exploitation rates ($u < u_{MSY}$) are maintained. Note that most stock assessments come from intensely managed fisheries in developed countries, and therefore our results may not apply to stocks in many developing countries, which are often not assessed but fished at high exploitation rates and low biomass. Full results are provided in table S2.

When we combined the biomass estimates of stocks assessed since 1977 ($n = 144$, Fig. 4A), we observed an 11% decline in total biomass. This trend is mostly driven by declines in pelagic (mid-water) species, whereas large declines in demersal (bottom-associated) fish stocks in the North Atlantic were offset by an increase in demersal biomass in the North Pacific after 1977. This shows how a global average can mask considerable regional variation. Although some ecosystems showed relative stability (e.g., the eastern Bering Sea, Fig. 4B), some experienced a collapse of biomass (e.g., eastern Canada, Fig. 4C), whereas others indicated rebuilding of some dominant target species (e.g., Northeast U.S. Shelf, Fig. 4D). These regional examples illustrate different stages of exploitation and rebuilding.

Research trawl surveys. The best sources of information to assess the state of fished communities are repeated scientific surveys that include both target and nontarget species. We analyzed research trawl survey data from 19 ecosystems where such data were available (see Fig. 1B for locations and fig. S2 and table S3 for full data set). We found that community trends averaged across all surveys (Fig. 4E) were broadly similar to the combined biomass trends seen in the recent assessments (Fig. 4A), with similar signatures of stability (Fig. 4F), collapse (Fig. 4G), and recovery (Fig. 4H) in selected regional ecosystems. Few of these surveys, however, reached back to the beginning of large-scale industrial exploitation in the 1950s and early 1960s. Where they did, for example, in the Gulf of Thailand and in Newfoundland, they revealed a rapid decline in total biomass within the first 15 to 20 years of fishing (fig. S2) as predicted by ecosystem models (Fig. 2). These declines were typically most pronounced for large predators such as gadoids (codfishes) and elasmobranchs (sharks and rays). Subsequent to the initial decline, total biomass and community composition have often remained relatively stable (fig. S2), although there may be substantial species turnover and collapses of individual stocks (see below). Across all surveys combined (*10*), we documented a 32% decline in total biomass, a 56% decline in large demersal fish biomass (species ≥90 cm maximum length), 8% for medium-sized demersals (30 to 90 cm), and 1% for small demersals (≤30 cm), whereas invertebrates increased by 23% and pelagic species by 143% (Fig. 4E). Increases are likely due to prey release from demersal predators (*17, 18*).

The trawl surveys also revealed changes in size structure that are consistent with model predictions: average maximum size (L_{max}) declined by 22% since 1959 when all communities were included (Fig. 4M). However, there were contrasting trends among our focal regions: L_{max} changed little in the eastern Bering Sea over the surveyed time period (Fig. 4N), dropped sharply in the southern Gulf of St. Lawrence, eastern Canada (Fig. 4O), as large demersal stocks collapsed, and increased because of rebuilding of large demersals (particularly haddock) on Georges Bank, Northeast U.S. Shelf (Fig. 4P). These trends included both target and nontarget species and show how changes in exploitation rates affect the broader community. Published analyses of the Gulf of St. Lawrence and adjacent areas in eastern Canada demonstrate that these community shifts involved large changes in predation regimes, leading to ecological surprises such as predator-prey reversals (*19*), trophic cascades (*17*), and the projected local extinction of formerly dominant species (*20*). Research on the Georges Bank closed area (*21*) and in marine protected areas worldwide (*22*) has shown how some of these changes may reverse when predatory fish are allowed to recover. This reveals top-down interactions cascading from fishers to predators and their multiple prey species as important structuring forces that affect community patterns of depletion and recovery (*18*).

Global fisheries catches. The benefits and costs involved in rebuilding depleted fisheries are demonstrated by an analysis of catch data. Global

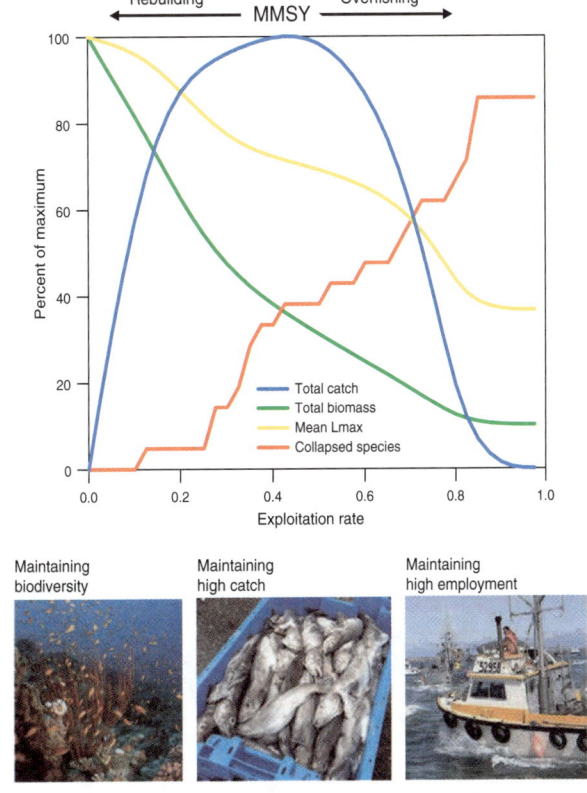

Fig. 2. Effects of increasing exploitation rate on a model fish community. Exploitation rate is the proportion of available fish biomass caught in each year. Mean L_{max} refers to the average maximum length that species in the community can attain. Collapsed species are those for which stock biomass has declined to less than 10% of their unfished biomass. This size-structured model was parameterized for 19 target and 2 nontarget species in the Georges Bank fish community (*13*). It includes size-dependent growth, maturation, predation, and fishing. Rebuilding can occur to the left, overfishing to the right, of the point of maximum catch. Three key objectives that inform current management are highlighted: biodiversity is maintained at low exploitation rate, maximum catch is maintained at intermediate exploitation rate, and high employment is often maintained at intermediate to high exploitation rate, because of the high fishing effort required.

catches have increased ~fivefold since 1950 as total biomass has been fished down (Fig. 4, A and E) then reached a plateau at ~80 million tons in the late 1980s (Fig. 4I). Catch composition with respect to the major species groups has remained relatively stable over time, with the exception of large demersal fishes, which have declined from 23 to 10% of total catch since 1950. Composition with respect to individual species, however, has fluctuated more widely owing to stock collapses (3) and expansion to new fisheries (6). Individual regions showed very different catch composition and trends, with large- and medium-sized demersal fish being historically dominant in the North Atlantic and North Pacific, small demersals being important in many tropical areas, and pelagic fish dominating the catch from oceanic and coastal upwelling systems (fig. S3). Among our focal regions, the eastern Bering Sea showed a high and stable proportion of large demersal fish (Fig. 4J), the Gulf of St. Lawrence displayed a collapse of the demersal catch and a replacement with small pelagic and invertebrate species (Fig. 4K), and Georges Bank (Fig. 4L) showed a large reduction in catch associated first with declining stocks and then with rebuilding efforts. These examples illustrate that the decline and rebuilding of fished stocks can incur significant costs because of lost catch, whereas sustained management for lower exploitation rates may promote greater stability with respect to both biomass and catches. Part of this stability may arise from the diversity of discrete populations and species that are more likely to persist in fisheries with low exploitation rates (3, 23).

Trends in species collapses. Theory suggests that increases in fishing pressure, even at levels below MMSY, cause an increasing number of target and non-target species to collapse (Fig. 2). Reductions in fishing pressure are predicted to reverse this trajectory, at least partially. By using biomass data from stock assessments compared to estimates of unfished biomass (B_0) (10), we found an increasing trend of stock collapses over time, such that 14% of assessed stocks were collapsed in 2007, that is, $B/B_0 < 0.1$ (Fig. 4M). This estimate is in the same range as figures provided by the United Nations Food and Agriculture Organization (FAO), which estimated that 19% of stocks were overexploited and 9% depleted or recovering from depletion in 2007 (24). Collapse trends vary substantially by region: The eastern Bering Sea had few assessed fish stocks collapsed (Fig. 4N), whereas collapses strongly increased to more than 60% of assessed stocks in eastern Canada (Fig. 4O) and more than 25% on the Northeast U.S. Shelf (Fig. 4P).

It appears that recent rebuilding efforts, although successful in reducing exploitation rates in several ecosystems (Fig. 3A), have not yet reversed a general trend of increasing depletion of individual stocks (Fig. 4M). This matches the model-derived prediction that reduction of exploitation rate to the level that produces MMSY will still keep a number of vulnerable species collapsed (Fig. 2). Rebuilding these collapsed stocks may require trading off short-term yields for conservation benefits or, alternatively, more selective targeting of species that can sustain current levels of fishing pressure while protecting others from overexploitation.

Small-scale fisheries. Fish or invertebrate stocks that are scientifically assessed ($n = 177$ in our analysis) or appear in research trawl surveys ($n = 1309$ taxa-by-survey combinations in fig. S2) constitute only a fraction of fisheries worldwide, which is an important caveat to the above discussion. Moreover they represent a nonrandom sample dominated by valuable industrial fisheries with some form of management in developed countries. The information on other fisheries, particularly small-scale artisanal and recreational fisheries is scarcer, less accessible, and more difficult to interpret. This is because small-scale fisheries are harder to track, with 12 million fishers compared with 0.5 million in industrialized

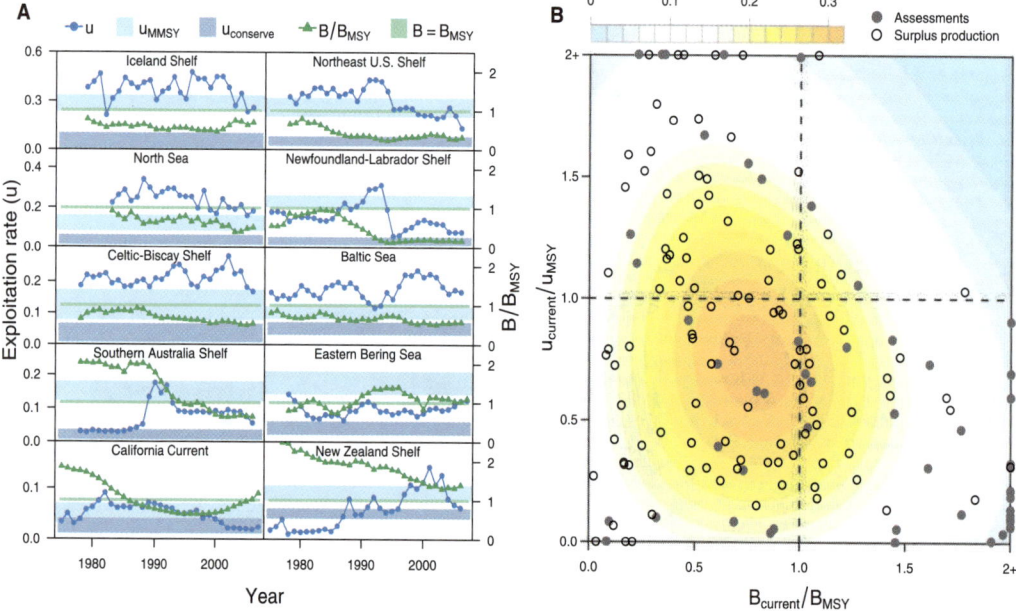

Fig. 3. Exploitation rate and biomass in large marine ecosystems and individual stocks. (**A**) Time trends of biomass (green triangles) are shown relative to the B_{MSY} (green band), exploitation rates (blue circles) relative to the u_{MMSY} (light blue band), and a hypothetical conservation objective at which less than 10% of species are collapsed ($u_{conserve}$, dark blue band). In each ecosystem, stock assessments were used to calculated average biomass relative to B_{MSY} and exploitation rate (total catch divided by total biomass) for assessed species. Reference points were calculated by using published ecosystem models; the width of the bands represents estimated uncertainty (10). (**B**) Current exploitation rate versus biomass for 166 individual stocks. Data are scaled relative to B_{MSY} and the exploitation rate (u_{MSY}) that allows for maximum sustainable yield. Colors indicate probability of occurrence as revealed by a kernel density smoothing function. Gray circles indicate that B_{MSY} and u_{MSY} estimates were obtained directly from assessments; open circles indicate that they were estimated from surplus production models (10).

Managing Fish Stocks

fisheries (25), and assessments or survey data are often lacking. Small-scale fisheries catches are also poorly reported; the best global estimate is about 21 million tons in 2000 (25). Conventional management tools used for industrial fisheries are generally unenforceable in small-scale fisheries when implemented in a top-down manner. More successful forms of governance have involved local communities in a co-management arrangement with government or nongovernmental organizations (26). An example is the rebuilding of depleted fish stocks on Kenyan coral reefs (Fig. 5A). A network of closed areas and the exclusion of highly unselective beach seines were implemented in cooperation with local communities and led to a recovery of the biomass and size of available fish (27). This translated into steep increases in fishers' incomes, particularly in

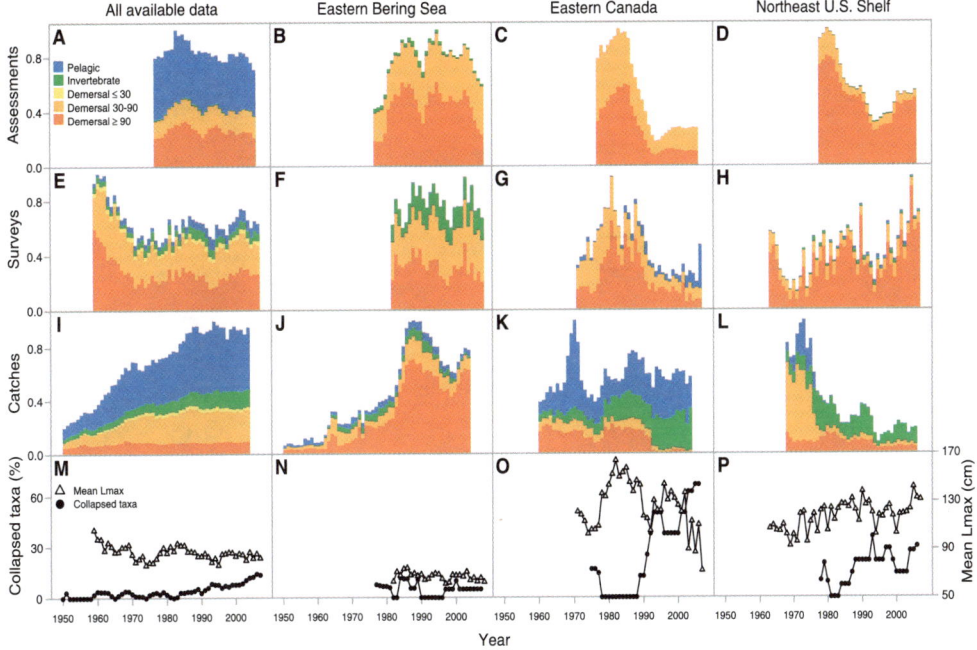

Fig. 4. Global and regional trends in fished ecosystems. Biomass trends computed from stock assessments (**A** to **D**), research surveys (**E** to **H**), as well as total catches (**I** to **L**) are depicted. Trends in the number of collapsed taxa (**M** to **P**, solid circles) were estimated from assessments, and changes in the average maximum size, L_{max} (M to P, open circles), were calculated from survey data (10). All data are scaled relative to the time series maximum. (G) and (K) represent the Southern Gulf of St. Lawrence (eastern Canada); (H) and (L), Georges Bank (Northeast U.S. Shelf) only. Collapsed taxa are defined as those where biomass declined to <10% of their unfished biomass. Colors refer to different species groups (demersal fish are split into small, medium, and large species based on the maximum length they can attain).

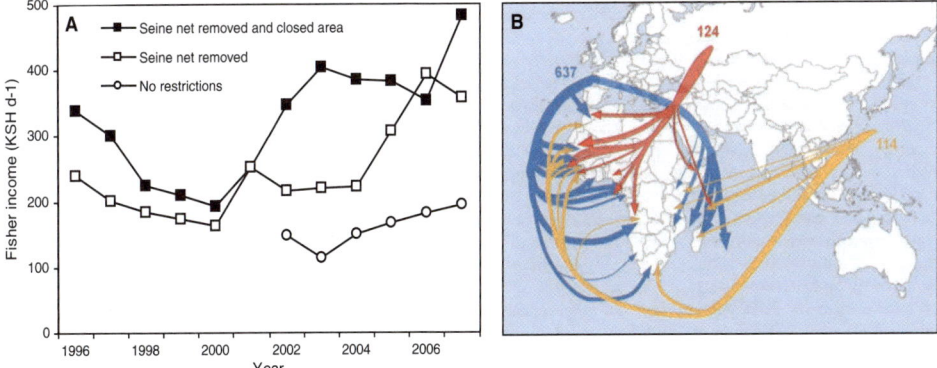

Fig. 5. Problems and solutions for small-scale fisheries. (**A**) Rebuilding of Kenyan small-scale fisheries through gear restrictions and closed area management. Updated, after (27). (**B**) Movement of fishing effort from developed nations to Africa in the 1990s. Data indicate total access years in distant-water fishing agreements. Updated, after (39).

Table 1. Management tools for rebuilding fisheries. Symbols indicate the contributions of a range of management tools to achieving reductions in exploitation rate: + tool contributed, ++ an important tool, or +++ an essential tool. Note that these examples are for industrialized fisheries, except Kenya, Chile, and Mexico. Ratings were supplied and checked by local experts.

Region	Gear restrictions	Capacity reduced	Total allowable catch reduced	Total fishing effort reduced	Closed areas	Catch shares	Fisheries certification	Community co-management
Bering Sea, Gulf of Alaska	+	++	+++		++	+++	+	+
California Current	+	++	+++		+++			
Northeast U.S. Shelf	+	++		+++	++			
North Sea, Celtic-Biscay	+	+	+++	++	+	+		+
Iceland	+	+	+++			+++		
Southeast Australian Shelf	+	+	+++		++	+++	+	
Northwest Australian Shelf	++				++			
New Zealand	+	+	+++			+++	+	
Kenya (Artisanal)	++				++			+++
Chile and Mexico (Artisanal)	+++				+			+++
Count	10	7	6	2	8	5	3	4
Total score	14	10	18	5	15	13	3	8

regions that had both closed areas and gear restrictions in place (Fig. 5A). Other examples of successful rebuilding come from Latin America, particularly Chile and Mexico, where open-access fisheries for valuable invertebrates were transformed by the establishment of spatial management units that had exclusive access by local fishing organizations (*26*). Despite these successes, rebuilding small-scale fisheries remains a significant challenge in developing countries where most fishers do not have access to alternative sources of food, income, and employment.

Tools for rebuilding. Management actions in a few ecosystems have prevented overfishing or, more commonly, reduced exploitation rates after a period of overfishing (Figs. 3 to 5). Diverse management tools have helped to achieve reductions in exploitation rates (Table 1). The most commonly used tools overall are gear restrictions, closed areas, and a reduction of fishing capacity, followed by reductions in total allowable catch and catch shares. Reductions in fishing capacity and allowable catch directly reduce the exploitation rate of target species by limiting catches. Gear modifications may be used to increase selectivity and reduce by-catch of nontarget species. Closed areas are either fully protected marine reserves (as in the Kenyan example discussed above) or are designed to exclude specific fisheries from certain areas. They can initiate recovery by providing refuge for overfished stocks (*21*, *28*), restoring community structure (*22*) and biodiversity (*3*), protecting important habitat features, and increasing ecosystem resilience (*29*). Assigning dedicated access privileges, such as catch shares or territorial fishing rights, to individual fishers or fishing communities has often provided economic incentives to reduce effort and exploitation rate (*30*) and may also improve compliance and participation in the management process (*31*). Likewise, the certification of sustainable fisheries is increasingly used as an incentive for improved management practices. Realigning economic incentives with resource conservation (rather than overexploitation) is increasingly recognized as a critical component of successful rebuilding efforts (*8*).

We emphasize that the feasibility and value of different management tools depends heavily on local characteristics of the fisheries, ecosystem, and governance system. For example, the most important element of small-scale fisheries success has been community-based management (Table 1), in which local communities develop context-dependent solutions for matching exploitation rates to the productivity of local resources (*26*). A combination of diverse tools, such as catch restrictions, gear modifications, and closed areas, is typically required to meet both fisheries and conservation objectives.

Here we have only identified the proximate tools, not the ultimate socioeconomic drivers that have enabled some regions to prevent or reduce overfishing while others remained overexploited. Yet it is generally evident that good local governance, enforcement, and compliance form the very basis for conservation and rebuilding efforts (*32*). Legislation that makes overexploitation illegal and specifies unambiguous control rules and rebuilding targets has also been critically important, for example, in the United States (*8*, *28*).

Most rebuilding efforts only begin after there is drastic and undeniable evidence of overexploitation. The inherent uncertainty in fisheries, however, requires that agencies act before it comes to that stage (*33*); this is especially true in light of accelerating global change (*34*). We found that only Alaska and New Zealand seemed to have acted with such foresight, whereas other regions experienced systemic overexploitation. The data that we have compiled cannot resolve why inherently complex fish-fisher-management systems (*35*) behaved differently in these cases; possible factors are a combination of abundant resources and low human population, slow development of domestic fisheries, and little interference from international fleets. It would be an important next step to dissect the underlying socioeconomic and ecological variables that enabled some regions to conserve, restore, and rebuild marine resources.

Problems for rebuilding. Despite local successes, it has also become evident that rebuilding efforts can encounter significant problems and short-term costs. On a regional scale, the reduction of quotas, fishing effort, and overcapacity eliminates jobs, at least in the short term. Initial losses may create strong resistance from fisheries-dependent communities through the political process. For instance in the United States, where 67 overfished stocks have rebuilding plans, 45% of those were still being overfished in 2006, whereas only 3 stocks had been rebuilt at that time (*36*). This problem is exacerbated by the fact that the recovery of depleted stocks can take years or even decades (*28*, *37*), and during this time catches may be dramatically reduced (e.g., Fig. 4L). Furthermore, government subsidies often promote overfishing and overcapacity and need to be reduced against the interests of those who receive them (*38*). Lastly, there is the problem of unreported and illegal fishing, which can seriously undermine rebuilding efforts (*11*). Illegal and unreported catches vary between regions, ranging between an estimated 3% of total catch in the Northeast Pacific to 37% in the East Central Atlantic, with a global average of 18% in 2000–2003 (*11*).

On a global scale, a key problem for rebuilding is the movement of fishing effort from industrialized countries to the developing world (Fig. 5B). This north-south redistribution of fisheries has been accelerating since the 1960s (*39*) and could in part be a perverse side effect of efforts to restore depleted fisheries in the developed world, as some fishing effort is displaced to countries with weaker laws and enforcement capacity. The situation is particularly well documented for West Africa (*39*) and more recently East Africa, where local fisheries have seen increasing competition from foreign fleets operating under national access agreements (Fig. 5B) and where illegal and unreported catches are higher than anywhere else (*11*). Almost all of the fish caught by foreign fleets is consumed in industrialized countries and may threaten regional food security (*39*) and biodiversity (*40*) in the developing world. Clearly, more global oversight is needed to ensure that rebuilding efforts in some regions do not cause problems elsewhere. For example, fishing vessels removed in effort-reduction schemes would ideally be prohibited from migrating to other regions and exacerbating existing problems with overcapacity and overexploitation.

Open questions. Rebuilding efforts raise a number of scientific questions. Recovery of depleted stocks is still a poorly understood process, particularly for demersal species (*37*). It is potentially constrained by the magnitude of previous decline (*37*), the loss of biodiversity (*3*, *23*), species life histories (*37*), species interactions (*17*, *18*, *20*), and climate (*28*, *34*). Yet, many examples of recovery exist, both in protected areas (*3*, *21*, *22*) and in large-scale ecosystems where exploitation was substantially reduced (Fig. 3A). A better understanding of how to predict and better manage for recovery will require insight into the resilience and productivity of individual populations and their communities.

This could be gained by more widespread spatial experimentation, involving proper controls, good monitoring, and adaptive management. Some of the most spectacular rebuilding efforts, such as those undertaken in California (*41*), the northeast United States (*21*), and northwest Australia (*42*) have involved bold experimentation with closed areas, gear and effort restrictions, and new approaches to catch allocation and enforcement. Science has a key role to play in guiding such policies, analyzing the effects of changes in management and advancing toward more general rules for rebuilding.

A second area of inquiry relates to the question of how to avoid contentious trade-offs between allowable catch and the conservation of vulnerable or collapsed species. Recovering these species while maintaining global catches may be possible through improved gear technology and a much more widespread use of ocean zoning into areas that are managed for fisheries benefits and others managed for species and habitat conservation. Designing appropriate incentives for fishers to avoid the catch of threatened species, for example, through tradable catch and by-catch quotas, has yielded good results in some regions (*16*). Temporary area closures can also be effective but require detailed mapping of the distribution of depleted populations and their habitats.

Conclusions. Marine ecosystems are currently subjected to a range of exploitation rates, resulting in a mosaic of stable, declining, collapsed, and rebuilding fish stocks and ecosystems. Management actions have achieved measurable reductions in exploitation rates in some regions, but a significant fraction of stocks will remain collapsed unless there are further reductions in exploitation rates. Unfortunately, effective controls on exploitation rates are still lacking in vast areas of the ocean, including those beyond national jurisdiction (*6*, *8*, *32*). Ecosystems examined in this paper account for less than a quarter of world fisheries area and catch, and lightly to moderately fished and rebuilding ecosystems (green and yellow areas in Fig. 1B) comprise less than half of those. They may best be interpreted as large-scale restoration experiments that demonstrate opportunities for successfully rebuilding marine resources elsewhere. Similar trajectories of recovery have been documented in protected areas around the world (*3*, *21*, *22*), which currently cover less than 1% of ocean area. Taken together, these examples provide hope that despite a long history of overexploitation (*1*, *2*) marine ecosystems can still recover if exploitation rates are reduced substantially. In fisheries science, there is a growing consensus that the exploitation rate that achieves maximum sustainable yield (u_{MSY}) should be reinterpreted as an upper limit rather than a management target. This requires overall reductions in exploitation rates, which can be achieved through a range of management tools. Finding the best management tools may depend on the local context. Most often,

it appears that a combination of traditional approaches (catch quotas, community management) coupled with strategically placed fishing closures, more selective fishing gear, ocean zoning, and economic incentives holds much promise for restoring marine fisheries and ecosystems. Within science, a new cooperation of fisheries scientists and conservation biologists sharing the best available data, and bridging disciplinary divisions, will help to inform and improve ecosystem management. We envision a seascape where the rebuilding, conservation, and sustainable use of marine resources become unifying themes for science, management, and society. We caution that the road to recovery is not always simple and not without short-term costs. Yet, it remains our only option for insuring fisheries and marine ecosystems against further depletion and collapse.

References and Notes
1. J. B. C. Jackson et al., *Science* **293**, 629 (2001).
2. H. K. Lotze et al., *Science* **312**, 1806 (2006).
3. B. Worm et al., *Science* **314**, 787 (2006).
4. R. Hilborn, *Ecosystems (N. Y., Print)* **10**, 1362 (2007).
5. S. Murawski et al., *Science* **316**, 1281b (2007).
6. D. Pauly et al., *Science* **302**, 1359 (2003).
7. T. H. Huxley, *Inaugural Meeting of the Fishery Congress: Address by Professor Huxley* (William Clowes and Sons, London, 1883).
8. J. R. Beddington, D. J. Agnew, C. W. Clark, *Science* **316**, 1713 (2007).
9. S. M. Garcia, A. Zerbi, C. Aliaume, T. Do Chi, G. Lasserre, "The ecosystem approach to fisheries" (FAO, Rome, 2003).
10. Details on methods and data sources are available as supporting online material on *Science* Online.
11. D. J. Agnew et al., *PLoS One* **4**, e4570 (2009).
12. E. A. Fulton, A. D. M. Smith, C. R. Johnson, *Mar. Ecol. Prog. Ser.* **253**, 1 (2003).
13. S. J. Hall et al., *Can. J. Fish. Aquat. Sci.* **63**, 1344 (2006).
14. R. Q. Grafton, T. Kompas, R. W. Hilborn, *Science* **318**, 1601 (2007).
15. S. Jennings, A. S. Revill, *ICES J. Mar. Sci.* **64**, 1525 (2007).
16. T. A. Branch, R. Hilborn, *Can. J. Fish. Aquat. Sci.* **65**, 1435 (2008).
17. K. T. Frank, B. Petrie, J. S. Choi, W. C. Leggett, *Science* **308**, 1621 (2005).
18. J. K. Baum, B. Worm, *J. Anim. Ecol.* **78**, 699 (2009).
19. D. P. Swain, A. F. Sinclair, *Can. J. Fish. Aquat. Sci.* **57**, 1321 (2000).
20. D. P. Swain, G. A. Chouinard, *Can. J. Fish. Aquat. Sci.* **65**, 2315 (2008).
21. S. A. Murawski, R. Brown, H.-L. Lai, P. J. Rago, L. Hendrickson, *Bull. Mar. Sci.* **66**, 775 (2000).
22. F. Micheli, B. S. Halpern, L. W. Botsford, R. R. Warner, *Ecol. Appl.* **14**, 1709 (2004).
23. R. Hilborn, T. P. Quinn, D. E. Schindler, D. E. Rogers, *Proc. Natl. Acad. Sci. U.S.A.* **100**, 6564 (2003).
24. FAO, *The State of World Fisheries and Aquaculture 2008* (FAO, Rome, 2009).
25. D. Pauly, *Marit. Stud.* **4**, 7 (2006).
26. O. Defeo, J. C. Castilla, *Rev. Fish Biol. Fish.* **15**, 265 (2005).
27. T. R. McClanahan, C. C. Hicks, E. S. Darling, *Ecol. Appl.* **18**, 1516 (2008).
28. J. F. Caddy, D. J. Agnew, *Rev. Fish Biol. Fish.* **14**, 43 (2004).
29. J. A. Ley, I. A. Halliday, A. J. Tobin, R. N. Garrett, N. A. Gribble, *Mar. Ecol. Prog. Ser.* **245**, 223 (2002).
30. C. Costello, S. D. Gaines, J. Lynham, *Science* **321**, 1678 (2008).
31. NRC, *Cooperative Research in the National Marine Fisheries Service* (National Research Council Press, Washington, DC, 2003).
32. C. Mora et al., *PLoS Biol.* **7**, e1000131 (2009).

33. D. Ludwig, R. Hilborn, C. Walters, *Science* **260**, 17 (1993).
34. K. M. Brander, *Proc. Natl. Acad. Sci. U.S.A.* **104**, 19709 (2007).
35. P. A. Larkin, *Annu. Rev. Ecol. Syst.* **9**, 57 (1978).
36. A. A. Rosenberg, J. H. Swasey, M. Bowman, *Front. Ecol. Environ* **4**, 303 (2006).
37. J. A. Hutchings, J. D. Reynolds, *Bioscience* **54**, 297 (2004).
38. U. R. Sumaila et al., *Fish. Res.* **88**, 1 (2007).
39. J. Alder, U. R. Sumaila, *J. Environ. Dev.* **13**, 156 (2004).
40. J. S. Brashares et al., *Science* **306**, 1180 (2004).
41. M. Dalton, S. Ralston, *Mar. Resour. Econ.* **18**, 67 (2004).
42. K. J. Sainsbury, R. A. Campbell, R. Lindholm, A. W. Whitelaw, in *Fisheries Management: Global Trends*, E. K. Pikitch, D. D. Huppert, M. P. Sissenwine, Eds. (American Fisheries Society, Bethesda, MD, 1997), pp. 107–112.
43. This work was conducted as part of the "Finding common ground in marine conservation and management" Working Group supported by the National Center for Ecological Analysis and Synthesis funded by NSF, the University of California, and the Santa Barbara campus. The authors acknowledge the Natural Sciences and Engineering Research Council (NSERC) and the Canadian Foundation for Innovation for funding database development, the Sea Around Us Project funded by Pew Charitable Trusts for compiling global catch data, and numerous colleagues and institutions around the world for sharing fisheries assessment, catch, access, and survey data, and ecosystem models (see SOM for full acknowledgments).

Supporting Online Material
www.sciencemag.org/cgi/content/full/325/5940/578/DC1
Materials and Methods
Figs. S1 to S6
Tables S1 to S7
References and Notes

27 April 2009; accepted 22 June 2009
10.1126/science.1173146

1.3 HONORABLE MENTION FULL CITATIONS AND ABSTRACTS

Conover, D., and S. Munch. 2002. Sustaining fisheries yields over evolutionary time scales. Science 297(5578):94–96.

> Fishery management plans ignore the potential for evolutionary change in harvestable biomass. We subjected populations of an exploited fish (*Menidia menidia*) to large, small, or random size-selective harvest of adults over four generations. Harvested biomass evolved rapidly in directions counter to the size-dependent force of fishing mortality. Large-harvested populations initially produced the highest catch but quickly evolved a lower yield than controls. Small-harvested populations did the reverse. These shifts were caused by selection of genotypes with slower or faster rates of growth. Management tools that preserve natural genetic variation are necessary for long-term sustainable yield.

Fournier, D. A., and C. P. Archibald. 1982. A general theory for analyzing catch at age data. Canadian Journal of Fisheries and Aquatic Sciences 39:1195–1207.

> We present a general theory for analyzing catch at age data for a fishery. This theory seems to be the first to address itself properly to the stochastic nature of the errors in the observed catch at age data. The model developed is very flexible and accommodates itself easily to the inclusion of extra information such as fishing effort data or information about errors in the aging procedure. An example is given to illustrate the use of the model.

Hoenig, J. M. 1983. Empirical use of longevity data to estimate mortality rates. US Fishery Bulletin 81:898–903.

> Various investigators have utilized compendia of life history parameters to develop equations for predicting values of difficult-to-estimate parameters from easily measured or estimated quantities. For example, Pauly (1979) developed multiple regressions to predict the natural mortality rate of fish from growth parameters and mean water temperature. Ohsumi (1979) developed linear regressions for estimating natural mortality of cetaceans from maximum length or maximum age. In this paper, a general regression equation is developed to predict the total mortality rate of fish, cetacean, and mollusk stocks from the maximum age.

Jackson, J. B. C., M. X. Kirby, W. H. Berger, K. A. Bjorndal, L. W. Botsford, B. J. Bourque, R. H. Bradbury, R. Cooke, J. Erlandson, J. A. Estes, T. P. Hughes, S. Kidwell, C. B. Lange, H. S. Lenihan, J. M. Pandolfi, C. H. Peterson, R. S. Steneck, M. J. Tegner, and R. R. Warner. 2001. Historical overfishing and the recent collapse of coastal ecosystems. Science 293(5530):629–638.

Ecological extinction caused by overfishing precedes all other pervasive human disturbance to coastal ecosystems, including pollution, degradation of water quality, and anthropogenic climate change. Historical abundances of large consumer species were fantastically large in comparison with recent observations. Paleoecological, archaeological, and historical data show that time lags of decades to centuries occurred between the onset of overfishing and consequent changes in ecological communities, because unfished species of similar trophic level assumed the ecological roles of overfished species until they too were overfished or died of epidemic diseases related to overcrowding. Retrospective data not only help to clarify underlying causes and rates of ecological change, but they also demonstrate achievable goals for restoration and management of coastal ecosystems that could not even be contemplated based on the limited perspective of recent observations alone.

Pauly, D. 1980. On the interrelationships between natural mortality, growth parameters, and mean environmental temperature in 175 fish stocks. Journal du Conseil, Conseil International pour l'Exploration de la Mer 39:175–192.

A compilation of values for the exponential coefficient of natural mortality (M) is given for 175 different stocks of fish distributed in 84 species, both freshwater and marine, and ranging from polar to tropical waters. Values of L_∞ (LT, cm), W_∞ (g, fresh weight), K (l/year) and T (°C, mean annual water temperature) were attributed to each value of M, and the 175 sets of values plotted such that:

1) $\log M = -0.2107 - 00824 \log W_\infty + 0.6757 \log K + 0.4627 \log T$

and

2) $\log M = -0.0066 - 0.279 \log L_\infty + 0.6543 \log K + 0.4634 \log T$

The multiple correlation coefficients are for 1) 0.845, and for 2) 0.847, while the critical value (171 d.f.) is 0.275 (for $P = 001$). All slopes are significantly * 0 (for $P = 0.001$). The standard deviation of estimates of $\log M$ are for 1) 0.247, and for 2) 0.245.

The equations provide highly reliable estimates of M for any given fish stock, given the values of W_∞ or L_∞ and K of the von Bertalanffy growth formula, and an estimate of the mean water temperature in which the stock in question lives. Only two groups have values of M generally differing from those obtained through the proposed equations: the Clupeidae, with generally lower and the polar fishes with generally higher values. Correction factors are given for both groups. Potential applications of the findings to population dynamics are discussed together with some ecological implications.

Section 2

Managing People

ROBERT ARLINGHAUS

2.1. SYNTHESIS

Background: Selection Does Not Guarantee Completeness

The articles in this section all relate to the social aspects of fisheries, particularly the management of people. The "human-dimensions of fisheries" are a diverse collection of disciplines involving economics, empirical social sciences, sociology, socio-psychology, political science, anthropology, human geography, and many more. For this reason, it was challenging to identify eight key articles in this area of fisheries science. Several sections in this book deal with specifics of the overarching discipline of fisheries biology, which reflects that fisheries biology remains the dominant approach to fisheries science. For space reasons, all social science-related articles have been combined into this section. This of course cannot give credit to the many seminal articles in the social science of fisheries literature that span the disciplines outlined above. Thus, it was decided to expand the Honorable Mention list to ten articles. While still not fully encompassing, it is believed that the eighteen articles cover the major streams of innovation that characterize the field of management-oriented human-dimensions studies in commercial, recreational, small-, and large-scale fisheries.

Note that the article list, due to space limitations, falls short on social science papers from fisheries sociology and anthropology. In particular, the list largely excluded articles that dealt with classical sociological and anthropological topics such as artisanal fishing communities, livelihoods, social networks, power relations, spiritual value of fisheries, religion, culture, ethics, and gender issues. I am aware of a number of key articles in these areas, some of which have accumulated very high citation rates (e.g., Callon 1988). However, many of these articles were not strongly focused on management-oriented issues, which is the core of all sections in this book. Naturally, the final choice also represented the literature knowledge of those involved in selecting the works, which was biased towards quantitative human-dimensions literature. Although advice from sociologists, political scientists, and anthropologists was solicited, it is most likely that some key publications have been missed.

Some of the key inspirations for understanding the core of the social science endeavor in fisheries—to unravel the rules of human behavior in relation to fish stocks and aquatic ecosystems—originated from political science in the context of natural resource management in general. Fisheries have often been featured as prominent cases in this literature (Ostrom

1990). Therefore, some of the articles reviewed in this section are not fisheries-specific, but instead have a more general theme in terms of how humans interact with renewable natural resources (e.g., Hardin 1968; Dietz et al. 2003).

Early Fisheries Economics

One important inquiry in the social sciences of fisheries has been economics. It is an unlikely coincidence that about the same time as the theory of fisheries biology became codified in textbooks (Beverton and Holt 1957; Ricker 1958), two key articles (Gordon 1954; Scott 1955) appeared that have largely set the research agenda in fisheries economics up to the present day. These articles have also strongly influenced fisheries biologists' thinking about the consequences of open access exploitation of fisheries resources.

The innovation of Gordon (1954) was to call attention to the role of property rights (institutions, or "rules-in-use," Ostrom 1990) for fisheries management. He illustrated that without a corrective hand, open-access and profit-maximizing behavior of the fishing industry constitute two conditions for attracting fishing effort until a so called bionomic equilibrium is reached (not to be confused with the dynamic bioeconomic equilibrium, see Clark 2006). In this steady state, no (economic) rents (surpluses) are produced to society [note that the concept of economic rent is not to be confused with resource rent, see Bromley 2009]. The bionomic equilibrium is defined as the point where the average revenue by the industry (or by any individual fishing firm) would equal their average (or individual) opportunity costs. Thus, while at equilibrium economic surpluses will be dissipated, the industry would usually still earn income (Bromley 2009).

In Gordon's (1954) model, the behavior of the fishing industry was represented very broadly. It was simply assumed that people enter the fishery or increase effort as long as economic rents are greater than the opportunity costs of fishing. Depending on the total cost structure of fishing, the equilibrium harvest level and corresponding fish biomass could occur at lower or greater biomass levels than the biomass that produces the maximum sustainable yield (Schaefer 1957). Hence, the bionomic equilibrium is not to be confused with overexploitation and collapse of fisheries. It is also important to realize that Gordon (1954) applied his formal analysis to a demersal fishery with two fishing grounds of differential productivity, noting that the better ground would be overfished and brought into balance with the inferior ground (which is the economic version of the ideal free distribution theory in ecology). These important aspects of the Gordon (1954) model have often been overlooked, instead assuming the bionomic equilibrium is economically wasteful and biologically unsustainable. Neither of this is necessarily true (Bromley 2009).

Gordon (1954) is credited with introducing the now famous concept of the maximum possible economic surplus, also referred to as the maximum (or net) economic yield (MEY) (Schaefer 1957). The MEY would, according to Gordon (1954), be expected to exist at the point where the marginal revenue (defined as the slope of the landings or revenue curve) would equal the marginal cost (defined as the slope of the total cost curve), graphically representing the maximum distance between the revenue and the total cost curves. A sole owner of a fishery would therefore tailor effort so as to reap the maximum possible economic rent, but in a restrictive market with few substitutes, this might come at the cost of high prices to the consumer (Bromley 2009). The idea that maximized economic rents would often correspond with much reduced effort is attractive for many researchers and stakeholders interested in reducing fishing mortality in overfished stocks because this can result in a "zone of new consensus" between fisheries and conservation interests (Hilborn 2007).

Whenever societies (or fisheries managers) wish for greater fish populations than the bionomic equilibrium provides, this desire becomes a motivating factor to change either the property rights or fishing mortality through management regulations. In this context, Gordon (1954) was the first to emphasize property right changes as a solution in fisheries calling for (demersal) fishing grounds to be owned individually or by the state, and he was soon echoed by another prominent article of the time. In Hardin's (1968) powerful herder metaphor, unavoidable ruin of natural resources was predicted to be caused by privately rational, self-interested harvesters seeking to maximize the net benefits from resource extraction, while externalizing the extraction costs to society. Stated differently, each harvester would reap the benefits of his/her additional boat put on the common fishing ground, but neither consider nor bear all of the corresponding costs of (cumulative) overfishing the fish stock. The failure of each individual to fully internalize (recognize) the costs of overfishing to society or—perhaps more forgiving—the perception of individuals that their own small take of fish will not affect the common good noticeably, would then result in a social dilemma of overuse of the commons (Hardin 1968). Individual interests may thus trump societal interest.

Gordon (1954), and particularly Hardin (1968), left the legacy that changes to property rights in fisheries implementing either state property or private property would be needed to avoid overharvest of renewable natural resources in open-access situations (see below). A foundation for an economic rationale for management of fisheries was laid by identifying the objective that ought to be pursued (MEY), by contrasting the implications of open-access resource use relative to this ideal, and by characterizing the (economic and biological) costs of not pursuing the maximization of economic rents.

Scott (1955) published the related economic analysis of optimal use of fisheries under sole ownership. Scott's (1955) innovation is his outline of the first dynamic theory of the privately owned (optimized) fishery referring to short and long term human decision-making regarding fisheries. Note that sole (or private) ownership here means sole control of harvest, which can entail an exclusive long-term harvest privilege or ownership of the fish stocks/grounds. Scott (1955) showed how the optimal state of exploitation of a fishery under sole ownership would balance marginal current profits against marginal user costs in the long term. His explicit innovation was the accounting of costly effects of intensive harvesting in terms of the impacts on future stock size and revenues. Under the condition of optimized sole ownership, an additional unit of input (effort) or output (landings) is worth investing (effort) or taking (landings) today only if its addition to current net revenue exceeds the present value of its long-term costs in reduced future profitability. This results, theoretically at least, in keeping future returns from the fishery as high as possible, while maximizing current income. A conservative harvesting strategy is the essential result. However, such effects strongly depend on the degree of discounting of future revenue (i.e., the opportunity cost of capital), and hence on human time preferences in terms of balancing current versus future revenue. It could be privately rational for a sole owner to continually overharvest a very valuable, but slow-growing resource, and drive it to extinction if the discount rates were high (Clark 1973).

Scott (1955) also added the important theoretical insight that one must consider the ability of the operator to control the access to the fishery resource in the long term; the less control one has, the more incentive there will be to fish. Viewed differently, when an owner of a fishery has a long time horizon for planning the fishing initiatives, he/she can plan the operations to maximize the net present value (where future benefits are discounted to a value in present day). Scott (1955) then argued that sole ownership may bring about the best use of the fishery from a social point of view (the above mentioned caveat related to high discount rates of course still applies).

Many subsequent studies have found strong empirical support for Scott's (1955) theoretical predictions. However, further research has also found that fisheries management success is not contingent on private or state ownership; having long-term secured harvesting rights—even group ownership and shared management of fisheries resources—can produce similar outcomes as those available under either private (sole) and state ownership (Ostrom 1990; Dietz et al. 2003), and this condition was not foreseen by Hardin (1968) who confused common-property with open access.

From Privatization to Common-property and the Power of Informal Institutions

Gordon (1954) and Hardin (1968), as well as many subsequent scholars, can be faulted in hindsight for having relied exclusively upon two oversimplifications in terms of the underlying reasons and solutions to combat overfishing from a social science perspective. The first is the claim that only two possible institutional arrangements—government control or private property—could sustain fisheries over the long term. The second is the assumption that harvesting and investment behavior is driven *only* by individualistic behavior by harvesters seeking to maximize expected revenue—or some other form of tangible utility. This assumes that harvesters are unable to create cooperative solutions to manage sustainable fisheries. Both simplifications have been found to be wrong in specific settings and circumstances, for example in many artisanal fisheries with co-management systems (Ostrom 1990; Gutiérrez et al. 2011).

Acheson (1975) was among the first to outline the power of voluntarily enforced territorial fishing rights, and the power of local social norms, to help curtail fishing effort and maintain fisheries in regimes of successful self-governance. Acheson (1975) described a regime of self-devised and enforced harvest rights—a form of common property—in the Maine lobster *Homarus americanus* fishery. These arrangements constitute informal institutions (Ostrom 1990) that have helped to conserve the resource and maintain viable fisheries without significant external input from fisheries agencies or any sort of formal institutions (laws).

Acheson (1975) identified two general types of lobster communities (provocatively called "harbor gangs"): (1) perimeter-defended; and (2) nucleus. Perimeter-defended territories exhibited the feature of more clearly delineated boundaries, which reduced encroaching on the territory by other fishers from different "harbor gangs." In addition, it was more difficult to become a member of "harbor gangs" in perimeter-defended territories. This rule kept communities stable and fostered communication. The system has generated a specific bioeconomic equilibrium reflecting a healthy resource base. Although Acheson (1975) maintained that the lobster system differed from Scott's (1955) "sole owner," there are similarities because the perimeter-defended community achieved unified control over harvests and involved *de-facto* (as opposed to *de-jure*) property rights. The important difference is one of "composition." Scott's (1955) sole owner seems to be a single decision maker, while Acheson's (1975) "gang" is a *cartel* of closely-related fishers. The conditions of long-term stable relationships and closed communities were later identified by others as promoting sustainable exploitation of common-pool-resources. Particularly noteworthy in this context is the work by Ostrom (1990), to which I will address further below.

It is therefore not privatization *per se* that is a necessary condition for sustainable fisheries because, under the right institutional arrangements, fishing communities can effectively sustain fisheries in a self-organized manner (Acheson 1975; Ostrom 1990). Some form of long-term tenure for the resources is however needed for proper incentives to sustain fisheries to develop (Dietz et al. 2003). Note that even in this situation users can organize to overexploit fisheries because it may be economically rational for an individual—or a group—to drive a biological resource to extinction (Clark 1973).

Acheson's (1975) article inspired many scholars who were not convinced that either privatization or state control was the only possible property regimes that would sustain fisheries. In the 1980s, the political scientist Elinor Ostrom and her colleagues started to conduct and synthesize many case studies like the one by Acheson (1975). Based on hundreds of case studies, she derived general insights into the conditions under which local self-governance of fisheries and other natural resources was possible by communities of resource users leading to sustainable outcomes (Ostrom 1990; a summary of the famous nine institutional design principles can be found in Dietz et al. 2003 reprinted in this book). Ostrom (1990) found effective governance of common-pool-resources, such as fishes, were easier to achieve overall when: (1) monitoring information (of fish and users) can be verified and understood at relatively low cost; (2) rates of change in resources, resource-user populations, technology, and economic and social conditions are moderate; (3) communities maintain frequent face-to-face communication and dense social networks; (4) outsiders can be excluded at relatively low cost from using the resource; and (5) users support effective monitoring and rule enforcement (Dietz et al. 2003).

Few settings in the world are characterized by all of these conditions. The famous case study by Johannes (1978) on island fishing communities in Oceania has clearly illustrated this point. The policy and management challenge is to devise arrangements that will help to establish such institutional conditions to achieve sustainable management in the absence of ideal conditions. Success in cooperative fisheries management is most likely when local leadership meets with social capital and strong incentives of local resource users to maintain resources (Guettiérez et al. 2011).

Co-management

In large-scale marine fisheries, fishing fleets are often internationally owned, meaning that community-based management in its pure form is often difficult to achieve. Indeed, many large-scale marine fisheries are controlled by governments. However, even under government control, there is a strong role for community involvement to produce better fisheries management outcomes. One of the milestone articles highlighting the many benefits of co-management between government and fisher organizations was published by Jentoft (1989).

Jentoft (1989) discussed the advantages and disadvantages of government-based fisheries management relative to more explicit co-management between government and fisher organizations. One of the key assumptions of Jentoft's (1989) work was that the legitimacy, and hence the expediency and quality, of fisheries regulations could be strongly improved by involving fisher organizations directly in the regulatory process. Up until around the 1970s, co-management was rarely accepted as a viable form of governance; this was the era of private or governmental control.

Jentoft's (1989) research blended theoretical insight with comparative analysis of co-management systems. Co-management, according to Jentoft (1989), means that fisher organizations not only have a say in the decision-making process, but also have the authority to make and implement regulatory decisions—importantly including enforcement—on their own. Usually, there is a shared system in which governments set the general framework (e.g., laws, quotas) and user organizations may then work out the details—or help in developing them with state officials. The most important contribution one can realistically hope for is that co-management will help to bolster the legitimacy of the regulatory process. This "legitimacy premium" will serve to make management more effective and less costly—especially in the realm of monitoring, compliance, and enforcement—in comparison to government control.

Co-management is a meeting point to blend a government's interest in sustainable resource use and protection along with the group's interest in maintaining equal opportunities, self-determination, and self-control. The essential key to success is the sharing of management functions among government and fisher organizations. Such co-management must be strictly formal, with mutually agreed and respected procedural rules. The factors of long-term success are: (1) formal acceptance of the procedures; (2) responsible and formal leadership; (3) a competent and honest executive staff; (4) a small scale of the organization (which increases relationships and produces fewer free riders); and (5) a homogenous membership to reduce conflicts and minimize varying perspectives. We are now seeing many international fisheries programs stressing the importance of a greater involvement by fishers and fisher organizations to improve procedures, legitimacy, and compliance with regulations. One unresolved problem is the decision on who to include in the process within user organizations and how to decide upon inclusiveness. There are also conditions under which co-management is likely to fail, which Jentoft (1989) clearly outlined in his milestone article.

Getting the Human Behavior Right

An economic rationale for fisheries management was established in the 1950s, while newer formulations simply elaborated—and sometimes misapplied (Bromley 2009)—the early models of economically optimal solutions to the resource management problem developed by Gordon (1954) and Scott (1955). However, up to the 1990s, economic and other social-scientific models rarely represented the explicit behavioral processes of human decision-making in terms of capital investments, spatial site choice, gear choice, or entry-exit behavior. Moreover, most economic models considered long-term solutions in the steady state (equilibrium), neglecting transitional dynamics. Forecasting future conditions must however account not only for the dynamics of the fish population and the technology and behavior of the fisher community, but also include the dynamics of the behavior and technology of the regulatory apparatus (Wilen 2000). In the 1970s, many nations declared exclusive economic zones (EEZs) resulting in many fisheries moving into some form of government-regulated, restricted-access condition. Under these conditions, Gordon's (1954) theory and predictions for open-access are expected to break down. Under regulated (managed) access, the nature of the dynamic equilibrium should be affected as much by public policy as it is by choices made by the fishing industry (Wilen 2000).

New models and theories were thus needed to represent the dynamic human behavior in fisheries-management models. Homans and Wilen (1997) were among the first (see also Smith 1968) to put forward economic models that accounted for human behavior under regulated conditions. The authors compared predictions from a model with a behavioral representation of regulatory actions with the standard Gordon (1954) open-access prediction. They showed how the Gordon (1954) model greatly underestimated effort (capacity), biomass, and harvest, which emerged from misrepresenting the role played by increasingly stringent regulations to control harvest. Relative to the Gordon (1954) model, the regulated model predicted higher biomass and harvest levels, a higher level of fishing capacity, and a substantially shorter fishing season, which was the regulatory tool used in the model. The short fishing season "stifled" the greater fishing capacity, which in turn, allowed biomass to be greater than in the Gordon (1954) model (Homans and Wilen 1997). To the extent that regulations were successful, they held the biomass at greater levels, which *ceteris paribus* generated higher economic rents and larger levels of fishing capacity. The latter point is worth mentioning because it explains why under regulated open access typical of many fisheries within the EEZ, so much redundant capital exists.

There are three key implications in the highly innovative paper by Homans and Wilen (1997). First, the technological and behavioral character of modern fisheries is intimately bound with nature, operation of the policy environment, and regulatory structure. This means that the technology and behavior will be affected by the regulatory structure, but will also be the attributes of the harvested populations. Second, regulated fisheries may look considerably different from what would emerge under pure open access—and they may, in fact, even be generating substantial economic rents. Finally, carefully accounting for the behavior of regulators will produce considerably different predictions than those from simple economic models. Taken together, Homans and Wilen (1997) left the legacy to better represent the interaction of policy, fisher behavior, and fish populations when attempting to generate insights of relevance to operational management. Their work placed an even greater burden on modelers to know and understand the particular features of the fisheries they are modeling, from an economic and a biological perspective. The community is still struggling with this task.

Smith and Wilen (2003) went one step further by incorporating an explicit random utility maximization (RUM) model of spatial fisher behavior into an economic-biological model. Using RUM to simulate discrete behavioral decisions was first introduced by McFadden (1974), and first applied to fisheries by Bockstael and Opaluch (1983) (see Carson et al. 2009 for an early example of a RUM from recreational fisheries, which was published in the primary literature many years after its first appearance in the gray literature). Smith and Wilen (2003) based their empirical application on the theoretical model of Sanchirico and Wilen (1999), showing how explicit modeling of fisher decisions in a spatial setting strongly altered predictions about the conservation value of marine protected areas. While this is an innovation by itself, the other key innovation is the integrated nature of the bioeconomic model by linking a calibrated model of fisher behavior to a sophisticated biological model of a sea urchin fishery. Explicit modeling of fisher behavior was accomplished much earlier than Smith and Wilen (2003), but the statistical modeling of spatial economic behavior in relation to an empirically calibrated resource model had not been done before.

Marine biologists had developed much faith in marine reserves as a management tool. However, most modeling of reserve performance had invoked strong simplifying assumptions about the behavior of fishers in response to spatial closures. Smith and Wilen (2003) showed that a realistic depiction of fisher behavior greatly altered the conclusions about the efficacy of reserves. The behavioral model showed how economic incentives determine participation and location choices of fishers. Simulations with behavioral response were compared to more traditional biological models that presumed that effort is spatially uniform and unresponsive to economic incentives. Through this approach, Smith and Wilen (2003) demonstrated that optimistic conclusions about reserves can be an artifact of simplifying assumptions that ignore economic behavior of harvesters. Economists had previously not paid much attention to incorporating realistic descriptions of spatial processes into resource dynamic models. As a result of this work, the importance of incorporating economic behavior of commercial harvesters into models intended to forecast the implications of reserves (or other fisheries management tools) has gained acceptance.

Recreational Fisheries

My discussions of key innovations on the (management-oriented) human-dimensions of fisheries have so far mainly dealt with commercial fisheries. World-wide, recreational fisheries constitute a very important use of wild living resources in coastal areas and inland ecosystems. From an economic perspective, the basic behavior of recreational anglers can be assumed to mirror the classical models by Gordon (1954) and others, with anglers seeking to maximize their individual net utility. Economists have modeled this process such that utility is a function of days fished and the quality of fishing minus the opportunity costs of lost income and time (Anderson 1993). Social-psychologists have used other constructs to understand angler behavior, such as angler motivations (Fedler and Ditton 1994) and satisfaction (Arlinghaus 2006); the latter can be interpreted as realized utility. From an economic perspective, open access recreational fisheries are also expected to move into a bionomic equilibrium where rents (consumer surpluses) are dissipated. Note there is, in principle, nothing wrong with surpluses being dissipated as long as biological sustainability is guaranteed through proper management (daily bag limits, length-based harvest limits, stocking, effort limitations on particular fisheries).

It is important to understand that the behavior of anglers is strongly different from the behavior of commercial fishers because the determinants of behavior differ starkly. For example, while commercial fishers are mainly profit-driven, the rewards sought by recreational anglers are multi-dimensional. They entail a range of catch-related and non-catch-related components of the fishing experience (Fedler and Ditton 1994). It is important to realize that the article by Anderson (1993)—claimed to represent the "complete theory of recreational fisheries"—conceptualizes the benefit of fishing to an angler as being comprised by number of days fishing and catch rate. It thus omitted any non-catch aspects of quality, such as crowding or distance to the fishing site, which are known to strongly affect utility of anglers (Hunt 2005). Further, the general motives of fishing involve non-catch aspects—for instance relaxation in nature or social relationships (Fedler and Ditton 1994), and these aspects are under greater control by the angler and thus most easily satisfied (Arlinghaus 2006). This means that some aspect of catch and travel distance (cost) seems to be the main constraint to satisfaction by many anglers (Arlinghaus 2006). If this is the case, then Anderson's (1993) approach to catch-rate driven fishing utility may offer a reasonable approximation. What will still vary strongly among angler types is which component of catch is most important; catch rate, size of fish, or harvest rate (Arlinghaus 2006, Johnston et al. 2010). The multi-dimensional nature of angler utility/satisfaction overall renders the prediction of the behavior of recreational anglers far more complex than the behavior of commercial fishers. Moreover, understanding the heterogeneity of angler preferences (e.g., catch-and-kill versus catch-and-release, types of sites choices, Hunt 2005) is essential if one attempts to predict the behavioral reactions of a full population of fishers to social or ecological change (for an application see Johnston et al. 2010).

Bryan (1977) put forward an influential framework—called recreational specialization—to understand the heterogeneity of recreational anglers (see also Ditton et al. 1992). Bryan (1977) organized his typology of anglers along a continuum of very general to very particular (specialized) interests and behavior. Reflected here are distinct site attribute preferences, consumptive orientation, preferences towards quantity and size of the catch, and management actions. One can think about specialization as a range of co-varying traits and attributes of an angler, resulting in sufficiently different angler types that behave consistently across contexts and have remarkably different attitudes and opinions. One can think of differently specialized anglers as functional groups in ecological jargon or personalities from a behavioral ecological perspective.

In the many empirical studies following Bryan (1977), much support for the predictive power of the specialization construct as an organizing framework to describe angler diversity has been accumulated. For example, it has been shown that commitment to fishing, a salient sub-dimension of specialization, correlates with many managerially relevant attitudes and preferences of anglers such as likelihood to accept regulations or to engage in catch-and-release conservation behavior. The challenge remains of how to accurately and reliably measure specialization of anglers in different cultures—and then to use this characterization to predict specific behaviors. Unfortunately, despite two decades of research, no accepted operationalization of the multidimensional specialization construct has been developed to be of general use in surveys among anglers. Similarly, there is still intense debate on the proper terminology for—and on the similarity and differences of—the specialization framework to other leisure frameworks. Notwithstanding this ongoing discussion, Johnston et al. (2010) demonstrated the importance of accounting for the differently specialized angler types in a coupled social-ecological model of recreational fisheries when attempting to derive socially optimal input and output regulations. In their model, not only did the impact of recreational angling on a fish stock depend on the frequency of different angler types fishing, but also the socially optimal regulations predicted by the model were found to vary depending upon the composition of angler types. Therefore, models that ignore angler heterogeneity will not identify regulations that provide the greatest level of angler well-being and they may also put populations of fish at risk of overfishing by inaccurately predicting local angling effort levels (Johnston et al. 2010).

Optimality Versus Adaptability in the Face of Uncertainty

Economists are "obsessed" with finding optimal regulatory policies. But how realistic is the achievement of this objective in complex, constantly changing ecosystems and fisheries? Ludwig et al. (1993) make a compelling case that we will never know with certainty all of the variables, processes, and mechanisms driving a fishery. With that being the case, the hope for some *a priori* ability to identify optimal management regulations may be naïve. Instead, suitable management outcomes may be achieved by some form of trial-and-error management where policies are treated as experiments. This call resonates strongly with what is known today as active adaptive management (Walters 1986). Ludwig et al.'s (1993) article stands as a challenge to the belief of Scott (1955), Anderson (1993), Johnston et al. (2010), and others that "optimal" management is possible in practical terms (but see Sethi et al. 2005 for optimal management accounting for uncertainty). Instead, perhaps we must be content with what we might call a "pretty good" (Hilborn 2010) outcome in the face of uncertainty. To this end, confronting the massive uncertainty inherent in most fisheries can be achieved by (Ludwig et al. 1993): (1) considering a variety of plausible hypotheses about the world; (2) considering a variety of possible strategies; (3) favoring actions that are robust to uncertainties and reversible; (4) probing and experimenting; and (5) updating assessments and policy accordingly.

One of the most dynamic and unpredictable sources of uncertainty is human behavior in fisheries. Hence, more innovations in the field of human-dimensions will be needed to better represent the behavior of fishers and policy makers in fisheries management. Perhaps in a future update of *Foundations of Fisheries Science* we will see an equal representation of sections on the biology and human dimensions, supplemented with some integrated sections where both fields are combined in the spirits of Smith and Wilen (2003) and Johnston et al. (2010). This is because most fisheries problems are essentially human problems—human-caused or human-mediated (Ludwig et al. 1993)—but they cannot be solved in isolation from the biological world and vice versa.

Acknowledgments

The section editor acknowledges the advice during the paper selection process provided by Dan Bromley, Eli Fenichel, Beth Fulton, Len Hunt, Stephen Sutton, Bonnie McCay, Jim Wilen, and Doug Wilson. Jim Wilen, Dan Bromley and Eli Fenichel also provided friendly reviews of this chapter. At no point did any of the mentioned colleagues exert any influence on the final choice of the articles. A portion of this summary was written while the section editor spent his sabbatical at the University of Florida in early 2013. Financial support by the School of Forest Resource and Conservation and the hospitality of Mike Allen and Mendy Willis are gratefully acknowledged. The section editor received funding by the German Ministry for Education and Research within the Besatzfisch project in the Program for Social-Ecological Research (grant # 01UU0907).

References

Anderson, L. G. 1993. Towards a complete economic theory of the utilization and management of recreational fisheries. Journal of Environmental Economics and Management 24:272–295.

Beverton, R. J. H., and S. J. Holt. 1957. On the dynamics of exploited fish populations. Fishery Investigations Series 2:Sea Fisheries 19:1–533.

Bockstael, N. E., and J. J. Opaluch. 1983. Discrete modeling of supply response under uncertainty: the case of the fishery. Journal of Environmental Economics and Management 10:125–137.

Bromley, D. W. 2009. Abdicating responsibility: the deceits of fisheries policy. Fisheries 34:280–290.

Callon, M. 1986. Some elements of a sociology of translation: domestication of the scallops and the fishermen of St Brieuc Bay. Pages 196–223 *in* J. Law, editor. Power, action, and belief: a new sociology of knowledge? Sociological Review Monograph, London, Routledge.

Clark, C. W. 2006. The worldwide crisis in fisheries: economic models and human behavior. Cambridge University Press, Cambridge.

Ditton, R. B., D. K. Loomis, and S. Choi. 1992. Recreational specialization: re-conceptualization from a social worlds perspective. Journal of Leisure Research 24:33–51.

Hilborn, R. 2007. Defining success in fisheries and conflicts in objectives. Marine Policy 31:153–158.

Hilborn, R. 2010. Pretty good yield and exploited fishes. Marine Policy 34:193–196.

Hunt, L. M. 2005. Recreational fishing site choice models: insights and future opportunities. Human Dimensions of Wildlife 10:153–172.

McFadden, D. 1974. Conditional logit analysis of qualitative choice behavior. Pages 105–142 *in* P. Zarembka, editor. Frontiers in econometrics. Academic Press, New York.

Ostrom, E. 1990. Governing the commons: the evolution of institutions for collective action. Cambridge University Press, Cambridge.

Ricker, W. E. 1958. Handbook of computations for biological statistics of fish populations. Bulletin of the Fisheries Research Board of Canada 119:1–300.

Sanchirico, J. N., and J. E. Wilen. 1999. Bioeconomics of spatial exploitation in a patchy environment. Journal of Environmental Economics and Management 37:129–150.

Sethi, G., C. Costello, A. Fisher, M. Hanemann, and L. Karp. 2005. Fishery management under multiple uncertainty. Journal of Environmental Economics and Management 50:300–318.

Smith, V. L. 1968. Economics of production from natural resources. The American Economic Review 58:409–431.

Walters, C. J. 1986. Adaptive management of renewable resources. MacMillan, New York.

Wilen, J. E. 2000. Renewable resource economists and policy: what differences have we made? Journal of Environmental Economics and Management 39:306–327.

2.2 REPRINTED ARTICLES

Acheson, J. M. 1975. The lobster fiefs: economic and ecological effects of territoriality in the Maine lobster industry. Human Ecology 3(3):183–207.

Bryan, H. 1977. Leisure value systems and recreational specialization: the case of trout fishermen. Journal of Leisure Research 9(3):174–187.

Dietz, T., E. Ostrom, and P. C. Stern. 2003. The struggle to govern the commons. Science 302:1907–1912.

Gordon, H. S. 1954. The economic theory of a common-property resource: the fishery. Journal of Political Economy 62(2):124–142.

Homans, F. R., and J. E. Wilen. 1997. A model of regulated open access resource use. Journal of Environmental Economics and Management 32:1–21.

Jentoft, S. 1989. Fisheries co-management: delegating government responsibility to fishermen's organizations. Marine Policy 13(2):137–154.

Scott, A. 1955. The fishery: the objectives of sole ownership. The Journal of Political Economy 63(2):116–124.

Smith, M. D., and J. E. Wilen. 2003. Economic impacts of marine reserves: the importance of spatial behavior. Journal of Environmental Economics and Management 46:183–206.

The Lobster Fiefs: Economic and Ecological Effects of Territoriality in the Maine Lobster Industry

James M. Acheson[1]

Received May 20, 1974; revised September 16, 1974

Lobstermen from each community along the coast of central Maine claim inshore fishing rights in particular areas. Although their claims are unrecognized by the state, they are well established and backed by surreptitious violence. Two kinds of lobstering territories exist, here termed "nucleated" and "perimeter-defended," which differ essentially in the extent to which exclusive fishing rights are maintained. These differences in territorial organization affect the fishing effort of lobstermen, which in turn has a strong biological and economic impact.

KEY WORDS: lobster fishing; territoriality; common property resources; Maine.

INTRODUCTION

Fisheries throughout the world are characterized by a persistent and, in some cases, disastrous tendency toward overexploitation. The most important and widely accepted cause is the common property or "open-access" nature of legal rights in the marine environment (Christy and Scott, 1965; Gordon, 1954: 132). In the absence of ownership, fishermen have no incentives to curtail fishing activities in response to declines in catches or increases in costs, because no property right guarantees that fish not taken today will be available in large quantity or at greater weight in the future. What one fisherman does not catch today simply goes to the other fishermen (Crutchfield, 1973: 116-117).

The research on which this paper is based was financed by the National Marine Fisheries Service (Contract No. N-043-30-72) and a Faculty Research Grant from the University of Maine in 1973.

[1] Department of Anthropology, University of Maine, Orono, Maine. (On leave 1974-1975 to serve with the National Marine Fisheries Service, Washington, D.C.)

Acheson

Accordingly, the fishing industry of the United States is having serious long-term difficulties, recently exacerbated by intense competition with foreign fishing fleets. In the New England area, cod, haddock, hake, and mackerel are far less plentiful than they were in the past, and the once flourishing sardine industry has been reduced to a shadow of its former self. In Maine, the site of our research and traditionally one of the most productive lobstering regions in the Atlantic, lobster landings have decreased very rapidly in recent years (Maine Sea and Shore Fisheries, 1973).

However, the lobster industry of Maine is an unusual fishery in that informal norms about territoriality and hence ownership do exist. Territorial arrangements have a substantial economic and ecological impact, with implications for the management of the Maine lobster fishing industry and for fisheries management as a whole. While the lobster catch in Maine is lower than in the past, the decrease appears to be highly differential: fishermen in some areas earn higher incomes and have a different effect on the biomass than those in others. Such differences can be explained in terms of variations in lobster fishing rights that exist in various parts of the Maine coast.[2]

GENERAL FEATURES OF THE MAINE LOBSTER INDUSTRY

The American lobster (*Homarus americanus*) is found in the waters off the Atlantic Coast of North America from Newfoundland to the Carolinas. However, Maine consistently produces far more lobsters than any other state, and our study area along the central Maine coast is one of the most productive areas of all.

Throughout the 1960s, there were approximately 6000 lobstermen in Maine. About 2800 of these men were full-time fishermen; the remainder were "part-timers" who earned most of their income "ashore."

The technology employed by lobstermen along the entire length of the Maine coast is relatively uniform. Lobsters are caught in wooden traps or "pots" about 3 ft long made of oak frames covered with hardwood lathes. Lathes are placed about 1.5 inches apart, allowing free circulation of sea water while retaining the larger, legal-sized lobsters (carapace length[3] of over 3 3/16 inches and under 5 inches). The open end of the trap is fitted with a funnel-shaped nylon

---

[2] Some biologists would argue that natural factors such as food supply, predation, competition, and water temperature have a more important effect on lobster stocks and landings than the social and economic factors I am here concerned with. Water temperature fluctuations in Maine are highly correlated with landings for the period 1904-1967 (Dow, 1969). However, I hold that property rights are also critical to differentials in lobster landings.

[3] It is standard practice in fisheries biology to measure lobsters (in millimeters) on the carapace, from the eye socket to the back of the body, where the carapace ends and the tail begins.

The Lobster Fiefs

net or "head," which lets lobsters climb in easily but makes it difficult for them to get out. Along the central part of the coast, the study area, one or sometimes two traps are attached to a small styrofoam buoy via a "warp" (nylon or hemp rope).[4] Distinctive sets of colors, registered with the state, mark the buoys belonging to each lobsterman. Fish remnants obtained from nearby processing plants constitute bait. The traps are usually placed in the water in "strings," or long rows, so that a man can see from one buoy to another in the fog. On a good day (usually the calm morning hours), a lobsterman in this area might pull about 200 traps. Typical lobstermen have about 400-600 traps.

Most lobstermen fish alone from gasoline- or diesel-powered boats 28–32 ft long, equipped with a depth sounder, hydraulic "pot" hauler, ship-to-shore radio, and compass. In the island areas, boats may be somewhat larger, more often diesel-powered, and also equipped with radar, so as to cope with the more violent offshore seas and the fog. Capital investment in boats constructed at present prices ranges from $13,000 to $17,000; besides the boat, a lobsterman may have from $8000 to $10,000 invested in traps and fishing equipment, a pickup truck, a dock, and some kind of workshop. Replacement values easily run over $40,000.

A lobsterman's activities vary greatly from season to season. The midwinter months are unquestionably the slowest time of year. During January, February, and March, when lobsters can be caught only in relatively warm, deep water, 3-10 miles offshore, lobstering is generally more dangerous and unprofitable. Bad weather and high winds increase trap losses and make the work more difficult. Some men stay ashore during this period to build lobster traps, while others use their boats for scalloping or shrimping. Those who persist in lobstering during the winter may pull their traps no more than six or seven times a month.

As the water warms in the spring, lobsters are available in shallower water closer to shore. Spring (April 15 to June 15) and fall (August 15 to November 15) are unquestionably the busiest months of the year, when men have a maximum number of traps in the water and pull them every chance they get. During the 3- or 4-week molting season (June 15 to August 15, depending on the area), traps are typically placed very close to shore — literally feet away from breaking surf. During this period, catches are so small that men bring many of their traps ashore and do maintenance work on their boats. In the fall, lobstermen begin to move their equipment into deeper water again.

While all lobstermen move their traps according to this general pattern, they are not all equally effective as fishermen. Skill and willingness to work

---

[4] In the more southern part of the state, around Casco Bay, it is customary to attach six to ten traps to one "warp"; this practice is called "fishing trawls," in contrast to "fishing singles." In the Casco Bay area, lobster fishermen have many more traps than in the central region of the Maine coast, and they fish with larger boats and two-man crews. On a good day, a man "fishing trawls" might easily pull 350 traps while a man "fishing singles" could pull about 200 traps.

greatly affect catches and incomes. I know one very experienced man who has 300 traps and had a net income in 1972 of approximately $21,000. Another man with only 5 years of experience, fishing in the same area with 900 traps, had a net income of only $12,000. The first carefully places his traps in areas known through experience to be productive; the second scatters them over the bay.

Any sizable harbor has at least one lobster dealer who buys from local lobstermen and sells to tourists or to one of the three or four large wholesale firms distributing lobsters in Maine and the nation. While the prices lobstermen receive do fluctuate seasonally, there is little price competition in the Maine lobster industry. On any given day, all the dealers in the state are paying approximately the same price. Dealers compete for a supply of lobsters by attempting to attach as many lobstermen to themselves as possible. They supply "their" lobstermen with gas, oil, and bait at low margins of profit, allow them to use their wharfs free of charge, and supply them with large amounts of credit. A few fishermen sell to two or more dealers, and periodically a man will change from one dealer to another in rapid succession. But typically a lobster fisherman maintains a longstanding relationship with only one dealer and sells his catch exclusively to that dealer.

The location of dealerships is not connected to lobster fishing areas. A lobsterman usually sells to a dealer in his own home harbor, but he may sell to any dealer — regardless of location.

The federal government has made no effort to regulate the lobster industry, and the state of Maine has passed only a few conservation laws. To be legally taken, a lobster must have a carapace length of over 3 3/16 inches and under 5 inches. Regardless of size, breeding females are protected by a law requiring any lobsterman who catches an egg-bearing female to cut two notches in her tail flipper and throw her back. "Notched-tail" lobsters may never be taken. In addition, the state has a licensing requirement, and lobstermen must mark their traps and buoys with their license number and assigned colors. Despite the fact that only 37 state Sea and Shore Fisheries wardens patrol some 2500 miles of coast, these laws are now almost universally obeyed. Violations can result in court action and suspension of fishing licenses, which effectively bars lobstermen from the fishery.

There are few data on the relative effectiveness of these formal, statewide measures. However, in other lobstering regions and in other fisheries, such restrictive conservation laws have been found to exacerbate the problem of economic inefficiency and to have questionable effects on fish population levels (Crutchfield and Pontecorvo, 1969; De Wolf, 1974). Certainly the Maine laws do not begin to solve the problems inherent in managing a common property resource. They do not limit entry into the fishery and do little to restrict fishing effort. However, the informal norms concerning territoriality certainly do

The Lobster Fiefs

operate to limit both. In all lobstering areas there are barriers to entry; and in some areas there are special local rules about the number of traps that may be fished or about local fishing seasons.

TERRITORIALITY IN THE MAINE LOBSTER INDUSTRY

From the legal view, anyone who has a license can go lobster fishing anywhere. In reality, far more is required. To go lobster fishing at all, one needs to be accepted by the men fishing out of one harbor; and once one has gained admission to a "harbor gang,"[5] one is ordinarily allowed to go fishing only in the traditional territory of that harbor. Interlopers are met with strong sanctions, sometimes merely verbal, but more often involving the destruction of lobstering gear. This entire territorial system is entirely the result of political competition between groups of lobstermen. It contains no "legal" or jural elements.

As a rule, the area fished by any one harbor gang is quite small. In the summer, a man will rarely fish more than 3 or 4 miles from his home harbor, and even in winter, when men are fishing in deeper water, they are rarely more than 10 miles from home. Territories fished by one harbor gang are rarely more than 100 square miles; most are far smaller. This means that a lobsterman spends his working life crossing and recrossing one small body of water.

The dividing lines between most lobstering territories are relatively minor features which would be familiar only to men intimately acquainted with the area (see Fig. 1). Along shore, a rock, ledge, cove, or perhaps a big pine tree may mark a boundary. Offshore, boundaries are usually marked by reference to landmarks onshore or on islands.

To some extent, delineation of boundaries varies with distance from shore. Close to shore, boundaries are known to the yard. Farther offshore, boundaries are less definite. Thus in the middle of the winter, when men are fishing far from shore, there is a good deal of "mixed fishing" between two harbor gangs. In part, this pattern can be explained by the fact that it is simply more difficult to establish exact boundaries far from shore. More important is the relative competition

---

[5] While the men fishing out of one harbor form very definite social entities, they do not have universally recognized names. One refers to the "Monhegan boys" or the "Friendship fishermen" or the "Pemaquid Harbor gang." I refer to these groups as "harbor gangs," although this term is used only rarely by the fishermen themselves. Usually the fishermen from one community form a harbor gang, but this is not always the case. For example, some of the men who live in the village of New Harbor fish out of Pemaquid Harbor and others out of New Harbor itself. Moreover, the men who fish the areas around some of the unoccupied islands in Penobscot Bay live onshore and commute daily to the islands to fish. Even though they may reside in different towns, they form some of the most cohesive harbor gangs along the entire coast.

Fig. 1. Map of lobster fishing territories along the coast of central Maine.

The Lobster Fiefs

for fishing bottom. In the winter, lobsters are available over a wider area and fewer men are fishing. In the summer, when more men are fishing shallow areas along shore, there is more competition and more interest in maintaining exact boundaries.

Violation of territorial boundaries meets with no set response. An older, well-established man from a large family might infringe on the territorial rights of others almost indefinitely, whereas a new man or a "part-timer" would almost certainly have trouble very quickly. Sooner or later, however, someone — usually one man acting completely on his own — will decide to sanction the interloper. First, the violator may be warned, usually by having his traps opened or by having two half-hitches tied around the spindle of his buoys. If he persists, some or all of his traps will be "cut off." That is, his traps will be pulled, the buoy, toggles, and warp cut off, and the trap pushed over in deep water where he has little chance of finding it.

Small-scale trap cutting is a constant problem all along the coast. There are, however, factors keeping trap cutting at a minimum. Most important is the knowledge that the destruction of traps will bring almost certain retaliation. Perhaps once a decade, these small incidents escalate into full "lobster wars," in which dozens of men thrust against each other in widespread forays, destroying large numbers of traps and even boats (Acheson, 1974).

All conflicts, regardless of how many men are involved, are kept very quiet. No matter how justified a man feels in "cutting off" another man's gear, he will rarely advertise his "skill with the knife." Trap cutting is illegal and can lead to loss of license; silence reduces the chances of retaliation by victims. Thus, when trap cutting occurs, people in coastal towns and state officials are apt to hear only vague rumors, but not the details, from the primary actors themselves. Even the victims remain silent because legal prosecution is usually unsuccessful and there is a strong feeling among fishermen that the "law should be kept at a distance." The result is what Bailey (1969: 144-147) calls an "encapsulated system" — one political system operating with its own set of rules within a larger system.

There are two different types of lobster fishing areas along this part of the central Maine coast. I am calling them here "nucleated" and "perimeter-defended" areas.[6] The lobstermen do not have terms for the different territorial arrangements, although they are aware of the differences, which essentially relate to the amount of mixed fishing allowed. In the western part of the study area (between Cape Newagan and Friendship), territorial arrangements are highly nucleated. That is, men from each harbor gang have a strong sense of territoriality close to the mouth of the harbor where they anchor their boats.

---

[6] In a previous article (Acheson, 1974), I referred to the differences in these areas in terms of "open" and "closed." I have abandoned this terminology as misleading and have adopted these new terms from Hockett (1973: 69).

But this sense of "ownership" grows progressively weaker the farther from the harbor mouth one goes. On the periphery there is almost no sense of territoriality and a good deal of mixed fishing takes place. In fact, in this area virtually all the deepwater fishing grounds exploited in the winter are fished by men from at least two harbors. (Areas of overlap may be fished jointly by men from particular harbor gangs but are not open to men from any harbor.) A stranger trying to fish in an area close to a harbor nucleus would almost certainly meet with violent opposition from a number of men. However, the reaction would be much less violent if he were to invade an area fished by multiple harbor gangs, where the sense of ownership is weaker.

Farther east on the mainland (see Fig. 1), the territories are still basically nucleated, although lobstermen say that boundaries become "harder," meaning that there is less area where mixed fishing is allowed and that larger areas are fished exclusively by men of a single harbor gang.

The island areas of Penobscot Bay are, however, perimeter-defended territories. Particularly in the cases of Monhegan, Matinicus, Green Island, Little Green Island, and Metinic Island, boundaries are sharply drawn and defended to a yard. Definition of the areas is in terms of their outermost boundaries, and the feeling of ownership does not sharply decrease with distance from the island harbors where the men anchor their boats. There is very little mixed fishing. Even in winter, when fishing offshore in deep water, men from these perimeter-defended territories rarely venture into areas they do not claim for their exclusive use. Periodically, a man may furtively and guiltily put a few traps outside his own area where he thinks no one will notice, but there is a strong feeling that if one is going to keep others on their side of a boundary one should stay on one's own side.

In both types of territories, but most clearly in the perimeter-defended areas, claims over ocean areas are tied up with formal ownership of land. Ownership of land on an island is held to mean ownership of "fishing rights" in nearby waters, despite the fact that legally the ocean areas are part of the public domain. On Matinicus, for example, no one is allowed to fish in the island's territory unless he owns land on the island. A major argument against selling land to "summer people" is that thereby an island family may lose its fishing rights. One man bought a very small island in the Muscle Ridge channel solely to gain fishing rights to adjacent waters. He does not live on the island, and does nothing else with it except to use the surrounding ocean for lobster fishing.

In perimeter-defended areas, ownership rights to the waters are not merely usufructuary. Even if the owner is not using his water territory, his fishing rights remain, and may be rented out. In these areas, ownership rights are so strong that men who own whole islands rent out "water areas" to men from nearby mainland harbors. Most of the lobstermen fishing in the Metinic Island area, for example, rent fishing rights from the two families who are legal owners of the

The Lobster Fiefs

island. The family members cannot completely exploit the island's water area, and so they outfit several men from the mainland with boats and gear and allow them to go fishing in their preserve. Arrangements vary considerably, but in some cases the families take half the gross income of these renters as return on the capital equipment and as rent on the "water area." These rental rights are traditionally held, and inherited patrilineally, as are land property rights.

A critical difference between nucleated and perimeter-defended territories is the extent to which entry is more severely limited and controlled in the latter. In nucleated areas it is relatively easy to gain acceptance to harbor gangs. If a man is a resident of the community and shows a willingness to abide by local fishing norms, he will eventually be accepted into the local harbor gang.[7] It is vastly more difficult to gain admission to harbor gangs that maintain perimeter-defended territories. This is to be expected, given the nature of such defensive boundary arrangements. The object of maintaining strict boundaries is to keep down the number of men fishing in an area. There is no sense in maintaining sharply defined boundaries by violence against other harbor gangs if anyone can join one's own gang. Thus, men who strongly defend the perimeters of their territories also limit membership in their own harbor gangs to a much greater extent than do men fishing in nucleated areas. As will be seen below, this results in more fishing area per fisherman than is evident in nucleated areas.

Entry to harbor gangs in perimeter-defended areas is essentially limited to men who belong to families owning the land and adjacent water areas, and to men who are accepted on a rental basis. The unoccupied islands (Green Island, Little Green Island, and Metinic) are each privately owned by one or two families, whose members reserve all fishing rights solely for themselves. It is virtually impossible for a newcomer to enter these harbor gangs, except in some instances as a renter. A few newcomers, however, have been able to establish themselves as fishermen on permanently occupied islands such as Matinicus and Monhegan. Usually such men are already known to the islanders; several come from families who have summered on an island for years, so that they grew up with island children. Normally they begin fishing as helpers, or "stern-men," and eventually

---

[7] Entry into harbor gangs is, however, by no means "open." Admission depends on a combination of factors. In all nucleated harbors a man will most easily gain admission if his family has long been resident in the town, if he began fishing as a boy and has gradually become a full-time fisherman, and if his father is a fisherman. Such a man inherits a place in his father's harbor gang. A man will have the most difficulty if he is not a long-term resident, if he began fishing as an adult, if he plans to fish part-time, and if his family has no connection with fishing. However, the most critical single factor is a man's willingness to abide by the local norms of the industry. One who gains a reputation for molesting other men's gear will not last long in the industry, regardless of family ties or long residence. The strong prejudice against part-time fishermen can be explained by their relative invulnerability to the effects of transgressing the norms; with little capital investment and with alternative means of employment, the part-time lobsterman may suffer little from trap cutting and other sanctions.

are permitted to purchase land (if their families do not already own land) and to become full-time fishermen. In the process these men are carefully scrutinized, and would be rejected if they made any trouble or showed any unwillingness to abide by local norms.

PROCESSES OF BOUNDARY MAINTENANCE AND CHANGE

Before 1920, all lobstering areas along the Maine coast were undoubtedly perimeter-defended areas. While most of the areas are now nucleated, the advent of such territories is a relatively recent phenomenon, and the change from perimeter-defended to nucleated areas is by no means complete.

Fifty years ago, lobstering was done only in the summer in very small territories held by small groups of men who defended them vigorously. In great part, this pattern was connected to the technology in use. Since lobstering was done from a sloop or dory, fishing in stormy winter waters was very difficult. Even in summer, the area that one could fish was very limited, since one could learn the bottom only by hand lead line and the travel radius was small. Since a man's income was dependent on a very small area, he zealously maintained its boundaries. These small territories were typically held by a small group of kinsmen, and were passed down to descendants patrilineally. Fishing areas were usually adjacent to legally owned coastal property, where it was felt the landowner had a right to the lobsters in the waters off his "very own property."

This pattern has persisted to the present day, as we have seen, around the small privately owned islands of Penobscot Bay, around the large permanently occupied islands farther out to sea (i.e., Matinicus and Monhegan), and to some extent near the mainland harbors in the eastern part of the study area. In the western part of the study area, this pattern has completely broken down. The small perimeter-defended territories have been amalgamated into larger nucleated areas where most ocean area is fished by men from at least two harbors.

The breakdown of small, perimeter-defended territories was made possible by technological change. As motors came into common use in the 1930s, the range a lobsterman could fish increased. In the 1950s, the use of depthfinding equipment, which made it relatively easy to learn the characteristics of the bottom, further increased the effective range a lobsterman could fish. While the new technology increased potential fishing area, one cannot explain why the older system of small perimeter-defended areas has persisted in the east and around certain islands but has broken down completely in the western part of the study area.[8] As we have seen, the technology is the same along the entire coast.

---

[8] The major criterion by which the lobstermen of this part of the Maine coast differentiate between what we call nucleated and perimeter-defended areas is a perceived difference in

The Lobster Fiefs

Several different sets of factors have produced the differential boundary breakdown. While these have been discussed in detail elsewhere (Acheson, 1974), several points bear repeating. First, boundary breakdown or maintenance is the result of conflict and political pressure. Second, the willingness to defend a boundary or invade another area depends on (1) the ability to form what Bailey would call "political teams" (Bailey, 1969: Chap. 3) and (2) the alternate income opportunities open to the men in question.

The technological and political factors, operating in different geographical contexts, have produced the different kinds of areas we have been discussing. In the western part of the study area, where the coast is strongly convoluted and formed into deep bays, the traditional areas of communities on open ocean have been under great pressure from men in towns farther up the rivers and bays. Before the advent of engines, men from bay communities, such as Bremen and Muscongus, had to restrict their fishing to the summer months when they could catch lobsters in the waters close to their home harbors. During the past few years, more and more men from such bay communities have been purchasing larger boats capable of fishing in open ocean. Since one cannot cover the costs of a $15,000 boat by using it only during the summer, these men must fish on a year-round basis. This means that they must invade what were formerly the exclusive territories of towns such as New Harbor and Friendship on open ocean. Obviously, such men are willing to sacrifice a great deal to get to open water. The alternative is to be bottled up in small traditional territories near their home harbors where they can fish only a few months a year.

Thus, it is not surprising that the traditional perimeter-defended boundaries in areas at the mouths of estuaries have broken down, resulting in nucleated fishing territories. Around Muscongus Bay and the deeply indented bays of the Boothbay region, harbors on the ends of peninsulas are apt to have very small areas for exclusive use. The rest are fished jointly with men from harbors farther inland. The amount of mixing is especially great in winter, when men from more inland harbors must come outside. However, perimeter-defended areas have been maintained farther up these rivers and bays.

For the men in the invaded areas, it is not worthwhile to repel the interlopers. It is true that invasion means that men from harbors on open ocean will have more competition in areas which they once held exclusively. This will ultimately mean a smaller yield for them. However, an attempt to stop the men

---

their defenders' propensity for violence. Their explanations for the difference are usually in cultural terms. Men from areas like Monhegan are believed to have traditional ideas about territoriality, to be very secretive and competitive, to cooperate rarely, and to be prone to use force and violence in defending their territories. Such "traditional" attitudes and behaviors are linked to factors such as older age, less education, and greater isolation from the outside world. None of these characteristics accurately describes the violence-prone men of perimeter-defended areas, who are, in fact, better educated and more sophisticated than most men fishing nucleated territories.

from upriver from their incursions would mean a full-scale "war," resulting in large financial losses. Even though there is some bitterness directed toward the men from bay communities, men from open-ocean harbor gangs feel it is better to mix than fight.

The men from ocean harbors feel this situation is very unfair. They would not mind yielding some of their exclusive areas if they could fish farther upriver during the summer. There is a good deal of hard talk about the "river rats who invade our areas but will not allow us to fish upriver." The invaders have a different point of view. As one of the older fishermen from a bay community phrased it, "If the old boundaries were maintained, we would have no place to fish much of the time. We would all have to be part-time fishermen." They clearly have no intention of reverting to part-time status again.

A combination of factors has allowed the small perimeter-defended areas to be maintained around some of the islands in our study areas.

First, men from mainland harbors have not been aggressive in pushing into these outer islands. Distance alone has had a dampening effect on the offensive ardor of the mainlanders. Islands such as Monhegan and Matinicus are over an hour from the nearest mainland harbor even for the fastest boats. Moreover, mainland harbors in this part of the coast, e.g., Pleasant Point, Tenant's Harbor, and Port Clyde, have both shoal water and deep water so that members of these harbor gangs can go fishing year round without pushing into areas controlled by other men. (In this respect, these areas are unlike upriver ones such as Round Pond and Bremen.) Second, and more important, these mainland harbor gangs are composed of larger numbers of men, who are not organized in ways that allow political effectiveness in the long run. Friendship, for example, has about 95 lobster boats owned by men who live in at least three towns and who have little in common besides occupation. The same is true of other mainland harbors in the area. Periodically, one man or a few men will attempt to put traps in areas controlled by island gangs, but their lack of organization prevents aggressive, long-term efforts which would result in change in boundary lines. In fact, I am aware of no instance where all the men of a nucleated territory have acted in concert to defend their own territory or to invade another.

"Owners" of island areas have long put up a spirited defense of their boundaries. Several factors have made them very effective.

These islands are owned and controlled by a small number of nuclear households strongly linked by ties of agnatic kinship and their need for mutual self-help. Their activities are customarily coordinated by older, experienced adult men, who rule their sons and nephews with an iron hand, and who have a great deal of influence over non-kin. Their small numbers, kinship ties, and a longstanding tradition of cooperation make it relatively easy to coordinate the defense of their lobstering area.[9]

---

[9] The social organization of Maine offshore islands is the topic of a future publication.

The Lobster Fiefs

The vehemence of this defense is bolstered by the fact that residents of islands do not have the alternative economic opportunities that men from mainland harbors have. Moreover, their long-term ownership of island property gives them special claim to the adjacent ocean areas. At least one family has owned an island since the mid-eighteenth century. In Maine, where virtue is attached to remaining in the same place for a long time, this family's claim to its traditional fishing area has strong moral overtones. Friends and enemies alike see the family as preservers of a valued tradition or admired *status quo*. To a great degree, this is true of all the perimeter-defended areas. The defensive posture of perimeter-defended areas is further enhanced by the fact that some prominent men from these areas have been able to use their wealth and long residence in the area to gain political eminence in the region as a whole. Thus, for the time being, there are several areas along the coast where perimeter-defended areas remain — little fiefs carved out of the public domain.

PERIMETER-DEFENDED VS. NUCLEATED AREAS: DIFFERENTIAL EFFECTS ON THE ECONOMICS AND ECOLOGY OF MAINE LOBSTERING

The remaining perimeter-defended territories are held by men who tend to form highly effective political groupings which serve both to severely restrict entry into the fishery of these areas and to enable to enactment of local conservation measures. Both these factors have produced great differences in fishing effort between nucleated and perimeter-defended areas.

Differences in ease of entry into harbor gangs fishing the two different kinds of areas have produced substantial variation in the number of boats (and fishermen) per square mile. Data comparing three perimeter-defended areas with three adjacent nucleated areas show that men from perimeter-defended areas have far more fishing area per boat than those from nucleated areas (Table I).

In reality, the difference between the nucleated and perimeter-defended areas is far greater than these figures indicate. First, in the nucleated areas, not all of the ocean bottom is productive of lobsters. An estimated 25% of the area fished by Friendship and New Harbor men is mud bottom. This rarely produces lobsters. More important, much of the area fished by men from nucleated harbors consists of zones of mixed fishing. Men from New Harbor, for example, fish over 44.7 square nautical miles of ocean. But only approximately 12.6 square nautical miles of that is fished exclusively by men from New Harbor. The men from Port Clyde fish some 30.4 square nautical miles, but 13.3 square nautical miles of that area they share with men from other harbors. By way of contrast, *all* of the area "owned" by men in perimeter-defended territories produces lobsters at some time in the seasonal cycle; and no part of these territories is fished with men from other harbor gangs. Thus, if the amount of productive

Table I. Square Miles per Boat in Nucleated vs. Perimeter-Defended Areas[a]

	Harbor	Number of boats	Total area in square nautical miles	Square nautical miles/boat
Nucleated areas	Port Clyde	39	30.4	0.78
	New Harbor	36	44.7	1.2
	Friendship	95	25.3	0.27
Perimeter-defended areas	Green Island	8	11	1.4
	Metinic (south end only)	7	10.8	1.5
	Monhegan	12	20	1.7

[a]Only the boats of full-time fishermen have been counted. Part-time (skiff) fishermen have been excluded. The number of boats, and not number of men, has been used since many boats carry more than one man.

fishing bottom and the amount of mixed fishing were taken into account, the ratio of square nautical miles per boat would be far lower than indicated in the case of these nucleated areas.

Since there are fewer boats per square mile in perimeter-defended areas, and since men fishing in those areas use no more capital equipment per man than those in nucleated areas, there is a great difference in the amount of fishing *effort* in these types of territories. Thus, in perimeter-defended areas, a higher proportion of the lobsters reaching the minimum legal size remain uncaught and grow to larger sizes.

Another difference between the two types of areas is that conservation efforts are practiced in the perimeter-defended territories. The men fishing off Matinicus Island, for example, have voluntarily agreed to limit the number of traps they fish. Such trap limits have two benefits. First, they increase a fisherman's profits by lowering his production costs. A man with a small number of traps does not spend as much for trap stock, and usually pays less for bait, fuel, and boat depreciation as well. Second, a man with fewer traps keeps better track of them, pulls them more frequently, and loses fewer. This reduces lobster mortality, since lobsters in lost traps are apt to be permanently incarcerated. Moreover, when traps are pulled frequently, molting lobster which would otherwise have been eaten by their brethren are released and given a better chance of survival.[10]

---

[10]State biologists are convinced that a very large number of lobsters are killed by being caught in untended or lost traps. In the past, bills have been unsuccessfully introduced requiring fishermen to pull their traps every 48 hr. At present, a "venting bill" is being introduced. This would require that one lathe on each trap be made of some kind of wood that rots quickly. Such traps, if lost, would automatically release their victims in time to prevent excessive mortality.

The Lobster Fiefs

Trap limits do not cut down the total catch. When trap limits are imposed, men dispense with the marginal traps — those not fishing well — and put the remainder to better use so that yield per trap is increased. Thus, with a trap limit, it may take men a little longer to catch the same volume of lobsters, but over the course of the annual cycle the same catch is harvested.[11]

Monhegan Island, another perimeter-defended territory, is attempting to conserve the lobster resource in its area by imposing a closed season. The inhabitants of Monhegan have persuaded the legislature to pass a law forbidding fishing in Monhegan waters from June 25 to January 1. Thus, Monhegan lobstermen are fishing in midwinter when the price for lobster is very high and when they have few other economic options. They put their traps "on the bank" in the summer when they have alternate employment in the tourist industry, leaving the defense of their territory to the state fish wardens. This closed season keeps anyone from setting traps during the critical months of July and August, when fishing for molting lobsters would result in very high mortality.

It is critical to note that one man cannot gain the benefits of a trap limit or closed season by unilaterally reducing the number of traps he fishes or the number of months he fishes. If he reduces his fishing effort (and others do not), he is obviously putting himself at a competitive disadvantage. But if everyone reduces effort simultaneously, then each man will still catch his "fair share" of lobsters. Such conservation measures will work only in areas where everyone agrees to them. So far this has occurred only in perimeter-defended areas, where harbor gangs are organized to enforce strong controls over the actions of their members and defend their area. Such voluntary controls are not characteristic of the relatively loosely organized harbor gangs fishing nucleated areas. These gangs are unable to control their own members, to defend effectively much area for their exclusive use, or to limit drastically the numbers of men they allow to fish in their area.

There is a persistent rumor along the coast that men fishing in areas such as Green Island, Monhegan, Matinicus, and other perimeter-defended areas earn larger incomes and catch larger lobsters than men fishing anywhere else along the coast. The reason, lobstermen explain, is that the "islanders reserve private fishing areas for themselves and keep everyone else out." A preliminary study indicated that these rumors had some substance to them.

We hypothesized that the differences observed were due to the differences in effort and fishing practices discussed above. To verify this set of hypotheses, we set up a project to gather data on the relationships among type of territoriality (nucleated or perimeter-defended), income, catch size, fishing effort, estimated stock density, and other measures. Two types of information were

---

[11]The Canadian government has instituted trap limits in a number of lobster fishing districts in New Brunswick and Nova Scotia with no noticeable decrease in annual total harvest (De Wolf, 1974).

collected. First, between June 1972 and May 1973, 28 lobstermen were persuaded to give detailed information on income and business expenses. Unfortunately, it was early in 1973 that the U.S. Internal Revenue Service began an intensive investigation of the men in the Maine lobster industry. A very high proportion of the fishermen in Maine were "overhauled" by the IRS. Many men were forced to pay additional tax money, and several have been charged with criminal fraud. Late in the spring of 1973, men had become so fearful and suspicious of anyone asking information about income that this part of the project had to be dropped. However, enough data had been collected in the preceding months to allow us to come to tentative conclusions concerning the economic effects of territoriality.

Second, 14 lobstermen from various harbors were asked to record data on their catches twice a month from March 1973 through January 1974. On every string of traps pulled, they recorded the date, location of the string, depth (in fathoms), number of "setover days" (number of days since the trap was last pulled and rebaited), and whether the string was in an area fished exclusively by men of one harbor or in an area of mixed fishing. On every trap pulled, they recorded another set of data: the carapace lengths of the legal-sized lobsters caught in that trap and the length of the proven breeding stock, i.e., female lobsters carrying eggs or with notched tails. To facilitate the recording of these data, the men were provided with metric calipers and data sheets. About a fourth of the total information was obtained by members of the University staff who accompanied the fishermen. Data were obtained on 9089 lobsters, of which 3327 were caught in nucleated areas and 5762 in perimeter-defended areas.

The lobstermen were chosen with a view toward controlling for several important variables. To control for ecological differences, all were picked from the same small study area. Seven men fish in nucleated areas, the other seven in perimeter-defended areas. All areas in this part of the study are adjacent to each other. In some cases, the men from nucleated areas are fishing only yards away from the men from perimeter-defended areas.

Moreover, the amount of capital used by the men studied is comparable. On this part of the coast, lobstermen use between 350 and 600 traps per man. The exact number a particular man uses depends less on the type of area he fishes than on factors such as age, skill, and ambition. More men fishing the offshore islands have larger diesel-powered boats with radar. But not all boats on islands are large, and many men on the mainland have comparable investments in boats. In short, there is no obvious difference in the amount of capital employed by the men in nucleated areas and by those in perimeter-defended areas.

The lobstermen claim that a difference in skills makes an enormous difference in the number and size of lobsters caught. While there has been no research on this point, I have little doubt that they are correct. To control for skills, all the men chosen were full-time fishermen, with a reputation for marked success.

The Lobster Fiefs

From the information gathered during the carapace-measuring project, there is unmistakable evidence that a reduction in fishing effort in the perimeter-defended areas has resulted in both biological and economic benefits.

Biological Benefits

First, the lobsters caught in the perimeter-defended areas are larger than those caught in nucleated areas. This can be demonstrated by analyzing the frequency distribution of lobsters caught.[12]

From Fig. 2, it can be seen that men catch a much higher percentage of small lobsters in the nucleated areas than in the perimeter-defended areas, and a higher percentage of larger lobsters in the perimeter-defended areas than in nucleated ones. For example, 8% of all lobsters caught in nucleated areas were 83 mm, but only 6.4% of those caught in perimeter-defended areas were. Additionally, 1.9% of the lobsters caught in perimeter-defended areas were 98 mm, whereas only 0.8% of those caught in the nucleated areas were this big. Since a high proportion of the lobsters which molt into the legal size range are immediately caught in nucleated areas, fewer survive to larger sizes compared to those in perimeter-defended areas. This is reflected in the fact that the mean carapace length of lobsters caught in the nucleated areas is shorter (87.8 mm) than that of those caught in the perimeter-defended areas (89.9 mm).

This difference in size undoubtedly has a profound influence on the numbers of eggs released in the water in these two types of areas and ultimately on the long-term prospects for the lobster fishery itself. Only 6% of female lobsters become sexually mature under 90 mm (carapace length), while nearly all females are mature by 105 mm (Krouse, 1973: 170-171). Thus, all but a very small proportion of female lobsters become sexually mature between 90 and 100 mm. (There are no figures available indicating when an *average* female becomes sexually mature.)

The percentage of lobsters in the critical size range (90-100 mm) is higher in the perimeter-defended areas than in the nucleated ones. There can be no question that a higher percentage of female lobsters in the perimeter-defended areas than in the nucleated areas are reaching a size where they can possibly extrude eggs.

This conclusion is further buttressed by the fact that the probability of a female reaching maturity in the perimeter-defended areas is 1.52 or nearly 50% higher than it is in the nucleated areas (Wilson, 1975). In our sample, 2.7% of

---

[12] The difference in the mean carapace length of lobsters caught in the nucleated vs. perimeter-defended areas is very small. However, since females become mature between 90 and 100 mm, these slight differences in size are very significant for the fishery, since they mean that there is a much higher proportion of mature females in the perimeter-defended areas.

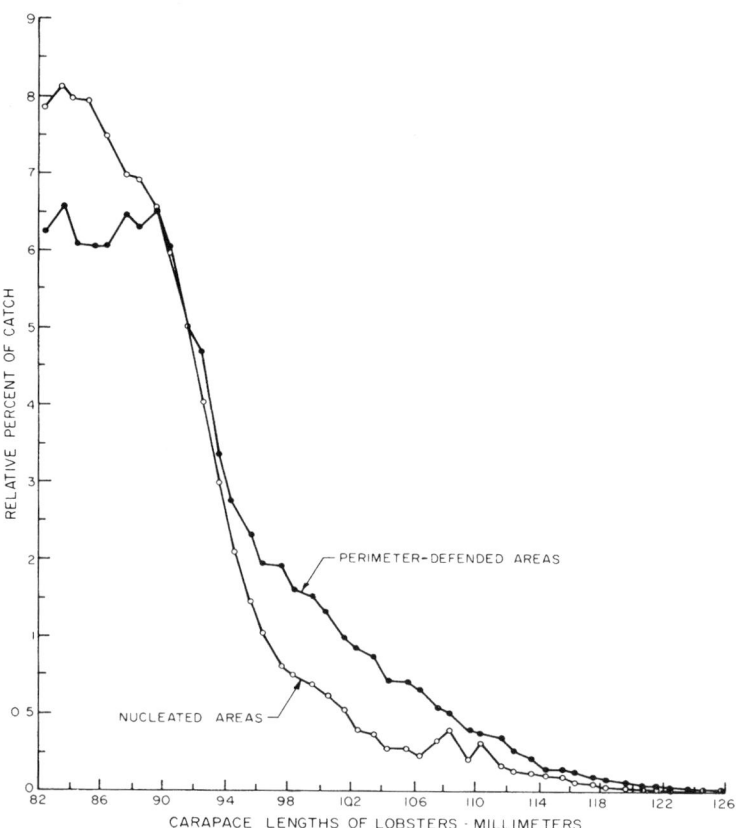

Fig. 2. Distribution of catch by length: nucleated and perimeter-defended areas. Perimeter-defended: mean, 89.98; SD, 7.23; N, 5762. Nucleated areas: mean, 87.89; SD, 5.88; N, 3327. The curves have been smoothed by using a 4-mm moving average. A t test (used to test whether or not there is a significant difference in two means) indicates that there is a less than 0.005 chance of the two samples being drawn from the same population. Calculated value of the t equals 12.55.

the lobsters caught in perimeter-defended areas were berried females as opposed to only 1.2% in the nucleated areas.

The weight of evidence suggests that the fishery can benefit if a larger number of females are allowed to survive to breeding size. For example, Thomas (1973: 43-56) argues that the relationship between recruitment (of lobsters into the fishery) and egg production is positive (as eggs increase, recruits will increase) and will remain so in the foreseeable future. Since the number of eggs is a function of the number of females, it can be argued that lobstermen in perimeter-defended areas are doing more to maintain the biomass than those in nucleated areas. However, the men in the perimeter-defended areas are not gaining

The Lobster Fiefs

all the benefits of their activities. There is no question that lobster larvae float and travel via currents during one growth stage and then settle down to a sedentary existence. The exact pattern of this movement is not known (Pontecorvo, 1962: 245). Thus, a lobster born by virtue of conservation practices carried out at Monhegan or Metinic may well end up living its entire adult life in an area many miles away.

There is, however, a good deal of evidence to suggest that the stock density of lobsters is much better in the perimeter-defended areas than in the nucleated ones. This means that the number of lobsters on any given unit of bottom in a perimeter-defended area is larger, with all that portends for future catches and the numbers of lobsters which can reproduce.

Stock density must be measured indirectly, since the use of divers has proven to be impractical. In the past, attempts have been made to assess the density of lobster stock by using figures for catch in numbers of pounds, or pounds caught per trap. Neither is a valid index, since neither takes into account the number of times a trap is pulled or the working time of the bait. Two areas producing 1 lb per trap do not have the same stock density if the traps in one are pulled twice as often as in the other. The best estimates of stock density can be obtained by figures on *catch per unit of effort*. Stock density is measured by catch in numbers per trap haul times setover days (TH SOD),[13] which can be figured by using the following equation (Thomas, 1973: 38):

$$\frac{\text{catch}}{\text{traps hauled} \times \text{setover days}} \quad \text{or} \quad \frac{\text{No.}}{\text{TH SOD}}$$

Thus, if a man caught 180 lobsters in a day out of 200 traps which had been pulled and rebaited 4 days before, the figures for No./(TH SOD) would be 180/(200 × 4) or 0.225.

Using the concept of numbers per trap hauled times setover days, we see an enormous difference between perimeter-defended and nucleated areas with regard to stock density (Table II). Three aspects of Table II call for comment. First, in seasons II and III there is a great difference in the lb/(TH SOD) between the exclusive territory of perimeter-defended areas and the overlapping areas of nucleated areas (compare lines A and D). Since most of the traps are located in these two categories, these figures indicate a great degree of difference in the stock densities.

Second, stock densities are relatively high in those parts of the perimeter-defended areas fished exclusively (and this would include most of such areas). As one moves toward the boundaries of regions fished with men from other harbor gangs, the stock densities drop (compare A and B regarding lb/(TH SOD) in perimeter-defended areas).

---

[13] As already mentioned, "setover days" indicates the number of days the trap has been in the water since it was pulled and rebaited.

Table II. Pounds/(Trap Haul × Setover Days) in Nucleated and Perimeter-Defended Areas[a]

		Season I[b]	Season II[c]	Season III[d]
Perimeter-defended areas	A. Exclusive	0.24409 (662)	0.26478 (2245)	0.47130 (2681)
	B. Overlapping	—	0.25465 (31)	0.33414 (98)
Nucleated areas	C. Exclusive	0.05353 (326)	0.11133 (313)	0.26735 (610)
	D. Overlapping	—	0.13782 (400)	0.27949 (1483)

[a] The first number in each square is the lb/(TH SOD), where TH SOD = trap haul × setover days; the number in parentheses represents the sample number of lobsters in each category. We did not begin to collect data on setover days until mid-April, so our information on season I is slight.
[b] August 1 to December 31.
[c] January 1 to April 30.
[d] May 1 to July 31.

Third, in the nucleated areas, the lb/(TH SOD) values for the small areas fished exclusively are lower than for those areas fished with men from other harbors (compare lines C and D). Since the small areas fished exclusively by men from nucleated harbors are close to the harbor mouths, they are apt to be the zones fished most heavily. These are areas that part-timers and high school boys can reach with their small skiffs. The low figures probably reflect this greater fishing effort.

Economic Benefits

Men in perimeter-defended areas are catching more and bigger lobsters, and they are catching them with less effort. Both factors translate directly into economic benefits. The evidence that the men from perimeter-defended areas get more lobsters can be seen by comparing mean numbers of lobsters caught per trap in the two types of areas (see Table III).

Table III. Catch Characteristics by Area and Season

	Season I		Season II		Season III	
	Nucleated	Perimeter-defended	Nucleated	Perimeter-defended	Nucleated	Perimeter-defended
Number of lobsters caught	366	767	710	2268	2093	2779
Number of lobsters caught per trap hauled	0.61986	1.294	0.4788	0.9569	0.7609	1.038
Mean pounds of lobster caught per trap hauled	0.72278	1.644	0.5654	1.209	0.9113	1.295
Pounds per lobster	1.1648	1.253	1.181	1.319	1.197	1.238

The Lobster Fiefs

In every season, the number of lobsters caught per trap is larger in perimeter-defended areas than in nucleated areas. Thus, if a man from one of these areas pulls as many traps as a man from a nucleated area, he obviously catches a larger number of lobsters to market. It is possible for men from nucleated areas to earn more than men from the perimeter-defended areas, but to do so they must pull vastly more traps and greatly increase their effort. There is, of course, evidence that some of the men in perimeter-defended areas are cutting down on numbers of traps hauled (e.g., informal trap limits). However, there is no evidence to suggest that such trap limits have so greatly cut down the number of traps any single man uses that incomes are jeopardized. Rather, such limits have stopped the escalation in numbers of traps.

The fact that men from perimeter-defended areas catch larger lobsters than men in nucleated areas is indicated by Fig. 2. This conclusion is buttressed by the data on pounds of lobster caught per trap (see Table III). At all times of year, the men from perimeter-defended areas catch more pounds per trap than men from nucleated areas.

The fact that lobsters are sold by the pound contributes to giving the men from perimeter-defended areas a higher economic return per trap. Moreover, larger lobsters which can be sold as "dinner lobsters" in high-priced restaurants bring a higher price per pound than the smaller lobsters or "culls." Since men from perimeter-defended areas catch a higher proportion of dinner lobsters, their income is boosted still higher. In this regard, one group of men who fish a perimeter-defended area have made arrangements with a dealer to receive $0.06/lb above the going price for all lobsters they catch. The dealer is more than happy to give it to them because of the large number of dinner lobsters they bring in.

There is an enormous difference between the gross income earned by lobstermen from nucleated areas and that of those from perimeter-defended areas. From October 1972 to March 1973, I interviewed eight lobstermen from perimeter-defended areas who made an average of $22,929 from the lobster industry during the preceding year.[14] The average gross income of fishermen from nucleated areas was $16,499.[15]

These differences in income are not merely due to the fact that men from perimeter-defended areas catch more lobsters than do men in nucleated areas; price factors play an important role as well. Men on Monhegan, for example, fish

---

[14] These are gross income figures, earned from lobster fishing alone. While net income figures would undoubtedly be more valuable, they are almost impossible to compute accurately since it is difficult to estimate depreciation on boats and equipment, and no records are kept on routine maintenance. In order to protect informants, the income figures were not broken down by variables such as age or harbor gang.

[15] While the number of fishermen in the sample was very small, the difference in mean income between nucleated and perimeter-defended areas is a surprising $6430.

only from January to June, and thus market most of their lobsters in months when the price is at an annual high.[16]

Summary of Biological and Economic Effects of Territoriality

Perimeter-defended lobster territories are characterized by more restricted entry than are nucleated ones. This shows up as fewer boats or fishermen per square mile, and implies lower levels of fishing effort in the perimeter-defended areas. Fishing effort has also been controlled by various local conservation measures, such as limits on the number of traps allowed or local closed seasons. These measures do not affect total catch, but do tend to reduce total lobster mortality, as well as increase the proportion of the more valuable large lobsters caught. Greater stock density in the perimeter-defended areas suggests that reduced effort has halted the process of overexploitation. Economic gains from the strategy of maintaining strong defense of traditional boundaries are shown in higher mean numbers of lobsters per trap, more large and hence higher-priced lobsters, and higher gross incomes.

MANAGEMENT IMPLICATIONS OF TERRITORIALITY

The state of Maine does not take advantage of the principle of having perimeter-defended areas, with all of the favorable results they bring. Moreover, the state has yet to institute any of the conservation measures that men from these areas have adopted for themselves, such as trap limits and closed seasons. At present, Maine is considering legislation concerning trap limits and other practices to limit fishing effort. However, the state has no interest in establishing legalized lobstering territories analogous to the traditional perimeter-defended areas. There are two reasons for this. The Commission of Sea and Shore Fisheries argues that the administrative costs involved in establishing and maintaining small fishing areas would be prohibitive (S. Appolonio, personal communication, 1973). This may be correct. In addition, since the existence of the territorial systems is hidden from official eyes, the state's officers have little information on possible beneficial effects.

---

[16]In recent years the price has been highest in February and March and during the height of the summer tourist season. It has been lowest in the fall, when lobsters are plentiful and demand is relatively low. For example, in October 1973 the price paid to lobstermen was $1.10 per pound; in January 1974 it was $1.65.

The Lobster Fiefs

CONCLUSION

One of the themes that run through the literature on fishing communities and fisheries economics is the competitive, predatory behavior of individual fishing firms, their inability and unwillingness to conserve the marine resources on which they depend, and their tendency to overcapitalize. These factors are related to the fact that, by law, oceans are common property resources. Since oceans are owned by no one and may be exploited by everyone, no one has any interest in maintaining the resources. Why should one man cut his fishing effort to conserve? The fish he does not catch today will be caught by someone else tomorrow. Under these conditions, a fisherman is only being rational when he expands the amount of capital equipment he owns and tries to catch all the fish he can as quickly as possible.

The result is what Hardin (1968: 1244) has called the "tragedy of the commons." Common property resources of all kinds — the air, waters, oceans, publicly-owned land — are subject to abuses and overexploitation that do not exist with privately owned resources (Hardin, 1968: 1245-1246). It is not only that common property resources are overexploited by a callous public; they are also subject to a kind of escalating abuse as people vie with each other to strip bare resources owned in common. As Hardin (1968: 1244) explains it, those exploiting a common property resource are locked into a system in which it is only "logical" that they increase their exploitation without limit. As far as fisheries are concerned, the "tragedy" takes the form of overexploitation, depletion of fish stocks, underutilization of capital resources, and, where opportunity costs are low, acceptance of low incomes (Crutchfield, 1964: 212).

At both the national and regional levels, attempts to regulate fisheries usually take the form of manipulating fishing seasons, fishing areas, and the type of fishing gear used. While such regulations may limit fish mortality, economists have pointed out that they are probably ineffective and certainly make fishing more inefficient (Pontecorvo and Vartdal, 1967; Crutchfield and Pontecorvo, 1969; De Wolf, 1974). Fishing quotas and seasons, for example, leave expensive boats tied to the dock much of the time and trained crews in the welfare line. They also increase competition for new equipment and for a "share of the catch" when fishing is allowed. Several economists have strongly argued that a much better management system would involve limiting entry into the fishery either by a licensing system (Pontecorvo, 1967; Christy, 1973) or by taxation (Pontecorvo and Vartdal, 1967), which would cut production by increasing marginal costs.

The lobstermen of Maine have already evolved a system that produces all the benefits these economists have envisioned. Given what we know about the theory of the common property resource, we can only deplore the breakdown of the traditional perimeter-defended areas. In perimeter-defended areas, access to the resource is highly controlled, and escalation of fishing effort has been prevented by the use of political power to enforce local conservation measures. Moreover, in these areas, fishermen are not caught in a self-defeating "competitive withdrawal" (Crutchfield and Pontecorvo, 1969), but can cooperate to conserve the lobster stock and raise their own income levels. In nucleated fishing areas, the traditional means of controlling escalating fishing effort are breaking down. The data we have on differences in average size of lobsters, size of catch, size of stock, and fishing incomes suggest that men fishing in nucleated areas are bearing the costs of increased participation in an industry that has many characteristics of a "common property" fishery.[17]

REFERENCES

Acheson, J. M. (1972). The territories of the lobstermen. *Natural History* 81: 60-69.
Acheson, J. M. (1974). Variations in inshore fishing rights in Maine lobstering communities. In Andersen, R. (ed.), *Northern Maritime Europeans,* Mouton, The Hague.
Bailey, F. G. (1969). *Stratagems and Spoils,* Schocken, New York.
Christy, F. T. (1973). Fishermen quotas: A tentative suggestion for domestic management. Occasional Paper 19, Law of the Sea Institute, University of Rhode Island, Kingston.
Christy, F. T., and Scott, A. (1965). *The Common-Wealth in Ocean Fisheries,* Johns Hopkins Press, Baltimore.
Crutchfield, J. (1964). The marine fisheries: A problem in international cooperation. *American Economic Review Proceedings* 54: 207-218.
Crutchfield, J. (1973). Resources from the sea. In English, T. S. (ed.), *Ocean Resources and Public Policy,* University of Washington Press, Seattle, pp. 105-133.
Crutchfield, J., and Pontecorvo, G. (1969). *The Pacific Salmon Fisheries,* Johns Hopkins Press, Baltimore.
De Wolf, G. (1974). *The Lobster Fishery of the Maritime Provinces: Economic Effects of Regulations,* Bulletin 187, Fisheries Research Board of Canada, Ottawa.

---

[17] In two respects, the theory of the common property resource does not quite fit the situation in the Maine lobster industry. First, the economists writing on the common property resource model are assuming that common property is completely accessible to everyone while private property has only one owner (e.g., Scott, 1955: 120, 121). In fact, no lobstering area is accessible to the general public. As we have pointed out, it is vastly more difficult to gain entry into harbor gangs fishing perimeter-defended areas than those fishing nucleated areas. But one must have certain characteristics to become a member of a harbor gang fishing even in a nucleated area. Moreover, there is no area along the coast owned by a single individual. All areas are "owned" by *groups*. Second, entrepreneurs exploiting "privately owned resources" cut down on the volume of production because of increased marginal costs, whereas those exploiting "common property resources" have no such constraint. In the case of the Maine lobster industry, there is no question that men fishing perimeter-defended areas try harder to cut down fishing effort than men fishing nucleated areas. In part, this reduction in effort is due to self-enforced "trap limits" and closed seasons. Even more important is the fact that they use political pressure to keep the number of fishermen in their areas small. This naturally decreases the number of traps being fished there. In short, decreased fishing effort is not so much due to the sacrifices of the individual men fishing in "perimeter-defended areas" as it is due to the fact that they keep other men out, and enforce the sacrifices on them. The effect is the same. Men fishing perimeter-defended areas restrict fishing effort, which has clear beneficial effects.

The Lobster Fiefs

Dow, R. L. (1969). Cyclic and geographic trends in seawater temperature and abundance of American lobster. *Science* 164: 1060-1063.

Gordon, H. S. (1954). The economic theory of a common property resource. *Journal of Political Economy* 62: 124-142.

Hardin, G. (1968). The tragedy of the commons. *Science* 162: 1243-1248.

Hockett, C. F. (1973). *Man's Place in Nature,* McGraw-Hill, New York.

Krouse, J. S. (1973). Maturity, sex ratio, and size competition of the natural population of American lobster, *Homarus americanus,* along the Maine coast. *Fishery Bulletin* 71: 165-173.

Maine Sea and Shore Fisheries (1973). *Maine Lobster Landings 1939 to Present,* Statistical Bulletin No. 1, Department of Sea and Shore Fisheries, Augusta.

Pontecorvo, G. (1962). Regulation of the North American lobster fishery. In Hamlisch, R. (ed.), *Economic Effects of Fishery Regulation,* F.A.O. Fisheries Reports No. 5, Rome, pp. 240-267.

Pontecorvo, G. (1967). Optimization and taxation in an open-access resource: The fishery. In Gaffney, M. (ed.), *Extractive Resources and Taxation,* University of Wisconsin Press, Madison, pp. 157-167.

Pontecorvo, G., and Vartdal, K., Jr. (1967). Optimizing resource use: The Norwegian winter herring fishery. *Statsøkonomisk Tidsskrift* 2: 65-87.

Scott, A. (1955). The objectives of sole ownership. *Journal of Political Economy* 63: 118-124.

Thomas, J. (1973). *An Analysis of the Commercial Lobster (Homarus americanus) Fishery Along the Coast of Maine, August 1966 Through December 1970,* National Oceanic and Atmospheric Administration Technical Report, National Marine Fisheries Service, Washington, D.C.

Wilson, J. A. (1975). The tragedy of the commons: A test. In Baten and Hardin, G. (eds.), *Managing the Commons,* Freeman, San Francisco.

Leisure Value Systems and Recreational Specialization: The Case of Trout Fishermen[1]

Hobson Bryan

ABSTRACT: *A conceptual framework of trout fishermen is developed around the concept "recreational specialization." This refers to a continuum of behavior from the general to the specialized. It is reflected by equipment, skills used, and preferences for specific recreation setting. Two hundred sixty-three on-site interviews with fishermen in Wyoming, Montana, and Idaho, supplemented by participant observation, yielded four types. They range from sportsmen with minimal interest and skill in the sport to those highly committed and specialized members of a leisure social world. Resulting propositions are: (1) Fishermen tend to become more specialized over time, (2) the most specialized comprise a leisure subculture with unique minority recreationist values, (3) increased specialization implies a shift from fish consumption to preservation and emphasis on the activity's nature and setting, and (4) as specialization increases, dependency on particular resource types increases. Management implications of these propositions are discussed.*

KEYWORDS: *Leisure social world, outdoor recreation, fishermen, specialization, values, management.*

AUTHOR: Hobson Bryan *is an Associate Professor of Sociology and Research Associate, National Resources Center, at the University of Alabama.*

---

[1] This report is a revised segment of a larger investigative effort presented to the 1974 American Fisheries Society Meetings in Honolulu. Modified versions were also presented to the 1975 Southwestern Sociological Association Meetings in San Antonio and the 1975 Southern Sociological Meetings in Washington, D.C. Acknowledgment is made to the U.S. Forest Service Southeastern Forest Experiment Station, Asheville, North Carolina, for support to report these findings and to research their applicability to other outdoor recreation activities.

Bryan, Hobson, *Leisure Value Systems and Recreational Specialization: The Case of Trout Fishermen*, Journal of Leisure Research, 9:3 (1977) p.174

Leisure Value Systems

Researchers are pointing to evidence that individuals can center their lives around leisure activities as well as work. In fact, as Roberts (1970: 25) notes, ". . . it can be argued that for many people leisure has now become such a central and dominant part of their lives that it is their behavior and attitudes towards work that are determined by their leisure rather than the other way around." He hypothesizes that self-concepts can be shaped by leisure activity. Related to this notion is Shibutani's (1955) concept of "social world" reference groups, which come into existence with the development of specialized communication channels. Devall (1973) elaborates on *leisure* social worlds in surfing and mountaineering. Leisure social groups are major sources of orientation and reward for members, just as the workplace meets the primary needs of others. What may well be significant about these groups is that they not only serve as standards of reference for leisure behavior, but may revolve around and influence central life interests and most other areas of life activity.

But all or even most of a given recreational group are not members of its social world segment. A broad range of orientations and behavior attends any recreational activity. In fact, a major weakness of past research efforts has been the assumption of sportsmen group homogeneity, with variations among individual sportsmen remaining largely unexplored (Bryan 1976a).

The research reported here explores a leisure social world of sportfishermen. Other types of fishermen who do not belong to this group are investigated for comparative purposes. The object is the development of a conceptual framework,[2] covering a broad spectrum of angler types, utilizing the variable "recreational specialization." The ramifications of leisure value systems and behavior are explored as they relate to the specialization variable among trout fishermen. More generally, the thesis is examined that the specialization dimension is a significant variable in understanding the behavior and attitudes of these sportsmen. A primary objective of this research is to develop explanatory principles of recreational behavior based on leisure commitment and specialization and to propose propositions for testing under controlled conditions.

Recreational Specialization. The term "recreational specialization" as used here refers to a continuum of behavior from the general to the particular, reflected by equipment and skills used in the sport and activity setting preferences. In other words, the dictionary definition of specialization applies. At one end of the continuum is the person who devotes or limits interest to some special branch of the sport. At the other end is the person who has more

---

[2] The conceptual framework is to be distinguished from a simple ad hoc classificatory system, or from a descriptive taxonomy. In classificatory systems, more or less arbitrary classes are constructed for the sake of summarizing data; in taxonomies, categories are formed to fit the data and there are interrelationships among categories, but they remain in the descriptive rather than the analytical realm (Zetterberg 1965: 24–28). Conceptual frameworks, on the other hand, logically direct empirical and theoretical activity around a core set of problems and, as such, offer the beginnings of systematic theory.

general recreational interests. In short, the research sets out to explore the idea that trout fishermen can be arranged along a specialization continuum which is linked to the diverse sportsmen preferences and behavior.

Though within-sport variability has been dealt with in the literature, particularly with regard to trends within a given sport, satisfactory treatment of the underlying processes accounting for the variability is lacking. Devall (1973: 57) recognizes this in his work in pointing to the need for theoretical elaboration in regard to the career patterns and social-psychological meanings of leisure activities. Thus, the emphasis here is on the exploration of the variation among sportsmen in terms of an activity's meaning to the individual and his behavior.[3]

Variable meanings of leisure activities have been treated by several researchers (Havighurst 1957; Faunce 1963; Kando and Summers 1971; Parker 1971). Among the distinctions is whether a sport involves active participants in an outdoor setting (McIntosh 1963: 44). Outdoor sports in general are characterized as "conquest sports." Challenge is provided by the environment or situation, not directly by other people opponents. Obvious examples would be backpacking, skiing, sailing, surfing. Devall (1973: 53) notes that in surfing and mountaineering ". . . the essence of the activity is that some people begin to define certain aspects of the natural environment as appropriate for expressive, play activities." But such sports may evolve into competitive activity with people opponents, at least for some of the participants. Skiing, sailing, and fishing[4] are cases in point.

The Case of Trout Fishermen. Turning specifically to the topic of this research, in accounting for variations within a leisure sphere, Kelly (1974) views recreational activity as a lifelong process of leisure socialization and advocates a "developmental approach to leisure careers." By inference, people approach their sports or hobbies differently, depending on their "stage of development" in the activity. In the research reported here the idea is explored that fishermen can be arranged along a continuum of experience and commitment to the sport, from the beginning recreationist to the specialist, that distinctively different preferences and behavior attend sportsmen at each level.

The research is on trout fishing due to the depth of activity and the wide range of orientation and behavior it presents. Thus, development of a conceptual framework potentially applicable to other types of sportsmen is made feasible. This contention is supported by Gingrich's (1965) research indicating that the writing on angling exceeds in extent and diversity all other works devoted to a single branch of sport. A full bibliography would go back almost five centuries and contain more than fifty thousand entries. It is reported that the trout is the most written about fish from the sport perspective and flyfishing the most written about technique.

A contention guiding this research is that "flyfishing" for trout (a technique

---

[3] This is in keeping with Kando's (1975: 93) statement that "the sociology of leisure deals with certain types of meanings that activity may have, rather than a specific category of activities."

[4] An example is tournament bass fishing, where the appeal of this sport has been analyzed as a status-achieving activity (Bryan 1974c).

Bryan, Hobson, *Leisure Value Systems and Recreational Specialization: The Case of Trout Fishermen*, Journal of Leisure Research, 9:3 (1977) p.174

Leisure Value Systems

by which imitations of insects on which the fish feed are cast with specialized equipment, i.e., fly rod, reel and line) represents the end-product of a progression of angling experiences leading to a more and more "mature" or specialized state. Popular notions of the sport generally support this thesis. For example, Gingrich (1965: 481) in reviewing the literature on fishing further observes:

> Since the kindergarten of angling is still fishing with a pole and a worm, and serious anglers generally agree that the progressive education of an angler culminates in stream fishing with a fly, it is only natural that the highest reaches of the literature should be concerned chiefly with this form of fishing.

Methodology

The bulk of the research consists of 263 on-site interviews with fishermen supplemented by participant observation techniques. Interviews and correspondence were also conducted with people in the sportfishing industry concerning their observations of fishing and fishermen in the initial development of the research. Fishermen were approached as they were leaving the stream and asked if they would be willing to "talk a while about their fishing" for a study on angling preferences (there were virtually no refusals). While interviews were informal and flexible, they followed a preplanned design to obtain information concerning beliefs, attitudes, values, and ideologies connected with the sport of fishing and its place in the individual's life. Questions were posed concerning: (1) Fishing preferences, (2) orientation toward the stream resource, (3) history of interest and activity in the sport, and (4) relationship of the leisure activity to other areas of life (family, career, other leisure activities).

Supplementary information was derived from participant observation on the streams and around such traditional fishermen hangouts as campgrounds, tackle stores, bars, and eating places. On-stream observations included skill displayed, techniques used, and social setting. Away-from-the-stream observations focused on interactions with friends, tackle purchasing habits, and related activity.

Interviews were portioned among eight sites over a 4-summer period: Silver Creek near Hailey, Idaho; Henry's Fork on the Snake River near Last Chance, Idaho; Big Spring Creek near Lewistown, Montana; Armstrong Spring Creek near Livingston, Montana; Poindexter Slough near Dillon, Montana; Firehole River in Yellowstone Park, Wyoming; Madison River near Ennis, Montana; and Yellowstone River near Gardiner, Montana.

The strategy was to select a variety of stream types and settings to attract a full range of fishermen types. To insure that the most specialized of fishermen would be selected, a large number of streams appealing to fly-fishermen was included. Other sites were selected on the basis of their general fishing fame (as established by McClane 1965 and Brooks 1966), thus appealing to a wider variety of fishermen. The sites were not intended to be proportionally representative of all waters available to trout fishermen. Nor was an attempt made to draw proportional samples of anglers. The purpose of the research was to develop a conceptual framework, rather than to determine from a probability sample, survey design how many of each type of recreationist there are in the sportsmen population, or to make other inferences about the angling population at large. This strategy stemmed from the contention that the virtual absence of conceptual frameworks to guide research is partially responsible for the many uncoordinated and sometimes unproductive studies of the human dimensions of fish and wildlife. (Bryan 1976a).

The intent of the discussion to follow is to compare and contrast fishermen categories and to detail trends in the data. The resulting "ideal types" serve as a basis for the conceptual framework from which propositions are proposed for controlled testing. Frequencies and percentages of response to specific areas of inquiry are presented in Table 1.

Findings

An overview of the results suggests a fishermen typology based on degree of specialization. This is reflected by amount of participation and technique and setting preferences. The types are:

1. Occasional Fishermen—those who fish infrequently because they are new to the activity and have not established it as a regular part of their leisure, or because it simply has not become a major interest.
2. Generalists—fishermen who have established the sport as a regular leisure activity and use a variety of techniques.
3. Technique Specialists—anglers who specialize in a particular method, largely to the exclusion of other techniques.
4. Technique-Setting Specialists—highly committed anglers who specialize in method and have distinct preferences for specific water types on which to practice the activity.

Equipment Preference. The proposed typology of anglers is based in part on equipment usage. Occasional fishermen are more likely than generalists not to have particular preferences (in 44% as opposed to 21% of the cases), while generalists are evenly divided (33% to 33%) in preferring spinning or spin-casting methods. Specialist fishermen, by definition, prefer flyfishing tackle.

On-stream observation[5] revealed that the "ideal typical" occasional angler can be identified by his green rubberized-cloth creel, small net, and—if the water is restricted to flyfishing—spin-casting outfit with a casting bubble for sufficient weight to cast an artificial fly (he neither owns nor knows how to use flyfishing tackle). Generalists are well equipped to "catch a limit," so their equipment is marked by its functional quality. Large wicker-basket creels, wide-diameter nets, and hip-waders mark this angler. Technique-specialists (flyfishermen in this case) typically own several rods to match fishing conditions. They wear chest-high waders with felt soles to prevent slipping and carry a large amount of tackle in their fishing vests for a variety of situations.

---

[5] Equipment enumeration per se is limited to rod type, because this was considered a key variable at the research's onset. As evidence accumulated and pointed to the typology, observations were directed to the several equipment items typifying each fishermen category. Conclusions, therefore, are based on impressions gleaned and field notes taken in the latter stages of the research.

Leisure Value Systems

TABLE 1

Fishermen Types and Responses to Specialization Variables

Fishermen Characteristics	Fishermen Types							
	Occasional (N = 48)		Generalists (N = 70)		Technique Specialists (N = 63)		Tech-Setting Specialists (N = 82)	
	%	(No.)	%	(No.)	%	(No.)	%	(No.)
Equipment Preference								
No preference	44	(21)	21	(15)	—	—	—	—
Spin-casting	23	(11)	33	(23)	—	—	—	—
Spinning	19	(9)	33	(23)	—	—	—	—
Flyfishing	15	(7)	13	(9)	100	(63)	100	(82)
Total*	101	(48)	100	(70)	100	(63)	100	(82)
Orientation to Fish								
Emphasize quantity	71	(32)	48	(30)	32	(20)	20	(16)
Emphasize size	20	(9)	48	(30)	52	(33)	40	(33)
Emphasize setting	9	(4)	5	(3)	16	(10)	40	(33)
Total*	100	(45)	101	(63)	100	(63)	100	(82)
Species Preference								
No preference	39	(17)	24	(17)	7	(4)	4	(3)
Trout	57	(25)	72	(48)	59	(34)	54	(41)
"Any fish caught on a fly"	5	(2)	3	(2)	34	(20)	42	(32)
Total*	101	(44)	100	(67)	100	(58)	100	(76)
Water Preference								
"Any water containing fish"	52	(25)	16	(11)	3	(2)	—	—
Lakes	25	(12)	31	(22)	21	(13)	—	—
Large streams	8	(4)	29	(20)	38	(24)	—	—
Small streams	15	(7)	24	(17)	38	(24)	100	(82)
Total*	100	(48)	101	(70)	100	(63)	100	(82)
Management Preference								
Stocking	57	(27)	47	(32)	32	(18)	20	(15)
Ease of access	23	(11)	16	(11)	21	(12)	14	(10)
Habitat management	19	(9)	37	(25)	47	(27)	66	(49)
Total*	99	(47)	100	(68)	100	(57)	100	(74)
Angling History								
Cumulative	60	(29)	65	(45)	66	(40)	76	(59)
Noncumulative	40	(19)	35	(24)	34	(21)	24	(19)
Total*	100	(48)	100	(69)	100	(61)	100	(78)
Social Setting								
Fishes alone	13	(6)	29	(20)	25	(15)	30	(23)
With family	63	(29)	29	(20)	21	(13)	14	(11)
With peers	24	(11)	43	(29)	54	(33)	55	(42)
Total*	100	(46)	101	(68)	100	(61)	99	(76)

TABLE 1 (*Continued*)

	Fishermen Types							
Fishermen Characteristics	Occasional (N = 48)		Generalists (N = 70)		Technique Specialists (N = 63)		Tech-Setting Specialists (N = 82)	
Distance Traveled								
Within 200 miles	27	(13)	59	(41)	27	(17)	23	(19)
Within geographic region†	35	(17)	30	(21)	14	(9)	15	(12)
Out of geographic region	38	(18)	10	(7)	59	(37)	63	(50)
Total*	100	(48)	99	(69)	100	(63)	101	(81)
Vacation Patterns								
Extended	15	(7)	26	(17)	42	(25)	49	(39)
Short	48	(22)	42	(27)	28	(17)	30	(24)
Seldom take vacations	37	(17)	32	(21)	30	(18)	20	(16)
Total*	100	(46)	100	(65)	100	(60)	99	(79)
Leisure Priority								
Career influenced by sport	5	(1)	15	(5)	39	(15)	54	(19)
Career not influenced by sport	95	(20)	85	(29)	61	(23)	46	(16)
Total*	100	(21)	100	(34)	100	(38)	100	(35)

* Percentages and numbers do not always add to total N's due to rounding and incomplete responses.
† But more than 200 miles.

Those individuals who specialize in the setting of the activity (spring streams in this research), as well as in the technique, are differentiated from other flyfishermen as much by what they do not carry with them as by what they do. Nets are not used, for this might indicate lack of expertise in playing the fish. Fish are landed with bare hands or beached. Nor is a creel carried, since fish are rarely kept, even large ones. Ownership of high quality rods is a characteristic of these fishermen. Anglers who started flyfishing with a $20 glass rod progress from "better quality" $50 rods to those in the $100 category. Or they may choose the traditional bamboo rod starting at $175. At the highest reaches of specialization, the flyfisherman is concerned with having exactly the right equipment for a particular fishing situation. Technique-setting specialists may own several custom-made bamboo rods in the $300 and above category and are among the first to purchase the new so-called space age graphite tackle.[6]

---

[6] Similar buying trends have been observed with regard to highly specialized bass fishermen in the Southeast (Bryan 1974c). But it is difficult to determine the dividing line between specialization and conspicuous consumption.

Bryan, Hobson, *Leisure Value Systems and Recreational Specialization: The Case of Trout Fishermen* , Journal of Leisure Research, 9:3 (1977) p.174

Leisure Value Systems

Orientation to Fish. Different types of anglers look for different things in the fishing experience. The emphasis among occasional and generalist anglers is on the number of fish caught (in 71% and 48% of the cases as opposed to 32% and 20% of the respective specialist types). But generalists equally emphasize size. For the specialist anglers, emphasis definitely changes from number of fish to size (for 52% of the technique specialists as opposed to 20% of the occasional fishermen) and then more to the setting of the activity (technique-setting specialists being evenly divided—40% to 40%—on preferring size and setting as compared to 9% and 5% of the occasional and generalist anglers).

Preferences regarding species of fish are also varied. Naturally, most anglers prefer trout to other species of fish. After all, individuals were interviewed on trout stream locations. But follow-up questioning revealed that the first concern of the occasional angler is to catch a fish, any fish. Thus, though these anglers prefer catching trout, a larger percentage (39%) than those in the other categories (7% and 4% of the respective specialists) express no species preference. Further, as the angler becomes more specialized, he is less likely to list trout as his first choice. The technique-setting specialist, as a matter of fact, often reported (in 42% of the cases as contrasted with 5% of the occasional and 3% of the generalist fishermen) that his first concern was whether the fish could be caught on flyfishing tackle. Follow-up questioning revealed that for specialist anglers technique preference begins to override species preference. In other words, the first concern of the flyfishing specialist is to be able to catch his quarry on fly tackle. Some flyfishermen prefer catching other species, particularly if they happen to be from a section of the country which does not have trout fishing.

Resource Orientation. Orientation to the setting of the fishing experience, a major component of the conceptual framework, differs by fisherman type. For the occasional fisherman, preference for a particular type of water seems to be overridden by his concern with the ease with which he might catch a fish, thus (in 52% of the cases as opposed to 7% and 4% of the specialists) he has no particular preference for lake or stream. Generalists are fairly evenly divided on their preferences (31% preferring lakes, 29% larger streams, and 24% smaller streams). Technique specialists (the flyfishermen) definitely prefer streams to lakes (in 76% of the cases if the two stream sizes are combined, as contrasted with the combined figures of 23% of the occasional and 53% of the generalist fishermen), but they are evenly divided on stream size (38% preferring large streams, 38% small streams). Technique-setting specialists, by definition, seek out the spring stream for their fishing experience.[8]

---

[7] Specialist fishermen were quick to distinguish between "free-stone" and limestone streams, the latter having chemical and biological properties contributing to the abundant insect life necessary for consistently good flyfishing. Limestone streams (which are often of spring origin) are usually relatively small in volume.

[8] For the purpose of this research, the spring stream is defined as one originating from springs and retaining the qualities of relatively constant temperature and volume, low gradient, high clarity, and supporting large quantities of insect and trout life.

Significantly, the more specialized the fisherman, the more his enjoyment and pursuit of the activity is inextricably linked to the nature and setting of the resource he fishes. One explanation is that the very nature of flyfishing, especially on the spring stream, implies a close tie to the qualities of the resource. The rationale is that a key attraction for any specialist, whether reference is to a recreationist or to one working at his profession, is the degree of control or manipulation which he can bring to the activity. For the fruits of his knowledge to have impact, he must have a monitoring capability. The spring stream offers the setting and predictability to allow for such control. The popular angling writer, Charles Waterman (1971: 40), writes:

> It is the difficulty, delicacy, and complexity of such fishing that obsesses the expert, but it is the predictability of the trout's feeding habits that adds to the stream's attraction, for the fisherman can see the fish feed and know they can be caught if his approach is carefully correct. His failures can hardly be attributed to bad luck.

Management Philosophy. Management concerns differ according to the fisherman's resource orientation and angling preference. Occasional fishermen and generalists favor an active stocking policy (in 57% and 47% of the cases respectively as opposed to 32% and 20% of the specialists). On-site observation revealed that a frequent concern of the occasional fisherman is whether the water is stocked with catchable sized fish. Ease of access is also a factor (the major concern of 23% of these anglers). For generalists, a "good return for the license dollar" is a large harvest of catchable-size fish during the year In contrast, technique and technique-setting specialists are more likely to favor habitat management (in 47% and 66% of the cases respectively as compared with 19% of occasional and 37% of generalist anglers). Streamside conversations revealed a concern that the wild fish population not be "contaminated" with hatchery-bred trout. These anglers frequently talk about "fishing quality" and "good management" in terms of harvesting policies to enhance the size of the fish. Catch-and-release policies are favored on certain streams if deemed necessary to maintain a healthy wild trout population.

Angling History. Popular conceptions of fishermen moving through stages in their "fishing careers" were supported. Socialization into the sport is cumulative—in other words, fishermen typically start with simple, easily mastered techniques which maximize chances of a catch, then move to more involved and demanding methods the longer they engage in the sport. Thus, a "cumulative response" is when the individual reports starting with rudimentary tackle (e.g., cane pole and worms) in his early experiences, progresses to lures cast with spinning or spin-cast tackle at a later stage, then progresses to flyfishing equipment still later. Technique-setting specialists were most likely to have a cumulative response (in 76% of the cases), while occasional, generalist, and technique specialists also reported such responses in the majority of cases (60%, 65%, and 66%, respectively). As in the case of any status attribute, one does not have to start at the "bottom" of the experiential sequence, nor, by the same token, does he have to move to the "top." But the tendency is to move toward the specialization end of the continuum.

Bryan, Hobson, *Leisure Value Systems and Recreational Specialization: The Case of Trout Fishermen*, Journal of Leisure Research, 9:3 (1977) p.174

Leisure Value Systems

Social Context. A major component of fishing as a leisure activity is its social context. The social setting of the fishing experience ranges from family outings (the most frequent situation among 63% of the occasional anglers) to fishing with peers (54% of the technique-setting specialists). In the case of occasional fishermen, angling is usually secondary to other activities, such as picnicking and sightseeing. But for those individuals who regularly incorporate fishing in their leisure time, the primary purpose of the trip is fishing, and the individual is more likely to engage in it with peers who have similar interests and skills.

Participant observation in the extensive fishermen friendship networks revealed that among the specialists the activity of fishing is much more than the casting of a fly. Tackle shop, bar, and campfire "bull sessions" are key ingredients to the experience. Specialist anglers are usually in the exclusive company of sportsmen having similar orientations to and interests in the sport. The fishermen peer group may serve as a reference group as well. The opinion leaders of this angling fraternity are the highly visable outdoor writers, the articulators of the specialist's value system. The cohesiveness of this system is cemented and reinforced through frequent contact among members of the friendship network. Ties are formed which transcend traditional occupational and class barriers to mold these fishermen into a true leisure social world. In addition to shared fishing experiences, contact with this far-flung network is maintained through attendance at meetings of organizations which promote the interests of specialist fishermen (e.g., Trout Unlimited, Federation of Fly-Fishermen), the magazines and newsletters of these organizations, and personal correspondence.

Vacation Patterns. Linked to the social context of the fishing experience are vacation patterns. These patterns may facilitate or restrict contact with fishing peers. In the case of distance traveled to the angling site, generalist anglers especially are likely to have been fishing either within relatively close proximity of their residences (in 59% of the cases), or at least within the geographic region (in 30% of the cases). The more specialized fishermen were likely to have traveled from outside the region (in 59% and 63% of the cases). Thus, generalist fishermen tend to fish with neighbors or friends from work. But more specialized anglers may travel all the way across the continent to "fish the circuit" each year. Having established the prime fishing times for certain streams within a region, they meet at these times with friends from all over the country.

Technique and technique-setting specialists are also more likely to take extended vacations (in 42% and 49% of the cases respectively) than the other anglers (15% of the occasional and 26% of the generalist fishermen). The latter either take short vacations (in 48% and 42% of the cases as compared with 28% and 30% of the specialists) or seldom take time off from the job (in 37% and 32% of the cases, compared with 30% and 20% of the specialists). Informal discussion with occasional anglers gave the impression that these individuals are little involved in any leisure activity. Such reasons were given as heavy family responsibilities, the press of work, and lack of free time.

In the case of generalists, since they tend to be from the local area, fishing is not strictly a vacation time proposition. The sport can be pursued after

Bryan, Hobson, *Leisure Value Systems and Recreational Specialization: The Case of Trout Fishermen*, Journal of Leisure Research, 9:3 (1977) p.174

work or during weekends. Vacations may be reserved for activities requiring larger blocks of time, such as hunting trips and family excursions to recreational areas some distance away.

Specialist fishermen are likely to center their leisure time, vacation and otherwise, around fishing. An obvious intervening variable here and throughout the study is socio-economic status. Though not examined directly, the amount of time available due to job or career factors seemed to be more crucial. Technique-setting specialists in particular had taken jobs for less pay and prestige to be close to exceptional fishing opportunities.[9] Those employed in other areas of the country sometimes made career choices on the basis of the free-time attributes of the job (e.g., the choice of a teaching career for the three months off in the summer, physicians specializing in radiology to permit extended vacations). Others manage to combine work and leisure elements, as in the case of teachers meeting the costs of a summer's fishing by working in tackle shops or guiding. Or work may be shaped to incorporate an aspect of fishing as in the case of a shop teacher who manufactures hand-crafted fishing equipment. There are those who have turned their sport into a full-time business enterprise (four tackle store owners and a rod maker). Included among the specialists are the "fishing bums." Four men in their early twenties had not established a steady career or job pattern after completing their education. But three individuals left middle to upper-level management positions to center their lives around fishing.[10]

Conclusions and Implications

The designation of fishermen types is a useful heuristic tool, providing points of comparison along a continuum of fishing specialization. It is not contended that the framework is definitive. The fact that most generalists originate within the region may indicate that their preference for trout over other species is part of the local value system. It is not so much that they have chosen to specialize in the trout, but more likely that they lack familiarity with other species. This contrasts with the specialists who have had a variety of angling experiences with different species in various locales.

Difficulties are encountered at times in distinguishing between technique specialists and technique-setting specialists and the former sometimes specialize in flyfishing as a technique almost to the degree of the latter. A critical dimension operating at all levels, but especially at the upper end of the typology, is degree of commitment to the activity—the extent of the individual's time and effort invested in the sport. Thus, distinctions between types are sometimes

---

[9] Sixty percent of the technique-setting specialists and 38 percent of the technique specialists indicated that career decisions had been "in some way influenced" by their interest in fishing. As this question was posed only after in-depth interviews revealed the variable's importance in the latter stages of the investigation, the N's are small.

[10] One works in a grocery store (he was formerly in a middle management position in the aerospace industry). Rather than opting for higher pay on his promotion from clerk to butcher, he negotiated for greater time flexibility and more days off for fishing. His custom-made fishing equipment is traded for such things as automobile repairs, a piece of furniture, or clothing. Existing on a semi-barter system, he pays little income tax, lives frugally, and manages to fish most days of the season.

Leisure Value Systems

TABLE 2

DEGREE OF ANGLING SPECIALIZATION AND FISHERMEN CHARACTERISTICS

Degree of Specialization	Fishing Orientation, Equipment	Resource Orientation, Management Philosophy	Social Setting, Leisure Orientation
Occasional Fishermen	Catching *a* fish, *any* fish on any tackle available.	Any water containing fish. Ease of access to water.	Fishing with family. Seldom take vacations.
Generalists	Catching a *limit* of *trout* on spinning or spincasting tackle.	Lakes, larger freestone streams. Stocking to supplement fish reproduced in streams.	Fishing with peers. Take short vacations within region.
Technique Specialists	Catching large fish on specialized equipment (fly-tackle).	Prefer stream fishing to lake. Harvesting policy to increase size of fish.	Fishing with peers. Take extended fishing vacations.
Technique-Setting Specialists	Catching fish under exacting conditions (on spring streams) with specialized equipment (fly-tackle).	Limestone spring streams. Habitat management, preservation of natural setting.	Fishing with fellow specialists (a reference group). May center lives around sport.

blurred by disproportionate commitment in a particular category. Further, fly-fishermen sometimes specialize in areas other than the spring stream (e.g., high mountain lakes) or they fish for species other than the trout for the greater part of the year.

Yet the consistency and direction of the findings lend validity to an angling specialization framework. A summary of the conclusions is presented in Table 2. What follows are propositions which found support in the research and form the basis for extending the fishermen typology into a conceptual framework. Since the study strategy precluded rigorous control and testing of variables and constitutes a first step for more systematic efforts, these inferences are to be considered tentative until subjected to more controlled testing under varied and representative conditions.

1. Fishermen tend to go through a predictable syndrome of angling experiences, usually moving into more specialized stages over time.[11] But increasing specialization does not necessarily imply narrowing or restriction of

---

[11] Yet some anglers reported that they had been "pushed" into the higher stages of specialization by participation in such activities as fly-fishing schools promoted by major fishing tackle companies. Thus, the promotion of recreational specialization by special interests may serve as an intervening variable in the socialization process.

activities outside the speciality. Instead, an ever-increasing commitment to the sport in general may be found. The more specialized fishermen tend to have high knowledge and commitment to a variety of angling pursuits as an outgrowth of high time and skill commitment to the sport generally.

2. The most specialized fishermen have in effect joined a leisure social world—a group of fellow sportsmen holding similar attitudes, beliefs, and ideologies, engaging in similar behavior, and having a sense of group identification. This leisure social world serves as a major reference group for its members. As a cohesive group it is effective in propounding the values of the so-called minority recreationist.

3. As level of angling specialization increases, attitudes and values about the sport change. Focus shifts from consumption of the fish to preservation and emphasis on the nature and setting of the activity. In short, for the most specialized fishermen the fish are not so much the object as the experience of fishing as an end in itself.

4. The values attendant to specialization are inextricably linked to the properties of the resource on which the sport is practiced. As level of angling specialization increases, resource dependency increases. What appeals to the specialist is a resource setting allowing for predictability and manipulation, a degree of control so as to be able to determine the difference between luck and skill.

The mere listing of leisure proclivities can be unfruitful unless the meaning of a particular activity for the individual and his involvement in it are ascertained. Thus, from the standpoint of this study, to say that someone fishes means little more than at some point in time in his life he has dangled a hook (or cast a net, used a spear, etc.) to capture fish. But the investigator should determine the degree of specialization brought to the activity. Such knowledge may well have predictive utility for orientation to and behavior in the sport. It may also be possible to infer the meaning and significance attached to the activity, leisure orientation in general, and relationship of the sport to occupation and life style. Further, the specialization dimension should have explanatory potential for a variety of other leisure activities.

Implications abound on the applied side of the ledger for recreation managers. What the "quality" experience is to one sportsman is not to another.[12] This implies variability in management strategies for resource utilization to meet variability in recreational orientations and needs (Hendee 1974). If the specialization concept can be applied to a variety of activities, the manager may be provided with a decision-making tool in matching the motivations of users with the appropriate resource. In the case of trout fishermen, the most direct implication is that the different water resources should be managed for those with the most specific motivations for using them. The technique-resource specialist is dependent on the unique attributes of certain types of water to find his satisfaction. Though other anglers may fish these waters, they are not bound to this type of resource for enjoyment of their sport and can be directed to areas better fitted to the more general user.

A reduction in the quality and availability of such unique resources as the spring stream is occurring due to more frequent environmental abuse and less public access.[13] To compound the problem, progression of fishermen

---

[12] Harry, et al. (1972), Hendee (1974), Knopf, et al. (1973), and Talhelm (1973) discuss this issue.

[13] For an elaboration of the problems in preserving this resource, see Bryan (1974b).

Leisure Value Systems

to specialist stages and attendant resource preferences has been hastened by tackle company promotions and other elements in the fishing industry of refined angling techniques and equipment. The fact that managers are faced with the problem of diminishing water resources which appeal to specialists in the face of increasing demand for them makes it all the more necessary to work out rational systems of resource allocation. A framework for prediction and interpretation of what different sportsmen constituencies are seeking in the outdoors is a first step. This can be followed by an inventory of the biological and social attributes of available resources. Subsequently, decisions can be made in matching user demand with resource supply.

References

Brooks, Joe. 1966. *Complete guide to fishing across North America.* New York: Harper and Row.

Bryan, Hobson. 1976a. "The sociology of fishing: a review and critique." In *Marine Recreational Fisheries.* Washington, D.C.: Sport Fishing Institute.

―――――. 1974b. Spring-stream flyfishermen: management implications of a specialized leisure subculture. Paper presented at the American Fisheries Society Meetings, Honolulu, Hawaii, September (mimeographed).

―――――. 1974c. Working at waste in leisure. *Journal of Environmental Education* 6 Fall: 20–23.

Devall, Bill. 1973. The development of leisure social worlds. *Humboldt Journal of Social Relations* 1 Fall: 53–59.

Faunce, William A. 1963. Automation and leisure. In Erwin O. Smigel, ed., *Work and Leisure—A Contemporary Social Problem.* New Haven, Conn.: College and University Press.

Gingrich, Arnold. 1965. Literature of angling. In A. J. McClane, ed., *McClane's Standard Fishing Encyclopedia.* New York: Holt, Rinehart and Winston.

Harry, Joseph, Hendee, John C., and Stein, Robert B. 1972. A sociological criterion for outdoor recreation resource allocation. Paper presented to the Annual Meetings of the American Sociological Association, New Orleans.

Havighurst, Robert J. 1957. The leisure activities of the middle-aged. *American Journal of Sociology* 63 September: 152–162.

Hendee, John C. 1974. A multiple-satisfaction approach to game management. *Wildlife Society Bulletin* 2(3): 104–112.

Kando, Thomas M. 1975. Leisure and popular culture in transition. St. Louis: C. V. Mosby Co.

Kelly, John R. 1974. Socialization toward leisure: a developmental approach. *Journal of Leisure Research* 6(3): 181–193.

Knopf, Richard C., Driver, B. L., and Bassatt, John R. 1973. Motivations for fishing. In John C. Hendee and Clay Schoenfeld, eds., *Human Dimensions in Wildlife Programs.* Rockville, Md.: Mercury Press.

McClane, A. J., ed. 1965. *McClane's standard fishing encyclopedia and international angling guide.* New York: Holt, Rinehart and Winston.

McIntosh, Peter C. 1963. *Sport in society.* London: C. A. Watts & Co. Ltd.

Parker, Stanley. 1971. *The future of work and leisure.* London: Praeger Publishers, Inc.

Roberts, Kenneth. 1970. *Leisure.* London: Longman.

Shibutani, Tamotsu. 1955. Reference groups as perspectives. *American Journal of Sociology* 60(May): 562–569.

Talhelm, Daniel R. 1973. Defining and evaluating recreation quality. In John C. Hendee and Clay Schoenfeld, eds., *Human Dimensions in Wildlife Programs.* Rockville, Md.: Mercury Press.

Waterman, Charles F. 1971. *The fisherman's world.* New York: Random House.

Zetterberg, Hans. 1965. *On theory and verification in sociology,* 2nd. ed. Totawa, N.J.: Bedminster Press.

TRAGEDY OF THE COMMONS?

REVIEW

The Struggle to Govern the Commons

Thomas Dietz,[1] Elinor Ostrom,[2] Paul C. Stern[3]*

Human institutions—ways of organizing activities—affect the resilience of the environment. Locally evolved institutional arrangements governed by stable communities and buffered from outside forces have sustained resources successfully for centuries, although they often fail when rapid change occurs. Ideal conditions for governance are increasingly rare. Critical problems, such as transboundary pollution, tropical deforestation, and climate change, are at larger scales and involve nonlocal influences. Promising strategies for addressing these problems include dialogue among interested parties, officials, and scientists; complex, redundant, and layered institutions; a mix of institutional types; and designs that facilitate experimentation, learning, and change.

In 1968, Hardin (1) drew attention to two human factors that drive environmental change. The first factor is the increasing demand for natural resources and environmental services, stemming from growth in human population and per capita resource consumption. The second factor is the way in which humans organize themselves to extract resources from the environment and eject effluents into it—what social scientists refer to as institutional arrangements. Hardin's work has been highly influential (2) but has long been aptly criticized as oversimplified (3–6).

Hardin's oversimplification was twofold: He claimed that only two state-established institutional arrangements—centralized government and private property—could sustain commons over the long run, and he presumed that resource users were trapped in a commons dilemma, unable to create solutions (7–9). He missed the point that many social groups, including the herders on the commons that provided the metaphor for his analysis, have struggled successfully against threats of resource degradation by developing and maintaining self-governing institutions (3, 10–13). Although these institutions have not always succeeded, neither have Hardin's preferred alternatives of private or state ownership.

In the absence of effective governance institutions at the appropriate scale, natural resources and the environment are in peril from increasing human population, consumption, and deployment of advanced technologies for resource use, all of which have reached unprecedented levels. For example, it is estimated that "the global ocean has lost more than 90% of large predatory fishes" with an 80% decline typically occurring "within 15 years of industrialized exploitation" (14). The threat of massive ecosystem degradation results from an interplay among ocean ecologies, fishing technologies, and inadequate governance.

Inshore fisheries are similarly degraded where they are open access or governed by top-down national regimes, leaving local and regional officials and users with insufficient autonomy and understanding to design effective institutions (15, 16). For example, the degraded inshore ground fishery in Maine is governed by top-down rules based on models that were not credible among users. As a result, compliance has been relatively low and there has been strong resistance to strengthening existing restrictions. This is in marked contrast to the Maine lobster fishery, which has been governed by formal and informal user institutions that have strongly influenced state-level rules that restrict fishing. The result has been credible rules with very high levels of compliance (17–19). A comparison of the landings of ground fish and lobster since 1980 is shown in Fig. 1. The rules and high levels of compliance related to lobster appear to have prevented the destruction of this fishery but probably are not responsible for the sharp rise in abundance and landings after 1986.

Resources at larger scales have also been successfully protected through appropriate international governance regimes such as the Montreal Protocol on stratospheric ozone and the International Commission for the Protection of the Rhine Agreements (20–24). Figure 2 compares the trajectory of atmospheric concentrations of ozone-depleting substances (ODS) with that of carbon dioxide since 1982. The Montreal Protocol, the centerpiece of the international agreements on ozone depletion, was signed in 1987. Before then, ODS concentrations were increasing faster than those of CO_2; the increases slowed by the early 1990s and the concentration appears to have stabilized in recent years. The international treaty regime to reduce the anthropogenic impact on stratospheric ozone is widely considered an example of a successful effort to protect the global commons. In contrast, international efforts to reduce greenhouse gas concentrations have not yet had an impact.

Knowledge from an emerging science of human-environment interactions, sometimes called human ecology or the "second environmental science" (25, 26), is clarifying the characteristics of institutions that facilitate or undermine sustainable use of environmental resources under particular conditions (6, 27). The knowledge base is strongest with small-scale ecologies and institutions, where long time series exist on many successes and failures. It is now developing for larger-scale systems. In this review, we address what science has learned about governing the commons and why it is always a struggle (28).

Why a Struggle?

Devising ways to sustain the earth's ability to support diverse life, including a reasonable quality of life for humans, involves making tough decisions under uncertainty, complexity, and substantial biophysical constraints as well as conflicting human values and interests. Devising effective governance systems is akin to a coevolutionary race. A set of rules crafted to fit one set of socioecological conditions can erode as social, economic, and

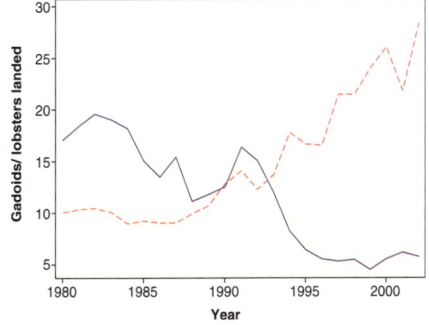

Fig. 1. Comparison of landings of ground fish (gadoids, solid blue line) and lobster (dashed red line) in Maine from 1980 to 2002. Measured in millions of kilograms of ground fish and lobsters landed per year. International fishing in these waters ended with the extended jurisdiction that occurred in 1977 (155).

[1]Environmental Science and Policy Program and Departments of Sociology and Crop and Soil Sciences, Michigan State University, East Lansing, MI 48824, USA. [2]Center for the Study of Institutions, Population, and Environmental Change and Workshop in Political Theory and Policy Analysis, Indiana University, Bloomington, IN 47408, USA. [3]Division of Social and Behavioral Sciences and Education, The National Academies, Washington, DC 20001, USA.

*To whom correspondence should be addressed. E-mail: pstern@nas.edu

Tragedy of the Commons?

technological developments increase the potential for human damage to ecosystems and even to the biosphere itself. Furthermore, humans devise ways of evading governance rules. Thus, successful commons governance requires that rules evolve.

Effective commons governance is easier to achieve when (i) the resources and use of the resources by humans can be monitored, and the information can be verified and understood at relatively low cost (e.g., trees are easier to monitor than fish, and lakes are easier to monitor than rivers) (29); (ii) rates of change in resources, resource-user populations, technology, and economic and social conditions are moderate (30–32); (iii) communities maintain frequent face-to-face communication and dense social networks—sometimes called social capital—that increase the potential for trust, allow people to express and see emotional reactions to distrust, and lower the cost of monitoring behavior and inducing rule compliance (33–36); (iv) outsiders can be excluded at relatively low cost from using the resource (new entrants add to the harvesting pressure and typically lack understanding of the rules); and (v) users support effective monitoring and rule enforcement (37–39). Few settings in the world are characterized by all of these conditions. The challenge is to devise institutional arrangements that help to establish such conditions or, as we discuss below, meet the main challenges of governance in the absence of ideal conditions (6, 40, 41).

Selective Pressures

The characteristics of resources and social interaction in many subsistence societies present favorable conditions for the evolution of effective self-governing resource institutions (13). Hundreds of documented examples exist of long-term sustainable resource use in such communities as well as in more economically advanced communities with effective, local, self-governing rights, but there are also many failures (6, 11, 42–44). As human communities have expanded, the selective pressures on environmental governance institutions increasingly have come from broad influences. Commerce has become regional, national, and global, and institutions at all of these levels have been created to enable and regulate trade, transportation, competition, and conflict (45, 46). These institutions shape environmental impact, even if they are not designed with that intent. They also provide mechanisms for environmental governance (e.g., national laws) and part of the social context for local efforts at environmental governance. Larger scale governance may authorize local control, help it, hinder it, or override it (47–52). Now, every local place is strongly influenced by global dynamics (48, 53–57).

The most important contemporary environmental challenges involve systems that are intrinsically global (e.g., climate change) or are tightly linked to global pressures (e.g., timber production for the world market) and that require governance at levels from the global all the way down to the local (48, 58, 59). These situations often feature environmental outcomes spatially displaced from their causes and hard-to-monitor, larger scale economic incentives that may not be closely aligned with the condition of local ecosystems. Also, differentials in power within user groups or across scales allow some to ignore rules of commons use or to reshape the rules in their own interest, such as when global markets reshape demand for local resources (e.g., forests) in ways that swamp the ability of locally evolved institutions to regulate their use (60–62).

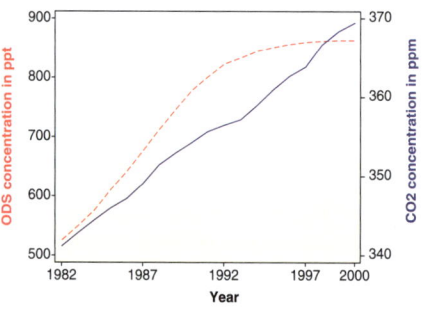

Fig. 2. Atmospheric concentration of CO_2 (solid blue line, right scale) and three principal ODS (dashed red line, left scale). The ODS are chlorofluorocarbons (CFCs) 11, 12, and 113 and were weighted based on their ozone-depleting potential (156). Data are from (157). ppt, parts per trillion; ppm, parts per million.

The store of governance tools and ways to modify and combine them is far greater than often is recognized (6, 63–65). Global and national environmental policy frequently ignores community-based governance and traditional tools, such as informal communication and sanctioning, but these tools can have significant impact (63, 66). Further, no single broad type of ownership—government, private, or community—uniformly succeeds or fails to halt major resource deterioration, as shown for forests in multiple countries (supporting online material text, figs. S1 to S5, and table S1).

Requirements of Adaptive Governance in Complex Systems

Providing information. Environmental governance depends on good, trustworthy information about stocks, flows, and processes within the resource systems being governed, as well as about the human-environment interactions affecting those systems. This information must be congruent in scale with environmental events and decisions (48, 67). Highly aggregated information may ignore or average out local information that is important in identifying future problems and developing solutions.

For example, in 2002, a moratorium on all fishing for northern cod was declared by the Canadian government after a collapse of this valuable fishery. An earlier near-collapse had led Canada to declare a 200-mile zone of exclusive fisheries jurisdiction in 1977 (68, 69). Considerable optimism existed during the 1980s that the stocks, as estimated by fishery scientists, were rebuilding. Consequently, generous total catch limits were established for northern cod and other ground fish, the number of licensed fishers was allowed to increase considerably, and substantial government subsidies were allocated for new vessels (70). What went wrong? There were a variety of information-related problems including: (i) treating all northern cod as a single stock instead of recognizing distinct populations with different characteristics, (ii) ignoring the variability of year classes of northern cod, (iii) focusing on offshore-fishery landing data rather than inshore data to "tune" the stock assessment, and (iv) ignoring inshore fishers who were catching ever-smaller fish and doubted the validity of stock assessments (70–72). This experience illustrates the need to collect and model both local and aggregated information about resource conditions and to use it in making policy at the appropriate scales.

Information also must be congruent with decision makers' needs in terms of timing, content, and form of presentation (73–75). Informational systems that simultaneously meet high scientific standards and serve ongoing needs of decision makers and users are particularly useful. Information must not overload the capacity of users to assimilate it. Systems that adequately characterize environmental conditions or human activities with summary indicators such as prices for products or emission permits, or certification of good environmental performance can provide valuable signals as long as they are attentive to local as well as aggregate conditions (76–78).

Effective governance requires not only factual information about the state of the environment and human actions but also information about uncertainty and values. Scientific understanding of coupled human-biophysical systems will always be uncertain because of inherent unpredictability in the systems and because the science is never complete (79). Decision makers need information that characterizes the types and magnitudes of this uncertainty, as well as the nature and extent of scientific ignorance and disagreement (80). Also, because every environmental decision requires tradeoffs,

TRAGEDY OF THE COMMONS?

knowledge is needed about individual and social values and about the effects of decisions on various valued outcomes. For many environmental systems, local and easily captured values (e.g., the market value of lumber) have to be balanced against global, diffuse, and hard-to-capture values (e.g., biodiversity and the capability of humans and ecosystems to adapt to unexpected events). Finding ways to measure and monitor the outcomes for such varied values in the face of globalization is a major informational challenge for governance.

Dealing with conflict. Sharp differences in power and in values across interested parties make conflict inherent in environmental choices. Indeed, conflict resolution may be as important a motivation for designing resource institutions as is concern with the resources themselves (*81*). People bring varying perspectives, interests, and fundamental philosophies to problems of environmental governance (*74, 82–84*), and their conflicts, if they do not escalate to the point of dysfunction, can spark learning and change (*85, 86*).

For example, a broadly participatory process was used to examine alternative strategies for regulating the Mississippi River and its tributaries (*87*). A dynamic model was constructed with continuous input by the Corps of Engineers, the Fish and Wildlife Service, local landowners, environmental groups, and academics from multiple disciplines. After extensive model development and testing against past historical data, most stakeholders had high confidence in the explanatory power of the model. Consensus was reached over alternative management options, and the resulting policies generated far less conflict than had existed at the outset (*88*).

Delegating authority to environmental ministries does not always resolve conflicts satisfactorily, so governments are experimenting with various governance approaches to complement managerial ones. They range from ballots and polls, where engagement is passive and participants interact minimally, to adversarial processes that allow parties to redress grievances through formal legal procedures, to various experiments with intense interaction and deliberation aimed at negotiating decisions or allowing parties in potential conflict to provide structured input to them through participatory processes (*89–93*).

Inducing rule compliance. Effective governance requires that the rules of resource use are generally followed, with reasonable standards for tolerating modest violations. It is generally most effective to impose modest sanctions on first offenders, and gradually increase the severity of sanctions for those who do not learn from their first or second encounter (*39, 94*). Community-based institutions often use informal strategies for achieving compliance that rely on participants' commitment to rules and subtle social sanctions. Whether enforcement mechanisms are formal or informal, those who impose them must be seen as effective and legitimate by resource users or resistance and evasion will overwhelm the commons governance strategy.

Much environmental regulation in complex societies has been "command and control." Governments require or prohibit specific actions or technologies, with fines or jail terms possible for punishing rule breakers. If sufficient resources are made available for monitoring and enforcement, such approaches are effective. But when governments lack the will or resources to protect "protected areas" (*95–97*), when major environmental damage comes from hard-to-detect "nonpoint sources," and when the need is to encourage innovation in behaviors or technologies rather than to require or prohibit familiar ones, command and control approaches are less effective. They are also economically inefficient in many circumstances (*98–100*).

Financial instruments can provide incentives to achieve compliance with environmental rules. In recent years, market-based systems of tradable environmental allowances (TEAs) that define a limit to environmental withdrawals or emissions and permit free trade of allocated allowances under those limits have become popular (*76, 101, 102*). TEAs are one of the bases for the Kyoto agreement on climate change.

Economic theory and experience in some settings suggest that these mechanisms have substantial advantages over command and control (*103–106*). TEAs have exhibited good environmental performance and economic efficiency in the U.S. Sulfur Dioxide Allowance Market intended to reduce the prevalence of acid rain (*107, 108*) and the Lead Phasedown Program aimed at reducing the level of lead emissions (*109*). Crucial variables that differentiate these highly successful programs from less successful ones, such as chlorofluorocarbon production quota trading and the early EPA emission trading programs, include: (i) the level of predictability of the stocks and flows, (ii) the number of users or producers who are regulated, (iii) the heterogeneity of the regulated users, and (iv) clearly defined and fully exchangeable permits (*110*).

TEAs, like all institutional arrangements, have notable limitations. TEA regimes tend to leave unprotected those resources not specifically covered by trading rules (e.g., by-catch of noncovered fish species) (*111*) and to suffer when monitoring is difficult (e.g., under the Kyoto protocol, the question of whether geologically sequestered carbon will remain sequestered). Problems can also occur with the initial allocation of allowances, especially when historic users, who may be called on to change their behavior most, have disproportionate power over allocation decisions (*76, 101*). TEAs and community-based systems appear to have opposite strengths and weaknesses (*101*), suggesting that institutions that combine aspects of both systems may work better than either approach alone. For example, the fisheries tradable permit system in New Zealand has added comanagement institutions to complement the market institutions (*102, 112*).

Voluntary approaches and those based on information disclosure have only begun to receive careful scientific attention as supplements to other tools (*63, 77, 113–115*). Success appears to depend on the existence of incentives that benefit leaders in volunteering over laggards and on the simultaneous use of other strategies, particularly ones that create incentives for compliance (*77, 116–118*). Difficulties of sanctioning pose major problems for international agreements (*119–121*).

Providing infrastructure. The importance of physical and technological infrastructure is often ignored. Infrastructure, including technology, determines the degree to which a commons can be exploited (e.g., water works and fishing technology), the extent to which waste can be reduced in resource use, and the degree to which resource conditions and the behavior of humans users can be effectively monitored. Indeed, the ability to choose institutional arrangements depends in part on infrastructure. In the absence of barbed-wire fences, for example, enforcing private property rights on grazing lands is expensive, but with barbed wire fences, it is relatively cheap (*122*). Effective communication and transportation technologies are also of immense importance. Fishers who observe an unauthorized boat or harvesting technology can use a radio or cellular phone to alert others to illegal actions (*123*). Infrastructure also affects the links between local commons and regional and global systems. Good roads can provide food in bad times but can also open local resources to global markets, creating demand for resources that cannot be used locally (*124*). Institutional infrastructure is also important, including research, social capital, and multilevel rules, to coordinate between local and larger levels of governance (*48, 125, 126*).

Be prepared for change. Institutions must be designed to allow for adaptation because some current understanding is likely to be wrong, the required scale of organization can shift, and biophysical and social systems change. Fixed rules are likely to fail because they place too much confidence in the current state of knowledge, whereas systems that guard against the low probability, high consequence possibilities and allow for change may be suboptimal in the short run but prove wiser in the long run. This is a principal lesson of adaptive management research (*31, 127*).

TRAGEDY OF THE COMMONS?

Strategies for Meeting the Requirements of Adaptive Governance

The general principles for robust governance institutions for localized resources (Fig. 3) are well established as a result of multiple empirical studies (*13, 39, 128–137*). Many of these also appear to be applicable to regional and global resources (*138*), although they are less well tested at those scales. Three of them seem to be particularly relevant for problems at larger scales.

Analytic deliberation. Well-structured dialogue involving scientists, resource users, and interested publics, and informed by analysis of key information about environmental and human-environment systems, appears critical. Such analytic deliberation (*74, 139, 140*) provides improved information and the trust in it that is essential for information to be used effectively, builds social capital, and can allow for change and deal with inevitable conflicts well enough to produce consensus on governance rules. The negotiated 1994 U.S. regulation on disinfectant by-products in water that reached an interim consensus, including a decision to collect new information and reconsider the rule on that basis (*74*), is an excellent example of this approach.

Nesting. Institutional arrangements must be complex, redundant, and nested in many layers (*32, 141, 142*). Simple strategies for governing the world's resources that rely exclusively on imposed markets or one-level, centralized command and control and that eliminate apparent redundancies in the name of efficiency have been tried and have failed. Catastrophic failures often have resulted when central governments have exerted sole authority over resources. Examples include the massive environmental degradation and impoverishment of local people in Indonesian Borneo (*95*), the increased rate of loss and fragmentation of high-quality habitat that occurred after creating the Wolong Nature Reserve in China (*143*), and the closing of the northern cod fishery along the eastern coast of Canada partly attributable to the excessive quotas granted by the Canadian government (*70*).

Institutional variety. Governance should employ mixtures of institutional types (e.g., hierarchies, markets, and community self-governance) that employ a variety of decision rules to change incentives, increase information, monitor use, and induce compliance (*6, 63, 117*). Innovative rule evaders can have more trouble with a multiplicity of rules than with a single type of rule.

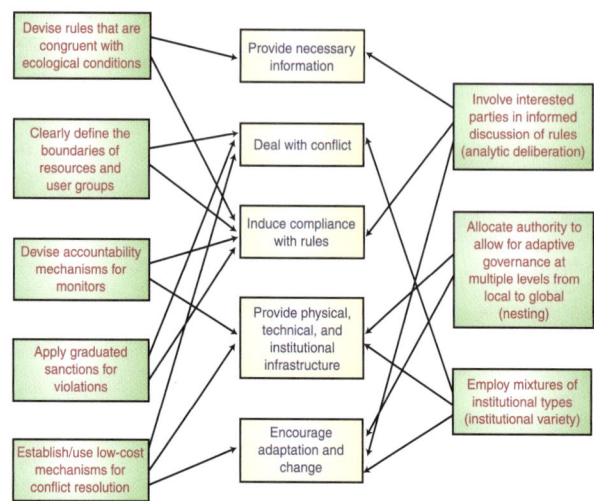

Fig. 3. General principles for robust governance of environmental resources (green, left and right columns) and the governance requirements they help meet (yellow, center column) (*13, 158*). Each principle is relevant for meeting several requirements. Arrows indicate some of the most likely connections between principles and requirements. Principles in the right column may be particularly relevant for global and regional problems.

Conclusion

Is it possible to govern such critical commons as the oceans and the climate? We remain guardedly optimistic. Thirty-five years ago it seemed that the "tragedy of the commons" was inevitable everywhere not owned privately or by a government. Systematic multidisciplinary research has, however, shown that a wide diversity of adaptive governance systems have been effective stewards of many resources. Sustained research coupled to an explicit view of national and international policies as experiments can yield the scientific knowledge necessary to design appropriate adaptive institutions.

Sound science is necessary for commons governance, but not sufficient. Too many strategies for governance of local commons are designed in capital cities or by donor agencies in ignorance of the state of the science and local conditions. The results are often tragic, but at least these tragedies are local. As the human footprint on the Earth enlarges (*144*), humanity is challenged to develop and deploy understanding of large-scale commons governance quickly enough to avoid the large-scale tragedies that will otherwise ensue.

References and Notes
1. G. Hardin, *Science* **162**, 1243 (1968).
2. See (*6, 145*). It was the paper most frequently cited as having the greatest career impact in a recent survey of biologists (*146*). A search performed by L. Wisen on 22 and 23 October 2003 on the Workshop Library Common-Pool Resources database (*147*) revealed that, before Hardin's paper, only 19 articles had been written in English-language academic literature with a specific reference to "commons," "common-pool resources," or "common property" in the title. Since then, attention to the commons has grown rapidly. Since 1968, a total of over 2300 articles in that database contain a specific reference to one of these three terms in the title.
3. B. J. McCay, J. M. Acheson, *The Question of the Commons: The Culture and Ecology of Communal Resources* (Univ. of Arizona Press, Tucson, 1987).
4. P. Dasgupta, *Proc. Br. Acad.* **90**, 165 (1996).
5. D. Feeny, F. Berkes, B. McCay, J. Acheson, *Hum. Ecol.* **18**, 1 (1990).
6. Committee on the Human Dimensions of Global Change, National Research Council, *The Drama of the Commons*, E. Ostrom *et al.*, Eds. (National Academy Press, Washington, DC, 2002).
7. J. Platt, *Am. Psychol.* **28**, 642 (1973).
8. J. G. Cross, M. J. Guyer, *Social Traps* (Univ. of Michigan Press, Ann Arbor, 1980).
9. R. Costanza, *Bioscience* **37**, 407 (1987).
10. R. McC. Netting, *Balancing on an Alp: Ecological Change and Continuity in a Swiss Mountain Community* (Cambridge Univ. Press, Cambridge, 1981).
11. National Research Council, *Proceedings of the Conference on Common Property Resource Management* (National Academy Press, Washington, DC, 1986).
12. J.-M. Baland, J.-P. Platteau, *Halting Degradation of Natural Resources: Is There a Role for Rural Communities?* (Clarendon Press, Oxford, 1996).
13. E. Ostrom, *Governing the Commons: The Evolution of Institutions for Collective Action* (Cambridge Univ. Press, New York, 1990).
14. R. A. Myers, B. Worm, *Nature* **423**, 280 (2003).
15. A. C. Finlayson, *Fishing for Truth: A Sociological Analysis of Northern Cod Stock Assessments from 1987 to 1990* (Institute of Social and Economic Research, Memorial Univ. of Newfoundland, St. Johns, Newfoundland, 1994).
16. S. Hanna, in *Northern Waters: Management Issues and Practice*, D. Symes, Ed. (Blackwell, London, 1998), pp. 25–35.
17. J. Acheson, *Capturing the Commons: Devising Institutions to Manage the Maine Lobster Industry* (Univ. Press of New England, Hanover, NH, 2003).
18. J. A. Wilson, P. Kleban, J. Acheson, M. Metcalfe, *Mar. Policy* **18**, 291 (1994).
19. J. Wilson, personal communication.
20. S. Weiner, J. Maxwell, in *Dimensions of Managing Chlorine in the Environment*, report of the MIT/Norwegian Chlorine Policy Study (MIT, Cambridge, MA, 1993).
21. U. Weber, *UNESCO Courier*, June 2000, p. 9.
22. M. Verweij, *Transboundary Environmental Problems and Cultural Theory: The Protection of the Rhine and the Great Lakes* (Palgrave, New York, 2000).
23. C. Dieperink, *Water Int.* **25**, 347 (2000).
24. E. Parson, *Protecting the Ozone Layer: Science and Strategy* (Oxford Univ. Press, New York, 2003).
25. E. Ostrom, C. D. Becker, *Annu. Rev. Ecol. Syst.* **26**, 113 (1995).

26. P. C. Stern, *Science* **260**, 1897 (1993).
27. E. Ostrom, J. Burger, C. B. Field, R. B. Norgaard, D. Policansky, *Science* **284**, 278 (1999).
28. We refer to adaptive governance rather than adaptive management (*31, 127*) because the idea of governance conveys the difficulty of control, the need to proceed in the face of substantial uncertainty, and the importance of dealing with diversity and reconciling conflict among people and groups who differ in values, interests, perspectives, power, and the kinds of information they bring to situations (*139, 148–151*). Effective environmental governance requires an understanding of both environmental systems and human-environment interactions (*26, 82, 152, 153*).
29. E. Schlager, W. Blomquist, S. Y. Tang, *Land Econ.* **70**, 294 (1994).
30. J. H. Brander, M. S. Taylor, *Am. Econ. Rev.* **88**, 119 (1998).
31. L. H. Gunderson, C. S. Holling, *Panarchy: Understanding Transformations in Human and Natural Systems* (Island Press, Washington, DC, 2001).
32. M. Janssen, *Complexity and Ecosystem Management* (Elgar, Cheltenham, UK, 2002).
33. R. Putnam, *Bowling Alone: The Collapse and Revival of American Community* (Simon and Schuster, New York, 2001).
34. A. Bebbington, *Geogr. J.* **163**, 189 (1997).
35. R. Frank, *Passions Within Reason: The Strategic Role of the Emotions* (Norton, New York, 1988).
36. J. Pretty, *Science* **302**, 1912 (2003).
37. J. Burger, E. Ostrom, R. B. Norgaard, D. Policansky, B. D. Goldstein, Eds., *Protecting the Commons: A Framework for Resource Management in the Americas* (Island Press, Washington, DC, 2001).
38. C. Gibson, J. Williams, E. Ostrom, in preparation.
39. M. S. Weinstein, *Georgetown Int. Environ. Law Rev.* **12**, 375 (2000).
40. R. Meinzen-Dick, K. V. Raju, A. Gulati, *World Dev.* **30**, 649 (2002).
41. E. L. Miles et al., Eds., *Environmental Regime Effectiveness: Confronting Theory with Evidence* (MIT Press, Cambridge, MA, 2001).
42. C. Gibson, M. McKean, E. Ostrom, Eds., *People and Forests* (MIT Press, Cambridge, MA, 2000).
43. S. Krech III, *The Ecological Indian: Myth and History* (Norton, New York, 1999).
44. For relevant bibliographies, see (*147, 154*).
45. D. C. North, *Structure and Change in Economic History* (North, New York, 1981).
46. R. Robertson, *Globalization: Social Theory and Global Culture* (Sage, London, 1992).
47. O. R. Young, Ed., *The Effectiveness of International Environmental Regimes* (MIT Press, Cambridge, MA, 1999).
48. O. R. Young, *The Institutional Dimensions of Environmental Change: Fit, Interplay, and Scale* (MIT Press, Cambridge, MA, 2002).
49. R. Keohane, E. Ostrom, Eds., *Local Commons and Global Interdependence* (Sage, London, 1995).
50. J. S. Lansing, *Priests and Programmers: Technologies of Power in the Engineered Landscape of Bali* (Princeton Univ. Press, Princeton, NJ, 1991).
51. J. Wunsch, D. Olowu, Eds., *The Failure of the Centralized State* (Institute for Contemporary Studies Press, San Francisco, CA, 1995).
52. N. Dolšak, E. Ostrom, Eds., *The Commons in the New Millennium: Challenges and Adaptation* (MIT Press, Cambridge, MA, 2003).
53. Association of American Geographers Global Change and Local Places Research Group, *Global Change and Local Places: Estimating, Understanding, and Reducing Greenhouse Gases* (Cambridge Univ. Press, Cambridge, 2003).
54. S. Karlsson, thesis, Linköping University, Sweden (2000).
55. R. Keohane, M. A. Levy, Eds., *Institutions for Environmental Aid* (MIT Press, Cambridge, MA, 1996).
56. O. S. Stokke, *Governing High Seas Fisheries: The Interplay of Global and Regional Regimes* (Oxford Univ. Press, London, 2001).
57. A. Underdal, K. Hanf, Eds., *International Environmental Agreements and Domestic Politics: The Case of Acid Rain* (Ashgate, Aldershot, England, 1998).

58. W. Clark, R. Munn, Eds., *Sustainable Development of the Biosphere* (Cambridge Univ. Press, New York, 1986).
59. B. L. Turner II et al., *Global Environ. Change* **1**, 14 (1991).
60. T. Dietz, T. R. Burns, *Acta Sociol.* **35**, 187 (1992).
61. T. Dietz, E. A. Rosa, in *Handbook of Environmental Sociology*, R. E. Dunlap, W. Michelson, Eds. (Greenwood Press, Westport, CT, 2002), pp. 370–406.
62. A. P. Vayda, in *Ecology in Practice*, F. di Castri et al., Eds. (Tycooly, Dublin, 1984).
63. Committee on the Human Dimensions of Global Change, National Research Council, *New Tools for Environmental Protection: Education, Information, and Voluntary Measures*, T. Dietz, P. C. Stern, Eds. (National Academy Press, Washington, DC, 2002).
64. M. Auer, *Policy Sci.* **33**, 155 (2000).
65. D. H. Cole, *Pollution and Property: Comparing Ownership Institutions for Environmental Protection* (Cambridge Univ. Press, Cambridge, 2002).
66. F. Berkes, J. Colding, C. Folke, Eds., *Navigating Social-Ecological Systems: Building Resilience for Complexity and Change* (Cambridge Univ. Press, Cambridge, 2003).
67. K. J. Willis, R. J. Whittaker, *Science* **295**, 1245 (2002).
68. Kirby Task Force on Atlantic Fisheries, *Navigating Troubled Waters: A New Policy for the Atlantic Fisheries* (Department of Fisheries and Oceans, Ottawa, 1982).
69. G. Barrett, A. Davis, *J. Can. Stud.* **19**, 125 (1984).
70. A. C. Finlayson, B. McCay, in *Linking Social and Ecological Systems*, F. Berkes, C. Folke, Eds. (Cambridge Univ. Press, Cambridge, 1998), pp. 311–338.
71. J. A. Wilson, R. Townsend, P. Kleban, S. McKay, J. French, *Ocean Shoreline Manage.* **13**, 179 (1990).
72. C. Martin, *Fisheries* **20**, 6 (1995).
73. Committee on Risk Perception and Communication, National Research Council, *Improving Risk Communication* (National Academy Press, Washington, DC, 1989).
74. Committee on Risk Characterization and Commission on Behavioral and Social Sciences and Education, National Research Council, *Understanding Risk: Informing Decisions in a Democratic Society*, P. C. Stern, H. V. Fineberg, Eds. (National Academy Press, Washington, DC, 1996).
75. Panel on Human Dimensions of Seasonal-to-Interannual Climate Variability, Committee on the Human Dimensions of Global Change, National Research Council, *Making Climate Forecasts Matter*, P. C. Stern, W. E. Easterling, Eds. (National Academy Press, Washington, DC, 1999).
76. T. Tietenberg, in *The Drama of the Commons*, Committee on the Human Dimensions of Global Change, National Research Council, E. Ostrom et al., Eds. (National Academy Press, Washington, DC, 2002), pp. 233–257.
77. T. Tietenberg, D. Wheeler, in *Frontiers of Environmental Economics*, H. Folmer, H. Landis Gabel, S. Gerking, A. Rose, Eds. (Elgar, Cheltenham, UK, 2001), pp. 85–120.
78. J. Thøgerson, in *New Tools for Environmental Protection: Education, Information, and Voluntary Measures*, T. Dietz, P. C. Stern, Eds. (National Academy Press, Washington, DC, 2002), pp. 83–104.
79. J. A. Wilson, in *The Drama of the Commons*, Committee on the Human Dimensions of Global Change, National Research Council, E. Ostrom et al., Eds. (National Academy Press, Washington, DC, 2002), pp. 327–360.
80. R. Moss, S. H. Schneider, in *Guidance Papers on the Cross-Cutting Issues of the Third Assessment Report of the IPCC*, R. Pachauri, T. Taniguchi, K. Tanaka, Eds. (World Meteorological Organization, Geneva, Switzerland, 2000), pp. 33–51.
81. B. J. McCay, in *The Drama of the Commons*, Committee on the Human Dimensions of Global Change, National Research Council, E. Ostrom et al., Eds. (National Academy Press, Washington, DC, 2002), pp. 361–402.
82. Board on Sustainable Development, National Research Council, *Our Common Journey: A Transition Toward Sustainability* (National Academy Press, Washington, DC, 1999).

83. Committee on Noneconomic and Economic Value of Biodiversity, National Research Council, *Perspectives on Biodiversity: Valuing Its Role in an Ever-changing World* (National Academy Press, Washington, DC, 1999).
84. W. M. Adams, D. Brockington, J. Dyson, B. Vira, *Science* **302**, 1915 (2003).
85. P. C. Stern, *Policy Sci.* **24**, 99 (1991).
86. V. Ostrom, *Public Choice* **77**, 163 (1993).
87. R. Costanza, M. Ruth, in *Institutions, Ecosystems, and Sustainability*, R. Costanza, B. S. Low, E. Ostrom, J. Wilson, Eds. (Lewis Publishers, Boca Raton, FL, 2001), pp. 169–178.
88. F. H. Sklar, M. L. White, R. Costanza, *The Coastal Ecological Landscape Spatial Simulation (CELSS) Model* (U.S. Fish and Wildlife Service, Washington, DC, 1989).
89. O. Renn, T. Webler, P. Wiedemann, Eds., *Fairness and Competence in Citizen Participation: Evaluating Models for Environmental Discourse* (Kluwer Academic Publishers, Dordrecht, Netherlands, 1995).
90. R. Gregory, T. McDaniels, D. Fields, *J. Policy Anal. Manage.* **20**, 415 (2001).
91. T. C. Beierle, J. Cayford, *Democracy in Practice: Public Participation in Environmental Decisions* (Resources for the Future, Washington, DC, 2002).
92. W. Leach, N. Pelkey, P. Sabatier, *J. Policy Anal. Manage.* **21**, 645 (2002).
93. R. O'Leary, L. B. Bingham, Eds., *The Promise and Performance of Environmental Conflict Resolution* (Resources for the Future, Washington, DC, 2003).
94. E. Ostrom, R. Gardner, J. Walker, Eds., *Rules, Games, and Common-Pool Resources* (Univ. of Michigan Press, Ann Arbor, 1994).
95. L. M. Curran et al., in preparation.
96. J. Liu et al., *Science* **300**, 1240 (2003).
97. R. W. Sussman, G. M. Green, L. K. Sussman, *Hum. Ecol.* **22**, 333 (1994).
98. F. Berkes, C. Folke, Eds., *Linking Social and Ecological Systems: Management Practices and Social Mechanisms* (Cambridge Univ. Press, Cambridge, 1998).
99. G. M. Heal, *Valuing the Future: Economic Theory and Sustainability* (Colombia Univ. Press, New York, 1998).
100. B. G. Colby, in *The Handbook of Environmental Economics*, D. Bromley, Ed. (Blackwell Publishers, Oxford, 1995), pp. 475–502.
101. C. Rose, in *The Drama of the Commons*, Committee on the Human Dimensions of Global Change, National Research Council, E. Ostrom et al., Eds. (National Academy Press, Washington, DC, 2002), pp. 233–257.
102. T. Yandle, C. M. Dewees, in *The Commons in the New Millennium: Challenges and Adaptation*, N. Dolšak, E. Ostrom, Eds. (MIT Press, Cambridge, MA, 2003), pp. 101–128.
103. G. Libecap, *Contracting for Property Rights* (Cambridge Univ. Press, Cambridge, 1990).
104. R. D. Lile, D. R. Bohi, D. Burtraw, *An Assessment of the EPA's SO_2 Emission Allowance Tracking System* (Resources for the Future, Washington, DC, 1996).
105. R. N. Stavins, *J. Econ. Perspect.* **12**, 133 (1998).
106. J. E. Wilen, *J. Environ. Econ. Manage.* **39**, 309 (2000).
107. A. D. Ellerman, R. Schmalensee, P. L. Joskow, J. P. Montero, E. M. Bailey, *Emissions Trading Under the U.S. Acid Rain Program* (MIT Center for Energy and Environmental Policy Research, Cambridge, MA, 1997).
108. E. M. Bailey, "Allowance trading activity and state regulatory rulings" (Working Paper 98-005, MIT Emissions Trading, Cambridge, MA, 1998).
109. B. D. Nussbaum, in *Climate Change: Designing a Tradeable Permit System* (OECD, Paris, 1992), pp. 22–34.
110. N. Dolšak, thesis, Indiana University, Bloomington, IN (2000).
111. S. L. Hsu, J. E. Wilen, *Ecol. Law Q.* **24**, 799 (1997).
112. E. Pinkerton, *Co-operative Management of Local Fisheries* (Univ. of British Columbia Press, Vancouver, 1989).
113. A. Prakash, *Bus. Strategy Environ.* **10**, 286 (2001).
114. J. Nash, in *New Tools for Environmental Protection: Education, Information and Voluntary Measures*, T. Dietz, P. C. Stern, Eds. (National Academy Press, Washington, DC, 2002), pp. 235–252.

TRAGEDY OF THE COMMONS?

115. J. A. Aragón-Correa, S. Sharma, *Acad. Manage. Rev.* **28**, 71 (2003).
116. A. Randall, in *New Tools for Environmental Protection: Education, Information and Voluntary Measures*, T. Dietz, P. C. Stern, Eds. (National Academy Press, Washington, DC, 2002), pp. 311–318.
117. G. T. Gardner, P. C. Stern, *Environmental Problems and Human Behavior* (Allyn and Bacon, Needham Heights, MA, 1996).
118. P. C. Stern, *J. Consum. Policy* **22**, 461 (1999).
119. S. Hanna, C. Folke, K.-G. Mäler, *Rights to Nature* (Island Press, Washington, DC, 1996).
120. E. Weiss, H. Jacobson, Eds., *Engaging Countries: Strengthening Compliance with International Environmental Agreements* (MIT Press, Cambridge, MA, 1998).
121. A. Underdal, *The Politics of International Environmental Management* (Kluwer Academic Publishers, Dordrecht, Netherlands, 1998).
122. A. Krell, *The Devil's Rope: A Cultural History of Barbed Wire* (Reaktion, London, 2002).
123. S. Singleton, *Constructing Cooperation: The Evolution of Institutions of Comanagement* (Univ. of Michigan Press, Ann Arbor, 1998).
124. E. Moran, Ed., *The Ecosystem Approach in Anthropology: From Concept to Practice* (Univ. of Michigan Press, Ann Arbor, 1990).
125. M. Janssen, J. M. Anderies, E. Ostrom, paper presented at the Workshop on Resiliency and Change in Ecological Systems, Santa Fe Institute, Santa Fe, NM, 25 to 27 October 2003.
126. T. Princen, *Global Environ. Polit.* **3**, 33 (2003).
127. K. Lee, *Compass and Gyroscope* (Island Press, Washington, DC, 1993).
128. C. L. Abernathy, H. Sally, *J. Appl. Irrig. Stud.* **35**, 177 (2000).
129. A. Agrawal, in *The Drama of the Commons*, Committee on the Human Dimensions of Global Change, National Research Council, E. Ostrom *et al.*, Eds. (National Academy Press, Washington, DC, 2002), pp. 41–85.
130. P. Coop, D. Brunckhorst, *Aust. J. Environ. Manage.* **6**, 48 (1999).
131. D. S. Crook, A. M. Jones, *Mt. Res. Dev.* **19**, 79 (1999).
132. D. J. Merrey, in *Irrigation Management Transfer*, S. H. Johnson, D. L. Vermillion, J. A. Sagardoy, Eds. (International Irrigation Management Institute, Colombo, Sri Lanka and the Food and Agriculture Organisation, Rome, 1995).
133. C. E. Morrow, R. W. Hull, *World Dev.* **24**, 1641 (1996).
134. T. Nilsson, thesis, Royal Institute of Technology, Stockholm, Sweden (2001).
135. N. Polman, L. Slangen, in *Environmental Co-operation and Institutional Change*, K. Hagedorn, Ed. (Elgar, Northampton, MA, 2002).
136. A. Sarker, T. Itoh, *Agric. Water Manage.* **48**, 89 (2001).
137. C. Tucker, *Praxis* **15**, 47 (1999).
138. R. Costanza *et al.*, *Science* **281**, 198 (1998).
139. T. Dietz, P. C. Stern, *Bioscience* **48**, 441 (1998).
140. E. Rosa, A. M. McWright, O. Renn, "The risk society: Theoretical frames and state management challenges" (Dept. of Sociology, Washington State Univ., Pullman, WA, 2003).
141. S. Levin, *Fragile Dominion: Complexity and the Commons* (Perseus Books, Reading, MA, 1999).
142. B. Low, E. Ostrom, C. Simon, J. Wilson, in *Navigating Social-Ecological Systems: Building Resilience for Complexity and Change*, F. Berkes, J. Colding, C. Folke, Eds. (Cambridge Univ. Press, New York, 2003), pp. 83–114.
143. J. Liu *et al.*, *Science* **292**, 98 (2001).
144. R. York, E. A. Rosa, T. Dietz, *Am. Sociol. Rev.* **68**, 279 (2003).
145. G. Hardin, *Science* **280**, 682 (1998).
146. G. W. Barrett, K. E. Mabry, *Bioscience* **52**, 282 (2002).
147. C. Hess, *The Comprehensive Bibliography of the Commons*, database available online at www.indiana.edu/~iascp/Iforms/searchcpr.html.
148. V. Ostrom, *The Meaning of Democracy and the Vulnerability of Democracies* (Univ. of Michigan Press, Ann Arbor, 1997).
149. M. McGinnis, Ed., *Polycentric Governance and Development: Readings from the Workshop in Political Theory and Policy Analysis* (Univ. of Michigan Press, Ann Arbor, 1999).
150. M. McGinnis, Ed., *Polycentric Games and Institutions: Readings from the Workshop in Political Theory and Policy Analysis* (Univ. of Michigan Press, Ann Arbor, 2000).
151. T. Dietz, *Hum. Ecol. Rev.* **10**, 60 (2003).
152. R. Costanza, B. S. Low, E. Ostrom, J. Wilson, Eds., *Institutions, Ecosystems, and Sustainability* (Lewis Publishers, New York, 2001).
153. Committee on the Human Dimensions of Global Change, National Research Council, *Global Environmental Change: Understanding the Human Dimensions*, P. C. Stern, O. R. Young, D. Druckman, Eds. (National Academy Press, Washington, DC, 1992).
154. C. Hess, *A Comprehensive Bibliography of Common-Pool Resources* (CD-Rom, Workshop in Political Theory and Policy Analysis, Indiana Univ., Bloomington, 1999).
155. Ground fish data were compiled by D. Gilbert (Maine Department of Marine Resources) with data from the National Marine Fisheries Service. Lobster data were compiled by C. Wilson (Maine Department of Marine Resources). J. Wilson (University of Maine) worked with the authors in the preparation of this figure.
156. United Nations Environment Programme, *Production and Consumption of Ozone Depleting Substances, 1986–1998* (United Nations Environment Programme Ozone Secretariat, Nairobi, Kenya, 1999).
157. World Resources Institute, *World Resources 2002–2004: EarthTrends Data CD* (World Resources Institute, Washington, DC, 2003).
158. P. C. Stern, T. Dietz, E. Ostrom, *Environ. Pract.* **4**, 61 (2002).
159. We thank R. Andrews, G. Daily, J. Hoehn, K. Lee, S. Levin, G. Libecap, V. Ruttan, T. Tietenberg, J. Wilson, and O. Young for their comments on earlier drafts; and G. Laasby, P. Lezotte, C. Liang, and L. Wisen for providing assistance. Supported in part by NSF grants BCS-9906253 and SBR-9521918, NASA grant NASW-01008, the Ford Foundation, and the MacArthur Foundation.

Supporting Online Material
www.sciencemag.org/cgi/content/full/302/5652/1907/DC1
SOM Text
Fig. S1 to S5
Table S1

Web Resources
www.sciencemag.org/cgi/content/full/302/5652/1907/DC2

THE ECONOMIC THEORY OF A COMMON-PROPERTY RESOURCE: THE FISHERY[1]

H. SCOTT GORDON

Carleton College, Ottawa, Ontario

I. INTRODUCTION

THE chief aim of this paper is to examine the economic theory of natural resource utilization as it pertains to the fishing industry. It will appear, I hope, that most of the problems associated with the words "conservation" or "depletion" or "overexploitation" in the fishery are, in reality, manifestations of the fact that the natural resources of the sea yield no economic rent. Fishery resources are unusual in the fact of their common-property nature; but they are not unique, and similar problems are encountered in other cases of common-property resource industries, such as petroleum production, hunting and trapping, etc. Although the theory presented in the following pages is worked out in terms of the fishing industry, it is, I believe, applicable generally to all cases where natural resources are owned in common and exploited under conditions of individualistic competition.

II. BIOLOGICAL FACTORS AND THEORIES

The great bulk of the research that has been done on the primary production phase of the fishing industry has so far been in the field of biology. Owing to the lack of theoretical economic research,[2] biologists have been forced to extend the scope of their own thought into the economic sphere and in some cases have penetrated quite deeply, despite the lack of the analytical tools of economic theory.[3] Many others, who have paid no specific attention to the economic aspects of the problem have nevertheless recognized that the ultimate question is not the ecology of life in the sea as such, but man's use of these resources for his own (economic) purposes. Dr. Martin D. Burkenroad, for example, began a recent article on fishery management with a section on "Fishery Management as Political Economy," saying that "the Management of fisheries is intended for the benefit of man, not fish; therefore effect of management upon fishstocks cannot be regarded as beneficial *per se*."[4] The

---

[1] I want to express my indebtedness to the Canadian Department of Fisheries for assistance and co-operation in making this study; also to Professor M. C. Urquhart, of Queen's University, Kingston, Ontario, for mathematical assistance with the last section of the paper and to the Economists' Summer Study Group at Queen's for affording opportunity for research and discussion.

[2] The single exception that I know is G. M. Gerhardsen, "Production Economics in Fisheries," *Revista de economía* (Lisbon), March, 1952.

[3] Especially remarkable efforts in this sense are Robert A. Nesbit, "Fishery Management" ("U.S. Fish and Wildlife Service, Special Scientific Reports," No. 18 [Chicago, 1943]) (mimeographed), and Harden F. Taylor, *Survey of Marine Fisheries of North Carolina* (Chapel Hill, 1951); also R. J. H. Beverton, "Some Observations on the Principles of Fishery Regulation," *Journal du conseil permanent international pour l'exploration de la mer* (Copenhagen), Vol. XIX, No. 1 (May, 1953); and M. D. Burkenroad, "Some Principles of Marine Fishery Biology," *Publications of the Institute of Marine Science* (University of Texas), Vol. II, No. 1 (September, 1951).

[4] "Theory and Practice of Marine Fishery Management," *Journal du conseil permanent international pour l'exploration de la mer*, Vol. XVIII, No. 3 (January, 1953).

THEORY OF A COMMON-PROPERTY RESOURCE

great Russian marine biology theorist, T. I. Baranoff, referred to his work as "bionomics" or "bio-economics," although he made little explicit reference to economic factors.[5] In the same way, A. G. Huntsman, reporting in 1944 on the work of the Fisheries Research Board of Canada, defined the problem of fisheries depletion in economic terms: "Where the take in proportion to the effort fails to yield a satisfactory living to the fisherman";[6] and a later paper by the same author contains, as an incidental statement, the essence of the economic optimum solution without, apparently, any recognition of its significance.[7] Upon the occasion of its fiftieth anniversary in 1952, the International Council for the Exploration of the Sea published a *Rapport Jubilaire*, consisting of a series of papers summarizing progress in various fields of fisheries research. The paper by Michael Graham on "Overfishing and Optimum Fishing," by its emphatic recognition of the economic criterion, would lead one to think that the economic aspects of the question had been extensively examined during the last half-century. But such is not the case. Virtually no specific research into the economics of fishery resource utilization has been undertaken. The present state of knowledge is that a great deal is known about the biology of the various commercial species but little about the economic characteristics of the fishing industry.

The most vivid thread that runs through the biological literature is the effort to determine the effect of fishing on the stock of fish in the sea. This discussion has had a very distinct practical orientation, being part of the effort to design regulative policies of a "conservation" nature. To the layman the problem appears to be dominated by a few facts of overriding importance. The first of these is the prodigious reproductive potential of most fish species. The adult female cod, for example, lays millions of eggs at each spawn. The egg that hatches and ultimately reaches maturity is the great exception rather than the rule. The various herrings (Clupeidae) are the most plentiful of the commercial species, accounting for close to half the world's total catch, as well as providing food for many other sea species. Yet herring are among the smallest spawners, laying a mere hundred thousand eggs a season, which, themselves, are eaten in large quantity by other species. Even in inclosed waters the survival and reproductive powers of fish appear to be very great. In 1939 the Fisheries Research Board of Canada deliberately tried to kill all the fish in one small lake by poisoning the water. Two years later more than ninety thousand fish were found in the lake, including only about six hundred old enough to have escaped the poisoning.

The picture one gets of life in the sea is one of constant predation of one species on another, each species living on a narrow margin of food supply. It reminds the economist of the Malthusian law of population; for, unlike man, the

---

[5] Two of Baranoff's most important papers—"On the Question of the Biological Basis of Fisheries" (1918) and "On the Question of the Dynamics of the Fishing Industry" (1925)—have been translated by W. E. Ricker, now of the Fisheries Research Board of Canada (Nanaimo, B.C.), and issued in mimeographed form.

[6] "Fishery Depletion," *Science*, XCIX (1944), 534.

[7] "The highest take is not necessarily the best. The take should be increased only as long as the extra cost is offset by the added revenue from sales" (A. G. Huntsman, "Research on Use and Increase of Fish Stocks," *Proceedings of the United Nations Scientific Conference on the Conservation and Utilization of Resources* [Lake Success, 1949]).

H. SCOTT GORDON

fish has no power to alter the conditions of his environment and consequently cannot progress. In fact, Malthus and his law are frequently mentioned in the biological literature. One's first reaction is to declare that environmental factors are so much more important than commercial fishing that man has no effect on the population of the sea at all. One of the continuing investigations made by fisheries biologists is the determination of the age distribution of catches. This is possible because fish continue to grow in size with age, and seasonal changes are reflected in certain hard parts of their bodies in much the same manner as one finds growth-rings in a tree. The study of these age distributions shows that commercial catches are heavily affected by good and bad brood years. A good brood year, one favorable to the hatching of eggs and the survival of fry, has its effect on future catches, and one can discern the dominating importance of that brood year in the commercial catches of succeeding years.[8] Large broods, however, do not appear to depend on large numbers of adult spawners, and this lends support to the belief that the fish population is entirely unaffected by the activity of man.

There is, however, important evidence to the contrary. World Wars I and II, during which fishing was sharply curtailed in European waters, were followed by indications of a significant growth in fish populations. Fish-marking experiments, of which there have been a great number, indicate that fishing is a major cause of fish mortality in developed fisheries. The introduction of restrictive laws has often been followed by an increase in fish populations, although the evidence on this point is capable of other interpretations which will be noted later.

General opinion among fisheries biologists appears to have had something of a cyclical pattern. During the latter part of the last century, the Scottish fisheries biologist, W. C. MacIntosh,[9] and the great Darwinian, T. H. Huxley, argued strongly against all restrictive measures on the basis of the inexhaustible nature of the fishery resources of the sea. As Huxley put it in 1883: "The cod fishery, the herring fishery, the pilchard fishery, the mackerel fishery, and probably all the great sea fisheries, are inexhaustible: that is to say that nothing we do seriously affects the number of fish. And any attempt to regulate these fisheries seems consequently, from the nature of the case, to be useless."[10] As a matter of fact, there was at this time relatively little restriction of fishing in European waters. Following the Royal Commission of 1866, England had repealed a host of restrictive laws. The development of steam-powered trawling in the 1880's, which enormously increased man's predatory capacity, and the marked improvement of the trawl method in 1923 turned the pendulum, and throughout the interwar years discussion centered on the problem of "overfishing" and "depletion." This was accompanied by a considerable growth of restrictive regula-

---

[8] One example of a very general phenomenon: 1904 was such a successful brood year for Norwegian herrings that the 1904 year class continued to outweigh all others in importance in the catch from 1907 through to 1919. The 1904 class was some thirty times as numerous as other year classes during the period (Johan Hjort, "Fluctuations in the Great Fisheries of Northern Europe," *Rapports et procès-verbaux, Conseil permanent international pour l'exploration de la mer*, Vol. XX [1914]; see also E. S. Russell, *The Overfishing Problem* [Cambridge, 1942], p. 57).

[9] See his *Resources of the Sea* published in 1899.

[10] Quoted in M. Graham, *The Fish Gate* (London, 1943), p. 111; see also T. H. Huxley, "The Herring," *Nature* (London), 1881.

THEORY OF A COMMON-PROPERTY RESOURCE

tions.[11] Only recently has the pendulum begun to reverse again, and there has lately been expressed in biological quarters a high degree of skepticism concerning the efficacy of restrictive measures, and the Huxleyian faith in the inexhaustibility of the sea has once again begun to find advocates. In 1951 Dr. Harden F. Taylor summarized the over-all position of world fisheries in the following words:

> Such statistics of world fisheries as are available suggest that while particular species have fluctuated in abundance, the *yield of the sea fisheries as a whole or of any considerable region has not only been sustained, but has generally increased with increasing human populations*, and there is as yet no sign that they will not continue to do so. No single species so far as we know has ever become extinct, and no regional fishery in the world has ever been exhausted.[12]

In formulating governmental policy, biologists appear to have had a hard struggle (not always successful) to avoid oversimplification of the problem. One of the crudest arguments to have had some support is known as the "propagation theory," associated with the name of the English biologist, E. W. L. Holt.[13] Holt advanced the proposition that legal size limits should be established at a level that would permit every individual of the species in question to spawn at least once. This suggestion was effectively demolished by the age-distribution studies whose results have been noted above. Moreover, some fisheries, such as the "sardine" fishery of the Canadian Atlantic Coast, are specifically for *immature* fish. The history of this particular fishery shows no evidence whatever that the landings have been in any degree reduced by the practice of taking very large quantities of fish of prespawning age year after year.

The state of uncertainty in biological quarters around the turn of the century is perhaps indicated by the fact that Holt's propagation theory was advanced concurrently with its diametric opposite: "the thinning theory" of the Danish biologist, C. G. J. Petersen.[14] The latter argued that the fish may be too plentiful for the available food and that thinning out the young by fishing would enable the remainder to grow more rapidly. Petersen supported his theory with the results of transplanting experiments which showed that the fish transplanted to a new habitat frequently grew much more rapidly than before. But this is equivalent to arguing that the reason why rabbits multiplied so rapidly when introduced to Australia is because there were no rabbits already there with which they had to compete for food. Such an explanation would neglect all the other elements of importance in a natural ecology. In point of fact, in so far as food alone is concerned, thinning a cod population, say by half, would not double the food supply of the remaining individuals; for there are other species, perhaps not commercially valuable, that use the same food as the cod.

Dr. Burkenroad's comment, quoted earlier, that the purpose of practical policy is the benefit of man, not fish, was not gratuitous, for the argument has at times been advanced that commercial fishing should crop the resource in such a way as to leave the stocks of fish in the sea completely unchanged. Baranoff was largely responsible for destroying this

---

[11] See H. Scott Gordon, "The Trawler Question in the United Kingdom and Canada," *Dalhousie Review*, summer, 1951.

[12] Taylor, *op. cit.*, p. 314 (Dr. Taylor's italics).

[13] See E. W. L. Holt, "An Examination of the Grimsby Trawl Fishery," *Journal of the Marine Biological Association* (Plymouth), 1895.

[14] See C. G. J. Petersen, "What Is Overfishing?" *Journal of the Marine Biological Association* (Plymouth), 1900–1903.

approach, showing most elegantly that a commercial fishery cannot fail to diminish the fish stock. His general conclusion is worth quoting, for it states clearly not only his own position but the error of earlier thinking:

> As we see, a picture is obtained which diverges radically from the hypothesis which has been favoured almost down to the present time, namely that the natural reserve of fish is an inviolable capital, of which the fishing industry must use only the interest, not touching the capital at all. Our theory says, on the contrary, that a fishery and a natural reserve of fish are incompatible, and that the exploitable stock of fish is a changeable quantity, which depends on the intensity of the fishery. The more fish we take from a body of water, the smaller is the basic stock remaining in it; and the less fish we take, the greater is the basic stock, approximating to the natural stock when the fishery approaches zero. Such is the nature of the matter.[15]

The general conception of a fisheries ecology would appear to make such a conclusion inevitable. If a species were in ecological equilibrium before the commencement of commercial fishing, man's intrusion would have the same effect as any other predator; and that can only mean that the species population would reach a new equilibrium at a lower level of abundance, the divergence of the new equilibrium from the old depending on the degree of man's predatory effort and effectiveness.

The term "fisheries management" has been much in vogue in recent years, being taken to express a more subtle approach to the fisheries problem than the older terms "depletion" and "conservation." Briefly, it focuses attention on the quantity of fish caught, taking as the human objective of commercial fishing the derivation of the largest sustainable catch. This approach is often hailed in the biological literature as the "new theory" or the "modern formulation" of the fisheries problem.[16] Its limitations, however, are very serious, and, indeed, the new approach comes very little closer to treating the fisheries problem as one of human utilization of natural resources than did the older, more primitive, theories. Focusing attention on the maximization of the catch neglects entirely the inputs of other factors of production which are used up in fishing and must be accounted for as costs. There are many references to such ultimate economic considerations in the biological literature but no analytical integration of the economic factors. In fact, the very conception of a *net economic yield* has scarcely made any appearance at all. On the whole, biologists tend to treat the fisherman as an exogenous element in their analytical model, and the behavior of fishermen is not made into an integrated element of a general and systematic "bionomic" theory. In the case of the fishing industry the large numbers of fishermen permit valid behavioristic generalization of their activities along the lines of the standard economic theory of production. The following section attempts to apply that theory to the fishing industry and to demonstrate that the "overfishing problem" has its roots in the economic organization of the industry.

III. ECONOMIC THEORY OF THE FISHERY

In the analysis which follows, the theory of optimum utilization of fishery re-

---

[15] T. I. Baranoff, "On the Question of the Dynamics of the Fishing Industry," p. 5 (mimeographed).

[16] See, e.g., R. E. Foerster, "Prospects for Managing Our Fisheries," *Bulletin of the Bingham Oceanographic Collection* (New Haven), May, 1948; E. S. Russell, "Some Theoretical Considerations on the Overfishing Problem," *Journal du conseil permanent international pour l'exploration de la mer*, 1931, and *The Overfishing Problem*, Lecture IV.

sources and the reasons for its frustration in practice are developed for a typical demersal fish. Demersal, or bottom-dwelling fishes, such as cod, haddock, and similar species and the various flatfishes, are relatively nonmigratory in character. They live and feed on shallow continental shelves where the continual mixing of cold water maintains the availability of those nutrient salts which form the fundamental basis of marine-food chains. The various feeding grounds are separated by deep-water channels which constitute barriers to the movement of these species; and in some cases the fish of different banks can be differentiated morphologically, having varying numbers of vertebrae or some such distinguishing characteristic. The significance of this fact is that each fishing ground can be treated as unique, in the same sense as can a piece of land, possessing, at the very least, one characteristic not shared by any other piece: that is, location.

(Other species, such as herring, mackerel, and similar pelagic or surface dwellers, migrate over very large distances, and it is necessary to treat the resource of an entire geographic region as one. The conclusions arrived at below are applicable to such fisheries, but the method of analysis employed is not formally applicable. The same is true of species that migrate to and from fresh water and the lake fishes proper.)

We can define the optimum degree of utilization of any particular fishing ground as that which maximizes the net economic yield, the difference between total cost, on the one hand, and total receipts (or total value production), on the other.[17] Total cost and total production can each be expressed as a function of the degree of fishing intensity or, as the biologists put it, "fishing effort," so that a simple maximization solution is possible. Total cost will be a linear function of fishing effort, if we assume no fishing-induced effects on factor prices, which is reasonable for any particular regional fishery.

The production function—the relationship between fishing effort and total value produced—requires some special attention. If we were to follow the usual presentation of economic theory, we should argue that this function would be positive but, after a point, would rise at a diminishing rate because of the law of diminishing returns. This would not mean that the fish population has been reduced, for the law refers only to the *proportions* of factors to one another, and a fixed fish population, together with an increasing intensity of effort, would be assumed to show the typical sigmoid pattern of yield. However, in what follows it will be assumed that the law of diminishing returns in this pure sense is inoperative in the fishing industry. (The reasons will be advanced at a later point in this paper.) We shall assume that, as fishing effort expands, the catch of fish increases at a diminishing rate but that it does so because of the effect of catch upon the fish population.[18] So far as the argument of the next few pages is concerned, all that is formally necessary is to assume that, as fishing intensity increases, catch will grow at a diminishing rate. Whether this reflects the pure law of diminishing returns or the reduction

---

[17] Expressed in these terms, this appears to be the monopoly maximum, but it coincides with the social optimum under the conditions employed in the analysis, as will be indicated below.

[18] Throughout this paper the conception of fish population that is employed is one of *weight* rather than *numbers*. A good deal of the biological theory has been an effort to combine growth factors and numbers factors into weight sums. The following analysis will neglect the fact that, for some species, fish of different sizes bring different unit prices.

H. SCOTT GORDON

of population by fishing, or both, is of no particular importance. The point at issue will, however, take on more significance in Section IV and will be examined there.

Our analysis can be simplified if we retain the ordinary production function instead of converting it to cost curves, as is usually done in the theory of the firm. Let us further assume that the functional relationship between average production (production-per-unit-of-fishing-effort) and the quantity of fishing effort is uniformly linear. This does not distort the

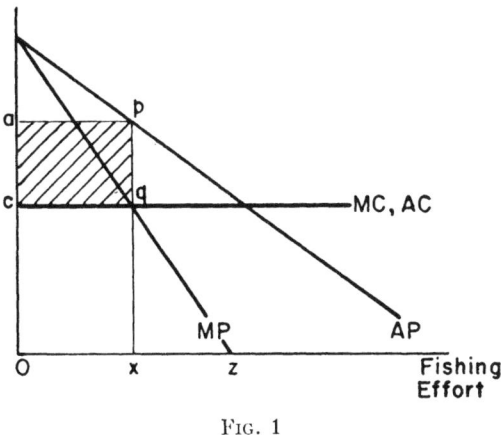

FIG. 1

results unduly, and it permits the analysis to be presented more simply and in graphic terms that are already quite familiar.

In Figure 1 the optimum intensity of utilization of a particular fishing ground is shown. The curves AP and MP represent, respectively, the average productivity and marginal productivity of fishing effort. The relationship between them is the same as that between average revenue and marginal revenue in imperfect competition theory, and MP bisects any horizontal between the ordinate and AP. Since the costs of fishing supplies, etc., are assumed to be unaffected by the amount of fishing effort, marginal cost and average cost are identical and constant, as shown by the curve MC, AC.[19] These costs are assumed to include an opportunity income for the fishermen, the income that could be earned in other comparable employments. Then Ox is the optimum intensity of effort on this fishing ground, and the resource will, at this level of exploitation, provide the maximum net economic yield indicated by the shaded area $apqc$. The maximum sustained physical yield that the biologists speak of will be attained when marginal productivity of fishing effort is zero, at Oz of fishing intensity in the chart shown. Thus, as one might expect, the optimum economic fishing intensity is less than that which would produce the maximum sustained physical yield.

The area $apqc$ in Figure 1 can be regarded as the rent yielded by the fishery resource. Under the given conditions, Ox is the best rate of exploitation for the fishing ground in question, and the rent reflects the productivity of that ground, not any artificial market limitation. The rent here corresponds to the extra productivity yielded in agriculture by soils of better quality or location than those on the margin of cultivation, which may produce an opportunity income but no more. In short, Figure 1 shows the determination of the intensive margin of utilization on an intramarginal fishing ground.

We now come to the point that is of greatest theoretical importance in understanding the primary production phase of the fishing industry and in distinguishing it from agriculture. In the sea fish-

---

[19] Throughout this analysis, fixed costs are neglected. The general conclusions reached would not be appreciably altered, I think, by their inclusion, though the presentation would be greatly complicated. Moreover, in the fishing industry the most substantial portion of fixed cost—wharves, harbors, etc.—is borne by government and does not enter into the cost calculations of the operators.

THEORY OF A COMMON-PROPERTY RESOURCE

eries the natural resource is not private property; hence the rent it may yield is not capable of being appropriated by anyone. The individual fisherman has no legal title to a section of ocean bottom. Each fisherman is more or less free to fish wherever he pleases. The result is a pattern of competition among fishermen which culminates in the dissipation of the rent of the intramarginal grounds. This can be most clearly seen through an analysis of the relationship between the fishermen are free to fish on whichever ground they please, it is clear that this is not an equilibrium allocation of fishing effort in the sense of connoting stability. A fisherman starting from port and deciding whether to go to ground *1* or *2* does not care for *marginal* productivity but for *average* productivity, for it is the latter that indicates where the greater total yield may be obtained. If fishing effort were allocated in the optimum fashion, as shown in Figure 2, with Ox on

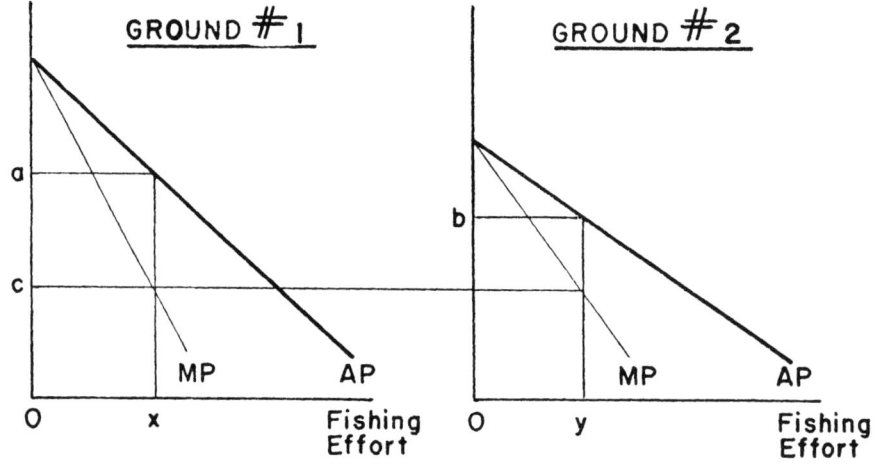

Fig. 2

intensive margin and the extensive margin of resource exploitation in fisheries.

In Figure 2, two fishing grounds of different fertility (or location) are shown. Any given amount of fishing effort devoted to ground *2* will yield a smaller total (and therefore average) product than if devoted to *1*. The maximization problem is now a question of the allocation of fishing effort between grounds *1* and *2*. The optimum is, of course, where the marginal productivities are equal on both grounds. In Figure 2, fishing effort of Ox on *1* and Oy on *2* would maximize the total net yield of $Ox + Oy$ effort if marginal cost were equal to Oc. But if under such circumstances the individual *1*, and Oy on *2*, this would be a disequilibrium situation. Each fisherman could expect to get an average catch of Oa on *1* but only Ob on *2*. Therefore, fishermen would shift from *2* to *1*. Stable equilibrium would not be reached until the average productivity of both grounds was equal. If we now imagine a continuous gradation of fishing grounds, the extensive margin would be on that ground which yielded nothing more than outlaid costs plus opportunity income—in short, the one on which average productivity and average cost were equal. But, since average cost is the same for all grounds and the average productivity of all grounds is also brought to equality by

H. SCOTT GORDON

the free and competitive nature of fishing, this means that the intramarginal grounds also yield no rent. It is entirely possible that some grounds would be exploited at a level of *negative* marginal productivity. What happens is that the rent which the intramarginal grounds are capable of yielding is dissipated through misallocation of fishing effort.

This is why fishermen are not wealthy, despite the fact that the fishery resources of the sea are the richest and most indestructible available to man. By and large, the only fisherman who becomes rich is one who makes a lucky catch or one who participates in a fishery that is put under a form of social control that turns the open resource into property rights.

Up to this point, the remuneration of fishermen has been accounted for as an opportunity-cost income comparable to earnings attainable in other industries. In point of fact, fishermen typically earn less than most others, even in much less hazardous occupations or in those requiring less skill. There is no effective reason why the competition among fishermen described above must stop at the point where opportunity incomes are yielded. It may be and is in many cases carried much further. Two factors prevent an equilibration of fishermen's incomes with those of other members of society. The first is the great immobility of fishermen. Living often in isolated communities, with little knowledge of conditions or opportunities elsewhere; educationally and often romantically tied to the sea; and lacking the savings necessary to provide a "stake," the fisherman is one of the least mobile of occupational groups. But, second, there is in the spirit of every fisherman the hope of the "lucky catch." As those who know fishermen well have often testified, they are gamblers and incurably optimistic. As a consequence, they will work for less than the going wage.[20]

The theory advanced above is substantiated by important developments in the fishing industry. For example, practically all control measures have, in the past, been designed by biologists, with sole attention paid to the production side of the problem and none to the cost side. The result has been a wide-open door for the frustration of the purposes of such measures. The Pacific halibut fishery, for example, is often hailed as a great achievement in modern fisheries management. Under international agreement between the United States and Canada, a fixed-catch limit was established during the early thirties. Since then, catch-per-unit-effort indexes, as usually interpreted, show a significant rise in the fish population. W. F. Thompson, the pioneer of the Pacific halibut management program, noted recently that "it has often been said that the halibut regulation presents the only definite case of sustained improvement of an overfished deep-sea fishery. This, I believe, is true and the fact should lend special importance to the principles which have been deliberately used to obtain this improvement."[21] Actually, careful study of the statistics indicates that the estimated recovery of halibut stocks could not have been due principally to the control measures, for the average catch was, in fact, greater during the recovery years than during the years of

---

[20] "The gambling instinct of the men makes many of them work for less remuneration than they would accept as a weekly wage, because there is always the possibility of a good catch and a financial windfall" (Graham, *op. cit.*, p. 86).

[21] W. F. Thompson, "Condition of Stocks of Halibut in the Pacific," *Journal du conseil permanent international pour l'exploration de la mer*, Vol. XVIII, No. 2 (August, 1952).

decline. The total amount of fish taken was only a small fraction of the estimated population reduction for the years prior to regulation.[22] Natural factors seem to be mainly responsible for the observed change in population, and the institution of control regulations almost a coincidence. Such coincidences are not uncommon in the history of fisheries policy, but they may be easily explained. If a long-term cyclical fluctuation is taking place in a commercially valuable species, controls will likely be instituted when fishing yields have fallen very low and the clamor of fishermen is great; but it is then, of course, that stocks are about due to recover in any case. The "success" of conservation measures may be due fully as much to the sociological foundations of public policy as to the policy's effect on the fish. Indeed, Burkenroad argues that biological statistics in general may be called into question on these grounds. Governments sponsor biological research when the catches are disappointing. If there are long-term cyclical fluctuations in fish populations, as some think, it is hardly to be wondered why biologists frequently discover that the sea is being depleted, only to change their collective opinion a decade or so later.

Quite aside from the *biological* argument on the Pacific halibut case, there is no clear-cut evidence that halibut fishermen were made relatively more prosperous by the control measures. Whether or not the recovery of the halibut stocks was due to natural factors or to the catch limit, the potential net yield this could have meant has been dissipated through a rise in fishing costs. Since the method of control was to halt fishing when the limit had been reached, this created a great incentive on the part of each fisherman to get the fish before his competitors. During the last twenty years, fishermen have invested in more, larger, and faster boats in a competitive race for fish. In 1933 the fishing season was more than six months long. In 1952 it took just twenty-six days to catch the legal limit in the area from Willapa Harbor to Cape Spencer, and sixty days in the Alaska region. What has been happening is a rise in the average cost of fishing effort, allowing no gap between average production and average cost to appear, and hence no rent.[23]

Essentially the same phenomenon is observable in the Canadian Atlantic Coast lobster-conservation program. The method of control here is by seasonal closure. The result has been a steady growth in the number of lobster traps set

---

[22] See M. D. Burkenroad, "Fluctuations in Abundance of Pacific Halibut," *Bulletin of the Bingham Oceanographic Collection*, May, 1948.

[23] The economic significance of the reduction in season length which followed upon the catch limitation imposed in the Pacific halibut fishery has not been fully appreciated. E.g., Michael Graham said in summary of the program in 1943: "The result has been that it now takes only five months to catch the quantity of halibut that formerly needed nine. This, *of course*, has meant profit, where there was none before" (*op. cit.*, p. 156; my italics). Yet, even when biologists have grasped the economic import of the halibut program and its results, they appear reluctant to declare against it. E.g., W. E. Ricker: "This method of regulation does not necessarily make for more profitable fishing and certainly puts no effective brake on waste of effort, since an unlimited number of boats is free to join the fleet and compete during the short period that fishing is open. However, the stock is protected, and yield approximates to a maximum if quotas are wisely set; as biologists, perhaps we are not required to think any further. Some claim that any mixing into the economics of the matter might prejudice the desirable biological consequences of regulation by quotas" ("Production and Utilization of Fish Population," in a Symposium on Dynamics of Production in Aquatic Populations, Ecological Society of America, *Ecological Monographs*, XVI [October, 1946], 385). What such "desirable biological consequences" might be, is hard to conceive. Since the regulatory policies are made by man, surely it is necessary they be evaluated in terms of human, not piscatorial, **objectives.**

by each fisherman. Virtually all available lobsters are now caught each year within the season, but at much greater cost in gear and supplies. At a fairly conservative estimate, the same quantity of lobsters could be caught with half the present number of traps. In a few places the fishermen have banded together into a local monopoly, preventing entry and controlling their own operations. By this means, the amount of fishing gear has been greatly reduced and incomes considerably improved.

That the plight of fishermen and the inefficiency of fisheries production stems from the common-property nature of the resources of the sea is further corroborated by the fact that one finds similar patterns of exploitation and similar problems in other cases of open resources. Perhaps the most obvious is hunting and trapping. Unlike fishes, the biotic potential of land animals is low enough for the species to be destroyed. Uncontrolled hunting means that animals will be killed for any short-range human reason, great or small: for food or simply for fun. Thus the buffalo of the western plains was destroyed to satisfy the most trivial desires of the white man, against which the long-term food needs of the aboriginal population counted as nothing. Even in the most civilized communities, conservation authorities have discovered that a bag-limit *per man* is necessary if complete destruction is to be avoided.

The results of anthropological investigation of modes of land tenure among primitive peoples render some further support to this thesis. In accordance with an evolutionary concept of cultural comparison, the older anthropological study was prone to regard resource tenure in common, with unrestricted exploitation, as a "lower" stage of development comparative with private and group property rights. However, more complete annals of primitive cultures reveal common tenure to be quite rare, even in hunting and gathering societies. Property rights in some form predominate by far, and, most important, their existence may be easily explained in terms of the necessity for orderly exploitation and conservation of the resource. Environmental conditions make necessary some vehicle which will prevent the resources of the community at large from being destroyed by excessive exploitation. Private or group land tenure accomplishes this end in an easily understandable fashion.[24] Significantly, land tenure is found to be "common" only in those cases where the hunting resource is migratory over such large areas that it cannot be regarded as husbandable by the society. In cases of group tenure where the numbers of the group are large, there is still the necessity of co-ordinating the practices of exploitation, in agricultural, as well as in hunting or gathering, economies. Thus, for example, Malinowski reported that among the Trobriand Islanders one of the fundamental principles of land tenure is the co-ordination of the productive activities of the gardeners by the person possessing magical leadership in the group.[25] Speaking generally, we may say that stable primitive cultures appear to have discovered the dangers of common-property tenure and to have de-

[24] See Frank G. Speck, "Land Ownership among Hunting Peoples in Primitive America and the World's Marginal Areas," *Proceedings of the 22nd International Congress of Americanists* (Rome, 1926), II, 323–32.

[25] B. Malinowski, *Coral Gardens and Their Magic*, Vol. I, chaps. xi and xii. Malinowski sees this as further evidence of the importance of magic in the culture rather than as a means of co-ordinating productive activity; but his discussion of the practice makes it clear that the latter is, to use Malinowski's own concept, the "function" of the institution of magical leadership, at least in this connection.

THEORY OF A COMMON-PROPERTY RESOURCE

veloped measures to protect their resources. Or, if a more Darwinian explanation be preferred, we may say that only those primitive cultures have survived which succeeded in developing such institutions.

Another case, from a very different industry, is that of petroleum production. Although the individual petroleum producer may acquire undisputed lease or ownership of the particular plot of land upon which his well is drilled, he shares, in most cases, a common pool of oil with other drillers. There is, consequently, set up the same kind of competitive race as is found in the fishing industry, with attending overexpansion of productive facilities and gross wastage of the resource. In the United States, efforts to regulate a chaotic situation in oil production began as early as 1915. Production practices, number of wells, and even output quotas were set by governmental authority; but it was not until the federal "Hot Oil" Act of 1935 and the development of interstate agreements that the final loophole (bootlegging) was closed through regulation of interstate commerce in oil.

Perhaps the most interesting similar case is the use of common pasture in the medieval manorial economy. Where the ownership of animals was private but the resource on which they fed was common (and limited), it was necessary to regulate the use of common pasture in order to prevent each man from competing and conflicting with his neighbors in an effort to utilize more of the pasture for his own animals. Thus the manor developed its elaborate rules regulating the use of the common pasture, or "stinting" the common: limitations on the number of animals, hours of pasturing, etc., designed to prevent the abuses of excessive individualistic competition.[26]

There appears, then, to be some truth in the conservative dictum that everybody's property is nobody's property. Wealth that is free for all is valued by none because he who is foolhardy enough to wait for its proper time of use will only find that it has been taken by another. The blade of grass that the manorial cowherd leaves behind is valueless to him, for tomorrow it may be eaten by another's animal; the oil left under the earth is valueless to the driller, for another may legally take it; the fish in the sea are valueless to the fisherman, because there is no assurance that they will be there for him tomorrow if they are left behind today. A factor of production that is valued at nothing in the business calculations of its users will yield nothing in income. Common-property natural resources are free goods for the individual and scarce goods for society. Under unregulated private exploitation, they can yield no rent; that can be accomplished only by methods which make them private property or public (government) property, in either case subject to a unified directing power.

IV. THE BIONOMIC EQUILIBRIUM OF THE FISHING INDUSTRY

The work of biological theory in the fishing industry is, basically, an effort to delineate the ecological system in which a particular fish population is found. In the main, the species that have been extensively studied are those which are subject to commercial exploitation. This is due not only to the fact that funds are forthcoming for such research but also because the activity of commercial fishing vessels provides the largest body of data upon which the biologist may work.

[26] See P. Vinogradoff, *The Growth of the Manor* [London, 1905], chap. iv; E. Lipson, *The Economic History of England* [London, 1949], I, 72.

H. SCOTT GORDON

Despite this, however, the ecosystem of the fisheries biologist is typically one that excludes man. Or, rather, man is regarded as an exogenous factor, having influence on the biological ecosystem through his removal of fish from the sea, but the activities of man are themselves not regarded as behaviorized or determined by the other elements of a system of mutual interdependence. The large number of independent fishermen who exploit fish populations of commercial importance makes it possible to treat man as a behavior element in a larger, "bionomic," ecology, if we can find the rules which relate his behavior to the other elements of the system. Similarly, in their treatment of the principles of fisheries management, biologists have overlooked essential elements of the problem by setting maximum physical landings as the objective of management, thereby neglecting the economic factor of input cost.

An analysis of the bionomic equilibrium of the fishing industry may, then, be approached in terms of two problems. The first is to explain the nature of the equilibrium of the industry as it occurs in the state of uncontrolled or unmanaged exploitation of a common-property resource. The second is to indicate the nature of a socially optimum manner of exploitation, which is, presumably, what governmental management policy aims to achieve or promote. These two problems will be discussed in the remaining pages.

In the preceding section it was shown that the equilibrium condition of uncontrolled exploitation is such that the net yield (total value landings *minus* total cost) is zero. The "bionomic ecosystem" of the fishing industry, as we might call it, can then be expressed in terms of four variables and four equations. Let P represent the population of the particular fish species on the particular fishing bank in question; L the total quantity taken or "landed" by man, measured in value terms; E the intensity of fishing or the quantity of "fishing effort" expended; and C the total cost of making such effort. The system, then, is as follows:

$$P = P(L), \quad (1)$$
$$L = L(P, E), \quad (2)$$
$$C = C(E), \quad (3)$$
$$C = L. \quad (4)$$

Equation (4) is the equilibrium condition of an uncontrolled fishery.

The functional relations stated in equations (1), (2), and (3) may be graphically presented as shown in Figure 3. Segment *1* shows the fish population as a simple negative function of landings. In segment *2* a map of landings functions is drawn. Thus, for example, if population were P_3, effort of Oe would produce Ol of fish. For each given level of population, a larger fishing effort will result in larger landings. Each population contour is, then, a production function for a given population level. The linearity of these contours indicates that the law of diminishing returns is not operative, nor are any landings-induced price effects assumed to affect the value landings graphed on the vertical axis. These assumptions are made in order to produce the simplest determinate solution; yet each is reasonable in itself. The assumption of a fixed product price is reasonable, since our analysis deals with one fishing ground, not the fishery as a whole. The cost function represented in equation (3) and graphed in segment *3* of Figure 3 is not really necessary to the determination, but its inclusion makes the matter somewhat clearer. Fixed prices of input

THEORY OF A COMMON-PROPERTY RESOURCE

factors—"fishing effort"—is assumed, which is reasonable again on the assumption that a small part of the total fishery is being analyzed.

Starting with the first segment, we see that a postulated catch of Ol connotes an equilibrium population in the biological ecosystem of Op. Suppose this population to be represented by the contour P_3 of segment 2. Then, given P_3, Oe is the effort required to catch the postulated landings Ol. This quantity of effort involves a total cost of Oc, as shown in segment 3 of the graph. In full bionomic equilibrium, $C = L$, and if the particular values Oc and Ol shown are not equal, other quantities of all four variables, L, P, E, and C, are required, involving movements of these variables through the functional system shown. The operative movement is, of course, in fishing effort, E. It is the equilibrating variable in the system.

The equilibrium equality of landings (L) and cost (C), however, must be a position of stability, and $L = C$ is a necessary, though not in itself sufficient, condition for stability in the ecosystem. This is shown by Figure 4. If effort-cost and effort-landings functions were both linear, no stable equilibrium could be found. If the case were represented by C and L_1, the fishery would contract to zero; if by C and L_2, it would undergo an infinite expansion. Stable equilibrium requires that either the cost or the landings function be nonlinear. This condition is fulfilled by the assumption that population is reduced by fishing (eq. [1] above). The equilibrium is therefore as shown in Figure 5. Now Oe represents a fully stable equilibrium intensity of fishing.

The analysis of the conditions of stable equilibrium raises some points of general theoretical interest. In the foregoing we

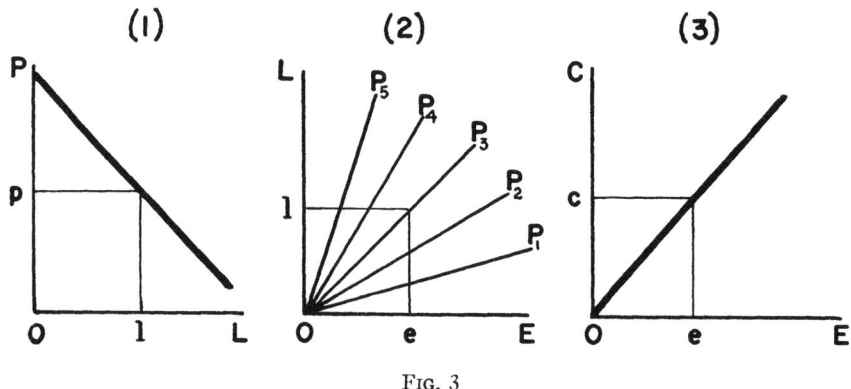

Fig. 3

have assumed that stability results from the effect of fishing on the fish population. In the standard analysis of economic theory, we should have employed the law of diminishing returns to produce a landings function of the necessary shape. Market factors might also have been so employed; a larger supply of fish, forthcoming from greater fishing effort, would reduce unit price and thereby produce a landings function with the necessary negative second derivative. Similarly, greater fishing intensity might raise the unit costs of factors, producing a cost function with a positive second derivative. Any one of these three—population effects, law of diminishing returns, or market effects—is alone sufficient to produce stable equilibrium in the ecosystem.

As to the law of diminishing returns, it has not been accepted per se by fisheries biologists. It is, in fact, a principle that becomes quite slippery when one applies it to the case of fisheries. Indicative of this is the fact that Alfred Marshall, in whose *Principles* one can find extremely little formal error, misinterprets the application of the law of diminishing returns to the fishing industry, arguing, in effect, that the law exerts its influence through the reducing effect of fishing on the fish population.[27] There have been some interesting expressions of the law or, rather, its essential varying-proportions-of-factors aspect, in the biological literature. H. M. Kyle, a German biologist, included it in 1928 among a number of reasons why catch-per-unit-of-fishing-effort indexes are not adequate measures of population change.[28] Interestingly enough, his various criticisms of the indexes were generally accepted, with the significant exception of this one point. More recently, A. G. Huntsman warned his colleagues in fisheries biology that "[there] may be a decrease in the take-per-unit-of-effort without any decrease in the total take or in the fish population.... This may mean that there has been an increase in fishermen rather than a decrease in fish."[29] While these statements run in terms of average rather than marginal yield, their underlying reasoning clearly appears to be that of the law of diminishing returns. The point has had little influence in biological circles, however, and when, two years ago, I advanced it, as Kyle and Huntsman had done, in criticism of the standard biological method of estimating population change, it received pretty short shrift.

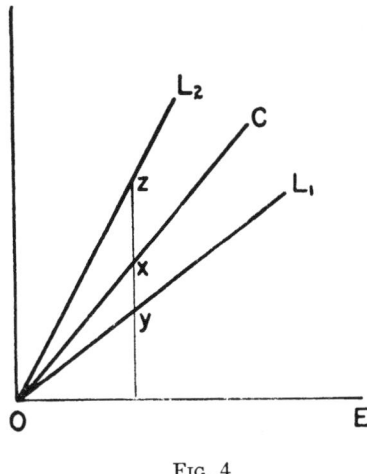

Fig. 4

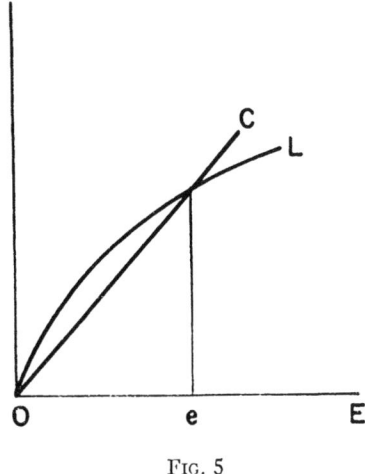

Fig. 5

[27] See H. Scott Gordon, "On a Misinterpretation of the Law of Diminishing Returns in Alfred Marshall's *Principles*," *Canadian Journal of Economics and Political Science*, February, 1952.

[28] "Die Statistik der Seefischerei Nordeuropas," *Handbuch der Seefischerei Nordeuropas* (Stuttgart, 1928).

[29] A. G. Huntsman, "Fishing and Assessing Populations," *Bulletin of the Bingham Oceanographic Collection* (New Haven), May, 1948.

THEORY OF A COMMON-PROPERTY RESOURCE

In point of fact, the law of diminishing returns is much more difficult to sustain in the case of fisheries than in agriculture or industry. The "proof" one finds in standard theory is not empirical, although the results of empirical experiments in agriculture are frequently adduced as subsidiary corroboration. The main weight of the law, however, rests on a *reductio ad absurdum*. One can easily demonstrate that, were it not for the law of diminishing returns, all the world's food could be grown on one acre of land. Reality is markedly different, and it is because the law serves to render this reality intelligible to the logical mind, or, as we might say, "explains" it, that it occupies such a firm place in the body of economic theory. In fisheries, however, the pattern of reality can easily be explained on other grounds. In the case at least of developed demersal fisheries, it cannot be denied that the fish population is reduced by fishing, and this relationship serves perfectly well to explain why an infinitely expansible production is not possible from a fixed fishing area. The other basis on which the law of diminishing returns is usually advanced in economic theory is the prima facie plausibility of the principle as such; but here, again, it is hard to grasp any similar reasoning in fisheries. In the typical agricultural illustration, for example, we may argue that the fourth harrowing or the fourth weeding, say, has a lower marginal productivity than the third. Such an assertion brings ready acceptance because it concerns a process with a zero productive limit. It is apparent that, ultimately, the land would be completely broken up or the weeds completely eliminated if harrowing or weeding were done in ever larger amounts. The law of diminishing returns signifies simply that such a zero limit is *gradually approached*, all of which appears to be quite acceptable on prima facie grounds. There is nothing comparable to this in fisheries at all, for there is no "cultivation" in the same sense of the term, except, of course, in such cases as oyster culture or pond rearing of fish, which are much more akin to farming than to typical sea fisheries.

In the biological literature the point has, I think, been well thought through, though the discussion does not revolve around the "law of diminishing returns" by that name. It is related rather to the fisheries biologist's problem of the interpretation of catch-per-unit-of-fishing-effort statistics. The essence of the law is usually eliminated by the assumption that there is no "competition" among units of fishing gear—that is, that the ratio of gear to fishing area and/or fish population is small. In some cases, corrections have been made by the use of the compound-interest formula where some competition among gear units is considered to exist.[30] Such corrections, however, appear to be based on the idea of an increasing catch-population ratio rather than an increasing effort-population ratio. The latter would be as the law of diminishing returns would have it; the idea lying behind the former is that the total population in existence represents the maximum that can be caught, and, since this maximum would be gradually approached, the ratio of catch to population has some bearing on the efficiency of fishing gear. It is, then, just an aspect of the population-reduction effect. Similarly, it has been pointed out that, since fish are recruited into the

[30] See, e.g., W. F. Thompson and F. H. Bell, *Biological Statistics of the Pacific Halibut Fishery, No. 2: Effect of Changes in Intensity upon Total Yield and Yield per Unit of Gear: Report of the International Fisheries Commission* (Seattle, 1934).

catchable stock in a seasonal fashion, one can expect the catch-per-unit-effort to fall as the fishing season progresses, at least in those fisheries where a substantial proportion of the stock is taken annually. Seasonal averaging is therefore necessary in using the catch-effort statistics as population indexes from year to year. This again is a population-reduction effect, not the law of diminishing returns. In general, there seems to be no reason for departing from the approach of the fisheries biologist on this point. The law of diminishing returns is not necessary to explain the conditions of stable equilibrium in a static model of the fishery, nor is there any prima facie ground for its acceptance.

Let us now consider the exploitation of a fishing ground under unified control, in which case the equilibrium condition is the maximization of net financial yield, $L - C$.

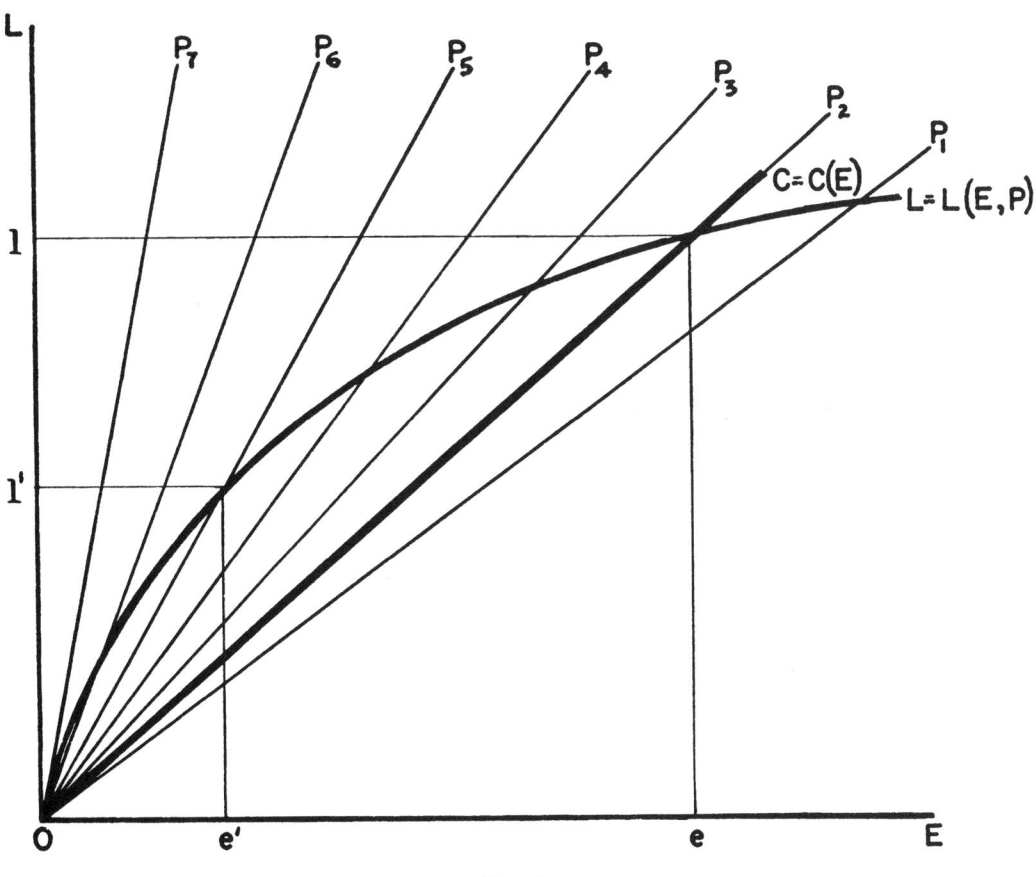

Fig. 6

The map of population contours graphed in segment 2 of Figure 3 may be superimposed upon the total-landings and total-cost functions graphed in Figure 5. The result is as shown in Figure 6. In the system of interrelationships we have to consider, population changes affect, and are in turn affected by, the amount of fish landed. The map of population contours does not include this roundabout effect that a population change has upon itself. The curve labeled L, however, is a landings function which accounts for the fact that larger landings reduce the population, and this is why it is shown to have a steadily diminishing slope. We may regard the landings function as moving progressively to lower population contours P_7, P_6, P_5, etc., as total landings increase in magnitude. As a consequence, while each population contour represents many hypothetical combinations of E, L, and P, only one such combination on each is actually compatible in this system of interrelationships. This combination is the point on any contour where that contour is met by the landings function L. Thus the curve labeled L may be regarded as tracing out a series of combinations of E, L, and P which are compatible with one another in the system.

The total-cost function may be drawn as shown, with total cost, C, measured in terms of landings, which the vertical axis represents.[31] This is a linear function of effort as shown. The optimum intensity of fishing effort is that which maximizes $L - C$. This is the monopoly solution; but, since we are considering only a single fishing ground, no price effects are introduced, and the social optimum coincides with maximum monopoly revenue. In this case we are maximizing the yield of a natural resource, not a privileged position, as in standard monopoly theory. The rent here is a social surplus yielded by the resource, not in any part due to artificial scarcity, as is monopoly profit or rent.

If the optimum fishing intensity is that which maximizes $L - C$, this is seen to be the position where the slope of the landings function equals the slope of the cost function in Figure 6. Thus the optimum fishing intensity is Oe' of fishing effort. This will yield Ol' of landings, and the species population will be in continuing stable equilibrium at a level indicated by P_5.

The equilibrium resulting from uncontrolled competitive fishing, where the rent is dissipated, can also be seen in Figure 6. This, being where $C = L$, is at Oe of effort and Ol of landings, and at a stable population level of P_2. As can be clearly seen, the uncontrolled equilibrium means a higher expenditure of effort, higher fish landings, and a lower continuing fish population than the optimum equilibrium.

Algebraically, the bionomic ecosystem may be set out in terms of the optimum solution as follows. The species population in equilibrium is a linear function of the amount of fish taken from the sea:

$$P = a - bL. \qquad (1)$$

In this function, a may be described as the "natural population" of the species—the equilibrium level it would attain if not commercially fished. All natural factors, such as water temperatures, food supplies, natural predators, etc., which affect the population are, for the purposes of the system analyzed, locked up in a. The magnitude of a is the vertical intercept of the population function graphed in segment 1 of Figure 3. The slope of this function is b, which may be described as the "depletion coefficient," since it indicates the effect of catch on population. The landings function is such that no landings are forthcoming with either zero effort or zero population; therefore,

$$L = cEP. \qquad (2)$$

---

[31] More correctly, perhaps, C and L are both measured in money terms.

H. SCOTT GORDON

The parameter c in this equation is the technical coefficient of production or, as we may call it simply, the "production coefficient." Total cost is a function of the amount of fishing effort.

$$C = qE.$$

The optimum condition is that the total net receipts must be maximized, that is,

$$L - C \text{ to be maximized}.$$

Since q has been assumed constant and equal to unity (i.e., effort is counted in "dollars-worth" units), we may write $L - E$ to be maximized. Let this be represented by R:

$$R = L - E, \quad (3)$$

$$\frac{dR}{dE} = 0. \quad (4)$$

The four numbered equations constitute the system when in optimality equilibrium. In order to find this optimum, the landings junction (2) may be rewritten, with the aid of equation (1), as:

$$L = cE(a - bL).$$

From this we have at once

$$L(1 + cEb) = cEa,$$

$$L = \frac{caE}{1 + cbE}.$$

To find the optimum intensity of effort, we have, from equation (3):

$$\frac{dR}{dE} = \frac{dL}{dE} - \frac{dE}{dE}$$

$$= \frac{(1 + cbE)(ca) - caE(cb)}{(1 + cbE)^2} - 1,$$

$$= \frac{ca}{(1 + cbE)^2} - 1;$$

for a maximum, this must be set equal to zero; hence,

$$ca = (1 + cbE)^2,$$

$$1 + cbE = \pm \sqrt{ca},$$

$$E = \frac{-1 \pm \sqrt{ca}}{cb}.$$

For positive E,

$$E = \frac{\sqrt{ca} - 1}{cb}.$$

This result indicates that the effect on optimum effort of a change in the production coefficient is uncertain, a rise in c calling for a rise in E in some cases and a fall in E in others, depending on the magnitude of the change in c. The effects of changes in the natural population and depletion coefficient are, however, clear, a rise (fall) in a calling for a rise (fall) in E, while a rise (fall) in b means a fall (rise) in E.

# A Model of Regulated Open Access Resource Use*

FRANCES R. HOMANS

Department of Applied Economics, University of Minnesota, St. Paul, Minnesota 55108

AND

JAMES E. WILEN

Department of Agricultural and Resource Economics, University of California at Davis, Davis, California 95616, and Giannini Foundation

Received October 5, 1994; revised August 24, 1995

This paper develops a model of regulated open access resource exploitation. The regulatory model assumes that regulators are goal oriented, choosing target harvest levels according to a safe stock concept. These harvest quotas are implemented by setting season lengths, conditioned on the industry fishing capacity. The industry enters until rents are dissipated, conditioned on season length regulations. Harvest levels, fishing capacity, season length, and biomass are determined jointly. Using parameter estimates from the long-regulated North Pacific Halibut fishery, predictions of these variables from the regulated open access model are compared to predictions that arise from the Gordon model. © 1997 Academic Press

1. INTRODUCTION

A survey of the literature in fisheries economics would leave one with the impression that fisheries fall into one of two institutional configurations: pure open access or optimized rent maximizing. This impression would be conveyed because these two paradigms totally dominate the published literature in fisheries economics. At the same time, if one were to examine the institutional configurations under which real world fisheries operate, one would find virtually no fisheries operating under either pure open access or rent maximizing conditions. Instead, most of the world's most important fisheries operate under what might best be termed *regulated open access*. In regulated open access fisheries, participants are free to enter but subject to certain regulations imposed by a management agency. These typically take the form of gear restrictions, area closures, and importantly, season length restrictions.[1]

Economists have not, of course, ignored regulations in discussions of fisheries. As early as the fifties, economists were hypothesizing about the economic impacts of regulations on fisheries, including output and input taxes, gear restrictions, and time and area closures, among others.[2] These somewhat speculative discussions arose in a particular institutional context in which most of the world's important

---

*The authors acknowledge helpful comments from three anonymous referees. This is Minnesota Agricultural Experiment Station Paper 21, 494.

[1] Some fisheries also operate under a more stringent form of regulation, in which regulations are imposed and enforced in a restricted access setting that uses a license limitation scheme or another form of closure to entry. These are perhaps best viewed as regulated restricted access fisheries.

[2] Cf. Turvey and Wiseman [22], Scott [18], and Crutchfield and Zellner [8].

fisheries were virtually unregulated and operating under open access conditions. H. S. Gordon's [10] important paper had just appeared, pointing out how open access fisheries inevitably dissipate potential rents and as a result, economists' early discussions about regulations focused on normative issues such as: how can regulations be designed in order to coax an open access fishery closer to a rent maximizing ideal? The extension of jurisdiction by coastal nations in 1976 changed the institutional context of fisheries dramatically since virtually all of the most important fisheries suddenly came under the authority of adjacent coastal nations. This important change set the stage for a new era of regulated fisheries use by legitimizing coastal nations' authority to explicitly manage the use of their fisheries.

As coastal nations have begun to exert more regulatory control over their adjacent fisheries, economists have continued to focus primarily on normative issues associated with regulatory design. There have been several important policy contributions emerging from this work. Economists were instrumental, for example, in framing management legislation (such as the Magnuson Act) in a manner that incorporated socioeconomic as well as biological goals and in promoting rationalization schemes including limited entry and transferable quota schemes. Interestingly, however, in their descriptive analytical and empirical work, economists have continued to depict fisheries as if they are still operating under pure open access conditions. This is in spite of the fact that most modern fisheries are heavily influenced by regulatory structures which proliferated especially after the extension of coastal jurisdictions at the end of the seventies.

A question which might be asked is, why not simply incorporate regulations as technological constraints and treat them as a minor modification of the basic open access model? The answer to that question is that the regulatory structure is fundamentally more important than this, particularly as a force affecting the character of fisheries in the long run. Moreover, regulatory constraints are not simply exogenous but instead are fluid and an outcome of purposeful behavior by institutions with goals and objectives. The menu of regulatory instruments is often extensive and regulators actively choose both the suite of instruments and their respective levels in a manner that reflects current and expected future conditions in each fishery. When external factors change, regulations also change in response and hence regulations are fundamentally endogenous and dynamic.

The implications of these observations are at least three. First, the technological and behavioral character of modern fisheries is intimately bound up with the nature and operation of the regulatory structure. Not only will technology and behavior be affected by the regulatory structure, but so also will the health and attributes of the harvested species. Second, discussions of policy alternatives need to measure the gains from rationalization relative to the appropriate status quo. Regulated fisheries may look considerably different from what would emerge under pure open access and they may, in fact, even be generating substantial economic rents. Finally, modeling the evolution of fisheries and particularly forecasting future conditions must account for not only the dynamics of the biomass and the technology and behavior of the industry but also the dynamics of the behavior and technology of the regulatory apparatus. This places an even greater burden on modelers to know and understand the particular features of the fisheries they are modeling.

In this paper we present a model of a regulated open access fishery which we believe more accurately captures some of the features of modern fisheries. It is

REGULATED OPEN ACCESS RESOURCE USE

important to highlight that this is a predictive rather than a normative model; we are interested in modeling and understanding the implications of structures which exist in real world setting rather than addressing questions about whether they are efficient or which institutions might be optimal. The model developed elevates the role of the regulatory structure to one on par with consideration of industry behavior and biological dynamics. We treat the regulatory sector as rational and purposeful (although not necessarily efficient), so that regulatory behavior and industry behavior jointly and endogenously determine the character of the fishery in question. We also view the process of regulator/industry interaction as dynamic so that when internal and external conditions change, the regulatory sector responds rather than being considered simply exogenous. In the next section the conceptual model is discussed and in Section 3 this model is used to generate some hypotheses about regulated open access fisheries. Section 4 discusses results of an empirical application of the model and the concluding section summarizes the differences between our model of regulated open access use and the more widely used pure open access model.

2. A MODEL OF A REGULATED OPEN ACCESS FISHERY

The model developed here has three fundamental components. First, in the industry component, it is assumed that the fishing industry commits a given amount of fishing capacity each season, based upon anticipated prices, costs, biomass level, and (importantly) the regulations set by the regulatory agency. Second, in the regulatory component, it is assumed that the regulatory agency selects regulations, based upon specific biologically oriented goals and the anticipated fishing capacity level of the industry. Thus there is a joint equilibrium established between the industry and the regulatory sector. Finally, in the biological component, we assume that the biomass evolves between seasons in a manner dependent upon how much has been harvested each season and the initial biomass level. The fishery is characterized by an equilibrium consisting of fishing capacity and regulation levels determined endogenously, and the biomass level.[3] The motivation for the specific characterizations of these components are as follows.

2.1. *Fishermen's Behavior*

H. S. Gordon's model of rent dissipation is a useful point of departure for considering industry behavior. We assume that fishermen behave as Gordon suggested, that is, they enter in response to rents and entry proceeds until effort is earning its opportunity cost. Rents will be assumed to be the difference between industry revenues and industry costs, defined over a given fishing season. Revenues are defined as total seasonal harvest multiplied by an exvessel price P per pound. Assume that there is an instantaneous Schaefer type harvest function defined by

$$h(t) = qEX(t), \qquad (1)$$

---

[3] For a more complete exposition of this model, see Homans [13]. An earlier attempt to model regulation in fisheries as endogenous is in Wilen [24].

where h is the harvest rate, q is the catchability parameter, E is a measure of fishing capacity or power, and X is the biomass level in period t of a given season. Assume also that the industry commits an amount of capacity E each season so that E can be considered variable between seasons but fixed within a season.

Let the level of biomass at the beginning of a particular season be designated X_0. Assume also that within the fishing season, the biomass declines by the fishing rate so that:

$$\dot{X}(t) = -qEX(t). \qquad (2)$$

Between fishing seasons we assume that the biomass grows in a density dependent fashion so that the beginning biomass level in one season is a function of the biomass level in the previous season. Natural mortality during the season is ignored here for analytical convenience. In Section 2.3 below we discuss the between-season dynamics. Under these assumptions, total cumulative harvest for the industry over a season of length T will be

$$H(T) = X_0 - X(T) = X_0(1 - e^{-qET}), \qquad (3)$$

which is determined by integrating (2) over the season length T, assuming that E is constant and X_0 is given.

With respect to costs we assume simple linear costs related to both the level of fixed fishing capacity and to the cumulative level of variable effort expended over the season. Consider capacity costs first. We assume fixed costs of f per season per unit of capacity E must be incurred to participate in the fishery. These costs may be assumed to be outfitting, repair, and preparation costs associated with gear, the opportunity cost associated with the investment, and other implicit rents associated with other inputs. We also assume that there are variable costs v per unit of capacity used per unit time associated with the (assumed constant) rate of input use over the season.

With both cost and revenue formulations as described, we may write total industry rents anticipated for a season of length T as:

$$\text{Rents} = \left[PX_0(1 - e^{-qET})\right] - [vET + fE]. \qquad (4)$$

Total revenues are given by the left hand term and are determined by multiplying total seasonal harvest in (3) by the exvessel price P. Total costs are given by integrating (ignoring the discount factor) variable flow costs vE over the season length T and adding fixed capacity costs of fE. Setting rents equal to zero yields an implicit equation for capacity E as a function of T, P, X_0, v, f, and q. This gives the rent dissipating level of capacity that would be expected to be attracted into a fishery in a given season. We are thus assuming that capacity enters each season until rent dissipation occurs.

The implicit equation $J(E,T) = 0$ derived by setting rents in (4) equal to zero describes industry behavior associated with capacity levels that dissipate rents, given a season length T. The functional forms used here, while standard practice and sensible representations of production and cost functions, create some analytical complexities because it is not possible to explicitly isolate the rent dissipating capacity E as a function of the other variables and parameters. The function $E = E(T; X_0, P, v, f, q)$ may be analyzed by indirect methods and Fig. 1 depicts the

REGULATED OPEN ACCESS RESOURCE USE

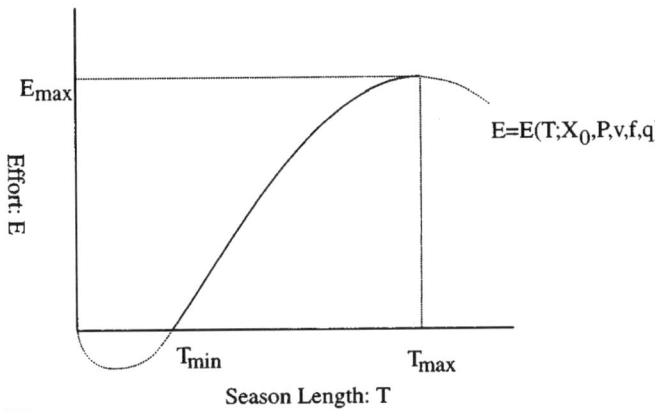

FIG. 1. Rent dissipating capacity.

shape of the function identified by setting (4) equal to zero (see Appendix A for details).

The function generally describes a monotonic relationship between E and T over the relevant range. Other things equal, longer season lengths will require larger amounts of fishing capacity to dissipate rents. Note that there is a minimum season length T_{min} below which no effort will be attracted, however. This is because there is a fixed cost f per unit of capacity and the season length must be long enough to generate sufficient variable profits to cover these costs for the first unit of positive capacity. In addition, there is a maximum capacity E_{max} associated with season length T_{max} which is the longest season length which the industry would voluntarily choose to use under any circumstance (see Appendix A).

2.2. Regulator Behavior

As discussed in the Introduction, economists have essentially ignored the fact that regulations are endogenous in modern fisheries. What hypotheses might we entertain to describe the motivations of a regulatory agency?[4] In this paper we assume a simplified goal structure for the regulatory body which captures the biological orientation of most real world fisheries regulatory bodies. In particular, we assume a two stage regulatory process where in the first stage, a target harvest quota is chosen to ensure stock safety. While there are several quota rules which regulators actually use, a common and analytically simple one is to assume that the

---

[4] One possibility is a rent seeking model, in which constituents are assumed to lobby regulators for actions that generate rents (Bhagwati [1], Buchanan et al. [2] and Rowley et al. [17]). A related possibility is a regulatory capture model in which regulatees "capture" the regulators through political or voting processes and manipulate outcomes (Karpoff [14]). These are plausible but at a level in the policy process once removed from that we are interested in. Fisheries in the United States and elsewhere are generally managed within broadly defined enabling legislation such as the Magnuson Fisheries Conservation and Management Act. Fisheries management acts such as this and related laws specify broad goals for management but not generally how to achieve these goals. Day to day implementation of fisheries regulations is usually left in the hands of local, regional, and fishery-familiar managers housed within biologically oriented field level agencies.

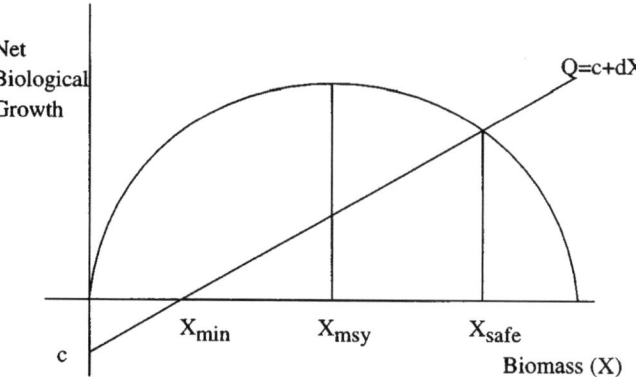

FIG. 2. Regulator quota rule.

allowable quota or exploitation rate is a linear function of the biomass over some range.[5] The quota would be set, then, according to the rule

$$Q = c + dX_0. \qquad (5)$$

In Fig. 2, for example, we show a quadratic yield function with a linear harvest quota rule superimposed. Whenever the biomass is below the desired long run "safe" stock level, the quota is set below yield so that the biomass grows and conversely if the biomass is above X_{safe}. This quota rule allows a gradual adjustment toward the equilibrium stock level whenever biomass is above or below it. Note that if $c < 0$, this rule would also call for a moratorium when the biomass falls to a level like X_{min}, hence a rule like $Q = \max(0, c + dX)$ better describes this situation. Note also that this simple rule encompasses the possibility that the safe stock is the maximum sustainable yield stock as well as others such as the $F_{0.1}$ strategy.[6] Obviously other variants of these types of rules are possible.

In the second stage of the regulatory process, we assume that regulatory instruments are selected to achieve the quota target determined from the quota rule. This distinction between targets and instruments is important and often ignored in descriptive analysis. In real fisheries, it is not enough to simply select an aggregate harvest level since there is nothing to ensure that the industry won't exceed the target. Typical regulatory instruments are the levels of various constraints on fishing technology and practices allowed by regulators, including mesh size restrictions, area and season length closures, engine size, and gear dimensions. By far the most commonly used instrument to ensure harvest quota adherence is the season length restriction. For the model discussed here, we assume that the

---

[5] Cf. the discussion in Hilborn and Walters [12], p. 22–43.
[6] The $F_{0.1}$ strategy prescribes a constant exploitation rate which leaves exploitation slightly less than the one which maximizes yield per recruit, cf. Hilborn and Walters [12].

REGULATED OPEN ACCESS RESOURCE USE

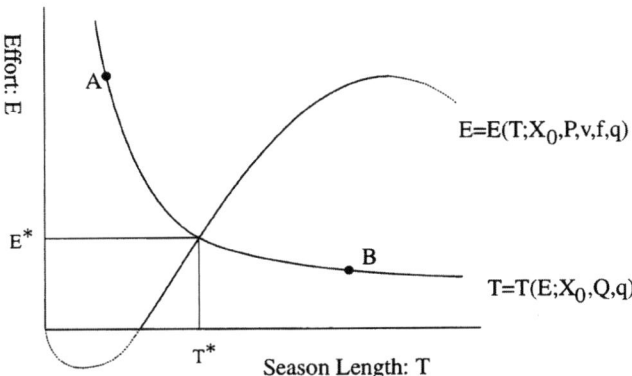

FIG. 3. Regulator season length choice and joint regulated equilibrium.

relevant instrument is the season length T, although it would be straightforward to generalize.[7]

Suppose that the biomass measured at the beginning of the fishing season is X_0. Then if the quota target is applied to the initial biomass, we have an allowable harvest quota of $Q = c + dX_0$, which, if realized, will leave a final end of season biomass of $X_T = X_0 - (c + dX_0)$. In the second stage, we thus assume that regulators select the season length instrument level which ensures that the target will be achieved. This is found by substituting the quota into Eq. (3), so that we have

$$Q = X_0(1 - e^{-qET}). \qquad (6)$$

This equation can be solved for the season length T as a function of capacity, beginning biomass, the harvest quota, and the catchability coefficient. In particular, the regulators are assumed to behave by choosing T so that

$$T = \frac{1}{qE} \ln\left[\frac{X_0}{X_0 - Q}\right]. \qquad (7)$$

Note that, once X_0 is given, this describes a rectangular hyperbola in T, E space as shown in Fig. 3. When fishing capacity is large as at point A, regulators will only allow fishing over a short season but when capacity is low as at point B, a longer season can be permitted.

A jointly determined regulated open access equilibrium occurs at a capacity level and season length as depicted at the intersection of the two curves. The industry

---

[7] For example, since aggregate fishing capacity is E in our model, we could disaggregate to incorporate input controls by defining the capacity function to be effective capacity (on the grounds) which is a function of various regulatable inputs. If effective capacity is assumed to be a function of a vector of inputs, regulators could further be assumed to be able to choose the attributes of those inputs or to regulate their levels directly. For example, fishing capacity is certainly a function of gear dimensions and other specifications and these are almost always directly regulated. The complication introduced by multiple inputs is that one must also develop a theory about instrument choice which answers questions like: if both season length and mesh size restrictions can be used to curtail aggregate effective fishing pressure, how would the mix of the two be chosen by regulators?

equilibrium depicts the industry's "choice" of a rent dissipating capacity, given the season length, and regulatory behavior is depicted as a choice of season length, given fishing capacity, the initial biomass, and the quota target established in the first stage.

2.3. *Between-Season Dynamics and the Harvest Quota Target*

The discussions above characterize an industry and regulator equilibrium during a single fishing season in which the initial biomass and hence quota are predetermined. If the quota is an equilibrium quota, the biomass will remain constant and the regulated open access fishery will be in a full steady state. If the biomass is changing, however, both the industry and regulator equilibrium curves will shift from season to season, since they are both functions of a changing X_0.[8] We can then think of the intersection of the two curves as determining a sequence of temporary equilibria, in which the industry and regulators attain successive within-season equilibria, each associated with the existing biomass level each period. Over time we would observe the entire system moving through a sequence of temporary seasonal equilibria as the biomass approached its long run level. In order to close the model and describe this long run equilibrium, we thus need to specify the dynamic relationships which determine the evolution of biomass levels between seasons.

The simplest way to specify this process is to utilize a hybrid Schaefer/Beverton-Holt model in which within season dynamics are simply governed by Eqs. (1) and (2) above and between season dynamics are governed by a density dependent mechanism. That is, during the fishing season, total biomass is assumed to evolve to reflect the harvest rate, falling from its initial level X_0 to X_T by the end of the season. Then, between seasons, it is assumed that the net biological growth depends in some density dependent way on previous biomass. Let $X_{0,t+1}$ be the biomass at the opening date of next year's season and let $X_{T,t}$ and $X_{0,t}$ be the ending and beginning biomass levels this season, respectively. Here we are indexing seasons by t and denoting the beginning and ending dates of any particular season with 0 and T.

Assume that additions to the biomass can be defined as $G(X_{0,t}) = aX_{0,t} - bX_{0,t}^2$.[9] Then between season dynamics can be represented by

$$X_{0,t+1} = X_{T,t} + G(X_{0,t}) = X_{T,t} + aX_{0,t} - bX_{0,t}^2. \tag{8}$$

---

[8] In particular, the industry equilibrium function shifts up with increases in biomass. The regulatory equilibrium function may shift in or out, depending upon the form of the quota rule. It can be shown that the regulatory equation will shift out (in) if the regulatory rule is elastic (inelastic) with respect to biomass. Thus for a linear rule, the rectangular hyperbola will shift out as biomass increases if c is positive and in if c is negative.

[9] Our use of the biomass growth function deserves some comment. Since halibut reach reproductive age in 7 to 8 years, it would be more realistic to employ an age-structured cohort model to capture biomass dynamics rather than a simple annual model. While the introduction of a cohort model may improve the accuracy of our representation of biomass dynamics and would be tractable for that purpose, it would introduce considerable analytical complexities into the complete model of season length and effort determination. Using a lumped parameter annual model for halibut has several precedents, see Cook [4], Stollery [21], and Conklin and Kolberg [5].

REGULATED OPEN ACCESS RESOURCE USE

That is, beginning biomass next season is equal to ending biomass this season plus net growth added to reproductive activities taking place or near the beginning of this season.[10]

With between season dynamics operating, under a linear exploitation rate rule, regulators allow a certain fraction $c + dX_0$ of the initial biomass to be taken each season. Thus the harvest target during any season t will be $Q_t = c + dX_{0,t}$ and we know that the end of the season biomass level will be $X_{T,t} = X_{0,t} - (c + dX_{0,t})$. As Fig. 2 shows, a linear exploitation rate leads the biomass to a long run equilibrium level X_{safe}. When the biomass is below X_{safe}, the linear rule results in a harvest level below the biological growth and the biomass approaches X_{safe} and similarly when the biomass is above X_{safe}. Along any path to a steady state, biomass dynamics may be expressed in terms of initial biomass levels, biological parameters a and b, and the exploitation rate parameters c and d, or

$$X_{0,t+1} = [(1-d)X_{0,t} - c] + aX_{0,t} - bX_{0,t}^2. \tag{9}$$

In a long run steady state equilibrium we have

$$X_{0,t+1} - X_{0,t} = 0 = [(1-d)X_{0,t} - c] + aX_{0,t} - bX_{0,t}^2 - X_{0,t}, \tag{10}$$

and if we let X_{safe} be the equilibrium value of the beginning of season biomass, we have, by substituting into (10) and solving, the expression

$$X_{\text{safe}} = \frac{a - d \pm \sqrt{(a-d)^2 - 4bc}}{2b}. \tag{11}$$

As expected, this is quadratic, representing the two intersections of the quota rule with the yield curve depicted in Fig. 2. We will assume $(a - d)$ is positive and that the desired steady state is the larger of the two roots, which would be considered the least vulnerable of the two stock levels.

Note that since the equilibrium biomass is a function of the regulatory parameters c and d, the notion of a safe biomass level is equivalent to a choice of the exploitation rule parameters c and d in the steady state. That is, we can view the regulatory goal as ensuring that a certain minimal long run biomass is maintained in the steady state, or alternatively, that the exploitation rule associated with the parameters is the one which achieves this goal.

---

[10] Alternatively, we could assume that biomass dynamics are governed by the alternative relationship $X_{0,t+1} = X_{T,t} + F(X_{T,t})$. This formulation would be appropriate if reproductive activities took place after the fishing season, while $X_{0,t+1} = X_{T,t} + G(X_{0,t})$ would reflect reproduction just prior to the season opening. The alternative relationship would change the corresponding expression for biomass to

$$X_{\text{safe}} = \frac{(1 + a + 2bc)(1-d) - 1 \pm \sqrt{[(1 + a + 2bc)(1-d) - 1]^2 - 4bc(1-d)^2(1 + a + bc)}}{2b(1-d)^2}.$$

Other expressions that embed biomass would change accordingly. Since halibut reproduce in the winter, just prior to the spring fishing season, the formulation used in the paper reflects halibut biomass dynamics more accurately than the alternative.

3. ANALYSIS OF REGULATED OPEN ACCESS BEHAVIOR

The three components described above characterize a regulated open access fishery. The industry is assumed to commit capacity E each season until rents are dissipated. Regulators are assumed to set a harvest target quota Q each season and then choose a season length T which ensures that the quota is achieved. A regulated open access equilibrium is achieved by the interaction of the industry and regulators each season. Biomass evolves between seasons according to whether the corresponding harvest is greater than, equal to, or less than biological growth. A long run steady state is achieved when the biomass is in equilibrium and when industry and regulatory behavior are constant.

In a steady state equilibrium, both beginning and ending biomass levels are given and constant from season and their levels depend upon c, d, a, and b. Since the harvest quota target is simply the difference between beginning and ending biomass levels, Q is predetermined also once c and d are chosen. Then a regulated open access equilibrium within the season is achieved when the industry commits a level of capacity E and the regulators choose a season length T such that

$$PX_0(1 - e^{-qET}) - vET - fE = 0$$
$$T = \frac{1}{qE}\ln\left[\frac{X_0}{X_T}\right] = \frac{1}{qE}\ln\left[\frac{X_0}{(1-d)X_0 - c}\right], \quad (12)$$

where X_0 satisfies (11) above. A regulated open access fishery is thus the joint outcome of the behavior of both the industry and regulatory agency, coupled with biological dynamics.

What happens to this equilibrium as economic, regulatory, and biological parameters change? This can be addressed by performing comparative statics computations on the equilibrium. Note first that this system is recursive in that the steady state biomass level is determined only by the biological parameters, a and b, and the regulatory parameters, c and d. The regulatory agency's instrument choice T depends only upon the exploitation rule parameters and the level of capacity chosen by the industry. Thus the two equations in (12) determine the equilibrium pair E, T which satisfy the open access regulatory equilibrium. For any arbitrary X_0, this pair of equations yields a temporary equilibrium $[E(T, X_0), T(E, X_0)]$ dependent upon X_0, and when X_0 satisfies (11), the system is in a full long run equilibrium with X_0 constant.

Comparative statics properties of the long run equilibrium are given in Table I and Appendix B. These are all plausible results. When prices P or harvest efficiency q rise (or costs v or f fall) the rent dissipating level of E rises. But if the biomass and quota are in long run equilibrium, this increase in potential fishing capacity must be stifled by corresponding reductions in season length. Since the biological system is recursive, price/cost parameters do not affect the steady state level of biomass which is determined only by biological and regulatory parameters (a, b, c, and d). On the other hand, changes in a, b, c, and d affect the levels T and E in potentially complicated ways via their effects on X and Q. For example, if the intrinsic growth rate a rises, then (*ceteris paribus*; holding the exploitation rule constant) both the allowable long run biomass and quota will rise. As it turns

REGULATED OPEN ACCESS RESOURCE USE

TABLE I
Comparative Statics

	Variable			
	E	T	X	Q
P	+	−	0	0
v	−	+	0	0
f	−	+	0	0
q	+	−	0	0
a	+	−	+	+
b	−	+	−	−
c	?	?	−	?
d	?	?	−	?

out, this higher quota is achieved with an increase in the equilibrium level of fishing capacity and a decrease in the regulated season length.

The only ambiguous results are those associated with the effects of changes in the regulatory parameters c and d on E and T. Recall that increasing the quota rule parameter d reduces the steady state biomass level while it increases the fraction of initial biomass targeted for harvest. Thus whether or not an increase in d will attract more entry depends upon the interplay between two forces in their effects on revenues. On the one hand, a lower steady state biomass level will reduce E ceteris paribus because there is a stock effect via the production function. On the other hand, increasing the exploitation rate increases the fraction of initial biomass allowed to be taken. If d is such that the steady state biomass is to the right of X_{msy}, then increasing d has opposing effects: a reduction in revenues via the stock effect of a reduced biomass level on the harvest function, and an increase in revenues via the fact that lower biomass levels actually increase the amount of growth and hence steady state yield. If d is relatively large, however, so that the equilibrium is to the left of X_{msy}, increases in d cause a negative stock effect and a reduction in allowable harvest, both of which lead to lower E and longer T. The bottom line is that for large enough values of d, the derivative of E with respect to d is negative and the derivative of T with respect to d is positive, whereas for smaller values these signs are indeterminate.

Note that the qualitative properties of the regulated open access model are different from those predicted by the pure open access Gordon model in fundamental ways. In the Gordon model, rents generate excess capacity, which in turn results in excessive harvest levels. These harvest levels, coupled with biological dynamics, determine an approach to a bioeconomic equilibrium. With higher prices and/or lower costs, the bioeconomic equilibrium will occur at lower, more vulnerable biomass levels. In the model presented here, the existence of the regulatory structure decouples the effects of economic parameters from impacts on the biomass. To the extent that the regulatory structure is effective, the biomass level will approach its safe level. Harvest quota targets ensure this, but the manner in which they are implemented depends upon the interplay between biological and economic factors. For example, when prices are high, potential rents attract more capacity which is stifled on the grounds with short seasons so that the targeted harvest quota is not exceeded. In the long run the regulated open access fishery will be characterized by higher biomass levels and generally even higher levels of inefficient input use than the Gordon model would predict.

4. AN APPLICATION—THE NORTH PACIFIC HALIBUT FISHERY

The North Pacific Halibut fishery provides a good case study with which to estimate and test some of the hypotheses generated from the model of regulated open access described above. In the first place, the fishery has a long history of regulation dating back to the 1930s. Over this whole period, an extensive data collection effort has been maintained in order to regulate the fishery. Long time series exist on measures of fishing capacity, total fishing effort expended, biomass estimates, quota targets, season lengths, and other economic variables. Second, the halibut fishery has been conducted with relatively simple fishing technology over the whole period, a longline gear with fairly standardized design and under relatively constant practices. Third, the halibut fishery has been managed over this whole period under a structure much like that described above. In particular, harvest quotas have been set by regulators annually and season lengths have been used to constrain the application of fishing pressure in order to achieve the desired harvest target.

We assembled data from sources published by the International Pacific Halibut Commission (IPHC) over the 1935–1977 period. Two separate series were constructed as a consistency check of the models, one for each of two management areas referred to as Area 2 (off British Columbia and up to Cape Spencer in Southeastern Alaska) and Area 3 north of Cape Spencer off Alaska but excluding waters in the Western Aleutians and Bering Sea).[11] The data used include biomass estimates from Quinn *et al.* [16] which were computed from logbook entries over the entire halibut program history. These are computed *ex post*, of course, but we assume they are unbiased representations of estimates used by regulators over the period for seasonal regulation decisions, as well as by the industry for capacity commitment decisions. Our measure of fishing capacity is also derived from logbook information. These data were used to calculate the total number of standard skate soaks (units of longline gear, approximately 1800 feet long with 100 hooks attached at 18 foot intervals, set in the water for 12 hours) in each area, which we divided by season length to estimate average fishing capacity. This measure thus assumes that effort intensity does not vary over the sample and that each unit of standardized capacity has costs proportional by their conversion factors to actual costs. Harvest and price data are also derived from published IPHC sources and prices are deflated by a wholesale price index with base year 1982. Quotas for each of the areas are published in annual reports by the IPHC. Season lengths were derived from published annual reports by the IPHC and from a summary in Skud [20]. These are expressed in days of season length and are essentially continuous over the period examined.

---

[11] The period of estimation was truncated at 1977 for several reasons. First, the jurisdiction extension by both Canada and the U.S. led to a separation of the waters of Area 2 into separate Canadian and U.S. waters, each with new data collection procedures and management methods. In addition, Canada instituted a limited entry program in 1979 which would have added estimation complications associated with expected change in industry and perhaps regulatory behavior.

REGULATED OPEN ACCESS RESOURCE USE

The econometric model consists of four structural equations: the biomass growth function (8), the quota rule (5), the entry/exit equation (4), and the season length determination function (7). Econometric error terms are appended to each of these equations, so that the model becomes

$$X_t + H_{t-1} = (1 + a)X_{t-1} - bX_{t-1}^2 + \epsilon_{1t} \tag{13}$$

$$Q_t = c + dX_t + \epsilon_{2t} \tag{14}$$

$$T_t = \frac{1}{qE_t} \ln\left[\frac{X_t}{X_t - Q_t}\right] + \epsilon_{3t} \tag{15}$$

$$0 = P_t X_t [1 - e^{-qE_t T_t}] - vE_t T_t - fE_t - f_w E_t D_{\text{WAR}} + \epsilon_{4t}. \tag{16}$$

Since the biomass level is predetermined each year, and the quota is set based upon the predetermined biomass level, Eqs. (13) and (14) are estimated individually using ordinary least-squares. The capacity level and the season length are determined simultaneously in equations (15) and (16) once the biomass and quota are known. We allow for a nonzero covariance between ϵ_{3t} and ϵ_{4t} and impose the cross equation restriction that the catchability coefficient q is the same in both equations. Because of the simultaneity, the correlation in the error terms, and the cross-equation restriction, Eqs. (15) and (16) are estimated jointly using nonlinear three-stage least-squares. All the predetermined variables in the system (P, Q, X, and D_{WAR}) as well as the regulatory variable ($\ln[X/(X - Q)]$) and lagged values for E, T, and X were used as instruments.

The entry/exit equation (16) is modified slightly by including a dummy variable for the war years. Since our estimation period spans the disruptive World War II years, we modify the fixed cost coefficient by including a dummy variable (D_{WAR}) for the years 1943–1945 when the war was active in the Pacific. Management reports suggest that fishermen were explicitly warned against fishing in the Bering Sea during the war due to dangers posed by Japanese submarines.

Results are presented in Table II. The biological dynamics and quota rule results are also shown in Fig. 4. The quota rule equation was corrected for autocorrelation, while the biological growth function was not.[12] The parameter estimates for these equations are all reasonable: the signs are as expected and all coefficients are significant at the 5% level, with most significant at the 1% level.

The implies maximum biomass levels are 318 and 416 million pounds for Areas 2 and 3, respectively. Maximum sustainable physical yields occur at half of these

---

[12] Durbin's h tests indicate that autocorrelation is present in equation (13). Our suspicion is that the positive test for autocorrelation, along with the recognition that the age of recruitment is 7–8 years, is an indication that a more complicated lag structure is in order. However, since our principal aim was to focus on the model of regulatory and effort choice rather than on biomass dynamics, we chose to keep our estimation (and our theoretical model) simple. Among population biologists who have paid serious attention to these modeling/estimation issues (see, for instance, Ludwig and Walters [15]), consensus seems to be that more accurate representations of biological dynamics do not necessarily lead to more accurate estimates of variables of interest such as optimum fishing effort or surplus yield. For example, simple autoregressive specifications often provide more accurate predictions of catch than more complicated age structured models, which demand more information out of limited data sets. Our corrections for higher order autocorrelation did not change parameter estimates appreciably, indicating bias associated with autocorrelation in the presence of a simple lagged dependent variable on the right-hand side is not too severe. Our estimates for the yield equations are also in accord with results from other studies, including Criddle [6], Criddle and Havenner [7], Deriso [9], and Capalbo [3].

TABLE II
Parameter Estimates

Parameter	Eq. (13)[a]		Eq. (14)[a]		Eqs. (16) and (15)[b]	
	Area 2	Area 3	Area 2	Area 3	Area 2	Area 3
$1 + a$	1.3786^c	1.3119^c	—	—	—	—
	(24.01)	(31.51)				
b	0.00119^c	0.00075^c	—	—	—	—
	(2.472)	(3.212)				
c	—	—	12.329^c	16.417^c	—	—
			(2.990)	(2.807)		
d	—	—	0.0897^c	0.0575^d	—	—
			(2.943)	(1.703)		
q	—	—	—	—	0.00114^c	0.000975^c
					(24.984)	(30.895)
v	—	—	—	—	0.0555^c	0.07848^c
					(8.4263)	(6.9705)
f	—	—	—	—	1.0318^d	2.0993
					(2.5569)	(1.8104)
f_W	—	—	—	—	0.79697	9.6556^d
					(1.2154)	(3.4284)
ρ	—	—	0.96	0.92	—	—
Durbin's h	6.21	5.76	—	—	—	—
D.W.	—	—	1.33	2.01	—	—
$\bar{R}^2$	0.95	0.97	0.94	0.84	—	—
Hansen's J	—	—	—	—	1.33	1.49
method	OLS		OLS		NL3SLS	

[a] t-statistics in parentheses.
[b] Asymptotic t-statistics in parentheses.
[c] Significant at the 1% level.
[d] Significant at the 5% level.

biomass levels and the implied levels are 159 and 208 million pounds. The four parameter estimates can also be used to compute the safe biomass level implied by regulators' choices of quota levels over this period. Using Eq. (11) above, the levels of X_{safe} implied by actual choices of harvest quota levels over the 1935–1977 period are 187 and 252 million pounds, respectively. Interestingly, the targeted or safe biomass levels implied by these equations are greater than the maximum sustainable yield biomass levels in both Areas 2 and 3, implying a conservative concern for stock safety.

Results from the estimation of the industry/regulator behavioral equations are also given in Table II. Parameter estimates are all of the correct sign and generally significant. Hansen's J test of overidentifying restrictions [11] indicates that the model is not misspecified.[13] With respect to magnitudes, the parameter estimates are also reasonable and similar across regions. We would expect higher costs in

---

[13] Hansen's test is for estimators in the class of Generalized Method of Moments estimators, of which non-linear three stage least-squares is one. It is calculated as the value of the criterion function multiplied by the number of observations, and has a Chi-square distribution where the degrees of freedom are the number of instruments (8) less the number of free parameters (4). Since Eq. (16) is an implicit equation where the dependent variable is zero, there is no adequate goodness of fit measure available.

REGULATED OPEN ACCESS RESOURCE USE

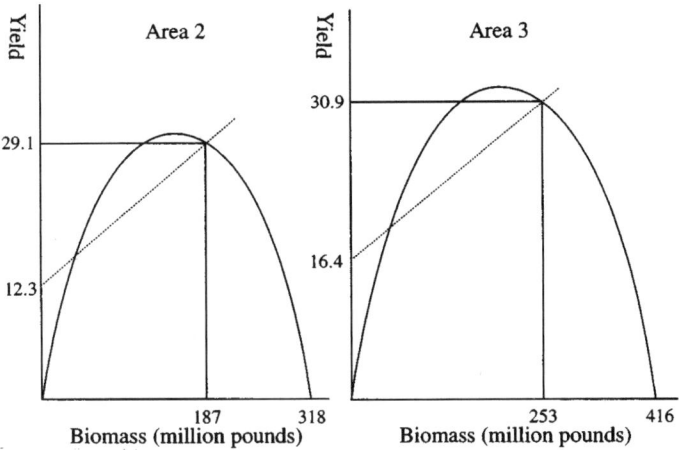

FIG. 4. Quota rule targets.

Area 3 due to its remoteness and this is the case for fixed costs. Fixed cost estimates suggest a twofold difference, with gear up cost estimates about $1,030 per unit of gear in Area 2 and $2,100 in Area 3. The War also had the expected effect of increasing the implicit opportunity cost of participating in the fishery. The implicit opportunity costs added by the hazards of war amount to an extra $790 per unit of gear in Area 2 and almost $9,700 in the Bering Sea. Variable costs are about $56 and $78 per gear set (skate soak) in Areas 2 and 3, respectively. This suggests that operating costs are lower in the more accessible region.

5. IMPLICATIONS AND ANALYSIS

The predictions generated from this model of regulated open access are significantly different from those from the standard open access paradigm. In view of the fact that most fisheries are regulated rather than pure open access, we should also expect that they are closer to what we observe in real fisheries. In this section we compare the predictions of our new model with those that would be generated using the basic H. S. Gordon model of open access equilibrium. The issue to be examined is: what are the implications of (correctly, we would argue) including the regulatory sector in a model of a fishery? To examine this question we calculate the steady state for two alternatives, using parameter estimates described above.

As discussed in the Introduction, the dominant paradigm in the renewable resource literature is the Gordon model of pure open access. We take this model to be represented by a zero rent condition together with a biological dynamics equation which describes how the species evolves. We thus assume instantaneous rent dissipation or fast dynamics for the entry/exit of fishing capacity, and slow dynamics for the biology. Since we have also brought into the analysis the concept of a season length, a question arises as to what to assume about $T(t)$. We assume that the industry ends up fishing for a period equal to $T_{\max}$, which is the rent dissipating season length also. In Fig. 1, the industry reaches a temporary equilibrium which depends upon the biomass level each year, in which $E(t) = E_{max}$ and $T(t) = T_{max}$. Then, depending upon the corresponding level of harvest, the biomass will either rise or fall in an approach to the open access equilibrium.

The Gordon open access model thus consists of three equations

$$P(t)X(t)[1 - e^{-qE(t)T(t)}] - vE(t)T(t) - fE(t) = 0$$

$$X(t+1) = (1+a)X(t) - bX(t)^2 - X(t)[1 - e^{qE(t)T(t)}]$$

$$T(t) = T_{max} = \left[\frac{1}{qE(t)}\right]\ln\left[\frac{PqX(t)}{v}\right]$$

in three unknowns (E, T, X). In the long run steady state, $X(t)$ is constant and the system is in a full open access equilibrium.

The regulated open access model also includes the rent dissipation equation and the biological dynamics equation as in the Gordon model above. The principal difference is that the season length is fixed by regulators in the second stage of the regulatory process, at a length dictated by a quota rule tied to the biomass level. Thus the regulated open access model consists of four equations

$$P(t)X(t)[1 - e^{-qE(t)T(t)}] - vE(t)T(t) - fE(t) = 0$$

$$X(t+1) = (1+a)X(t) - bX(t)^2 - X(t)[1 - e^{qE(t)T(t)}]$$

$$T(t) = \left[\frac{1}{qE(t)}\right]\ln\left[\frac{X(t)}{X(t) - Q(t)}\right]$$

$$Q(t) = c + dX(t)$$

which determine four unknowns (E, T, X, Q).

We simulated these two systems using parameter estimates from Table II. We fixed the value of the exvessel price at its value in 1977. Table III shows the equilibrium values predicted by the alternative models for each of Areas 2 and 3. The main differences in predictions from the two models are significant. The regulated model predicts higher biomass and harvest levels, a higher level of capacity and a shorter fishing season. The reason for these differences are, of course, associated with the explicit inclusion of the regulatory sector. To the extent that regulations are successful, they hold the biomass at larger levels, which *ceteris paribus* generate higher rents and larger levels of capacity, which in turn must be

TABLE III
Regulated vs Unregulated Open Access: Steady State Predictions

	Area 2		Area 3	
	Gordon model	Regulated open access model	Gordon model	Regulated open access model
Capacity	4.5	45.9	2.1	23.6
Season length	79.6	3.3	153.1	5.7
Biomass	37.7	183	56.8	253
Quota/yield	12.6	28.7	15.3	30.9

REGULATED OPEN ACCESS RESOURCE USE

stifled in order to hold the harvest at the targeted quota. For example, in Area 2 the regulated open access equilibrium has ten times the capacity of the unregulated open access equilibrium. However, instead of fishing over 80 days as in the unregulated case, the regulated season length is only 3 days. The actual values predicted by the regulated open access model are also relatively close to what has been observed in the actual fishery recently. For example, total exploitable biomass in Areas 2 and 3 combined reached about 275 million pounds in the late 1980s. Area 2 and 3 total harvest levels during the same period averaged about 24 and 55 million pounds, respectively. Season lengths off Alaska in both Areas 2 and 3 have fallen to about 3–5 days.[14]

The predictions from the regulated open access model point to and explain several interesting facts about modern fisheries. First, regulated fisheries are likely to attract even more redundant capital than was predicted by Gordon's unregulated open access model. The degree to which this is true depends, in fact, directly on the degree to which regulations are successful in holding biomass close to a safe level. At higher biomass levels, more potential rents exist, generating more entry pressure which must be controlled with efficiency decreasing policies. Second, as real prices rise (due to population growth), the industry equilibrium curve in Fig. 3 shifts upward, causing the equilibrium to slide along the regulatory equilibrium hyperbola. Thus it is almost inevitable that over the long run, more capacity is attracted and regulations are tightened. This explains the process we observe whereby in many fisheries, seasons have been reduced to a few weeks, days, and even hours. Finally, this model suggests some important but sometimes counterintuitive links between regulations, the output market, and rent dissipation. In particular, as seasons and other regulations are tightened, a likely market consequence is that exvessel prices are lower than they would otherwise be. For example, short seasons result in poorly delivered and handled product, and high storage costs. As prices are affected by regulatory actions, rent dissipating pressures are actually mitigated somewhat. Thus the direct effects of the regulatory process are to draw in more capacity than the Gordon model would suggest, but dampened somewhat due to indirect effects of regulations on the market. This in turn explains recent observations about fisheries that have become rationalized with individual transferable quota (ITQ) programs. In many of these, a surprising outcome has been that most of the immediate rent gains have seemed to emerge on the revenue rather than the cost side.[15] But this is what we would expect: as the system is released from its restrictive regulatory structure which reduces revenues, the first easy gains come from new marketing opportunities engendered both the unconstrained system and also by the new incentives generated under property rights based regulations.

APPENDIX A

Rent Dissipating Capacity Function

This appendix discusses the shape of the equation depicting industry equilibrium in a regulated open access setting. Industry behavior is assumed to be driven by

---

[14] Cf. U.S. Department of Commerce [23].
[15] Cf. Wilen and Homans [25], Wilen and Homans [26].

rent dissipation in a manner similar to that first described by Gordon in 1954. In particular, with a Schaefer production function and with biomass dynamics within the season governed by harvesting mortality, undiscounted rents for any season are simply

$$\text{Rents} = PX_0(1 - e^{-qET}) - vET - fE. \tag{A1}$$

This equation cannot be solved explicitly for capacity E as a function of other variables. Let $J(E, T; X_0, P, q, v, f)$ describe the function obtained by setting rents in (A1) above equal to zero. Then we can use the implicit function theorem to get

$$\frac{dE}{dT} = -\frac{J_T}{J_E} = \frac{E[v - PqX_0 E^{-qET}]}{T[PqX_0 e^{-qET} - v] - f}. \tag{A2}$$

The denominator of this expression is the net value of marginal physical product of an additional unit of capacity. If capacity enters until rents are dissipated, the net value of average physical product will be zero. Since marginal product is less than average product, and since average product is zero, the denominator must be negative.

The numerator of (A2) is the (negative of the) value of marginal physical product of an additional unit of season length. The sign of the numerator depends upon where in the domain of $E(T)$ we are operating. As it turns out, in the relevant region, the value of marginal product is positive and hence the numerator is negative. This can be seen as follows. Let $\Pi(ET)$ be variable profits associated with arbitrary levels of ET before fixed costs fE are subtracted, or

$$\Pi = PX_0(1 - e^{-qET}) - vET. \tag{A3}$$

Note first that Π is strictly concave in E, T, and ET. In addition, the average variable profit function has a finite limit as E approaches zero, namely

$$\lim_{E \to 0} (\Pi/E) = PqX_0T - vT. \tag{A4}$$

This leads to:

PROPOSITION 1. *A necessary and sufficient condition for positive levels of capacity is that the season length allowed by regulators T be larger than some $T_{\min}$ where*

$$T_{\min} = \frac{f}{PX_0q - v}. \tag{A5}$$

This can easily be seen by noting that capacity will enter only if average variable profits cover fixed entry costs. Since average variable profits are decreasing in E, there will be entry only if the first unit finds it profitable or if

$$\lim_{E \to 0} (\Pi/E) = PqX_0T - vT > f. \tag{A6}$$

But this is true only if T is greater than $T_{\min}$ defined in (A5). This defines a minimum season length below which no positive capacity is feasible.

There is also a maximum season length $T_{\max}$, beyond which the industry would not voluntarily choose to fish. This can be seen by noting that the numerator of

REGULATED OPEN ACCESS RESOURCE USE

(A2) approaches zero at some season length defined by $PqX_0 e^{-qET} - v = 0$. Solving this simultaneously with the rent equation (A1) set equal to zero for T defines what we refer to as T_{max} and we have:

PROPOSITION 2. *The longest season length that the industry would voluntarily operate over is defined by*

$$T_{max} = \frac{f \ln[(PqX_0)/v]}{PqX_0 - v(1 + \ln[(PqX_0)/v])}. \quad (A7)$$

At season lengths longer than T_{max}, the industry would actually be incurring negative variable profits. This can be seen by noting that $X_0 e^{-qET}$ is equal to the terminal stock level X_T. Moreover, since the harvest rate is $h = qEX(t)$, $PqX_T - v$ is the average daily variable profit associated with the last unit of capacity utilized on the last day of the season. As the season progresses, average variable profits per day decline from their initial maximum of of $PqX_0 - v$ to their minimum of $PqX_T - v$ on the last day. Thus Proposition 2 ensures nonnegative variable profits throughout the season. Season lengths larger than T_{max} would be forcing the industry to fish the ending biomass down to levels that generate variable profit losses at season end and the industry would always choose not to incur such losses by truncating fishing at T_{max}. Hence season lengths where regulations are binding are limited to those between T_{min} and T_{max}.

If $T_{min} < T < T_{max}$, then the numerator of (A2) will be negative and the sign of the derivative will be positive. The second derivative of the entry function with respect to T can be shown to be

$$\frac{d^2 E}{dT^2} = E \left\{ 2 \left[\frac{v - PqX_0 e^{-qET}}{T(PqX_0 e^{-qET} - v) - f} \right]^2 + \frac{f^2 Pq^2 EX_0 e^{-qET}}{[T(PqX_0 e^{-qET} - v) - f]^3} \right\}. \quad (A8)$$

At T_{max}, the $(PqX_0 e^{-qET} - v)$ terms vanish and hence the second derivative is negative. Hence, over the relevant range, as T increases E also increases monotonically, reaching a maximum at $E(T_{max})$ as shown in Fig. 1.

APPENDIX B

Comparative Statics and Existence of a Regulated Equilibrium

Comparative statics properties of the regulated open access model are tedious but easily derived. The basic model is as denoted in the text in equation system (12) along with the quota rule (5) and the biomass dynamics equation (8) with the biological equation assumed to be in a steady state. As discussed above, the system is recursive and the parameters a, b, c, and d determine equilibrium values for the biomass level X and the quota Q. Comparative statics properties of the equilibrium levels of E and T in the industry/regulator block can be computed most simply by determining the direct effects of parameters and the indirect effects of parameters operating on the biomass/quota block. (Comparative statics calculations are available upon request from the authors.)

HOMANS AND WILEN

The comparative statics properties summarized in Table I assume an "interior" equilibrium in the sense that the fishery is characterized by a joint equilibrium where regulations are binding. Circumstances where active regulation is necessary exist where there is an equilibrium intersection of the two equations representing the industry and regulator behavior. The circumstances which will lead to a regulated open access equilibrium can be described as follows. Begin with the relationship which describes the level of cumulative effort $ET_{\max}$ defined as

$$ET_{\max} = \frac{1}{q}\ln\left[\frac{PqX_0}{v}\right] \tag{B1}$$

for some arbitrary level of biomass X_0. Note that this defines a rectangular hyperbola in E, T space and hence can be compared with the regulatory agency's choice which is also defined by a rectangular hyperbola, namely

$$ET_{\text{reg}} = \frac{1}{q}\ln\left[\frac{X_0}{(1-d)X_0 - c}\right]. \tag{B2}$$

A joint or binding temporary regulatory equilibrium occurs whenever the maximum level of effort which dissipates unregulated rents or $ET_{\max}$ is greater than the amount of total fishing effort desired by the regulatory agency ET_{reg}. Thus the condition for existence of a temporary regulatory equilibrium is

$$ET_{\max} > ET_{\text{reg}} \Rightarrow \frac{PqX_0}{v} > \frac{X_0}{(1-d)X_0 - c}. \tag{B3}$$

Thus a regulated equilibrium is likely to occur when P, q, or X_0 are large, or when v is small, or whenever d or c are small. Note that d small implies that regulators are particularly concerned about high exploitation rates.

The above refers to the (temporary equilibrium) case where X_0 is some arbitrary value which could of course be its long run open access steady state value. In general, however, a full long run equilibrium must not only have the industry earning zero rents, but the biomass must also be in equilibrium at that state. The long run steady state biomass level that would evolve in an unregulated (Gordon) system is

$$X_G = \frac{a - 1 + \sqrt{(1-a)^2 - (4bv/Pq)}}{2b}, \tag{B4}$$

while the biomass level that would emerge in a regulated open access system is given by Eq. (11) in the text. The complete characterization of necessary conditions for a binding long run regulated open access equilibrium is given when these are inserted into Eq. (B3) above, or

$$\frac{PqX_G}{v} > \frac{X_{\text{safe}}}{(1-d)X_{\text{safe}} - c}. \tag{B5}$$

REGULATED OPEN ACCESS RESOURCE USE

REFERENCES

1. J. Bhagwati, Directly unproductive profit seeking activities, *J. Polit. Econom.* **90**, 988–1002 (1982).
2. J. Buchanan, R. Tollison, and G. Tullock (Eds.), "Toward a Theory of a Rent Seeking Society," Texas A & M Univ. Press, College Station (1980).
3. S. Capalbo, "Bioeconomic Supply and Imperfect Competition: The Case of the North Pacific Halibut Industry," Ph.D. dissertation, University of California at Davis (1982).
4. B. Cook, "Optimal Harvest Levels for Canada's Pacific Halibut Fishery," Ph.D. dissertation, Simon Fraser University (1983).
5. J. Conklin and W. Kolberg, Chaos for the halibut?, *Mar. Resour. Econom.* **9**, 159–182 (1994).
6. K. Criddle, "Modeling Dynamic Nonlinear Systems," Ph.D. dissertation, University of California at Davis (1989).
7. K. Criddle and A. Havenner, Forecasting halibut biomass using systems theoretic time series, *Amer. J. Agr. Econom.* **71**, 422–431 (1989).
8. J. Crutchfield and A. Zellner, "Economic Aspects of the Pacific Halibut Fishery," U.S. Department of the Interior, Fishery Industrial Research, Washington, DC (1962).
9. R. Deriso, Risk averse harvesting strategies, *in* "Resource Management: Proceedings of the Second Ralf Yorque Workshop, Lecture Notes in Biomathematics" (M. Mangel, Ed.), Springer-Verlag, Berlin (1985).
10. H. S. Gordon, The economic theory of a common property resource: The fishery, *J. Polit. Econom.* **62**, 124–142 (1954).
11. L. P. Hansen, Large sample properties of generalized method of moments estimators, *Econometrica* **50**, 1029–1054 (1982).
12. R. Hilborn and C. J. Walters, "Quantitative Fisheries Stock Assessment: Choice, Dynamics, and Uncertainty," Chapman and Hall, New York (1992).
13. F. R. Homans, "Modeling Regulated Open Access Resource Use," Ph.D. dissertation, University of California at Davis, (1993).
14. J. Karpoff, Suboptimal controls in common resource management: The case of the fishery, *J. Polit. Econom.* **95**, 179–194 (1987).
15. D. Ludwig and C. J. Walters, A robust method for parameter estimation from catch and effort data, *Canad. J. Fish. Aquat. Sci.* **46**, 137–144 (1989).
16. T. Quinn, R. Deriso, and S. Hoag, "Methods of Population Assessment of Pacific Halibut," International Pacific Halibut Commission Scientific Report No. 72 (1985).
17. C. Rowley, R. Tollison, and G. Tullock (Eds.), "The Political Economy of Rent Seeking," Kluwer Academic, Boston (1988).
18. A. Scott, The fishery: The objectives of sole ownership, *J. Polit. Econom.* **63**, 116–124 (1955).
19. "SHAZAM User's Reference Manual Version 7.0," McGraw–Hill, New York (1993).
20. B. Skud, "Revised Estimates of Halibut Abundance and the Thompson–Burkenroad Debate," International Pacific Halibut Commission Scientific Report No. 63 (1975).
21. K. Stollery, A short-run model of capital stuffing in the Pacific Halibut Fishery, *Mar. Resour. Econom.* **9**, 137–153 (1986).
22. R. Turvey, and J. Wiseman (Eds.), "The Economics of Fisheries," FAO, Rome (1957).
23. U.S. Department of Commerce, National Ocean and Atmospheric Administration, Draft Environmental Impact Statement for Proposed Individual Quota Management Alternatives for the Halibut Fisheries in the Gulf of Alaska and Bering Sea/Aleutian Islands, July (1991).
24. J. E. Wilen, Towards a theory of the regulated fishery, *Mar. Resour. Econom.* **1**, 369–388 (1985).
25. J. E. Wilen and F. R. Homans, Marketing losses in regulated open access fisheries, *in* "Fisheries Economics and Trade: Proceedings of the Sixth Conference" (J. Catanzano, Ed.) IFREMER-SEM (1994).
26. J. E. Wilen and F. R. Homans, Unraveling rent losses in modern fisheries: Production, market, or regulatory inefficiencies? "Fisheries Management—Global Trends" (D. Huppert, E. Pikitch, and M. Sissenwine, Eds.) Univ. of Washington Press, Seattle, 1997.

Fisheries co-management

Delegating government responsibility to fishermen's organizations

Svein Jentoft

'The only kind of coercion I recommend is mutual coercion, mutually agreed up by the majority of the people affected'

Garret Hardin, 'The tragedy of the commons', *Science*, Vol 162, 1968, pp 1243–1248.

This article addresses the role of cooperative organizations in fisheries management and the extent to which fishermen's organizations are capable of handling regulatory functions. What are the problems inherent in the cooperative management approach, and what may be the benefits compared to other regulatory systems? Which circumstances may be beneficial for the success of co-management? The author draws on comparative international experiences to form conclusions regarding the efficacy of a cooperative management regime.

The author is at the Institute of Social Sciences, University of Tromso, Breivika, PO Box 1040, N 9001 Tromso, Norway.

This article is a revised version of a paper presented to a symposium on 'Gulf Coast Maritime Utilization', Mobile, Alabama, 4–6 May 1988, organized by the University of South Alabama. The paper was also presented on 23 June 1988 at a staff meeting in the US Department of Commerce, National Marine Fisheries Services, Washington, DC. Preparation of this paper was supported by the Norwegian Fisheries Research Council and written while in residence at the Department of Agricultural Economics and Rural Sociology at Auburn University, Alabama. The paper has benefitted from constructive critiques made by Conner Bailey, Soren Christensen, Petter Holm, Helge O. Larsen, Leigh Mazany,
continued on p 138

In order to ensure sustainable harvests of fisheries resources and avoid what is generally known as the 'Tragedy of the Commons', strict management practices are needed. Generally, it is assumed that fisheries management is a government responsibility. Various management mechanisms have been used including licensing systems, catch quotas and other control measures. However, the experience from most countries shows that very often these management systems have met mixed success.[1]

One focus for this debate has been what kind of regulatory means should the government use. For instance, should input regulations (licences) be replaced by output regulations (fish-quotas)? A more recent issue for debate, which will be the focus in this article, has been the division of responsibility between the government and the fishing industry. Should the government take full responsibility for all management functions, including the establishment of quotas, deciding which fishermen should be allowed access into the fishery, promulgating detailed rules for the conduct of the fishery, and monitoring the fishery to see that all the rules are being obeyed? Or could some, perhaps all, of these functions be more efficiently carried out by fishermen's cooperative organizations? If the answer to the second question is affirmative, why is this so?

These are questions which will be addressed in this article. More specifically I will discuss: what explains the failure of government regulations in the fisheries? What exactly makes fishermen's organizations suitable instruments for fisheries management? What are the organizational implications of delegating responsibility? What are the possible negative effects of delegating management tasks to fishermen's organizations? What circumstances may be beneficial for a successful result?

Fisheries co-management
continued from p 137
Mike Skladany and Jim Stallings. Responsibility for accuracy of fact, interpretations and analysis lies with the author.

[1] L. Chatterton and B. Chatterton, 'How much political compromise can fisheries management stand? Premiums and politics in closed coastal fisheries', *Marine Policy*, Vol 5, No 2, 1981, pp 114–134; P. Copes, 'A critical review of the individual quota as a device in fisheries management', *Land Economics*, Vol 62, No 3, 1986, pp 278–291; A. Davis and V. Thiessen, 'Public policy and social control in the Atlantic fisheries', *Canadian Public Policy*, Vol XIV, No 1, 1988, pp 66–77; R. Hannesson, 'Inefficiency through government regulation: the case of Norway's fishery policy', *Marine Resource Economics*, Vol 2, 1985, pp 115–141; and K.H. Mikalsen, *Limits to Limited Entry? Entry Limitation as a Fishery Management Tool*, paper presented at the 20th Annual Meeting of the Atlantic Association of Sociologists and Anthropologists, University of Prince Edward Island, 14–16 March 1985.
[2] J. Kearney, 'The transformation of the Bay of Fundy herring fisheries in 1976–1978: an experiment in fishermen–government co-management', in C. Lamson and A. Hansson (eds), *Atlantic Fisheries and Coastal Communities: Fisheries Decision-making Case Studies*, Institute of Resource and Environmental Studies, Dalhousie University, Halifax, 1985; B.J. McCay and J.A. Acheson, *The Question of the Commons: The Culture and Ecology of Communal Resources*, The University of Arizona Press, Tucson, 1987; E. Pinkerton, 'Co-operative management of local fisheries: a route to development', in J.W. Bennett and J.R. Bowen (eds), *Production and Autonomy: Anthropological Perspectives on Development*, Society for Economic Anthropology and University Press of America, Lanham, Maryland, 1987.
[3] H. Scott Gordon 'The economic theory of a common property resource: the fishery', *Journal of Political Economy*, Vol 62, No 2, 1954, pp 124–42; G. Munro, 'Fisheries, extended jurisdiction and the economics of common property resources', *Canadian Journal of Economics*, Vol XV, No 3, 1982, pp 405–425.
[4] G. Calabresi and P. Bobbitt, *Tragic Choices: The Conflicts Society Confronts in the Allocation of Tragically Scarce Resources*, W.W. Norton & Co, New York, 1978.
[5] S. Jentoft, 'Fiskeripolitikk som nullsumspill', in B. Hersoug (ed), *Kan Fiskerinaeringa styres?* Novus Forlag, Oslo, 1983; L. Thurow, *The Zero Sum Society*. Basic Books, New York, 1980.

Experiences with delegating management responsibility to fishermen's organizations in different countries will be used as a basis from which to address these questions and draw conclusions regarding the efficacy of co-operative regimes as a management tool. Fishermen's organizations take an active part in designing, implementing and enforcing fisheries regulations have by various authors been termed 'co-management'.[2] This concept will also be used here.

Expediency and legitimacy

The rationale for government action in fisheries management is at least threefold. First, it is argued that the government should get involved for efficiency reasons. Fish as a common pool resource introduces externalities which, with open access, frequently leads to depletion of the resource base and dissipation of the potential resource rent.[3] To prevent this from happening, the state is called on to exercise strict control over harvesting capacity and the total volume of catches. Second, it is argued, the state must be involved for *equity reasons*. It has a role in securing a fair distribution of fishing opportunities and incomes among participant groups. In many countries development policy is closely connected to management schemes. Thus, one motivation for government control is to allow marginal regions and small-scale fisheries a chance to survive. Third, it is argued that the state must be involved for *administrative reasons*. Only the state is seen to have authority and resources sufficient to implement management schemes. And only the state has at its disposal the means of force to ensure that the rules are followed.

These arguments have motivated extensive government involvement in fisheries management in most industrialized countries. However, in performing the management role, governments have faced 'tragic choices'. Keeping the industry viable and profitable while at the same time securing equitable income distribution may be mutually exclusive goals. Somehow these goals must be balanced. Also, for a given quota, the fishermen's race for fish is a zero-sum game. The government can influence the outcome of this game, but there will still be losers as well as winners. It is a general experience, not only in the fishing industry, that solving such conflicts is a political process, requiring hard decisions.[5]

In fisheries management, governments usually choose between two general options: *indirect regulation* and *direct regulation*. Indirect regulations try to control the total harvesting effort by regulating the number of participant fishermen, the size of their boats, and/or the number and type of gear. Territorial and seasonal regulations, which restrict fishermen's access to certain fishing grounds at certain periods of time also belong to this category. While indirect regulations try to control the inputs of manpower and/or capital, direct regulations seek to limit output. Fixing a level for a total allowable catch (TAC) is one way. Dividing the TAC into individual quotas (per man or per boat) is another.

Experiences with indirect regulations are primarily negative: they have scarcely obtained the intended results and often produced unintended consequences. For instance, such regulations fail to cope with overcapitalization and resource depletion because they stimulate the adoption of more efficient technology. They close the door to new

Fisheries co-management

entrants, and, as a consequence, they establish privileges which make the fishery a 'rich man's club'. Indirect regulations are difficult to administer and enforce. They also create a very inflexible regulatory system: once adopted they are hard to change.[6]

Today, fisheries economists contend that indirect regulations should be replaced by direct regulations.[7] They argue that the introduction of individual quotas will simplify the regulatory system dramatically. Fishermen should receive quotas, free or for some price, and they should be allowed to trade their quotas. Transferability, it is argued, will help to increase economic efficiency.

There are some promising reports on successful management systems based on output control, from countries such as New Zealand, Canada and Iceland.[8] However, as has been the lesson from many years of input regulations, there have been unintended effects. As Copes has demonstrated, individual quota management also has its pitfalls. For instance, it has proved difficult to ensure that fishermen do not exceed their quotas. Fishermen will often misreport their catches.[9] Thus, Copes finds reasons to conclude that fisheries regulations are exceptionally vulnerable to Murphy's Law: 'If anything can go wrong with a new fisheries management scheme ... it will'.[10] Regulations, both indirect and direct, mean by definition that the government imposes restrictions on fishermen. Fishermen almost always have an immediate economic interest in finding ways to bypass them. 'There is no reason to assume that fishermen, when confronted with the rules of individual quota management, will lose either their ingenuity at circumvention or their incentive to promote individual interests at the expense of collective interest'.[11]

The crucial question for the success of any management scheme is what measures are needed to get fishermen voluntarily to advance their collective interests at the expense of their private ones. In other words, what could motivate fishermen to adhere loyally to the regulations? A keyword here is 'legitimacy'; ie to what extent fishermen willingly accept the regulations as appropriate and consistent with their persisting values.[12] If fishermen find the regulatory scheme legitimate, there is more reason to believe that they will follow the rules. Then, how could legitimacy be improved?

We suggest that the legitimacy of a regulatory scheme is related to at least four general hypotheses: 1) *Content of the regulations*: the more that regulations coincide with the way fishermen themselves define their problems, the greater will be their legitimacy. 2) *Distributional effects*: the more equitably are restrictions imposed, the more legitimate will the regulations be regarded. 3) *Making of the regulations*: the more fishermen are involved in the decision-making process, the more legitimate the regulatory process will be perceived. 4) *Implementation of the regulations*: the more directly involved fishermen are in installing and enforcing the regulations, the more the regulations will be accepted as legitimate.

Thus, there may be at least four ways to improve the legitimacy of fisheries regulations and to increase their prospects of success; each requires taking the fishermen's point of view into closer consideration. In the first two hypotheses, the content and quality of the regulations per se are the focal points. The last two hypotheses concern the organization of the decision-making process.

In this article we are particularly interested in hypothesis three and

[6]Mikalsen, *op cit*, Ref 1.
[7]F.T. Christy, *Fishermen Quotas: A Tentative Suggestion for Domestic Management*, occasional paper no 19 of the Law of the Sea Institute, University of Rhode Island, Kingston, 1973; Hannesson, *op cit*, Ref 1; O. Flaten, 'Fiskeriplanlegging og biooekonomisk teori', *op cit*, Ref 5; and A. Scott and P.A. Neher *The Public Regulation of Commercial Fisheries in Canada*, Economic Council of Canada, Ottawa, 1981.
[8]R. Hannesson, *Fishermen's Organizations and their Role in Fisheries Management*, a study undertaken for the FAO, The Norwegian School of Economics and Business Administration, Bergen, 1987.
[9]R. Arnason, *Management of the Icelandic Demersal Fisheries*, a report for the workshop on North Pacific fishery management options, April 21–24 1986; J.A. Gulland *Control of the Amount of Fishing by Catch Limits*, FAO Fisheries Report No 289, Supplement 2, Rome, 1983; and R.L. Stokes 'Limitation of fishing effort: an economic analysis of options', *Marine Policy*, Vol 3, No 4, 1979, pp 289–301.
[10]Copes, *op cit*, Ref 1.
[11]*Ibid*.
[12]'Legitimacy refers to the degree of acceptance which the political regime enjoys among the community', G. Ponton and P. Gill, *Introduction to Politics*, Martin Peterson, Oxford, 1982. J.C. Plano and R.E. Riggs *Dictionary of Political Analysis*, The Dryden Press, Minsdale, II, 1973, provide a more elaborate definition. They see legitimacy as the quality 'of being justified or willingly accepted by subordinates that converts the exercise of political power into rightful authority'. The classical treatise of the foundations of legitimacy can be found in M. Weber, *The Theory of Social and Economic Organization*, The Free Press, New York, 1964.

Fisheries co-management

four above. How can the legitimacy, and hence the expediency, of fisheries regulations be improved by involving fishermen's organizations directly in the regulatory making process? At best, one should expect both a direct and an indirect effect. Participation would in itself tend to advance legitimacy, but in addition, participation should also improve the quality of the regulations as such. In other words, by organizing the regulatory process (hypothesis three and four), the content as well as the distributional effects of the regulations (hypothesis one and two) should be improved. This argument will be outlined in the following sections. We start, however, by describing some international experiences with fisheries co-management.

International experiences

The existence of locally organized informal fisheries management systems have been well documented by social anthropologists with interest in fisheries and maritime communities.[13] These regulations usually take the form of territorial use rights. Here, fishermen from a certain community share tacit agreements on the conduct of the fishery within waters which they consider as 'theirs', and which they actively protect from 'intruders'. Sometimes these regulations are established for reasons of resource protection. Very often their main rationale is to create order and avoid gear conflicts or to ensure fair distribution of access opportunities to the fishing grounds.

Compared to the many studies of informal regulations by fishermen, there are few reports on regulations by formal fishermen's cooperative organizations. But those which are available give a clear picture of a fisheries management system which can not be discarded as Utopian or irrelevant, not even in industrialized fisheries. These reports demonstrate that fishermen, if properly organized, can handle management functions and that they are able to solve their conflicts of interest even if they take the form of zero-sum games.

In some cases, management by cooperatives has developed spontaneously and exists in addition to central government regulations. McCay's study of a fishermen's cooperative in the New York Bight Region of the Mid-Atlantic coast, can be classified here. The cooperative performs management tasks, based on exclusive control of dock facilities, restriction of access of newcomers as members, and the imposition of catch quotas among its members. This is done primarily for the purpose of controlling the price on the products the cooperative is selling. Nevertheless, its success in regulating the fishery leads McCay to draw the conclusion that this is a way of fisheries management with a much wider potential.

Berkes[14] also regards cooperatives as positive tools in fisheries management, particularly as they relate to small-scale fisheries. In a Turkish case study he describes several examples of cooperatives actively taking part in management functions. Berkes argues that effective local-level management is impossible 'without the existence of institutions and mechanisms suitable for achieving consensus among fishermen participating in the fishery.'

This conclusion is also supported by a Norwegian case study of the Lofoten Fishery.[15] It describes an example of fishermen's cooperative management which has been in existence, codified by law, for more than 90 years. Fishing takes place from January to April off the Lofoten Islands in north Norway where the arctic cod has its spawning grounds.

[13]J.M. Acheson, 'The lobster fiefs: economic and ecological effects of territoriality in the Maine lobster industry', *Human Ecology*, Vol 3, No 3, 1975, pp 183–207; F. Berkes 'Fishermen and the tragedy of the commons', *Environmental Conservation*, Vol 12, No 3, 1985, pp 199–205; C. Dahl 'Traditional marine tenure: a basis for artisanal fisheries management', *Marine Policy*, Vol 12, No 1, 1988, pp 40–48; A. Davis, 'Property rights and access management in the small boat fishery: a case study from South Western Nova Scotia', in C. Lamson and A. Hanson (eds), *Atlantic Fisheries and Coastal Communities: Fisheries Decision-making Case Studies*, Institute for Resource and Environmental Studies, Dalhouse University, Halifax, 1985; E.P. Durrenberger and G. Palsson 'Ownership at sea: fishing territories and access to sea resources', *American Ethnologist*, Vol 14, No 3, 1987, pp 508–522; B.J. McCay 'A fishermen's cooperative limited: indigenous resource management in a complex society', *Anthropological Quarterly*, Vol 53, 1980, pp 29–38; J.R. McGoodwin 'Some examples of self-regulatory mechanisms in unmanaged fisheries', FAO Fisheries Report No 289, Supplement 2, Rome, 1983.
[14]F. Berkes 'Local level management and the commons problem', *Marine Policy*, Vol 10, No 3, 1986, pp 215–229.
[15]S. Jentoft and T. Kristoffersen *Fishermen's Self Management: The Case of the Lofoten Fishery*, forthcoming.

Fisheries co-management

For hundreds of years the Lofoten fishery has attracted fishermen from north to south in the country. The high number of participant fishermen caused enormous crowding problems on the fishing grounds which led to frequent conflicts, particularly between fishermen using different kinds of gear. During the 19th century, various kinds of regulatory systems were tried, but none of them seemed to be able to solve the regulatory problems; not until co-management principles were introduced in the late 1890s. The Norwegian Government enacted special legislation for the Lofoten Fishery which actually delegated responsibility for the regulation of the fishery to the fishermen themselves. Special district committees of fishermen representing different gear groups were set up to make the rules for the fishery, such as: allowable fishing times; which gear is allowed on which fishing grounds; and how much space should be reserved for certain gears such as handlines, gillnets, longlines, seines. In addition to elected fishermen inspectors, a public enforcement agency was established to assure that the rules initiated by the fishermen committees were being obeyed. This system still prevails today. Some minor changes have been initiated, but the co-management principles are intact.

An example of fisheries co-management that failed is reported by Kearney[16] in the Bay of Fundy herring fishery on the east coast of Canada. A cooperative was established in the mid-1970s. In addition to fisheries regulations, the cooperative also had a marketing function. Thus, it was able to strengthen the bargaining position of the fiß vis-a-vis the fish processors. The control over the harvesting operations given by allocating quotas among the member fishermen from a total fleet quota reserved for the cooperative by the government, contributed to this strong bargaining position. A further contributing factor was that the cooperative was authorized to organize 'over-the-side' sales to foreign vessels. This gave the fishermen an alternative sales outlet to the private local fish processors. The cooperative was also responsible for policing the vessel quotas, allocating nightly markets, distributing surplus quotas among the fleet, and collecting statistical information for the government. Thus, according to Kearney, the cooperative 'assumed many administrative functions normally performed by the government, and in its day-to-day control of harvesting effort in relation to market availability, the AHFMC had taken on a decision making function usually associated with government regulation of a common property resource'.

However, the cooperative failed after a few years. A general decline in the fishery made it difficult to enforce its regulatory scheme. Gear conflicts and tensions over the contribution of resource benefits among traditional small-scale fishermen and fishermen using modern capital intensive fishing technology had a similar impact. As a consequence, some fishermen left and established individual marketing arrangements. After a couple of years, and as a result of intense lobbying by the processors who also grew dissatisfied with the cooperative, the Government withdrew its authority to negotiate contracts for over-the-side sales. This was the final blow.

The most successful example of fishermen's cooperatives playing a prominent role in fisheries regulations occurs in Japan. While the cases of fisheries co-management referred to above are exceptions in the regulatory system of those countries, this is not the case in Japan where co-management is the main principle in coastal waters.[17]

[16]Kearney, *op cit*, Ref 2.
[17]The Japanese management system is well documented in the academic literature, see for example, K. Shima, *The Role of Cooperatives in the Exploration and Management of Coastal Resources in Japan*, FAO Fisheries Report, No 295, Supplement, Rome. This author visited fisheries cooperatives in Japan in June 1987.

Fisheries co-management

The management function of the cooperatives has roots in feudal times, and was, until the turn of this century, largely administered by village guilds. In 1901, a new fisheries law was promulgated. Inspired by the famous Rochdale Pioneers' Society in England in 1844, in which the original cooperative principles were formulated, these guilds were redesigned as fisheries cooperatives and granted their legal status. They started as organizations to administer fisheries regulations, but gradually expanded into other areas, such as marketing, processing, leasing out fishing equipment, purchasing supplies, education and the like. Today, there are close to 5 000 fisheries cooperatives scattered all around the coast.[18] On the regional and national level these cooperatives form federations and an umbrella organization. In addition there are supportive cooperative institutions for finance, insurance and the like.

The Japanese fisheries management system is based on two pillars: fishery rights and fishing licences. Fisheries rights concern fixed gears and fish or marine plants which are relatively stable. Thus, fisheries rights are mainly confined to the inshore waters. Fishing licences concern offshore fisheries and fishermen that operate throughout a wider area with non-stationary fishing gear like the trawl and purse seine. Fishery rights are defined by territory. Each cooperative has exclusive ownership to the area outside their port, extending as far as 10 km out to sea. Depending on the type of fishery or aquaculture, the cooperatives have either a monopoly or priority over private individuals or companies. Fishing licences are seldom held by cooperatives in the offshore or distant water fishery, even though they are eligible to do so. In the inshore fishery, however, they apply to government for licences which they distribute among their members.

The high percentage of organizational coverage of fisheries cooperatives in Japan is because they have been authorized to regulate fishing rights, and fishermen have to be members of a cooperative in order to engage in fishing.[19] Member fishermen which do not abide by the rules established by the cooperative risk being expelled from the cooperative by the general membership.

While the principle of co-management in Japan is primarily restricted to the inshore fisheries, this is not the case in the UK fisheries. Here, co-management also is introduced in the offshore fisheries. Another difference is worth noticing. In Japan, inshore regulations have a territorial basis: by reserving a limited area at sea for members of a certain cooperative, fishermen cannot expand into another cooperative territory. In the UK, on the other hand, regulations are enforced through quota allocations.

In the early 1970s, when the UK joined the EEC, producers' organizations were set up all around the country. Their function was to organize raw fish sales and to administer the EEC price support scheme. (In 1986 there were 14 such organizations.) Fisheries regulations were a government responsibility, and quota allocations were a matter between government and individual fishermen. However, in 1984 the Government decided to decentralize the management function by transferring the regulatory responsibility to the producers' organizations. Instead of dividing the TAC among individual fishermen, the Government now allocated sectoral quotas to the producers' organizations. Thus, these organizations became responsible for the distribution of quotas among their members. Rules for fishing operations and

[18]Zengoryen, *Fisheries Cooperative Associations in Japan*, National Federation of Fisheries Cooperative Associations, Tokyo, 1984.
[19]*Ibid.*

Fisheries co-management

enforcement of the quotas also became a task for the organizations.

How this system works has not yet been closely studied, but apparently it works well.[20] There has been little opposition among the fishermen to the arrangement. John Goodlad, chief executive of the Shetland Fish Producers' Organisation Limited, concludes that it has been 'a successful experiment in the devolution of fisheries management responsibility from National Government to the fishermen'.[21] Other chief executives in the Scottish fish producers' organizations voiced similar opinions when interviewed by this author. A problem underlined by all, however, was that these organizations do not have any monopoly power. Membership is voluntary and fishermen outside the organizations can get individual quotas directly from the Government. According to the same chief executives, this tends to undermine the system. (In 1984, 65% of the UK quota was administered by the producers' organizations.) Another problem is that different producers' organizations, even when located in the same port, may have different regulations. This creates tensions between fishermen belonging to different organizations. A positive factor voiced by several representatives of the producers' organizations was the fact that they, in addition to management, also are responsible for fish-marketing. The market situation could be taken into account when regulatory decisions were made, thereby ensuring a stable fish price. Also the other cooperatives described above, with the exception of the Norwegian case, are multi-purpose organizations in this respect, and similar effects are obtained.

The general rules applying to all producers' organizations in Scotland and the Islands (Shetland, Orkneys, Hebrides) are outlined in a consultation paper from the Department of Agriculture and Fisheries of Scotland. For instance, the calculation of sectoral quotas is based on the track record of particular vessels. The track record calculation is applied to the vessels currently holding membership of a producers' organization. If, during the course of the year, it becomes clear that a producer organization will not catch its quota allocation, reallocation to other organizations is made. Account is then taken of the 'need' of the different groups for additional quotas. When a producers' organization overfishes its quota, a tonne for tonne reduction is made from the group's corresponding quota in the following year.

Summing up, these examples show that delegating responsibility for fisheries regulations to fishermen has been carried out in various countries, but with mixed results. Co-management systems have been introduced in both inshore and offshore fisheries, for stationary as well as highly mobile fleets. In some cases co-management takes the form of territorial regulations, in other cases quota allocation is the tool. We will return to these examples of fisheries co-management later for a closer analysis of what may explain the variable success. In the next section, however, we will discuss the co-management concept. What does co-management really mean? What are the implications of delegating responsibility to fishermen's cooperative organizations.

Delegating responsibility

By definition, fisheries co-management means that government agencies and fishermen, through their cooperative organizations, are sharing responsibility for management functions.[22] The point of departure for

[20] This author interviewed representatives of several producers' organizations in Scotland and Shetland in April 1987.
[21] J. Goodlad, *Regional Fisheries Management: The Shetland Experience*, notes prepared for the Norwegian/Canadian Fisheries Management Workshop, Tromsø, 16–21 June 1986.
[22] C. Bailey, 'Managing an open access resource: the case of coastal fisheries', in D.C. Corten and R. Klauss (eds), *People-Centered Development: Contributions Toward Theory and Planning Framework*, Kumarian Press, West Hartford, CT, 1984; Kearney, *op cit*, Ref 2; Pinkerton *op cit*, Ref 2.

initiating co-management agreements as part of a political process can vary from country to country. In one case it can mean that the government formally recognizes regulations which are already being enforced in an informal manner by the fishermen themselves. In another, the actual regulatory power is transferred from the government to fishermen's organizations. This would normally be the situation in fisheries where the government already plays a prominent management role.

Organizational conditions affecting the deregulation of regulatory responsibility differ from country to country. In the UK system, organizations suitable for fisheries management were already in place when the Government decided to introduce sectoral quota allocations. If such organizations had been absent prior to the decision to introduce co-management, they would have had to be formed as part of the process of introducing co-management. This, for instance, happened in the Lofoten Fishery of Norway, described by Jentoft and Kristoffersen.[23] When organizational formation becomes a component of the new regulatory strategy, co-management becomes a more ambitious, and certainly a more complicated process. The prospects of success of co-management will largely depend on whether or not such organizations can function as viable institutions. That was indeed one of the problems which caused the failure of the Canadian experience.[24]

Co-management also means that fishermen's organizations are granted authority by law to enforce regulations on member fishermen. In some cases, as in the Japanese, this authority is based on legislated ownership rights to fishing territories. In Lofoten regulations follow a somewhat similar principle in that different gear types are allocated different territories. In other cases, as in the UK, each fishermen's organization gets quotas for its own discretional disposal. In all three cases, the organizations have the right to exclude non-members from sharing the territory or the quota and to sanction members who violate the rules.

Co-management is to be distinguished from 'consultative' arrangements which, for instance, have been in existence for several years in Norway as well as in many other countries like Canada[25] and the USA.[26] Such arrangements usually involve an advisory board, in which representatives of the fishing industry are consulted by the government before regulations are introduced. In contrast, co-management means that fishermen's organizations not only have a say in the decision-making process, but also have the authority to make and implement regulatory decisions on their own. Thus, in Norway, the regulation of the Lofoten Fishery is an exception from the general rule. The law which delegates regulatory responsibility to the fishermen's committees has nothing to say on the content of the decisions per se, only on how the decision-making process is to be organized.

How, then, is co-management to be distinguished from other common property management systems, such as government regulations or community initiated regulations? Co-management takes a middle course. It is a meeting point between overall government concerns for efficient resource utilization and protection, and local concerns for equal opportunities, self-determination and self-control. The responsibility for initiating regulations is shared. The government's responsibility may be to provide the general framework for operation of the cooperatives such as: the general legislation to install co-management principles; fixing total allowable catch; allocation of quotas between

---

[23]Jentoft and Kristoffersen, *op cit*, Ref 15.
[24]Kearney, *op cit*, Ref 2.
[25]*Ibid.*
[26]P. Fricke 'Use of sociological data in the allocation of common property resources: a comparison of practices, *Marine Policy*, Vol 9, No 1, 1985, pp 39–52.

Fisheries co-management

different fishermen's organizations; and, perhaps, also deciding the general framework for the organization of the regulatory process as in the Lofoten case. The government could also retain control over the total catching capacity through a licensing system. This is, for instance, the situation in the UK. However, the practical use of the licence is very much influenced by the fishermen's organizations. The producers' organization controls who is to become a member and thereby obtaining a share of its quota. In Japan, when a licensing system is installed for a fishery, the cooperative may receive a number of licences leaving it to the cooperative to distribute them among its members at its own discretion.[27]

Such overall rules could also be worked out in cooperation between government agencies and fishermen's organizations, as suggested by Chatterton and Chatterton[28] in the case of Australia:

> Negotiations could take place between fishermen in a particular fishery and government for a contract that would lay down important principles of ownership, participation and conservation. Once the contract had been negotiated the government could hand the day-to-day management of the fishery over to a cooperative board of fishermen elected from among the fishermen themselves.

Co-management is formal in the sense that regulations are made explicit and public and that the decision-making process itself has to follow certain procedures which ensure active participation from the affected interests. Importantly, fishermen are not necessarily the only affected group. Co-management allows the rules to be less detailed and comprehensive, and decisions can be made in a more ad hoc fashion. In comparison, local community regulations often result from a process of mutual adjustment, taking the form of unwritten norms and carried out through informal sanctions.[29] Co-management requires formal leadership and an executive staff. Leaders are elected from among the membership, and an executive staff has administrative responsibility for ensuring that regulatory decisions are implemented.

The essential characteristics of co-management as distinguished from government management systems and informal community based management systems are summarized in Table 1.

Why co-management?

The fishing industry is extremely complex, characterized by a wide range of social conditions and technological processes. Furthermore, fishing operations may vary over the years, seasons and places. There is no simple management solution appropriate for integrating all the different needs, demands, and interests within the sector. Thus, co-management agreements will hardly be a panacea for solving all the problems of fisheries management. However, when benefits and costs

[27]Y. Hirasawa, *Coastal Fishery and Fishery Rights*, paper delivered at the Symposium on Development and Management of Small-scale Fisheries, Indo-Pacific Fishery Commission, Kyoto, May 1980.
[28]Chatterton *et al*, *op cit*, Ref 1.
[29]Acheson, *op cit*, Ref 13.

Table 1. Main characteristics of fisheries management systems.

Characteristic	Fisheries management systems		
	Government	Cooperative	Community
Initiative	Central	(De) central	Local
Organization	Formal	Formal	Informal
Leadership	Hierarchy	Participant	Mutual adj
Control	Central	(De) central	Decentral
Autonomy	No	Some	Yes
Participation	No	Yes	Yes

Fisheries co-management

are taken into account, co-management must be considered a viable option in comparison to other management alternatives.

A central argument in this article is that the expediency of fisheries regulations hinges on their legitimacy. From international experiences one may conclude that there is little chance fisheries regulations can succeed unless they have the active support of affected interests, particularly the fishermen. Without the active support of fishermen they will find ways to bypass the regulatory measures. The legitimacy of fisheries regulations is largely contingent on the decision-making process itself. The distributive effects on incomes of fisheries regulations are important, but so also is the distribution of influence in the decision-making process.

The distribution of influence is an organizational matter, and that is basically what co-management is all about. Co-management entails 'mutual coercion, mutually agreed upon by the majority of the people affected'.[30] In contrast, government management is management from the top down. If the decision-making process is fair and just, which is co-management at its best, the majority rule is more likely to be followed by all. Jentoft and Kristoffersen contend[34] that this has been the effect of the co-management system in the Lofoten Fishery, where violations of the regulations are few. In the UK case, Goodlad, argues[12] that regulations by fishermen's organizations 'are generally more "respected" than regulations by Government'.

Another key point in this article is that content of the decisions and organizational form are closely related: what comes out of the decision-making process is heavily dependent on how it is organized. In other words, as hypothesized earlier, when co-management is introduced the quality of the regulations will improve, and this will also increase legitimacy.

Perhaps the most common argument is that fishermen's cooperative organizations are in a position to make more equitable regulations than are governments.[33] Not only are fishermen's organizations better able to determine what the relevant equity considerations are, they are also more capable of responding adequately to the special needs, demands and interests of individual fishermen or fishermen groups. Governments tend to follow principles of 'universalism' when dealing with client fishermen. This may guarantee neutral, but not necessarily fair, treatment.[34] Fishermen's organizations, on the other hand, can be more 'particularistic', which is sometimes needed to ensure fairness and equal opportunities. For instance, an accident may hinder a fisherman from catching his quota. It would therefore be fair if his quota was increased next year to compensate for his loss. Representatives of the UK producers' organizations when interviewed, pinpointed this as one of the important improvements of co-management. In Japan, lottery systems combined with a rotation principle are used to ensure equal opportunities. When fishermen obtain through a lottery certain use rights, they may be excluded from participating in lotteries for other fisheries rights. Needs of individual fishermen are also taken in consideration. For instance, the number of nets a fisherman operates can be determined by the size and age bracket of his family.[35]

Following Pinkerton, information about the resource base should be improved as a consequence of direct fishermen involvement in fisheries management. Fishermen have more detailed information based on their practical experience than do governments. Hence, more fine-grained

---

[30] G. Hardin, 'The tragedy of the commons', *Science*, Vol 162, 1968, pp 1243–1248.
[31] Jentoft *et al*, *op cit*, Ref 15.
[32] Goodlad, *op cit*, Ref 21.
[33] See Pinkerton, *op cit*, Ref 2; and Zengyoren, *op cit*, Ref 18. Hannesson sees equity improvement as the main contribution of fisheries co-management: 'The arguments for collective solutions of the common property problem are arguments of equity and social justice rather than efficiency, while pseudo-market solutions based on transferable catch quotas or fishing licences held by individuals or firms seem more likely to promote efficiency'. The same equity effects could also, he argues, be obtained through leasing or taxing licenses or quotas, and by making transferability subject to certain conditions. He thus concludes that 'In theory the case for this type of solution (ie co-management) is not entirely convincing'. As pointed out in this article, co-management has in several cases proved viable in practice. The empirical evidence for the efficiency and workability of market solutions in fisheries management is still very scarce.
[34] The neutrality of government may be questionable. Government agencies are often exposed to lobbying and political pressure from powerful economic interests within the fishing industry. This may, for instance, result in favouritism of large-scale operators at the expense of small-scale operators when regulations are implemented. Cf. Barret and Davis (1984) for a discussion of the Canadian case, and Bailey (1988) for the case of many Third World countries. Also the rationality of government agencies may be questioned. In the case of Norway Orebech (1982) finds that fisheries authorities are sloppy in following their own rules for the distribution of licences.
[35] Hirazawa, *op cit*, Ref 27.

Fisheries co-management

decisions can be made. Also co-management, according to Pinkerton, has the potential to 'increase the responsible sharing of information . . . with consequent reduction in conflicts between state and fishermen'. The willingness to share local catch information is a function of being trusted as responsible participants in management schemes – and not adversaries who have to be controlled by government. Pinkerton argues that fishermen's behaviour and attitudes alter as a result of the changes in their role with the introduction of co-management principles.

Government bureaucracies have a limited capacity to oversee the many local and seasonal variations within different regions and sectors of the fishery. For regulations to be efficiently carried out they must be fair, and to be fair this diversity must be taken into account. This, however, requires a large amount of detailed knowledge of local circumstances in the fishing industry and the ecological conditions which exist in various fisheries. Government agencies usually do not have this stock of knowledge, and if they try to get it, the costs are prohibitive. This was an important reason why the UK Government, after several years of centralized fisheries management, decided to delegate responsibility for fisheries regulations to the producers' organizations. The increased management effort which was needed caused an 'overload' on government agencies which was eased by transferring regulatory functions to the producers' organizations. In Lofoten, the importance of local knowledge of the conduct of the fishery and the natural conditions on each fishing ground, was the main reason for introducing co-management in the 1890s.

Variations entailed in the nature of the fisheries require flexible management systems. A central argument for introducing co-management is that government bureaucracies are less flexible than fishermens' organizations in enforcing management schemes. Goodlad for instance, argues that UK producers' organizations 'are generally more able to react to a situation more quickly than National Governments'.[36] This was also an important factor in the Lofoten fishery leading to the institutionalization of co-management. Decisions to change the rules of fishery could be reached much more quickly by the fishermen's committees than by the government.

Delegating responsibility to fishermen's cooperatives means that fishermen become active and responsible individuals in the decision-making process. By definition, cooperatives rely on membership participation, which is reflected in the internal structure of the organization. Member fishermen form the general assembly and the board of directors which make the strategic decisions. Transferring responsibility for management functions should therefore indicate that more democracy is introduced in the regulatory process. This should not only result in better management solutions, as suggested above, but it would also be a valuable societal benefit in its own right.

What are the problems of co-management?

The responsibility for fisheries regulations can become a heavy burden for government agencies, as seen, for example, in the UK case. The same is true for fishermen's organizations. It requires sophisticated administrative resources and skills to handle fisheries regulations. This can be an obstacle for some organizations. On the other hand, some organizations have adopted the resources required. By the means of a

[36]Goodlad, *op cit*, Ref 21.

Fisheries co-management

computer used for keeping control of how much each member is fishing, the UK producers' organizations have managed well and with minimal administrative costs.

A more serious problem for cooperative organizations are internal conflicts and disputes which may arise among members or groups. Delegating management responsibility does not alter the conflict nature of fisheries regulations. Co-management simply represents another way of handling such conflicts. Fishermen's cooperative organizations are usually established to provide various benefits to their members.[37] Assuming additional responsibility for fisheries regulations means that restrictions on members behaviour have to be enforced. While the benefits obtained by cooperative members are experienced as good, the same is not necessarily the case with fisheries regulations. Importantly, a classic cooperative principle is that they are fundamentally voluntary organizations. However, co-management involves the non-voluntary imposition of restrictions on the membership. By enforcing strict regulations, the members get easily frustrated, and as a consequence, the cooperative risks that members leave. Thus, the political costs of regulating fishing behaviour can be high.

The legitimacy of regulations enforced by the cooperative can be challenged by member fishermen for other reasons. There may be conflicting views among members concerning the rules for the fishery. This is especially true when the membership is heterogenous, for instance, according to boat size, gear types, capital costs, and ownership. Even if the membership is homogenous, there may still be variations in skills and, consequently, catch results. How, then, should skills be accounted for when regulations are defined? Should variations in skills be reflected in the distribution of quotas? In fact, the explanation of variations in catches, and the effects of the skill-factor, is among the most controversial issues among fishermen.[38]

A common cooperative principle, usually stated in the charter of the organization, is open membership and a low entrance fee. In order to keep the total fishing effort under control fisheries management requires limited entry, which means that this principle is abandoned, as illustrated in the US case reported by McCay.[39] Therefore, it should not come as a surprise if fishermen's organizations are sceptical of carrying out the responsibility for management functions. They may prefer to act as a pressure group vis-a-vis government authorities. Inevitably, someone will be blamed when fisheries regulations are implemented and enforced. To have the state targetted for the blame lessens the local impact on cooperative decision makers.

The Japanese case may provide a solution to this problem. Certain exclusive rights accompany the delegation of responsibility for fisheries regulations. The cooperatives have ownership rights to fishing territories and a fisherman must be a member of a cooperative to be granted access. Ownership rights are a main reason for the success of Japanese fisheries cooperatives. The UK producers' organizations do not have similar monopoly rights. A fisherman can obtain an individual quota directly from the government if he feels uncomfortable with the regulations of the producers' organizations. This seriously weakens the organizations' ability to enforce restrictions.

If co-management is going to have any real effects, the fishermen's organizations must have a certain amount of autonomy. This concerns the relation between the cooperative and its environment. As to fisher-

[37] S. Jentoft, 'Fisheries co-operatives: lessons drawn from international experiences', *Canadian Journal of Development Studies*, Vol VII, No 2, 1986, pp 198–209.
[38] G. Palsson, *Representations and Reality: Cognitive Models and Social Relations among the Fishermen of Sandigerdi, Iceland*, PhD dissertation, University of Manchester, 1982.
[39] McCay, *op cit*, Ref 13.

Fisheries co-management

ies management, the environment tends to be rather turbulent. As Foreman points out,[40] fisheries are particularly difficult to manage because accurate data on the state of the fish stocks is hard to provide. However, when it is provided it is often, and unexpectedly, portraying the fish stocks on the brink of depletion. This calls for immediate protective action. In the New England case which he studied, the Regional Management Council (see below) had to make frequent changes in its management responses, which put the Council under heavy stress. After having put much effort in working out a compromise solution, the decision-making process would have to start all over again. There were other implications as well. Regulatory decisions were often made under great uncertainty. Also, the Council became very dependent on external expertise and information provided by resource biologists.

The autonomy of fishermen's organizations in fisheries management is also determined by the division of responsibility; ie how many regulatory functions which are actually delegated from the government to the fishermen's organization. The greater the number of functions delegated, the greater the autonomy. In the Japanese and UK cases presented above, the government has retained the responsibility for fixing the size of the TAC. In the Norwegian case, the Government in the early 1900s withdrew the fishermen committees' authority to decide what kind of fishing gear is to be allowed on the Lofoten grounds. The government claimed that the fishermen were too conservative in letting in new and more efficient gear.

Thus, there may be reasons to exclude some functions from being delegated. There are limitations on what functions can or should be transferred to fishermen's organizations. In general, these limitations are influenced by the number of organizations involved. The higher the number of organizations involved, the fewer the functions which can be delegated. Competition among organizations of fishermen can be quite as devastating for the resource base as competition among individual fishermen. On the other, if there was only one organization, all the regulatory functions, including the decision of deciding the TAC, could, in theory, be delegated. Competition would be replaced by internal command within the organization. In the fisheries management literature, this is defined as the 'sole ownership option'.[41]

Whether co-management will in all cases promote a more democratic process with proper consideration of equity and fairness is an open question. A crucial variable pertains to the social dynamics of the participatory process. Even though cooperative organizations entail participatory decision making, in practice this could be more formal than real. Participation of members in a real sense could be limited to just casting votes. Democratic organizations are often victims of oligarchic tendencies, group rivalry, conspiracy, and elite expropriation. Consequently, instead of advancing participant democracy, delegating responsibility can be a contribution to the consolidation of rigid, inequitable power structures.[42] If this is the case, a government agency may be preferred as a mediator in conflicts and may be a more democratic institution than a cooperative organization.

Fishermen are not the only group with an interest in how the fishery is regulated. Other groups within the fishing industry, such as processors and fish plant workers, are also affected by the regulations. Groups external to the fishing industry may have an interest as well.

---

[40]C.H. Foreman Jr, *Fisheries and the Search for Regulatory Consensus: Lessons from New England*, paper prepared for the meeting of the American Fisheries Society, Cornell University, Ithaca, New York, 12–16 August 1984.
[41]E.A. Keen, 'Common property in fisheries: is sole ownership an option?', *Marine Policy*, Vol 7, No 3, 1983, pp 197–211.
[42]C. Bailey 'Optimal development of Third World fisheries', in M. Morris (ed), *North South Perspectives on Marine Policy*, Westview Press, Boulder, forthcoming.

Fisheries co-management

Internationally, environmental groups have become increasingly concerned with fisheries management practices. In many countries, recreational fishermen struggle for more influence on management decisions which they claim exclusively benefit commercial fishermen. In Norway, for instance, the effects of fisheries regulations on the settlement structure is a major concern among the public at large. Consequently, while delegation of responsibility for fisheries regulations to fishermen's organizations may improve legitimacy among fishermen, the opposite may be the result among the other groups. Fishermen's organizations are often powerful relative to such other groups,[43] and delegating responsibility for fisheries regulations would strengthen their power base even further. Thus, one may expect external opposition to the co-management concept.

A solution may be to create organizations with a broader representation allowing all affected interests to take part in the decision-making process. The regional fisheries management councils in the USA, established under the Magnuson Act of 1976, have such a broad representation. They include, in addition to fishermen representatives, public officials, processors, consumers, recreationalists and environmentalists. The councils also arrange public hearings in various communities to ensure participation from the public at large. The councils, however, have their responsibility restricted to making recommendations to the Government concerning fisheries management.

However, allowing additional affected interests a say in the decision-making process may lead to other problems. The organizations become more complex and internal conflicts are more likely to arise. As Foreman argues in the New England case:[44] 'In important respects the council has proved to be less an "organization" (with the sense of coherence and mission that term implies) than an "arena" where diverse fishing constituencies contest with one another'.

As Dahl[45] has pointed out, one concern here should be the extra costs of decision making that the broader representation will lead to. Another is the fact that groups are affected disproportionately by fisheries regulations. They may have more or less economic risk at stake. How then should this be reflected in the decision-making process and the voting? A third problem is the question of competence. Fisheries management requires special knowledge of the fishery; but, such competence may be unevenly distributed among the participant decision makers. One of the alleged advantages of allowing affected interest into the decision-making process is the special competence which they will bring with them: and consequently, this will result in more qualified decisions. But, when there are conflicts of interest, special competence may be an obstacle rather than a help in the decision-making process. As contended by Foreman:

Indeed, one is often left with the sense that greater knowledge of management technique on the part of 'generalist' council representatives could prove a double-edged sword in the search for concensus; such sophistication could result in nothing more than more elegant (but undiminished) conflict.

Our final concern with the co-management solution is hypothesis four outlined earlier, which argues that legitimacy will be improved if fishermen are involved in implementation and enforcement. Peer group pressure to adhere to the rules can obviously be very effective, but also

[43]S. Jentoft and K.H. Mikalsen, 'Government subsidies in the Norwegian fisheries: regional development or political favouritism?', *Marine Policy*, Vol 11, No 3, pp 217–228.
[44]Foreman, *op cit*, Ref 40.
[45]R.A. Dahl, *After the Revolution: Authority in a Good Society*, Yale University Press, New Haven, 1970.

Fisheries co-management

quite as intimidating and repressive as government control. Nothing is worse than losing face among colleagues. Fishermen would also have to be each others' policemen, and reporting may be another way to lose face. In the Lofoten case, in addition to serving as ombudsmen for fellow fishermen, the elected fishermen inspectors are supposed to report other fishermen if they discover that regulations are broken. The fact that the fishermen know that inspectors are fishing next to them, and may follow their actions, restrains them from rule-breaking. However, the fishermen inspectors usually find it difficult to carry out the role as 'informers' and will rarely report other fishermen. Therefore, the public enforcement agency, which has inspection vessels on the fishing grounds, is, in practice, solely carrying out this function. If there is a general lesson to be learned from this it is that enforcement is one of the regulatory functions which seems better handled by government than by a fishermen's organization.

Conclusion

This article has addressed the division of responsibility between the state and the fishing industry in fisheries management. Some of the achievements and problems of assigning more responsibility to fishermen's cooperative organizations have been discussed. When strengths and weaknesses are considered, what conclusions can be drawn for the potential success of introducing co-management arrangements? Is co-management to be recommended? The answer is conditional. Experience shows that while some co-management systems have persisted, others have failed. From the case studies presented, some generalizations regarding critical variables can be identified.

The importance of legislation which gives fishermen's organizations not only the responsibility but also the authority to implement and enforce restrictions on fishermen's behaviour, should not be underestimated. The Canadian experience failed because of reluctant support from the Government. The UK co-management system is vulnerable because fishermen can escape the collective regulations by obtaining quotas directly from the government.[46] The two most long-lasting and successful examples of fisheries co-management are the Norwegian and Japanese cases. In both countries, fishermen's organizations are, by law, given exclusive rights which rule out an exit option. If fishermen are dissatisfied with the regulations, they have to use their voice and vote. It is noteworthy to point out that this does not necessarily entail a less democratic process than the exit option.[47]

Successes which have been noted in this article to a large degree reflect the scale of the organizations. A common feature of all co-management systems described is the limited scale of the cooperatives, both in terms of membership and regional jurisdiction. In general, participant democracy seems to flourish in smaller rather than larger organizations. Small organizations allow direct, personal participation. Large organizations must rely on indirect, intermediary representation in the decision-making processes. The problem of free riders is found to occur more often in large organizations. In small organizations, free riders breaking the rules are easier to identify and control by informal sanctions. This problem is also a question of fishermen's sense of belonging to an organization. Members tend to feel a stronger identification with a small organization rather than a large organization.

[46]The producers' organizations have made complaints to the government and asked for revision. However, by April 1987, no corrective action had been taken.
[47]A.O. Hirschman, *Exit, Voice and Loyalty; Responses to Decline in Firms, Organizations, and States*, Harvard University Press, Cambridge, MA, 1970.

Fisheries co-management

After evaluating international experiences with fisheries cooperatives in developing countries, Pollnac argues:[48] 'There are cases where fishermen's organizations failed because they were made so large that members no longer felt that the group was their own'. Another consequence of organizational scale is that, when organizations grow in membership, regulations will affect a larger number of individuals. Hence, it is more likely that some individuals will find that the regulations are contrary to their interests. As pointed out by Young:[49] 'Assuming that actor preferences are distributed normally, every increase in the size of a regime will lower the probability that programs chosen will conform precisely to the preferences of any individual member of the beneficiary group'.

Consequently, when organizations grow in scale, dissatisfaction, frustration and internal conflicts within the membership are more likely to arise.

Organizations with a relatively homogeneous socioeconomic membership will have less internal conflicts of interest and this will make decision-making easier. In the Canadian case, conflicts arose between small-scale and large-scale fishermen, and this was a contributing factor to the failure of the cooperative. The success of co-management is contingent on fair and equal distribution of resource benefits. When the membership is homogeneous, equal distribution will also be fair distribution. On the other hand, when the membership is heterogeneous as in the Canadian case, fair distribution is not necessarily the same as equal distribution. For instance, quota allocations may have to be made relative to capital invested in boats and gear. What is fair would then have to be negotiated among the member groups. If this is the case, conflicts are likely to arise which may threaten the organization.

The answer to the problem of heterogeneity, as well as the problem of participation in large organizations, may be as suggested by Robert A. Dahl:[50]

> In an association where members are competent but greatly in conflict, it may make sense to dissolve the association into more harmonious groups that will be able to honor political equality and majority rule. But this solution rarely is completely attainable. For a broader association (which may be that peculiarly important association known as the state) may be necessary to regulate conflict among the smaller, more homogenous associations.

At first glance, Dahl's solution seems to indicate a dilemma. More participant democracy, enhanced by dissolving fishermen into smaller organizations, necessitates state intervention which leads to less self-management. However, a state agency is not the only possible external mediator. A cooperative umbrella organization may serve a similar role. In Poland, for instance, the national quota is divided by the government between the state corporate sector, the cooperative sector, and the private sector. Thereafter, The National Union of Fishery Cooperatives allocates the cooperative quota among its members.[51]

As Berkes argues in the Turkish case, the traditions of cooperation among fishermen may be important. The Japanese are well known for their strong commitment to collective values and participatory decision making in business management.[52] Undoubtedly, this is an important factor in explaining their success of fisheries co-management. Cooperation is in itself a learning process, and collective values are reinforced through such a process. If fishermen lack a positive exper-

[48]R. Pollnac, *Evaluating the Potential of Fishermen's Organizations in Developing Countries*, ICMRD report, University of Rhode Island, Kingston, 1988.
[49]O.R. Young, *Resource Regimes: Natural Resources and Social Institutions*, University of California Press, Berkeley, 1982.
[50]Dahl, *op cit*, Ref 45.
[51]Anonymous, *Fisheries and Related Industries in Poland*; Office of Maritime Economy and Central Union of Work Cooperatives in Poland, prepared for the First UNIDO consultation on fisheries industry, Gdansk, 1–5 June 1987. This author visited Polish fisheries cooperatives and had interviews with the chief executive of the National Union of Fishery Cooperatives in October 1987.
[52]W.G. Ouchi, *Theory Z: How American Business can Meet the Japanese Challenge*, Addison-Wesley Publishing Company, Reading, MA, 1981; C. Pegels *Japan vs. the West: Implications for Management*, Kluwer-Nijhof Publishers, Boston, 1984.

Fisheries co-management

ience of cooperation and collective action, introducing co-management have less chance of becoming successful.

The fact that the UK Government could delegate management functions to already existing cooperative organizations, eased the transition period required for assuming full management responsibility. If such organizations do not exist, they will have to be established before delegation of responsibility for management functions can take place. However, delegating responsibility to existing organizations may be regarded negatively by fishermen. Fishermen do not always trust cooperatives more than government. In fact, internationally, scepticism among fishermen of cooperative models is widespread.[53] Important for a successful result, therefore, are the factors which produce trust in organizations and whether or not these factors are prevalent in the existing organizations.[54]

In part, trust is dependent on fishermen's previous relations with these organizations. Trust develops over time and through experience. For instance, the regulatory system in the Lofoten fishery of Norway has worked well for so long that fishermen take it for granted. The concrete regulations are often questioned among fishermen, but not the co-management principle itself. Trust is also based on the quality of the social relations fishermen have to each other. Some of the main findings of organizational research are the existence of informal organization within formal organizational structures, and informal rules and relationships among members which have crucial impacts on organizational behaviour. Some of these relationships are developed within the organization; others stem from outside interaction.[55] The argument here would be that trust is crucial for the workability of fisheries co-management. However, trust is not only a product of formal organization, but also of informal organization. Informal organization develops through long-term interaction among members inside and/or outside the organization. This leads to the proposition that the more long lasting and multifaceted relations among fishermen, the more likely is the success of co-management. This is also why community self-management often works well in small-scale inshore fisheries, in contrast to large-scale offshore fisheries where the mobility of the fleet is much higher. Consequently, relationships of trust have less chances of being developed in offshore fisheries. The UK case, however, suggests that mobility is not an insurmountable obstacle.

The UK producers' organizations, as well as the Japanese, the Canadian and the US cooperatives described above, are multi-purpose organizations. They combine fisheries management with fish marketing, as well as other important functions (eg credit, supplies, gas etc). These are tasks which should be coordinated for the attainment of economic and social objectives,[56] and transaction costs can be saved if such coordination takes place within the same organization rather than among several independent organizations.[57] The fact that these cooperatives have other functions reinforces the management function. The costs and burdens fishermen experience because of the regulations installed by the cooperative can be compensated for by the various benefits of belonging to the same organization.

The long-term effect of introducing co-management agreements is hard to predict, as it is with most major institutional reforms.[58] In particular this is pertinent in the special case of the fishing industry. The short-term effects may be quite different from the long-term effects.

[53]Jentoft, *op cit*, Ref 37; and Pollnac, *op cit*, Ref 48.
[54]M. Granovetter, 'Economic action and social structure: the problem of embeddedness', *American Journal of Sociology*, Vol 91, No 3, 1985, pp 481–510; L.G. Zucker, 'Production of trust: institutional sources of economic structure, 1840–1920', *Research in Organizational Behaviour*, Vol 8, 1985, pp 53–111.
[55]C. Perrow, *Complex Organizations: A Critical Essay*, Random House, New York, 1986.
[56]S. Jentoft, 'Models of fishery development: the cooperative approach', *Marine Policy*, Vol 8, No 4, 1985, pp 322–331; I. MacSween, *The Interaction Between Fisheries Management and the Marketing of Fish*, FAO Fisheries Report No 289, Supplement 2, Rome, 1983.
[57]O.E. Williamson, *Markets and Hierarchies: Analysis and Antitrust Implications*, The Free Press, New York, 1975.
[58]J. Elster, 'Skeptiske tanker om samfunnsplanlegging', *Forskning og framtid*, Nr 1, Resource Policy Group, Oslo, 1984, pp 27–36.

There may be transitional problems. The history of fisheries management schemes tells us that unexpected effects will occur. Moreover, the prospects of success will be contingent on the way co-management is introduced, for instance, depending on whether it is introduced 'incrementally' or as a 'grand scheme'. Co-management in small enclaves, as in the Canadian case, may have different possibilities of success than if co-management was made the system for the whole sector, as in Japan. When co-management is implemented incrementally it must operate within an environment which may be disfunctional or even hostile to the experiment. When co-management is introduced as a macro reform such environmental factors will be minimized. Importantly, however, one cannot uncritically draw conclusions from how co-management works in a small enclave to how it will work as a global solution.[59]

The most important contribution one can realistically hope for is that co-management will imbue the regulatory process with legitimacy. This will tend to make management both more effective and less costly compared with government control. Legitimacy, we have argued, will be improved for both procedural and substantive reasons; procedurally because co-management introduces participatory decision making; and substantively because fishermen's organizations will be more inclined to base their regulatory decisions on considerations of fairness and equity. In view of the devastating effects the absence of legitimacy has had on management schemes in the past, this would be no small achievement.

Crucial to the success of co-management is the actual division of responsibility between government and industry. The context into which co-management is introduced varies from country to country and from fishery to fishery, and there is no single model for implementing such management principles. The context should therefore be taken into account when co-management schemes are designed. Nevertheless, we have argued that the government has a role in overall planning, total quota management, in solving distributional conflicts among various co-operative organizations, in providing sufficient legal support for the cooperatives, as well as in enforcing the regulatory decisions. Apart from that, when it comes to fishing practices, access control, and making distributional decisions among individual fishermen or boats, fishermen's organizations in general are well suited for the task.

[59]*Ibid*.

THE FISHERY: THE OBJECTIVES OF SOLE OWNERSHIP[1]

ANTHONY SCOTT

University of British Columbia

> The rights of property, as such, have not been venerated by those master minds who have built up economic science; but the authority of the science has been wrongly assumed by some who have pushed the claims of vested rights to extreme and antisocial uses. It may be well therefore to note that the tendency of careful economic study is to base the rights of private property not on any abstract principle, but on the observation that in the past they have been inseparable from economic progress. . . .—ALFRED MARSHALL, *Principles of Economics* (8th ed.), p. 48.

IT IS *a* commonplace to observe that for natural resources—as for other types of wealth—"everybody's property is nobody's property." No one will take the trouble to husband and maintain a resource unless he has a reasonable certainty of receiving some portion of the product of his management; that is, unless he has some property right in the yield. Yet the mere existence of the institution of private property is not sufficient to insure the efficient management of natural resources; the property must be allocated on a *scale* sufficient to insure that one management has complete control of the asset. In this paper, for example, I shall show that private property in fishing boats is not a sufficient condition for efficiency; sole ownership of the fishery is also necessary. Some assets, such as oil fields, fisheries, and watersheds, occur on an immense scale, and it is a very real problem to know whether the efficiency gained from unified management provides a social gain sufficient to offset the possible dangers of the creation of some immense sole-ownership organization (such as a cooperative, a government board, a private corporation, or an international authority).

This paper continues the discussion of the economics of private and common property undertaken in "The Economic Theory of a Common-Property Resource: The Fishery," by H. Scott Gordon, which appeared in the *Journal of Political Economy* for April, 1954 (pp. 124–42). Gordon's contribution was a most stimulating, original, and important study of the advantages of sole ownership, which seem practically to have escaped theoretical discussion since Marshall's time.[2] While the economics of the farm and the forest are continually under revision, the earlier economists' insistence[3] that efficiency in production depends upon scarce wealth

---

[1] In writing this paper, I have had interesting and helpful discussions with Messrs. D. C. Corbett, Stuart Jamieson, W. J. Anderson, O. R. Reischer, and E. E. Snyder.

[2] See Marshall, *Principles*, pp. 166–67 and 369–72; see al→ Gordon's article in the *Canadian Journal of Economics and Political Science*, February, 1952.

[3] See J. S. Mill, *Principles* (5th ed.; New York, 1897), Book II, chap. ii, sec. 5, and the citation to Sismondi's *Étude sur l'économie politique* (n.d.) in the footnote.

THE FISHERY: THE OBJECTIVES OF SOLE OWNERSHIP

being "appropriated" has been relegated to the introductory chapters of principles textbooks and is scarcely considered in them again. This is all the more reason for welcoming Gordon's bringing some "political economy" back into economics.

In this paper I wish to compare the use of a fishery by competing fishermen with the mode of management that would be most profitable to a "sole owner" of the same fishery. In particular, I wish to show that *long-run* considerations of efficiency suggest that sole ownership is a much superior regime to competition but that in the *short run* in the ordinary case there is little difference between the efficiency of common and of private property.[4]

I do not wish to dispute in any way the facts about fisheries presented by Gordon (except those about opportunity costs) but to welcome them and to use them myself. Particularly interesting was the point made about the ignorance among fisheries biologists as to whether recent conservation measures have ever produced changes in demersal (deep-water, sea-bottom) fish populations. It is remarkable that many popular books advocating the conservation of natural resources, after describing the undoubted success of conservation in some pelagic (surface) and in-shore fisheries, then—perhaps unintentionally—give the reader to understand that the same conservation might be achieved by controlling demersal fisheries, an assertion that is apparently as yet unverified.

I

In many ways the central part of Gordon's paper is Section IV, "The Bionomic Equilibrium of the Fishing Industry." In this section he sets out to suggest the nature of the equilibrium of this common-property industry as it occurs in the state of uncontrolled or unmanaged exploitation. Later, he undertakes to indicate the nature of a socially optimum manner of exploitation, which is presumably what governmental management policy seeks to achieve or promote.[5] In a subsequent section of the present paper I shall discuss this optimum, but I should like first to recapitulate his exposition of the equilibrium of a given fishery under conditions of competitive exploitation by individual fishermen.

Gordon first argues that, because there is no sole owner to capture for himself whatever gain there may be from using the fishery conservatively, it will pay every fisherman to enter the industry so long as he can earn something above his cash expenses plus his opportunity costs. As long as fishermen do this, the tendency will be for exploitation to continue beyond the point where the marginal product of fishing effort equals its marginal cost, to the point where the average product of effort just covers the marginal cost of effort (where, in fact, every fisherman just covers his opportunity costs, and average cost is equal to price). There tends, it is argued, to be no "surplus" earned in the industry—the dollar value of the catch exactly equals

---

[4] "Sole ownership" is not monopoly but merely complete appropriation of all of a natural resource in a particular location. Putting a resource into sole ownership is sometimes called making a resource "specific" to one owner (see my forthcoming *Natural Resources: The Economics of Conservation* [Toronto: University of Toronto Press]).

[5] These two aims have been paraphrased from Gordon, *op. cit.*, p. 136.

the dollar cost of landing the catch.

Under such a regime, the argument continues, the possible equilibrium of effort, fish population, and income may be described in terms of a system of four variables, which may be drawn in four figures. These diagrams can be conveniently studied in one four-quadrant diagram (Fig. 1).

If we start with the two upper quadrants, we see that output (or landings) L depends upon both the size of the input (or effort) E and the size of the capital stock, or population, P. The size of the capital stock, or population, is in turn itself assumed to be dependent upon the effort.[6] In the southeast quadrant is a cost function, C, showing constant marginal costs of effort, and in the southwest quadrant is a simple transition function of 45° which confronts the cost of various kinds of effort with the revenue from various sizes of landing. The condition of equilibrium is that the total costs must equal the total

---

[6] Although Gordon admits that this is not always the case, the analysis holds only when population *is* affected by effort.

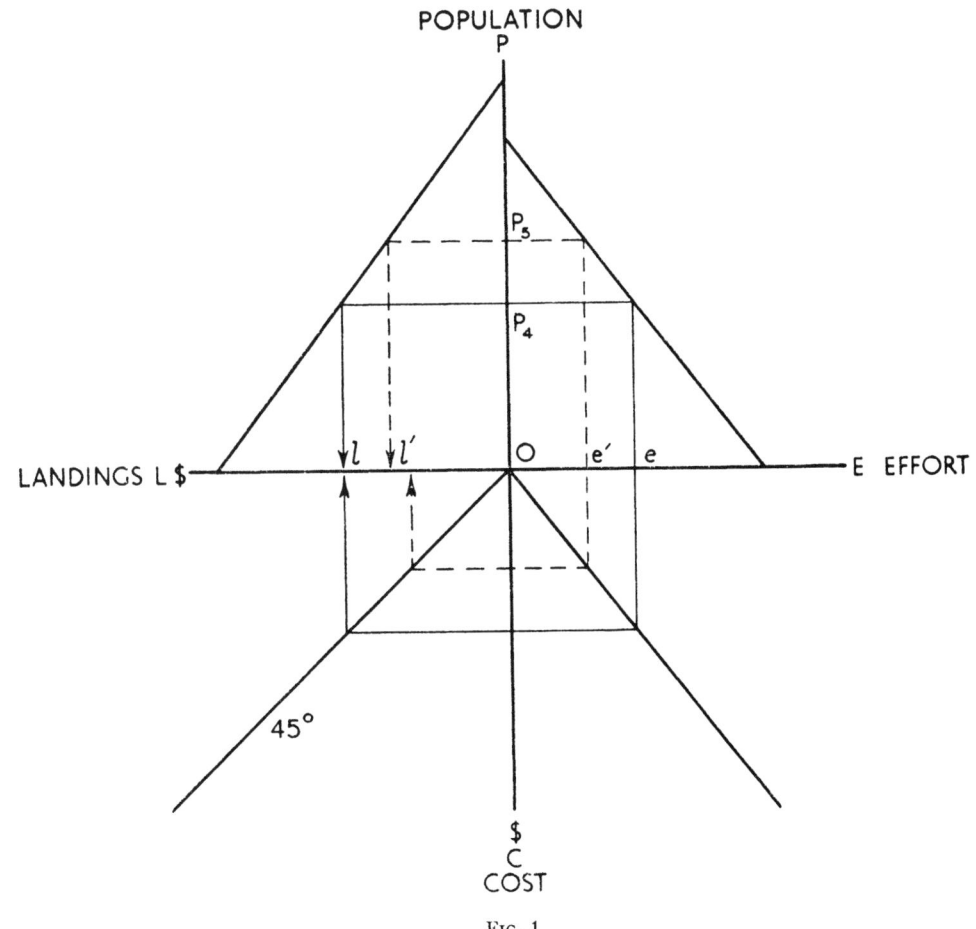

Fig. 1

THE FISHERY: THE OBJECTIVES OF SOLE OWNERSHIP

revenues, Ol. In diagrammatic terms there must be an inscribed rectangle in the four quadrants. Hence the dotted lines do not represent an equilibrium, but the solid lines do.

Assume that, in equilibrium, a certain total "effort," E (that is, of fishermen-plus-equipment), is being expended in the fishery. This effort, expended annually, is compatible with a certain size of fish population, shown on the P-axis, and there is an equilibrium annual catch that can be achieved by this effort with this population, shown on the L-axis. But, as was argued above, the fishery will not be in equilibrium unless the cost of expending that effort (shown on the C-axis) is exactly equal to the revenue, L. When L and C coincide, the fishery is in equilibrium.

A condition for achieving a positive, stable equilibrium is that there must be one and only one rectangle that will fit into the four quadrants. There may be no unique equilibrium possible, given the three functions, unless one of them is curvilinear, or (as here) unless the E- and L-functions have different intercepts on the P-axis.[7]

It is also possible to show the suggested equilibrium position of a sole owner. The dotted lines may now be used to represent this equilibrium: the difference at l' between the cost and the landings revenue is assumed to have been maximized. Gordon's description of the equilibrium applies to this diagram:

> The optimum intensity of fishing effort is that which maximizes $L - C$. This is the monopoly solution; but, since we are considering only a single fishing ground, no price effects are introduced, and the social optimum coincides with maximum monopoly revenue. In this case we are maximizing the yield of a natural resource, not a privileged position, as in standard monopoly theory. The rent here is a social surplus yielded by the resource, not in any part due to artificial scarcity, as is monopoly profit or rent.

> If the optimum fishing intensity is that which maximizes $L - C$, . . . the optimum fishing intensity is Oe' of fishing effort. This will yield Ol' of landings, and the species population will be in continuous stable equilibrium at a level indicated by P_5.[8]

II

Gordon calls these two situations "positions of equilibrium." It is true that in the first, the competitive exploitation of the fishery, the equilibrium indicates values of C, P, and E that are compatible with one another, though we are not told just how the fisherman arrives at this equilibrium. It should be noted that it is quite possible that a competitive fishery in the process of production—which might be described diagrammatically as a trial-and-error procedure of searching for the inscribed rectangle—might easily miss an unstable equilibrium and gravitate rapidly between the alternate extremes of zero output and zero population. Furthermore, in those fisheries there may be no stable equilibrium short of zero landings and full natural population. It is not unlikely that fished-out lakes —which are, after all, demersal fisheries in the *economic* sense of Gordon's paper—would be examples of this impossibility of finding a competitive equilibrium position.

"Would be" because the fundamental assumption in Gordon's paper (apart from the admittedly unknown

---

[7] For those readers who would like to identify this diagram with the algebraic system given in Gordon's paper, we can say that $P = a - bL$; $P = a/bCE + 1$; $C = qE$; $C = L$ (Gordon, *op. cit.*, pp. 141–42).

[8] *Ibid.*, p. 141.

ANTHONY SCOTT

nature of the biological relationship between landings and demersal populations) is that there are in fishing no diminishing returns and hence no increasing costs and no incentive to stop operations short of the equality of total costs and landings. Surely this fundamental assumption is incorrect; surely *in the short run* (with population and equipment fixed) each fishing boat will experience increasing costs as it attempts to increase its landings.

Gordon's analysis, which I have followed in Figure 1, relies upon the depletion of the population to produce a species of "diminishing returns" effect that will explain, with price given, why the competitive fishery does not expand indefinitely. But this explanation applies only to the long run and cannot hold within a single season, when the fish population is one of the fixed inputs. In the short run, fishermen do not expand their catch indefinitely because they *do* experience increasing costs in attempting to increase their landings. Gordon depends upon the omnibus variable "effort" to cover the changeable combinations of men, boats, and other equipment used by individual fishermen. But, if we look through this omnibus variable, we see that in fact the short-run situation in a fishery exploited by competing fishermen will be very like the standard situation in pure competition. The supply curve of this fishery (with the price given by the world market situation) will be made up by the addition of the relevant portions of the supply curves of the individual fishermen. These curves will slope upward because, with fixed equipment and a fixed number of boats, there will be some number of landings per boat which has a least cost; if the crew is worked long hours, or the boat is kept running without time for maintenance or repair, the cost per landing will begin to rise. Each boat will increase its landings until its supply price (marginal cost) is equal to the going price. The "surplus" that might be captured in this situation is the usual quasi-rent, available to each boat by operating at the point where marginal costs are equal to marginal revenue.

Now (if we continue to consider only short-run decisions), would a sole owner select a different rate of output than that which was determined under competition? There are two possible situations here: (1) the sole owner may take over an existing competitive fishery, boats, canneries, and crews. (2) The sole owner may reorganize the fishery in the most efficient way; this is not the *same* short-run situation but an alternative situation.

1. If the sole owner were taking over *for a season only* a fishery that had been equipped in the manner suitable for operation by competing fishermen, he would operate it in exactly the same way as they had, that is, at the output for which the marginal cost of fishing equaled the price of the product. There is, however, one qualification of this assertion. If it were the case that competing fishermen were so numerous that boats got in each other's way, then the sole owner would rationally lay off some of the boats (and perhaps canneries and collecting boats) for the season. In this way he could reduce the external diseconomies of fishing. But, apart from this qualification (which is really a matter of the long run), the sole owner and competitive fisherman would in the short run oper-

THE FISHERY: THE OBJECTIVES OF SOLE OWNERSHIP

ate the fleet identically, so that marginal cost equaled price and so that the marginal product of labor equaled the price of labor.

2. However, if a sole owner expected to have permanent tenure, then even in the short run his organization of the fishery would probably be quite different from that of small competing fishermen. For instance, it has been suggested that on the West Coast the sole owner of a salmon fishery would rely more on traps than on vessels; doubtless economically similar techniques are known in the demersal fisheries. Not only would a sole owner prevent the wasteful interference of competing fishermen with each other, but he would also design his fleet and his transport and packing facilities so as to take advantage of the economies of integration and scale. When he had worked this out, assuming that he was in competition with the owners of other fisheries, he would still tend to operate where short-run marginal cost equaled price. Whether this rule would result in his using more or less variable factors, and whether his catch would be larger or smaller, it is impossible to guess a priori. There is no reason to believe that it would be significantly different, although the productivity of all inputs would almost certainly be higher, since the sole owner has the choice of a wider range of techniques.[9]

Hence, we can say that, as a general rule, the mere fact of sole ownership does not bring about a significant change in the exploitation of the fishery in the short run. Both the sole owner and the competing fisherman will operate at an output which is theoretically similar (in its equality of marginal cost and marginal revenue) to that in other industries. Only if there is an opportunity for adopting alternative fishing techniques that reduce the investment necessary for a given output is there an argument in favor of sole ownership. Some efficient techniques may be profitable only on the assumption that a very large fishing operation can adopt them. But why cannot there be large-scale efficient operations under competition among fishermen? Perhaps because of the danger of diminishing the population or of the fear of its diminution; but these are long-run dangers which I shall discuss below.

While I am on the short-run part of the argument, it is relevant to comment on the subject of the cost of the variable factors. These can be divided into cash costs and opportunity costs of the fishermen. The smaller the opportunity costs—perhaps because of the immobility of the fishermen—the greater the use of factors in fishing, regardless of whether the industry is competitive or typified by sole ownership. The low opportunity costs do not provide a basic explanation of the inefficiency of competitive exploitation of fisheries; it is the inability to control the size of the fish population in the long run which does that. Hence even in areas where relevant opportunity costs are high, as they are in the West Coast industry, we find more men and more rigs employed than would be employed in a "monopolized" fishery. The price system, when it works well, does not depend only upon high opportu-

---

[9] Another external diseconomy arises from the shortage of really skilled labor. A sole owner could either plan to economize on the use of labor by adopting labor-saving techniques (which, to an extent, is a method also open to competing boats) or to act as a monopsonist in the purchase of local labor services, or both.

ANTHONY SCOTT

nity costs to draw factors into the most productive employment. It also relies on employers dispensing with factors that are not needed; and our subject here is really the alleged failure of competitive fisheries to do this. Low opportunity costs are not relevant to the immediate problem. Where Gordon brings in the low opportunity costs

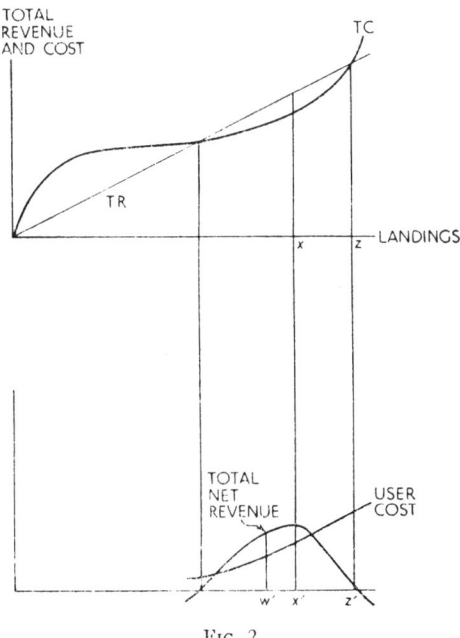

Fig. 2

of the industry, he drags in a red herring.[10]

III

But it is when we come to the long period that we see where the four-variable analysis fails as a description of the sole owner's optimum. What is now needed is an indication of the best use of the factors of production and of the fishery over time. It is not to be concluded, for example, that the ra-

tional owner would even wish to find an "equilibrium" size for the fish population. His most profitable action might be instead to deplete the fishery, gradually, over time; or, alternatively, to build it up over time. *As long as the user of a fishery is sure that he will have property rights over the fishery for a series of periods in the future,* he can plan the use of the fishery in such a way as to maximize the present value (future net returns discounted to the present) of his enterprise. From the social point of view it can be said that he will bring about the "best" use of the fishery and of all other factors invested in it over future periods by thus allocating outputs and outlays over time in accordance with the current rate of discount.

I can best illustrate the nature of this decision-making process by using the following diagrams (Fig. 2).[11] The first diagram indicates the total costs and total revenues of a fishery in a period. The total revenue curve is shown as linear to indicate that the output (landings) of this one fishery has no effect on the price at which it may be sold. The total cost curve is shaped to suggest diminishing marginal returns (increasing costs) in the short run.

Under competitive fishing conditions, or if the fishery is taken into sole ownership, the tendency is to maximize net returns from the fishery by producing x, where TC is parallel to TR (that is, where marginal cost equals price). This holds only in the short run.[12] However, if the catch today has an influence on the population and so

[10] *Op. cit.*, p. 132.

[11] See r→ "Notes on User Cost," *Economic Journal*, June, 1953, p. 372.

[12] The output z is the competitive equilibrium (no-profit) output suggested by Gordon.

THE FISHERY: THE OBJECTIVES OF SOLE OWNERSHIP

on the catch tomorrow, the sole owner will wish not only to maximize current returns but also to arrange for the optimum series of landings through the ensuing future periods. He will, in fact, wish to maximize the present value of his property. This he will do by investigating the effect of his marginal current output on the present value (or sum of the discounted net returns of all future periods) and by fixing current output where marginal current net revenue is equal to marginal user cost. In such a position, since a unit of output is produced only if its addition to current net revenue exceeds its cost in diminished present value, the sole owner succeeds in keeping the future returns from the fishery as high as possible while maximizing current income. This position is to be found at w' where the total net revenue curve is parallel to the user-cost curve. The total net revenue curve is derived directly from the TC and TR curves and shows the difference between them. The user-cost curve shows the effect of succeeding units of current output on the "present value" of the enterprise. The greater the rate of interest (or the personal rate of time preference of the owner), the lower the valuation put on landings in the remote future, and the lower the user cost.

If increased output tends to diminish the population and so to reduce the net revenues that could be earned in other periods had output been restrained today, the user-cost curve will slope upward, marginal user cost will equal marginal net revenue at less than the maximum total net revenue, and sole ownership will result in a still greater reduction of desired output than would be the case if short-run considerations only were at stake. This slope of the UC curve is presumed to describe the situation in those fisheries that are exhaustible. Pelagic fisheries such as salmon and seal might be suggested.

If, on the other hand, increased output should tend to increase the population and the net revenues to be earned in future periods, the user-cost curve would slope downward. This, it has been said, is true up to a point of some fisheries: effort today not only produces a catch today but also improves conditions for increase of the fishery. But this too may be truer of pelagic than of demersal fisheries. In these special circumstances the current rate of output of a sole owner would be somewhat larger than that which yielded the maximum current return, and perhaps even more than that of Gordon's competing fishermen.

If landings have no effect on population (or, more precisely, on future landings), there is no user cost, no user-cost curve, and the output x is most profitable even when long-run effects are taken into consideration. This seems to be the thoroughgoing "demersal" case.

IV

What has been left out of this picture of the sole owner's planning for the long run?

In the first place, following Gordon, I have not given much attention to the nature of the fixed equipment used in the industry. It may not be possible to replan the duration of the fishery (that is, the size of the population) from period to period with complete freedom because the type of equipment needed to bring in small, unusually large, or postponed landings may not

ANTHONY SCOTT

be available. In actual practice it is necessary to plan the "scale" of a private fishery at the outset, to fit into the complementary provision of canneries, transport, etc. Once this "scale" has been established, it is not easy to change the general range of output per season if user cost changes. The problem here could be illustrated diagrammatically by showing a variety of short-run cost curves corresponding to the choice among "scales" of fishing industry. The net revenue curves so established would have to be confronted with a series of user-cost curves corresponding to alternative future long-run scales of the industry.[13]

In the second place, I have assumed that the sole owner is not the monopolist of his product. If he were a monopolist and could influence the price by his output, he would be confronted by a nonlinear total revenue curve in Fig. 2, and it is conceivable that his landings per period would be even smaller than those of a corresponding solely owned fishery competing with many other fisheries. Also, if he were a monopolist, he would be able to influence the future price and even the trend of taste and demand for his output. There are so many possible consequences of this power that it is impossible to generalize about them. One

→ Professor Donald Carlisle in his recent "The Economics of a Fund Resource: Mining," *American Economic Review*, September, 1954, p. 609, has suggested a three-dimensional cost surface, with rate of mining on one axis, level (or scale) of the mine on another, and costs on the vertical axis. Instead of the usual U-shaped cost curve, a saucer-shaped surface is developed. It is not easy to see how the "present value optimum," which corresponds to our long-run equality of marginal user cost and marginal net revenue, is determined —its position is merely drawn into the diagram, apparently arbitrarily. However, the diagram is a useful reminder that scale and current output cannot be dissociated; each must be constantly reappraised.

important possibility, however, is that the uncertainty surrounding the price and sales in each period would be somewhat less than the uncertainty borne by competitive producers. Such monopolists might well increase the investment in the scale of the fishery. This, in turn, would tend to keep the fish population high and to promote a sustained yield rather than the gradually decreasing yield that is a likely outcome of sole ownership.

Finally, I have asserted that the equilibrium of the sole owner who maximized the present value of the fishery would correspond more closely to the social optimum than would the competitive equilibrium. This is true only if the other enterprises in the economy are run by purely competitive businessmen who attempt to maximize their profits and to maximize the present value of their enterprises in terms of the market rate of discount. If these assumptions are satisfied by equalizing marginal user cost and marginal net return, we have a situation where the marginal productivity of each factor is the same wherever it is used, and where the allocation of production over time corresponds to the rate of discount determined by the marginal rate of time preference of the community. In such circumstances, given the endowment of the economy with factors and resources, and given the tastes of savers and consumers, it would not be possible to increase the value of output of any product without reducing some other output by a greater amount. In this sense, the social optimum in both the long run and the short run would demand that common-property resources be allocated to maximizing owners, associations, co-operatives, or governments.

JOURNAL OF
ENVIRONMENTAL
ECONOMICS AND
MANAGEMENT

http://www.elsevier.com/locate/jeem

Economic impacts of marine reserves: the importance of spatial behavior

Martin D. Smith[a,*] and James E. Wilen[b]

[a] *Nicholas School of the Environment and Earth Sciences, Duke University, Box 90328, Durham, NC 27708, USA*
[b] *Department of Agricultural & Resource Economics, University of California, Davis, USA*

Received 15 March 2002

Abstract

Marine biologists have shown virtually unqualified support for managing fisheries with marine reserves, signifying a new resource management paradigm that recognizes the importance of spatial processes in exploited systems. Most modeling of reserves employs simplifying assumptions about the behavior of fishermen in response to spatial closures. We show that a realistic depiction of fishermen behavior dramatically alters the conclusions about reserves. We develop, estimate, and calibrate an integrated bioeconomic model of the sea urchin fishery in northern California and use it to simulate reserve policies. Our behavioral model shows how economic incentives determine both participation and location choices of fishermen. We compare simulations with behavioral response to biological modeling that presumes that effort is spatially uniform and unresponsive to economic incentives. We demonstrate that optimistic conclusions about reserves may be an artifact of simplifying assumptions that ignore economic behavior.
© 2003 Elsevier Science (USA). All rights reserved.

JEL classification: Q22

1. Introduction

An important paradigm shift is underway in marine policy that will profoundly alter the future management of our coastal resources. The shift is toward the use of spatial zoning measures that will effectively partition the ocean into a system of areas with regulated exploitation and areas protected with marine reserves.[1] A confluence of emerging science and political interest in marine

---
*Corresponding author.
 E-mail address: marsmith@duke.edu (M.D. Smith).
[1] The emerging literature on marine reserves uses a variety of terms that are sometimes synonymous and sometimes not. We will use marine protected areas, marine reserves, and no-take zones interchangeably to mean areas in which no exploitation is permitted, including especially commercial and recreational fisheries exploitation.

0095-0696/03/$ - see front matter © 2003 Elsevier Science (USA). All rights reserved.
doi:10.1016/S0095-0696(03)00024-X

protected areas provides the underpinnings of this new vision. In the sciences, there is growing consensus among marine ecologists, biologists, and many fisheries managers that conventional season length and gear restriction management methods have failed, are bound to fail in the future, and that a new approach is therefore needed. In the political arena, there is a growing view among NGOs and influential environmental lobbying groups that a network of protected areas similar to our terrestrial park system will best achieve long-term biodiversity and conservation goals.

Up to this point, biologists have promoted this new spatial zoning view with little input solicited from economists and other policy analysts. But as real proposals emerge for specific systems of protected areas, there will be increasing calls for economic analyses of actual policy options. Economists are only beginning to think about how to introduce space into conventional models of renewable resources and how spatially differentiated policies might compare with second best undifferentiated policies. With a case study of the northern California red sea urchin fishery, we present a comprehensive empirical investigation of how marine reserves would perform as a fishery management tool. We analyze both economic and biological consequences of implementing reserves. To this end, we focus particularly on harvester spatial behavior, and we assess the degree to which accounting for economically driven behavior alters the conclusions of recent biological investigations of marine reserves. We find, somewhat to our surprise, that the main simplifying modeling assumptions made by biologists to handle harvester behavior do not actually "cancel out" in the final analysis of reserves as a fishery policy instrument. Instead, virtually every economic simplification in biological modeling biases the case in favor of using reserves to manage fisheries. Our results thus call into question whether the optimism displayed for reserves as a fisheries management tool is warranted.

In Section 2 we review both the biological and economic literature on marine reserves. We then describe the sea urchin fishery case study in Section 3 and focus particularly on its spatial character and its short- and long-term dynamics. In Section 4 we discuss and estimate a model of spatial behavior and then use it, in Section 5, to simulate the impacts of marine reserves with an integrated bioeconomic model. In Section 6 we draw some broad conclusions for both research and policy analysis of marine protected areas.

2. Related biological and economic work

The notion of using closed areas as a fishery management tool emerged over a decade ago among marine ecologists and conservation biologists. The first proposals were modest and focused on reserves as laboratories, calling for small areas off coastal research institutes in which ecologists could study unexploited systems in order to gauge the ecological impacts of exploitation. By the early 1990s, the idea had morphed into a grander vision that called for significant areas to be set aside, often on the order of 20–30% of the coastline. The transformation in the scale of the proposals coincided with several important papers on fisheries management. Most of these studies concluded that the world's fisheries were in a state of crisis, that conventional methods were to blame, and that a new approach to management was needed.[2]

---

[2] See for example the oft-cited paper by Ludwig et al. [13], which argues that reductionist approaches cannot overcome the natural variation and irreducible uncertainty inherent in natural systems in order to guide us toward

Early modeling papers thus focused on the impact that permanent spatial closures might have on maintaining fisheries at safe levels of exploitation.

The first modeling work on marine reserves is Polacheck [20], which uses a Beverton and Holt [1] model to examine how a spatial closure might affect an exploited fishery. By assuming exogenous larval recruitment in the fishery and exogenously determined fishing effort, Polacheck shows that marine reserves always increase spawning biomass within the reserve itself and under some circumstances, but not many, reserves increase fishing yield. Empirical studies of reserves proceeded to confirm the finding that spawning biomass increases *within* closed areas [7].

One may not need voluminous empirical work to believe that removing exploitation from an ecosystem will increase biomass and broaden the age distribution by providing protection for larger and more fecund fish. Less certain, however, is whether the increase in spawning stock biomass and its composition within a reserve can provide a net increase to the fishery *outside* the reserve.[3] The first generation modeling work identifies mechanisms that are germane to this question, namely the mobility of adults and the level of pre- and post-reserve exploitation in the fishery. These studies show that reserves most likely provide a net yield increase from migration into the open area when adult mobility is neither too low nor too high. The intuition is that, for a given size reserve, adults must not move so widely that the reserve does not afford them protection, but there must be some movement of adults (or juveniles) to increase the fishable abundance in the remaining open area.

The second generation of papers generalizes the earlier approaches by closing the relationship between spawning biomass and recruits to the fishery. This is an important addition to understanding because, in addition to spillover of adults, protected areas may also produce eggs and larvae that subsequently redistribute to exploitable populations in the remaining open areas. A few studies from the late 1990s assume that adults within and outside the reserve produce larvae, which disperse in some manner and then recruit into both the fishery and the reproductive population.[4] A significant and frequently cited second generation paper is by two economists, Holland and Brazee [10]. In a detailed age-structured two-patch population model, they depict sophisticated biological mechanisms, including density-dependent stock/recruitment relationships in both the reserve and open area, migration of adults according to a density-dependent mechanism, and (uniform) larval dispersal. Perhaps more importantly, their model incorporates economic variables and is fully dynamic so that it computes the present values of transition paths. Holland and Brazee confirm the Polacheck results that spawning stock biomass will always

---

(footnote continued)
anything like sustainable yield. They urge that managers "confront uncertainty" by explicitly accounting for it in common sense ways including taking actions that are robust to uncertainty, that allow monitoring and learning, that are reversible, that hedge risks and that incorporate scientific principles from decision making under uncertainty. Walters [29] provides more detail on these ideas in his important book on adaptive management.

[3] Some of the fisheries literature does not seem aware that it is a net increase overall and not just a gross increase in yield that is needed in the remaining open areas to make a reserve worthwhile as a fisheries enhancement measure. That is, from the fishing industry perspective, it is not enough that a reserve increases harvest outside the reserve, but rather that the increase be large enough to compensate for the area removed from fishing. In general, this is a significant hurdle, and as the reserve size gets larger, ceteris paribus, this hurdle rises.

[4] Most models assume a common pool dispersal process whereby total system larvae are distributed instantaneously and uniformly over the whole system.

increase with reserves. They also find that whether this increase creates conditions to generate a net increase in the present value of economic benefits depends importantly on the discount rate and the pre-reserve exploitation level, as well as bioeconomic parameters. The role of discounting is intuitive because reserves always decrease harvests initially and then increase harvests as spillovers begin to emerge. At high discount rates it may not be worth the sacrifice necessary to rebuild sustainable harvests to higher levels. Thus, the discount rate is an essential determinant of whether reserves generate net economic benefits. In contrast, the first generation modeling ignored transition paths and hence missed this important characteristic of reserves. Holland and Brazee find that pre-reserve exploitation rates are the other important determinant of whether a reserve pays off; dramatically, overharvested fisheries are more likely to be worth an investment in rebuilding the overall biomass by using a reserve.[5]

One key simplifying assumption in virtually all of the early analyses of reserves is that effort is fixed both before and after reserve formation. Under a spatial closure, most analysis presumes that effort simply displaces to the remaining open area. This analysis thus does not account for the fact that economic conditions will, in part, determine pre-reserve fishing effort and that the reserve itself will alter relative profitabilities and hence subsequent effort decisions by fishermen. Sanchirico and Wilen [23,24] relax this assumption in a conceptual analysis that determines spatially explicit effort endogenously.[6] They find that, under open access, most reserve scenarios produce a biological benefit but that there are very few combinations of biological and economic parameters that give rise to both a harvest increase and a biological benefit. In particular, they find that harvest increases are likely only when the designated reserve patch has been severely overexploited in the pre-reserve setting.

Taken as a whole, the conceptual biological and economic literature warrants a serious empirical investigation of reserves that accounts for spatial and dynamic aspects of the biological system and the harvest sector. In this paper we consider how incorporating realistic depictions of harvester behavior affects the implications of marine spatial closures. A priori, economically motivated harvester behavior ought to matter in several ways. First, since the pre-reserve status quo is important to the net effect of a policy change, it should be of interest to know how economic variables condition the initial circumstances in a spatial bioeconomic system. Second, since a spatial closure will affect the subsequent spatial distribution of relative economic returns, we expect that effort redistribution will have complicated spatial and intertemporal effects, both in the short run and in the long run. Third, since most real spatial systems embody complicated biological and economic heterogeneities that affect profit differentials over space, we expect these to color the conclusions about the economic impacts of reserves. For all of these reasons, simplified assumptions about effort distribution and its determinants are likely to confuse the debate about these new forms of marine policy instruments.

---

[5] At the same time, what this shows more fundamentally is that it can be a sound investment to rebuild an overexploited stock by reducing effort. Closing a fraction of a fishery's area is one way to reduce overall fishing mortality (as Polacheck [20] concluded), but another way is to simply crank down conventional methods of effort and fishing mortality control. Hastings and Botsford [9] have established that area controls are equivalent to conventional effort controls under certain reasonable circumstances. Some fisheries observers point out, however, that closed areas may be easier to enforce than conventional effort control measures such as mesh size, days at sea, etc.

[6] Hannesson [8] also allows for endogenous effort in an open access model of marine reserves but incorporates less generality in the biological and economic models than Sanchirico and Wilen [23,24].

3. Case study: the northern California red sea urchin fishery

The northern California red sea urchin fishery is an ideal case study with which to examine empirically marine reserves and the dynamics of harvester spatial behavior. Urchin population dynamics are consistent with a biological structure that is favorable for reserve formation, and regulators are currently considering spatial management in the urchin fishery.

Red sea urchins, *Strongylocentrutus franciscanus*, are found along the Pacific Coast in rocky inter-tidal kelp forests from southern California to Alaska.[7] Urchins have hard spiny shells and are harvested for the gonads (roe) found inside their shells. Divers harvest urchins on 1-day trips to fishing grounds. Their vessels are designed especially to travel fast to the fishing locations. Once on a site, a diver begins harvesting while connected to the surface by a line that supplies compressed air from the boat. A tender aboard the vessel monitors the air compressor and the diver's line. Harvesters use rakes to remove urchins from the rocky bottom, filling mesh bags that are then loaded onto the vessel by the tender. Weather conditions are critical determinants of when and where divers choose to dive, particularly in northern California where a trip is typically made only 14% of available open days.

At the end of a fishing day, harvesters deliver the whole urchins to processors. At the processing plant, workers split the shells, scoop out the gonads, and then wash and bathe the gonads in alum solutions to firm the roe. Roe skeins are then carefully packed in special wooden trays holding 250 g, and the product is shipped overnight to Japan. In Japan, the roe is sold mainly at the Tokyo Central Wholesale Market, in competition with roe from Japan and other North American fisheries. The roe is relatively high valued, with current Tokyo Wholesale prices in the range of $25 per pound, translating into a markdown to divers of approximately $1 per pound of whole urchin. A typical day trip currently brings in 750 pounds of urchin, grossing $750 per trip for the diver/tender team.[8]

Sea urchins have been harvested in southern California since the early 1970s and in northern California since 1988.[9] Regulatory restrictions include closed seasons, minimum size limits, and a limited entry program. Current regulations require that northern California urchins be at least 3.5 inches in diameter,[10] a size reached at approximately 6 years of age. A full closure is in effect throughout all of July, and 1-week closures are operative from May to September. Within each open week, 3-day per week openings are in force in June and August, and 4-day openings prevail in April and October. In total, the number of potential open days per year is approximately 240 in northern California. The limited entry program was introduced into the California-wide fishery in

---

[7] We draw the factual background in this section from Kato and Schroeter [12], Kalvass and Hendrix [11], the log book and landings ticket data described below, and the website of the Sea Urchin Harvesters Association of California, http://www.seaurchin.org.

[8] The recovery of roe from a whole urchin is about 10%. Hence, ten pounds of whole urchins are needed to generate a pound of roe, mostly explaining the markdown.

[9] Ironically, urchins were considered pests by abalone divers in the 1960s because they competed for the lucrative abalone for habitat and food. Abalone divers often poured quicklime on urchins in order to remove them and open up habitat for the commercially valuable abalone. In the 1970s, a market for California red sea urchins developed as the Japanese stocks were overharvested [22]. The fishery was harvested mainly in southern California by ex-abalone divers after the abalone fishery collapsed. The continued growth in the Japanese market led to opening the Northern California fishery in the late 1980s.

[10] Sea urchin diameter is measured across the test, i.e. the urchin shell from which the spines extend.

1989, grandfathering all existing permit holders (roughly 850) into the fishery. The program has tightened over the intervening years in order to steer participation to a long-term goal of 300 divers. Each diver must now land 300 pounds per trip for a minimum of 20 trips during either the current or preceding year in order to be granted a license.[11] The license holder must do the fishing, and licenses may not be leased or sold.

Sea urchin biological characteristics are consistent with those identified by biological modelers as holding promise for successful use of reserves. First, urchins are "patchy" in that they are found in multiple discrete areas where substrate and habitat characteristics are suitable for the species. Second, adult movement within each patch is relatively low (an average of 7–15 cm per day) and hence closed areas promise protection of spawning biomass. Urchins reach sexual maturity approximately at age 5, and egg production increases with age at an increasing rate.[12] Third, larvae are redistributed considerable distances from spawning areas by currents, winds, and sea surface changes and hence protected urchins have the capability of replenishing and sustaining remaining open areas.

In addition to favorable biological characteristics, the urchin fishery is also an ideal case study for examining spatial behavior of harvesters. Most importantly, urchin trips are day trips that are made repeatedly by the same individuals under conditions that vary with biological and economic variables. This means that it is possible to assemble a large panel data set with substantial intertemporal variation in determinants of behavior. The data set that we use to examine spatial harvester behavior in the northern California sea urchin fishery is unusually rich. It consists of over 57,000 actual individual dives made by up to 358 divers in each year over the period between 1988 and 1997. Divers record information in mandatory logbooks and on landings tickets. Each landings ticket reports port of landing, processor code, quantity landed, price paid, and diver code. Each logbook entry contains dive location (latitude), dive duration, average depth, and divers per vessel. We also collected daily weather data on wind speed, wave height, and wave period from weather buoy records, averaged it over the period preceding a typical trip decision, and linked it to each potential dive trip decision. The fact that each trip is a day trip is convenient since we can model decisions as repeated nested discrete choices.[13] The model that we report here estimates parameters of a repeated choice structure for several hundred divers making decisions over ten years and amounts to over 400,000 choice occasions.[14] During this period, the urchin fishery was harvested from a virgin fishery to the present level approaching a steady state in which returns to fishing equilibrate across space. Prices vary over the period as a result of exchange rate changes, quality changes, and landings variability. Weather conditions vary on a daily basis and with some inter-annual variation as well.

---

[11] Provisions are also made to allow one new entrant for each ten licenses retired.

[12] Brown and Roughgarden [5] make a similar returns to scale argument to show that harvesting one patch from a metapopulation can be the optimal strategy.

[13] A repeated decision structure is more tractable than trying to model situations with a fixed end point. In addition, fisheries in which multi-day and multi-area trips are made must contend with the complications introduced by decisions that are essentially searching decisions made along the journey to a targeted destination.

[14] Total choice occasions are days for which the urchin fishery is open to harvesting. Because weather conditions so frequently keep fishermen ashore, the actual number of trips is only a fraction of total choice occasions, about 14% of open days. In addition, Sunday closures in the Tokyo Central Wholesale Market induce lagged responses that reduce diving activity on Fridays and Saturdays.

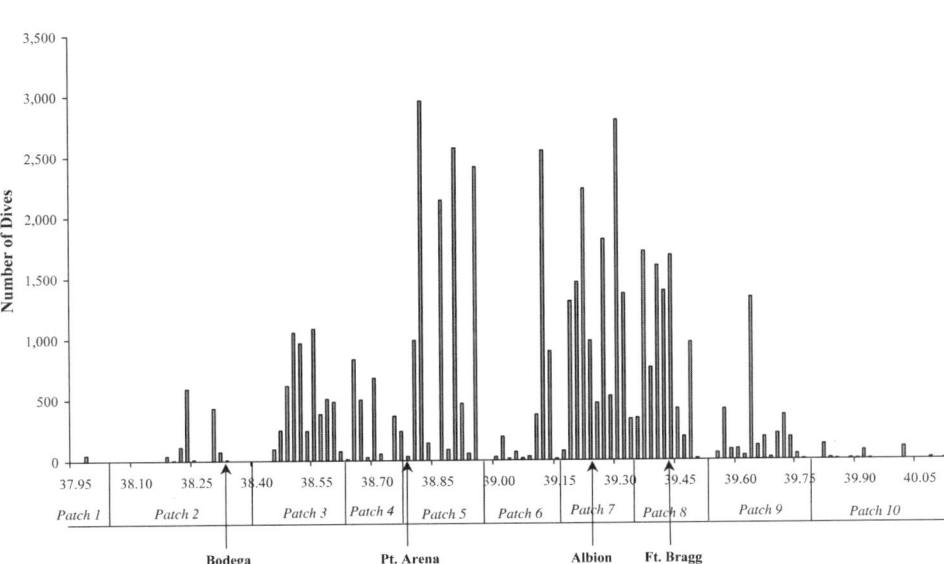

Fig. 1. Diving activity coastal histogram. Trips in North-Central California 1988–1997.

In northern California, harvesters land sea urchins in six ports. From south to north, the ports are Half Moon Bay, Bodega, Point Arena, Albion, Fort Bragg, and Crescent City. Crescent City near the Oregon border and Half Moon Bay on the San Francisco peninsula contribute less than 10% of the total harvest, while the remaining harvest is landed at the four ports in the middle, all within 200 road miles north of San Francisco International Airport.

To begin our analysis of the urchin fishery, we identify and define patches. In our definition, a patch is a bioeconomic concept, not purely a biological one. Thus, patches are discrete subpopulations of *harvested* sea urchins. Before settling on patch definitions, we first looked at the spatial distribution of fishing effort generally. Not surprisingly, fishing effort fans out in a manner that declines geometrically with distance from the port. Because the four main ports are relatively close, when aggregating effort across ports, there is considerable overlap. That is, some areas are fished from multiple ports. Still, there are areas close to ports in which no exploitation occurs. Using these effort data and knowledge that there are breaks in suitable sea urchin habitat, we partitioned the entire data set into eleven patches along the coast of northern California. Fig. 1 is a histogram over the whole period with diving activity characterized by "patch" number.[15] With the exception of the Farallon Islands, patches correspond to latitudinal corridors that extend from the shore out to sea. Reducing the patches to a single latitude-based spatial dimension is sensible in this fishery because urchins only live close to shore. The x-axis shows degrees of latitude (converted to decimals rather than degrees and minutes). The figure shows the large concentrations of effort located near the four ports with highest landings, together with breaks between patches associated with substrate and habitat variation. Throughout the fleet, divers

---

[15] We do not show the eleventh, which is off the Farallon Island off San Francisco Bay. See [27] for details.

Table 1
Diver spatial mobility across patches

No. of patches active in	Number of active divers											
	1988	1989	1990	1991	1992	1993	1994	1995	1996	1997	1998	1999
1	50	60	146	139	139	99	107	49	37	36	43	43
2	44	51	72	72	62	65	67	48	44	45	33	39
3	20	19	59	60	52	46	45	30	25	33	27	24
4	10	8	36	32	36	32	25	21	21	22	18	9
5	4	0	20	23	38	18	13	15	0	7	5	9
6	1	0	6	12	13	15	13	3	4	3	3	1
7	0	0	2	8	15	9	2	1	0	0	1	0
8	0	0	0	4	3	0	0	1	0	0	0	0
9	0	0	1	3	0	0	0	0	0	0	0	0
10	0	0	0	0	0	0	0	0	0	0	0	0
11	0	0	0	0	0	0	0	0	0	0	0	0
Total divers	129	138	342	353	358	284	272	168	131	146	130	125
Weighted average no. patches	2.05	1.82	2.25	2.53	2.68	2.60	2.33	2.54	2.35	2.51	2.40	2.24

show considerable spatial mobility across patches, as Table 1 demonstrates. In any given year, individuals visit between one and nine patches with fleet annual average patches visited generally ranging between 2 and 3 over the sample period.

In Table 1, one can also see the open access implications of opening up the previously unexploited northern California fishery starting in 1988 before regulators began to act. From zero activity in 1987, harvesters from the pool of divers in the southern California fishery and elsewhere were drawn rapidly into the new fishery, with over 100 new divers participating in 1988. The northern California part of the fishery peaked in 1992 with 358 total divers. During the peak period, some divers fished in as many as 6–8 northern California patches during the year, although the majority concentrated on one or two patches from a single port. As the northern California fishery matured and as the unexploited stocks were drawn down and profitability reduced, regulations were implemented, and fishermen exited, some returning to the southern California fishery and others dropping out entirely in response to the stringent limited entry requirements. Total California-wide numbers dropped from 826 in 1994 to 440 in 1998, whereas in northern California, numbers dropped from 358 to about 125 in recent years. In the last couple of years, there has been a slight reduction in average mobility across northern California ports, and the fishery appears to have settled into a pattern approximating a bioeconomic equilibrium, with harvesters fishing more intensively to maintain a living under reduced abundance conditions. The average number of trips per year has increased from about 26 per year in 1992/93 to a current level of about 34 trips per year. Average depth per dive (mean 35.45 ft over all years/locations) has similarly increased at a rate of about 1 foot per year as the extensive margins expanded out from all port locations.

4. Modeling spatial behavior

In this section, we report results using the logbook/landings ticket-based data set on divers' daily spatial behavior to estimate a model of individual choice. It is convenient to begin with the hypothesis that individual divers make daily choices that maximize some random utility function:

$$U_{ijt} = v_{ijt} + \varepsilon_{ijt} = f(\mathbf{X}_{it}, \mathbf{Z}_{i1t}, \mathbf{Z}_{i2t}, \ldots, \mathbf{Z}_{iMt}; \boldsymbol{\theta}) + \varepsilon_{ijt}, \qquad (1)$$

where $\mathbf{X}_{it}$ includes harvester-specific and time-specific characteristics that are constant across choices, $\mathbf{Z}_{ijt}$ includes choice-specific characteristics such as travel costs and expected resource abundance that may also be harvester- and time-specific, $\boldsymbol{\theta}$ is a parameter vector, and ε_{ijt} is a random component that is unobservable to the analyst. This random utility model posits that, given M possible dive locations and the possibility of not diving, diver i in period t will choose location k if the utility of choice k is higher than that of the other $M - 1$ location choices as well as the choice of not to dive in period t.

The nesting structure for daily choices is straightforward. On any given open day, divers choose to go or not go fishing. If they choose to go fishing, they choose among the set of locations. For identification, we normalize the utility of not going to zero. Note, however, that the utility of not diving captures the utility of leisure, work opportunities outside of fishing, and the value of avoiding exposure to unsafe diving conditions.

Although there are numerous discrete formulations that can represent this choice structure, we adopt a Repeated Nested Logit (RNL) framework similar to Morey et al. [17] and based on McFadden [14].[16] Smith [28], which to our knowledge is the first analysis to model discrete participation and location choices jointly in commercial fisheries, identifies reasons for taking this approach in modeling spatial decisions of commercial fishing fleets. Here we summarize these reasons with an eye toward building a bioeconomic simulation model. First, a forward-looking model is unnecessary because even under limited entry, there is some element of non-excludability in harvest, and fishers do not perceive that they will realize future gains from restraining present harvest. Second, the RNL is relatively easy to estimate with the large size data set that we have. A more advanced technique that relies on simulation-based estimation, e.g. Mixed Multinomial Logit [16], could require a super computer to estimate over this 400,000 observation data set. Third, we need an estimation scheme with a closed form solution in order to embed the estimated behavioral choice model into the bioeconomic simulation model discussed in the next section.[17] Finally, the RNL also has well-known properties that are desirable in general, including the flexibility to admit different variances at different decision nodes.[18]

---

[16] The choice structure is basically the same as the RNL proposed by Morey et al. [17], which deals with repeated participation and site choice in recreation demand. The key difference in our setting is that we do not explicitly model income effects, and our site choice is conditional on the diver's port. In contrast, Morey et al., in dealing with recreation demand, do model income effects and include separate decision branches for region choice followed by specific site choice.

[17] In contrast, a simulation-based estimation approach would require nesting complicated numerical integrations within each time step of the bioeconomic simulation model below and would lead to computational infeasibility.

[18] The drawbacks of RNL are equally well known and include assumptions of independence across both time and individuals.

Following McFadden [14], we assume that the ε_{ijt} is independently and identically distributed generalized extreme value. We further assume that utility is linear in individual- and choice-specific variables so that the following model characterizes individual choices:

$$Pr(Go\ to\ j) = \frac{\exp\{\frac{z'_{jt}\gamma}{(1-\sigma)} + \mathbf{x}'_t\boldsymbol{\beta} + (1-\sigma)I\}}{\sum_{k=0}^{10}[\exp\{\frac{z'_{kt}\gamma}{(1-\sigma)}\} + \exp\{\frac{z'_{kt}\gamma}{(1-\sigma)} + \mathbf{x}'_t\boldsymbol{\beta} + (1-\sigma)I\}]} \quad (2)$$

and

$$Pr(Do\ not\ go) = 1 - \sum_{k=0}^{10} Pr(Go\ to\ k)$$

$$= \frac{1}{1 + \exp[\mathbf{x}'_t\boldsymbol{\beta} + (1-\sigma)I]}, \quad (3)$$

where

$$I = \ln\left[\sum_{k=0}^{10} \exp\left\{\frac{z'_{kt}\gamma}{(1-\sigma)}\right\}\right]. \quad (4)$$

In the above equations, the i subscripts for the individuals are suppressed since the form of the model is the same for each individual in the data set. Some characteristics $\mathbf{x}$ and $\mathbf{z}$ could, in principle, vary across individuals. Here $\boldsymbol{\beta}$ denotes the parameter vector for characteristics that vary across individuals and/or choice occasions but not across choices, γ is the parameter vector for choices that vary across choices, and $(1-\sigma)$ is the coefficient on the nested logit inclusive value.

Table 2 reports the results of a parsimonious specification using the entire 401,151 observation data set.[19] Variables that are not choice-specific include three coast-wide weather variables and a day-of-week dummy variable. The weather variables are WP (wave period), WS (wind speed), and WH (wave height), all computed as averages over weather buoy data for the 12-h period preceding noon of the day in question. The day of the week dummy (DWEEK) is one on Friday, Saturday, and Sunday, reflecting reduced propensity to dive on weekends and on days that precede the Sunday closure of the Tokyo Central Wholesale Market. Variables that are location-specific are distance to the center of the patch in question (DISTANCE), and expected revenue (ER) in each patch.[20] The inclusive value is computed from the lower level utility branch.

---

[19] Our strategy for estimating the model involved first drawing a random sample of 30 divers followed over the entire period. This relatively smaller sample of 27,000 choice occasion observations was used for preliminary specification testing. We used the smaller subset to determine the effects of using different backward lags for our expectations variables, different averaging methods for prices, and the impacts of various diver-specific variables computed from the data set. See [28] for empirical results from this smaller data set and comparisons to a more aggregated econometric approach.

[20] Some of the methodological issues associated with computing expected revenues are explored in [26]. Expected revenues reported here are calculated as the product of expected price and expected catch per trip. The former is a rolling 1-month backward looking average across the entire northern California fishery, so that for each day the expected price is based on the average price for the previous 30.4 days. Expected catch is a patch-specific rolling 1-month backward looking average. These proved to be best fit in our preliminary specification tests. In addition, 1-month averages are convenient for simulating the model in the next section.

Table 2
Nested logit estimates

Variable	Coefficient	Standard error	Z-statistic
Not location-specific			
Constant	1.06	0.048	22.21
WP	−0.18	0.005	−34.69
WS	−0.11	0.003	−36.69
WH	−0.74	0.011	−70.36
DWEEK	−0.74	0.012	−60.02
Location-specific			
DISTANCE	−7.27	0.036	−203.72
ER	0.08	0.001	65.17
σ	0.22	0.027	8.34
Log-likelihood		−189,878	
Observations		401,151	
Pseudo-$R^2(1)$		0.21	
Pseudo-$R^2(2)$		0.81	

Pseudo-$R^2(1)$ is based on the log-likelihood in a conditional logit model with choice-specific constants. Pseudo-$R^2(2)$ is based on the log-likelihood of $n\ln(1/J)$, where $J = 12$ possible choices.

As Table 2 shows, all variables have signs as expected and all are highly significant as one would expect.[21]

These estimated RNL equations may be used to generate simulated effects of changes in economic variables, such as a change in expected revenues. An increase in expected revenues anticipated in patch k causes a *spatial substitution effect* as effort is drawn from other patches. However, the increase in expected revenues in patch k will also have a *participation effect* because urchin fishing becomes a more attractive use of time overall. This participation effect will cause more effort to be distributed over all possible patches. The upshot is that the own effect of an increase in expected revenues will always be positive. The cross effects may be positive or negative, depending upon whether the participation effect outweighs the substitution effect. In general, the signs of cross effects will hinge on all of the variable values, and the substitution effect will be larger for patches that are visited more frequently (because $Pr(k)$ is higher). Table 3 shows some of the computed elasticities of patch choice with respect to revenues.[22] A 10% change in expected

---
[21] Since σ is significantly different from zero (and from one), the inclusive value coefficient is between zero and one. This has two implications. First, it suggests that the model is globally consistent with stochastic utility maximization [6,15]. Second, it suggests that variances of utility for participation and for location choices are different. This implies that choices across branches of the decision tree are less similar than choices within each branch of the tree. For example, choosing between patches 7 and 8 on a given day is more similar than choosing between 8 and not going fishing on that day.

[22] These are computed assuming DWEEK = 0. In addition, since there are 11 travel distances from each port, we average across probabilities associated with these actual distances rather than using the sample mean distance. Using a sample mean would effectively act as if all divers came from a fictive port in the middle of the coast rather than from their actual ports. We also compute elasticities with respect to the weather variables, finding elasticities

Table 3
Nested logit cross-revenue elasticities. Mid-week distance adjusted

		1% Change in ER_j										
	j	0	1	2	3	4	5	6	7	8	9	10
% Change in P_k	k											
	0	1.515	0.042	0.085	0.080	0.086	0.065	0.058	0.047	0.043	0.058	0.112
	1	0.130	0.598	0.093	0.097	0.104	0.080	0.071	0.057	0.052	0.071	0.136
	2	0.117	0.045	1.212	0.088	0.094	0.072	0.064	0.051	0.047	0.064	0.122
	3	0.119	0.046	0.094	1.145	0.095	0.073	0.065	0.052	0.048	0.065	0.124
	4	0.117	0.045	0.093	0.088	1.221	0.071	0.064	0.051	0.047	0.064	0.122
	5	0.124	0.048	0.098	0.093	0.099	0.938	0.067	0.054	0.050	0.067	0.129
	6	0.126	0.049	0.100	0.094	0.101	0.077	0.836	0.055	0.051	0.069	0.131
	7	0.129	0.050	0.102	0.097	0.103	0.079	0.070	0.672	0.052	0.070	0.134
	8	0.130	0.050	0.103	0.097	0.104	0.079	0.071	0.057	0.620	0.071	0.135
	9	0.126	0.049	0.100	0.094	0.101	0.077	0.069	0.055	0.051	0.837	0.131
	10	0.104	0.041	0.083	0.078	0.084	0.064	0.057	0.046	0.042	0.057	1.580
Net change in P_{GO}		2.736	1.063	2.163	2.051	2.190	1.675	1.491	1.196	1.103	1.492	2.857

revenues off patch 8, for example, will result in an increase in total participation of 11.03%. This increase in effort will be distributed mostly to patch 8 which will experience a 6.2% increase, and in small amounts averaging 0.5% to each of the other patches. We use this model of spatial choice to predict the dynamic and spatial distribution of effort to be used in the biological model discussed next.

5. An integrated bioeconomic model of reserve formation

In this section we outline a spatially explicit and dynamic bioeconomic model of the sea urchin fishery. The model is innovative in several respects. First, the model is a true bioeconomic model in that it integrates a population model of sea urchins with a behavioral model of the harvesting sector and generates joint bioeconomic equilibria. Second, the biological model is explicitly spatial and dynamic. We depict the sea urchin population as a metapopulation of 11 discrete patches, each with its own natality/mortality and growth parameters. The populations are linked with a dispersal matrix capable of characterizing any type of qualitative dispersal pattern. We parameterize the dispersal matrix with parameters calibrated to mimic field observations of larval settlement along the northern California coast. Third, the economic model is also explicitly spatial and dynamic. We use the model discussed above to depict industry behavior as an aggregation of individual choices made by divers, each of which is presumed responsive to the relative expected profitability of participation and location. Fourth, the economic and population

---

(footnote continued)
of: WP(-1.1), WS(-0.51), and WH(-1.42). These suggest that wave height is regarded as more risky than wind speed, a finding that seems sensible when one considers the process of trying to hold a vessel steady in high waves as a diver climbs in an out of the boat and the tender unloads urchins.

models explicitly account for size limits, season restrictions, and limited entry. Finally, the economic model of harvester behavior and the biological model are linked and integrated over both time and space. This allows us to experiment with different spatially explicit policies, change economic and biological parameters, and trace out both short run impacts, long run steady state impacts, and the dynamic and spatial adjustments that take place in transition to steady states.

5.1. The metapopulation model

The metapopulation model developed to examine spatial management policies in the red sea urchin fishery consists of 11 discrete age- and size-structured subpopulations linked by a dispersal matrix.[23] Each separate subpopulation has a size structure described by a von Bertalanffy equation, so that the size of an individual of age a in patch j is given by

$$Size_{j,a} = L_\infty^j (1 - e^{-k_j a}), \tag{5}$$

where a is a monthly time index from 1 to 360 and L_∞^j and k_j are patch-specific growth parameters. Note that L_∞^j is the terminal size of an individual organism, and this parameter ultimately dictates the maximum amount of biomass per individual. The model begins computations with a set of initial abundance matrices for each site. To generate these initial abundance matrices, we simulate a non-harvested steady state that mimics the situation in 1988 when the northern California resource first came under exploitation. The populations are then aged by advancing the abundance values for each month to the next older month so that $A_{i,a} = A_{i,a-1}$, where A denotes the number of organisms in the cohort. After the populations are aged, the numbers surviving in the population are computed, along with the catch. Survival is determined by a Beverton–Holt mortality relationship, which embeds both patch-specific natural mortality rates m_j as well as fishing mortality rates f_j if the size is above the minimum size limit L_{limit}. We link the economic model of diver behavior to the population model by making monthly fishing mortality rates a function of predicted diver trips. Accounting for both natural and fishing mortality, survival of the number of individuals to age a becomes

$$A_{j,a} = \begin{cases} A_{j,a} e^{-m_j} & \text{if } Size_{j,a} < L_{\text{limit}}, \\ A_{j,a} e^{-m_j - f_j} & \text{if } Size_{j,a} > L_{\text{limit}} \end{cases} \tag{6}$$

and total catch (C) consists of the sum of harvests of all sizes greater than the minimum size over all patches, which is

$$C = \sum_{j=0}^{10} \sum_{a=0}^{360} \frac{f_j}{m_j + f_j} (1 - e^{-(f_j + m_j)}) w \, Size_{j,a}^b A_{j,a}, \quad \forall \, Size_{j,a} > L_{\text{limit}}, \tag{7}$$

where w and b are allometric parameters relating weight and urchin test diameter. These parameters essentially convert number of organisms of each size to an aggregate measure of biomass, and harvest is a function of that biomass based on the fishing mortality parameters f_j. Note that for a given organism terminal size, L_∞^j, there is a corresponding terminal biomass. This, in turn, implies a maximum possible catch for a given number of organisms. The allometric

---

[23] The metapopulation model is more fully described in [2].

parameters give rise to the possibility of an increasing returns production technology because $b>1$ is the usual case, which means that the second derivative of catch with respect to size is positive.[24]

The metapopulation model also computes egg production, larval dispersal, settlement and survival. Egg production is computed after survival has been calculated for each month. If the month is a spawning month, then egg production in patch j is computed with

$$e_j = \sum_{a=0}^{a=360} \alpha x^\beta A_{j,a}, \quad \text{where } x = \begin{cases} Size_{j,a} & \text{if } Size_{j,a} > L_{\text{maturity}}, \\ 0 & \text{if } Size_{j,a} < L_{\text{maturity}}. \end{cases} \quad (8)$$

This equation sums the egg production from each size class, where there is only positive production for sizes greater than the size at reproductive maturity. The exponent on size (β) is greater than one, since egg production is known to be increasing and convex in organism size. Thus, the egg production relationship gives rise to another dimension of increasing returns production technology due to a combination of the minimum sexual maturity size and the positive second derivative of the egg production with respect to size.

After eggs are produced, they are distributed spatially over the system, using a dispersal matrix which can take on a number of different qualitative forms. During the months in which larval dispersal is assumed to take place, settlement of larvae is calculated. For each month of the egg production period, a fraction of egg production is presumed to survive and this is distributed via the dispersal matrix from each of the patches to each individual patch according to

$$\mathbf{s}^{\text{in}} = p\mathbf{D}\mathbf{e}. \quad (9)$$

This 11×1 vector gives the array of settlement associated with the array of egg production from the system, modified by the survival probability p, and distributed by the dispersal matrix $\mathbf{D}$. If all of the patches cover all possible dispersal sites, then the rows of D sum to one. Beyond that possible restriction, $\mathbf{D}$ is general and can characterize a variety of dispersal mechanisms. The most commonly assumed dispersal mechanism is uniform dispersal, in which the production of larvae in each location redistributes uniformly over the entire system, often referred to as a common larval pool assumption.

The number that actually settle successfully follows the following stock-recruitment function:

$$s_j^{\text{out}} = \frac{s_j^{\text{in}}}{a^{-1} + c^{-1} s_j^{\text{in}}}. \quad (10)$$

This specification enables the model to simulate various density-dependent larval survival mechanisms in the system, all of which may be patch specific. Once the settlement is calculated for any given site, the successful settlers (s_j^{out}) become the next period's age zero entry and the growth process starts again. Appendix A contains baseline values of the parameters that we used to

---

[24] In our case, $b = 2.68$ for sea urchins. The parameter b, in the absence of empirical work, is often assumed to be 3. The idea is that diameter (or length) is a one-dimensional measure and to get to mass, one goes through an approximation of volume first, and then mass is proportional to volume. The extent to which the cube of an organism's one-dimensional measure is a good approximation for its volume will vary by species, but it is typical for b to be greater than 1.

calibrate the spatial biological model. The parameters are based on ongoing fieldwork being conducted off the coast of northern California by colleagues investigating the sea urchin in a joint long-term research program.[25] Raw data on growth increments and adult size distributions have been gathered in dive transects at several exploited and unexploited sites along the coast. These have been used to compute growth and natural mortality coefficients using maximum likelihood techniques.

The dispersal process used in our model is a special feature and does not rely on the common larval pool assumption. Instead, we specify the dispersal matrix to reflect current understanding arising out of our joint work with biologists. That work has collected sea urchin larvae on a continuous basis and correlated larval settlement patterns with sea surface, wind, and current patterns to understand dispersal processes. The current thinking is that during the upwelling season from April to July, there is a strong offshore and equator-directed flow of current that sweeps larvae southward. The larvae then collect in "retention zones" or gyres to the south of two promontories of Point Reyes and Point Arena. Then, during relaxation of the upwelling events, a reversal of the process creates a nearshore pole-directed flow of currents that sweep and distribute larvae back along the coast. We thus find relatively more settlement in areas just north of the two gyres. The dispersal matrix used here reflects this realistic system that incorporates oceanographic influences on larval dispersal.

5.2. Calibrating the bioeconomic model

To simulate the implications of spatial closures, we combine the spatially explicit biological model with the model of spatial behavior estimated with the logbook/landings ticket database. The key link in the integrated model is the connection between monthly fishing mortality in each patch f_{jt} (in Eqs. (6) and (7)) and monthly trips, or

$$f_{jt} = (Trips_{jt})hq = \left(o_t \sum_{p=1}^{4} d_p p_{pjt}\right)hq, \qquad (11)$$

where h is diving hours per trip, q is the scaling or catchability coefficient (assumed constant across time), o_t is the number of choice occasions in that month,[26] d_p is the number of divers in port p, and p_{pjt} is the probability of a trip from port p to patch j in month t based on the RNL estimates. We presume that h is fixed across all ports and time periods, and we use q to calibrate model forecasts to actual harvests aggregated across space on an annual basis.[27] That is, we adjust the catchability coefficient so that the integrated model (1) appears to be on the same trajectory to a steady state as the actual data at the end of the sample period and (2) tracks the initial draw-down phase of the actual data. We place more weight on the first criterion than on the second because we are particularly interested in evaluating the steady-state implications for marine

---

[25] See [2–4,18,19,21,25,30] for further details on biological and oceanographic parameters.
[26] Choice occasions are determined by the season closure regulations, which are the same across all northern California ports.
[27] On average, harvesters spend 3.5 h diving per trip, and though there is considerable variation in dive hours, we have not found any variables that explain substantial amounts of that variation.

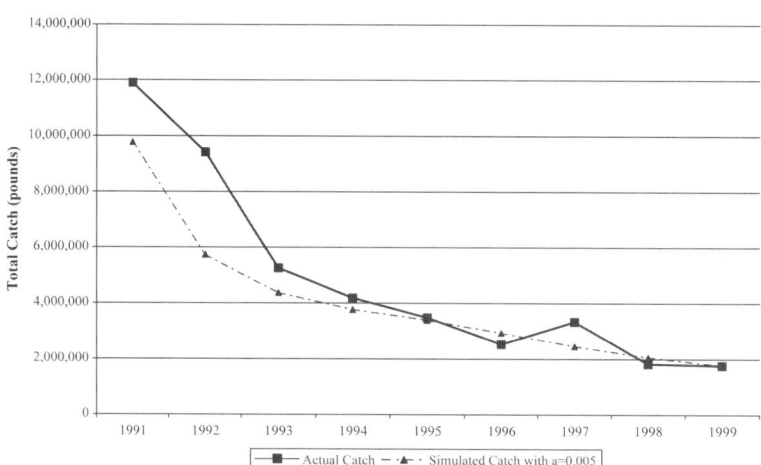

Fig. 2. Calibration of bioeconomic simulation model.

reserves.[28] In all simulations, since expected revenues are rolling 1 month backward averages, we use lagged catch per trip multiplied by an exogenous price to predict trips and shares of divers in each location. To simplify matters, we use a reduced-dimension model that includes only the four main northern California ports because fishing from the southern- and northern-most ports is erratic and accounts for less than 5% of the total fishing trips in the data set. Fig. 2 shows how the integrated model calibrates with the actual harvest path during the draw-down of the northern California urchin population over the past decade.

To understand how economic behavior affects forecasted impacts of marine reserves, we compare simulations of the integrated bioeconomic model (the ECON model) with a standard biology-only model that we dub the NOECON model. Both models use the dual gyre dispersal pattern that mirrors current thinking about coastal oceanography off northern California, and both incorporate the minimum size limit of 3.5 in. The NOECON model assumes, consistent with the biological literature, that total fishing effort before and after reserve formation is constant and is distributed uniformly across space. In contrast, the ECON model is based on our RNL behavioral model and incorporates seasonal and weekly closures by varying choice occasions over time.

Although the ECON model is the correct model, we need to calibrate the NOECON model also in order to evaluate the impacts of our behavioral model. We first calibrate NOECON to the draw-down phase of the actual system-wide harvest to mirror what the biological modeler would do. However, NOECON by nature of its behavioral simplification is unable to both track the draw-down phase and end up at the same steady state as ECON, which we believe to be the true steady state. Thus, we provide a second calibration of NOECON to the ECON steady state. This latter calibration allows us to explore the implications of ignoring spatial behavior without the confounding effect of inter-temporal effort dynamics.

---

[28] Attempts to use an "objective" calibration criterion, e.g. minimize sum of squared errors, tended to do well for early draw-down phase but poorly for the steady state trajectory.

Table 4
Marine reserves and economic behavior the northern California red sea urchin fishery

	Steady-state harvest (1000 pounds)	Steady-state egg production (billions)	Discounted[a] revenues ($1000)
With discrete choice behavioral model—ECON ($a = 0.005$)			
No closure	830	1316	17,440
Close patch 8	752	1441	15,074
With no economic model—NOECON ($a = 0.005$) approach path calibration			
No closure	386[b]	267	8096
Close patch 8	545	397	8204
With no economic model—NOECON ($a = 0.005$) steady-state calibration			
No closure	829[c]	434	17,400
Close patch 8	868	553	16,423

[a] Uses a 5% constant discount rate and assumes $1 per pound of sea urchin.
[b] Calibrated approach path catch to actual catch.
[c] Calibrated steady-state harvest to behavioral model.

5.3. The importance of behavior

With the calibrated models we examine how a reserve placed in various patches might affect the whole fishery. Table 4 reports simulation results of closing patch 8, the heavily fished area off Fort Bragg.[29] This table demonstrates the importance of incorporating spatial behavior. First, the NOECON model dramatically overstates the extent of decline when calibrated to the draw-down phase of early fishery exploitation. The intuition for this is obvious once stated. Early in a fishery's development, abundance is high and revenues are high, drawing large amounts of effort into the fishery. This results in a draw-down phase of high fishing mortality, and if one assumes that this level of effort continues, that ultimately overpredicts the amount of effort, overstates the extent of overfishing, and thus underpredicts steady-state harvest. For example, calibrated to the draw-down phase, the NOECON model predicts a steady state harvest of only 386,000 pounds and a dramatically reduced egg production, reflecting the assumed low reproductive potential of the overexploited biomass. In contrast, the ECON model continuously adjusts effort to profitability. As the fishery is drawn down, effort exits and fishing mortality falls, generating steady-state biomass and harvest levels that are much larger. Under the ECON scenario, the steady-state harvest is more than double the NOECON prediction, at 830,000 pounds. Second, perhaps equally important, the predicted egg production in the ECON model is almost five times that

[29] We use patch 8 to demonstrate the impacts of closures mainly because the findings by Holland and Brazee [10] and Sanchirico and Wilen [24] suggest that reserves are likely to be beneficial to fisheries production in two instances: (a) either when the closed patch is heavily exploited; or (b) when the closed patch operates as a source rather than a sink. The gyre pattern and north–south pattern of larval flow makes patch 8 a type of source, although the definition is not always clear in the literature.

predicted with the NOECON model. The higher egg production emerges from a magnitude effect and a distribution effect. Since overall exploitation is lower, total reproductive biomass and egg production are larger. But the ECON model also predicts a spatial distribution of effort that depends upon relative spatial profits rather than assuming a uniform distribution. In this more realistic setting, areas that are high cost (e.g. more distant) will be lightly exploited and hence will serve as de facto "reserves", even without explicit spatial closures.[30] In contrast, the NOECON model essentially assumes uniform overexploitation in these de facto reserves.

Table 4 also shows how misleading biological models without behavior may be about the impacts of reserve formation. For example, the NOECON model calibrated to the approach path predicts that a closure of patch 8 will produce a 40% increase in steady-state harvest, coupled with a 48% increase in system-wide egg production. The NOECON model assumes, as is typical in the literature, that displaced effort adjusts immediately and uniformly to the remaining open areas. Thus total harvest falls to zero in the closed area but increases initially in all open areas. In general, whether the present value of the loss over the whole adjustment period is made up by the corresponding gain depends on initial conditions and adjustment speeds. In the NOECON simulation, the model predicts that the initial loss is compensated for over the whole transition path since the overall present value of a spatial closure is positive. Contrast these results with the ECON case, which makes the transition path and ultimate steady state dependent upon relative profits. In the ECON simulation, a closure of patch 8 results in a 10% loss in steady-state harvests, in addition to the transition losses associated with the closure. The result is that discounted revenues fall by 14% compared with no spatial closure, rather than rising as predicted by the NOECON model. The importance of incorporating economic behavior into models intended to forecast the implications of reserves is thus profound. Importantly, the assumption of uniformly distributed and unresponsive effort used for simplicity in the biological literature biases predictions toward overly pessimistic status quo harvest and egg production predictions, and toward overly optimistic predictions of harvest gains and the net economic costs of reserve formation.

While so far we have compared our integrated model with what the biologists would do, it is interesting to isolate the impacts of ignoring spatial behavior without the confounding effect of dynamic changes in the overall level of fishing effort. Thus, the third section of Table 4 calibrates the pure biological model to the same steady-state harvest as ECON. In this case, the NOECON model still predicts a steady state harvest gain, although much smaller than under the approach path calibration. These harvest gains are not enough to compensate for initial harvest losses, and hence the system-wide present value of revenues falls. A more important difference may be in predicted egg production from the system. Because the NOECON model assumes uniform effort distribution, the whole system's reproductive biomass is predicted to be drawn down relatively uniformly to establish the status quo. In contrast, the ECON model predicts a heterogeneous distribution of effort reflecting distance from ports and other costs of remote patches. These act as de facto sources, hence contributing to predictions of overall egg production that are larger than the NOECON model. Fig. 3 clarifies how this arises, depicting the steady-state size distribution

---

[30] The steady-state spatial coefficient of variation for NOECON is 0.28 whereas for ECON it is 0.43. We know that spatial variation in the NOECON model must be due to net larval dispersal differences. In the ECON model, spatial variation is also due to differential fishing effort responding to differential rents.

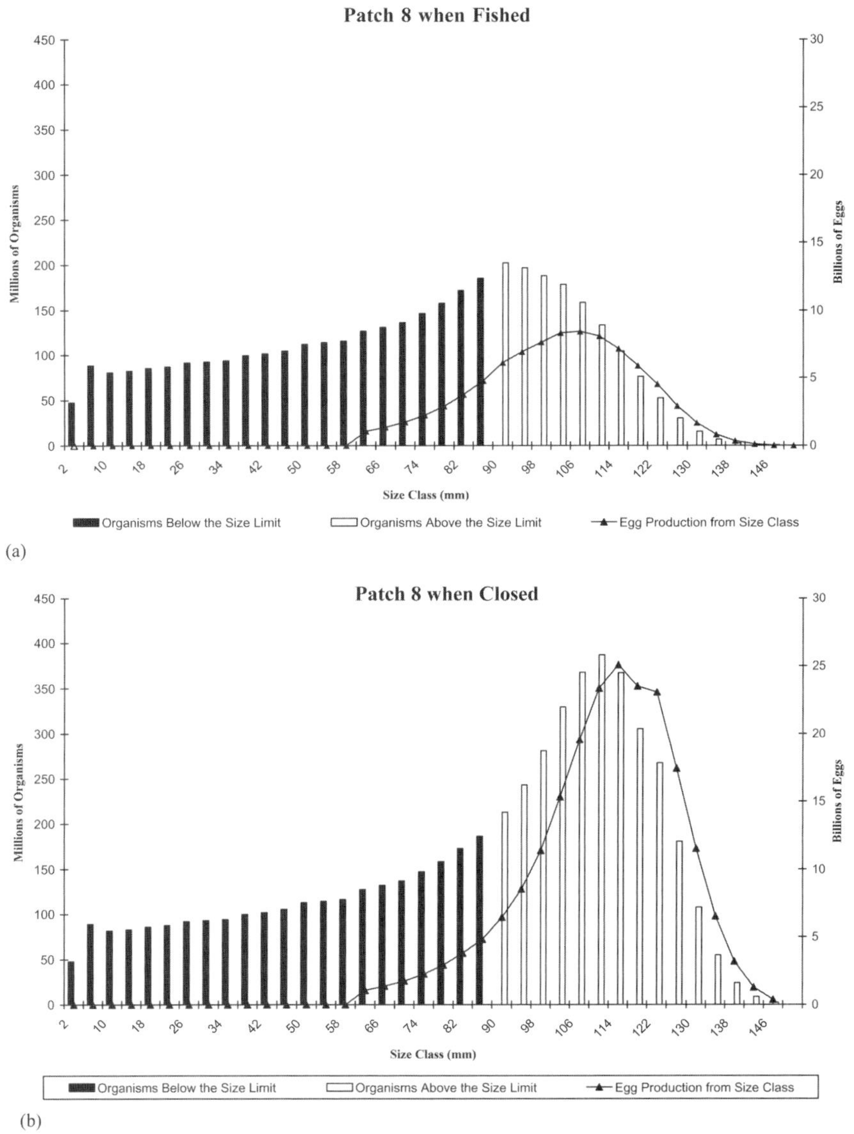

Fig. 3. Steady-state size distribution and egg production.

and egg production from a heavily exploited patch and from a closed patch. The closed patch has a wider size distribution with more large individuals that produce more total eggs. Egg production peaks beyond the size class with the most number of organisms both because egg production is

Table 5
Impacts of dispersal and spatial behavior

	System-wide totals for the NOECON model		System-wide totals for the ECON model	
	SS harvest (1000 pounds)	SS eggs (billions)	SS harvest (1000 pounds)	SS eggs (billions)
No closure	828.6	433.6	830.5	1316.0
Separate patches				
Close Farralons	670.6	566.5	819.1	1322.5
Close patch 1	749.4	580.3	827.1	1317.5
First gyre				
Close patch 2	807.5	584.1	755.2	1395.4
Close patch 3	812.4	571.3	765.3	1382.3
Close patch 4	816.3	559.3	761.2	1385.3
Gyre border				
Close patch 5	911.6	559.7	755.0	1425.6
Second gyre				
Close patch 6	868.3	594.7	743.6	1446.9
Close patch 7	869.1	572.6	746.7	1447.5
Close patch 8	868.0	553.3	752.4	1440.6
Close patch 9	862.3	522.2	787.3	1374.3
Close patch 10	839.5	470.8	829.6	1316.0

increasing in organism size and is convex in size.[31] These depict how the de facto closures associated with relatively uneconomic patches can contribute in important ways to a system's egg production. Ignoring this possibility causes the pre-reserve status quo predictions to bias reproductive biomass predictions low.

Table 5 shows that the received wisdom about reserve siting based on oceanographic dispersal may not be robust when policy makers account for economic behavior. This table shows the implications of siting reserves in various different patches in order to achieve different objectives such as harvest gains and system reproductive capacity gains. We calibrate the NOECON model to the same steady-state harvest level as the ECON model but with a uniform effort distribution. Again, the ECON model predicts a drop in steady-state harvests under any option chosen, whereas the NOECON model predicts increases in over half of the patch closure options. The NOECON model predicts harvest decreases from closing the southern-most patches and patches in the southern gyre around Point Arena. This would lead one to favor other areas as candidates for closure. In contrast, in the ECON model, closing those southern-most patches is *least* costly in term of harvest loss because they are most lightly exploited before the reserve. In terms of total system-wide egg production, the gains predicted by the ECON model are not as large proportionately as those predicted with the NOECON model. The NOECON model predicts relatively large egg production gains in patches 1 and 2, primarily because the model overpredicts pre-reserve harvest pressure in those relatively unprofitable patches. The ECON model, however, predicts relatively large egg production gains from closing patch 2 because it is lightly exploited pre-reserve and a large contributor to the first gyre. Large egg production gains are also predicted

---

[31] In viewing this figure, total egg production in each panel is the area under each egg production line.

from closing patches 6–8, but for different reasons. Patch 8 production gains come from closing a heavily exploited area, but patch 6 gains come from its critical role as a source feeding the second gyre. Overall, then, ranking the best sites depends upon both economic and biological dispersal. Ignoring economic dispersal profoundly affects what would appear to be wise siting choices, often missing the true configuration of the pre-reserve status quo, or incorrectly predicting eventual larval dispersal and harvest adjustments by failing to anticipate the behavioral response to spatial closures.

6. Summary and conclusions

This paper predicts some economic and biological impacts of marine reserve creation with particular attention to how economically motivated behavior determines outcomes. We address a potentially important shortcoming of the vast marine reserves literature, namely the assumption that fishing effort is fixed and uniformly distributed. Instead, we presume that effort in realistic settings responds to economic incentives, particularly differential profit opportunities that are dynamic and spatial. To address the importance of behavior, we construct and estimate a spatial choice model using a comprehensive database from logbooks and landings tickets. It is important to realize that fisheries scientists collect these data to guide regulatory decisions rather than to aid economic research per se. In spite of this, we show how this type of data can inform analyses of economic choices. The RNL econometric model confirms what we expect, namely that divers respond to differences in expected returns across different patches. The RNL model also shows that fishermen respond negatively to weather risk and travel distance and positively to expected returns, which are composed of both relative abundance and expected price. Although divers actually choose to participate only about 14% of the time, even our parsimonious model successfully explains a reasonable fraction of their variation in behavior.

We link the spatial model of choice behavior to a biological model intended to capture the most important features of our case study. The biological model is detailed and is the first to depict a multi-patch system with explicit empirically driven hypotheses about larval dispersal. Most existing literature uses two patch models; ours is an 11-patch system calibrated with parameters derived from field data. A unique feature of the biological model is its incorporation of the "dual gyre" nature of coastal circulation in northern California. This feature represents current thinking about dispersal processes that affect urchin larvae, and it adds detail and complexity that generate new hypotheses about reserve impacts and reserve siting questions. We simulate the model on a monthly time scale and the output from the model consists of aggregates such as system-wide egg production, total harvest, and present values of revenues. Only a handful of studies consider the role of larval dispersal analytically; this paper is among the first to study fishery policy with an empirically driven representation of dispersal.

Our results confirm that economic behavior is a critical determinant of the predicted impacts of reserves. Moreover, we show that the typical assumptions made by biologists for analytical tractability consistently bias the predicted impacts in a manner that makes reserves look more favorable than they actually might be. For example, a model that calibrates fishing mortality during a draw-down phase overpredicts the extent of overexploitation compared with an economically based model that incorporates the natural decline in profitability as the steady state is approached. Since aggregate harvest is more likely to be increased in overexploited fisheries, results based on mis-calibrated fishing mortality optimistically favor reserve creation. However, we also show that even without draw-down phase mis-calibration, the assumptions of uniform effort rather than economically motivated effort produce mistaken characterizations of reserve siting in realistic settings. We show, for example, that some remote or high cost or high risk areas are naturally less profitable and hence exploited less by fishermen. These patches form de facto reserves and hence contribute to overall egg production in ways not revealed by simpler uniform effort distribution models. However, the manner in which they bolster both harvest and system reproduction is complicated because their role depends upon economic, biological, and oceanographic factors. And the remoteness of particular patches is relative and dependent upon fishery independent factors such as port location and roadways.

For reasons discussed above, our overall assessment of reserves as a fisheries policy tool is more ambivalent than the received wisdom in the biological literature. Although our model incorporates a rich depiction of biological and physical oceanographic processes, the conclusions are more in accord with the very small amount of conceptual economic analysis that has been devoted to the topic. We find, as did Holland and Brazee [10], that reserves can produce harvest gains in an age-structured model but only when the biomass is severely overexploited. We also find, as Holland and Brazee did, that even when steady state harvests are increased with a spatial closure, the discounted returns are often negative, reflecting slow biological recovery relative to the discount rate. We find results that confirm the Sanchirico and Wilen [23,24] analytical work that shows that easily exploited (low bioeconomic ratio) patches are most likely to be the best sites to produce both harvest and reproductive potential gains. Their modeling implicitly assumes that harvest gains accrue from emigrating adults responding to density buildups in the reserve patches. The model used in this paper depicts a different process in which larvae numbers depend upon spawning stock biomass, but larval spatial distribution is driven by oceanographic transport of larvae. Many marine biologists believe that larval transport mechanisms may be more important determinants of dispersal in marine spatial systems than adult migration. We show that some of the sink/source notions about site selection currently in vogue appear to be based on naïve views of spatial dynamics that fail to incorporate realistic detail about either biophysical processes or fisher behavior. In particular, whether a particular patch is a source or sink depends on its relative level of exploitation as well as its physical placement in an oceanographic system. Patches with high intrinsic productivity in an unexploited system or a lightly exploited system may be less productive to the system as a whole when differential harvesting pressure (driven by relative economic opportunities) affects the spatial distribution of abundance and hence larval production. These results suggest, in contrast to the tone of recent dialogue among marine scientists about reserves, that there are still unanswered questions about whether they can deliver what they promise based on the simple modeling done to date. At the very least, our integrated bioeconomic model raises new questions about whether oceanographic dispersal is the key driver of spatial closure impacts, or whether harvester dispersal may be equally important.

Acknowledgments

The authors thank Loo Botsford, Dan Holland, Doug Larson, David Layton, Bill Provencher, Jeffrey Williams and two anonymous referees for helpful comments and suggestions and thank

Table 6

Parameter	Description	Value
k	Growth	0.24
m	Natural mortality	0.09
L_{inf}	Terminal size (mm)	118
L_{limit}	Min. size limit (mm)	89
L_{mature}	Min. size of sexually mature organism	60
f	Fishing mortality	0.29
w	1st allometric weighting parm.	0.001413
b	2nd allometric weighting parm.	2.68
α	1st egg production parm.	5.47E–06
β	2nd eggs production parm.	3.45
p	Survival probability	1.0
a	Resiliency settlement parm.	0.005–0.05
c	Carrying capacity settlement parm.	1.2E+07–2.4E+07

Dale Lockwood for programming assistance. This research is funded in part by a grant from the National Sea Grant College Program, National Oceanic and Atmospheric Administration, US Department of Commerce, under Grant NA06RG0142 project number R/F-179 through the California Sea Grant College System, and in part by the California State Resources Agency. The views expressed herein are those of the authors and do not necessarily reflect the views of NOAA or any of its sub-agencies, or the Resources Agency.

Appendix A

Baseline parameter values for bioeconomic simulations are given in Table 6.

References

[1] R.J.H. Beverton, S.J. Holt, On the Dynamics of Exploited Fish Populations, Chapman & Hall, London (1957, reprinted in 1993).
[2] L.W. Botsford, D. Lockwood, L. Morgan, J.E. Wilen, Marine reserves and management of the northern California red sea urchin fishery, CALCOFI Report No. 40, 1999, pp. 87–93.
[3] L.W. Botsford, J.F. Quinn, S.R. Wing, J.G. Brittnacher, Rotating spatial harvest of a benthic invertebrate, the red sea urchin, *Strongylocentrotus franciscanus*, in: Proceedings of the International Symposium on Management Strategies for Exploited Fish Populations, University of Alaska Sea Grant College Program, 1993.
[4] L.W. Botsford, B. Smith, J.F. Quinn, Bimodality in size distributions: the red sea urchin *Strongylocentrutus franciscanus* as an example, Ecolog. Appl. 4 (1) (1994) 42–50.
[5] G.M. Brown, J. Roughgarden, A metapopulation model with private property and a common pool, Ecolog. Econom. 22 (1997) 65–71.
[6] A. Daly, S. Zachary, Improved multiple choice models, in: D. Henscher, Q. Dalvi (Eds.), Identifying and Measuring the Determinants of Mode Choice, Teakfield, London, 1979, pp. 335–357.
[7] B. Halpern, The impact of marine reserves: does size matter?, Ecolog. Appl. (2002), forthcoming.
[8] R. Hannesson, Marine reserves: what should they accomplish?, Marine Resource Econom. 13 (1998) 159–170.

[9] A. Hastings, L.W. Botsford, Equivalence in yield from marine reserves and traditional fisheries management, Science 284 (1999) 1537–1538.
[10] D.S. Holland, R.J. Brazee, Marine reserves for fisheries management, Marine Resource Econom. 11 (1996) 157–171.
[11] P.E. Kalvass, J.M. Hendrix, The California red sea urchin, *Stongylocentrotus franciscanus*, fishery: catch, effort, and management trends, Marine Fisheries Rev. 59 (2) (1997) 1–17.
[12] S. Kato, S.C. Schroeter, Biology of the red sea urchin, Marine Fisheries Rev. 47 (3) (1985) 1–20.
[13] D. Ludwig, R. Hilborn, C. Walters, Uncertainty, resource exploitation, and conservation: lessons from history, Science 260 (1993) 17–18.
[14] D. McFadden, Modeling the choice of residential location, in: A. Karlqvist, L. Lundvist, F. Snickbars, J.W. Weibull (Eds.), Spatial Interaction Theory and Planning Models, North-Holland, Amsterdam, 1978, pp. 75–96.
[15] D. McFadden, Quantitative methods for analyzing travel behavior of individuals: some recent developments, in: D. Hensher, P. Stopher (Eds.), Behavioral Travel Modeling, Croom Helm, London, 1979, pp. 279–318.
[16] D. McFadden, K.E. Train, Mixed MNL models for discrete response, J. Appl. Econometrics 15 (5) (2000) 447–470.
[17] E.R. Morey, R.D. Rowe, M. Watson, A repeated nested-logit model of Atlantic salmon fishing, Amer. J. Agric. Econom. 75 (1993) 578–592.
[18] L. Morgan, Spatial variability in growth, mortality and recruitment in the Northern California Red Sea Urchin Fishery, Ph.D. Dissertation, University of California, Davis, 1997.
[19] L. Morgan, S.R. Wing, L.W. Botsford, C. Lundquist, J. Diehl, Spatial variability in red sea urchin recruitment in northern California, Fisheries Oceanogr. 9 (1) (2000) 83–98.
[20] T. Polacheck, Year around closed areas as a management tool, Nat. Resource Modeling 4 (2) (1990) 327–353.
[21] J.F. Quinn, S.R. Wing, L.W. Botsford, Harvest refugia in marine invertebrate fisheries: models and applications to the red sea urchin, *Strongylocentrotus franciscanus*, Amer. Zoologist 33 (1993) 537–550.
[22] J.A. Reynolds, J.E. Wilen, The sea urchin fishery: harvesting, processing, and the market, Marine Resource Econom. 15 (2) (2000) 115–126.
[23] J.N. Sanchirico, J.E. Wilen, Bioeconomics of spatial exploitation in a patchy environment, J. Environ. Econom. Management 37 (1999) 129–150 doi:10.1006/jeem.1998.1060.
[24] J.N. Sanchirico, J.E. Wilen, Bioeconomics of marine reserve creation, J. Environ. Econom. Management 42 (2001) 257–276 doi:10.1006/jeem.2000.1162.
[25] B. Smith, L.W. Botsford, S.R. Wing, Estimation of growth and mortality parameters from size frequency distributions lacking age patterns: the red sea rchin (*Strongylocentrotus franciscanus*) as an example, Canad. J. Fisheries Aquatic Sci. 55 (1998) 1236–1247.
[26] M.D. Smith, Spatial search and fishing location choice: methodological challenges of empirical modeling, Amer. J. Agric. Econom. 82 (5) (2000) 1198–1206.
[27] M.D. Smith, Spatial behavior, marine reserves, and the Northern California Sea Urchin Fishery, Ph.D. Dissertation, University of California, Davis, 2001.
[28] M.D. Smith, Two econometric approaches for predicting the spatial behavior of renewable resource harvesters, Land Econom. 78 (4) (2002) 522–538.
[29] C. Walters, Adaptive Management of Renewable Resources, Macmillan Publishing Company, New York (1986, reprinted by Fisheries Centre, University of British Columbia, 1997).
[30] J.E. Wilen, M.D. Smith, D. Lockwood, L.W. Botsford, Avoiding surprises: incorporating fishermen behavior into management models, Bull. Marine Sci. 70 (2) (2002) 553–575.

2.3 HONORABLE MENTION FULL CITATIONS AND ABSTRACTS

Arlinghaus, R. 2006. On the apparently striking disconnect between motivation and satisfaction in recreational fishing: the case of catch orientation of German anglers. North American Journal of Fisheries Management 26:592–605.

In this study, three distinct segments of German anglers differing with respect to degree of catch orientation as the main fishing motive were identified in a nationwide telephone survey ($N = 474$). Noncatch aspects of the fishing experience played a major role in the motivations of anglers: about 80% of the sample was classified as anglers with a low, or minimal, catch orientation. Angler satisfaction and its determinants were examined across degrees of catch orientation to improve understanding of the link between angler motivation and satisfaction. Highly catch-oriented anglers were significantly less satisfied with the previous angling season than were minimally catch-oriented anglers. An exclusivity of activity-specific, mainly catchrelated, satisfaction components as predictors of overall angling year satisfaction was found in all angler segments, irrespective of catch orientation. Satisfaction was unrelated to actual catch or harvest rates, and no significant differences in catch and harvest were found across the three catch orientation groups. This suggested that catch expectation was the primary driver of angler satisfaction. This study revealed that there are anglers, most often the majority within the population, who can be characterized as attaching relatively little importance to catch motives but whose satisfaction is still mainly catch dependent. It is not warranted to conclude that there is a striking disconnect in this finding. The reasons for the apparently striking inconsistency between motivation and satisfaction are related to (1) the fundamental conceptual differences in meaning and definition of motivation and satisfaction and (2) the differential ease in satisfying activity-general and activity-specific aspects of the fishing experience. Care must be taken not to draw overly simplified management implications from motivational information. However, by knowing the determinants of angler satisfaction, the manager's ability to plan future management actions is improved, and satisfaction rather than motivation is the ultimate product of the fishing experience.

Carson, R. T., W. M. Hanemann, and T. C. Wegge. 2009. A nested logit model of recreational fishing demand in Alaska. Marine Resource Economics 24:101–129.

This paper (and the large study behind it) was in some ways a labor of love. We set out to build a discrete choice model to predict recreational fishing behavior that would extend the literature along several dimensions. Our interest in doing so stemmed from earlier work that we were individually or collectively involved in. Hanemann's dissertation (1978) suggested the power of the discrete choice random utility model (RUM) to look at a large set of recreational alternatives, in this case beaches in the Boston area, and identify the role that attributes like water quality played in consumer choice behavior. Carson worked as a research assistant on a large recreational demand project for Jeff Vaughan and Cliff Russell at Resources for the Future (Vaughan and Russell 1982a,b). This project attempted to expand the travel cost framework to dealing with water quality issues and different types of fishing on a large spatial scale using the 1975 Survey of Hunting, Fishing, and Wildlife-Associated Recreation and a much smaller survey undertaken by the research team aimed at gathering specific information for placing a monetary value on different types of fishing days. The project used a discrete choice framework and had bumped up against both computational limits and limits of what could be estimated using data from government surveys being implemented at the time. Hanemann had a fruitful collaboration with Nancy Bockstael and Ivar Strand on the EPA-funded study of the Chesapeake Bay, where he was involved mainly in the theoretical formulation of demand models rather than data collection or estimation. He also collaborated with Ivar Strand and Thomas Wegge of Jones & Stokes, a Sacramento-based environmental consulting company, in all phases of a study of marine recreational fishing in Southern California for the National Marine Fisheries Service (Wegge, Hanemann, and Strand 1986). These studies whetted our appetite for the opportunity to engage in a large-scale data collection effort for modeling discrete/continuous choice behavior.

By the mid-1980s, the original zonal travel cost model as suggested by Hotelling and implemented by Clawson and Knetsch (1966), while still a workhorse of applied work for government agencies, had largely run out of steam from an academic perspective. The properties of the model had been well explored (e.g., Gum and Martin 1975; Dwyer, Kelly, and Bowes 1977), and its problems loomed large once one moved away from very simple situations and the limited origin-destination data that was typically available in secondary datasets (e.g., Smith and Kopp 1980). Interest in moving to a discrete choice RUM framework was driven by the desire to value marginal changes in the characteristics of sites and in valuing the opening and closing of sites.

What we wanted to do required much more extensive data than had been collected in the past. This put such a study out of the reach of the pure academic realm and even beyond what most state or federal agencies were used to funding. By happenstance, we did find an agency with a serious problem, a willingness to find the resources needed to resolve it, and, perhaps most importantly, one that was willing to consider the use of economic values in making its decisions. The Alaska Department of Fish and Game (ADFG) was wrestling with politically controversial issues related to managing the stocks of the major salmon species, determining funding priorities for different hatchery programs, and allocating the catch among commercial and recreational fishermen. ADFG felt it needed a valuation model for sport fishing with considerable spatial, temporal, and species resolution. Because of the distinctively compressed timing of salmon runs at particular spawning sites, ADFG needed to be able to evaluate decisions involving actions on the scale of a particular week, a particular species of salmon, and a narrowly defined location. ADFG also recognized the importance of site substitution; it needed to know how much closing one site to sport fishing would increase fishing pressure at other sites. ADFG put out a request for proposals to develop such a valuation model, and we ended up winning the contract. Jones & Stokes would be in charge of the large data collection effort and would also conduct a companion analysis of sport fishing expenditures and the resultant impact on the Alaskan economy.

From our perspective, Alaska was ideal. Geographically, the area of analysis was self contained. Alaskans fish at a greater rate and more often than any other state. Much of the population lives in a few locations. A fairly sparse road system coupled with an emphasis on catching different species of salmon near the mouths of rivers helped to define fishing sites. Mike Mills, our project officer at AFDG, had an encyclopedic knowledge of fisheries and fishing in Alaska.

The first major decision that we made was to put most of the money from the contract into data collection in order to be sure to get the right data, knowing that this meant that we would largely be working for free on the data analysis. The study was one of the first travel cost studies to use focus groups to understand how recreational fishing decisions are made. This gave us substantial insight into what variables to collect and how they might be used in modeling anglers' choice behavior.

Some of the things we learned are as follows. First, it became clear that the size of vehicle matters greatly; some anglers have small cars while others have large trucks or even RVs. This led us to collect information on the type of vehicle owned by respondents so that we could employ individualized travel costs. Second, we found out that for people who have a cabin, loosely defined as anything from a nicely appointed cottage to a lean-to shack, this had a major impact on where (and how frequently) they chose to go fishing. The possession of a cabin entails a lower cost, better knowledge, and a type of habit formation somewhat different from what was then generally considered in the literature. Without this variable, the modeling would not have gone nearly as well. Third, we had to develop a definition of leisure time availability that fit the reality that at any single point in time any Alaskan seemed to be able to take off for a couple of days of non-stop fishing, although most were not able to do this week in and week out. Fourth, there is a relatively high degree of angler awareness of current fishing opportunities in Alaska because ADFG puts out a weekly announcement describing fishing conditions and stating where the salmon are expected to be running. We got a sense that anglers' perceptions are based partly on this information and partly on their knowledge of what happened last season. The combination of harvest estimates from the previous year plus the ADFG announcements allowed us to create variables which avoided the endogeneity of the respondent's current catch as a proxy for fishing quality. Fourth, we also saw that there was likely to be a strong temporal pattern to recreational fishing, especially in the shoulder portion of the fishing season (May and September), influenced by when particular species of salmon are running and how cold it is outdoors. It seemed like everyone who fished in Alaska went somewhere to fish over the Fourth of July. This temporal dimension ultimately led us to build a model whereby respondents made recreational fishing choices week by week over a 22-week fishing season. The propensity to fish was allowed to vary weekly, and site choice varied weekly with changes in crowding and an index of fishing quality that took considerable effort to develop.

In the focus groups we were also struck by the tremendous heterogeneity in fishing behavior and preferences. Some people target a specific fish species and then look for the best site, while others pick a site and fish for whatever is there. Some people choose sites to avoid crowds, while others actively seek out places with many other people. This observation influenced how we structured the model that was eventually estimated.

We attempted to avoid sample selection problems by first locating people who might fish during the coming season using a first-stage screening survey of the general Alaskan population. We attempted to avoid recollection problems by sending respondents a diary in which to record their trips and collected that information in two waves during the fishing season. The first wave covered fishing in May, June, and July, and the second covered August and September. We made an effort to minimize non-response by sending out a composite survey for the whole fishing season to those who did not respond to the first wave. We attempted to avoid problems with identifying geographic locations by providing maps and lists of sites in the survey instrument. Examples of how to record trip information were provided to help reduce respondent recording errors.

Given the data, the major challenge in formulating and estimating a model of choice behavior was the sheer complexity of the potential angler choices. In each of the 22 weeks, an angler could potentially choose among 29 sites, and, at each site, among up to 13 species. An angler also could choose how many times to fish that week. In all, there were almost 30 million potential choice alternatives. It was clear to us that this structure called for a nested logit model, and we ended up developing a model with four levels of nesting. At that time, no one had ever estimated a model with so many levels of nests.

Programming the estimation of this complex model and securing the computational resources to conduct the estimation were monumental challenges. At the time, statistical packages with conditional logit models, such as LIMDEP, were just becoming available. However, they were too limited for our needs. Our estimation problem was irregular in the sense that the choice set varied week by week (e.g., if there were no king salmon at a site in a particular week that site could not be chosen for king salmon fishing). Beyond this, our choice problem was just too big given the usual way of trying to estimate the model in the computer's RAM memory. We were very fortunate that Dan Steinberg, a colleague of Carson's, had a software company, Salford Systems, which was developing a LOGIT package at the time. Steinberg was willing to make many of the modifications we needed because he saw what we were doing as the wave of the future. A number of these enhancements were incorporated into the next version of LOGIT and eventually into other packages for estimating conditional logit models. Steinberg had a long-standing working relationship with Scott Cardell, who is credited with first proposing the nested logit model, and the two of them were an invaluable resource in dealing with the statistical properties of the model we were estimating.

We used just over 2,000 hours on Digital Equipment Corporation (DEC) VAX minicomputers and six hours on what at the time was one of the world's fastest supercomputers, a CRAY XMP/48, at the San Diego Super Computer Center. The VAX minicomputers we commandeered comprised the bulk of UCSD's instructional computers. They were being used lightly and sporadically for classes during the summer and, to use them, we had to structure the programs so that they would essentially go into background and out of memory if the machines were being used by students. The VAX computers had eight megabytes of memory, which was enormous relative to PCs of the day. For much of our project, this eight megabytes of memory (minus the space taken by the operating system) was the critical limitation on what could be estimated. Now the PC sitting on your desk is likely to have two plus gigabytes of memory and 5,000+ times the raw processing speed. Even the $20 million Cray XMP/48, which was several hundred times faster and had double the memory of the VAX, pales in comparison with the common desktop PC of today. With software packages that are now readily available, today's research would face no computational challenge in getting our entire model to run in hours rather than the months it required.

This paper languished with regard to publication because other events intervened, and we were never quite satisfied that all of the loose ends had been tied up. ADFG was pleased with the study, and as soon as we submitted our final report, they commissioned a similar study covering the southeastern part of the state. This broke little new ground methodologically and consumed time that would otherwise have been spent refining the Southcentral study for publication as a journal article, but we felt obliged to accommodate ADFG. While the Southeast study was under way, the *Exxon Valdez* oil spill occurred, and we were approached by the federal government and by the Alaska Attorney General to work on this (Carson et al. 1992).

With regard to this study, there was more that we wanted to do, including: *i)* moving from a sequential limited information maximum likelihood approach to a full information likelihood approach; *ii)* testing the sensitivity of the results to alternative nesting structures; *iii)* allowing for more interactions between respondent characteristics and site attributes along the lines of what we did with crowding as a way for allowing for more heterogeneity in respondent preferences; and *iv)* allowing explicitly for temporal variation in some of the parameters. One modification that did get done later was the estimation of a truncated count data model to predict the number of trips taken for those who took a trip in a particular week where the inclusive value from the lower branches of the tree served as predictor (Grogger and Carson 1991). This effectively would have modeled the top branch of the tree in terms of a "fish/no fish this week" decision and then, if fish, how many times in the week, so that the resulting tree had five—not four—levels. We also wanted to explore issues related to non-linearity of welfare measures in terms of the estimated parameters and the implications of how the specification of the error component influenced estimates of those welfare measures. The complexity of the model in conjunction with the computational limits we faced largely forced us into considering a representative agent formulation. Indeed, we had to develop a PC program based on the model parameters that would run on a PC of the day so that the ADFG could use it to predict the response of anglers to regulatory actions, such as closing down sites for king salmon fishing in a particular week. This program was used for many years and must have been one of the early interactive bioeconomic modeling tools used by a government agency.

After the *Exxon Valdez* study, two of us, Carson and Hanemann, became heavily involved in the contentious debate over the use of contingent valuation surveys, while Wegge had moved on to open his own firm as a private consultant. Over time, the study took on a life of its own in the grey literature. Since it was not widely distributed or posted on the web, the study's main influence up to now was limited to researchers who had access to a hard copy of this paper or the larger report.

Clark, C. W. 1973. The economics of overexploitation. Science 181:630–634.

Renewable resources, by definition, possess self-regeneration capacities and can provide man with an essentially endless supply of goods and services. But man, in turn, possesses capacities both for the conservation and for the destruction of the renewable resource base.

Indeed, man's increasing capacity to seriously deplete the world's natural resources appears to be reaching a critical stage (1); if this is not imminent for the nonrenewable resources (2), it certainly appears so for many of the renewable ones (3). The problems of environmental pollution that loom so large today, for example, often result from a process of overexploitation of the regenerative capacity of our atmospheric and water resources. Economists lately have devoted much attention to environmental questions (4), and most are agreed that "externalities"- that is, effects not normally accounted for in the cost-revenue analyses of producers-are the leading economic cause of pollution and the destruction of natural beauty.

Animate resources, or biological resources, are also subject to serious misuse by man. An accelerating decline has been observed in recent years in the productivity of many important fisheries (5), particularly the great whale fisheries and the famous Grand Banks fisheries of the western Atlantic, as well as the spectacularly productive Peruvian anchovy fishery (6). As technology improves and demand increases, so the pressure on renewable resources grows more severe. The long-recognized need for effective international regulation of fisheries has never been so pressing as it is today.

A prerequisite for effective regulation is a clear understanding of the basic reasons for overexploitation, and in this regard the outstanding article by Hardin (7) on "The tragedy of the commons" has been a positive asset, even though economists have long been aware of the common property problem in fisheries (8). Indeed, in concentrating their attention on the problems of competitive overexploitation of fisheries, economists appear to have largely overlooked the fact that a corporate owner of property rights in a biological resource might actually prefer extermination to conservation, on the basis of maximization of profits (9). In this article I argue that overexploitation, perhaps even to the point of actual extinction, is a definite possibility under private management of renewable resources.

The implications of this argument for successful international regulation would seem to be that, if it is assumed that society wishes to preserve the productivity of the oceans and to prevent the extermination of valuable commercial species, control of the physical aspects of exploitation is essential. In particular the popular idea of maximum sustainable yield should be generally adopted, at least in the sense of setting an upper limit on the allowable degree of exploitation. Only a dire emergency in local food supply should be considered as a valid reason for temporarily running down the basic stock of a biological resource.

Fedler, A. J., and R. B. Ditton. 1994. Understanding angler motivations in fisheries management. Fisheries 19(4):6–13.

Motivation studies provide insight into what anglers seek in a fishing experience. Seventeen comparable angler studies were examined to identify motivational characteristics of angler populations and subpopulation groups. Five categories of motivations were used: general psychological and physiological, natural environment, social, fishery resource, and skill and equipment. Motivational profiles of angler populations were found to differ little between freshwater and saltwater environments or between the two states examined. Greater diversity was found in the importance of individual motivations among subpopulation groups, based on mode of fishing or target species, and between subpopulation groups and statewide populations. Results discourage extending motivational characteristics from a general sample of fishermen to subpopulation angler groups and vice-versa. Implications for using motivations to better meet angler needs and build supportive constituencies are discussed, and suggestions for further angler motivation research are presented.

Gutiérrez, N. L., R. Hilborn, and O. Defeo. 2011. Leadership, social capital and incentives promote successful fisheries. Nature 470:386–389

One billion people depend on seafood as their primary source of protein and 25% of the world's total animal protein comes from fisheries. Yet a third of fish stocks worldwide are overexploited or depleted. Using individual case studies, many have argued that community-based co-management should prevent the tragedy of thecommons4 because cooperativemanagement by fishers, managers and scientists often results in sustainable fisheries. However, general and multidisciplinary evaluations of co-management regimes and the conditions for social, economic and ecological success within such regimes are lacking. Here we examine 130 comanaged fisheries in a wide range of countries with different degrees of development, ecosystems, fishing sectors and type of resources. We identified strong leadership as the most important attribute contributing to success, followed by individual or community quotas, social cohesion and protected areas. Less important conditions included enforcement mechanisms, long-term management policies and life history of the resources. Fisheries were most successful when at least eight co-management attributes were present, showing a strong positive relationship between the number of these attributes and success, owing to redundancy in management regulations. Our results demonstrate the critical importance of prominent community leaders and robust social capital, combined with clear incentives through catch shares and conservation benefits derived from protected areas, for successfully managing aquatic resources and securing the livelihoods of communities depending on them. Our study offers hope that co-management, the only realistic solution for the majority of the world's fisheries, can solve many of the problems facing global fisheries.

Hardin, G. 1968. The tragedy of the commons. Science 162:1243–1248.

At the end of a thoughtful article on the future of nuclear war, Wiesner and York concluded that: "Both sides in the arms race are…confronted by the dilemma of steadily decreasing national security. *It is our considered professional judgment that this dilemma has no technical solution.* If the great powers continue to look for solutions in the area of science and technology only, the result will be to worsen the situation."

I would like to focus your attention not on the subject of the article (national security in a nuclear world) but on the kind of conclusion they reached, namely that there is no technical solution to the problem. An implicit and almost universal assumption of discussions published in professional and semipopular scientific journals is that the problem under discussion has a technical solution. A technical solution may be defined as one that requires a change only in the techniques of the natural sciences, demanding little or nothing in the way of change in human values or ideas of mortality.

In our day (though not in earlier times) technical solutions are always welcome. Because of previous failures in prophecy, it takes courage to assert that a desired technical solution is not possible. Wiesner and York exhibited this courage; publishing in a science journal, they insisted that the solution to the problem was not to be found in natural sciences. They cautiously qualified their statement with the phrase, "It is our considered professional judgment,…" Whether they were right or not is not the concern of the present article. Rather, the concern here is with the important concept of a class of human problems which can be called "no technical solution problems," and, more specifically, with the identification and discussion of one of these.

It is easy to show that the class is not a null class. Recall the game of tick-tack-toe. Consider the problem, "How can I win the game of tick-tack-toe?" It is well knowing that I cannot, if I assume (in keeping with the conventions of game theory) that my opponent understands the game perfectly. Put another way, there is no "technical solution" to the problem. I can win only by giving a radical meaning to the word "win." I can hit my opponent over the head; or I can drug him; or I can falsify the records. Every way in which I "win" involves, in some sense, an abandonment of the game, as we intuitively understand it. (I can also, of course, openly abandon the game—refuse to play it. This is what most adults do.)

The class of "No technical solution problems" has members. My thesis is that the "population problem," as conventionally conceived, is a member of this class. How it is conventionally conceived needs some comment. It is fair to say that most people who anguish over the population problem are trying to find a way to avoid the evils of overpopulation without relinquishing any of the privileges they now enjoy. They think that farming the seas or developing new strains of wheat will solve the problem—technologically. I try to show here that the solution they seek cannot be found. The population problem cannot be solved in a technical way, any more than can the problem of winning the game of tick-tack-toe.

Johannes, R. E. 1978. Traditional marine conservation methods in Oceania and their demise. Annual Review of Ecology and Systematics 9:349–364.

Understanding a conservation system means understanding not only the nature of what is being conserved, but also the viewpoint of the conserver. Knowledge of this second element is essential if we are to comprehend a system of resource management employed by a people whose perception of their environment differs from our own. Watt (83) has said that a prudent civilization should take seriously the ideas of other civilizations about resource use. "Over the short term," he states, "the ideas of civilization A might appear vastly superior to those of civilization B. But over the long term it could turn out that the apparently 'primitive' practices of civilization B were based on millenia of trial and error and incorporated deep wisdom that was unintelligible to civilization A." The following is an account of the rise and decline of a millenia-old system of controlled exploitation of marine resources that incorporates a wisdom Westerners are only now beginning to appreciate after having brought about its widespread decay.

The inhabitants of Oceania [defined here as the islands of Polynesia (excluding New Zealand), Melanesia (excluding New Guinea), and Micronesia] traditionally obtained the bulk of their protein from the sea. They often had no alternative. Population densities commonly reached several hundred people per square mile and sometimes climbed to more than one thousand per square mile. On some islands the land (often consisting of calcareous soil with little humus) barely supplied their vegetable needs.

Terrestrial food supplies were not only limited, but also precarious. On many islands typhoons, droughts, and tsunamis periodically destroyed them. Warm, humid climates tended to discourage the long-term storage of coconuts, sweet potatoes, breadfruit, or taro as insurance against hard times. Some islanders had pigs, but even on larger islands with sufficient land to support considerable livestock, they were raised indifferently and often sufficed only for feasts or the enjoyment of royalty.

But the supply of seafood was relatively substantial and dependable. And what the islanders lacked as animal husbandrymen they compensated for as fishermen and students of marine life. Of Tahitians, for example, Ellis (30) said, "In no other part of the world, perhaps, are the inhabitants better fishermen." Ichthyologists Gosline & Brock (36) state, "It is probable that the Hawaiians of Captain James Cook's time knew more about the fishes of their islands than is known today."

The sea's produce was dependable but not unlimited. In some island groups extensive reef, mangrove, and seagrass communities produced more fish and shellfish than the population could use. But more often these islands-the tips of submerged mountains-plunged steeply into abyssal depths, and productive shallow waters were limited to a narrow band of coral reef. Offshore waters were not only hazardous much of the time but also far less productive than the waters extending from the island to the outer reef slope. And although those who lived on atolls had sheltered lagoons, these also were much less productive of food than the narrow strip of reef that encircled them (21, 43, 53).

Possessing a clearly limited fishery on which they depended for about 90% of their animal protein, these people viewed marine resources in a way different from that of continental peoples with abundant terrestrial food sources and wide continental shelves. Until recently, Westerners have looked upon the sea's supply of fish as virtually unlimited. T. H. Huxley, for example, once proclaimed, "I believe that probably all the great sea fisheries are inexhaustible; that is to say, nothing we do seriously affects the number of fish." In contrast, the natives of Oceania, knowing that their precious fisheries could easily be depleted, devised centuries ago a variety of measures designed to guard against this eventuality.

Johnston, F. D., R. Arlinghaus, and U. Dieckmann. 2010. Diversity and complexity of angler behaviour drive socially optimal regulations in a bioeconomic recreational-fisheries model. Canadian Journal of Fisheries and Aquatic Sciences 67:1507–1531. Erratum published in same volume, p. 1897–1898.

In many areas of the world, recreational fisheries are not managed sustainably. This might be related to the omission or oversimplification of angler behaviour and angler heterogeneity in fisheries-management models. We present an integrated bioeconomic modelling approach to examine how differing assumptions about angler behaviour, angler preferences, and composition of the angler population altered predictions about optimal recreational-fisheries management, where optimal regulations were determined by maximizing aggregated angler utility. We report four main results derived for a prototypical northern pike (*Esox lucius*) fishery. First, accounting for dynamic angler behaviour changed predictions about optimal angling regulations. Second, optimal input and output regulations varied substantially among different angler types. Third, the composition of the angler population in terms of angler types was important for determining optimal regulations. Fourth, the welfare measure used to quantify aggregated utility altered the predicted optimal regulations, highlighting the importance of choosing welfare measures that closely reflect management objectives. A further key finding was that socially optimal angling regulations resulted in biological sustainability of the fish population. Managers can use the novel integrated modelling framework introduced her to account, quantitatively and transparently, for the diversity and complexity of angler behaviour when determining regulations that maximize social welfare and ensure biological sustainability.

Ludwig, D., R. Hilborn, and C. Walters. 1993. Uncertainty, resource exploitation, and conservation: lessons from history. Science 260:17–36.

There are currently many plans for sustainable use or sustainable development that are founded upon scientific information and consensus. Such ideas reflect ignorance of the history of resource exploitation and misunderstanding of the possibility of achieving scientific consensus concerning resources and the environment. Although there is considerable variation in detail, there is remarkable consistency in the history of resource exploitation: resources are inevitably overexploited, often to the point of collapse or extinction. We suggest that such consistency is due to the following common features: (i) Wealth or the prospect of wealth generates political and social power that is used to promote unlimited exploitation of resources. (ii) Scientific understanding and consensus is hampered by the lack of controls and replicates, so that each new problem involves learning about a new system. (iii) The complexity of the underlying biological and physical systems precludes a reductionist approach to management. Optimum levels of exploitation must be determined by trial and error. (iv) Large levels of natural variability mask the effects of overexploitation. Initial overexploitation is not detectable until it is severe and often irreversible.

Schaefer, M. B. 1957. Some considerations of population dynamics and economics in relation to the management of commercial marine fisheries. Journal of the Fisheries Research Board of Canada 14:669–681.

Fishing is one of man's oldest occupations; some of the sea fisheries pre-date recorded history. So long, however, as men relied on oar and sail to reach the fishing grounds, and on simple, hand-operated gear to catch the fish, the intensity of fishing on the high seas remained low, so that the amount of the catch had apparently little effect on the magnitude of the fish stocks. There were great variations in the harvests, of course, but these were due to fluctuations in the fish populations quite independent of the amount of fishing. A fisherman's success depended on uncontrollable natural factors, and was not much affected by whether the number of fishermen was many or few.

With the industrialization of sea fishing in the latter part of the last century, bringing steam and later diesel power to the vessels, and bringing new and more efficient types of fishing gear, and machinery to handle it, the sea fisheries near northern Europe and in some other parts of the world began to show signs of diminishing return per unit of fishing effort. Near the turn of the century, there was considerable controversy as to whether or not the amount of a given kind of fish which man is able to take from the sea is sufficient to have any noticeable effect on the supply, This matter was discussed at some length, for example, by McIntosh (1899), Garstang (1900) and others. Alfred Marshall was preparing the first edition of his famous *Principles of Economics* (1890) at the time when the industrialization of the British trawl fishery was proceeding rapidly and this controversy was going on. It was, therefore, yet a moot question whether his Law of Diminishing Returns applied to the sea fisheries (Marshall, 8th edition, 1938, p. 166).

Subsequent history of the North Sea demersal species, the haddock of Iceland and the Northwest Atlantic, the Pacific halibut, and of numerous other fisheries, leaves little room to doubt that a modern commercial fishery can so affect the stock of fish in the sea that the return per unit of fishing effort is thereby diminished, and can even become so intense that the *total* harvest is also reduced. It may be noted here, although the matter will be developed in more detail later, that the law of diminishing returns as applied by Marshall and others to agriculture is somewhat different than the application to the sea fisheries. As originally developed, the law holds that the increased application of other factors of production to the land results in a decreased *rate* of return, but so long as the fundamental fertility of the land is not reduced, the *total* return would not diminish, but would increase, at a falling rate, toward some upper limit determined by the fertility of the land, the rainfall, amount of solar radiation, etc. (Ricardo's "original and indestructible powers of the soil".)

Experience having shown that the stock of commercial sizes of a sea-fish species, and the annual harvest obtainable from that stock, is related to the amount of fishing effort applied, there arises the irnportant question of how the amount of fishing should be managed in order to provide the greatest benefits to mankind. This, of course, is a socio-economic problem which will have unique solutions only if it can be specified what situation among possible alternatives is to be regarded as most beneficial. Fundamental to rational consideration of the matter, however, is knowledge respecting what are the possibilities, which must depend on the dynamic relationships between amount of fishing and amount and yield of the fish stocks, and corollary economic implications.

Attempts to systematize some of the significaut biological and economic facts bearing on this problem have been made by a number of persons in recent years, among which may be cited Russel (1931), Graham (1935, 1939, 1953), Beverton (1953), Gordon (1953, 1954), Burkenroad (1951), and Schaefer (1954a, b). The main result has been a considerable advance in our understanding of the biological and economic principles involved. There is, however, a fairly large degree of confusion, resulting from the biologists' inadequate consideration of economic principles, and, in part, from economists' failure to fully consider the properties of a self-renewing natural resource, the rate of renewal of which is dependent on the magnitude of the stock of the resource, which importantly distinguish such a resource from other classes of natural resources.

It seems worthwhile, therefore, to consider together some significant aspects both of the population dynamics of commercial fish stocks and of the economics of commercial fishing in order to arrive at a rational basis of considering the social problem of fisheries management.

In considering the bionomic properties of a fishery rve shall be concerned with the "long run" relationships. That is, we are interested in the average annual harvests that will be *sustained* by the fish population indefinitely at different levels of fishing effort, the monetary value of the harvests, and the monetary cost of the fishing effort. This approach has been admirably and carefully applied by Gordon (1954), but some further consideration appears to be necessary, because (1) he has not made the necessary distinction between the self-regulating, density dependent fish resources and other common-property resources of a different nature, (2) the mathematical model in Section IV of his paper is not consistent with the assumptions in Section III (and is not quite in accord with some dynamic properties of fish populations), and (3) he has (p. 129) defined the optimum degree of utilization of any particular fish stock as that which maximizes the net econlomic yield, the difierence between total cost, on the one hand, and total receipts (or total value production), on the other. This is one possible choice, of course, but it is not immediately obvious that it is the social optimum; indeed other possibilities have been explicitly chosen both for particular fisheries and as a general objective of fishery management.

Section 3

Managing Fish Habitat

DANIEL E. SCHINDLER

3.1. SYNTHESIS

No rational fisheries scientist or manager would disagree that habitat is critically important for sustaining fish populations and the fisheries they support. That said, the survey results that informed the contents of this book produced a somewhat idiosyncratic set of articles about fish habitat that should be considered classics of fisheries science. Articles that garnered the most support focused on river and lake ecosystems, and those that focused on marine ecosystems were rare and produced little consensus. The selection of articles deemed classics certainly reflects the emphasis that scientists working on freshwater ecosystems have placed on habitat. Further, the intensity of connections between human populations and freshwater ecosystems puts these at substantially higher risk of human-caused habitat degradation than their marine counterparts. However, the habitat requirements of freshwater fishes cannot be any more important than the habitat requirements of marine fishes. Many of the same issues that freshwater ecologists have grappled with have gained some traction in the marine literature, especially associated with how habitat features affect risk-sensitive foraging and species interactions among marine taxa. The growing interest in marine protected areas (MPAs) as a management tool has also renewed interest in habitat-mediated ecological interactions in marine ecosystems, though it is not clear that the classic research in freshwater habitats is fully appreciated by scientists working on habitat related issues in the marine realm.

Three articles are included here that provide the foundational basis of understanding the spatial and temporal organization of habitat in river ecosystems. The River Continuum Concept (RCC) of Vannote et al. (1980) was by far the most widely recognized habitat paper in our survey. The RCC is a framework to account for how hydrologic energy is dissipated along the downstream gradient in river systems, how these changes control transport and deposition of organic matter, and thus translate into fundamental changes in biological communities. The RCC is a powerful framework because it provided a testable set of predictions, based on established physical principles, about how river systems are organized from the smallest headwater streams to the largest rivers flowing into oceans. The RCC provided a conceptual baseline against which many management actions that affect ecosystem functioning can be established.

Although the RCC emphasized the predictable changes that characterize rivers along their downstream paths to oceans, in the late 1980s, there was increasing emphasis placed on the more stochastic processes linking rivers laterally to their flood plains. Junk et al. (1989) developed the "Flood Pulse Concept" (FPC) to describe the key organizing features of flood plain ecosystems. Unlike the RCC that stressed the continuous dissipation of hydrologic energy and its effects on sediment transport and storage along the downstream axis of rivers, the FPC described how floodplains were characterized by processes that were less dependent on location along the upstream-downstream gradient. Instead, the FPC described how floodplains were characterized by lateral fluxes of sediment and nutrients between the main channel and "off channel" habitat. Although the FPC is often offered as an alternative to the RCC, the reality is that both models contribute in distinct and substantial ways to the current understanding of spatial and temporal organization of habitat in lotic ecosystems.

The FPC identified that episodic high flows were critical for transporting nutrients and organisms between the main channel and the floodplains of rivers. The importance of flow dynamics were more critically examined in rivers in the western United States that had been subjected to massive changes in their hydrologic dynamics due to impoundments and flow regulations. Poff et al. (1997) discussed how the spatial organization of rivers, particularly the lateral connections between the channel and the floodplain, were intimately dependent on their temporal organization (i.e., their flow regime). Poff et al. (1997) pointed out that most successes in the management of United States rivers had focused on improving water quality by controlling pollutants, and little attention had been paid to hydrologic dynamics. Further, efforts to manage hydrology had focused singularly on minimum flows and how these were affected by diversions and withdrawals. The overall temporal organization of river flows, that could be characterized in terms of several statistical properties characterizing the natural flow regime of rivers, were fundamentally missing from river management strategies.

Poff et al. (1997) also described how a wide variety of human effects on rivers and watersheds have substantially changed the natural flow regime of rivers. Dams built for flood control have obvious effects on the intensity and frequency of large flow events. However, dams built for generating hydroelectricity or for irrigation have similar effects. Other human activities, ranging from urbanization to groundwater withdrawals, can also change the natural flow regime as defined by the magnitude and timing of high and low flow events. The flow regime concept of Poff et al. (1997) has become a key organizing principle for river restoration. Much of this restoration effort is aimed at removing dams that interfere with the natural flow regimes that define specific rivers and maintain the habitat requirements for lotic communities and the fisheries they support. There have been parallel efforts to more explicitly consider the natural flow regime when developing spill schedules over dams, to more closely approximate the hydrology that characterizes the river in the absence of human effects. How this story plays out in a future of new climate regimes and growing human needs for water remains unclear. Rapidly growing demands for energy and water in the developing world have led to increased damming of rivers. The realities of maintaining the natural hydrologic variability of rivers, while simultaneously meeting the demands of humans for reliable energy and water, will continue to produce difficult policy choices about how to manage river flows and habitats that support fisheries.

In lentic systems, ecologists have tended to emphasize the three-dimensional structural complexity of the environment as the key determinant of fish habitat. Many of the theoretical underpinnings of aquatic habitats can be traced to G.E. Hutchinson and his conceptualization of the niche requirements of aquatic species (Hutchinson 1957). However, Crowder and Cooper (1982) were particularly influential because they were among the first to translate this abstract concept to a tangible set of features of the environment that affected the ecology of fishes. Crowder and Cooper (1982) experimentally assessed the effects of habitat structural complexity (as determined by the density of aquatic macrophytes) on predator-prey interactions in fish communities. Their experiments demonstrated that increasing habitat complexity caused declines in predator search efficiencies. Subsequently, the diversity and abundance of prey increased with habitat complexity because of the predation refuges provided by structures and reduced predator search efficiencies. This contribution put the growing appreciation for the fact that predators could control the abundance and composition of prey communities (sensu Paine 1966) within a habitat context, whereby the physical characteristics of the environment often controlled the intensity of species interactions.

Concurrent with the experimental work of Crowder and Cooper (1982), Earl Werner and colleagues were developing and testing simple theoretical models to describe the trade-offs in foraging opportunity and predation risk that fishes encountered in different habitats of lakes. Werner et al. (1983) experimentally demonstrated why optimal foraging models often did not accurately capture the foraging behavior of fishes in the field, in terms of their habitat use and prey selection. Fish often showed what appeared to be sub-optimal behaviors from an optimal foraging perspective. Werner et al. (1983) demonstrated that fishes are faced with trade-offs between growth opportunity and predation risk, which are modulated by habitat complexity. A critical component of this model emphasized that the predation risk–growth opportunity trade-off among habitats changed substantially as fish proceeded through their ontogeny. As fish grew larger and invulnerable to predation, their behavior conformed more to expectations from optimal foraging models. The complexities of the ontogenetic implications of habitat trade-offs were developed more extensively in the noteworthy contribution of Werner and Gilliam (1984), who described the ontogenetic niche that fishes pass through as they grow from larvae into mature adults. Together, this body of research was a critical foundation for understanding behaviorally-mediated indirect effects, and the complexities of size-structured interactions among species in community ecology (de Roos and Persson 2013).

Most conceptions of habitat describe the physical characteristics of ecosystems as they are organized across *space*, and have inferred how biological processes respond to these changes in the environment. As shown in previous examples, the trade-offs in growth and mortality of different habitats often vary in response to the physical attributes of the habitat in which fishes live. Fisheries scientists have also inferred changes in habitat conditions by evaluating how biological features, such as fishery productivity, change through *time*. This approach has been particularly common in efforts to understand how climate forcing affects the productivity and abundance of fishes; such changes through time likely reflect changes to habitat, broadly defined. It has become widely appreciated that marine fisheries are often organized into distinct production regimes (Vert-pre et al. 2013) that likely reflect shifts in habitat conditions of the ecosystem.

Mantua et al. (1997) provided a particularly compelling example of how subtle, but persistent changes, in the physical conditions in oceans translate into sustained production regimes of fish stocks. Mantua et al. (1997) described the Pacific Decadal Oscillation (PDO) as characterized by distinct production regimes of Pacific Salmon *Oncorhynchus* spp., each of which last for several decades. There is a distinct spatial fingerprint of PDO dynamics defined by high biological productivity in the northeastern Pacific Ocean during periods when productivity in the California Current ecosystem is low, and vice-versa. The Mantua et al. (1997) article is foundational because it provided the first, clear-cut demonstration of how climatically-driven changes in the physical structure of the ocean could be amplified into large shifts in the biological productivity of marine ecosystems that could last for extended periods of time. While studies of paleo-records suggested that long-term regimes punctuated by distinct shifts are common in marine fish stocks (Soutar and Isaacs 1974; Rogers et al. 2013), the Mantua et al. (1997) research remains the most convincing example of this from direct observations of fish abundance. Although understanding the mechanisms through which regime dynamics are expressed in fisheries continues to be a motivation for much of fisheries oceanography, the more important contribution is to illustrate the non-stationary dynamics of ecosystems. From a management perspective, understanding the mechanisms that produce non-stationarity is far less important than adequately preparing for shifts in productivity by maintaining responsive assessment and management systems (Walters and Parma 1996).

Two other articles focused on the habitat of marine fisheries garnered sufficient attention to include in this book. Duggins et al. (1989) is one of a set of articles demonstrating that habitat conditions for marine fishes can be generated through ecological interactions. Duggins et al. (1989) described the responses of kelp forests to loss of Sea Otters *Enhydra lutris* along the west coast of North America, showing that grazing by sea urchins mediated indirect effects of changes in otter abundance on kelp productivity. In locations where otters were rare (where they had not recovered from intense hunting pressure), sea urchin populations exploded because of a release from otter predation, and grazed kelp forests into barrens. In locations where sea otters had recovered from hunting, they exerted sufficient predation pressure on urchins to functionally eliminate them from coastal ecosystems, thereby allowing kelp forests to flourish. Organic matter formed by kelp primary production could be traced via stable isotopes of carbon into fishes and other consumers in the ecosystem.

Although there are a number of articles that provide complementary descriptions of this habitat-forming species interaction in benthic marine ecosystems, Duggins et al. (1989) demonstrated that strong trophic cascades could occur in large marine ecosystems and that they were substantial enough to fundamentally change habitat structure for other species that do not necessarily interact directly with otters and urchins. Furthermore, they showed that changes in habitat that result in structural differences can translate into changes in the dominant energy flows through ecosystems.

Finally, in marine ecosystems, establishment of MPAs has arguably been the most widely advocated and implemented form of habitat management. Here, discrete geographic areas of marine ecosystems are protected to reduce a variety of human impacts on the ecosystem, including the reduction or outright ban of fishing. The intention of establishing MPAs varies from restoring and conserving biodiversity to enhancing regional fisheries through supposed "spill-over" effects of recruitment deriving from protection of spawning stocks within MPAs. MPAs have come to be regarded by many as the panacea for problems associated with marine ecosystems, with little consideration of the ecological and economic trade-offs associated with their implementation. Hilborn et al. (2004) provided a clear overview of the trade-offs associated with the establishment of MPAs for sustaining and managing fisheries. Although benefits to biodiversity appear to be the clearest, the effects on fisheries are mixed and dependent upon spatial overlap between fisheries management, habitat units, and human communities. Those advocating for MPAs as a fisheries management tool would be well-served to consider the complexity of trade-offs associated with their effects on ecosystems, as described by Hilborn et al. (2004).

The collection of articles included here do not necessarily provide as cohesive of a foundation to the topic of habitat and its relationship to fisheries management as can be traced to the topic of population management (Walters, Section 1). However, they do provide a set of useful entry points to the diverse and scattered literature on fish habitat and its roles in fisheries management.

Acknowledgment

I thank Tim Cline for providing comments on this manuscript.

References

de Roos, A. M., and L. Persson. 2013. Population and community ecology of ontogenetic development. Princeton University Press, New Jersey.

Hutchinson, G. E. 1957. Concluding remarks. Cold Spring Harbor Symposium 22:415–427.

Paine, R. T. 1966. Food web complexity and species diversity. American Naturalist 100(910):65–75.

Rogers, L. A., D. E. Schindler, P. J. Lisi, G. W. Holtgrieve, P. R. Leavitt, L. Bunting, B. P. Finney, D. T. Selbie, G. J. Chen, I. Gregory-Eaves, M. J. Lisac, and P. B. Walsh. 2013. Centennial-scale fluctuations and regional complexity characterize Pacific salmon population dynamics over the past five centuries. Proceedings of the National Academy of Sciences 110:1750–1755.

Vert-pre, K. A., R. O. Amoroso, O. P. Jensen, and R. Hilborn. 2013. Frequency and intensity of productivity regime shifts in marine fish stocks. Proceedings of the National Academy of Sciences 110:1779–1784.

Walters, C., and A. M. Parma. 1996. Fixed exploitation rate strategies for coping with effects of climate change. Canadian Journal of Fisheries and Aquatic Sciences 53:148–158.

3.2 REPRINTED ARTICLES

Crowder, L. B., and W. E. Cooper. 1982. Habitat structural complexity and the interaction between bluegills and their prey. Ecology 63:1802–1813.

Duggins, D. O., C. A. Simenstad, and J. A. Estes. 1989. Magnification of secondary production by kelp detritus in coastal marine ecosystems. Science 245(4914):170–173.

Hilborn, R., K. Stokes, J.-J. Maguire, T. Smith, C. W. Botsford, M. Mangel, J. Orensanz, A. Parma, J. Rice, J. Bell, K. L. Cochrane, S. Garcia, S. J. Hall, G. P. Kirkwood, K. Sainsbury, G. Stefansson, and C. Walters. 2004. When can marine reserves improve fisheries management? Ocean & Coastal Management 47:197–205.

Junk, W. J., P. B. Bayley, and R. E. Sparks. 1989. The flood pulse concept in river-floodplain systems. Pages 110–127 *in* D. P. Dodge, editor. Proceedings of the international large river symposium. Canadian Special Publication of Fisheries and Aquatic Sciences 106.

Mantua, N. J., S. R. Hare, Y. Zhang, J. M. Wallace, and R. C. Francis. 1997. A Pacific interdecadal climate oscillation with impacts on salmon production. Bulletin of the American Meteorological Society 78(6):1069–1079.

Poff, N. L., J. D. Allan, M. B. Bain, J. R. Karr, K. L. Prestegaard, B. D. Richter, R. E. Sparks, and J. C. Stromberg. 1997. The natural flow regime. BioScience 47:769–784.

Vannote, R. L., G. W. Minshall, K. W. Cummins, J.R. Sedell, and C. E. Cushing. 1980. The river continuum concept. Canadian Journal of Fisheries and Aquatic Sciences 37:130–137.

Werner, E. E., J. F. Gilliam, D. J. Hall, and G. G. Mittelbach. 1983. An experimental test of the effects of predation risk on habitat use in fish. Ecology 64:1540–1548.

HABITAT STRUCTURAL COMPLEXITY AND THE INTERACTION BETWEEN BLUEGILLS AND THEIR PREY[1]

LARRY B. CROWDER[2] AND WILLIAM E. COOPER
Department of Zoology, Michigan State University, East Lansing, Michigan 48824 USA

Abstract. Structural complexity of the habitat often reduces predatory efficiency by reducing prey capture rates. Prey density is often positively correlated with habitat structure because it provides food and substrate to the prey as well as a relative refuge from predators. Dense structure inhibits foraging, allowing abundant, highly profitable prey to coexist with predators. Sparse structure allows efficient foraging and generally contains few highly profitable prey. This suggests that feeding rates of predators may be maximized at intermediate structure. If this is true, we might also expect predator growth rates to be higher in intermediate structure habitats. Since diet breadth is thought to be related to rates of encounter with profitable prey, we also expect diets of predators to be narrower at intermediate structure than in either sparsely or densely structured habitats.

Bluegill sunfish (*Lepomis macrochirus*) restricted to experimental ponds varying in vegetation density grew better and consumed more prey at intermediate macrophyte density than fish held at either low or high macrophyte densities. Fish at low macrophyte density had narrower diets than expected due to high initial prey availability relative to prey available at intermediate and high macrophyte density. Fish at high macrophyte density ate fewer, but larger, prey and thus had a narrower diet than expected. Fish predation reduced total prey biomass as well as mean prey size and altered the prey community structure by removing large active invertebrate predators and herbivores with subsequent release of smaller invertebrate predators and herbivores. These changes in prey community structure were also mediated by habitat structure. Habitat structure–food density interactions may be added to temperature and presence of predators as variables that influence the use of resources by fishes.

Key words: benthos; community structure; foraging behavior; habitat complexity; keystone predator; Lepomis; *macrophytes; predator-prey; refuge; resource use; structural complexity.*

INTRODUCTION

Habitat structural complexity may have a profound effect on ecological interactions. Increased physical structure in the habitat creates more microhabitat types—a greater total niche space—which may allow the coexistence of competitors and the persistence of both predators and their prey (Smith 1972, Crowley 1978). Littoral zones of lakes and ponds are often both physically complex and spatially patchy. Thus, the influence of habitat structural complexity on the interactions between littoral zone fishes and their prey should be examined. We consider this to be an appropriate model system which may suggest some general characteristics of predator-prey interactions in structurally complex habitats.

Physical structure acts in several ways to reduce predator efficiency. It may provide complete refuges for the prey (e.g., oatmeal sediment in Gause's [1934] *Paramecium-Didinium* system) or simply provide partial protection for prey due to reduced predator efficiency in portions of the habitat (Huffaker 1958, Smith 1972). Because prey capture rates are reduced in these relative refuges, one expects strong selection pressure for prey to occupy these regions preferentially (Huffaker 1958, Smith 1972). Reduction of predator efficiency via physical complexity of the habitat tends to stabilize predator-prey interactions (Smith 1972, Crowley 1978).

Habitat complexity also influences resource use, which is apparently closely tied to the utility or profitability of a particular resource relative to alternatives. The growing body of literature on optimal foraging theory (Pyke et al. 1977) and recent work on foraging in fishes suggests that fishes tend to employ optimal feeding strategies. Studies on prey size selection (Werner and Hall 1974, Eggers 1977, Werner 1977), swimming speed (Ware 1975, 1978), and habitat use in fishes (Werner and Hall 1979, Mittelbach 1981, Werner et al. 1981) all suggest that fishes behave in ways that maximize net energy intake. Overall habitat profitability to fishes is likely dependent on several variables including habitat structure, prey size and density (Werner and Hall 1979, Mittelbach 1981), temperature (Magnuson et al. 1979), predator risk (Hall and Werner 1977, Mittelbach 1981), and higher order interactions among these variables (Cooper and Crowder 1979, Mittelbach 1981, Crowder and Magnuson 1982).

Little field or experimental work exists on the effects of habitat structural complexity on fish-prey interactions. Apparently, structural complexity can alter

---

[1] Manuscript received 14 August 1981; revised 4 February 1982; accepted 8 February 1982.

[2] Present address: Department of Zoology, North Carolina State University, P.O. Box 5577, Raleigh, North Carolina 27650 USA.

the outcome of fish-prey interactions (Glass 1971, Ware 1973, Stein and Magnuson 1976) but few attempts have been made to assess the role of habitat complexity in predator-prey interactions in the field (Colwell and Fuentes 1975).

In fish-prey systems, per capita prey capture rates generally decline monotonically with increasing structural complexity due to increased search and pursuit times (Glass 1971, Ware 1973, Stein and Magnuson 1976). But prey density and diversity is often positively correlated with structural complexity (Macan 1949, Gerking 1957, Hruska 1961). Larger, higher utility prey and those most vulnerable to predation tend to be associated most closely with dense structure or other refuges (Crossman 1959, Charnov et al. 1976, Stein 1977, Van Dolah 1978). These observations suggest that predator feeding rates and thus growth rates may be maximized at intermediate structure density (Crowder and Cooper 1979). Because of relatively greater search and pursuit times for high utility prey at low and high structure densities relative to intermediate structure density, we expected fish diets to be narrower in habitats of intermediate structural complexity (Pyke et al. 1977, Cooper and Crowder 1979, Pyke 1979). Therefore, the impact of fish predation on the prey community (prey size and taxon) also should be dependent on habitat structural complexity.

In this paper we examine the interaction of habitat structural complexity and prey density on the diet and growth of bluegill sunfish. Fish were experimentally confined to a single habitat patch of known structural complexity so that overall habitat profitability could be inferred from feeding rates and growth. We also show that both habitat structural complexity and fish predation influence prey community structure.

METHODS

In early June 1978, three experimental ponds were selected from among 18 circular experimental ponds (28 m diameter, 1.7 m deep) at the Kellogg Biological Station, Michigan State University, Hickory Corners, Michigan, USA. Ponds were selected based on uniformity of macrophyte development, macrophyte species present, and similarity of the invertebrate prey community. *Ceratophyllum demersum* was the dominant macrophyte in all ponds; small stands of *Myriophyllum spicatum, Elodea canadensis, Potamogeton* spp., *Utricularia* spp., and *Chara* spp. also occurred. Cattails (*Typha angustifolia, T. latifolia*) around the perimeter of the ponds were cut to substrate level and removed. In mid-June, each pond was subdivided into four quadrants with seine cloth fencing (4.8 mm Ace-style netting) which was suspended by polypropylene rope and weighted into the substrate with steel chain sewn to the bottom of the net. Nets were quickly colonized by algae and movement by invertebrates through the fencing was assumed to be minimal.

On 29 and 30 June, submerged macrophytes were harvested by hand with sickle and shears and all cut vegetation removed. One pond was harvested throughout to an average low density of macrophytes (36 ± 5 [SE]stems/m^2), which extended from the substrate to within 0.5 m of the surface. Another pond was harvested to an intermediate density (111 ± 7 stems/m^2) and a third was unharvested (high density, 177 ± 10 stems/m^2). Vegetation extended to the surface in both the intermediate- and high-density ponds. Stem counts were made in ≈40 randomly placed 0.25-m^2 quadrats in each pond at 1 m depth. Macrophyte species were mapped on several occasions during the experiment; maps were verified by direct observation while snorkeling. Weekly stem counts and harvesting were done to maintain these macrophyte densities throughout the experimental period.

On 6 July, 200 bluegill sunfish (8.6 ± 0.2 [SE] cm standard length [SL, from the tip of the snout to the end of the vertebral column], ≈18.5 g) were introduced into each of two of the four quadrants in each pond. This fish density (~240 kg/ha) is well within the range of observed bluegill densities in natural lakes (Carlander 1977). Because of a shortage of fish, slightly smaller fish (≈7–8 cm) were stocked into one of the quadrants at intermediate macrophyte density. Because we predicted better growth at intermediate macrophyte density, the effect was conservative. The other two quadrants in each pond served as controls for fish effects on prey community structure.

Invertebrate prey species were quantitatively sampled about weekly through the experimental period. Collections were made with 23 cm diameter blind-end plankton nets (153-μm mesh) which were collapsed below a clump of vegetation (≈10 stems of *C. demersum* 1 m long, tied and weighted into the net). We left the vegetation clumps in the pond for 1 wk before sampling. Based on pilot samples, we found that colonization takes place quickly (within 3 d) and stabilizes within 1 wk. Samples were then recovered by pulling the net up around the clump of vegetation via a monofilament harness attached to the net on the bottom and to floats on the pond surface. Three samples were taken in each quadrant on each date (12 samples per pond per date). Invertebrates were gently washed from the vegetation samples into buckets. Vegetation was checked for clinging invertebrates under a dissecting microscope, dried, and weighed. Invertebrate samples were washed onto a 600-μm mesh sieve (number 30 United States Standard) and preserved in 10% formalin. Samples chosen from four representative dates to reflect prey dynamics both early and later in the season were counted into 25 taxonomic categories (Appendix) and all whole individuals were measured. Zooplankton from vegetation samples were not counted. Length-mass regressions derived for these pond species (E. Werner, D. Hall, and D. Laughlin, *personal communication*) were used to convert counts to masses.

LARRY B. CROWDER AND WILLIAM E. COOPER

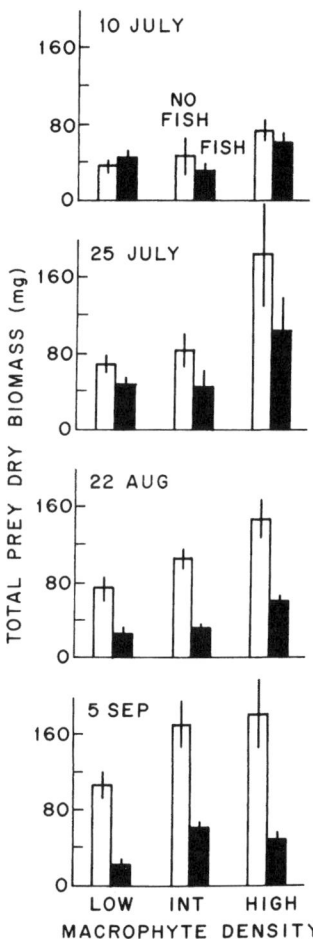

FIG. 1. Total dry biomass of invertebrates per vegetation sample ($\bar{x} \pm 1$ SE, $n = 6$) in ponds containing high, intermediate, and low density vegetation. Data are for four dates in pond quadrants with and without fish.

Fish were sampled about weekly by hook and line for stomach analysis, as other collection methods, such as seining, would disrupt the vegetation. An average of five fish taken from each quadrant on each date were replaced with marked fish of a similar size. Since bluegills are diurnal feeders and diets may differ with time of day (Keast and Welsh 1968, Sarker 1977), fish were collected during daily feeding peaks (Sarker 1977) both morning (AM, 0830–1030 EDT) and evening (PM, 2000–2200 EDT), immediately measured, and sacrificed. Their stomachs were then removed and placed in vials of 10% formalin. Morning fish diets likely reflect feeding beginning at dawn; evening diets reflect feeding in afternoon and near dusk. On most dates, fish were sampled from all ponds and stomachs preserved within 1 h. Ponds were fished in random order and individual fish were usually sacrificed within 15 min of capture. Over 400 stomachs were examined and all prey identified to the lowest taxonomic level possible (62% to family, 38% to genus or species, Appendix). On each date an average of 10.4 fish ($\pm$ 0.5, SE) characterize the diet. The contents of each stomach were counted and whole prey were measured. Length-mass regressions were used to convert counts to masses.

Fish diet breadths were calculated using Petraitis' (1979) method, which takes into account the differences in proportional prey representation in each pond. Diet breadths were calculated based on both prey size and taxon for both morning (AM) and evening (PM) sample on dates nearest (mean 3 d, range 0–6 d) to the four prey sample dates examined. Only benthic prey were included in the diet analysis because zooplankton were not counted in our prey samples. Zooplankton accounted for <3% of the diet biomass in both the low and high macrophyte ponds. *Simocephalus* and large ostracods which were abundant only in the intermediate macrophyte pond accounted for an average 10.7% of the biomass in fish diets.

Daily maximum temperatures measured with mercury thermometers at both the surface and 1 m depth varied between 23° and 28°C throughout the experiment. No distinct thermal stratification with depth was detected, though temperature 1 m depth was as much as 2° cooler on still, hot days. Dissolved oxygen was also assessed regularly at the surface and 1 m depth during the oxygen minimum ($\approx$0430 EDT). At no time were any of the ponds found to have less O_2 than 2 mg/L. Since oxygen levels were near saturation during the day, we doubt that fish were oxygen limited during the period of intense feeding.

On 7 September, vegetation was removed and the ponds were seined repeatedly to recapture stocked fish. On subsequent days, ponds were drained and all fish collected. Fish from the original stocking (unmarked fish) were measured, weighed, and preserved in 10% neutral formalin. Marked fish were counted and preserved.

RESULTS

Prey biomass dynamics

Total dry biomass of benthic invertebrates in the vegetation was clearly reduced by fish in all three macrophyte densities (Fig. 1). Prey biomass in quadrants without fish exceeded prey biomass with fish in all three ponds on both 22 August and 5 September ($P < .05$, t test). Prey biomass increased significantly ($P < .05$, Tukey's test, Box et al. 1978) between the first and final sample dates in no-fish quadrants at both low and intermediate macrophyte densities. As expected, mean prey biomass correlates well with

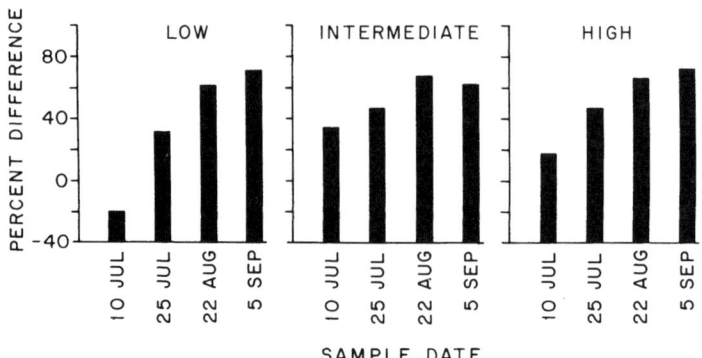

FIG. 2. Percent difference [100(NF − F)/NF] in dry biomass of invertebrates during the experiment in pond quadrants without fish (NF) and quadrants with fish (F) at three macrophyte densities.

macrophyte density in the quadrants without fish, though these patterns are generally not significant ($P > .05$, Tukey's test). On 22 August, significantly higher prey biomass occurred in fish quadrants at high macrophyte density than at either low or intermediate macrophyte densities ($P < .05$, Tukey's test). On 5 September, both high and intermediate quadrants with fish had higher prey biomass than in the low macrophyte fish quadrants ($P < .05$, Tukey's test). No other significant differences in prey biomass were noted among ponds.

We examined the temporal change in the difference between total prey biomass in quadrants without fish and prey biomass in quadrants with fish (Fig. 2), and found that the percent difference increased through the season in all three ponds. Rate of change was most dramatic in the low macrophyte pond (-20% to $+70\%$) and slowest in the intermediate macrophyte pond ($+35\%$ to 68% by 22 August). This suggests a greater fish effect at low macrophyte density compared to the other densities. Percent difference appeared to level off at 60–70% in all three ponds by the end of the experiment. The impact of fish on prey biomass was obvious in all three ponds by the 3rd wk of the experiment. Total prey numbers (excluding zooplankton) also tended to correlate with macrophyte density in quadrants without fish. Prey were often more numerous in fish quadrants than in the no-fish controls. This was particularly true in the low and intermediate macrophyte ponds, which averaged 29% and 51% more prey where fish were present than in controls on the final two sampling dates. Prey density did not appear to increase in the high macrophyte treatment. Given the observed dramatic declines in biomass (Fig. 1) we infer that most of the observed increase in prey numbers must have been contributed by small prey (Fig. 3). After the 1st wk, prey were significantly smaller in quadrants with fish ($P < .01$, t test).

The stomachs of bluegills collected in the morning (AM) tended to contain a greater biomass of prey than those collected in the evening (PM, Fig. 4) though stomach masses were somewhat variable. This pattern is correlated with vegetation density; fish at low macrophyte density contained significantly more prey in the morning than in the evening beginning 3 August ($P < .05$, t test). At intermediate macrophyte density this difference was significant beginning 15 August ($P < .05$, t test). No differences occurred at high macrophyte density. Mean stomach masses did not differ systematically among macrophyte densities in either the morning or the evening.

To clarify these patterns we calculated the ratio of mean mass of fish stomach contents to mean prey biomasses in vegetation samples from the quadrants in which the fish were collected (Fig. 5). This ratio suggests that fish at low marophyte density not only fed more in the morning than in the evening, but that they consumed a much higher proportion of available benthic resources (ratio = 20–50) in the morning than fish in the intermediate or high macrophyte pond (ratio ≤ 20).

Prey dynamics by taxon

In quadrants without fish, 8–10 taxonomic groups were abundant enough to account for >1% of the total biomass on each date (Fig. 6). Without fish, relatively large organisms (odonates, *Hyalella*) predominated in all three ponds, but total biomass tended to increase with increasing vegetation (Fig. 1), so that even though the proportional composition of odonates or *Hyalella* was similar among ponds the actual biomass contribution may differ.

In the presence of fish predators, several of these groups became uncommon in the vegetation samples. *Hyalella* and large damselflies (Coenagrionidae) were proportionally much reduced over the no-fish controls. This, of course, implies that biomass of these organisms was reduced because overall prey biomass

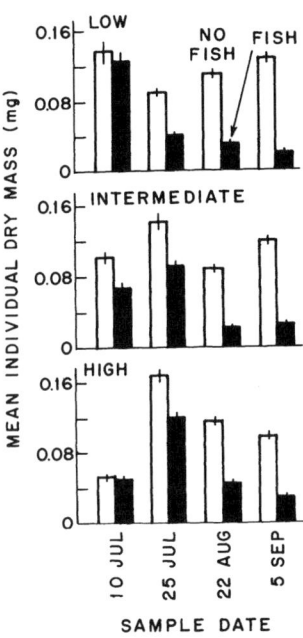

FIG. 3. Mean mass (±SE) of an individual benthic invertebrate in ponds at three macrophyte densities. Data are for four dates in pond quadrants with and without fish.

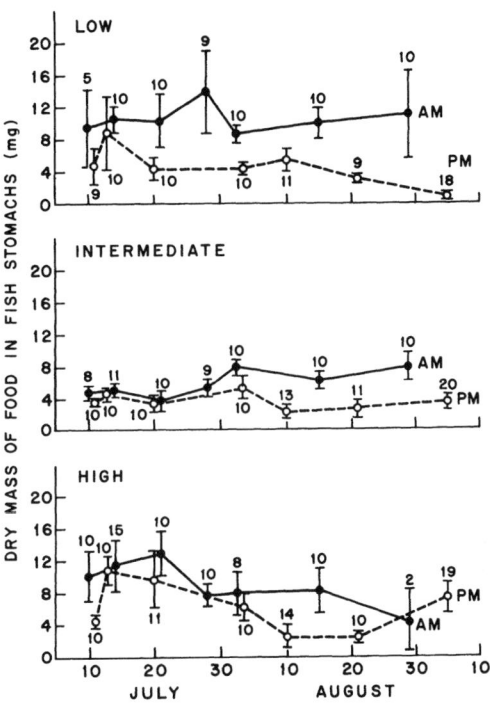

FIG. 4. Average dry mass of stomach samples (±SE) from fish collected in the morning (AM) and in the evening (PM) in ponds at three macrophyte densities.

was reduced in quadrants with fish (Fig. 1). Corixids and *Anax* were never significant components of the prey community biomass in fish quadrants of any pond. For example, in the low vegetation pond, the prey biomass in quadrants without fish was 47% odonates at the end of the experiment; in quadrants with fish, odonates only accounted for 20% of the much reduced biomass. At the end of the experiment, *Hyalella* was reduced from an average 30% of the total biomass without fish to 8% of total biomass in the presence of fishes. In fish quadrants, coenagrionids never contributed significantly to total biomass at low vegetation. Their proportional contribution was basically unchanged, though, at high vegetation density.

We also noted shifts in invertebrate trophic structure, taxon composition, and size of individuals. Over the season, invertebrate predators (Fig. 6) accounted for an average of $33 \pm 3\%$ (SE; $n = 8$) of the biomass in low density vegetation, $32 \pm 11\%$ ($n = 8$) in intermediate density vegetation and $44 \pm 4\%$ ($n = 8$) in the high vegetation pond. No differences in percent biomass of invertebrate predators occurred between fish and no-fish quadrants. What was most striking was the shift in species and size composition of the benthic invertebrates from large mobile forms to smaller, more sedentary species with the addition of fish. As coenagrionids and libellulids declined, a concomitant increase occurred in the smaller invertebrate predator group, the Tanypodinae. A clear shift occurred in invertebrate predators to smaller individuals, though trophic structure (percent invertebrate predators) was fairly constant with or without fish.

A similar shift from large mobile prey to small, more sessile prey occurred among the herbivore/detritivores. *Hyalella* was rapidly reduced in all three ponds and mayflies (Ephemeroptera) and midges (Chironomidae) increased. Baetid mayflies increased most in the low macrophyte pond. Caenid mayflies increased dramatically in all ponds but reached peak abundance at intermediate macrophyte densities. The midges and *Tanytarsus* also were most abundant at intermediate macrophyte densities.

We conclude, then, that fishes can significantly reduce prey biomass below that observed in similar pond habitats without fish, though prey numbers may actually increase with fish. Fish also appear to prey upon larger or more active invertebrates, leading to shifts in invertebrate community structure toward smaller or less active forms.

Fish diets: prey taxa

In morning (AM) fish diets, 8–9 taxonomic groups were abundant enough to account for >1% of the total stomach biomass (Fig. 7). In evening (PM), fish diets expanded to include 11-12 items. Relatively large prey (odonates, mayflies) accounted for 60-80% of total prey biomass in stomachs in all three ponds, both morning and evening. But more prey types were included in evening diets. Diet breadths calculated based on biomass, using Petraitis' (1979) method, were significantly ($P < .01$) narrower in the morning on the first (10 July) and fourth (5 September) dates in the low macrophyte pond. In the intermediate macrophyte pond, morning diets were significantly ($P < .01$) narrower only on the final date (5 September). Fish on the first (10 July) and third (22 August) dates had narrower diets ($P < .01$) at high macrophyte densities. In no case were diets narrower in the PM samples. This decrease in AM diet breadth along with the observation that average stomach masses tended to be higher in the morning (Fig. 4) suggests that encounter rates with prey were reduced in evening relative to morning.

Diet breadths based on prey taxa also differed among ponds (Table 1). Fish in the low macrophyte pond had narrower AM diets than fish in either intermediate or high macrophyte densities, especially early in the experiment. Low macrophyte fish also had narrower PM diets than fish at high macrophyte densities during the second and third dates examined. Fish at intermediate macrophyte densities had significantly narrower diets only on one occasion: the PM sample on the final date examined.

Fish at low macrophyte density, however, consumed many *Ceriodaphnia* and other small zooplankton for the last month of the experiment. On the final 2 wk of the experiment, bluegills in the low macrophyte pond ate 79.4% zooplankton (by numbers) in the morning (AM) and 91.7% zooplankton in the evening (PM). Fish in the final 2 wk at intermediate macrophyte density averaged 59.1% (by numbers) (AM) and 64.4% (PM) zooplankton, most of which were *Simocephalus*, which are larger than *Ceriodaphnia*. At high macrophyte density over the same period, fish ate (by numbers) 73.1% (AM) and 62.0% (PM) zooplankton, which were predominantly *Ceriodaphnia*. This suggests that zooplankton were substantial parts of the diet, particularly at low macrophyte density, but because these zooplankton were small, they contributed little to total prey biomass eaten.

Fish diets: prey size

Fishes ate more in amount (Fig. 4) and fewer kinds of benthic prey (Fig. 7) in morning than in evening. On average, fishes consumed larger prey than the average prey available in the benthos (Table 2). Larger prey were consumed more frequently (9 of 12 cases) in morning than in evening (Table 2). On average, fish at high macrophyte density consumed fewer but larger prey than fish at either low or intermediate macrophyte densities both in morning and in evening.

Diet breadths (on a $\log_2$ prey mass basis) were cal-

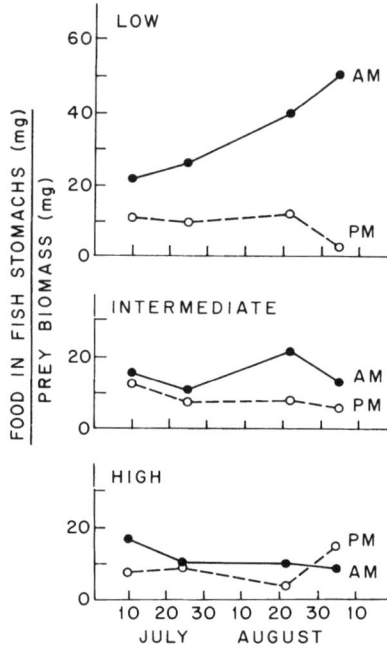

FIG. 5. Ratio of dry mass of fish stomach contents and dry mass of prey in vegetation samples for fish caught both morning (AM) and evening (PM) in ponds at three macrophyte densities.

culated using the approach outlined in Petraitis (1979). Bluegill diets tended to be rather broad in all three ponds. Significant differences in diet width occurred on only two dates. During the 1st wk of the experiment fish in the intermediate density pond had significantly narrower diets in the evening than fish in the low macrophyte pond ($P < .05$). On the week of 22 August, fish in the low macrophyte pond had narrower diets in the morning than did fish in the high macrophyte pond ($P < .05$). Fish diet breadths tended to be wider in evening samples relative to morning samples, but significant differences ($P < .05$) occurred only on one date. In general, diet breadths on a prey size axis did not differ with time of day or macrophyte density.

Fish growth

The mean size (centimetres SL ± SE) and mass (grams wet mass ± SE) of fish recovered from both quadrants in each pond at the end of the experimental period were as follows: low macrophyte density: 9.47 ± 0.07 cm (24.5 ± 0.6 g); intermediate macrophyte density: 9.71 ± 0.09 cm (26.3 ± 0.8 g); high macrophyte density: 9.37 ± 0.08 cm (25.0 ± 0.5 g), $N > 100$. Fish were largest (ANOVA, $P < .01$) at intermediate macrophyte densities despite the fact that slightly smaller fish were stocked in one quadrant of

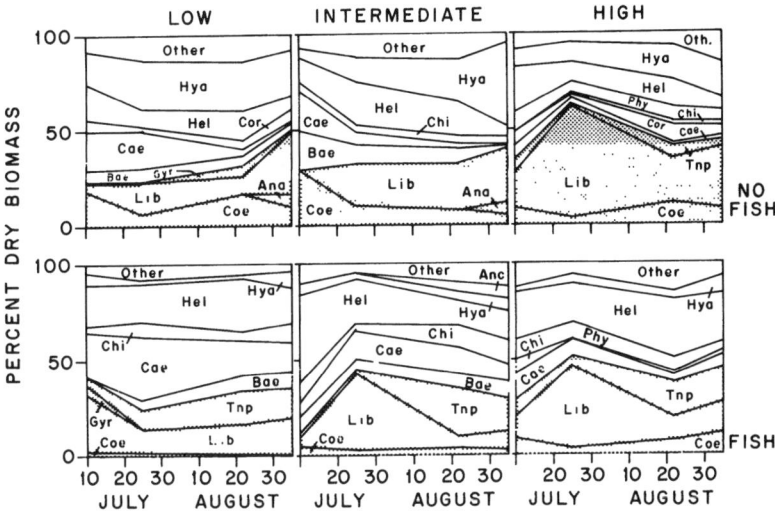

FIG. 6. Percent dry biomass of benthic invertebrate taxa in each pond with and without fish predators. Taxa in "Other" prey category each accounted for <1% of total prey biomass/sample. Shaded area indicates invertebrate predators. Taxon codes are as follows: Coe-Coenagrionidae; Lib-Libellulidae; Ana-*Anax*; Gyr-Gyrinidae; Tnp-Tanypodinae; Bae-Baetidae; Cae-Caenidae; Cor-Corixidae; Chi-*Chironomus*; Phy-*Physa*; Hel-*Helisoma*; Hya-*Hyalella*; Anc-Ancylidae; Oth-Other taxa. Relationships among these taxa are outlined in the Appendix.

the intermediate macrophyte pond. Survival varied among pond quadrants but averaged 89% in the low macrophyte pond and 91 and 86% in the intermediate and high macrophyte ponds, respectively (Crowder and Cooper 1979).

We used a bioenergetics-growth model for bluegill (Kitchell et al. 1977, Breck and Kitchell 1979) to back-calculate prey consumption for fish in each quadrant of each pond. The model calculates an energy balance equation so that if one knows fish size and numbers, growth, and water temperature, then food consumption can be estimated for bluegills in each pond. Cal-

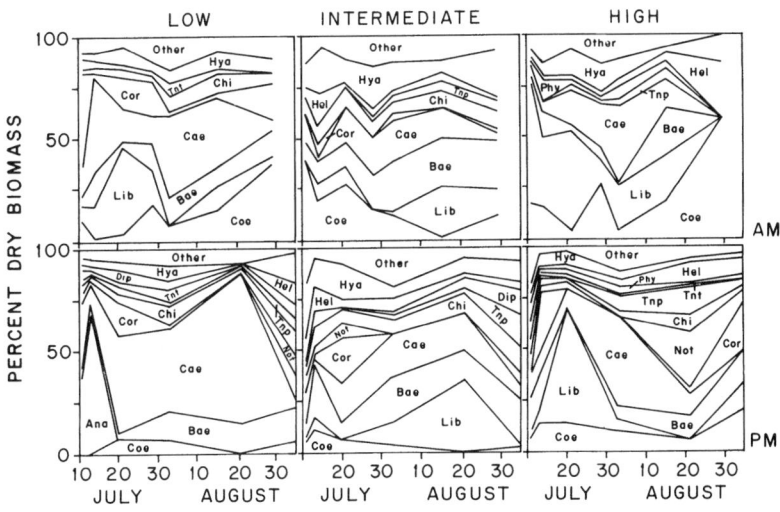

FIG. 7. Percent dry mass of prey taxa in fish stomachs from both morning (AM) and evening (PM) samples. Taxa in "Other" prey category each accounted for <1% of the mean prey biomass/stomach. Prey taxon codes as follows: Not-Notonectidae; Tnt-*Tanytarsus*; Dip-miscellaneous small Diptera. Other taxon codes as in Fig. 6.

TABLE 1. Diet breadth (method of Petraitis 1979) comparisons among ponds based on biomass in various prey taxa. "<" indicates a significantly narrower diet ($P < .05$).

10 July	AM	Low < intermediate, high
	PM	No significance differences ($P > .05$)
25 July	AM	Low < intermediate, high
	PM	Low < high
22 August	AM	Low < intermediate
	PM	Low < high
5 September	AM	Low, high < intermediate
	PM	Intermediate < low

culations were made separately for each quadrant because survival and growth varied slightly among quadrants. Based on model calculations, fish at intermediate macrophyte densities had the highest daily consumption rates (Table 3) when differences among ponds in temperature, fish size, and survival were taken into account.

We also used the model to estimate the average initial size of fish in the southeast (SE) quadrant of the intermediate density pond which was stocked with slightly smaller fish than the other ponds. We assumed the same consumption rates for fish in the SE quadrant of the intermediate macrophyte pond as was estimated for fish in the NW quadrant. This was conservative because maximum ration decreases with fish size. From the model, initial fish size in the SE quadrant was 8.0 cm SL (13.8 g). This agrees well with our qualitative estimate of 7–8 cm SL. Fish were significantly larger at the end of the experiment at intermediate macrophyte density despite slightly smaller fish being planted in one of the quadrants.

DISCUSSION

Bluegill sunfish, restricted to different vegetation densities, experienced higher overall habitat profitability at intermediate macrophyte density. At this density fish grew best, and model calculations which adjust for differences in survival, fish size, and temperature among ponds also indicated that their average daily rations were higher. By comparison, low macrophyte density and high macrophyte density appear to be less profitable habitats.

Fish diets (based on prey taxa) tended to be narrower at low macrophyte density than at either intermediate or high macrophyte density. Because prey biomass is often positively correlated with increasing macrophyte density (Macan 1949, Gerking 1957, Hruska 1961) and predator efficiency is often negatively correlated with increasing structure (Glass 1971, Ware 1973, Stein and Magnuson 1976), we expected higher prey capture rates and thus a narrower diet at intermediate macrophyte density (Cooper and Crowder 1979). Because prey biomass available at various macrophyte densities did not differ significantly until the last half of the experiment, and predators are often

TABLE 3. Bluegill bioenergetics model (Breck and Kitchell 1979) estimate of daily ration, accumulative consumption during the experiment and daily growth rates for fish in this field experiment. Actual mass gains, survivorship, and pond temperatures were used to make the estimates.

			Model estimates		
Macrophyte density	Pond quadrant	Average actual mass gain (g)	Proportion of maximum daily ration	Cumulative consumption per fish (g)	Growth (% body mass per d)
Low	SE	5.7	0.434	43.8	0.44
	NW	7.5	0.460	47.6	0.54
Intermediate	SE*	4.6†	0.508*	45.3	0.82
	NW	10.8	0.508	54.8	0.73
High	SE	6.5	0.436	43.4	0.48
	NW	5.3	0.416	40.6	0.40

* Stocked with slightly smaller fish than other treatments. Proportion of maximum ration was assumed equal to NW quadrant to make consumption and growth estimates.
† Conservative model estimate for average mass gain in SE quadrant, intermediate macrophyte density, is 9.3 g.

TABLE 2. Mean mass of benthic prey organisms in ponds and fish stomachs, morning (AM) and evening (PM), on four sampling dates.

		Date			
Vegetation density	Sample	10 July	25 July	22 August	5 September
		Mean mass of organisms (mg)			
Low	Pond	0.13	0.04	0.03	0.02
	Fish AM	0.19	0.12	0.11	0.25
	PM	0.13	0.08	0.10	<0.01
Intermediate	Pond	0.07	0.09	0.02	0.03
	Fish AM	0.13	0.12	0.10	0.07
	PM	0.09	0.13	0.06	0.04
High	Pond	0.05	0.12	0.05	0.03
	Fish AM	0.32	0.14	0.20	*
	PM	0.06	0.29	0.09	0.14

* Mean mass based on only two fish.

more efficient in a low structure habitat, the observed narrower diets at low macrophyte density might be expected. Fish at intermediate macrophyte densities had narrower diets only on the final date examined, on which prey biomass available was also significantly higher than at low macrophyte density. Apparently, fish at high macrophyte densities experienced reduced prey capture rates (Table 3) and slower growth despite the presence of a higher biomass of prey available throughout most of the experiment. Fish in high macrophyte density likely experienced reduced encounter rates and thus reduced attempts to capture prey (Glass 1971). Fish in the high macrophyte pond ate fewer but larger prey and did not expand their diets to include low utility items. Fish diets at high macrophyte density did not broaden as we expected; the fish simply consumed a lower ration and grew more poorly as a result.

Bluegills probably respond to benthic prey based on prey behavior and movement more than prey size per se (cf. Ware 1973). Fish in all three ponds rapidly removed actively moving *Hyalella*. But no differences were apparent among ponds in fish diets based on prey size. The fish were size selective in all three ponds, but prey behavior is probably critically important to determining prey availability and prey choice.

In the field, small bluegills of the size studied here are often closely associated with vegetation (Werner et al. 1977, Keast et al. 1978). This association would reduce intraspecific competition among size classes (Keast 1978), but would also reduce predation pressure on small bluegills (Hall and Werner 1977, Mittelbach 1981). Although no piscine predators were included in our experiments, their presence would likely reduce bluegill densities most rapidly in the low macrophyte pond. Because intermediate and dense vegetation provide increasing protection from predators and dense vegetation apparently inhibits foraging, small bluegills might be found most commonly in association with intermediate to dense vegetation. This appears to be the case (Keast et al. 1978) but bluegill abundance relative to quantified vegetation densities is not well established.

The behavioral interactions of bluegills and their prey are also very interesting. Though we made no direct observations on bluegill foraging behavior, evidence exists for resource depression (Hall et al. 1970, Charnov et al. 1976, Stein and Magnuson 1976) among prey species during daylight. Fish contained a higher biomass of prey as a result of feeding in the morning than in the evening. Diet breadth (taxon groups) increased and mean prey size eaten declined in evening relative to morning. This suggests that encounter rates with prey may have been higher in the morning. However, too little is known regarding the diel prey behavior patterns and how they might influence encounter rates to evaluate this hypothesis. Fish foraging behaviors or abilities may also be altered on a diel basis. Further evaluation of these observations and diel aspects of predator-prey interactions might well prove useful.

Little is known regarding the mechanisms underlying the reduction, with increasing structural complexity, of invertebrate prey capture rates by fish. Prey refuging behavior may reduce captures and capture rates of large active prey may decline with increasing structure due to increased search or pursuit times. Largemouth bass (*Micropterus salmoides*) often capture fewer fish prey with increasing structure levels (Glass 1971), but this also depends on dynamics in prey behavior (Savino and Stein 1982). In laboratory experiments with largemouth bass as predators, small bluegill change behaviors from schooling to dispersed distributions with increasing structure; capture rates decline to zero at high structure levels. Fathead minnows (*Pimephales promelas*) do not alter their behavior; they continue to school even at high structure levels and structural levels do not influence capture success (J. F. Savino and R. A. Stein, *personal communication*). Thus, differences in prey behavior may alter their availability and vulnerability to predators in structured environments.

Our results run counter to the hypothesis proposed by Thorp and Bergey (1981) that "individual species of keystone benthic predators do not occur in the littoral zone of freshwater lentic environments with soft bottoms." Bluegills in our experiment significantly reduced the total biomass and size of benthic invertebrates as compared to controls, though this effect was tempered by structure levels. Fish also altered the invertebrate community composition by removing large invertebrate predators (odonates) and herbivores (*Hyalella*) to the benefit of smaller forms.

Thorp and Bergey's (1981) results on predation effects in the littoral zone may be misleading. They erected cages to exclude predators but did not in any way document predation intensity or predator diets in cages open to predators. It is conceivable, then, that no obvious differences were found between predator exclusion cages and predator access cages because predation intensity in open cages was light. Also, their comparisons were based on total invertebrate density or prey density aggregated into functional guilds and across seasons. This would tend to mask changes in species composition within these aggregate groups. As we have shown, total prey density may actually increase in the presence of fish. But some groups are dramatically reduced while others increase. Prey biomass reductions and changes in mean invertebrate size are probably more appropriate response variables for tests of these hypotheses. Thorp and Bergey (1981) give no information on invertebrate biomass or size dynamics, except for chironomids, which were not reduced in size when predators had access to the cages.

Shifts in the invertebrate prey community in our experiments resulted from both changes in the density of vegetation and the presence of predators. Invertebrate biomass correlates with vegetation density in the absence of predators. Without fish, the taxon composition of each pond was similar; higher density vegetation simply scaled up the densities of various taxa. Large odonate predators predominated; active *Hyalella azteca* was the most common detritivore. With fish, large invertebrates were cropped with a concomitant increase in smaller forms.

Invertebrate predators accounted for 33–50% of the total benthic invertebrate biomass throughout the experiment, but taxon shifts occurred; smaller Tanypodinae increased as the large odonates were removed. Benke (1976, 1978) has explained how the relatively large biomass of invertebrate predators can be supported in littoral ecosystems; major prey species of large odonates such as chironomids and other midges have high population turnover rates relative to the odonates. The cropping of odonate predators was dependent on vegetation density; fewer odonates persisted at low macrophyte density than at high macrophyte densities.

The major increases in chironomids, caenid mayflies, and Tanypodinae we observed were probably responses to reduced invertebrate predation. Midges and caenids are the major prey of odonates (Lawton 1970, Benke 1978, Baker 1980). Fish remove the large invertebrate predator, releasing smaller species from predation and perhaps reducing competition between large invertebrate predators and small invertebrate predators. We cannot, of course, rule out the effects of competition in the increase of small invertebrates, but the generally low total biomass of invertebrates in fish quadrants relative to the no-fish controls and the striking effects of the invertebrate predators makes competition a less likely explanation for many of the prey community changes.

Habitat structural complexity may also influence the long-term stability of predator-prey interactions (Glass 1971, Benke 1976, Stenseth 1977, 1980, Crowley 1978, Baker 1980). At low levels of structural complexity, predators are efficient, deplete the high profitability prey, and the system tends to become unstable. Increased structure reduces predatory efficiency (Glass 1971), increases the "effective size" of the system (Crowley 1978) and may tend to stabilize the interaction. But extremely dense structure may so reduce predation that prey tend to increase uncontrolled and destabilize the interaction. Glass (1971) argued that some intermediate level of structure may tend to stabilize systems where both predator and prey are fish, and may also produce the greatest biomass of predators. We cannot directly address the question of stability of the bluegill invertebrate interaction explored here. But we do have evidence for greater production of bluegills at intermediate structure levels. Reduced prey biomass in the low structure pond and generally low feeding rates and growth of fish at high macrophyte densities suggest that in the long term, intermediate macrophyte density may be the best habitat both in terms of foraging and growth of the predator and stability of the fish-prey interaction.

We limited fish to one habitat type to assess profitability of habitats of various vegetation densities. In lakes, of course, fish may choose from a variety of habitat patches. We may expect patch choice to be dependent on several seasonally dynamic variables including food density, habitat structure, temperature and predators. In this paper, we have demonstrated the interaction of prey density and habitat structure in foraging and growth of bluegill sunfish. The interaction of food availability and temperature may be important in terms of net energy available for growth and thus habitat profitability (Crowder and Magnuson 1982). Finally as Mittelbach (1981) has recently shown, seasonal shifts in food availability and presence of predators may also influence the foraging behavior and habitat choice of small fishes.

Structural complexity of the habitat mediates fish-prey interactions through behavioral changes of both fish and their prey. An optimal level of structure does exist which can be characterized by higher feeding rates and better growth in the predators and perhaps by a tendency for more stable long-term interaction with their prey. The importance of considering multiple variables (food, habitat structure, temperature, predators) and assessing resource availability is increasingly obvious if we are to understand resource use in spatially complex environments.

Acknowledgments

We thank Dennis Laughlin, Karen Patterson, Brad Pease, Mike Pease, and Scott Cooper for their assistance in data collection and analysis. Don Hall, Jim Kitchell, and Roy Stein provided useful comments on an earlier draft of the manuscript. Cheryle Hughes drew the figures. This research was supported by National Science Foundation grant DEB 77-04818.

Literature Cited

Baker, R. L. 1980. Use of space in relation to feeding areas by Zygopteran nymphs in captivity. Canadian Journal of Zoology **58**:1060–1065.

Benke, A. C. 1976. Dragonfly production and prey turnover. Ecology **57**:915–927.

———. 1978. Interactions among coexisting predators—a field experiment with dragonfly larvae. Journal of Animal Ecology **47**:335–350.

Box, G. E. P., W. G. Hunter, and J. S. Hunter. 1978. Statistics for experimenters. John Wiley and Sons, New York, New York, USA.

Breck, J. E., and J. F. Kitchell. 1979. Effects of macrophyte harvesting on simulated predator-prey interactions. Pages 211–228 *in* J. E. Breck, R. T. Prentki, and O. L. Loucks, editors. Aquatic plants, lake management and ecosystem consequences of lake harvesting. Institute of Environmental Studies, University of Wisconsin, Madison, Wisconsin, USA.

Carlander, K. D. 1977. Handbook of freshwater fishery biology. Volume 2. Iowa State University Press, Ames, Iowa, USA.

Charnov, E. L., G. H. Orians, and K. Hyatt. 1976. Ecological implications of resource depression. Ameican Naturalist 110:247–259.

Colwell, R. K., and E. R. Fuentes. 1975. Experimental studies of the niche. Annual Review of Ecology and Systematics 6:281–310.

Cooper, W. E., and L. B. Crowder. 1979. Patterns of predation in simple and complex environments. Pages 257–267 in R. H. Stroud and H. Clepper, editors. Predator-pre systems in fisheries management. Sport Fishing Institute, Washington, D.C., USA.

Crossman, E. D. 1959. Distribution and movements of a predator, the rainbow trout, and its prey, the redside shiner, in Paul Lake, British Columbia. Journal of the Fisheries Research Board of Canada 16:247–267.

Crowder, L. B., and W. E. Cooper. 1979. Structural complexity and fish-prey interactions in ponds: a point of view. Pages 2–10 in D. L. Johnson and R. A. Stein, editors. Response of fish to habitat structure in standing water. Special Publication Number 6, North Central Division, American Fisheries Society, Bethesda, Maryland, USA.

Crowder, L. B., and J. J. Magnuson. 1982, in press. Cost-benefit analysis of temperature and food resource use: a synthesis with examples from the fishes. In W. P. Aspey and S. I. Lustick, editors. Behavioral energetics: the cost of survival in vertebrates. Ohio State University Biosciences Colloquia, Number 7, Ohio State University Press, Columbus, Ohio, USA.

Crowley, P. H. 1978. Effective size and the persistence of ecosystems. Oecologia 35:185–195.

Eggers, D. M. 1977. The nature of prey selection by planktivorous fish. Ecology 58:46–59.

Gause, G. F. 1934. The struggle for existence. Williams and Wilkins, Baltimore, Maryland, USA.

Gerking, S. D. 1957. A method of sampling the littoral macrofauna and its application. Ecology 38:219–225.

Glass, N. R. 1971. Computer analysis of predation energetics in the largemouth bass. Pages 325–363 in B. C. Patten, editor. Systems analysis and simulation in ecology. Volume 1. Academic Press, New York, New York, USA.

Hall, D. J., W. E. Cooper, and E. E. Werner. 1970. An experimental approach to the production dynamics and structure of freshwater animal communities. Limnology and Oceanography 15:829–928.

Hall, D. J., and E. E. Werner. 1977. Seasonal distribution and abundance of fishes in the littoral zone of a Michigan lake. Transactions of the American Fisheries Society 106:545–555.

Hruska, V. 1961. An attempt at a direct investigation of the influence of the carp stock on the bottom fauna of two ponds. Internationale Vereinigung für Theoretische und Angewandte Limnologie, Verhandlungen 14:732–736.

Huffaker, C. B. 1958. Experimental studies on predation: dispersion factors and predator-prey oscillations. Hilgardia 27:343–383.

Kitchell, J. F., D. J. Stewart, and D. Weininger. 1977. Applications of a bioenergetics model to yellow perch (*Perca flavescens*) and walleye (*Stizostedion vitreum vitreum*). Journal of the Fisheries Research Board of Canada 34:1922–1935.

Keast, A. 1978. Trophic and spatial relationships in the fish species of an Ontario temperate lake. Environmental Biology of Fishes 3:7–31.

Keast, A., J. Harker, and D. Turnbull. 1978. Nearshore fish habitat utilization and species associations in Lake Opinicon (Ontario, Canada). Environmental Biology of Fishes 3:173–184.

Keast, A., and L. Welsh. 1968. Daily feeding periodicities, food uptake rates, and dietary changes with hour of day in some lake fishes. Journal of the Fisheries Research Board of Canada 25:1133–1144.

Lawton, J. H. 1970. Feeding and food energy assimilation in larvae of the damselfly *Pyrrhosoma nymphula* (Sulz.) (Odonata: Zygoptera). Journal of Animal Ecology 39:669–689.

Macan, T. T. 1949. Survey of a moorland fishpond. Journal of Animal Ecology 18:160–186.

Magnuson, J. J., L. B. Crowder, and P. A. Medvick. 1979. Temperature as an ecological resource. American Zoologist 19:331–343.

Mittelbach, G. G. 1981. Foraging efficiency and body size: a study of optimal diet and habitat use by bluegills. Ecology 62:1370–1386.

Petraitis, P. 1979. Likelihood measures of niche breadth and overlap. Ecology 60:703–710.

Pyke, G. H. 1979. Optimal foraging in fish. Pages 199–202 in R. H. Stroud and H. Clepper, editors. Predator-prey systems in fisheries management. Sport Fishing Institute, Washington, D.C., USA.

Pyke, G. H., H. R. Pulliam, and E. L. Charnov. 1977. Optimal foraging: a selective review of theory and tests. Quarterly Review of Biology 52:137–154.

Sarker, A. L. 1977. Feeding ecology of the bluegill, *Lepomis macrochirus*, in two heated reservoirs of Texas. III. Time of day and patterns of feeding. Transactions of the American Fisheries Society 106:596–601.

Savino, J. F., and R. A. Stein. 1982. Predator-prey interaction between largemouth bass and bluegills as influenced by simulated submersed vegetation. Transactions of the American Fisheries Society 111:255–266.

Smith, F. E. 1972. Spatial heterogeneity, stability and diversity in ecosystems. Transactions of the Connecticut Academy of Arts and Sciences 44:309–335.

Stein, R. A. 1977. Selective predation, optimal foraging, and the predator-prey interaction between fish and crayfish. Ecology 58:1237–1253.

Stein, R. A., and J. J. Magnuson. 1976. Behavioral response of crayfish to a fish predator. Ecology 57:751–761.

Stenseth, N. C. 1977. Evolutionary aspects of demographic cycles: the relevance of some models for microtine fluctuations. Oikos 29:525–538.

———. 1980. Spatial heterogeneity and population stability: some evolutionary consequences. Oikos 35:165–184.

Thorp, J. H., and E. A. Bergey. 1981. Field experiments on responses of a freshwater benthic macroinvertebrate community to vertebrate predators. Ecology 62:365–375.

VanDolah, R. F. 1978. Factors regulating the distribution and population dynamics of the amphipod (*Gammarus palustris*) in an intertidal salt marsh community. Ecological Monographs 48:191–217.

Ware, D. M. 1973. Risk of epibenthic prey to predation by rainbow trout (*Salmo gairdneri*). Journal of the Fisheries Research Board of Canada. 30:787–797.

———. 1975. Growth, metabolism and optimal swimming speed of a pelagic fish. Journal of the Fisheries Research Board of Canada 32:33–41.

———. 1978. Bioenergetics of pelagic fish: theoretical change in swimming speed and ration with body size. Journal of the Fisheries Research Board of Canada 35:220–228.

Werner, E. E. 1977. Species packing and niche complementarity in three sunfishes. American Naturalist 111:553–578.

Werner, E. E., and D. J. Hall. 1974. Optimal foraging and the size selection of prey by the bluegill sunfish (*Lepomis macrochirus*). Ecology 55:1042–1052.

Werner, E. E., and D. J. Hall. 1979. Foraging efficiency and habitat switching in competing sunfishes. Ecology 60:256–264.

Werner, E. E., D. J. Hall, D. R. Laughlin, D. J. Wagner, L. A. Wilsmann, and F. C. Funk. 1977. Habitat partitioning in a freshwater fish community. Journal of the Fisheries Research Board of Canada **34**:360–370.

Werner, E. E., G. G. Mittelbach, and D. J. Hall. 1981. The role of foraging profitability and experience in habitat use by the bluegill sunfish. Ecology **62**:116–125.

APPENDIX

Groups of invertebrates identified from vegetation samples and fish stomachs. Species codes are included as used in Figs. 6 and 7.

Class	Order	Family	Subfamily	Genus	Species code
Crustacea	Copepoda*				
	Ostracoda*				
	Cladocera*			*Chydorus*	
				Bosmina	
				Ceriodaphnia	
				Simocephalus	
				Daphnia	
				Alona	
	Amphipoda			*Hyalella*	Hya
Annelida	Hirudinea				
Arachnoidea	Hydracarina				
Insecta	Collembola				
	Ephemeroptera	Baetidae			Bae
		Caenidae			Cae
	Odonata	Aeschinidae		*Anax*	Ana
		Libellulidae			Lib
		Coenagrionidae			Coe
	Hemiptera	Corixidae			Cor
		Notonectidae			Not
	Trichoptera	Leptoceridae			
		Hydroptilidae			
		Polycentropidae			
	Coleoptera	Gyrinidae			Gyr
		Dytiscidae			
		Haliplidae			
	Diptera	Chaoboridae		*Chaoborus*	
		Ceratopogonidae			
		Chironomidae			
			Tanypodinae		Tnp
			Tanytarsini	*Tanytarsus*	Tnt
				Chironomus	Chi
		Miscellaneous small			Dip
Mollusca	Gastropoda	Physidae		*Physa*	Phy
		Planorbidae		*Heliosoma*	Hel
		Ancylidae			Anc
	Pelecypoda				

* Collected but not quantified in vegetation samples.

Magnification of Secondary Production by Kelp Detritus in Coastal Marine Ecosystems

D. O. DUGGINS, C. A. SIMENSTAD, J. A. ESTES

Kelps are highly productive seaweeds found along most temperate latitude coastlines, but the fate and importance of kelp production to nearshore ecosystems are largely unknown. The trophic role of kelp-derived carbon in a wide range of marine organisms was assessed by a natural experiment. Growth rates of benthic suspension feeders were greatly increased in the presence of organic detritus (particulate and dissolved) originating from large benthic seaweeds (kelps). Stable carbon isotope analysis confirmed that kelp-derived carbon is found throughout the nearshore food web.

ALTHOUGH PHYTOPLANKTON IS UNdoubtedly the primary source of organic carbon in much of the world's oceans, benthic plants are thought to be important contributors to food webs in estuarine and coral reef habitats (1). In the early 1970s, Mann and others (2–4) showed exceptionally high productivity in benthic macrophytes belonging to the order Laminariales (kelps) and inferred that kelp-derived organic carbon could play a significant role in temperate coastal (nearshore) secondary production. We assessed the significance of kelp-derived organic carbon to secondary production by a natural experiment involving islands in the Aleutian archipelago (Alaska) with and without sea otters, and thus with and without extensive kelp forests (5). We show that growth rates of benthic suspension feeders are two to five times as high at kelp-dominated islands as at those without kelp beds. Stable carbon isotope ($\delta^{13}C$) analyses show that kelp-derived carbon contributes significantly to the carbon assimilated by secondary consumers at these islands.

Kelps are a dominant feature of many exposed and semiexposed temperate coastlines, where they frequently form dense stands from the low intertidal zone to depths approaching 40 m. Individual kelps can achieve large biomass and rapid growth even at high density, thus forming one of the world's most productive habitats (3, 6). Benthic marine herbivores such as sea urchins (Echinoidea) can retard the growth of kelp populations and occasionally decimate extant populations, but most kelp biomass is not consumed directly (7). This has led to speculation that kelp biomass enters the nearshore food web through indirect (detrital) routes. By releasing particulate as well as dissolved organic matter (POM and DOM, respectively) as they grow and senesce, kelps could provide a significant organic carbon source for the diverse and abundant assemblages of nearshore suspension feeders, pelagic as well as benthic. Even the considerable quantity of kelp biomass deposited on beaches adjacent to kelp stands ultimately may reenter the nearshore food web as POM and DOM after decomposition.

The reestablishment of sea otters in the Aleutians after their near extinction in the 19th century and the subsequent resurgence of otter predation upon sea urchins allow us to compare secondary production between areas that differ greatly in kelp biomass but are otherwise similar. Oceanographic data indicate that the pervasive influence of the westward-flowing Alaskan Stream accounts for relatively uniform physical conditions among the central and western Aleutian Islands (8). Neither the few prior studies nor our surveys provide evidence for significant differences in species composition (including phytoplankton) along this segment of the archipelago (9).

The mid- and low-intertidal zones throughout the Aleutian Islands are dominated by kelps belonging to the genera

---

D. O. Duggins, Friday Harbor Laboratories, Friday Harbor, WA 98250.
C. A. Simenstad, Fisheries Research Institute, WH-10, University of Washington, Seattle, WA 98195.
J. A. Estes, U.S. Fish and Wildlife Service, Institute of Marine Sciences, University of California, Santa Cruz, CA 95064.

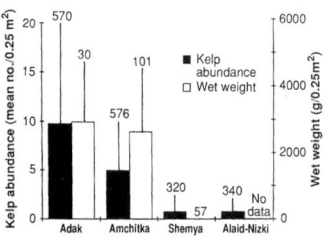

Fig. 1. Kelp abundance and wet weight (means and 1 SD of quadrat counts and sample weights) from islands with sea otters (Adak and Amchitka) and without otters (Shemya and Alaid-Nizki). Biomass at Shemya = 0. Sample size is given above each error bar.

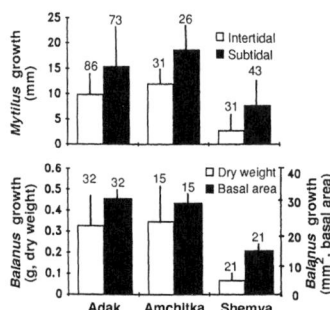

Fig. 2. Results of translocation experiments; means and 1 SD of growth in length (mussels, *Mytilus*) or dry weight and basal area (barnacles, *Balanus*). Sample size is given above each error bar.

Laminaria and *Alaria*, which exist in a refuge above the foraging range of sea urchins. Rocky subtidal habitats at islands with sea otters are characterized by low urchin biomass and large stands of both understory and surface canopy kelps; however, islands where sea otters have not become reestablished are characterized by large urchin biomasses and few kelps (5). This nearshore community variation allowed us to assess (by comparison) the importance of kelp carbon to the production of nearshore consumers. Between 1985 and 1987 we conducted benthic surveys, experiments, and $\delta^{13}C$ analyses at Adak and Amchitka Islands (sea otters abundant) and at Shemya and Alaid-Nizki Islands (no sea otters) to determine whether growth (as an indicator of production) and $\delta^{13}C$ of consumer organisms were related positively to kelp biomass.

Systematic benthic surveys verified differences in kelp abundance and biomass between islands. At each island, transects were established at 16 to 30 randomly selected sites along the 6-m contour. All sites were on the Bering Sea side of each island and were judged subjectively to be of similar wave exposure. Along each transect, 20 randomly selected 0.25-m² quadrats were censused by divers for kelp abundance. The resulting data (Fig. 1) show orders of magnitude differences in subtidal kelp abundance and biomass between islands with and without sea otters. Although abundance and biomass are not synonymous with production, research on similar kelp assemblages indicates that benthic production should be vastly different among these islands (2, 10).

If kelp-derived organic carbon is available to nearshore suspension feeders, detritivores, and (indirectly) their predators, and if phytoplankton alone is a limiting resource (11), then secondary production should be significantly different between island groups. We used two analyses to determine whether suspension-feeding organisms grow faster in kelp-dominated than in urchin-dominated environments. First, we translocated two species of suspension feeders (the mussel *Mytilus edulis* and the barnacle *Balanus glandula*) from a common source (Puget Sound, Washington) into cages at six intertidal and six subtidal sites at Adak, Amchitka, and Shemya islands. Mussels were translocated in 1985 and 1986 and barnacles in 1986. Mussels were individually tagged and measured (maximum valve length) (12); barnacles were allowed to settle on fiberglass plates, thinned to minimize competition, and measured in four dimensions (length and width of operculum and base). Barnacle locations were mapped for individual identification and the plates were placed in the same cages with the mussels. All animals were remeasured after 1 year. Mussel valve elongation, barnacle final dry weight (13), and change in barnacle basal plate area were the parameters used to compare growth among islands.

Growth rates were significantly different (14) among islands. Mussels in kelp-dominated habitats (Adak and Amchitka) grew approximately two (subtidal) to four (intertidal) times as fast as mussels in urchin-dominated habitats (Fig. 2). Likewise, barnacles (intertidal) grew up to five times as fast in kelp-dominated environments (Fig. 2) (barnacles from subtidal cages did not survive).

As a second, independent verification of the translocation data, age-size relations were analyzed for intertidal mussels collected from six sites at each of four islands (15). Mussel valves were sectioned and age determined by the methods of Lutz (16). For year classes 2 to 5, mussels were significantly larger at islands with substantial subtidal kelp forests (Fig. 3) (17).

Our results do not exclude the possibility that a carbon source other than that derived from kelps accounted for differences in mussel and barnacle growth among islands. We employed $\delta^{13}C$ analyses to determine the extent to which consumers were using kelp-derived organic carbon and if such use differed among islands as predicted (18). The Aleutian nearshore food web lends itself well to such analyses for several reasons. Unlike other habitats such as estuaries where diverse autotrophs (phytoplankton, benthic algae, and marine and terrestrial angiosperms), each with a distinctive $\delta^{13}C$ signature, contribute POM and DOM to nearshore waters, the Aleutians have only two principal sources of organic carbon: benthic algae and phytoplankton (19). Furthermore, at high latitudes, phytoplankton is more deplete in ^{13}C than at low latitudes (20) and thus the difference in signatures between kelps and phytoplankton is large, leading to less ambiguous interpretation of the origins of organic carbon in nearshore consumers. In addition, because islands differ greatly in the potential input of kelp carbon, we could incorporate the measured consumer $\delta^{13}C$ enrichment at kelp-dominated islands in a simple mixing model to assess quantitatively the magnitude of kelp carbon input to the nearshore food web.

To characterize predominant carbon sources, $\delta^{13}C$ was determined from samples of the dominant kelps (*Laminaria groenlandica*, *L. longipes*, and *Alaria fistulosa*) collected at Adak and Amchitka islands. Six suspension feeders, two detritivores, and three

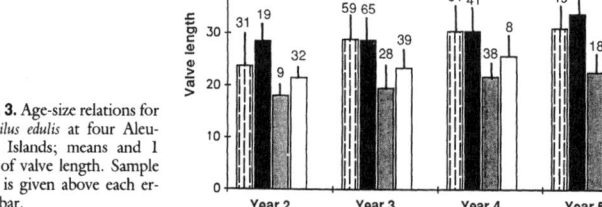

Fig. 3. Age-size relations for *Mytilus edulis* at four Aleutian Islands; means and 1 SD of valve length. Sample size is given above each error bar.

predator taxa were collected at each island as indicators of the islands' nearshore consumers (21). Phytoplankton isotopic values were derived from cultures incubated on board the R.V. *Alpha Helix*. Cultures inoculated from ambient seawater were used rather than net tow samples because of the inevitable problem of contamination by non-phytoplankton carbon in plankton net tows. These cultures were made up primarily of *Chaetoceros* and *Thalassiosira*, which were the dominant genera in net tows at all islands as well as the most common genera reported for the region (9). The mean phytoplankton isotopic value (-24.0 per mil $\pm$ 1.0 SD, $n = 5$) corresponds with published values (-22.9 to -24.4 per mil) for this region and surface water temperatures. Kelps were considerably enriched in ^{13}C relative to phytoplankton, and values were relatively consistent among taxa and islands (overall mean of -17.7 per mil $\pm$ 2.3 SD, $n = 162$).

Differences in consumer $\delta^{13}C$ between kelp-dominated islands and urchin-dominated islands support the hypothesis that carbon fixed by kelp is found throughout the nearshore food web and may even be consequential at islands with comparatively low kelp abundance. For each of 11 consumers tested, with the single exception of *Mytilus edulis* (22), mean $\delta^{13}C$ values for animals from kelp-dominated islands (Adak and Amchitka pooled) were more enriched than mean values for animals from islands without kelps (Shemya and Alaid-Nizki pooled)

(23). On the basis of a simple mixing model (24), primary consumers at kelp-dominated islands average, conservatively, 58.3% kelp-derived carbon (Fig. 4), whereas at urchin-dominated islands they average only 32.0%. This was the case despite all consumers being collected in midsummer during periods of peak phytoplankton abundance, when $\delta^{13}C$ values should be indicative of maximum phytoplankton influence. The moderate percentage of kelp-derived carbon in consumers at urchin-dominated islands is probably the result of input from intertidal kelps and kelps existing in subtidal refuges from urchin grazing. The relatively consistent enrichment of consumers at Adak and Amchitka, regardless of feeding type or trophic level, argues for the pervasive occurrence throughout the nearshore food web of organic carbon originally derived from kelp photosynthesis.

Both the transplant translocation experiments and age-size analysis show that suspension feeders grew at a significantly higher rate at islands with extensive subtidal kelp forests than at those without. The $\delta^{13}C$ data indicate that this difference in growth rate most likely results from the use of organic carbon photosynthesized by kelps, rather than from differences among islands in phytoplankton production or some other variable. Isotopically enriched signatures of organisms such as mysids, rock greenling, and pelagic cormorants at kelp-dominated islands further indicate that use of kelp-derived organic carbon is not restricted to benthic organisms.

The ecological role of kelps in the nearshore region is multifaceted. Strong evidence exists that they provide habitat (substratum and canopy) for a wide range of benthic, epibenthic, and pelagic organisms and alter the hydrodynamic environment of the nearshore region (25). Here we report a strong trophic link between kelps and a wide range of organisms of varied feeding strategies and trophic levels, extending beyond the obvious kelp-grazer-predator food chain. The common occurrence of extensive, highly productive kelp forests along most temperate to subpolar coastlines suggests that our results are not specific to the Aleutian Islands. The relative trophic contributions of kelp detritus and phytoplankton may vary with latitude, habitat, or season, particularly given the punctuated nature of phytoplankton production. The role of kelp production may actually be greatest in winter, when phytoplankton production is at a minimum, and kelp standing stock is either senescing (annual species) or being physically degraded during storms (annual and perennial species).

Our data support quantitatively the suggestions of Mann (2, 3) and others that kelps contribute significantly to coastal secondary production, perhaps ultimately establishing limits to the abundances of food-limited populations. Factors affecting kelp occurrence and production such as the recolonization or extinction of sea otters, sea urchin disease, catastrophic storms, nutrient depletion during El Niño events, or oil spills and other pollution are thus likely to have wide-ranging and often long-lasting influences on the productivity of coastal ecosystems.

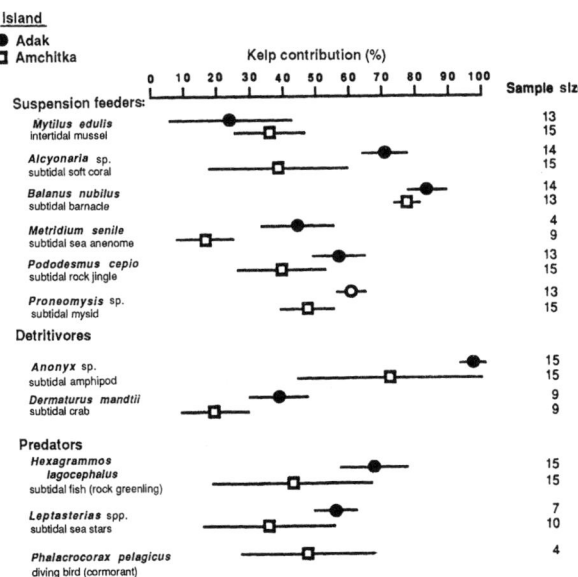

Fig. 4. Percentage of carbon photosynthesized by kelps found in tissues of consumers at islands with extensive subtidal kelp beds. Values are means and 1 SD. Three to five samples were collected and analyzed from each of three sites at each island.

REFERENCES AND NOTES

1. R. C. Newell, in *Flows of Energy and Materials in Marine Ecosystems*, M. J. R. Fasham, Ed. (Plenum, New York, 1984), pp. 317–343; C. A. Simenstad and R. C. Wissmar, *Mar. Ecol. Prog. Ser.* **22**, 141 (1985); R. Carpenter, *Ecol. Monogr.* **56**, 345 (1986).
2. K. H. Mann, *Mar. Biol.* **14**, 199 (1972).
3. ———, *Science* **182**, 975 (1973).
4. D. O. Duggins, *Ecology* **61**, 447 (1980).
5. J. A. Estes and J. F. Palmisano, *Science* **185**, 1058 (1974); J. A. Estes et al., *Ecology* **59**, 822 (1974).
6. K. H. Mann, *Ecology of Coastal Waters: A Systems Approach* (Univ. of California Press, Berkeley, CA, 1982).
7. C. Harrold and J. S. Pearse, in *Echinoderm Studies*, M. Jangoux and J. M. Lawrence, Eds. (Balkema, Rotterdam, 1987), pp. 137–233; V. A. Gerard, thesis, University of California, Santa Cruz (1976).
8. F. Favorite, *Occas. Publ. Inst. Mar. Sci. Univ. Alaska* **2** (1974), pp. 3–37; W. B. McAlister and F. Favorite, in *The Environment of Amchitka Island, Alaska*, M. L. Merritt and R. G. Fuller, Eds. (Energy Research and Development Administration, Oak Ridge, TN, 1977), pp. 331–352.
9. S. Motoda and T. Minoda, *Occas. Publ. Inst. Mar. Sci. Univ. Alaska* **2** (1974), pp. 207–241; P. A. Lebednik and J. F. Palmisano, in *The Environment of*

Amchitka Island, Alaska, M. L. Merritt and R. G. Fuller, Eds. (Energy Research and Development Administration, Oak Ridge, TN, 1977), pp. 353–393; N. J. Wilimovsky, *Proceedings 14th Alaska Science Conference* (1964), pp. 172–190.
10. D. O. Duggins, thesis, University of Washington, Seattle (1980).
11. Phytoplankton production in the Aleutians is comparatively low (38 to 243 mg of carbon per square meter per day) concentrated during a relatively short growing season and probably restricted to bays (rather than the exposed coast) with vertical water stability (9).
12. Mussels were initially of equal size and were assigned to cages randomly. We could not replicate the "no kelp" treatment of this experiment because we were unable to visit Alaid-Nizki Island until the summer of 1987.
13. All barnacles recruited to the settling plates within a span of several weeks and were of similar size. Barnacle physical dimensions were measured within 48 hours of placement in cages; at that time there was no statistically significant difference in mean basal plate area among islands ($n = 4$) for those individuals that survived to the experiment's conclusion (Kruskal-Wallis, $P > 0.1$). Final dry weight was measured on animals removed from the plates and dried at 50°C for 24 hours.
14. Kruskal-Wallis ANOVA on mean growth of individuals at a site (cage) were mussel intertidal, $\chi^2(2) = 19.3$, $P < 0.0001$; mussel subtidal, $\chi^2(2) = 17.2$, $P < 0.0002$; barnacle dry weight, $\chi^2(2) = 6.4$, $P < 0.05$; barnacle basal plate area, $\chi^2(2) = 6.2$, $P < 0.05$.
15. Mussels were collected from equivalent tidal heights at sites of equivalent wave exposure. Spatially isolated mussels were chosen to reduce the possible confounding consequences of intraspecific competition.
16. R. A. Lutz, *J. Mar. Biol. Assoc. U.K.* **56**, 723 (1976).
17. Complete random block ANOVA for age-size analysis with age (2 to 5 years) as the blocking factor; $F(1,11) = 203$, $P < 0.001$. Such a random block analysis allows us to examine island effects (kelp-dominated compared to urchin-dominated) for all four age classes simultaneously.
18. See B. Fry and E. B. Sherr [*Contrib. Mar. Sci.* **27**, 15 (1984)] for comprehensive discussion and critique of the application of $\delta^{13}C$ techniques to ecological studies. The ratio of ^{13}C to ^{12}C is fixed at the time (and according to the pathway) of photosynthesis. With minor modification (+0.5 to 1.5 per mil per trophic level), this ratio is maintained through consumer trophic levels. Thus consumer signatures reflect those of key primary producers.
19. Terrestrial input of organic matter to nearshore coastal waters was presumed to be insignificant primarily because most terrestrial vegetation is maritime tundra of grasses and lichens, which degrade in situ; there is no woody vegetation that would produce large amounts of exportable detrital matter.
20. T. McConnaughey and C. P. McRoy, *Mar. Biol.* **53**, 257 (1979); M. R. Fontugne and J.-C. Duplessy, *Oceanol. Acta* **4**, 85 (1981); G. H. Rau *et al.*, *Deep Sea Res.* **29**, 1035 (1982).
21. Five whole (above holdfast) kelp specimens per species were collected at each of three randomly selected sites at each island and subsampled systematically by taking a number of plugs uniformly along the length of the blade. Three to five specimens of each consumer taxon were collected at the same sites as the kelps and were used whole (mysid, amphipod, sea anemone) or subsampled (muscle tissue of others).
22. The single exception to the pattern of greater consumer enrichment at islands with kelps was *Mytilus edulis*, which was the only consumer we collected from the intertidal zone, where kelps are abundant at all four islands. *Mytilus* $\delta^{13}C$ values showed no pattern between kelp and no kelp islands.
23. Differences between kelp and no kelp islands were significant in a random-block ANOVA (with species as blocks, thus allowing analysis of all species simultaneously) considering all subtidal consumers pooled [$F(1,27) = 7.96$, $P < 0.0001$] or only suspension feeders [$F(1,14) = 11.64$, $P < 0.005$].
24. A simple mixing model based upon that of T. McConnaughey and C. P. McRoy [*Mar. Biol.* **53**, 263 (1979)] was possible because of the two-carbon source system. Percentage contribution from kelp is calculated as $[\delta^{13}C \text{ sample} - \delta^{13}C \text{ phytoplankton} - I]/[\delta^{13}C \text{ kelp} - \delta^{13}C \text{ phytoplankton}] \times 100$, where I represents a post-photosynthetic isotope fractionation and was empirically derived for each species by calculating the difference in $\delta^{13}C$ between the most deplete sample of that species ("pure" phytoplankton diet) and the mean phytoplankton value (-24.0). In cases where the most $\delta^{13}C$ deplete value for a species was less than our phytoplankton value, the mean of our measured enrichment values (2.5 per mil per trophic level) was used. This method makes our model conservative in favor of phytoplankton (reducing the percentage of carbon from kelp) in that our calculations indicate that even the most isotopically deplete consumers incorporate some kelp-derived carbon.
25. D. O. Duggins, in *The Community Ecology of Sea Otters,* G. R. VanBlaricom and J. A. Estes, Eds. (Springer-Verlag, Berlin, 1988), pp. 192–201.
26. Supported by NSF grant DPP-8421362. We sincerely appreciate the discussions and comments of our colleagues M. Dethier, B. Fry, T. Michaels, and G. VanBlaricom, and the assistance provided by a small army of divers plus the U.S. Coast Guard (17th District), the U.S. Navy (Adak Naval Station), the U.S. Air Force (Shemya Air Force Base), and the U.S. Fish and Wildlife Service (Alaska Maritime Refuge), during our surveys, collections, and experiments in the Aleutians. We thank P. Hassett, L. Johnson, A. Sewell, and J. Watson who were particularly helpful in the field or laboratory.

28 November 1988; accepted 16 May 1989

Recent developments

When can marine reserves improve fisheries management?

Ray Hilborn[a,*], Kevin Stokes[b], Jean-Jacques Maguire[c], Tony Smith[d], Louis W. Botsford[e], Marc Mangel[f], José Orensanz[g], Ana Parma[h], Jake Rice[i], Johann Bell[j], Kevern L. Cochrane[k], Serge Garcia[l], Stephen J. Hall[m], G.P. Kirkwood[n], Keith Sainsbury[o], Gunnar Stefansson[p], Carl Walters[q]

[a] *School of Aquatic and Fishery Sciences, P.O. Box 355020, University of Washington, Seattle, WA 98195, USA*
[b] *Seafood Industry Council, Private Bag 24-901, Manners St. PO, Wellington, New Zealand*
[c] *Halieutikos Inc., 1450 Godefroy, Sillery, Quebec, G1T 2E4, Canada*
[d] *CSIRO Division of Marine Research, GPO Box 1538, Hobart, Tasmania, 7001, Australia*
[e] *Department of Wildlife, Fish, and Conservation Biology, University of California, Davis, CA 95616, USA*
[f] *Department of Environmental Studies, University of California, Santa Cruz, CA 95064, USA*
[g] *Centro Nacional Patagonico, CONICET, 9120 Puerto Madryn, Chubut, Argentina*
[h] *Centro Nacional Patagonico, CONICET, 9120 Puerto Madryn, Chubut, Argentina*
[i] *DFO Science Advisory Secretariat, 200 Kent Street, Ottawa, Ontario, K1A 0E6, Canada*
[j] *World Fish Center, GPO Box 500, 10670 Penang, Malaysia*
[k] *Fishery Resources Division, FAO Fisheries Department, Viale delle Terme di Caracalla, 00100, Rome, Italy*
[l] *Fishery Resources Division, FAO Fisheries Department, Viale delle Terme di Caracalla, 00100, Rome, Italy*
[m] *Australian Institute of Marine Science, PMB No. 3, Townsville, Qld 4810, Australia*
[n] *Department of Environmental Science and Technology, Faculty of Life Sciences, Imperial College, London SW7 2BP, UK*
[o] *CSIRO Division of Marine Research, GPO Box 1538, Hobart Tasmania 7001, Australia*
[p] *University of Iceland/Marine Research Institute, Dunhaga 7, 101 Reykjavik, Iceland*
[q] *Fisheries Centre, University of British Columbia, Vancouver, B.C., Canada V6T 1Z4*

Abstract

Marine reserves are a promising tool for fisheries management and conservation of biodiversity, but they are not a panacea for fisheries management problems. For fisheries that

*Corresponding author.
E-mail address: rayh@u.washington.edu (R. Hilborn).

target highly mobile single species with little or no by-catch or habitat impact, marine reserves provide few benefits compared to conventional fishery management tools. For fisheries that are multi-species or on more sedentary stocks, or for which broader ecological impacts of fishing are an issue, marine reserves have some potential advantages. Their successful use requires a case-by-case understanding of the spatial structure of impacted fisheries, ecosystems and human communities. Marine reserves, together with other fishery management tools, can help achieve broad fishery and biodiversity objectives, but their use will require careful planning and evaluation. Mistakes will be made, and without planning, monitoring and evaluation, we will not learn what worked, what did not, and why. If marine reserves are implemented without case by case evaluation and appropriate monitoring programs, there is a risk of unfulfilled expectations, the creation of disincentives, and a loss of credibility of what potentially is a valuable management tool.
© 2004 Elsevier Ltd. All rights reserved.

1. Introduction

Globally, there is a wave of environmental groups, politicians and ecologists pushing for the large-scale implementation of Marine Protected Areas (MPAs),[1] with many calls for protecting 20–30% of the oceans [1]. The establishment of MPAs does not automatically require an outright banning of fishing activities in the designated area which may accommodate fishing and other economic activities under specific management regimes. However, it is often proposed to simply eliminate all consumptive uses (particularly fishing) from those areas, turning all or part of a traditional fishing ground into a no-take MPA or marine reserve.[2] Proponents argue that by eliminating all fishing from an area, marine reserves protect biodiversity, serve as an insurance policy, and benefit ecosystem and fisheries management. Initially, there was a clear distinction between establishing marine reserves for protection of biodiversity and establishing them for fisheries management. Most current calls for large scale implementation of marine reserves argue that they will provide both biodiversity and fishery benefits, whilst potential costs are seldom mentioned [1,2].

While the potential value of marine reserves for the protection of habitat and biodiversity is clear, their potential for improving fisheries management and particularly fisheries yields will be limited unless the roots of fisheries management failures are addressed. The same holds for other management tools. The major problems in fisheries management and conservation stem from improper incentives and institutional structures [3–5] that fail to control the race for fish leading to over-capacity, over-fishing and economic loss. Once overfishing becomes chronic, the socio-economic and political costs of the tough decisions needed for significant

---

[1] The term MPA is used here to mean areas that are closed to fishing, the meaning that is more widely used by the public. In the scientific literature, these areas are more commonly referred to as marine reserves (i.e., 9).

[2] The difference between marine protected area and a marine reserve is not always clearly made, generating confusion.

improvement represent a major impediment to change. Marine reserves are a tool for specifying the location of fishing; they do not affect the incentives, nor the institutional structures responsible for over-fishing [6]. Furthermore, imposition of ill-considered marine reserves may in fact be detrimental, and it is misleading to promote them as devices always likely to result in improved yields.

Area closures are just one tool of fisheries management and marine reserves implementation needs to be guided by the scientific principles of adaptive management: experimental treatments, controls and evaluation [7]. For marine reserves to be an effective fishery management tool, they need to be considered case by case in light of the objectives and the current state of the fishery. They need to be evaluated and compared to viable alternative fisheries management tools, and used, where appropriate, as one element in a broader package of measures. Planned programs are needed for testing the effectiveness of marine reserves for fisheries management. The utility of marine reserves in relation to alternative tools will likely be very different for different types of fisheries, as discussed below.

In the following sections, the knowledge available regarding the potential role of marine reserves specifically in fisheries management is reviewed.

2. Potential of marine reserves

There are several well-defined ways in which marine reserves may be expected to have merit as a fisheries management tool. These are examined briefly below.

2.1. Increases in yield

The empirical evidence that marine reserves enhance fish yields is sparse [8]. Setting aside a marine reserve initially reduces the area that can be fished, thus reducing yield. The question then is whether the yield in the area remaining open will increase enough to make up for losses from the closed area. We know that in many marine reserves, the abundance and size of fish increases [9]. This is expected. Yield from the fished open area can increase in two ways: (1) bigger fish can swim out of the closed area and be caught, and (2) the larger fish in the closed area can contribute more eggs and ultimately more larvae to the fished open area. However, neither result is guaranteed. If the fish or invertebrates species of concern are sessile they will not move into the fished open area. Conversely, if they are too mobile, virtually all will move into the fished open area, thus removing the anticipated benefit [10,11]. Also, larval dispersal patterns must be such that enough larvae are transported to the open areas [12], and (compensatory) density-dependent growth does not negate benefits within the closed areas [13]. Benefits will accrue only if recruitment to the fished area before its closure was less than the maximum possible. Thus, marine reserves can, subject to the conditions just described being met, increase yields only in fisheries in which heavy fishing mortality has substantially reduced recruitment [14–17]. This is a corollary of a formal result: management based on marine reserves and conventional management are analytically equivalent [18–20] with respect to the yield of the target species.

2.2. Buffer against uncertainty

Conventional management through catch or effort controls can fail due to stock assessment errors and inadequate institutional frameworks. To the extent that marine reserves may be effective at protecting breeding stock, they may help to buffer the impact of such failures [3,21,22]. However, persistence of populations in marine reserves, and their ability to replenish surrounding areas, depends on the reserve configuration and larval dispersal patterns, which are poorly known [23]. Thus, while MPAs have the potential to reduce uncertainty in the effects that fishing regulations will have, lack of relevant biological knowledge adds uncertainty. It should also be mentioned that other methods (e.g. seasonal closures to protect juveniles) can potentially have similar or even stronger effects than marine reserves in that respect [24].

2.3. Reduced collateral ecological impacts

Fishing has wider impacts on marine ecological systems, not just on target species [25]. Marine reserves can reduce impacts of fishing on benthic habitats, by-catch and protected species, and ecosystem structure and function. To the extent that the objectives of fisheries management have been broadened to include concern for such impacts [26,27], reserves are potentially an important tool in meeting such specified objectives.

2.4. Stocks of sedentary organisms

The term "sedentary", as used here, does not mean immobile. Sedentary organisms are those whose movements are short-range when compared to the spatial scale of the fishing process (fleet displacements) and/or pelagic larval dispersal. Marine reserves are one form of spatial management. For sedentary species, it has long been recognized that spatial management can be more easily understood, accepted and implemented than catch limits [28,29]. In the case of many fisheries targeting relatively small stocks of sedentary organisms, conventional stock assessment and catch regulation are unlikely to be affordable or effective. Instead, locally supported regulations, including spatial management such as marine reserves, have been shown to provide significant benefits in some cases [30,31]. In addition, global catch controls may be inappropriate for many sedentary invertebrates in terms of their population biology. For example, broadcast spawners require high-density concentrations in order to reproduce successfully, and these high-density concentrations are the first ones targeted by a fishery regulated by catch or effort limits. Spatial management may achieve larger reproductive outputs than global controls for comparable harvest rates.

2.5. Multispecies fisheries

When a fishery targets a multispecies complex, existing catch and net size limits may be poor management tools for some species. For example, in many fisheries the chief management tool currently used is ITQs/TACs.[3] These apply to a few species, whilst the fisheries may land dozens or even hundreds of species and discard many more.[4] Extending quota management to all species in such multispecies cases is not practicable. Even if sufficient data were available, such fisheries are rarely profitable enough to afford the assessment costs. Prohibiting landings of some protected species or sizes may simply force dumping. Setting catch limits on every species would practically close the fishery because at any time at least one species would likely need protection. Properly designed marine reserves may be a cost-effective management tool for such fisheries.

2.6. Improved knowledge

Marine reserves may provide valuable scientific reference areas to serve as controls (in the absence of take) on trends in fish production, age, size and sex structure of the stock, as well as on impacts of fishing on habitats [32,33]. Closed areas may provide the best basis for understanding the broader impacts of fishing on ecological systems. The spatial scale of the reserves would need to be appropriate to the life history of the species, but stock assessments that include data from an unfished control site would be highly informative. Such reference areas are particularly appropriate during the development of new fisheries, when sustainable exploitation rates of newly exploited species are highly uncertain, so that there is risk of over-fishing [34]. Carving out marine reserves from conventional fishing grounds, however adds on uncertainty concerning the induced behaviour of fishers and resulting fishing and societal costs [19].

3. Potential and actual problems with marine reserves

Conversely, marine reserves present problems under a number of circumstances which are reviewed briefly below.

3.1. Effects of spatial shifts in fishing effort

A consequence of closing an area to fishing is for the fishing effort to move elsewhere, which may have a number of undesirable consequences [35][5] that in most

---

[3] The total allowable catch (TAC) is the catch limit for a whole stock. The way in which that limit is allocated and managed will vary between management regimes. Individual transferable quotas (ITQs) are one was of allocating and managing TACs.

[4] Australia's south east trawl fishery, for example, catches well over 100 species, of which up to 80 are sometimes landed, but only 18 are currently managed by quotas [37].

[5] Rijnsdorp et al. [35] for example, showed that a closed area for protection of cod in the North Sea led to unintended transfer of effort to areas where skates and long lived benthic species were more vulnerable.

cases remain un-analyzed. If a reserve were large relative to the dispersal of adults and juveniles, protecting 30% of the area would lead to a 30% reduction in potential yield. Unless the quota or effort were reduced by 30% outside of the reserved area, the sedentary stock outside would be severely over-fished. If catch limits were reduced proportionally, the conservation benefits would come primarily from having reduced the overall catch, not from having closed the area to fishing. The spatial re-allocation of effort that occurs when areas are closed can have detrimental impacts on target species, non-target species and habitat in the areas that remain open. The impact of effort re-allocation must always be considered when planning the deployment of marine reserves.

3.2. Stocks of highly mobile organisms

Many of the species caught in industrialized and some artisanal fisheries are so mobile that marine reserves would have to be very large to effectively protect breeding stock. With mobile stocks, closing some areas imposes economic inefficiencies, forcing the catch to be taken at other times and places. The stock would not be protected without additional measures, but economic costs would be imposed [19].

3.3. Better options may be available

When existing fisheries systems protect the breeding stock through catch, size or area limits, it is unclear that imposing reserves will provide additional yield benefits. Where conventional fisheries management systems have not protected breeding stocks, such as New England groundfish and in many European fisheries, scientific recommendations have not been implemented. Similar problems may befall marine reserves. Marine reserves may also increase costs and overcapitalization, potentially defeating conservation purposes [19]. Many countries have attempted to impose top–down catch or size regulations on local fishermen with little success. Top–down imposition of reserves is equally unlikely to work; what is needed, as for any management measure, is bottom-up support of fishery stakeholders and communities. In addition, the possibility of using particular regulations of fishing operations in marine protected areas should also be carefully considered as an alternative to outright banning of the fishery.

3.4. Hardship to fishing communities

Fishing communities, just as many fish stocks, may have complex spatial structure and limited mobility. Marine reserves may cause extreme hardship to fishing communities, shortening fishing seasons, forcing fishers to travel much farther to unfamiliar grounds, increasing risk to the smaller vessels and to people [19]. Indeed, marine reserves that are large enough to protect some widely spread species may exclude local people from any form of fishing. The spatial structure of the fish and the human community must be considered in the analysis of marine reserves.

4. How should we proceed?

The empirical evidence of the positive effects for fisheries attributed to MPAs and marine reserves is scarce [36] and it is obvious that marine reserves have benefits (and costs) beyond fishing. However, as with terrestrial national parks, for example, they are proposed not only to prohibit fishing, mining, or dumping, but to preserve ecosystem functions and processes, and to provide opportunities for numerous other forms of human enjoyment. It can be argued that many of the short-term costs of marine reserves to fishing could be offset by other, long-term benefits to society, but this is also likely to vary from case to case. In this paper, we have intentionally considered marine reserves from a fisheries angle and agree that an integrated, multiple use perspective (as in an Integrated Coastal Areas Management framework) would be necessary to reach broader conclusions.

Marine reserves can be appropriate as a tool for the conservation of identified habitat, species and community biodiversity. However, to minimize the yield losses to fisheries, and to achieve the desired conservation benefits, reserves need to be evaluated in the context of: (1) clear biodiversity, ecosystem and fisheries objectives; (2) the social and institutional ability to maintain and enforce the closures; (3) existing fisheries management actions they could complement under certain conditions; and (4) the ability to monitor and evaluate success. Unqualified advocacy for no-take marine reserves, sometimes hidden under advocacy for MPAs in general, ignores the need for their scientific evaluation and the potential negative impacts to stocks, yields, and communities.

We need to learn how marine reserves (and MPAs in general) might be used to improve fisheries yields, and this will need careful experimental design and evaluation using the principles of adaptive management. Reserves of different sizes need to be set up in different environments with replicates and controls. Long-term evaluation needs to be in place and criteria for success need to be determined a priori. As the lack of scientific studies and inadequate sampling will be a major impediment to the successful implementation and evaluation of marine reserves, the appropriate scientific frameworks for their placement and evaluation are critical.

References

[1] Roberts CM, Hawkins JP. Fully protected marine reserves: a guide 2003. World Wildlife Fund, United States, Washington, DC, available at http://www.panda.org/resources/publications/water/mpreserves/mar_dwnld.htm

[2] NCEAS. Scientific consensus statement on marine reserves, marine protected areas. University of California. Statement submitted at the Annual Meeting of the American Association for the Advancement of the Sciences (AAAS), 17 February 2001. National Center for Ecological Analysis and Synthesis (NCEAS), 2003. Http://www.nceas.uscb.edu/Consensus

[3] Botsford LW, Castilla JC, Peterson CH. The management of fisheries and marine ecosystems. Science 1997;277(5325):509–15.

[4] Ludwig D, Hilborn R, Walters CJ. Uncertainty, resource exploitation and conservation: lessons from history. Science 1993;260(5104):17–36.

[5] Heinz Center. Fishing grounds: defining a new era for American fisheries management. John Heinz III center for science, economics and the environment. Washington, DC: Island Press; 2000. 241pp.
[6] Hannesson R, Fraser D, Garcia S, Kurien J, Makuch Z, Sissenwine M, Valdimarsson G, Williams M. Governance for a sustainable future: II fishing for the future. A Report by the World Humanity Action Trust, 2000. 67pp.
[7] Walters CJ. Adaptive management of renewable resources. New York: MacMillan Publishing; 1986.
[8] National Research Council. Marine protected areas: tools for sustaining ocean ecosystems. Washington, DC: National Academy Press; 2001. 272pp.
[9] Halpern B, Warner RR. Marine reserves have rapid and long lasting effects. Ecological Letters 2002;5:361–6.
[10] Polacheck T. Year around closed areas as a management tool. Natural Resource Modelling 1990;4(3):327–53.
[11] DeMartini EE. Modeling the potential of fishery reserves for managing Pacific coral reef fishes. Fisheries Bulletin 1993;91(3):414–27.
[12] Hastings A, Botsford LW. Comparing designs of marine reserves for fisheries and for biodiversity. Ecological Applications 2003; 13(1)(Suppl.): 65–70.
[13] Parrish R. Marine reserves for fisheries management: why not. Symposium of the CalCOFI Conference: a continuing dialogue on no-take reserves for resource management, Asilomar, CA, USA; 4 November 1998. California Cooperative Oceanic Fisheries Investigations Report 1999;40: 77–86.
[14] Quinn JF, Wing SR, Botsford LW. Harvest refugia in marine invertebrate fisheries: models and applications to the Red Sea urchin, *Strongylocentrotus franciscanus*. Annual meeting of the American Society of Zoologists and the Canadian Society of Zoologists, Vancouver, BC, Canada, 27–30 December 1992. American Zoologist 1993;33:537.
[15] Holland DS, Brazee RJ. Marine reserves for fisheries management. Marine Resources Economics 1996;11(3):157–71.
[16] Sladek-Nowlis J, Roberts CM. Fisheries benefits and optimal design of marine reserves. Fisheries Bulletin 1999;97(3):604–16.
[17] Botsford LW, Morgan LE, Lockwood DR, Wilen JE. Marine reserves and management of the northern California Red Sea urchin fishery. Symposium of the CalCOFI Conference: a continuing dialogue on no-take reserves for resource management, Asilomar, CA, USA, 4 November 1998. California Cooperative Oceanic Fisheries Investigation Report 1999;40:87–93.
[18] Mangel M. On the fraction of habitat allocated to marine reserves. Ecological Letters 2000;3(1): 15–22.
[19] Hannesson R. Marine reserves: what would they accomplish? Marine Resources Economics 1998;13:159–70.
[20] Hastings A, Botsford LW. Equivalence in yield from marine reserves and traditional fisheries management. Science 1999;284(5419):1537–8.
[21] Lauck T, Clark CW, Mangel M, Munro GR. Implementing the precautionary principle in fisheries management through marine reserves. Ecological Applications 1998;8(1):S72–8.
[22] Mangel M. Irreducible uncertainties, sustainable fisheries and marine reserves. Evolutionary Ecology Research 2000;2(4):547–57.
[23] Botsford LW, Hastings A, Gaines SD. Dependence of sustainability on the configuration of marine reserves and larval dispersal distance. Ecological Letters 2001;4:144–50.
[24] Garcia SM. Seasonal trawling ban can be very successful in heavily overfished areas: the Cyprus effect. Fishbyte 1986;4(1):7–12.
[25] Hall SJ. The effects of fishing on marine ecosystems and communities. Oxford, UK: Blackwell Science; 1999. 296pp.
[26] Sainsbury KJ, Punt AE, Smith ADM. Design of operational management strategies for achieving fishery ecosystem objectives. ICES Journal of Marine Science 2000;57:731–41.
[27] FAO. The ecosystem approach to fisheries. FAO technical guidelines for responsible fisheries. (Suppl. 2), Vol. 4. Rome: FAO; 2003. 112pp.

[28] Caddy JF. Recent developments in research and management for wild stocks of bivalves and gastropods. In: Caddy JF, editor. Marine invertebrate fisheries: their assessment and management. New York: Wiley; 1989. p. 665–700.

[29] Orensanz JM, Jamieson GS. The assessment and management of spatially structured stocks: an overview of the North Pacific symposium on invertebrate Stock assessment and management. In: Jamieson GS, Campbell A, editors. Proceedings of the North Pacific Symposium on Invertebrate Stock Assessment and Management. Canadian Special Publication on Fisheries and Aquatic Sciences 1998;125:441–59.

[30] Castilla JC, Manriquez P, Alvarado J, Rosson A, Pino C, Espoz C, Soto R, Oliva D, Defeo O. Artisanal "caletas" as units of production and co-managers of benthic invertebrates in Chile. In: Jamieson GS, Campbell A, editors. Proceedings of the North Pacific Symposium on Invertebrate Stock Assessment and Management. Canadian Special Publication on Fisheries and Aquatic Sciences 1998;125:407–13.

[31] Castilla JC. Coastal marine communities: trends and perspectives from human exclusion experiments. Trends in Ecology and Evolution 1999;14:280–3.

[32] Smith B, Botsford LW, Wing SR. Estimation of growth and mortality parameters from size frequency distributions lacking age patterns: the Red Sea urchin (*Strongylocentrotus franciscanus*) as an example. Canadian Journal of Fisheries and Aquatic Sciences 1998;55(5):1236–1247.

[33] Castilla JC, Defeo O. Latin American benthic shellfisheries: emphasis on co-management and experimental practices. Reviews in Fish Biology and Fisheries 2001;11(1):1–30.

[34] Perry RI, Walters CJ, Boutillier JA. A framework for providing scientific advice for the management of new and developing invertebrate fisheries. Reviews in Fish Biology and Fisheries 1999;9(2):125–50.

[35] Rijnsdorp AD, Piet GJ, Poos JJ. Effort allocation of the Dutch beam trawl fleet in response to a temporarily closed area in the North Sea. International Council for the Exploration of the Sea, Copenhagen, Denmark; 2001. ICES CM 2001/N:01.

[36] Willis TJ, Millar RB, Babcock RC, Tolimieri N. Burdens of evidence and the benefits of marine reserves: putting Descartes before des horse? Environmental conservation 2003;30(2):97–103.

[37] Smith ADM, Smith DC. A complex quota-managed fishery: science and management in Australia's South East fishery. Marine and Freshwater Research 2000;52:353–9.

The Flood Pulse Concept in River–Floodplain Systems

Wolfgang J. Junk

*Max Planck Institut für Limnologie, August Thienemann Strasse 2,
Postfach 165, D-2320 Plön, West Germany*

Peter B. Bayley and Richard E. Sparks

Illinois Natural History Survey, 607 E. Peabody Dr., Champaign, IL 61820, USA

Abstract

JUNK, W. J., P. B. BAYLEY, AND R. E. SPARKS. 1989. The flood pulse concept in river-floodplain systems, p. 110–127. *In* D. P. Dodge [ed.] Proceedings of the International Large River Symposium. Can. Spec. Publ. Fish. Aquat. Sci. 106.

The principal driving force responsible for the existence, productivity, and interactions of the major biota in river-floodplain systems is the flood pulse. A spectrum of geomorphological and hydrological conditions produces flood pulses, which range from unpredictable to predictable and from short to long duration. Short and generally unpredictable pulses occur in low-order streams or heavily modified systems with floodplains that have been leveed and drained by man. Because low-order stream pulses are brief and unpredictable, organisms have limited adaptations for directly utilizing the aquatic/terrestrial transition zone (ATTZ), although aquatic organisms benefit indirectly from transport of resources into the lotic environment. Conversely, a predictable pulse of long duration engenders organismic adaptations and strategies that efficiently utilize attributes of the ATTZ. This pulse is coupled with a dynamic edge effect, which extends a "moving littoral" throughout the ATTZ. The moving littoral prevents prolonged stagnation and allows rapid recycling of organic matter and nutrients, thereby resulting in high productivity. Primary production associated with the ATTZ is much higher than that of permanent water bodies in unmodified systems. Fish yields and production are strongly related to the extent of accessible floodplain, whereas the main river is used as a migration route by most of the fishes.

In temperate regions, light and/or temperature variations may modify the effects of the pulse, and anthropogenic influences on the flood pulse or floodplain frequently limit production. A local floodplain, however, can develop by sedimentation in a river stretch modified by a low head dam. Borders of slowly flowing rivers turn into floodplain habitats, becoming separated from the main channel by levees.

The flood pulse is a "batch" process and is distinct from concepts that emphasize the continuous processes in flowing water environments, such as the river continuum concept. Floodplains are distinct because they do not depend on upstream processing inefficiencies of organic matter, although their nutrient pool is influenced by periodic lateral exchange of water and sediments with the main channel. The pulse concept is distinct because the position of a floodplain within the river network is not a primary determinant of the processes that occur. The pulse concept requires an approach other than the traditional limnological paradigms used in lotic or lentic systems.

Résumé

JUNK, W. J., P. B. BAYLEY, AND R. E. SPARKS. 1989. The flood pulse concept in river-floodplain systems, p. 110–127. *In* D. P. Dodge [ed.] Proceedings of the International Large River Symposium. Can. Spec. Publ. Fish. Aquat. Sci. 106.

Les inondations occasionnées par la crue des eaux dans les systèmes cours d'eau-plaines inondables constituent le principal facteur qui détermine la nature et la productivité du biote dominant de même que les interactions existant entre les organismes biotiques et entre ceux-ci et leur environnement. Ces crues passagères, dont la durée et la prévisibilité sont variables, sont produites par un ensemble de facteurs géomorphologiques et hydrologiques. Les crues de courte durée, généralement imprévisibles, surviennent dans les réseaux hydrographiques peu ramifiées ou dans les réseaux qui ont connu des transformations importantes suite à l'endiguement et au drainage des plaines inondables par l'homme. Comme les crues survenant dans les réseaux hydrographiques d'ordre inférieur sont brèves et imprévisibles, les adaptations des organismes vivants sont limitées en ce qui a trait à l'exploitation des ressources de la zone de transition existant entre le milieu aquatique et le milieu terrestre (ATTZ), bien que les organismes aquatiques profitent indirectement des éléments transportés dans le milieu lotique. Inversement, une crue prévisible de longue durée favorise le développement d'adaptations et de stratégies qui permettent aux organismes d'exploiter efficacement l'ATTZ. Une telle crue s'accompagne d'un effet de bordure dynamique qui fait en sorte que l'ATTZ devient un « littoral mobile ». Dans ces circonstances, il n'y a pas de stagnation prolongée et le recyclage de la matière organique et des substances nutritives se fait rapidement, ce qui donne lieu à une productivité élevée. La production primaire dans l'ATTZ est beaucoup plus élevée que celle des masses d'eau permanentes dans les réseaux hydrographiques non modifiés. Le rendement et la production de poissons sont étroitement reliés à l'étendue de la plaine inondable, tandis que le cours normal de la rivière est utilisé comme voie de migration par la plupart des poissons.

production de poissons sont étroitement reliés à l'étendue de la plaine inondable, tandis que le cours normal de la rivière est utilisé comme voie de migration par la plupart des poissons.

Dans les régions tempérées, les variations de l'ensoleillement et/ou de la température peuvent modifier les effets de la crue, et l'action de l'homme sur la crue des eaux et sur les plaines inondables limite souvent la production. Une plaine inondable peut cependant se former localement par sédimentation dans un tronçon de cours d'eau modifié par un barrage de basse chute. Aussi, les rives des cours d'eau à faible débit se transforment en plaines inondables suite à la formation de levées alluviales qui les séparent du canal principal.

Les crues sont des phénomènes qui se manifestent par à-coups. Cette situation est différente de celles prises en compte par les concepts qui mettent l'accent sur les processus continus intervenant dans les eaux courantes, tel que le concept du continuum appliqué aux cours d'eau. Les plaines inondables constituent un cas particulier car elles ne sont pas tributaires de la transformation inefficace de la matière organique en amont, même si leur réserve d'éléments nutritifs dépend en partie des échanges latéraux périodiques d'eau et de sédiments avec le canal principal. La crue est un phénomène particulier par rapport aux conditions normales parce que la position d'une plaine inondable dans le réseau fluvial n'est pas un facteur qui détermine de façon fondamentale les processus observés dans ce type de milieu. Les questions soulevées par le phénomène des crues ne peuvent pas être résolues à l'aide des concepts traditionnels de la limnologie utilisés pour étudier les systèmes lotiques et lénitiques.

Hydrologists think of rivers as links in the hydrological cycle, which transport runoff water from the continents to the sea or to the center of endorheic basins (Curry 1972). Since water is a good solvent and flowing water provides kinetic energy, water transport by rivers is linked with the transport of dissolved and solid substances. However, precipitation and river discharge typically vary significantly during the annual cycle. At low discharge rates, rivers flow in well-defined channels, but at high water in natural systems wide floodplains are recurrently inundated.

River-floodplain systems provide important habitats for biota, and ecologists have tried to link the biota of river systems with local environmental conditions and to adopt existing paradigms from other aquatic systems. These attempts have met with two problems: (1) the division of ecology into terrestrial ecology and limnology; and (2) the classification of water bodies into more or less closed, lentic systems with accumulating characteristics (lakes, ponds) as outlined in traditional limnology texts (Ruttner 1952) and open, lotic systems with discharging characteristics (streams, rivers) (Hynes 1970). The transient nature of aquatic habitats in floodplains resulted in biased treatment or in their omission. When studying rivers, most limnologists restricted themselves to river channels; when studying floodplains, they concentrated on floodplain lakes, often treating them as classical lakes.

One recent theoretical construct in river ecology, the river continuum concept (RCC) (Vannote et al. 1980), is based on the hypothesis that a continuous gradient of physical conditions exists from headwater to mouth. Analogous to the energy equilibrium theory of fluvial geomorphologists, the RCC states that structural and functional characteristics of stream communities are adapted to conform to the most probable position or mean state of the physical system. Producer and consumer communities establish themselves in harmony with the dynamic physical conditions of a given river reach, and downstream communities are fashioned to capitalize on the inefficiencies of upstream processing. Both upstream inefficiency (leakage) and downstream adjustment seem predictable. Therefore the RCC purports to provide a framework that permits us to integrate predictable and observable biological features of lotic systems (Vannote et al. 1980).

In our view, the RCC suffers from two basic limitations: (1) it was developed on small temperate streams but has been extrapolated to rivers in general; and (2) it was based on a concept that had been elaborated for the river basin in a geomorphological sense but was in fact restricted to habitats that are permanent and lotic.

Most papers that discuss the RCC recognize these limitations (Winterbourn et al. 1981; Barmuta and Lake 1982; Minshall et al. 1983; Minshall et al. 1985; Statzner and Higler 1985; Sedell et al. 1989) but fail to consider the biological significance of processes within the seasonal, aquatic habitats of floodplains. It may prove acceptable to modify the RCC to account for brief and unpredictable floods in low-order streams, even for catastrophic floods which change the physical environment and "reset" systems (Cummins 1977; Fisher 1983). However, as the size of a floodplain increases, usually along with increasing river discharge, the frequency of floods decreases, and their duration and predictability increase. These changes result in a distinct geomorphological and hydrological system with an increasing ratio of periodically lentic to lotic areas. This system results in adaptations of biota that are distinct from those in systems dominated by stable lotic or lentic habitats.

Recently, the importance of river-floodplains to fish populations in temperate, subtropical, and tropical regions has been shown by Lambou (1959), Holčík and Bastl (1976, 1977), Bryan and Sabins (1979), Welcomme (1979, 1985, 1989), Bayley (1980, 1981a, 1983), Junk (1980, 1984), and Littlejohn et al. (1985). These studies have signaled a renewed appreciation of pioneer work by Antipa (1911, 1928) and Richardson (1921). The status of the forest in subtropical river-floodplain systems has been summarized by Gosselink et al. (1981) and Wharton et al. (1981). The biases and inadequacies of limnological paradigms when applied to floodplain systems were recently discussed by Bayley (1980, 1983), Junk (1980, 1984), and Junk and Welcomme (1989) based on their experience in tropical systems. Amoros et al. (1986) and Bravard et al. (1986), who analysed the impact of flood regulation on plant and animal communities of the Rhône R. floodplain, stressed the importance of lateral and vertical dimensions of the river-floodplain system. Davies and Walker (1985) emphasized that considerable modification of the RCC was required before it could be applied to large river systems.

In this paper we synthesize evidence that suggests a complementary concept, the "flood pulse", that attempts to explain the relationship between the biota and the environment of an unmodified, large river–floodplain system. This concept is based on our experiences in relatively pristine systems in the neotropics and Southeast Asia and in the Upper Mississippi R. We derive this concept from the known ecology of typical biota that have adapted to the geomorphology and hydrology of large river-floodplain systems.

The Flood Pulse Concept

We propose that the pulsing of the river discharge, the flood pulse, is the major force controlling biota in river-floodplains. Lateral exchange between floodplain and river channel, and nutrient recycling within the floodplain have more direct impact on biota than the nutrient spiralling discussed in the RCC (Vannote et al. 1980). We postulate that in unaltered large river systems with floodplains in the temperate, subtropical, or tropical belt, the overwhelming bulk of the riverine animal biomass derives directly or indirectly from production within the floodplains and not from downstream transport of organic matter produced elsewhere in the basin.

The effect of the flood pulse on biota is principally hydrological. We postulate that if no organic material except living animals were exchanged between floodplain and channel, no qualitative and, at most, limited quantitative changes would occur in the floodplain (Bayley 1989). The relative importance of imported versus recycled inorganic nutrients in floodplains is not clear and probably varies between systems. Given similar hydrological conditions, the longitudinal position of a floodplain in the drainage network is of little importance with respect to the biota.

The Highway Analogy

Faunal life histories in unaltered large river–floodplains can be viewed as analogous to vehicles on a highway network. Were non-terrestrials to investigate this network, they would observe numerous bodies traveling in opposite directions and might well surmise that resources for those bodies were derived from the highways. If funds permitted a detailed study, it would reveal that four-wheeled creatures need to leave highways periodically for sustenance, along with their apparently symbiotic occupants. Eventually, major sources of production would be identified in farms, oil fields, and mines, vehicles consuming and distributing resources via the highway network as a response to production cycles and long-term economic changes.

The life histories of major plant and animal groups, in particular fish, in large river-floodplains are beginning to be understood sufficiently to contribute to the theory that the river network in a river-floodplain system is in many ways analogous to a highway network with the vehicles corresponding to the fish. Detritivores, herbivores, and/or omnivores support large fisheries in the main channel (Petrere 1978, 1982; Welcomme 1979; Quirós and Baigún 1985), but the highest yields are associated with adjoining floodplains (Richardson 1921; Lowe-McConnell 1964; Petrere 1983) and most of their production is derived from floodplain habitats (Welcomme 1979; Bayley 1983). The main channel is used principally as a route for gaining access to adult feeding areas, nurseries, spawning grounds, or as a refuge at low water or during winter in temperate zones. An analogous situation is found in large north-temperate and arctic rivers where most of the ichthyomass is anadromous; here the main feeding grounds are found in the delta area or in the sea (Grainger 1953; Andrews and Lear 1956; Foerster 1968; Roy 1989).

We will describe the functions of the floodplain and main channel in large river–floodplain systems with respect to the biota and evaluate the links between them and the nonfloodable watershed in the light of recent data.

Definition of a Floodplain

Terms applied to classical limnological and terrestrial systems can be inappropriate for explaining concepts in river-floodplains. This is not merely a semantic discussion because the classical terms are understood to define features and functions in their respective systems.

The "active floodplain" of a river is defined by North American hydrologists as the area flooded by a 100-year flood (Bhowmik and Stall 1979). This period is arbitrary, longer than most existing records, and has little ecological meaning. Bayley (1981b) noted that huge areas of shallow, very acidic, largely deoxygenated swamp occur in the Peruvian Amazon. These areas are distant from the main channels and inhospitable to the bulk of aquatic animals. He proposed an active floodplain that excluded these peripheral swamps in order to compare fish production and fishery yields among systems.

We define floodplains as "areas that are periodically inundated by the lateral overflow of rivers or lakes, and/or by direct precipitation or groundwater; the resulting physicochemical environment causes the biota to respond by morphological, anatomical, physiological, phenological, and/or ethological adaptations, and produce characteristic community structures". This ecological definition recognizes that flooding causes a perceptible impact on biota and that biota display a defined reaction to flooding. Furthermore, it implies that the impact of water level pulsing on biota is independent of the nature of its source and that there are many ecological similarities between floodplains adjacent to, for example, pulsing lakes or reservoirs and pulsing rivers. The definition encompasses a wide hydrological spectrum from short- to long-duration floods and from unpredictable to predictable timing. Our examples from large river systems exhibit predictable flood pulses of long duration.

We have termed the floodplain area the "aquatic/terrestrial transition zone" (ATTZ) because it alternates between aquatic and terrestrial environments. We use this term to stress our more specific definition of floodplain, because 'floodplain' has often been defined to include permanent lentic and lotic habitats. The inshore edge of the aquatic environment that traverses the floodplain (ATTZ) we have termed the "moving littoral". The floodplain or ATTZ has unique properties that have been considered to comprise a specific ecosystem (Junk 1980; Odum 1981).

Hydrologists consider the river and its floodplain as one unit since they are inseparable with respect to the water, sediment, and organic budgets. We term this unit the "river-floodplain system". Therefore, this system comprises permanent lotic habitats (main channels), permanent lentic habitats, and the floodplain (ATTZ). Many limnologists have difficulty defining floodplains viz a viz other aquatic systems, and they have defined artificial, stable borders between land and water. Conversely, floodplains are ecosystems with water boundaries that recurrently traverse large areas. The environmental change from the aquatic to the terrestrial phase at a specific point in a floodplain (ATTZ) may be as severe as the change from a lake to a desert. Classical limnological terms describing morphological features of lakes or rivers (e.g., shoreline, littoral, profundal, size, depth) are unsuitable and must be redefined or qualified, because they have become time-dependent in the floodplain. This time dependency is important because it affects the productive processes and the life cycles of plants and animals. Pieczyńska's (1972) definition of eulittoral appears to have functional parallels with our definition of a floodplain; however, the eulittoral occupied a very small part ($\pm 5\%$) of the nonfloodplain lakes in her study and responded to a pulse amplitude of only about 40 cm. Also, we are cautious about drawing close parallels with the intertidal zone because the time scale of the tidal pulse is so much shorter, and is brief compared with the generation times of the higher biota.

Distinctions between aquatic and terrestrial organisms and processes have proved useful in studies of rivers and lakes with well-defined borders. The ecologist's view of floodplains, however, may vary according to the group of organisms being studied. Many of the organisms colonizing floodplains have developed adaptations that enable them to survive during an adverse period of drought or flood and even to benefit from it; thus neither a purely aquatic nor a wholly terrestrial view is appropriate.

Fisheries biologists tend to consider main channels and their floodplains as a single unit, because both are essential for the survival of fish stocks (Holčík and Bastl 1976; Welcomme 1979; Bayley 1980, 1981a, 1983). Conversely, studies of floodplains linked to African rivers or reservoirs show that they are also important for terrestrial game animals in adjacent nonflooded savannas, because the floodplains determine survival rates during the dry period (Sheppe and Osborne 1971; Davies 1985).

Were we to follow the arguments of hydrologists, all plant and animal material produced in a river–floodplain system would be autochthonous because it derives from riverine sediments and dissolved nutrients. Allochthonous would refer to the material introduced from outside the river–floodplain system. In limnological literature, however, the term autochthonous is applied to biota produced in the aquatic environment, and all terrestrial material is thereby classified as allochthonous. Oscillation between aquatic and terrestrial phases in floodplains makes the limnological differentiation of organic material according to its origin misleading. Similarly, the riparian zone, as understood in temperate areas, is difficult to define in a river–floodplain system. Consequently, we avoid unqualified references to these terms.

We have defined floodplain (ATTZ), river–floodplain system, and moving littoral, and explained why traditional limnological and hydrological paradigms are not appropriate from an ecologist's view. We now use examples to describe the effects of the flood pulse on biotic and abiotic components of the river–floodplain system.

Hydrology

The hydrological regime of rivers reflects the climate of its upstream catchment area. Low order streams have an irregular flood pattern with numerous peaks because they are strongly influenced by local precipitation. This influence generally diminishes with increasing size of the watershed and is almost imperceptible in the hydrograph of very large rivers.

The hydrological buffering capacity of a large catchment area results in a rather smooth and predictable flood curve. In mainly tropical or subtropical systems with large watersheds, the hydrograph reflects seasonality in precipitation, and typically shows only one pronounced flood peak per year. A few tropical rivers, e.g., the Zaire R., show two flood peaks due to two rainy seasons in their catchment areas. In temperate and cold climates, the impact of precipitation on the hydrograph is modified by the temperature regime. For example, minor flooding occurred in autumn in the Upper Mississippi R. prior to dam construction (Grubaugh and Anderson 1989a) because evapotranspiration rates decrease as temperature drops. Also, water accumulates as snow and ice in winter, which then contribute to the spring flood by melting.

Due to the size of large river basins, the effects of seasonal climatic changes may be felt downstream only after several weeks or even months. This time lag can be of ecological importance in downstream parts of large river systems. In the central Amazon the river is still rising at Manaus after the termination of the major rains; the flood peak follows the rainy season by 4-6 weeks. On the lower Mississippi R., cold water from melting snow in the head waters passes when the temperature in the backwaters of the floodplain is already much higher (Bryan et al. 1976; Holland et al. 1983).

The shape of the hydrograph depends not only on the discharge characteristics of the river, but also on valley slope, floodplain size, and vegetation. Although the Illinois R. has a mean discharge of only 627 $m^3 \cdot s^{-1}$ (Fitzgerald et al. 1986), it has protacted floods characteristics of a much larger river because it occupies a wide river valley carved by the ancestral Mississippi and Teays rivers. Because the valley has filled with alluvium, its gradient is very flat and the river drops only 1.6 $cm \cdot km^{-1}$.

At a given rate of discharge increase, the water level rises more slowly as the floodplain begins to fill. In larger floodplains the rate of rise is slower, the period of inundation increases, and more lentic habitats develop. As the water recedes, processes in the floodplain become less dependent on the river channel and more subject to local climatic events. During the terrestrial phase, the amount and distribution of local rains greatly affects the composition and productivity of plant communities as well as the life cycles of many animals. When local precipitation at low water is high, floodplains are forested, e.g., in the middle and upper Amazon, Zaire, and Mississippi rivers. Conversely, when local precipitation is low, savannas with gallery forest develop, e.g., in the floodplains of the lower Nile, Zambezi, and Volta rivers. Some lakes and swamps are isolated from the main channel for many months or even years. Their hydrological regimes are therefore independent of the main channel except during periods of high water.

Nutrients

According to hydrologists, a river's chemistry reflects its catchment area. This holistic view has been applied successfully to streams with respect to their nutrient budgets (Hynes 1975; Vannote et al. 1980). Nutrients can roughly be divided into inorganic and organic fractions; these in turn can be subdivided into gaseous compounds, dissolved solids, and particulate matter. The floodplain receives all classes of nutrients directly from the main channel, and its basic nutrient status would be expected to correspond to that of the river. Floodplains, however, tend to establish their own cycles since organisms and environmental conditions that influence the biogeochemical cycles differ considerably from those in the main channel. The effects of rain, runoff, groundwater, and input from floodplain tributaries may also be important.

The Inorganic Fraction

Gaseous Compounds

Gases such as CO_2, O_2, H_2S, CH_4, and N_2 are produced and/or consumed in the floodplain independently of processes in the main channel in systems with slow, regular flood pulses. Residence time of floodplain water and temperature modify concentrations. The lack of persistent thermal and chemical stratification in most Atchafalaya floodplain lakes is due to the short period of lentic conditions during warm weather (Bryan et al. 1974). In contrast, the water column becomes chemically stratified over large areas soon after entering the Amazon floodplain; the daily thermocline with a temperature difference of 1–3 °C is sufficient to inhibit circulation deeper than 2–6 m during periods of several weeks or even months. Large amounts of organic material under decomposition at high temperatures result in high rates of oxygen consumption and CO_2 release near the bottom. Hypoxic, or even anoxic conditions accompanied by H_2S and CH_4 production, are often found at a few metres depth (Schmidt 1973a; Melack and Fisher 1983; Junk et al. 1983).

In addition to nitrogen input from the river, high nitrogen fluxes to and from the atmosphere occur. These fluxes are related to oxygen levels and to organisms in water and soils, both of which change drastically between flood and dry periods. Denitrification in wetlands is well documented (Kemp and Day 1984) and has even been used in the treatment of wastewater (Dierberg and Breszonic 1984). Various nitrogen-fixing organisms, e.g., cyanophytes and bacteria, that are often associated with higher plants such as Leguminosae counteract denitrification by fixing atmospheric nitrogen (Heller 1969; Richey et al. 1985). Despite the high potential for denitrification, Brinson et al. (1980) consider tupelo-cypress swamps to be nitrogen sinks due to high nitrogen levels in the litter.

Dissolved Solids

River water is the major source for dissolved inorganic compounds, including plant nutrients. Abiotic and biotic processes in the floodplain, however, may considerably alter the total amount and ionic composition of dissolved materials. Increased evaporation may raise salinity in backwaters above the levels found in the river, in particular in arid climatic zones. Biogenic modifications are reported from Amazonian floodplain lakes where ten to twentyfold increases in total salinity have been measured in small pools at low water (Furch et al. 1983). A major change in ionic composition, such as an increase in potassium, has been principally associated with leaching of decomposing aquatic and terrestrial macrophytes (Furch 1984a, 1984b; Furch et al. 1983).

Further changes in ionic composition result from dilution by local rains or by mixing with lateral inflows of surface and ground water from nonflooded areas. During low river stages in the Amazon, water seeping through floodplain sediments has an electric conductance up to 200 times that of the Amazon R. water, with high levels of iron and manganese (Irion and Junk, unpublished data).

Levels of dissolved nutrients are seldom limiting factors for primary production in the main channels of large rivers. In the floodplain, however, phosphorous and/or nitrogen often limit productivity, and inflowing river water replenishes the nutrient levels, as shown for phytoplankton production in Amazonian floodplain lakes (Fisher 1979). In lake and swamp habitats receiving minimal influence from the Atchafalaya R., heterotrophic phytoplankters (flagellated euglenophytes and pyrrophytes) predominated during low water levels in association with minimal inorganic nutrients (Bryan et al. 1976; Seger and Bryan 1981).

Little is known concerning the amount of dissolved inorganic compounds released from the floodplain into the main channel, and findings are contradictory for phosphorous (Yarbro 1983) and nitrogen (Brinson et al. 1983). Release and storage may be related to the flood cycle and to vegetation cover, and in temperate regions, to the growth cycle of the vegetation (Klopatek 1978; Brinson et al. 1980). Because large floodplains represent a mosaic of habitats with different physical and chemical conditions supporting diverse biotic communities, they may act either as a sink, or as a source with respect to each nutrient, depending on the circumstances.

Particulate Matter

Particulate inorganic matter in suspension is normally considered an unimportant source of plant nutrients in the river channel. Conversely, such particles hinder growth of phytoplankton and submersed aquatic macrophytes due to shading. In floodplains, however, they become a basic part of the nutrient pool available to primary producers in the dry phase and during part of the wet phase. Fertility of floodplains depends largely upon the quality of deposited sediments. Irion (1983) states that transport and deposition of sandy and kaolinitic material produce an infertile floodplain (e.g., Rio Negro in Brazil), whereas the montmorillonite and illite of the Amazon and Mississippi rivers result in high floodplain fertility. However, an impoverishment of some mobile elements (Fe, Mn, Zn) was detected in the upper 10 m-layer of Amazon sediments, which are only a few hundred years old (Irion et al., unpublished data). Conversely, weathering of the sediments, which is accelerated in tropical climates, adds dissolved inorganic materials.

The Organic Fraction

According to the RCC, aquatic animal communities of low-order streams depend mainly upon material from the nonflooded watershed. Medium-order streams have an increased instream production. Fauna of high-order rivers lacking floodplains depend mainly on organic material from upstream areas because primary production in the main channel is very low (Vannote et al. 1980).

Practically all litter must be processed by microorganisms if it is to become attractive to higher consumers. A considerable portion continues to be practically indigestible, such as fine particulate organic material in the Amazon main channel (Hedges et al. 1986). Ertel et al. (1986) reported that humic materials comprised 60% of the dissolved organic carbon of the Amazon main channel; this carbon in turn made up about 50% of the total organic carbon. The comparatively low BOD of the water from the main channel of the Amazon itself contrasts sharply with values in its floodplain (Junk, unpublished data).

Part of the organic carbon transported in the main channel passes on to the floodplain. This amount, however, is negligible in comparison with in situ production of organic material in the floodplains of rivers (Bayley 1989). Estimates of the productivity of the Amazon floodplain show that annual primary production is of the same order of magnitude as the total amount of carbon transported by the river to the Atlantic Ocean (Richey et al. 1980; Junk 1985a).

The direct impact of floodplains on the carbon budget of main channels is not well known. Some evidence suggests that floodplains can be a source for particulate and dissolved carbon (Chowdhury et al. 1982; Martins 1982; Junk 1985a; Furch and Junk 1985; Grubaugh and Anderson 1989b). Conversely, retention mechanisms, such as settling of particulates, uptake by organisms, and retention of most macrophytes by stranding or trapping during falling water (Junk 1980) contribute to the recycling of most carbon in the floodplain and strongly reduce leakage to the river channel. Carbon export from floodplains also depends on hydroperiod, flushing rate, and in temperate regions, on the growth cycle of floodplain vegetation. Data from floodplains are limited, but Odum and de la Cruz (1967) estimated that the rate of export of organic material from a Georgia tidal marsh was directly proportional to volumetric flow rates.

Gosslink et al. (1981) assumed that flooding during winter and spring provides more detritus to main channels than during summer in temperate regions. In the tropics, consistently high temperatures favor high production and rapid processing of organic material throughout the year.

Biota in the River Channel

The channel is well defined in large, pristine rivers, and is delineated from the floodplains by natural levées and/or a marked increase in water velocity. In rivers modified by navigation dams, such as the Mississippi, broad, slow-flowing main channel borders are found on either side of the narrow main channel, which is defined by the thalweg (Fremling et al. 1989). These borders, which constitute a developing floodplain, are discussed separately below; however, the main channels of modified rivers have much in common with those in more pristine systems.

Plants

Great water depth, high suspensoid load, considerable turbulence, and strong current make the main channel unfavorable for primary production. Aquatic macrophytes and periphyton normally colonize shores and, in some transparent tropical rivers, rocky substrates (Podostemaceae). In slow-flowing tropical and subtropical rivers floating macrophytes may become important. Phytopotamoplankton density increases with stream order, transparency, and decreasing current velocity, but absolute values are low (e.g., Berner 1951). In most large rivers, physical factors, in particular light, rather than mineral nutrients limit primary production (Fisher 1979). Average primary production per unit area in the main stems of large turbid river systems such as the Amazon, Mekong, Ganges, and Mississippi can be only a small fraction of that in their floodplains.

The extent to which floodplain water bodies contribute to populations of potamoplankton and floating macrophytes in large rivers is unknown. The considerable increase of potamoplankton downstream of reservoirs, e.g., in the Nile (Brook and Rzóska 1954; Talling and Rzóska 1967; Hammerton 1976) and the increase of floating macrophytes in the Amazon main channel at rising and high water (Junk 1970) are due to high production of these plants in associated lentic habitats.

Invertebrates

Little information is available about colonization by animals of the bottoms of large rivers. The bed loads of large rivers in alluvial plains, e.g., the Mississippi, are sandy (Schumm 1977). Large river channels mostly consist of a monotonous sequence of slowly moving sand dunes unsuitable for benthic organisms. The Amazon R., for example, transports its bed load of coarse sand as dunes 6-8 m high (Sioli 1984).

High suspensoid loads hinder benthic and epizoic animals (Hynes 1970). Junk (1973) found a decrease in number and biomass of principally filter-feeding perizoon in floating macrophyte vegetation as amounts of inorganic suspensoids increased.

Although some invertebrates can live in the dominant sandy substrates of main channels (e.g., the chironomids *Gillotia*, *Cyphonella*, *Robackia*, and *Saetheria* [Coffman and Ferrington 1984]), densities are low. Berner (1951) and Morris et al. (1968) indicated average fresh invertebrate biomasses in the main channel of only 0.001 $g \cdot m^{-2}$ and 0.007-0.048 $g \cdot m^{-2}$, respectively, for the Missouri R., and attributed these low values to shifting substrates, siltation, fluctuating water levels, swift current, and absence of aquatic vegetation. In the Atchafalaya distributary, which receives 80% of the Mississippi R. discharge, Bryan et al. (1976) reported a mean quantity of 327 benthic individuals per m^2 in riverine habitats compared with densities up to ten times greater in floodplain habitats.

Logs and rocks provide stable substrates for organisms in a channel environment that is otherwise dominated by shifting alluvium. Over 10^6 logs were pulled from channels in the lower 1600 km of the Mississippi during a 5-year period (Harmon et al. 1986). The average fresh animal biomass colonizing logs in the Kaskaskia R., Illinois, varied between 0.57 and 1.65 $g \cdot m^{-2}$ (Nilsen and Larimore 1973). Nord and Schmulbach (1973) reported a range of 0.2–3.2 $g \cdot m^{-2}$ dry weight in the Missouri R. Assuming an average surface area per log of 5 m^2, a dry biomass density of 2 $g \cdot m^{-2}$ of log, and an average width of the lower Mississippi channel of 900 m, the overall biomass density of this fauna would be only 0.007 $g \cdot m^{-2}$.

Vertebrates

Vertebrates, particularly fish, are important consumers in the main channel. In subtropical and tropical rivers, freshwater dolphins, capybaras, manatee, hippos, turtles, and crocodiles may contribute considerably to the main channel biomass. White whales and seals occur in arctic rivers; beavers, muskrats, and otters in temperate rivers; and waterfowl and shorebirds in both. However, few higher animals have adapted to utilize main channel habitats exclusively. Those that do tend to be predators whose prey depends largely on production in floodplain habitats, such as large, piscivorous catfishes (Goulding 1981), to some extent river dolphins (Ferreira da Silva 1983), and fish that consume aquatic invertebrates (Lundberg et al. 1987). In the main channels of the Mississippi and Missouri rivers, pallid sturgeon (*Scaphirynchus albus*), blue sucker (*Cycleptus elongatus*), blue catfish (*Ictalurus furcatus*), and several chubs (*Hybopsis* spp.) feed largely on invertebrates, and, with respect to large pallid sturgeons and blue catfish, on fish (Pflieger and Grace 1987).

Most vertebrates use the main channel temporarily as migration routes, for spawning, as refuge during droughts or freeze-up, or for hibernation. Tropical rivers are famous for large-scale migrations of fish for dispersal and/or spawning in the main channel or floodplain, that result in large biomass densities in the main channel during falling or low-water periods (Godoy 1967; Bonetto et al. 1969a; Bayley 1973; Ribeiro 1983). Large channel catfish (*Ictalurus punctatus*), flathead catfish (*Pylodictis olivaris*), and freshwater drum (*Aplodinotus grunniens*) use drop-offs, scour holes or obstructions in or along the main channel of the Upper Mississippi R. for a winter refuge (Hawkinson and Grunwald 1979).

Except for limited amounts of potamoplankton, benthos, and predators, the biota of the main channel concentrate close to the river shoreline, to islands, or in the main channel border areas described below, areas where habitat diversity increases and food supply improves (edge effect). Therefore the "bank coefficient" (Sedell et al. 1989) is an index of the productivity potential of a river channel in the absence of a floodplain. Conversely, when a regularly inundated floodplain is present, most of the vertebrates found in the main channel depend to a great extent directly or indirectly on primary production in the laterally linked floodplain habitats.

Biota in the Floodplains

Flood Pulsing and Life Cycles

Life cycles of biota utilizing floodplain habitats are related to the flood pulse in terms of its annual timing, duration, and the rate of rise and fall. Timing is important in temperate rivers where seasonal temperature and light cycles also regulate productivity.

Because the ATTZ has pronounced aquatic and terrestrial phases, there are strong selective pressures on aquatic organisms to colonize it at rising or high water because of the feeding opportunities (Bonetto et al. 1969b; Welcomme 1979; Bayley 1983, 1988). Conversely, terrestrial organisms that occupy nonflooded habitats along the floodplain borders are adapted to exploit the ATTZ at low water levels (Sheppe and Osborne 1971; Fredrickson 1979; Davies 1985).

In low-order streams, the level of adaptation to flooding is rather low. For many organisms, unpredictable floods correspond to catastrophic events that periodically "reset" the physical and biotic environment (Cummins 1977; Fisher 1983). Obligate aquatic organisms concentrate mostly in the main channel because flood periods are too short and irregular to develop profitable strategies for occupying the ATTZ.

The predictable and prolonged flood pulse typical of large rivers favors the development of anatomical, morphological, physiological, and/or ethological adaptations of terrestrial and aquatic organisms in order to colonize the ATTZ as shown by Adis (1979) and Irmler (1981) for Amazonian terrestrial invertebrates and by Uetz et al. (1979) and Wharton et al. (1981) for N. American floodplains.

In the humid tropics, regular flooding and drying of floodplains provoke a pronounced seasonality in an otherwise unseasonal environment. Many Amazonian floodplain trees show distinct annual growth rings, because inundation causes a "physiological winter" through oxygen stress (Worbes 1985, 1986). Seed production is timed with the flood for dispersal by water or by fish (Gottsberger 1978; Goulding 1980). Terrestrial arthropods from central Amazonian floodplain forests show a defined reproduction period (Adis and Mahnert 1986; Irmler 1986) but are polyvoltine in neighboring dryland forests (Adis and Sturm 1989). The flood cycle has been hypothesized as the driving force behind species selection ("taxon pulse", Erwin and Adis 1982) and the acquisition of an annual seasonality that enabled tropical insects to colonize temperate zones (Paarmann et al. 1982; Adis et al. 1986). The regular pulsing of large rivers may have been as important for the development of biorhythms in the tropics as was the pulsing of the light/temperature regime in temperate regions or the change between dry and wet periods in the arid and semiarid tropics.

Because many vertebrates living in the main channel depend on the floodplain for food supply, spawning, and shelter, they have developed strategies to utilize periodically available habitats. High mobility is required, as witnessed by the extensive migrations referred to earlier. Such strictly aquatic animals as fish and manatees depend on the flood cycle of the river, which controls access to the floodplain. Others less strictly aquatic, such as hippos, beavers, or capybaras, make feeding trips out of the water.

The importance of lateral migration of animals between the floodplain and main channel of large river systems has been underestimated because modern civilization has substantially modified the hydrograph and separated floodplains from main channels. These modifications dominate large temperate river systems. The biologist's typical view of fish in temperate rivers has been that they complete their life cycles within the river channel. Indeed, fish have no alternative in sections of some highly altered systems such as major stretches of the Mississippi R. Their persistence in these areas attests to their great plasticity in coping with habitat change.

Fishes that depend on seasonal colonization of floodplain habitats dominate the fisheries, the biomass, and the production in river-floodplain systems (Bonetto et al. 1969a; Welcomme 1979; Bayley 1981a; Goulding 1981; Bayley 1983; Littlejohn et al. 1985). Spawning of many species occurs at the beginning or during some period of the rising flood, resulting in timely colonization of the floodplains for feeding and shelter (Bayley 1983, 1988; Holland et al. 1983; Welcomme 1985). Conversely, when the water recedes, fish find refuge in main channels, in residual floodplain water bodies, or in permanent tributaries (Welcomme 1979).

Adults of many species show seasonality in food uptake related to flood cycles, as shown for the Rupununi R. by Lowe-McConnell (1964) and for the large rivers of the Amazon basin by Goulding (1980, 1981) and Ribeiro (1983). Periods of fasting coincide with low or falling water levels and are associated with decreases in seasonal fat content in many adult fish (Junk 1985b). Studies of diets at rising and high water show that many species directly use pollen, fruits, seeds, and the small portion of terrestrial insects that drop into the water from the canopy of the forest (Goulding 1980).

Detritus plays a major part in the food webs in floodplains (Welcomme 1985). Fish are major detritivores in the tropics. For example, fine particulate organic matter (FPOM) is consumed directly by the highly specialized Prochilodontidae and Curimatidae in South America, and by Citharinidae and *Labeo* species in Africa (Bowen 1984; PBB, pers. obs.). Coarse particulate organic matter (CPOM) features in the diet of many omnivores in the Amazon (Almeida 1980; Santos 1981).

FPOM is also an important feature of the gut contents of large catostomids and *Dorosoma* in large N. American rivers, but its nutritional importance has only recently been indicated (Ahlgren 1988). Most of the commercially important fishes are bottom feeders utilizing macroinvertebrates, which in turn ingest detritus (Fremling et al. 1989).

The importance of remnant floodplain areas in the Mississippi and its tributaries was indicated by Risotto and Turner (1985), who found that 55 % of the variation in average fish catch was explained by bottomland hardwood area (as a proxy to floodplain area), fishing effort, and latitude. Because some bottomland forest is now cut off by manmade levees and not all floodplains are forested, the relationship might be improved with direct measurements of the active floodplain areas.

Adaptations to survive hypoxic conditions favor the colonization of periodically stagnant waters typical of many floodplains. Air breathing and other adaptations to low oxygen concentrations are frequently found in neotropical fishes (Carter and Beadle 1931; Kramer et al. 1978; Junk et al. 1983), other tropical floodplain rivers (Welcomme 1979), and in fish of the Mississippi drainage (e.g., gars, *Lepisosteus* spp. and bowfin, *Amia calva*; see also Marvin and Heath 1968).

In the temperate Upper Mississippi R. floods can reduce the overwinter survival of young-of-the-year freshwater drum (*Aplodinotus grunniens*) by the influx of channel water at 0°C into backwater thermal refuges where the temperature is 4°C (Bodensteiner and Sheehan, in press). The winter biology of fishes in large North American rivers has been little studied, and the recruitment of other species may be strongly affected by winter temperatures and flood patterns. From spring through summer, the timing and duration of the flood is critical to species which gain access to the ATTZ and permanent backwaters for feeding and spawning. Ideal conditions for spring spawners occur during years in which the flood and temperature rise are coupled; conversely, recruitment is poor if the flood retreats too soon during the warm growing season (Fig. 1). Finger and Stewart (1987) found that the timing and duration of flooding controlled the year-class dominance of spring- versus summer-spawners in Missouri floodplain forests.

In polar, sub-arctic, and taiga rivers the timing of the flood is predictable because of massive snow melt in the spring. However, the flood is accompanied by ice that scours the floodplains and subsequently recedes rapidly,

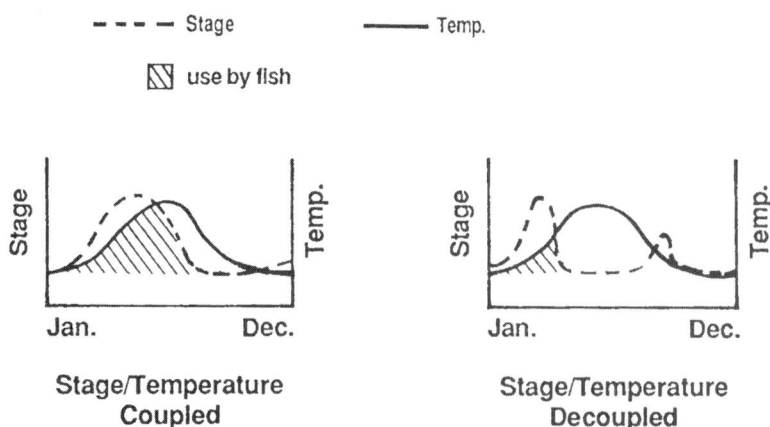

FIG. 1. Schematic of combinations of river stage and water temperature in temperate river-floodplain systems (see text).

creating an inhospitable environment for fishes (Roy 1989). The severe springtime conditions may explain why fish in high latitudes avoid the flood by spawning in the fall (R.A. Ryder, personal communication).

Tree growth is mainly retarded by floods because the rhizosphere becomes deoxygenated (Huffman 1980; Huffman et al. 1981). Gosselink et al. (1981) postulated that floods during winter or spring have a positive effect on the floodplain forest because they distribute nutrients and water to the soil before plant growth commences. Data on flood tolerance of tree species often appear to be contradictory because the timing of floods relative to growth and resting periods is not stated (Dister 1980).

The Mississippi R. is a major migratory flyway for waterfowl, shorebirds, gulls, and eagles. The dabbling ducks (mallard, pintails, greenwing and bluewing teal, black duck) utilize mast in floodplain forests, waste grain in adjacent harvested fields, and invertebrates associated with macrophytes in shallow water bodies, as well as the seeds, tubers, and plant leaves in the floodplain (Bellrose 1941). The diving ducks (canvasback, lesser scaup) utilize submerged macrophytes and macroinvertebrates that grow in deeper water (Thompson 1973). Aquatic and moist-soil vegetation in the Illinois and Upper Mississippi floodplains requires a period of shallow, stable water levels during the summer growing season (Bellrose et al. 1979). The summer's primary production is made more accessible to migratory waterfowl by the autumn rise in water levels. If an autumn flood does not occur, managers of refuges and duck clubs create one by pumping water from the river into the floodplain. They also pump water out of the same impoundments if the flood is too slow to retreat in the summer, so they can sow millet or allow native plants to grow (Bellrose et al. 1979).

Flood Pulsing and Plant Community Structure

Under given climatic conditions, plant communities become established in the ATTZ of large rivers according to the flood regime. Every place in this zone can be considered a point on a gradient reflecting the degree of annual flooding. Every plant has its optimum position on this gradient. The optimum, however, can be modified by such factors as stability, structure, and fertility of the substrate, groundwater level, and biogenic processes (e.g., accumulation of organic material, nitrogen fixation, and interspecific competition) (Lindsey et al. 1961; Bedinger 1979; Burgess et al. 1973; Johnson and Bell 1976; Bell 1980; Dister 1980, 1983; Gosselink et al. 1981; McKnight et al. 1981). Distributions of animals are also affected by this gradient in spite of their mobility (Wharton et al. 1981; Larson et al. 1981).

Basic changes in plant community structure occur mainly through a shift of the gradient, such as a rise of the floodplain surface due to additional inorganic or organic sediment deposition (allogenic or autogenic succession), a lowering by erosion, or a change in the hydrograph due to climatic change, tectonic movement, or human influence such as the construction of a dam or lateral dikes.

Plant communities, however, are characterized by smaller changes. There is strong pressure on communities to proceed to a later successional stage when the period of the flood pulse is reduced. The shape of the pulse often varies within large limits, thereby causing communities to respond. Annual plants react to annual differences whereas forest communities are affected by extreme annual floods, droughts, or even periods of successive years of extreme flood events that may occur every 10, 20, or 100 years. Establishment of tree seedlings in low-lying areas requires a period of exceptionally low water for several years, as Demaree (1932) found for *Taxodium distichum*. Aquatic communities tend to fill up periodically isolated water bodies with organic debris, thereby causing autogenic succession to marsh and swamp vegetation when the flood pulse fails. This process has been estimated to require about 200 years in the temperate Rhône R. (Amoros et al. 1986). Extreme floods clean these water bodies and "reset" communities to earlier successional stages. Resets can be especially severe when floods occur during the ice season in temperate rivers because trees and channels can be scoured by wind- or water-driven ice (Sigafoos 1964). Consequently, the observed community structure in floodplains is a result of short-, medium-, and/or long-term effects of the flood pulse. Shelford (1954) estimated that about 600 years were required to develop the late subclimax tulip-deer-oak communities on the lower Mississippi R. Most communities receiving the full amplitude of the flood pulse can be viewed as being in a dynamic equilibrium at an early successional level (pulse-stability, *sensu* Odum 1959; see also Margalef 1968).

Flood Pulsing and Production

Primary and secondary production in the river–floodplain system is the sum of production during terrestrial and aquatic phases. As indicated previously, the basic fertility of the floodplain depends on the nutrient status of the water and on the sediments deriving from the river. This fertility, however, may be modified by tributaries and by runoff from the local catchment area of the floodplain. Length, amplitude, frequency, timing, and predictability of the flood pulse determine occurrences, life cycles, and abundances of primary and secondary producers and decomposers, abundances which affect the level of exploitation and regeneration of the nutrient pool as well as its supply.

Gosselink and Turner (1978) proposed a classification of wetland systems according to a hydrodynamic energy gradient. They suggested that a positive relationship existed between productivity and water flow. Their theory may be valid within limits in a river–floodplain system; however, short-duration pulsing can flush out considerable organic matter and nutrients into the main channel (or into the estuary from a salt marsh as shown by Teal [1962]) and limit in situ productive processes and access by aquatic animals. In such systems, the aquatic biologist studying production is concerned with how the ATTZ benefits the river or the permanent lentic areas in the floodplain. Conversely, slow inundation of the same floodplain allows sufficient time for in situ processes along the moving littoral (Fig. 2), which traverses the ATTZ with each pulse. Aquatic and terrestrial biologists studying production in river–floodplain systems with slow pulsing should be concerned with how the river benefits the floodplain.

The flooding phase of the moving littoral (Fig. 2) finds its closest parallel to a reservoir in the process of being flooded (Wood 1951), with mineralized products from any preceding aquatic cycle and the current terrestrial one being

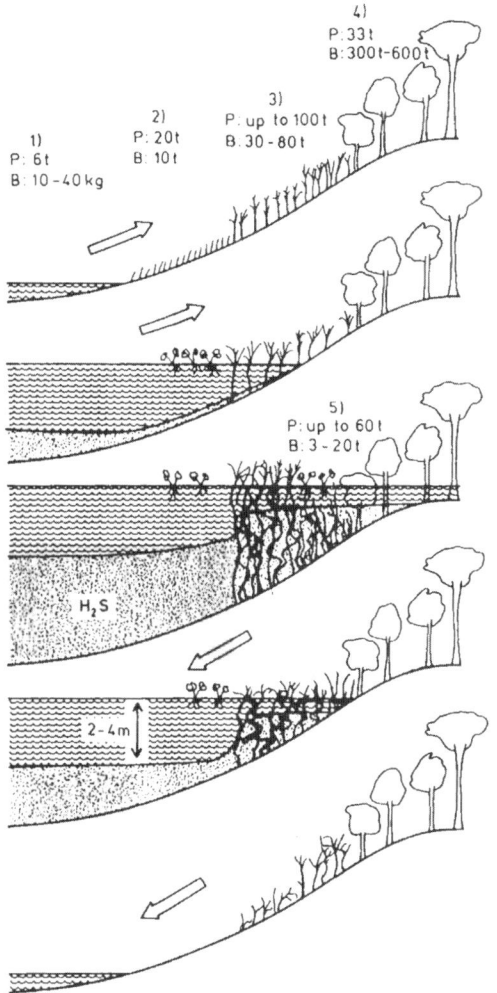

FIG. 2. The moving littoral in the transition zone (ATTZ) of a river-floodplain system in the central Amazon, with estimates of annual production (P) and biomass (B). Estimates are as dry weight per hectare. The H_2S zone has no dissolved O_2. The indicated zones are as follows: (1) Phytoplankton C14 (Schmidt 1973b), (2) annual terrestrial plants, (3) perennial grasses, (4) floodplain (várzea) forest, and (5) emergent macrophytes (from Junk 1985c and unpubl. data). Periphyton are not included, but preliminary data of periphyton on macrophytes from T. R. Fisher (pers. comm.) indicate a total productivity in the floodplain of the same order as phytoplankton (Bayley 1989).

released into the water. The various sources of primary production have high values (Fig. 2) but varying production to biomass ratios. When integrated over areas appropriate for each season in the floodplain, phytoplankton contributed less than 6% of the total carbon production in the central Amazon várzea floodplain (Junk 1985a; Bayley 1989).

Most carbon sources, including considerable detrital biomass, are important to some animals at some time (Welcomme 1979, 1985; Junk 1984), but their quantitative importance is unknown. Organic material produced in floodplains varies considerably with respect to consistency, protein content, digestibility, and availability, that result in large differences in decomposition time and in the types of organisms involved in decomposition processes. Phytoplankton and periphyton are easily decomposed in only a few hours or days. In the Amazon, aquatic and terrestrial herbaceous plants lose about 50% of their weight after 2-3 weeks in water (Howard-Williams and Junk 1976). Tree leaves vary widely according to species; some are as quickly decomposed as herbaceous plants whereas others remain little modified throughout months and even years. Softwood plants are destroyed in a few years, whereas hardwood plants may remain little modified for years and even decades (Junk, unpublished data).

Strong evidence suggests that the change between terrestrial and aquatic phases accelerates the decomposition of organic material, as the circumstantial evidence of Wood (1951) indicated. Terrestrial arthropods play an important role in the decomposition of leaf litter and wood as shown for cockroaches by Irmler and Furch (1979) and for termites by C. Martius (pers. comm. to WJJ). Oxygenation of sediments during dry periods promotes processing of organic material; later, when reflooding occurs, plant nutrients are recycled into the water, thereby enhancing productivity. This effect, sometimes in combination with a crop plantation or fallow period for an entire year, has been utilized for many years in European fish culture. Wood (1951) proposed the management of water levels in impoundments by changing them seasonally to increase fish production. Lambou (1959) suggested that the processes described by Wood explain the high productivity of backwater lakes due to natural water fluctuations in the Mississippi floodplain. In the Amazon floodplain during the period of rising water, mean growth increments by weight of 12 common fish species were 60% higher than during the remainder of the year (Bayley 1988).

Food supply in fertile floodplains during the flood phase can be so abundant that factors other than food may limit individual growth and population density of fish and other aquatic animals. Limitations during the flood phase include spawning success, lack of habitats with sufficient dissolved oxygen (Junk et al. 1983), and predation (Bayley 1983). Limitations at low water include higher levels of predation, a probable reduction in food supply, or even death by drought. Bayley (1988) found that growth of juveniles of 11 abundant fish species tested did not indicate a density-dependent relationship with potentially competing species guilds during the period of rising water. Only two out of eight species indicated density-dependency at $P < .05$ during the shorter falling-water period.

The preceding ideas have very little to do with traditional concepts of productive processes in rivers. The RCC predicts that lower reaches of river systems have low ratios of production to respiration (P/R) due to processing of material from upstream and reduced in situ production. Wissmar et al. (1981) noted that Amazon floodplain lakes have high respiration rates, and Melack and Fisher (1983) noted that carbon loss due to respiration exceeds the carbon contributed by phytoplankton. However, these are limnological perspectives that describe only part of the system. The evidence offered here for the lower reaches of the river-floodplain system indicates high in situ production and low importation of organic matter from upstream. Therefore, we predict high P/R ratios for large river–floodplain systems.

Flood Pulsing and Diversity of Habitats and Species

Sediments, which are deposited in the floodplain in well-defined geomorphological units, form bars, levees, swales, oxbows, backwaters, and side channels. Flowing water grades sediments according to grain size. The floodplain soils are stratified horizontally and vertically in a small scale pattern (Irion et al. 1983; Amoros et al. 1986), but the wind-induced transport of sediment may modify the water-induced sediment pattern.

The main river and its connecting channels represent the lotic part of the river–floodplain system; oxbow lakes, abandoned channels, and backwaters represent the lentic one. Both harbour sets of organisms which colonize the much more extensive, periodically flooded ATTZ and increase species numbers occurring in the floodplain.

Differences in the duration of flooding, in soil structure, and in vegetation result in small-scale habitats in the form of narrow, roughly parallel zones. This arrangement multiplies the edge effect far beyond that represented by the main channel and its islands. In addition to these topological edges, there are many physico-chemical edges in the form of sharp vertical and horizontal boundaries in oxygen, temperature, dissolved or suspended matter; in the main channel these are encountered only at confluences with tributaries or near the substrate. In the Amazon, oxygen levels in surface water may drop from about 5 mg·L^{-1} in the main channel to 0.5 mg·L^{-1} in the floodplain 50 m away (Junk et al. 1983).

Habitats shift horizontally and vertically according to the waterlevel (Fig. 2). In addition to this instability due to the moving littoral is another instability caused by sediment deposition and erosion by the river. Depending on the position of the river channel and its dynamics, habitats may be ephemeral or rather stable over decades or centuries. This affects such stationary organisms as trees.

Nonflooded areas inside and adjacent to the floodplain perimeter, as well as emergent vegetation or the floodplain forest canopy, can be termed terrestrial habitats. All of them harbour an abundance of plants and animals that colonize the ATTZ, increasing considerably the total number of plants and animals occurring in the system.

No attempt to explain the total diversity in all habitats has been made; however, studies on specific plant and animal groups show some tendencies and some apparent inconsistencies. Species diversity would be expected to be limited in aquatic and terrestrial taxa that are sedentary and experience the full impact of the physiological stress resulting from the change between the aquatic and terrestrial phase. Worbes (1983) showed that the central Amazon floodplain forest has a much lower plant species diversity than the nonflooded forest. Salo et al. (1986), however, state that high diversity in tree species characterizing the upper Amazon lowland forests occurs in existing and relict floodplains, but they did not present species numbers or diversity indices. They describe a mosaic of small habitats created by large-scale, continuous disturbance by lateral erosion and sedimentation from the river channel, with high diversity between habitats. They reason that the high diversity in the relatively short-lived habitats of the present floodplains was due to insufficient time to allow competitive exclusion, supporting Connell's (1978) intermediate disturbance hypothesis. In the former floodplain formations that are about 5 000–10 000 years old, habitats are very stable, and the high species diversity between habitats was attributed by them to allopatric speciation.

Diversity would be expected to increase with the ability of organisms to avoid the physiological stress in the ATTZ. High diversity in floodplains occurs in mobile groups, such as fish (Lowe-McConnell 1975; Welcomme 1985) and nonaquatic birds (Remsen and Parker 1983).

The drastic change between terrestrial and aquatic phases results in high seasonal losses for most plant and animal populations, but these losses tend to be recovered by quick growth, early maturity, high reproduction rates for r-strategy organisms (Pianka 1970), and fast dispersal. Many of the most persistent and productive tropical aquatic weeds (e.g., *Eichhornia crassipes, Salvinia auriculata, Ceratopteris pteridoides,* and *Alternanthera philoxeroides*) are endemic to neotropical river–floodplains. In floodplains they are periodically decimated during the dry phase, allowing coexistence of many plant species with similar habitat requirements. In hydrologically stable conditions, they become dominant due to their strong competitive ability. Conversely, many persistent weeds in agricultural crops dominate in the early successional stages of floodplain vegetation at low water due to their r-strategy traits and recurrent disturbance of the ATTZ by the flood pulse (Seidenschwarz 1986; WJJ, unpublished data).

Many plants and animals show an impressive resilience with respect to short-term catastrophic events; an example is the rapid response of fishes following extreme drought, overfishing, or poisoning. Due to their highly effective reproduction strategies and to their mobility which allows access to dispersed low-water refuges, fish recover quickly when the flood pulse returns (Welcomme 1979). An r-strategy is effective only when sufficient nutrient and food resources are available to fully utilize the growth potential. Floodplains of extremely low nutrient status may therefore favor K-selection (Pianka 1970), such as Magalhães and Walker (1989) have indicated for Amazonian freshwater shrimps.

If we consider the total number of species in a river–floodplain system, circumstantial evidence suggests that a physical factor, the flood pulse, produces and maintains a highly diverse and dynamic habitat structure, thereby allowing a high species diversity despite stresses in the ATTZ. This is consistent with the intermediate disturbance hypothesis of Connell (1978) and parallels the observations of Statzner and Higler (1986) and Statzner (1987) who noted that physical factors (stream hydraulics) affected zonation patterns of benthic invertebrates, and that longitudinal zones of transition were associated with higher species richness.

Man-Made River-Floodplains

Dams have altered the hydrology and created artificial sedimentation basins covering thousands of square kilometres in rivers worldwide. Dam construction continues. For example, about 100 large reservoirs totalling 100 000 km^2 are projected to utilize the hydroelectric potential of Amazon R. tributaries (Junk and Melo 1987).

The hydrological changes often remove the flood pulse from floodplains downstream and sometimes permanently inundate floodplains upstream.

In the longer term, sedimentation and the modified flood pulse produce man-made river–floodplains. The 26 mainstem navigation dams on the Upper Mississippi R. downstream from Minneapolis, Minnesota, divide the river into reaches where the entire floodplain width immediately upstream of the dam is currently inundated, but where sedimentation is creating shallows that will become levées, side channels, or backwaters, and eventually floodplains (Fig. 3A to H). Of course, former floodplains now behind manmade levées will remain isolated from the river, assuming no long-term changes in flood stages or flood protection policy. The new floodplains upstream from some of these dams will experience the full amplitude of the flood cycle because the dams maintain water depths for navigation only during low flows but have little effect on flood levels. Indeed, the gates are raised completely out of the water and the relatively low earthen weirs that connect the locks and gates to the bluffs are overtopped during floods. The extent to which these developing floodplains contribute to secondary production, fish yield, and waterfowl utilization should be measurable during the next 50 years, assuming that other factors (e.g., pollution) remain constant or are taken into account. Thus the flood pulse concept can be investigated by measuring changes in one system through time since the navigation dam construction.

Even now, in an intermediate stage of succession in the Mississippi pools, the channel borders, not the main channel, are centers of production. Concentrations of particulate and dissolved organic carbon, plankton, and microbes are higher closer to the fringing plant beds and diminish toward the channel (Fig. 3C, E, and D). The greatest biomass of benthic macroinvertebrates are the burrowing filterers and collectors (mayflies of the genus *Hexagenia* and sphaeriid clams, *Musculium* and *Sphaerium*), which occur in beds just offshore of the macrophytes (Fig. 3F). These invertebrates apparently did not appear in high densities (up to 100 000 clams·m^{-2}) in the oldest pooled reach of the Mississippi R., the Keokuk Pool, until the 1960's (Gale 1969; Sandusky et al. 1979), when sedimentation raised the channel border bottom to the 1-m euphotic zone, thereby triggering autochthonous production by macrophytes. Diving ducks, which feed on concentrations of these invertebrates, only began using this pool in substantial numbers in the mid 1960's (Mills et al. 1966; Thompson 1973; F.C. Bellrose, pers. comm.). If phytoplankton or upstream sources had fueled the clams and mayflies, dense populations of these invertebrates should have been present in Keokuk Pool (but evidently were not) when Ellis (1931a, 1931b) made his biological surveys 18 years after the dam was closed, which was sufficient time for the accumulation of substrate suitable for burrowers. Organic matter was not being trapped behind upstream dams before it could enter the pool because these dams were not constructed until the late 1930's and early 1940's. The historical evidence from the Upper Mississippi R. thus supports the idea that a high level of secondary production requires a nearby center of primary production, rather than long-distance transport of organic matter from upstream sources via the main channel.

Conclusions

From a hydrological aspect, floodplains are part of the drainage system of rivers and are periodically affected by transport of water and dissolved and particulate material. From an ecological point of view, they represent transition zones (ATTZ) that alternate between aquatic and terrestrial states and link river channels with permanent lentic bodies and permanently dry land. Most large river systems have geomorphological settings that produce floodplains that are large relative to the lotic surface area (Welcomme 1985), and, in unmodified watersheds, produce a pulse of long duration that results in extensive but temporary lentic areas covering the ATTZ. Conversely, flood pulses of short duration, which are typical of low-order streams or of some modified systems, are associated with ATTZ's that are frequently covered by flowing water for short periods.

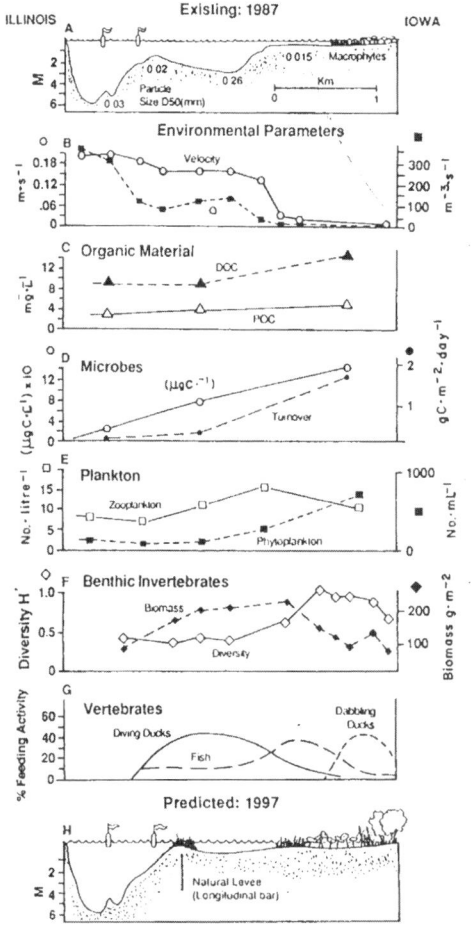

FIG. 3. A section of lower Keokuk Pool on the Upper Mississippi (A-G) with a projection of the stabilized system by the end of the century (H) (unpublished data from R. V. Anderson, R. E. Sparks, J. W. Grubaugh, K. S. Lubinski, and R. W. Gorden).

The flood pulse is the driving force for river-floodplain systems and maintains them in dynamic equilibrium. The system responds to the rate of rise and fall and to the amplitude, duration, frequency, and regularity of the pulses. Unpredictable pulses generally impede the adaptation of organisms and are counterproductive for many of them. Conversely, a regular pulse allows organisms to develop adaptations and strategies for efficient utilization of habitats and resources within the ATTZ, rather than depend solely on permanent water bodies or permanent terrestrial habitats. In temperate regions, the light and/or temperature regime may modify the biological effects of the pulse; timing of the pulse becomes important. In polar, sub-arctic, and taiga rivers where ice scouring occurs, the contribution to productivity from the ATTZ is not realized. In semiarid regions, local precipitation has a strong influence on the floodplain biota during the dry phase.

A variety of physical structures in combination with the flood pulse results in great habitat diversity. This diversity is coupled with the dynamic effect of the moving littoral, which extends the edge effect of the littoral over the entire ATTZ, thereby rendering channel banks bordering lotic zones insignificant by comparison. Organisms tend to invade the ATTZ from the terrestrial side also. Regular pulsing coupled with habitat diversity favors high diversity of aquatic and terrestrial plants and animals, despite considerable stress that results from the change between terrestrial and aquatic phases.

Aquatic and terrestrial productivity of river-floodplain systems depend mainly on the nutrient status of the water and sediments, on the climate, and on the flood pulse. Cycles specific to the floodplain, however, are decoupled to some extent from the nutrient status of the main channel. The moving littoral prevents permanent stagnation, thereby allowing the rapid recycling of organic matter and nutrients and resulting in a productivity that we predict to be greater than if the ATTZ were either permanently inundated or dry. Primary production associated with the ATTZ is much higher than that of permanent water bodies in unmodified systems and can often exceed that of permanent terrestrial habitats.

Transport of organic carbon from upstream catchment areas into the floodplain (spiralling) is of little importance to the productivity of the system. Conversely, primary and secondary production of the floodplains is essential to fauna in the main channels. A major component of energy transfer between floodplains and main channels is effected by animal migration, in particular of fish that also migrate upstream for considerable distances. Some bird species transfer nutrients from terrestrial areas or flooded mudflats, where they feed, to floodplain lakes, where they rest and defecate; other species do the reverse. The main function of the river channel in relation to plants and animals in the river-floodplain system is that of a migration route and dispersal system to access resources and refuges.

In conclusion, for those interested in the principal driving forces responsible for the structure, function, and evolutionary history of the biota in river-floodplain systems, we believe that the concept offered here will prove of heuristic rather than merely descriptive value. There is a fundamental dichotomy in the river-floodplain system: both continuous (e.g., the RCC) and batch processes occur. The latter, represented by the flood pulse concept, is dominant in systems with floodplains (ATTZs), in particular when the pulse is regular and of long duration. It is distinct because processes in floodplains do not depend on inefficient processing of organic matter upstream, although their inorganic nutrient pool may be replenished with periodic lateral inflows of water and sediments from the main channel. The pulse concept differs in that the position of the floodplain in the system relative to the river network is not a primary determinant of the processes that occur, although hydrological circumstances do not normally favor floodplain development in extreme upper reaches. However, examples do occur in upper reaches, such as the Pantanal of the Paraná system and the extensive Bolivian and Peruvian floodplains in the Amazon.

This concept implies an approach to studying the system different from the traditional limnological paradigms for either lotic or lentic systems. The space and time scales appropriate for understanding the mechanisms differ from those related to longitudinal processes in lotic channels. We hope that the flood pulse concept will help ecologists improve the design of studies and frame hypotheses that will lead more directly to a better understanding of river-floodplain systems. This is an urgent goal considering the modifications that continue to be proposed and that are sometimes put into practice in many tropical and temperate systems.

Acknowledgments

The following gave valuable suggestions: J.R. Adams, J. Adis, R.V. Anderson, C.F. Bryan, W.R. Edwards, R.W. Gorden, J.W. Grubaugh, M. Grubb, R.W. Larimore, L.L. Osborne, K. Robertson, S.K. Robinson, and M.J. Wiley. Amazon work was supported by CNPq, Brasilia and INPA of the Brazilian Government, and the Max-Planck-Institute for Limnology of West Germany. Research on the Upper Mississippi R. was supported by a National Science Foundation grant for long-term ecological research (LTER), No. BSR-8114563 and BSR-8612107, and by the loan of equipment from the Upper Mississippi River Basin Association.

References

(Addresses of personal communications follow references)

ADIS, J. 1979. Vergleichende ökologische Studien an der terrestrischen Arthropodenfauna zentralamerikanischer überschwemmungswälder. Ph.D. thesis, Ulm Univ., West Germany. 99 p.

ADIS, J., AND V. MAHNERT. 1986. On the natural history and ecology of Pseudoscorpiones (Arachnidae) from an Amazonian blackwater inundation forest. Amazoniana 9: 297-314.

ADIS, J., W. PAARMANN, AND T. L. ERWIN. 1986. On the natural history and ecology of small terrestrial ground beetles (Col.: Bembidiini: Tachyina: Polyderis) from an Amazonian blackwater inundation forest, p. 413-427. *In* P.J. Den Boer, M. L. Luff, D. Mossakowski, and F. Weber [ed.] Carabid beetles: their adaptations, dynamics and evolution, G. Fisher, Stuttgart, West Germany.

ADIS, J., AND H. STURM. 1989. Flood-resistance of eggs and life-cycle adaptation, a survival strategy of *Neomachilellus scandens* (Meinertellidae, Archaeognatha) in Central Amazonian inundation forest. Insect Sci. Appl. (In press)

AHLGREN, M. O. 1988. Diet selection and the seasonal contribution of detritus to the diet of the white sucker. Poster session of the 1988 American Fisheries Society Conference, Toronto, Canada.

ALMEIDA, R. G. 1980. Aspectos taxonómicos e hábitos alimentares de três espécies de *Triportheus* (Pisces: Characoidei, Characidae), do lago do Castanho, Amazonas. M.S. thesis, Instituto Nacional de Pesquisas da Amazônia, Manaus, Brazil. 104 p.

AMOROS, C., A. L. ROUX, J. L. REYGROBELLET, J. P. BRAVARD, AND G. PAUTOU. 1986. A method for applied ecological studies of fluvial hydrosystems. Regulated Rivers 1: 17-36.

ANDREWS, C. W., AND E. LEAR. 1956. The biology of arctic char (*Salvelinus alpinus*) in northern Labrador. J. Fish. Res. Board Can. 13: 843-860.

ANTIPA, G. P. 1911. Fischerei und Flussregulierung. Allgem. Fischerei-Zeitung, München 16/17: 1-5.

—— 1928. Die biologischen Grundlagen und der Mechanismus der Fischproduktion in den Gewässern der unteren Donau Academie Roumaine, Bull. de la Section Scientifique 11: 1-20.

BARMUTA, L. A., AND P. S. LAKE. 1982. On the value of the river continuum concept. N. Z. J. Mar. Freshwater Res. 16: 227-231.

BAYLEY, P. B. 1973. Studies on the migratory characin *Prochilodus platensis* Holmberg 1889 (Pisces, Characoidei) in the Rio Pilcomayo, S. America. J. Fish Biol. 5: 25-40.

—— 1980. The limits of limnological theory and approaches as applied to river-floodplain systems and their fish production, p. 739-746. *In* J.I. Furtado [ed.] Tropical ecology and development. Proceedings of the Vth International Symposium of Tropical Ecology. International Society of Tropical Ecology, Kuala Lumpur.

—— 1981a. Fish yield from the Amazon in Brazil: comparisons with African river yields and management possibilities. Trans. Am. Fish. Soc. 110: 351-359.

—— 1981b. Características de inundación en los rios y áreas de captación en la Amazonia Peruana: una interpretación basada en imagenes del 'LANDSAT' e informes de 'ONERN'. Inst. Mar. Peru(Callao) Inf. 81: 245-303.

—— 1983. Central Amazon fish populations: biomass, production and some dynamic characteristics. Ph.D. thesis, Dalhousie Univ., Nova Scotia, Canada. 330 p.

—— 1988. Factors affecting growth rates of young tropical fishes: seasonality and density-dependence. Env. Biol. Fishes 21:127-142.

—— 1989. Aquatic environments in the Amazon Basin, with an analysis of carbon sources, fish production, and yield, p. 399-408. *In* D. P. Dodge [ed.] Proceedings of the International Large River Symposium. Can. Spec. Publ. Fish. Aquat. Sci. 106.

BEDINGER, M. S. 1979. Relation between forest species and flooding, p. 427-435. *In* P. E. Greeson, P. E. Clark, and J. E. Clark [ed.] Wetland functions and values: the state of our understanding. American Water Resources Association, Anthony Falls Hydraulic Laboratory, Minneapolis, MN.

BELL, D. T. 1980. Gradient trends in the streamside forest of central Illinois. Bull. Torrey Bot. Club 107: 172-180.

BELLROSE, F. C. 1941. Duck food plants of the Illinois River valley. Ill. Nat. Hist. Surv. Bull. 21: 237-280.

BELLROSE, F. C., F. L. PAVEGLIO, Jr., AND D. W. STEFFECK. 1979. Waterfowl populations and the changing environment of the Illinois River valley. Ill. Nat. Hist. Surv. Bull. 32: 1-54.

BERNER, L. M. 1951. Limnology of the lower Missouri River. Ecology 32: 1-12.

BHOWMIK, N. G., AND J. B. STALL. 1979. Hydraulic geometry and carrying capacity of floodplains. Univ. of Illinois Water Resources Center. Research Report 145 UI LU-WRC-79-0145, Champaign, IL. 147 p.

BODENSTEINER, L. R., AND R. J. SHEEHAN. 1989. Implications of backwater habitat management strategies to fish populations. Proceedings of the 44th Annual Meeting of the Upper Mississippi River Conservation Committee. 8-10 March 1988, Peoria, Il. (In press)

BONETTO, A. A., W. DIONI, AND C. PIGNALBERI. 1969a. Limnological investigations on biotic communities in the Middle Parana River Valley. Int. Ver. Theor. Angew. Limnol. Verh. 17: 1035-1050.

BONETTO, A. A., E. CORDIVIOLA DE YUAN, C. PIGNALBERI, AND O. OLIVEROS. 1969b. Ciclos hidrológicos del Rio Paraná y las poblaciones de peces contenidas en las cuencas temporarias de su valle de inundación. Physis (Buenos Aires) 29: 213-223.

BOWEN, S. H. 1984. Detritivory in neotropical fish communities, p. 59-66. *In* T. Zaret [ed.] Evolutionary ecology of neotropical freshwater fishes. Dr W. Junk, The Hague, Netherlands.

BRAVARD, J. P., C. AMOROS, AND G. PAUTOU. 1986. Impact of civil engineering works on the succession of communities in a fluvial system. Oikos 47: 92-111.

BRINSON, M. M., H. D. BRADSHAW, AND J. B. ELKINS, Jr. 1980. Litterfall, stemflow, and throughfall nutrient fluxes in an alluvial swamp forest. Ecology 61: 827-835.

BRINSON, M. M., H. D. BRADSHAW, AND R. N. HOMES. 1983. Significance of the floodplain sediments in nutrient exchange between a stream and its floodplain, p. 199-221. *In* T. D. Fontaine, and S. M. Bartell [ed.] Dynamics of lotic ecosystems. Ann Arbor Science, Ann Arbor, MI.

BROOK, A. J., AND J. RZÓSKA. 1954. The influence of the Gebel Aulyia Dam on the development of Nile plankton. J. Anim. Ecol. 23: 101-114.

BRYAN, C. F., F. M. TRUESDALE, D. S. SABINS, AND C. R. DEMAS. 1974. Annual report on a limnological survey of the Atchafalaya Basin. Louisiana Cooperative Fishery Research Unit, School of Forestry and Wildlife Management, Louisiana State Univ., 1974. 208 p.

BRYAN, C. F., D. J. DEMONT, D. S. SABINS, AND J. P. NEWMAN, Jr. 1976. Annual report on a limnological survey of the Atchafalaya Basin. Louisiana Cooperative Fishery Research Unit, School of Forestry and Wildlife Management, Louisiana State Univ., 1976. 285 p.

BRYAN, C. F., AND D. S. SABINS. 1979. Management implications in water quality and fish standing stock information in the Atchafalaya River Basin, Louisiana, p. 293-316. *In* J. W. Day Jr., D. D. Culley Jr., R. E. Turner, and A. J. Mumphrey Jr. [ed.] Proceedings from the Third Coastal Marsh and Estuary Symposium, Louisiana State Univ., Division of Continuing Education, Baton Rouge, Louisiana.

BURGESS, R. L., W. C. JOHNSON, AND W. R. KEAMMERER. 1973. Vegetation of the Missouri River floodplain in North Dakota. North Dakota Water Resource Research Institute Report WI-221-018-07: 162 p.

CARTER, G. S., AND L. C. BEADLE. 1931. The fauna of the swamps of the Paraguayan Chaco in relation to its environment. II. Respiratory adaptations in the fishes. J. Linn. Soc. Lond. (Zool.) 37: 327-368.

CHOWDHURY, M. J., S. SAFIULLAH, S. M. IQBAL ALI, M. MOFIZUDDIN, AND S. E. KABIR. 1982. Carbon transport in the Ganges and the Brahmaputra: preliminary results. Mitt. Geol. Paläont. Inst. Univ. Hamburg, SCOPE/UNEP Sonderbd. 52: 457-468.

COFFMAN, W. P., AND L. C. FERRINGTON, Jr. 1984. Chironomidae, p. 551-652. *In* R. W. Merrit, and K. W. Cummings [ed.] An introduction to the aquatic insects of North America. 2nd. Edition, Kendall/Hunt Publishing Co., Dubuque, IA.

CONNELL, J. H. 1978. Diversity in tropical rain forests and coral reefs. Science 199: 1302-1310.

CUMMINS, K. W. 1977. From headwater streams to rivers. Am. Biol. Teach. 39: 305-312.

CURRY, R. R. 1972. Rivers — a geomorphic and chemical overview, p. 9-31. *In* R. T. Oglesby, C. A. Carlson, and J. A. McCan [ed.] River ecology and man. Academic Press, New York, NY.

DAVIES, B. R. 1985. The Zambezi river system, p. 225-267. *In* B. R. Davies, and K. F. Walker [ed.] The ecology of river systems. Dr W. Junk, The Hague, Netherlands.

DAVIES, B. R., AND K. F. WALKER. 1985. River systems as ecological units. An introduction to the ecology of river systems, p. 1-23. *In* B. R. Davies, and K. F. Walker [ed.] The ecology of river systems. Dr W. Junk, The Hague, Netherlands.

DEMAREE, D. 1932. Submerging experiment with *Taxodium*. Ecology 13: 258-262.

DIERBERG, F. E., AND P. L. BRESZONIC. 1984. The effect of wastewater on the surface water and ground water quality of cypress domes, p. 83-101. *In* K. C. Ewel, and H. T. Odum [ed.] Cypress swamps. Univ. of Florida Press, Gainesville, FL.

DISTER, E. 1980. Geobotanische Untersuchungen in der hessischen Rheinaue als Grundlage für die Naturschutzarbeit. Ph.D. thesis, Göttingen Univ., West Germany. 170 p.
1983. Zur Hochwassertoleranz von Auwaldbäumen an lehmigen Standorten. Verhandlungen der Gesellschaft für ökologie 11: 325-336.

ELLIS, M. M. 1931a. Some factors affecting the replacement of the commercial fresh-water mussels. Bureau of Fisheries Circular, U.S. Dep. of Commerce 7: 1-10.
1931b. A survey of conditions affecting fisheries in the Upper Mississippi River. Bureau of Fisheries Circular, U.S. Dep. of Commerce 5: 1-18.

ERTEL, J. R., J. I. HEDGES, A. H. DEVOL, J. E. RICHEY, AND M. N. G. RIBEIRO. 1986. Dissolved humic substances of the Amazon River system. Limnol. Oceanogr. 31: 739-754.

ERWIN, T. L., AND J. ADIS. 1982. Amazonian inundation forests, richness and taxon pulses, p. 358-371. *In* G. T. Prance [ed.] Biological diversification in the tropics. Proc. Fifth Int. Symp. Assoc. Trop. Biol. Columbia Univ. Press, New York, NY.

FERREIRA DA SILVA, V. M. 1983. Ecología alimentar dos dolfinhos da Amazônia. M.S. thesis, Instituto Nacional de Pesquisas da Amazônia, Manaus, Brazil. 110 p.

FINGER, T. R., AND E. M. STEWART. 1987. Response of fishes to flooding regime in lowland hardwood wetlands, p. 86-92. *In* W. J. Mathews and D. C. Heins [Ed.] Community and evolutionary ecology of North American stream fishes. University of Oklahoma Press, Norman, OK.

FISHER, S. G. 1983. Succession in streams, p. 7-27. *In* J. R. Barnes, and G. W. Minshall [ed.] Stream ecology: application and testing of general ecological theory. Plenum Press, N.Y.

FISHER, T. R. 1979. Plankton and primary production in aquatic systems of the central Amazon Basin. Comp. Biochem. Physiol. 62A: 31-38.

FITZGERALD, K. K., P. D. HAYES, T. E. RICHARDS, AND R. L. STAHL. 1986. Water resources data, Illinois, Water Year 1985. Vol. 2. Illinois River Basin. U.S. Geological Survey Water Data Report IL-85-2, USGS, Urbana, IL. 397 p.

FOERSTER, R. E. 1968. The sockeye salmon, *Oncorhynchus nerka*. Fish. Res. Board Canada Bull. 162: 422 p.

FREDERICKSON, L. H. 1979. Lowland hardwood wetlands: current status and habitat values for wildlife, p. 296-311. *In* P. E. Greeson, P. E. Clark, and J. E. Clark [ed.] Wetland functions and values: the state of our understanding. American Water Resources Association, Anthony Falls Hydraulic Laboratory, Minneapolis, MN.

FREMLING, C. R., J. L. RASMUSSEN, R. E. SPARKS, S. P. COBB, C. F. BRYAN, and T. O. CLAFLIN. 1989. Mississippi River fisheries: a case history, 309-351. *In* D. P. Dodge [ed.] Proceedings of the International Large River Symposium. Can. Spec. Publ. Fish. Aquat. Sci. 106.

FURCH, K. 1984a. Interanuelle Variation hydrochemischer Parameter auf der Ilha de Marchantaria. Biogeographica 19: 85-100.
1984b. Seasonal variations of the major cation content of the várzea-lake Lago Camaleão, middle Amazon, Brazil, in 1981 and 1982. Int. Ver. Theor. Angew. Limnol. Verh. 22: 1288-1293.

FURCH, K., AND W. J. JUNK. 1985. Dissolved carbon in a floodplain lake of the Amazon and in the river channel. Mitt. Geol. Paläont. Inst. Univ. Hamburg, SCOPE/UNEP Sonderbd. 58: 285-298.

FURCH, K., W. J. JUNK, J. DIETERICH, AND N. KOCHERT. 1983. Seasonal variation in the major cation (Na, K, Mg and Ca) content of the water of Lago Camaleão, an Amazonian floodplain lake near Manaus, Brazil. Amazonia 8: 75-89.

GALE, W. F. 1969. Bottom fauna of Pool 19, Mississippi River, with emphasis on the life history of the fingernail clam *Sphaerium transversum*. Ph.D. thesis, Iowa State Univ., Ames. 234 p.

GODOY, M. P. de. 1967. Dez anos de observações sôbre periodicidade migratoria de peixes do Rio Mogi Guassu. Rev. Brasil. Biol. 27: 1-12.

GOSSELINK, J. G., AND R. E. TURNER. 1978. The role of hydrology in freshwater wetland ecosystems, p. 63-78. *In* R. E. Good, D. F. Whigham, and R. L. Simpson [ed.] Freshwater wetlands: ecological processes and management potential. Academic Press, New York, NY.

GOSSELINK, J. G., S. E. BAYLEY, W. H. CONNER, AND R. E. TURNER. 1981. Ecological factors in the determination of riparian wetland boundaries, p. 197-219. *In* J. R. Clark and J. Benforado [ed.] Wetlands of bottomland hardwood forests. — Developments in agricultural and managed forest ecology 11, Elsevier Scientific Publishing Company, Amsterdam, Oxford, New York, NY.

GOTTSBERGER, G. 1978. Seed dispersal by fish in the inundated regions of Humaita, Amazonia. Biotropica 10: 170-183.

GOULDING, M. 1980. The fishes and the forest: explorations in Amazonia natural history. California Univ. Press, Berkeley. 280 p.
1981. Man and fisheries on an Amazon frontier. Dr. W. Junk, The Hague, Netherlands. 137 p.

GRAINGER, E. H. 1953. On the age, growth, migration, reproductive potential and feeding habits of the Arctic char (*Salvelinus alpinus*) of Frobisher Bay, Baffin Island. J. Fish. Res. Board Can. 10: 326-370.

GRUBAUGH, J. W., AND R. V. ANDERSON. 1989a. Long-term effects of navigation dams on a segment of the Upper Mississippi River. Regulated Rivers. (In press)
1989b. Seasonal fluxes and the influences of floodplain forest on organic matter dynamics in the Upper Mississippi River. Hydrobiologia. (In press)

HAMMERTON, D. 1976. The Blue Nile in the plains, p. 243-256. *In* J. Rzóska [ed.] The Nile, biology of an ancient river. Dr. W. Junk, The Hague, Netherlands.

HARMON, M. E., J. R. FRANKLIN, F. J. SWANSON, J. D. LATTIN, S. V. GREGORY, N. H. ANDERSON, S. P. CLINE, N. G. AUMEN, J. R. SEDELL, G. W. LIENKAEMPER, K. CROMACK Jr., AND K. W. CUMMINS. 1986. Ecology of course woody debris in temperate ecosystems. Adv. Ecol. Res. 15: 133-302.

HAWKINSON, B., AND G. GRUNWALD. 1979. Observations of a wintertime concentration of catfish in the Mississippi River. Fisheries Investigation Report No. 365, Minnesota Dep. of Natural Resources. 9 p.

HEDGES, J. I., W. A. CLARK, P. D. QUAY, J. E. RICHEY, A. H. DEVOL, AND U. M. SANTOS. 1986. Compositions and fluxes of particulate organic material in the Amazon river. Limnol. Oceanogr. 31: 717-738.

HELLER, H. 1969. Lebensbedingungen und Abfolge der Flussauevegetation in der Schweiz. Schweiz. Anst. Forstl. Versuchswes. Mitt. 45: 1-124.

HOLČÍK, J., AND I. BASTL. 1976. Ecological effects of water level fluctuation upon the fish populations in the Danube River floodplain in Czechoslovakia. Acta Sci. Natur. Acad. Scient. Bojemoslov. Brno 10: 1–46.
———. 1977. Predicting fish yield in the Czechoslovakian section of the Danube River based on the hydrological regime. Int. Rev. Gesamten Hydrobiol. 62: 523–532.
HOLLAND, L. E., C. F. BRYAN, AND J. P. NEWMAN, Jr. 1983. Water quality and the rotifer population in the Atchafalaya River Basin. Hydrobiologia 98: 55–69.
HOWARD-WILLIAMS, C., AND W. J. JUNK. 1976. The decomposition of aquatic macrophytes in the floating meadows of a Central Amazonian várzea lake. Biogeographica 7: 115–123.
HUFFMAN, R. T. 1980. The relation of flood timing and duration to variation in bottomland hardwood community structure in the Quachita River Basin of Southeastern Arkansas. U. S. Army Engineer Waterways Experiment Station Miscellaneous Paper E-80-4, Vicksburg, Mississippi: 22 p.
HUFFMAN, R. T., AND S. W. FORSYTHE. 1981. Bottomland hardwood forests and their relation to anaerobic soil conditions, p. 187–196. In J. R. Clark and J. Benforado [ed.] Wetlands of bottomland hardwood forests. — Developments in agricultural and managed forest ecology 11, Elsevier Scientific Publishing Company, Amsterdam, Oxford, New York, NY.
HYNES, H. B. N. 1970. The ecology of running waters. Liverpool Univ. Press, England. 555 p.
———. 1975. The stream and its valley. Int. Ver. Theor. Angew. Limnol. Verh. 19: 1–15.
IRION, G. 1983. Tonmineralien in der Schwebfracht von Flüssen des Amazonasgebietes und von Papua Neuguinea. Zentralblatt für Geologie und Paläontologie, Stuttgart 1: 502–515.
IRION, G., J. ADIS, W. J. JUNK, AND F. WUNDERLICH. 1983. Sedimentological studies of the "Ilha da Marchantaria" in the Solimões/Amazon River near Manaus. Amazoniana 8: 1–18.
IRMLER, U. 1981. Überlebensstrategien von Tieren im saisonal überschwemmtem amazonischen überschwemmungswald. Zool. Anz. 206: 26–38.
———. 1986. Temperature dependent generation cycle for the cicindelid beetle *Pentacomia egregia* Chaud. (Coleoptera, Carabidae, Cicindelidae) of the Amazon valley. Amazoniana 9: 431–439.
IRMLER, U., AND K. FURCH. 1979. Production, energy and nutrient turnover of the cockroach *Epilampra irmleri* Rocha e Silva and Aguiar in a Central Amazonian inundation forest. Amazoniana 6: 497–520.
JOHNSON, F. L., AND D. T. BELL. 1976. Plant biomass and net primary production along a flood-frequency gradient in the streamside forest. Castanea 41: 156–165.
JUNK, W. J. 1970. Investigations on the ecology and production biology of the floating meadows (Paspalo-Echinochloetum) on the Middle Amazon, Part 1: the floating vegetation and its ecology. Amazoniana 2: 449–495.
———. 1973. Investigations on the ecology and production biology of the floating meadows (Paspalo-Echinochloetum) on the Middle Amazon, Part 2: the aquatic fauna in the root zone of floating vegetation. Amazoniana 4: 9–102.
———. 1980. Areas inundáveis — Um desafío para Limnología. Acta Amazonica 10: 775–795.
———. 1984. Ecology of the várzea, floodplain of Amazonian whitewater rivers, p. 215–244. In H. Sioli [ed.] The Amazon (Monographiae biologicae, Vol 56). Dr. W. Junk, The Hague, Netherlands.
———. 1985a. The Amazon floodplain — a sink or source for organic carbon?. Mitt. Geol. Paläont. Inst. Univ. Hamburg, SCOPE/UNEP Sonderbd. 58: 267–283.
———. 1985b. Temporary fat storage, an adaptation of some fish species to the waterlevel fluctuations and related environmental changes of the Amazon system. Amazoniana 9: 315–351.
———. 1985c. Aquatic plants of the Amazon system, p. 319–337. In B. R. Davies and K. F. Walker [ed.] The ecology of river systems. Dr W. Junk, The Hague, Netherlands.
JUNK, W. J., AND J. A. S. N. de MELLO. 1987. Impactos ecológicos das represas hidroelétricas na bacía amazonica brasileira. p. 367–385. In G. Kohlhepp and A. Schrader [ed.] Homem e natureza na Amazônia. Tübinger Geographische Studien Vol. 95.
JUNK, W. J., G. M. SOARES, AND F. M. CARVALHO. 1983. Distribution of fish species in a lake of the Amazon river floodplain near Manaus (Lago Camaleão), with special reference to extreme oxygen conditions. Amazoniana 7: 397–431.
JUNK, W. J., AND R. L. WELCOMME. 1989. Management of floodplains. In B. C. Patten [ed.] Wetlands and shallow continental water bodies, Vol. 1. SPB Academic Publishing, The Hague, Netherlands. (In press)
KEMP, G. P., AND J. W. DAY. 1984. Nutrient dynamics in a Louisiana swamp receiving agricultural runoff, p. 286–293. In K. C. Ewel and H. T. Odum [ed.] Cypress swamps. Univ. of Florida Press, Gainesville, FL.
KLOPATEK, J. M. 1978. Nutrient dynamics of freshwater riverine marshes and the role of emergent macrophytes, p. 195–216. In R. E. Good, D. F. Whigham, and R. L. Simpson [ed.] Freshwater wetlands. Ecological processes and management potential. Academic Press, New York, NY.
KRAMER, D. L., C. C. LINDSEY, G. E. E. MOODRE, AND E. D. STEVENS. 1978. The fishes and the aquatic environment of the Central Amazon Basin, with particular reference to respiratory patterns. Can. J. Zool. 56: 717–729.
LAMBOU, V. W. 1959. Fish populations of backwater lakes in Louisiana. Trans. Am. Fish. Soc. 88: 7–15.
LARSON, J. S., M. S. BEDINGER, C. F. BRYAN, S. BROWN, R. T. HUFFMAN, E. L. MILLER, D. G. RHODES, AND B. A. TUCHET. 1981. Transition from wetlands to uplands in southeastern bottomland hardwood forests, p. 225–273. In J. R. Clark and J. Benforado [ed.] Wetlands of bottomland hardwood forests. — Developments in agricultural and managed forest ecology 11, Elsevier Scientific Publishing Company, Amsterdam, Oxford, New York.
LINDSEY, A. A., R. D. PETTY, D. K. STERLING, AND W. VAN ASDALL. 1961. Vegetation and environment along the Wabash and Tippecanoe Rivers. Ecol. Monogr. 31: 105–156.
LITTLEJOHN, S., L. E. HOLLAND, R. JACOBSON, M. HUSTON, AND T. HORNUNG. 1985. Habits and habitats of fish in the upper Mississippi River. U.S. Fish and Wildlife Resource Publication. June 1985, LaCrosse, WI. 20 p.
LOWE-MCCONNELL, R. H. 1964. The fishes of the Rupununi savanna district of British Guiana, South America. Part 1. Ecological groupings of fish species and effects of the seasonal cycle on the fish. J. Linn. Soc. Zool. 45(304): 1–103.
———. 1975. Fish communities in tropical freshwaters. Their distribution, ecology, and evolution. Longman, London. 337 p.
LUNDBERG, J. G., W. M. LEWIS, Jr., J. F. SAUNDERS III, AND F. MAGO-LECCIA. 1987. A major food web component in the Orinoco River Channel: evidence from planktivorous fishes. Science 237: 81–83.
MAGALHÃES, C., AND I. WALKER. 1989. Larval development and ecological distribution of Central Amazon palaemonid shrimps. Crustaceana. (In press)
MARGALEF, R. 1968. Perspectives in ecological theory. The Univ. of Chicago Press, Chicago, IL. 111 p.
MARTINS, O. 1982. Geochemistry of the Niger River. Mitt. Geol. Paläont. Inst. Univ. Hamburge SCOPE/UNEP Sonderbd. 52: 357–364.
MARVIN, D. E., AND A. G. HEATH. 1968. Cardiac and respiratory response to gradual hypoxia in three ecologically distinct species of freshwater fish. Comp. Biochem. Physiol. 27: 349–355.
MCKNIGHT, J. S., D. D. HOOK, O. G. LANGDON, AND R. L.

JOHNSON. 1981. Flood tolerance and related characteristics of trees of the bottomland forest of the Southern United States. p. 29-69. In J. R. Clark and J. Benforado [ed.] Wetlands of bottomland hardwood forests. — Developments in agricultural and managed forest ecology 11, Elsevier Scientific Publishing Company, Amsterdam, Oxford, New York, NY.

MELACK, J. M., AND T. R. FISHER. 1983. Diel oxygen variations and their ecological implications in Amazon floodplain lakes. Arch. Hydrobiol. 98: 422-442.

MILLS, H. B., W. C. STARRETT, AND F. C. BELLROSE. 1966. Man's effect on the fish and wildlife of the Illinois River. Ill. Nat. Hist. Surv. Biol. Notes 57: 1-24.

MINSHALL, G. W., K. W. CUMMINS, R. C. PETERSEN, C. E. CUSHING, D. A. BRUNS, J. R. SEDELL, AND R. L. VANNOTE. 1985. Developments in stream ecosystem theory. Can. J. Fish. Aquat. Sci. 42: 1045-1055.

MINSHALL, G. W., R. C. PETERSEN, K. W. CUMMINS, T. L. BOTT, J. R. SEDELL, C. E. CUSHING, AND R. L. VANNOTE. 1983. Interbiome comparison of stream ecosystem dynamics. Ecol. Monogr. 53: 1-25.

MORRIS, L. A., R. N. LANGMEIER, T. R. RUSSELL, AND A. WITT, Jr. 1968. Effects of main stem impoundments and channelization upon the limnology of the Missouri River, Nebraska. Trans. Am. Fish. Soc. 97: 380-388.

NILSEN, H. C., AND R. W. LARIMORE. 1973. Establishment of invertebrate communities on log substrates in the Kaskaskia River, Illinois. Ecology 54: 366-374.

NORD, A. E., AND J. C. SCHMULBACH. 1973. A comparison of the macroinvertebrate aufwuchs in the unstabilized and stabilized Missouri River. Proc. S. D. Acad. Sci. 52: 127-139.

ODUM, E. P. 1959. Fundamentals of ecology. W. B. Saunders Co., London. 574 p.
1981. Foreward. In J. R. Clark and J. Benforado [ed.] Wetlands of bottomland hardwood forests. — Development in agricultural and managed forest ecology 11, Elsevier Scientific Publishing Company, Amsterdam, Oxford, New York, NY.

ODUM, E. P., AND A. A. DE LA CRUZ. 1967. Particulate organic detritus in a Georgia salt marsh-estuarine system, p. 383-388. In G. H. Lauff [ed.] Estuaries. Publ. 83 of the American Association for the Advancement of Science, Washington D.C.

PAARMANN, W., U. IRMLER, AND J. ADIS. 1982. *Pentacomia egregia* Chaud. (Carabidae, Cicindelidae), an univoltine species in the Amazonian inundation forest. Coleopterists Bull. 36: 183-188.

PETRERE, M., Jr. 1978. Pesca e esforço da pesca no Estado do Amazonas. II Locais, aparelhos de captura e estatísticas de desembarque. Acta Amazonica 8: 1-54.
1982. Ecology of the fisheries in the River Amazon and its tributaries in the Amazonas State (Brazil). Ph.D. thesis, Univ. of East Anglia, UK. 96 p.
1983. Relationships among catches, fishing effort and river morphology for eight rivers in Amazonas State (Brazil), during 1976-1978. Amazoniana 8: 281-296.

PFLIEGER, W. L., AND T. B. GRACE. 1987. Changes in the fish fauna of the Lower Missouri River, 1940-1983, p. 166-177. In W. J. Mathews and D. C. Heins [ed.] Community and evolutionary ecology of North American stream fishes. University of Oklahoma Press, Norman, OK.

PIANKA, E. R. 1970. On r- and K-selection. Am. Nat. 104: 592-597.

PIECZYŃSKA, E. 1972. Production and decomposition in the eulittoral zone of lakes, p. 271-285. In Z. Kajak and A. Hillbricht-Ilkowska [ed.] Proceedings of the IBP-UNESCO Symposium on Productivity Problems of Freshwaters. Kazimierz Dolny, Poland.

QUIRÓS, R., AND C. BAIGÚN. 1985. Fish abundance related to organic matter in the Plata River Basin, South America. Trans. Am. Fish. Soc. 114: 377-387.

REMSEN, J. V., Jr., AND T. A. PARKER. 1983. Contribution of river-created habitats to bird species richness in Amazonia. Biotropica 15: 223-231.

RIBEIRO, M. C. L. B. 1983. As migrações dos jaraquis (Pisces, Prochilodontidae) no Rio Negro, Amazonas, Brasil. M.S. thesis, Instituto Nacional de Pesquisas da Amazônia, Manaus, Brazil. 192 p.

RICHARDSON, R. E. 1921. The small bottom and shore fauna of the middle and lower Illinois River and its connecting lakes, Chillicothe to Grafton; its valuation; its sources of food supply; and its relation to the fishery. Ill. Nat. Hist. Surv. Bull. 13: 363-522.

RICHEY, J. E., J. T. BROCK, R. J. NAIMAN, R. C. WISSMAR, AND R. F. STALLARD. 1980. Organic carbon: oxidation and transport in the Amazon River. Science 207: 1348-1351.

RICHEY, J. E., E. SALATI, AND U. SANTOS. 1985. Biochemistry of the Amazon River: an update. Mitt. Geol. Paläont. Inst. Univ. Hamburg SCOPE/UNEP Sonderbd. 58: 245-257.

RISOTTO, S. P., AND R. E. TURNER. 1985. Annual fluctuation in abundance of the commercial fisheries of the Mississippi River and tributaries. N. Am. J. Fish Manage. 5: 557-574.

ROY, D. 1989. Physical and biological factors controlling the distribution and abundance of fish in Hudson/James Bay rivers, p. 159-171. In D. P. Dodge [ed.] Proceedings of the International Large River Symposium. Can. Spec. Publ. Fish. Aquat. Sci. 106.

RUTTNER, F. 1952. Fundamentals of limnology. Univ. Toronto Press, Toronto, Ont. 242 p.

SALO, J., R. KALLIOLA, I. HÄKKINEN, Y. MÄKINEN, P. NIEMELÄ, M. PUHAKKA, AND P. D. COLEY. 1986. River dynamics and the diversity of Amazon lowland forest. Nature 322: 254-258.

SANDUSKY, M. J., R. E. SPARKS, AND A. A. PAPARO. 1979. Investigations of declines in fingernail clam (*Musculium transversum*) populations in the Illinois River and Pool 19 of the Mississippi River. Bulletin of the American Malacological Union 1979: 11-15.

SANTOS, G. M. 1981. Estudos de alimentação e hábitos alimentares de *Schizodon fasciatus* Agassiz, 1829, *Rhytiodus microlepis* Kner, 1859 e *Rhytiodus argenteofuscus* Kner, 1859, do lago Janauacá -AM. (Osteichthyes, Characoidei, Anostomidae). Acta Amazonica 11: 267-283.

SCHMIDT, G. W. 1973a. Primary production of phytoplankton in the three types of Amazonian waters. II. The limnology of a tropical flood-plain lake in Central Amazônia, Lago do Castanho. Amazonas, Brazil. Amazoniana 4: 139-203.
1973b. Primary production of phytoplankton in the three types of Amazonian waters. III. Primary productivity of phytoplankton in a tropical flood-plain lake of Central Amazonia, Lago do Castanho. Amazonas, Brazil. Amazoniana 4: 379-404.

SCHUMM, S. A. 1977. The fluvial system. John Wiley and Sons, New York, NY. 338 p.

SEDELL, J. R., J. E. RICHEY, AND F. J. SWANSON. 1989. The river continuum concept: a basis for the expected ecosystem behavior of very large rivers? p. 49-55. In D. P. Dodge [ed.] Proceedings of the International Large River Sysmposium. Can. Spec. Publ. Fish. Aquat. Sci. 106.

SEGER, D. R., AND C. F. BRYAN. 1981. Temporal and spatial distribution of phytoplankton in the Lower Atchafalaya River Basin, Louisiana, p. 91-101. In L. A. Krumholz [ed.] Proceedings of Warmwater Streams Symposium. Allen Press, Lawrence, KS.

SEIDENSCHWARZ, F. 1986. Pionervegetation im Amazonasgebeit Perus. Ein pflanzensoziologischer Vergleich von vorandinem Flussufer und Kulturland. Monographs on Agriculture and Ecology of Warmer Climates, Vol. 3, Margraf, Tropical Scientific Books, Gaimersheim, West Germany. 226 p.

SHELFORD, V. E. 1954. Some lower Mississippi Valley floodplain biotic communities: their age and elevation. Ecology 15: 126-142.

SHEPPE, W., AND T. OSBORNE. 1971. Patterns of use of a floodplain by Zambian mammals. Ecol. Monogr. 41: 179-205.

SIGAFOOS, R. S. 1964. Botanical evidence of floods and floodplain deposition. U.S. Geological Survey Profesionnal Paper 485-A: 1-35.

SIOLI, H. 1984. The Amazon and its main affluents: hydrography, morphology of the river courses, and river types, p. 127-166. In H. Sioli [ed.] The Amazon (Monographiae biologicae, Vol 56). Dr. W. Junk, The Hague, Netherlands.

STATZNER, B. 1987. Characteristics of lotic ecosystems and consequences for future research directions, p. 365-390. In E. D. Schulze and H. Zwolfer [ed.] Potentials and limitations of ecosystem analysis. Ecological Studies 61. Springer-Verlag, Berlin.

STATZNER, B., AND B. HIGLER. 1985. Questions and comments on the River Continuum Concept. Can. J. Fish. Aquat. Sci. 42: 1038-1044.

1986. Stream hydraulics as a major determinant of benthic invertebrate zonation patterns. Freshwater Biol. 16: 127-139.

TALLING, J. F., AND J. RZÓSKA. 1967. The development of plankton in relation to hydrological regime in the Blue Nile. J. Ecol. 55: 637-662.

TEAL, J. M. 1962. Energy flow in the salt marsh system of Georgia. Ecology 43: 614-624.

THOMPSON, J. D. 1973. Feeding ecology of diving ducks on Keokuk Pool, Mississippi River. J. Wildl. Manage. 37: 367-381.

UETZ, G. W., K. L. VAN DER LAAN, G. F. SUMMERS, P. A. GIBSON, AND L. L. GETZ. 1979. The effects of flooding on floodplains in arthropod distribution, abundance, and community structure. Am. Midl. Nat. 101: 286-299.

VANNOTE, R. L., G. M. MINSHALL, K. W. CUMMINS, J. R. SEDELL, AND C. E. CUSHING. 1980. The river continuum concept. Can. J. Fish. Aquat. Sci. 37: 130-137.

WELCOMME, R. L. 1979. Fisheries ecology of floodplain rivers. Longman, London. 317 p.

1985. River fisheries. FAO Fish. Tech. Pap. 262: 330.

1989. Review of the present state of knowledge of fish stocks and fisheries of African rivers, p. 515-532. In D. P. Dodge [ed.] Proceedings of the International Large River Symposium. Can. Spec. Publ. Fish. Aquat. Sci. 106.

WHARTON, C. H., V. W. LAMBOU, J. NEWSOM, P. V. WINGER, L. L. GADDY, AND R. MANCKE. 1981. The fauna of bottomland hardwoods in Southeastern United States. p. 87-160. In J. R. Clark and J. Benforado [ed.] Wetlands of bottomland hardwood forests. — Developments in agricultural and managed forest ecology 11, Elsevier Scientific Publishing Company, Amsterdam, Oxford, New York, NY.

WINTERBOURN, M. J., J. S. ROUNICK, AND B. COWIE. 1981. Are New Zealand stream ecosystems really different?. N. Z. J. Mar. Freshwater Res. 15: 321-328.

WISSMAR, R. C., J. E. RICHEY, R. F. STALLARD, AND J. M. EDMOND. 1981. Plankton metabolism and carbon processes in the Amazon river, its tributaries, and floodplain waters, Peru-Brazil, May-June 1977. Ecology 62: 1622-1633.

WOOD, R. 1951. The significance of managed water levels in developing the fisheries of large impoundments. J. Tenn. Acad. Sci. 26: 214-235.

WORBES, M. 1983. Vegetationskundkiche Untersuchungen zweier überschwemmungswälder in Zentralamazonien. Amazoniana 8: 47-65.

1985. Structural and other adaptations to longterm flooding by trees in central Amazonia. Amazoniana 9: 459-484.

1986. Lebensbedingungen und Holzwachstum in zentralamazonischen überschwemmungswäldern. Scripta Geobotanica 17: 112.

YARBRO, L. A. 1983. The influence of hydrological variations on phosphorus cycling and retention in a swamp stream ecosystem, p. 199-221. In T. D. Fontaine and S. M. Bartell [ed.] Dynamics of ecosystems. Ann Arbor Science, Ann Arbor, MI.

Addresses of persons referred to as "pers. comm." or "unpublished data"

ANDERSON, R. V. Department of Biological Sciences, Western Illinois University, Macomb, IL 61455, USA.

BELLROSE, F. C., River Research Laboratory, Box 599, Havana, IL 62644, USA.

FISHER, T. R., Center for Environmental and Estuarine Studies, Horn Point, University of Maryland, Box 775, Cambridge, MD 21613, USA.

IRION, G., Forschungsinstitut Senckenberg, Abteilung für Meersgeologie und Meeresbiologie, Schleusenstr. 39a, D-2940 Wilhemshaven, W. Germany.

MARTIUS, C., Max-Planck-Inst. für Limnologie, AG Tropenökologie, D-2320 Plön, August Thienemannstr. 2, W. Germany.

RYDER, R. A., Ontario Ministry of Natural Resources, Box 2089, Thunder Bay, Ont. P7B 5E7, Canada.

A Pacific Interdecadal Climate Oscillation with Impacts on Salmon Production*

Nathan J. Mantua,+ Steven R. Hare,# Yuan Zhang,+
John M. Wallace,+ and Robert C. Francis@

ABSTRACT

Evidence gleaned from the instrumental record of climate data identifies a robust, recurring pattern of ocean–atmosphere climate variability centered over the midlatitude North Pacific basin. Over the past century, the amplitude of this climate pattern has varied irregularly at interannual-to-interdecadal timescales. There is evidence of reversals in the prevailing polarity of the oscillation occurring around 1925, 1947, and 1977; the last two reversals correspond to dramatic shifts in salmon production regimes in the North Pacific Ocean. This climate pattern also affects coastal sea and continental surface air temperatures, as well as streamflow in major west coast river systems, from Alaska to California.

September 1915 (*Pacific Fisherman* 1915)

Never before have the Bristol Bay [Alaska] salmon packers returned to port after the season's operations so early.

The spring [chinook salmon] fishing season on the Columbia River [Washington and Oregon] closed at noon on August 25, and proved to be one of the best for some years.

1939 Yearbook (*Pacific Fisherman* 1939)

The Bristol Bay [Alaska] Red [sockeye salmon] run was regarded as the greatest in history.

---

*JISAO Contribution Number 379.
+Joint Institute for the Study of the Atmosphere and Oceans, University of Washington, Seattle, Washington.
#International Pacific Halibut Commission, University of Washington, Seattle, Washington.
@Fisheries Research Institute, University of Washington, Seattle, Washington.
Corresponding author address: Nathan Mantua, Joint Institute for the Study of the Atmosphere and Oceans, University of Washington, Box 354235, Seattle, WA 98195-4235.
E-mail: mantua@atmos.washington.edu
In final form 6 January 1997.

The [May, June and July chinook] catch this year is one of the lowest in the history of the Columbia [Washington and Oregon].

August/September 1972 (*National Fisherman* 1972)

Bristol Bay [Alaska] salmon run a disaster.

Gillnetters in the Lower Columbia [Washington and Oregon] received an unexpected bonus when the largest run of spring chinook since counting began in 1938 entered the river.

1995 Yearbook (*Pacific Fishing* 1995)

Alaska set a new record for its salmon harvest in 1994, breaking the record set the year before.

Columbia [Washington and Oregon] spring chinook fishery shut down; west coast troll coho fishing banned.

1. Introduction

Pacific salmon production has a rich history of confounding expectations. For much of the past two decades, salmon fishers in Alaska have prospered while those in the Pacific Northwest have suffered. Yet, in the 1960s and early 1970s, their fortunes were essentially reversed. Could this pattern of alternating fishery production extremes be connected to climate changes in the Pacific basin?

In this article we present a synthesis of results derived from the analyses of climate records and data describing biological aspects of variability in the large marine ecosystems of the northeast Pacific Ocean. Our goal is to highlight the widespread connections between interdecadal climate fluctuations and ecological variability in and around the North Pacific basin.

A considerable body of literature has been devoted to the discussion of persistent widespread changes in Pacific basin climate that took place in the late 1970s (Namias 1978; Trenberth 1990; Ebbesmeyer et al. 1991; Graham 1994; Trenberth and Hurrell 1994). Several studies have also documented interdecadal climate fluctuations in the Pacific basin, of which the changes that took place in the late 1970s are but a single realization (Ebbesmeyer et al. 1989; Francis and Hare 1994 and Hare and Francis 1995, hereafter FH–HF; Latif and Barnett 1994, 1996; Ware 1995; Hare 1996; Zhang 1996; Zhang et al. 1997, hereafter ZWB).

Widespread ecological changes related to interdecadal climate variations in the Pacific have also been noted. Dramatic shifts in an array of marine and terrestrial ecological variables in western North America coincided with the changes in the state of the physical environment in the late 1970s (Venrick et al. 1987; Ebbesmeyer et al. 1991; Brodeur and Ware 1992; Roemmich and McGowan 1995; Francis et al. 1997). Rapid changes in the production levels of major Alaskan commercial fish stocks have been connected to interdecadal climate variability in the northeast Pacific (Beamish and Boullion 1993; Hollowed and Wooster 1992; FH–HF), and similar climate–salmon production relationships have been observed for some salmon populations in Washington, Oregon, and California (Francis and Sibley 1991; J. Anderson 1996, personal communication).

Our results add support to those of previous studies suggesting that the climatic regime shift of the late 1970s is not unique in the century-long instrumental climate record, nor in the record of North Pacific salmon production. In fact, we find that signatures of a recurring pattern of interdecadal climate variability are widespread and detectable in a variety of Pacific basin climate and ecological systems. This climate pattern, hereafter referred to as the Pacific (inter) Decadal Oscillation, or PDO (following coauthor S.R.H.'s suggestion), is a pan-Pacific phenomenon that also includes interdecadal climate variability in the tropical Pacific.

2. Data and methodology

We analyze a wide collection of historical records of Pacific basin climate and selected commercial salmon landings. Specifically, this study examines records of (i) tropical and Northern Hemisphere extratropical sea surface temperature (SST) and sea level pressure (SLP); (ii) wintertime North American land surface air temperatures and precipitation; (iii) wintertime Northern Hemisphere 500-mb height fields; (iv) SST along the west coast of North America; (v) selected streamflow records from western North America; and (vi) salmon landings from Alaska, Washington, Oregon, and California.

Monthly mean SST data for the period of record 1900–93 were obtained from an updated version of the quality-controlled U.K. Meteorological Office Historical SST Dataset (HSSTD) provided by the Climatic Research Unit, University of East Anglia (Folland and Parker 1990, 1995). These data are on a 5° lat × 5° long grid. The monthly mean, 1° lat × 1° long gridded data of the Optimally Interpolated SST (OISST, Reynolds and Smith 1995) are averaged into 5° boxes and used to extend the HSSTD through the January 1994–May 1996 period of record. We also use 2° lat × 2° long Comprehensive Ocean–Atmosphere Data Set (COADS, Fletcher et al. 1983) SST for the period of record 1900–92 in the construction of Fig. 2.

Monthly mean SLP data were obtained from two sources: first, 5° lat × 5° long gridded fields from the Data Support Section/Computing Facility at the National Center for Atmospheric Research (NCAR) for the period of record 1900–May 1996 (Trenberth and Paolino 1980); and second, 2° lat × 2° long gridded surface marine observations from COADS for the period of record 1900–92, which are used to construct the station-based Southern Oscillation Index (SOI) shown in Fig. 1 and the SLP map in Fig. 2.

For the period of record 1900–92, the COADS-based SOI used here was constructed following

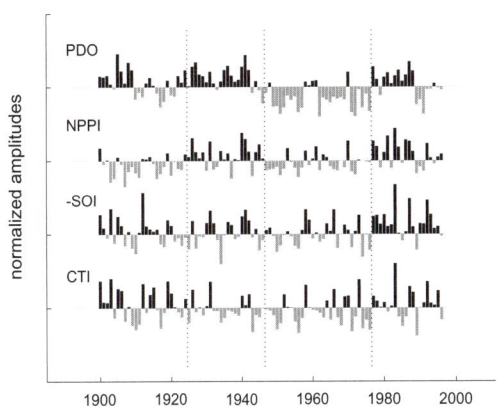

FIG. 1. Normalized winter mean (November–March) time histories of Pacific climate indices. Dotted vertical lines are drawn to mark the PDO polarity reversal times in 1925, 1947, and 1977. Positive (negative) values of the NPPI correspond to years with a deepened (weakened) Aleutian low. The negative SOI is plotted so that it is in phase with the tropical SST variability captured by the CTI. Positive value bars are black, negative are gray.

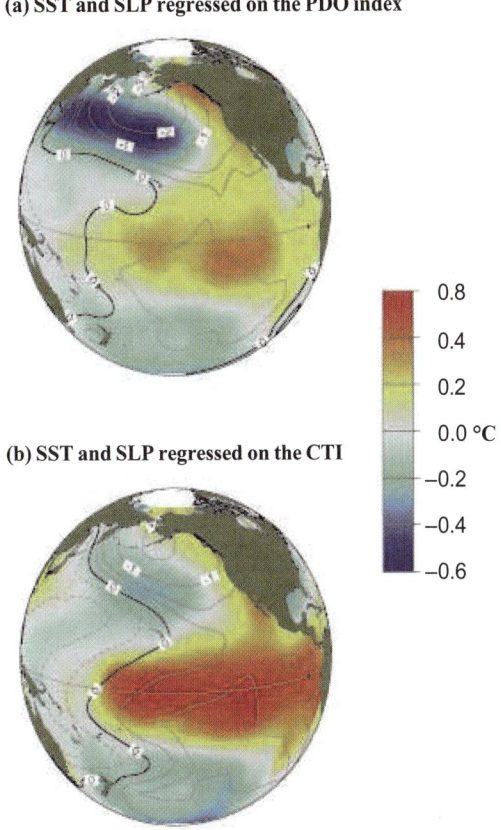

FIG. 2. COADS SST (color shaded) and SLP (contoured) regressed upon (a) the PDO index and (b) the CTI for the period of record 1900–92. Contour interval is 1 mb, with additional contours drawn for +/−0.25 and 0.50 mb. Positive (negative) contours are dashed (solid).

ZWB. The Tahiti pole is defined as the average SLP anomaly from 20°N to 20°S latitude from the international date line to the coast of South and Central America, while the Darwin pole is defined as the average SLP anomaly over the remainder of the global tropical oceans within the same range of latitudes. Missing SOI values for the period of record 1913–20 and 1993–May 1996, were estimated from a linear regression with the traditional Tahiti–Darwin SOI based on the common period of record 1933–90, obtained from the National Oceanic and Atmospheric Administration/National Centers for Environmental Prediction (NOAA/NCEP) Climate Prediction Center. For an early description of the Southern Oscillation the reader is referred to Walker and Bliss (1932).

Gridded, global, land surface air temperature and precipitation anomalies for the period of record 1900–92, based on station data, were obtained from the Carbon Dioxide Information Analysis Center in Oak Ridge, Tennessee. The air temperature data are provided as monthly anomalies on a 5° lat × 10° long grid, over land only (Jones et al. 1985). We used "cold-season" means (November–March) for Fig. 3a. The precipitation anomalies are provided as (3 month) seasonal mean anomalies on a 4° lat × 5° long grid, over land only (Eischeid et al. 1991). We used the December–February seasonal mean anomalies in constructing Fig. 3b.

Gridded, Northern Hemisphere 500-mb height fields were obtained from NMC (National Meteorological Center, now NCEP) operational analysis fields, as described by Kushnir and Wallace (1989). November through March mean anomalies were used in constructing Fig. 4.

Monthly mean streamflow records for the Kenai River at Cooper's Landing, Alaska; the Skeena River at Usk, British Columbia, Canada and the Fraser River at Hope, British Columbia, Canada; and the Columbia River at The Dalles, Oregon, were obtained from the National Water Data Exchange, which is part of the United States Geological Survey (USGS). The monthly records were used to generate annual water year (October–September) flow indices for each stream. The time series labelled BC/Columbia Streamflow in Fig. 5 is a composite of the normalized Skeena, Fraser, and Columbia river water year streamflow anomalies.

Coastal SST time series for British Columbia stations were obtained from the Institute of Ocean Sciences in Sidney, British Columbia, Canada. The time series for coastal BC SST shown in Fig. 5 is a composite of eight individual time series from the following coastal observing stations: Amphitrite Point, Departure Bay, Race Rocks, Langara Island, Kains Island, McInnes Island, Entrance Island, and Pine Island. We use a composite index in an attempt to emphasize regional-scale nearshore SST variability over the finescale variability that exists in that topographically diverse region.

Monthly mean values for Scripps Pier SST were obtained from the Scripps Institution of Oceanography in La Jolla, California. Scripps Pier SST variability is well correlated with that along the Alta and Baja California coastline (J. McGowan 1996, personal communication).

Coastal Gulf of Alaska cold season air temperatures were obtained from the National Climate Data Center. The November–March mean Gulf of Alaska air temperatures shown in Fig. 5 are a composite of Kodiak, King Salmon, and Cold Bay, Alaska, station records.

Prior to compositing, each individual SST, streamflow, and air temperature time series was normalized with respect to the 1947–95 period of record, a period for which data are available for all the time series used in the construction of Fig. 5. The mean for the available period of record was then removed from the composite time series before plotting in Fig. 5.

Alaska salmon landings for the period of record 1925–91 were provided by the Alaska Department of Fish and Game (1991). Catch data for 1992 through 1995 were obtained from Pacific Fishing magazine (1994, 1995). We focus on the catch records of sockeye salmon in western and central Alaska, and that of pink salmon in central and southeast Alaska (shown in Fig. 6). These four regional stocks account for about 75% of Alaska's annual salmon catch. The period of record from 1920 through the 1930s represents a "fishing-up" period while the industry was experiencing rapid growth. Subsequent to the late 1930s, fisheries for these stocks have been fully developed, and the catch records are good indicators of stock abundance (Beamish and Bouillon 1993; FH–HF).

Additionally, the record of chinook salmon catch from the Columbia River for the period of record 1938–93 and coho landings from Washington–Oregon–California (WOC) for the period of record 1925–93 are also shown. These records were obtained from the Washington Department of Fisheries (WDF), the Oregon Department of Fish and

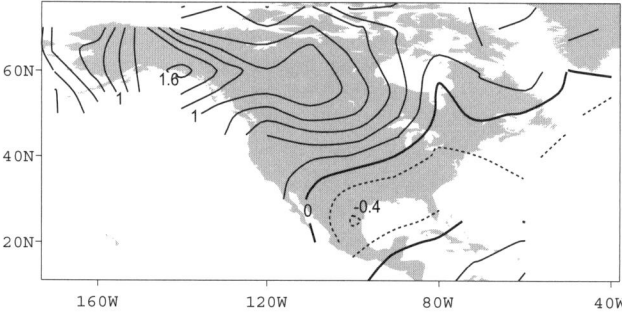

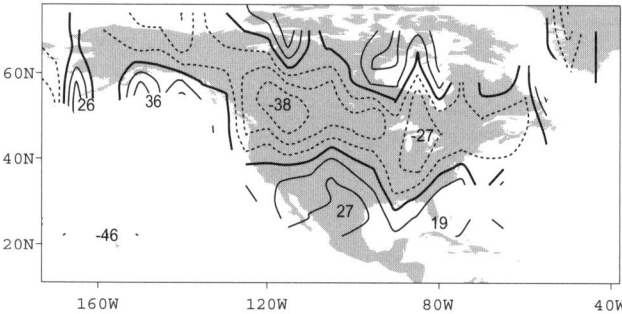

FIG. 3. Maps of PDO regression and correlation coefficients: (a) November–March surface air temperature regressed upon the PDO index shown in Fig. 1; contour interval is 0.2°C. (b) Correlation coefficients (× 100) between December–February precipitation and the PDO index shown in Fig. 1; contour interval is 10. Positive (negative) contours are solid (dashed).

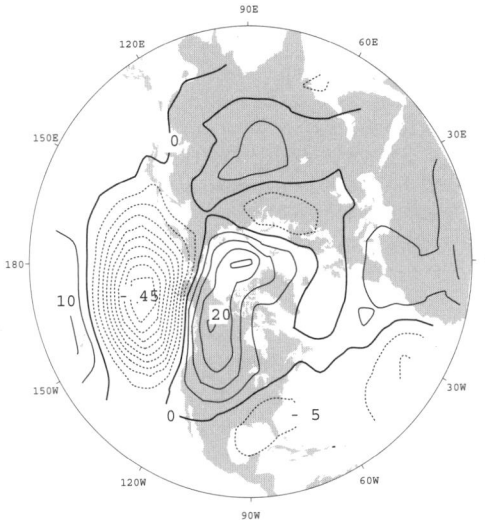

FIG. 4. Wintertime Northern Hemisphere 500-mb heights regressed upon the PDO index for the period of record 1951–90. Contour interval is 5 m, positive (negative) contours are solid (dashed).

Wildlife (ODFW), and the California Department of Fish and Game (WDF and ODFW 1992).

Parallel EOF/PC analyses of the monthly SST and SLP anomaly fields, carried out independently by two of the present authors, were based on the temporal covariance matrix from the 1900–93 period of record. For SST, we used the covariance matrix created from monthly HSSTD anomalies poleward of 20°N in the Pacific basin (Zhang 1996). For SLP, we used the covariance matrix created from monthly NCAR SLP anomalies poleward of 20°N and between 110°E and 110°W (Hare 1996). The resulting November–March mean PCs were normalized prior to plotting in Fig. 1. The leading PC for SLP in the North Pacific sector is labelled NPPI, while that for SST is labelled PDO.

3. Characteristics of the PDO

Of particular interest to this study is the fact that, since at least the 1920s, interdecadal fluctuations in the dominant pattern of North Pacific SLP (NPPI) have closely paralleled those in the leading North Pacific SST pattern (PDO) (Fig. 1; Zhang 1996; ZWB; Latif and Barnett 1996). It is this coherent, interdecadal timescale ocean–atmosphere covariability that we see as the essence of the PDO climate signature. For convenience, throughout the remainder of this report we refer to the time history of the leading eigenvector of North Pacific SST as an index for the state of the PDO.

Also shown in Fig. 1 are the SOI and the Cold Tongue Index (CTI, which is the average SST anomaly from 6°N to 6°S, 180° to 90°W), indices commonly used to monitor the atmospheric and oceanic aspects of ENSO, respectively. The SOI and CTI are correlated with the PDO (see Table 1) such that warm- (cold-) phase ENSO-like conditions tend to coincide with the years of positive (negative) polarity in the PDO. Interestingly, fluctuations in the CTI are mostly interannual, while those in the PDO are predominantly interdecadal (ZWB).

Interdecadal and interannual timescales are both apparent in the indices of atmospheric variability at low and high northern latitudes over the Pacific. The NPPI and SOI are correlated such that the mean wintertime Aleutian Low tends to be more (less) intense during winters with weakened (intensified) easterly winds near the equator in the Pacific.

Correlations between the atmospheric and oceanic climate indices shown in Fig. 1 within respective high- and low-latitude ranges are relatively strong. The NPPI is moderately well correlated with that of the extratropical SST, while at tropical latitudes the SOI and CTI are very well correlated (see Table 1).

By regressing the records of wintertime SST and SLP upon the PDO index, the spatial patterns typically associated with a positive unit standard deviation of the PDO are generated (Fig. 2a). The largest PDO-related SST anomalies are found in the

TABLE 1. Correlation coefficients for the Pacific basin climate indices, shown in Fig. 1, for the period of record 1900–92. Correlation coefficients have been adjusted to reflect the effective degrees of freedom, as a function of autocorrelation, in each time series.

	PDO	NPPI	SOI	CTI
PDO	—	0.50	−0.35	0.38
NPPI		—	−0.39	0.42
SOI			—	−0.82
CTI				—

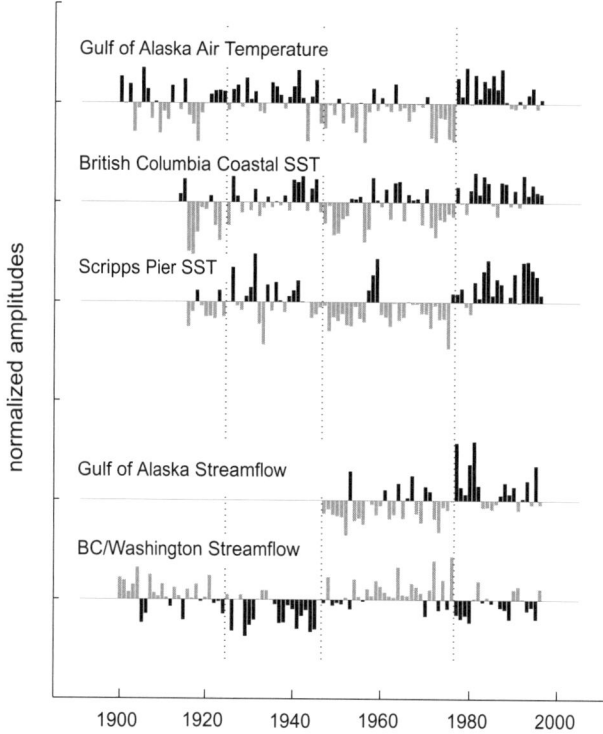

FIG. 5. Selected regional climate time series with PDO signatures. Dotted vertical lines are drawn to mark the PDO polarity reversal times in 1925, 1947, and 1977. Bars are shaded as in Fig. 1, with the shading convention reversed for the BC/Washington streamflow index.

Shown in Fig. 2b are the SST and SLP fields regressed upon the CTI, thus this map shows anomalies typically associated with a unit standard deviation ENSO index. Comparing Fig. 2a with Fig. 2b, it is evident that the tropical PDO-spatial signatures are in many ways reminiscent of canonical warm-phase ENSO SST and SLP anomalies (Rasmussen and Carpenter 1982). However, the PDO amplitudes in the tropical fields are weaker than those obtained by regressing the surface fields upon the CTI. Likewise, the PDO-regression amplitudes in the Northern Hemisphere extratropics are stronger than those obtained from regressions upon the CTI (ZWB).

To establish the significance and consistency of polarity reversals in time—referred to by some authors as regime shifts—FH–HF and Hare (1996) utilized a technique known as intervention analysis (Box and Tiao 1975), which is an extension of Autoregressive Integrated Moving Average (ARIMA) modeling (Box and Jenkins 1976). We applied this analysis to each of the time series shown in Fig. 1. Intervention analysis is essentially a two sample t test that can be applied to autocorrelated data, which is a common feature of environmental time series. While interventions can take many forms, we tested only step interventions. The implicit model, therefore, for each variable is a sequence of abruptly shifting levels, accounting for a significant portion of the total variance, around which occurs residual variability, either random or autocorrelated.[1]

central North Pacific Ocean, where a large pool of cooler than average surface water has been centered for much of the past 20 yr. The peak amplitude of the SST regression coefficients in the cold pool are on the order of −0.5°C. The narrow belt of warmer than average SST that, in the past two decades, has prevailed in the nearshore waters along the west coast of the Americas is also a distinctive feature of this pattern. Note also that the Southern Hemisphere midlatitude SST signature is very similar to that in the northern extratropics. The SLP anomalies that are typical of the positive PDO are characterized by basin-scale negative anomalies between 20° and 60°N. The peak amplitude of the midlatitude wintertime SLP signature is about 4 mb, which represents an intensification of the climatological mean Aleutian low. This SLP pattern is very similar to the dominant pattern of wintertime North Pacific SLP variability. It is noteworthy that there are no strong PDO signatures in the Atlantic or Indian Ocean SST and SLP fields.

---

[1]We followed the standard three-step process in fitting the intervention models. First we identify a model. For all time series, the initial model consisted of five parameters: Three interventions, a lag-1 autoregressive term and a constant. The three interventions (phase reversals) we used were 1925, 1947, and 1977. The timing of the interventions was derived independently in earlier studies by several of the authors in this study (FH–HF; ZWB). In the second step, parameters are estimated for significance. If any parameters are statistically insignificant, the least significant is dropped and the remaining parameters reestimated. This sequence is repeated as necessary. The model is then accepted if the final step, a white noise test for model residuals, is passed.

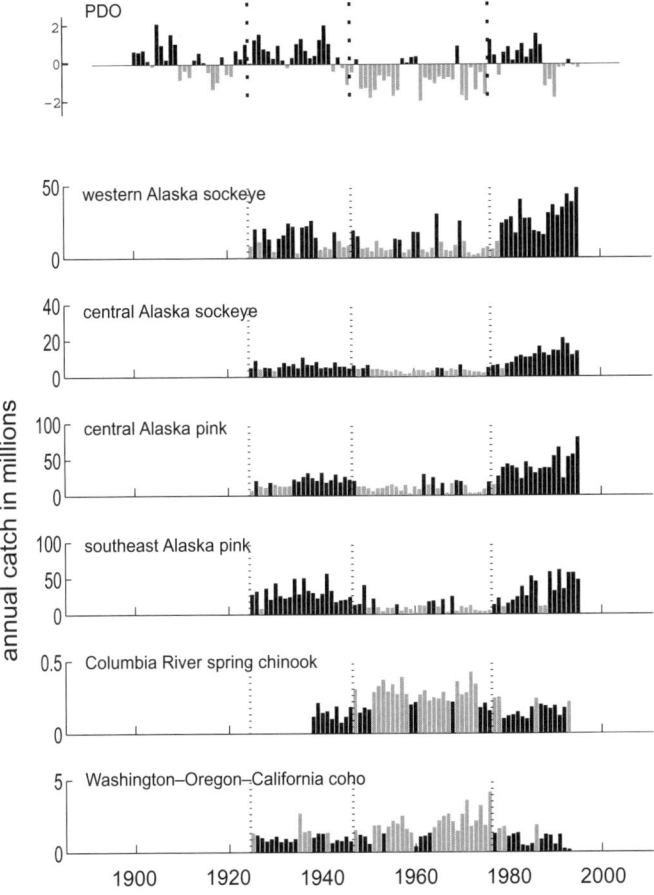

FIG. 6. Selected Pacific salmon catch records with PDO signatures. For Alaska catch, black (gray) bars denote values that are greater (less) than the long-term median. The shading convention is reversed for WOC coho and Columbia River spring chinook catch. Dotted vertical lines are drawn in each panel to mark the PDO polarity reversal times in 1925, 1947, and 1977. At the top, the PDO index is repeated from Fig. 1.

The statistical significance of the intervention model parameters are shown in Table 2. Excluding the CTI, polarity reversals in 1977 are supported in each of the time series shown in Fig. 1. Additional sign reversals in 1925 and 1947 are supported by the PDO and NPPI time series but not for the SOI or CTI.

The implications of this statistical exercise are as follows. We have identified an interdecadal climate signal that is evident in the oceanic and atmospheric climate record. We attribute these signatures to the PDO. During this century, using the North Pacific SST pattern time series as the indicator of polarity, the PDO was predominantly positive between 1925 and 1946, negative between 1947 and 1976, and positive since 1977. Note that these multidecade epochs contain intervals of up to a few years in length in which the polarity of the PDO is reversed e.g., the positive PDO values in 1958–61, and the strongly negative PDO values in 1989–91).

4. Coastal and continental signatures of the PDO

The signature of the PDO is clearly evident in the wintertime surface climate record for much of North America but not for that of the other continents. The strongest coefficients of wintertime air temperature regressed upon the PDO index are located in northwestern North America (Fig. 3a; cf. Latif and Barnett 1994, Fig. 5b), with local maxima of opposing centers over south central Alaska–western Canada and the southeastern United States. The PDO is positively correlated[2] with wintertime precipitation along the coast of the central Gulf of Alaska and over northern Mexico and south Florida, and negatively correlated with that over much of the interior of North America and over the Hawaiian Islands.

The continental PDO surface climate signatures are consistent with PDO-related circulation anomalies on the hemispheric scale. The Pacific–North America (PNA) (Wallace and Gutzler 1981) pat-

---
[2]To highlight the regional patterns of the PDO Dec–Feb precipitation signal over the North American continent, the correlation map is shown instead of the regression map. The regression coefficients are skewed toward extreme values in the Pacific Northwest and central Gulf of Alaska. Typical precipitation anomalies for a unit standard deviation positive PDO are about +20 to +30 mm for the central Gulf of Alaska, −20 to −30 mm for western Washington state, −40 mm for the Hawaiian Islands, +5 mm over northern Mexico, and −10 mm over the Great Lakes.

TABLE 2. P values for tests of step-changes in the mean level of the Pacific basin climate indices shown in Fig. 1. The four time periods tested for changes in the mean level were 1900–24, 1925–46, 1947–76, 1977–96. P-values greater than 0.05 are labeled "ns" (not significant).

Climate Index	Intercept	1925 step	1947 step	1977 step
PDO	ns	0.001	0.000	0.000
NPPI	0.005	0.001	0.000	0.000
SOI	ns	ns	ns	0.001
CTI	ns	ns	ns	ns

tern emerges when the cold season (November–March) 500-mb height fields are regressed upon the PDO index for the period of record 1951–90 (Fig. 4). This relationship suggests that during epochs in which the PDO is in its positive polarity, coastal central Alaska tends to experience an enhanced cyclonic (counterclockwise) flow of warm, moist air, which is consistent with heavier than normal precipitation. Washington state and British Columbia also tend to be subject to an increased flow of relatively warm humid air, but in their case it is within an area of enhanced anticyclonic circulation that is dynamically unfavorable for heavier than normal precipitation.

In an analysis of springtime (1 April) snowcourse data for the western United States, Cayan (1996) finds that the leading eigenvector of snowpack variability, what he calls the Idaho pattern, is centered in the Pacific Northwest. Cayan's time series for the Idaho pattern has tracked our PDO index since at least 1935 (when his data begins). This pattern of snowpack variability is consistent with the PDO-related wintertime air temperature and precipitation patterns shown in Fig. 3: relatively warm (cool) winter air temperatures and anomalously low (high) precipitation during positive (negative) PDO years contribute to reduced (enhanced) snowpack in the Pacific Northwest. Furthermore, Cayan's composite wintertime 700-mb height fields for the extreme years reveal that variability in the Idaho snowpack pattern is largely controlled by PNA circulation anomalies (cf. Cayan's Figs. 3 and 6 with our Figs. 3b and 4).

We used the PDO correlation and regression maps (Figs. 2 and 3) as guides to search for the local and regional instrumental records of PDO-driven climate variability shown in Fig. 5. Wintertime surface air temperature along the Gulf of Alaska, and SST near the coast from Alaska to southern California, varies in phase with the PDO. During positive PDO years the annual water year discharge in the Skeena, Fraser, and Columbia Rivers is on average 8%, 8%, and 14% lower, respectively, than that during negative PDO years. In contrast, positive-PDO-year discharge from the Kenai River in the central Gulf of Alaska region is on average about 18% higher than that during the negative polarity PDO years. Cayan and Peterson (1989) also noted that this dipole pattern in west coast streamflow fluctuations is related to the favored pattern of SLP variability in the North Pacific.

5. The PDO and salmon production in the northeast Pacific

Commercial fisheries for Alaskan pink and sockeye salmon are among the most lucrative in the United States (U.S. Department of Commerce 1994, 1995). The unique life history of salmon, which begins and ends in freshwater streams and involves an extensive period of feeding in the ocean pasture, makes them vulnerable to a variety of environmental changes. A growing body of evidence suggests that many populations of Pacific salmon are strongly influenced by marine climate variability (Pearcy 1992; Beamish and Bouillon 1993; FH–HF; Beamish et al. 1995; Francis et al. 1997).

A remarkable characteristic of Alaskan salmon abundance over the past half-century has been the large fluctuations at interdecadal timescales that resemble those of the PDO (Fig. 6, see also Table 3) (FH–HF; Hare 1996). Time series for WOC coho and Columbia River spring chinook landings tend to be out of phase with the PDO index (Fig. 6), though the correspondence is less compelling than that with Alaskan salmon. The weaker connections between the WOC and Columbia River salmon populations and the PDO may be a result of differing environmental–biological interactions. On the other hand, climatic influences on salmon in their southern ranges may also be masked or overwhelmed by anthropogenic impacts: Alaskan stocks are predominantly wild spawners in pristine watersheds, while the WOC coho and Columbia River spring chinook are mostly of hatchery origin and originate in watersheds that have been significantly altered by human activities.

The best-fit interventions for the Alaskan sockeye stocks occur 2 and 3 yr after those identified in the PDO history, while the best-fit interventions for the Alaskan pink salmon stocks occur 1 yr following the climate shifts (FH–HF). It is believed that sockeye and pink salmon abundances are most significantly impacted by marine climate variability early in the ocean phases of their life cycles (Hare 1996). If this is true, the key biophysical interactions are likely taking place in the nearshore marine and estuarine environments where juvenile salmon are generally found.

Recent work suggests that the marine ecological response to the PDO-related environmental changes starts with phytoplankton and zooplankton at the base of the food chain and works its way up to top-level predators like salmon (Venrick et al. 1987; FH–HF; Roemmich and McGowan 1995; Hare 1996; Brodeur et al. 1996; Francis et al. 1997). This "bottom-up" enhancement of overall productivity appears to be closely related to upper-ocean changes that are characteristic of the positive polarity of the PDO. For example, some phytoplankton–zooplankton population dynamics models are sensitive to specified upper-ocean mixed-layer depths and temperatures. For the decade following the 1960–76 period of record, such models have successfully simulated aspects of the observed increases in Gulf of Alaska productivity as a response to an observed 20%–30% shoaling and 0.5° to 1°C warming of the mixed layer (Polovina et al. 1995).

To the extent that high streamflows favor high survival of juvenile salmon, PDO-related streamflow variations are likely working in concert with the changes to the near-shore marine environment in regard to impacts on salmon production. For Alaskan salmon, the typical positive PDO year brings enhanced streamflows and nearshore ocean mixed-layer conditions favorable to high biological productivity. Generally speaking, the converse appears to be true for Pacific Northwest salmon.

6. Discussion

Our synthesis of climate and fishery data from the North Pacific sector highlights the existence of a very large-scale, interdecadal, coherent pattern of environmental and biotic changes. It has recently come to our attention that Minobe (1997) has compiled a complementary study of North Pacific climate variability that includes SST indices from the coastal Japan and Indian Ocean–Maritime Continent regions. Especially relevant to our work is the fact that Minobe used instrumental records to independently identify the same dates we promote for climatic regime shifts (1925, 1947, and 1977). Also intriguing is Minobe's analysis of (tree ring) reconstructed continental surface temperatures that suggest PDO-like climate variability has a characteristic recurrance interval of 50–70 yr and that these fluctuations are evident throughout the past 3 centuries.

It is clear from a visual inspection of the time series shown in Figs. 1, 5, and 6 that not all changes in our PDO index are indicative of interdecadal regime shifts that are equally apparent in the other indices. The difficulties inherent in real-time assessment of the state of the PDO are illustrated by the recent period of record: Alaskan salmon catches and coastal SSTs have remained above average since the late 1970s, while, in contrast, the PDO index dipped well below average from 1989–91 and has hovered around normal since this time. Without the benefit of hindsight it is virtually impossible to characterize such periods and to recognize long-lived regime shifts at the time they occur.

The ENSO and PDO climate patterns are clearly related, both spatially and temporally, to the extent that the PDO may be viewed as ENSO-like interdecadal climate variability (Tanimoto et al. 1993; ZWB). While it may be tempting to interpret interdecadal climatic shifts as responses to individual (tropical) ENSO events, it seems equally

TABLE 3. Percent change in mean catches of four Alaskan salmon stocks following major PDO polarity changes in 1947 and 1977. Mean catch levels were estimated from intervention models fitted to the data and incorporating a 1-yr lag for both pink salmon stocks, a 2-yr lag for western sockeye, and a 3-yr lag for central sockeye.

Salmon stock	1947 step	1977 step
Western AK sockeye	−37.2%	+242.2%
Central AK sockeye	−33.3%	+220.4%
Central AK pink	−38.3%	+251.9%
Southeast AK pink	−64.4%	+208.7%

conceivable that the state of the interdecadal PDO constrains the envelope of interannual ENSO variability.

To our knowledge, there are no documented robust relationships between Pacific salmon abundance and indices of ENSO. The slowly varying time series of salmon catches examined in this study are much more coherent with the interdecadal aspects of the PDO than the higher frequency fluctuations in tropical ENSO indices. In the future it seems very likely that the PDO will continue to change polarity every few decades, as it has over the past century, and with it the abundance of Alaskan salmon and other species sensitive to environmental conditions in the North Pacific and adjacent coastal waters.

This climatic regime–driven model of salmon production has broad implications for fishery management (Hare 1996; Adkison et al. 1996). The most critical implication concerns periods of low productivity, such as currently experienced by WOC salmon. Management goals, such as the current legislative mandate to double Washington State salmon production[3] (Salmon 2000 Technical Report 1992), may simply not be attainable when environmental conditions are unfavorable. Conversely, in a period of climatically favored high productivity, managers might be well advised to exercise caution in claiming credit for a situation that may be beyond their control.

Acknowledgments. We thank Ileana Bladé and Nick Bond for carefully reading an early draft of this article and offering constructive critiques. This study was prompted by the University of Washington's interdisciplinary project, "An Integrated Assessment of the Dynamics of Climate Variability, Impacts, and Policy Response Strategies for the Pacific Northwest," and was funded by NOAA's cooperative agreement #NA67RJ0155, Washington Sea Grant, and The Hayes Center.

References

Adkison, M. D., R. M. Peterman, M. P. Lapointe, D. M. Gillis, and J. Korman, 1996: Alternative models of climatic effects on sockeye salmon (Oncorhynchus nerka) productivity in Bristol Bay, Alaska and Fraser River, British Columbia. *Fish. Oceanogr.,* **5,** 137–152.

---

[3]"The [Washington State] legislature hereby establishes a production goal to double the state-wide salmon catch by the year 2000" (Salmon 2000 Technical Report 1992).

Alaska Department of Fish and Game (ADFG), 1991: Alaska commercial salmon catches, 1878–1991. Division of Commercial Fish Regional Information Rep. 5J91-16, Juneau, AK, 88 pp. [Available from Alaska Department of Fish and Game, P.O. Box 25526, Juneau, AK 99802.]

Beamish, R. J., and D. R. Bouillon, 1993: Pacific salmon production trends in relation to climate. *Can. J. Fish. Aquat Sci.,* **50,** 1002–1016.

——, G. E. Riddell, C.-E. M. Neville, B. L. Thomson, and Z. Zhang, 1995: Declines in chinook salmon catches in the Strait of Georgia in relation to shifts in the marine environment. *Fish. Oceanogr.,* **4,** 243–256.

Box, G. E. P., and G. C. Tiao, 1975: Intervention analysis with applications to economic and environmental problems. *J. Amer. Stat. Assoc.,* **70,** 70–79.

——, and G. M. Jenkins, 1976: *Time Series Analysis: Forecasting and Control.* Holden-Day, 575 pp.

Brodeur, R. D., and D. M. Ware, 1992: Interannual and interdecadal changes in zooplankton biomass in the subarctic Pacific Ocean. *Fish. Oceanogr.,* **1,** 32–38.

——, B. W. Frost, S. R. Hare, R. C. Francis, and W. J. Ingraham Jr. 1996: Interannual variations in zooplankton biomass in the Gulf of Alaska and covariation with California Current zooplankton. *Calif. Coop. Oceanic Fish. Invest. Rep.,* **37,** 80–99.

Cayan, D. R., 1996: Interannual climate variability and snowpack in the western United States. *J. Climate,* **9,** 928–948.

——, and D. H. Peterson, 1989: The influence of the North Pacific atmospheric circulation and streamflow in the west. *Aspects of Climate Variability in the Western Americas, Geophys. Monogr.,* No. 55, Amer. Geophys. Union, 375–397.

Ebbesmeyer, C. C., C. A. Coomes, C. A. Cannon, and D. E. Bretschneider, 1989: Linkage of ocean and fjord dynamics at decadal period. *Climate Variability on the Eastern Pacific and Western North America, Geophys. Monogr.,* No. 55, Amer. Geophys. Union, 399–417.

——, D. R. Cayan, D. R. McLain, F. H. Nichols, D. H. Peterson, and K. T. Redmond, 1991: 1976 step in the Pacific climate: Forty environmental changes between 1968–1975 and 1977–1985. *Proc. Seventh Annual Pacific Climate Workshop,* Asilomar, CA, California Dept. of Water Research, 115–126.

Eischeid, J. K., H. F. Diaz, R. S. Bradley, and P. D. Jones, 1991: A comprehensive precipitation data set for global land areas. U.S. Dept. of Energy, Carbon Dioxide Research Program DOE/ER-69017T-H1, TR051, Washington, DC, 81 pp. [Available from CDIAC, Oak Ridge National Laboratory, P.O. Box 2008, Oak Ridge, TN 37831-6335.]

Fletcher, J. O., R. J. Slutz, and S. D. Woodruff, 1983: Towards a comprehensive ocean–atmosphere dataset. *Trop. Ocean–Atmos. Newslett.,* **20,** 13–14.

Folland, C. K., and D. E. Parker, 1990: Observed variations of sea surface temperature. *Climate-Ocean Interaction,* M. E. Schlesinger, Ed., Kluwer, 21–52.

——, and ——, 1995: Correction of instrumental biases in historical sea surface temperature data. *Quart. J. Roy. Meteor. Soc.,* **121,** 319–367.

Francis, R. C., and T. H. Sibley, 1991: Climate change and fisheries: What are the real issues? *NW Environ. J.,* **7,** 295–307.

——, and S. R. Hare, 1994: Decadal-scale regime shifts in the large marine ecosystems of the north-east Pacific: A case for historical science. *Fish. Oceanogr.,* **3,** 279–291.

——, ——, A. B. Hollowed, and W. S. Wooster, 1997: Effects of interdecadal climate variability on the oceanic ecosystems of the northeast Pacific. *J. Climate,* in press.

Graham, N. E., 1994: Decadal-scale climate variability in the 1970s and 1980s: Observations and model results. *Climate Dyn.,* **10,** 135–159.

Hare, S. R., 1996: Low-frequency climate variability and salmon production. Ph.D. thesis, University of Washington, Seattle, 306 pp. [Available from University Microfilms, 1490 Eisenhower Place, P.O. Box 975 Ann Arbor, MI 48106.]

——, and R. C. Francis, 1995: Climate change and salmon production in the Northeast Pacific Ocean. *Can. Spec. Publ. Fish. Aquat. Sci.,* **121,** 357–372.

Hollowed, A. B., and W. S. Wooster, 1992: Variability of winter ocean conditions and strong year classes of Northeast Pacific groundfish. *ICES Mar. Sci. Symp.,* **195,** 433–444.

Jones, P. D., and Coauthors, 1985: A grid point temperature data set for the Northern Hemisphere. U.S. Dept. of Energy, Carbon Dioxide Research Division Tech. Rep. TR022, 251 pp. [Available from CDIAC, Oak Ridge National Laboratory, P.O. Box 2008, Oak Ridge, TN 37831-6335.]

Kushnir, Y., and J. M. Wallace, 1989: Low-frequency variability in the Northern Hemisphere winter: Geographical distribution, structure and timescale. *J. Atmos. Sci.,* **46,** 3122–3142.

Latif, M., and T. P. Barnett, 1994: Causes of decadal climate variability over the North Pacific and North America. *Science,* **266,** 634–637.

——, and ——, 1996: Decadal climate variability over the North Pacific and North America: Dynamics and predictability. *J. Climate,* **9,** 2407–2423.

Minobe, S., 1997: A 50–70 year climatic oscillation over the North Pacific and North America. *Geophys. Res. Lett.,* **24,** 683–686.

Namias, J., 1978: Multiple causes of the North American abnormal winter of 1976–77. *Mon. Wea. Rev.,* **106,** 279–295.

National Fisherman, 1972: August/September 1972. M. Freeman Publications, 88 pp.

Pacific Fisherman, 1915: September 1915. M. Freeman Publications, 50 pp.

——, 1939: *1939 Yearbook.* M. Freeman Publications, 312 pp.

Pacific Fishing, 1994: *1994 Yearbook.* Vol. XV, Pacific Fishing Partnership, 116 pp.

——, 1995: *1995 Yearbook.* Vol. XVI, Pacific Fishing Partnership, 90 pp.

Pearcy, W. G., 1992: *Ocean Ecology of North Pacific Salmonids.* University of Washington Press, 179 pp.

Polovina, J. J., G. T. Mitchum, and C. T. Evans, 1995: Decadal and basin-scale variation in mixed layer depth and the impact on biological production in the central and North Pacific, 1960–88. *Deep-Sea Res.,* **42,** 1701–1716.

Rasmussen, E. M., and T. H. Carpenter, 1982: Variations in tropical sea surface temperature and surface wind fields associated with the Southern Oscillation/El Niño. *Mon. Wea. Rev.,* **110,** 354–384.

Reynolds, R. W., and T. M. Smith, 1995: A high-resolution global sea surface temperature climatology. *J. Climate,* **8,** 1571–1583.

Roemmich, D., and J. McGowan, 1995: Climatic warming and the decline of zooplankton in the California Current. *Science,* **267,** 1324–1326.

Salmon 2000 Legislation, 1988: Washington State Senate Bill 6647, Olympia, Washington, 339 pp. [Available from Washington Department of Fisheries and Wildlife, 115 General Administration Bldg., Olympia, WA 98504.]

Tanimoto, Y., N. Iwasaka, K. Hanawa, and Y. Toba, 1993: Characteristic variations of sea surface temperature with multiple time scales in the North Pacific. *J. Climate,* **6,** 1153–1160.

Trenberth, K. E., 1990: Recent observed interdecadal climate changes in the Northern Hemisphere. *Bull. Amer. Meteor. Soc.,* **71,** 988–993.

——, and D. A. Paolino Jr., 1980: The Northern Hemisphere sea-level pressure data set: Trends, errors and discontinuities. *Mon. Wea. Rev.,* **108,** 855–872.

——, and J. W. Hurrell, 1994: Decadal atmosphere-ocean variations in the Pacific. *Climate Dyn.,* **9,** 303.

U.S. Department of Commerce, 1994: Fisheries of the United States. Current Fisheries Statistics Rep. 9400, 113 pp. [Available from Fisheries Statistics Div., (F/RE1), National Marine Fisheries Service, NOAA, 1315 East-West Highway, Silver Spring, MD 20910-3282.]

——, 1995: Current Fisheries of the United States. Current Fisheries Statistics 9400, 126 pp. [Available from Fisheries Statistics Div., (F/RE1), National Marine Fisheries Service, NOAA, 1315 East-West Highway, Silver Spring, MD 20910-3282.]

Venrick, E. L., and J. A. McGowan, D. R. Cayan, and T. L. Hayward, 1987: Climate and chlorophyll a: Long-term trends in the central North Pacific Ocean. *Science,* **238,** 70–72.

WDF, and ODFW, 1992: Status report: Columbia River fish runs and fisheries, 1938–91. Washington Dept. of Fisheries and Oregon Dept. of Fish and Wildlife, Portland, Oregon, 224 pp. [Available from Oregon Dept. of Fish and Wildlife, 2501 S.W. First Ave., P.O. Box 59, Portland, OR 97207.]

Walker, G. T., and E. W. Bliss, 1932: World Weather V, Mem. *Quart. J. Roy. Meteor. Soc.,* **4,** 53–84.

Wallace, J. M., and D. S. Gutzler, 1981: Teleconnections in the geopotential height field during the Northern Hemisphere winter. *Mon. Wea. Rev.,* **109,** 784–812.

Ware, D. M., 1995: A century and a half of change in the climate of the NE Pacific. *Fish. Oceanogr.,* **4,** 267–277.

Zhang, Y., 1996: An observational study of atmosphere-ocean interactions in the Northern Oceans on interannual and interdecadal time-scales. Ph.D. thesis, University of Washington. [Available from University Microfilms, 1490 Eisenhower Place, P.O. Box 975 Ann Arbor, MI 48106.]

——, J. M. Wallace, and D. S. Battisti, 1997: ENSO-like interdecadal variability: 1900–93. *J. Climate,* **10,** 1004–1020.

The Natural Flow Regime

A paradigm for river conservation and restoration

N. LeRoy Poff, J. David Allan, Mark B. Bain, James R. Karr, Karen L. Prestegaard,
Brian D. Richter, Richard E. Sparks, and Julie C. Stromberg

Humans have long been fascinated by the dynamism of free-flowing waters. Yet we have expended great effort to tame rivers for transportation, water supply, flood control, agriculture, and power generation. It is now recognized that harnessing of streams and rivers comes at great cost: Many rivers no longer support socially valued native species or sustain healthy ecosystems that provide important goods and services (Naiman et al. 1995, NRC 1992).

N. LeRoy Poff is an assistant professor in the Department of Biology, Colorado State University, Fort Collins, CO 80523-1878 and formerly senior scientist at Trout Unlimited, Arlington, VA 22209. J. David Allan is a professor at the School of Natural Resources & Environment, University of Michigan, Ann Arbor, MI 48109-1115. Mark B. Bain is a research scientist and associate professor at the New York Cooperative Fish & Wildlife Research Unit of the Department of Natural Resources, Cornell University, Ithaca, NY 14853-3001. James R. Karr is a professor in the departments of Fisheries and Zoology, Box 357980, University of Washington, Seattle, WA 98195-7980. Karen L. Prestegaard is an associate professor in the Department of Geology, University of Maryland, College Park, MD 20742. Brian D. Richter is national hydrologist in the Biohydrology Program, The Nature Conservancy, Hayden, CO 81639. Richard E. Sparks is director of the River Research Laboratories at the Illinois Natural History Survey, Havana, IL 62644. Julie C. Stromberg is an associate professor in the Department of Plant Biology, Arizona State University, Tempe, AZ 85281. © 1997 American Institute of Biological Sciences.

The ecological integrity of river ecosystems depends on their natural dynamic character

The extensive ecological degradation and loss of biological diversity resulting from river exploitation is eliciting widespread concern for conservation and restoration of healthy river ecosystems among scientists and the lay public alike (Allan and Flecker 1993, Hughes and Noss 1992, Karr et al. 1985, TNC 1996, Williams et al. 1996). Extirpation of species, closures of fisheries, groundwater depletion, declines in water quality and availability, and more frequent and intense flooding are increasingly recognized as consequences of current river management and development policies (Abramovitz 1996, Collier et al. 1996, Naiman et al. 1995). The broad social support in the United States for the Endangered Species Act, the recognition of the intrinsic value of noncommercial native species, and the proliferation of watershed councils and riverwatch teams are evidence of society's interest in maintaining the ecological integrity and self-sustaining productivity of free-flowing river systems.

Society's ability to maintain and restore the integrity of river ecosystems requires that conservation and management actions be firmly grounded in scientific understanding. However, current management approaches often fail to recognize the fundamental scientific principle that the integrity of flowing water systems depends largely on their natural dynamic character; as a result, these methods frequently prevent successful river conservation or restoration. Streamflow quantity and timing are critical components of water supply, water quality, and the ecological integrity of river systems. Indeed, streamflow, which is strongly correlated with many critical physicochemical characteristics of rivers, such as water temperature, channel geomorphology, and habitat diversity, can be considered a "master variable" that limits the distribution and abundance of riverine species (Power et al. 1995, Resh et al. 1988) and regulates the ecological integrity of flowing water systems (Figure 1). Until recently, however, the importance of natural streamflow variability in maintaining healthy aquatic ecosystems has been virtually ignored in a management context.

Historically, the "protection" of river ecosystems has been limited in scope, emphasizing water quality and only one aspect of water quantity: minimum flow. Water resources management has also suffered from the often incongruent perspectives and fragmented responsibility of agencies (for example, the US Army Corps of Engineers and Bureau of Reclamation are responsible for water supply and flood control, the US Environmental Protection Agency and state environmental agencies for water quality, and the US Fish &

December 1997

Figure 1. Flow regime is of central importance in sustaining the ecological integrity of flowing water systems. The five components of the flow regime—magnitude, frequency, duration, timing, and rate of change—influence integrity both directly and indirectly, through their effects on other primary regulators of integrity. Modification of flow thus has cascading effects on the ecological integrity of rivers. After Karr 1991.

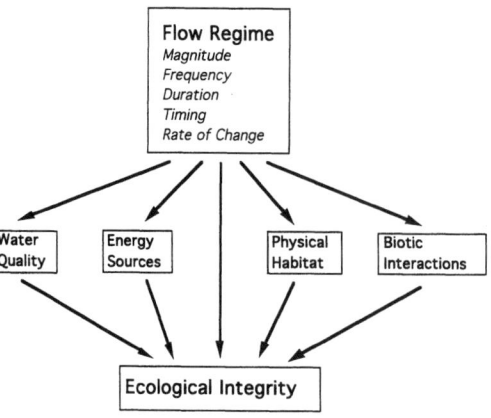

Wildlife Service for water-dependent species of sporting, commercial, or conservation value), making it difficult, if not impossible, to manage the entire river ecosystem (Karr 1991). However, environmental dynamism is now recognized as central to sustaining and conserving native species diversity and ecological integrity in rivers and other ecosystems (Holling and Meffe 1996, Hughes 1994, Pickett et al. 1992, Stanford et al. 1996), and coordinated actions are therefore necessary to protect and restore a river's natural flow variability.

In this article, we synthesize existing scientific knowledge to argue that the natural flow regime plays a critical role in sustaining native biodiversity and ecosystem integrity in rivers. Decades of observation of the effects of human alteration of natural flow regimes have resulted in a well-grounded scientific perspective on why altering hydrologic variability in rivers is ecologically harmful (e.g., Arthington et al. 1991, Castleberry et al. 1996, Hill et al. 1991, Johnson et al. 1976, Richter et al. 1997, Sparks 1995, Stanford et al. 1996, Toth 1995, Tyus 1990). Current pressing demands on water use and the continuing alteration of watersheds require scientists to help develop management protocols that can accommodate economic uses while protecting ecosystem functions. For humans to continue to rely on river ecosystems for sustainable food production, power production, waste assimilation, and flood control, a new, holistic, ecological perspective on water management is needed to guide society's interactions with rivers.

The natural flow regime

The natural flow of a river varies on time scales of hours, days, seasons, years, and longer. Many years of observation from a streamflow gauge are generally needed to describe the characteristic pattern of a river's flow quantity, timing, and variability—that is, its natural flow regime. Components of a natural flow regime can be characterized using various time series (e.g., Fourier and wavelet) and probability analyses of, for example, extremely high or low flows, or of the entire range of flows expressed as average daily discharge (Dunne and Leopold 1978). In watersheds lacking long-term streamflow data, analyses can be extended statistically from gauged streams in the same geographic area. The frequency of large-magnitude floods can be estimated by paleohydrologic studies of debris left by floods and by studies of historical damage to living trees (Hupp and Osterkamp 1985, Knox 1972). These historical techniques can be used to extend existing hydrologic records or to provide estimates of flood flows for ungauged sites.

River flow regimes show regional patterns that are determined largely by river size and by geographic variation in climate, geology, topography, and vegetative cover. For example, some streams in regions with little seasonality in precipitation exhibit relatively stable hydrographs due to high groundwater inputs (Figure 2a), whereas other streams can fluctuate greatly at virtually any time of year (Figure 2b). In regions with seasonal precipitation, some streams are dominated by snowmelt, resulting in pronounced, predictable runoff patterns (Figure 2c), and others lack snow accumulation and exhibit more variable runoff patterns during the rainy season, with peaks occurring after each substantial storm event (Figure 2d).

Five critical components of the flow regime regulate ecological processes in river ecosystems: the magnitude, frequency, duration, timing, and rate of change of hydrologic conditions (Poff and Ward 1989, Richter et al. 1996, Walker et al. 1995). These components can be used to characterize the entire range of flows and specific hydrologic phenomena, such as floods or low flows, that are critical to the integrity of river ecosystems. Furthermore, by defining flow regimes in these terms, the ecological consequences of particular human activities that modify one or more components of the flow regime can be considered explicitly.

- The *magnitude* of discharge[1] at any given time interval is simply the amount of water moving past a fixed location per unit time. Magnitude can refer either to absolute or to relative discharge (e.g., the amount of water that inundates a floodplain). Maximum and minimum magnitudes of flow vary with climate and watershed size both within and among river systems.
- The *frequency* of occurrence refers to how often a flow above a given magnitude recurs over some specified time interval. Frequency of occurrence is inversely related to flow magnitude. For example, a 100-year flood is equaled or exceeded on average once every 100 years (i.e., a chance of 0.01 of occurring in any given year). The average (median)

---

[1]Discharge (also known as streamflow, flow, or flow rate) is always expressed in dimensions of volume per time. However, a great variety of units are used to describe flow, depending on custom and purpose of characterization: Flows can be expressed in near-instantaneous terms (e.g., ft^3/s and m^3/s) or over long time intervals (e.g., acre-ft/yr).

flow is determined from a data series of discharges defined over a specific time interval, and it has a frequency of occurrence of 0.5 (a 50% probability).

• The *duration* is the period of time associated with a specific flow condition. Duration can be defined relative to a particular flow event (e.g., a floodplain may be inundated for a specific number of days by a ten-year flood), or it can be a defined as a composite expressed over a specified time period (e.g., the number of days in a year when flow exceeds some value).

• The *timing*, or *predictability*, of flows of defined magnitude refers to the regularity with which they occur. This regularity can be defined formally or informally and with reference to different time scales (Poff 1996). For example, annual peak flows may occur with low seasonal predictability (Figure 2b) or with high seasonal predictability (Figure 2c).

• The *rate of change*, or *flashiness*, refers to how quickly flow changes from one magnitude to another. At the extremes, "flashy" streams have rapid rates of change (Figure 2b), whereas "stable" streams have slow rates of change (Figure 2a).

Hydrologic processes and the flow regime. All river flow derives ultimately from precipitation, but in any given time and place a river's flow is derived from some combination of surface water, soil water, and groundwater. Climate, geology, topography, soils, and vegetation help to determine both the supply of water and the pathways by which precipitation reaches the channel. The water movement pathways depicted in Figure 3a illustrate why rivers in different settings have different flow regimes and why flow is variable in virtually all rivers. Collectively, overland and shallow subsurface flow pathways create hydrograph peaks, which are the river's response to storm events. By contrast, deeper groundwater pathways are responsible for baseflow, the form of delivery during periods of little rainfall.

Variability in intensity, timing, and duration of precipitation (as rain or as snow) and in the effects of terrain, soil texture, and plant evapotranspiration on the hydrologic cycle combine to create local and regional flow patterns. For example, high flows due to rainstorms may occur over periods of hours (for permeable soils) or even minutes (for impermeable soils), whereas snow will melt over a period of days or weeks, which slowly builds the peak snowmelt flood. As one proceeds downstream within a watershed, river flow reflects the sum of flow generation and routing processes operating in multiple small tributary watersheds. The travel time of flow down the river system, combined with nonsynchronous tributary inputs and larger downstream channel and floodplain storage capacities, act to attenuate and to dampen flow peaks. Consequently, annual hydrographs in large streams typically show peaks created by widespread storms or snowmelt events and broad seasonal influences that affect many tributaries together (Dunne and Leopold 1978).

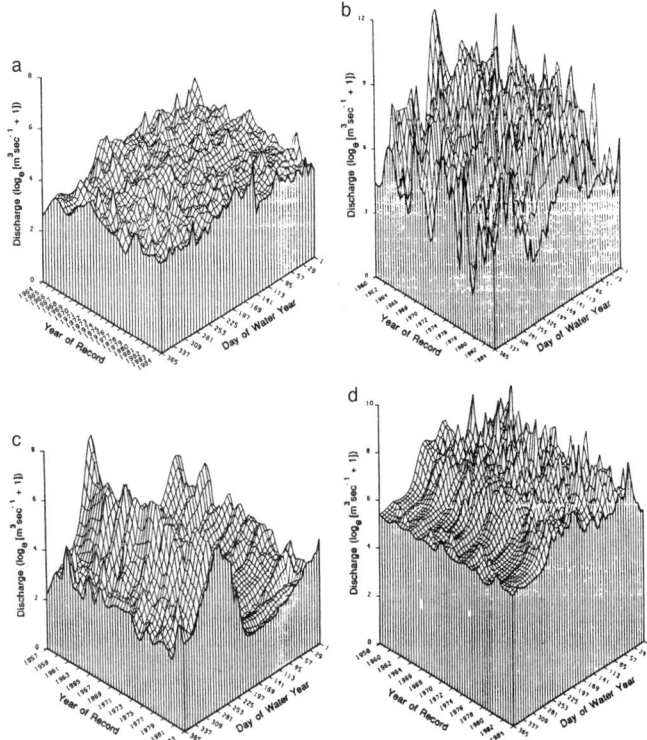

Figure 2. Flow histories based on long-term, daily mean discharge records. These histories show within- and among-year variation for (**a**) Augusta Creek, MI, (**b**) Satilla River, GA, (**c**) upper Colorado River, CO, and (**d**) South Fork of the McKenzie River, OR. Each water year begins on October 1 and ends on September 30. Adapted from Poff and Ward 1990.

The natural flow regime organizes and defines river ecosystems. In rivers, the physical structure of the environment and, thus, of the habitat, is defined largely by physical processes, especially the movement of water and sediment within the channel and between the channel and floodplain. To understand the biodiversity, production, and sustainability of river ecosystems, it is necessary to appreciate the central organizing role played by a dynamically varying physical environment.

The physical habitat of a river includes sediment size and heterogeneity, channel and floodplain morphology, and other geomorphic features. These features form as the available sediment, woody debris, and other transportable materials are moved and deposited by flow. Thus, habitat conditions associated with channels and floodplains vary among

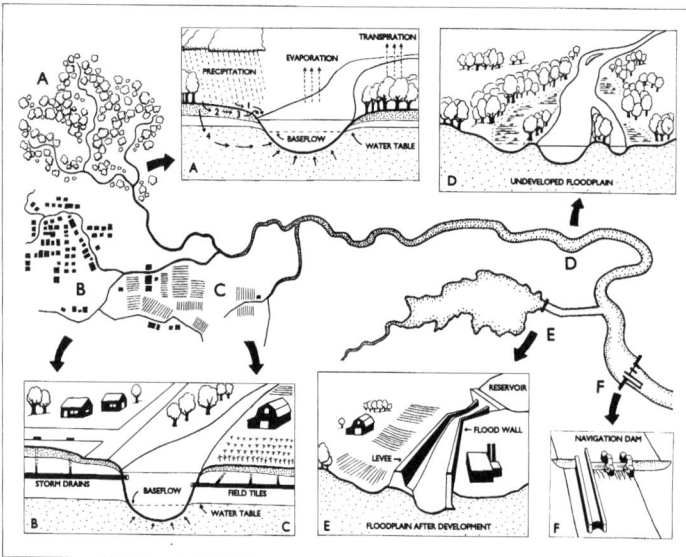

Figure 3. Stream valley cross-sections at various locations in a watershed illustrate basic principles about natural pathways of water moving downhill and human influences on hydrology. Runoff, which occurs when precipitation exceeds losses due to evaporation and plant transpiration, can be divided into four components (**a**): overland flow (1) occurs when precipitation exceeds the infiltration capacity of the soil; shallow subsurface stormflow (2) represents water that infiltrates the soil but is routed relatively quickly to the stream channel; saturated overland flow (3) occurs where the water table is close to the surface, such as adjacent to the stream channel, upstream of first-order tributaries, and in soils saturated by prior precipitation; and groundwater flow (4) represents relatively deep and slow pathways of water movement and provides water to the stream channel even during periods of little or no precipitation. Collectively, overland and shallow subsurface flow pathways create the peaks in the hydrograph that are a river's response to storm events, whereas deeper groundwater pathways are responsible for baseflow. Urbanized (**b**) and agricultural (**c**) land uses increase surface flow by increasing the extent of impermeable surfaces, reducing vegetation cover, and installing drainage systems. Relative to the unaltered state, channels often are scoured to greater depth by unnaturally high flood crests and water tables are lowered, causing baseflow to drop. Side-channels, wetlands, and episodically flooded lowlands comprise the diverse floodplain habitats of unmodified river ecosystems (**d**). Levees or flood walls (**e**) constructed along the banks retain flood waters in the main channel and lead to a loss of floodplain habitat diversity and function. Dams impede the downstream movement of water and can greatly modify a river's flow regime, depending on whether they are operated for storage (**e**) or as "run-of-river," such as for navigation (**f**).

or other features that are left by infrequent high-magnitude floods (e.g., Miller 1990).

Over periods of years to decades, a single river can consistently provide ephemeral, seasonal, and persistent types of habitat that range from free-flowing, to standing, to no water. This predictable diversity of in-channel and floodplain habitat types has promoted the evolution of species that exploit the habitat mosaic created and maintained by hydrologic variability. For many riverine species, completion of the life cycle requires an array of different habitat types, whose availability over time is regulated by the flow regime (e.g., Greenberg et al. 1996, Reeves et al. 1996, Sparks 1995). Indeed, adaptation to this environmental dynamism allows aquatic and floodplain species to persist in the face of seemingly harsh conditions, such as floods and droughts, that regularly destroy and re-create habitat elements.

From an evolutionary perspective, the pattern of spatial and temporal habitat dynamics influences the relative success of a species in a particular environmental setting. This habitat template (Southwood 1977), which is dictated largely by flow regime, creates both subtle and profound differences in the natural histories of species in different segments of their ranges. It also influences species distribution and abundance, as well as ecosystem function (Poff and Allan 1995, Schlosser 1990, Sparks 1992, Stanford et al. 1996). Human alteration of flow regime changes the established pattern of natural hydrologic variation and disturbance, thereby altering habitat dynamics and creating new conditions to which the native biota may be poorly adapted.

Human alteration of flow regimes

Human modification of natural hydrologic processes disrupts the dynamic equilibrium between the movement of water and the movement of sediment that exists in free-flowing rivers (Dunne and Leopold 1978). This disruption alters both gross- and fine-scale geomorphic features that constitute habitat for aquatic and riparian species (Table 1). After

rivers in accordance with both flow characteristics and the type and the availability of transportable materials.

Within a river, different habitat features are created and maintained by a wide range of flows. For example, many channel and floodplain features, such as river bars and riffle–pool sequences, are formed and maintained by dominant, or bankfull, discharges. These discharges are flows that can move significant quantities of bed or bank sediment and that occur frequently enough (e.g., every several years) to continually modify the channel (Wolman and Miller 1960). In many streams and rivers with a small range of flood flows, bankfull flow can build and maintain the active floodplain through stream migration (Leopold et al. 1964). However, the concept of a dominant discharge may not be applicable in all flow regimes (Wolman and Gerson 1978). Furthermore, in some flow regimes, the flows that build the channel may differ from those that build the floodplain. For example, in rivers with a wide range of flood flows, floodplains may exhibit major bar deposits, such as berms of boulders along the channel,

Table 1. Physical responses to altered flow regimes.

Source(s) of alteration	Hydrologic change(s)	Geomorphic response(s)	Reference(s)
Dam	Capture sediment moving downstream	Downstream channel erosion and tributary headcutting	Chien 1985, Petts 1984, 1985, Williams and Wolman 1984
		Bed armoring (coarsening)	Chien 1985
Dam, diversion	Reduce magnitude and frequency of high flows	Deposition of fines in gravel	Sear 1995, Stevens et al. 1995
		Channel stabilization and narrowing	Johnson 1994, Williams and Wolman 1984
		Reduced formation of point bars, secondary channels, oxbows, and changes in channel planform	Chien 1985, Copp 1989, Fenner et al. 1985
Urbanization, tiling, drainage	Increase magnitude and frequency of high flows	Bank erosion and channel widening	Hammer 1972
		Downward incision and floodplain disconnection	Prestegaard 1988
	Reduced infiltration into soil	Reduced baseflows	Leopold 1968
Levees and channelization	Reduce overbank flows	Channel restriction causing downcutting	Daniels 1960, Prestegaard et al. 1994
		Floodplain deposition and erosion prevented	Sparks 1992
		Reduced channel migration and formation of secondary channels	Shankman and Drake 1990
Groundwater pumping	Lowered water table levels	Streambank erosion and channel downcutting after loss of vegetation stability	Kondolf and Curry 1986

such a disruption, it may take centuries for a new dynamic equilibrium to be attained by channel and floodplain adjustments to the new flow regime (Petts 1985); in some cases, a new equilibrium is never attained, and the channel remains in a state of continuous recovery from the most recent flood event (Wolman and Gerson 1978). These channel and floodplain adjustments are sometimes overlooked because they can be confounded with long-term responses of the channel to changing climates (e.g., Knox 1972). Recognition of human-caused physical changes and associated biological consequences may require many years, and physical restoration of the river ecosystem may call for dramatic action (see box on the Grand Canyon flood, page 774).

Dams, which are the most obvious direct modifiers of river flow, capture both low and high flows for flood control, electrical power generation, irrigation and municipal water needs, maintenance of recreational reservoir levels, and navigation. More than 85% of the inland waterways within the continental United States are now artificially controlled (NRC 1992), including nearly 1 million km of rivers that are affected by dams (Echeverria et al. 1989). Dams capture all but the finest sediments moving down a river, with many severe downstream consequences. For example, sediment-depleted water released from dams can erode finer sediments from the receiving channel. The coarsening of the streambed can, in turn, reduce habitat availability for the many aquatic species living in or using interstitial spaces. In addition, channels may erode, or downcut, triggering rejuvenation of tributaries, which themselves begin eroding and migrating headward (Chien 1985, Petts 1984). Fine sediments that are contributed by tributaries downstream of a dam may be deposited between the coarse particles of the streambed (e.g., Sear 1995). In the absence of high flushing flows, species with life stages that are sensitive to sedimentation, such as the eggs and larvae of many invertebrates and fish, can suffer high mortality rates.

For many rivers, it is land-use activities, including timber harvest, livestock grazing, agriculture, and urbanization, rather than dams, that are the primary causes of altered flow regimes. For example, logging and the associated building of roads have contributed greatly to degradation of salmon streams in the Pacific Northwest, mainly through effects on runoff and sediment delivery (NRC 1996). Converting forest or prairie lands to agricultural lands generally decreases soil infiltration and results in increased overland flow, channel incision, floodplain isolation, and headward erosion of stream channels (Prestegaard 1988). Many agricultural areas were drained by the construction of ditches or tile-and-drain systems, with the result that many channels have become entrenched (Brookes 1988).

These land-use practices, combined with extensive draining of wetlands or overgrazing, reduce retention of water in watersheds and,

A controlled flood in the Grand Canyon

Since the Glen Canyon dam first began to store water in 1963, creating Lake Powell, some 430 km (270 miles) of the Colorado River, including Grand Canyon National Park, have been virtually bereft of seasonal floods. Before 1963, melting snow in the upper basin produced an average peak discharge exceeding 2400 m^3/s; after the dam was constructed, releases were generally maintained at less than 500 m^3/s. The building of the dam also trapped more than 95% of the sediment moving down the Colorado River in Lake Powell (Collier et al. 1996).

This dramatic change in flow regime produced drastic alterations in the dynamic nature of the historically sediment-laden Colorado River. The annual cycle of scour and fill had maintained large sandbars along the river banks, prevented encroachment of vegetation onto these bars, and limited bouldery debris deposits from constricting the river at the mouths of tributaries (Collier et al. 1997). When flows were reduced, the limited amount of sand accumulated in the channel rather than in bars farther up the river banks, and shallow low-velocity habitat in eddies used by juvenile fishes declined. Flow regulation allowed for increased cover of wetland and riparian vegetation, which expanded into sites that were regularly scoured by floods in the constrained fluvial canyon of the Colorado River; however, much of the woody vegetation that established after the dam's construction is composed of an exotic tree, salt cedar (*Tamarix* sp.; Stevens et al. 1995). Restoration of flood flows clearly would help to steer the aquatic and riparian ecosystem toward its former state and decrease the area of wetland and riparian vegetation, but precisely how the system would respond to an artificial flood could not be predicted.

In an example of adaptive management (i.e., a planned experiment to guide further actions), a controlled, seven-day flood of 1274 m^3/s was released through the Glen Canyon dam in late March 1996. This flow, roughly 35% of the pre-dam average for a spring flood (and far less than some large historical floods), was the maximum flow that could pass through the power plant turbines plus four steel drainpipes, and it cost approximately $2 million in lost hydropower revenues (Collier et al. 1997). The immediate result was significant beach building: Over 53% of the beaches increased in size, and just 10% decreased in size. Full documentation of the effects will continue to be monitored by measuring channel cross-sections and studying riparian vegetation and fish populations.

instead, route it quickly downstream, increasing the size and frequency of floods and reducing baseflow levels during dry periods (Figure 3b; Leopold 1968). Over time, these practices degrade in-channel habitat for aquatic species. They may also isolate the floodplain from overbank flows, thereby degrading habitat for riparian species. Similarly, urbanization and suburbanization associated with human population expansion across the landscape create impermeable surfaces that direct water away from subsurface pathways to overland flow (and often into storm drains). Consequently, floods increase in frequency and intensity (Beven 1986), banks erode, and channels widen (Hammer 1972), and baseflow declines during dry periods (Figure 3c).

Whereas dams and diversions affect rivers of virtually all sizes, and land-use impacts are particularly evident in headwaters, lowland rivers are greatly influenced by efforts to sever channel–floodplain linkages. Flood control projects have shortened, narrowed, straightened, and leveed many river systems and cut the main channels off from their floodplains (NRC 1992). For example, channelization of the Kissimmee River above Lake Okeechobee, Florida, by the US Army Corps of Engineers transformed a historical 166 km meandering river with a 1.5 to 3 km wide floodplain into a 90 km long canal flowing through a series of five impoundments, resulting in great loss of river channel habitat and adjacent floodplain wetlands (Toth 1995). Because levees are designed to prevent increases in the width of flow, rivers respond by cutting deeper channels, reaching higher velocities, or both.

Channelization and wetland drainage can actually increase the magnitude of extreme floods, because reduction in upstream storage capacity results in accelerated water delivery downstream. Much of the damage caused by the extensive flooding along the Mississippi River in 1993 resulted from levee failure as the river reestablished historic connections to the floodplain. Thus, although elaborate storage dam and levee systems can "reclaim" the floodplain for agriculture and human settlement in most years, the occasional but inevitable large floods will impose increasingly high disaster costs to society (Faber 1996). The severing of floodplains from rivers also stops the processes of sediment erosion and deposition that regulate the topographic diversity of floodplains. This diversity is essential for maintaining species diversity on floodplains, where relatively small differences in land elevation result in large differences in annual inundation and soil moisture regimes, which regulate plant distribution and abundance (Sparks 1992).

Ecological functions of the natural flow regime

Naturally variable flows create and maintain the dynamics of in-channel and floodplain conditions and habitats that are essential to aquatic and riparian species, as shown schematically in Figure 4. For purposes of illustration, we treat the components of a flow regime individually, although in reality they interact in complex ways to regulate geomorphic and ecological processes. In describing the ecological functions associated with the components of a flow regime, we pay particular attention to high- and low-flow events, because they often serve as ecological "bottlenecks" that present critical stresses and opportunities for a wide array of riverine species (Poff and Ward 1989).

The magnitude and frequency of high and low flows regulate numerous ecological processes. Frequent, moderately high flows effectively transport sediment through the channel (Leopold et al. 1964). This sediment movement, combined with the force of moving water, exports organic resources, such as detritus and attached algae, rejuvenating the biological community and allowing many species with fast life cycles and good colonizing ability to reestablish (Fisher 1983). Consequently, the composition and relative abundance of species that are present in a stream or river often reflect the frequency and intensity of high flows (Meffe and Minckley 1987, Schlosser 1985).

High flows provide further ecological benefits by maintaining ecosystem productivity and diversity. For example, high flows remove and transport fine sediments that would otherwise fill the interstitial spaces in productive gravel habitats (Beschta and Jackson 1979). Floods import woody debris into the channel (Keller and Swanson 1979), where it creates new, high-quality habitat (Figure 4; Moore and Gregory 1988, Wallace and Benke 1984). By connecting the channel to the floodplain, high overbank flows also maintain broader productivity and diversity. Floodplain wetlands provide important nursery grounds for fish and export organic matter and organisms back into the main channel (Junk et al. 1989, Sparks 1995, Welcomme 1992). The scouring of floodplain soils rejuvenates habitat for plant species that germinate only on barren, wetted surfaces that are free of competition (Scott et al. 1996) or that require access to shallow water tables (Stromberg et al. 1997). Flood-resistant, disturbance-adapted riparian communities are maintained by flooding along river corridors, even in river sections that have steep banks and lack floodplains (Hupp and Osterkamp 1985).

Flows of low magnitude also provide ecological benefits. Periods of low flow may present recruitment opportunities for riparian plant species in regions where floodplains are frequently inundated (Wharton et al. 1981). Streams that dry temporarily, generally in arid regions, have aquatic (Williams and Hynes 1977)

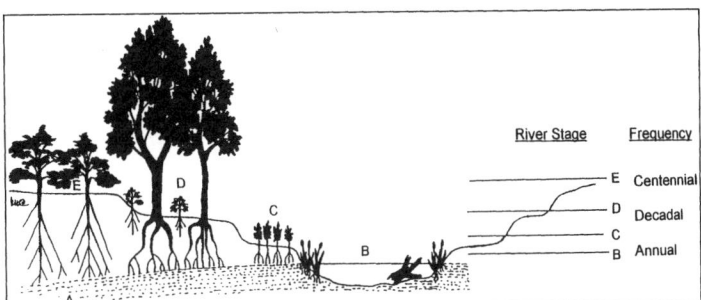

Figure 4. Geomorphic and ecological functions provided by different levels of flow. Water tables that sustain riparian vegetation and that delineate in-channel baseflow habitat are maintained by groundwater inflow and flood recharge (A). Floods of varying size and timing are needed to maintain a diversity of riparian plant species and aquatic habitat. Small floods occur frequently and transport fine sediments, maintaining high benthic productivity and creating spawning habitat for fishes (B). Intermediate-size floods inundate low-lying floodplains and deposit entrained sediment, allowing for the establishment of pioneer species (C). These floods also import accumulated organic material into the channel and help to maintain the characteristic form of the active stream channel. Larger floods that recur on the order of decades inundate the aggraded floodplain terraces, where later successional species establish (D). Rare, large floods can uproot mature riparian trees and deposit them in the channel, creating high-quality habitat for many aquatic species (E).

and riparian (Nilsen et al. 1984) species with special behavioral or physiological adaptations that suit them to these harsh conditions.

The duration of a specific flow condition often determines its ecological significance. For example, differences in tolerance to prolonged flooding in riparian plants (Chapman et al. 1982) and to prolonged low flow in aquatic invertebrates (Williams and Hynes 1977) and fishes (Closs and Lake 1996) allow these species to persist in locations from which they might otherwise be displaced by dominant, but less tolerant, species.

The timing, or predictability, of flow events is critical ecologically because the life cycles of many aquatic and riparian species are timed to either avoid or exploit flows of variable magnitudes. For example, the natural timing of high or low streamflows provides environmental cues for initiating life cycle transitions in fish, such as spawning (Montgomery et al. 1983, Nesler et al. 1988), egg hatching (Næsje et al. 1995), rearing (Seegrist and Gard 1978), movement onto the floodplain for feeding or reproduction (Junk et al. 1989, Sparks 1995, Welcomme 1992), or migration upstream or downstream (Trépanier et al. 1996). Natural seasonal variation in flow conditions can prevent the successful establishment of nonnative species with flow-dependent spawning and egg incubation requirements, such as striped bass (*Morone saxatilis*; Turner and Chadwick 1972) and brown trout (*Salmo trutta*; Moyle and Light 1996, Strange et al. 1992).

Seasonal access to floodplain wetlands is essential for the survival of certain river fishes, and such access can directly link high wetland productivity with fish production in the stream channel (Copp 1989, Welcomme 1979). Studies of the effects on stream fishes of both extensive and limited floodplain inundation (Finger and Stewart 1987, Ross and Baker 1983) indicate that some fishes are adapted to exploiting floodplain habitats, and these species decline in abundance when floodplain use is restricted. Models indicate that catch rates and biomass of fish are influenced by both maximum and minimum wetland area (Power et al. 1995, Welcomme and Hagborg 1977), and empirical work shows that the area of floodplain water bodies during nonflood periods influences the species richness of those wetland habitats (Halyk and Balon 1983). The timing of floodplain inundation is important for some fish because migratory and reproductive behaviors must coincide with access to and avail-

Table 2. Ecological responses to alterations in components of natural flow regime.[a]

Flow component	Specific alteration	Ecological response	Reference(s)
Magnitude and frequency	Increased variation	Wash-out and/or stranding	Cushman 1985, Petts 1984
		Loss of sensitive species	Gehrke et al. 1995, Kingsolving and Bain 1993, Travnichek et al. 1995
		Increased algal scour and wash-out of organic matter	Petts 1984
		Life cycle disruption	Scheidegger and Bain 1995
	Flow stabilization	Altered energy flow	Valentin et al. 1995
		Invasion or establishment of exotic species, leading to:	
		Local extinction	Kupferberg 1996, Meffe 1984
		Threat to native commercial species	Stanford et al. 1996
		Altered communities	Busch and Smith 1995, Moyle 1986, Ward and Stanford 1979
		Reduced water and nutrients to floodplain plant species, causing:	
		Seedling desiccation	Duncan 1993
		Ineffective seed dispersal	Nilsson 1982
		Loss of scoured habitat patches and secondary channels needed for plant establishment	Fenner et al. 1985, Rood et al. 1995, Scott et al. 1997, Shankman and Drake 1990
		Encroachment of vegetation into channels	Johnson 1994, Nilsson 1982
Timing	Loss of seasonal flow peaks	Disrupt cues for fish:	
		Spawning	Fausch and Bestgen 1997, Montgomery et al. 1993, Nesler et al. 1988
		Egg hatching	Næsje et al. 1995
		Migration	Williams 1996
		Loss of fish access to wetlands or backwaters	Junk et al. 1989, Sparks 1995
		Modification of aquatic food web structure	Power 1992, Wootton et al. 1996
		Reduction or elimination of riparian plant recruitment	Fenner et al. 1985
		Invasion of exotic riparian species	Horton 1977
		Reduced plant growth rates	Reily and Johnson 1982
Duration	Prolonged low flows	Concentration of aquatic organisms	Cushman 1985, Petts 1984
		Reduction or elimination of plant cover	Taylor 1982
		Diminished plant species diversity	Taylor 1982
		Desertification of riparian species composition	Busch and Smith 1995, Stromberg et al. 1996
		Physiological stress leading to reduced plant growth rate, morphological change, or mortality	Kondolf and Curry 1986, Perkins et al. 1984, Reily and Johnson 1982, Rood et al. 1995, Stromberg et al. 1992
	Prolonged baseflow "spikes"	Downstream loss of floating eggs	Robertson 1997
	Altered inundation duration	Altered plant cover types	Auble et al. 1994
	Prolonged inundation	Change in vegetation functional type	Bren 1992, Connor et al. 1981
		Tree mortality	Harms et al. 1980
		Loss of riffle habitat for aquatic species	Bogan 1993
Rate of change	Rapid changes in river stage	Wash-out and stranding of aquatic species	Cushman 1985, Petts 1984
	Accelerated flood recession	Failure of seedling establishment	Rood et al. 1995

[a]Only representative studies are listed here. Additional references are located on the Web at http://lamar.colostate.edu/~poff/natflow.html.

ability of floodplain habitats (Welcomme 1979). The match of reproductive period and wetland access also explains some of the yearly variation in stream fish community composition (Finger and Stewart 1987).

Many riparian plants also have life cycles that are adapted to the seasonal timing components of natural flow regimes through their "emergence phenologies"—the seasonal sequence of flowering, seed dispersal, germination, and seedling growth. The interaction of emergence phenologies with temporally varying environmental stress from flooding or drought helps to maintain high species diversity in, for example, southern floodplain forests (Streng et al. 1989). Productivity of riparian forests is also influenced by flow timing and can increase when short-duration flooding occurs in the growing season (Mitsch and Rust 1984, Molles et al. 1995).

The rate of change, or flashiness, in flow conditions can influence species persistence and coexistence. In many streams and rivers, particularly in arid areas, flow can change dramatically over a period of hours due to heavy storms. Non-native fishes generally lack the behavioral adaptations to avoid being displaced downstream by sudden floods (Minckley and Deacon 1991). In a dramatic example of how floods can benefit native species, Meffe (1984) documented that a native fish, the Gila topminnow (*Poeciliopsis occidentalis*), was locally extirpated by the introduced predatory mosquitofish (*Gambusia affinis*) in locations where natural flash floods were regulated by upstream dams, but the native species persisted in naturally flashy streams.

Rapid flow increases in streams of the central and southwestern United States often serve as spawning cues for native minnow species, whose rapidly developing eggs are either broadcast into the water column or attached to submerged structures as floodwaters recede (Fausch and Bestgen 1997, Robertson in press). More gradual, seasonal rates of change in flow conditions also regulate the persistence of many aquatic and riparian species. Cottonwoods (*Populus* spp.), for example, are disturbance species that establish after winter–spring flood flows, during a narrow "window of opportunity" when competition-free alluvial substrates and wet soils are available for germination. A certain rate of floodwater recession is critical to seedling germination because seedling roots must remain connected to a receding water table as they grow downward (Rood and Mahoney 1990).

Ecological responses to altered flow regimes

Modification of the natural flow regime dramatically affects both aquatic and riparian species in streams and rivers worldwide. Ecological responses to altered flow regimes in a specific stream or river depend on how the components of flow have changed relative to the natural flow regime for that particular stream or river (Poff and Ward 1990) and how specific geomorphic and ecological processes will respond to this relative change. As a result of variation in flow regime within and among rivers (Figure 2), the same human activity in different locations may cause different degrees of change relative to unaltered conditions and, therefore, have different ecological consequences.

Flow alteration commonly changes the magnitude and frequency of high and low flows, often reducing variability but sometimes enhancing the range. For example, the extreme daily variations below peaking power hydroelectric dams have no natural analogue in freshwater systems and represent, in an evolutionary sense, an extremely harsh environment of frequent, unpredictable flow disturbance. Many aquatic populations living in these environments suffer high mortality from physiological stress, from wash-out during high flows, and from stranding during rapid dewatering (Cushman 1985, Petts 1984). Especially in shallow shoreline habitats, frequent atmospheric exposure for even brief periods can result in massive mortality of bottom-dwelling organisms and subsequent severe reductions in biological productivity (Weisberg et al. 1990). Moreover, the rearing and refuge functions of shallow shoreline or backwater areas, where many small fish species and the young of large species are found (Greenberg et al. 1996, Moore and Gregory 1988), are severely impaired by frequent flow fluctuations (Bain et al. 1988, Stanford 1994). In these artificially fluctuating environments, specialized stream or river species are typically replaced by generalist species that tolerate frequent and large variations in flow. Furthermore, life cycles of many species are often disrupted and energy flow through the ecosystem is greatly modified (Table 2). Short-term flow modifications clearly lead to a reduction in both the natural diversity and abundance of many native fish and invertebrates.

At the opposite hydrologic extreme, flow stabilization below certain types of dams, such as water supply reservoirs, results in artificially constant environments that lack natural extremes. Although production of a few species may increase greatly, it is usually at the expense of other native species and of systemwide species diversity (Ward and Stanford 1979). Many lake fish species have successfully invaded (or been intentionally established in) flow-stabilized river environments (Moyle 1986, Moyle and Light 1996). Often top predators, these introduced fish can devastate native river fish and threaten commercially valuable stocks (Stanford et al. 1996). In the southwestern United States, virtually the entire native river fish fauna is listed as threatened under the Endangered Species Act, largely as a consequence of water withdrawal, flow stabilization, and exotic species proliferation. The last remaining strongholds of native river fishes are all in dynamic, free-flowing rivers, where exotic fishes are periodically reduced by natural flash floods (Minckley and Deacon 1991, Minckley and Meffe 1987).

Flow stabilization also reduces the magnitude and frequency of overbank flows, affecting riparian plant species and communities. In rivers with constrained canyon reaches or multiple shallow channels, loss of high flows results in increased cover of plant species that would otherwise be removed by flood scour (Ligon et al. 1995, Williams and Wolman 1984). Moreover, due to other related effects of flow regulation, including increased water salinity, non-native vegetation often dominates, such as the salt cedar (*Tamarix* sp.) in the semiarid western United States (Busch and Smith 1995). In alluvial valleys, the loss of overbank flows can greatly modify riparian communities by causing plant desiccation, reduced growth, competitive exclusion, ineffective seed dispersal, or failure of seedling establishment (Table 2).

The elimination of flooding may also affect animal species that depend on terrestrial habitats. For example, in the flow-stabilized Platte River of the United States Great Plains, the channel has narrowed dramatically (up to 85%) over a period of decades (Johnson 1994). This narrowing has been facilitated by vegetative colonization of sandbars that formerly provided nesting habitat for the threatened piping plover (*Charadius melodius*) and endangered least tern (*Sterna antillarum*; Sidle et al. 1992). Sandhill cranes (*Grus canadensis*), which made the Platte River famous, have abandoned river segments that have narrowed the most (Krapu et al. 1984).

Changes in the duration of flow conditions also have significant biological consequences. Riparian plant species respond dramatically to channel dewatering, which occurs frequently in arid regions due to surface water diversion and groundwater pumping. These biological and ecological responses range from altered leaf morphology to total loss of riparian vegetation cover (Table 2). Changes in duration of inundation, independent of changes in annual volume of flow, can alter the abundance of plant cover types (Auble et al. 1994). For example, increased duration of inundation has contributed to the conversion of grassland to forest along a regulated Australian river (Bren 1992). For aquatic species, prolonged flows of particular levels can also be damaging. In the regulated Pecos River of New Mexico, artificially prolonged high summer flows for irrigation displace the floating eggs of the threatened Pecos bluntnose shiner (*Notropis sinius pecosensis*) into unfavorable habitat, where none survive (Robertson in press).

Modification of natural flow timing, or predictability, can affect aquatic organisms both directly and indirectly. For example, some native fishes in Norway use seasonal flow peaks as a cue for egg hatching, and river regulation that eliminates these peaks can directly reduce local population sizes of these species (Næsje et al. 1995). Furthermore, entire food webs, not just single species, may be modified by altered flow timing. In regulated rivers of northern California, the seasonal shifting of scouring flows from winter to summer indirectly reduces the growth rate of juvenile steelhead trout (*Oncorhyncus mykiss*) by increasing the relative abundance of predator-resistant invertebrates that divert energy away from the food chain leading to trout (Wootton et al. 1996). In unregulated rivers, high winter flows reduce these predator-resistant insects and favor species that are more palatable to fish.

Riparian plant species are also strongly affected by altered flow tim-

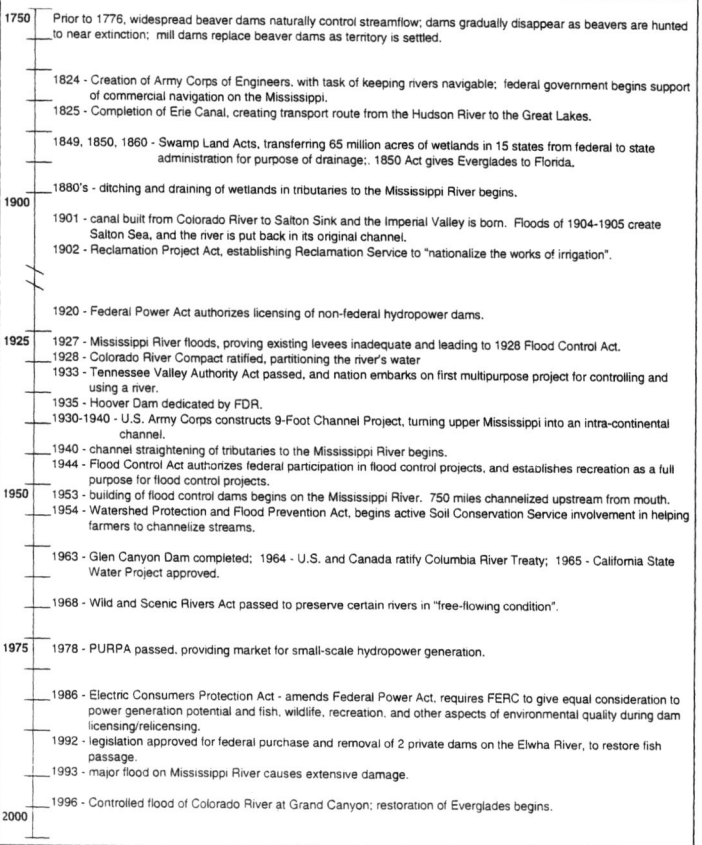

Figure 5. A brief history of flow alteration in the United States.

ing (Table 2). A shift in timing of peak flows from spring to summer, as often occurs when reservoirs are managed to supply irrigation water, has prevented reestablishment of the Fremont cottonwood (*Populus fremontii*), the dominant plant species in Arizona, because flow peaks now occur after, rather than before, its germination period (Fenner et al. 1985). Non-native plant species with less specific germination requirements may benefit from changes in flood timing. For example, salt cedar's (*Tamarix* sp.) long seed dispersal period allows it to establish after floods occurring any time during the growing season, contributing to its abundance on floodplains of the western United States (Horton 1977).

Altering the rate of change in flow can negatively affect both aquatic and riparian species. As mentioned above, loss of natural flashiness threatens most of the native fish fauna of the American Southwest (Minckley and Deacon 1991), and artificially increased rates of change caused by peaking power hydroelectric dams on historically less flashy rivers creates numerous ecological problems (Table 2; Petts 1984). A modified rate of change can devastate riparian species, such as cottonwoods, whose successful seedling growth depends on the rate of groundwater recession following floodplain inundation. In the St. Mary River in Alberta, Canada, for example, rapid drawdowns of river stage during spring have prevented the recruitment of young trees (Rood and Mahoney 1990). Such effects can be reversed, however. Restoration of the spring flood and its natural, slow recession in the Truckee River in California has allowed the successful establishment of a new generation of cotton-

Table 3. Recent projects in which restoration of some component(s) of natural flow regimes has occurred or been proposed for specific ecological benefits.

Location	Flow component(s)	Ecological purpose(s)	Reference
Trinity River, CA	Mimic timing and magnitude of peak flow	Rejuvenate in-channel gravel habitats; restore early riparian succession; provide migration flows for juvenile salmon	Barinaga 1996[a]
Truckee River, CA	Mimic timing, magnitude, and duration of peak flow, and its rate of change during recession	Restore riparian trees, especially cottonwoods	Klotz and Swanson 1997
Owens River, CA	Increase base flows; partially restore overbank flows	Restore riparian vegetation and habitat for native fishes and non-native brown trout	Hill and Platts in press
Rush Creek, CA (and other tributaries to Mono Lake)	Increase minimum flows	Restore riparian vegetation and habitat for waterfowl and non-native fishes	LADWP 1995
Oldman River and tributaries, southern Alberta, Canada	Increase summer flows; reduce rates of postflood stage decline; mimic natural flows in wet years	Restore riparian vegetation (cottonwoods) and cold-water (trout) fisheries	Rood et al. 1995
Green River, UT	Mimic timing and duration of peak flow and duration and timing of nonpeak flows; reduce rapid baseflow fluctuations from hydropower generation	Recovery of endangered fish species; enhance other native fishes	Stanford 1994
San Juan River, UT/NM	Mimic magnitude, timing, and duration of peak flow; restore low winter baseflows	Recovery of endangered fish species	—[b]
Gunnison River, CO	Mimic magnitude, timing, and duration of peak flow; mimic duration and timing of nonpeak flows	Recovery of endangered fish species	—[b]
Rio Grande River, NM	Mimic timing and duration of floodplain inundation	Ecosystem processes (e.g., nitrogen flux, microbial activity, litter decomposition)	Molles et al. 1995
Pecos River, NM	Regulate duration and magnitude of summer irrigation releases to mimic spawning flow "spikes"; maintain minimum flows	Determine spawning and habitat needs for threatened fish species	Robertson 1997
Colorado River, AZ	Mimic magnitude and timing	Restore habitat for endangered fish species and scour riparian zone	Collier et al. 1997
Bill Williams River, AZ (proposed)	Mimic natural flood peak timing and duration	Promote establishment of native trees	USCOE 1996
Pemigewasset River, NH	Reduce frequency (i.e., to no more than natural frequency) of high flows during summer low-flow season; reduce rate of change between low and high flows during hydropower cycles	Enhance native Atlantic salmon recovery	FERC 1995
Roanoke River, VA	Restore more natural patterning of monthly flows in spring; reduce rate of change between low and high flows during hydropower cycles	Increased reproduction of striped bass	Rulifson and Manooch 1993
Kissimmee River, FL	Mimic magnitude, duration, rate of change, and timing of high- and low-flow periods	Restore floodplain inundation to recover wetland functions; reestablish in-channel habitats for fish and other aquatic species	Toth 1995

[a] J. Polos, 1997, personal communication. US Fish & Wildlife Service, Arcata, CA.
[b] F. Pfeifer, 1997, personal communication. US Fish & Wildlife Service, Grand Junction, CO.

wood trees (Klotz and Swanson 1997).

Recent approaches to streamflow management

Methods to estimate environmental flow requirements for rivers focus primarily on one or a few species that live in the wetted river channel. Most of these methods have the narrow intent of establishing minimum allowable flows. The simplest make use of easily analyzed flow data, of assumptions about the regional similarity of rivers, and of professional opinions of the minimal flow needs for certain fish species (e.g., Larson 1981).

A more sophisticated assessment of how changes in river flow affect aquatic habitat is provided by the Instream Flow Incremental Methodology (IFIM; Bovee and Milhous

1978). IFIM combines two models, a biological one that describes the physical habitat preferences of fishes (and occasionally macroinvertebrates) in terms of depth, velocity, and substrate, and a hydraulic one that estimates how the availability of habitat for fish varies with discharge. IFIM has been widely used as an organizational framework for formulating and evaluating alternative water management options related to production of one or a few fish species (Stalnaker et al. 1995).

As a predictive tool for ecological management, the IFIM modeling approach has been criticized both in terms of the statistical validity of its physical habitat characterizations (Williams 1996) and the limited realism of its biological assumptions (Castleberry et al. 1996). Field tests of its predictions have yielded mixed results (Morehardt 1986). Although this approach continues to evolve, both by adding biological realism (Van Winkle et al. 1993) and by expanding the range of habitats modeled (Stalnaker et al. 1995), in practice it is often used only to establish minimum flows for "important" (i.e., game or imperiled) fish species. But current understanding of river ecology clearly indicates that fish and other aquatic organisms require habitat features that cannot be maintained by minimum flows alone (see Stalnaker 1990). A range of flows is necessary to scour and revitalize gravel beds, to import wood and organic matter from the floodplain, and to provide access to productive riparian wetlands (Figure 4). Interannual variation in these flow peaks is also critical for maintaining channel and riparian dynamics. For example, imposition of only a fixed high-flow level each year would simply result in the equilibration of in-channel and floodplain habitats to these constant peak flows.

Moreover, a focus on one or a few species and on minimum flows fails to recognize that what is "good" for the ecosystem may not consistently benefit individual species, and that what is good for individual species may not be of benefit to the ecosystem. Long-term studies of naturally variable systems show that some species do best in wet years, that other species do best in dry years, and that overall biological diversity and ecosystem function benefit from these variations in species success (Tilman et al. 1994). Indeed, experience in river restoration clearly shows the impossibility of simultaneously engineering optimal conditions for all species (Sparks 1992, 1995, Toth 1995). A holistic view that attempts to restore natural variability in ecological processes and species success (and that acknowledges the tremendous uncertainty that is inherent in attempting to mechanistically model all species in the ecosystem) is necessary for ecosystem management and restoration (Franklin 1993).

Managing toward a natural flow regime

The first step toward better incorporating flow regime into the management of river ecosystems is to recognize that extensive human alteration of river flow has resulted in widespread geomorphic and ecological changes in these ecosystems. The history of river use is also a history of flow alteration (Figure 5). The early establishment of the US Army Corps of Engineers is testimony to the importance that the nation gave to developing navigable water routes and to controlling recurrent large floods. However, growing understanding of the ecological impacts of flow alteration has led to a shift toward an appreciation of the merits of free-flowing rivers. For example, the Wild and Scenic Rivers Act of 1968 recognized that the flow of certain rivers should be protected as a national resource, and the recent blossoming of natural flow restoration projects (Table 3) may herald the beginning of efforts to undo some of the damage of past flow alterations. The next century holds promise as an era for renegotiating human relationships with rivers, in which lessons from past experience are used to direct wise and informed action in the future.

A large body of evidence has shown that the natural flow regime of virtually all rivers is inherently variable, and that this variability is critical to ecosystem function and native biodiversity. As we have already discussed, rivers with highly altered and regulated flows lose their ability to support natural processes and native species. Thus, to protect pristine or nearly pristine systems, it is necessary to preserve the natural hydrologic cycle by safeguarding against upstream river development and damaging land uses that modify runoff and sediment supply in the watershed.

Most rivers are highly modified, of course, and so the greatest challenges lie in managing and restoring rivers that are also used to satisfy human needs. Can reestablishing the natural flow regime serve as a useful management and restoration goal? We believe that it can, although to varying degrees, depending on the present extent of human intervention and flow alteration affecting a particular river. Recognizing the natural variability of river flow and explicitly incorporating the five components of the natural flow regime (i.e., magnitude, frequency, duration, timing, and rate of change) into a broader framework for ecosystem management would constitute a major advance over most present management, which focuses on minimum flows and on just a few species. Such recognition would also contribute to the developing science of stream restoration in heavily altered watersheds, where, all too often, physical channel features (e.g., bars and woody debris) are re-created without regard to restoring the flow regime that will help to maintain these re-created features.

Just as rivers have been incrementally modified, they can be incrementally restored, with resulting improvements to many physical and biological processes. A list of recent efforts to restore various components of a natural flow regime (that is, to "naturalize" river flow) demonstrates the scope for success (Table 3). Many of the projects summarized in Table 3 represent only partial steps toward full flow restoration, but they have had demonstrable ecological benefits. For example, high flood flows followed by mimicked natural rates of flow decline in the Oldman River of Alberta, Canada, resulted in a massive cottonwood recruitment that extended for more than 500 km downstream from the Oldman Dam. Dampening of the unnatural flow fluctuations caused by hydroelectric generation on the Roanoke River in

Virginia has increased juvenile abundances of native striped bass. Mimicking short-duration flow spikes that are historically caused by summer thunderstorms in the regulated Pecos River of New Mexico has benefited the reproductive success of the Pecos bluntnose shiner.

We also recognize that there are scientific limits to how precisely the natural flow regime for a particular river can be defined. It is possible to have only an approximate knowledge of the historic condition of a river, both because some human activities may have preceded the installation of flow gauges, and because climate conditions may have changed over the past century or more. Furthermore, in many rivers, year-to-year differences in the timing and quantity of flow result in substantial variability around any average flow condition. Accordingly, managing for the "average" condition can be misguided. For example, in human-altered rivers that are managed for incremental improvements, restoring a flow pattern that is simply proportional to the natural hydrograph in years with little runoff may provide few if any ecological benefits, because many geomorphic and ecological processes show nonlinear responses to flow. Clearly, half of the peak discharge will not move half of the sediment, half of a migration-motivational flow will not motivate half of the fish, and half of an overbank flow will not inundate half of the floodplain. In such rivers, more ecological benefits would accrue from capitalizing on the natural between-year variability in flow. For example, in years with above-average flow, "surplus" water could be used to exceed flow thresholds that drive critical geomorphic and ecological processes.

If full flow restoration is impossible, mimicking certain geomorphic processes may provide some ecological benefits. Well-timed irrigation could stimulate recruitment of valued riparian trees such as cottonwoods (Friedman et al. 1995). Strategically clearing vegetation from river banks could provide new sources of gravel for sediment-starved regulated rivers with reduced peak flows (e.g., Ligon et al. 1995). In all situations, managers will be required to make judgments about specific restoration goals and to work with appropriate components of the natural flow regime to achieve those goals. Recognition of the natural flow variability and careful identification of key processes that are linked to various components of the flow regime are critical to making these judgments.

Setting specific goals to restore a more natural regime in rivers with altered flows (or, equally important, to preserve unaltered flows in pristine rivers) should ideally be a cooperative process involving river scientists, resource managers, and appropriate stakeholders. The details of this process will vary depending on the specific objectives for the river in question, the degree to which its flow regime and other environmental variables (e.g., thermal regime, sediment supply) have been altered, and the social and economic constraints that are in play. Establishing specific criteria for flow restoration will be challenging because our understanding of the interactions of individual flow components with geomorphic and ecological processes is incomplete. However, quantitative, river-specific standards can, in principle, be developed based on the reconstruction of the natural flow regime (e.g., Richter et al. 1997). Restoration actions based on such guidelines should be viewed as experiments to be monitored and evaluated—that is, adaptive management—to provide critical new knowledge for creative management of natural ecosystem variability (Table 3).

To manage rivers from this new perspective, some policy changes are needed. The narrow regulatory focus on minimum flows and single species impedes enlightened river management and restoration, as do the often conflicting mandates of the many agencies and organizations that are involved in the process. Revisions of laws and regulations, and redefinition of societal goals and policies, are essential to enable managers to use the best science to develop appropriate management programs.

Using science to guide ecosystem management requires that basic and applied research address difficult questions in complex, real-world settings, in which experimental controls and statistical replication are often impossible. Too little attention and too few resources have been devoted to clarifying how restoring specific components of the flow regime will benefit the entire ecosystem. Nevertheless, it is clear that, whenever possible, the natural river system should be allowed to repair and maintain itself. This approach is likely to be the most successful and the least expensive way to restore and maintain the ecological integrity of flow-altered rivers (Stanford et al. 1996). Although the most effective mix of human-aided and natural recovery methods will vary with the river, we believe that existing knowledge makes a strong case that restoring natural flows should be a cornerstone of our management approach to river ecosystems.

Acknowledgments

We thank the following people for reading and commenting on earlier versions of this paper: Jack Schmidt, Lou Toth, Mike Scott, David Wegner, Gary Meffe, Mary Power, Kurt Fausch, Jack Stanford, Bob Naiman, Don Duff, John Epifanio, Lori Robertson, Jeff Baumgartner, Tim Randle, David Harpman, Mike Armbruster, and Thomas Payne. Members of the Hydropower Reform Coalition also offered constructive comments. Excellent final reviews were provided by Greg Auble, Carter Johnson, an anonymous reviewer, and the editor of *BioScience*. Robin Abell contributed to the development of the timeline in Figure 5, and graphics assistance was provided by Teresa Peterson (Figure 3), Matthew Chew (Figure 4) and Robin Abell and Jackie Howard (Figure 5). We also thank the national offices of Trout Unlimited and American Rivers for encouraging the expression of the ideas presented here. We especially thank the George Gund Foundation for providing a grant to hold a one-day workshop, and The Nature Conservancy for providing logistical support for several of the authors prior to the workshop.

References cited

Abramovitz JN. 1996. Imperiled waters, impoverished future: the decline of freshwater ecosystems. Washington (DC): Worldwatch Institute. Worldwatch paper nr 128.

Allan JD, Flecker AS. 1993. Biodiversity conservation in running waters. BioScience 43: 32–43.

Arthington AH, King JM, O'Keefe JH, Bunn SE, Day JA, Pusey BJ, Bluhdorn DR, Thame R. 1991. Development of an holistic approach for assessing environmental flow requirements of riverine ecosystems. Pages 69–76 in Pigram JJ, Hooper BA, eds. Water allocation for the environment: proceedings of an international seminar and workshop. University of New England Armidale (Australia): The Centre for Water Policy Research.

Auble GT, Friedman JM, Scott ML. 1994. Relating riparian vegetation to present and future streamflows. Ecological Applications 4: 544–554.

Bain MB, Finn JT, Booke HE. 1988. Streamflow regulation and fish community structure. Ecology 69: 382–392.

Baringa M. 1996. A recipe for river recovery? Science 273: 1648–1650.

Beschta RL, Jackson WL. 1979. The intrusion of fine sediments into a stable gravel bed. Journal of the Fisheries Research Board of Canada 36: 207–210.

Beven KJ. 1986. Hillslope runoff processes and flood frequency characteristics. Pages 187–202 in Abrahams AD, ed. Hillslope processes. Boston: Allen and Unwin.

Bogan AE. 1993. Freshwater bivalve extinctions (Mollusca: Unionida): a search for causes. American Zoologist 33: 599–609.

Bovee KD, Milhous R. 1978. Hydraulic simulation in instream flow studies: theory and techniques. Ft. Collins (CO): Office of Biological Services, US Fish & Wildlife Service. Instream Flow Information Paper nr 5, FWS/OBS-78/33.

Bren LJ. 1992. Tree invasion of an intermittent wetland in relation to changes in the flooding frequency of the River Murray, Australia. Australian Journal of Ecology 17: 395–408.

Brookes A. 1988. Channelized rivers, perspectives for environmental management. New York: John Wiley & Sons.

Busch DE, Smith SD. 1995. Mechanisms associated with decline of woody species in riparian ecosystems of the Southwestern US. Ecological Monographs 65: 347–370.

Castleberry DT, et al. 1996. Uncertainty and instream flow standards. Fisheries 21: 20–21.

Chapman RJ, Hinckley TM, Lee LC, Teskey RO. 1982. Impact of water level changes on woody riparian and wetland communities. Vol. 10. Kearneysville (WV): US Fish & Wildlife Service. Publication nr OBS-82/83.

Chien N. 1985. Changes in river regime after the construction of upstream reservoirs. Earth Surface Processes and Landforms 10: 143–159.

Closs GP, Lake PS. 1996. Drought, differential mortality and the coexistence of a native and an introduced fish species in a south east Australian intermittent stream. Environmental Biology of Fishes 47: 17–26.

Collier M, Webb RH, Schmidt JC. 1996. Dams and rivers: primer on the downstream effects of dams. Reston (VA): US Geological Survey. Circular nr 1126.

Collier MP, Webb RH, Andrews ED. 1997. Experimental flooding in the Grand Canyon. Scientific American 276: 82–89.

Connor WH, Gosselink JG, Parrondo RT. 1981. Comparison of the vegetation of three Louisiana swamp sites with different flooding regimes. American Journal of Botany 68: 320–331.

Copp GH. 1989. The habitat diversity and fish reproductive function of floodplain ecosystems. Environmental Biology of Fishes 26: 1–27.

Cushman RM. 1985. Review of ecological effects of rapidly varying flows downstream from hydroelectric facilities. North American Journal of Fisheries Management 5: 330–339.

Daniels RB. 1960. Entrenchment of the willow drainage ditch, Harrison County, Iowa. American Journal of Science 258: 161–176.

Duncan RP. 1993. Flood disturbance and the coexistence of species in a lowland podocarp forest, south Westland, New Zealand. Journal of Ecology 81: 403–416.

Dunne T, Leopold LB. 1978. Water in Environmental Planning. San Francisco: W. H. Freeman and Co.

Echeverria JD, Barrow P, Roos-Collins R. 1989. Rivers at risk: the concerned citizen's guide to hydropower. Washington (DC): Island Press.

Faber S. 1996. On borrowed land: public policies for floodplains. Cambridge (MA): Lincoln Institute of Land Policy.

Fausch KD, Bestgen KR. 1997. Ecology of fishes indigenous to the central and southwestern Great Plains. Pages 131–166 in Knopf FL, Samson FB, eds. Ecology and conservation of Great Plains vertebrates. New York: Springer-Verlag.

[FERC] Federal Energy Regulatory Commission. 1995. Relicensing the Ayers Island hydroelectric project in the Pemigewasset/Merrimack River Basin. Washington (DC): Federal Energy Regulatory Commission. Final environmental impact statement, FERC Project nr 2456–009.

Fenner P, Brady WW, Patten DR. 1985. Effects of regulated water flows on regeneration o Fremont cottonwood. Journal of Range Management 38: 135–138.

Finger TR, Stewart EM. 1987. Response of fishes to flooding in lowland hardwood wetlands. Pages 86–92 in Matthews WJ, Heins DC, eds. Community and evolutionary ecology of North American stream fishes. Norman (OK): University of Oklahoma Press.

Fisher SG. 1983. Succession in streams. Pages 7–27 in Barnes JR, Minshall GW, eds. Stream ecology: application and testing of general ecological theory. New York: Plenum Press.

Franklin JF. 1993. Preserving biodiversity: species, ecosystems, or landscapes? Ecological Applications 3: 202–205.

Friedman JM, Scott ML, Lewis WM. 1995. Restoration of riparian forest using irrigation, artificial disturbance, and natural seedfall. Environmental Management 19: 547–557.

Gehrke PC, Brown P, Schiller CB, Moffatt DB, Bruce AM. 1995. River regulation and fish communities in the Murray-Darling river system, Australia. Regulated Rivers: Research & Management 11: 363–375.

Greenberg L, Svendsen P, Harby A. 1996. Availability of microhabitats and their use by brown trout (*Salmo trutta*) and grayling (*Thymallus thymallus*) in the River Vojman, Sweden. Regulated Rivers: Research & Management 12: 287–303.

Halyk LC, Balon EK. 1983. Structure and ecological production of the fish taxocene of a small floodplain system. Canadian Journal of Zoology 61: 2446–2464.

Hammer TR. 1972. Stream channel enlargement due to urbanization. Water Resources Research 8: 1530–1540.

Harms WR, Schreuder HT, Hook DD, Brown CL, Shropshire FW. 1980. The effects of flooding on the swamp forest in Lake Oklawaha, Florida. Ecology 61: 1412–1421.

Hill MT, Platts WS. In press. Restoration of riparian habitat with a multiple flow regime in the Owens River Gorge, California. Journal of Restoration Ecology.

Hill MT, Platts WS, Beschta RL. 1991. Ecological and geomorphological concepts for instream and out-of-channel flow requirements. Rivers 2: 198–210.

Holling CS, Meffe GK. 1996. Command and control and the pathology of natural resource management. Conservation Biology 10: 328–337.

Horton JS. 1977. The development and perpetuation of the permanent tamarisk type in the phreatophyte zone of the Southwest. USDA Forest Service. General Technical Report nr RM-43: 124–127.

Hughes FMR. 1994. Environmental change, disturbance, and regeneration in semi-arid floodplain forests. Pages 321–345 in Millington AC, Pye K, eds. Environmental change in drylands: biogeographical and geomorphological perspectives. New York: John Wiley & Sons.

Hughes RM, Noss RF. 1992. Biological diversity and biological integrity: current concerns for lakes and streams. Fisheries 17: 11–19.

Hupp CR, Osterkamp WR. 1985. Bottomland vegetation distribution along Passage Creek, Virginia, in relation to fluvial landforms. Ecology 66: 670–681.

Johnson WC. 1994. Woodland expansion in the Platte River, Nebraska: patterns and causes. Ecological Monographs 64: 45–84.

Johnson WC, Burgess RL, Keammerer WR. 1976. Forest overstory vegetation and environment on the Missouri River floodplain in North Dakota. Ecological Monographs 46: 59–84.

Junk WJ, Bayley PB, Sparks RE. 1989. The flood pulse concept in river-floodplain systems. Canadian Special Publication of Fisheries and Aquatic Sciences 106: 110–127.

Karr JR. 1991. Biological integrity: a long-neglected aspect of water resource management. Ecological Applications 1: 66–84.

Karr JR, Toth LA, Dudley DR. 1985. Fish communities of midwestern rivers: a history of degradation. BioScience 35: 90–95.

Keller EA, Swanson FJ. 1979. Effects of large organic material on channel form and fluvial processes. Earth Surface Processes and Landforms 4: 351–380.

Kingsolving AD, Bain MB. 1993. Fish assemblage recovery along a riverine disturbance gradient. Ecological Applications 3: 531–544.

Klotz JR, Swanson S. 1997. Managed instream flows for woody vegetation recruitment, a case study. Pages 483–489 in Warwick J, ed. Symposium proceedings: water resources education, training, and practice: opportunities for the next century. American Water Resources Association, Universities Council on Water Resources, American Water Works Association; 29 Jun–3 Jul; Keystone, CO.

Knox JC. 1972. Valley alluviation in southwestern Wisconsin. Annals of the Association of American Geographers 62: 401–410.

Kondolf GM, Curry RR. 1986. Channel erosion along the Carmel River, Monterey County, California. Earth Surface Processes and Landforms 11: 307–319.

Krapu GL, Facey DE, Fritzell EK, Johnson DH. 1984. Habitat use by migrant sandhill cranes in Nebraska. Journal of Wildlife Management 48: 407–417.

Kupferberg SK. 1996. Hydrologic and geomorphic factors affecting conservation of a river-breeding frog (*Rana boylii*). Ecological Applications 6: 1332–1344.

Larson HN. 1981. New England flow policy. Memorandum, interim regional policy for New England stream flow recommendations. Boston: US Fish & Wildlife Service, Region 5.

Leopold LB. 1968. Hydrology for urban land planning: a guidebook on the hydrologic effects of land use. Reston (VA): US Geological Survey. Circular nr 554.

Leopold LB, Wolman MG, Miller JP. 1964. Fluvial processes in geomorphology. San Francisco: W. H. Freeman & Sons.

Ligon FK, Dietrich WE, Trush WJ. 1995. Downstream ecological effects of dams, a geomorphic perspective. BioScience 45: 183–192.

[LADWP] Los Angeles Department of Water and Power. 1995. Draft Mono Basin stream and channel restoration plan. Los Angeles: Department of Water and Power.

Meffe GK. 1984. Effects of abiotic disturbance on coexistence of predator and prey fish species. Ecology 65: 1525–1534.

Meffe GK, Minckley WL. 1987. Persistence and stability of fish and invertebrate assemblages in a repeatedly disturbed Sonoran Desert stream. American Midland Naturalist 117: 177–191.

Miller AJ. 1990. Flood hydrology and geomorphic effectiveness in the central Appalachians. Earth Surface Processes and Landforms 15: 119–134.

Minckley WL, Deacon JE, ed. 1991. Battle against extinction: native fish management in the American West. Tucson (AZ): University of Arizona Press.

Minckley WL, Meffe GK. 1987. Differential selection by flooding in stream-fish communities of the arid American Southwest. Pages 93–104 in Matthews WJ, Heins DC, eds. Community and evolutionary ecology of North American stream fishes. Norman (OK): University of Oklahoma Press.

Mitsch WJ, Rust WG. 1984. Tree growth responses to flooding in a bottomland forest in northern Illinois. Forest Science 30: 499–510.

Molles MC, Crawford CS, Ellis LM. 1995. Effects of an experimental flood on litter dynamics in the Middle Rio Grande riparian ecosystem. Regulated Rivers: Research & Management 11: 275–281.

Montgomery WL, McCormick SD, Naiman RJ, Whoriskey FG, Black GA. 1983. Spring migratory synchrony of salmonid, catostomid, and cyprinid fishes in Rivière á la Truite, Québec. Canadian Journal of Zoology 61: 2495–2502.

Moore KMS, Gregory SV. 1988. Response of young-of-the-year cutthroat trout to manipulations of habitat structure in a small stream. Transactions of the American Fisheries Society 117: 162–170.

Morehardt JE. 1986. Instream flow methodologies. Palo Alto (CA): Electric Power Research Institute. Report nr EPRIEA-4819.

Moyle PB. 1986. Fish introductions into North America: patterns and ecological impact. Pages 27–43 in Mooney HA, Drake JA, eds. Ecology of biological invasions of North America and Hawaii. New York: Springer-Verlag.

Moyle PB, Light T. 1996. Fish invasions in California: do abiotic factors determine success? Ecology 77: 1666–1669.

Næsje T, Jonsson B, Skurdal J. 1995. Spring flood: a primary cue for hatching of river spawning Coregoninae. Canadian Journal of Fisheries and Aquatic Sciences 52: 2190–2196.

Naiman RJ, Magnuson JJ, McKnight DM, Stanford JA. 1995. The freshwater imperative: a research agenda. Washington (DC): Island Press.

[NRC] National Research Council. 1992. Restoration of aquatic systems: science, technology, and public policy. Washington (DC): National Academy Press.

_____. 1996. Upstream: salmon and society in the Pacific Northwest. Washington (DC): National Academy Press.

Nesler TP, Muth RT, Wasowicz AF. 1988. Evidence for baseline flow spikes as spawning cues for Colorado Squawfish in the Yampa River, Colorado. American Fisheries Society Symposium 5: 68–79.

Nilsen ET, Sharifi MR, Rundel PW. 1984. Comparative water relations of phreatophytes in the Sonoran Desert of California. Ecology 65: 767–778.

Nilsson C. 1982. Effects of stream regulation on riparian vegetation. Pages 93–106 in Lillehammer A, Saltveit SJ, eds. Regulated rivers. New York: Columbia University Press.

Perkins DJ, Carlsen BN, Fredstrom M, Miller RH, Rofer CM, Ruggerone GT, Zimmerman CS. 1984. The effects of groundwater pumping on natural spring communities in Owens Valley. Pages 515–527 in Warner RE, Hendrix KM, eds. California riparian systems: ecology, conservation, and productive management. Berkeley (CA): University of California Press.

Petts GE. 1984. Impounded rivers: perspectives for ecological management. New York: John Wiley & Sons.

_____. 1985. Time scales for ecological concern in regulated rivers. Pages 257–266 in Craig JF, Kemper JB, eds. Regulated streams: advances in ecology. New York: Plenum Press.

Pickett STA, Parker VT, Fiedler PL. 1992. The new paradigm in ecology: implications for conservation biology above the species level. Pages 66–88 in Fiedler PL, Jain SK, eds. Conservation biology. New York: Chapman & Hall.

Poff NL. 1996. A hydrogeography of unregulated streams in the United States and an examination of scale-dependence in some hydrological descriptors. Freshwater Biology 36: 101–121.

Poff NL, Allan JD. 1995. Functional organization of stream fish assemblages in relation to hydrological variability. Ecology 76: 606–627.

Poff NL, Ward JV. 1989. Implications of streamflow variability and predictability for lotic community structure: a regional analysis of streamflow patterns. Canadian Journal of Fisheries and Aquatic Sciences 46: 1805–1818.

_____. 1990. The physical habitat template of lotic systems: recovery in the context of historical pattern of spatio-temporal heterogeneity. Environmental Management 14: 629–646.

Power ME. 1992. Hydrologic and trophic controls of seasonal algal blooms in northern California rivers. Archiv für Hydrobiologie 125: 385–410.

Power ME, Sun A, Parker M, Dietrich WE, Wootton JT. 1995. Hydraulic food-chain models: an approach to the study of food-web dynamics in large rivers. BioScience 45: 159–167.

Prestegaard KL. 1988. Morphological controls on sediment delivery pathways. Pages 533–540 in Walling DE, ed. Sediment budgets. Wallingford (UK): IAHS Press. International Association of Hydrological Sciences Publication nr 174.

Prestegaard KL, Matherne AM, Shane B, Houghton K, O'Connell M, Katyl N. 1994. Spatial variations in the magnitude of the 1993 floods, Raccoon River Basin, Iowa. Geomorphology 10: 169–182.

Reeves GH, Benda LE, Burnett KM, Bisson PA, Sedell JR. 1996. A disturbance-based ecosystem approach to maintaining and restoring freshwater habitats of evolutionarily significant units of anadromous salmonids in the Pacific Northwest. American Fisheries Society Symposium 17: 334–349.

Reily PW, Johnson WC. 1982. The effects of altered hydrologic regime on tree growth along the Missouri River in North Dakota. Canadian Journal of Botany 60: 2410–2423.

Resh VH, Brown AV, Covich AP, Gurtz ME, Li HW, Minshall GW, Reice SR, Sheldon AL, Wallace JB, Wissmar R. 1988. The role of disturbance in stream ecology. Journal of the North American Benthological Society 7: 433–455.

Richter BD, Baumgartner JV, Powell J, Braun DP. 1996. A method for assessing hydrologic alteration within ecosystems. Conservation Biology 10: 1163–1174.

Richter BD, Baumgartner JV, Wigington R, Braun DP. 1997. How much water does a river need? Freshwater Biology 37: 231–249.

Robertson L. In press. Water operations on the Pecos River, New Mexico and the Pecos bluntnose shiner, a federally-listed minnow. US Conference on Irrigation and Drainage Symposium.

Rood SB, Mahoney JM. 1990. Collapse of riparian poplar forests downstream from dams in western prairies: probable causes and prospects for mitigation. Environmental Management 14: 451–464.

Rood SB, Mahoney JM, Reid DE, Zilm L. 1995. Instream flows and the decline of riparian cottonwoods along the St. Mary River, Alberta. Canadian Journal of Botany 73: 1250–1260.

Ross ST, Baker JA. 1983. The response of fishes to periodic spring floods in a southeastern stream. American Midland Naturalist 109: 1–14.

Rulifson RA, Manooch CS III, eds. 1993. Roanoke River water flow committee report for 1991–1993. Albemarle-Pamlico estuarine study. Raleigh (NC): US Environmental Protection Agency. Project nr APES 93-18.

Scheidegger KJ, Bain MB. 1995. Larval fish in natural and regulated rivers: assemblage composition and microhabitat use. Copeia 1995: 125–135.

Schlosser IJ. 1985. Flow regime, juvenile abundance, and the assemblage structure of stream fishes. Ecology 66: 1484–1490.

_____. 1990. Environmental variation, life history attributes, and community structure in stream fishes: implications for environmental management assessment. Environmental Management 14: 621–628.

Scott ML, Friedman JM, Auble GT. 1996. Fluvial processes and the establishment of bottomland trees. Geomorphology 14: 327–339.

Scott, ML, Auble GT, Friedman JM. 1997 Flood dependency of cottonwood establishment along the Missouri River, Montana, USA. Ecological Applications 7: 677–690.

Sear DA. 1995. Morphological and sedimentological changes in a gravel-bed river following 12 years of flow regulation for hydropower. Regulated Rivers: Research & Management 10: 247–264.

Seegrist DW, Gard R. 1972. Effects of floods on trout in Sagehen Creek, California. Transactions of the American Fisheries Society 101: 478–482.

Shankman D, Drake DL. 1990. Channel migration and regeneration of bald cypress in western Tennessee. Physical Geography 11: 343–352.

Sidle JG, Carlson DE, Kirsch EM, Dinan JJ. 1992. Flooding mortality and habitat renewal for least terns and piping plovers. Colonial Waterbirds 15: 132–136.

Southwood TRE. 1977. Habitat, the templet for ecological strategies? Journal of Animal Ecology 46: 337–365.

Sparks RE. 1992. Risks of altering the hydrologic regime of large rivers. Pages 119–152 in Cairns J, Niederlehner BR, Orvos DR, eds. Predicting ecosystem risk. Vol XX. Advances in modern environmental toxicology. Princeton (NJ): Princeton Scientific Publishing Co.

_____. 1995. Need for ecosystem management of large rivers and their floodplains. BioScience 45: 168–182.

Stalnaker CB. 1990. Minimum flow is a myth. Pages 31–33 in Bain MB, ed. Ecology and assessment of warmwater streams: workshop synopsis. Washington (DC): US Fish & Wildlife Service. Biological Report nr 90(5).

Stalnaker C, Lamb BL, Henriksen J, Bovee K, Bartholow J. 1995. The instream flow incremental methodology: a primer for IFIM. Ft. Collins (CO): National Biological Service, US Department of the Interior. Biological Report nr 29.

Stanford JA. 1994. Instream flows to assist the recovery of endangered fishes of the Upper Colorado River Basin. Washington (DC): US Department of the Interior, National Biological Survey. Biological Report nr 24.

Stanford JA, Ward JV, Liss WJ, Frissell CA, Williams RN, Lichatowich JA, Coutant CC. 1996. A general protocol for restoration of regulated rivers. Regulated Rivers: Research & Management 12: 391–414.

Stevens LE, Schmidt JC, Brown BT. 1995. Flow regulation, geomorphology, and Colorado River marsh development in the Grand Canyon, Arizona. Ecological Applications 5: 1025–1039.

Strange EM, Moyle PB, Foin TC. 1992. Interactions between stochastic and deterministic processes in stream fish community assembly. Environmental Biology of Fishes 36: 1–15.

Streng DR, Glitzenstein JS, Harcombe PA. 1989. Woody seedling dynamics in an East Texas floodplain forest. Ecological Monographs 59: 177–204.

Stromberg JC, Tress JA, Wilkins SD, Clark S. 1992. Response of velvet mesquite to groundwater decline. Journal of Arid Environments 23: 45–58.

Stromberg JC, Tiller R, Richter B. 1996. Effects of groundwater decline on riparian vegetation of semiarid regions: the San Pedro River, Arizona, USA. Ecological Applications 6: 113–131.

Stromberg JC, Fry J, Patten DT. 1997. Marsh development after large floods in an alluvial, arid-land river. Wetlands 17: 292–300.

Taylor DW. 1982. Eastern Sierra riparian vegetation: ecological effects of stream diversion. Mono Basin Research Group Contribution nr 6, Report to Inyo National Forest.

[TNC] The Nature Conservancy. 1996. Troubled waters: protecting our aquatic heritage. Arlington (VA): The Nature Conservancy.

Tilman D, Downing JA, Wedin DA. 1994. Does diversity beget stability? Nature 371: 257–264.

Toth LA. 1995. Principles and guidelines for restoration of river/floodplain ecosystems—Kissimmee River, Florida. Pages 49–73 in Cairns J, ed. Rehabilitating damaged ecosystems. 2nd ed. Boca Raton (FL): Lewis Publishers/CRC Press.

Travnichek VH, Bain MB, Maceina MJ. 1995. Recovery of a warmwater fish assemblage after the initiation of a minimum-flow release downstream from a hydroelectric dam. Transactions of the American Fisheries Society 124: 836–844.

Trépanier S, Rodríguez MA, Magnan P. 1996. Spawning migrations in landlocked Atlantic salmon: time series modelling of river discharge and water temperature effects. Journal of Fish Biology 48: 925–936.

Turner JL, Chadwick HK. 1972. Distribution and abundance of young-of-the-year striped bass, *Morone saxatilis*, in relation to river flow in the Sacramento-San Joaquin estuary. Transactions of the American Fisheries Society 101: 442–452.

Tyus HM. 1990. Effects of altered stream flows on fishery resources. Fisheries 15: 18–20.

[USCOE] US Army Corps of Engineers, Los Angeles District. 1996. Reconnaissance report, review of existing project: Alamo Lake, Arizona.

Valentin S, Wasson JG, Philippe M. 1995. Effects of hydropower peaking on epilithon and invertebrate community trophic structure. Regulated Rivers: Research & Management 10: 105–119.

Van Winkle W, Rose KA, Chambers RC. 1993. Individual-based approach to fish population dynamics: an overview. Transactions of the American Fisheries Society 122: 397–403.

Walker KF, Sheldon F, Puckridge JT. 1995. A perspective on dryland river ecosystems. Regulated Rivers: Research & Management 11: 85–104.

Wallace JB, Benke AC. 1984. Quantification of wood habitat in subtropical coastal plains streams. Canadian Journal of Fisheries and Aquatic Sciences 41: 1643–1652.

Ward JV, Stanford JA. 1979. The ecology of regulated streams. New York: Plenum Press.

Weisberg SB, Janicki AJ, Gerritsen J, Wilson HT. 1990. Enhancement of benthic macroinvertebrates by minimum flow from a hydroelectric dam. Regulated Rivers: Research & Management 5: 265–277.

Welcomme RL. 1979. Fisheries ecology of floodplain rivers. New York: Longman.

_____. 1992. River conservation—future prospects. Pages 454–462 in Boon PJ, Calow P, Petts GE, eds. River conservation and management. New York: John Wiley & Sons.

Welcomme RL, Hagborg D. 1977. Towards a model of a floodplain fish population and its fishery. Environmental Biology of Fishes 2: 7–24.

Wharton CH, Lambou VW, Newsome J, Winger PV, Gaddy LL, Mancke R. 1981. The fauna of bottomland hardwoods in the southeastern United States. Pages 87–160 in Clark JR, Benforado J, eds. Wetlands of bottomland hardwood forests. New York: Elsevier Scientific Publishing Co.

Williams JG. 1996. Lost in space: minimum confidence intervals for idealized PHABSIM studies. Transactions of the American Fisheries Society 125: 458–465.

Williams DD, Hynes HBN. 1977. The ecology of temporary streams. II. General remarks on temporary streams. Internationale Revue des gesampten Hydrobiologie 62: 53–61.

Williams GP, Wolman MG. 1984. Downstream effects of dams on alluvial rivers. Reston (VA): US Geological Survey. Professional Paper nr 1286.

Williams RN, Calvin LD, Coutant CC, Erho MW, Lichatowich JA, Liss WJ, McConnaha WE, Mundy PR, Stanford JA, Whitney RR. 1996. Return to the river: restoration of salmonid fishes in the Columbia River ecosystem. Portland (OR): Northwest Power Planning Council.

Wolman MG, Gerson R. 1978. Relative scales of time and effectiveness of climate in watershed geomorphology. Earth Surface Processes and Landforms 3: 189–208.

Wolman MG, Miller JP. 1960. Magnitude and frequency of forces in geomorphic processes. Journal of Hydrology 69: 54–74.

Wootton JT, Parker MS, Power ME. 1996. Effects of disturbance on river food webs. Science 273: 1558–1561.

PERSPECTIVES

The River Continuum Concept[1]

ROBIN L. VANNOTE

Stroud Water Research Center, Academy of Natural Sciences of Philadelphia, Avondale, PA 19311, USA

G. WAYNE MINSHALL

Department of Biology, Idaho State University, Pocatello, ID 83209, USA

KENNETH W. CUMMINS

Department of Fisheries and Wildlife, Oregon State University, Corvallis, OR 97331, USA

JAMES R. SEDELL

Weyerhauser Corporation, Forestry Research, 505 North Pearl Street, Centralia, WA 98531, USA

AND COLBERT E. CUSHING

Ecosystems Department, Battelle-Pacific Northwest Laboratories, Richland, WA 99352, USA

VANNOTE, R. L., G. W. MINSHALL, K. W. CUMMINS, J. R. SEDELL, AND C. E. CUSHING. 1980. The river continuum concept. Can. J. Fish. Aquat. Sci. 37: 130–137.

From headwaters to mouth, the physical variables within a river system present a continuous gradient of physical conditions. This gradient should elicit a series of responses within the constituent populations resulting in a continuum of biotic adjustments and consistent patterns of loading, transport, utilization, and storage of organic matter along the length of a river. Based on the energy equilibrium theory of fluvial geomorphologists, we hypothesize that the structural and functional characteristics of stream communities are adapted to conform to the most probable position or mean state of the physical system. We reason that producer and consumer communities characteristic of a given river reach become established in harmony with the dynamic physical conditions of the channel. In natural stream systems, biological communities can be characterized as forming a temporal continuum of synchronized species replacements. This continuous replacement functions to distribute the utilization of energy inputs over time. Thus, the biological system moves towards a balance between a tendency for efficient use of energy inputs through resource partitioning (food, substrate, etc.) and an opposing tendency for a uniform rate of energy processing throughout the year. We theorize that biological communities developed in natural streams assume processing strategies involving minimum energy loss. Downstream communities are fashioned to capitalize on upstream processing inefficiencies. Both the upstream inefficiency (leakage) and the downstream adjustments seem predictable. We propose that this River Continuum Concept provides a framework for integrating predictable and observable biological features of lotic systems. Implications of the concept in the areas of structure, function, and stability of riverine ecosystems are discussed.

Key words: river continuum; stream ecosystems; ecosystem structure, function; resource partitioning; ecosystem stability; community succession; river zonation; stream geomorphology

VANNOTE, R. L., G. W. MINSHALL, K. W. CUMMINS, J. R. SEDELL, AND C. E. CUSHING. 1980. The river continuum concept. Can. J. Fish. Aquat. Sci. 37: 130–137.

De la tête des eaux à l'embouchure, un réseau fluvial offre un gradient continu de conditions physiques. Ce gradient devrait susciter, chez les populations habitant dans le réseau, une série de réponses aboutissant à un continuum d'ajustements biotiques et à des schémas uniformes de charge, transport, utilisation et emmagasinage de la matière organique sur tout le

---

[1]Contribution No. 1 from the NSF River Continuum Project.

Printed in Canada (J5632)
Imprimé au Canada (J5632)

PERSPECTIVES

parcours d'une rivière. Faisant appel à la théorie de l'équilibre énergétique des spécialistes de la géomorphologie fluviale, nous avançons l'hypothèse que les caractéristiques structurales et fonctionnelles des communautés fluviatiles sont adaptées de façon à se conformer à la position ou condition moyenne la plus probable du système physique. Nous croyons que les communautés de producteurs et de consommateurs caractéristiques d'un segment donné de la rivière se mettent en harmonie avec les conditions physiques dynamiques du chenal. Dans des réseaux fluviaux naturels, on peut dire que les communautés biologiques forment un continuum temporel de remplacements synchronisés d'espèces. Grâce à ce remplacement continu, il y a répartition dans le temps de l'utilisation des apports énergétiques. Ainsi, le système biologique vise à un équilibre entre une tendance vers l'utilisation efficace des apports d'énergie en partageant les ressources (nourriture, substrat, etc.), d'une part, et une tendance opposée vers un taux uniforme de transformation de l'énergie durant l'année, d'autre part. A notre avis, les communautés biologiques habitant dans des cours d'eau naturels adoptent des stratégies de transformation comportant une perte minimale d'énergie. Les communautés d'aval sont organisées de façon à tirer profit de l'inefficacité de transformation des communautés d'amont. On semble pouvoir prédire à la fois l'inefficacité (fuite) d'amont et les ajustements d'aval. Nous suggérons ce concept d'un continuum fluvial comme cadre dans lequel intégrer les caractères biologiques prévisibles et observables des systèmes lotiques. Nous analysons les implications du concept quant à la structure, fonction et stabilité des écosystèmes fluviaux.

Received May 14, 1979
Accepted September 19, 1979

Reçu le 14 mai 1979
Accepté le 19 septembre 1979

Statement of the Concept

Many communities can be thought of as continua consisting of mosaics of integrading population aggregates (McIntosh 1967; Mills 1969). Such a conceptualization is particularly appropriate to streams. Several workers have visualized streams as possessing assemblages of species which respond by their occurrences and relative abundances to the physical gradients present (Shelford 1911; Thompson and Hunt 1930; Ricker 1934; Ide 1935; Burton and Odum 1945; Van Deusen 1954; Huet 1954, 1959; Slack 1955; Minshall 1968; Ziemer 1973; Swanston et al. 1977; Platts 1979). Expansion of this idea to include functional relationships has allowed development of a framework, the "River Continuum Concept," describing the structure and function of communities along a river system. Basically, the concept proposes that understanding of the biological strategies and dynamics of river systems requires consideration of the gradient of physical factors formed by the drainage network. Thus energy input, and organic matter transport, storage, and use by macroinvertebrate functional feeding groups may be regulated largely by fluvial geomorphic processes. The patterns of organic matter use may be analogous to those of physical energy expenditure proposed by geomorphologists (Leopold and Maddock 1953; Leopold and Langbein 1962; Langbein and Leopold 1966; Curry 1972). Further, the physical structure coupled with the hydrologic cycle form a templet (Southwood 1977) for biological responses and result in consistent patterns of community structure and function and organic matter loading, transport, utilization, and storage along the length of a river.

Derivation of the Concept

As the cyclic theory for explaining the evolution of land forms and streams (young, mature, ancient) proved unsatisfactory, the concepts gradually were replaced by a principle of dynamic equilibrium (Curry 1972). The concept of the physical stream network system and the distribution of watersheds as open systems in dynamic ("quasi") equilibrium was first proposed by Leopold and Maddock (1953) to describe consistent patterns, or adjustments, in the relationships of stream width, depth, velocity, and sediment load. These "steady state" systems are only rarely characterized by exact equilibria and generally the river and its channel tend toward a mean form, definable only in terms of statistical means and extremes (Chorley 1962); hence, the idea of a "dynamic" equilibrium. The equilibrium concept was later expanded to include at least nine physical variables and was progressively developed in terms of energy inputs, efficiency in utilization, and rate of entropy gain (Leopold and Langbein 1962; Leopold et al. 1964; Langbein and Leopold 1966). In this view, equilibration of river morphology and hydraulics is achieved by adjustments between the tendency of the river to maximize the efficiency of energy utilization and the opposing tendency toward a uniform rate of energy use.

Based upon these geomorphological considerations, Vannote initially formulated the hypothesis that structural and functional characteristics of stream communities distributed along river gradients are selected to conform to the most probable position or mean state of the physical system. From our collective experience with a number of streams, we felt it was possible to translate the energy equilibrium theory from the physical system of geomorphologists into a biological analog. In this analysis, producer and consumer communities characteristic of a given reach of the river continuum conform to the manner in which the river system utilizes its kinetic energy in achieving a dynamic equilibrium. Therefore, over extended river reaches, biological communities should become established which approach equilibrium with the dynamic physical conditions of the channel.

Implications of the Concept

It is only possible at present to trace the broad outlines of the ways the concept should apply to stream ecosystems and to illustrate these with a few examples for which reasonably good information is available. From headwaters to downstream extent, the physical variables within a stream system present a continuous gradient of conditions including width, depth, velocity, flow volume, temperature, and entropy gain. In developing a biological analog to the physical system, we hypothesize that the biological organization in rivers conforms structurally and functionally to kinetic energy dissipation patterns of the physical system. Biotic communities rapidly adjust to any changes in the redistribution of use of kinetic energy by the physcial system.

STREAM SIZE AND ECOSYSTEM STRUCTURE AND FUNCTION

Based on considerations of stream size, we propose some broad characteristics of lotic communities which can be roughly grouped into headwaters (orders 1–3), medium-sized streams (4–6), and large rivers (>6) (Fig. 1). Many headwater streams are influenced strongly by the riparian vegetation which reduces autotrophic production by shading and contributes large amounts of allochthonous detritus. As stream size increases, the reduced importance of terrestrial organic input coincides with enhanced significance of autochthonous primary production and organic transport from upstream. This transition from headwaters, dependent on terrestrial inputs, to medium-sized rivers, relying on algal or rooted vascular plant production, is thought to be generally reflected by a change in the ratio of gross primary productivity to community respiration (P/R) (Fig. 2). The zone through which the stream shifts from heterotrophic to autotrophic is primarily dependent upon the degree of shading (Minshall 1978). In deciduous forests and some coniferous forests, the transition probably is approximately at order 3 (Fig. 1). At higher elevations and latitudes, and in xeric regions where riparian vegetation is restricted, the transition to autotrophy may be in order 1. Deeply incised streams, even with sparse riparian vegetation, may be heterotrophic due to side slope ("canyon") shading.

Large rivers receive quantities of fine particulate organic matter from upstream processing of dead leaves and woody debris. The effect of riparian vegetation is insignificant, but primary production may often be limited by depth and turbidity. Such light attenuated systems would be characterized by P/R < 1. Streams of lower order entering midsized or larger rivers (e.g. the 3rd order system shown entering the 6th order

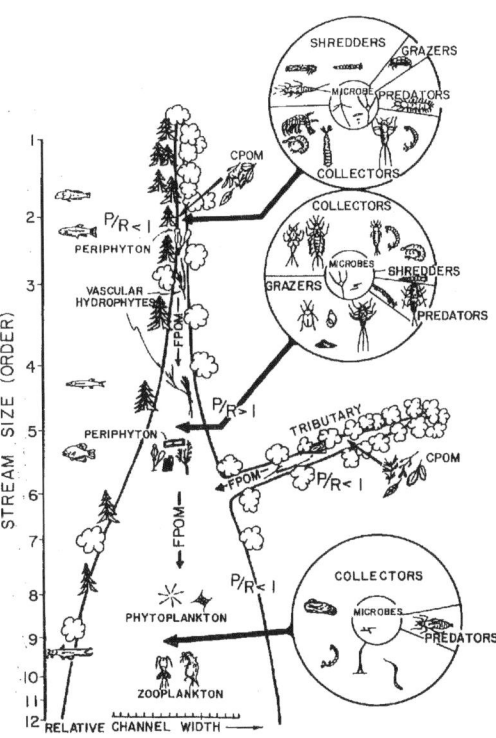

FIG. 1. A proposed relationship between stream size and the progressive shift in structural and functional attributes of lotic communities. See text for fuller explanation.

river in Fig. 1) have localized effects of varying magnitude depending upon the volume and nature of the inputs.

The morphological–behavioral adaptations of running water invertebrates reflect shifts in types and locations of food resources with stream size (Fig. 1). The relative dominance (as biomass) of the general functional groups — shredders, collectors, scrapers (grazers), and predators are depicted in Fig. 1. Shredders utilize coarse particulate organic matter (CPOM, >1 mm), such as leaf litter, with a significant dependence on the associated microbial biomass. Collectors filter from transport, or gather from the sediments, fine and ultrafine particulate organic matter (FPOM, 50 μm–1 mm; UPOM 0.5–50 μm). Like shredders, collectors depend on the microbial biomass associated with the particles (primarily on the surface) and products of microbial metabolism for their nutrition. Scrapers are adapted primarily for shearing attached algae from surfaces. The proposed dominance of scrapers follows shifts in primary production, being maximized in midsized rivers

PERSPECTIVES

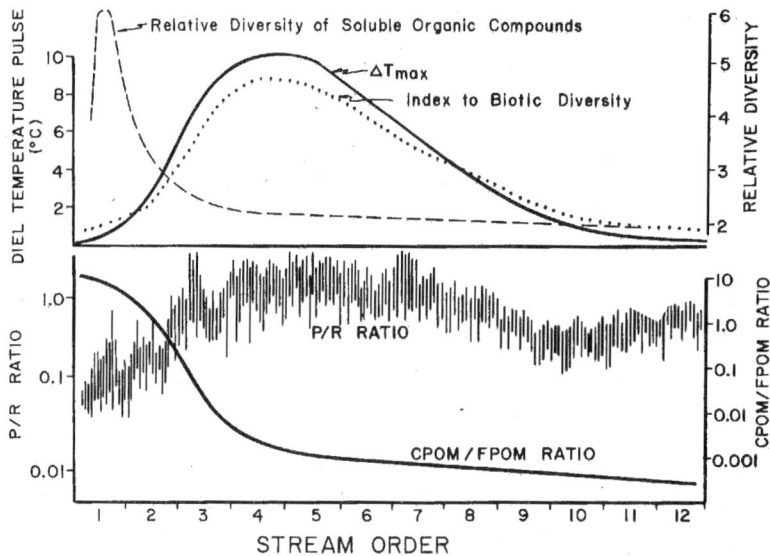

FIG. 2. Hypothetical distribution of selected parameters through the river continuum from headwater seeps to a twelfth order river. Parameters include heterogeneity of soluble organic matter, maximum diel temperature pulse, total biotic diversity within the river channel, coarse to fine particulate organic matter ratio, and the gross photosynthesis/respiration ratio.

with P/R > 1. Shredders are hypothesized to be codominant with collectors in the headwaters, reflecting the importance of riparian zone CPOM and FPOM–UPOM derived from it. With increasing stream size and a general reduction in detrital particle size, collectors should increase in importance and dominate the macroinvertebrate assemblages of large rivers (Fig. 1).

The predatory invertebrate component changes little in relative dominance with stream order. Fish populations (Fig. 1) show a shift from cool water species low in diversity to more diverse warm water communities (e.g. Huet 1954). Most headwater species are largely invertivores. Piscivorous and invertivorous species characterize the midsized rivers and in large rivers some planktivorous species are found — reflecting the semi-lentic nature of such waters.

The expected diversity of soluble organic compounds through the continuum is shown in Fig. 2 (dashed line). Headwater streams represent the maximum interface with the landscape and therefore are predominantly accumulators, processors, and transporters of materials from the terrestrial system. Among these inputs are heterogeneous assemblages of labile and refractory dissolved compounds, comprised of short- and long-chain organics. Heterotrophic use and physical absorption of labile organic compounds is rapid, leaving the more refractory and relatively high molecular weight compounds for export downstream. The relative importance of large particle detritus to energy flow in the system is expected to follow a curve similar to that of the diversity of soluble organic compounds; however, its importance may extend further downstream.

Thus the river system, from headwaters to mouth, can be considered as a gradient of conditions from a strongly heterotrophic headwater regime to a seasonal, and in many cases, an annual regime of autotrophy in midreaches, and then a gradual return to heterotrophic processes in downstream waters (Fisher 1977). Major bioenergetic influences along the stream continuum are local inputs (allochthonous litter and light) and transport from upstream reaches and tributaries (Fig. 1). As a consequence of physical and biological processes, the particle size of organic material in transport should become progressively smaller down the continuum (reflected by CPOM:FPOM ratio in Fig. 2, except for localized input of lower order tributaries) and the stream community response reflect progressively more efficient processing of smaller particles.

RIVER ECOSYSTEM STABILITY

Stability of the river ecosystem may be viewed as a tendency for reduced fluctuations in energy flow, while community structure and function are maintained, in the face of environmental variations. This implicitly couples community stability (sensu Ricklefs 1979) to the instability ("noise") of the physical system. In highly stable physical systems, biotic contribution to ecosystem stability may be less critical. However, in widely fluctuating environments (e.g. stream reaches with lage fluctuations in temperature), the biota may assume critical importance in stabilizing the entire system. In this interpretation, ecosystem stability is achieved by a dynamic balance between forces contributing to stabilization (e.g. debris dams, filter feeders, and other retention devices; nutrient cycling) and those contributing to its instability (e.g. floods, temperature fluctuations, microbial epidemics). In systems with a highly stable physical structure, biotic diversity may be low and yet total stability of the stream ecosystem still be maintained. In contrast, systems with a high degree of physical variation may have high species diversity or at least high complexity in species function which acts to maintain stability.

For example, in stream zones experiencing wide diel temperature changes, organisms may be exposed to suboptimum temperatures for significant portions of the day, but over some range in the diel cycle each organism encounters a favorable or optimum temperature range. Under these conditions an optimum temperature will occur for a larger number of species than if the thermal regime displayed minimum variance. Also, in the thermally fluctuating system, many populations have an opportunity to process energy, and as temperatures oscillate around a mean position, various populations may increase or decrease their processing rates. Thus, an important aspect of the predictably fluctuating physical system is that it encompasses optimum conditions for a large number of species. This interplay between physical and biological components can be seen in terms of ecosystem stability by considering the response of total biotic diversity in the river channel as balanced against the maximum diel temperature range (ΔT max) (Fig. 2). Headwater streams in proximity to groundwater supply or infiltration source areas exhibit little variation in ΔT max. With increased distance from subsurface sources and separation of the forest canopy, ΔT max will attain its widest variance because of increased solar input. The ΔT max amplitude is greatly diminished in high order streams due to the buffering effect of the large volume of water in the channel (Ross 1963). In headwater springs and brooks, diversity may be low because biological communities are assembled from those species which can function within a narrow temperature range on a restricted nutritional base; the stability of the system may be maintained by the low amplitude of diel and annual temperature regimes. Total community diversity is greatest in medium-sized (3rd to 5th order in Fig. 2) streams where temperature variations tend to be maximized. The tendency to stabilize energy flow in midsized streams may be aided by high biotic diversity which mitigates the influence of high variance in the physical system as characterized by ΔT max; i.e. variation due to fluctuating thermal regimes should be offset by a high diversity of biota. In large rivers, stability of the system should be correlated with reduction in variance of diel temperature. We wish to emphasize that temperature is not the only factor responsible for the change in community structure; it is simply one of the easiest to visualize. Other factors such as riparian influence, substrate, flow, and food also are important and change in predictable fashion downstream both absolutely and in terms of the relative heterogeneity of each.

Temporal Adjustments in Maintaining an Equilibrium of Energy Flow

Natural stream ecosystems should tend towards uniformity of energy flow on an annual basis. Although the processing rates and efficiencies of energy utilization by consumer organisms are believed to approach equilibrium for the year, the major organic substrates shift seasonally. In natural stream systems, both living and detrital food bases are processed continuously, but there is a seasonal shift in the relative importance of autotrophic production vs. detritus loading and processing. Several studies (Minshall 1967; Coffman et al. 1971; Kaushik and Hynes 1971; MacKay and Kalff 1973; Cummins 1974; Sedell et al. 1974) have shown the importance of detritus in supporting autumn-winter food chains and providing a fine particle base for consumer organisms during other seasons of the year. Autotrophic communities often form the major food base, especially in spring and summer months (Minshall 1978).

Studies on headwater (order 1–3) streams have shown that biological communities in most habitats can be characterized as forming a temporal sequence of synchronized species replacement. As a species completes its growth in a particular microhabitat, it is replaced by other species performing essentially the same function, differing principally by the season of growth (Minshall 1968; Sweeney and Vannote 1978; Vannote 1978; Vannote and Sweeney 1979). It is this continuous species replacement that functions to distribute the utilization of energy inputs over time (e.g. Wallace et al. 1977). Individuals within a species will tend to exploit their environment as efficiently as possible. This results in the biological system (composite species assemblage) tending to maximize energy consumption. Because some species persist through time and because new species become dominant, and these too are exploiting their environment as efficiently as possible, processing of energy by the changing biological system tends to result in uniform energy processing over time. Thus, the biological system moves towards equilibrium by a trade-off between a tendency to make most efficient use of energy inputs through resource partitioning of food, substrate, temperature, etc. and tendency toward a uniform rate of energy processing throughout the year. From strategies observed on small to medium-sized streams (orders 1–5), we propose that biological communities, developed in natural streams

PERSPECTIVES

in dynamic equilibrium, assume processing strategies involving minimum energy loss (termed maximum "spiraling" by Webster 1975).

ECOSYSTEM PROCESSING ALONG THE CONTINUUM

The dynamic equilibrium resulting from maximization of energy utilization and minimization of variation in its use over the year determines storage or leakage of energy. Storage includes production of new tissue and physical retention of organic material for future processing. In stream ecosystems, unused or partially processed materials will tend to be transported downstream. This energy loss, however, is the energy income, together with local inputs, for communities in downstream reaches. We postulate that downstream communities are structured to capitalize on these inefficiencies of upstream processing. In every reach some material is processed, some stored, and some released. The amount released in this fashion has been used in calculating system efficiency (Fisher 1977). Both the upstream inefficiency (leakage) and the downstream adjustments seem predictable. Communities distributed along the river are structured to process materials (specific detrital sizes, algae, and vascular hydrophytes) thereby minimizing the variance in system structure and function. For example, materials prone to washout, such as flocculant fine-particle detritus, might be most efficiently processed either in transport or after deposition in downstream areas. The resistivity of fine particle detritus to periodic washout is increased by sedimentation in depositional zones or by combination in a matrix with the more cohesive silt and clay sediments. Thus, enhanced retention results in the formation of a distinct community adapted to utilize this material. The minimization of the variance of energy flow is the outcome of seasonal variations of energy input rates (detritus and autotrophic production), coupled with adjustments in species diversity, specialization for food processing, temporal expression of functional groups, and the erosional–depositional transport and storage characteristics of flowing waters.

TIME INVARIANCE AND THE ABSENCE OF SUCCESSION IN STREAM COMMUNITIES

A corollary to the continuum hypothesis, also arising from the geomorphological literature (Langbein and Leopold 1966), is that studies of biological systems established in a dynamically balanced physical setting can be viewed in a time independent fashion. In the context of viewing adaptive strategies and processes as continua along a river system, temporal change becomes the slow process of evolutionary drift (physical and genetic). Incorporation of new functional components into the community over evolutionary time necessitates an efficiency adjustment towards reduced leakage. In natural river systems, community structure gains and loses species in response to low probability cataclysmic events and in response to slow processes of channel development.

The concept of time invariance allows integration of community structure and function along the river without the illusion that successional stages are being observed at a given location in a time-dependent series. The concept of biological succession (Margalef 1960) is of little use for river continua, because the communities in each reach have a continuous heritage rather than an isolated temporal composition within a sequence of discrete successional stages. In fact, the biological subsystems for each reach are in equilibrium with the physical system at that point in the continuum. The concept of heritage implies that in natural river systems total absence of a population is rare, and biological subsystems are simply shifting spatially (visualize a series of overlapping normal species-abundance curves in which all species are present at any point on the spatial axis but their abundance differs from one point to the next) and not in the temporal sense typical of plant succession.

On an evolutionary time scale, the spatial shift has two vectors: a donwstream one involving most of the aquatic insects and an upstream one involving molluscs and crustaceans. The insects are believed to have evolved terrestrially and to be secondarily aquatic. Since the maximum terrestrial–aquatic interface occurs in the headwaters, it is likely that the transition from land to water first occurred here with the aquatic forms then moving progressively downstream. The molluscs and crayfish are thought to have developed in a marine environment and to have moved through estuaries into rivers and thence upstream. The convergence of the two vectors may explain why maximum species diversity occurs in the midreaches.

Conclusion

We propose that the River Continuum Concept provides a framework for integrating predictable and observable biological features of flowing water systems with the physical–geomorphic environment. The model has been developed specifically in reference to natural, unperturbed stream ecosystems as they operate in the context of evolutionary and population time scales. However, the concept should accommodate many unnatural disturbances as well, particularly those which alter the relative degree of autotrophy:heterotrophy (e.g. nutrient enrichment, organic pollution, alteration of riparian vegetation through grazing, clear-cutting, etc.) or affect the quality and quantity of transport (e.g. impoundment, high sediment load). In many cases, these alterations can be thought of as reset mechanisms which cause the overall continuum response to be shifted toward the headwaters or seaward depending on the type of perturbation and its location on the river system.

A concept of dynamic equilibrium for biological communities, despite some difficulties in absolute definition, is useful because it suggests that community structure and function adjust to changes in certain geomorphic, physical, and biotic variables such as stream flow, channel morphology, detritus loading, size of particulate organic material, characteristics of autotrophic production, and thermal responses. In developing a theory of biological strategies along the river continuum, it also should be possible to observe a number of patterns that describe various processing rates, growth strategies, metabolic strategies, and community structures and functions. Collection of extensive data sets over the long profile of rivers are needed to further test and refine these ideas.

Acknowledgments

The ideas presented here have been refined through discussions with our associates in the River Continuum Project, especially T. L. Bott, J. D. Hall, R. C. Petersen, and F. J. Swanson; and other colleagues including S. A. Fisher, C. A. S. Hall, L. B. Leopold, F. J. Triska, and J. B. Webster. We are grateful to J. T. Brock, A. B. Hale, C. P. Hawkins, J. L. Meyers, and J. R. Moeller for their constructive criticism of a draft version of the manuscript. This work was supported by the U.S. National Science Foundation, Ecosystems Studies program under grant numbers BMS-75-07333 and DEB-7811671.

BURTON, G. W., AND E. P. ODUM. 1945. The distribution of stream fish in the vicinity of Mountain Lake, Virginia. Ecology 26: 182–194.

CHORLEY, R. J. 1962. Geomorphology and general systems theory. U.S. Geol. Surv. Prof. Pap. 500-B: 10 p.

COFFMAN, W. P., K. W. CUMMINS, AND J. C. WUYCHECK. 1971. Energy flow in a woodland stream ecosystem. I. Tissue support trophic structure of the autumnal community. Arch. Hydrobiol. 68: 232–276.

CUMMINS, K. W. 1974. Structure and function of stream ecosystems. BioScience 24: 631–641.

CURRY, R. R. 1972. Rivers — a geomorphic and chemical overview, p. 9–31. In R. T. Oglesby, C. A. Carlson, and J. A. McCann [ed.] River ecology and man. Academic Press, N.Y. 465 p.

FISHER, S. G. 1977. Organic matter processing by a stream-segment ecosystem: Fort River, Massachusetts, U.S.A. Int. Rev. Hydrobiol. 62: 701–727.

HUET, M. 1954. Biologie, prifils en long et en travers des eaux courantes. Bull. Fr. Piscic. 175: 41–53.

――――― 1959. Profiles and biology of Western European streams as related to fish management. Trans. Am. Fish. Soc. 88: 153–163.

IDE, F. P. 1935. The effect of temperature on the distribution of the mayfly fauna of a stream. Publ. Ont. Fish. Res. Lab. 50: 1–76.

KAUSHIK, N. K., AND H. B. N. HYNES. 1971. The fate of dead leaves that fall into streams. Arch. Hydrobiol. 68: 465–515.

LANGBEIN, W. B., AND L. B. LEOPOLD. 1966. River meanders — theory of minimum variance. U.S. Geol. Surv. Prof. Pap. 422-H: 15 p.

LEOPOLD, L. B., AND T. MADDOCK JR. 1953. The hydraulic geometry of stream channels and some physiographic implications. U.S. Geol. Surv. Prof. Pap. 252: 57 p.

LEOPOLD, L. B., AND W. B. LANGBEIN. 1962. The concept of entropy in landscape evolution. U.S. Geol. Surv. Prof. Pap. 500-A: 20 p.

LEOPOLD, L. B., M. G. WOLMAN, AND J. P. MILLER. 1964. Fluvial processes in geomorphology. W. H. Freeman, San Francisco, Calif. 522 p.

MACKAY, R. J., AND J. KALFF. 1973. Ecology of two related species of caddisfly larvae in the organic substrates of a woodland stream. Ecology 54: 499–511.

MCINTOSH, R. P. 1967. The concept of vegetation. Bot. Rev. 33: 130–187.

MARGALEF, R. 1960. Ideas for a synthetic approach to the ecology of running waters. Int. Rev. Gesamten Hydrobiol. 45: 133–153.

MILLS, E. L. 1969. The community concept in marine zoology, with comments on continua and instability in some marine communities: a review. J. Fish. Res. Board Can. 26: 1415–1428.

MINSHALL, G. W. 1967. Role of allochthonous detritus in the trophic structure of a woodland springbrook community. Ecology 48: 139–149.

――――― 1968. Community dynamics of the benthic fauna in a woodland springbrook. Hydrobiologia 32: 305–339.

――――― 1978. Autotrophy in stream ecosystems. BioScience 28: 767–771.

PLATTS, W. S. 1979. Relationships among stream order, fish populations, and aquatic geomorphology in an Idaho river drainage. Fisheries 4: 5–9.

RICKER, W. E. 1934. An ecological classification of certain Ontario streams. Univ. Toronto Stud. Biol. 37: 1–114.

RICKLEFS, R. E. 1979. Ecology. Chiron Press, Inc. New York, N.Y. 966 p.

ROSS, H. H. 1963. Stream communities and terrestrial biomes. Arch. Hydrobiol. 59: 235–242.

SEDELL, J. R., F. J. TRISKA, J. D. HALL, N. H. ANDERSON, AND J. H. LYFORD JR. 1974. Sources and fate of organic inputs in coniferous forest streams, p. 57–69. In R. H. Waring and R. L. Edmonds [ed.] Integrated research in the Coniferous Forest Biome. Bull. Coniferous Forest Biome Ecosystem Analysis Studies. Univ. Washington, Seattle, Wash. 78 p.

SHELFORD, V. E. 1911. Ecological succession. I. Stream fishes and the method of physiographic analysis. Biol. Bull. 21: 9–35.

SLACK, K. V. 1955. A study of the factors affecting stream productivity by the comparative method. Invest. Indiana Lakes Streams 4: 3–47.

SOUTHWOOD, T. R. E. 1977. Habitat, the templet for ecological strategies? J. Anim. Ecol. 46: 337–365.

SWANSTON, D. N., W. R. MEEHAN, AND J. A. MCNUTT. 1977. A quantitative geomorphic approach to predicting productivity of pink and chum salmon in Southeast Alaska. Publ. Pac. N. W. Forest Range Exp. Stn.

SWEENY, B. W., AND R. L. VANNOTE. 1978. Size variation and the distribution of hemitabolous aquatic insects: two thermal equilibrium hypotheses. Science 200: 444–446.

THOMPSON, D. H., AND F. D. HUNT. 1930. The fishes of Champaign County: a study of the distribution and abundance of fishes in small streams. Bull. Ill. Nat. Hist. Surv. 19: 5–101.

VAN DEUSEN, R. D. 1954. Maryland freshwater stream classification by watersheds. Contr. Chesapeake Biol. Lab. 106: 1–30.

VANNOTE, R. L. 1978. A geometric model describing a quasi-equilibrium of energy flow in populations of stream insects. Proc. Natl. Acad. Sci. U.S.A. 75: 381–384.

PERSPECTIVES

VANNOTE, R. L., AND B. W. SWEENEY. 1979. Geographic analysis of thermal equilibria: a conceptual model for evaluating the effect of natural and modified thermal regimes on aquatic insect communities. Am. Nat. 14: (In press)

WALLACE, J. B., J. R. WEBSTER, AND W. R. WOODALL. 1977. The role of filter feeders in flowing waters. Arch. Hydrobiol. 79: 506–532.

WEBSTER, J. R. 1975. Analysis of potassium and calcium dynamics in stream ecosystems on three southern Appalachian watersheds of contrasting vegetation. Ph.D. thesis, Univ. Georgia, Athens, Ga. 232 p.

ZIEMER, G. L. 1973. Quantitative geomorphology of drainage basins related to fish production. Alaska Fish Game Dep. Info. Leafl. 162: 1–26.

AN EXPERIMENTAL TEST OF THE EFFECTS OF PREDATION RISK ON HABITAT USE IN FISH[1]

Earl E. Werner, James F. Gilliam[2], Donald J. Hall, and Gary G. Mittelbach[3]

Kellogg Biological Station and Department of Zoology, Michigan State University, Hickory Corners, Michigan 49060 USA

Abstract. We present an experiment designed to test the hypothesis that fish respond to both relative predation risk and habitat profitability in choosing habitats in which to feed. Identical populations of three size-classes of bluegill sunfish (*Lepomis macrochirus*) were stocked on both sides of a divided pond (29 m in diameter), and eight piscivorous largemouth bass (*Micropterus salmoides*) were introduced to one side. Sizes of both species were chosen such that the small class of bluegills was very vulnerable to the bass, whereas the largest class was invulnerable to bass predation. We then compared mortality, habitat use, and growth of each size-class in the presence and absence of the bass.

Only the small size-class suffered significant mortality from the bass (each bass consumed on average about one small bluegill every 3.8 d); the two larger size-classes exhibited similar mortality rates on both sides of the pond. In the absence of the bass, we found that habitat use of all size-classes was similar and that the pattern of habitat use maximized foraging return rates (Werner et al. 1983). In the presence of the bass the two larger size-classes chose habitats to maximize return rates, but the small size-class obtained a greater fraction of its diet from the vegetation habitat, where foraging return rates were only one-third of those in the more open habitats. The small size-class further exhibited a significant depression in individual growth in the presence of the bass; the growth increment during the experiment was 27% less than that for small bluegills in the absence of the bass. Because of the reduced utilization of more open habitats by the small fish in the presence of bass, resources in these habitats were released to the larger size-classes, which showed greater growth in the presence of the bass than in its absence. We develop methods to predict the additional mortality expected on a cohort due to a reduction in growth rate (because individuals are spending a longer time in vulnerable sizes), and discuss the potential for predation risk to enforce size-class segregation, which leads de facto to resource partitioning.

Key words: foraging efficiency; habitat use; Lepomis; *Michigan;* Micropterus; *Osteichthyes; predation risk; predator avoidance; size-class interactions.*

Introduction

Recently, optimality models have been usefully applied to problems in animal behavior and evolution. In general, the costs and benefits associated with particular behaviors are described and solutions derived which minimize a postulated cost/benefit function. When this approach is used to study the evolution of morphological structures or life histories, serious questions arise concerning genetic constraints and the existence of tradeoffs (e.g., Gould and Lewontin 1979). When applying the approach to the study of behavior, we are often able to measure directly the costs and benefits associated with a tradeoff and simply ask: does the individual organism over the short term have the capabilities to assess changes in its environment and have the flexibility to respond to these changes as the model predicts? A second related question, especially germane when such models are to be tested under relatively uncontrolled field situations, is: are the costs and constraints conceived by the investigator sufficiently accurate and inclusive to account for the major selective forces that have molded the behavior(s) of interest?

We have experimentally demonstrated that fish have the capability to respond to changes in resource levels in the environment by modifying their selection of food particle size in approximate accordance with optimal foraging models (Werner and Hall 1974, Mittelbach 1981, Werner et al. 1983). We have further demonstrated that fish have the flexibility to shift habitats as relative resource levels in these habitats change and that we can predict these shifts, using the foraging models in small experimental ponds (Werner 1982, Werner et al. 1983). However, testing the predictions of such models under less controlled conditions is more difficult due to additional constraints which might be expected to modify optimal behavior, i.e., the second question above. For example, we have noted that in natural lakes small fish are restricted to weedbeds and do not conform to the predictions of models specifying optimal habitat use from the standpoint of foraging rates (Hall and Werner 1977, Mittelbach 1981). We postulate that predation risk due to piscivorous fish is responsible for this deviation from predicted behavior.

---

[1] Manuscript received 19 April 1982; revised 22 December 1982; accepted 29 December 1982.
[2] Present address: Department of Biological Sciences, State University of New York, Albany, New York 12222 USA.
[3] Present address: Department of Zoology, Ohio State University, Columbus, Ohio 43210 USA.

Often in natural communities a richer habitat, from the standpoint of potential foraging rate, is also one in which a forager experiences higher predation risk. Thus decisions on where to feed presumably involve some weighting of these factors according to their relative impact on fitness, i.e., there is a foraging rate/mortality risk tradeoff. It is therefore important to test whether animals assess predation risk and modify their foraging behavior accordingly, and to build this constraint into our models of optimal habitat use if they do.

A large literature documents the qualitative effects of predators on prey behavior (see Stein [1979] and Curio [1976] for reviews), but surprisingly few studies have quantified this effect, especially in the context of methods which predict the optimal behavior in the predator's absence (but see Milinski and Heller 1978, Caraco et al. 1980, Sih 1980). Clearly such studies are required if a quantitative theory of how animals adaptively balance these two conflicting demands is to be constructed.

In this paper we present an experiment designed to test the hypothesis that fish modify their habitat use under risk of predation and examine the consequences of such changes in behavior on individual growth rate, which is a major component of fitness and population dynamics in fish. Further, the magnitude of the growth depression when predators are present provides some index of the magnitude of the foraging rate/mortality risk tradeoff. In particular, we demonstrate that the presence of the largemouth bass (*Micropterus salmoides*, hereafter simply the bass) causes vulnerable sizes of the bluegill sunfish (*Lepomis macrochirus*) to utilize less profitable but safer habitats. We do this by first measuring resource levels and estimating habitat-specific foraging rates, which are used to predict optimal habitat use successfully in the absence of the bass (i.e., the control situation; see Werner et al. [1983] for details). We then contrast this case with habitat use by the bluegill in an identical environment in the presence of the bass. We further show that the changes in habitat use in the presence of the bass result in a significant decrease in growth rates of the vulnerable bluegill sizes. These results suggest that fish can balance the conflicting demands of foraging and predation risk but that this behavioral response occurs at some significant cost in terms of growth rate. We discuss the implications of these results for optimal foraging theory and the theory of species interactions.

EXPERIMENTAL DESIGN AND METHODS

The experiment was performed in a circular pond (29 m diameter, 1.8 m deep) at the Kellogg Biological Station. All macrophytes were removed from the pond except for a 3 m wide border of cattails (*Typha* spp.). This manipulation yielded three very discrete habitats: a ring of dense vegetation and an unstructured pond center of open water and bare sediments. The pond was then divided in half by a 0.6-cm mesh nylon partition, which was suspended from ropes and anchored to the pond bottom.

The experiment was initiated on 15 July 1979 and terminated by draining the pond on 28 September. Each half of the pond was stocked with identical bluegill populations: 500 small (35.5 ± 0.4 mm, average standard length measured from the tip of the snout to the posterior of the vertebral column ±1 SE), 300 medium (52.9 ± 0.4 mm), and 100 large (73.0 ± 0.8 mm) bluegills. These size-classes and relative proportions were similar to those found in bluegill populations of local lakes (Hall and Werner 1977). One-half of the pond was also stocked with eight bass (198.8 ± 2.9 mm in length). The size of the bass was carefully chosen to set up a gradient in predation risk for the different size-classes of bluegills. Using laboratory data on largemouth bass feeding (Lawrence 1957, Werner 1977) as a guide, we chose a bass size, such that the small bluegills would be extremely vulnerable and the large bluegills would be too large for the bass to catch and swallow. Bass of this size can swallow the medium size-class of bluegills in the laboratory, but it is doubtful that they could very easily capture bluegills of this size in the field. All surviving fish were recovered in September by draining the pond.

At intervals of 1 wk (or more frequently, in July), 10–20 small, 10 medium, and 10 large bluegills were seined and removed from each half of the pond for stomach analyses and determination of growth rates. We replaced the sampled fish with bluegills of identical length from a nearby holding pond. Because we wanted to make comparisons of bluegill growth in the presence and absence of the bass, we were concerned about the potentially confounding effects of reduced bluegill densities in the one pond-half due to predation by the bass. In an attempt to minimize this difference, we assumed that the bass would be eating primarily the small size-class and estimated that initially a bass would consume one small bluegill every 3 d. We further adjusted this estimate as the size of the bluegills and bass increased during the experiment. Thus on the bass side we initially added ≈20 small bluegills in addition to the replacements of the fish sampled for stomach analyses. Replacement for bass predation tapered to <10 individuals per sample date at the end of the season. Over the entire experimental period, we added an additional 144 small bluegills to the pond-half with the bass.

Habitat utilization by the fish was determined by classifying prey in the stomachs according to the habitat from which those prey originated (open water, sediments, vegetation). The vast majority of the prey in all size classes across the season could be unambiguously assigned to one of these habitat types (>90% of the diets on average across the season). Further details concerning prey sampling and the generation of predictions of optimal diet and habitat use for the fish can be found in the companion paper (Werner et al. 1983).

RESULTS

Mortality

Mortality rates were similar for the two larger size-classes of bluegills across treatments. When we recovered populations in the fall, the cumulative mortality of the medium size-class was 10% on both sides, and for the large size-class, 11% in the presence of the bass and 19% in the control half of the pond. No mortality occurred in the bass population.

As explained earlier, we added more small bluegills to the bass side to compensate for expected predation losses. Our estimates of bass feeding rates were appropriate since the numbers of small fish recovered in the fall were similar: 348 on the bass side and 359 on the control side. Thus growth data were not confounded by different bluegill densities on the two sides. From these data we estimated a mortality of 28% for small fish on the control side, which is higher than that experienced by the two larger size-classes. The mortality rate of the small bluegills in the presence of the bass, however, was 59% of the original 500 fish. (Including the 144 fish added during the experiment, 296 fish died on the bass side.) Assuming the same non-bass mortality rate (28%) on each side, we estimate that each bass consumed a small fish every 3.8 d. The small size-class, therefore, did incur significant mortality due to the presence of the bass. In the Discussion and the Appendix we further examine how this increased mortality might be apportioned between the "direct" and "indirect" effects of the presence of the bass.

Habitat use

The design of this experiment was predicated on the assumption that predation risk for the bluegill would be much reduced in the cattails compared to the unstructured water column and bare sediments. Accordingly we chose a pond with very high resource levels in the open water and bare sediments, relative to the vegetation. Only if the more profitable habitats were also more dangerous could we test whether the fish were capable of responding to a predation risk/foraging rate tradeoff.

It is generally accepted that complex habitats are safer for prey, but this has not often been quantitatively demonstrated (see, e.g., Huffaker 1958). We know of three sources of direct experimental inference that the cattail habitat should interfere with the bass' predatory efficiency. First, Glass (1971) found in laboratory pools with different densities of vertical wooden dowels that the capture rate of bass preying on guppies (*Poecilia reticulata*) decreased monotonically with an increase in dowel density (0–370 dowels/m^2). Second, Savino and Stein (1982) have found that predatory success of bass feeding on bluegills declined with increasing simulated plant density (0–1000 stems/m^2). In small wading pools (2.4–3 m in diameter) containing strands of polypropylene rope, success of the bass declined most sharply between 50 and 250 stems/m^2. Third, in a similar set of experiments we contrasted the success of bass (100–270 mm) feeding on bluegills (20–75 mm) in open and vegetated habitats (0 and 500 stems/m^2 polypropylene rope). Preliminary analyses of these data indicate that the small bluegills in the present study ($\approx$35 mm), if pursued by a 200-mm bass, would be at least twice as vulnerable in the open habitat. The density of cattails in the pond study reported here averaged 176 stems/m^2, but individual samples ranged as high as 400 stems/m^2. Thus we are confident that the small bluegills incurred much less risk in the cattail habitat of the pond.

In the companion paper (Werner et al. 1983) we examined habitat use by the fish in the absence of predation risk from the bass. We generated foraging rates for each size-class of fish in the three habitats (open water, sediments, and vegetation), using an optimal foraging model where costs and benefits were estimated from laboratory feeding experiments. Examining the return rates in the three habitats across the season, we predicted that to maximize return rates, all size-classes should begin feeding in the open water and then shift to feeding from the sediments when the profitabilities of these two habitats crossed in late July. Thus, the more open habitats were indeed the more profitable, as we had anticipated.

In the absence of the bass, the behavior of all three size-classes was in excellent accord with the model predictions (Werner et al. 1983: Fig. 5). The fish consumed >80% plankton initially and between 21 and 25 July switched dramatically to a diet of >80% sediment-dwelling prey. Utilization of sediments remained high for the remainder of the experiment except for the small size-class, which switched back to plankton in late September when profitability of the open-water habitat was again highest for this size-class (Werner et al. 1983).

While the medium and large bluegills exhibited very similar patterns of habitat use in the presence and absence of the bass, the small bluegills behaved very differently. Several lines of evidence demonstrate the effects of the bass on the foraging behavior of small bluegills. Early in the experiment *Daphnia pulex* was extremely abundant (up to 73 individuals/L) and as a consequence the profitability of the open water was 7- to 27-fold greater than that of either the vegetation or the sediments (Werner et al. 1983). All fish fed extensively on *D. pulex*, and this species very quickly disappeared. Reduced utilization of this very profitable resource by the small bluegills in the presence of the bass can be shown by comparing their use with that of the larger size-classes. (Comparisons of the small class across treatments cannot be made due to the large disparity in *D. pulex* abundance which devel-

TABLE 1. Average percent composition (±1 SE) of the diet by habitat for the three bluegill size-classes (6 August–6 September). Row sums do not add to 100% because a small fraction (≤3%) of prey could not be assigned to a specific habitat (see text for details).

		Vegetation	Plankton	Benthos
No predator	Small	9 ± 2	19 ± 5	69 ± 7
	Medium	14 ± 4	2 ± 0.5	81 ± 4
	Large	11 ± 3	trace	86 ± 3
Predator	Small	34 ± 10	17 ± 5	46 ± 9
	Medium	9 ± 3	16 ± 6	74 ± 7
	Large	14 ± 5	6 ± 4	78 ± 6

oped on the two sides of the pond; see below.) As *D. pulex* densities declined, its contribution to the diets declined most rapidly in the small size-class. By 30 July the small class had nearly ceased feeding on *D. pulex* (4% of diet), whereas the diet of the large and medium classes still contained 67 and 74% of this species, respectively. Over the first four dates the diet of the large size-class averaged 78% *D. pulex* and that of the small size-class only 53%. This contrasts with the control side where large, medium, and small size-classes averaged 90, 73, and 85% *D. pulex* in the diet, respectively, on those dates when they were feeding on plankton.

The reduced utilization of *D. pulex* on the side with the bass was also clearly reflected in the dynamics of the daphnids. On the control side, *D. pulex* abundances declined from 73 to 1 individual/L in 10 d, and by 25 July no *D. pulex* were found in the fish from this half of the pond. In contrast, on the side with the bass, *D. pulex* remained abundant for >20 d into the experiment and was a major part of the diet of the two larger size-classes through 6 August. Clearly the predation pressure on *D. pulex* in the open water was much reduced on this side, evidently due to the reduced utilization by the small size-class discussed above.

Direct comparisons of habitat use in the presence and absence of the bass are best made when resource levels in the three habitats are similar on the two sides. Following the demise of *D. pulex*, resource levels in each habitat were similar on the two sides for the period 6 August–6 September. Over this period then, we can compare habitat utilization of size-classes in the presence and absence of the bass, unconfounded by large differences in resource levels across treatments. Further, a large fraction of the seasonal growth occurred during this period, and the small size-classes of bluegills began to diverge in size on the two sides (see later).

Between 6 August and 6 September, a distinct pattern of habitat use emerged (Table 1). All three size-classes on the control side and the two larger size-classes on the bass side had switched to feeding from the sediments and exhibited similar diets. The small class on the bass side, however, averaged 36% vege-

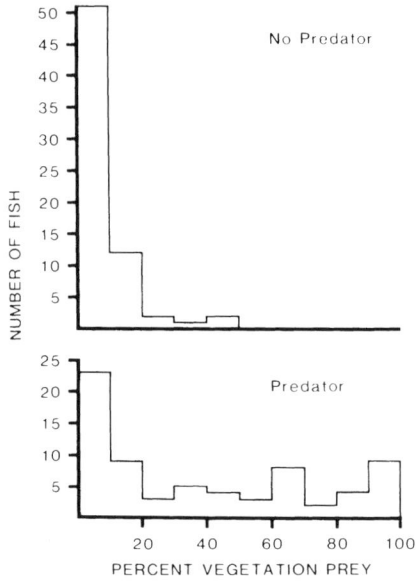

FIG. 1. Number of fish containing different fractions of prey derived from the vegetation. Data are for the small size-class in the presence and absence of largemouth bass. Numbers are for fish from 6 August and 6 September combined. The distributions differ at the $P < .001$ level (Kolmogorov-Smirnov test).

tation-dwelling prey in the diet, as compared to 9–15% for all other fish. This increased use of prey from the vegetation by the small fish was accompanied by a reduced use of prey from the sediments; all other fish were taking at least 70% benthic prey as opposed to 46% in the small class with the bass (Table 1). The increased use of vegetation-dwelling prey is clearly exhibited in Fig. 1, which indicates that use of the vegetation was not uniform among individuals in the population. In the absence of the bass, 93% of the small fish contained <20% prey from the vegetation, and no fish contained >50%. In the presence of the bass, however, 54% of the individuals contained >20% vegetation-dwelling prey. There was no relation between the fraction of vegetation-dwelling prey used and fish size, within the small size-class (regression of percent prey from vegetation against body size over range of 35–55 mm, $r^2 = 0.05$, $n = 70$, $P > .5$).

The small fish on both sides consumed more zooplankton than the large and medium classes (the high value for the medium class on the bass side is due to one anomalous date [Table 1]). The vast majority of the zooplankton eaten during this period was *Ceriodaphnia*, a form found in the vegetation as well as in the open water, and therefore it is possible that the small class in the presence of bass was also obtaining these prey from the vegetation. Thus we conclude that all fish were using habitats in very similar ways, except

TABLE 2. Mean individual dry mass (g) ± 1 SE for the three bluegill size-classes in the presence and absence of the predator. Final values are for the entire population recovered in the fall. Differences between means between predator and no predator treatments were determined by t test.

	Size-class					
	Small		Medium		Large	
	Predator	No predator	Predator	No predator	Predator	No predator
Initial mass (g)	0.28 ± 0.01 (n = 44)		1.35 ± 0.03 (n = 27)		3.64 ± 0.12 (n = 30)	
Final mass (g)	0.90 ± 0.02 (n = 348)	1.13 ± 0.02**† (n = 359)	4.45 ± 0.05 (n = 270)	4.35 ± 0.05 (n = 270)	9.17 ± 0.15 (n = 89)	8.64 ± 0.15** (n = 81)
Increment (g)	0.62	0.85	3.10	3.00	5.53	5.00
Population increment (g)	191.0	269.5	768.8	750.0	392.6	230.0
Difference in population increment (g)	−78.5		18.8		72.6	

** $P < .01$.
**† $P < .005$.

for the small size-class in the presence of the bass, which exhibited increased use of the vegetation.

The expected foraging profitabilities provide a measure of the cost of foraging more in the vegetation (Werner et al. 1983). The predicted return rate for the small class feeding in the vegetation averaged only 32 ±6% of that for the sediments from 6 August through 6 September. We show below that, as a consequence, growth was slower in the presence of the bass.

In late August, *Daphnia ambigua* (an open-water species) appeared and increased to ≈100 individuals/L by late September. Similarly, *Bosmina* and *Ceriodaphnia* increased in late September. These three species are much smaller than adult *D. pulex* (by at least an order of magnitude in mass), and, in general, only the small bluegills utilized these forms to any extent. If we compare the last two dates when the small size-class on both sides was taking predominantly *D. ambigua*, we find that 77% of the diet of the small fish on the control side was plankton, whereas only 52% of the diet of the small fish on the bass side consisted of plankton, although resource levels in the open water and vegetation were similar on both sides of the pond. Thus, the effect of the bass was again to reduce the small bluegills use of the open water, as was the case when the small bluegills were feeding on *D. pulex* early in the experiment.

Growth

The presence of the predator clearly caused a shift in the habitat use of the small size-class; the question now is whether this habitat shift had any effect on growth of the surviving fish. In the absence of the bass, the three size-classes each grew markedly, but at very different rates (Table 2). Larger size-classes exhibited progressively higher growth rates. However, in the presence of the bass, the small fish exhibited a significant depression in growth, whereas the medium and large classes grew larger than in the absence of the bass (Table 2). The average growth increment of small fish was 27% less when with the bass, while those of the medium and large fish were 3 and 11% more than when no bass were present. These differences are highly significant for the small and large classes (Table 2). Examination of the size-frequency distributions of the small class on the two sides indicated no evidence of selective predation within this class. The distributions were very similar in shape, but smaller fish were recovered on the predator side, indicating a growth response. The bass increased in length from 198.8 ± 2.9 to 228 ± 3.8 mm during the experiment.

Thus, the presence of the bass significantly depressed growth rates of the small fish in accord with their increased use of the poorer habitat (vegetation). Further, because the small fish spent more time in the less profitable vegetation, this apparently released resources for the larger fish in the sediments and open water. The large class especially benefited from this release in resources (Table 2). Indeed, there was nearly equal compensation by the larger classes for the total production lost to the small class in the bass' presence. A crude estimate of total fish production (number surviving × growth increment) on the two sides only differed by 13 g or ≈1% of the production of either side (Table 2).

The resource samples also indicated the effect of the predator. We have already noted that *D. pulex* lasted nearly 2 wk longer in the predator's presence. Though not as dramatic, the effect was also apparent in the sediment habitat. *Chironomus* densities were always higher in the predator's presence after July, when the fish began feeding on this species (with the exception of two dates when they were equal). Though these differences were not large, the trend to higher midge densities in the predator's presence was very consistent, apparently due to reduced foraging pressure by the small fish. In both the open water and the sediments prey densities prior to the introduction of fish were actually slightly higher in the pond-half without the bass. Clearly, the presence of the bass had striking effects on the distribution of resources among size-classes of the bluegill.

Discussion

The commonly observed fact that habitats vary temporally and spatially in foraging profitability and predation risk suggests that many animals need to balance gains and risks in their decisions on where and when to forage. Can animals assess these gains and risks, and are these factors weighted or balanced in such a way that tends to maximize fitness? This is an especially complex and critical question in the context of species that exhibit strong ontogenetic niche shifts due to the fact that relative foraging abilities and risks to predators change markedly with body size over the life history. Thus, decisions on where and when to forage must be made not only in the face of changing resource and predator dynamics, but also as these relations change with increases in body size.

We have demonstrated that the bluegill is able to assess changes in both foraging profitability and predation risk. In the companion paper (Werner et al. 1983) we showed that temporal habitat shifts by all size-classes occurred when foraging rates in another habitat became greater than those of the habitat currently used. In this paper, we further demonstrated that small, vulnerable size-classes of the bluegill showed a marked shift in foraging behavior in the presence of the bass. These data provide experimental support for the hypothesis of Hall and Werner (1977) and Mittelbach (1981) that small bluegills in natural lakes are confined to weedbeds because of a behavioral response to the greater predation risk in more open habitats. The quantitative predictions of habitat profitabilities also indicate that the small bluegills were evaluating this risk in the face of threefold greater foraging rates in the more open habitats. Whether this response also maximizes fitness remains to be tested.

The response of the small bluegills was not of an all or none nature, i.e., they did not use the unstructured or vegetated habitats exclusively on any given day. This must in part be due to the proximity of these habitats in the ponds; the small fish could feed in the open water or sediments and yet be only a matter of a metre or two from the vegetation refuge. In natural lakes the spatial separation of these habitats is usually much greater and consequently precludes this possibility. Mittelbach (1981) found that small bluegills in a natural lake did not feed on the very profitable offshore plankton prey at all except on one date when zooplankton were found within several metres of the shore. Thus the effect of the presence of the bass may be expected to be even stronger and more sharply defined in natural lakes where large areas of open habitats intervene between those in which the fish feed. Of course, decreased foraging rates due to the presence of a predator may be generated not only by shifts in habitat use, but also by the necessity of greater wariness, or escape responses, which can decrease feeding rates in a particular habitat (Milinski and Heller 1978, Caraco et al. 1980). It is not known how the fish evaluate predation risk, but laboratory studies indicate that bluegills seemingly pay little attention to bass in a tank, until the bass shows subtle inclinations to begin to feed (R. Stein and J. O'Brien, *personal communication*).

The question also arises as to why the small fish exhibit such individual variation in their use of the vegetation in the presence of the predator (Fig. 1). We do not know if certain fish consistently spend more time in the vegetation than do others, i.e., if individuals tend to be risk averse or risk prone, or if this variation simply represents short-term (days, weeks) changes in habitat use by all individuals. This is an important problem, as these two hypotheses lead to very different ideas concerning individual behavior and fitness. We would expect that if risk-prone individuals existed, they would incur higher mortality rates but would also grow faster because of their use of the richer habitats. Thus, there should be a relation between size and habitat use. We noted earlier that there was no relation between body size among the small fish (ranging from ≈35 to 55 mm in length) and the fraction of their diet that came from the vegetation. The question of this individual variation in behavior deserves more detailed study.

In this study we were able to quantify the effects of predator-restricted habitat use on the growth rates of the fish. The small fish exhibited a 27% reduction in growth over part of one growing season. A growth reduction of this magnitude would certainly have far-reaching effects on the dynamics and population structure of these fish. Fish are indeterminant growers, and it is well recognized that fecundity is a direct function of size (e.g., Bagenal 1978) and that mortality rate is an inverse function of size, at least during the early part of the life history (Ricker 1979). Thus, lower growth rates protract the time spent in vulnerable stages, lower survivorship, and increase the time to reproductive maturity. Where the probability of death per day is a function of size, some of the demographic consequences of increased daily mortality rates and decreased growth rates can be assessed by examining the survivorship of fish through a size- (not age-) interval. Assume that for all sizes in the interval the presence of a predator multiplies the daily mortality rate by a factor c_μ and the growth rate by a factor c_g ($c_g = 1$ indicates no effect). It can then be shown (see Appendix) that the survivorship from size s_1 to size s_2 in the presence of a predator is given by

$$l_p(s_1, s_2) = [l_{np}(s_1, s_2)]^{c_\mu/c_g}, \qquad (1)$$

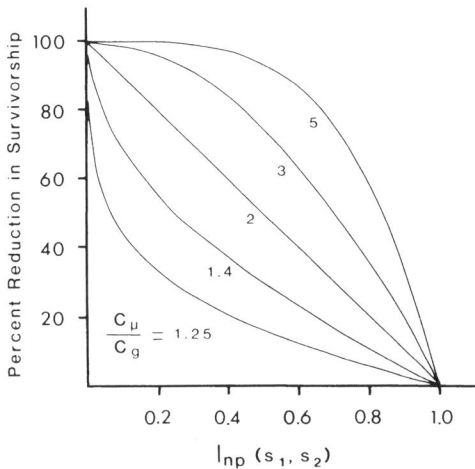

FIG. 2. Percent reduction in survivorship through a size-interval due to the presence of a predator. $l_{np}(s_1, s_2)$ is the survivorship in the absence of the particular predator (or before a predator's density is increased). c_μ is the factor by which the predator's presence (or increase) multiplies the size-specific daily mortality rate. c_g is the factor by which the predator's presence (or increase) multiplies the forager's daily (individual) growth rate.

where $l_p(s_1, s_2)$ = survivorship in the presence of a predator, and $l_{np}(s_1, s_2)$ = survivorship in the absence of the predator. Thus, halving the growth rate (setting $c_g = \frac{1}{2}$, holding $c_\mu = 1$) has the same effect on survivorship through a size-interval as doubling the daily mortality rate (setting $c_\mu = 2$, holding $c_g = 1$); both result in $c_\mu/c_g = 2$. In both cases the survivorship through the size-interval is squared. The decrease in growth rate has the further consequence of increasing the time required to reach reproductive size. It appears that a predator's "indirect" effect on the prey, inducing a lowered growth rate measured by c_g, might have greater effects on population demography than the "direct" effect of raising daily mortality rates (measured by c_μ).

Fig. 2 illustrates the effect of various values of c_μ/c_g on survivorship through a size-interval. The crucial observation here is that for a given value of c_μ/c_g the impact of the presence of the predator is much stronger if the predator affects a size-interval which already exhibits low survivorship, which is the case for small fish. For example, in the case of $c_\mu/c_g = 2$, a size-interval with $l_{np}(s_1, s_2) = 0.8$ initially will have a new survivorship of $l_p(s_1, s_2) = 0.64$ by application of Eq. 1. This is a reduction of 20%. However, if $l_{np}(s_1, s_2)$ were 0.1, this would yield $l_p(s_1, s_2) = 0.01$, a reduction of 90%. Thus, unless density-dependent processes compensate in other size-intervals (which is likely), the number of fish reaching a given adult size will be reduced by 20 or by 90%, respectively, for the two cases.

We can also use this approach to derive a crude estimate of the relative effect of the "direct" effect of adding more predators to a system (i.e., raising the probability of death/day) and the "indirect" effect of inducing lowered growth rates (which causes the prey to be in a size-interval longer and therefore accrue greater mortality through the size-interval; see Appendix). In the experiment presented here we partition the effects over the size-interval of 0.28 g (the initial mass of the small size-class of bluegills on both sides) to 0.90 g (the final mass on the predator side). We estimate that over the size-range the total reduction in survivorship in the presence of the bass was 37%. If only the "direct" effect of increased daily mortality rates (c_μ) had occurred, we estimate the reduction in survivorship would have been 23%. If only the "indirect" effect of decreased growth rates (and hence a longer time in the interval) had occurred, we estimate that the reduction would have been 10%. Thus the total reduction of 37% can be partitioned into 23% due to the c_μ alone, 10% due to c_g alone, and an additional 4% when both factors acted concurrently. Our estimates of the value of c_μ/c_g ranged from 2.34 to 3.89.

A further consequence of the predator-restricted habitat use of the small fish was the significant compensatory increase in growth of the larger size-classes. Thus, predation risk enforced a degree of intraspecific resource partitioning which exacerbated the differences in growth rates between size-classes (Table 2). This habitat segregation obviously has important consequences to population size-structure and intraspecific competition. Resources in open habitats, which often would be the preferred prey of all size-classes, may become exclusive resources for large fish in the presence of predators. The open-water plankton in particular is likely an important exclusive resource for larger bluegills in natural lakes with extensive limnetic zones. In the absence of predators, bluegills invariably develop "stunted" populations of uniformly small individuals (e.g., Swingle and Smith 1940, Wenger 1972). In such cases foraging demands evidently drive down resource levels (and mean prey size [Hall et al. 1970]), such that intense intraspecific competition prevents sustained growth. The presence of predators, which provides larger fish with exclusive resources, may be a major factor enabling them to continue to grow. Thus predation risk and its effect on habitat use may be an important mechanism mediating intraspecific competition in the field that has been largely overlooked (but see Jackson 1961). We feel that predation risk is most likely the cause of the patterns in size-class segregation noted in the bluegill, rather than an evolutionary response to intraspecific competition (e.g., Keast 1977).

We have noted that predation risk tends to concentrate the young of many species in the vegetation of natural lakes (Hall and Werner 1977, Laughlin and

Werner 1980, Mittelbach 1984). The observation that juvenile diets of several sunfishes are more similar than adult diets (Laughlin and Werner 1980, Mittelbach 1984) is likely the result of this habitat restriction of juveniles. Thus if vegetation is relatively rare and/or resources in it low, risk of predation can create significant competitive bottlenecks for these species at this point in their life histories. Identifying these bottlenecks is obviously central to considerations of niche packing in such systems, but the effects of these sorts of bottlenecks have been virtually unexplored. Our understanding of community structure in fish and other organisms with size-structured populations will be very limited until we systematically explore the consequences of these sorts of interactions.

Our experimental work suggests that competition and predation may interact in subtle but critical ways in fish communities. Shifts in competitive advantages between species due to the presence of a predator may be a great deal more subtle than simply accounting for the differential removal of each species. If bottlenecks of the sort we have hypothesized are important, the mere presence of a predator population could result in extinction of prey that are not even eaten by the predator due largely to the predator's indirect effects of increasing competition in protected habitats. These ideas connote very different mechanisms of "predator-mediated" coexistence than those ordinarily considered. Clearly the effects of a predator on the spatial distribution of its prey and the resultant changes in the strength of interactions within and between prey species may have profound effects that would not be evident from measures of prey removal rates or energy flows.

Acknowledgments

We thank Julie Bebak and Laura Riley for their help with endless samples and the data analysis, and Scott Gleeson, Carol Folt, Craig Osenberg, David Hart, Tim Ehlinger, and James Wetterer for comments on an earlier draft. Ed Turanchik and Beth Hutchison assisted in the field. John Gorentz did the computer programming. This work was generously supported by the National Science Foundation (DEB-7824271 and DEB-8119258 to E. E. Werner and D. J. Hall and DEB-8104697 to G. G. Mittelbach). Contribution number 489 of the Kellogg Biological Station.

Literature Cited

Bagenal, T. B. 1978. Aspects of fish fecundity. Pages 75–101 *in* S. D. Gerking, editor. Ecology of freshwater fish production. J. Wiley and Sons, New York, New York, USA.

Caraco, T., S. Martindale, and H. R. Pulliam. 1980. Avian flocking in the presence of a predator. Nature **285**:400–401.

Curio, E. 1976. The ethology of predation. Zoophysiology and ecology. Volume 7. Springer-Verlag, Berlin, Germany.

Glass, N. R. 1971. Computer analysis of predation energetics in the largemouth bass. Pages 325–363 *in* B. C. Patten, editor. Systems analysis and simulation in ecology. Academic Press, New York, New York, USA.

Gould, S. J., and R. C. Lewontin. 1979. The spandrels of San Marco and the Panglossian paradigm: a critique of the adaptationists programme. Proceedings of the Royal Society of London B Biological Sciences **205**:581–598.

Hall, D. J., W. E. Cooper, and E. E. Werner. 1970. An experimental approach to the production dynamics and structure of freshwater animal communities. Limnology and Oceanography **15**:829–928.

Hall, D. J., and E. E. Werner. 1977. Seasonal distribution and abundance of fishes in the littoral zone of a Michigan lake. Transactions of the American Fisheries Society **106**:545–555.

Hassell, M. P. 1978. The dynamics of arthropod predator-prey systems. Princeton University Press, Princeton, New Jersey, USA.

Huffaker, C. B. 1958. Experimental studies on predation: dispersion factors and predator-prey oscillations. Hilgardia **27**:343–383.

Jackson, P. B. N. 1961. The impact of predation especially by the tiger fish (*Hydrocynus vittatus* Cast) on African freshwater fishes. Proceedings of the Zoological Society of London **136**:603–622.

Keast, A. 1977. Mechanisms expanding niche width and minimizing intraspecific competition in two centrarchid fishes. Pages 333–395 *in* M. K. Steere and B. Wallace, editors. Evolutionary biology. Volume 10. Plenum Press, New York, New York, USA.

Laughlin, D. R., and E. E. Werner. 1980. Resource partitioning in two coexisting sunfish: pumpkinseed (*Lepomis gibbosus*) and northern longear sunfish (*Lepomis megalotis peltastes*). Canadian Journal of Fisheries and Aquatic Sciences **37**:1411–1420.

Lawrence, J. M. 1957. Estimated sizes of various forage fishes largemouth bass can swallow. Proceedings of the Southeastern Association of Game and Fish Commission **11**:220–226.

Milinski, M., and R. Heller. 1978. Influence of a predator on the optimal foraging behaviour of sticklebacks (*Gasterosteus aculeatus* L.). Nature **275**:642–644.

Mittelbach, G. G. 1981. Foraging efficiency and body size: a study of optimal diet and habitat use by bluegills. Ecology **62**:1370–1386.

———. 1984, *in press*. Predation and resource partitioning in two sunfishes (Centrarchidae). Ecology.

Ricker, W. E. 1979. Growth rates and models. Pages 677–743 *in* W. S. Hoar, D. J. Randall, and J. R. Brett, editors. Fish physiology. Volume VIII. Academic Press, New York, New York, USA.

Savino, J. F., and R. A. Stein. 1982. Predator-prey interactions between largemouth bass and bluegills as influenced by simulated, submersed vegetation. Transactions of the American Fisheries Society **111**:255–266.

Sih, A. 1980. Optimal behavior: can foragers balance two conflicting demands? Science **210**:1041–1043.

Stein, R. A. 1979. Behavioral response of prey to fish predators. Pages 343–353 *in* R. H. Stroud and H. Clepper, editors. Black bass biology and management. Sport Fishing Institute, Washington, D.C., USA.

Swingle, H. S., and E. V. Smith. 1940. Experiments on the stocking of fish ponds. Transactions of the North American Wildlife Conference **5**:267–276.

Van Sickle, J. 1977. Analysis of a distributed-parameter population model based on physiological age. Journal of Theoretical Biology **64**:571–586.

Wenger, A. 1972. A review of the literature concerning largemouth bass stocking techniques. Technical Series Number 13, Texas Parks and Wildlife Department, Sheldon, Texas, USA.

Werner, E. E. 1977. Species packing and niche complementarity in three sunfishes. American Naturalist **111**:553–578.

———. 1982, *in press*. The mechanisms of species interactions and community organization in fish. *In* D. Simberloff and D. Strong, editors. Ecological communities: conceptual issues and evidence. Princeton University Press, Princeton, New Jersey, USA.

Werner, E. E., and D. J. Hall. 1974. Optimal foraging and the size selection of prey by the bluegill sunfish (*Lepomis macrochirus*). Ecology 55:1042–1052.

Werner, E. E., G. G. Mittelbach, and D. J. Hall. 1981. Foraging profitability and the role of experience in habitat use by the bluegill sunfish. Ecology 62:116–125.

Werner, E. E., G. G. Mittelbach, D. J. Hall, and J. F. Gilliam. 1983. Experimental tests of optimal habitat use in fish: the role of relative habitat profitability. Ecology 64: 1525–1539.

APPENDIX

The survivorship of a fish from age x_1 to age x_2 can be described by

$$l(x_1, x_2) = \exp\left[-\int_{x_1}^{x_2} \mu(x)\, dx\right], \quad (A1)$$

where x = age, and $\mu(x)$ = instantaneous mortality rate at age x (see, e.g., Hassell 1978: Appendix I). If the mortality rate is explicitly a function of size rather than age, we can rewrite $\mu(x)$ as $\mu[s(x)]$. We can then change the variable of integration from age to size to obtain an expression for survivorship from size s_1 to size s_2. Eq. A1 can thus be rewritten as:

$$l[x(s_1), x(s_2)] = \exp\left\{-\int_{x(s_1)}^{x(s_2)} \mu[s(x)]\, dx\right\}. \quad (A2)$$

Changing the variable of integration yields:

$$l(s_1, s_2) = \exp\left[-\int_{s_1}^{s_2} \mu(s)/(ds/dx)\, ds\right]. \quad (A3)$$

Since ds/dx is the growth rate, $g(s)$, we have

$$l(s_1, s_2) = \exp\left[-\int_{s_1}^{s_2} \mu(s)/g(s)\, ds\right]. \quad (A4)$$

Van Sickle (1977) has derived this equation by a different method. Intuitively, Eq. A4 represents survivorship across a size-interval because $\mu(s)/g(s)$ is the instantaneous probability of death at a particular size, since that probability is the product of $\mu(s)$ (i.e., deaths/time) and the inverse of the growth rate (a measure of the time spent at that size).

For heuristic purposes, assume that over some size-interval (s_1, s_2) the presence of the bass uniformly multiplies $\mu(s)$ by a factor c_μ and multiplies $g(s)$ by a factor c_g. Then the survivorship through a size-interval in the presence of the predator (l_p) is:

$$l_p(s_1, s_2) = \exp\left\{-\int_{s_1}^{s_2} [c_\mu \cdot \mu(s)]/[c_g \cdot g(s)]\, ds\right\}$$
$$= \exp\left[-(c_\mu/c_g)\int_{s_1}^{s_2} \mu(s)/g(s)\, ds\right]$$
$$= \left\{\exp\left[-\int_{s_1}^{s_2} \mu(s)/g(s)\, ds\right]\right\}^{(c_\mu/c_g)}. \quad (A5)$$

Since the expression in the braces is just the survivorship without the predator, $l_{np}(s_1, s_2)$, this yields:

$$l_p(s_1, s_2) = [l_{np}(s_1, s_2)]^{c_\mu/c_g}. \quad (A6)$$

This is Eq. 1. The percent reduction in survivorship is thus (the arguments of l_p and l_{np} are dropped for convenience):

$$(100)(l_{np} - l_p)/l_{np} = (100)(l_{np} - l_{np}^{c_\mu/c_g})/l_{np}$$
$$= (100)[1 - l_{np}^{(c_\mu/c_g - 1)}]. \quad (A7)$$

By estimating c_μ and c_g in this experiment, we can somewhat crudely estimate the relative intensities of the predator's "direct" effect (increased daily mortality rates) and its "indirect" effect (reduced growth rate) on the bluegill's survivorship through a size-interval. Here we consider survivorship over the size-interval from 0.28 g (the initial mass on both pond sides) to 0.90 g (the final mass on the predator side). We take $c_g = 0.729$ (Table 2). By specifying l_p (0.28, 0.90) and l_{np} (0.28, 0.90), c_μ can be calculated from Eq. A6 and the relative effects of c_μ and c_g assessed. Below, we calculate values for l_p (0.28, 0.90) and l_{np} (0.28, 0.90) and then partition the total reduction in survivorship into components due to the direct effect (c_μ), the indirect effect (c_g), and their interaction when both occur simultaneously.

The value of l_p (0.28, 0.90) cannot be calculated exactly because fish were replaced at various times and sizes on the predator side, and we have no data on the shape of $\mu(s)$. However, we can bracket the possible values for l_p (0.28, 0.90) by calculating the survivorship as if all the replacements were added on (1) the first day of the experiment, or (2) the last day of the experiment (equivalently, not added at all). The first calculation yields a survivorship of 348/644 = 0.540 (348 fish recovered from 644 total fish). The second calculation yields 204/500 = 0.408 (348 recovered minus 144 replacements, divided by the 500 original fish). Thus we take l_p (0.28, 0.90) to lie somewhere between the extreme limits of 0.408 and 0.540. The parameter l_{np} (0.28, 0.90) also cannot be calculated exactly since we know only the survivorship to 1.13 g (the final size with no predator); l_{np} (0.28, 1.13) was 359/500 = 0.718. We can bracket the possible values of l_{np} (0.28, 0.90) by performing the calculation as if (1) no fish died after reaching 0.90 g (i.e., the mortality rate was a very strongly declining function of fish size), and (2) the mortality rates were independent of fish size. The first case yields l_{np} (0.28, 0.90) = l_{np} (0.28, 1.13) = 0.718. The second case yields l_{np} (0.28, 0.90) = 0.787. This was calculated by letting μ represent the size-independent daily mortality rate. Then $N_t = N_0 e^{-\mu t}$, where N_t = survivors at time t, N_0 = initial number of fish, t = time. Taking $t = 76$ d (the length of the experiment), $N_{76} = 359$, and $N_0 = 500$ yields $\mu = 0.004359$. The fish reached a mean size of 0.90 g after 55 d; $N_{55} = N_0 e^{-\mu(55)}$, which yields $N_{55} = 393.4$. Thus l_{np} (0.28, 0.90) = 393.4/500 = 0.787.

These survivorship ranges can now be used to estimate the relative impact of c_μ and c_g on survivorship. First, we calculate our "best estimate" of the relative effects by taking the midpoints of the survivorship ranges. Thus, we take l_{np} (0.28, 0.90) = 0.753 and l_p (0.28, 0.90) = 0.474, a reduction in survivorship of 37.1%. Application of Eq. A6 with $c_g = 0.729$ yields $c_\mu = 1.92$. If the predator had not affected the prey's growth rate (take $c_g = 1$), this model would predict survivorship to have been l_p (0.28, 0.90) = $l_{np}^{c_\mu}$ (0.28, 0.90) = $(0.753)^{(1.92)} = 0.580$, a reduction of 23.0%. Similarly, if the predator had affected only the growth rate, the model would predict l_p (0.28, 0.90) = l_{np}^{1/c_g} (0.28, 0.90) = $(0.753)^{(1/0.729)} = 0.678$, a decrease of 10.0%. Thus, we estimate that the total of 37.1% reduction in survivorship can be partitioned into 23.0% due to the "direct" effect of increased daily mortality rates acting alone, 10.0% due to the "indirect" effect of decreased growth rates acting alone, and the remainder, 4.1%, to their interaction when both factors acted simultaneously.

Finally, there are four combinations of the endpoints of the ranges of l_{np} (0.28, 0.90) and l_p (0.28, 0.90). Taking the highest estimate of l_{np} (0.28, 0.90) and the lowest estimate of l_p (0.28, 0.90) yields l_{np} (0.28, 0.90) = 0.787 and l_p (0.28, 0.90) = 0.408, a reduction of 48.2%. Taking $c_g = 0.729$ and solving Eq. A6 yields $c_\mu = 2.73$. This results in a partitioning of the 48.2% into 33.9, 8.5, and 5.8% attributable to c_μ, c_g, and their interaction, respectively. Taking the lowest estimate of l_{np} (0.28, 0.90) and the highest estimate of l_p (0.28, 0.90) yields l_{np} (0.28, 0.90) = 0.718 and l_p (0.28, 0.90) = 0.540. In this case, $c_\mu = 1.36$, and the total reduction of 24.8% can be partitioned into 11.3, 11.5, and 2.0% attributable to c_μ, c_g, and their interaction. The results of the other two combinations lie within the range of the above combinations.

3.3 HONORABLE MENTION FULL CITATIONS AND ABSTRACTS

Soutar, A., and J. D. Isaacs. 1974. Abundance of pelagic fish during the 19th and 20th centuries as recorded in anaerobic sediment off the coast of the Californias. Fishery Bulletin 72(2):257–273.

> Anaerobic sediment preserves a chronographic record of the bioclimatological conditions in coastal seas. Of the myriad elements within this record, the accumulation of pelagic fish debris is of particular interest. The deposition of scales of the Pacific sardine, the northern anchovy, the Pacific hake, the Pacific saury, and the Pacific mackerel in the sediment of the Santa Barbara Basin, Alta California, and the Soledad Basin, Baja California, is generally in accord with available population estimates. The relation between scale deposition and population, when applied to the sedimentary record over the past 150 yr, suggests that major pelagic-fish productivity between 1925 and 1970 was substantially below pre-1925 levels.

Werner, E. E., and J. F. Gilliam. 1984. The ontogenetic niche and species interactions in size-structured populations. Annual Review of Ecology and Systematics 15:393–425.

> Body size is manifestly one of the most important attributes of an organism from an ecological and evolutionary point of view. Size has a predominant influence on an animal's energetic requirements, its potential for resource exploitation, and its susceptibility to natural enemies. A large literature now exists on how physiological, life history, and population parameters scale with body dimensions.
> The ecological literature on species interactions and the structure of animal communities also stresses the importance of body size. Differences in body size are a major means by which species avoid direct overlap in resource use, and size-selective predation can be a primary organizing force in some communities. Size thus imposes important constraints on the manner in which an organism interacts with its environment and influences the strength, type, and symmetry of interactions with other species.
> Paradoxically, ecologists have virtually ignored the implications of these observations for interactions among species that exhibit size-distributed populations. For instance, it has been often suggested that competing species using the same habitats must differ in size by a factor of 2 in weight in order for them to coexist. In many taxa, however, the body weight of individuals *within* species commonly spans 1 order of magnitude and sometimes even 4 or more orders of magnitude (e.g. fish, reptiles). Thus, the body dimensions experienced ontogenetically often transcend those limits purported to isolate strongly competing species.
> Given that resource utilization abilities and predation risk are generally related to body size, many species will undergo extensive ontogenetic shifts in food or habitat use. Such shifts create a complex fabric of ecological interactions in natural communities. Individuals face different competitors and/or predators as they grow, and in many cases even the sign of the interaction between species changes with size (see below). The size- or stage-specific nature of these interactions is critically important in shaping species life histories, the dynamics of species interactions, and the structure of the communities in which they are imbedded.

In this review, we first document the widespread existence of ontogenetic shifts in diet and habitat and then explore the consequences of such shifts for species interactions and community structure. The majority of our examples are from the lower vertebrates and invertebrates in freshwater communities, in part because this reflects our experience and in part because the phenomena are well documented in aquatic communities. Such stage-specific interactions should be abundantly represented in other taxa, however, since most exhibit strongly size-distributed populations (e.g. most invertebrate phyla, fish, amphibians, reptiles). Birds and mammals may be exceptions, but even in these taxa some have precocial young or adults and juveniles that differ in resource use. We have not attempted to review such interactions in plants or other sessile organisms, though they are clearly important in these groups.

In the second half of the paper, we offer a conceptual framework for predicting ontogenetic shifts and suggest some preliminary approaches for exploring their ecological and evolutionary consequences. We indicate how such life histories may be incorporated into a population dynamics framework and review the relevant findings from studies of structured population models. These are only tentative suggestions about how we might begin to develop a theory of species interactions in size-structured populations. We hope that this review will serve to highlight some important questions and stimulate broader treatments of the problem.

Section 4

Managing Fish Communities and Ecosystems

James F. Kitchell

4.1. SYNTHESIS

This set of articles includes a wide range of topics selected by the AFS membership and other fisheries societies around the world. Use of the label "communities" can include things that range from results focused on characteristics for an assemblage of fishes in similar habitats to those based on assemblages across a wide range of habitat types and systematic groups. A diversity of analytical approaches have developed. Some are based on empirical evidence. Some include theoretical tenets based on predator–prey interactions. Some include the direct and indirect effects of fishery practices as influences on community structure. Some emphasize trophic interactions expressed in ecosystem structure and function. Articles selected for presentation herein include pioneering contributions arranged largely by the scope of the ideas represented. Some, not all, are presented chronologically.

The history of community ecology is oft-clouded by arguments over what is a "community," what rules or definitions apply, and why they do or do not differ from one another. The article by Forney (1974) reflects a mix of that. Forney (1974) focused on a set of species that dominate ideas about population dynamics for trophic competitors and their effects on predator–prey interactions. One of the major merits of this research is the long-term focus on Yellow Perch *Perca flavescens* and Walleye *Sander vitreus* populations. This article offered one of the classic examples of how recruitment variability is expressed at population levels for individual species and their interactions owing to predation by adults on juvenile members of their own and those of other species. Yes, weather effects on recruitment is a part of that and the results are then expressed as strong or weak year classes that proceed through trophic ontogeny over the course of many years. John Forney and his colleagues offered decades of rigorous research dedicated to the study of these dynamics in Oneida Lake, New York, where fishery management efforts had been maintained in a relatively uniform way and served as a guide to stocking policies. More recently, Oneida Lake has changed substantially in response to invasions by non-native species. Results of the extensive early work and those reported after Forney's leadership efforts serve as baseline and guidance to expectations that might develop elsewhere.

Tonn and Magnuson (1982) provided a synthesis of physical–chemical conditions and the trophic interactions that can determine fish community structure. Their results offer broadly applicable insights for lakes at northern latitudes in North America, Europe, and Asia where winter conditions create oxygen limitations that "filter" species composition by preventing or allowing sustained populations of large piscivores. The mix of sampling methods and statistical analyses offer broadly applicable guidance to answers about why fish communities may vary greatly in what appear to be otherwise similar settings. Karr (1981) developed analogous approaches for streams of the midwestern United States based on the United States Environmental Protection Agency's Clean Water Act criteria for "fishable waters." This approach was adapted to systems elsewhere and drew locally modified applications in many kinds of aquatic habitats.

Winemiller and Rose (1992) tackled a much greater challenge in seeking explanations for common principles of fish life histories. This article assembled massive and diverse data sets for members of a wide array of fish communities. These data represent 216 species, 57 families, and 7 orders; the result of many, many hours of digging in the literature mentioned above, including the miscellany of books on "*Fishes of* ____." Using an impressive set of univariate and multivariate methods, they tested for principles that could derive from phylogenetic backgrounds, life history characteristics, marine and freshwater habitat types, trophic position, patterns of reproduction, and responses to natural or anthropogenic disturbances. Many of the positive and negative correlations in life history traits create insightful trade-offs. These offered a framework for predicting population responses to environmental variability and effects of altered habitat conditions, including fishery effects. The article included some particularly interesting discussion on evolutionary features of life history strategies. The analyses and ideas of Winemiller and Rose (1992) are a gold mine for future research that can help set new results in a larger framework.

A huge literature has developed for individual fish species based on descriptive and comparative results for studies of diets, growth rates, fecundity, habitats, and life history attributes. Library shelves bulge with outcomes including the extensive collection of bound volumes as graduate theses and dissertations. Results have been used to develop inferences and speculation about ecological interactions involving competition and predation effects in a community context. Werner and Hall (1977) conducted a pioneering experimental study relevant to competition effects on congeneric members of the sunfish community (Centrachidae). When isolated in pond ecosystems, Green Sunfish *Lepomis cyanellus* and Bluegill *L. macrochirus* preferred vegetated habitats and had similar diets. When sympatric, Green Sunfish occupied vegetation, ate larger prey, and evidenced competitive superiority; Bluegill switched to pelagic habitats where diets contained more small zooplankton and benthic prey. The main result was an obvious segregation along the gradient of habitats and an asymmetry of competition effects. The principle of habitat segregation in response to competition gained substantial momentum in subsequent studies and generally extended to selection for and evidence of predator avoidance in a diversity of more recent studies. Analogous efforts spread to include effects on invertebrate communities as part of an extensive development of experimental pond systems. As developed below (Carpenter et al. 1985), these approaches served as a basis for the role of food web interactions in an ecosystem context.

Recruitment variability remains a complexity for many kinds of resource management questions. In fisheries science, stock assessments and harvest policies depend heavily on estimations of recruitment success. There are general models of stock-recruit relationships (e.g., Ricker or Beverton-Holt), but they prove to be idiosyncratic and too often, surprisingly wrong. Weather plays a major role in the "match-mismatch" ideas presented by Cushing (1990) that ascribes variability in early life history success to relative synchrony of prey availability and larval development. Walters and Kitchell (2001) addressed different issues developing during juvenile life stages and provided some mechanistic explanations for frequent and unexpected outcomes. Recruitment depensation (i.e., failure) was evaluated using modeling approaches that included positive and negative effects of adult abundances set in the context of trophic interactions involving competition and predation effects. One outcome was the "cultivation hypothesis" that expressed how adult fish increase survival prospects for their juveniles by preying on members of other species that compete with their own juveniles. A second modeling approach described a "foraging arena" explanation for depensatory effects. These can derive from risk-sensitive predator avoidance behaviors that alter the outcomes of conventional functional response models used in characterizing density-dependence of predator-prey interactions. The central lesson in these analyses was that non-linear outcomes derive from theory that rises above conventional wisdom expressed in most textbooks. Those lessons offered explanatory principles of value to harvest strategies and research efforts. In fact, analogs of the foraging arena concepts are developing rapidly in many ecological applications.

Setting ecological interactions in an ecosystem context was energized by the development of the International Biological Program (IBP) launched (i.e., widely funded) during the late 1960s and early 1970s. The role of fishes was welcomed to that view. Aquatic ecosystem wisdom of the time had largely depended on the principles of Lindeman (1942) and data reflected in bottom-up reverence for regression analyses on nutrients limiting primary production. Yet, only about half of the observed variance in primary productivity could be explained by nutrient concentrations and the precision became even less predictable at higher trophic levels. The trophic cascade idea (Carpenter et al. 1985) argued that much of the unexplained variability could be due to different nutrient cycling rates owing to contrasts in food web structure. This article outlined that hypothesis based on a history of results from lakes with very different food webs. Carpenter et al. (1985) presented a theoretical picture of outcomes for the time course of ecosystem responses to events such as a strong piscivore year class or a winterkill effect, such as those described by Tonn and Magnuson (1982). The key trophic change was expressed in size-selective predation on zooplankton and their effects on phytoplankton productivity. In this case, fishes are the apex predators and feedback down into the food web is a major cause of differences in ecosystem processes. Analogous studies make the case for managing fishes as a part of a "biomanipulation" intended to improve water quality. Many assertions have followed about tracing "top down" effects on food webs in aquatic and terrestrial systems. Of particular relevance to fisheries interests, Stein et al. (1995) offered an interesting example in freshwater systems where the rapidly growing, omnivorous Gizzard Shad *Dorosoma cepedianum* creates "middle out" food web effects.

Modeling approaches cover a wide range of topics in fisheries science. Among the most recently developed are ecosystem-scale models that combine life history characteristics, trophic interactions, and population dynamics in simulating a range of environmental and management applications. The EcoPath/EcoSim (EwE) models developed at the University of British Columbia have proven of substantial and growing value. The Walters et al. (2005) application of these models is used to test ideas about single species versus an ecosystem context in the widely practiced management goal of maximum sustainable yield (MSY) for individual fishery stocks. In this case, pursuit of MSY harvest policies for one species can be evaluated for its effects on food web interactions. Analyses derived from applications through models for eleven different large, marine ecosystems with long histories of fishery exploitation. Two main lessons emerged: 1) MSY management can, in some cases, cause severe negative effects on ecosystem structure; and 2) management to protect major forage species has positive effects on higher trophic levels. These answers were unique for each of the systems simulated and very significant to the issues of escalating public and professional concern about effects of fishery exploitation at the ecosystem scale.

Effects of fisheries drew national and international attention when Pauly et al. (1998) presented evidence of global changes in fish populations owing to increasing exploitation in marine and freshwater systems. Because large fishes typically occupy higher trophic levels and are commonly a primary target of fisheries, many of these apex predator species have exhibited major declines in abundance over recent decades. Based on patterns of declining mean trophic level for fishery landings, Pauly et al. (1998) asserted that we were "fishing down marine food webs." Claims of dire outcomes followed. Media attention blossomed. A huge public outcry developed. Scientific controversy was fueled within professional fisheries circles to the extent that one primary journal refused to continue publishing elements of the debate. The result should, at a minimum, convince people to read the primary paper! This example draws attention to the prospect that overexploitation of fish stocks and consequent effects on food webs are important issues at an international scale and that the public has awakened to these issues. Popularity of the original article fueled conservationist fund raising efforts. That became a major source of grants and contracts for analogous work. The results increased dire forecasts for the demise of fisheries and more scientific controversy about their adequacy as guidance to research. That is clear justification for greater attention to increased support for research and management.

Although Pauly et al. (1998) and its kin have been challenged in many ways, among the most instructive analyses and interpretations was the more recent article by Essington et al. (2006). This article provided evidence that the majority of the outcomes presented in Pauly et al. (1998) actually derived from "fishing through" (not "down") the food webs. The arithmetic changed because mean trophic level declined in a few systems (e.g., the North Atlantic) due to expansion of fisheries at intermediate trophic levels. The majority of marine ecosystems did not show declines in trophic level for fisheries. In fact, the most common response was complex compensation like the trophic cascade described in Carpenter et al. (1985). Nevertheless, overexploitation effects can and do develop. Continued media attention seems to focus on the catastrophes of collapsing fisheries, which therefore, becomes a dominant public view. In fact, the majority of large fisheries are well managed. The core problem is lack of support for stock assessment research on the many smaller fisheries that represent the majority of catches. *Time Magazine* identified that problem as among the top 100 scientific revelations for 2012. Correcting the management problems could increase global fisheries yield by more than 50%. Young scientists should note that Essington et al. (2006) emerged from a graduate student seminar! Clearly, there is room for important and highly relevant growth in this discipline. Feed your mind and sharpen your quantitative skills!

Honorable Mention Contributions

Articles selected as Honorable Mention for the Managing Fish Communities and Ecosystems section have been cited above as part of reference to their value and/or important complements to the articles selected by our survey of AFS members and other fisheries societies around the world. Those are Lindeman (1942), Karr (1981), Cushing (1990), Stein et al. (1995), and Essington et al. (2006). Two others (Kitchell et al. 1977; Gabelhouse 1984) are briefly described in the following because neither of them fit neatly into the general classification system adopted. Both offer utility in a more general arena. A third addition differs from all others in that it represents a particularly interesting view from Swingle (1951) on the history of fisheries management. Swingle (1951) is an epistemological lesson about some continuing debates.

As fishery exploitation develops, species- and size-selective harvesting effects are commonly imposed on those of greatest economic value (i.e., fishes representing higher trophic levels and the larger adult members). Major changes can follow. Those can be manifested in age and/or size structure, reproductive output, intra-specific interactions (e.g., competition and cannibalism), and density-dependent growth rates. Full development of stock assessment methods may include many or all as representative of population dynamics, but correlates and indicators are often taken as a surrogate. Size distributions and growth rates are among those.

Size distributions are among the most frequently and readily used metrics of fish stock assessment and fishery effects. A diversity of efforts has developed ways to use population size structure as a guide to management activities. Many of these have focused on regional and local guidelines for management based on proportional stock densities (PSD) (more recently, proportional size distributions) of individual species (Anderson 1978). Gabelhouse (1984) offered a review of comparative outcomes for relative stock density (RSD) metrics for 69 fish species ranging across habitats from freshwater to marine and from tropical to coldwater species. Size distributions are arranged based on categories for stock density. Those include "minimum, quality, preferred, memorable, and trophy" as size spectra reflecting opportunities for management alternatives. Accomplishing a selected management goal is clearly challenging, but the empirical characteristics of size distributions can serve as guidelines.

How much does a fish consume? That question emerged in many aspects of basic and applied issues in fisheries science. We commonly collect data on diets and growth. Those serve as estimators of competition and predation effects in field studies. Energy allocation to reproduction and growth are key components for estimating production, population dynamics, and life history strategies. Aquaculture practices require direct measures of energy budgets and effects of different feeding practices. Each of those above calls for an understanding of bioenergetics. The Kitchell et al. (1977) modeling approach has served as a tool applied in pursuit of answers. Kitchell et al. (1977) offered examples of applications in field and laboratory studies. A key merit of this approach was its capacity for including the roles of temperature effects on metabolic rates, feeding rates, and consequent growth outcomes. It is a modeling method that, rightly so, invites challenges to the assumptions and modifications for improvement. It is a flexible tool in that each species and application has potentially different input parameters. The American Fisheries Society has conducted two international symposia and scores of training workshops based on this modeling method. As of this writing, bioenergetics parameter tables have been developed for >50 fish species.

Homer Swingle is a name well known to fisheries science. His view on artificial propagation was published in 1951 and is among the Honorable Mention list. First, his research offers a general lesson on the history of fisheries management that led to the development of aquaculture and stocking practices:

"…untouched by common sense or the systematizing guidance of research," and

"…the hatchery movement reached its greatest absurdity…"

Swingle was among the heretical voices advocating a greater breadth for management based on research. His career developed accordingly in pioneering the development of predator-prey and habitat management principles practiced for farm ponds and reservoirs commonly used in the southern United States. Over time, the lessons from Swingle and his colleague's begat now-familiar practices embodied in PSD and the kinds of ideas presented by Gabelhouse (1984). Alternatively, controversies over hatchery and aquaculture practices continue many decades later.

4.2 REPRINTED ARTICLES

Carpenter, S. R., J. F. Kitchell, and J. R. Hodgson. 1985. Cascading trophic interactions and lake productivity. BioScience 35(10):634–639.

Forney, J. L. 1974. Interactions between yellow perch abundance, walleye predation, and survival of alternate prey in Oneida Lake, New York. Transactions of the American Fisheries Society 103(1):15–24.

Pauly, D., V. Christensen, J. Dalsgaard, R. Froese, and F. Torres. 1998. Fishing down marine food webs. Science 279(5352):860–863.

Tonn, W. M., and J. J. Magnuson. 1982. Patterns in species composition and richness of fish assemblages in northern Wisconsin lakes. Ecology 63(4):1149–1166.

Walters, C. J., and J. F. Kitchell. 2001. Cultivation/depensation effects on juvenile survival and recruitment: implications for the theory of fishing. Canadian Journal of Fisheries and Aquatic Sciences 58:39–50.

Walters, C. J., V. Christensen, S. J. Martell, and J. F. Kitchell. 2005. Possible ecosystem impacts of applying MSY policies from single-species assessment. ICES Journal of Marine Sciences 62:558–568.

Werner, E. E., and D. J. Hall. 1977. Competition and habitat shift in two sunfishes (Centrarchidae). Ecology 58:869–876.

Winemiller, K. O., and K. A. Rose. 1992. Patterns of life-history diversification in North American fishes: implications for population regulation. Canadian Journal of Fisheries and Aquatic Sciences 49:2196–2218.

Cascading Trophic Interactions and Lake Productivity

Fish predation and herbivory can regulate lake ecosystems

Stephen R. Carpenter, James F. Kitchell, and James R. Hodgson

Limnologists have been studying patterns in lake primary productivity for more than 60 years (Elster 1974). More recently, concern about eutrophication has focused attention on nutrient supply as a regulator of lake productivity. However, nutrient supply cannot explain all the variation in the primary productivity of the world's lakes. Schindler (1978) analyzed a sample of 66 lakes that were likely to be limited in productivity by phosphorus because their nitrogen/phosphorus ratios exceeded five. Phosphorus supply, corrected for hydrologic residence time, explained only 48% of the variance in primary production, and lakes with similar phosphorus supply rates differed nearly a thousandfold in productivity. Phosphorus loading explains 79–95% of the variance in chlorophyll *a* concentration (Dillon and Rigler 1974, Oglesby 1977, Schindler 1978), but chlorophyll *a* concentration is a poor predictor of primary production (Brylinsky and Mann 1973, Oglesby 1977).

The concept of cascading trophic

Stephen R. Carpenter is an associate professor, Department of Biology, and assistant director of the Environmental Research Center, University of Notre Dame, Notre Dame, IN 46556. James F. Kitchell is professor of zoology and associate director of the Center for Limnology, University of Wisconsin, Madison, WI 53706. James R. Hodgson is associate professor of biology and chairman of the Division of Sciences, St. Norbert College, De Pere, WI 54115. © 1985 American Institute of Biological Sciences.

> **Altering food webs by altering consumer populations may be a promising management tool**

interactions, on the other hand, explains differences in productivity among lakes with similar nutrient supplies but contrasting food webs. The concept reflects an elaboration of long-standing principles of fishery management based on logistic models (Larkin 1978). Simply put, a rise in piscivore biomass brings decreased planktivore biomass, increased herbivore biomass, and decreased phytoplankton biomass (Figure 1). Specific growth rates at each trophic level show the opposite responses. Productivity at a given trophic level is maximized at an intermediate biomass of its predators. Productivity at all trophic levels, and energy flow through the food web, are highest where intensities of predation are intermediate at all trophic levels (Kitchell 1980). Although this simple conceptual model is heuristically useful, real ecosystems exhibit nonequilibrium dynamics that result from different life histories and variable interactions among the major species.

Cascading trophic interactions and nutrient loading models are complementary, not contradictory. Potential productivity at all trophic levels is set by nutrient supply. Actual productivity depends on the recycling of nutrients and their allocation among populations with different growth rates. The phosphorus availability to phytoplankton, for example, is determined by processes that operate over a wide range of temporal and spatial scales (Harris 1980, Kitchell et al. 1979). Nutrient excretion by zooplankton is a major recycling process (Lehman 1980) that is strongly influenced by

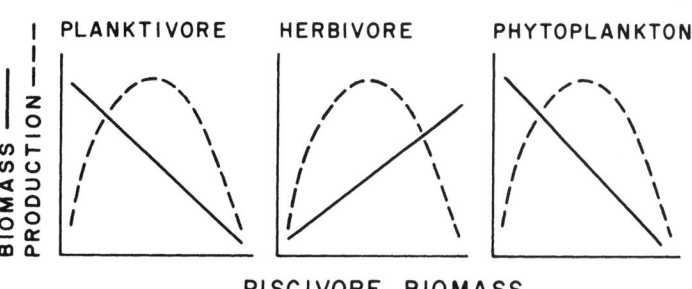

Figure 1. Piscivore biomass in relation to biomass (solid line) and production (dashed line) of vertebrate zooplanktivores, large herbivores, and phytoplankton.

selective predation on zooplankton by fishes (Bartell and Kitchell 1978). Thus, by regulating recycling rates, consumers regulate primary production.

The trophic cascade

To explain the details of cascading trophic interactions, we consider a lake food web that includes limiting nutrients and four trophic levels: piscivores such as bass, pike, or salmon, zooplanktivores, herbivorous zooplankton, and phytoplankton (Figure 2). Invertebrate planktivores like insect larvae and predaceous copepods take smaller prey than vertebrate planktivores like minnows. (Even though the rotifers include herbivores and predators, we will treat them collectively as a size class of zooplankton that includes grazers and is preyed on most heavily by invertebrates.) Small crustacean zooplankton include grazers, such as *Bosmina*, which remain small throughout their life cycle, and the young of large crustacean grazers, such as *Daphnia pulex*, and invertebrate planktivores. We divide the phytoplankton into three functional groups: nannoplankters subject to grazing by all herbivores, edible net phytoplankters like *Scenedesmus* that are grazed only by larger zooplankton, and inedible algae.

Examples of consumers controlling species composition, biomass, and productivity are available for each trophic level. Changes in the density of large piscivorous fishes result in changes in density, species composition, and behavior of zooplanktivorous fishes. In Wisconsin lakes containing bass or pike, spiny-rayed planktivorous fishes replace softrayed minnows, which are common in the absence of piscivores (Tonn and Magnuson 1982). The depletion of prey fishes by salmonids stocked in Lake Michigan (Stewart et al. 1981) and in European reservoirs (Benndorf et al. 1984) shows how piscivores can regulate zooplanktivorous fishes. Prey fish biomass declines as their predators increase in density; in contrast, prey fish productivity reaches a maximum at intermediate predator densities (Larkin 1978).

High planktivory by vertebrates is associated with low planktivory by invertebrates as well as high densities of rotifers and small crustaceans. Where planktivorous fishes are absent, invertebrate planktivores and large crustacean zooplankton predominate. Planktivorous fishes select the largest available prey and can rapidly reduce the density of zooplankters larger than about 1 mm (Hall et al. 1976). In contrast, planktivorous invertebrates select and deplete herbivores smaller than 0.5–1 mm. Lynch (1980) concludes that contrasting planktivore pressures have led to two distinct types of life history in cladoceran herbivores. Heavy planktivory by invertebrates favors large cladocerans that grow rapidly until they cannot be taken by the planktivores. At this size, these cladocerans shift energy allocation from growth to producing many small offspring. Planktivorous fishes, which consume large zooplankton (including invertebrate planktivores), promote dominance of small cladocerans that grow continually, reproduce at an early age, and have small clutches of large offspring.

Differences in size structure among herbivorous zooplankton communities lead to pronounced differences in grazing and nutrient recycling rates. Effects of zooplankton on phytoplankton biomass and productivity are not intuitively clear because they result from countervailing processes (grazing vs. nutrient recycling) and potentially compensatory allometric relationships. Larger zooplankters can ingest larger algae (Burns 1968). Absolute grazing rate (cells · animal^{-1} · t^{-1}) increases with grazer size, but mass-specific grazing rate (cells · mg animal^{-1} · t^{-1}) declines with grazer size (Peters and Downing 1984). Similarly, absolute excretion rate increases with grazer size, and mass-specific excretion rate decreases with grazer size (Ejsmont-Karabin 1983, Peters and Rigler 1973). Herbivorous zooplankton alter phytoplankton species composition and size structure directly by selective grazing and indirectly through nutrient recycling (Bergquist 1985, Carpenter and Kitchell 1984, Lehman and Sandgren 1985). Changes in phytoplankton size structure imply substantial changes in chlorophyll concentration and productivity because of several allometric relationships. In-

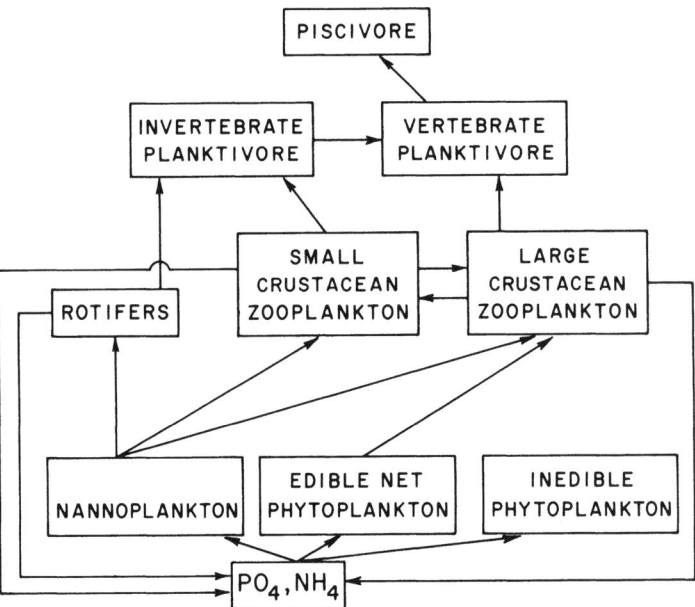

Figure 2. Conceptual model of trophic structure in a typical lake. See text for further details.

creasing algal cell size is accompanied by decreases in maximum growth rate, susceptibility to grazing, cell quotas for N and P, and intracellular chlorophyll concentrations and by increases in sinking rate and half-saturation constants for nutrient uptake (Reynolds 1984).

We investigated the complex interactions among zooplankton and phytoplankton using simulation models that yielded response surfaces of algal biomass and productivity as functions of zooplankton biomass and zooplankter body size (Carpenter and Kitchell 1984). Chlorophyll a concentrations were highest at low biomasses of small herbivores, declining smoothly as both biomass and grazer mass increased. The response of primary production to zooplankton biomass was unimodal, with maximum production at intermediate zooplankton biomass. The zooplankton biomass that maximized primary production declined as herbivore mass increased. At low herbivore biomass, productivity was limited by recycling; it increased as grazer biomass increased. When herbivore biomass was high, productivity was restrained by grazing and declined as grazer biomass rose. Although the model was far more complex than the familiar logistic equation of population biology, the phytoplankton as a whole behaved logistically in two respects: Productivity was related parabolically to chlorophyll a, with maximum productivity at intermediate chlorophyll a, and specific productivity decreased as chlorophyll a increased.

An increase in piscivore density cascades through the food web in the following way. Vertebrate zooplanktivores are reduced while planktivory by invertebrates increases, shifting the herbivorous zooplankton community toward larger zooplankters and higher biomass. Chlorophyll a concentration declines.

A decrease in piscivore density has the reverse effects. Vertebrate zooplanktivory rises at the expense of invertebrate zooplanktivores, and small zooplankters dominate the herbivore assemblage. Chlorophyll a concentration rises. A change in piscivore density can increase or decrease primary production, which is a unimodal function of zooplankton biomass.

Rates of cascading responses

In natural systems, sequences of cascading trophic interactions will propagate from stochastic fluctuations in piscivore year-class strength and mortality. Fish stocks, reproduction rates, and mortality rates in turn exhibit enormous variance (Peterson and Wroblewski 1984, Steele and Henderson 1984).

Fluctuations in piscivore reproduction do not cascade instantaneously through lake food webs. Rather, lags in ecosystem response occur because generation times differ among trophic levels. In temperate lakes, piscivores and many invertebrate and vertebrate planktivores reproduce annually. Crustacean herbivores and rotifers, which go through a generation in several days, pass through many generations in a summer. Phytoplankton generation times are shorter still, ranging from hours to a few days. Inorganic nutrients turn over in only a few minutes to a few hours. Because of this hierarchy of generation and turnover times, ecosystem components respond at different rates to changes in piscivore abundance.

The longest lags in the trophic cascade result from predatory ontogeny and predatory inertia. Predatory ontogeny occurs when a piscivore cohort develops, and the fish act first as zooplanktivores and then as piscivores (Figure 3, solid lines). As zooplanktivores, the fish drive the ecosystem toward small zooplankton and higher chlorophyll concentrations. These trends reverse as the fish grow and increase the proportion of planktivorous fish in their diet. Predatory inertia refers to the persisting effect of older age classes despite reproductive failure in any one year (Stewart et al. 1981). It takes several consecutive year-class failures to reduce piscivory enough for vertebrate planktivores to increase, with associated shifts in zooplankton and phytoplankton.

Cascading trophic interactions can be reversed by increasing or decreasing the intensity of piscivory. Because of lags, however, responses to increased piscivory involve transitions among food web configurations that do not occur during responses to decreased piscivory. Hysteresis will therefore occur when a change in piscivory is reversed: The sequence of ecosystem states and the rate of transition among states in the reverse pathway will differ from those of the forward pathway. The hysteresis effect is illustrated by two contrasting disturbances (Figure 3). Solid lines show the results of an unusually strong piscivore year class, which could occur naturally or through stocking young fish. Dashed lines show the results of a reduction in piscivores, such as those caused by winter kill or human exploitation. In each case, the system returns to the same state, but the pathways are very different.

Lake ecosystems are buffeted at irregular intervals by variations in fish recruitment and mortality rates. The system responses are nonequilibrium, transient phenomena that are difficult to detect using long-term averages. Finer-grained time course data are needed.

Correlations vs. experiments

Ecologists have been urged to develop theories based on multiple regression analyses of data from the literature (Peters 1980). We doubt that this approach can be used successfully to analyze relationships between food web structure and productivity. Correlations among trophic levels reflect nutrient supply effects, which influence biomass at each trophic level in an essentially stoichiometric fashion (cf. McCauley and Kalff 1981). The effects of food web structure are independent of those due to nutrient supply. Therefore, the appropriate statistical procedure is to first remove nutrient effects by regression, and then seek food web effects in the residuals of the regressions. Such a study would be subject to the pitfalls of interpreting regressions pointed out by Box et al. (1978, pp. 487–498). Common statistical problems in data from the literature relevant to cascading trophic interactions are dependencies among predictor variables and lack of control or precise measurements of predictor variables.

Literature data have serious shortcomings, in addition to purely statistical problems, which make them unsuitable for regression analyses of cascading trophic interactions. Frequently, data on biomasses of trophic levels do not distinguish between ed-

1983). Similar unimodal curves of productivity versus grazer biomass occur in grasslands grazed by ungulates (McNaughton 1979).

Management implications

In sum, enhanced piscivory can decrease planktivore densities, increase grazer densities, and decrease chlorophyll concentrations. Stocking piscivores therefore has promise as a tool for rehabilitating eutrophic lakes. Shapiro was among the first to recognize the potential of food web alteration as a management tool and has termed the approach biomanipulation (Shapiro and Wright 1986). A recent review has advocated stocking piscivores and/or harvesting zooplanktivores as a practical approach toward enhanced fishery production and mitigation of water quality problems (Kitchell et al. 1986). The approach has been successfully used to control eutrophication in European reservoirs (Benndorf et al. 1984).

Limnology and fisheries biology have developed independently and remain largely separate professions (Larkin 1978, Rigler 1982). An analogous distinction persists between water quality management and fisheries management. The concept of cascading trophic interactions links the principles of limnology with those of fisheries biology and suggests a biological alternative to the engineering techniques that presently dominate lake management. Variation in primary productivity is mechanistically linked to variation in piscivore populations. Piscivore reproduction and mortality control the cascade of trophic interactions that regulate algal dynamics. Through programs of stocking and harvesting, fish populations can be managed to regulate algal biomass and productivity.

Acknowledgments

This article is a contribution from the University of Notre Dame Environmental Research Center, funded by the National Science Foundation through grant BSR 83 08918. We thank David Lodge, Ann Bergquist, and the referees for their constructive comments on the manuscript and Carolyn Robinson for word processing.

References Cited

Bartell, S. M., and J. F. Kitchell. 1978. Seasonal impact of planktivory on phosphorus release by Lake Wingra zooplankton. Verh. Int. Ver. Theoret. Angew. Limnol. 20: 466–474.

Benndorf, J., H. Kneschke, K. Kossatz, and E. Penz. 1984. Manipulation of the pelagic food web by stocking with predacious fishes. Int. Rev. Gesamten. Hydrobiol. 69: 407–428.

Bergquist, A. M. 1985. Effects of herbivory on phytoplankton community composition, size structure, and primary production. Ph.D. dissertation, University of Notre Dame, Notre Dame, IN.

Box, G. E., W. G. Hunter, and W. S. Hunter. 1978. Statistics for Experimenters. John Wiley & Sons, New York.

Brylinsky, M., and K. H. Mann. 1973. An analysis of factors governing productivity in lakes and reservoirs. Limnol. Oceanogr. 18: 1–14.

Burns, C. W. 1968. The relationship between body size of filter-feeding Cladocera and the maximum size particle ingested. Limnol. Oceanogr. 13: 675–678.

Carpenter, S. R., and J. F. Kitchell. 1984 Plankton community structure and limnetic primary production. Am. Nat. 124: 159–172.

Cooper, D. C. 1973. Enhancement of net primary productivity by herbivore grazing in aquatic laboratory microcosms. Limnol. Oceanogr. 18: 31–37.

DiBernardi, R. 1981. Biotic interactions in freshwater and effects on community structure. Boll. Zool. 48: 353–371.

Dillon, P. J., and F. H. Rigler. 1974. The phosphorus-chlorophyll relationship in lakes. Limnol. Oceanogr. 20: 767–773.

Ejsmont-Karabin, J. 1983. Ammonia, nitrogen, and inorganic phosphorus excretion by the planktonic rotifers. Hydrobiologia 104: 231–236.

Elliott, E. T., L. G. Castanares, D. Perlmutter, and K. G. Porter. 1983. Trophic-level control of production and nutrient dynamics in an experimental planktonic community. Oikos 41: 7–16.

Elster, H.-J. 1974. History of limnology. Mitt. Int. Ver. Theoret. Angew. Limnol. 20: 7–30.

Flint, R. W., and C. R. Goldman. 1975. The effects of a benthic grazer on the primary productivity of the littoral zone of Lake Tahoe. Limnol. Oceanogr. 20: 935–944.

Gregory, S. V. 1983. Plant-herbivore interactions in stream systems. Pages 157–189 in J. R. Barnes and G. W. Minshall, eds. Stream Ecology. Plenum Press, New York.

Hall, D. J., S. T. Threlkeld, C. W. Burns, and P. H. Crowley. 1976. The size-efficiency hypothesis and the size structure of zooplankton communities. Annu. Rev. Ecol. Syst. 7: 177–203.

Harris, G. P. 1980. Temporal and spatial scales in phytoplankton ecology: mechanisms, methods, models, and management. Can. J. Fish. Aquat. Sci. 37: 877–900.

Henrikson, L., H. G. Nyman, H. G. Oscarson, and J. A. E. Stenson. 1980. Trophic changes, without changes in the external nutrient loading. Hydrobiologia 68: 257–263.

Hrbacek, J., M. Dvorakova, V. Korinek, and L. Prochazkova. 1961. Demonstration of the effect of the fish stock on the species composition of zooplankton and the intensity of metabolism of the whole plankton assemblage. Verh. Int. Ver. Theoret. Angew. Limnol. 14: 192–195.

Kitchell, J. F. 1980. Fish dynamics and phosphorus cycling in lakes. Pages 81–91 in D. Scavia and R. Moll, eds. Nutrient cycling in the Great Lakes: a summarization of the factors regulating cycling of phosphorus. NOAA Spec. Rep. 83. Great Lakes Environmental Research Laboratory, Ann Arbor, MI.

Kitchell, J. F., H. F. Henderson, E. Grygierek, J. Hrbacek, S. R. Kerr, M. Pedini, T. Petr, J. Shapiro, R. A. Stein, J. Stenson, and T. Zaret. 1986. Management of lakes by food-chain manipulation. FAO Fish. Tech. Pap. UN Food and Agricultural Organization, Rome, in press.

Kitchell, J. F., R. V. O'Neill, D. Webb, G. Gallepp, S. M. Bartell, J. F. Koonce, and B. S. Ausmus. 1979. Consumer regulation of nutrient cycling. BioScience 29: 28–34.

Korstad, J. E. 1980. Laboratory and field studies of phytoplankton-zooplankton interactions. Ph.D. dissertation, University of Michigan, Ann Arbor.

Larkin, P. A. 1978. Fisheries management—an essay for ecologists. Annu. Rev. Ecol. Syst. 9: 57–74.

Lehman, J. T. 1980. Release and cycling of nutrients between planktonic algae and herbivores. Limnol. Oceanogr. 25: 620–632.

Lehman, J. T., and C. D. Sandgren. 1985. Species-specific rates of growth and grazing loss among freshwater algae. Limnol. Oceanogr. 30: 34–46.

Lynch, M. 1980. The evolution of cladoceran life histories. Q. Rev. Biol. 55: 23–42.

McCauley, E., and J. Kalff. 1981. Empirical relationships between phytoplankton and zooplankton biomass in lakes. Can. J. Fish. Aquat. Sci. 38: 458–463.

McNaughton, S. J. 1979. Grazing as an optimization process: grass-ungulate relationships in the Serengeti. Am. Nat. 113: 691–703.

Oglesby, R. T. 1977. Phytoplankton summer standing crop and annual productivity as functions of phosphorus loading and various physical factors. J. Fish. Res. Board Can. 34: 2255–2270.

Peters, R. H. 1980. Useful concepts for predictive ecology. Synthèse 43: 215–228.

Peters, R. H., and J. A. Downing. 1984. Empirical analysis of zooplankton filtering and feeding rates. Limnol. Oceanogr. 29: 763–784.

Peters, R. H., and F. H. Rigler. 1973. Phosphorus release by Daphnia. Limnol. Oceanogr. 13: 821–839.

Peterson, I., and J. S. Wroblewski. 1984. Mortality rate of fishes in the pelagic ecosystem. Can. J. Fish. Aquat. Sci. 41: 1117–1120.

Reynolds, C. S. 1984. The Ecology of Freshwater Phytoplankton. Cambridge University Press, London.

Rigler, F. H. 1982. The relation between fisheries management and limnology. Trans. Am. Fish. Soc. 111: 121–132.

Schindler, D. W. 1978. Factors regulating phytoplankton production and standing crop in the world's lakes. Limnol. Oceanogr. 23: 478–486.

Seale, D. B. 1980. Influence of amphibian larvae on primary production, nutrient flux, and competition in a pond ecosystem. *Ecology* 61: 1531–1550.

Shapiro, J. 1980. The importance of trophic-level interactions to the abundance and species composition of algae in lakes. Pages 105–115 in J. Barica and L. R. Mur, eds. *Hypertrophic Ecosystems*. Dr. W. Junk Publishing Co., The Hague, Netherlands.

Shapiro, J., and D. I. Wright. 1984. Lake restoration by biomanipulation. *Freshwater Biol.* 14: 371–383.

Steele, J. H., and E. W. Henderson. 1984. Modeling long-term fluctuations in fish stocks. *Science* 224: 985–987.

Stewart, D. J., J. F. Kitchell, and L. B. Crowder. 1981. Forage fishes and their salmonid predators in Lake Michigan. *Trans. Am. Fish. Soc.* 110: 751–763.

Tonn, W. M., and J. J. Magnuson. 1982. Patterns in the species composition and richness of fish assemblages in northern Wisconsin lakes. *Ecology* 63: 1149–1166.

Interactions Between Yellow Perch Abundance, Walleye Predation, and Survival of Alternate Prey in Oneida Lake, New York[1]

JOHN L. FORNEY

*Department of Natural Resources, Cornell University
Ithaca, New York 14850*

ABSTRACT

Species of forage fish in stomachs of walleye and their abundance in trawl catches were compared in 1968-71. Young yellow perch were the predominant species in trawls and were consistently selected by walleyes. Consumption of young white perch and walleyes by older walleyes increased during periods of low yellow perch abundance which suggested that young yellow perch might act as a buffer controlling intensity of predation. This possibility was assessed by comparison of relative survival of white perch and walleye cohorts between the first and second year of life with indices of yellow perch density between 1959 and 1970. Close correlations between these variables support the conclusion that abundance of young perch governs intensity of predation on other forage size fish and indirectly controls the size of the walleye population by regulating cannibalism.

Piscivorous species are of fundamental importance in molding species and size composition of fish populations. This has been demonstrated by manipulation of predator-prey combinations in ponds (Swingle 1950) and by introduction of predators into natural waters (Gammon and Hasler 1965). Although these and other studies illustrate the importance of predators in regulating abundance of prey, the interactions between predator and prey in multi-species populations have remained obscure.

The object of this study was to determine how changes in density of young (0+) yellow perch (*Perca flavescens*) affect the diet of yearling (I+) and older walleye (*Stizostedion vitreum*) and the survival of alternate prey. Abundance of forage species was monitored from 1959-1971, and their occurrence in walleye stomachs was compared with their abundance in trawl catches in 1968-71. Young perch were the predominant species eaten in all four years, and young walleyes and white perch (*Morone americana*) were of secondary importance. Other forage fish of lesser importance are the tessellated darter (*Etheostoma olmstedi*), trout-perch (*Percopsis omiscomaycus*), mottled sculpin (*Cottus bairdi*), logperch (*Percina caprodes*) and young pumpkinseeds (*Lepomis gibbosus*), burbot (*Lota lota*) and ciscoes (*Coregonus artedii*).

In most years young yellow perch are more abundant in Oneida Lake than all other forage-sized fish combined. However, yellow perch cohorts vary in size and 10-fold differences in density are common in late summer of some years (Forney 1971). The walleye is the most abundant predator and the standing crop has averaged about 22 kg/ha during the past decade (Forney 1967). Growth is relatively slow and male walleyes attain a length of about 34 cm at age IV and females 36 cm. Annual growth increments are directly correlated with abundance of young perch which suggests that the population is food-limited (Forney 1965).

Oneida is a large (207 km²) eutrophic lake located in central New York State. Mean depth is 6.8 m and the maximum depth 16.8 m. The lake is usually homothermal but brief periods of temperature and oxygen stratification occasionally develop in summer (Greeson and Meyers 1969). Much of the shoreline is exposed to vigorous wave action and submergent plants are restricted to the protected bays. The bottom is rubble or sand in shallow areas grading into mud at depths over 4 m. Walleyes are widely distributed, occupying both shoal and deeper mud bottom areas during the summer and fall.

---

[1] A contribution from Federal Aid in Fish Restoration Project F-17-R, New York.

METHODS

Stomach contents of walleyes over 20 cm total length caught between late April and December 1968–1971 were examined to establish seasonal and annual changes in diet. Size and age composition of samples varied but age III to VI walleyes predominated each year. Most walleyes examined in spring and fall were captured in bottom trawls fished at depths of 6 to 12 m. Samples in June through August were generally taken in gill nets of graded mesh set overnight at depths of 4 to 12 m. No qualitative differences could be detected between stomach contents of walleyes caught in gill nets and trawls fished simultaneously in the same area, so samples taken with both gears were combined for analysis.

Stomach contents were removed either by dissection or by insertion of a straight glass tube from the mouth into the stomach. Pressure from the inserted tube and the distended stomach walls forced most food into the tube. As the tube was withdrawn the end was closed with the thumb to prevent loss of organisms. The tube was then reinserted to visually check the stomach for any remaining food items.

Fish removed from stomachs were counted and identified to species when possible. Occurrence of other food organisms was recorded. Figures reported for frequency of occurrence and number of fish per stomach are based on total number of walleyes in the sample examined.

Abundance of forage-sized fish was monitored by trawling. Fish were collected with a 5.9-m otter trawl with a 13-mm stretch mesh (½-in) cod end towed for 5 min at a speed of 3.4 km/hour. Hauls were made during daylight hours. In 1959 and 1960 replicate hauls were made at four sites, three times a week for four consecutive weeks beginning in mid-August. In later years the number of stations was increased to 10 and samples were taken at approximately weekly intervals from late July through October and at less frequent intervals in the spring. All fish were counted. Young and older yellow perch, walleye and white perch were segregated by size and larger individuals were aged. Counts of young and older trout-perch, tessellated darters and other small forage species were combined since all ages were vulnerable to predation.

Relative size of yellow perch, walleye and white perch year-classes in the first year of life was estimated either from the mean catch per unit effort (c/f) in mid-August to mid-September or from linear regressions fitted to the log of weekly catches from July through October. The latter method was well adapted to estimating abundance of young perch and walleye since the catches decreased exponentially during the summer.

It was necessary to determine if trawl catches accurately portrayed species composition of the forage population since proportions of each forage species in trawl catches and in stomachs of walleyes were used to evaluate food selection. Consequently, paired day-night hauls were made in 1969–1971. Catches of young yellow perch and walleyes were highest in hauls made during the day while most other species were taken in greater numbers at night (Table 1). Daytime trawl catches were adjusted by the ratio of night/day catches to provide a refined but still imprecise measure of the relative abundance of forage species.

I compared the proportions of various forage fish in stomachs of walleyes with their ratio in trawl catches using Ivlev's (1961) electivity index (E):

$$E = (r - p)/(r + p)$$

where r = relative amount of any ingredient in the ration expressed as a percentage of the whole ration and p = relative value of the

TABLE 1.—*Total catch of young-of-the-year fish in 23 paired day-night trawl hauls, 1969–71, and the ratio of day to night catches*

Species	Catch		Ratio $\frac{Night}{Day}$
	Day	Night	
Yellow perch	45,795	19,548	0.43
Walleye	342	140	0.41
White perch	308	956	3.10
Tessellated darter	226	2,790	12.34
Trout-perch	870	815	0.94
White bass	28	2	0.07
Sculpin	15	80	5.33
Logperch	9	8	0.88
Pumpkinseed	153	744	4.86
Burbot	0	7	—

FORNEY—YELLOW PERCH AND WALLEYE INTERACTIONS

TABLE 2.—*Frequency of occurrence (%), average number of fish (in parentheses) and frequency of occurrence (%) of invertebrates in stomachs of walleyes, 1968–71*

Length class (cm)	Number examined	Percent empty	Yellow perch 0+	Yellow perch 1+	Walleye 0+	Walleye 1+	White perch	Tessellated darter	Trout-perch	Other fish Identified	Other fish Unidentified	Invertebrates
20–24	470	38.3	27.4 (.94)	0 0	0.6 (.01)	0 0	0.4 (<.01)	0.6 (.01)	0.4 (<.01)	1.6 (.02)	12.3 (.25)	20.2
25–29	916	37.5	32.1 (1.82)	2.8 (.04)	0.2 (<.01)	0 0	0.9 (.01)	1.1 (.01)	0.4 (.01)	2.2 (.03)	18.7 (.81)	16.2
30–34	1572	39.0	32.9 (2.54)	2.7 (.05)	0.2 (<.01)	0.1 (<.01)	1.0 (.01)	1.7 (.02)	0.2 (<.01)	2.8 (.04)	26.3 (1.57)	9.0
35–39	1880	38.9	40.0 (3.61)	1.4 (.03)	0.5 (<.01)	0 0	1.1 (.01)	1.4 (.02)	0 0	2.0 (.03)	27.1 (1.85)	6.5
>40	1631	42.5	34.8 (3.80)	3.0 (.04)	1.8 (.02)	0.5 (.01)	0.8 (.01)	1.3 (.02)	0.4 (<.01)	1.9 (.04)	22.1 (2.10)	8.4

same ingredient in the food complex of the environment. Electivity values were used to examine the effects of changing perch density on consumption of alternate prey. These electivity values are not a true measure of food preference since trawl catches are not a precise measure of relative forage abundance. However seasonal and annual changes in electivity should reflect changes in preference.

FOOD OF WALLEYE

Relation to Walleye Size

Young perch were the most important item in the diet of all size classes of walleyes between 20 and 65 cm long (Table 2). Average number of perch per stomach increased from less than one in small walleyes to over three in large walleyes. Invertebrates, principally chironomid pupae, occurred most frequently in stomachs of walleyes under 30 cm.

Most young perch in the stomachs of walleyes were 30 to 80 mm long and the majority of other forage fish in stomachs fell within this length range. Parsons (1971) found the rapidly growing 1959 year class of walleyes in Lake Erie consumed first yellow perch and then spottail shiners, emerald shiners, and alewives in sequence. Failure of walleyes in Oneida Lake to show a similar shift to larger forage fish with increasing size probably reflects a paucity of intermediate-sized forage. Catches of yearling perch usually comprised less than 1% of the forage and other potential forage species in the 100- to 150-mm range were even less numerous.

Seasonal and Annual Changes in Diet

Seasonal changes in diet followed a predictable pattern. Stomachs of walleyes examined in May through mid-June contained mostly yearling perch and an occasional yearling walleye (Table 3). The number of fish in stomachs was low during this period but invertebrates were common in the diet. In late June walleyes began feeding on young perch. At this time perch were about 20 mm long and weighed less than 0.1 g. Walleyes fed almost exclusively on young perch during July and August. During the fall, incidence of other fish in the diet increased but perch remained the primary food through the following spring. Thus, the walleye population was largely dependent on a single cohort of perch for food from June to June. Numerical size of perch year classes at the time predation began and growth of young during the summer must largely determine the forage available to the walleye population during the following 12 mo. The number of young perch in stomachs of walleye declined rapidly after mid-July 1969 and the decrease paralleled a precipitous drop in adjusted trawl catches (Table 4). A similar decline occurred in 1970 when trawl catches indicated a relatively low density of young perch in late summer. Walleye stomachs contained much larger numbers of perch in 1968 and 1971 when the c/f in trawls remained high in the fall.

Although trawl catches indicated a higher density of perch in 1968 than in 1971, the mean number per stomach was consistently

TABLE 3.—*Mean number of fish and frequency of occurrence (%) of invertebrates in stomachs of walleyes examined between April and November, 1968–1971*

	Number examined	Percent empty	Yellow perch		Walleye		White perch	Tessellated darter	Trout-perch	Other fish		Invertebrates
			0+	I+	0+	I+				Identified	Unidentified	
1 May–14 June												
1968	105	51.4	—	.59	—	.01	—	—	.01	.05	.23	37.1
1969	122	52.5	—	.24	—	.01	—	—	—	.01	.07	27.9
1970	611	49.9	.13	.01	—	.01	—	—	.01	<.01	.10	49.3
1971	475	29.7	—	.30	—	.01	—	.02	.01	<.01	.15	22.9
15 June–14 July												
1968	193	61.1	3.56	.06	<.01	—	—	—	—	<.01	2.84	8.3
1969	156	37.2	6.41	—	.01	—	—	—	—	.03	3.00	22.4
1970	355	43.6	2.38	—	.01	—	—	.01	—	.20	2.05	10.7
1971	255	27.8	14.29	<.01	<.01	—	—	—	—	.06	16.78	—
15 July–14 Aug												
1968	381	40.4	6.94	<.01	—	—	—	—	—	.03	.22	4.5
1969	279	54.8	1.75	.01	.04	—	—	.02	—	—	.30	3.6
1970	265	33.9	4.28	—	.01	—	.02	.07	—	.01	.63	4.5
1971	309	28.5	8.19	.01	<.01	—	—	.03	—	.02	6.98	4.8
15 Aug–14 Sept												
1968	435	30.0	3.58	<.01	<.01	—	—	—	—	.01	.36	1.1
1969	276	50.0	1.48	<.01	.03	—	.01	.02	.02	—	.36	1.8
1970	251	60.5	1.25	—	.01	—	.01	.06	—	.06	.18	1.6
1971	165	15.8	8.36	—	.01	—	—	.02	—	.01	4.54	0.6
15 Sept–14 Oct												
1968	190	37.9	1.76	.03	<.01	—	—	.01	—	.01	.20	—
1969	86	66.3	.19	—	.02	—	.04	.09	.01	.05	.10	3.5
1970	144	43.0	.46	—	.01	—	.05	.11	—	.06	.29	1.4
1971	265	29.0	2.86	—	<.01	—	.01	.03	—	<.01	1.54	—
15 Oct–Nov												
1968	228	34.6	1.53	.02	.04	—	<.01	<.01	—	.04	.55	—
1969	346	71.4	.16	.01	.03	—	.03	.01	.01	.01	.13	0.3
1970	534	51.1	.62	<.01	.02	—	.09	.06	.01	.12	.55	—
1971	370	28.4	3.53	—	—	—	.04	.02	—	.02	.93	0.3

greater in 1971. Small size of individual perch probably contributed to the high consumption in 1971. Young perch averaged 1.4 g on 1 September 1971 compared to 2.6 g in 1968. Therefore, to maintain the same food intake about twice as many perch were consumed in 1971 as in 1968.

A few yearling perch and a variety of other species appeared in walleye stomachs, but none contributed significantly to the diet. More young walleyes, white perch and tessellated darters were consumed in 1969 and 1970 when density of young perch was low. Generally the contribution of these and other alternate forage fish to the diet increased slightly in the fall when perch density was

TABLE 4.—*Adjusted catch per unit effort of young-of-the-year yellow perch, walleye and white perch and catch of other forage sized fish in trawls during monthly periods beginning 15 July in 1968–71*[1]

Year	Period	Number of hauls	Catch					
			Yellow perch	Walleye	White perch	Tessellated darter	Trout-perch	Other species
1968	15 July–14 Aug	40	3304.8	24.4	0.7	20.9	31.1	0.3
	15 Aug–14 Sept	40	1758.3	21.5	5.0	129.4	70.2	45.8
	15 Sept–14 Oct	40	1029.2	20.9	3.7	136.8	64.4	40.0
	15 Oct–Nov	20	558.6	22.5	5.9	135.5	72.6	4.0
1969	15 July–14 Aug	30	1040.6	10.7	40.7	46.7	36.7	4.2
	15 Aug–14 Sept	40	286.4	16.4	39.6	156.1	66.9	51.4
	15 Sept–14 Oct	40	45.8	5.9	22.1	101.8	31.8	39.8
	15 Oct–Nov	20	25.6	4.4	20.0	46.2	24.3	4.0
1970	15 July–14 Aug	50	1836.3	7.8	22.0	24.7	4.7	14.0
	15 Aug–14 Sept	40	540.0	2.3	51.0	32.4	10.6	127.0
	15 Sept–14 Oct	40	151.8	3.0	72.5	65.2	14.8	83.9
	15 Oct–Nov	20	90.0	1.1	21.4	30.8	6.8	2.4
1971	15 July–14 Aug	50	1706.3	3.3	46.9	144.2	14.0	12.3
	15 Aug–14 Sept	50	1552.4	5.1	25.9	185.3	17.9	10.2
	15 Sept–14 Oct	40	400.2	3.0	38.1	273.4	13.8	9.5
	15 Oct–Nov	20	362.9	4.3	29.2	326.4	12.8	5.2

[1] Catches of white perch, darters, sunfish, and sculpins increased by the ratio of night/day catches shown in Table 1.

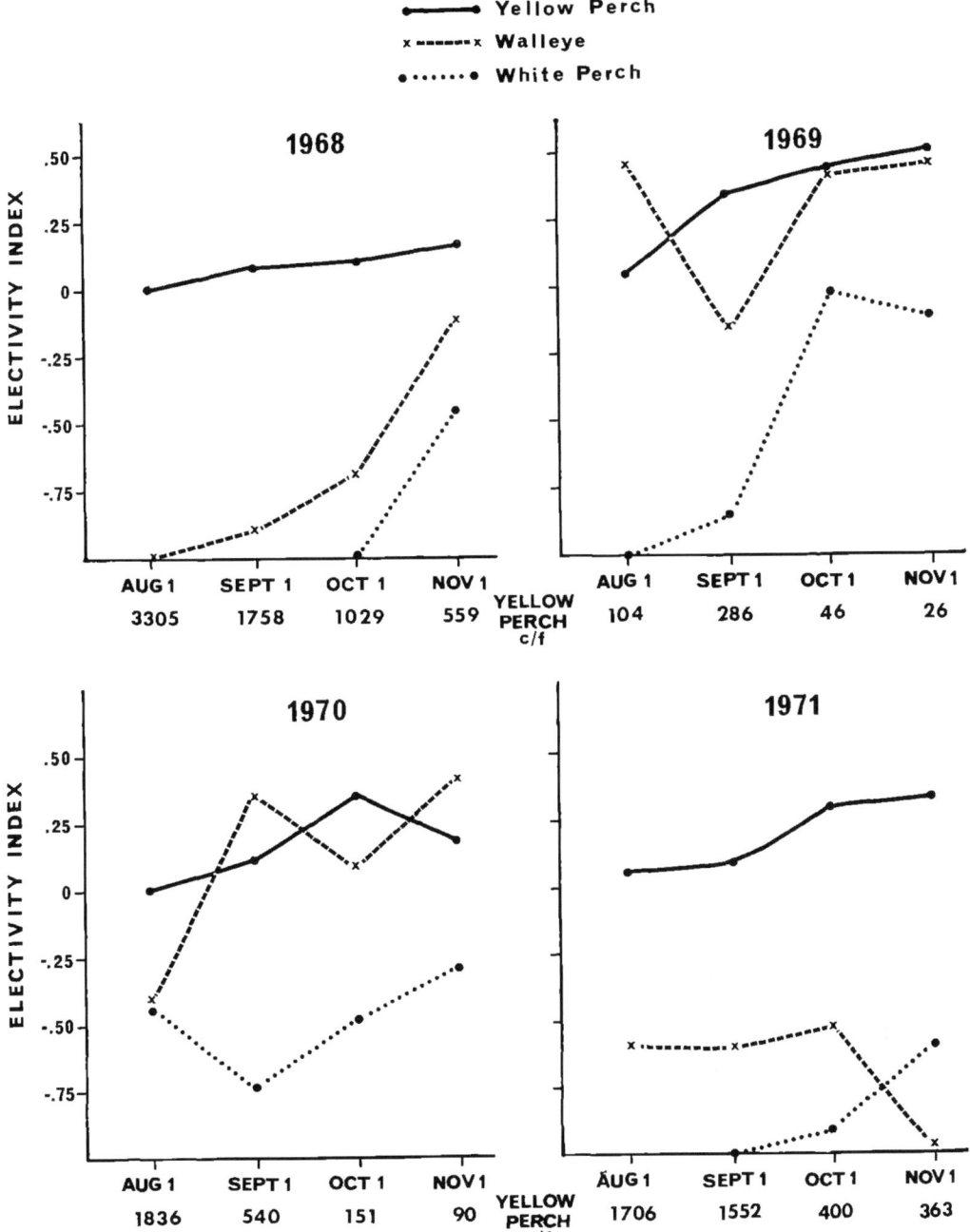

FIGURE 1.—Electivity indices of prey fishes consumed by older walleyes in Oneida Lake calculated from ratios of young fishes in stomachs and their ratios in trawl catches.

low. This suggests that a high density of young perch might act as a buffer, thus reducing predation on other forage species and cannibalism on young walleyes. Electivity values were calculated to evaluate this possibility.

FOOD SELECTION

Contribution of forage species to the ration of walleyes over 20 cm long (Table 3) and their contribution to adjusted trawl catches (Table 4) were compared to evaluate food selection. Ciscoes, burbot and yearling perch and walleye were not incorporated in estimates of electivity since they were not fully vulnerable to trawls.

Electivity values for yellow perch were consistently positive and increased as perch density declined (Fig. 1). Young perch dominated the forage population in early summer but their numbers decreased rapidly (Table 3). By October in 1969–71 yellow perch comprised less than half the forage population as measured by adjusted trawl catches. Increased selection for yellow perch as forage density decreased during the summer is interpreted as evidence that yellow perch were the preferred species in this forage complex. Shorygin as quoted by Ivlev (1961, p. 73) stated, "In the event of a reduction in the biomass of any of the organisms preferred by the fish, there will be an increase in the degree to which it is selected by this fish; but simultaneously there will also begin to be a slight increase in the selectivity indices of other, less favored organisms." The increase in selection of less favored organisms is evident from electivity values for white perch and walleye (Fig. 1).

Consumption of white perch generally increased during the summer as abundance of young yellow perch declined. Although electivity values for white perch were usually negative (Fig. 1), values increased seasonally as yellow perch density decreased, and values were highest in 1969 and 1970 when yellow perch density was low (Table 4). These trends imply that intensity of walleye predation on white perch was a function of yellow perch abundance. Trawl catches of white perch did not show a consistent decline during the summer which suggests predation was low until late fall as indicated by electivity values.

Young walleyes also appeared more vulnerable to predation in 1969 and 1970 when perch density was low than in 1968 and 1971 when perch were more abundant. High selectivity for young walleyes in 1969 and 1970 was consistent with the rapid decrease in trawl c/f in both years (Table 4). Electivity values for young walleyes increased during the summer of 1968 as perch density decreased and were high in the fall of 1968–70 when yellow perch abundance was low but electivity values declined in the fall of 1971. Rapid growth of walleyes in 1971 may have reduced their vulnerability to predation in the fall. Chevalier (1971) found young walleyes in the stomachs of adults were significantly smaller than the population mean length and few young over 150 mm were eaten. Mean length of young captured in trawls on 15 October 1971 was 172 mm compared to 145 mm in 1968, 146 mm in 1969 and 136 mm in 1970.

EFFECT OF SELECTION ON
YEAR-CLASS SURVIVAL

Reduced selection of young white perch and walleyes by older walleyes in years of high yellow perch density should enhance white perch and young walleye survival. I assessed the importance of young yellow perch as a buffer by comparing abundance of several year classes of white perch and walleyes in their first year and in later years of life. From these data the relative survival of several year classes was estimated and compared to abundance of young yellow perch in the year of hatching.

White Perch

Relative density of young white perch in 1959 through 1966 was estimated from trawl catches at four stations at weekly or more frequent intervals between mid-August and mid-September (Table 5). Mean c/f of young white perch was compared to the total contribution of these year classes to the catch in gill nets at ages I through V in 1959 through 1971 (Table 6). Gillnetting effort was con-

FORNEY—YELLOW PERCH AND WALLEYE INTERACTIONS

TABLE 5.—*Mean catch of the 1959–1966 year classes of young white perch and yellow perch in trawl tows during mid-August to mid-September (number of hauls in parentheses) and contribution of white perch to the catch in gill nets*

Year class	White perch c/f 0+	White perch c/f Age I–V	Relative survival of white perch (Age I–V/ 0+ × 100)	Yellow perch c/f
1959	9.3 (96)	27	.0290	57 (96)
1960	224.3 (96)	131	.0058	814 (96)
1961	69.6 (48)	27	.0039	516 (48)
1962	28.0 (36)	159	.0568	1062 (36)
1963	6.8 (32)	22	.0324	542 (32)
1964	3.6 (28)	79	.2194	1126 (28)
1965	96.2 (28)	62	.0064	688 (28)
1966	89.9 (24)	28	.0031	81 (24)

stant and sets were made overnight at the same 15 sites each year.

Based on gill net catches, the dominant year classes of white perch were those produced in 1960, 1962, and 1964. However, young white perch were most abundant in 1960, 1965 and 1966. A low correlation (r = .35) between density of young white perch and their contribution to the catch as adults suggests mortality between the fingerling and adult stages substantially altered year class size. The possible importance of young yellow perch as a buffer in this system was indicated by the strong correlation (r = .74) between density of yellow perch and size of adult white perch year classes.

The catch of yearling and older white perch was divided by the c/f of young to obtain a quantitative estimate of the relative survival of year classes between the fingerling and adult stage. Graphic analysis suggested an exponential decrease in survival with increasing density of young white perch and an exponential increase in survival of white perch with increasing young yellow perch abundance. To examine the relationship further, logarithmic transformation of the three variables was performed and a multiple regression calculated for the 8 yr with survival as the dependent variable.

Differences in abundance of young white perch and yellow perch accounted for 86% ($R^2 = .86$) of the variation in survival of the 1959–66 year-class between the fingerling and adult stages. The reduction in variance due to regression was highly significant (F = 15.34, d.f. = 5). The partial correlation coefficient of survival on abundance of young yellow perch was .65 indicating survival of white perch was enhanced in years of high yellow perch density. However the partial correlation between density of young white perch and survival was negative (−.91). This suggests the loss of young white perch between the first and second year of life would be compensatory with yellow perch abundance held constant.

Walleye

I divided the c/f for age I+ walleyes by the catch of 0+ on 1 August (Table 7) to determine relative survival of walleye year classes produced in 1961 through 1970. Yearling c/f was based on numbers taken in 130 to 180 trawl hauls between late July through October. Catch of young walleyes on 1 Au-

TABLE 6.—*Number of age I to V white perch caught in 15 gill net sets made annually in Oneida Lake, 1960–1971*

Year class	Year of capture												Total
	1960	1961	1962	1963	1964	1965	1966	1967	1968	1969	1970	1971	
1959	0	5	2	9	11								27
1960		1	10	26	43	51							131
1961			0	5	6	8	8						27
1962				0	63	54	27	15					159
1963					0	7	6	5	4				22
1964						6	18	28	12	15			79
1965							0	2	10	10	40		62
1966								0	0	5	20	3	28

TABLE 7.—*Relative abundance of the 1961–1970 year classes of walleye at age 0+ and 1+ and size of yellow perch year classes estimated from the c/f in trawls*

Year class	Number of walleye (c/f)		Relative survival of walleye (1+/0+)	Number 0+ yellow perch 15 Oct (c/f)
	0+ 1 Aug	1+ July–Oct		
1961	31.8	0.66	.0208	285
1962	19.6	1.12	.0571	435
1963	26.7	1.21	.0453	78
1964	12.3	1.28	.1041	496
1965	13.0	0.96	.0738	260
1966	3.6	0.08	.0222	15
1967	9.6	0.72	.0750	210
1968	22.4	3.89	.1737	670
1969	10.8	0.03	.0029	21
1970	6.4	0.16	.0250	90

TABLE 8.—*Mean total length of young walleyes taken in trawls on 1 October 1961–1971 and relative year class survival between age 0+ and 1*

Year class	Mean length (mm)	Relative survival
1961	152	.0208
1962	144	.0571
1963	132	.0453
1964	139	.1041
1965	154	.0738
1966	142	.0222
1967	130	.0750
1968	145	.1737
1969	146	.0029
1970	136	.0250
1971	172	—

gust was predicted from regressions fitted to trawl catches made during the same period. Survival of the 1959 and 1960 year classes was not estimated since samples in these years were limited to the period mid-August to mid-September.

Relative survival of walleye year classes was closely correlated ($r = .86$) with catch of young yellow perch in October. I compared survival with abundance of perch in October since analysis of the diet suggested cannibalism was most intense in the fall. However, walleye survival between the first and second year was correlated with other indices of yellow perch abundance including c/f in mid-August to mid-September ($r = .92$), mean biomass per haul during the same period ($r = .86$), and biomass on 1 August plus production (Forney 1971) from August to October ($r = .83$). Agreement between correlation coefficients would be expected since rankings of yellow perch year classes seldom changed between age 0+ and 1+ and biomass and production are dependent on numerical size of the year class.

Low survival of young walleyes in years of low perch density agrees well with occurrence of young walleyes in the diet of older walleyes. Walleyes began feeding on their young earlier in the summers of 1969 and 1970 when perch density was low than in 1968 and 1971 which were years of high perch density (Table 3). Electivity values of young walleyes were also higher in 1969 and 1970 than in 1968 and 1971 indicating more intense selection for walleyes in years of low perch density (Fig. 1). Survival of the 1969 and 1970 year classes between the first and second year of life was low while survival of the 1968 year class was exceptionally high (Table 7).

Relative survival of the 1961–70 walleye year classes was not correlated ($r = -.02$) with mean lengths of young in October (Table 8), although in an earlier section of this paper I suggested the unusually rapid growth of the 1971 year class may have reduced their vulnerability to cannibalism. It does not appear that fluctuations in perch density mask the expected relation between size and vulnerability to predation. The partial correlation coefficient between walleye survival and number of yellow perch on 15 October was 0.88 and −.36 between survival and length of walleye year classes on 15 October.

The negative relationship between survival and length is probably a spurious correlation since it is unlikely slow growth would enhance survival. Probably in 1961 to 1970 when mean lengths in October ranged from 130 to 154 mm (Table 8), most young were vulnerable to predation and small differences in first-year growth did not have a detectable effect on survival. However, in 1971, young walleyes attained a length of 172 mm which may have afforded some degree of protection from predation.

DISCUSSION

High mortality of white perch and walleye cohorts produced in years when density of young yellow perch was low was consistent with observed changes in diet of walleyes over 20 cm long in 1968 to 1971. Selection for young white perch and walleye increased in the fall as density of yellow perch declined and was highest in 1969 and 1970 when young yellow perch were scarce. Since the walleye was the most abundant predator, it is reasonable to attribute differences in survival of white perch and walleye cohorts to predation by the walleye population.

Standing crop of age III and older walleyes fluctuated within a relative narrow range of 21 to 34 kg/ha between 1959 and 1970 (Forney, 1967 and unpublished data). A stable predator population should feed more intensively on smaller than larger year classes of prey (Ivlev, 1961), and in Oneida Lake small year classes of young yellow perch experienced higher mortality than large year classes produced in 1961 to 1968 (Forney 1971). Under these conditions I would expect that other forage-sized fish associated with yellow perch would also be subject to more intense predation in years of low perch density, and this is supported by the close correlation of relative survival rates of both white perch and walleye cohorts with density of young yellow perch.

In lakes where yellow perch are an important component in the forage complex, synchronous fluctuation in year class size of yellow perch and walleyes should be evident when age structure of the adult population is examined. Heyerdahl and Smith (1971) detected such parallel fluctuations in yellow perch year classes over a 22-yr period in the Red Lakes, Minnesota. In Oneida Lake year class strength of adult perch and walleye generally coincided over a period of 12 yr (Chevalier 1971).

The unique position of young yellow perch as the only important forage fish in Oneida Lake and intense utilization of young by the walleye population may explain the close relationship between success of yellow perch year classes and survival of young white perch and walleye. Young yellow perch in 1968-1971 comprised over 80% of trawl catches in August in terms of both numbers and weight. Rapid exploitation of this food resource is indicated by the decline in c/f from about 2,000 in August to a few hundred in late October. By May 1969 the catch of the 1968 year class had declined to 18 per haul while catches of the 1969 and 1970 cohorts dropped to under 3 per haul. Based on area swept by trawls, perch density was only 30 to 178/ha (12–72/acre). Despite the low density, yellow perch at age I remained the predominant fish in the diet of walleyes during the spring (Table 3) which suggests that few alternate forage fish survived the winter. This was supported by very low catches of all forage species in trawls during the spring except for trout-perch which showed only a modest decline in numbers overwinter. Trout-perch may be less palatable than other forage fish since they were seldom eaten by walleyes although the two species inhabit the deeper areas of the lake during much of the year.

Chevalier (1973) demonstrated that mortality of young walleyes in Oneida Lake was probably attributable to cannibalism. I have shown in this paper that intensity of cannibalism is partly a function of young perch density. From these two observations the outlines of a simple, compensatory predator-prey system emerge.

Production of large year classes of young yellow perch should enhance survival of young walleye in the same year. Growth of these walleyes and the associated increase in capacity to consume forage fish will increase predation on subsequent year classes of perch and simultaneously increase cannibalism. The effect of these responses over a period of years may adjust the size of the walleye population to some mean level of forage production. Since the response of the predator population to an increase in prey abundance involves a time lag and density-independent factors may affect prey abundance, fluctuations in both predator and prey year classes would be expected and are generally observed in walleye-yellow perch associations.

ACKNOWLEDGMENTS

I am indebted to R. L. Noble for reviewing the manuscript and to numerous colleagues and students who have assisted in the collection of data over the past decade.

LITERATURE CITED

Chevalier, J. R. 1971. Cannibalism as a factor in first year survival of walleye in Oneida Lake. M. S. thesis. Cornell Univ., Ithaca, N. Y. 44 p. (Unpublished).

———. 1973. Cannibalism as a factor in first year survival of walleye in Oneida Lake. Trans. Amer. Fish. Soc. 102(4): 739–744.

Forney, J. L. 1965. Factors affecting growth and maturity in a walleye population. N. Y. Fish Game J. 13(2): 146–167.

———. 1967. Estimates of biomass and mortality rates in a walleye population. N. Y. Fish Game J. 14(2): 176–192.

———. 1971. Development of dominant year classes in a yellow perch population. Trans. Amer. Fish. Soc. 100: 739–749.

Gammon, J. R., and A. D. Hasler. 1965. Predation by introduced muskellunge on perch and bass, I: Years 1–5. Wisconsin Acad. Sci., Arts, Letters 54: 249–272.

Greeson, P. E., and G. S. Meyers. 1969. The limnology of Oneida Lake, an interim report. N. Y. Conserv. Dep., Water Resour. Comm. Invest. RI-8.

Heyerdahl, E. G., and L. L. Smith, Jr. 1971. Annual catch of yellow perch from Red Lakes, Minnesota, in relation to growth rate and fishing effort. Univ. Minn. Agr. Exp. Sta. Bull. 285. 51 p.

Ivlev, V. S. 1961. Experimental ecology of the feeding of fishes. Yale Univ. Press, New Haven, Conn. 302 p.

Parsons, J. W. 1971. Selective food preferences of walleyes of the 1959 year class in Lake Erie. Trans. Amer. Fish. Soc. 100: 474–485.

Swingle, H. S. 1950. Relationships and dynamics of balanced and unbalanced fish populations. Alabama Agr. Exp. Sta. Bull. 274. 74 p.

Fishing Down Marine Food Webs

Daniel Pauly,* Villy Christensen, Johanne Dalsgaard,
Rainer Froese, Francisco Torres Jr.

The mean trophic level of the species groups reported in Food and Agricultural Organization global fisheries statistics declined from 1950 to 1994. This reflects a gradual transition in landings from long-lived, high trophic level, piscivorous bottom fish toward short-lived, low trophic level invertebrates and planktivorous pelagic fish. This effect, also found to be occurring in inland fisheries, is most pronounced in the Northern Hemisphere. Fishing down food webs (that is, at lower trophic levels) leads at first to increasing catches, then to a phase transition associated with stagnating or declining catches. These results indicate that present exploitation patterns are unsustainable.

Exploitation of the ocean for fish and marine invertebrates, both wholesome and valuable products, ought to be a prosperous sector, given that capture fisheries—in contrast to agriculture and aquaculture—reap harvests that did not need to be sown. Yet marine fisheries are in a global crisis, mainly due to open access policies and subsidy-driven over-capitalization (1). It may be argued, however, that the global crisis is mainly one of economics or of governance, whereas the global resource base itself fluctuates naturally. Contradicting this more optimistic view, we show here that landings from global fisheries have shifted in the last 45 years from large piscivorous fishes toward smaller invertebrates and planktivorous fishes, especially in the Northern Hemisphere. This may imply major changes in the structure of marine food webs.

Two data sets were used. The first has estimates of trophic levels for 220 different species or groups of fish and invertebrates, covering all statistical categories included in the official Food and Agricultural Organization (FAO) landings statistics (2). We obtained these estimates from 60 published mass-balance trophic models that covered all major aquatic ecosystem types (3, 4). The models were constructed with the Ecopath software (5) and local data that included detailed diet compositions (6). In such models, fractional trophic levels (7) are estimated values, based on the diet compositions of all ecosystem components rather than assumed values; hence, their precision and accuracy are much higher than for the integer trophic level values used in

D. Pauly and J. Dalsgaard, Fisheries Centre, 2204 Main Mall, University of British Columbia, Vancouver, British Columbia, Canada V6T 1Z4.
V. Christensen, R. Froese, F. Torres Jr., International Center for Living Aquatic Resources Management, M.C. Post Office Box 2631, 0718 Makati, Philippines.

*To whom correspondence should be addressed. E-mail: pauly@fisheries.com

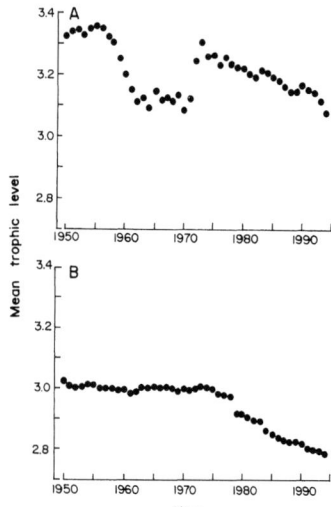

Fig. 1. Global trends of mean trophic level of fisheries landings, 1950 to 1994. (**A**) Marine areas; (**B**) inland areas.

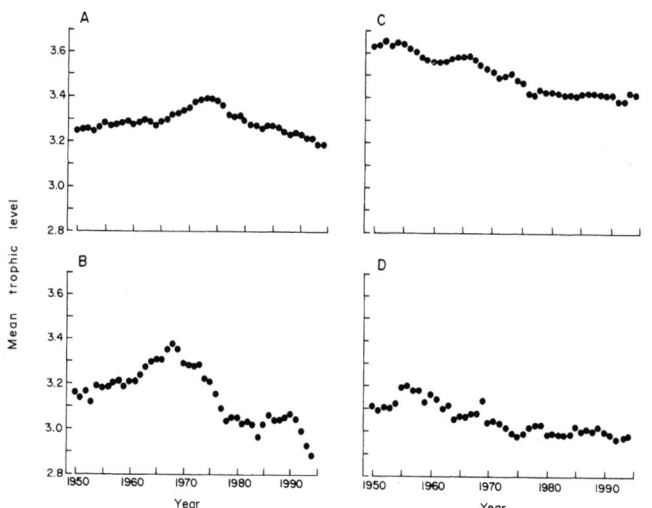

Fig. 2. Trends of mean trophic level of fisheries landings in northern temperate areas, 1950 to 1994. (**A**) North Pacific (FAO areas 61 and 67); (**B**) Northwest and Western Central Atlantic (FAO areas 21 and 31); (**C**) Northeast Atlantic (FAO area 27); and (**D**) Mediterranean (FAO area 37).

earlier global studies (8). The 220 trophic levels derived from these 60 Ecopath applications range from a definitional value of 1 for primary producers and detritus to 4.6 ($\pm$ 0.32) for snappers (family Lutjanidae) on the shelf of Yucatan, Mexico (9). The second data set we used comprises FAO global statistics (2) of fisheries landings for the years from 1950 to 1994, which are based on reports submitted annually by FAO member countries and other states and were recently used for reassessing world fisheries potential (10). By combining these data sets we could estimate the mean trophic level of landings, presented here as time series by different groupings of all FAO statistical areas and for the world (11).

For all marine areas, the trend over the past 45 years has been a decline in the mean trophic level of the fisheries landings, from slightly more than 3.3 in the early 1950s to less than 3.1 in 1994 (Fig. 1A). A dip in the 1960s and early 1970s occurred because of extremely large catches [>12 $\times$ 10^6 metric tons (t) per year] of Peruvian anchoveta with a low trophic level (12) of 2.2 ($\pm$ 0.42). Since the collapse of the Peruvian anchoveta fishery in 1972–1973, the global trend in the trophic level of marine fisheries landings has been one of steady decline. Fisheries in inland waters exhibit, on the global level, a similar trend as for the marine areas (Fig. 1B): A clear decline in average trophic level is apparent from the early 1970s, in parallel to, and about 0.3 units below, those of marine catches. The previous plateau, from 1950 to 1975, is due to insufficiently detailed fishery statistics for the earlier decades (10).

In northern temperate areas where the fisheries are most developed, the mean trophic level of the landings has declined steadily over the last two decades. In the North Pacific (FAO areas 61 and 67; Fig. 2A), trophic levels peaked in the early 1970s and have since then decreased rapidly in spite of the recent increase in landings of Alaska pollock, *Theragra chalcogramma*, which has a relatively high trophic level of 3.8 ($\pm$ 0.24). In the Northwest Atlantic (FAO areas 21 and 31; Fig. 2B), the fisheries were initially dominated by planktivorous menhaden, *Brevoortia* spp., and other small pelagics at low trophic levels. As their landings decreased, the average trophic level of the fishery initially increased, then in the 1970s it reversed to a steep decline. Similar declines are apparent throughout the time series for the Northeast Atlantic (FAO area 27; Fig. 2C) and the Mediterranean (FAO area 37; Fig. 2C), although the latter system operates at altogether lower trophic levels.

The Central Eastern Pacific (FAO area 77; Fig. 3A), Southern and Central Eastern Atlantic (FAO areas 41, 47, and 34; Fig. 3B), and the Indo-Pacific (FAO areas 51, 57, and 71; Fig. 3C) show no clear trends over time. In the southern Atlantic this is probably due to the development of new fisheries, for example, on the Patagonian shelf, which tends to mask declines of trophic levels in more developed fisheries. In the Indo-Pacific area, the apparent stability is certainly due to inadequacies of

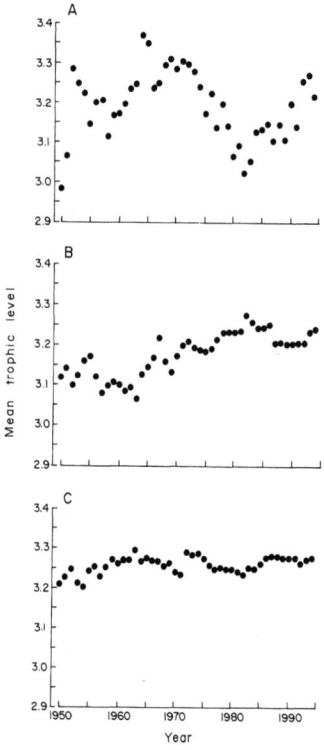

Fig. 3. Trends of mean trophic levels of fisheries landings in the intertropical belt and adjacent waters. (**A**) Central Eastern Pacific (FAO area 77); (**B**) Southwest, Central Eastern, and Southeast Atlantic (FAO areas 41, 34, and 47); and (**C**) Indo (west)-Pacific (FAO areas 51, 57, and 71).

the statistics, because numerous accounts exist that document species shifts similar to those that occurred in northern temperate areas (13).

The South Pacific areas (FAO areas 81 and 87; Fig. 4A) are interesting in that they display wide-amplitude fluctuations of trophic levels, reflecting the growth in the mid-1950s of a huge industrial fishery for Peruvian anchoveta. Subsequent to the anchoveta fishery collapse, an offshore fishery developed for horse mackerel, *Trachurus murphyi*, which has a higher trophic level (3.3 ± 0.21) and whose range extends west toward New Zealand (14). Antarctica (FAO areas 48, 58, and 88; Fig. 4B) also exhibits high-amplitude variation of mean trophic levels, from a high of 3.4, due to a fishery that quickly depleted local accumulations of bony fishes, to a low of 2.3, due to *Euphausia superba* (trophic level 2.2 ± 0.40), a large krill species that dominated the more recent catches.

The gross features of the plots in Figs. 2 through 4, while consistent with previous knowledge of the dynamics of major stocks, may provide new insights on the effect of fisheries on ecosystems. Further interpretation of the observed trends is facilitated by plotting mean trophic levels against catches. For example, the four systems in Fig. 5 illustrate patterns different from the monotonous increase of catch that may be expected when fishing down food webs (15). Each of the four systems in Fig. 5 has a signature marked by abrupt phase shifts. For three of the examples, the highest landings are not associated with the lowest trophic levels, as the fishing-down-the-food-web theory would predict. Instead, the time series tend to bend backward. The exception (where landings continue to increase as trophic levels decline) is the Southern Pacific (Fig. 5C), where the westward expansion of horse mackerel fisheries is still the dominant feature, thus masking more local effects.

The backward-bending feature of the plots of trophic levels versus landings, which also occurs in areas other than those in Fig. 5, may be due to a combination of the following: (i) artifacts due to the data, methods, and assumptions used; (ii) large and increasing catches that are not reported to FAO; (iii) massive discarding of bycatches (16) consisting predominantly of fish with low trophic levels; (iv) reduced catchability as a result of a decreasing average size of exploitable organisms; and (v) fisheries-induced changes in the food webs from which the landings were extracted. Regarding item (i), the quality of the official landing statistics we used may be seen as a major impediment for analyses of the sort presented here. We know that considerable under- and misreporting occur (16). However, for our analysis, the overall accuracy of the landings is not of major importance, if the trends are unbiased. Anatomical and functional considerations support our assumption that the trophic levels of fish are conservative attributes and that they cannot change much over time, even when ecosystem structure changes (17). Moreover, the increase of young fish as a proportion of landings in a given species that result from increasing fishing pressure would strengthen the reported trends, because the young of piscivorous species tend to be zooplanktivorous (18) and thus have lower trophic levels than the adults. Items (ii) and (iii) may be more important for the overall explanation. Thus, for the Northeast Atlantic, the estimated (16) discard of 3.7×10^6 t year^{-1} of bycatch would straighten out the backward-bending curve of Fig. 5B.

Item (iv) is due to the fact that trophic levels of aquatic organisms are inversely related to size (19). Thus, the relation between trophic level and catch will always break down as catches increase: There is a lower size limit for what can be caught and marketed, and zooplankton is not going to be reaching our dinner plates in the foreseeable future. Low catchability due to small size or extreme dilution (<1 g m^{-3}) is, similarly, a major reason why the huge global biomass ($\approx 10^9$ t) of lanternfish (family Myctophidae) and other mesopelagics (20) will continue to remain latent resources.

If we assume that fisheries tend to switch from species with high trophic levels to species with low trophic levels in response to changes of their relative abundances, then the backward-bending curves in Fig. 5 may be also due to changes in ecosystem structure, that is, item (v). In the North Sea, Norway pout, *Trisopterus esmarkii*, serves as a food source for most of the important fish species used for human consumption, such as cod or saithe. Norway pout is also the most important predator on euphausiids (krill) in the North Sea (3).

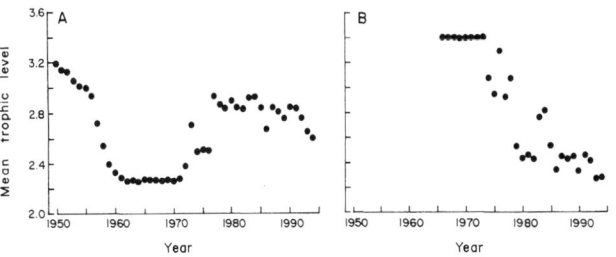

Fig. 4. High-amplitude changes of mean trophic levels in fisheries landings. (**A**) South Pacific (FAO areas 81 and 87); (**B**) Antarctica (FAO areas 48, 58, and 88).

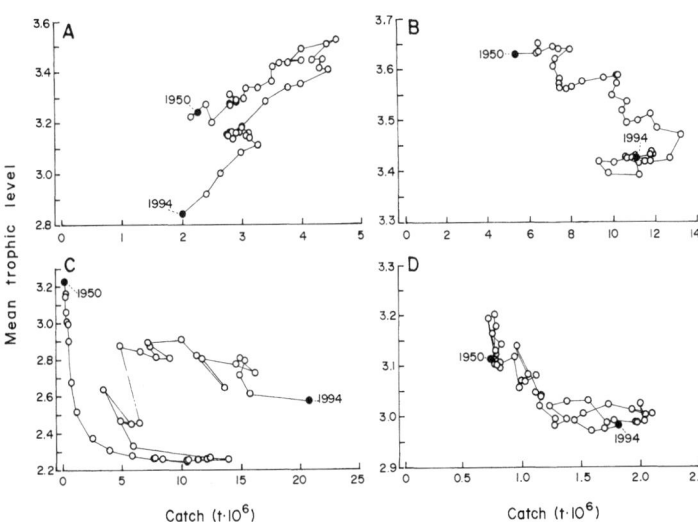

Fig. 5. Plots of mean trophic levels in fishery landings versus the landings (in millions of metric tons) in four marine regions, illustrating typical backward-bending signatures (note variable ordinate and abcissa scales). (**A**) Northwest Atlantic (FAO area 21); (**B**) Northeast Atlantic (FAO area 27); (**C**) Southeast Pacific (FAO area 87); (**D**) Mediterranean (FAO area 37).

We must therefore expect that a directed fishery on this small gadoid (landings in the Northeast Atlantic are about 3×10^5 t year^{-1}) will have a positive effect on the euphausiids, which in turn prey on copepods, a much more important food source for commercial fish species than euphausiids. Hence, fishing for Norway pout may have a cascading effect, leading to a build-up of nonutilized euphausiids. Triangles such as the one involving Norway pout, euphausiids, and copepods, and which may have a major effect on ecosystem stability, are increasingly being integrated in ecological theory (21), especially in fisheries biology (22).

Globally, trophic levels of fisheries landings appear to have declined in recent decades at a rate of about 0.1 per decade, without the landings themselves increasing substantially. It is likely that continuation of present trends will lead to widespread fisheries collapses and to more backward-bending curves such as in Fig. 5, whether or not they are due to a relaxation of top-down control (23). Therefore, we consider estimations of global potentials based on extrapolation of present trends or explicitly incorporating fishing-down-the-food-web strategies to be highly questionable. Also, we suggest that in the next decades fisheries management will have to emphasize the rebuilding of fish populations embedded within functional food webs, within large "no-take" marine protected areas (24).

REFERENCES AND NOTES

1. S. M. Garcia and C. Newton, in *Global Trends in Fisheries Management*, E. Pikitch, D. D. Hubert, M. Sissenwine, Eds. (American Fisheries Society Symposium 20, Bethesda, MD, 1997), pp. 3–27.
2. FAO, *FISHSTAT-PC: Data Retrieval, Graphical and Analytical Software for Microcomputers* (FAO, Rome, 1996).
3. V. Christensen, *Dana* **11**, (1995).
4. The bulk of the 60 published models are documented in (25); D. Pauly and V. Christensen, in *Large Marine Ecosystems: Stress, Mitigation and Sustainability Stratified*, K. Sherman, L. M. Alexander, B. D. Gold, Eds. (AAAS Publication, Washington, DC, 1993), pp. 148–174; D. Pauly and V. Christensen, *Nature* **374**, 255 (1995). References to the remaining models are given in FishBase 97 (9).
5. V. Christensen and D. Pauly, *Ecol. Model.* **61**, 169 (1992).
6. The documentation of the Ecopath models in (3) and (4) includes sources of diet compositions of all consumer groups in each ecosystem. These diet compositions are rendered mutually compatible when mass-balance within each model is established.
7. As initially proposed by W. E. Odum and E. J. Heald, in *Estuarine Research*, L. E. Cronin, Ed. (Academic Press, New York, 1975), vol. 1, pp. 265–286.
8. J. H. Ryther, *Science* **166**, 72 (1969).
9. All trophic level estimates are fully documented on the home pages of the Fisheries Centre, University of British Columbia, (www.fisheries.com) and of the FishBase project (www.fishbase.org), and on the FishBase 97 CD-ROM (R. Froese and D. Pauly, *FishBase 97, Concepts, Design and Data Sources* [International Center for Living Aquatic Resources Management (ICLARM), Manila, Philippines, 1997]), where references to the 60 published Ecopath applications are given as well. FishBase 97 also includes the FAO statistics (2), so Figs. 1 through 5 can be reproduced straightforwardly. To estimate the standard error (SE) we used the square root of the variance of the estimate of trophic level, in agreement with S. Pimm (21), who defined an omnivore as "a species which feeds on more than one trophic level." Thus, our estimates of SE do not necessarily express uncertainty about the exact values of trophic level estimates; rather, they reflect levels of omnivory. We do not present SE for the trophic levels of fisheries landings, as fisheries are inherently "omnivorous."
10. R. J. R. Grainger and S. M. Garcia, *FAO Fish. Tech. Pap. No. 359* (1996).
11. Mean trophic level, $\overline{TL}_i$, for year i is estimated by multiplying the landings (Y_j) by the trophic levels of the individual species groups j, then taking a weighted mean, that is, $\overline{TL}_i = \Sigma_{ij} TL_{ij} Y_{ij} / \Sigma Y_{ij}$.
12. A. Jarre, P. Muck, D. Pauly, *ICES Mar. Sci. Symp.* **193**, 171 (1991).
13. J. R. Beddington and R. M. May, *Sci. Am.* **247**, 42 (November 1982); P. Dalzell and D. Pauly, *Neth. J. Sea Res.* **24**, 641 (1989); contributions in G. Silvestre and D. Pauly, Eds., *Status and Management of Tropical Coastal Fisheries in Asia* (Conf. Proc. 53, ICLARM, Manila, Philippines, 1997).
14. R. Parrish, in *Peruvian Upwelling Ecosystem: Dynamics and Interactions*, D. Pauly, P. Muck, J. Mendo, I. Tsukayama, Eds. (Conf. Proc. 18, ICLARM, Manila, Philippines, 1989).
15. V. Christensen, *Rev. Fish Biol. Fish.* **6**, 417 (1996).
16. D. L. Alverson, M. H. Freeberg, S. A. Murawski, J. G. Pope, *FAO Fish. Tech. Paper 339* (1994).
17. We refer here to gill rakers, whose spacing determines the sizes of organisms that may be filtered, the length of the alimentary canal, which determines what may be digested, or the caudal fin aspect ratio, which determines attack speed and, hence, which prey organisms that may be consumed. See S. J. de Groot [*Neth. J. Sea Res.* **5**, 121 (1981)] for an example for flatfish (order Pleuronectiformes).
18. A. R. Longhurst and D. Pauly, *Ecology of Tropical Oceans* (Academic Press, San Diego, CA, 1987); A. P. Robb and J. R. G. Hislop, *J. Fish. Biol.* **16**, 199 (1980).
19. Contributions in (25) document the strong correlation between size and trophic level in aquatic ecosystems, a case also made for the North Sea by J. Rice and H. Gislason [*ICES J. Mar. Sci.* **53**, 1214 (1996)].
20. J. Gjøsaeter and K. Kawaguchi, *FAO Fish. Tech. Paper No. 193* (1980).
21. S. Pimm, *Food Webs* (Chapman & Hall, London, 1982).
22. R. Jones, in *Theory and Management of Tropical Fisheries*, D. Pauly and G. I. Murphy, Eds. (Conf. Proc. 9, ICLARM, Manila, Philippines, 1982), pp. 195–240; E. Ursin, *Dana* **2**, 51 (1982).
23. M. E. Power, *Ecology* **73**, 733 (1992).
24. A. C. Alcala and G. R. Russ, *J. Cons. Cons. Int. Explor. Mer* **46**, 40 (1990); M. H. Carr and D. C. Reed, *Can. J. Fish. Aquat. Sci.* **50**, 2019 (1993); J. E. Dugan and G. E. Davis, *ibid.*, p. 2029; C. M. Roberts and N. V. C. Polunin, *Ambio* **6**, 363 (1993).
25. V. Christensen and D. Pauly, Eds., *Trophic Models of Aquatic Ecosystems* (Conf. Proc. 26, ICLARM, Manila, Philippines, 1993).
26. D.P. acknowledges a Canadian (National Science and Engineering Council of Canada) and V. Christensen a Danish (Danish International Development Agency) grant for the development of Ecopath. R.F. thanks the European Commission (Directorate-General VIII) for successive grants to FishBase. We also thank H. Vatlysson and A. Laborte for a discussion and the FishBase programming, respectively. This is ICLARM contribution number 1401.

22 August 1997; accepted 10 December 1997

PATTERNS IN THE SPECIES COMPOSITION AND RICHNESS OF FISH ASSEMBLAGES IN NORTHERN WISCONSIN LAKES[1]

WILLIAM M. TONN AND JOHN J. MAGNUSON
Laboratory of Limnology, Department of Zoology, University of Wisconsin–Madison, Madison, Wisconsin 53706 USA

Abstract. Fish assemblage structure, and factors and mechanisms appearing important in the ecological maintenance of these structures, were examined for 18 small lakes in northern Wisconsin during summer and winter. The study was focused around the following questions. Are there discrete, repeatable groups of fish assemblages? If so, are they temporally stable? What are the relations between fish assemblage structure and habitat complexity, physical disturbance, biotic interactions, and the insular nature of small lakes? A comparative approach was used to generate hypotheses and propose explanations concerning the roles of these factors in structuring the assemblages.

Multivariate classification, ordination, and discriminant analyses helped discern two assemblage types: *Umbra*-cyprinid and centrarchid-*Esox*. Each had a distinctive species composition and seasonal change in composition. Environmental characteristics of the lakes occupied by each assemblage type also differed consistently.

The type of assemblage present in a lake appeared related to oxygen concentrations in winter, interacting with the availability of refuges from either a severe physical environment (low oxygen during winter) or from large piscivores. Centrarchid-*Esox* assemblages occurred in lakes with high winter oxygen levels, and also in lakes with low oxygen levels if a stream or connecting lake could provide a refuge from these conditions in winter. When no refuge was present, low winter oxygen lakes lacked piscivorous fishes, but contained *Umbra*-cyprinid assemblages.

The relationships between species richness in summer and environmental factors were generally similar for the two assemblage types, but the relative importance of individual factors differed. In winter, richness relationships in centrarchid-*Esox* assemblages for most environmental factors were reversed from those of summer. No significant seasonal change occurred in the *Umbra*-cyprinid assemblages.

Habitat complexity factors, particularly vegetation diversity, were significantly related to summer species richness in both assemblage types. Lake area was also related to summer richness for both types, but the slope of the species-area regression was much steeper for *Umbra*-cyprinid assemblages than for those in centrarchid-*Esox* lakes. Species richness relationships with winter oxygen concentration were negative in both seasons in *Umbra*-cyprinid lakes, but the relationship was positive for centrarchid-*Esox* assemblages in winter. A measure of lake connectedness was related to summer richness in centrarchid-*Esox* lakes. These patterns suggest that centrarchid-*Esox* assemblages are in ecological equilibrium but that a disturbance-induced disequilibrium occurs in *Umbra*-cyprinid assemblages.

*Key words: centrarchid-*Esox*; disturbance; fish assemblages; habitat complexity; insular; migration; multivariate analysis; predation; productivity; refuges;* Umbra*-cyprinid; Wisconsin lakes.*

INTRODUCTION

Our paper examines patterns in the species composition, richness (diversity), and seasonal dynamics of the fish assemblages in 18 small lakes in northern Wisconsin, and discusses factors and mechanisms which appear important in the ecological maintenance of assemblage structure. We addressed the following questions:

1) Are there discrete types of fish assemblages that are repeatable in many lakes?
2) If so, are they temporally stable in the face of seasonally (and probably unpredictably) harsh environmental conditions?
3) What are the relations between fish assemblage structure and habitat complexity, physical disturbance and biotic interactions?

4) To what extent does the insular nature of small lakes contribute to the composition and structure of their fish assemblages?

Species diversity theories

The ecological literature provides a plethora of theories and hypotheses to explain differences in species diversity among communities. Although all hypotheses are somewhat distinct, they can be grouped together into two major theories; one proposes mechanisms based on equilibrium conditions, the other on the absence of equilibrium (see discussions in Mac Arthur 1972, Connell 1978, Huston 1979).

Equilibrium-based mechanisms are inseparably linked to niche structure in communities. In saturated communities, richness is proposed to be a function of the number of discrete resources available, the tolerable niche overlap and minimum niche size possible

[1] Manuscript received 21 January 1981; revised 15 September 1981; accepted 29 September 1981.

along a resource gradient, or both (Pianka 1975, 1978, Menge and Sutherland 1976, Connell 1978). If the tolerable niche overlap and minimum niche size are relatively constant (Roughgarden 1974, Werner 1977), species richness should depend upon habitat complexity. Similarly, more productive habitats allow greater dietary specialization and should support more species (MacArthur 1972). Equilibrium theories also predict higher diversities in more stable and/or predictable environments (Slobodkin and Sanders 1969).

Some ecologists have questioned how often communities meet the assumptions and conditions of equilibrium and propose that nonequilibrium mechanisms dominate (e.g., Wiens 1977, Connell 1978, Huston 1979). Under nonequilibrium conditions, less efficient or poorly adapted species can persist without competitive exclusion, thereby increasing diversity. Competition-induced specialization cannot evolve over ecological time, so if competition does occur under equilibrium conditions, diversity will be reduced by competitive exclusion. At intermediate levels of disturbance a "dynamic equilibrium" of increased diversity can be maintained (Connell 1978, Huston 1979). At these levels, disturbances are sufficiently frequent or intense to prevent the community from reaching equilibrium, but still allow some populations to recover.

A severe environment that might otherwise be expected to produce a depauperate community may instead produce a more diverse community if refuges are present, reducing the severity. Refuges can also reduce the impact of predation and competition (Dodson 1970, Dayton 1971, Thomson and Lehner 1976), resulting in increased diversity. Woodin (1978) identified five categories of spatial and temporal refuges and argued that communities could be viewed as combinations of species successfully exploiting these refuges.

Lakes as islands

Another set of factors that should be considered in the examination of fish assemblage structure in small lakes is their insular nature (Barbour and Brown 1974, Magnuson 1976, Browne 1981). One important insular parameter of small lakes is isolation. Relative isolation of an island depends on the likely mode of colonization (flight, drift in ocean currents, etc.). Certainly, seepage lakes without permanent inlets or outlets are more isolated for fishes than are drainage lakes. Measuring the degree of isolation (or connectedness) in drainage lakes is not as straightforward as for oceanic islands. Factors to be considered include the length of the interconnecting waterways and size of the watershed, the degree of differences in habitat between the lakes and connecting streams, and the presence of marked barriers (e.g., waterfalls) between lakes (Magnuson 1976).

As with oceanic islands, population size in lakes should be a function of area. Larger lakes should have larger populations, lower probabilities of local extinction, and therefore more species than should smaller lakes (Magnuson 1976).

Other components of assemblage structure

Knowledge of species diversity alone is insufficient for understanding the organization, dynamics, and controlling mechanisms of assemblages. Assemblages may differ in species composition, reflecting differences in the seasonal responses, environmental regimes, dominance relationships, and controlling factors (Coull and Fleeger 1977). Compositional differences can also result from differential dispersal abilities and extinction probabilities among the potentially available species (Simberloff and Connor 1981). Dynamic properties of the species structures are also important. Cycling of species suites (Coull and Fleeger 1977), differences in seasonal fluctuations of populations (Thomson and Lehner 1976) or historical components (Osman 1977, 1978) can also contribute information about the structure and function of assemblages.

A lake system and an approach

Our study lakes possess a variety of sizes, shapes, and environmental conditions. During the winter, the extent of low dissolved oxygen conditions varies among the lakes, providing a gradient of environmental disturbance. Different lakes have different morphometries, substrates, and macrophyte vegetation so that the role of habitat complexity in structuring the fish assemblages can be evaluated. Trophic types range from eutrophic and mesotrophic to dystrophic. The lakes are relatively close together and exposed to the same species pool but differ in surface area and degree of connectedness, providing a perspective for considering the insular biogeography of fish assemblages.

The investigation of how environmental factors (physical and biotic) determine the structure of natural assemblages has benefitted greatly from the "natural experiments" of comparative studies (e.g., Cody 1974, Diamond 1978, Werner et al. 1978). This method can relatively quickly generate and test hypotheses, assess mechanisms, and produce acceptable explanations for community-level problems under a wide variety of conditions.

STUDY AREA

The 18 study lakes (Table 1) are in Vilas County, Wisconsin, USA, the center of the Northern Highlands Lake District of Wisconsin and Michigan (Juday and Birge 1930; Fig. 1). This area is one of the most concentrated lake districts in the world (Vilas County alone has over 1300 lakes) and is well suited for comparative studies in aquatic ecology.

In choosing a lake we considered the following major factors: a history of low oxygen concentration in

TABLE 1. Morphometric and limnological characteristics of the 18 study lakes in Vilas County, Wisconsin (from Black et al. 1963).

Lake	Water source*	Area (ha)	Watershed (km²)	Maximum depth (m)	Length (km)	Shoreline development factor†	Alkalinity (CaCO₃ mg/L)	Conductivity (μS/cm) at 20°C	pH	History of low winter oxygen‡	Predominant substrate types§
1. Apeekwa	D	76.1	33.7	3.0	1.6	1.5	22	57	7.2	0	M
2. Aurora	D	38.0	4.7	1.2	1.4	2.2	40	89	6.8	+	M, S, G
3. Blueberry	S	4.9	0.5	8.2	0.5	1.5	2	15	5.8	0	M, S
4. Camp 2	S	5.7	0.5	1.5	0.3	1.1	2	16	5.9	+	M
5. Gateway	S	3.2	0.3	2.4	0.3	1.5	36	145	7.5	+	?
6. Grassy	Spr	42.9	1.6	1.2	1.6	2.5	38	85	7.0	+	M, S
7. Johnson	Spr	9.7	9.1	3.6	0.3	1.1	23	52	7.2	0	M
8. Landing	D	89.0	12.9	3.3	1.8	3.0	35	79	7.5	+	S, G, M
9. Little Rice	Spr	23.9	176.1	2.1	0.6	1.0	21	53	7.0	0	M
10. Maple	S‖	19.0	1.3	4.3	1.3	2.3	4.5	18	6.0	0	M
11. Mill	D	53.0	9.1	1.2	1.1	1.9	47.5	113	8.0	+	S, M, G
12. Mystery	D	8.1	1.8	2.1	0.3	1.0	15	30	7.1	+	M
13. Nixon	D	44.5	19.4	1.5	1.1	1.4	41.5	23	7.0	0	M
14. Spruce	S‖	6.1	0.4	4.9	0.3	1.0	8	16	6.2	0	M, G
15. Whitney	Spr	89.8	2.6	2.4	0.8	1.1	22	53	7.3	+	M, S
16. Whynot	S	3.2	0.3	5.8	0.2	1.5	3	4	5.2	0	?
17. 33-6	S	2.4	2.6	3.3	0.2	1.2	3	13	5.1	?	?
18. 33-13	S	2.8	2.6	3.0	0.2	1.8	12	25	6.0	?	?

* D = drainage lake, having an inlet and outlet; Spr = spring-fed lake, having an outlet; S = seepage lake, having no inlet or outlet.
† Shoreline development factor = $S/2\sqrt{a\pi}$, where S = length of shoreline and a = area of lake.
‡ + = known or suspected history of winterkill; 0 = no history of winterkill; ? = no information on winterkill history.
§ M = muck; S = sand; G = gravel; ? = no information.
‖ Maple and Spruce have intermittent outlets.

winter, water source (drainage or spring-fed vs. seepage), surface area, maximum depth, and predominant substrate type. Eight lakes had a history of low winter oxygen conditions. Ten lakes were drainage or spring-fed (and thus had inlets and/or outlets) and eight lakes were seepage (no inlets or outlets), though two of the seepage lakes had intermittent outlets (P. Brenner, *personal communication*; W. M. Tonn, *personal observation*). An attempt was made to select lakes spanning a variety of surface areas while keeping maximum depths and predominant substrate types of all study lakes similar.

MATERIALS AND METHODS

We sampled during two 9-wk periods in 1978, one during January–March (winter) and one during June–August (summer). Each lake was sampled once per season. Preliminary sampling had occurred during June–August 1977. Two lakes (Grassy and 33–13) were re-sampled in August 1978, and one (Mill) in August 1979, to examine sampling replicability.

Fish sampling

Winter.—In each lake, 15 regularly spaced stations along 3–4 transects were selected. A 25 cm diameter hole was drilled through the ice and a pair of Gee's wire minnow traps (44.5 cm long, 23 cm at largest diameter, 2.5 cm funnel diameter, 6 mm square mesh), baited with bread and liver, was placed in the water. One trap was set just under the ice-water interface, the other 1.5 m below the first or on the bottom, whichever was shallower. Two small fyke nets (4.6 m leads, four 0.76 m diameter hoops per net, 10 cm throat diameter, 5 mm square mesh) were also used in most lakes, at water depths under the ice of 1–2 m. Four lakes (Aurora, Mill, Little Rice, and Camp 2) were too shallow for fyke nets. Traps and nets were set in each lake for approximately 48 h, usually concurrently, but always within 24 h of each other.

Summer.—Sampling in summer was similar to the winter. Minnow traps were placed in approximately the same locations and fyke nets were set perpendicularly from shore, immediately shoreward from where they were set in winter. All lakes were sampled with fyke nets during the summer. A trammel net (30.5 × 1.2 m; 18 and 2.5 cm square mesh) was added to increase the variety of sampling gear, and thus to examine the thoroughness of the methods. All gear were set concurrently for 48 h. In August 1979, Mill was resampled, doubling the number of fyke and trammel nets, also to examine sampling thoroughness.

Fishes were identified to species and counted in the field; a subsample was preserved to verify identification. Identification followed keys of Eddy and Underhill (1974) and Becker and Johnson (1970).

Environmental sampling

In addition to data from Black et al. (1963; Table 1), several physical, chemical, and habitat complexity factors were measured for each lake during the 48-h

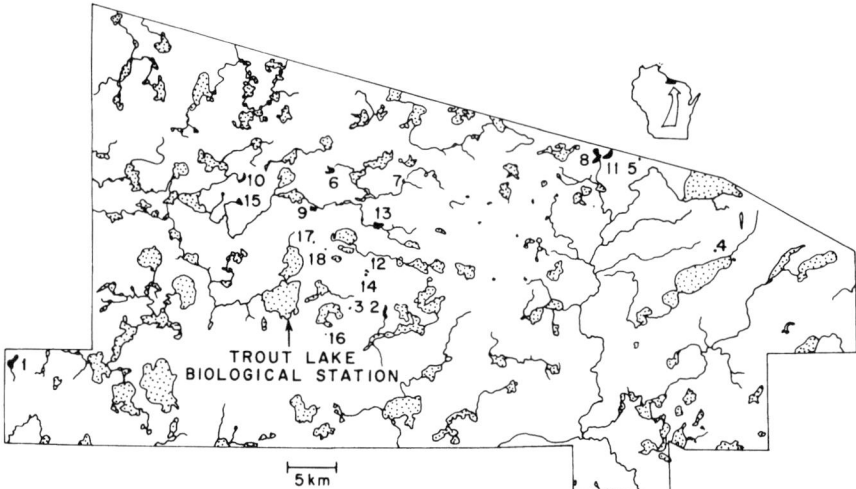

FIG. 1. Map of Vilas County, Wisconsin. Vilas County is the center of the Northern Highlands Lake District, where all 18 study lakes are located. Study lakes (solid black areas) are numbered as in Table 1.

fish sampling periods. Each factor was measured in one season only, but all measurements occurred at midday, 1100–1300. Some measurements were made in the alternate season to check for seasonal variation.

In winter, an inshore and an offshore station were selected in each lake. At each station, a hole was carefully drilled through the ice so as not to disturb the water below (the last bit of ice was gently tapped out with an ice chisel). Water samples were collected 1–3 cm below the ice-water interface with a siphon sampler (Magnuson and Stuntz 1970), and with a Kemmerer bottle centered at 1 m below the interface, if depth permitted. For oxygen analyses, duplicate samples from each depth at each station were fixed immediately in the field and titrated (Winkler method, azide modification, American Public Health Association 1976) in the laboratory later the same day. Conductivity measurements were also made on the same day in the laboratory, using a Hach conductivity meter, Model #2510, (Hach Chemical Company, Loveland, Colorado) on samples taken at each of the two stations, warmed to 20°C. Total dissolved solids were measured on 100-mL water samples, filtered through 0.45-μm filters and evaporated at 103°.

In summer, habitat structure was measured for three variables: depth, substrate type, and macrophytes. Measurements were made along the same transects used for the minnow traps, with 15–25 stations in small and/or structurally simple lakes, and 25–35 stations in larger, more complex lakes. At each station, water depth was measured to the nearest 1 cm, later grouped into five categories (Table 2). A substrate sample was taken with an Ekman dredge, and the bottom material was classified on a five-point scale, based on the proportion of muck, sand, gravel, and litter that was present, estimated by visual and tactile inspection (Table 2). Within an estimated 5 m radius of each sampling station, the macrophytes were classified on a presence/absence basis for submergent, floating, and emergent forms (Table 2), from visual inspection. Where the lake bottom was not visible, presence/absence of submergent plants was determined by 2–4 dredge samples. Macrophyte measurements were not made for Aurora until June 1980.

Preliminary analyses

Several preliminary analyses were performed on the data to identify unknown biases prior to analyzing for fish assemblage structure. For these and all subsequent analyses, $P \leq .05$ was used as the level of statistical significance. Rank correlations (Siegel 1956) were performed on the sampling sequence of the lakes against dissolved oxygen, surface area, watershed size, pH and alkalinity. These correlations revealed no significant trends. Thus sampling sequence should not bias conclusions relating these variables to fish assemblage data.

Preliminary analysis of the fish data addressed the question: did fish samples accurately represent the available assemblages? Evidence for the adequacy of sampling comes from the two lakes (Grassy and 33-13) sampled twice during the summer, 1978, and from Mill, where in 1979 a resampling with twice the effort was done. Czekanowski's similarity coefficients, S_c, were calculated from presence/absence data by the formula:

TABLE 2. Descriptions of the five depth, five substrate, and seven vegetation categories used in calculating habitat diversity measurements.

Habitat variable	Habitat category						
	1	2	3	4	5	6	7
Depth (m)	0.00–0.50	0.51–1.00	1.01–2.00	2.01–4.00	4.01–8.00		
Substrate (% of each type)	>50 litter	1–50 litter; 50–99 muck	0–49 sand; 51–100 muck	1–50 muck; 50–99 sand	0–49 gravel; 51–100 sand		
Vegetation (type present)*	e	f	s	e, f	e, s	f, s	e, f, s

* e = emergent (e.g., *Sagittaria*); f = floating (e.g., *Nuphar*); s = submergent (e.g., *Anacharis*).

$$S_c = \frac{2x_{jk}}{x_j + x_k}$$

where x_j is the number of species in the first sample, x_k is the number of species in the second sample, and x_{jk} is the number of species common to both samples (Bray and Curtis 1957). S_c for Grassy, 33–13, and Mill were 0.90, 1.00, and 0.86 respectively. Since this replicate sampling involved a doubling or tripling of the effort but yielded a total change of only one species in Grassy, one in Mill, and none in 33–13, we concluded that our level of effort was sufficient to obtain almost all susceptible species. Thus our assumption was that any sampling inadequacy was relatively minor, similar for all lakes, and did not significantly bias comparisons among lakes.

The numbers of individuals caught did not appear to influence species richness in our sampling. When we correlated species richness against the numbers of individuals and ln (individuals) for each season only the correlation with ln (winter individuals) was significant. If four lakes with 0 or 1 individuals (and thus no degrees of freedom for species richness) were excluded, this one correlation lost its significance.

Finally, to test for a time trend in sample species richness, a one-sample runs test and a Spearman rank correlation (Siegel 1956) of sample richness with sampling sequence were performed. Both tests showed that there was no significant time trend for either sampling season.

Fish assemblage analyses

We used multivariate techniques of classification and ordination to detect assemblage patterns in the fish data. In the matrices, the lakes were rows (entities) and the species were columns (attributes). Classifications were from the CLUSTAN 1C package (Wishart 1975). Association analysis (Williams and Lambert 1959), with sum of chi-square as the maximum attribute sum on which cluster division was based, was used. Ordinations were Bray-Curtis types (Bray and Curtis 1957, Post et al. 1973). $1.0-S_c$ (Czekanowski's coefficient) was employed as the distance measure; regression was used for endpoint selection (Post et al. 1973). Hypotheses generated by the classification-ordination analyses were examined using discriminant analysis programs (Schlater and Learn 1974, Dixon and Brown 1979).

Czekanowski's similarity coefficients, S_c, between winter and summer fish assemblages were calculated for each lake. These were used as measurements of the seasonal stability of the species compositions. To standardize sampling effort for these summer–winter comparisons, data from gear not used in both seasons were excluded: summer fyke net catches in four lakes (see Fish sampling, above) and all summer trammel net catches.

Habitat diversities were calculated for each lake with the Shannon-Wiener formula, $H' = -\Sigma p_i \ln (p_i)$ (Shannon and Weaver 1948), where p_i is the proportion of all stations in the ith habitat category (from Table 2). Diversities were calculated for the three habitat factors (depth, substrate, vegetation) separately, all two-factor combinations of the three factors, and for the three factors combined, using the components of diversity method (Pielou 1977).

We derived a connectedness parameter for the lake "islands." For seepage lakes, these values were equal to the lakes' watershed areas (Table 1). For drainage lakes, the values also included the watershed areas of the next adjacent lake both upstream and downstream, obtained from Black et al. (1963). For Maple and Spruce, seepage lakes with intermittent outlets, we added only one-half of the downstream watershed area. We believe that these connectedness values, while admittedly somewhat arbitrary, are meaningful ecologically because fishes in drainage lakes have the connecting lakes and streams as both potential refuges during severe conditions, and as source areas for potential immigrants.

To examine the relationships between fish species richness and the lake environments, single and stepwise multiple linear regressions were performed on richness vs. the morphometric, limnological, and habitat diversity factors across all lakes, for each season. Multiple regressions were performed on 15 available independent variables. These included lake area, max-

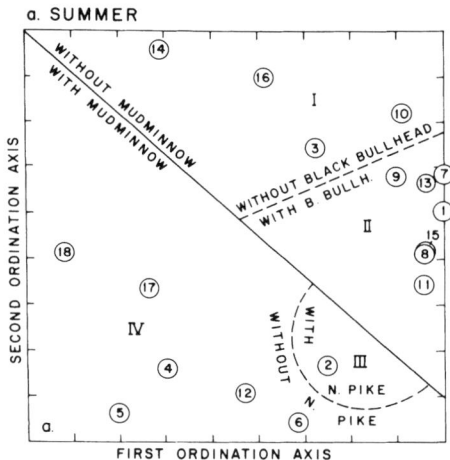
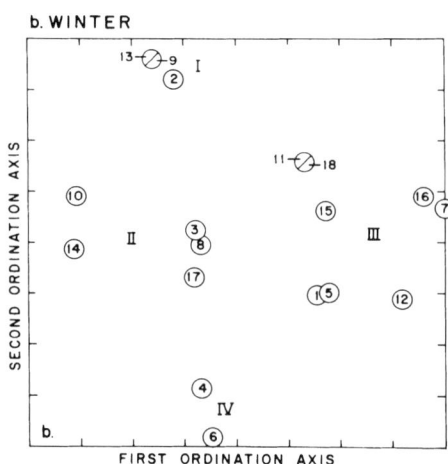
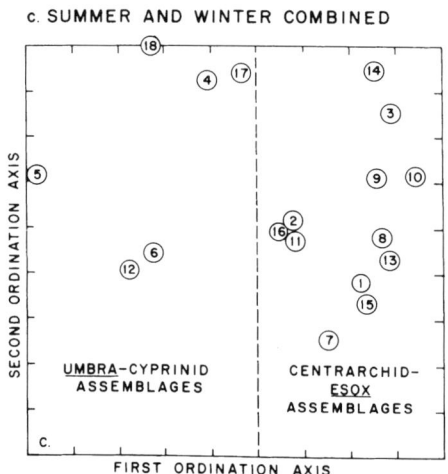

FIG. 2. Bray-Curtis ordinations of fish assemblages for the 18 study lakes in 1978. Lakes are numbered as in Table 1. (a) Summer. Key species, on whose presence or absence the first three divisions of the association analysis are based, are given for the four groups distinguished; (b) Winter; and (c) Summer and winter combined.

imum depth, alkalinity, and pH from Black et al. (1963; Table 1), and all factors measured in this study (Table 3), except mean depth. Because some factors (pH and the habitat diversity measures) were in a logarithmic form, and because many different units of measurements were used, a $\log_{10}$ transformation was used on all other environmental data. A log-log model was also used on the richness vs. lake area regressions, since this is the most standard way of presenting species-area relationships (Connor and McCoy 1979). A $P = .05$ level was used as the entering and leaving criterion for all multiple regressions.

In addition to the standard stepwise procedure, in which the independent variable that is most highly correlated with the dependent variable is always first to enter the model, the multiple regressions were also run by starting with each independent variable as a base variable (Allen and Learn 1973). For this, the regression began with the base variable already included in the model, regardless of its correlation with the dependent variable. In this way, a combination of independent variables more successful (in terms of the probability level of the model's F ratio) than that chosen by the standard route could be identified. Correlations were also calculated among the environmental factors (Appendix I).

RESULTS

Fish assemblage composition

We caught a total of 23 species, 18 in winter and 22 in summer (Appendix II).

Summer.—In the ordination of summer fish assemblages (Fig. 2a), the first two axes accounted for 79% of the variation. An association analysis, which divided the assemblages hierarchically into progressively more similar groups, based on the presence and absence of key species, corresponded closely to the results of the ordination. The classification results at the four-group level are demarcated in Fig. 2a, and the "key" and abundant species of each group are listed in Table 4.

The first division of the association analysis was based on the presence of the central mudminnow. (Scientific names for all species are provided in Appendix II.) The 7 assemblages with mudminnow had richnesses of 1–11 species ($\bar{x} = 5.6$), while the 11 assemblages without mudminnow ranged from 2–10 ($\bar{x} = 6.1$) species. The second division (Fig. 2a) split assemblages without mudminnows into Group I (four assemblages without black bullhead), and Group II (seven

TABLE 3. Morphometric, limnological, and habitat diversity measurements of the 18 study lakes (from the present study).

				Total dis-		Diversity (H')‡						
				solved								Depth-sub-
	Lake connect-	Winter	Conduc-	solids	Mean				Depth-	Depth-	Sub-	strate-
	edness*	oxygen	tivity (μS/cm)	(TDS)	depth		Sub-	Vegeta-	sub-	vegeta-	strate-	vegeta-
Lake	(km²)	(mg/L)	at 20°C	(mg/L)	(m)†	Depth	strate	tion	strate	tion	vegetation	tion
1. Apeekwa	117.8	2.36	107	98	1.14	0.90	0.95	1.69	1.71	1.88	2.20	2.20
2. Aurora	35.7	0.28	130	95	0.66	0.91	0.55	1.62	1.21	2.45	1.52	2.52
3. Blueberry	0.5	9.14	36.5	8	2.63	1.16	1.25	0.37	2.02	1.51	1.53	2.13
4. Camp 2	0.5	2.44	58	38	1.08	0.69	0.51	1.08	1.00	1.58	1.41	1.83
5. Gateway	0.3	0.07	141	114	1.09	1.02	0.50	0.93	1.38	1.73	1.48	2.03
6. Grassy	10.1	0.52	112.5	68	0.64	0.88	0.58	1.47	1.37	2.10	1.53	2.34
7. Johnson	185.2	6.48	92	39	1.54	1.20	0.00	0.58	1.20	1.61	1.09	1.61
8. Landing	22.0	1.58	130.5	74	1.72	0.83	1.30	0.78	1.89	1.51	1.54	2.03
9. Little Rice	255.1	0.00	157	92	0.88	0.76	1.19	0.91	1.71	1.71	2.12	2.28
10. Maple	33.0§	8.34	43	20	2.00	0.61	1.03	0.50	1.45	1.03	1.16	1.57
11. Mill	38.6	0.00	209	148	0.99	0.83	1.39	0.96	2.06	1.51	1.79	2.38
12. Mystery	8.3	0.56	67	53	1.18	0.82	0.84	0.65	1.46	1.43	1.41	1.87
13. Nixon	197.6	0.57	92	84	1.44	0.56	0.84	1.55	0.98	1.61	1.98	2.09
14. Spruce	1.3§	12.26	40	37	2.16	0.79	0.68	0.00	1.23	0.79	0.68	1.23
15. Whitney	339.3	12.52	76	30	1.63	0.43	1.04	0.91	1.28	1.17	1.57	1.77
16. Whynot	0.3	10.11	37.5	29	4.18	0.86	0.62	0.35	1.12	1.03	1.01	1.24
17. 33-6	2.6	2.73	53	34	1.64	1.21	0.86	0.40	1.34	1.42	1.03	1.51
18. 33-13	2.6	5.78	47.5	41	1.58	0.98	0.53	0.32	1.38	1.31	0.85	1.66

* For seepage lakes, these values are equal to the lakes' watershed areas (Table 1). For drainage lakes, the values also included the watershed areas of the next adjacent lake both upstream and downstream.

† Mean depth was determined by averaging the depths measured at 15–35 sampling stations in each lake (see text). It was not used in the multiple regression analyses.

‡ These were calculated using the general formula $H' = -\Sigma p_i \ln(p_i)$, where p_i is the proportion of a habitat category described in Table 2 (see text).

§ Maple and Spruce have intermittent outlets. One-half of the adjacent downstream watershed was added to their watershed area from Table 1.

assemblages with the bullhead, Table 4). The third division (Fig. 2a) split the assemblages with mudminnows into Group III (one assemblage with northern pike) and Group IV (six lakes without pike, Table 4).

All four Group I assemblages (Table 4) contained largemouth bass, three contained yellow perch, and three the bluegill. Three Group I lakes are dystrophic, small, seepage, and relatively deep, with much of their shorelines formed from sphagnum mat. They contained 2–4 species. The fourth lake (Maple) also has a considerable proportion of sphagnum shoreline, but is larger, somewhat shallower and connected to a relatively large watershed via an intermittent outlet. With 7 species, Maple's summer assemblage was richer than the other three Group I lakes.

The seven Group II assemblages (Table 4), without mudminnow but with black bullhead, had summer richnesses ranging from 5–10 species. Other abundant species were northern pike, white sucker, yellow perch, and pumpkinseed sunfish. All seven lakes are drainage or spring-fed lakes with relatively high levels of conductivity, pH, and other edaphic-productivity related characteristics of mesotrophic to eutrophic lakes.

The Group IV assemblages had richnesses of 1–11 species. All six contained mudminnow, four had yellow perch, and four contained at least two minnow species (Cyprinidae; Table 4). In contrast with the other groups, these assemblages contained neither large piscivores (pike, bass) nor sunfishes (Centrarchidae). Group IV lakes included four seepage lakes and two drainage lakes with small watersheds. All were shallow, with low to moderate winter dissolved oxygen levels (Tables 1, 3).

The single Group III assemblage with both mudminnow and pike had a relatively rich summer assemblage of eight species (Table 4). It shared compositional similarities both with the other lakes with mudminnows (Group IV, e.g., two cyprinids) and the lakes without mudminnows (especially Group II, e.g., pike, black bullhead, pumpkinseed).

Winter.—The results of the classification and the ordination of the winter assemblages showed little agreement. We felt that the ordination resulted in better patterns (Fig. 2b). Still, within-group compositional similarity was not high; only 51% of the variation was accounted for by the first two axes. From the ordination, four assemblage groups were subjectively distinguished (Fig. 2b). Their characteristic species are summarized in Table 4.

The four winter groups did not correspond well to the four summer groups. Taken alone, the pattern of winter assemblages yielded few insights. However, as with the species richness patterns (see below), they did show that the relatively clear assemblage relationships of the summer broke down during winter.

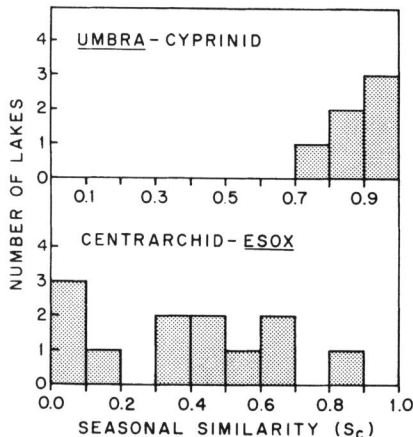

FIG. 3. Seasonal (summer/winter) similarities (S_c) for *Umbra*-cyprinid (above) and centrarchid-*Esox* (below) fish assemblages, based on an equal sampling effort (see Methods).

Summer-winter (combined).—Although the two single-season analyses differed greatly, we were interested in obtaining a more complete, year-round picture of the fish assemblages. To do this, the classification and ordination analyses were run with summer and winter data sets combined. The classification results with the combined data were identical to those of the summer. The ordination (Fig. 2c), however, did show influence of the winter assemblages (e.g., Aurora, Maple, and Mill were not especially similar in either season alone, but, showing a similar summer–winter response, they were grouped closely together). Two assemblage types were distinguished from the ordination (Fig. 2c).

The six assemblages with mudminnow but without pike during summer (Group IV, Fig. 2a) maintained their integrity as a group especially well on the first axis of the combined ordination. This axis accounted for 36% of the variation, the first two axes 56%. The presence of mudminnow in both seasons appeared to be a key factor. These six lakes appeared to form a distinct fish assemblage type. We call them "*Umbra*-cyprinid assemblages" in Fig. 2c, after their characteristic component species.

The other 12 lakes made up the second major group, which we call "centrarchid-*Esox* assemblages" in Fig. 2c. The two groups without mudminnow in summer (Groups I and II, Fig. 2a) were not as distinct when winter data were included. The single Group III lake in Fig. 2a also grouped with these 11 lakes when the winter data were included (Fig. 2c).

Seasonal similarity.—The within-lake similarities in species composition between winter and summer (based on equal sampling; see Methods) were greater in *Umbra*-cyprinid assemblages ($\bar{S}_c = .90$) than in centrarchid-*Esox* assemblages ($\bar{S}_c = .35$, Fig. 3); the two assemblage types were significantly different (Wilcoxon rank sum test). Thus, *Umbra*-cyprinid assemblages were not only similar in their species compositions, but also in the seasonal stability of their compositions. The significantly lower seasonal similarities of centrarchid-*Esox* assemblages generally resulted from reduced species richness during the winter.

Discriminant analyses.—To describe quantitatively the separation of the two assemblage types identified above (*Umbra*-cyprinid and centrarchid-*Esox*) we applied discriminant analysis to the combined summer-winter data on species presence/absence. Lakes were assigned to one of the two assemblage types from the classification-ordination analyses, and the percentages

TABLE 4. A summary of species composition of the four fish assemblage groups from the summer sampling (upper) and the winter sampling (lower) identified by the ordination-classification analyses of Figs. 2a and 2b, respectively. Species listed include those denoted as "key" species by the association analysis and those that were numerically abundant in a majority of lakes of each group.

Group I (2–7 species)	Group II (5–10 species)	Group III (8 species)	Group IV (1–11 species)
Summer			
Largemouth bass	Black bullhead	Mudminnow	Mudminnow
Yellow perch	Northern pike	Northern pike	Yellow perch
Bluegill	White sucker	Black bullhead	Golden shiner
	Yellow perch	Yellow bullhead	Redbelly dace
	Pumpkinseed	Pumpkinseed	

Group I (0–1 species)	Group II (1–4 species)	Group III (1–7 species)	Group IV (4–6 species)
Winter			
Pumpkinseed or no species present	Yellow perch	Mudminnow	Mudminnow
		Black bullhead	Pearl dace
			Golden shiner
			Yellow perch

TABLE 5. Correlation coefficients (r), statistical significance of r, and linear regression values for summer and winter species richness (y) vs. each of 15 environmental factors (x) for all 18 assemblages (top), the 12 centrarchid-*Esox* assemblages (middle), and the 6 *Umbra*-cyprinid assemblages (bottom). Multiple regressions for each season and assemblage are given below the linear regression set.

	Summer				Winter			
			$y = a + bx$				$y = a + bx$	
Independent variable	r	($P \leq .05$)	a	b	r	($P \leq .05$)	a	b
All lakes ($N = 18$)								
1. Log (lake area)	.69	*	1.86	3.50	−.08	NS	3.14	−0.26
2. Log (maximum depth)	−.47	*	8.25	−5.50	.04	NS	2.69	0.34
3. Log (connectedness + 1)	.60	*	3.58	1.96	−.30	NS	3.64	−0.67
4. Log (alkalinity)	.66	*	1.58	3.87	−.02	NS	2.94	−0.09
5. Log (conductivity)	.60	*	−7.70	7.20	−.06	NS	3.82	−0.52
6. pH	.70	*	−9.98	2.39	.14	NS	0.58	0.34
7. Log (total dissolved solids)	.42	NS	−0.58	3.85	−.07	NS	3.59	−0.45
8. Log (winter oxygen + 1)	−.42	NS	7.48	−2.83	.02	NS	2.78	0.11
9. Substrate diversity	−.08	NS	6.48	−0.66	−.27	NS	4.04	−1.48
10. Vegetation diversity	.69	*	2.66	3.93	.00	NS	2.83	0.00
11. Depth diversity	−.12	NS	7.33	−1.61	.19	NS	1.36	1.72
12. Depth and substrate	.08	NS	4.92	0.72	−.02	NS	2.98	−0.10
13. Depth and vegetation	.58	*	−0.47	4.22	.07	NS	2.30	0.35
14. Substrate and vegetation	.50	*	0.73	3.67	−.16	NS	3.98	−0.81
15. Depth, substrate, and vegetation	.57	*	−2.16	4.25	−.08	NS	3.58	−0.39

Summer richness = 3.75 + 4.56 log area −3.84 substrate diversity
$R^2 = .67 \quad P \leq .05$

Winter richness = −3.15 + 1.14 pH − 1.30 log (watershed + 1)
$R^2 = .24 \quad P > .05$

	Summer				Winter			
Independent variable	r	($P \leq .05$)	a	b	r	($P \leq .05$)	a	b
Centrarchid-*Esox* ($N = 12$)								
1. Log (lake area)	.62	*	2.36	2.92	−.13	NS	2.80	−0.40
2. Log (maximum depth)	−.52	NS	8.53	−4.72	.59	*	0.66	3.42
3. Log (connectedness + 1)	.73	*	3.25	2.03	−.28	NS	2.99	−0.49
4. Log (alkalinity)	.67	*	2.28	3.44	−.23	NS	3.13	−0.74
5. Log (conductivity)	.58	*	−3.84	5.32	−.29	NS	5.44	−1.67
6. pH	.67	*	−7.37	2.02	−.06	NS	3.01	−0.11
7. Log (total dissolved solids)	.46	NS	1.23	3.04	−.32	NS	4.50	−1.33
8. Log (winter oxygen + 1)	−.41	NS	7.66	−2.16	.59	*	1.01	2.01
9. Substrate diversity	−.24	NS	7.64	−1.45	−.25	NS	3.12	−0.97
10. Vegetation diversity	.66	*	3.81	2.96	−.31	NS	3.00	−0.88
11. Depth diversity	.09	NS	5.52	0.99	.44	NS	−0.23	3.03
12. Depth and substrate	−.02	NS	6.49	−0.11	.11	NS	1.59	0.45
13. Depth and vegetation	.60	*	1.48	3.27	−.21	NS	3.33	−0.73
14. Substrate and vegetation	.37	NS	3.23	2.08	−.46	NS	4.69	−1.64
15. Depth, substrate, and vegetation	.47	NS	1.29	2.62	−.40	NS	5.00	−1.43

Summer richness = −1.18 + 2.46 log (watershed + 1) + 4.59 depth diversity
$R^2 = .69 \quad P \leq .05$

Winter richness = −7.28 + 9.94 log maximum depth + 4.16 log alkalinity
$R^2 = .72 \quad P \leq .05$

	Summer				Winter			
Independent variable	r	($P \leq .05$)	a	b	r	($P \leq .05$)	a	b
***Umbra*-cyprinid ($N = 6$)**								
1. Log (lake area)	.90	*	−0.25	7.02	.73	NS	1.24	3.57
2. Log (maximum depth)	−.80	NS	10.7	−17.0	−.70	NS	7.04	−9.31
3. Log (connectedness + 1)	.51	NS	2.40	4.86	.50	NS	2.32	2.94
4. Log (alkalinity)	.64	NS	0.67	4.37	.41	NS	2.22	1.73
5. Log (conductivity)	.74	NS	−21.4	14.2	.51	NS	−7.32	6.07
6. pH	.72	NS	−13.5	2.90	.68	NS	−7.00	1.71
7. Log (total dissolved solids)	.59	NS	−13.8	11.0	.41	NS	−4.19	4.76
8. Log (winter oxygen + 1)	−.81	*	8.94	9.67	−.78	NS	6.28	−5.84
9. Substrate diversity	−.06	NS	6.01	−1.33	.19	NS	2.32	2.63
10. Vegetation diversity	.84	*	−0.48	6.98	.63	NS	1.36	3.27
11. Depth diversity	−.38	NS	12.3	−7.65	−.56	NS	10.6	−7.08
12. Depth and substrate	.22	NS	−1.42	4.98	.16	NS	1.00	2.27
13. Depth and vegetation	.86	*	−12.3	10.9	.52	NS	−2.65	4.17
14. Substrate and vegetation	.85	*	−9.22	11.2	.85	*	−4.98	6.99
15. Depth, substrate, and vegetation	.93	*	−16.8	11.7	.69	NS	−6.17	5.43

Summer richness = 3.47 + 5.23 log lake area −6.01 log (winter oxygen + 1)
$R^2 = .99 \quad P \leq .05$

Winter richness = −4.98 + 6.99 substrate and vegetation
$R^2 = .72 \quad P \leq .05$

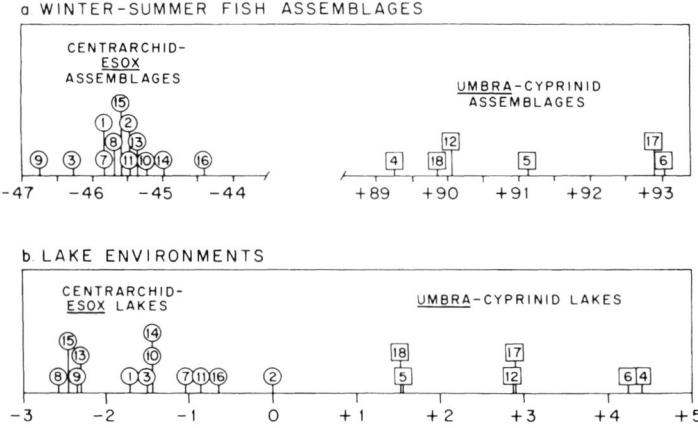

FIG. 4. (a) A discriminant analysis of the fish assemblages in the 18 study lakes from combined data for winter and summer, 1978. Assemblages were assigned to one of two groups (*Umbra*-cyprinid or centrarchid-*Esox*) based on the classification/ordination analyses of Fig. 2. Lakes are numbered as in Table 1. (b) A discriminant analysis of the study lakes using 15 morphometric, limnological, and habitat diversity measurements from Black et al. (1963) and the present study.

of correctly classified lakes were calculated, after Dixon and Brown (1979). All species were used in the analysis (i.e., a stepwise procedure with a critical entering and leaving level was not used).

Separation of the two groups was distinct in the plot of the 18 assemblages along the discriminant function (Fig. 4a). All *Umbra*-cyprinid and centrarchid-*Esox* assemblages were correctly classified. The summer occurrences of mudminnow and redbelly dace and the winter presence of mudminnow best defined the *Umbra*-cyprinid assemblage type, while the summer presence of northern pike, pumpkinseed sunfish, and white sucker best defined the centrarchid-*Esox* assemblage type, as indicated by F ratios from univariate F tests (Tonn 1980). The summer presence of mudminnow, golden shiner, and bluntnose minnow provided the greatest discriminatory power to the discriminant function.

A discriminant analysis also was applied to the log-transformed environmental data on the *Umbra*-cyprinid and the centrarchid-*Esox* lakes. Our purpose was to evaluate the environmental distinctness between the two groups of lakes, and to help identify environmental factors contributing to their separation. Lakes were plotted in the reduced discriminant space, after Green and Vascotto (1978). Lakes were classified by the type of fish assemblage present, from the previous analyses. A clear separation of the two groups resulted (Fig. 4b), with all lakes correctly classified. Individual factors which best defined the two groups included our lake connectedness measurement, lake area, and substrate diversity. The three-component habitat complexity variable (depth, substrate, and vegetation) depth-vegetation diversity and depth-substrate diversity contributed most to the discriminatory power of the discriminant function.

Thus, we found two groups of lakes having both distinctive fish species compositions and environments.

Fish species richness

Summer.—The relationships between species richness and some environmental factors were similar in summer for the *Umbra*-cyprinid and centrarchid-*Esox* assemblages (Table 5). For example, vegetation diversity appeared as the most important single component of habitat structure in both assemblage types. Lake area was also significantly related to richness for both assemblage types.

Other environmental factors were significantly related to richness in one assemblage type but not the other. Lake connectedness was significantly related to summer richness in centrarchid-*Esox* assemblages but not for the *Umbra*-cyprinid type (Table 5). Similarly, pH, conductivity, and alkalinity were also significantly related to richness in the centrarchid-*Esox* lakes, but not for *Umbra*-cyprinid assemblages (Table 5). The three-component habitat complexity factor and winter oxygen were significantly related to richness for the *Umbra*-cyprinid assemblage type, but they were not for centrarchid-*Esox* assemblages (Table 5). However, except for depth and depth-substrate diversity, the correlation coefficients between species richness and each environmental factor had the same sign in both assemblage types during summer.

Multiple linear regressions between summer richness and environmental factors also revealed differences between the two assemblage types. For *Umbra*-cyprinid assemblages lake area and winter oxygen levels accounted for the most variation (Table 5). Species richness increased with lake area and decreased with increasing winter oxygen levels. For centrarchid-*Esox* assemblages lake connectedness and depth diversity explained the most variation (Table 5). Species richness increased with lake connectedness and with depth diversity.

Winter.—The winter relationships in centrarchid-*Esox* assemblages were quite different from their summer patterns (Table 5). Species richness increased significantly with maximum depth and with winter dissolved oxygen levels, instead of decreasing as it did in summer (Table 5). Although the regressions for other factors were not statistically significant, 13 of 15 were opposite in sign from their summer patterns (Table 5). When we correlated the differences in richness within a lake between summer and winter (corrected for equal sampling) with the environmental variables, significant correlations were found with vegetation diversity, the three-factor habitat complexity, alkalinity, lake connectedness, and winter oxygen levels. In the multiple linear regression, maximum depth and alkalinity were included in the model (Table 5). Richness was greater in centrarchid-*Esox* lakes with greater depths and higher alkalinities.

For *Umbra*-cyprinid assemblages, relationships between species richness in winter and environmental factors were similar to those for the summer, although only substrate-vegetation diversity was statistically significant (Table 5). Richness increased with substrate-vegetation diversity. The within-lake differences in species richness between summer and winter were not significantly related to any environmental factor. In the multiple regression analysis only substrate-vegetation diversity was entered (Table 5).

Thus the two assemblage types, identified initially by differences in species composition, also differ in their species richness patterns, particularly in the seasonal changes in richness. These species richness differences might easily have gone unnoticed if only summer patterns in richness had been investigated. These richness differences undoubtedly reflect differences in the assemblage structuring mechanisms and/or show that the same mechanisms can have opposite consequences in two different assemblage types of the same region.

DISCUSSION

The ecological maintenance of the assemblage types

We found two discrete fish assemblage types, "*Umbra*-cyprinid" and "centrarchid-*Esox*," each having broadly repeatable patterns of species composition and seasonal stability. Both species composition and richness were seasonally stable in *Umbra*-cyprinid assemblages, but were seasonally dissimilar in centrar-

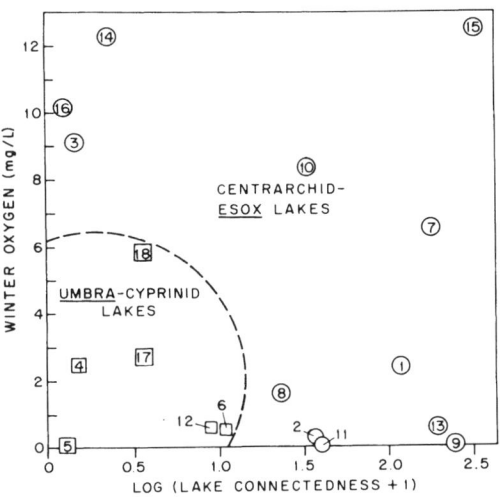

FIG. 5. A direct gradient ordination of the 18 study lakes. Lakes with *Umbra*-cyprinid fish assemblages are those with low levels of oxygen during winter and low connectedness. Lakes with centrarchid-*Esox* assemblages either have high winter oxygen levels or are connected to large watersheds via inlet or outlet streams which can serve as refuges from low oxygen conditions.

chid-*Esox* assemblages, especially in those occurring in productive, low winter oxygen, drainage lakes.

What causes the occurrence of these two discrete fish assemblage types? We believe that a combination of winter oxygen concentration and lake connectedness is most clearly related to the type of fish assemblage that was present. This can be seen in a "direct gradient analysis" ordination (Whittaker 1973), plotting the lakes in winter oxygen vs. lake connectedness space (Fig. 5).

Where winter oxygen levels are high, centrarchid-*Esox* assemblages occur, with largemouth bass as the usual top predator. In lakes with low oxygen levels, the type of fish assemblage present appears to depend on the presence or absence of a connection to a large watershed, whose streams or lakes can act as refuges from the low oxygen conditions. Lakes having direct connections to a stream or lake had centrarchid-*Esox* assemblages in summer. Northern pike tended to be the top predator in these lakes. During winter, as oxygen levels dropped, we hypothesize that most fish migrate out of these lakes into stream or lake refuges. In low oxygen lakes without such a refuge, *Umbra*-cyprinid assemblages occur.

Species inhabiting the *Umbra*-cyprinid lakes are, in general, better able to survive in lakes with low winter oxygen levels than species restricted to centrarchid-*Esox* lakes (Moore 1942, Cooper and Washburn 1946, Petrosky and Magnuson 1973, Gee et al. 1978, Klinger et al. 1982, J. J. Magnuson, *personal observation*). In our study, all 12 species caught in *Umbra*-cyprinid lakes during the summer were also caught in these lakes during winter, at oxygen concentrations <1.0 mg/L.

Several centrarchid-*Esox* lakes also had low levels of dissolved oxygen during winter, but they were distinguished by having greatly reduced, or even nonexistent, winter fish assemblages. Although sampling biases may have contributed to the depauperate winter assemblages, we believe that seasonal emigration accounts for most cases. The evidence, though indirect, comes from several independent lines. Many of the species missing in these centrarchid-*Esox* drainage lakes during the winter were caught with the same gear in *Umbra*-cyprinid lakes and other centrarchid-*Esox* lakes. Thus they were susceptible to the gear when present. For these centrarchid-*Esox* drainage lakes, similarities (S_c) between seasons were significantly lower than for centrarchid-*Esox* seepage lakes (Wilcoxon rank sum test).

The redistribution of fish under the ice in lakes of reduced oxygen levels has previously been observed (Moyle and Clothier 1959, Mills 1972, J. J. Magnuson, *personal observation*), including the aggregation around inlet/outlet streams (Cooper and Washburn 1946, Johnson and Moyle 1969). The actual migration of fish out of a low oxygen lake into the outlet has also been observed (Johnson and Moyle 1969, J. J. Magnuson, *personal observation*). Johnson and Moyle (1969) observed the migration of northern pike, which was a prominent summer component in all of our centrarchid-*Esox* drainage lake assemblages, but which was never caught in these lakes during winter.

A similar pattern, involving a switching of assemblage composition from one type to another, was noted by Jones (1973). When a large creek was disconnected from a shallow, heavily vegetated lake which often experienced low winter oxygen levels, the fish assemblage in the lake was reduced from 11 to 4 species. The species composition changed from one similar to our centrarchid-*Esox* assemblages to one like our *Umbra*-cyprinid pattern. Jones (1973) attributed the change to the removal of the creek refuge as a source for annual repopulation of the lake by the centrarchid-*Esox* assemblage after winter. Other changes in species composition as a result of severe winterkill have been reported in Michigan (Beckman 1948) and Illinois lakes (Bennett 1948).

While this low winter oxygen disturbance/stream refuge hypothesis can explain why centrarchid-*Esox* species are absent from *Umbra*-cyprinid lakes (and from their low-oxygen drainage lakes during winter), it does not address the complementary pattern: why are *Umbra*-cyprinid species rare or absent from centrarchid-*Esox* lakes? A number of factors may be involved, including predation and/or competition, working together with differences in habitat complexity.

One of the most readily apparent ecological differences between the two assemblage types is the presence of large piscivorous species (largemouth bass and/or northern pike) in all 12 centrarchid-*Esox* assemblages, and their complete absence in *Umbra*-cyprinid lakes (Appendix II). Do these top predators eliminate the minnows, mudminnows, and sticklebacks from centrarchid-*Esox* lakes? Are *Umbra*-cyprinid lakes refuges from predation just as centrarchid-*Esox* lakes appear to provide refuges from low winter oxygen conditions? Perhaps predators lowered population levels to the point of local extinction (Zaret and Paine 1973). Possibly the habitat in most of the centrarchid-*Esox* lakes offers little refuge from predation. The few centrarchid-*Esox* lakes in which species characteristic of *Umbra*-cyprinid assemblages occurred had the most rooted macrophytes, and this dense cover may have provided sufficient refuge to allow the coexistence of small populations of cyprinids or mudminnows.

With the exception of brook stickleback, species found primarily in *Umbra*-cyprinid assemblages are small, soft-rayed forms while their "replacements," those species restricted to centrarchid-*Esox* lakes, are chiefly larger, spiney-rayed forms. Most species present in both assemblage types also have spines (yellow perch and black bullhead) or become large (white sucker). Because spines are antipredator devices (Hoogland et al. 1957), they could promote coexistence with large predators. Likewise, if both spiney and spineless species co-occurred with piscivores, the spineless species should be selected by the predators (Hoogland et al. 1957, Lewis et al. 1961) and would more likely be eliminated from the lake.

As an alternative hypothesis, centrarchids might competitively exclude the *Umbra*-cyprinid species from centrarchid-*Esox* lakes. These sunfishes are generalized foragers, taking a wide variety of invertebrates from the sediments, vegetation, and open water, while the cyprinids often specialize on prey in the plankton (e.g., golden and blacknose shiners), on vegetation (e.g., redbelly dace), or even detritus (fathead minnow) (Keast and Webb 1966, Keast 1970, 1978, Werner and Hall 1976, 1977, 1979, Gascon and Leggett 1977, Werner et al. 1977, Hall et al. 1979). Sunfishes should also eat a wider variety of prey sizes than the smaller cyprinids, mudminnow, or stickleback (Werner 1979). This generalization of foraging site, prey type, and prey size has contributed to the general success of sunfishes in small glacial lakes (Werner et al. 1977, Werner and Hall 1979). The particular combination of habitat structure and prey types and sizes present in centrarchid-*Esox* lakes may be well suited to the sunfishes, and resulting competition may lead to the exclusion of the *Umbra*-cyprinid specialists. However, many of these hypothesized cyprinid-centrarchid competitors successfully coexist in other types of lakes (Werner et al. 1977, Keast 1978, Hall et al. 1979).

The predation hypothesis appears to offer the simplest explanation as to why the *Umbra*-cyprinid species were rarely found in the centrarchid-*Esox* lakes. Thus, we feel that combinations of physical and biological disturbances and refuges from these disturbance agents are the major factors responsible for the ecological maintenance of these two assemblage types. Specifically, disturbances come in the forms of low oxygen levels during winter, and predation. Refuges are provided by connections to well oxygenated streams or by the absence of predators due to low winter oxygen conditions. In lakes with predators, heavy densities of macrophytes may provide limited refuges from predation, allowing small populations of some *Umbra*-cyprinid species to persist. Because of these combinations of disturbances and refuges, we agree with Woodin (1978) that communities can be viewed as "compilations of species successfully exploiting refuges in space and/or time."

Species richness in the two assemblage types

Productivity and habitat complexity.—Summer species richness in both assemblage types was highly correlated with measurements related to habitat complexity, particularly vegetation diversity (Table 5). Summer richness in centrarchid-*Esox* assemblages, but not in the *Umbra*-cyprinid lakes, was significantly related to factors related to productivity (pH, alkalinity, conductivity, total dissolved solids; Table 5).

Habitat complexity has often been implicated as an important determinant of species richness in aquatic habitats. Werner et al. (1978), in a comparison of centrarchid lakes with similar structures from two different regions (Michigan and Florida), suggested the assemblages were "saturated," and that habitat structure and morphometry strongly influenced the numbers of fish species that could coexist. Keast (1978) came to similar conclusions about many of the smaller, glacier-formed lakes in North America. Niche segregation and complementarity have been observed in centrarchid-dominated assemblages (Werner et al. 1977, 1978, Keast et al. 1978). Since most species' niches proved distinct with regard to one or more habitat factors, the number of coexisting fish species should increase with increased habitat complexity and heterogeneity. Species diversity in several stream fish assemblages is also closely related to habitat complexity (Sheldon 1968, Tramer and Rogers 1973, Gorman and Karr 1978).

If the habitat complexity: niche complementarity: species richness relationship applies in our assemblages, vegetation diversity should be identified as a major factor related to species richness. Because we limited the range of substrates and depths by our selection of lakes, ranges of diversity were relatively small for these habitat factors. Thus, if habitat complexity contributed to species richness, vegetation represents the primary habitat dimension along which niche segregation and species packing could be demonstrated in our study. Also, we noted previously that vegetation may provide refuges from predation in centrarchid-*Esox* lakes and contribute to higher species richness.

More productive habitats should allow for greater dietary specialization under conditions of evolutionary equilibrium (Mac Arthur 1972). Certain resources in productive habitats may be able to support a species when they would be unable to do so in unproductive habitats (Mac Arthur 1965). Productivity, particularly associated with increased benthic and planktonic food levels, has been related to fish species diversity elsewhere (Nakashima et al. 1977). The reason(s) why productivity-related factors appear important in centrarchid-*Esox* assemblages but not in *Umbra*-cyprinid assemblages is not known, but might be related to a dichotomy between equilibrium and nonequilibrium assemblage types. This idea will be discussed below.

Environmental disturbance.—In *Umbra*-cyprinid lakes higher species richness in summer was associated with lower levels of dissolved oxygen in winter. This may imply a disturbance-related mechanism. The lowest oxygen levels (0.07 mg/L in Gateway, 0.52 mg/L in Grassy, 0.56 mg/L in Mystery) are usually considered "severe," capable of killing many species (Moore 1942, Cooper and Washburn 1946). Yet these lakes were the richest of the *Umbra*-cyprinid lakes.

Environmental disturbance, including severity, instability, and unpredictability, has been associated with both increased and decreased diversity in many aquatic and terrestrial systems. Kushlan (1976), Mahon and Balon (1977) and Horwitz (1978) all found lower fish species diversity in unstable environments. Werner et al. (1978) felt that some of the differences in fish assemblages between their lakes derived from fluctuations in water level. Gorman and Karr (1978) found that their significantly positive relationships between habitat diversity and fish species diversity broke down in stream environments that were unstable due to flooding or human activities.

At "intermediate" levels of frequency or intensity, environmental disturbance can promote species richness (Connell 1978, Huston 1979). An excellent example is the work on a marine epifaunal community by Osman (1977). Diversity was highest on intermediate-sized rocks because of their "optimal" frequency of disturbance. Thomson and Lehner (1976) indicated that environmental instability may have favored increased diversity in an intertidal fish assemblage. In spite of the growing theoretical discussions and field evidence from a variety of communities, we are unaware of any studies that demonstrate, or even implicate, environmental disturbance as a major mechanism promoting species richness in freshwater fish assemblages. This cannot be due to the absence of these conditions. Disequilibria actually or potentially occur due, for example, to floods and droughts (Starrett 1951, Larimore et al. 1959, Kushlan 1976, Gorman and Karr 1978, Harrell 1978, Horwitz 1978), low winter oxygen levels in "winterkill" lakes (Greenbank 1945, Cooper and Washburn 1946, Schneberger 1970), seasonal fluctuations in the abundance and distribution of critical resources in temperate lakes (Hall and Werner 1977, Keast 1978) or the effects of human activities (e.g., Gorman and Karr 1978).

Low winter oxygen levels might increase species richness simply by acting as a rarefying agent, reducing population levels below saturation so that species coexistence is possible at less intense competition. This would particularly be likely if the most susceptible species to winterkill are the dominant predators or competitors of the assemblage (W. M. Tonn, *personal observation*). Beckman (1948) and Bennett (1948) found that growth increased in fish surviving population reductions caused by a severe winterkill and attributed this to increased food per fish. Apparently, competition for food had been reduced.

In winter, richness in centrarchid-*Esox* assemblages was higher in lakes with higher winter oxygen levels (Table 5). Thus, richer winter assemblages of this type occurred in the more "benign" environments in terms of oxygen concentrations, the opposite of that found in *Umbra*-cyprinid assemblages. Winter oxygen levels were as low in *Umbra*-cyprinid lakes as in the depauperate centrarchid-*Esox* lakes, though they were apparently not as "severe" to the more tolerant *Umbra*-cyprinid species, and thus did not supress winter richness. This also implicates an "intermediate" disturbance mechanism operating to increase richness in *Umbra*-cyprinid assemblages.

Lake environments with low oxygen levels may be "severe" for the centrarchid-*Esox* species, reducing winter richness in the lakes themselves. However, the availability of stream/lake refuges may effectively eliminate any significant "disturbance" to the populations, so that when the fish return to the lake in the spring, the summer species richness of the lake returns to its relatively high level. Thus, although the seasonal richness patterns in the lakes themselves were measured to be unstable, the equilibria of the populations might be maintained.

Insular factors.—The second factor included in the multiple regression analysis of richness in *Umbra*-cyprinid assemblages was surface area (Table 5). Barbour and Brown (1974) were the first to look at fish species richness in lakes as a problem of island biogeography. Their analysis of species-area curves, primarily in large lakes, yielded slopes of the log-log regressions that tended to be lower for lake fishes than for plants and animals on oceanic islands. They attributed this to either the relative homogeneity of lake environments as compared to isolated terrestrial habitats, and/or to historical events that may tend to prevent large lakes from acquiring as many species as they can support ecologically.

The summer species-area slope for the *Umbra*-cyprinid assemblages was 0.62, much higher than the range discussed by Barbour and Brown (1974) for assemblages with greater than equilibral numbers of species. For centrarchid-*Esox* assemblages, the slope was 0.29, in the middle of the range noted by Barbour and Brown for lakes whose fish assemblages are in equilibrium between colonization and extinction. While the cause(s) of these different species-area relationships are not known, the values are consistent with the hypothesis that centrarchid-*Esox* assemblages are in ecological equilibrium, while disturbance-induced disequilibrium characterizes *Umbra*-cyprinid assemblages.

Summer species richness in the centrarchid-*Esox* assemblages was also significantly related to the biogeographically important factor of insular connectedness (Table 5). A lake which has greater insular connectedness should have an increased immigration rate and a richer assemblage at equilibrium (Mac Arthur and Wilson 1967, Magnuson 1976).

Summary and Conclusions

We summarize the structural characteristics of the fish assemblages in our small lakes by a list of assembly patterns. Some were directly supported by our results (as indicated by a "D" following the pattern). Others received only partial or indirect support (as indicated by an "I") and require further investigation for direct confirmation, modification, or refutation.

1) Large piscivorous fishes are absent from lakes with low concentrations of dissolved oxygen in winter and no stream refuges from these conditions (D).
2) In lakes with high concentrations of dissolved oxygen in winter, or with refuges from low oxygen conditions provided by streams or connecting lakes, large piscivores are present in the summer (D).
3) Small species tolerant to low oxygen, such as the mudminnow and several cyprinids, form important year-round components of the fish assemblages in lakes without piscivores (D).
4) In those lakes containing large piscivores, the remaining fishes are dominated by medium-sized, spiney-rayed species such as centrarchids, bullheads and yellow perch. The small, soft-rayed species of the piscivore-free assemblages are either rare or absent in lakes with large piscivores, just as centrarchids, along with piscivores, are absent from low winter oxygen lakes (D).
5) In small seepage lakes with high winter oxygen levels, the top predator tends to be largemouth bass. The fish assemblages are similar in summer and winter (D).
6) Conversely, in larger drainage lakes with low winter oxygen levels, the top predator tends to be northern pike. The fish assemblages are seasonally unstable, being much reduced in richness during

winter. Most populations apparently emigrate from these drainage lakes during winter, obtaining refuge from the low oxygen conditions (1).

7) Winter oxygen level is an important determinant of species richness in "*Umbra*-cyprinid" lakes in both summer and winter. The relationship is negative and suggests a disturbance-related mechanism operating to increase richness. The species-area slope for these assemblages also is consistent with the hypothesis that disturbance-induced rarefaction is maintaining greater than equilibrial numbers of species in the richer of these assemblages (1).

8) Productivity, habitat complexity, and lake connectedness are significantly related to summer species richness in "centrarchid-*Esox*" lakes. These are basic components of equilibrium theories of diversity. The species-area slope of these assemblages is in the range of lakes whose fish assemblages are hypothesized to be at equilibrium (1).

The identification of these assembly patterns describes what we feel are ecologically striking fish assemblage structures which appear to result from deterministic mechanisms of assemblage maintenance. Only now that these patterns have been described can meaningful, specific hypotheses be tested by intensive autecological or experimental studies.

Acknowledgments

This research is based on a thesis (W. M. Tonn) submitted to the Department of Zoology, University of Wisconsin–Madison in partial fulfillment of the requirements for an M.S. degree. We are grateful to the many friends and colleagues who helped us out with field work at Trout Lake: J. Capelli, J. Elias, B. Horns, B. Javenkoski, J. Lorman, S. Lozano, P. Medvick, P. Rasmussen, and D. Rondorf. Special thanks go to R. Evans, L. Kitchel, S. Lester, D. Riege, and K. Webster for their many-faceted assistance. Technical and administrative assistance were generously provided by G. Chipman and D. Egger. A. Forbes helped with the ordination program. C. Hughes prepared the figures.

We would like to express our appreciation to Professor J. A. Jones of Macalester College for kindly loaning us a copy of his dissertation. He also provided early encouragement and enthusiasm and introduced the senior author to the special qualities of the mudminnow. The manuscript benefitted from the comments and criticisms of E. Beals, G. Capelli, S. Dodson, L. Fraser, T. Frost, J. Kushlan, E. Werner, and an anonymous reviewer. Special recognition and thanks goes to W. Haag for his care, comments, and criticisms concerning all phases of work and life at Trout Lake, and to C. Paszkowski for her continual help in the collection, analysis, and interpretation of the data and overall support of the project.

Financial support from a Wisconsin Alumni Research Foundation graduate fellowship and a Zoology Department Teaching Assistantship to W. M. Tonn, and grants from the Brittingham Foundation, Wisconsin Department of Natural Resources (Dingell-Johnson Project F-83-R) and National Science Foundation (DEB 7912337) to J. J. Magnuson is gratefully acknowledged.

Literature Cited

Allen, J., and J. Learn. 1973. REGAN 3: stepwise linear regression analysis. Academic Computing Center. University of Wisconsin–Madison, Madison, Wisconsin, USA.

American Public Health Association. 1976. Standard methods for the examination of water and wastewater. 14th edition. American Public Health Association, Washington, D.C., USA.

Barbour, C. D., and J. H. Brown. 1974. Fish species diversity in lakes. American Naturalist **108**:473–489.

Becker, G. C., and T. R. Johnson. 1970. Illustrated key to the minnows of Wisconsin. Department of Biology, University of Wisconsin–Stevens Point, Stevens Point, Wisconsin, USA.

Beckman, W. C. 1948. Changes in growth rates of fishes following reduction in population densities by winterkill. Transactions of the American Fisheries Society **78**:82–90.

Bennett, G. W. 1948. Winterkill of fishes in an Illinois lake. Lake Management Reports, Biological Note Number 19, Illinois State Natural History Survey, Urbana, Illinois, USA.

Black, J. J., L. M. Andrews, and C. W. Threinen. 1963. Surface water resources of Vilas County. Wisconsin Department of Natural Resources, Madison, Wisconsin, USA.

Bray, J. R., and J. T. Curtis. 1957. An ordination of the upland forest communities of southern Wisconsin. Ecological Monographs **27**:325–349.

Browne, R. A. 1981. Lakes as islands: biogeographic distribution, turnover rates, and species composition in the lakes of central New York. Journal of Biogeography **8**:75–83.

Cody, M. L. 1974. Competition and the structure of bird communities. Princeton University Press, Princeton, New Jersey, USA.

Connell, J. H. 1978. Diversity in tropical rain forests and coral reefs. Science **199**:1302–1310.

Conner, E. F., and E. D. McCoy. 1979. The statistics and biology of the species-area relationship. American Naturalist **113**:791–833.

Cooper, G. P., and G. N. Washburn. 1946. Relation of dissolved oxygen to winter mortality of fish in Michigan lakes. Transactions of the American Fisheries Society **76**:23–33.

Coull, B. C., and J. W. Fleeger. 1977. Long-term temporal variation and community dynamics of meiobenthic copepods. Ecology **58**:1136–1143.

Dayton, P. K. 1971. Competition, disturbance, and community organization: the provision and subsequent utilization of space in a rocky intertidal community. Ecological Monographs **41**:351–389.

Diamond, J. M. 1978. Niche shifts and the rediscovery of interspecific competition. American Scientist **66**:322–331.

Dixon, W. J., and M. B. Brown. 1979. BMDP Biomedical Computer Programs. University of California Press, Berkeley, California, USA.

Dodson, S. I. 1970. Complementary feeding niches sustained by size-selective predation. Limnology and Oceanography **15**:131–137.

Eddy, S., and J. C. Underhill. 1974. Northern fishes. Third edition. University of Minnesota Press, Minneapolis, Minnesota, USA.

Gascon, D., and W. C. Leggett. 1977. Distribution, abundance, and resource utilization of littoral zone fish in Lake Memphremagog. Journal of the Fisheries Research Board of Canada **34**:1105–1117.

Gee, J. H., R. F. Tallman, and H. J. Smart. 1978. Reactions of some great plains fishes to progressive hypoxia. Canadian Journal of Zoology **56**:1962–1966.

Gorman, O. T., and J. R. Karr. 1978. Habitat structure and stream fish communities. Ecology **59**:507–515.

Green, R. H., and G. L. Vascotto. 1978. A method for the analysis of environmental factors controlling patterns of species composition in aquatic communities. Water Research **12**:583–590.

Greenbank, J. T. 1945. Limnological conditions in ice-cov-

ered lakes, especially as related to winter-kill of fish. Ecological Monographs 15:343–392.
Hall, D. J., and E. E. Werner. 1977. Seasonal distribution and abundance of fishes in the littoral zone of a Michigan lake. Transactions of the American Fisheries Society 106:545–555.
Hall, D. J., E. E. Werner, J. F. Giliam, G. G. Mittelbach, D. Howard, C. G. Doner, J. A. Dickermann, and A. J. Stewart. 1979. Diel foraging behavior and prey selection in the golden shiner (*Notemigonus crysoleucas*). Journal of the Fisheries Research Board of Canada 36:1029–1039.
Harrell, H. L. 1978. Response of the Devil's River (Texas) fish community to flooding. Copeia 1978:60–68.
Hoogland, R., D. Morris, and N. Tinbergen. 1957. The spines of sticklebacks (*Gasterosteus* and *Pygosteus*) as a means of defense against predators (*Perca* and *Esox*). Behaviour 10:205–236.
Horwitz, R. J. 1978. Temporal variability patterns and the distributional patterns of stream fishes. Ecological Monographs 48:307–321.
Huston, M. 1979. A general hypothesis of species diversity. American Naturalist 113:81–101.
Johnson, M. C., and J. B. Moyle. 1969. Management of a large shallow winter-kill lake in Minnesota for the production of pike. Transactions of the American Fisheries Society 98:691–697.
Jones, J. A. 1973. The ecology of the mudminnow, *Umbra limi*, in Fish Lake (Anoka County, Minnesota). Dissertation. Iowa State University, Ames, Iowa, USA.
Juday, C., and E. A. Birge. 1930. The highland lake district of northeastern Wisconsin and the Trout Lake Limnological Laboratory. Transactions of the Wisconsin Academy of Science, Arts and Letters 25:337–352.
Keast, A. 1970. Food specializations and bioenergetic interrelations in the fish faunas of some small Ontario waterways. Pages 377–411 *in* J. H. Steele, editor. Marine food chains. Oliver and Boyd, Edinburgh, Scotland.
———. 1978. Trophic and spatial interrelationships in the fish species of an Ontario temperate lake. Environmental Biology of Fishes 3:7–31.
Keast, A., J. Harker, and D. Turnbull. 1978. Nearshore fish habitat utilization and species associations in Lake Opinicon, Ontario. Environmental Biology of Fishes 3:173–184.
Keast, A., and D. Webb. 1966. Mouth and body form relative to feeding ecology in the fish fauna of a small lake, Lake Opinicon, Ontario. Journal of the Fisheries Research Board of Canada 23:1845–1874.
Klinger, S. A., J. J. Magnuson, and G. W. Gallepp. 1982, *in press*. Survival mechanisms of the central mudminnow (*Umbra limi*), fathead minnow (*Pimephales promelas*) and brook stickleback (*Culaea inconstans*) for low oxygen in winter. Environmental Biology of Fishes 7.
Kushlan, J. A. 1976. Environmental stability and fish community diversity. Ecology 57:821–825.
Larimore, R. W., W. F. Childers, and C. Heckrote. 1959. Destruction and reestablishment of stream fish and invertebrates affected by drought. Transactions of the American Fisheries Society 88:261–285.
Lewis, W. M., G. E. Gunning, E. Lyles, and W. L. Bridges. 1961. Food choice of largemouth bass as a function of availability and vulnerability of food items. Transactions of the American Fisheries Society 90:277–280.
Mac Arthur, R. H. 1965. Patterns of species diversity. Biological Reviews 40:510–533.
———. 1972. Geographical ecology. Harper and Row, New York, New York, USA.
Mac Arthur, R. H., and E. O. Wilson. 1967. The theory of island biogeography. Princeton University Press, Princeton, New Jersey, USA.
Magnuson, J. J. 1976. Managing with exotics—a game of chance. Transactions of the American Fisheries Society 105:1–9.
Magnuson, J. J., and W. E. Stuntz. 1970. A siphon water sampler for use through the ice. Limnology and Oceanography 15:156–158.
Mahon, R., and E. K. Balon. 1977. Fish community structure in lakeshore lagoons on Long Point, Lake Erie, Canada. Environmental Biology of Fishes 2:71–82.
Menge, B. A., and J. P. Sutherland. 1976. Species diversity gradients: synthesis of the roles of predation, competition and temporal heterogeneity. American Naturalist 110:351–359.
Mills, K. H. 1972. Distribution of fishes under the ice in relation to dissolved oxygen, temperature, and free, dissolved carbon dioxide in Mystery Lake, Wisconsin. Thesis. University of Wisconsin–Madison, Madison, Wisconsin, USA.
Moore, W. G. 1942. Field studies on the oxygen requirements of certain freshwater fishes. Ecology 23:319–329.
Moyle, J. B., and W. D. Clothier. 1959. Effects of management and winter oxygen levels on the fish population of a prairie lake. Transactions of the American Fisheries Society 88:178–185.
Nakashima, B. S., D. Gascon, and W. C. Leggett. 1977. Species diversity of littoral zone fishes along a phosphorus-production gradient in Lake Memphremagog, Quebec-Vermont. Journal of the Fisheries Research Board of Canada 34:167–170.
Osman, R. W. 1977. The establishment and development of a marine epifaunal community. Ecological Monographs 47:37–63.
———. 1978. The influence of seasonality and stability on the species equilibrium. Ecology 59:383–399.
Petrosky, B. R., and J. J. Magnuson. 1973. Behavioral responses of northern pike, yellow perch, and bluegill to oxygen concentrations under simulated winterkill conditions. Copeia 1973:124–133.
Pianka, E. R. 1975. Niche relations of desert lizards. Pages 292–314 *in* M. L. Cody and J. Diamond, editors. Ecology and evolution of communities. Belknap Press, Cambridge, Massachusetts, USA.
———. 1978. Evolutionary ecology. Second Edition. Harper and Row, New York, New York, USA.
Pielou, E. C. 1977. Mathematical ecology. John Wiley and Sons, New York, New York, USA.
Post, W., E. Beals, and T. Allen. 1973. BCORD documentation. Departments of Botany and Zoology, University of Wisconsin–Madison, Madison, Wisconsin, USA.
Roughgarden, J. 1974. Species packing and the competition function with illustrations from coral reef fish. Theoretical Population Biology 5:163–186.
Schlater, J., and J. Learn. 1974. DISCRIM1: discriminant analysis reference manual. Academic Computing Center, University of Wisconsin–Madison, Madison, Wisconsin, USA.
Schneberger, E. 1970. A symposium on the management of midwestern winterkill lakes. Special Publication, North Central Division, American Fisheries Society, FAS Publishing, Madison, Wisconsin, USA.
Shannon, C. E., and W. Weaver. 1949. The mathematical theory of communication. University of Illinois Press, Urbana, Illinois, USA.
Sheldon, A. L. 1968. Species diversity and longitudinal succession in stream fishes. Ecology 49:193–198.
Siegel, S. 1956. Nonparametric statistics for the behavioral sciences. McGraw-Hill, New York, New York, USA.
Simberloff, D., and E. F. Connor. 1981. Missing species combinations. American Naturalist 118:215–239.
Slobodkin, L. B., and H. L. Sanders. 1969. On the contribution of environmental predictability to species diversity. Brookhaven Symposia in Biology 22:82–95.

Starrett, W. C. 1951. Some factors affecting the abundance of minnows in the Des Moines River, Iowa. Ecology 32:13–27.

Thomson, D. A., and C. E. Lehner. 1976. Resilience of a rocky intertidal fish community in a physically unstable environment. Journal of Experimental Marine Biology and Ecology 22:1–29.

Tonn, W. M. 1980. Patterns in the assembly and diversity of fish communities in northern Wisconsin lakes. Thesis. University of Wisconsin–Madison, Madison, Wisconsin, USA.

Tramer, E. J., and P. M. Rogers. 1973. Diversity and longitudinal zonation in fish populations of two streams entering a metropolitan area. American Midland Naturalist 90:366–375.

Werner, E. E. 1977. Species packing and niche complementarity in three sunfishes. American Naturalist 111:553–578.

———. 1979. Niche partitioning by food size in fish communities. Pages 311–322 in H. Clepper, editor. Predator-prey systems in fisheries management. Sport Fishing Institute, Washington, D.C., USA.

Werner, E. E., and D. J. Hall. 1976. Niche shifts in sunfishes: experimental evidence and significance. Science 191:404–406.

Werner, E. E., and D. J. Hall. 1977. Competition and habitat shift in two sunfishes (Centrarchidae). Ecology 58:869–876.

Werner, E. E., and D. J. Hall. 1979. Foraging efficiency and habitat switching in competing sunfishes. Ecology 60:256–264.

Werner, E. E., D. J. Hall, D. R. Laughlin, D. J. Wagner, L. A. Wilsmann, and F. C. Funk. 1977. Habitat partitioning in a freshwater fish community. Journal of the Fisheries Research Board of Canada 34:360–370.

Werner, E. E., D. J. Hall, and M. D. Werner. 1978. Littoral zone fish communities of two Florida lakes and a comparison with Michigan lakes. Environmental Biology of Fishes 3:163–172.

Whittaker, R. H. 1973. Direct gradient analysis: techniques. Pages 7–32 in R. H. Whittaker, editor. Ordination and classification of communities. Volume 5. Handbook of vegetation science. Dr. W. Junk, The Hague, The Netherlands.

Wiens, J. A. 1977. On competition and variable environments. American Scientist 65:590–597.

Williams, W. T., and J. M. Lambert. 1959. Multivariate methods in plant ecology. I. Association analysis in plant communities. Journal of Ecology 47:83–101.

Wishart, D. 1975. CLUSTAN IC user's manual. University College, London, England.

Woodin, S. A. 1978. Refuges, disturbance, and community structure: a marine soft-bottom example. Ecology 59:274–284.

Zaret, T. M., and R. T. Paine. 1973. Species introduction in a tropical lake. Science 182:449–455.

Appendix I

Correlation coefficients (r) among the 15 environmental variables for the 18 study lakes. Statistically critical value is 0.47 for $P = .05$.

	Log lake area	Log maximum depth	Log (watershed + 1)	Log alkalinity	Log conductivity	pH	Log total dissolved solids	Log (winter oxygen + 1)	Substrate	Vegetation	Depth	Depth and substrate	Depth and vegetation	Substrate and vegetation	Depth, substrate and vegetation
									\multicolumn{7}{c|}{Diversity}							
Log lake area	1.00	−.46	.77	.65	.61	.69	.45	−.29	.44	.66	−.53	.28	.36	.66	.59
Log maximum depth		1.00	−.35	−.62	−.70	−.54	−.74	.77	.06	−.73	.33	.17	−.62	−.51	−.64
Log (connectedness + 1)			1.00	.60	.53	.58	.36	−.19	.20	.49	−.40	.08	.25	.58	.37
Log alkalinity				1.00	.84	.88	.80	−.61	.02	.58	−.18	.13	.50	.50	.59
Log conductivity					1.00	.85	.87	−.81	.18	.66	−.07	.32	.65	.70	.76
pH						1.00	.69	−.56	.18	.52	−.23	.34	.39	.60	.62
Log total dissolved solids							1.00	−.81	.00	.62	−.13	.05	.52	.50	.54
Log (winter oxygen + 1)								1.00	−.13	−.62	−.02	−.19	−.70	−.65	−.75
Substrate diversity									1.00	.01	−.31	.74	−.16	.49	.32
Vegetation diversity										1.00	−.27	−.07	.80	.74	.77
Depth diversity											1.00	.23	.24	−.32	−.03
Depth and substrate diversity												1.00	.07	.36	.44
Depth and vegetation diversity													1.00	.56	.84
Substrate and vegetation diversity														1.00	.82
Depth, substrate, and vegetation diversity															1.00

Appendix II

Winter and summer assemblages, and assemblage type designation, for the 18 study lakes, based on the 1978 fish sampling. W = present in the winter sampling only; S = present in summer only; * = present in both seasons; C-E = centrarchid-*Esox* assemblage type; U-C = *Umbra*-cyprinid assemblage type.

Lake	Central mudminnow (*Umbra limi*)	Northern pike (*Esox lucius*)	Pearl dace (*Semotilus margarita*)	Redbelly dace (*Chrosomus eos*)	Finescale dace (*C. neogaeus*)	Golden shiner (*Notemigonus crysoleucas*)	Bluntnose minnow (*Pimephales notatus*)	Fathead minnow (*P. promelas*)	Common shiner (*Notropis cronutus*)	Blacknose shiner (*N. heterolepis*)	Redhorse (*Moxostoma* sp.)
1. Apeekwa	W	S									S
2. Aurora	S	S				S				S	
3. Blueberry											
4. Camp 2	*		*			*					
5. Gateway	*			S	*	S		*			
6. Grassy	*		*		S	*		*		*	
7. Johnson	W	S				S					
8. Landing		S						*	S		
9. Little Rice		S									
10. Maple		S									
11. Mill	W	S				S	S				
12. Mystery	*			*	*					*	
13. Nixon		S									
14. Spruce											
15. Whitney		S				W					
16. Whynot	W										
17. 33-6	*										
18. 33-13	*										

Appendix II
Continued.

Lake	White sucker (*Catostomus commersoni*)	Black bullhead (*Ictalurus melas*)	Yellow bullhead (*I. natalis*)	Brook stickleback (*Culaea inconstans*)	Largemouth bass (*Micropterus salmoides*)	Rock bass (*Ambloplites rupestris*)	Bluegill (*Lepomis macrochirus*)	Pumpkinseed (*L. gibbosus*)	Black crappie (*Pomoxis nigromaculatus*)	Yellow perch (*Perca flavescens*)	Iowa darter (*Etheostoma exile*)	Mottled sculpin (*Cottus bairdi*)	Assemblage type
1. Apeekwa	S	*				S	*	S		*	S		C-E
2. Aurora		S	S				*			*	S		C-E
3. Blueberry			*		S			*		*			U-C
4. Camp 2				*									U-C
5. Gateway	S			S						*	S		U-C
6. Grassy	S	*			S	S	*	S	S	S		W	C-E
7. Johnson	S	*					*			*			C-E
8. Landing	S		S			S				S			C-E
9. Little Rice	S	S					S	S		*			C-E
10. Maple	S	S							S		S		C-E
11. Mill	W	*								S		*	U-C
12. Mystery	S	S	S			S	S		S	S			C-E
13. Nixon	*				*					*			C-E
14. Spruce													C-E
15. Whitney	S			W	S		S			S			C-E
16. Whynot										*			C-E
17. 33-6													U-C
18. 33-13													U-C

Cultivation/depensation effects on juvenile survival and recruitment: implications for the theory of fishing[1]

Carl Walters and James F. Kitchell

Abstract: Large, dominant fish species that are the basis of many fisheries may be naturally so successful due partly to "cultivation effects," where adults crop down forage species that are potential competitors/predators of their own juveniles. Such effects imply a converse impact when adult abundance is severely reduced by fishing: increases in forage species may then cause lagged, apparently depensatory decreases in juvenile survival. Depensatory effects can then delay or prevent stock rebuilding. Cultivation effects are apparently common in freshwater communities and may also explain low recruitment success following severe declines of some major marine stocks such as Newfoundland Atlantic cod (*Gadus morhua*). Risk of depensatory effects should be a major target of recruitment research, and management policies should aim for considerably higher spawning abundances than has previously been assumed necessary based on recruitment data collected during adult stock declines associated with fishery development.

Résumé : Le succès des grosses espèces dominantes de poissons qui sont à la base de nombreuses pêches peut être dû en partie à un « effet cultural », les adultes récoltant les espèces fourrage qui sont des concurrents ou des prédateurs potentiels de leurs propres juvéniles. Cet effet a par contre un impact inverse lorsque l'abondance des adultes est fortement réduite par la pêche : les augmentations chez les espèces fourrage peuvent causer des baisses décalées dans le temps, à caractère apparemment dépensatoire, de la survie des juvéniles. Les effets dépensatoires peuvent alors retarder ou empêcher le rétablissement des stocks. L'effet cultural semble courant dans les communautés dulcicoles et peuvent aussi expliquer le faible succès de recrutement qui suit les déclins graves de certains grands stocks marins comme la morue franche (*Gadus morhua*) de Terre-Neuve. Le risque d'effets dépensatoires devrait être un thème majeur de la recherche sur le recrutement, et les politiques de gestion devraient fixer pour objectifs des abondances de géniteurs considérablement plus élevées que ce qu'on jugeait jusqu'ici nécessaire en se fondant sur les données de recrutement recueillies pendant les déclins de stocks d'adultes associés au développement des pêches.

[Traduit par la Rédaction]

Introduction

Single-species stock assessments and harvest policy development generally assume either that recruitment is independent of stock size or that recruitment varies around some compensatory relationship described by a simple saturating or dome-shaped curve. Compensatory effects are assumed to arise through reductions in intraspecific competition and (or) cannibalism when abundance is reduced by fishing. There is now broad empirical support for limits to compensatory response, so that low parental stock sizes can indeed result in lower mean recruitment (Myers and Barrowman 1996; Myers et al. 1999), so most assessments include a recruitment relationship that at least recognizes some risk of recruitment overfishing. But there has been little empirical support for the existence of "depensatory" or "recruitment failure" effects where recruitment decline with stock size is even more rapid than expected from a decrease in egg production combined with high juvenile survival (Myers et al. 1995*a*, 1995*b*; Liermann and Hilborn 1997). However, such effects are likely to be difficult to detect in typical data sets that have few observations at very low stock size (Shelton and Healey 1999). Depensatory effects are not routinely incorporated in assessments except via risk management tactics such as setting arbitrary minimum population size goals. The approach of assuming "stationary" mean stock–recruitment relationships has been criticized on grounds that it does not account for effects of either persistent environmental change or changes in trophic relationships (juvenile predation risk, food) that might accompany overfishing (e.g., see Walters 1987; Walters and Korman 1999; Hall 1999), but criticism has focussed mainly on our inability to forecast short-term recruitment changes. In particular, we have paid little attention to the risk of very severe nonstationarity, in the form of persistent depensatory effects (low juvenile survival) that develop with some time lag following periods of adult stock depletion.

Using Ecosim II (Walters et al. 1997, 2000), we have been conducting exploratory simulations to detect possible depensatory recruitment effects due to trophic interactions.

Received December 14, 1999. Accepted April 19, 2000. Published on the NRC Research Press web site on November 10, 2000.
J15485

C. Walters.[2] Fisheries Centre, University of British Columbia, Vancouver, BC V6T 1Z4, Canada.
J.F. Kitchell. Center for Limnology, University of Wisconsin, Madison, WI 53706, U.S.A.

[1]Invited perspective for this 100th Anniversary Issue.
[2]Corresponding author (e-mail: walters@fisheries.ubc.ca).

Fig. 1. We envision recruitment depensation at low stock sizes as arising from a trophic triangle: prey organisms of adult fishes can respond positively to reductions in adult fish abundance by fishing and then causing reductions in juvenile survival via competition and (or) predation interactions with the juveniles.

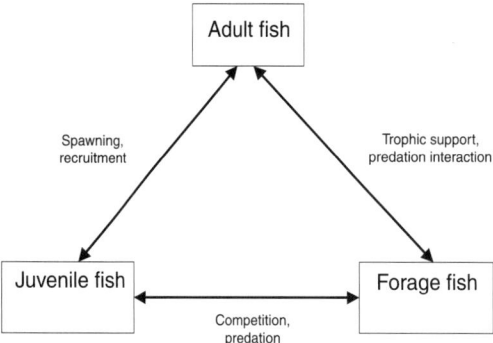

This model combines simple biomass dynamics models for some ecosystem components with age-structured (delay-difference) models for selected species that have strong trophic ontogeny and (or) size-selective fishing impacts. It links recruitment to trophic changes by explicit (and reciprocal, predator–prey) representation of how dynamic changes in food availability and predation risk affect juvenile mortality rates and how juvenile fish may moderate these rates through risk-sensitive changes in foraging times (Walters and Juanes 1993). Most often, the "emergent" stock–recruit relationships predicted by Ecosim II look like the classic Beverton–Holt, broken stick, or Ricker relationships, i.e., we predict mainly strong compensatory (stabilizing) effects due to the usual mechanisms thought to result in increased juvenile survival at low densities (more food, less cannibalism, etc.). But in some cases, we see a catastrophic pattern: as simulated fishing mortality rate is increased over time, recruitment initially appears to be stable or declining along a Beverton–Holt or Ricker relationship with declining spawning stock. But then, juvenile mortality rates "suddenly" increase (over a few simulated years) after some time delay to result in delayed depensatory effects that may result in extinction even if fishing is stopped.

Here, we describe the mechanism that causes models like Ecosim II to predict depensatory recruitment changes that strongly contradict classic compensatory stock–recruitment theory. We propose a "cultivation hypothesis" to suggest why this mechanism could in fact be quite common, especially for large, predatory fish species, and discuss factors that may prevent it from operating in some circumstances. We review case examples where it may have occurred. We conclude that the mechanism is plausible enough, and supported by enough circumstantial case evidence, to warrant immediate policy response in the form of higher spawning stock (lower exploitation rate) goals than would be estimated from single-species population theory.

How juvenile trophic interactions can cause depensatory dynamics

Delayed depensatory effects arise in Ecosim II models (Appendix) through the following sequence of events. As fishing reduces the adult population size of a fish species, the total number of juveniles produced per time decreases. In the absence other trophic effects, this results in increased food density in the localized "foraging arena" habitats (usually near predation refuges; see Walters and Juanes 1993) where juvenile feeding is concentrated. Juveniles respond to increased food density by reducing feeding time and hence time at risk to predation (or total time spent at body sizes small enough to be vulnerable to high predation risk). Juvenile mortality rate then decreases, so net recruitment at first stays nearly constant despite fewer juveniles entering the juvenile life stage per time. But if adult abundance is severely reduced, one or more smaller "forage fish" species are "released" to increase in abundance. Then, one or two negative effects can occur. First, the forage fish may directly (even if incidentally) prey on the juveniles, causing increased predation risk per time spent foraging and hence higher juvenile mortality rate. Second, if the forage and juvenile fish share at least some foods (e.g., zooplankton, benthic invertebrates) and use overlapping foraging arenas and tactics for reducing predation risk, increased forage fish abundance leads to reduced food density and hence to increased juvenile foraging time and general predation risk. A simple way to visualize this dynamic is as a "trophic triangle" (Ursin 1982; Cohen et al. 1993; He et al. 1993; Rudstam et al. 1994), "competitive juvenile bottleneck" (Bystroem et al. 1998), or "predator–prey role reversal" (Barkai and McQuaid 1988). The triangle adds prey/competitor dynamics to the usual juvenile/adult dynamic linkage that has traditionally been emphasized in population dynamics modeling (Fig. 1).

Figure 2 presents a graphical model for the elements of this mechanism, in terms of patterns that should be observable in the field if it is operating: (*i*) negative relationship between forage fish abundance and adult abundance ("spawning stock" biomass), (*ii*) depensatory increase in juvenile foraging time (and (or) reduced juvenile growth rate) as the forage fish become more abundant, and (*iii*) declining juvenile survival rate when adult abundance has been low for long enough for the forage fish increase to occur. Additionally, we should be able to observe (*iv*) diet and habitat use overlap between the juvenile fish and the forage fishes and (or) (*v*) direct evidence of predation by the forage fish on juveniles, in stomach contents sampling. Of these observations, we should not be surprised if direct evidence of predation is not found. Forage fish are likely to be much more abundant than the juveniles (and to have high food consumption rates) and may thus cause a high juvenile mortality rate (eat a large total number of juveniles) even if only a very tiny percentage of their diet is juveniles.

Note that this mechanism for causing decreased reproductive performance at low stock size is quite different from the common concern that fishing too hard on a dominant species may allow competitors to increase and "take over" its niche. We are talking not about competitors in general, but very specifically about other small fish (and some invertebrates like squid) that are likely to be directly impacted in abundance and distribution by adults of the fish species in question. These species may be direct competitors and (or) predators of the juveniles of the species during a life history stage where we know from recruitment experience that the

Fig. 2. Elements of a hypothesis for depensatory recruitment changes at low stock sizes. (*a*) Increase in abundance of small "forage" fishes/invertebrates if predatory stock size decreases; (*b*) increasing rather than decreasing juvenile foraging time when adult abundance is low due to competition with forage fishes; (*c*) decreased juvenile survival rate at low adult population size due to increased foraging time and (or) direct predation by forage species.

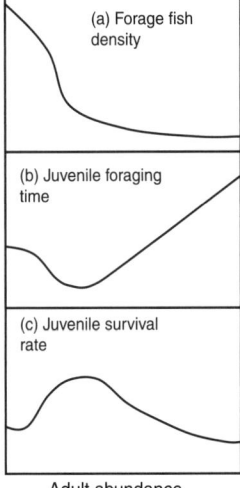

juveniles are likely to be "sensitive" (to have high and strongly density-dependent mortality rates). In part, this is not a new idea or concern: biologists have long speculated about how predatory fish species are able to achieve large body sizes, given that their juveniles must grow through a predation–competition "bottleneck" involving the very species that will be their prey later in life (Crowder et al. 1992; Wooten 1994). It is familiar in management procedures designed to create "balanced" predator–prey interactions such as those for bluegill (*Lepomis macrochirus*) – largemouth bass (*Micropterus salmoides*) systems (Gutreuter and Anderson 1985), as guidance for size at stocking procedures (Madenjian et al. 1992), and to ecologists engaged in evaluating "size-structured" or "trait-mediated" interactions (Persson and Eklov 1995; Werner 1998).

We can of course envision more complex mechanisms by which impacts of adult abundance on trophic structure may modify survival conditions for their juveniles. For example, we found in early Ecosim tests that invasion of Nile perch (*Lates niloticus*) in Lake Victoria was possibly slowed initially by competition/predation from the natural fish community of the lake. Population growth rate then apparently increased as the perch became abundant enough to depress this community and allow increases in an invertebrate (*Caridina*) and a fish (*Rastrinebola*) that later became its dominant foods (Kitchell et al. 1996; Walters et al. 1997). Another example would be the possibility of trophic "quadrangles" in zooplanktivores: if adults feed selectively on larger zooplankters, reduction in adult abundance may allow an increase in abundance of these larger forms with an attendant negative impact on abundance of smaller zooplankters

that are needed by smaller juveniles. A third example is argued to account for the progressively increased juvenile mortality rates observed for stocked lake trout (*Salvelinus namaycush*) in Lake Superior. After sea lamprey (*Petromyzon marinus*) and fisheries mortality were reduced, natural reproduction allowed a gradual increase in adult abundance of a deepwater trout race (siscowet). This created a predator population that imposed increased mortality on the stocked juveniles of the shallow water trout race (lean) and may be responsible for the lack of successful reproduction by the latter (Hansen et al. 1995).

The depensatory mechanism described in Fig. 2 is fundamentally different from models for direct depensatory predation effects based on the form of predator responses to prey densities (Fig. 3) (e.g., Collie and Spencer 1993; Spencer and Collie 1997*a*, 1997*b*) or models based on reproductive failure at low population size. In classical "reaction vat" models of predator–prey interaction, decreasing prey mortality rate with increasing prey density is caused by increased handling time or satiation of predators, such that the proportion of the prey population killed by each predator decreases with increasing prey (juvenile fish) density (Fig. 4). This may occur in a few circumstances where prey are particularly vulnerable to predation and predators can be "overwhelmed", for example, during downstream migrations of salmon fry from small streams (Neave 1954), but it is probably not common. In Ecosim, we assume that predation takes place largely in spatial patches or "foraging arenas" where juveniles are forced to accept predation risk in order to forage and where predation rates are limited not by predator satiation but rather by juvenile movement rates into and out of (or time spent in) behavioral refuges and by predation risk per time spent foraging (see Appendix; also see fig. 1 in Walters and Juanes 1993). We think that this model for spatial organization is a much better description of general natural history experience in aquatic ecology (stomach contents data rarely show predator satiation, juvenile fish distributions obviously dominated by tactics to reduce predation risk) than is the classic reaction vat model. Further, it better explains the ecosystem-scale observation that trophic cascade effects are relatively weak and suggestive of ratio dependence in predator–prey interactions (McCarthy et al. 1995; Scheffer and De Boer 1995; Brett and Goldman 1996).

Why perverse interactions could be common: the cultivation hypothesis

Most fisheries develop at least initially to take the largest, most abundant, ecologically "dominant" fishes. This may be precisely the suite of species most vulnerable to depensatory responses because a reversal of these responses may be why such species are dominant in the first place. That is, ecological dominance may well be due at least partly to "cultivation effects": dominants may be species that are fortuitously capable of being especially good at capturing (and otherwise suppressing) the particular smaller forage fishes that could cause the worst competition/predation effects on their own juveniles. Note that this is not a group or population selection argument about selection favoring adults that consume particular forage species so as to protect their own juveniles.

© 2001 NRC Canada

Fig. 3. Contrasting assumptions and predictions in models that explain depensatory effects by the form of predator functional responses in random search (reaction vat) environments versus models that assume spatial organization of predation interactions in patchy "foraging arenas."

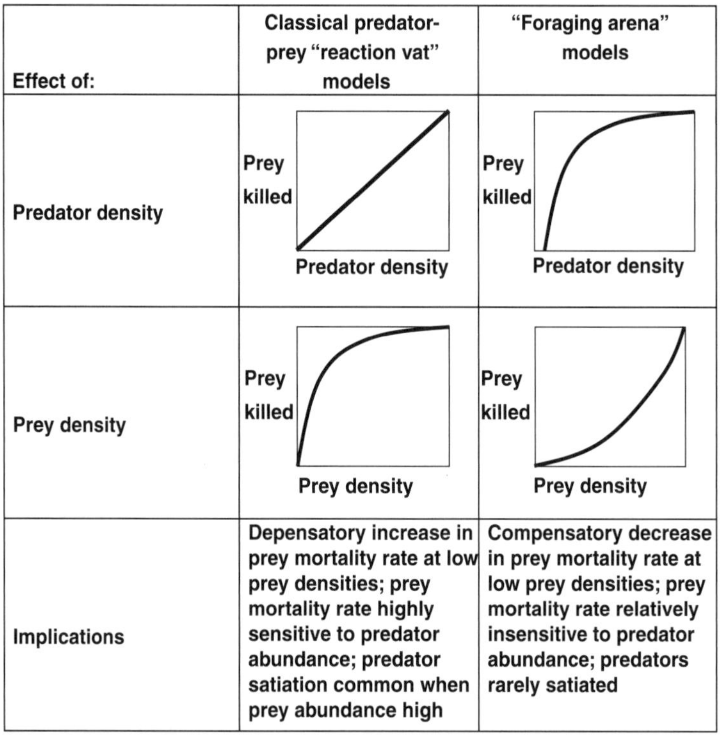

Effect of:	Classical predator-prey "reaction vat" models	"Foraging arena" models
Predator density	Prey killed increases linearly with Predator density	Prey killed saturates with Predator density
Prey density	Prey killed saturates with Prey density	Prey killed increases convexly with Prey density
Implications	Depensatory increase in prey mortality rate at low prey densities; prey mortality rate highly sensitive to predator abundance; predator satiation common when prey abundance high	Compensatory decrease in prey mortality rate at low prey densities; prey mortality rate relatively insensitive to predator abundance; predators rarely satiated

It simply says that if there are several large species in a system, with varied diets as both juveniles and adults, the dominant large species should end up being that one that happens to cultivate the best survival conditions for its juveniles by having particularly large impacts on its juveniles' competitors/predators.

We usually think of dominant fishes as those capable of best using trophic (food) production and physical habitat and of being long-lived enough to accumulate large unfished population sizes. But when we make this assumption, we ignore the large body of evidence that abundance is generally "limited," not at the adult stage but rather at the juvenile stage (recruitment most often observed to be independent of or flat across a wide range of adult abundance). Dominance may well require adult feeding patterns that efficiently use production by lower trophic levels, but it certainly also requires relatively good conditions for juvenile survival and growth. The cultivation hypothesis is that dominance is a result of not only being able to acquire trophic resources but also to insure the best possible trophic conditions for juveniles.

Factors that mitigate against cultivation/depensation effects

An obvious and immediate objection to the cultivation/depensation arguments presented above is that they offer no mechanism by which large, dominant species could become abundant enough to have cultivation effects in the first place. Why is the world not dominated by small forage species that successfully prevent larger predatory species from ever becoming abundant through impacts on juvenile survival of the predatory species? If depensatory effects are common, they must not be so strong as to entirely prevent large predatory species from invading ecosystems, at least when there is no fishing.

At least four factors likely mitigate against very strong depensatory effects: (i) niche specialization, (ii) limitation of predation impacts in forage species via risk-sensitive behaviors by the forage species, (iii) diffuse predation impacts that act to prevent strong population responses by forage fishes, and (iv) spatial propagation effects. Niche specialization is an obvious possibility: successful large predatory species may be ones whose juveniles are competent at acquiring particular food resources, relative to forage fish competitors (i.e., competition may not be all that severe). In terms of the graphical model in Fig. 2, such niche specialization would imply a "failure" in Fig. 2b: juvenile foraging time not increasing with increases in abundance of forage species competitors.

The second and third factors involve mitigation of the forage fish numerical response to predator abundance, i.e., a less dramatic response than shown in Fig. 2a. If the forage fishes have severely restricted habitat use/foraging activities

Invited perspectives and article

Fig. 4. Risk-sensitive behaviors by juvenile fish imply a deep reversal of predictions about predation impact. Small increases in the space–time scale of experimentation and modeling can result in a reversal in form of the functional response observed, from a type II response for a reaction vat experiment to type III for an experiment where prey can hide from predators unless prey density is high enough to force the prey to spend more time foraging.

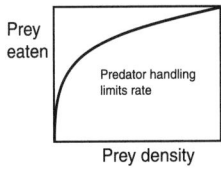

due to predation risk in general (risk-sensitive foraging), they may simply not be limited in abundance in the first place by the predatory species in question. In fisheries terms, their natural mortality rate M and (or) recruitment may not be sensitive to changes in abundance of any particular predator that might be reduced through fishing. Alternatively, any temporary increase in their abundance (due to a decrease in their mortality rate) may be reversed by numerical responses of a variety of other predators (diffuse predation impacts).

In physically large ecosystems with a diversity of habitats, juveniles of large predators may find refuges for persistence in particular sites where predation/competition effects are relatively weak. Such sites may then act as "epicenters" for spatial population expansion, as adults produced from the centers gradually move in enough numbers to other sites so as to generate cultivation effects in these sites. That is, cultivation effects may be critical in the range expansion/contraction dynamics often observed for large fish populations (MacCall 1990).

We likewise would not expect strong cultivation/depensation effects for species that show large-scale ontogenetic habitat shifts (large physical separation between juvenile nursery areas and adult feeding areas), possible mainly in marine ecosystems. In cases like anadromous salmon, it is difficult to see how adults could have much direct effect on competition/predation conditions faced by juveniles (although they may have other indirect effects such as fertilization of rearing areas with carcasses).

It should be noted that large ecosystem size per se does not imply that predation/competition effects should be weaker because predation is more "dilute," as suggested by Verity and Smetacek (1996) to explain differences between freshwater and marine systems. Since interactions can be spatially patchy (highly localized) in both environments, it is irrelevant that average densities of predators are much lower in the ocean environment.

Examples?

The most obvious examples of apparent cultivation/depensation effects have been in freshwater ecosystems. There has long been a concern about how to establish "balance" in centrarchid communities, which have a nasty propensity to shift toward dominance by stunted sunfish populations when basses are heavily exploited (Swingle 1950a, 1950b; Hackney 1979; Olson 1996). In these systems, it is obvious how sunfish forage species impact recruitment of bass via both competition and direct predation. There is a south–north cline toward increasing risk of sunfish dominance to the north, most likely related to the impact of growing season length and the duration of the competition/predation window faced by juvenile basses.

Under heavy exploitation, walleye (*Stizostedion vitrium*) populations in Alberta, Canada, have shown persistent recruitment failure, accompanied by dramatic increases in minnow populations that are thought to prey heavily on walleye larvae (M. Sullivan, Alberta Department of Natural

© 2001 NRC Canada

Fig. 5. Patterns in stock–recruitment data expected under alternative hypotheses. In the regular compensation case, juvenile survival rate increases smoothly as spawning stock size decreases. In the apparent (prior) depensation case, recruitment decline precedes (and causes) the spawning stock decline, giving the appearance of depensation unless enough observations are available for the recovery portion of the recruitment "hook." In the delayed depensation case, recruitment may remain high as the stock declines and then finally collapse.

Regular compensatory case:

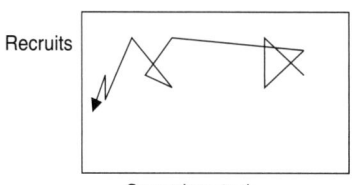

Apparent depensation due to prior factor impacting recruitment (spawning stock decline due to recruitment decline):

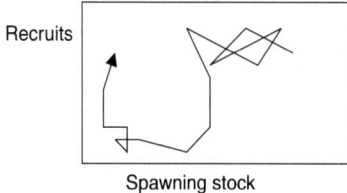

Delayed depensatory effect due to reduced spawning stock size:

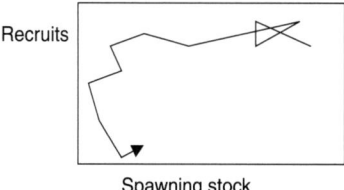

Resources, Edmonton, Atla., personal communication). Similar walleye recruitment failures may have occurred in Wisconsin lakes, and walleye have not become established in some lakes with apparently excellent habitat and forage conditions (D. Beard, Wisconsin Department of Natural Resources, Madison, Wis., personal communication). However, most Wisconsin lakes have maintained strong recruitment despite heavy fishing (e.g., Escanaba; Hansen et al. 1998). The striking difference between these regions supports the possibility mentioned above that "diffuse predation" may prevent delayed depensatory effects: Alberta lakes lack the centrarchids (basses, sunfish) that are prominent in the littoral zones of most Wisconsin Lakes.

The northern (2J3KL) Atlantic cod (*Gadus morhus*) stock off Newfoundland showed declines in apparent juvenile survival rate (recruits per spawning biomass) and also body growth rate during the 1980s, following severe stock reduction during the 1970s (Myers et al. 1996; Anderson and Dalley 1997; Shelton and Healey 1999). That stock has not recovered as expected following fishery closure in 1992, and in particular, there is little evidence of recruitment to the larger, offshore migratory component of the stock. If the cultivation/depensation hypothesis applies in this case, it should be possible to demonstrate substantial increases in some smaller species that are competitors/predators on juvenile cod in nursery areas and were preferred prey of adult cod (note that "preferred" is in the technical sense: a high proportion in the diet compared with the proportion in the environment). One possible candidate species is Arctic cod (*Boreogadus saida*). However, note that for the 10 North Atlantic cod stocks that have undergone severe decline (80% or more) in recent years, 6 have not shown the survival decline predicted by the cultivation/depensation hypothesis (Myers et al. 1996). Bax (1998) suggested that predation on juvenile cod by clupeids could exaggerate fishing effects and help lead to a "planktivore-dominated" ecosystem in the Baltic.

The large number of stock–recruitment data sets assembled by Myers and colleagues (Myers et al. 1995a, 1995b, 1999; www.mscs.dal.ca/~myers/data.html) provides an opportunity to broadly examine the frequency of occurrence of depensatory effects. Because we were concerned about possible transitory and delayed effects that might not be detected by simply fitting stationary stock–recruitment curves to the data (Fig. 5), we had three independent scientists examine the data sets and provide a visual assessment of whether delayed depensation might be present. Of 330 stock–recruitment data sets excluding anadromous salmonid cases, we found (Table 1) that 44–112 could be interpreted as showing some sort of depensatory response, but these represent almost a third of the cases where there are observations at relatively low (20% or less of maximum) spawning stock sizes. Liermann and Hilborn (1997) also suggested that these data sets might contain more examples of depensation than were detected by Myers et al. (1995a, 1995b). However, only a small number (17–45) of these possible depensatory cases show the delayed response expected under the cultivation/depensation hypothesis, characterized by a downward "hook" (Walters 1987) in the stock–recruit time series (Fig. 5). Far more common, especially for clupeoids, are hooks of the reverse shape where recruitment initially declines, then the spawning stock declines, and then recruitment begins to recover; these cases should not be interpreted as evidence for depensation. Fisheries scientists have generally interpreted these cases as "bad luck": poor environmental conditions leading to recruitment failure, and persistence of the poor conditions for at least some time following implementation of measures aimed at protecting spawning stocks. While such cases might be due to trophic effects (e.g., increase in predators leading to recruitment decline and then predator collapse allowing recruitment to recover; likewise for food supply

Table 1. Visual characterization of 330 worldwide stock–recruit data sets assembled by Myers and colleages, exlcuding anadromous salmonid cases.

Taxonomic group	No evidence	No depensation	Possible depensation	
			Prior	Delay
Classification by three scientists of data at www.mscs.dal.ca/~myers/data.html				
Clupeiformes	24–32	11–23	14–25	3–7
Gadiformes	36–52	6–18	6–15	7–14
Perciformes	38–47	6–17	1–10	2–11
Pleuronectiformes	27–30	1–11	1–7	1–3
Salmoniformes	16–19	11–16	3–4	2–5
Miscellaneous	11–22	3–8	2–6	2–5
Total	152–202	38–93	27–67	17–45
Classification by Walters of data in Myers et al. 1995a and 1995b				
Clupeiformes	21	28	3	2
Gadiformes	36	18	4	11
Perciformes	12	5	2	1
Pleuronectiformes	21	7	2	0
Salmoniformes	2	3	1	2
Miscellaneous				
Total	92	61	12	16

Note: Ranges are for three independent scientists. "No evidence" means no observations at low enough spawning stock size (<20% of maximum) to expect depensatory effects, "no depensation" means recruitment relatively high at lowest stock sizes and (or) an upward hook in the recruitment time series, and "possible depensation" means at least a few observations of relatively low recruitment per spawner (juvenile survival rate) at low stock size. In depensation cases, "prior" means that survival decline preceded stock decline, suggesting an agent other than delayed depensation likely responsible for apparent depensation.

dynamics), we would not interpret them as evidence for persistent depensatory effects, delayed or otherwise.

The taxonomic distribution of possible depensation cases in the Myers and colleagues database (Table 1) provides direct support for using foraging arena assumptions in ecosystem models and recruitment analysis (Figs. 3 and 4). Were predator–prey interactions mainly of the mass action or reaction vat functional form, we would expect depensation effects to be most common in taxonomic groups dominated by smaller "forage" species (Clupeiformes fishes) and least common in groups dominated by piscivores (Gadiformes fishes). In fact, we see the opposite: the incidence of depensatory cases is highest in Gadiformes fishes, and the two delayed depensation cases in Salmoniformes fishes are for a large piscivore (northern pike (*Esox lucius*) in the two basins of Lake Windermere, Great Britain. This is just what we expect from foraging arena theory, assuming that there has been strong selection in smaller species for distributional/ behavioral tactics to limit predation risk. Also, we generally predict unrealistically violent predator–prey oscillations in Ecosim models and unrealistically large temporal variation in natural mortality rates of forage species unless we assume such tactics.

It is likely that available stock–recruitment data sets provide an underestimate of the risk of cultivation/depensation effects. Most of the data have been collected since the middle of the twentieth century, and many stock collapses due to depensatory effects could have occurred much earlier in world fishery development so that what we have left to study today are mainly the most productive and resilient stocks. It is not unusual to hear laments by older, experienced fisheries observers about the disappearance of various species and stock components, before anyone had the time or resources to study them. Data collected for research purposes are typically among the last things to occur in the developmental sequence of most fisheries. On an even longer time scale, the patterns of mortality imposed by the industrial fisheries of this century are unlike anything in the evolutionary history of most fish species (Frank and Leggett 1994).

Implications for harvest management and research

The arguments presented in this paper imply need for a very particular third step in the evolution of the theory of fishing. The first step in this evolution was the development of simple catch–effort relationships (Baranov, Graham, Scahefer, Gulland) that did not explicitly use any ecological variables for prediction; this approach "worked" in a world of slow fisheries development that did not cause either rapid population size/structure transients or severe depletion. The second step, heralded by Schaefer's (1957) method for fitting logistic population models to time series data, was to recognize population size and structure as dynamic variables. Most of the elaborate machinery of modern fisheries assessment has really just added detail to the population state representation and statistical analysis, allowing better interpretation of data from rapidly changing populations under modern, more violent exploitation regimes. It is noteworthy that we generally do not obtain much better fits to population time series data (or predictions of harvest impact) with the elaborate models than we can with simple logistic or delay-difference models; the really big step was to recognize population size as a critical state variable. Single-species models served us well until very severe stock depletions began to occur.

We argue that in this "new" domain of fisheries system states, where severe depletions and risk of recruitment overfishing are common, that the single-species recruitment models are no longer reliable. While we understand and agree with the call for an "ecosystem approach" to fisheries management (Mooney 1998), we also recognize that this approach is complex and will require a substantial effort before successes can reinforce its value. In the interim, we believe that we must at least try to extend the theory of fishing so as represent some other, particularly important variables (predators/competitors of juvenile fish) in order to predict the pathological dynamics that sometimes accompany severe depletion. That is, we need to take a third basic step toward inclusion of more variables in predictive models, but in a very particular way. Just as population dynamics modelers had much of the modeling and statistical machinery already available when they began the step into modern stock assessment, so do we have the machinery largely in place to begin the next step, via ecosystem analysis tools like multispecies virtual population analysis (Pope 1991; Sparre 1991) and Ecopath/Ecosim (Walters et al. 1997, 2000).

As for any depensatory mechanism, the primary policy implication of cultivation/depensation effects is the existence of a critical population size, defining a division between two qualitatively different domains of population behavior. In the high-abundance domain, traditional single-species assessment procedures and prescriptions should work reasonably well. But should abundance be driven into the lower domain, we expect to see accelerating population collapse toward some low equilibrium or extinction. Reduction in exploitation rates after entering this domain may or may not allow recovery, depending in a quite unpredictable way on the *quantitative* details of how the juvenile survival rate is impacted.

Can we say anything in general about the probability of there being such a critical population size, or what this size might be relative to reference points like unfished abundance? We think not: the development of depensatory effects as population size is reduced depends on the quantitative pattern of forage (competitor/predator) response, i.e., the specific form of the numerical response in Fig. 2a. Models like Ecosim II can be used to define a range of possible responses, depending on assumptions about factors ranging from predator feeding rates to behavioral characteristics of the forage fish that may limit their vulnerability to predation. But the parameters that define the "correct" response obviously cannot be estimated reliably from historical data where the response is not yet evident, and these parameters summarize a very complex set of direct and indirect impacts of predation (Bax 1998). Comparative analysis of stock–recruitment data (Table 1; Myers et al. 1995a, 1995b; Liermann and Hilborn 1997) hints that at least some delayed depensation effects may occur in up to 10–20% of severely overfished cases. But the available data sets do not have enough observations at low stock sizes to make a convincing quantitative case about risk. We suspect that 10% may be a considerable underestimate of the risk for freshwater piscivores and possibly also for larger marine piscivores (particularly Gadiformes). We note, too, that many of the data sets in hand derive from populations that are components of multispecies fisheries. Therefore, the observed responses to exploitation include some degree of ecological change greater than that of simple mortality rates for single stocks. Ecosim may be a particularly valuable tool in evaluating those mixed impacts.

From a biological perspective, we might be willing to treat relatively rare instances (e.g., 10–20% of stocks; Table 1) of delayed depensation as biological curiosities rather than a matter for considerable research investment. But from a social and economic perspective, a 10% risk of stock collapse and (or) delayed recovery can be a very serious matter, especially where a substantial community of people is deeply dependent on the stocks. Would any fishery manager in Canada be willing to step forward and admit to having knowingly accepted a 10% chance that the Newfoundland cod stocks would collapse and show long delays in recovery? We think not, and we conclude from such examples that the risk should be taken very seriously indeed.

Depensation and physical "environmental effects" may interact, making prediction of critical population size even more difficult (Collie and Spencer 1993). Physical changes that impact productivity and size of juvenile nursery areas may move the response curves in Fig. 2 in complex ways, depending on the details of how both juveniles and their competitors/predators are impacted. Outcomes can range from mitigation of effects if there is a differential negative effect on the other species to severe reinforcing of negative effects if the other species are differentially enhanced. Indeed, the cultivation/depensation hypothesis offers an explanation for why correlations between recruitment and environmental factors are so prone to break down over time (Drinkwater and Myers 1987; Myers et al. 1997). Strong immediate responses to physical change are likely to be followed by dampening of the effect as trophic structure adjusts to the change.

If we cannot predict the critical population size in advance, what might we monitor in order to provide the earliest possible management reaction should depensatory effects start to develop? Here there are two obvious recommendations. First, develop survey methods to provide direct, immediate measures of juvenile survival rate and recruitment performance. Age-structured methods based on surveys and harvest of older fish do not provide reliable recruitment/survival estimates for each cohort until that cohort has been in the fishery for at least a few years. Second, also develop survey methods (or use the juvenile survey methods themselves) for abundance trends of a suite of potential competitor/predators in major juvenile nursery and rearing areas. In these surveys, routinely monitor diet compositions of juveniles and these species. Although multispecies surveys may fail to detect potential depensatory effects if such effects occur in concentrated space–time windows (e.g., seasonal impact on a particular size range of juveniles), their failure is the essential next step toward discovering this type of "critical period" or "bottleneck" effect.

As of 1990, assessments based on goals such as $F_{0.1}$, along with general belief that recruitment is poorly correlated with spawning biomass, led to a common view that the spawning biomass for most fish can be safely reduced at least 60–80% from natural levels without a substantial risk of recruitment failure. This view has been strongly challenged in the last decade (Mace 1994; Myers et al. 1994), thanks particularly to the comparative recruitment studies by Myers and his colleagues along with empirical studies of long-term population change (Patterson 1992). It is now rare to see suggestions that spawning biomass can be safely reduced by more than 60–70%. We suggest that even these more conservative goals are based on very limited temporal experience with initial recruitment responses to stock size reduction and over the long term may be dangerously optimistic. To insure against lagged depensation effects, we suggest that spawning stock abundance goals should generally be no less than 50% of unfished spawning biomass, which in any case should usually produce yields not much less than at the more dangerously low levels (e.g., 30%) often recommended.

Where is the burden of proof?

Ecosystem models have drawn considerable criticism from proponents of single-species assessment methods. Ecosystem models have not been "proven" to work and have not been "tested" by fitting them to available time series data, they have many parameters whose effects are not easily seen in the data, and there is no proof that representations of species interaction effects are really necessary for policy formulation (e.g., see Hilborn and Walters 1992, p. 448: "We believe that the food web modeling approach is hopeless as an aid to formulating management advice"). In short, defenders of single-species assessment have argued that (*i*) you cannot do it and (or) (*ii*) we do not need it. We think that these arguments are deeply misleading and in fact represent a bizarre reversal of the burden of proof: what really demands justification is not attempts to understand consequences of trophic interactions but rather continuing the pretense that we can get away with not doing so!

Consider the "you cannot do it" argument. By appropriate choice of vulnerability parameters in functions for predicting predation mortality rates and foraging time/predation risk responses (e.g., Ecosim eq. A3, Appendix), we can turn ecosystem models into a collection of "independent" single-species models with essentially the same response dynamics as the single-species models now used for most assessment. If we then follow the standard assessment practice of including many nuisance parameters to account for unexplained recruitment variation ("process errors," "recruitment anomalies"), we can then fit these population "submodels" just as well as we can fit their single-species analogs and make the same predictions about policy parameters like maximum sustainable yield. If we then vary parameters so as to strengthen trophic interaction effects, we are almost bound (given many covarying time series) to "explain" at least some of the variation initially attributed to nuisance parameters. Such exercises prove nothing (correlations could be spurious), just as do exercises showing that recruitment anomalies are correlated with environmental factors. At this point, ecosystem model predictions about impacts of extreme abundance changes and impacts of policy changes such as fishing at the bottom of the food chain will begin to depart from the predictions of single-species models. Should we trust such predictions? Of course we should not, any more than we should trust the predictions of single-species assessments: the only "proof" is to see which predictions stand the test of time, and we cannot obtain such proof if we resist making the predictions in the first place.

It is even more misleading to argue that we do not "need" ecosystem models. Most of the apparent success of single-species assessment approaches has come from three tactics employed by experienced assessment scientists like the senior author. First, we take considerable care in choice of case populations and data sets to use as test cases in reporting methods development, when possible avoiding uninformative and (or) perverse data sets (which in fact make up a clear majority in the Myers synthesis of stock–recruitment data sets). Second, we shrug off much of the interesting variation, by calling it "anomalies" or "environmental effects," saying only that we need to perform risk assessments under various alternative hypotheses about future patterns of variation. This tactic leads us directly away from recognizing serious policy issues such as the risk of delayed depensation. Third, we restrict ourselves to asking only the most menial of policy questions, and in this, we do deep disservice to fisheries management by encouraging the use of correspondingly myopic policy approaches (e.g., my model cannot tell you anything about the efficacy of marine protected areas because it does not account for the spatial and trophic effects of such a policy, so let us talk about next year's allowable catch instead). Assessment scientists who use these tactics may soon find themselves left behind by both the science and fisheries decision-making.

Acknowledgements

We are particularly grateful to members of the Apex Predators Working Group, National Center for Ecological Analysis and Synthesis, for helping clarify the ideas presented here: Kerim Aydin, Chris Boggs, Bob Francis, Bob Olson, Jeff Polovina, Tim Essington, and George Watters. Further encouragement and support was provided by Villy Christensen and Daniel Pauly, University of British Columbia. Rob Ahrens and Sean Cox spent long hours examining stock–recruitment data. Financial support for C. Walters was provided by a Natural Sciences and Engineering Research Council of Canada operating grant and for J.F. Kitchell by the U.S. National Science Foundation and Wisconsin Sea Grant Program.

References

Anderson, J.T., and Dalley, E.L. 1997. Year-class strength of northern cod (2J3KL) estimated from pelagic juvenile fish surveys in the Newfoundland region, 1994, 1995, 1996. NAFO Sci. Counc. Res. Doc.

Barkai, A., and McQuaid, C. 1988. Predator–prey role reversal in a marine benthic ecosystem. Science (Washington, D.C.), **242**: 62–64.

Bax, N.J. 1998. The significance and prediction of predation in marine fisheries. ICES J. Mar. Sci. **55**: 997–1030.

Brett, M.J., and Goldman, C.R. 1996. A meta-analysis of the freshwater trophic cascade. Proc. Natl. Acad. Sci. U.S.A. **93**: 7723–7726.

Bystroem, P., Persson, L., and Wahlstrom, E. 1998. Competing predators and prey: juvenile bottlenecks in whole-lake experiments. Ecology, **79**: 2153–2167.

Cohen, J.E., Pimm, S.L., Yodzis, P., and Saldana, J. 1993. Body sizes of animal predators and animal prey in food webs. J. Anim. Ecol. **62**: 67–78.

Collie, J.S., and Spencer, P.D. 1993. Management strategies for fish populations subject to long term environmental variability and depensatory predation. *In* Proceedings of the International Symposium on Management Strategies for Exploited Fish Populations. *Edited by* G. Kruse, D.M. Eggers, R.J. Maresco, C. Pautzke, and T.J. Quinn II. Lowell Wakefield Fisheries Symposium, October 1992, Anchorage, Alaska. Alaska Sea Grant Program, Anchorage. pp. 629–650.

Crowder, L.B., Rice, J.A., Miller, T.J., and Marshall, E.A. 1992. Empirical and theoretical approaches to size-based interactions and recruitment variability in fishes. *In* Individual-based approaches in ecology: concepts and individual models. *Edited by* D. DeAngelis and L. Gross. Routledge, Chapman and Hall, New York. pp. 237–255.

Drinkwater, K.F., and Myers, R.A. 1987. Testing predictions of marine fish and shellfish landings from environmental variables. Can. J. Fish. Aquat. Sci. **44**: 1568–1573.

Frank, K.T., and Leggett, W.C. 1994. Fisheries ecology in the context of ecological and evolutionary theory. Annu. Rev. Ecol. Syst. **25**: 401–422.

Gutreuter, S.J., and Anderson, R.O. 1985. Importance of body size to the recruitment process in largemouth bass populations. Trans. Am. Fish. Soc. **114**: 317–327.

Hackney, P.A. 1979. Influence of piscivorous fish on fish community structure of ponds. *In* Proceedings of an International Symposium on Predator–Prey Systems in Fish Communities and Their Role in Fisheries Management, 24 July 1978, Atlanta, Ga. *Edited by* H. Clepper. Sport fishing Institute, Washington, D.C. pp. 11–121.

Hall, S.J. 1999. The effects of fishing on marine ecosystems and communities. Blackwell Science Ltd., Oxford, U.K.

Hansen, M.J., and 11 coauthors. 1995. Lake trout (*Salvelinus namaycush*) populations in Lake Superior and their restoration in 1959–1993. J. Gt. Lakes Res. **21**(Suppl. 1): 152–175.

Hansen, M.J., Bozek, M.A., Newby, J.R., Newman, S.P., and Staggs, M.D. 1998. Factors affecting recruitment of walleyes in Escanaba Lake, Wisconsin, 1958–1996. N. Am. J. Fish. Manage. **18**: 764–774.

He, X., Wright, R.A., and Kitchell, J.F. 1993. Fish behavioral and community responses to manipulation. *In* The trophic cascade in lakes. *Edited by* S.R. Carpenter and J.F. Kitchell. Cambridge University Press, Cambridge, U.K. pp. 69–84.

Hilborn, R., and Walters, C. 1992. Quantitative fisheries stock assessment: choice, dynamics, and uncertainty. Chapman and Hall, New York.

Kitchell, J.F., Schindler, D.E., Ogatu-Ohwayo, R., and Reinthal, P.M. 1996. The Nile perch in Lake Victoria: interactions between predation and fisheries. Ecol. Appl. **7**: 653–664.

Liermann, M., and Hilborn, R. 1997. Depensation in fish stocks: a hierarchic Bayesian meta-analysis. Can. J. Fish. Aquat. Sci. **54**: 1976–1984.

MacCall, A.D. 1990. Dynamic geography of marine fish populations. Washington Sea Grant Program, University of Washington Press, Seattle, Wash.

Mace, P.M. 1994. Relationships between common biological reference points used as thresholds and targets of fisheries management strategies. Can. J. Fish. Aquat. Sci. **51**: 110–122.

Madenjian, C.P., Johnson, B.M., and Carpenter, S.R. 1992. Individual-based modeling: application to walleye stocking. *In* Food web management: a case study of Lake Mendota. *Edited by* J.F.Kitchell. Springer-Verlag, New York. pp. 493–505.

McCarthy, M.A., Ginzburg, L.R., and Akcakaya, H.R. 1995. Predator interference across trophic chains. Ecology, **76**: 1310–1319.

Mooney, H.A. (*Editor*). 1998. Ecosystem management for sustainable fisheries. Ecol. Appl. **8**(Suppl. 1).

Myers, R.A., and Barrowman, N.J. 1996. Is fish recruitment related to spawner abundance? Fish. Bull. U.S. **94**: 707–724.

Myers, R.A., Rosenberg, A.A., Mace, P.M., Barrowman, N., and Restrepo, V.R. 1994. In search of thresholds for recruitment overfishing. ICES J. Mar. Sci. **51**: 191–205.

Myers, R.A., Barrowman, N.J., Hutchings, J.A., and Rosenberg, A.A. 1995*a*. Population dynamics of exploited fish stocks at low population levels. Science (Washington, D.C.), **269**: 1106–1108.

Myers, R.A., Bridson, J., and Barrowman, N.J. 1995*b*. Summary of worldwide spawner and recruitment data. Can. Tech. Rep. Fish. Aquat. Sci. No. 2024.

Myers, R.A., Hutchings, J.A., and Barrowman, N.J. 1996. Hypotheses for the decline of cod in the North Atlantic. Mar. Ecol. Prog. Ser. **138**: 293–308.

Myers, R.A., Mertz, G., and Bridson, J. 1997. Spatial scales of interannual recruitment variations of marine, anadromous, and freshwater fish. Can. J. Fish. Aquat. Sci. **54**: 1400–1407.

Myers, R.A., Bowen, K.G., and Barrowman, N.J. 1999. Maximum reproductive rate of fish at low population sizes. Can. J. Fish. Aquat. Sci. **56**: 2404–2419.

Neave, F. 1954. Principles affecting the size of pink and chum salmon populations in British Columbia. J. Fish. Res. Board Can. **9**: 450–491.

Olson, M.H. 1996. Predator–prey interactions in size-structured fish communities: implications of prey growth. Oecologia, **108**: 757–763.

Patterson, K. 1992. Fisheries for small pelagic species: an empirical approach to management targets. Rev. Fish Biol. Fish. **2**: 321–338.

Persson, L., and Eklov, P. 1995. Prey refuges affecting interactions between piscivorous perch and juvenile perch and roach. Ecology, **76**: 70–81.

Pope, J.G. 1991. The ICES Multispecies Assessment Working Group: evolution, insights and future problems. ICES Mar. Sci. Symp. **193**: 22–33.

Rudstam, I., Aneer, G., and Hilden, M. 1994. Top-down control in the pelagic Baltic ecosytem. Dana, **10**: 105–129.

Schaefer, M.B. 1957. A study of the dynamics of the fishery for yellowfin tuna in the eastern tropical Pacific Ocean. Inter-Am. Trop. Tuna Comm. Bull. **2**: 247–285.

Scheffer, M., and De Boer, R.J. 1995. Implications of spatial heterogeneity for the paradox of enrichment. Ecology, **76**: 2270–2277.

Shelton, P.A., and Healey, B.P. 1999. Should depensation be dismissed as a possible explanation for the lack of recovery of the northern cod (*Gadus morhua*) stock? Can. J. Fish. Aquat. Sci. **56**: 1521–1524.

Sparre, P. 1991. Introduction to multispecies virtual population analysis. ICES Mar. Sci. Symp. **193**: 12–21.

Spencer, D.D., and Collie, J.S. 1997*a*. Patterns of population variability in marine fish stocks. Fish. Oceanogr. **6**: 188–204.

Spencer, D.D., and Collie, J.S. 1997*b*. Effect of nonlinear predation rates on rebuilding the Georges Bank haddock (*Melanogrammus aeglefinus*) stock. Can. J. Fish. Aquat. Sci. **54**: 2920–2929.

Swingle, H.S. 1950*a*. Relationships and dynamics in balanced and unbalanced fish populations. Ala. Polytech. Inst. Agric. Exp. Stn. Bull. No. 274.

Swingle, H.S. 1950*b*. Experiments with various rates of stocking bluegills, *Lepomis macrochirus* Rafinesque, and largemouth bass, *Micropterus salmoides* (Lacepede), in ponds. Trans. Am. Fish. Soc. **80**: 218–230.

Ursin, E. 1982. Stability and variability in the marine ecosystem. Dana, **2**: 51–65.

Invited perspectives and article

Verity, P.G., and Smetacek, V. 1996. Organism life cycles, predation, and the structure of marine pelagic ecosystems. Mar. Ecol. Prog. Ser. **130**: 277–293.
Walters, C. 1987. Nonstationarity of production relationships in exploited populations. Can. J. Fish. Aquat. Sci. **44**(Suppl. 2): 156–165.
Walters, C.J., and Juanes, F. 1993. Recruitment limitation as a consequence of natural selection for use of restricted feeding habitats and predation risk taking by juvenile fishes. Can. J. Fish. Aquat. Sci. **50**: 2058–2070.
Walters, C., and Korman, J. 1999. Linking recruitment to trophic factors: revisiting the Beverton–Holt recruitment model from a life history and multispecies perspective. Rev. Fish Biol. Fish. **9**: 187–202.
Walters, C., Christensen, V., and Pauly, D. 1997. Structuring dynamic models of exploited ecosystems from trophic mass-balance assessments. Rev. Fish Biol. Fish. **7**: 1–34.
Walters, C., Pauly, D., Christensen, V., and Kitchell, J. 2000. Representing density dependent consequences of life history strategies in aquatic ecosystems: Ecosim II. Ecosystems, **3**: 70–83.
Werner, E.E. 1998. Ecological experiments and a research program in community ecology. *In* Experimental ecology: issues and perspectives. *Edited by* W. Resetarits and J. Bernardo. Oxford University Press, Oxford, U.K. pp. 3–26.
Wooten, J.T. 1994. The nature and consequences of indirect effects in ecological communities. Annu. Rev. Ecol. Syst. **25**: 443–466.

Appendix. "Foraging arena" concept as represented in Ecosim

In trophic models, we need to predict consumption rates Q_{ij} of prey types i by predator types j. Suppose that at some moment in time the total prey population is N_i and the total predator population is N_j. Simple reaction vat or mass action encounter models predict Q_{ij} from encounter rate arguments as $Q_{ij} = a_{ij} N_i N_j$ or as $Q_{ij} = f(N_i) N_j$, where $f(N_i)$ is the predator functional response to overall prey density. But in reality, if we look at any collection N_i of prey, we will find these individuals at any moment to be in a wide variety of "vulnerability states" with respect to N_j, depending on spatial position (e.g., in hiding places) and activity (e.g., resting versus actively feeding). Ecosim attempts to model this vulnerability distribution by treating the prey as being in one of two behavioral states, "invulnerable" and "vulnerable," with exchange between these states possibly representing both behavioral and physical mixing processes. Animals in the vulnerable state are said to be "in the foraging arena for i–j interaction," and we assume that there are V_{ij} of these at any moment. The dynamics of V_{ij} are modeled as having three components: (1) movement of individuals into the vulnerable (foraging arena) state at rate $v_{ij}(N_i - V_{ij})$, (2) movement of individuals out of this state at rate $v'_{ij} V_{ij}$, and (3) consumption of vulnerable individuals at mass action rate $Q_{ij} = a_{ij} V_{ij} N_j$. Note that we ignore predator handling time/satiation in the attack rate component 3, following the observation that predators with full stomachs are not a common field observation (D. Schindler, Department of Zoology, University of Washington, Seattle, Wash., unpublished data). We then assume that the dynamics of V are very fast compared with the dynamics of the Ns, so V quickly adjusts so that the three rates 1–3 remain near balance (dV/dt stays near zero). This variable speed-splitting assumption (similar to speed-splitting arguments used to derive classic type II functional response equations from handling time considerations) leads to the prediction

(A1) $\quad V_{ij} = v_{ij} N_i / (v_{ij} + v'_{ij} + a_{ij} N_j).$

Combining this prediction with assumption 3 above leads to

(A2) $\quad Q_{ij} = a_{ij} v_{ij} N_i N_j / (v_{ij} + v'_{ij} + a_{ij} N_j).$

This model implies Q_{ij} proportional to N_i and saturating in N_j (Fig. 3). In traditional predator–prey terminology, the denominator term $a_{ij} N_j$ represents a "ratio dependence" or localized "predator interference competition" effect. From eq. A2, instantaneous prey mortality rate Z_{ij} due to predator j is predicted to vary as

(A3) $\quad Z_{ij} = Q_{ij}/N_i = a_{ij} v_{ij} N_j / (v_{ij} + v'_{ij} + a_{ij} N_j).$

That is, Z_{ij} is not simply proportional to predator abundance but rather increases asymptotically toward maximum rate v_{ij} as predator abundance increases; for low v_{ij} and high predator search rate a_{ij}, Z_{ij} is predicted to be nearly constant (constant natural mortality M assumption).

In Ecosim, we allow users to introduce additional variation into predation mortality rates Z_{ij} by assuming changes in v_{ij} due to intraspecific competition among prey i and risk-sensitive behavioral responses. We monitor simulated food consumption rates by N_i and increase/decrease v_{ij} as feeding rate decreases/increases; the "target" feeding rate can be made inversely proportional to predation risk per time foraging. A basic implication of these time adjustment hypotheses is that for fixed predator abundance (all N_j constant) and N_i representing juveniles of some fish species, Z_{ij} will vary linearly with N_i so as to produce the widely observed Beverton–Holt form (flat-topped) of stock–recruitment curve (Walters and Korman 1999).

We see four common mechanisms that can decrease the vulnerability parameters v_{ij} so as to create stabilizing effects (Abrams and Walters 1996) and the appearance of "ratio-dependent" or "bottom-up" control of consumption rates (Q_{ij} limited to maximum $v_{ij} N_i$ no matter how many predators are present).

(1) Risk-sensitive prey behaviors

Prey may spend only a small proportion of their time in foraging arenas where they are subject to predation risk, otherwise taking refuge in schools, deep water, littoral refuge sites, etc.

(2) Risk-sensitive predator behaviors (the "three to tango" argument)

Especially if the predator is a small fish, it may severely restrict its own range relative to the range occupied by the prey, so that only a small proportion of the prey move or are mixed into the habitats used by it per unit time; in other words, its predators may drive it to behave in ways that make its own prey less vulnerable to it.

(3) Size-dependent graduation effects

If N_i represents an aggregate of different prey sizes, and predator j can take only some limited range of sizes, v_{ij} can represent a somewhat slower process of prey graduation into and out of the vulnerable size range due to growth. Size effects may of course also be associated with distribution (predator–prey spatial overlap) shifts.

(4) Passive, differential spatial depletion effects

Even if neither prey nor predator shows active behaviors that create foraging arena patches, any physical or behavioral processes that create spatial variation in encounters between i and j will lead to local depletion of i in high-risk areas and concentrations of i in partial predation "refuges" represented by low-risk areas. "Flow" between low- and high-risk areas (v_{ij}) is then created by any processes that move organisms.

These mechanisms are so ubiquitous that any reader with aquatic natural history experience might wonder why modelers have ever chosen to assume mass action, random encounter models (or infinite v_{ij}) in the past.

For readers who might think it practical to avoid simplifications like eqs. A1 and A2 by explicitly modeling the full space–time structure of individual predator–prey encounters, be warned that the foraging arena structure arises from biology and physics operating at very small scales indeed. Foraging and movement dynamics generally take place at time scales of minutes to hours and involve complicated spatial movements at scales of a few metres to a few hundred metres. Indeed, it is probably because we have not thought carefully about heterogeneity at these very difficult scales that we have been willing to use mass action models in the past. Note further that just complicating the trophic model state representation by including details of size structure and macroscale spatial overlaps of predators and prey (e.g., Stefansson and Palsson 1998) does not solve the microscale representation problem at all and could still be completely misleading if only simple mass action interaction rates are used in the detailed calculations.

Appendix references

Abrams, P.A., and Walters, C. 1996. Invulnerable prey and the paradox of enrichment. Ecology, **77**: 1125–1133.

Stefansson, G., and Palsson, O. 1998. A framework for multispecies modeling of Arcto-boreal systems. Rev. Fish Biol. Fish. **8**: 101–104.

Walters, C., and Korman, J. 1999. Linking recruitment to trophic factors: revisiting the Beverton–Holt recruitment model from a life history and multispecies perspective. Rev. Fish Biol. Fish. **9**: 187–202.

Possible ecosystem impacts of applying MSY policies from single-species assessment

Carl J. Walters, Villy Christensen, Steven J. Martell, and James F. Kitchell

Walters, C. J., Christensen, V., Martell, S. J., and Kitchell, J. F. 2005. Possible ecosystem impacts of applying MSY policies from single-species assessment. – ICES Journal of Marine Science, 62: 558–568.

Ecosim models have been fitted to time-series data for a wide variety of ecosystems for which there are long-term data that confirm the models' ability to reproduce past responses of many species to harvesting. We subject these model ecosystems to a variety of harvest policies, including options based on harvesting each species at its maximum sustainable yield (MSY) fishing rate. We show that widespread application of single-species MSY policies would in general cause severe deterioration in ecosystem structure, in particular the loss of top predator species. This supports the long-established practice in fisheries management of protecting at least some smaller "forage" species specifically for their value in supporting larger piscivores.

© 2005 Published by Elsevier Ltd on behalf of International Council for the Exploration of the Sea.

Keywords: Ecosim, ecosystem models, MSY, multispecies production, overharvesting.

Received 1 April 2004; accepted 13 September 2004.

C. J. Walters, V. Christensen and S. J. Martell: Fisheries Centre, University of British Columbia, Vancouver, British Columbia V6T1Z4, Canada. J. F. Kitchell: Center for Limnology, University of Wisconsin, Madison, WI 53706, USA. Correspondence to C. J. Walters: tel: +1 604 8226320; fax: +1 604 8228934; e-mail: c.walters@fisheries. ubc.ca.

Introduction

Single-species assessments and management controls may produce misleading predictions and pathological changes in ecosystems (Hollowed et al., 2000). Obvious examples include the appropriation of production by fisheries that would otherwise support valued predators such as marine mammals, and the appropriation of production needed to support valued piscivore species by reduction fisheries that target small planktivores. Application at ecosystem scale of harvest rates calculated from single-species assessments (e.g. F_{MSY}) might result in widespread degradation in ecosystem function, and in considerably smaller overall yield and value than would be predicted from the sum of corresponding single-species yields. However, there is a counter-argument to this concern: in complex foodwebs, fishing a variety of species at their species-specific F_{MSY} might reduce both competition among and within these species, as well as predation mortality rates (owing to harvesting predators, so reducing the carrying capacity for unharvested predators).

The ecological basis of sustainable net production and harvest is compensatory response, especially of juvenile survival rates, to reductions in stock size by fishing. In single-species assessment, these responses are seen and measured mainly through stock-recruitment relationships, where "flat" curves (no effect of reduced spawner abundance on recruitment) are caused by compensatory changes in juvenile survival rate. Compensatory responses may be direct and immediate (for example, behavioural changes in foraging time and predation risk), or indirect and delayed owing to ecosystem-scale changes in predation risk and food production. Simple food-chain models predict stronger compensatory responses for any species when interactions with the rest of the system are accounted for, owing to the negative impact of fishing on its predators, and the positive impact on its food organisms. Figure 1 shows Ecosim estimates of equilibrium surplus production plotted against stock size for two fish species and two cases: (i) community structure and abundance (except the target species) held fixed; and (ii) community structure allowed to reach equilibrium in response to changes caused by fishing on the target species. In virtually all such cases that have been examined with Ecosim, regardless of what trophic level the target species occupies, the same pattern is

Ecosystem impacts from applying MSY policies from single-species assessments

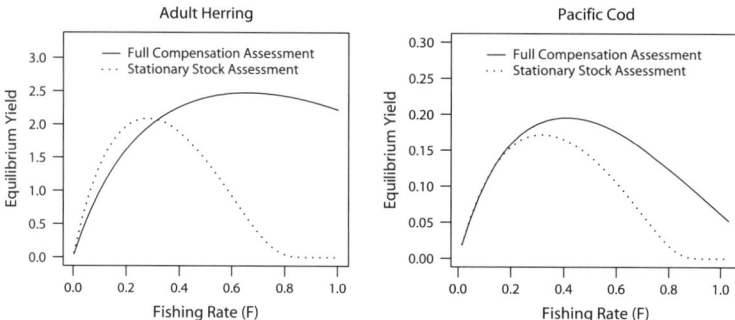

Figure 1. Typical Ecosim predictions of the equilibrium relationship between fishing mortality rate and yield for two species, a top predator (cod; left panel) and a mid-foodweb planktivore (herring; right panel) when holding all other biomass pools at Ecopath base biomasses (stationary assessment), and when allowing for ecosystem interactions (full compensation assessment).

predicted: indirect food-chain effects increase, rather than decrease, the productivity of the harvested species.

If chain-like interactions were important, the main task in ecosystem-scale harvest policy design would be to seek policies that avoid uneconomical or socially unacceptable impacts of fisheries at different trophic levels on one another (and on other valued species), caused by appropriation of potential production and hence weakening of natural compensatory responses. Unfortunately, the web-like structure of interactions, along with the prevalence of trophic ontogeny (most species change trophic position as they grow), can cause much more complex and perverse trophic changes. It is critical to have at least some basic recognition of potentially perverse changes when we engage in developing ecosystem models. There are at least three ways that more complex interactions can cause apparently depensatory rather than compensatory responses to fishing, all involving ways that decreases in the abundance of one species may lead to increases in other species that affect it negatively:

(i) *Cultivation-depensation effects* (Walters and Kitchell, 2001). Adults of a dominant species may be abundant *in the first place* because they exert control on competitors and predators of their own juveniles (trophic triangles). Reducing adult abundance of the dominant species by fishing could result in increased abundance of the competitors and predators, and in turn to reduced juvenile survival, and consequently adult abundance, of the harvested species.

(ii) *Competition-predation trade-offs among species at the same trophic level*. If the suite of species feeding at approximately the same trophic level varies in competitive ability vs. their ability to avoid predation, the dominant species in the suite is likely to be the one that is less vulnerable to predation. Targeted fishing on that dominant species may enhance its competitors (and hence weaken its compensatory responses). Further, harvesting its predator populations may lead to even stronger negative effects on the target species (Yodzis, 2001).

(iii) *Predators feeding on multiple trophic levels*. If some piscivore species of harvest interest is fed upon by a predator that is also happy to feed on the smaller fish species preferred by the piscivore (for example, seals and sea lions often feed on herring as well as on piscivores like cod and pollock, that also feed on herring), harvesting the piscivore should enhance (or at least stabilize) the food supply for the predator, which may in turn exert depensatory effects on the piscivore, should its abundance be reduced by fishing (higher predation mortality rate caused by a type II feeding response of the predator).

Fortunately, there is not much field evidence of severe disruptions of compensatory responses (as evidenced by non-stationarity or apparent depensatory patterns in single-species stock-recruitment relationships), except in cases where foodweb structure has been severely disrupted by overfishing (e.g. Newfoundland cod). Still, depensatory effects represent risks that need to be included in prudent designs for ecosystem harvest policy.

Therefore, it is not at all obvious whether we should expect higher or lower yields from an entire ecosystem than would be predicted from application of single-species harvest control policies. Pitcher and Cochrane (2002), Christensen and Walters (2004), and Okey and Wright (2004) used Ecosim (Walters et al., 1997) to explore policy options for maximizing total ecosystem yield, net economic value, and other ecosystem-scale measures of total value from fishing. These optimization exercises revealed complex trade-offs between fisheries and management objectives, and one simple answer to such complexity is to argue that a "pretty good" ecosystem value might be obtained just by managing each species to its potential maximum sustainable yield (MSY). We evaluate this policy option using simulation rather than optimization procedures, through Ecopath/Ecosim biomass dynamics and size-structured foodweb models for a variety of marine ecosystems. The models have been used to assemble and summarize a wide variety of recent information on abundances and foodweb interaction patterns from stock assessments and diet studies, mostly for ecosystems that historically have been highly disturbed by fishing.

Methods

Ecosim cases

We selected 11 case studies (Table 1) for which Ecosim has been shown to reproduce reasonably well the past patterns of change in relative abundance of major species (and sometimes catches and total mortality rate), given historical disturbance patterns (fishing mortality rates and/or effort over time, and in some cases changes in oceanographic or nutrient-loading indices of relative primary productivity). In the absence of such response data, ecosystem models might be constructed that either over- or underestimated the importance of various trophic interactions, and the need to account for these in harvest policy design.

As indicated by the number of biomass pools included (Table 1), the models vary in complexity. All include one or more species for which the basic Ecosim differential equations for biomass dynamics are replaced by age-structured population accounting, using either delay-difference models ("split pools", Walters et al., 2000) or full size—age "multistanza" accounting, with multiple life history stanzas representing major ontogenetic changes in mortality, feeding, and vulnerability to fishing (Walters and Martell, 2004). Typically, Ecosim model behaviour is dominated not by size—age accounting details, but rather by the "foraging arena" (Walters and Martell, 2004) limitations assumed for trophic interaction rates (food supplies, predation rates), owing to behavioural processes that organize fine-scale overlap patterns between predators and their prey.

For each case model, an informal procedure was used to fit the model to the abundance (and in a few cases, catch and total mortality rate) time-series. Possible causes of model-data discrepancies were first identified by examining modelled time-series patterns in components of change (food consumption, growth, mortality rate), then Ecopath base parameters (biomass, mortality rate, diet composition) and Ecosim dynamic response parameters (prey vulnerability exchange, foraging time adjustment) were varied to improve fit. Promising estimates were then examined more carefully using time-series fitting (maximum likelihood assuming lognormal errors in relative abundance measurement) methods similar to those used for single-species assessments. In no case has this stepwise process produced a "fully validated" model. Rather, the fitting procedure is a developing process, subject to changes over time as new information, and ideas about important interactions that may have been missed, become available.

Mainly, two types of patterns have been encountered during the fitting process: (i) "one-way-trip" (Hilborn and Walters, 1992) declines in abundance under fishery development; and (ii) more complex cyclic or dome-shaped patterns indicative of both more complex fishing regimes, and, in some cases, apparently large changes in species productivity. Fitting one-way-trip data has been largely a matter of adjusting the Ecosim vulnerability exchange parameters that play the same role as compensatory response parameters (particularly recruitment curve slopes) in single-species models. For such species, Ecosim typically predicts similar dynamic response patterns to future fishing policies to those from single-species assessments. Fitting the more complex patterns has typically required examination of multiple alternative hypotheses that might equally well explain the data (e.g. environmental forcing vs. fishing vs. trophic interaction effects).

Key results from fitting process

The following subsections review a few key results from the fitting exercises, particularly in relation to the role of trophic interactions in shaping historical response patterns.

Top-down vs. bottom-up effects

There has been much debate in the field of ecology about the relative importance of top-down (predation) vs. bottom-up (food supply) effects of trophic interactions (Power, 1992; Brett and Goldman, 1996). We commonly found strong evidence for top-down effects of harvesting (or protecting) predators on the productivity of their prey, but we did not see correspondingly strong evidence for bottom-up effects of harvesting prey on the productivity of their predators. In other words, a given trophic interaction can have strongly asymmetric effects, with typically stronger effects on the prey than on the predator.

Examples of strong top-down effects involved both marine mammals and piscivores. In the Georgia Strait case, increases in the numbers of seals and sea lions following marine mammal protection, since the mid-1970s, appear to be partly responsible (in conjunction with changes in primary productivity) for declines in survival rate and recruitment of a variety of fish species, ranging from Pacific salmon to demersal cod and flatfish. In the central North Pacific and eastern tropical Pacific, the exploitation of large piscivores (tuna, marlin) has apparently resulted in increased productivity of smaller piscivores (skipjack tuna, mahi mahi). In the Baltic Sea, Bering Sea, and North Sea cases, gadoids are predicted to have had strong predation impacts on small planktivores. Few would doubt the positive impact that cod declines around the Atlantic have had on shrimp productivity (Worm and Myers, 2003), and this interaction is important in the North American case of the West Coast of Vancouver Island (Martell, 2002).

Interestingly, only the relatively simple Baltic Sea model shows strong bottom-up impact of fishing on lower trophic levels. In the Georgia Strait and West Coast of Vancouver Island cases, herring were nearly eliminated by reduction fisheries during the 1960s, but there is no signal of this in the abundance time-series for piscivores. In the Bering Sea case, we have been unable to explain declines in Steller sea lions as having been caused directly by the groundfish fisheries (NRC, 2003). In the case of the West Coast of Vancouver Island, development (and early overfishing) of the shrimp fishery had no apparent impact on cod and other benthic predators. Ecosim predicts that, in each of these cases, predators were able to adjust diet compositions so as to feel little impact of reductions in dominant prey species caused by fishing, and/or that productivity of alternate prey increased owing to reduced competition/predation with/by the harvested prey.

In contrast to the apparent lack of response to depletion of prey, there is good evidence for ecosystem-scale bottom-up effects apparently as a consequence of temporal changes in primary productivity. For several cases (Georgia Strait, West Coast of Vancouver Island, Bering Sea), roughly half the total variance in relative abundance can be explained by fitting a single temporal pattern of relative primary production anomalies. Considering the number of abundance time-series included, the variance reduction is far larger than would be expected by chance alone if each series were subject to an independent anomaly pattern, owing to species-scale environmental influence.

Cryptic predation effects

Ecopath base diet compositions have been estimated mainly from stomach studies aimed at identifying the main prey items that support predators. Unfortunately, an abundant predator may have large predation impacts on a low-biomass prey, without that prey making up a recognizable proportion of its diet. Such cryptic predation impacts may be the most common reason for model "failure" to represent important species interactions.

In the Georgia Strait case, visual observations of marine mammals feeding on juvenile salmon shortly after ocean entry support the assumption that mammal predation has an impact on juvenile survival, but diet studies have failed to sample at the space/time scales that could confirm this assumption. In several cases, calculated juvenile fish biomasses are low compared with adult biomasses, and both cannibalism by the adults and incidental predation by other (temporal) piscivores could have large impacts not evidenced in adult diets. In cases involving gadoids, lack of response to changes in prey availability may well represent recruitment control by cannibalism, rather than food resource availability.

Competition effects

The case models display two types of competition effect, neither particularly common nor dominating overall system behaviour: (i) release (increase) of relatively rare species when dominant species are reduced by fishing or predation; (ii) cultivation—depensation patterns (discussed above).

Examples of simple competitive release effects include predicted (and observed) increases in jellyfish in the Bering Sea following declines in small pelagic fish biomass. Both are planktivores. The observed increases in smaller pelagic predators following reduction in large tuna and billfish appear to reflect a mixed predation—competition effect. In a few cases, competitive effects were stronger than appeared reasonable on the basis of time-series patterns, and these model "errors" were solved by increasing the detail representing prey populations along with diet composition changes to reflect less severe diet overlaps than had been assumed initially.

Cultivation—depensation effects have been noticed for only a few dominant piscivore species, and then only under more extreme fishing scenarios than observed historically. Both pelagic system models show some reduction in juvenile survival rate of yellowfin tuna when this species is severely reduced, owing to increased abundance and predation by skipjack tuna and mahi mahi. However, fisheries for these smaller species may have prevented the effects from being as large as predicted in extreme model scenarios, and we may also have overestimated predation impacts by the smaller species, and diet overlaps.

Bycatch effects

The case studies include several instances where non-target species have suffered relatively severe fishing mortality. The most prominent examples involve billfish and shark populations in the pelagic ecosystem models, which have been reduced by longline fishing directed at tuna. In the Gulf of Thailand, where a wide variety of species are landed, it is difficult to classify species as bycatch or target. Responses in bycatch species have involved mainly simple one-way-trip depletions, with evidence of the usual compensatory response. In the West Coast of Vancouver Island case, a complex interaction exists between fisheries, where the shrimp fishery takes juvenile cod as bycatch while the finfish trawl fishery takes mainly larger cod; Ecosim explains persistence of the shrimp fishery partly by the negative impact on cod by the shrimp fishery itself.

Trophic mediation effects

There has been concern in Ecosim model development about indirect (Dill et al., 2003; Werner and Peacor, 2003) or trophic mediation effects, where interaction rates (or spatial overlap patterns) between predators and prey are influenced by abundances of some third, "mediating" type of organism (e.g. cover provided by seagrass and macro-algae in the Tampa Bay case). None of the cases show enough change in the mediating species to provide a clear test of the importance of such effects, particularly because the main effects might have taken place before observations started.

C. J. Walters et al.

Table 1. Biomass pools represented in Ecopath/Ecosim case examples used in MSY comparisons (pools in italics were included in MSY analyses; species code numbers in later figures correspond to row numbers; N/A, no reference available; YOY, young of year).

	Ecosystem model, area, reference and pool				
	Central North Pacific	Benguela	Baltic	Chesapeake Bay	Eastern Bering Sea
Source:	Cox et al.(2002a, b)	Shannon et al. (2004)	Harvey et al. (2003)	N/A	NRC (2003)
Group #					
1	*Large bigeye*	Phytoplankton	Spring phytoplankton	Piscivorous birds	*Baleen whales*
2	*Small bigeye*	Benthic producers	Other phytoplankton	Non-piscivorous seabirds	Toothed whales
3	*Large yellowfin*	Microzooplankton	Bacteria	*Spot*	Sperm whales
4	*Small yellowfin*	Mesozooplankton	Microzooplankton	Striped bass YOY	Beaked whales
5	*Large albacore*	Macrozooplankton	Mesozooplankton	*Striped bass resident*	*Walrus and bearded seals*
6	*Small albacore*	Gelatinous zooplankton	Mysids/Invertebrates	*Striped bass migratory*	Seals
7	*Large blue sharks*	Anchovy	Meiofauna	Reef fish	*Steller sea lions*
8	*Small blue Sharks*	Sardine	Macrofauna	Bluefish YOY	Piscivorous birds
9	*Blue marlin*	Round herring	*Juvenile sprat*	*Bluefish adult*	*Adult pollock 2+*
10	*Large sharks*	Other small pelagics	*Juvenile herring*	Weakfish YOY	Juvenile pollock 0–1
11	*Brown sharks*	Chub mackerel	*Juvenile cod*	*Weakfish adults*	*Other demersal fish*
12	*Swordfish*	Juvenile horse mackerel	*Adult sprat*	Summer flounder	*Large flatfish*
13	*Other billfish*	Adult horse mackerel	*Adult herring*	Atlantic croaker	*Small flatfish*
14	*Mahi mahi*	*Mesopelagic fish*	*Adult cod*	Menhaden YOY	Shallow pelagics
15	*Small scombrids*	*Snoek*	Salmon	*Menhaden adults*	Deep pelagics
16	*Flying squid*	Other large pelagics	Seals	Black drum	Deepwater fish
17	*Skipjack*	Cephalopods	Detritus	Littoral forage fish	Jellyfish
18	*Large leatherback*	Small shallow-water hake		*Alewife and herring*	Cephalopods
19	*Small leatherback*	Large shallow-water hake		American eel	Crustaceans
20	*Large loggerhead*	Small deepwater hake		*American shad*	Infauna
21	*Small loggerhead*	Large deepwater hake		Bay anchovy	Epifauna
22	Jellyfish	Pelagic-feeding demersals		Channel and other catfish	Large Zooplankton
23	Lance	Benthic-feeding demersals		Other flatfish	Herbivorous zooplankton
24	Squid	Pelagic sharks		White perch YOY	Phytoplankton
25	Flying fish	Benthic sharks		*White perch adults*	Discards
26	Mesopelagic fish nekton	Apex sharks		Other elasmobranchs	Detritus
27	Epipelagic fish nekton	Seals		Gizzard shad	
28	Epipelagic micronekton	Cetaceans		Blue catfish	
29	Mesopelagic micronekton	Seabirds		Non-reef-associated fish	
30	Phytoplankton	Meiobenthos		Sandbar shark	
31	Detritus	Macrobenthos		Hard clam	
32		Detritus		Other suspension feeders	
33				Other in/epifauna	
34				Mesozooplankton	
35				Microzooplankton	
36				Ctenophores	
37				Sea nettles	
38				Blue crab YOY	
39				*Blue crab adults*	
40				Oysters YOY	
41				*Oysters 1+*	
42				Soft clam	
43				Phytoplankton	
44				Subaquatic vegetation	
45				Benthic algae	
46				Detritus	
47					
48					
49					
50					
51					
52					

Ecosystem impacts from applying MSY policies from single-species assessments

		Ecosystem model, area, reference and pool			
Eastern tropical Pacific	Georgia Strait	Gulf of Thailand	North Sea	Tampa Bay	West Coast of Vancouver Island
Olson and Watters (2003)	Martell et al. (2002)	FAO/FISHCODE (2001)	Christensen et al. (2002)	N/A	Martell (2002)
Pursuit birds	Transient orcas	*Rastrelliger* spp.	Cod	0–3 Snook	Lingcod
Grazing birds	Dolphins (resident orcas)	*Scomberomorus*	Haddock	3–12 Snook	*Juvenile lingcod*
Baleen whales	Seals and sea lions	Carangidae	Herring	12–48 Snook	Dogfish
Toothed whales	Halibut	Pomfret	Mackerel	48–90 Snook	Adult Pacific cod
Spotted dolphin	Lingcod	Small pelagics	Norway pout	90+ Snook	Juvenile Pacific cod
Meso dolphin	Dogfish shark	False trevally	Plaice	0–3 Red drum	*Demersal fish*
Sea turtles	Adult hake	Large piscivores	Saithe	3–8 Red drum	*Adult herring*
Large yellowfin	Juvenile hake	Sciaenidae	Sandeel	8–18 Red drum	*Juvenile herring*
Large bigeye	Adult coho	*Saurida* spp.	Sole	*18–36 Red drum*	*Eulachon*
Large marlin	Juvenile coho	Lutianidae	Whiting	*36+ Red drum*	*Euphausiids*
Large sailfish	Adult chinook	Plectorhynchidae	Birds	0–3 Sea trout	*Pink shrimp*
Large swordfish	Juvenile chinook	*Priacanthus* spp.	Gurnards	3–18 Sea trout	Copepods
Large dorado	Demersal fish	*Sillago*	Horse mackerel	*18+ Sea trout*	Herbivorous Zooplankton
Large wahoo	Seabirds	*Nemipterus* spp.	Other predators	0–3 Sand trout	Phytoplankton
Large sharks	Small pelagics	Ariidae	Raja	3–12 Sand trout	Detritus
Rays	Eulachon	Rays	Seals	*12+ Sand trout*	
Skipjack	Adult herring	Sharks	West mackerel	0–6 Mullet	
Albacore	Juvenile herring	Cephalopods	Other invertebrates	*6–18 Mullet*	
Auxis	Jellyfish	Shrimps	Juvenile cod	*18+ Mullet*	
Bluefin	Predatory invertebrates	Crabs and lobsters	Juvenile haddock	0–3 Mackerel	
Small yellowfin	Shellfish	Trashfish	Juvenile saithe	*3+ Mackerel*	
Small bigeye	Grazing invertebrates	Small demersals	Juvenile whiting	0–10 Ladyfish	
Small marlin	Carnivorous zooplankton	Demersal piscivores	Sprat	*10+ Ladyfish*	
Small sailfish	Herbivorous zooplankton	Demersal benthivores	Dab	Jacks	
Small swordfish	Kelp/seagrass	Shellfish	Copepods	Bay anchovy	
Small dorado	Phytoplankton	Jellyfish	Euphausiids	*Pin fish*	
Small wahoo	Detritus	Sea cucumber	Other crustaceans	Spot	
Small sharks		Seaweeds	Echinoderms	*Silver perch*	
Miscellaneous piscivores		Coastal tuna	Polychaetes	Scaled sardine	
Flying fish		Sergestid shrimps	Other macrobenthos	Mojarra	
Miscellaneous epipelagic fish		Mammals	Phytoplankton	Threadfin herring	
Miscellaneous mesopelagic fish		Pony fish	Detritus	*Menhaden*	
Cephalopods		Benthos		Menidia (silverside)	
Crabs		Zooplankton		Catfish	
Mesozooplankton		Juvenile pelagics		Bumper	
Microzooplankton		Juvenile Caranx		Caridean shrimp	
Large phytoplankton		Juvenile Saurida		*Shrimp*	
Small phytoplankton		Juvenile Nemipterus		Stone crab	
Detritus		Phytoplankton		*Blue crab*	
		Detritus		Cyprinodontids	
				Poecilids	
				Pigfish	
				Gobies	
				Rays	
				Benthic invertebrates	
				Macrozooplankton	
				Microzoolplankton	
				Infauna	
				Attached microalgae	
				Phytoplankton	
				Detritus	

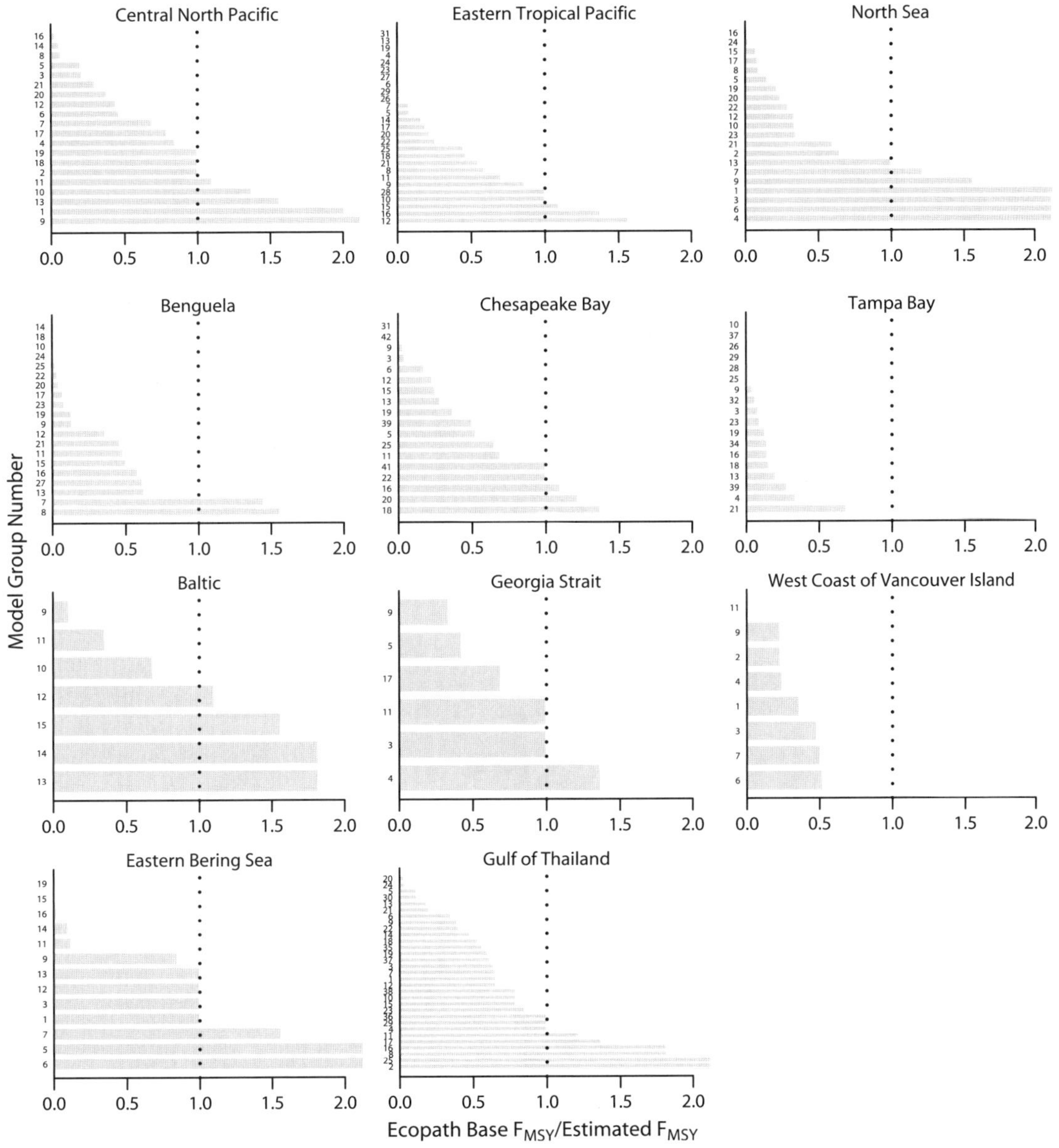

Figure 2. Ratios of Ecopath base fishing rate F to estimated F_{MSY} (full compensation assessment) for all fished species in the case ecosystems (species codes are the rows in Table 1).

Estimating F_{MSY} and equilibrium system responses

We used a simple two-step procedure to evaluate the ecosystem-scale impact of widespread application of single-species MSY policies for each case. First, for each harvested species (each group with non-zero catch and/or bycatch), we ran a long simulation (1000+ years), where fishing mortality rate (F) of that species was incremented or decremented slowly, while holding all other F constant at Ecopath base values. F_{MSY} for the species was taken to be the F that resulted in maximum average catch. The simulation for each species was repeated for two alternative assumptions about ecosystem change in response to changes in fishing on the one species: (i) no response (food and predator populations held constant, so that the species is buffered from predator/prey effects); (ii) with full ecosystem-scale dynamic response effects in all species (but only

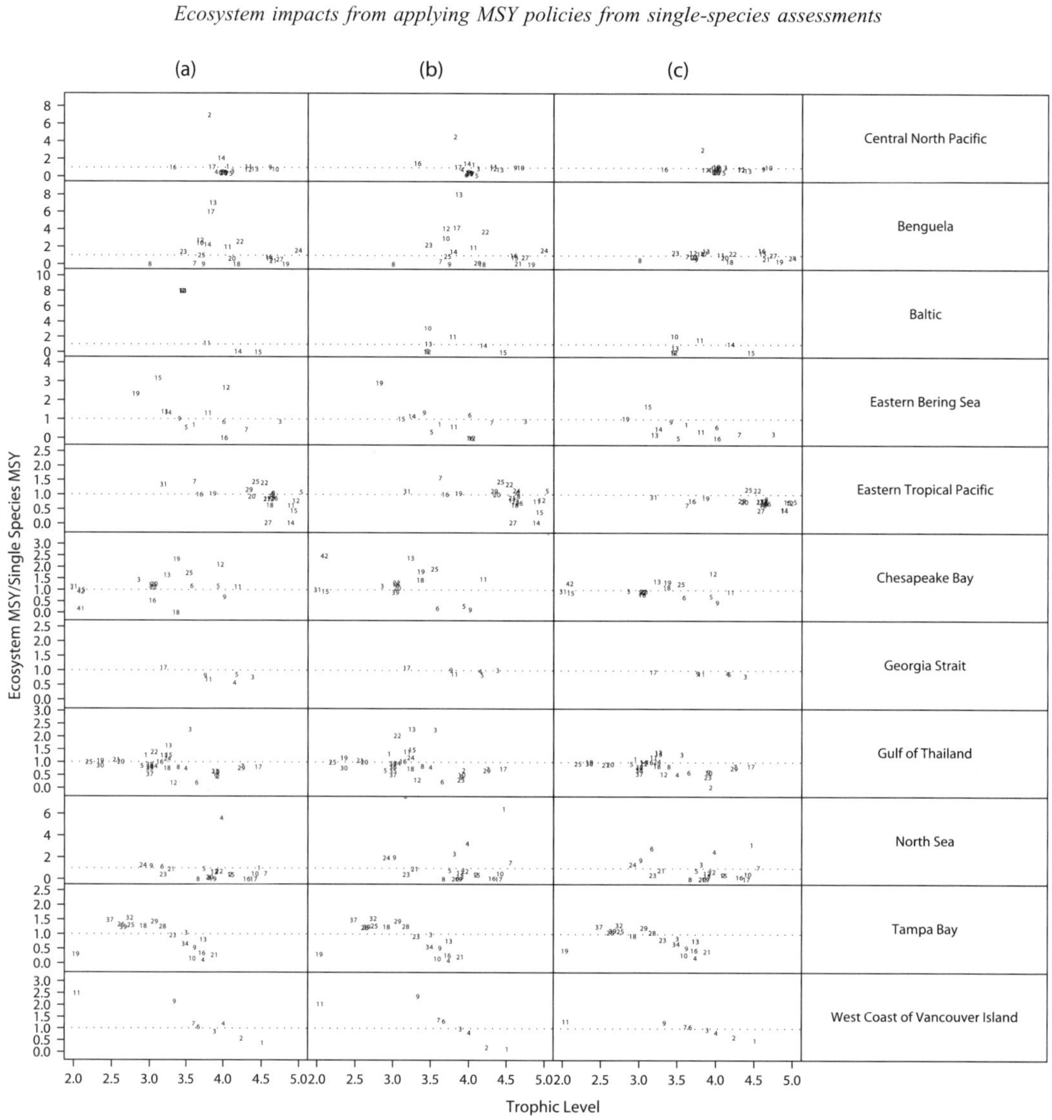

Ecosystem impacts from applying MSY policies from single-species assessments

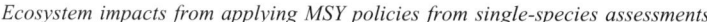

Figure 3. Predicted ratios of equilibrium yield when all species are fished simultaneously at k × F_{MSY}, to equilibrium yield predicted for each species when only that species is fished at F_{MSY} and all others are fished at Ecopath base F (species code numbers are the rows in Table 1): (a) k = 1, applying the F_{MSY} estimated for each species while holding all other species constant; (b) k = 1, applying the F_{MSY} estimated for each species while allowing abundances (but not Fs) of other species to vary; (c) k = 0.7, as case (b) except applying 0.7F_{MSY} for all species, to reduce impact.

to changes in F for the species being tested). As noted in the Introduction, these estimates of F_{MSY} and MSY are generally higher than those predicted by holding all other biomasses constant, because it includes indirect compensatory responses throughout the ecosystem to changes in abundance of the target species.

Second, for each of the two methods of estimating single-species F_{MSY}, we ran one additional long-term simulation, with the F for each harvested species set to its F_{MSY} estimated from the long simulation for that species. Biomasses and catches at the end of this simulation represent Ecosim predicted equilibrium values under the all-species F_{MSY} policy (not to be confused with a policy aimed at maximizing some multispecies MSY). This simulation amounts to pretending that long-term experience from managing each species under a variable harvesting regime could be applied successfully to the ecosystem as a whole, without (i) uncertainty arising from

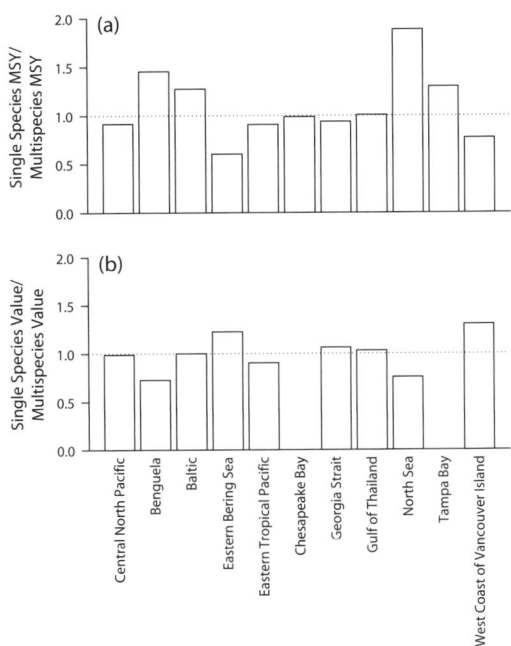

Figure 4. Ratios of predicted total ecosystem (a) yield (biomass), and (b) landed value (based on price multipliers), estimated as the sum of single-species MSY assessments, to equilibrium yield/landed value predicted by Ecosim when all species are fished at F_{MSY} (landed values only for a subset where reasonable price information was available for all species).

incomplete response information for each species, (ii) implementation errors associated with uncertain relationships between target Fs and policy instruments used to try and achieve the targets, and (iii) any adaptive learning after implementation about changes in productivity caused by applying F_{MSY} to other species. Assumption (iii) represents a conservative assessment of our ability to detect changes in productivity in single-species assessment situations.

Results

Estimated F_{MSY} values vary widely over species compared with the Ecopath base F estimates for all case systems (Figure 2). Every system appears to have at least a few lightly fished species ($F \ll F_{MSY}$), and most have at least one species that has been overfished historically ($F > F_{MSY}$). In most cases, the simulation to test widespread application of F_{MSY} over species involves quite a large change in F, compared with Ecopath base values, and such changes might be expected to cause large changes in ecosystem structure whether or not they result in sustainable harvests for all species.

Figure 3 summarizes Ecosim predictions of the impact of widespread application of single-species MSY policies, using ratios of MSYs predicted when all species are harvested at their F_{MSY} rates, to the MSYs predicted for each species when all other species are fished at Ecopath base rates as indicators of relative sustainable (equilibrium) performance, for the two options (interactions held constant and full ecosystem response), respectively. Species are ordered by Ecopath estimates of mean trophic level (calculated from Ecopath base diet compositions). Only in the Georgia Strait case are the predicted performance ratios almost the same for all harvested species, whether they are examined individually or in combination. For all other cases, trophic interactions in combination with widespread F_{MSY} fishing rates are predicted to cause considerable change in community structure and to MSY performance very different for at least some species than would be predicted from single-species assessment.

We expected to see strong negative relationships, with yields increasing at lower trophic levels at the expense of yields at higher trophic levels, as a consequence of appropriation by fishing of production at the lower levels. This effect is evident in most cases, but is also quite variable. We also expected the pattern to be stronger for the simpler models that do not include many species with complex trophic ontogeny, but surprisingly the pattern is most clear in the Tampa Bay case, one of the most complex models. For most models, the main conclusions from the equilibrium yield comparisons are that (i) yields under a many-species F_{MSY} policy can diverge grossly from single-species predictions, and (ii) the direction of divergence is not consistently related to trophic level.

Comparison of the yield ratios in Figure 3 reveals that including the full range of ecosystem effects in the estimates of F_{MSY} can cause quite large differences in predicted patterns of variation among species in the predicted relationship between trophic level and yield ratio. For several models, ratios are more often >1.0, and there are also some drastic shifts in ratios for intermediate trophic level species. Overall, however, the predicted patterns are similar for the two methods of MSY estimation, and for some cases are virtually identical.

Despite predicting large changes in the fish community under widespread F_{MSY} policy application, for most cases Ecosim predicts equilibrium total ecosystem yield from such policies to be quite similar to the sum of the individual MSY estimates (Figure 4a). This surprising result is not just due to the total system yields being dominated by one or a few lower trophic level species, although such species generally do make up the bulk of total yield. Changes in the predicted total landed value of catch for the cases where good data on prices are available (Figure 4b) do not show the expected decrease in landed value if the all-species F_{MSY} policy only caused a shift towards catching lower priced, small fish.

It is possible that the large performance changes shown in Figure 3a and b could be mitigated considerably by use of precautionary F policies, rather than F_{MSY}. To test this possibility, we did additional long-term simulations with the F for every species set to $k \times F_{MSY}$, where $k < 1$ represents a single ecosystem-scale precautionary factor in setting target F. Generally, $k < 1$ would be expected to produce equilibrium yields of around $k(2 - k)$ times MSY (assuming equilibrium biomass decreases linearly with increasing F), e.g. $k = 0.7$ implies a sustainable yield of around 0.9 MSY. Applying a precautionary $k = 0.7$ value in setting all target Fs is indeed predicted to have little impact on total yield, and to have the additional benefit of reducing the departures of catch from those predicted from single-species analysis (Figure 3c).

Discussion

The results in Figure 3 support previous analyses that warn of the need in ecosystem management to provide explicit protection for species whose value derives in part from support of other species ("forage fish"; Baxter, 1997) as well as from harvesting. It is not true that successful regulation of all harvesting to F_{MSY} would "solve" the ecosystem harvest management problem without creating severe conflicts among fisheries. Further, in cases such as the Gulf of Thailand, where most of the catch is taken by non-selective gear (trawling), implementation of F_{MSY} policies that differ widely across species would require major changes in fishing methods, towards more selective practices.

Predicted severity of impacts and trade-offs among different fisheries and values would not have been apparent from simple analysis of pairwise predator—prey relationships. It is fishing whole ecosystems with F_{MSY} policies for every species that leads to Ecosim predictions of severe erosion, not applying MSY policies to just a few selected species, while managing others more conservatively. When trade-offs between fisheries/species are examined pairwise, Ecosim typically predicts a much less severe reduction (or improvement) in MSY for each species, caused by "buffering" effects, such as availability of alternative prey for piscivores when some subset of the prey field is harvested.

The difference between whole-system vs. pairwise responses very likely explains the apparent contradiction between the finding from Ecosim data fitting about asymmetric effects (bottom-up effects not visible), and the predictions that most often warn of decreased yields from top trophic levels (Figure 3), should there be more intense fishing on lower trophic levels. In most cases, whole-system application of F_{MSY} would involve fishing a variety of lower trophic level species much harder than in the past.

With growing pressure for future fishery development, and with most opportunities involving lower trophic level species (Pauly et al., 1998), there will be increasing demand for more precise calculations of just how far towards F_{MSY} policies it would be possible to move without impacting other fishery and existence values. The results presented will be questioned, perhaps with justification, as overestimating the effects of fisheries on one another, and we cannot claim that results are robust to uncertainties in inputs such as diet composition, assumed trophic ontogeny pattern, and productivity change driven by environmental factors. In some cases, the ratio results can be changed drastically just by changing a few diets or by making adjustments to avoid harvesting younger fish in age-structured populations. In cases involving large changes in predicted trophic level, there is no way to be sure that species currently too rare to be included in the models would not increase to take over the trophic roles of others impacted by fishing. Such possibilities are ecosystem-scale analogues of the "black hole" arguments that fishers often raise in response to assessments that indicate a need for reduced fishing ("there are still plenty of fish, just over the horizon, not included in the assessment calculations"), and they are particularly credible in cases where fisheries practices are selective enough not to depress community abundance in general, through simple swept-area impacts.

In addition to reasonable arguments about why Ecosim may overestimate the effects of trophic interaction, there are also arguments for stronger interaction effects than are currently represented in the structure. In particular, switching behaviour by large predators may cause unexpected, sequential depletion in prey type following depletion of any one type by fishing; for example, Springer et al. (2003) argue that killer whale predation may have caused declines in several marine mammal species in the Bering Sea, following the depletion of other whales.

Despite the large number of case studies and relatively long-term monitoring data now available, there simply is not yet enough empirical experience under widely contrasting fishing patterns to allow precise assessment of whether Ecosim (and other ecosystem and multispecies models) can successfully predict the impacts of complex ecosystem harvest management policies. However, we can conclude that there are significant dangers of adopting single-species management approaches that may be myopic, and that these are large enough to justify continuation and improvement in monitoring programmes aimed at detecting unexpected consequences, should they begin. There is also clear justification when uncertainty is so large for treating all fisheries development policies as adaptive management experiments, with a requirement to monitor trophic interaction and fish community effects along with the usual fishery performance indicators.

Acknowledgements

We specially thank Niels Daan, Poul Degnbol, and Lynne Shannon for their valued comment on the draft manuscript. Financial support was provided by a NSF grant to JFK for Apex Predators in Pelagic Ecosystems. VC acknowledges support from the few Charitable Trusts.

References

Baxter, B. S. (Ed). 1997. Forage Fishes in Marine Ecosystems. Proceedings of the International Symposium on the Role of Forage Fishes in Marine Ecosystems, 13–16 November 1996. University of Alaska Sea Grant Report, 97–01.

Brett, M., and Goldman, C. 1996. A meta-analysis of the freshwater trophic cascade. Proceedings of the National Academy of Sciences of the United States of America, 93: 7723–7726.

Christensen, V., Beyer, J. E., Gislason, H., and Vinter, M. 2002. A comparative analysis of the North Sea based on Ecopath with Ecosim and multispecies virtual population analysis. In Proceedings of the INCO-DC Conference Placing Fisheries in their Ecosystem Context, Galápagos Islands, Ecuador, 4–8 December 2000, p. 39. Ed. by V. Christensen, G. Reck, and J. L. Maclean. ACP-EU Fisheries Research Report, 12: vii + 79 pp.

Christensen, V., and Walters, C. J. 2004. Trade-offs in ecosystem-scale optimization of fisheries management policies. Bulletin of Marine Science, 74: 549–562.

Cox, S. P., Martell, S. J. D., Walters, C., Essington, T. E., Kitchell, J. F., Boggs, C., and Kaplan, I. 2002a. Reconstructing ecosystem dynamics in the central Pacific Ocean, 1952–1998. 1. Estimating population biomass and recruitment of tunas and billfishes. Canadian Journal of Fisheries and Aquatic Sciences, 59: 1724–1735.

Cox, S. P., Martell, S. J. D., Walters, C., Essington, T. E., Kitchell, J. F., Boggs, C., and Kaplan, I. 2002b. Reconstructing ecosystem dynamics in the central Pacific Ocean, 1952–1998. 2. A preliminary assessment of the trophic impacts of fishing and effects on tuna dynamics. Canadian Journal of Fisheries and Aquatic Sciences, 59: 1736–1747.

Dill, L. M., Heithaus, M. R., and Walters, C. J. 2003. Behaviorally mediated indirect interactions in marine communities and their conservation implications. Ecology, 84: 1151–1157.

FAO/FISHCODE. 2001. Report of a bio-economic modelling workshop and a policy dialogue meeting on the Thai demersal fisheries in the Gulf of Thailand held at Hua Hin, Thailand, 31 May–9 June 2000. FI: GCP/INT/648/NOR: Field Report F-16 (En). Rome, FAO. 104 pp.

Harvey, C. J., Cox, S. P., Essington, T. E., Hansson, S., and Kitchell, J. F. 2003. An ecosystem model of food web and fisheries interactions in the Baltic Sea. ICES Journal of Marine Science, 60: 939–950.

Hilborn, R., and Walters, C. J. 1992. Quantitative Fisheries Assessment and Management: Choice, Dynamics, and Uncertainty. Chapman & Hall, New York.

Hollowed, A. B., Bax, N., Beamish, R., Collie, J., Fogarty, M., Livingston, P., Pope, J., and Rice, J. C. 2000. Are multispecies models an improvement on single-species models for measuring fishing impacts on marine ecosystems? ICES Journal of Marine Science, 57: 707–719.

Martell, S. J., Beattie, A. I., Walters, C. J., Nayar, T., and Briese, R. 2002. Simulating fisheries management strategies in the Gulf of Georgia ecosystem using Ecopath with Ecosim. In The Use of Ecosystem Models to Investigate Multispecies Management Strategies for Capture Fisheries, pp. 16-23. Ed. by T. Pitcher and K. Cochrane. University of British Columbia, Fisheries Centre Research Reports 10(2). Vancouver, B.C.

Martell, S. J. D. 2002. Variation in pink shrimp populations off the west coast of Vancouver Island: oceanographic and trophic influences. PhD dissertation, University of British Columbia, Vancouver, B.C.

NRC. 2003. The Decline of the Steller Sea Lion in Alaskan Waters: Untangling Food Webs and Fishing Nets. National Academies Press, Washington, D.C. 216 pp.

Okey, T. A., and Wright, B. 2004. Toward ecosystem-based extraction policies for Prince William Sound, Alaska: integrating conflicting objectives and rebuilding pinnipeds. Bulletin of Marine Science, 74: 727–747.

Olson, R., and Watters, G. 2003. A model of the pelagic ecosystem in the eastern tropical Pacific Ocean. Bulletin of the Inter-American Tropical Tuna Commission, 22(3): 91 pp.

Pauly, D., Christensen, V., Dalsgaard, J., Froese, R., and Torres, F. 1998. Fishing down marine food webs. Science, 279(5352): 860–863.

Pitcher, T., and Cochrane, K. (Eds). 2002. The use of ecosystem models to investigate multispecies management strategies for capture fisheries. University of British Columbia. Fisheries Centre Research Reports 10(2). Vancouver, B.C.

Power, M. E. 1992. Top-down and bottom-up forces in food webs: do plants have primacy? Ecology, 73: 733–746.

Shannon, L. J., Christensen, V., and Walters, C. J. 2004. Modelling stock dynamics in the southern Benguela ecosystem for the period 1978–2002. African Journal of Marine Science, 26: 179–196.

Springer, A. M., Estes, J. A., van Vliet, G. B., Williams, T. M., Doak, D. F., Danner, E. M., Forney, K. A., and Pfister, B. 2003. Sequential megafaunal collapse in the North Pacific Ocean; an ongoing legacy of industrial whaling? Proceedings of the National Academy of Sciences of the United States of America, 100(21): 12223–12228.

Walters, C., and Kitchell, J. 2001. Cultivation/depensation effects on juvenile survival and recruitment: implications for the theory of fishing. Canadian Journal of Fisheries and Aquatic Sciences, 58: 39–50.

Walters, C., Christensen, V., and Pauly, D. 1997. Structuring dynamic models of exploited ecosystems from trophic mass-balance assessments. Reviews in Fish Biology and Fisheries, 7: 139–172.

Walters, C., and Martell, S. 2004. Fisheries Ecology and Management. Princeton University Press, Princeton, New Jersey.

Walters, C., Pauly, D., Christensen, V., and Kitchell, J. F. 2000. Representing density dependent consequences of life history strategies in aquatic ecosystems: EcoSim II. Ecosystems, 3: 70–83.

Werner, E. E., and Peacor, S. D. 2003. A review of trait-mediated indirect interactions in ecological communities. Ecology, 84: 1083–1100.

Worm, B., and Myers, R. A. 2003. Meta-analysis of cod–shrimp interactions reveals top-down control in oceanic food webs. Ecology, 84: 162–173.

Yodzis, P. 2001. Must top predators be culled for the sake of fisheries? Trends in Ecology and Evolution, 16: 78–84.

COMPETITION AND HABITAT SHIFT IN TWO SUNFISHES (CENTRARCHIDAE)[1]

EARL E. WERNER AND DONALD J. HALL
*W. K. Kellogg Biological Station and Department of Zoology, Michigan State University,
Hickory Corners, Michigan 49060 USA*

Abstract. The bluegill sunfish (*Lepomis macrochirus*) in small ponds feeds on relatively large prey associated with the vegetation. However, in the presence of the green sunfish (*L. cyanellus*) it shifts to feeding on smaller, less preferred prey in the open water column. The mechanisms responsible for this habitat shift were examined by experimentally confining both species, alone and together, in homogeneous patches of the preferred habitat (vegetation).

When confined together in the vegetation the green sunfish exhibited higher survivorship, growth rates, and amount of food in the stomachs than the bluegill. The bluegill fed on smaller items and consumed more benthic prey than did the green sunfish. The presence of the congener did not alter the food habits of either species or the growth rates of the green sunfish in relation to species stocked alone. Presence of the congener did affect the growth rates of the bluegill. Overlap in the diet was 70% when these species were confined to the vegetation as compared to 44% in an earlier study where habitat separation was permitted.

The green sunfish is more of a sit-and-wait predator and is able to utilize a wider food size spectrum than the bluegill. This results in a strong asymmetry in the competition function favoring the green sunfish in the vegetation. However, in the open water column the distribution of food sizes is truncated and this provides a competitive refuge for the bluegill which handles small foods more efficiently. The bluegill appears to be more flexible in its habitat use while the green sunfish is more aggressive and limited in the habitats utilized.

Comparisons with studies of habitat shifts in salmonids suggest that the competitive mechanisms outlined are of general relevance in fish communities. Consideration of the relation between habitat structure, the correlated distribution of food sizes, and species morphology provides a framework for specifying the occurrence of habitat shifts and which species of the interactive set will shift.

Key words: Asymmetrical competitive effects; competition; fish; food size; foraging strategies; habitat shift; Lepomis; Michigan.

INTRODUCTION

Segregation by habitat is one of the most important means by which ecologically similar species partition resources (Schoener 1974). Shifts in habitat use by species when similar forms are absent thus provide some of the strongest evidence for the action of competition in structuring communities. Such niche shifts have often been documented, particularly in the study of island and mainland faunas or other "natural experiments" (Schoener 1975), but the mechanisms that underlie these habitat shifts have not received careful study. In general, mechanisms have been inferred from niche comparisons that are confounded by different resource levels between study sites as well as ecological vs. evolutionary time scales.

In this study we attempt to determine experimentally the factors responsible for habitat shifts by the bluegill sunfish (*Lepomis macrochirus*) in the presence of a congener, the green sunfish (*Lepomis cyanellus*). These species coexist in natural lakes and streams over much of central North America and segregate on the habitat dimension (Werner et al. 1977). The bluegill is generally found in the water column of the deeper littoral zone (1–6 m deep) and the green sunfish in shallow areas (usually <1 m deep) near the shore, often in areas of emergent vegetation.

Ecological segregation between the bluegill and green sunfish in small ponds has been shown to parallel that observed in natural lakes (Werner and Hall 1976). When these species were stocked together in small ponds (with a third, bottom feeding species, *L. gibbosus*), the bluegill fed to a large extent on open water zooplankton and occupied the deeper areas of the pond. The green sunfish fed predominantly on prey associated with vegetation and inhabited the border of cattails (*Typha*) that surrounded the ponds. However, when stocked in the absence of congeners, the bluegill fed largely on vegetation-dwelling prey (Werner and Hall 1976) suggesting that, in the presence of the green sunfish, the bluegill is forced to shift to the open water habitat and forage less preferred prey.

In order to identify the competitive mechanisms involved in the habitat shift by the bluegill, we confined each species alone and with the congener in homogeneous patches of the vegetation (cattail) habitat. This permitted us to contrast the relative abilities of these species to harvest the (preferred) resources in this habitat with and without direct effects due to the presence of the congener. The nature and magnitude of these effects have implications concerning the segregation of these species along the habitat

[1] Manuscript received 18 October 1976; accepted 12 January 1977.

dimension and asymmetries in the competition function (cf. Roughgarden 1972).

EXPERIMENTAL DESIGN AND METHODS

The experiment was performed in one of the 18 circular ponds (29 m diam, 1.8 m deep) at the Kellogg Biological Station of Michigan State University. The pond contained a uniform and continuous border of cattails (*Typha* spp.) extending 3–4 m from the shore, and growing to a depth of ≈0.9–1.3 m. The cattails were interspersed with a few bulrushes (*Scirpus* sp.) near the shore and *Chara* and pondweeds (*Potamogeton* spp.) in deeper water. Exposed sediments constituted <25% of the bottom, the rest being obscured by a mat of cattail stems, roots, and debris.

The cattails were partitioned from the open water of the pond by a 3.2-mm mesh nylon fence sewn to a continuous piece of flexible polyvinyl chloride plastic pipe. Concrete blocks and gravel anchored the pipe to the bottom. The partition was suspended 15.2 cm above the water surface from floats. This enclosure was then divided into three compartments of equal habitat volume (≈50 m^3).

Six randomly placed 0.5 × 0.5-m quadrats were censused in each compartment for the density of cattails. The mean number of stems ranged from 8 to 10 per 0.25 m^2 and did not differ among compartments.

On 9 July 1974, 500 bluegills were introduced into one compartment (hereinafter Section B), 500 green sunfish into another (Section G), and 250 of each species in the third compartment (Section B × G). Both species were in their second growing season and were obtained from brood ponds on site. The fish ranged from 24 to 37 mm standard length; the mean size (± standard error) for the bluegill was 30 ± 0.2 mm and for the green sunfish, 29 ± 0.2 mm ($n > 150$ in both cases).

Fish were sampled for stomach analyses at intervals ranging from 4 days to 2 wk throughout the 3-mo duration of the experiment. Samples consisted of 10 individuals of each species from each of the 3 sections. The fish were trapped in 20-cm diam acrylic tubes with a hardware cloth funnel attached at each end. To hold densities constant among sections, 5 individuals of each species of sizes equivalent to those removed were replaced in Section B × G on each date. Most of the samples were taken between 0800–1000 h. Stomach contents were enumerated according to 40 taxonomic prey groups (more than 50% identified to genus or species). Intact prey were measured and length–weight regressions were calculated for prey categories which permitted conversion of the stomach counts to dry weight. More than 500 stomachs were examined.

An accurate estimate of fish size was obtained near the midpoint of the experiment (22 August) by trapping 75–175 individuals of each species in the three sections. The fish were anesthetized with MS-222™

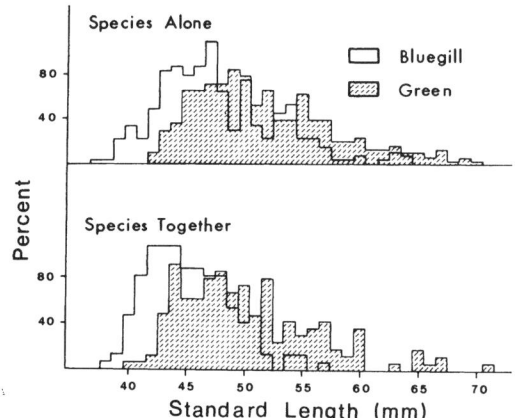

FIG. 1. Size-frequency distributions of the populations recovered in October. The upper panel presents the bluegill and green sunfish when stocked alone; lower panel shows the two species when stocked together.

(tricaine methanesulfonate), measured, and returned to the appropriate section. The experiment was terminated on 10 October by draining the pond. All fish were then counted and measured.

RESULTS

Survivorship and growth

The green sunfish exhibited higher survivorship than the bluegill under all conditions. The number of green sunfish sampled plus those recovered in October accounted for 85% of the initial 250 in Section B × G and 80% of the initial 500 in Section G. Survivorships in the bluegill populations were 75% (B × G) and 70% (B).

Growth in fishes is a very sensitive index of resource availability and is generally positively related to fitness (Bagenal 1967; Hall et al. 1970; Werner and Hall 1976). Reproductive activity in the small fish we used was nil, and survivorship among the populations quite similar; thus, growth rates serve as a measure of the success of these species in the cattail habitat.

The green sunfish grew faster than the bluegill under all conditions (Fig. 1). When stocked as single species populations (Sections B and G), the green sunfish were 24% larger than the bluegill by October (cf. Table 1). When stocked together (Section B × G) the disparity was even greater; the green sunfish were 44% larger than the bluegill. The green sunfish in Section B × G were slightly smaller than those in Section G but this difference is not statistically significant (t-test, $.1 > P > .05$). All other comparisons are significantly different. Thus, the presence of the congener had little effect on the growth of the green sunfish but had a marked effect on the growth of the bluegill.

The differences in size between species were estab-

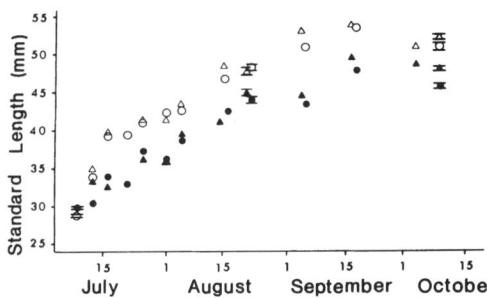

FIG. 2. Seasonal growth pattern of the green sunfish and bluegill when stocked alone and together. The initial, MS-222™, and final samples represent >75 animals per species population and are provided with standard errors. Open symbols represent the green sunfish and solid symbols the bluegill; triangles are populations alone, circles are populations with the congener.

lished very early in the experiment (Fig. 2). However, the populations continued to diverge throughout the summer as seen by comparing the large MS-222™ sample taken in August to the final sample of October. In August the average green sunfish was 0.24 g larger than the bluegill in Section B × G and 0.14 g larger in Sections B and G. In October these differences had increased to 0.31 and 0.22 g, respectively. The observed growth differences were not due to genetic factors; both species exhibited more rapid growth at equivalent sizes in other experiments (Werner and Hall 1976).

Food habits

The food habits of the fish provide clues concerning their foraging behavior in the cattails. Prey were assigned to three categories reflecting their microdistribution: zooplankton or forms that use the open water column, leaf- or stem-dwelling prey, and the benthic infauna or epifauna. Prey that were not restricted to one of these microenvironments were placed in the "other" category. The following analyses are based on dry weight of the prey in the diet averaged over the experimental period.

The major differences in the food of the green sunfish and bluegill occurred in the vegetation-dwelling and benthic categories (Table 2). The proportion of benthic infauna and epifauna in the diet of the bluegill was twice as large (37% and 33%) as that in the green sunfish (18% and 18%) whereas the latter exhibited a greater proportion of vegetation-dwelling prey (39% and 32%) than the bluegill (19% and 19%). The open water prey, of course, are not important to either species in the cattail habitat. These patterns did not vary over the season with the exception of the first sample date when the pre-stocking high resource levels were still evident (*personal observation*) and the diets of both species contained 70%–90% vegetation-dwelling prey. Again, these prey appear to be preferred by both species (Werner and Hall 1976).

The two most important prey contributing to the above differences in the diets were Coenagrionidae and Chironominae which are very different in life habit. Coenagrionidae (damselfly) nymphs are larger and intimately associated with plant stems. The Chironominae (midge) larvae are much smaller and virtually all of the forms encountered were strictly tube-building, sediment-dwelling species (R. King, *personal communication*).

We further compared the diets by computing the percent overlap (α) according to Schoener (1970).

$$\alpha = 100\left[1 - \tfrac{1}{2} \sum_i |P_{x,i} - P_{y,i}|\right],$$

where $P_{x,i}$ is the proportion of the ith prey in species x and $P_{y,i}$ the proportion of the ith prey in species y. Date-by-date comparisons indicated that overlap changed little across the season. Thus, overlap was computed on the basis of the average dry weight of each prey in the diet over the season (utilizing all 40 prey groups).

Overlap between the green sunfish and bluegill was

TABLE 2. Percent contribution of prey to the diets of the bluegill and green sunfish according to microhabitat of the prey. Based on dry wt of prey in the stomachs averaged over the experimental period

Section	Species	Benthic in- and epi-fauna (%)	Vegetation-dwellers (%)	Open water zooplankton (%)	Other (%)
B × G	Bluegill	36.8	19.2	4.2	40.0
	Green sunfish	18.0	38.6	1.4	42.0
B	Bluegill	33.2	19.4	3.9	43.5
G	Green sunfish	17.9	31.8	2.4	47.9

TABLE 1. Average dry wt biomass (± SE) of individual bluegill and green sunfish at the beginning and termination of the experiment. Final weights were computed on the basis of the entire population recovered in October. Dry weights were obtained from length–weight regressions

Weight parameter	Green sunfish G	Bluegill B	Green sunfish B × G	Bluegill B × G
Final weight (g)	1.08 ± 0.02	0.87 ± 0.02	1.01 ± 0.03	0.70 ± 0.017
Initial weight (g)	0.17 ± 0.004	0.16 ± 0.003	0.17 ± 0.004	0.16 ± 0.003
Weight increment (g	0.91	0.71	0.84	0.54

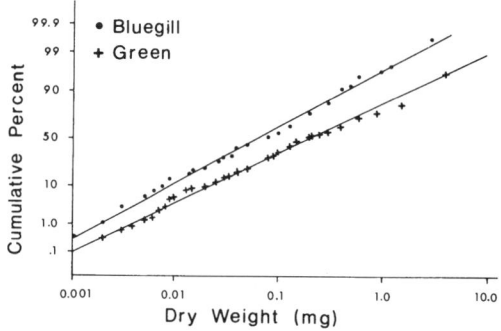

FIG. 3. Cumulative percent of the diet by weight plotted against food size on log probability paper for the green sunfish and bluegill when together.

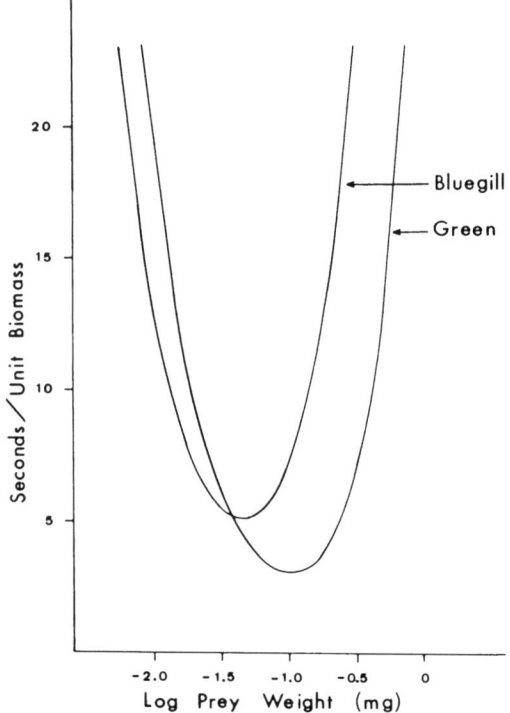

FIG. 4. Cost curves for the green sunfish and bluegill of a common size (0.6 g) plotted against the log of prey weight (after Werner 1977).

72% in the absence of the congener and 70% where the species were together. In a previous study (Werner and Hall 1976) the green sunfish and bluegill were stocked together in a pond but were free to segregate by habitat and the overlap in diet was only 44%. Thus, forced confinement to the cattails increased overlap in foods utilized by a factor of ≈1.6. Note that the food types utilized in the cattails are not influenced by the presence of the congener (see also Table 2); this supports arguments concerning the influence of competitors on food and habitat use (cf. MacArthur and Pianka 1966) which state that the presence of a competitor should not change the kinds of foods consumed within a habitat patch, but may influence the amount of time spent foraging in that habitat relative to other habitats.

Food size

The mean prey size (milligrams dry weight) for each species was computed on every sampling date; these data were then averaged across the season (Table 3). The average food size for the green sunfish was 3.2× that of the bluegill in Section B × G and 2.3× between Sections G and B. Average food sizes of the conspecific populations in the different sections were not statistically different (t-test, in both cases $.4 > P > .2$).

The distributions of food size in the diets of both species were essentially lognormal. Frequency distributions were constructed according to individual prey weight over all prey species for Section B × G, normalized, and the cumulative percent plotted on logarithmic probability paper (Fig. 3). The more gradual slope of the relationship for the green sunfish indicates the greater variance (range) in food sizes utilized by this species. The comparision of Sections B and G gave identical results.

The differences in food size (Table 3) are not simply a result of the relative proportions of prey types in the diet for even within a prey type the green sunfish usually consumed individuals that were larger than those eaten by the bluegill. The average size of Chironominae in the bluegill stomachs in both sections was 0.02 mg but in the green sunfish was 0.04 mg (B × G) and 0.03 mg (G). Average size of Coenagrionidae in the bluegill was 0.15 mg (B × G) and 0.33 mg (B) but 0.46 mg (B × G) and 0.45 mg (G) in the green sunfish. Most other prey were also larger in the green sunfish.

The fact that the green sunfish consistently consumed larger food sizes probably contributes to its greater growth rate in the cattail habitat. Mouth size and other morphological features enable it to handle larger food than a similar-sized bluegill (Werner 1977). However, the green sunfish also quickly grew larger

TABLE 3. Mean food size (milligrams dry weight ± SE) for the bluegill and green sunfish. Data are averages over all sample dates

Species	Section	
	B × G	B and G
Bluegill	0.023 ± 0.0025	0.025 ± 0.003
Green sunfish	0.074 ± 0.015	0.057 ± 0.012

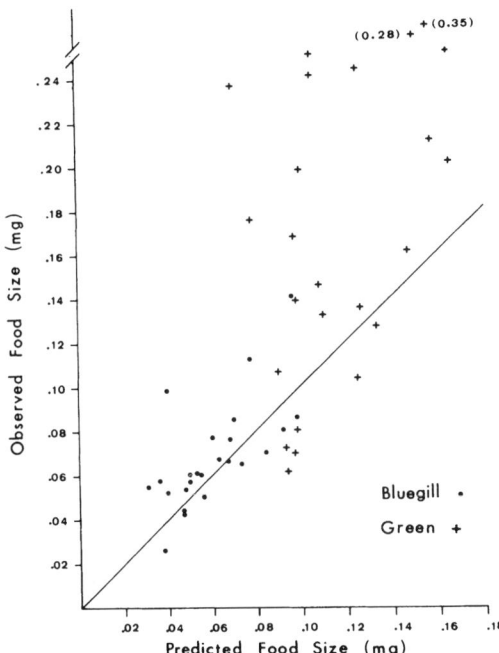

FIG. 5. Observed mean food size for the green sunfish and bluegill plotted against the expected food size predicted by the cost curve analysis.

than the bluegill and remained so throughout the experiment. Consequently, the observed food sizes in the diets must be considered relative to factors of both size and foraging efficiency in order to draw more precise conclusions.

We approached this problem by comparing the mean food size in the diet with a standard derived from independent measurements of the relative abilities of these species to consume different food sizes. The "costs" of handling and pursuing foods of different sizes have been measured as a function of body size for these fish in the laboratory and used to construct "cost curves" ranking prey by their profitability to the predator (Werner 1974, 1977). An example of these curves is provided in Fig. 4 for the green sunfish and bluegill at an identical size (0.6 g). Thus, any change in diet breadth, e.g., expansion, should include prey of equal cost both larger and smaller than the optimum. In addition, we have estimates of the size-frequency distribution of prey available from a similar pond situation (Werner 1977). Once provided with an upper and lower boundary on the diet from the cost curves (i.e., prey of equal cost), we can numerically integrate the area under the resource distribution between these limits, renormalize, and compute an expected prey size for the bluegill and the green sunfish of any given body size. These procedures are outlined more fully in Werner (1977).

As a first idea of the performance of these species relative to their morphological capabilities we computed the expected food size for each species on every sample date based on the mean weight of the fish sampled on that date and a "maximum" diet (i.e., a diet whose limits were the maximum food size the species could handle and the minimum prey size of equal cost). Thus, this procedure takes into account both the differences in body size and morphology between the species. For each date, the observed mean food size was plotted against the predicted value (Fig. 5). Because we do not have estimates of search times, we cannot be more precise about optimal diets. However, the comparison of observed and predicted food sizes can be used as an index of the foraging success of each species.

Mean food sizes of the bluegill did not deviate greatly from those predicted using a diet of maximal width (Fig. 5). On the other hand, mean food sizes of the green sunfish deviated considerably from the predicted, and nearly always in the direction of larger average food size. The green sunfish was therefore selecting larger (higher yield) prey than the bluegill and prey that were larger than predicted on the basis of a diet of maximum width. Although the bluegill was capable of consuming larger prey (based on maximum prey sizes found in the stomachs on the first sample date), it rarely did so. We cannot as yet interpret these results directly in terms of cost/benefit ratios; however, it seems clear that the green sunfish was more capable of obtaining the higher yield resources from the cattail environment.

The apparent differences in foraging success of the green sunfish was further corroborated by the fact that it consistently contained greater absolute amounts of food in the stomach. The mean amount per stomach was computed for each species, pooling the data for all dates and from different sections. The seasonal average stomach content for the green sunfish (3.67 ± 0.45 mg dry wt) was considerably larger than that for the bluegill (2.75 ± 0.2 mg dry wt). These values are significantly different (t-test, $.05 > P > .025$).

Discussion

In an earlier paper (Werner and Hall 1976), we argued that the bluegill largely abandoned habitats containing its preferred prey items due to the presence of the green sunfish. From the present experiment we now know that under identical densities in the preferred habitat (or at least a habitat containing the preferred prey) the green sunfish exhibits significantly greater growth and therefore presumably greater fitness. We now examine the mechanisms that this experiment suggests are responsible for the habitat shift by the bluegill.

Werner (1977) argued that the green sunfish makes broader use of the food size spectrum than the bluegill but that it is more restricted in its habitat use. A constellation of morphological and behavioral attributes are involved. The green sunfish is well suited to the nearshore vegetation habitat due to: (1) a larger mouth in relation to body size which enables it to feed on large prey found there, (2) the moderately fusiform body which increases capture efficiency on (usually larger) prey that require pursuit, and (3) strong homing tendencies (e.g., Kudrna 1967) and aggressiveness (which suggest that the green sunfish is a more sedentary species and is closely tied to particular spatial-structural features of its habitat). In contrast, the bluegill has: (1) a smaller, more protrusible mouth which permits efficient exploitation of small prey, especially zooplankton, but restricts the range of prey sizes consumed, (2) a more laterally compressed, gibbose body and relatively larger fins which enable more efficient and precise maneuvering for prey that must be gleaned in large numbers from vegetation, bottom surfaces, or by turning motions in open water, and (3) a relatively strong tendency to school which suggests greater mobility and utilization of more open habitats. These morphological and behavioral contrasts indicate that the bluegill is more specialized for small foods but is able to utilize a broader array of habitats, points which are further supported by comparative food (Keast 1970; Werner and Hall 1976; Werner 1977) and habitat studies (Werner et al. 1977).

The foraging procedure of these fish has not been systematically studied but anecdotal information points to substantial differences. Observations in ponds and lakes indicate that the green sunfish is more of a sit-and-wait predator and the bluegill a searcher, although both are far from the pure extremes of this dichotomy. The difference in absolute amount of food in the stomachs when confined to the same physical space suggests that the effective density of food in the vegetation is higher for the green sunfish. A fish which spends more time sitting-and-waiting in the dense vegetation may actually encounter more prey, particularly large insect larvae which are known to retreat to cover in the presence of active fish (Charnov et al. 1976; see also Goss-Custard 1970). By contrast, an active, searching predator such as the bluegill may depress the level of prey available to it (Charnov et al. 1976). Apparently the effective resource levels in the cattails for the bluegill were not sufficient for it to obtain quantities similar to the green sunfish or to concentrate on the preferred resources (e.g., large insects); consequently, this species consumed more benthos (Table 2), particularly small, tube-dwelling midge larvae (predominantly Chironominae).

The green sunfish then is capable of consuming larger items than the bluegill and in the vegetation consumes on the average more food per unit time. Roughgarden (1972) and Wilson (1975) point out that these relations produce an asymmetry in the competition function. The fact that the green sunfish was larger in body size after the first few weeks of the experiment simply exacerbates these two differences in feeding relations producing an even greater potential asymmetry.

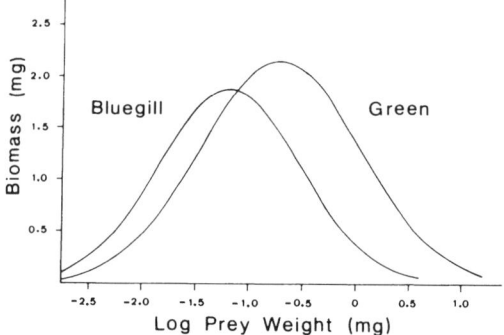

FIG. 6. Distribution of biomass in the diets of the bluegill and green sunfish according to food size. The area under the curves corresponds to the average amount of food in the stomachs of each species. Overlap of the green sunfish on the bluegill is 83% and the bluegill on the green sunfish is 64%.

The data on the size-frequency distributions (Fig. 3) and amounts of food in the stomachs can be used to provide an idea of the magnitude of this asymmetry. The means and standard deviations of the food size distributions in Fig. 3 were estimated by graphic techniques (e.g., Aitchison and Brown 1957); i.e., green sunfish, $\bar{X} = -0.699$ and SD = 0.74, and bluegill, $\bar{X} = -1.18$ and SD = 0.66. Normal curves were then constructed on a logarithmic food size axis using these parameters and weighted such that the total area under the respective curves corresponded to the average amount of food in the stomachs of the two species (Fig. 6). The reciprocal areas of overlap were determined using tabular values of the area under the normal curve knowing the absolute area under each distribution. The overlap of the green sunfish on the bluegill, α_{BG}, was 83% whereas that of the bluegill on the green sunfish, α_{GB}, was 64%. These values, derived from the data, demonstrate that the asymmetry in overlap on the food size dimension is considerable and would confer a decided competitive advantage to the green sunfish in the cattail environment. This may explain why growth of the bluegill was additionally suppressed in the presence of the green sunfish.

The interaction between these species may also involve interference competition. The green sunfish is a very aggressive fish even in the small prereproductive sizes (Greenburg 1947) and clearly dominates bluegills of the same size in aquama to the point that cohabitation is not possible (*personal observation*; Janssen 1974). We have also observed green sunfish defending small areas in ponds (see also Janssen 1974) and exhibiting aggression toward congeners. To the extent that interference competition is operating, it would appear that the green sunfish would dominate the interaction. Such interference would augment the asymmetry in the competitive relations between these species in the cattails.

The differences in aggression between these two species reflect the nature of the habitats that they normally utilize. In shallow, highly structured environments resources are more defendable and aggression is advantageous. In open environments, with dispersed prey, aggression would be maladaptive. There are, of course, costs involved in interference activities (e.g., chasing the actively searching bluegills) which may explain why growth of the green sunfish was not greater in Section B × G than in Section G. However, if this is the case it is clear that interspecific aggression is advantageous in the sense that relative fitness is even greater for the green sunfish when with the bluegill than in the comparison of species alone.

The magnitude of the competitive advantage of the green sunfish is evidently sufficient to cause the bluegill to shift from the vegetation to the open water when not confined to the cattails. In a previous study we showed that this shift occurred and resulted in a 3-fold decrease in average food size for the bluegill (Werner and Hall 1976). Consequently, this habitat shift would entail a marked increase in handling time per unit return of the prey (Werner 1974). Evidently foraging on the small, abundant plankton also reduced search costs as the growth of the bluegill was nearly identical to that of the green sunfish (Werner and Hall 1976), indicating that the relative fitness of the bluegill was indeed improved by this habitat shift.

The open water environment appears to be a competitive refuge for the bluegill. The distribution of prey sizes in open water is truncated such that no advantage is provided to a species able to take larger food (except for piscivory which requires considerable morphological specialization). Also, the dispersed nature of the prey and the less structured environment preclude the effective use of aggression for defending resources. Thus, the asymmetry in the competition relations between the green sunfish and the bluegill so evident in the cattails would not be possible in the open water. In fact, the asymmetry noted above may be reversed (see Wilson 1975).

The open water refuge and relative flexibility in habitat use of the bluegill may explain its success in natural associations where it is often the dominant species (Brown and Ball 1942; Keast 1970; Werner et al. 1977). The deeper, more open habitats are volumetrically the largest in the littoral zones of lakes and thus would be expected to support a much larger species population. Flexibility in habitat use would permit the bluegill to capitalize on the extensive seasonal fluctuations in prey that occur in the different habitats in lakes. This flexibility would also appear to account for the results of our earlier experiment (Werner and Hall 1976) where both species were stocked alone in identical ponds and the bluegill exhibited a significantly greater growth rate than the green sunfish. The bluegill fed predominantly on vegetation-dwelling prey utilizing both the cattails and the areas of abundant submerged vegetation (*Potamogeton foliosus*) found in deeper water. By contrast, 80%–90% of the green sunfish remained in the cattails. These results serve to strengthen our conclusion that habitat flexibility is greater in the bluegill and is an important component of its success. Thwarting this flexibility, as we did in the present experiment, limits its success. Thus, this experiment underscores the nature of the compromises involved in generalizing on the habitat and prey dimensions.

The general relevance of the mechanisms responsible for habitat shift among the centrarchids is supported by a number of excellent studies of salmonids. Allopatric populations of the brown trout (*Salmo trutta*) and arctic char (*Salvelinus alpinus*) in Sweden exhibit a common preferred prey: amphipods, snails, mayfly nymphs, and terrestrial insects found in shallow weedy areas (Nilsson 1960, 1963). When sympatric in similar lakes the char shifts to offshore prey, primarily zooplankton, whereas the trout continues to feed on the preferred prey items. Nilsson ascribes this niche shift by the char to both exploitation and interference competition. The trout is more effective than the char in exploiting the preferred prey items and is much more territorial and aggressive in interspecific encounters. By contrast, the char is more plastic in its feeding behavior, switching readily to zooplankton (even when rare) upon which it feeds with greater efficiency than the trout. A similar interaction occurs between a trout (*Salmo clarki*) and char (*Salvelinus malma*) in sympatric and allopatric populations in Canada (Andrusak and Northcote 1971; Schutz and Northcote 1972).

The parallel nature of species interactions and habitat shifts in the centrarchids and salmonids suggest a general pattern in the evolution of fish communities. Segregation on the habitat dimension is probably the most important means of niche separation in freshwater fishes (Werner et al. 1977). However, habitat type is not independent of resource distribution; the two are usually strongly correlated. Optimal specialization of trophic morphology and feeding behavior is closely linked with the types of habitat utilized and the associated food size distributions (Werner 1977). Cast in this framework, it should be possible to examine the competitive relations between species and predict asymmetries in the competition function which depend on the given habitat, its associated resource distribution, and the morphology of the competing species. We may then begin to forecast the occurrence of habitat shifts and the feasibility of habitat partitioning in order to gain further insight into how communities are assembled.

Acknowledgments

We thank Martin Werner and Thomas Shuba for assistance throughout the study. Robert King and Kenneth Cummins kindly provided information and identifications for several prey groups. Donald Beaver, Dennis Laughlin, Donald Wagner, Patricia Werner, and Leni Wilsmann offered helpful comments on an earlier draft. This research was supported by NSF grants GB-35988 and BMS 74-09013 AO1. Contribution #319 of the W. K. Kellogg Biological Station.

Literature Cited

Aitchison, J., and J. A. C. Brown. 1957. The lognormal distribution. Cambridge Univ. Press. 176 p.

Andrusak, H., and T. G. Northcote. 1971. Segregation between adult cutthroat trout (*Salmo clarki*) and dolly varden (*Salvelinus malma*) in small coastal British Columbia lakes. J. Fish. Res. Board Can. **28**:1259–1268.

Bagenal, T. B. 1967. A short review of fish fecundity, p. 89–111. *In* S. D. Gerking [ed.] The biological basis of freshwater fish production. Wiley and Sons, Inc., New York.

Brown, C. J. D., and R. C. Ball. 1942. A fish population study of Third Sister Lake. Trans. Am. Fish. Soc. **72**:177–186.

Charnov, E. L., G. H. Orians, and K. Hyatt. 1976. Ecological implications of resource depression. Am. Nat. **110**:247–259.

Goss-Custard, J. D. 1970. Feeding dispersion in some overwintering wading birds. p. 3–35. *In* J. H. Crook [ed.] Social behavior in birds and mammals. Academic Press, New York.

Greenburg, B. 1947. Some relations between territory, social hierarchy, and leadership in the green sunfish, (*Lepomis cyanellus*). Physiol. Zool. **20**:267–294.

Hall, D. J., W. E. Cooper, and E. E. Werner. 1970. An experimental approach to the production dynamics and structure of freshwater animal communities. Limnol. Oceanogr. **15**:829–928.

Janssen, J. A. 1974. Competition and growth of bluegills, green sunfish, and their hybrids and the possible role of aggressive behavior. Ph. D. thesis, Michigan State Univ., East Lansing. 66 p.

Keast, A. 1970. Food specialization and bioenergetic interrelations in the fish faunas of some small Ontario waterways, p. 377–411. *In* J. H. Steele [ed.] Marine food chains. Oliver and Boyd, Edinburgh.

Kudrna, J. J. 1967. Movement and homing of sunfishes in Clear Lake. Proc. Iowa Acad. Sci. **72**:263–271.

MacArthur, R. H., and E. Pianka. 1966. On optimal use of a patchy environment. Am. Nat. **100**:603–609.

Nilsson, N. 1960. Seasonal fluctuations in the food segregation of trout, char and whitefish in 14 North-Swedish lakes. Rep. Inst. Freshwater Res. Drottningholm **41**:185–205.

———. 1963. Interaction between trout and char in Scandinavia. Trans. Am. Fish. Soc. **92**:276–285.

Roughgarden, J. 1972. Evolution of niche width. Am. Nat. **106**:683–718.

Schoener, T. W. 1970. Nonsynchronous spatial overlap of lizards in patchy habitats. Ecology **51**:408–418.

———. 1974. Resource partitioning in ecological communities. Science **185**:27–39.

———. 1975. Presence and absence of habitat shift in some widespread lizard species. Ecol. Monogr. **45**:233–258.

Schutz, D. C., and T. G. Northcote. 1972. An experimental study of feeding behavior and interaction of coastal cutthroat trout (*Salmo clarki clarki*) and dolly varden (*Salvelinus malma*). J. Fish. Res. Board Can. **29**:555–565.

Werner, E. E. 1974. The fish size, prey size, handling time relation in several sunfishes and some implications. J. Fish. Res. Board Can. **31**:1531–1536.

———. 1977. Species packing and niche complementarity in three sunfishes. Am. Nat. **111**:553–578.

Werner, E. E., and D. J. Hall. 1976. Niche shifts in sunfishes: Experimental evidence and significance. Science **191**:404–406.

Werner, E. E., D. J. Hall, D. R. Laughlin, D. J. Wagner, L. A. Wilsmann, and F. C. Funk. 1977. Habitat partitioning in a freshwater fish community. J. Fish. Res. Board Can. **34**:360–370.

Wilson, D. S. 1975. The adequacy of body size as a niche difference. Am. Nat. **109**:769–784.

Patterns of Life-History Diversification in North American Fishes: Implications for Population Regulation

Kirk O. Winemiller[1] and Kenneth A. Rose

Environmental Sciences Division, Oak Ridge National Laboratory, Oak Ridge, TN 37831-6036, USA

Winemiller, K. O., and K. A. Rose. 1992. Patterns of life-history diversification in North American fishes: implications for population regulation. Can. J. Fish. Aquat. Sci. 49: 2196–2218.

Interspecific patterns of fish life histories were evaluated in relation to several theoretical models of life-history evolution. Data were gathered for 216 North American fish species (57 families) to explore relationships among variables and to ordinate species. Multivariate tests, performed on freshwater, marine, and combined data matrices, repeatedly identified a gradient associating later-maturing fishes with higher fecundity, small eggs, and few bouts of reproduction during a short spawning season and the opposite suite of traits with small fishes. A second strong gradient indicated positive associations between parental care, egg size, and extended breeding seasons. Phylogeny affected each variable, and some higher taxonomic groupings were associated with particular life-history strategies. High-fecundity characteristics tended to be associated with large species ranges in the marine environment. Age at maturation, adult growth rate, life span, and egg size positively correlated with anadromy. Parental care was inversely correlated with median latitude. A trilateral continuum based on essential trade-offs among three demographic variables predicts many of the correlations among life-history traits. This framework has implications for predicting population responses to diverse natural and anthropogenic disturbances and provides a basis for comparing responses of different species to the same disturbance.

Les caractéristiques du cycle biologique communes à plusieurs espèces de poissons ont été évaluées par rapport à plusieurs modèles théoriques de l'évolution à l'intérieur du cycle biologique. On a recueilli des données sur 216 espèces nord-américaines de poissons (57 familles) afin d'explorer les rapports entre différentes variables et afin de classer les espèces. Des tests multivariés, faits sur des matrices correspondant aux eaux douces, aux eaux de mer et aux deux, ont régulièrement fait ressortir un gradient qui associe les poissons à maturation lente à une fertilité élevée, à la petitesse des oeufs et au nombre restreint de périodes d'activité sexuelle au cours d'une brève saison de fraie, et qui associe les traits opposés aux poissons de petite taille. Un deuxième gradient marqué a indiqué des associations positives entre les soins des parents, la grosseur des oeufs et l'existence de saisons de fraie prolongées. La phylogénie a des effets sur chacune des variables, et certains groupes taxonomiques supérieurs sont associés à des stratégies particulières du cycle biologique. Il tendait à exister un rapport entre la fertilité élevée et l'aire de distribution des grosses espèces en milieu marin. L'âge à maturité, la vitesse de croissance des adultes, la durée de vie et la grosseur des oeufs étaient tous en corrélation positive avec l'anadromie. Les soins des parents étaient en corrélation inverse avec la latitude médiane. Un ensemble trilatéral de données fondées sur des compromis essentiels entre trois variables démographiques, permet de prédire beaucoup de corrélations avec les caractéristiques du cycle biologique. Ce cadre d'examen est utile à la prévision des réponses de populations à différentes perturbations naturelles et d'origine anthropique, et il procure la base pour la comparaison des réponses de différentes espèces à une même perturbation.

Received October 29, 1991
Accepted May 8, 1992
(JB286)

Reçu le 29 octobre 1991
Accepté le 8 mai 1992

Balon (1975) listed the requirements for a comparative framework useful for predicting the response of fish populations to different kinds of environments and disturbances. Such a framework should contain few categories and allow researchers "to build from bits and pieces of available information about reproductive strategies" (Balon 1975). Moreover, it should group similar species irrespective of phylogenetic origin. In other words, adaptive convergences should be stressed over phylogenetic affiliations. Balon's (1975; Balon et al. 1977) reproductive guild framework was based on the premise that environmental requirements and adaptations of early life stages are likely to account for a large amount of the variance in densities and geographical distributions of fish populations. Reproductive guilds permit researchers and resource managers to identify common ecological features and problems in different geographical locations involving different fish faunas.

Because it is qualitative and emphasizes the physiological ecology of early life stages, the reproductive guild concept is limited in its application to many practical problems. Much of population biology and fisheries science is founded in mathematical formulations, and a framework based on developmental physiology does not easily yield quantitative predictions. Hence, a general, yet quantitative comparative framework that could interface with both qualitative schemes, like reproductive guilds, and quantitative population models is desirable. To provide a further step toward a conceptual framework of fish ecological strategies, we examine patterns of life-history variation among North American fresh water and marine fishes and evaluate gradients of variation in relation to several earlier models of life-history evolution.

Because life-history traits are also the fundamental determinants of population performance, the investigation of life-

---

[1]Current address: Department of Wildlife and Fisheries Science, Texas A&M University, College Station, TX 77843-2258, USA.

strategies is central to both theoretical ecology and resource management. Life-history theory deals with constraints among demographic variables and traits associated with reproduction and the manner in which these constraints, or trade-offs, shape strategies for dealing with different kinds of environments. Life-history trade-offs may have a primarily physiological basis (e.g. clutch size and investment per offspring; Smith and Fretwell 1974), a demographic basis (e.g. intrinsic rate of increase and mean generation time; Birch 1948; Smith 1954), an ecological basis (e.g. provision of parental care and clutch size; Sargent et al. 1987; Nussbaum and Schultz 1989), or a phylogenetic basis (Gotelli and Pyron 1991). Of course, organisms consist of complex suites of life-history traits, so that genetic correlations between coevolved traits exhibiting a strong trade-off can indirectly result in correlations with other traits (Pease and Bull 1988). For example, Roff (1981) reevaluated Murphy's (1968) findings for clupeid life histories and concluded that variation in reproductive life span could be explained by its correlation with age at maturity, rather than as a direct evolutionary response to variation in reproductive success.

Insights into the evolutionary response of life-history parameters to different environmental conditions and spatiotemporal changes usually come from two approaches: theoretical models of life-history evolution (e.g. Cole 1954; Cohen 1967; Goodman 1974; Schaffer 1974; Green and Painter 1975; Boyce 1979; Roff 1984; Sibly and Calow 1985, 1986) and analyses of empirical patterns (e.g. Kawasaki 1980; Stearns 1983; Dunham and Miles 1985; Roff 1988; Winemiller 1989; Paine 1990). Both of these approaches have relied, to a large degree, on the r–K continuum (Pianka 1970) or similar unidimensional schemes (e.g. bet-hedging (Murphy 1968) or iteroparity–semelparity gradients (Cole 1954; Schaffer 1974)) as the basis for comparing alternative life-history strategies. Triangular continua containing three endpoint strategies (r-, K-, and stress- or adversity-resistance) have been adopted to interpret patterns and consequences of observed life-history variation in plants and insects (Grime 1977, 1979; Southwood 1977, 1988; Greenslade 1983). Studying fishes in very different environments, Kawasaki (1980, 1983), Baltz (1984), and Winemiller (1989; Winemiller and Taphorn 1989) independently identified three similar strategies as endpoints of a triangular continuum. Following Winemiller (1992), these can be classified as (1) small, rapidly maturing, short lived fishes (opportunistic strategists), (2) larger, highly fecund fishes with longer life spans (periodic strategists), and (3) fishes of intermediate size that often exhibit parental care and produce fewer but larger offspring (equilibrium strategists).

Here, we further evaluate the trilateral continuum model of fish life-history strategies by analyzing data from 216 North American freshwater and marine fish species. Even though many North American species are not included here, the breadth and evenness of phylogenetic coverage should be sufficient to identify major axes of life-history variation and to ordinate species into a framework of basic ecological and demographic strategies. We adopt broad interspecific comparisons under the assumption that consistent intercorrelations among life-history features across widely divergent taxa are likely to reveal both fundamental constraints and adaptive responses to environmental conditions. Life-history traits and strategies are then examined with respect to phylogeny, and observed patterns are evaluated in reference to several models of life-history evolution and population regulation. Finally, we argue that greater understanding of population regulation in fisheries can be achieved by contrasting alternative life-history strategies in relation to different scales of variation in resources and sources of mortality.

Materials and Methods

Life-History Data Set[2]

Estimates of fish life-history traits were obtained from literature sources that summarize large amounts of quantitative data for individual species (e.g. Carlander 1969, 1977; Hart 1973; and synopses of biological data published by Food and Agriculture Organization, Rome). In some instances, we consulted the original studies cited in the species synopses to allow better judgement of the method of estimation and reliability of data. Most fish species exhibit considerable interdemic variation in life-history traits over their geographical ranges. Therefore, we determined the average or modal value of traits by using data from populations located near the center of species' ranges. For example, if a freshwater species ranged from central Canada to the Tennessee River, we sought studies conducted near the latitudes of the Great Lakes. When limited data were available near the center of the range, we sought estimates from peripheral populations in an incremental fashion (working outward from the center of the range). When no reliable data were found for a given trait, that cell in the species by life-history trait matrix was left blank and all calculations calling for the trait eliminated the species from the analysis. Whenever maturation and growth data were reported for the sexes separately, we used the estimates for females. In some instances, total lengths were calculated from standard lengths or fork lengths using published conversion equations. Because no conversion equations were available for North American cavefishes (Amblyopsidae), we estimated conversion ratios from measurements of photographs.

Data were obtained for the following 16 life-history traits.
(1) Age at maturation — the mean age at maturation in years, or when estimates were in summarized form, the modal age of maturation.
(2) Length at maturation — the modal length at maturation in millimetres total length (TL), or if not reported, either the median or minimum length at maturation.
(3) Maximum length — the maximum length reported in millimetres TL.
(4) Longevity — maximum age in years.
(5) Maximum clutch size — the largest batch fecundity reported.
(6) Mean clutch size — the mean batch fecundity for a local population, i.e. data from a specific location or ecosystem, calculated as

$$(1) \quad E = \frac{\sum_{i=1}^{n} N_i F_i}{\sum_{i=1}^{n} N_i}$$

where E is the mean clutch, N_i is the number of individuals in age or size class i, F_i is the number of mature eggs per clutch, and n is the number of age or size classes in the population.
(7) Egg size — the mean diameter of mature (fully yolked) ovrian oocytes (to nearest 0.01 mm).

---
[2]A database listing our principal literature sources, life history traits, and numerical estimates is available, for a nominal fee, from the Depository of Unpublished Data, CISTI, National Research Council of Canada, Ottawa, Ont. K1A 0S2, Canada.

(8) Range of egg sizes — the range of diameters for mature ovarian oocytes reported for a local population (0.01 mm).

(9) Duration of spawning season — number of days that spawning or early larvae were reported.

(10) Number of spawning bouts per year — the mean number of times an individual female was reported to spawn during a year. Because multiple spawning bouts are difficult to document in wild fishes, many of the estimates used in this analysis are probably underestimates. Several recent studies of small cyprinids and percids indicate that multiple clutches may be common in small fishes having small clutches (e.g. Heins and Rabito 1986; Heins and Baker 1988; James et al. 1991). When two fairly distinctive size classes of ova were reported in mature ovaries and other evidence was consistent with a hypothesis of repeat spawning, we recorded the species as having two bouts per year. Following Hubbs (1985), we used an average interbrood interval of 10 d for estimates of spawning bouts for several darters (*Etheostoma*, Percidae) that exhibited strong evidence of multiple clutches.

(11) Parental care — following Winemiller (1989), quantified as Σx_i for $i = 1$ to 3 ($x_1 = 0$ if no special placement of zygotes, 1 if zygotes are placed in a special habitat (e.g. scattered on vegetation, or buried in gravel), and 2 if both zygotes and larvae are maintained in a nest; $x_2 = 0$ if no parental protection of zygotes or larvae, 1 if a brief period of protection by one sex (<1 mo), 2 if a long period of protection by one sex (>1 mo) or brief care by both sexes, and 4 if lengthy protection by both sexes; and $x_3 = 0$ if no nutritional contribution to larvae (yolk sac material is not considered here), 2 if brief period of nutritional contribution to larvae (= brief gestation (<1 mo) with nutritional contribution in viviparous forms), 4 if long period of nutritional contribution to larvae or embryos (= long gestation (1–2 mo) with nutritional contribution), or 8 if extremely long gestation (>2 mo). We reason that, in terms of benefits received by offspring, gestation with nutritional contribution is approximately equivalent to biparental brood guarding during an equivalent time period. Parental care values (Σx_i) ranged between 0 (no care) and 8 (long gestation in some embiotocid fishes) in the North American fish data set.

(12) Time to hatch — the mean time to hatch within the range of values for average midseason temperatures, or when not reported, the mean, modal, or midrange time to hatch at the highest temperature reported within a reasonable range (relative to local ambient temperatures) for a given locality.

(13) Larval growth rate — mean increment in millimetres TL during the first month following hatching. We subtracted the length of larvae at hatching from the mean length attained after the first month. For the few species with larval stage duration <1 mo, we converted mean daily growth rates to mean increments for 30 d.

(14) Young of the year (YOY) growth rate — mean increment in millimetres TL during the first year following hatching or independent life for viviparous fishes. We subtracted the length of larvae at hatching from the mean or modal length attained after the first growing season.

(15) Adult growth rate — mean increment in millimetres TL per year of life over an average adult life span (in this case, data were not weighted by sample size of age cohorts).

(16) Fractional adult growth — mean fraction of millimetres TL gained per year in a normal adult life span, calculated from

$$G = \frac{\sum_{i=2}^{n} I_i/L_{i-1}}{n}$$

where I_i is the annual length gain for adults in age class i, L_i is TL for an adult entering age class i, and n is the number of adult age classes.

Within local populations and size classes, individual fish exhibit considerable variation in life-history traits. Our analysis assumes that measures of central tendency for populations near the center of their species distributions allow the investigation of relationships among life-history variables within a broad inter-specific context.

Phylogeny and Ecological Data Set

Phylogeny of North American fishes follows Lee et al. (1980) for freshwater fishes and Nelson (1984) for marine fishes. We coded family and order as categorical variables for statistical analyses of phylogenetic influences on life-history traits. Each species was classified as either freshwater or marine depending on where the greatest fraction of the life cycle occurred. Because the majority of the estuarine populations extend into other coastal marine habitats, estuarine fishes were included in the marine category. We obtained data for marine populations of *Oncorhynchus mykiss* (steelhead), *Menidia beryllina* (inland silverside), and *Gasterosteus aculeatus* (threespine stickleback) and data for a landlocked freshwater population of *Oncorhynchus nerka kennerlyi* (kokanee salmon).

For each species, the ecological data set consisted of variables characterizing its geographical range, general habitat, and general ecological niche. We recorded the midrange latitude and total range in latitude for each species based on range maps or verbal accounts of species' ranges. Each species was classified as either benthic (1), epibenthic (2), or pelagic (3) based on accounts of the normal depth distribution of adult fishes. The basic adult habitat was classified as either caves or springs (0), small cold-water streams (1), small warmwater streams (2), river channels (3), river backwaters and lakes (4), estuaries (5), marine benthic (6), or marine pelagic (7). Fishes that are common in two or more categories or occupy intermediate habitats were assigned fractional values (e.g. habitat = 2.5 for *Ictiobus bubalus* (smallmouth buffalo) which commonly occurs in both rivers and lakes). Based on summarized diet information for adults, trophic status was classified as either detritivore/ herbivore (1), omnivore (2), invertebrate-feeder (3), or piscivore (4). Fishes that consume large quantities of both invertebrates and fishes formed a fifth intermediate category (e.g. trophic status = 3.5 for *Micropterus dolomieu* (smallmouth bass)). The relative migratory behavior of each species was classified under the headings anadromous (1), sedentary (0), and catadromous (-1). Fishes exhibiting spawning runs from lakes into rivers or from rivers into affluent tributaries (potamodromy) were classified as 0.5, and fishes exhibiting spawning migrations from nearshore to offshore were classified as -0.5.

Data Analysis and Comparisons

Analyses based on broad interspecific comparisons yield patterns that have resulted from many generations and thousands of years of evolution. As a consequence, comparisons involving many taxa and large phylogenetic breadth should contain fewer idiosyncracies due to genetic correlations carried along within a particular phylogenetic lineage (= phylogenetic constraints). If divergent lineages are not given fairly equivalent representation, taxonomic bias can enter into interspecific comparisons (Pagel and Harvey 1988). Some have even argued against the use of species in comparisons (e.g., Ridly 1989).

In essence, the comparative approach requires that a pattern be repeated consistently within a variety of taxa (i.e. conservation or convergence of pattern), or the effect of phylogeny be held constant (or adjusted for statistically), if hypotheses of adaptation are to be tested. To maximize the likelihood that emergent patterns reflect adaption, we examined life-history patterns based on a variety of different combinations of life-history traits and various ecological and phylogenetic subsets of the overall data set.

Because some distributions of raw life-history traits were lognormal, data were ln-transformed for parametric statistical tests. All statistics were calculated using SAS (SAS Institute Inc. 1987). Univariate comparisons between various subsets of the data set used either t-tests (two-sample, two-tailed) or Mann–Whitney U-tests when data were interval or did not approximate a normal distribution. Bivariate relationships among four categorical ecological variables and parental care (ordinal scale) were also analyzed using Spearman's rank correlation coefficient. Chi-square tests for goodness of fit were used to compare frequency distributions of life-history traits for marine and freshwater fishes.

Nested analysis of covariance was used to test for effects of phylogeny and body size on life-history traits. Order and family were used as independent variables in tests of phylogenetic effects (family nested within order). Mean TL at maturation (ln-transformed) was used as an index of body size. For additional insights into the effects of phylogenetic affiliation on life-history patterns, bivariate relationships among several key life-history variables were analyzed within several orders that contained many species.

To explore patterns of association among life-history traits and ordination of species, a series of principal components analyses (PCA) was performed on ln-transformed life-history data. All PCAs were calculated from the correlation matrix to standardize for the influence of unequal variances. Because data were lacking for some traits for some species, analyses that involved more life-history variables resulted in ordination of fewer species. Therefore, separate PCAs were performed on marine, freshwater, and combined fish data sets: one using 12 relatively nonredundant life-history variables (analyses omitting maximum length (redundant with size at maturity), maximum clutch (redundant with average clutch), range of egg size (redundant with egg size), and either length or age at maturation (highly correlated with each other)) and another using only five life-history variables (length at maturation, mean clutch, egg size, spawning bouts, parental care). The five life-history variables retained for the five-variable analyses were selected based on their dominant influence in the 12-variable PCA models, except that length at maturation was substituted for age at maturity in the five-variable data set in order to increase the number of species retained in the analysis.

Although values for parental care were ordinal, we included ln-transformed parental care values in multivariate analyses because they approximated a normal distribution within most groupings and were derived from an algorithm that combined several independent attributes into a single numeric value. Because some life-history researchers have identified an influence of body size on other life-history traits, some PCAs were done both with and without length partialled-out of the other variables (i.e. multivariate analyses are based on relationships among the residuals from regressions of variables with length). To test for potential biases resulting from disproportionate inclusion of some taxa in the global data set (e.g. *Lepomis* species, $N = 8$), PCA was also performed on a data set containing only one species per genus. The species used to represent genera were selected in alphabetical order to reduce experimenter bias.

To further test relationships between life-history patterns and the species' environmental biology, we performed canonical discriminant function (CDF) based on nine life-history variables and three ecological variables (habitat, trophic status, migration) and parental care recorded as classification variables. CDF derives canonical variables from the set of life-history variables in a manner that maximizes multiple correlations of the original variables within groups. To show general associations between ecological groupings and suites of life-history traits, we plotted the means and standard deviations of each class on the first two CDF axes.

Results

Univariate Comparisons

Life-history parameters showed large variation both within the entire data matrix and within orders containing the largest number of species. Standard deviations approached, and in many instances exceeded, the magnitude of the mean values for life-history traits (Table 1). The pygmy sunfish, *Elassoma zonatum* (Centrarchidae), was the smallest fish in the overall North American data set (minimum size at maturation = 25.0 mm). The largest fishes were the Atlantic sturgeon, *Acipenser oxyrhynchus* (Acipenseridae) (length at maturation = 2.5 m), Pacific halibut, *Hippoglossus stenolepis* (Pleuronectidae) (maximum length = 2.7 m), and ocean sunfish, *Mola mola* (Molidae) (maximum length = 3 m). The smallest maximum clutches (batch fecundities) were recorded for two live bearing surfperches (Embiotocidae), *Hyperprosopon argenteum* (12) and *Cymatogaster aggregata* (20). The largest clutch size estimates were for the ocean sunfish (average = 300×10^6), Atlantic cod, *Gadus morhua* (Gadidae) (maximum = 12×10^6), and tarpon, *Megalops atlanticus* (Elopidae) (maximum = 12.2×10^6). Reported estimates of average egg sizes ranged from a minimum of 0.45 mm in diameter for the bay anchovy, *Anchoa mitchilli* (Engraulidae), to a maximum of 20.5 mm for the mouthbrooding gafftopsail catfish, *Bagre marinus* (Ariidae). Average larval growth rates ranged from a minimum estimate of 1.3 mm TL/mo for lake whitefish, *Coregonus clupeaformis* (Salmonidae), to a high of 69.9 mm TL/mo for longnose gar, *Lepisosteus osseus* (Lepisosteidae).

Comparisons between marine and freshwater fishes showed differences in the distributions of life-history attributes (notable examples illustrated in Fig. 1). Statistically significant mean differences are reported with the following notation: t = value from two-sample t-test, z = value from Mann–Whitney U-test. Because both data sets were biased somewhat in favor of larger, commercial species, our interpretations of these univariate comparisons are tentative. As a group, marine fishes matured later (mean marine (m) = 3.38 yr, mean freshwater (f) = 2.74 yr, t = 1.97, df = 194, $P < 0.05$; Fig. 1a), matured at larger sizes (m = 320 mm, f = 186 mm, t = 4.03, df = 202, $P < 0.0001$), lived longer (m = 13.0 yr, f = 9.7 yr, t = 2.19, df = 185, $P < 0.05$), had larger mean clutches (m = 1 554 400, f = 113 376, t = 4.34, df = 189, $P < 0.0001$; Fig. 1b), had longer spawning seasons (m = 103 d, f = 59 d, t = 5.55, df = 213, $P < 0.0001$; Fig. 1c), had larger YOY growth rates (m = 131.2 mm/yr, f = 98.4 mm/yr, z = 2.10, $P < 0.0025$), and had larger adult growth rates (m = 50.8 mm/yr, f = 30.5 mm/yr, z = 3.12, $P < 0.001$) than freshwater

TABLE 1. Mean values (standard deviation in parentheses) for 16 life-history variables based on the entire data matrix (all species) and based on seven major fish orders.

Variable	All species	Clupeif.	Salmonif.	Cyprinif.	Silurif.	Percif.	Scorpaenif.	Pleuronectif.
No. of species	216	12	28	30	12	71	16	11
Maximum length (mm TL)	541 (551)	323 (131)	662 (438)	318 (259)	414 (409)	519 (551)	561 (341)	894 (657)
Age at maturity (yr)	3.0 (2.5)	2.3 (1.4)	3.4 (1.3)	2.7 (1.6)	2.5 (1.3)	2.2 (1.0)	4.7 (1.1)	4.4 (3.0)
Length at maturity (mm TL)	250 (261)	208 (104)	339 (212)	165 (128)	188 (139)	199 (176)	270 (172)	294 (74)
Longevity (yr)	11.1 (11.4)	7.7 (4.3)	11.2 (9.5)	7.0 (4.2)	7.6 (5.1)	9.8 (6.5)	24.4 (15.1)	19.7 (14.2)
Mean clutch size (mature oocytes)	234 000 (679 000)	129 000 (160 000)	13 500 (2 600)	38 000 (86 000)	2 390 (3 330)	356 000 (941 000)	229 000 (312 000)	818 000 (711 000)
Maximum clutch (mature oocytes)	598 000 (1 683 000)	258 500 (274 000)	36 500 (67 500)	74 500 (163 500)	8 840 (19 826)	747 500 (1 746 000)	591 000 (709 000)	2 026 000 (1 487 000)
Egg size (mm diameter)	2.14 (2.23)	1.26 (0.66)	3.4 (1.8)	1.81 (1.06)	6.24 (6.43)	1.20 (0.42)	1.37 (0.68)	1.68 (1.08)
Range egg size (mm diameter)	0.69 (1.53)	0.48 (0.23)	0.93 (0.54)	0.38 (0.25)	2.86 (4.35)	0.29 (0.20)	0.33 (0.22)	1.70 (3.77)
Spawn season (d)	80 (49)	108 (49)	50 (17)	50 (31)	59 (25)	91 (52)	56 (19)	118 (63)
Spawn bouts (no./yr)	3.4 (14.5)	18.1 (51.2)	1.1 (0.4)	2.4 (3.4)	1.7 (0.9)	2.7 (5.5)	1.2 (0.4)	1.4 (1.0)
Parental care (scale 0–8)	1.6 (1.7)	0.2 (0.4)	0.8 (0.6)	1.1 (1.0)	3.4 (0.5)	1.8 (2.2)	3.6 (1.0)	0 (—)
Hatch time (h)	1111 (4084)	88 (68)	1551 (1271)	150 (88)	253 (190)	510 (1522)	994 (637)	165 (120)
Larval growth (mm TL/mo)	14.9 (10.6)	19.5 (16.9)	10.8 (8.9)	14.2 (8.2)	13.2 (3.0)	14.0 (7.6)	12.0 (—)	13.8 (—)
YOY growth (mm TL/yr)	109.9 (76.2)	91.7 (32.7)	133.1 (65.6)	83.6 (56.9)	86.5 (36.0)	115.4 (75.5)	190.8 (61.0)	97.6 (35.7)
Adult growth (mm TL/yr)	38.5 (29.3)	23.7 (11.6)	58.8 (44.5)	29.4 (13.2)	34.9 (185)	36.7 (24.7)	52.9 (19.1)	30.8 (10.2)
Fraction growth (% TL/yr)[a]	0.21 (0.12)	0.13 (0.08)	0.23 (0.12)	0.25 (0.16)	0.24 (0.16)	0.21 (0.09)	0.17 (0.07)	(0.22 (0.08)

[a] Mean percentage of initial TL gained per year of adult life.

fishes. Freshwater fishes tended to have more highly developed parental care than marine fishes ($f = 1.8$, $m = 1.4$, $z = 3.33$, $P < 0.01$; Fig. 1d). Primarily due to skewed distributions, mean maximum clutch sizes ($m = 1\ 102\ 792$, $f = 113{,}376$), mean within-species range of egg diameters ($m = 2.69$, $f = 0.49$), mean hatching times ($m = 786$ h, $f = 1381$ h), and mean larval growth rates ($m = 12.5$ mm/mo, $f = 16.3$ mm/mo) did not differ statistically between groups of fishes classified as marine versus freshwater. Mean fractional adult growth rates were also nearly the same for the two groups ($m = 0.20$, $f = 0.21$). Due to a disproportionate influence of two marine catfishes (Ariidae), mean egg diameters were nearly the same for freshwater and marine fishes ($f = 2.10$ mm, $m = 2.17$ mm), yet distributions of egg size intervals differed significantly (Kolomorov–Smirnov test (K–S), $P < 0.05$; Fig. 1e), and medians differed ($f > m$) based on the nonparametric statistic ($z = 3.78$, $P < 0.0001$). Similarly, after the disproportionate influence of two marine anchovies was reduced through use of the nonparametric test, freshwater fishes tended to have more spawning bouts per year than marine fishes (K–S, $P < 0.05$; $z = 2.15$, $P < 0.05$; Fig. 1f).

Bivariate Life-History Relationships

Correlations between all pairwise combinations of life-history (ln-transformed) and ecological variables are given in Table 2. Some of the high correlations reflect measurement of similar traits or traits that would be expected to covary on a physiological basis (e.g. mean clutch and maximum clutch (0.98), maximum length and length at maturity (0.92), length at maturity and age at maturity (0.77)). Other high correlations need not follow from physiological mechanisms and reflect a diversity of life-history and ecological constraints (e.g. mean clutch and parental care (-0.54), age at maturity and spawning bouts per year (-0.61), longevity and spawning bouts per year (-0.51), length at maturity and YOY growth rate (0.77), mean clutch and YOY growth rate (0.65)).

Figure 2 compares marine and freshwater fishes while illustrating a wide range of bivariate relationships among several of the attributes frequently examined in the life-history literature. The bivariate relationship between ln length at maturity and ln clutch size was nearly the same for fishes categorized as freshwater and marine (Fig. 2a). Neither category showed a strong relationshp between egg size and ln mean clutch size or egg size and ln larval growth rate (Fig. 2b, 2d). The slope was greater for freshwater fishes in the linear relationship between ln mean clutch size and ln YOY growth rate (Fig. 2c), and this difference was due, in part, to the influence of slow YOY growth in cavefishes (*Amblyopsis rosae*, 8.8 mm/yr; *A. spelaea*, 10.6 mm/yr; Poulson 1963).

Correlations between five life-history variables, each measuring an essentially different life-history trait (length at maturity, mean clutch size, egg size, spawning bouts, and parental care), were performed for the seven orders that had the most species with recorded data (Table 3). Overall, matrices of life-history intercorrelations show fairly large deviations among

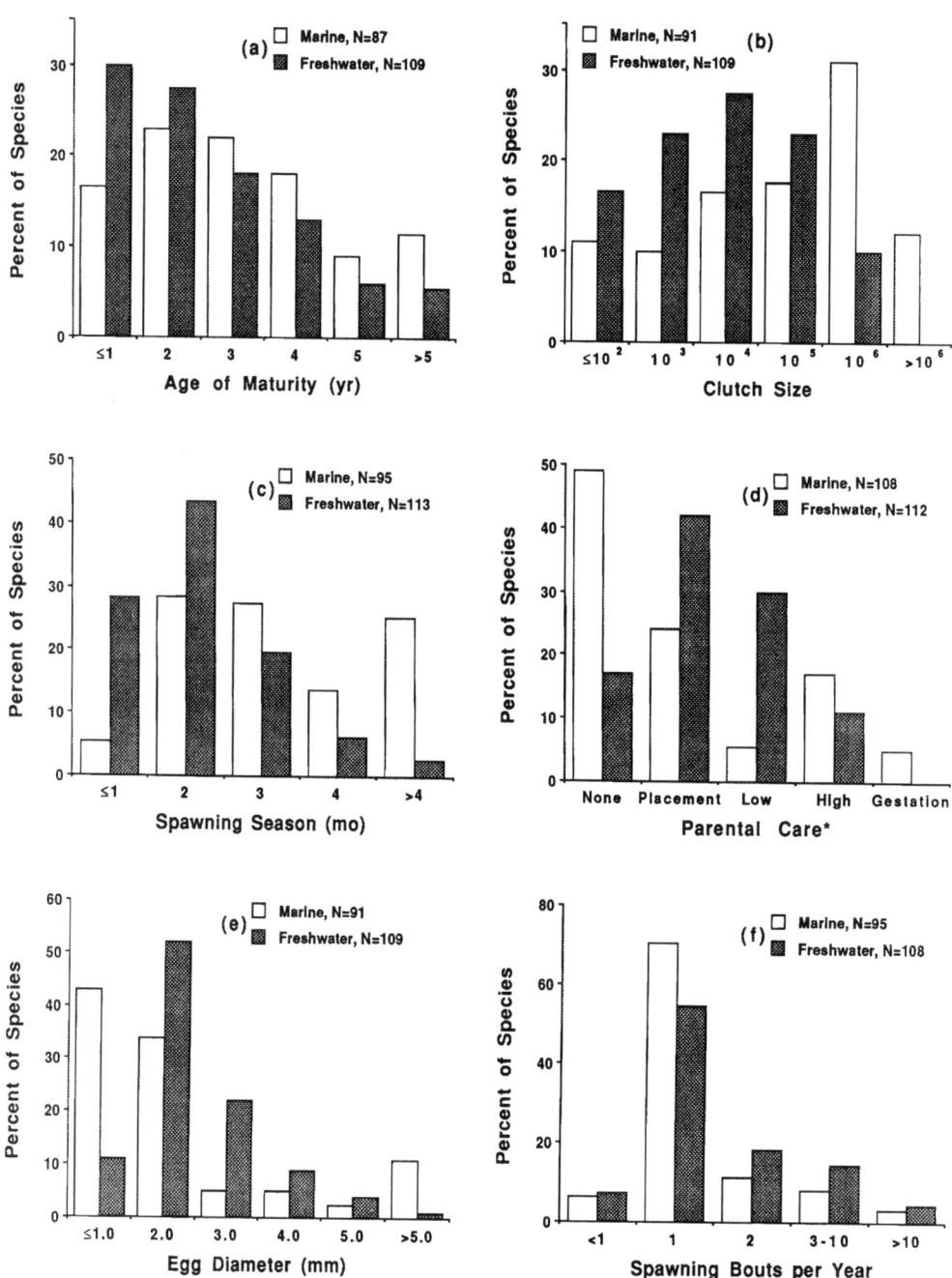

FIG. 1. Frequency distributions of (a) average age at maturation, (b) average clutch size, (c) average duration of the spawning season, (d) parental care (defined under Materials and Methods), (e) average diameter of mature eggs, and (f) average number of spawning bouts per year for fish species classified as either predominantly freshwater or marine.

TABLE 2. Correlation matrix of life-history (ln-transformed) and ecological variables (Spearman's rank correlation; otherwise, correlations are Pearson's product-moment). Correlations are based on all available data for freshwater and marine species (mean N = 171, range of N = 51–216). $*P < 0.05$; $**P < 0.0001$.

Variable	2	3	4	5	6	7	8	9	10	11	12	13	14	15	16	17	18	19	20	21	22
1. Maximum length	0.71**	0.92**	0.78**	0.75**	0.74**	0.19*	0.17	−0.10	−0.49**	−0.32**	−0.12	0.25*	0.77**	0.76**	−0.32*	−0.10	0.36**	0.14^s	0.40**	0.58**	0.16**
2. Age at maturity		0.77**	0.80**	0.53**	0.55**	0.37**	0.17	−0.33**	−0.61**	0.03	0.15	0.22*	0.42**	0.39**	−0.38**	0.30**	0.18*	0.01^s	0.31**	0.30**	0.25**
3. Length at maturity			0.75**	0.73**	0.72**	0.24*	0.23*	−0.21*	−0.51**	−0.30**	−0.04	0.26*	0.77**	0.71**	−0.39**	−0.01	−0.36**	0.18*	0.51**	0.51**	0.25**
4. Longevity				0.66**	0.67**	0.16	0.11	−0.17*	−0.51**	−0.17*	−0.01	0.23*	0.52**	0.44**	−0.43**	0.07	0.16*	0.02^s	0.34**	0.39**	0.06
5. Mean clutch size					0.98**	−0.29*	−0.24*	0.01	−0.37**	−0.54**	−0.45**	0.26*	0.65**	0.58**	−0.34**	−0.18*	0.40**	0.17**	0.44**	0.43**	0.01
6. Maximum clutch						−0.30**	−0.26*	0.01	−0.47**	−0.49**	−0.39**	0.27*	0.62**	0.55**	−0.36**	−0.19*	0.33**	0.17**	0.49**	0.51**	0.03
7. Egg size							0.76**	−0.35**	−0.19*	0.34**	0.52**	−0.13	0.09	0.19*	−0.01	0.34**	−0.13	−0.06^s	−0.21*	0.02^s	0.21**
8. Range egg size								−0.33*	−0.10	0.23*	0.37*	−0.04	0.13	0.13	0.13	0.26**	−0.21*	0.06^s	−0.03^s	0.20**	0.24**
9. Spawning season									0.29**	−0.26**	−0.21*	−0.15	−0.14	−0.06	0.13	−0.41**	0.09	0.04^s	0.22**	−0.04^s	−0.22**
10. Spawning bouts										0.00	0.00	−0.15	−0.26*	−0.24*	0.22*	0.26**	−0.18	0.01^s	−0.21**	−0.20**	−0.20**
11. Parental care											0.37**	−0.07	−0.27*	−0.28*	0.06	0.22*	−0.34**	−0.21*	−0.29**	−0.10^s	0.01^s
12. Hatching time												0.26*	−0.27**	−0.21*	−0.42**	−0.23*	−0.19*	−0.04^s	−0.08^s	−0.02^s	0.25**
13. Larval growth													0.27*	0.17	−0.11	−0.11	0.10	0.03^s	0.17^s	0.06^s	−0.06^s
14. YOY growth														0.69**	−0.07	−0.07	0.38**	0.20**	0.39**	0.39**	0.14
15. Adult growth															0.16	0.04	0.28*	0.27**	0.40**	0.47**	0.24**
16. Fractional growth																	−0.15	0.05^s	−0.16^s	−0.03^s	−0.03^s
17. Median latitude																	−0.35**	−0.27**	0.04^s	0.01^s	0.32**
18. Range latitude																		0.31**	0.46**	0.25**	0.05
19. Water column																			0.18^s	0.36**	0.12^s
20. Habitat																				0.30**	0.45**
21. Trophic status																					0.04^s
22. Migration																					

orders, indicating evolutionary divergences in life-history strategies among higher taxa. Yet, some orders demonstrated high degrees of concordance in the signs of the 10 bivariate relationships, indicating evolutionary conservation or convergences of life-history strategies. Clupeiforms, and siluriforms showed only two of 10 differences in the sign of correlations, and salmoniforms and cypriniforms showed three sign differences, two of which involved $r^2 \leq 0.10$. Except for pleuronectiforms (no statistically significant relationship), length at maturity was always positively associated with mean clutch size, and length at maturity was negatively associated with spawning bouts per year. Except for scorpaeniforms (statistically nonsignificant positive relationship), clutch size was always negatively associated with spawning bouts per year (five of six cases were statistically significant; Table 3).

Large positive correlations between life-history and ecological variables (Table 2) were obtained for maximum length with trophic status (0.58), length at maturity with habitat and trophic status (both 0.51), mean clutch size with habitat (0.44), maximum clutch size with habitat (0.49) and trophic status (0.51), hatch time with median latitude (0.49), and adult growth rate with trophic status (0.47). The largest negative correlations between life-history and ecological variables were for length at maturity with range of latitude (-0.36) and parental care with range of latitude (-0.34). The largest positive intercorrelations among ecological variables (Table 2) were for habitat with range of latitude (0.46) and relative migration (0.45). Fishes at higher median latitudes tended to be associated with smaller ranges in latitude (-0.35), and this trend was heavily influenced by marine species.

Multivariate Life-History Patterns

Results of PCA were nearly the same for the data sets involving five life-history variables with all fish species ($N = 147$) and the data set using only one species to represent each genus ($N = 83$; Table 4). In each case, the first three PCs modeled 91% of the total variation in the data set and resulted in a first axis with endpoints contrasting species with large body size, large clutches, small eggs, few bouts of reproduction per year, and little parental care against those with small body size, small clutches, multiple reproductive bouts per year, and more parental care. In each case, high scores on the second axis reflect associations of large body size, large eggs, well-developed parental care, and few spawning bouts per year (Table 4). High species scores on the third axis reflect large egg size in association with multiple spawning bouts and little parental care.

When maximum length was partialled-out of the data set containing five life-history variables and 147 species, the first three PC axes modeled 85% of the total variation. The first axis reflected an association of clutches (eigenvector = -0.599), egg size (0.593), and parental care (0.509), and the second axis reflected groupings based on maturation size (0.741) in association with number of bouts of reproduction (-0.626) and clutch size (0.219). The suite of characteristics described by the first orthogonal axis from the length-adjusted analysis was similar to that described by the second axis from the unadjusted analysis. Similarly, the second PC axis from length-adjusted data approximated the primary associations predicted by the first axis from the unadjusted analysis.

With 12 life-history variables, the data set consisting of one species per genus ($N = 31$) yielded a PCA in which the first three axes modeled 71% of the total variation. The largest variable loadings (eigenvectors) on the first axis were longevity (0.437), age at maturity (0.375), mean clutch (0.375), YOY

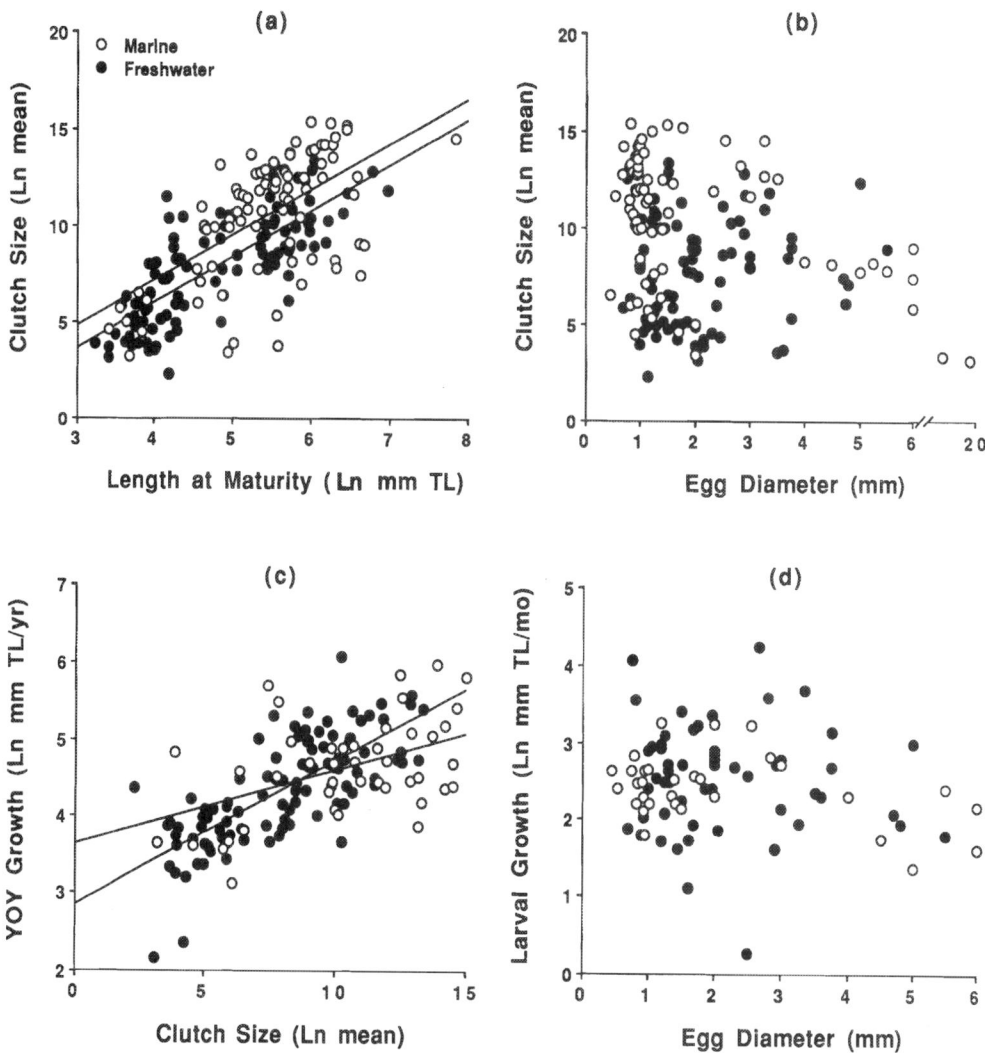

FIG. 2. Bivariate relationships among five life-history traits (open circles = freshwater fishes, closed circles = marine fishes). (a) Freshwater $r^2 = 0.62$, ln clutch = 2.36 (ln length) − 3.48; marine $r^2 = 0.35$, ln clutch = 2.34 (ln length) − 2.24. (b) Freshwater $r^2 < 0.01$; marine $r^2 = 0.17$. (c) Freshwater $r^2 = 0.55$, ln YOY growth = 0.19 (ln clutch) + 2.83; marine $r^2 = 0.35$, ln YOY growth = 0.10 (ln clutch) + 3.63. (d) Freshwater $r^2 < 0.01$; marine $r^2 = 0.20$.

growth (0.348), and spawning bouts (−0.311), and the second axis was dominated by egg size (eigenvector = 0.509), hatch time (0.474), spawning season (−0.407), and parental care (0.397). Based on 12 life-history variables, the PCA using all available fish species ($N = 64$) yielded a PCA that modeled 67% of the total variation. The first axis of this PCA was also dominated by longevity (0.422), age at maturity (0.417), spawning bouts (−0.368), YOY growth (0.346), and mean clutch (0.339), and the second axis was dominated by hatch time (eigenvector = 0.529), egg size (0.509), and mean clutch (−0.340). When maximum length was partialled-out of 12 life-history variables of the all-species data set, the first three PC axes modeled 59% of the total variation. The first axis was dominated by hatch time (eigenvector = 0.439), egg size (0.434), mean clutch (−0.434), and adult growth rate (0.382), and the second axis of the PCA (12 variables with length partialled-out) was dominated by age at maturity (eigenvector = −0.441), longevity (−0.389), spawning bouts (0.359), and YOY growth rate (0.340).

When separate PCAs were performed on data sets that grouped species as either freshwater or marine (unadjusted for length), the variable loadings on the first three PC axes predicted suites of life-history characteristics very similar to those revealed by the combined species data sets. Each of four analyses (freshwater versus marine, 12 versus five life-history variables; based on unadjusted values) identified a continuum of

TABLE 3. Pearson product-moment correlations among five life-history variables by fish order. All variables were ln-transformed. *$P < 0.05$; **$P < 0.0001$; ×, no variation recorded in one variable.

Variable	Clutch size	Egg size	Spawn bouts	Parental care
Clupeiformes (N = 12)				
Length at maturity	0.87*	0.74*	−0.81*	0.54
Mean clutch size	—	0.41	−0.83*	0.50
Egg size	—	—	−0.45	0.26
Spawn bouts	—	—	—	−0.38
Cypriniformes (N = 27)				
Length at maturity	0.78**	0.71*	−0.54*	−0.35
Mean clutch size	—	0.55*	−0.65*	−0.54*
Egg size	—	—	−0.38	−0.10
Spawn bouts	—	—	—	0.36
Perciformes (N = 58)				
Length at maturity	0.88**	−0.18	−0.67**	−0.45*
Mean clutch size	—	−0.48*	−0.54**	−0.65**
Egg size	—	—	0.38*	0.59**
Spawn bouts	—	—	—	0.19
Pleuronectiformes (N = 11)				
Length at maturity	−0.08	−0.34	0.14	×
Mean clutch size	—	−0.27	−0.68*	×
Egg size	—	—	−0.24	×
Salmoniformes (N = 28)				
Length at maturity	0.28	0.71**	−0.31	−0.18
Mean clutch size	—	−0.24	−0.16	−0.44*
Egg size	—	—	−0.43*	0.07
Spawn bouts	—	—	—	−0.04
Siluriformes (N = 12)				
Length at maturity	0.78*	0.26	−0.67*	0.47
Mean clutch size	—	−0.46	−0.72*	0.19
Egg size	—	—	0.41	0.43
Spawn bouts	—	—	—	−0.28
Scorpaeniformes (n = 14)				
Length at maturity	0.57*	0.06	−0.29	0.11
Mean clutch size	—	−0.44	0.40	−0.19
Egg size	—	—	−0.01	0.05
Spawn bouts	—	—	—	−0.40

life-history strategies with large size at maturity (or late maturity), large clutches, few spawning bouts per year, and little parental care on one end and small size at maturity (or early maturity), small clutches, multiple spawning bouts, and more parental care on the other (Table 5). This continuum was associated with the first PC for the freshwater group (based on both 12 and five variables) and the marine group based on five variables, but was associated with the second PC for the marine group based on 12 life-history variables (Table 5). The separate freshwater and marine analyses each identified a second basic life-history continuum that revealed an association between smaller clutches, larger eggs, and more parental care. This axis was identified by the second PC in the freshwater group (both 12 and five variables) and the first PC (12 variables) and second PC (five variables) in the marine group (Table 5).

The ordination of species based on the first two axes of PCA (based on five life-history variables; unadjusted for length) shows a large range of life-history strategies for most orders within both the freshwater (Fig. 3) and marine (Fig. 4) groups. Relative to other freshwater fishes, lake sturgeon (*Acipenser fulvescens*) and paddlefish (*Polyodon spathula*) exhibit an extreme strategy (i.e. extreme positive values on abscissa in Fig. 3) in which age and size at maturation are large, eggs are numerous and small in relation to body size, and spawning is episodic and seasonal. Salmon and trout (*Oncorhynchus, Salmo, Salvelinus* spp.) display a variation of this strategy that involves smaller clutches and greater egg size in relation to body size. Cavefishes (Amblyopsidae) and madtom catfishes (*Noturus* spp.) exhibit a different strategy (upper left region in Fig. 3) that involves large egg size relative to body size and parental care (branchial brooding in cavefishes, nest guarding in certain madtoms). Certain species of darters (*Etheostoma, Percina* spp.) and minnows (Cyprinidae) exhibit a strategy of early maturation and multiple spawning of small clutches consisting of small eggs. A life-history strategy consisting of large maturation size and episodic or seasonal spawning of large clutches of small eggs (lower right region in Fig. 3) is displayed by a taxonomically diverse group of fishes, including the gizzard shad (*Dorosoma cepedianum*), muskellunge (*Esox masquinongy*), burbot (*Lota lota*), and suckers (*Catostomus, Moxostoma* spp.)

The pattern of ordination of marine fishes on the first two PCs (Fig. 4) followed a pattern very similar to that shown by freshwater fishes. Sturgeons again represented an extreme example of the delayed-maturation, large-clutch, periodic-spawning strategy that involves large eggs in absolute sense, but small eggs relative to body size (points on the right-hand half of Fig. 4). Again, salmon exhibited a strategy of low-frequency spawning but larger relative egg sizes and much smaller clutches. The large-clutch, episodic-spawning strategy involving small eggs can be seen in a phylogenetically diverse mixture of fishes, including the Atlantic cod, striped bass (*Morone saxatilis*), cobia (*Rachycentron canadum*), skipjack tuna (*Katsuwonus pelamis*), red snapper (*Lutjanus campechanus*), and winter founder (*Pleuronectes americanus*). Relatively few of the marine fishes fell within the region of large egg size and parental care (upper left region in Fig. 4), but an extreme form of this life-history strategy is seen in the mouthbrooding sea catfishes (*Arius felis, B. marinus*). The strategy of rapid maturation at small sizes and production of multiple small clutches of small eggs (lower left region in Fig. 4) is seen in the bay anchovy, mummichog (*Fundulus heteroclitus*), and inland silverside. Small fishes with parental care lie intermediate between the parental care strategy and early-maturation, multiple-clutch strategy and include the threespine stickleback, gulf pipefish (*Syngnathus scovelli*), and the blackbelly eelpout (*Lycodopsis pacifica*).

Phylogenetic and Ecological Correlates of Life-History Strategies

Mean values for most life-history parameters (length, longevity, clutch size, egg size, parental care, hatch time) varied considerably among fish orders (Table 1). Relative to other life-history variables, larval growth rates (12.0–19.5 mm/mo) and average fractional adult growth rate (proportion of adult TL gained per year = 0.13–0.25) showed the least variability among orders. Phylogenetic affiliations are also apparent in the general pattern of ordination of species within orders in the plots of species scores on the first two PC axes (Fig. 3, 4). Based on ANCOVA, statistically significant effects of phylogeny at the ordinal and family levels were obtained for 15 ln-

TABLE 4. PCA statistics for the North American fish data matrix based on five life-history variables and using all species ($N = 147$ species) and one representative per genus ($N = 83$ genera). Variable loadings (eigenvectors) on the first three principal axes that were between -0.250 and 0 are listed as $-$, and those between 0 and 0.250 are listed as $+$.

	All species			One species per genus		
	PC1	PC2	PC3	PC1	PC2	PC3
Eigenvalue	2.346	1.496	0.688	2.417	1.459	0.696
% variance	46.9	29.9	13.8	48.3	29.2	13.9
Variable						
Length at maturity	0.558	0.290	+	0.498	0.384	0.347
Mean clutch size	0.602	−	−	0.613	−	−
Egg size	−	0.706	0.561	−0.253	0.625	0.582
Spawning bouts	−0.372	−0.465	0.604	−	−0.601	0.652
Parental care	−0.425	0.419	−0.494	−0.507	0.318	−0.334

TABLE 5. PCA statistics for North American freshwater fish data matrices based on 12 and five ln-transformed life-history variables. Data on the left-hand side are for freshwater fishes, and data on the right-hand side are for marine fishes. Variable loadings (eigenvectors) on the first three principal axes that were between -0.250 and 0 are listed as $-$, and those between 0 and 0.250 are listed as $+$.

	PC1	PC2	PC3	PC1	PC2	PC3
	Freshwater species $N = 44$			Marine species $N = 20$		
Eigenvalue	4.898	2.059	1.061	3.979	3.450	1.873
% variance	40.8	17.2	8.8	33.2	28.7	15.6
Variable						
Age at maturity	0.372	+	+	−	0.504	−
Longevity	0.400	−	+	−	0.464	−
Mean clutch size	0.306	−0.397	−	−0.270	0.357	+
Egg size	+	0.531	+	0.426	+	−
Spawning season	−	−	+	−0.346	−0.274	+
Spawning bouts	−0.320	−	−	−	−0.431	−
Parental care	−	−	0.854	0.288	+	−0.402
Hatching time	+	0.580	−0.255	0.463	−	−
Larval growth	+	−0.389	−	−0.305	+	+
YOY growth	0.368	−	+	+	+	0.648
Adult growth	0.347	−	−	0.319	+	0.497
Fractional growth	−0.268	+	−	0.316	−	0.264
	Freshwater species $N = 82$			Marine species $N = 65$		
Eigenvalue	2.457	1.142	0.823	2.157	1.810	0.584
% variance	49.1	22.9	16.4	43.1	36.2	11.7
Variable						
Length at maturity	0.596	+	+	0.450	0.466	0.371
Mean clutch size	0.526	−0.361	−0.281	0.641	−	−
Egg size	0.267	0.696	0.551	−0.256	0.601	0.517
Spawning bouts	−0.421	−0.363	−0.511	−0.400	−0.460	0.516
Parental care	−0.345	0.502	−0.597	−0.401	0.449	−0.566

transformed life-history variables (Table 6). Except for the lack of a significant family effect on adult growth rate, significant effects for the nested ANCOVA type III sum of squares show that phylogenetic effects are present even after the influence of length (maximum TL) as a covariate and the effects of ordinal affiliations for families were removed (Table 6). Relationships between maximum length and three variables (range of egg size, hatching time, duration of spawning season) were not statistically significant within the total fish data set (Tables 2, 6).

Each of the four discriminant function analyses (CDF) resulted in a first canonical axis in which high values corresponded with larger size at maturity, larger clutches, longer life span, and faster growth (Fig. 5; Table 7). When migration was used as the class variable, the first axis was also associated with larger egg size. The second canonical axis was more discordant between tests involving different class variables (Table 7).

The CDF using habitat as the class variable (Fig. 5a) showed a general pattern of larger, longer-lived fishes with large clutches and short spawning seasons in association with the marine environment, estuaries, river backwaters, and lakes versus small fishes with small clutches, slow YOY growth, and long spawning seasons in association with headwater habitats. River fishes were intermediate between headwater and estuarine/marine fishes in the multivariate life-history space defined

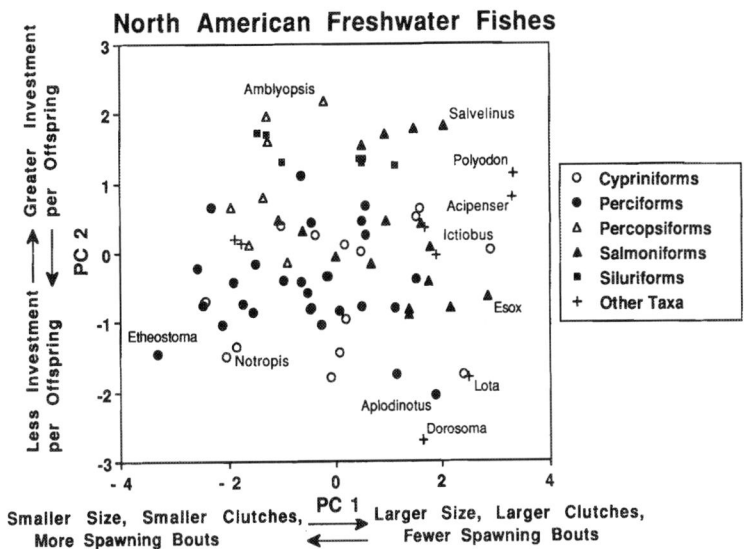

FIG. 3. Scores for freshwater fish species on the first two principle components axes based on five life-history variables (length at maturity, average clutch, egg size, bouts per year, parental care). The two axes are interpreted based on correlations of the original life-history variables (statistics associated with the PCA are given in Table 5).

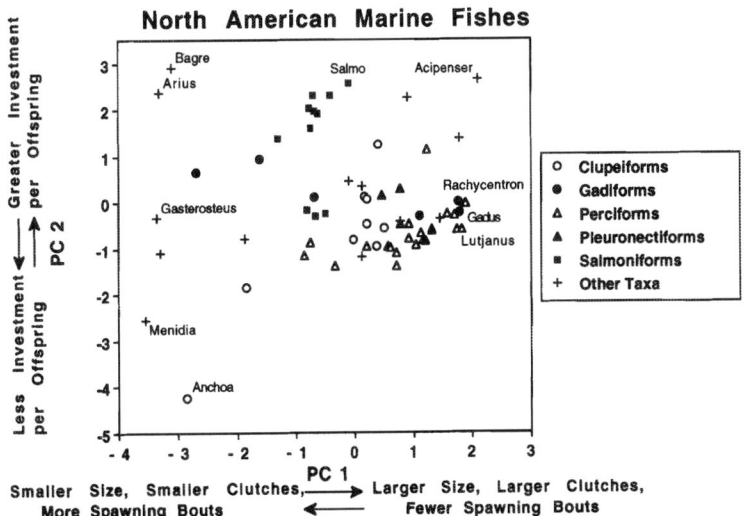

FIG. 4. Scores for marine fish species on the first two principle components axes based on five life-history variables (length at maturity, average clutch, egg size, bouts per year, parental care). The two axes are interpreted based on correlations of the original life-history variables (statistics associated with the PCA are given in Table 6).

by CDF. The CDF based on trophic status (Fig. 5b) showed a general pattern of larger, longer-lived fishes with large clutches, intermediate-sized eggs, and little parental care in association with piscivory versus small fishes with small clutches, slow YOY growth, larger eggs, and more parental care in association with feeding on invertebrates and omnivory. In addition, detritivores tended to be associated with small size, small clutches, small eggs, and little or no parental care (Fig. 5b).

Highly migratory fishes (Fig. 5c) were associated with large body size, long life spans, and large clutches. Catadromous fishes (represented by the American eel, *Anguilla rostrata*) exhibited a lower spawning frequency (i.e. one semelparous

TABLE 6. Results of nested ANCOVA for life-history variables (ln-transformed) with family nested within order and maximum total length as a covariate. Type I sum of squares (SS) are for main effects of variables without adjustment for size effects; Type III SS are for effects of family and order after adjustment for covariation due to size.

Variable		Type I SS		Type III SS		Model r^2
		F	P	F	P	
Age at maturity	Order	16.7	0.0001	5.8	0.0001	0.80
	Family (order)	4.0	0.0001	2.7	0.0001	
	Max. length	110.0	0.0001			
Length at maturity	Order	32.7	0.0001	1.5	0.082	0.91
	Family (order)	16.2	0.0001	1.7	0.015	
	Max. length	317.4	0.0001			
Longevity	Order	14.3	0.0001	2.4	0.0013	0.81
	Family (order)	6.0	0.0001	2.3	0.0012	
	Max. length	91.9	0.0001			
Mean clutch size	Order	27.8	0.0001	7.8	0.0001	0.91
	Family (order)	19.6	0.0001	9.8	0.0001	
	Max. length	149.3	0.0001			
Maximum clutch	Order	26.5	0.0001	8.7	0.0001	0.92
	Family (order)	22.9	0.0001	10.7	0.0001	
	Max. length	171.3	0.0001			
Egg size	Order	14.9	0.0001	11.2	0.0001	0.80
	Family (order)	5.3	0.0001	5.1	0.0001	
	Max. length	8.5	0.0042			
Range egg size	Order	4.7	0.0001	5.4	0.0001	0.70
	Family (order)	4.1	0.0001	4.0	0.0001	
	Max. length	3.8	0.055			
Spawning season	Order	5.5	0.0001	6.2	0.0001	0.57
	Family (order)	2.5	0.0003	2.7	0.0001	
	Max. length	1.9	0.16			
Spawning bouts	Order	8.8	0.0001	5.9	0.0001	0.70
	Family (order)	4.4	0.0001	2.8	0.0001	
	Max. length	36.2	0.0001			
Parental care	Order	12.2	0.001	9.5	0.0001	0.79
	Family (order)	8.7	0.0001	7.5	0.0001	
	Max. length	16.2	0.0001			
Hatching time	Order	14.6	0.0001	5.0	0.0001	0.83
	Family (order)	9.5	0.0001	9.1	0.0001	
	Max. length	1.2	0.26			
Larval growth	Order	2.0	0.028	1.4	0.15	0.47
	Family (order)	1.5	0.14	1.1	0.34	
	Max. length	4.3	0.042			
YOY growth	Order	10.4	0.0001	2.7	0.0006	0.79
	Family (order)	7.2	0.0001	1.8	0.026	
	Max. length	36.7	0.0001			
Adult growth	Order	8.8	0.0001	2.7	0.0009	0.75
	Family (order)	5.6	0.0001	1.3	0.18	
	Max. length	75.3	0.0001			
Fractional adult growth	Order	2.9	0.0005	2.1	0.013	0.46
	Family (order)	2.1	0.011	1.6	0.058	
	Max. length	6.3	0.013			

bout per 14 yr of adult life on average) and smaller egg size when compared with anadromous fishes (primarily salmon) with earlier maturation. Less-migratory fishes tended to be smaller with relatively smaller clutches and smaller eggs than migratory species; however, highly sedentary fishes tended to have longer spawning seasons and larger eggs than migratory fishes.

When CDF was performed with level of parental care as the class variable (none, placement of eggs into a special habitat, low guarding, high guarding/viviparity), fishes with highly developed parental care were associated with the small size, short life spans, small clutches, slow YOY and adult growth, and spawning seasons of intermediate length (Fig. 5d). Fishes with no parental care tended to be large with long lifespans,

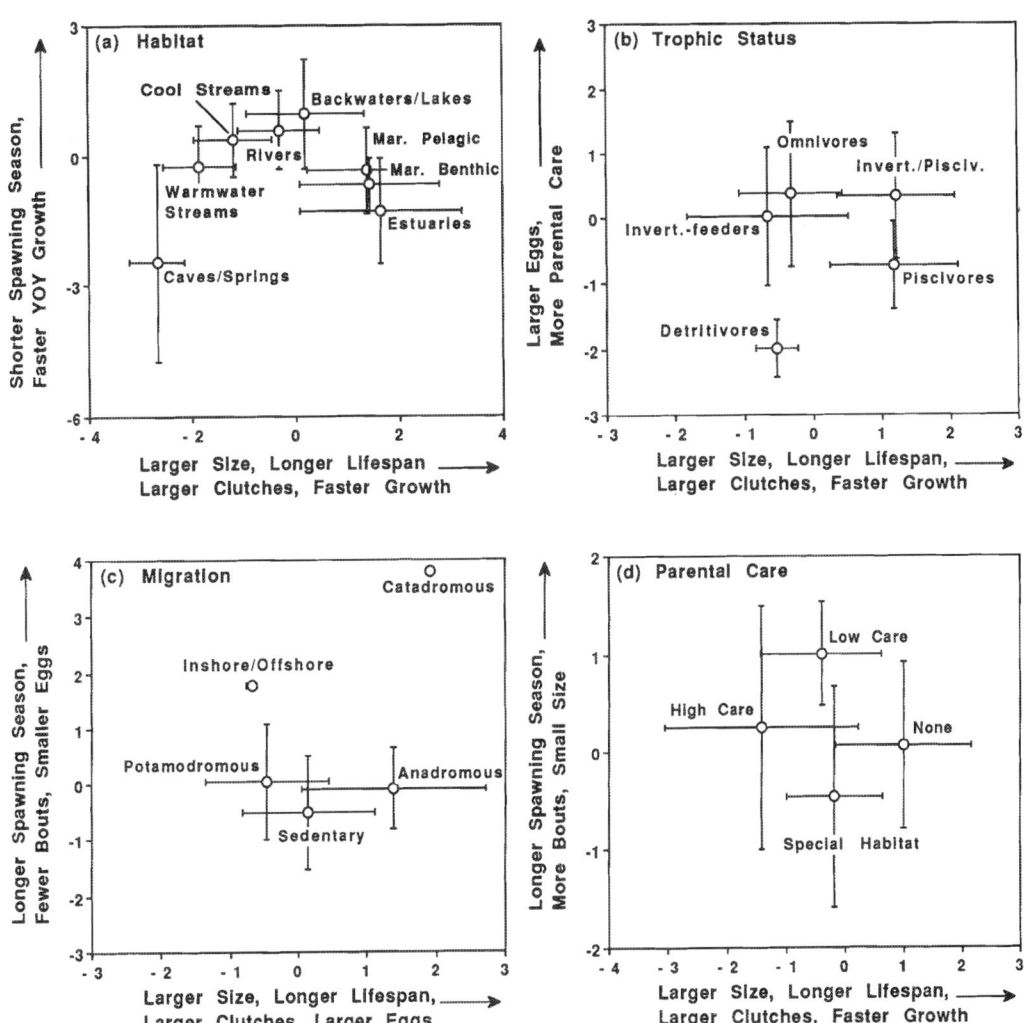

FIG. 5. Average species scores on the first two discriminant function axes by ecological and life-history groupings. Discriminant functions for a–c were based on nine life-history variables (length at maturity, average clutch, egg size, longevity, spawning season, bouts per year, YOY growth, adult growth, parental care), and plot d was based on eight variables with parental care as the grouping variable. The x-axis represents the first canonical axis, the y-axis is the second canonical axis, and verbal interpretations of axes are based on the eigenvectors of the original life-history variables. Statistics associated with each plot are summarized in Table 7.

large clutches, fast YOY and adult growth, and intermediate breeding seasons. Fishes exhibiting no parental care other than special placement of eggs (e.g. scattering over vegetation or gravel beds) tended to be smaller and less fecund with shorter spawning seasons than the group with no parental care. Fishes with low levels of parental care (e.g. brief period of nest guarding by the male) tended toward small body size, intermediate clutches, intermediate YOY growth rates, and longer spawning seasons involving multiple bouts.

Discussion

Life history trade-offs frequently are sought by means of bivariate regression analysis of attributes having a hypothesized functional relationship. For example, one way to achieve larger clutches is to partition reproductive biomass into smaller individual packages, which yields the negative correlation between clutch size and egg size frequently observed in fishes (Wootton 1984; Duarte and Alcarez 1989; Elgar 1990; Fleming and Gross 1990; Paine 1990). In the current study, egg size showed a significant, but weak, inverse relationship with clutch size when the comparison was based on the entire data set ($r = -0.29$, $P < 0.05$). This pattern is not due to a clutch size – body size interaction because many fishes (e.g. *B. marinus*) have much smaller clutches than species of much smaller body size (e.g. *Rhinichthys atratulus* (blacknose dace)). A negative relationship between clutch size and egg size was obtained for

TABLE 7. Statistics associated with canonical discriminant function analyses based on three ecological and one life-history classification variables: habitat, trophic status, migration, and parental care. P values for discriminant function axes represent the probability that canonical correlation for the axis and all that follow are zero (probability that Wilks' lambda $> F$); coefficients for variables are based on total canonical structure.

	Habitat		Trophic status		Migration		Parental care	
	Axis 1	Axis 2	Axis 1	Axis 2	Axis 1	Axis 2	Axis 1	Axis 2
Eigenvalue	2.054	1.142	0.869	0.221	0.555	0.219	0.563	0.289
Variance	0.42	0.23	0.67	0.17	0.60	0.24	0.60	0.31
P	0.0001	0.0001	0.0001	0.14	0.0001	0.09	0.0001	0.0023
Variable								
Length maturity	0.727	0.252	0.702	–	0.757	–	0.513	−0.743
Longevity	0.508	0.328	0.387	–	0.395	–	0.391	−0.537
Mean clutch	0.891	+	0.595	−0.347	0.447	0.265	0.770	−0.352
Egg size	–	0.285	0.232	0.655	0.419	−0.466	−0.386	−0.351
Spawn season	0.324	−0.677	–	–	−0.271	0.684	0.435	0.548
Spawn bouts	–	+	−0.260	–	−0.505	−0.464	+	0.477
Parental care	−0.692	–	+	0.525	+	+		
YOY growth	0.658	0.552	0.753	−0.335	0.326	–	0.623	−0.308
Adult growth	0.663	+	0.890	–	0.628	–	0.428	−0.551

perciforms, pleuronectiforms, salmoniforms, siluriforms, and scorpaeniforms, but the relationship was actually positive for clupeiforms and cypriniforms. Our data set, which involved a broader taxonomic and ecological survey than that used by Duarte and Alcarez (1989), indicates that larger clutches may indeed be produced by either delaying reproduction until achieving a large body size or packaging reproductive biomass into smaller eggs or both. However, not all of the narrower ecological or taxonomic groupings of fishes were consistent with these simple functional trade-offs. Simultaneous trade-offs with other life-history attributes can account for low correlations in instances in which strong functional relationships are hypothesized. Greater insights into the potential adaptive significance of individual attributes often can be gained by examining their interrelationships by multivariate methods. Much of the variance around a bivariate regression often can be explained by simultaneous trade-offs by one or both of the two life-history attributes with other attributes.

Life-History Patterns as Adaptive Strategies

Multivariate methods identified two general gradients of variation that were fairly consistent among the various subsets of life-history variables and species. In one form or another, a principal association was found between larger adult body size with delayed maturation, longer life span, larger clutches, smaller eggs, and fewer annual spawning bouts. In most data sets, a second orthogonal gradient contrasted fishes having more parental care, larger eggs, longer spawning seasons, and multiple bouts against fishes with the opposite suite of traits. When body length was partialled-out of the other life history traits, most of the strong associations were retained. When species are ordered simultaneously on the two primary gradients, three fairly distinctive life-history strategies are identified as the endpoints of a trilateral continuum (Fig. 3, 4). In most instances, the addition of a life-history gradient derived from a third or fourth axis produced little modification in the general pattern of species ordination derived from the first two principal axes.

The three primary strategies (i.e. endpoint strategies) of North American fishes have some striking similarities with earlier patterns presented in the empirical and theoretical life-history literature. We observed (1) species with delayed maturation, intermediate or large size at maturation, large clutches, small eggs, rapid larval and YOY growth rates, and short reproductive seasons, (2) species with early maturation, small size at maturation, small eggs, rapid larval growth, and long reproductive seasons with multiple spawning bouts, and (3) small- or medium-size species with large eggs, small clutches, well-developed parental care, slow YOY and adult growth, and long reproductive seasons. Freshwater fishes have a more restricted range of strategies within life-history space than marine fishes when the two groups are viewed jointly. When we plotted the first two PC coordinates of freshwater and marine fishes together (species loadings associated with Table 4), the orientation and shape of the scatterplot were very similar to Fig. 4, and each of the three endpoints was a marine representative (i.e. bay anchovy, gafftopsail catfish, Atlantic sturgeon).

We observe great consistency among the gradients of life-history variation here and among those derived from comparisons of commerical stocks of marine fishes (Kawasaki 1980, 1983), Pacific surfperches (Baltz 1984), neotropical freshwater fishes (Winemiller 1989), and North American darters (Paine 1990). Wootton (1984) clustered on Canadian freshwater fishes based on five variables reported in Scott and Crossman (1973). He discussed three prinicpal life-history groupings: (1) salmonid fishes with fall/winter spawning, large eggs, large body size, and low relative fecundity, (2) small species with low fecundities, short life spans, and spring/summer spawning, and (3) medium and large species with high fecundities, small eggs, and spring spawning. Based on fishes from a Canadian and a Polish river system, Mahon (1984) interpreted the pattern of species ordination on the first axes of PCA as a gradient of reproductive strategies correlated with a gradient of fluvial habitats (i.e. small, early-maturing, sedentary fishes with parental care and few large eggs in headwaters versus larger migratory fishes with the opposite suite of characteristics in large rivers). Mahon interpreted the second PC gradient as a trade-off between egg size and fecundity.

Given that populations exhibiting these divergent strategies are persistent, insights into population regulation can be gained by comparing suites of life-history traits in the context of adaptations for alternative environmental conditions (Southwood 1988). Next we discuss some hypothesized relationships between primary strategies, life-history trade-offs, and selection caused by different scales of environmental variation.

Periodic strategy

Following Winemiller's (1992) terminology, a "periodic" strategy identifies fishes that delay maturation in order to attain a size sufficient for production of a large clutch and adult survival during periods of suboptimal environmental conditions (e.g. winter, dry season, periods of reduced food availability). Species with large clutches frequently reproduce in synchronous episodes of spawning, and this trend yielded the negative association between clutch size and number of spawning bouts per year. This synchronous spawning often coincides either with movement into favorable habitats or with favorable periods within the temporal cycle of the environment (e.g. spring). Extreme forms of the high-fecundity strategy are often seen among marine species with pelagic eggs and larvae (e.g. cod, cobia, tuna, ocean sunfish). These marine species appear to cope with large-scale spatial heterogeneity of the marine pelagic environment by producing huge numbers of tiny offspring, at least some of which are bound to thrive once they encounter favorable areas or patches within zones and strata. Yet, on average, larval survivorship is extremely low among highly fecund fishes in the marine environment (Houde 1987). Miller et al. (1988) argued that the average larval fish dies during the first week of life, and greater understanding of recruitment may be achieved by seeking greater understanding of the unique features of survivors. Within a local population, some spawners probably contribute disproportionately large numbers of survivors to subsequent generations (relative to conspecifics with similar clutch sizes) based on purely stochastic aspects of larval movement into favorable zones. Despite the fact that egg size tends to be small in periodic fishes, both larval and YOY growth rates tended to be relatively fast. We presume that these fast growth rates for early life stages reflect assimilation from exogeneous feeding by the early survivors that encounter relatively high prey densities. This is consistent with Houde's (1989) assumption that higher growth rates for marine fish larvae at higher temperatures are supported by increased food consumption rather than increased growth efficiency.

At temperate latitudes, large-scale temporal variation in environmental conditions may be as influential as spatial variation on the timing of reproduction. Temporal variation at high latitudes is large and cyclic, hence to some extent predictable. In theory, highly fecund fishes can exploit predictable patterns in time or space by releasing massive numbers of small progeny in phase with periods in which environmental conditions are most favorable for larval growth and survival (Cohen 1967; Boyce 1979). Natural selection should strongly favor physiological mechanisms that enhance a fish's ability to detect cues that predict the moment of a periodic cycle (e.g. photoperiod, ambient temperature, solute concentrations). In the marine pelagic environemnt at low latitudes, large-scale variation in space may represent a periodic signal as strong as seasonal variation at temperate latitudes (i.e. patchily distributed physical parameters, primary production, zooplankton, etc., due to upwellings, gyres, convergence zones, and other predictable currents; Sinclair 1988; MacCall 1990).

Periodic strategists are among the most migratory of North American fishes, an association also revealed by Roff's (1988) analysis of marine fishes and observed among South American freshwater fishes (Winemiller 1989). In California, anadromous threespine stickleback were more periodic in their characteristics (e.g. large clutches, larger size at maturity) when compared with freshwater populations (Snyder 1990). Migration to favorable habitats for spawning is a means by which fishes can reduce uncertainty in their attempts to exploit large-scale temporal and spatial environmental variation. For example, American shad (*Alosa sapidissima*) are more iteroparous and devote a greater fraction of energy to migration at middle and higher latitudes where environments are less stable and less predictable (Leggett and Carscadden 1978). Rothschild and DiNardo (1987) viewed anadromy as a means by which adults seek favorable environments for larval development whereas the reproductive success of marine broadcast spawners may depend on rates of encounters by larvae with suitable zones or patches. Massive clutches of small eggs undoubtedly enhance dispersal capabilities of wide-ranging marine fishes during the early life stages. In a stable population, losses due to advection ultimately are balanced by the survival benefits derived from the passage of some fraction of larval cohorts into suitable regions or habitats (Sinclair 1988).

Opportunistic strategy

The "opportunistic" life-history strategy in fishes appears to place a premium on early maturation, frequent reproduction over an extended spawning season, rapid larval growth, and rapid population turnover rates, all leading to a large intrinsic rate of population increase (Winemiller 1989, 1992). Opportunistic fishes differ markedly from the r-strategists of Pianka (1970) and others in having among the smallest rather than largest clutches. The strong inverse relationship between the rate of population growth and generation time has been appreciated for a long time (Birch 1948; Smith 1954; Lewontin 1965; Pianka 1970; Michod 1979). Small fishes with early maturation, small eggs, small clutches (yet high relative reproductive effort), and continuous spawning are well equipped to repopulate habitats following disturbances or in the face of continuous high mortality in the adult stage (Lewontin 1965). This suite of life-history traits permits efficient recolonization of habitats over relatively small spatial scales. Extreme examples of the opportunistic strategy are seen in the bay anchovy, silversides, killifishes (*Fundulus* spp.), and mosquitofishes (*Gambusia* spp). These small fishes frequently maintain dense populations in marginal habitats (e.g. ecotones, constantly changing habitats) and frequently experience high predation mortality during the adult stage.

Equilibrium strategy

The "equilibrium" strategy in fishes is largely consistent with the suite of characteristics often associated with the traditional K-strategy of adaptation to life in resource-limited or density-dependent environments (Pianka 1970). Large eggs and parental care result in the production of relatively small clutches of larger or more advanced juveniles at the onset of independent life. Our equilibrium strategy differs from the traditional K-strategy model in that equilibrium strategists tended to rank among the smallest fishes rather than largest (the largest North American fishes were periodic strategists). Within the North American fish data set, marine ariid catfishes (egg diameters 16.0–20.5 mm, oral brooding of eggs and larvae) and amblyopsid cavefishes (branchial brooding of small clutches of relatively large eggs) represent extreme forms of the equilibrium life-history strategy. Cavefishes probably inhabit the most temporally stable and resource limited of the aquatic environments covered in our survey.

Intermediate strategies

Three endpoint life-history strategies are fairly distinctive, but intermediate strategies are also recognized near the center and along the boundaries of a trilateral gradient. Some of the largest periodic strategists (e.g. lake sturgeon, paddlefish) have relatively large eggs, which compromises the attainment of a theoretical maximum clutch size. Salmon and trout possess much larger eggs and smaller clutches than fishes exhibiting the extreme periodic life-history configuration. An inverse relationship between egg size and clutch size was found within the salmoniforms (Table 3). Among populations of coho salmon (*Oncorhynchus kisutch*), this inverse relationship is observed over a latitudinal gradient of declining egg size and increasing clutch size with increasing latitude (Fleming and Gross 1990). Fleming and Gross suggested that selection favors local optima in egg size with clutch size adjustments resulting from physiological constraints and ecological performance. Relative to periodic strategists with larger clutches and smaller eggs, salmon and trout appear to have adopted a more equilibrium strategy of larger investment in fewer offspring and larger offspring at the time of independent life. Migration to special spawning habitats and burial of zygotes (brood hiding) by salmon, char, and trout could be viewed as forms of parental investment that carry large energetic and survival costs in relation to future reproductive effort.

A number of medium-size fishes have seasonal spawning, moderately large clutches, and male nest guarding (e.g. *Ameiurus* spp., *Lepomis* spp.). Another intermediate group has large clutches, small eggs, and viviparity (e.g. *Sebastes* spp.). These fishes are in between the periodic and equilibrium endpoints of the gradient. Small fishes with rapid maturation, small clutches, large eggs relative to body size, and a degree of parental care (e.g. *Pimephales* spp., *Noturus* spp., *Etheostoma* spp., *G. aculeatus*, *S. scovelli*, *Cottus* spp.) lie between the opportunistic and equilibrium strategists. Similarly, small fishes with seasonal spawning, moderately large clutches, small eggs, and only one or a few bouts of reproduction per season (e.g. *Osmerus mordax* (rainbow smelt), *Notemigonus crysoleucas* (golden shiner), *Notropis* spp., *Percopsis omiscomaycus* (trout-perch)) lie between the opportunistic and periodic strategists on the gradient. Determinations of multiple spawning in fishes have been difficult (Heins and Rabito 1986; Heins and Baker 1988), and we suspect that some of the small North American fishes categorized as single spawners (and some species conservatively coded as two bouts per year here) may actually spawn several times each season. With improved estimates of multiple spawning, some of the small fishes intermediate between opportunistic and periodic strategists may actually cluster nearer the opportunistic endpoint.

Life-History Strategies and Population Regulation

Life-history theory attempts to explain patterns of covariation in demographic parameters and reproductive traits in relation to alternative environmental conditions (Lack 1954; Stearns 1976; Whittaker and Goodman 1979; Southwood 1988). For example, Murphy (1968) and Kawasaki (1980, 1983) each concluded that delayed maturation, iteroparity, and high fecundity increase the probability of recruitment to the adult population in the face of variable preadult mortality in marine environments. Armstrong and Shelton (1990) used a Monte Carlo model to show that within-season serial spawning reduces variation in clupeoid brood strength when within-season variation is large. Several other studies have explored the potential influence of alternative life-history strategies on fish population responses to both natural and anthropogenic disturbances (Adams 1980; Saunders and Finn 1983; Garrod and Horwood 1984; Ware 1984; Rago and Goodyear 1987; Schaaf et al. 1987; Barnthouse et al. 1990; Beverton 1990; Leaman 1991).

Following Winemiller (1992), we propose that the essential features of the three primary life-history strategies can be captured by the interrelationships among three basic demographic parameters: survival, fecundity, and onset and duration of reproductive life. In terms of life-history strategies, fitness can be estimated by either V_x, the reproductive value of an individual or age class (Fischer 1958; Pianka 1976; Leaman 1991), or r, the intrinsic rate of natural increase of a population or genotype (Birch 1948; Cole 1954; Southwood et al. 1974; Roff 1984; Stearns and Crandall 1984), or λ, the finite rate of growth from the Euler equation (and see Ware (1982, 1984) for discussions of surplus power as a measure of fitness). Each of these fitness measures can be expressed as a function of three essential components: survivorship, fecundity, and the onset and duration of reproductive life. In the case of reproductive value:

$$(2) \quad V_x = m_x + \sum_{t=x+1}^{\omega} \frac{l_t m_t}{l_x}$$

where for a stable population, m_x is age-specific fecundity, l_x is age-specific survivorship, and ω is the last age class of active reproduction. In this form of the equation, two components of reproductive value are added together: the current investment in offspring (m_x) and the expectation of future offspring (residual reproductive value). Reproductive value is therefore equivalent to the lifetime expectation of offspring (i.e. provided that x is equal to α, the age at first reproduction) and contains survivorship, fecundity, and timing components.

The intrinsic rate of population increase can be approximated as

$$(3) \quad r \simeq \frac{\ln R_0}{T}$$

where R_0 is the net replacement rate, T is the mean generation time ($T \simeq \sum_{x=\alpha}^{\omega} x\, l_x\, m_x$), and

$$(4) \quad R_0 = \Sigma\, l_x\, m_x$$

$$(5) \quad r \simeq \frac{\ln (\Sigma\, l_x\, m_x)}{T}.$$

The finite rate of growth (λ) is computed from the intrinsic rate of increase using $\lambda = e^{r\Delta t}$, where Δt is the time interval over which population change is measured. Therefore, the relative rate of population increase (or relative reproductive success among genotypes within a population) is directly dependent on fecundity, timing of reproduction, and survivorship during both the immature and adult stages (Lewontin 1965; Southwood et al. 1974; Southwood 1988; Itô 1978; Kawasaki 1980; Roff 1984; Taylor and Williams 1984; Sutherland et al. 1986; Winemiller 1992). Over long time periods and on average, the three parameters (l_x, m_x, T) must balance or the population declines to extinction or grows to precariously high densities and crashes.

Three primary life-history strategies result from trade-offs among age at maturation (α positively correlated with T), fecundity, and survivorship and are illustrated as the endpoints of a trilateral surface in Fig. 6. Here, we choose to focus attention on juvenile survivorship, the fraction of individuals surviving from the zygote stage until first reproduction. Plotting demographic trade-offs in relation to juvenile survivorship (l_α) results in separation of the equilibrium strategy of higher juvenile survivorship from the opportunistic and periodic strategies, whereas an axis of adult survivorship (the mean expectation of future life, $\bar{E}_x$, for all adult age classes, $x = a$ to ω, where $E_x = [\Sigma l_y]/l_x$ for $y = x$ to ω) would separate the opportunistic strategy

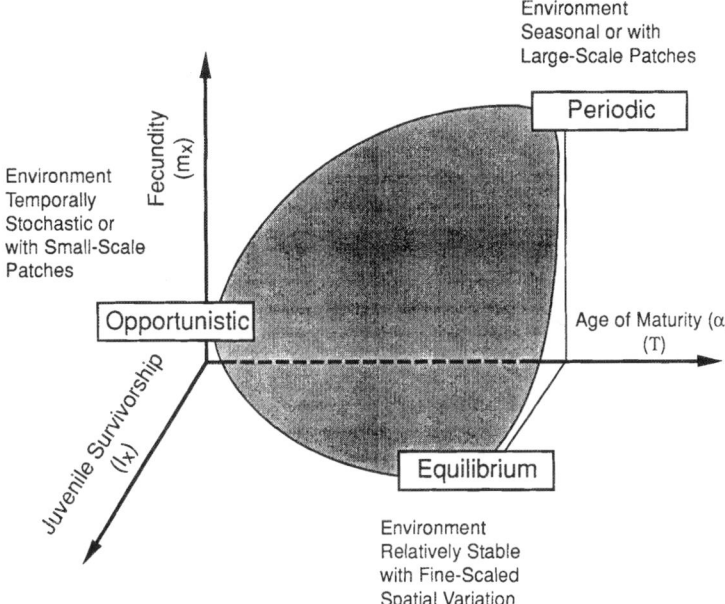

FIG. 6. Model for an adaptive surface of fish life-history strategies based on fundamental demographic trade-offs and selection in response to different kinds of environmental variation. The opportunistic strategy (small T, small m_x, small l_x) maximizes colonizing capability in environments that change frequently or stochastically on relatively small temporal and spatial scales. The periodic strategy (large m_x, large T, small l_x) is favored in environments having large-scale cyclic or spatial variation. The equilibrium strategy (large l_x, large T, small m_x) is favored in environments with low variation in habitat quality and strong direct and indirect biotic interactions. Curvilinear edges of the surface portray diminishing returns in the theoretical upper limits of bivariate relationships between adult body size and clutch size, adult body size and parental investment/offspring (a correlate of juvenile l_x), and clutch size and juvenile survivorship.

of low adult survivorship from the other two strategies. The suites of traits predicted by this adaptive surface appear similar to those described in the trichotomous comparative frameworks proposed for plants (Grime 1977, 1979) and other animal groups (Walters 1975; Allan 1976; Greenslade 1983). Our periodic strategy corresponds to high values on both the fecundity and age at maturity axes (the latter a correlate of population turnover rate) and a low value on the juvenile survivorship axis. Our opportunistic strategy (optimization of population turnover rate via a reduction in developmental time) corresponds to low values on all three axes. The equilibrium strategy is defined by low values on the fecundity axis and high values on the age at maturity and juvenile survivorship axes.

Figures 7 (freshwater) and 8 (marine) plot the positions of North American fishes in relation to three life-history variables used as surrogates (strong correlates) for the demographic axes in the trilateral gradient life-history model. Because data on size at maturation were available for more species, and size and age at maturation are highly correlated ($r = 0.77$), we used ln maturation size as a surrogate for age at maturation (x-axis). We used ln mean clutch size for fecundity (y-axis) and investment per progeny (calculated as the sum of ln parental care value and ln egg diameter) to reflect differences in the probability of juvenile survivorship (z-axis). The basic form of each empirical plot conforms well with the triangular surface predicted by the demographic model of primary life-history strategies (Fig. 6). Figures 7 and 8 strongly imply that natural selection eliminates certain combinations of life-history traits, such as late maturation/small clutches/small investment per offspring. Other combinations of life history traits, such as the "Darwinian superorganism" (early maturation/large clutches/large investment per offspring/long life span), are prohibited by direct physical and physiological constraints.

Response of the periodic strategy

The periodic strategy maximizes age-specific fecundity (clutch size) at the expense of optimizing turnover time (turnover times are lengthened by delayed maturation) and juvenile survivorship (maximum fecundities are attained by producing smaller eggs and larvae). Several theoretical models predict maximization of fecundity in response to predictable (= seasonal) environmental variation (Cohen 1967; Boyce 1979). If conditions favorable for growth and survival of immatures are periodic and occur at frequencies smaller than the normal life span (Southwood 1977), selection will favor the strategy of synchronous reproduction in phase with the periodicity of optimal conditions and production of large numbers of small offspring that require little or no parental care. In effect, the periodic strategy represents an iteroparous "bet-hedging tactic" on a scale of interannual variation.

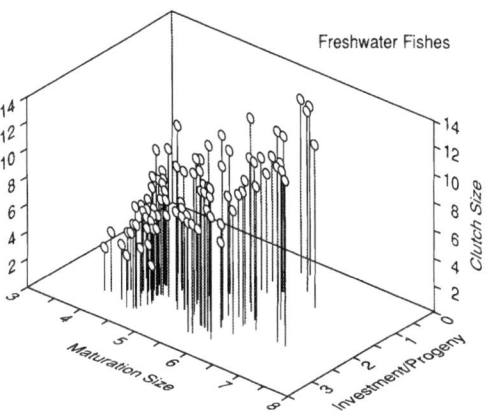

FIG. 7. Three-dimensional plot of ln maturation length (correlate of age at maturity), ln mean clutch size (fecundity), and relative investment per progeny (a surrogate for juvenile survivorship that was equal to ln ((egg diameter + 1) (parental care + 1))) for 88 freshwater fishes. The three axes permit visualization of the basic pattern of life-history variation in relation to demographic trade-offs and strategies (Fig. 6).

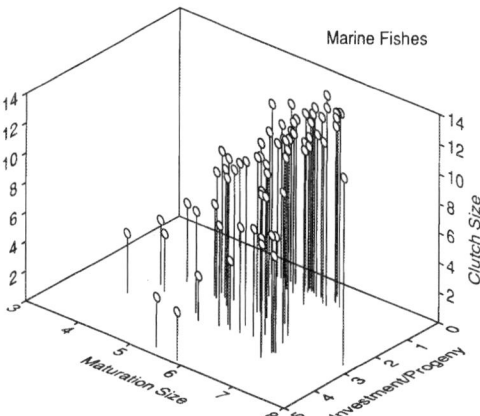

FIG. 8. Three-dimensional plot of ln maturation length, ln mean clutch, and investment per progeny for 68 marine fishes. The life-history pattern generated by marine fishes is similar to that shown by freshwater fishes (Fig. 7) and reflects essential demographic trade-offs and strategies illustrated in Figure 6.

According to this model, large body size enhances adult survivorship during suboptimal conditions and permits storage of energy and nutrients for future bouts of reproduction. The periodic production of offspring allows the organism to repeatedly sample the environment (for most perennial species, this occurs only on an annual basis) until, sooner or later, reproduction coincides with favorable conditions and the fitness payoff is collected (Murphy 1968; Armstrong and Shelton 1990). All environments exhibit either spatial or temporal variation that is to some degree predictable. This must be especially true in both freshwater and marine environments because the periodic strategy of large clutch sizes is predominant among bony fishes in both temperate and tropical settings. At arctic and temperate latitudes, the seasonal cycle appears to have selected for fish life-history cycles calibrated to annual periodicity. In marine pelagic environments, large-scale spatial patchiness of both larval food resources and predation mortality have been estimated by a variety of means (e.g. Sherman et al. 1984; McGurk 1986; Smith et al. 1989; Fortier and Gagné 1990; Brander and Hurley 1992). Large variation in the recruitment of annual cohorts appears to be the rule among periodic fishes in both marine and freshwater habitats, and the effects of climate on the distribution of larval food and predators have been implicated as principal determinants of both spatial and temporal variance in larval growth and survival (Sissenwine 1984; Fletcher and Deriso 1988; Shepherd and Cushing 1990). Variability in recruitment of several marine fishes was shown to be greater in populations closer to the latitudinal extremes of species' ranges (Myers 1992). Based on 21 marine fishes, Pepin and Myers (1991) found correlations between recruitment variability and both change in body length and duration of the larval stage. This finding is consistent with our view that the periodic strategy results in rapid growth and enhanced survival in a small fraction of initial larvae that somehow encounter zones or time periods in which food resources are abundant.

Response of the opportunistic strategy

As stated earlier, the opportunistic life-history configuration maximizes the intrinsic rate of population growth (r) through a reduction in the mean generation time (T). This reduction in T is achieved via early maturation, which in turn diminishes the capability to produce large clutches and large eggs. Owing to their small size, the relative reproductive effort of opportunistic strategists is continuously high, despite the fact that absolute clutch size and egg size are small. The production of multiple batches sometimes results in annual fecundities (mass) that greatly exceed female body mass in these small species (Wootton 1973; Hubbs 1976; Burt et al. 1988). We noted a tendency toward larger eggs and parental care in some small benthic fishes, within both freshwater and marine environments. As Mahon (1984) suggested, increasing juvenile survivorship by means of parental care may be a more viable tactic than increasing fecundity or egg size, given the material constraints imposed by small body size in these fishes. Of course, this view implies that other ecological factors (e.g. adult feeding ecology or microhabitat requirements) constrain the evolution of larger body size.

Birch (1948) was among the first to stress the strong negative correlation between population turnover time (strongly correlated with α and T) and the potential for rapid population growth in density-independent settings (high r). The advantages of early maturation for the achievement of a rapid population growth rate, and hence colonizing ability, were clearly demonstrated by Lewontin (1965), Taylor and Williams (1983), and Sibly and Calow (1986) among others. In effect, the opportunistic life history can be viewed as a sort of null life-history strategy that is favored in density-independent settings or environments that vary unpredictably on small scales of space and time. Fishes exhibiting the opportunistic strategy seem to be associated with very shallow marginal habitats: killifishes on the fringes of salt marshes and mosquitofishes in small headwater tributaries, pool edges, or swamp fringes. These shallow edge habitats (terrestrial/aquatic ecotones) are the kinds of environments that experience the largest and most unpredictable spatial and temporal changes on scales measured in days or hours. Changes in precipitation and temperature induce major alterations in water depth, substrate characteristics, and productivity in shallow aquatic habitats. The opportunistic suite of life-history charcteristics allows fishes to rebound from local disturbances in the absence of intense predation and resource limitation.

The opportunistic strategy of repeated local recolonizations through continual and rapid population turnover is well exemplified in the bay anchovy and marsh-dwelling killifishes like the mummichog. Over much of its range, the bay anchovy ranks among the most numerically dominant fishes (Morton 1989). Bay anchovies are also one of the principal food resources for a variety of fishes, birds, and invertebrates. We postulate that early maturation and frequent spawning permit the bay anchovy to sustain large numbers in regions where local subpopulations are continually cropped by predators. For bay anchovy, silversides, and other small pelagic fishes, predation pressure may partially override the signal from environmental seasonality. The opportunistic strategy appears to be more common in tropical freshwaters than in the temperate zone (Burt et al. 1988; Winemiller 1989).

Response of the equilibrium strategy

Following the basic premise of the K-selection theory of life histories (Pianka 1970), we postulated that the equilibrium strategy in fishes optimizes juvenile survivorship by apportioning a greater amount of material into each individual egg and/or the provision of parental care. Trade-offs between body size, egg size, and clutch size in fishes are probably inherent in the requirement for an adult body size sufficient to permit successful implementation of parental care behavior (e.g. nest defense, brooding, and gestation until larvae or embryos have attained sufficient size to escape predation or to forage efficiently). The tactic of providing parental care for a small number of offspring is of no advantage in highly seasonal habitats or environments dominated by strong density-independent selection. We believe that the equilibrium configuration of life-history traits is best understood in relation to the traditional model of density dependence and resource limitation (e.g. Pianka 1970; Roughgarden 1971; Goodman 1974). For example, some stream-dwelling darters and madtoms may experience resource limitation within shallow riffle microhabitats during periods of reduced stream flow. Microhabitats that provide refuge from predators could also be considered a resource that is either limiting or causes fishes to compete for depleted food supplies on the margins of the refuge (Fraser and Cerri 1982). In the marine environment, parental care appears to be most frequently associated with small benthic fishes (e.g. pipefishes, seahorses, some gobies, eelpouts). Among North American fishes, parental care was positively correlated with median latitude, but this trend was due to a major influence of high-fecundity marine fishes with pelagic larvae from low latitudes. Small benthic marine fishes were not well represented in the data set, and the relationship between parental care and latitude is inverse among freshwater fishes ($r = -0.36, P < 0.0001$). On a global basis, parental care appears to be most advanced and common within tropical ichthyofaunas in both freshwater (Winemiller 1989) and shallow, marine benthic environments (e.g. coral reefs; Miller 1979).

Salmon and trout life histories seem to represent variations of the equilibrium strategy that involve anadromous spawning migrations, egg hiding, and slow larval development rates rather than parental care in the traditional sense of nest guarding or brooding. The growing season at northern latitudes may be so short that large fishes are prohibited from adopting a brood guarding tactic in oligotrophic systems. Data on Arctic char (*Salvelinus alpinus*) indicated that the growing season at high latitudes probably constrains age at maturity and the frequency of spawning (Dutil 1986). Senescence associated with semelparity in Pacific salmon might have evolved as a consequence of the survival cost of returning to the sea after energetically costly upstream runs to fluvial habitats that enhance larval survivorship. By comparison, freshwater whitefishes (*Coregonus, Prosopium*) exhibit a perennial periodic strategy that involves large clutches, small eggs, and annual spawning bouts (Hutchings and Morris 1985).

Phylogenetic Trends

All life-history variables used in this study showed highly significant effects of phylogeny (Table 6). Yet, overall, the pattern of basic life-history strategies was highly repeatable within data sets containing different life-history traits and species. The high concordance between patterns revealed within this comparative study and others offers evidence that essential trade-offs produce numerous convergences in life-history strategies. For example, mouth brooding has evolved in numerous widely divergent equilibrium-type fishes worldwide (e.g. Osteoglossidae, Ariidae, Cichlidae), and miniature, multiple-clutching opportunistic-type fishes are found within many large families, especially in shallow environments in the tropics (e.g. Characidae, Cyprinidae, Cyprinodontidae, Gobiidae). Formal phylogentic studies of life-history evolution would contribute immensely to our understanding of the scenarios favoring the evolution of specific life-history strategies (Gotelli and Pyron 1991).

The restriction of some phylogenetic clades to subregions of life-history space results from historical events shaping evolution, morphological and physiological design constraints, and the interaction of these with contemporary ecology. Despite the reality of phylogentic constraints, wide divergence within several higher taxa is apparent in plots of species scores in relation to the life-history continua that were reflected in the principal axes from PCA (Fig. 3, 4). Perciforms are a widely divergent group, particularly within the freshwater data set. Cypriniforms exhibit a wide range of strategies primarily along an axis linking the periodic and opportunistic endpoints. Salmoniforms show a range of strategies along an axis between a periodic strategy and an intermediate strategy located midway between the equilibrium and periodic extremes. In the marine group, clupeiforms span an axis between opportunistic and periodic strategies, and gadiforms span an axis between the periodic and equilibrium strategy endpoints. Several other orders show much greater restrictions in their ranges of life-history strategies (e.g. percopsiforms tended to be equilibrium strategists, acipenseriforms and pleuronectiforms were relative periodic strategists).

Life-History Strategies and Fisheries Management

Life-history theory can provide some general insights into where attention might be most profitably focused in monitoring and research (Adams 1980; Garrod and Horwood 1984; Ware 1984; Schaaf et al. 1987; Barnthouse et al. 1990; Leaman 1991). For example, the periodic strategy is designed to exploit differences in environmental quality on temporal and spatial scales that are relatively large and to some degree regular or predictable. We have noted that the periodic strategy is predominant among commercial fish stocks worldwide. At arctic and temperate latitudes where spawning is largely annual and synchronous, generations are often recognizable as fairly discrete age cohorts. Correlations between parental stock densities and densities of YOY recruits are frequently weak or negligible in periodic strategists (Garrod 1983; Fletcher and Deriso 1988; Shepherd and Cushing 1990). Consequently, fisheries projections rely on juvenile cohort estimates as short-term forecasters of catchable stocks.

We view the periodic strategy as the perennial tactic of spreading reproductive effort over many years (or over a large area), so that high larval/juvenile survivorship during one year (or in one spatial zone) compensates for the many bad years (or zones). For example, anadromous female striped bass live up to 17 yr on average and produce an average clutch of 4×10^6 eggs every year or two, which translates to an egg to maturation survivorship of roughly 3×10^{-8} to maintain an equilibrium population. We speculate that most years probably result in a larval survivorship approaching zero for most females. From the standpoint of an individual female, it is likely that the fitness payoff only comes during one or two spawning acts over the course of a normal life span. These extremely low expectations for larval survivorship are far smaller than measurement errors involved with field estimates of mortality. In species like striped bass, the variance in larval survivorship that serves as input for population projections lies well beyond our ability to measure differences in the field (Koslow 1992). The maintenance of some critical density of adult stocks and the protection of spawners and spawning habitats during the short reproductive period should be crucial in the management of long-lived periodic fishes. If most larvae never recruit into the adult population even under pristine conditions, it follows that spawning must proceed unimpeded each year in fairly undegraded habitats if a fitness payoff is to be collected during the exceptional year.

Because they tend to be small and occur in shallow shoreline habitats, opportunistic strategists are not usually exploited commercially. Some species intermediate between opportunistic and periodic are very important commercial stocks (e.g. gulf menhaden (*Brevoortia patronus*)). Yet, opportunistic species are often the most important food resources for larger piscivorous species. By virtue of their small size and rapid turnover rates, these species should be efficient colonizers of frequently disturbed habitats like intermittent streams and salt marshes. Some opportunistic strategists, like the bay anchovy and silversides, inhabit more stable habitats and sustain dense populations in the face of intense predation. Given the innate ability of opportunistic strategists to sustain losses during all stages of life spans that are typically rather short, one of the keys to their management might be protection from large-scale or chronic perturbations that eliminate key refugia in space or time. Obviously, what may be perceived by humans to be a minor disturbance (e.g. a few degrees centigrade, or a few parts per thousand salinity) might pose a major impact from the perspective of a small fish that must maintain high reproductive output over a short life span.

Because equilibrium strategists produce small numbers of offspring, larval and juvenile survivorship must be comparatively high if these populations are to maintain themselves around some average density. Parental care is often well developed in equilibrium strategists, so that survivorship of eggs and larvae is dependent on the condition of adults and the integrity of the adult habitat (e.g. threatened cavefishes and some species of darters and madtoms). These fishes ought to conform to stock–recruit fishery models to a greater extent than periodic and opportunistic strategists (and see Koslow (1992) for a discussion of the effect of fecundity on the stock–recruit relationship). Yet, relatively few equilibrium strategists are commercially exploited in North American marine and freshwaters.

Several species exploited by sport fisheries exhibit brood guarding and appear to be intermediate between equilibrium and periodic strategies, including lingcod (*Ophiodon elongatus*) and the black basses (*Micropterus* spp.). In general terms, management of these larger equilibrium-type species should be aimed at maintenance of a productive environment that promotes surplus yields (which can be harvested and replaced via compensation) and the maintenance of healthy adult stocks. We suggested that anadromous salmonids could represent a variation on the equilibrium strategy, in which parental investment in the form of costly migrations to oligotrophic habitats take the place of guarding, brooding, or bearing. Following this view, degradation of spawning habitats or impedement of access to spawning sites would have the same net impact as removing nest-guarding or brooding adults during the spawning season.

Finally, we point out that a variety of fishes with divergent life-history strategies frequently coexist in the same habitats. The feeding niche probably determines a large proportion of the total environmental variance experienced by an organism. Inherited design constraints, including the morphological features required for trophic function in particular microhabitats, place restrictions on the evolution of life-history features. A diversity of life-history strategies are consequently observed among species that perceive the same environment very differently from another. As a consequence, management efforts designed to abate a problem for a given species may sometimes have unanticipated effects on sympatric species that exhibit alternative strategies.

Acknowledgements

We thank J. S. Mattice, W. Van Winkle, R. G. Otto, and other members of the Electic Power Research Institute's COMPMECH program for encouraging comparative research on fish life histories and population regulation. T. J. Miller, R. C. Chambers, W. Van Winkle, and an anonymous reviewer provided valuable comments on earlier drafts of the manuscript. J. H. Cowan contributed several references on fish life history. A. L. Brenkert and D. D. Schmoyer assisted in data management. The study was performed under the sponsorship of the Electric Power Research Institute under contract No. RP2932-2 (DOE No. ERD-87-672) with the Oak Ridge National Laboratory (ORNL). ORNL is managed by Martin Marietta Energy Systems, Inc. under contract No. DE-AC05-84OR21400 with the U.S. Department of Energy. This is Publication No. 3970 of the Environmental Sciences Division, ORNL.

References

ADAMS, P. B. 1980. Life history patterns in marine fishes and their consequences for fisheries management. Fish. Bull 78: 1–12.
ALLAN, J. D. 1976. Life history patterns in zooplankton. Am. Nat. 110: 165–180.
ARMSTRONG, M. J., AND P. A. SHELTON. 1990. Clupeoid life-history styles in variable environments. Environ. Biol. Fishes. 28: 77–85.
BALON, E. K. 1975. Reproductive guilds of fishes: a proposal and definition. J. Fish. Res. Board Can. 32: 821–864.
BALON, E. K., W. T. MOMOT, AND H. A. REGIER. 1977. Reproductive guilds of percids: results of the paleogeographical history and ecological succession. J. Fish. Res. Board Can. 34: 1910–1921.
BALTZ, D. M. 1984. Life history variation among female surfperches (Perciformes: Embiotocidae). Environ. Biol. Fishes 10: 159–171. (B3)
BARNTHOUSE, L. W., G. W. SUTER, II, AND A. E. ROSEN. 1990. Risks of toxic contaminants to exploited fish populations: influence of life history, data uncertainty and exploitation intensity. Environ. Toxicol. Chem. 9: 297–311.
BEVERTON, R. J. H. 1990. Small marine pelagic fish and the threat of fishing: are they endangered? J. Fish Biol. 37(Suppl. A): 5–16.

BIRCH, L. C. 1948. The intrinsic rate of natural increase of an insect population. J. Anim. Ecol. 17: 15–26.
BOYCE, M. S. 1979. Seasonality and patterns of natural selection for life histories. Am. Nat. 114: 569–583.
BRANDER, K., AND P. C. F. HURLEY. 1992. Distribution of early-stage Atlantic cod (*Gadus morhua*), haddock (*Melanogrammus aeglefinus*), and witch flounder (*Glyptocephalus cynoglossus*) eggs on the Scotian Shelf: a reappraisal of evidence on the coupling of cod spawning and plankton production. Can. J. Fish. Aquat. Sci. 49: 238–251.
BURT, A., D. L. KRAMER, K. NAKATSURU, AND C. SPRY. 1988. The tempo of reproduction in *Hyphessobrycon pulchripinnis* (Characidae), with a discussion on the biology of 'multiple spawning' in fishes. Environ. Biol. Fishes 22: 15–27.
CARLANDER, K. D. 1969. Handbook of freshwater fisheries biology. Vol. 1. Iowa State University Press, Ames, IA.
1977. Handbook of freshwater fisheries biology. Vol. 2, Iowa State University Press, Ames, IA.
COHEN, D. 1967. Optimizing reproduction in a randomly varying environment when a correlation may exist between the conditions at the time a choice has to be made and the subsequent outcome. J. Theor. Biol. 16: 1–14.
COLE, L. C. 1954. The population consequences of life history phenomena. Q. Rev. Biol. 29: 103–137.
DUARTE, C. M., AND M. ALCAREZ. 1989. To produce many small or few large eggs: a size-independent reproductive tactic of fish. Oecologia 80: 401–404.
DUNHAM, A. E., AND D. B. MILES. 1985. Patterns of covariation in life history traits of squamate reptiles: the effects of size and phylogeny reconsidered. Am. Nat. 126: 231–257.
DUTIL, J. D. 1986. Energetic constraints and spawning interval in the anadromous arctic char (*Salvelinus alpinus*). Copeia 1986: 945–955.
ELGAR, M. A. 1990. Evolutionary compromise between a few large and many small eggs: comparative evidence in teleost fish. Oikos 59: 283–287.
FISHER, R. A. 1958. The genetical theory of natural selection. 2nd ed. Dover, New York, NY.
FLEMING, I. A., AND M. R. GROSS. 1990. Latitudinal clines: a trade-off between egg number and size in Pacific salmon. Ecology 71: 1–11.
FLETCHER, R. I., AND R. B. DERISO. 1988. Fishing in dangerous waters: remarks on a controversial appeal to spawner–recruit theory for long-term impact assessment. Am. Fish. Soc. Monogr. 4: 232–244.
FORTIER, L., AND J. A. GAGNÉ. 1990. Larval herring (*Clupea harengus*) dispersion, growth, and survival in the St. Lawrence estuary: match/mismatch or membership/vagrancy. Can. J. Fish. Aquat. Sci. 47: 1898–1912.
FRASER, D. F., AND R. D. CERRI. 1982. Experimental evaluation of predator–prey relationships in a patchy environment: consequences for habitat use in minnows. Ecology 63: 307–313.
GARROD, D. J. 1983. On the variability of yearclass strength. J. Cons. Perm. Int. Explor. Mer 41: 63–66.
GARROD, D. J., AND J. W. HORWOOD. 1984. Reproductive strategies and the response to exploitation, p. 367–384. *In* G. W. Potts and R. J. Wootton [ed.] Fish reproduction. Academic Press, London and New York.
GOODMAN, D. 1974. Natural selection and a cost ceiling on reproductive effort. Am. Nat. 108: 247–268.
GOTELLI, N. J., AND M. PYRON. 1991. Life history variation in North American freshwater minnows: effects of latitude and phylogeny. Oikos 62: 30–40.
GREEN, R., AND P. R. PAINTER. 1975. Selection for fertility and development time. Am. Nat. 109: 1–10.
GREENSLADE, P. J. M. 1983. Adversity selection and the habitat template. Am. Nat. 122: 352–365.
GRIME, J. P. 1977. Evidence for the existence of three primary strategies in plants and its relevance to ecological and evolutionary theory. Am. Nat. 111: 1169–1194.
1979. Plant strategies and vegetation process. Wiley, New York, NY.
HART, J. L. 1973. Pacific fishes of Canada. Bull. Fish. Res. Board Can. 180: 740 p.
HEINS, D. C., AND J. A. BAKER. 1988. Egg sizes in fishes: do mature oocytes accurately demonstrate statistics of ripe ova? Copeia 1988: 238–240.
HEINS, D. C., AND F. G. RABITO, JR. 1986. Spawning performance in North American minnows: direct evidence of the occurrence of multiple clutches in the genus *Notropis*. J. Fish Biol. 28: 343–357.
HOUDE, E. D. 1987. Fish early life dynamics and recruitment variability. Am. Fish. Soc. Symp. 2: 17–29.
1989. Comparative growth, mortality, and energetics of marine fish larvae: temperature and implied latitudinal effects. Fish. Bull. 87: 471–495.
HUBBS, C. 1976. The diel reproductive pattern and fecundity of *Menidia audens*. Copeia 1976: 386–388.
1985. Darter reproductive seasons. Copeia 1985: 56–68.

HUTCHINGS, J. A., AND D. W. MORRIS. 1985. The influence of phylogeny, size and behavior on patterns of covariation in salmonid life histories. Oikos 45: 118–124.
ITÔ, Y. 1978. Comparative ecology. Cambridge University Press, Cambridge.
JAMES, P. W., O. E. MAUGHAN, AND A. V. ZALE. 1991. Life history of the leopard darter, *Percina pantherina* in Glover River, Oklahoma. Am. Midl. Nat. 125: 173–179.
KAWASAKI, T. 1980. Fundamental relations among the selections of life history in the marine teleosts. Bull. Jpn. Soc. Sci. Fish. 46: 289–293.
1983. Why do some pelagic fishes have wide fluctuations in their numbers? Biological basis of fluctuation from the viewpoint of evolutionary ecology. FAO Fish. Rep. 291(1): 1065–1080.
KOSLOW, J. A. 1992. Fecundity and the stock–recruitment relationship. Can. J. Fish. Aquat. Sci. 49: 210–217.
LACK, D. 1954. The natural regulation of animal numbers. Oxford University Press, New York, NY.
LEAMAN, B. M. 1991. Reproductive styles and life history variables relative to exploitation and management of *Sebastes* stocks. Environ. Biol. Fishes 30: 253–271.
LEE, D. S., C. R. GILBERT, C. H. HOCUTT, R. E. JENKINS, D. E. MCALLISTER, AND J. R. STAUFFER JR. 1980. Atlas of North American freshwater fishes. North Carolina State Museum of Natural History, Raleigh, NC.
LEGGETT, W. C., AND J. E. CARSCADDEN. 1978. Latitudinal variation in reproductive characteristics of American shad (*Alosa sapidissima*): evidence for population specific life history strategies in fish. J. Fish. Res. Board Can. 35: 1469–1478.
LEWONTIN, R.C. 1965. Selection for colonizing ability, p. 79–94. *In* H. G. Baker and G. L. Stebbins [ed.] The genetics of colonizing species. Academic Press, New York, NY.
MACCALL, A. D. 1990. Dynamic geography of marine fish populations. University of Washington Press, Seattle, WA. 153 p.
MAHON, R. 1984. Divergent structure in fish taxocenes of north temperate streams. Can. J. Fish. Aquat. Sci. 41: 330–350.
MCGURK, M. D. 1986. Natural mortality of marine pelagic fish eggs and larvae: role of spatial patchiness. Mar. Ecol. Prog. Ser. 34: 227–242.
MICHOD, R. E. 1979. Evolution of life histories in response to age specific mortality factors. Am. Nat. 113: 531–550.
MILLER, P. J. 1979. Adaptiveness and implications of small size in teleosts. Symp. Zool. Soc. Lond. 44: 263–306.
MILLER, T. J., L. B. CROWDER, J. A. RICE, AND E. A. MARSCHALL. 1988. Larval size and recruitment mechanisms in fishes: towards a conceptual framework. Can. J. Fish. Aquat. Sci. 45: 1657–1670.
MORTON, T. 1989. Species profiles: life histories and environmental requirements of coastal fishes and invertebrates (mid-Atlantic): Bay anchovy. U.S. Fish Wildl. Serv. Biol. Rep. 82(11.97): 13 p.
MURPHY, G. I. 1968. Patterns in life history and the environment. Am. Nat. 102: 391–403.
MYERS, R. A. 1992. Population variability of juvenile fish and the range of cod, haddock and herring in the North Atlantic. J. Anim. Ecol. (Inpress)
NELSON, J. S. 1984. Fishes of the world. 2nd ed. Wiley, New York, NY.
NUSSBAUM, R.A., AND D. L. SCHULTZ 1989. Coevolution of parental care and egg size. Am. Nat. 133: 591–603.
PAGEL, M. D., AND P. H. HARVEY. 1988. Recent developments in the analysis of comparative data. Q. Rev. Biol. 63: 413–440.
PAINE, M. D. 1990. Life history tactics of darters (Percidae: Etheostomatiini) and their relationship with body size, reproductive behavior, latitude and rarity. J. Fish Biol. 37: 473–488.
PEASE, C. M., AND J. J. BULL. 1988. A critique of methods for measuring life history trade-offs. J. Evol. Biol. 1: 293–303.
PEPIN, P., AND R. A. MYERS. 1991. Significance of egg and larval size to recruitment variability of temperate marine fish. Can. J. Fish. Aquat. Sci. 48: 1820–1828.
PIANKA, E. R. 1970. On *r*- and *K*-selection. Am. Nat. 104: 592–597.
1976. Natural selection of optimal reproductive tactics. Am. Zool. 16: 775–784.
POULSON, T. L. 1963. Cave adaptations in amblyopsid fishes. Am. Midl. Nat. 70: 257–290.
RAGO, P. L., AND C. P. GOODYEAR. 1987. Recruitment mechanisms of striped bass and Atlantic salmon: comparative liabilities of alternative life histories. Am. Fish. Soc. Symp. 1: 402–416.
RIDLY, M. 1989. Why not to use species in comparative tests. J. Theor. Biol. 136: 361–364.
ROFF, D. A. 1981. Reproductive uncertainty and the evolution of iteroparity: why don't flatfish put all their eggs in one basket? Can. J. Fish. Aquat. Sci. 38: 968–977.
1984. The evolution of life history parameters in teleosts. Can. J. Fish. Aquat. Sci. 41: 989–1000.

1988. The evolution of migration and some life history parameters in marine fishes. Environ. Biol. Fishes 22: 133–146.
ROTHSCHILD B. J., AND G. T. DiNARDO. 1987. Comparison of recruitment variability and life history data among marine and anadromous fishes. Am. Fish. Soc. Symp. 1: 531–546.
ROUGHGARDEN, J. 1971. Density-dependent natural selection. Ecology 52: 453–468.
SARGENT, R. C., P. D. TAYLOR, AND M. R. GROSS 1987. Parental care and the evolution of egg size in fishes. Am. Nat. 129: 32–46.
SAS INSTITUTE INC. 1987. SAS/STAT guide for personal computers, version 6. SAS Institute Inc., Cary, NC.
SAUNDERS, W. P. JR., AND J. T. FINN. 1983. Stability analysis of a fish population undergoing life history changes, p. 345–354. In W. K. Lauenroth, G. V. Skogerboe, and M. Flug [ed.] Analysis of ecological systems: state of the art in ecological modelling. Elsevier Scientific, Amsterdam, The Netherlands.
SCHAAF, W. E., D. S. PETERS, D. S. VAUGHN, L. COSTON-CLEMENTS, AND C. W. KROUSE. 1987. Fish population responses to chronic and acute pollution: the influence of life history strategies. Estuaries 10: 267–275.
SCHAFFER, W. M. 1974. Optimal reproductive effort in fluctuating environments. Am. Nat. 108: 783–790.
SCHAFFER, W. M. 1974. Optimal reproductive effort in fluctuating environments. Am. Nat. 108: 783–790.
SCOTT, W. B., AND E. J. CROSSMAN. 1973. Freshwater fishes of Canada. Bull. Fish. Res. Board Can. 184: 966 p.
SHEPHERD, J. G., AND D. H. CUSHING. 1990. Regulation in fish populations: myth or mirage? Philos. Trans. R. Soc. Lond. B 330: 151–164.
SHERMAN, K., W. SMITH, W. MORSE, M. BERMAN, J. GREEN, AND L. EJSYMONT. 1984. Spawning strategies of fishes in relation to circulation, phytoplankton production, and pulses in zooplankton off the northeastern United States. Mar. Ecol. Prog. Ser. 18: 1–19.
SIBLY, R., AND P. CALOW. 1985. The classification of habitats by selection pressures: a synthesis of life-cycle and r/K theory, p. 75–90. In R. M. Sibly and R. H. Smith [ed.] Behavioural ecology: ecological consequences of adaptive behavior. Blackwell, Oxford.
1986. Why breeding earlier is always worthwhile J. Theor. Biol. 123: 311–319.
SINCLAIR, M. 1988. Marine populations: an essay on population regulation and speciation. University of Washington Press, Seattle, WA. 252 p.
SISSENWINE, M. P. 1984. Why do fish populations vary?, p. 59–94. In R. M. May [ed.] Exploitation of marine communities. Report of the Dalhem workshop on exploitation of marine communities, Berlin 1984. Springer-Verlag, Berlin.
SMITH, F. E. 1954. Quantitative aspects of population growth, p. 277–294. In E. Boell [ed.] Dynamics of growth processes. Princeton University Press, Princeton, NJ.
SMITH, C. C., AND S. D. FRETWELL. 1974. The optimal balance between size and number of offspring. Am. Nat. 108: 499–506.
SMITH, P. E., M. D. OHMAN, AND L. E. EBER. 1989. Analysis of the patterns of distribution of zooplankton aggregations from an acoustic doppler current profiler. CalCOFI Rep. 30: 88–103.
SNYDER, R. J. 1990. Clutch size of anadromous and freshwater threespine sticklebacks: a reassessment. Can. J. Zool. 68: 2027–2030.
SOUTHWOOD, T. R. E. 1977. Habitat, the templet for ecological strategies? J. Anim. Ecol. 46: 337–365.
1988. Tactics, strategies and templates. Oikos 52: 3–18.
SOUTHWOOD, T. R. E., R. M. MAY, M. P. HASSELL, AND G. R. CONWAY. 1974. Ecological strategies and population parameters. Am. Nat. 108: 791–804.
STEARNS, S. C. 1976. Life-history tactics: a review of the ideas. Q. Rev. Biol. 51: 3–47.
1983. The influence of size and phylogeny on patterns of covariation among life-history traits in mammals. Oikos 41: 173–187.
STEARNS, S. C., AND R. E. CRANDALL. 1984. Plasticity for age and size at sexual maturity: a life history response to unavoidable stress, p. 13–33. In G. W. Potts and R. J. Wootton [ed.] Fish reproduction. Academic Press, London and New York.
SUTHERLAND, W. J., A. GRAFEN, AND P. H. HARVEY. 1986. Life history correlations and demography. Nature (Lond.) 320: 88.
TAYLOR, P. D., AND G. C. WILLIAMS. 1983. A geometric model for optimal life history, p. 91–97. In H. I. Freedman and C. Strobeck [ed.] Population biology, Lect. Notes Biomath. 52. Springer-Verlag, Berlin.
1984. Demographic parameters in evolutionary equilibrium. Can. J. Zool. 62: 2264–2271.
WALTERS, C. J. 1975. Dynamic models and evolutionary strategies, p. 68–82. In S. A. Levin [ed.] Ecosystem, analysis and prediction. Society for Industrial and Applied Mathematics. Philadelphia, PA.

WARE, D. M. 1982. Power and evolutionary fitness of teleosts. Can. J. Fish. Aquat. Sci. 39: 3–13.
1984. Fitness of different reproductive strategies in teleost fishes, p. 1349–366. In G. W. Potts and R. J. Wootton [ed.] Fish reproduction. Academic Press, London and New York.
WHITTAKER, R. H., AND D. GOODMAN. 1979. Classifying species according to their demographic strategy: I. Population fluctuations and environmental heterogeneity. Am. Nat. 113: 185–200.
WINEMILLER, K. O. 1989. Patterns of variation in life history among South American fishes in seasonal environments. Oecologia 81: 225–241.
1992. Life-history strategies and the effectiveness of sexual selection. Oikos. (In press)
WINEMILLER, K. O., AND D. C. TAPHORN. 1989. La evolución de las estrategias de la vida en los peces de los llanos occidentales de Venezuela. Biollania 6: 77–122.
WOOTTON, R. J. 1973. Fecundity of the three-spined stickleback, *Gasterosteus aculeatus* (L.). J. Fish Biol. 5: 683–688.
1984. Introduction: strategies and tactics in fish reproduction, p. 1–12. In G. W. Potts and R. J. Wootton [ed.] Fish reproduction. Academic Press, London and New York.

Appendix: List of Fishes Used in Statistical Analyses

Freshwater

Lake sturgeon, *Acipenser fulvescens*; paddlefish, *Polyodon spathula*; longnose gar, *Lepisosteus osseus*; bowfin, *Amia calva*; goldeye, *Hiodon alosoides*; gizzard shad, *Dorosoma cepedianum*; threadfin shad, *Dorosoma petenense*; cisco/lake herring, *Coregonus artedi*; lake whitefish, *Coregonus clupeaformis*; bloater, *Coregonus hoyi*; sockeye salmon, *Oncorhynchus nerka kennerlyi*; round whitefish, *Prosopium cylindraceum*; mountain whitefish, *Prosopium williamsoni*; brook trout,[3] *Salvelinus fontinalis*; lake trout,[3] *Salvelinus namaycush*; Arctic grayling, *Thymallus arcticus*; Alaska blackfish, *Dallia pectoralis*; central mudminnow, *Umbra limi*; redfin pickerel, *Esox americanus americanus*; northern pike, *Esox lucius*; muskellunge, *Esox masquinongy*; chain pickerel, *Esox niger*; central stoneroller, *Campostoma anomalum*; redside dace, *Clinostomus elongatus*; spotfin shiner, *Cyprinella spiloptera*; Utah chub, *Gila atraria*; Mississippi silvery minnow, *Hybognathus nuchalis*; hornyhead chub, *Nocomis biguttatus*; golden shiner, *Notemigonus crysoleucas*; emerald shiner, *Notropis atherinoides*; spottail shiner, *Notropis hudsonius*; rosyface shiner, *Notropis rubellus*; sand shiner, *Notropis stramineus*; redfin shiner, *Lythrurus umbratilis*; bluntnose minnow, *Pimephales notatus*; fathead minnow, *Pimephales promelas*; northern squawfish, *Ptychocheilus oregonensis*; blacknose dace, *Rhinichthys atratulus*; longnose dace, *Rhinichthys cataractae*; redside shiner, *Richardsonius balteatus*; creek chub, *Semotilus atromaculatus*; river carpsucker, *Carpoides carpio*; longnose sucker, *Catostomus catostomus*; white sucker, *Catostomus commersoni*; mountain sucker, *Catostomus platyrhynchus*; lake chubsucker, *Erimyzon sucetta*; smallmouth buffalo, *Ictiobus bubalus*; bigmouth buffalo, *Ictiobus cyprinellus*; silver redhorse, *Moxostoma anisurum*; black redhorse, *Moxostoma duquesnei*; golden redhorse, *Moxostoma erythrurum*; shorthead redhorse, *Moxostoma macrolepidotum*; black bullhead, *Ameiurus melas*; yellow bullhead, *Ameiurus natalis*; brown bullhead, *Ameiurus nebulosus*; channel catfish, *Ictalurus punctatus*; Ozark madtom, *Noturus albater*; stonecat, *Noturus flavus*; tadpole madtom, *Noturus gyrinus*; brindled madtom, *Noturus miurus*; freckled madtom, *Noturus nocturnus*; flathead catfish, *Pylodictis olivaris*; Ozark cavefish, *Amblyopsis rosae*; northern cavefish, *Amblyopsis spelaea*; spring cavefish, *Chologaster agassizi*;

---
[3]The authors prefer the term "char."

swampfish, *Chologaster cornuta*; southern cavefish, *Typhlichthys subterraneus*; pirate perch, *Aphredoderus sayanus*; trout-perch, *Percopsis omiscomaycus*; burbot, *Lota lota*; banded killifish, *Fundulus diaphanus*; plains killifish, *Fundulus zebrinus*; western mosquitofish, *Gambusia affinis*; freshwater drum, *Aplodinotus grunniens*; white bass, *Morone chrysops*; rock bass, *Ambloplites rupestris*; flier, *Centrarchus macropterus*; banded pygmy sunfish, *Elassoma zonatum*; redbreast sunfish, *Lepomis auritus*; green sunfish, *Lepomis cyanellus*; pumpkinseed, *Lepomis gibbosus*; warmouth, *Lepomis gulosus*; orangespotted sunfish, *Lepomis humilis*; bluegill, *Lepomis macrochirus*; longear sunfish, *Lepomis megalotis*; redear sunfish, *Lepomis microlophus*; smallmouth bass, *Micropterus dolomieu*; spotted bass, *Micropterus punctulatus*; largemouth bass, *Micropterus salmoides*; white crappie, *Pomoxis annularis*; black crappie, *Pomoxis nigromaculatus*; naked sand darter, *Ammocrypta beani*; greenside darter, *Etheostoma blennioides*; rainbow darter, *Etheostoma caeruleum*; Iowa darter, *Etheostoma exile*; fantail darter, *Etheostoma flabellare*; yoke darter, *Etheostoma juliae*; greenthroat darter, *Etheostoma lepidum*; johnny darter, *Etheostoma nigrum*; tessellated darter, *Etheostoma olmstedi*; orangethroat darter, *Etheostoma spectabile*; variegate darter, *Etheostoma variatum*; banded darter, *Etheostoma zonale*; yellowperch, *Perca flavescens*; logperch, *Percina caprodes*; blackside darter, *Percina maculata*; slenderhead darter, *Percina phoxocephala*; dusky darter, *Percina sciera*; sauger, *Stizostedion canadense*; walleye, *Stizostedion vitreum*; slimy sculpin, *Cottus cognatus*.

Marine

Shortnose sturgeon, *Acipenser brevirostrum*; Atlantic sturgeon, *Acipenser oxyrhynchus*; ladyfish, *Elops saurus*; tarpon, *Megalops atlanticus*; American eel, *Anguilla rostrata*; blueback herring, *Alosa aestivalis*; alewife, *Alosa pseudoharengus*; American shad, *Alosa sapidissima*; gulf menhaden, *Brevoortia patronus*; Atlantic menhaden, *Brevoortia tyrannus*; Atlantic herring, *Clupea harengus*; Pacific sardine, *Sardinops sagax*; bay anchovy, *Anchoa mitchilli*; anchoveta, *Cetengraulis mysticetus*; northern anchovy, *Engraulis mordax*; cutthroat trout, *Oncorhynchus clarki*; pink salmon, *Oncorhynchus gorbuscha*; chum salmon, *Oncorhynchus keta*; coho salmon, *Oncorhynchus kisutch*; rainbow trout/steelhead, *Oncorhynchus mykiss*; chinook salmon, *Oncorhynchus tshawytscha*; Atlantic salmon, *Salmo salar*; Arctic char, *Salvelinus alpinus*; Dolly Varden, *Salvelinus malma*; capelin, *Mallotus villosus*; rainbow smelt, *Osmerus mordax*; longfin smelt, *Spirinchus thaleichthys*; eulachon, *Thaleichthys pacificus*; hardhead catfish, *Arius felis*; gafftopsail catfish, *Bagre marinus*; plainfin midshipman, *Porichthys notatus*; oyster toadfish, *Opsanus tau*; Pacific cod, *Gadus macrocephalus*; Atlantic cod, *Gadus morhua*; Pacific hake, *Merluccius productus*; Atlantic tomcod, *Microgadus tomcod*; red brotula, *Brosmophycis marginata*; blackbelly eelpout, *Lycodopsis pacifica*; California grunion, *Leuresthes tenuis*; inland silverside, *Menidia beryllina*; Atlantic silverside, *Menidia menidia*; mummichog, *Fundulus heteroclitus*; tubesnout, *Aulorhynchus flavidus*; threespine stickleback, *Gasterosteus aculeatus*; ninespine stickleback, *Pungitius pungitius*; gulf pipefish, *Syngnathus scovelli*; skipjack tuna, *Katsuwonus pelamis*; chub mackerel, *Scomber japonicus*; Atlantic mackerel, *Scomber scombrus*; Atlantic threadfin, *Polydactylus octonemus*; Pacific barracuda, *Sphyraena argentea*; cobia, *Rachycentron canadum*; American sand lance, *Ammodytes americanus*; northern sand lance, *Ammodytes dubius*; jack mackerel, *Trachurus symmetricus*; bluefish, *Pomatomus saltatrix*; tomtate, *Haemulon aurolineatum*; white grunt, *Haemulon plumieri*; pigfish, *Orthopristis chrysoptera*; spotted seatrout, *Cynoscion nebulosus*; weakfish, *Cynoscion regalis*; spot, *Leiostomus xanthurus*; Atlantic croaker, *Micropogonias undulatus*; red drum, *Sciaenops ocellatus*; pinfish, *Lagodon rhomboides*; scup, *Stenotomus chrysops*; black grouper, *Mycteroperca bonaci*; red grouper, *Epinephelus morio*; Nassau grouper, *Epinephelus striatus*; black sea bass, *Centropristis striata*; red snapper, *Lutjanus campechanus*; tautog, *Tautoga onitis*; common snook, *Centropomus undecimalis*; white perch, *Morone americana*; striped bass, *Morone saxatilis*; shiner perch, *Cymatogaster aggregata*; striped seaperch, *Embiotoca lateralis*; walleye surfperch, *Hyperprosopon argenteum*; rubberlip seaperch, *Rhacochilus toxotes*; pile perch, *Rhacochilus vacca*; high cockscomb, *Anoplarchus purpurescens*; arrow goby, *Clevelandia ios*; clown goby, *Microgobius gulosus*; coastrange sculpin, *Cottus aleuticus*; showy snailfish, *Liparis pulchellus*; Pacific ocean perch, *Sebastes alutus*; brown rockfish, *Sebastes auriculatus*; copper rockfish, *Sebastes caurinus*; splitnose rockfish, *Sebastes diploproa*; yellowtail rockfish, *Sebastes flavidus*; chilipepper, *Sebastes goodei*; shortbelly rockfish, *Sebastes jordani*; black rockfish, *Sebastes melanops*; bocaccio, *Sebastes paucispinis*; canary rockfish, *Sebastes pinniger*; stripetail rockfish, *Sebastes saxicola*; shortspine thornyhead, *Sebastolobus alascanus*; lingcod, *Ophiodon elongatus*; gulf flounder, *Paralichthys albigutta*; California halibut, *Paralichthys californicus*; summer flounder, *Paralichthys dentatus*; southern flounder, *Paralichthys lethostigma*; flathead sole, *Hippoglossoides elassodon*; Pacific halibut, *Hippoglossus stenolepis*; rock sole, *Pleuronectes bilineatus*; Dover sole, *Microstomus pacificus*; English sole, *Pleuronectes* starry flounder, *Platichthys stellatus*; winter flounder, *Pleuronectes americanus*; ocean sunfish, *Mola mola*.

4.3 HONORABLE MENTION FULL CITATIONS AND ABSTRACTS

Anderson, R. O. 1978. New approaches to recreational fishery management. Pages 73–78 *in* G. D. Novinger and J. G. Dillard, editors. New approaches to the management of small impoundments. American Fisheries Society, Special Publication 5, Bethesda, Maryland.

This concluding paper develops perspectives of recreational fishery management and the need for new approaches. New approaches proposed include the use of an index of length-frequency distribution, Proportional Stock Density (PSD), and a new condition factor, Relative Weight (W_r). The data obtained from fish populations in small impoundments provide a basis for the identification of problems and proposed solutions. The problems in fish populations are related to unfavorable rates of reproduction, growth, or mortality which result in unsatisfactory fish population or community structure. Solutions proposed include more effective coordination and cooperation of fishery researchers, managers, and anglers, and more effective regulation of harvest.

Cushing, D. H. 1990. Plankton production and year-class strength in fish populations: an update of the match-mismatch hypothesis. Advances in Marine Biology 26:249–293.

The degree of match and mismatch in the time of larval production and production of their food has been put forward as an explanation of part of the variability in recruitment to a stock of fish (Cushing, 1974, 1975, 1982). The magnitude of recruitment is not completely determined until the year-class finally joins the adult stock, and the processes involved probably begin early in the life-history of the fish when both their growth and mortality rates are high (Ricker, 1954). Hjort (1914) thought the level of recruitment was established during this period between hatching and first-feeding (the critical period according to Marr, 1956; May, 1974). However, I extended the match/mismatch hypothesis to cover the subsequent development through larval life up to metamorphosis, and possibly just beyond. An essential part of my hypothesis was the suggestion of Ricker and Foerster (1948) that under adventitious predation, i.e. without aggregation, well-fed larvae grow quickly and experience less predatory mortality than poorly fed ones at a given stage of development (Cushing and Harris, 1973; Shepherd and Cushing, 1980). Hence, one would expect growth and mortality to be inversely related. Almost as a consequence of match and mismatch, fish in temperate waters should release their larvae during the spring or autumn peaks in the production cycle, when more food is available. Anderson (1988) listed a number of hypotheses regarding the survival of pre-recruits, and he came to the conclusion that the growth/mortality hypothesis (as he named it) was most significant. As well as extending my match/mismatch hypothesis to cover a longer period in the life-history of fish, it was also more explicitly related to climatic factors and to the Sverdrup (1958) model than Hjort's first hypothesis.

Hjort's (1914) second hypothesis, i.e. that recruitment may be determined by the loss of larvae through advective processes, was revived by Bailey (1981), who found recruitment to the hake stock off the west coast of the U.S.A. to be negatively correlated with an index of upwelling. I decided to examine the effect of upwelling on recruitment in waters equatorward of 40" latitude and, subsequently, I re-examined the data in support of my original hypothesis together with more recent information, including the member/vagrant hypothesis.

The match/mismatch hypothesis has developed over a number of years in several publications, and I thought it desirable to bring all of the evidence together here. However, it should be noted that although my original hypothesis applied only to recruitment to fish stocks, I now also include spiny lobsters and Dungeness crabs.

Essington, T. E., A. H. Beaudreau, and J. Wiedenmann. 2006. Fishing through marine food webs. Proceedings of the National Academy of Sciences of the United States of America 103:3171–3175.

A recurring pattern of declining mean trophic level of fisheries landings, termed "fishing down the food web," is thought to be indicative of the serial replacement of high-trophic-level fisheries with less valuable, low-trophic-level fisheries as the former become depleted to economic extinction. An alternative to this view, that declining mean trophic levels indicate the serial addition of low-trophic-level fisheries ("fishing through the food web"), may be equally severe because it ultimately leads to conflicting demands for ecosystem services. By analyzing trends in fishery landings in 48 large marine ecosystems worldwide, we find that fishing down the food web was pervasive (present in 30 ecosystems) but that the sequential addition mechanism was by far the most common one underlying declines in the mean trophic level of landings. Specifically, only 9 ecosystems showed declining catches of upper-trophic-level species, compared with 21 ecosystems that exhibited either no significant change ($n = 6$) or significant increases ($n = 5$) in upper-trophic-level catches when fishing down the food web was occurring. Only in the North Atlantic were ecosystems regularly subjected to sequential collapse and replacement of fisheries. We suggest that efforts to promote sustainable use of marine resources will benefit from a fuller consideration of all processes giving rise to fishing down the food web.

(Copyright 2014 National Academy of Sciences, U.S.A.)

Gabelhouse, D. W., Jr. 1984. A length-categorization system to assess fish stocks. North American Journal of Fisheries Management 4:273–285.

A length-categorization system was developed to assess structure of fish stocks with greater precision than is possible using Proportional Stock Density (PSD). Three new size categories—"preferred," "memorable," and "trophy"—were developed to accompany previously established "stock" and "quality" lengths. Like minimum stock and quality lengths, minimum lengths for the new categories are defined as percentag lengths of the all-tackle, world-record fish. Length ranges from or near which minimum stock, quality, preferred, memorable, and trophy lengths should be selected were computed for all freshwater fish species having a world-record length listed by the International Game Fish Association in 1982. Minimum lengths corresponding to each of the five size categories are proposed for several species. By arraying samples of fish population data or angler catch data according to the five size-group categories, a length-frequency distribution can be easily assessed and verbalized. Relative Stock Density (RSD) or models for catch rates also can be developed to set management objectives that are easily understandable, yet reflect recruitment, mortality, and growth functions of fish populations and communities. Desirable percentages and catch rates for size-group categories may differ among individual waters or geographic regions depending upon management objectives and the capacity to produce the species of interest.

Karr, J. R. 1981. Assessment of biotic integrity using fish communities. Fisheries 6(6):21–27.

> Man's activities have had profound, and usually negative, influences on freshwater fishes from the smallest streams to the largest rivers. Some negative effects are due to contaminants, while others are associated with changes in watershed hydrology, habitat modifications, and alteration of energy sources upon which the aquatic biota depends. Regrettably, past efforts to evaluate effects of man's activities on fishes have attempted to use water quality as a surrogate for more comprehensive biotic assessment. A more refined biotic assessment program is required for effective protection of freshwater fish resources. An assessment system proposed here uses a series of fish community attributes related to species composition and ecological structure to evaluate the quality of an aquatic biota. In preliminary trials this system accurately reflected the status of fish communities and the environment supporting them.

Kitchell, J. F., D. J. Stewart, and D. Weininger. 1977. Applications of a bioenergetics model to yellow perch (*Perca flavescens*) and walleye (*Stizostedion vitreum vitreum*). Journal of the Fisheries Research Board of Canada 34:1910–1921.

> A simple energy budget equation is developed to yield bioenergetics model designed to simulate fish growth. Parameters for the model are estimated from the literature for application to yellow perch (*Perca flavescens*) and walleye (*Stizostedion vitreum vitreum*). Simulations are presented that demonstrate model output as functions of body size, activity level, ration level, food quality, and environmental temperature. Sensitivity analyses identify the importance of food consumption, activity, and excretion as biological processes represented in the parameters. On the basis of temperature conditions in selected lakes and specified feeding levels, simulations are presented to quantify the importance of year-to-year variation of temperature in determining growth. In heterothermal systems, temperature selection by percids can have a significant effect on growth. For walleye on fixed rations, annual growth can vary from zero to twofold increments due entirely to differences in summer temperatures. Variations in food quality have lesser effects.

Lindeman, R. L. 1942. The trophic-dynamic aspect of ecology. Ecology 23(4):157–176.

> Recent progress in the study of aquatic food-cycle relationships invites a re-appraisal of certain ecological tenets. Quantitative productivity data provide a basis for enunciating certain trophic principles, which, when applied to a series of successional stages, shed new light on the dynamics of ecological succession.

Stein, R. A., D. R. DeVries, and J. M. Dettmers. 1995. Food-web regulation by a planktivore: exploring the generality of the trophic cascade hypothesis. Canadian Journal of Fisheries and Aquatic Sciences 52:2518–2526.

The trophic cascade hypothesis currently being tested in north temperate systems may not apply to open-water communities in lower latitude U.S. reservoirs. These reservoir communities differ dramatically from northern lakes in that an open-water omnivore, gizzard shad (*Dorosoma cepedianum*), often occurs in abundance. Neither controlled by fish predators (owing to high fecundity and low vulnerability) nor by their zooplankton prey (following the midsummer zooplankton decline, gizzard shad consume detritus and phytoplankton), gizzard shad regulate community composition rather than being regulated by top-down or bottom-up forces. In experiments across a range of spatial scales (enclosures, 1–9 m^2; ponds, 4–5 ha; and reservoirs, 50–100 ha), we evaluated the generality of the trophic cascade hypothesis by assessing its conceptual strength in reservoir food webs. We reviewed the role of gizzard shad in controlling zooplankton populations and hence recruitment of bluegill, *Lepomis macrochirus* (via exploitative competition for zooplankton), and largemouth bass, *Micropterus salmoides* (by reducing their bluegill prey). Reservoir fish communities, owing to the presence of gizzard shad, appear to be regulated by complex weblike interactions among species than by the more chainlike interactions characteristic of the trophic cascade.

Swingle, H. S. 1951. Experiments with various rates of stocking bluegills, *Lepomis macrochirus* Rafinesque, and largemouth bass, *Micropterus salmoides* (Lacepede) in ponds. Transactions of the American Fisheries Society 80:218–230.

Results from stocking adult bluegills and largemouth bass, fingerling bass and adult bluegills, and fingerling bluegills and fingerling largemouth bass are compared. Increase in rates of stocking bluegills from 8 to 1,500 per acre in bluegill-bass combinations gave corresponding increases in pounds of harvestable fish per acre and corresponding decreases in the cost per pound. The calculated A_T value (percentage harvestable fish) could be used to predict the success of various stocking rates. Unbalanced populations resulted from stocking with insufficient numbers of either largemouth bass or bluegills. When normally adequate numbers of each species were stocked, unbalanced populations occasionally resulted from extremely high mortality among the stocked fish. Where no fish were removed by fishing, an average of 25.6 percent of the stocked bass died during the first 6 months and an additional 20.4 percent during the following year; similarly an average of 15.4 percent of the stocked bluegills died during the first year and an additional 19.1 percent during the second year.

Section 5

Managing Fisheries Enhancements

Kai Lorenzen

5.1. SYNTHESIS

Articles in this section deal with the use of hatchery programs in fisheries enhancement and restoration—the third approach to managing fisheries after harvest and habitat management, and possibly the most controversial. Hatchery programs have been used successfully to maintain fisheries where natural recruitment of target species is low or absent, to enhance certain wild fisheries, and to conserve or restore threatened or endangered fish populations. At the same time, many hatchery programs have been associated with deleterious ecological or genetic impacts on wild fish populations and fisheries. In addition to biological interactions, hatchery programs have brought about varied and often significant human responses. Some have provided the impetus for fish conservation and habitat restoration initiatives, while others have encouraged overexploitation of wild stock components in mixed fisheries or masked fisheries impacts of habitat loss and thereby reduced incentives for restoration.

Even this brief introduction suggests that evaluating hatchery programs and using them effectively where potential exists for them to improve fisheries outcomes is a complex endeavor—quite the opposite of the "quick fix" that hatcheries are sometimes believed to offer. Among the issues that need to be considered are the dynamics of the fish population enhanced or created by stocking, hatchery techniques and their implications for post-stocking survival, strategies for releasing hatchery fish successfully into natural environments, genetic management, and the behavior of stakeholders and governance systems. Moreover, these facets need to be integrated into a coherent enhancement system framework to assess whether a hatchery program may meet its intended fisheries management goals and to design, implement, or reform the program where this is the case.

Hatchery programs have been used in fisheries management for well over a century, yet much of our current understanding of their potentials and limitations has emerged only over the past few decades. The articles assembled in this section are milestones that have advanced our understanding of hatchery programs through visionary and critical reviews that have defined the place of hatchery programs in fisheries management and the critical issues that need to be considered, and through primary research on these issues. This section aims to provide a unifying context for the selected, seminal articles and to point the reader to key subsequent studies that may have confirmed or challenged the conclusions of the selected articles.

Setting the Scene

The two articles in this section ask "what are hatchery programs useful for from a fisheries management perspective" and "what needs to be considered in order to make it work?" Both articles appeared around the same time and were motivated by a prevailing sense that many hatchery programs operated without a clear rationale, without consideration of key factors likely to be crucial to outcomes, and without evaluation. Cowx (1994) draws mostly on experience and examples from European freshwater systems where hatchery programs have been long established, while Blankenship and Leber (1995) focus on marine hatchery programs with a much shorter history. Cowx (1994) emphasized decision making frameworks such as flow charts, while Blankenship and Leber (1995) outlined a set of broad recommendations. The articles independently arrive at many of the same conclusions—the need for a strategic approach with defined objectives, targeted program design, and rigorous evaluation being the overarching one. Others include the need to consider stocking/release strategies, ecological interactions with wild fish, genetic management, and disease control. Some differences in approach are evident that may be traced to differences between freshwater and marine hatchery programs and the degree to which they have become part of operational management rather than research. Cowx (1994) differentiates between uses of hatchery programs (for mitigation, enhancement, restoration or creation of new fisheries) and emphasizes the need to quantitatively assess the status of the fishery (e.g., abundance relative to carrying capacity, size and age structure) to identify the need and scope for enhancement. At the time, such assessments were more practical in freshwater systems where empirical yield and stocking models had been developed from comparative studies across multiple lakes or streams than in marine systems where stocks were assessed individually using mathematical models that were not set up to deal with the issues surrounding stocking. This and other aspects of the disciplinary elements of hatchery programs are further explored below. As for overarching, strategic approaches to hatchery programs, the articles by Cowx (1994) and Blankenship and Leber (1995) have remained important points of reference. In 2010, Lorenzen, Leber, and Blankenship published a comprehensive update of the responsible approach that integrates and expands on key recommendations from both articles. The updated responsible approach has fifteen key elements arranged in three stages as follows: (Phase I) initial appraisal and goal setting; (Phase II) research and technology development including pilot studies; and (Phase III) operational implementation and adaptive management. Stages are ordered in this sequence to ensure that broad-based and rigorous appraisal of enhancement contributions to fisheries management goals is conducted prior to more detailed research and technology development and operational implementation.

Population Dynamics

The central aim of most hatchery programs is to increase the abundance of fish populations to enhance, rebuild, and/or conserve small fisheries. Hence, understanding and predicting the dynamics of fish populations subject to stocking is crucial to managing hatchery programs effectively.

Experimental stocking studies have been conducted in freshwater systems since the early 20th century, often for the dual purpose of studying fundamentals of production ecology and developing stocking strategies. Of these studies, Homer Swingle's experiments on the management of fisheries in farm ponds in the southeastern United States are the most well-known and influential. In the article reproduced here, Swingle (1951) designed a stocking regime that quite reliably produced a good annual crop of harvestable size fish from a predator-prey community of Largemouth Bass *Micropterus salmoides* and Bluegill *Lepomis macrochirus* stocked into small impoundments. The study systematically applied ecological concepts and quantitative indicators Swingle had developed earlier through a large number of pond experiments (Swingle 1950), and which are well laid out in the introduction of the article reproduced here. The article makes several important contributions to the development of fisheries enhancement science. First, it sets out to use stocking for a specific fisheries management goal (a balanced fish community yielding good catches) that, in small impoundments, cannot reliably be achieved by harvest or habitat management alone. That is a far cry from the many *ad hoc* hatchery programs that have motivated Cowx (1994) and Blankenship and Leber (1995) to call for systematic and responsible approaches to enhancement. Secondly, in addition to using his community indices to guide stocking strategies, Swingle carefully analyzed survival and the effects of stocking regimes and inter-specific interactions upon it, making this an early empirical study of population dynamics of stocked fisheries. Swingle's studies, with some extensions and modification, have continued to inform the management of small impoundments in the southeastern United States to this day.

Pacific Salmon *Oncorhynchus* spp. stocks have been subjected to the world's largest and longest-running hatchery programs. The relative ease of quantifying abundance of juveniles and spawners during their migrations out of and into their natal rivers means that some of the best quantitative data on enhanced populations are available for these stocks. Hilborn and Eggers (2000) took advantage of such long-term data for Alaskan Pink Salmon *O. gorbuscha* stocks and most importantly, evaluated the impact of hatchery programs by comparing long-term variation between enhanced stock and non-enhanced controls. While the focal hatchery program in Prince William Sound was associated with a substantial increase in salmon catches, similar increases were observed over the same period in stocks not enhanced with hatchery fish—suggesting that catch increases were due to large-scale changes in ocean conditions and that a substantial contribution of hatchery fish to catches in Prince William Sound signified displacement of wild by hatchery fish rather than a net positive contribution to catches. The study sparked some debate (Wertheimer et al. 2001; Hilborn and Eggers 2001), which served to further highlight the risk of displacing the wild stock component in enhanced fisheries, and the importance of adopting a sound experimental design with non-enhanced controls and replication when evaluating enhancements experimentally.

While experimental and comparative observation studies have played a major role in informing the management and enhancement of freshwater and anadromous fisheries, such approaches are less suited to marine fisheries and those in larger freshwater systems which rely on fewer, larger stocks and offer only limited opportunities for replicated experiments. Population dynamics modeling, therefore, is the primary tool for assessing management options in such fisheries. The same approach holds promise for assessing the potential or actual contribution of hatchery releases to fisheries management goals. However, the dynamic pool models commonly used in fisheries assessment are based on a simplified representation of population dynamics that is appropriate to the assessment of harvesting of recruited fish, but precludes the evaluation of enhancements. Lorenzen (2005) extended the dynamic pool theory of fishing to stock enhancement by unpacking recruitment, incorporating regulation in the recruited stock, and accounting for biological differences between wild and hatchery fish. Lorenzen (2005) then used the extended model to analyze the dynamics of stock enhancement and restocking and its potential role in fisheries management. He showed that due to multiple density-dependent processes in the life histories of fishes, enhancements can be designed to increase total yield and stock abundance, but will almost inevitably reduce abundance of the naturally recruited stock component; that is, re-stocking of overfished populations is likely to be beneficial only in combination with fishing restrictions and only when populations have been very severely depleted. Releasing hatchery fish of compromised fitness will be most deleterious to wild stocks if fitness is only moderately compromised. Along with the enhanced fishery model of Walters and Martell (2004), this study paved the way for fisheries enhancement and restoration through hatchery programs to be evaluated quantitatively alongside conventional fishing regulations and, with other extensions to commonly used dynamic pool models, habitat management.

Impacts of Hatchery Rearing on the Biology of Stocked Fish

Rearing of fish in hatcheries and other culture facilities subjects the organisms to an inadvertent or intentional process of domestication. Domestication involves plastic developmental responses to the culture environment, an altered selection regime, and has strong, almost always negative impacts on the capacity of fish to survive, grow, and reproduce in the wild (Lorenzen et al. 2012). The review by Olla et al. (1998) synthesized experimental evidence for domestication effects on the most plastic aspect of fish biology: behavior. Evidence suggested that being reared in a simple, psycho-sensorily deprived hatchery environment tends to lessen the innate capabilities of fish to avoid predation and to forage for prey. The authors also outline a number of approaches to improve post-release behavioral capabilities in hatchery-reared fish, including exposure to predators or predatory stimuli, alteration of spatial and temporal distribution of food, mitigation of rearing and transport stress, and control of the social environment. Olla et al.'s (1998) review was the first synthesis of this research area which has seen a great deal of activity since. A later, much cited review (Brown and Dey 2002) focused on life-skills training of hatchery fish.

Post-release Ecology and Release Strategies

Hatchery fish are typically stocked as juveniles, at a life stage characterized by fairly specific food and habitat requirements and high vulnerability to predation. As discussed in Olla et al. (1998), hatchery juveniles are also often deficient in life skills compared to their wild conspecifics. Not surprisingly, the size, time, and habitat of release can have major impacts on post-release survival of hatchery fish and systematic studies on release strategies can yield very substantial improvements. Santucci and Wahl's (1993) study on release strategies for Walleye *Sander vitreus* is remarkable in not only testing for the effect of stocking size on survival, but elucidating ecological mechanisms underlying the observed patterns. It shows how stocking of hatchery fish can be used to gain ecological insights through manipulations that are otherwise difficult to undertake. At the same time, by analyzing economic returns for different release sizes, the study provided very practical information for management.

Genetic Management

Three main sets of issues are associated with the genetic management of hatchery programs: (1) potential disruption of neutral and adaptive spatial population structure due to translocation; (2) impacts of hatchery spawning and rearing on genetic diversity of stocked fish and the enhanced, mixed stock; and (3) impacts of hatchery rearing on the fitness of released fish and their naturally recruited offspring. The issue of disruption of spatial genetic populations structure is discussed in the article by Cowx (1994) reproduced in this volume, while Blankenship and Leber's (1995) section on genetic resource management emphasized maintenance of genetic diversity in hatchery and mixed populations.

By far the best known study on the implications of mixing wild and hatchery populations of different genetic diversity is the short note by Ryman and Laikre (1992) (Honorable Mention). Ryman and Laikre (1992) developed a simple model for predicting the genetically effective population size of an admixture of populations with different effective population size. Results showed how stocking of large numbers of fish from a population of small effective size risks lowering the effective size of the combined population, while supplementing small natural populations with hatchery fish of larger effective population size (note here that techniques such as factorial or minimum kinship mating can be used in hatcheries to raise the ratio of genetically effective to census population size) can have the opposite effect.

While genetic diversity implications of hatchery programs have received much attention in research, they are arguably more tractable through appropriate sourcing and management of brood stock than the third issue; loss of fitness in hatchery-reared fish. Hatchery populations experience regimes that relax selection pressure on many traits, while exerting pressure in others that result in adaptation to the hatchery environment. Both these changes result in loss of fitness in the wild. Reisenbichler and McIntyre (1978) reported the first rigorous study on loss of fitness related to hatchery rearing. Comparing fitness in the wild, in a pond of hatchery fish, hatchery-wild hybrids, and wild fish all derived from the same local population, they showed that wild fish outperform hatchery fish in natural streams, while the reverse is true for hatchery ponds. The performance of hybrids was intermediate in both cases. Overall, this demonstrated a loss of fitness in hatchery fish with possible implications for the productivity of wild populations if hatchery fish interact with them ecologically or genetically. Reisenbichler and McIntyre's (1978) work has been broadly confirmed by subsequent studies summarized in Araki et al. (2008).

Ford (2002) (Honorable Mention) used a combined quantitative genetic and demographic model to explore consequences of the loss of fitness due to captive rearing for wild populations supplemented or enhanced by hatchery programs. Assuming that the hatchery and wild environments select for different optimal trait values, the model showed that when the captive population is closed to gene flow from the wild population, even low levels of gene flow from the captive population to the wild population will shift the wild population's mean phenotype so that it approaches the optimal phenotype in captivity. If the captive population receives gene flow from the wild, the shift in the wild population's mean phenotype becomes less pronounced. He also showed that a decline in fitness of around 30% can occur over a broad range of scenarios. For a recent re-evaluation (which has broadly confirmed Ford 2002 and other previous results) see Baskett and Waples (2013).

Human Dimensions

Considering human dimensions of hatchery programs is crucial for several reasons. First, individual and collective responses of fishers are intended outcomes of many hatchery programs (e.g., those aimed at increasing recreational fishing participation), but may also have unintended consequences such as an increase in fishing pressure on wild stock components. Secondly, hatchery programs involve active replenishments of common pool resources and are likely to be initiated and sustained only where effective governance arrangements allow for regulation of resource use and ensure that benefits of enhancements accrue to those bearing the costs. Three articles included here as "Honorable Mentions" explore these human dimensions of hatchery programs.

Loomis and Fix (1998) (Honorable Mention) conducted an empirical analysis of the effects of fish stocking on license sales and the fishing effort expended in lakes and streams in Colorado. Their results showed that total license sales were unresponsive to fish stocking, suggesting that a reduction in the state's stocking efforts would not result in a reduction in fishing participation or license income. Fishing effort in individual water bodies was found to be responsive to stocking of catchable fish, but only moderately so (with a 1% increase in stocking causing effort to increase by 0.43% in lakes and 0.23% in streams). Stocking in recreational fisheries is often intended to yield an increase in fishing effort, either to generate economic benefits or to divert effort away from pristine or vulnerable fisheries. Conversely, effort increases may negatively affect wild components of mixed stock fisheries or dissipate benefits from stocking in commercial fisheries (where effort is associated with economic costs rather than benefits). In either case, quantifying the effort response to fish stocking is important to understanding and managing the outcomes of a hatchery program. The moderate, less-than-proportional responsiveness found by Loomis and Fix (1998) is not surprising, because to most recreational fishers, expectation or experience of a higher catch rate associated with stocking is only one of several factors influencing the decision on how much and where to fish.

Anderson (2002) (Honorable Mention) considered the interrelationship between the strength of property rights and the degree of control exercised over biological production and product marketing in a fishery. He showed that fisheries enhanced by hatchery programs tended to occupy an intermediate position along the continuum from traditional capture fisheries (typically weak property rights and weak control over production) and aquaculture (strong property rights and control over production). This result can likely be generalized from "property rights" to "governance arrangements" (effective community-based, governmental, or cooperative governance arrangements may substitute for individual property rights where appropriate) and from commercial to recreational fisheries (where the product marketed is the fishing experience). The interaction between governance and production control may work both ways; strengthening of governance arrangements (as is happening in many fisheries) can provide incentives for the development of hatchery programs and other forms of production enhancements, while the availability of promising hatchery technologies may provide incentives for strengthening governance arrangements in order to take advantage of the technical opportunities.

Complementing Anderson's (2002) comparative analysis, Pinkerton (1994) (Honorable Mention) provided a detailed case study of the technical and governance interactions in the Prince William Sound, Alaska hatchery program that has become effectively integrated into the fisheries management framework. She showed how the salmon hatchery program has facilitated the emergence of cooperative fishery management involving the state and fishing communities and how the greater control over production achieved through the hatchery program resulted in economic benefits related to the more consistent volume and quality of product and collective marketing arrangements. Note that many of the social, economic, and political benefits of the hatchery program discussed here are related to qualitative changes in management and marketing rather than an overall production enhancement (the occurrence of which has been challenged for this fishery in the article by Hilborn and Eggers 2000).

Closing Remarks

The articles and citations reproduced in this section provide the scientific foundations for our understanding of hatchery programs and their role in fisheries management. At the same time, the articles are testament to the advances and insights that research on hatchery programs has made to the fundamentals of fisheries science. This research has enhanced our understanding of size and density-dependent processes in fish populations (Lorenzen 2005), the role of foraging and predator avoidance behavior in the fitness of wild fish (Olla et al. 1998), the role of continuous natural selection in maintaining fitness and the rapidity with which such fitness can be lost in altered selection regimes, and the close connection between use rights, resource stewardship, and enhancement (Pinkerton 1994; Anderson 2002). Continued development of hatchery technologies for more fish and invertebrate species, the expansion of rights-based governance systems that provide incentives for active resource enhancement and replenishment, and impacts of global environmental change that may motivate increasingly interventionist approaches to resource conservation and management suggest that the role of hatchery programs in fisheries management is unlikely to diminish. Applying the insights from the seminal studies reprinted here and from the studies they have motivated will be crucial to the responsible development of hatchery programs.

References

Araki, H., B. A. Berejikian, M. J. Ford, and M. S. Blouin. 2008. Fitness of hatchery-reared salmonids in the wild. Evolutionary Applications 1:342–355.

Baskett, M. L., and R. S. Waples. 2013. Evaluating alternative strategies for minimizing unintended fitness consequences of cultured individuals on wild populations. Conservation Biology 27:83–94.

Brown, C., and R. L. Day. 2002. The future of enhancements: lessons for hatchery practice from conservation biology. Fish and Fisheries 3:79–94.

Hilborn, R., and D. Eggers. 2001. A review of the hatchery programs for pink salmon in Prince William Sound and Kodiak Island, Alaska: response to comment. Transactions of the American Fisheries Society 130:720–724.

Lorenzen, K., K. M. Leber, and H. L. Blankenship. 2010. Responsible approach to marine stock enhancement: an update. Reviews in Fisheries Science 18:189–210.

Lorenzen, K., M. C. M. Beveridge, and M. Mangel. 2012. Cultured fish: integrative biology and management of domestication and interactions with wild fish. Biological Reviews 87:639–660.

Swingle, H. S. 1950. Relationships and dynamics in balanced and unbalanced fish populations. Alabama Polytechnic Institute, Agricultural Experimental Station Bulletin 274.

Walters, C. J., and S. J. D. Martell. 2004. Fisheries ecology and management. Princeton University Press, Princeton, New Jersey.

Wertheimer, A. C., W. W. Smoker, T. L. Joyce, and W. R. Heard. 2001. Comment: a review of the hatchery programs for pink salmon in Prince William Sound and Kodiak Island, Alaska. Transactions of the American Fisheries Society 130:712–720.

5.2 REPRINTED ARTICLES

Blankenship, H. L., and K. M. Leber. 1995. A responsible approach to marine stock enhancement. Pages 167–175 *in* H. L. Schramm, Jr., and R. G. Piper, editors. Uses and effects of cultured fishes in aquatic ecosystems. American Fisheries Society, Symposium 15, Bethesda, Maryland.

Cowx, I. G. 1994. Stocking strategies. Fisheries Management and Ecology 1:15–30.

Hilborn, R., and D. Eggers. 2000. A review of the hatchery programs for pink salmon in Prince William Sound and Kodiak Island, Alaska. Transactions of the American Fisheries Society 129:333–350.

Loomis, J., and P. Fix. 1998. Testing the importance of fish stocking as a determinant of the demand for fishing licenses and fishing effort in Colorado. Human Dimensions of Wildlife 3:46–61.

Lorenzen, K. 2005. Population dynamics and potential of fisheries stock enhancement: practical theory for assessment and policy analysis. Philosophical Transactions of the Royal Society 360:171–189.

Olla, B. L., M. W. Davis, and C. H. Ryer. 1998. Understanding how the hatchery environment represses or promotes the development of behavioral survival skills. Bulletin of Marine Science 62:531–550.

Reisenbichler, R. R., and J. D. McIntyre. 1977. Genetic differences in growth and survival of juvenile hatchery and wild steelhead trout, *Salmo gairdneri*. Journal of the Fisheries Research Board of Canada 34:123–128.

Santucci, V. J., and D. H. Wahl. 1993. Factors influencing survival and growth of stocked walleye (*Stizostedion vitreum*) in a centrarchid dominated impoundment. Canadian Journal of Fisheries and Aquatic Sciences 50:1548–1558.

Swingle, H. S. 1951. Experiments with various rates of stocking bluegills, *Lepomis macrochirus* Rafinesque and largemouth bass, *Micropterus salmoides* (Lacepede), in ponds. Transactions of the American Fisheries Society 80:218–230.

A Responsible Approach to Marine Stock Enhancement

H. LEE BLANKENSHIP

Washington Department of Fish and Wildlife
600 Capitol Way North, Mail Stop 43149
Olympia, Washington 98501, USA

KENNETH M. LEBER

The Oceanic Institute, Makapuu Point
Waimanalo, Hawaii 96795, USA

Abstract.—Declining marine fish populations worldwide have rekindled an interest in marine fish enhancement. Recent technological advances in fish tagging and marine fish culture provide a basis for successful hatchery-based marine enhancement. To ensure success and avoid repeating mistakes, we must take a responsible approach to developing, evaluating, and managing marine stock enhancement programs. A responsible-approach concept with several key components is described. Each component is considered essential to control and optimize enhancement. The components include the need to (1) prioritize and select target species for enhancement; (2) develop a species management plan that identifies harvest opportunity, stock rebuilding goals, and genetic objectives; (3) define quantitative measures of success; (4) use genetic resource management to avoid deleterious genetic effects; (5) use disease and health management; (6) consider ecological, biological, and life-history patterns when forming enhancement objectives and tactics; (7) identify released hatchery fish and assess stocking effects; (8) use an empirical process for defining optimum release strategies; (9) identify economic and policy guidelines; and (10) use adaptive management. Developing case studies with Atlantic cod *Gadus morhua*, red drum *Sciaenops ocellatus*, striped mullet *Mugil cephalus*, and white seabass *Atractoscion nobilis* are used to verify that the responsible approach to marine stock enhancement is practical and can work.

Marine fish populations are declining worldwide. In the United States, current abundance trends are known for only 15 of the most important marine stocks; about half of them are declining (NOAA 1991, 1992). Current harvest rates on most declining stocks are far in excess of exploitation levels needed to maintain the high long-term average yields that could be achieved through contemporary fishery management practices. Projected increases in human population size worldwide suggest this trend will continue into the future (FAO 1991).

Three principal tactics are available to fishery managers to replenish depleted stocks and manage fishery yields: regulating fishing effort; restoring degraded nursery and spawning habitats; and increasing recruitment through propagation and release (stock enhancement). The first two methods form the basis for the current federal approach to managing marine fisheries in the United States. The potential of the third method has not been convincingly documented with marine fishes.

Marine stock enhancement is not a new concept. In fact, hatchery-based stock enhancement was the principal technique used in an attempt to restore marine fisheries during the last part of the nineteenth century and early decades of the twentieth century. However, stock enhancement fell out of favor among fishery biologists after a half century of hatchery releases produced no evidence of an increased yield. Atlantic cod *Gadus morhua*, haddock *Melanogrammus aeglefinus*, pollock *Pollachius virens*, winter flounder *Pleuronectes americanus*, and Atlantic mackerel *Scomber scombrus* were stocked. Regrettably, when the last of the early marine hatcheries in the United States closed in 1948, after 50 years of stocking marine fishes, the technology had progressed no further than the stocking of unmarked, newly hatched fry. This was partly a result of the early approach to assessment, in which the success of hatchery programs was judged by numbers of fry stocked rather than by numbers of adults surviving to enter the fishery (Richards and Edwards 1986).

A New and Responsible Approach

Two general problems have restricted development of marine stock enhancement technology this century. Lack of an evaluation capability to determine whether hatchery releases were successful has been a major obstacle. Before the development of modern marking methods, fish-tagging systems were not applicable to the small, early life history stages released by hatcheries. The other impediment to development of marine enhancement has been the inability to culture marine fishes beyond early larval stages to the juvenile stage (fingerlings and larger sizes).

A new approach to marine stock enhancement is long overdue. Faced with declining stocks and an expanding world population, managers around the globe are looking at marine enhancement with renewed interest. To develop and evaluate stock enhancement's full potential, a process is needed for designing and refining stock enhancement tactics based on the combined effects of managing the resource (i.e., the interactive effects of hatchery practices, release strategies, harvest regulations, and habitat restoration on the condition of the managed stock).

Recent advances in both tagging technology and marine fish culture provide basic tools for a new approach to marine enhancement. We now have the technology for benign tagging of fish from juvenile through adult life stages (Bergman et al. 1992). Such tagging provides the basis for a quantitative assessment of stock enhancement success. Several marine fishes can be cultured to provide a wide range of life stages for release (e.g., McVey 1991; Honma 1993). Together, these tools allow an empirical evaluation of survival of cultured fish in the wild, and feedback on hatchery-release effects can be used to refine enhancement strategies. Release effects on wild stocks, and the fisheries based on them, can be quantified and evaluated. Survival can be examined over a range of hatchery practices and release variables (such as culture practices, fish size at release, release magnitude, release site, and season) to identify optimum combinations of hatchery and release strategies.

These new tools provide the basis for significantly increasing wild stock abundances. To ensure their successful use and avoid repeating past mistakes experienced in both marine and freshwater enhancement, we must use a careful approach in developing marine stock enhancement programs. The expression "a responsible approach to marine stock enhancement" embraces a logical and conscientious strategy for applying aquaculture technology to help conserve and expand natural resources. This approach prescribes several key components as integral parts in developing, evaluating, and managing marine stock enhancement programs. Each component is considered essential to control and to optimize the results of enhancement. The components include the need to (1) prioritize and select target species for enhancement; (2) develop a species management plan that identifies harvest opportunity, stock rebuilding goals, and genetic objectives; (3) define quantitative measures of success; (4) use genetic resource management to avoid deleterious genetic effects; (5) use disease and health management; (6) consider ecological, biological, and life-history patterns when forming enhancement objectives and tactics; (7) identify released hatchery fish and assess stocking effects; (8) use an empirical process for defining optimum release strategies; (9) identify economic and policy guidelines; and (10) use adaptive management. Combining new marine fish culture and tagging technologies with these ten principles is gaining support as a responsible approach to marine stock enhancement.

Empirical data suitable for accurately assessing the effect of hatchery releases on wild populations are often lacking. Partly because of this uncertainty, there is an increasing division of conservationists into two camps—one adamantly favoring increased fishing regulations and habitat protection and restoration in preference to hatchery releases, the other supporting propagation and release as an additional tool to manage fisheries and restore declining stocks. This split must be reconciled. Is stock enhancement of marine fishes a powerful, yet undeveloped technology for rebuilding depleted wild stocks and increasing fishery yields? Or are emerging marine enhancement programs merely futile attempts at recovering precious resources, thus diverting money and attention away from habitat restoration and the regulations needed to control overfishing? To realize the full potential of marine enhancement for the conservation and rapid replenishment of declining marine stocks, we must develop the technology to supplement and replenish marine stocks responsibly and quickly.

We must act now to assess the potential of marine stock enhancement through carefully planned research programs. Using strong inference (Platt 1964), which is essentially the scientific method, and addressing all of the components of the responsible approach concept, research programs will either document the value of marine enhancement or reveal that enhancement is not a useful concept. Without determined and careful attention to the 10 points listed above, marine hatchery releases in the 1990s may serve only to fuel divisiveness between the two conservationist camps, with little or no positive effect on natural resources.

Applying the responsible approach concept to new stock enhancement initiatives is straightforward. Existing enhancement programs may find it useful to review the 10 components discussed here. Incorporating those components expanded upon below, that are not already part of ongoing enhancement programs should provide a measurable increase in the realized effectiveness of replenishment efforts.

Prioritize and Select Target Species for Enhancement

In the absence of a candid and straightforward method, targeting species for stock enhancement can become a difficult and biased process. Unless attention is focused on the full spectrum of criteria that can be used to prioritize species, consideration of an immediate need by an advocacy group or simply the availability of aquaculture technology can become the driving factors in species selection. Commercial and recreational demand are obviously important criteria, but should they take precedence over other factors?

To reduce the bias inherent in selecting species, a semiquantitative approach was developed in Hawaii to identify selection criteria and prioritize species for stock enhancement research (Leber 1994). This approach involved four phases: (1) an initial workshop, where selection criteria were defined and ranked in order of importance; (2) a community survey, which was used to solicit opinions on the selection criteria and generate a list of possible species for stock enhancement research; (3) interviews with local experts to rank each candidate species with regard to each selection criterion; and (4) a second workshop, in which the results of the quantitative species selection process were discussed and a consensus was sought. This decision-making process focused discussions, stimulated questions, and quantified participants' responses. Panelists' strong endorsement of the ranking results and selection process used in Hawaii demonstrate the potential for applying formal decision making to species selection in other regions.

A critical step in removing bias from the species selection process lies in the type of numerical analysis used. The relative importance of the various criteria can be used in the analysis by factoring the degree to which each fish meets each criterion by the criterion weight. This produces a score for each species. This same concept is used to determine dominance in ecological studies of species assemblages (i.e., relative abundance times frequency of occurrence in samples). Using a trained facilitator to conduct the workshops also reduces bias by focusing activities on achieving results and by encouraging participation by all present.

Formal decision-making tools have been used effectively to prepare comprehensive plans for fisheries research (Bain 1987). Mackett et al. (1983) discuss the interactive management system for the Southwest Fisheries Center of the National Marine Fisheries Service. Similar processes have been used for research on North Pacific pelagic fisheries, in strategic planning for Hawaii's commercial fishery for skipjack tuna *Katsuwonus pelamis* (Boggs and Pooley 1987), and for a 5-year scientific investigation of marine resources of the main Hawaiian Islands (Pooley 1988).

Develop a Species Management Plan

A management plan identifies the context into which enhancement fits into the total strategy for managing stocks. The goals and objectives of stock enhancement programs should be clearly defined and understood prior to implementation. The genetic structure of wild stocks targeted for enhancement should be identified and managed according to objectives of the enhancement program. What is the population being enhanced? Can it be geographically defined? Clearly, in the interest of both production aquaculture and conservation, effort must be made to maintain genetic diversity (Kapuscinski and Jacobson 1987; Shaklee et al. 1993a, 1993b).

Assumptions and expectations about the performance and operation of the enhancement program necessary to make it successful should be identified (such as postrelease survival, interactions with wild stocks, long-term fitness, and disease). Critical uncertainties about basic assumptions that would affect the choice of production and management strategies should likewise be identified and prioritized. Evaluation of these uncertainties should be an integral part of the species management plan, and a feedback loop to evaluate and change production and management objectives should be included.

Define Quantitative Measures of Success

Without a definition of success, how do you know if or when you have it? Explicit indicators of success are clearly needed to evaluate stock enhancement programs. The objectives of enhancement programs need to be stated in terms of testable hypotheses. To be testable, a hypothesis must be falsifiable (Popper 1965). Depending on enhancement objectives, multiple indicators of success may be needed. These could include statements such as

> Hatchery releases will provide at least a 20% increase in annual landings of *Polydactylus sexfilis* in the Kahana Bay recreational fishery by the third year of the project.

Monitoring will show less than 3% change in the frequency of rare alleles (frequency less than 0.05) after 5 years of hatchery releases (this assumes that a control for the effects of environmentally induced change in allele frequencies is possible).

Numerous indicators should be identified to track progress over time. Although simplistic, indicators like the two examples above could be linked to success and would provide a basis for evaluating enhancement efforts during the initial period of full-scale releases. Clearly, to examine such indicators requires a reliable, quantitative marking and assessment system for tracking hatchery fish.

Use Genetic Resource Management

The need for genetic resource management in stock enhancement programs is currently the subject of intense public debate, and its importance cannot be over-rated. Responsible guidelines are now becoming available to aid resource managers in revitalizing stocks without loss of genetic fitness that could follow from inbreeding in the hatchery and subsequent outbreeding depression in the wild (Kapuscinski and Jacobson 1987; Shaklee et al. 1993a, 1993b). Once the genetic status of the target stock and the genetic goals of the enhancement program are identified, the approach for managing genetic resources is similar to the approach for managing other enhancement objectives (e.g., controlling the level of impact of stocked fish on abundances of the target population). This approach includes (1) identifying the genetic risks and consequences of enhancement; (2) defining an enhancement strategy; (3) implementing genetic controls in the hatchery and a monitoring and evaluation program for wild stocks; (4) outlining research needs and objectives; and (5) developing a feedback mechanism. These points are discussed in detail by Kapuscinski and Jacobson (1987) and Shaklee et al. (1993a, 1993b).

A genetic resource management plan should encompass genetic monitoring prior to, during, and after enhancement, as well as proper use of a sufficiently large and representative broodstock population and spawning protocols, to maintain adequate effective broodstock population size. Prior to enhancement, a comprehensive genetic baseline evaluation of the wild population should be developed to describe the level and distribution of genetic diversity. This baseline evaluation should at least include the geographical range of the particular stock targeted for enhancement. The monitoring should take place over a long enough period to observe possible short-term fluctuation or long-term change. The baseline can be used as a basis to determine an effective population or broodstock size to minimize the undesirable genetic effects of inbreeding, changes in allele frequencies, and loss of alleles. Genetic monitoring of the broodstock and its released progeny should be undertaken to measure success. Long-term genetic monitoring of the wild stock after enhancement should also occur to measure possible loss of genetic diversity, which might be attributed to enhancement efforts.

Maintenance and proper use of a sufficient broodstock population may be one of the toughest and most expensive components of marine stock enhancement. It is also one of the most important. The typically high fecundity rate of marine fish provides the opportunity for a greatly reduced effective population size in a hatchery environment because relatively few adults could potentially contribute a large number of eggs. Fortunately however, marine fish are genetically more homogeneous than freshwater and anadromous species on a relative scale, and genetic studies show relatively little stock separation due to geographic, clinal, or temporal factors (Gyllensten 1985; Waples 1987; Bartley and Kent 1990; King et al. 1995, this volume). In vagile marine species gene flow is often sufficient to homogenize the genetic structures over broad areas. Regardless, sufficient numbers of broodstock must be used so that the genetic diversity (including rare alleles) of the fish being released is the same over time as their wild counterparts.

Hubbs-Sea World Research Institute (Hubbs) has been an early promoter of a responsible genetic management plan. Hubbs leads a consortium of California researchers who are evaluating the feasibility of enhancement of white seabass *Atractoscion nobilis*. Although the genetic profile of progeny from an individual spawn may differ from wild spawns, use of multiple hatchery spawns can approximate the genetic variability observed in the wild. Bartley and Kent (1990) successfully used this concept with white seabass and showed that over 98% of the genetic variability observed in the wild could be maintained with an effective population of 100 broodfish.

Texas Parks and Wildlife Department's (Texas) enhancement program for red drum *Sciaenops ocellatus* provides a good example of maintaining a large broodstock with yearly replenishment (McEachron et al. 1995, this volume). Texas has 140–170 adult broodstock for its program, with an annual replacement of at least 25%. In Norway, studies of allele frequencies are being used to com-

MARINE STOCK ENHANCEMENT

pare broodstock and their progeny with wild populations of Atlantic cod (Svasand et al. 1990).

Use Disease and Health Management

Disease and health guidelines are important to both the survival of the fish being released and the wild populations of the same species or other species with which they interact. Florida Department of Environmental Protection (Florida) has developed an aggressive and responsible approach in this area in association with their red drum enhancement project (Landsberg et al. 1991). Florida's policy requires that all groups of fish pass a certified inspection for bacterial and viral infections and parasites prior to release. Maximum acceptable levels of infection and parasites in the hatchery populations are established based on the results of screening healthy wild populations.

Form Enhancement Objectives and Tactics

During the design phase of enhancement programs ecological factors that can contribute to the success or failure of hatchery releases should be considered. Predators, food availability, accessibility of critical habitat, competition over food and space, environmental carrying capacity, and abiotic factors, such as temperature and salinity, are all key variables that can affect survival, growth, dispersal, and reproduction of cultured fish in the wild. Predatory losses and food availability have long been thought to be among the principal variables that mediate recruitment success in wild populations (Lasker 1987; Houde 1987).

Habitat degradation in marine environments can also affect recruitment success. For example, seagrass meadows are important nursery habitats for fishes and crustaceans (see Kikuchi 1974). In vegetated aquatic environments, habitat availability and habitat quality (e.g., structural complexity) have been shown to mediate survival from predators (Crowder and Cooper 1982; Stoner 1982; Main 1987). In some cases, habitat degradation in marine environments may be so complete that certain habitats are unsuitable for stock enhancement (Stoner 1994). To enhance fisheries in some locales, restoration of coastal habitat may be the first priority.

The authors feel strongly that marine stock enhancement should never be used as mitigation to justify loss of habitat. However, we also feel that enhancement efforts with cultured fishes can fill a void where critically important habitats such as coastal wetlands and estuaries, which provide nurseries for early life stages, are irretrievably lost or degraded.

In addition to ecological factors, there may be physiological and behavioral deficits in hatchery-reared fish that strongly reduce survival in the wild (e.g., swimming ability, feeding behavior, predator avoidance, agonism, schooling, and habitat selection). In Japan, Tsukamoto (1993) has evaluated the effect of behavior on survival of cultured madai *Pagrus major* (called red sea bream by Tsukamoto) released into the sea. Tsukamoto's results indicate that a predator-avoidance behavior (tilting), in which wild fish lay flat against the substratum, may be reduced or absent in cultured fish during the first few days after release into the sea. Abnormal tilting behavior was directly correlated with mortality rate. For certain learned behaviors, exposure to behavioral cues and responses by wild fish in hatchery microcosms may be needed to overcome behavioral deficits (Olla and Davis 1988).

A solid understanding of the ecological and biological mechanisms mediating target species abundances can require exhaustive field studies for each species considered for enhancement. Whole careers have been dedicated to understanding mechanisms behind animal distributions and abundance; it does not seem practical to hold off on stock enhancement research until the ecological mechanisms are completely understood. However, failure to consider such factors can result in poor performance of released fish at best and at worst have negative impacts on natural stocks (Murphy and Kelso 1986).

Our viewpoint is that preliminary, pilot-scale experimental releases with subsequent monitoring of cultured fish afford a direct method for evaluating assumptions about the effects of uncontrolled environmental factors. For example, assumptions about carrying capacity in particular release habitats can and should be evaluated through pilot releases conducted prior to full-scale enhancement at those sites (Leber et al. 1995, this volume). This approach is elaborated below.

Identify Released Hatchery Fish and Assess Stocking Effects

One of the most critical components of any enhancement effort is the ability to quantify success or failure. Without some form of assessment, one has no idea to what degree the enhancement was effective or, more critically, which approaches were totally successful, partially successful, or a downright failure. Natural fluctuations in marine stock abundance can mask successes and failures. Maximization of benefits cannot be realized without the proper monitoring and evaluation system.

Tagging or marking systems that are benign and satisfy the basic assumption that identified fish are representative of untagged counterparts are essential, but weren't available until relatively recently. The detrimental effects of external tags are well documented (Isaksson and Bergman 1978; Hansen 1988; McFarlane and Beamish 1990), and few fishery managers or researchers defend their use today, especially with juvenile fish. Useful information retrieved from external tags is usually restricted to migration and growth rates of relatively large fish (Scott et al. 1990; Trumble et al. 1990).

In recent years, a few identification systems (e.g., coded wire tags, passive integrated transponder tags, genetic markers, and otolith marks) have been developed that meet the requirements that identified fish are representative of the species with regard to behavior, biological functions, and mortality factors, and thus provide unbiased data (Buckley and Blankenship 1990). The story of the development and now widespread use of the coded wire tag (Jefferts et al. 1963) is well known, and it is fair to say that it has revolutionized the approach to stock enhancement (Soloman 1990).

With an unbiased tag or mark, quantitative assessment of the effects of release is possible. In developing enhancement programs, evaluation of hatchery contributions can be partitioned into at least four distinct stages: initial survival, survival through the nursery stage, survival to adulthood (entry into the fishery), and successful contribution to the breeding pool. In Hawaii, the percent of hatchery fish in field samples taken after pilot releases of striped mullet *Mugil cephalus* has been as high as 80% in initial collections, 50% in some nursery habitats through the tenth month after release, and (in a recreational fishery in Hilo, Hawaii) as high as 20% of the catch (Leber, in press; Leber et al. 1995; Leber et al., in press). In Norway, genetic markers are beginning to show that released Atlantic cod produce viable offspring in the wild (Jorstad 1994).

Assessment of the effects of release should go further than evaluation of survival and contribution rates of hatchery fish. Evaluation of hatchery fish interactions with wild stocks is also critical. Clearly, evaluation of genetic impact is important. It is equally important to understand whether hatchery releases increase abundances in the wild or simply displace the wild stocks targeted for enhancement. At least one experimental study in Hawaii has documented that released hatchery fish can indeed increase abundances in a principal nursery habitat, without displacing wild individuals (Leber et al. 1995).

Use an Empirical Process to Define Optimum Release Strategies

Just as preliminary releases can be used to evaluate ecological assumptions, pilot release experiments afford a means of quantifying and controlling the effects of release variables and their influence on the performance of cultured fish in coastal environments (Tsukamoto et al. 1989; Svasand and Kristiansen 1990; Leber, in press; Willis et al. 1995, this volume).

Experiments to evaluate fish size at release, release season, release habitat, and release magnitude should always be conducted prior to launching full-scale enhancement programs. These experiments are a critical step in identifying enhancement capabilities and limitations and in determining release strategy. They also provide the empirical data needed to plan enhancement objectives, test assumptions about survival and cost effectiveness, and model enhancement potential. The lack of monitoring to assess survival of the fish released by marine enhancement programs early in this century (through the 1940s) was the single greatest reason for the failure of those programs to increase stock abundances and fishery yields (Richards and Edwards 1986).

Based on the results of pilot experiments by The Oceanic Institute in Hawaii, hatchery-release variables were steadily refined to maximize striped mullet enhancement potential. This resulted in an increase in recapture rates by at least 400% over a 3-year period (Leber et al. 1995; Leber et al., in press.) During the third year of pilot studies in Kaneohe Bay, hatchery fish provided at least 50% of the striped mullet in net samples during the entire 10-month collection period after releases. An understanding of how fish size at release and release habitat affected survival were the primary factors needed to increase recapture rates. However, understanding the interaction of release season with size at release and release habitat also had significant effect on refinement of release strategies. The apparent doubling effect on abundances in the third year was achieved with a release of only 80,000 juveniles into the principal striped mullet nursery habitat in Kaneohe Bay, the largest estuary in Hawaii. A subsequent study documented that mullet releases did not displace wild juveniles from that nursery habitat (Leber et al. 1995). Thus, hatchery

MARINE STOCK ENHANCEMENT

releases in Kaneohe Bay appear to be increasing population size in the primary nursery habitat. Clearly, these pilot experiments are crucial for managing enhancement impact.

Identify Economic and Policy Objectives

Initially, costs and benefits can be estimated and economic models developed to predict the value of enhancement. This information can be used to generate funding support through reprioritization, legislation, or user fees. The information can contribute to an explicit understanding with policy makers and the general public on the time frame that is needed for components such as adaptation of culture technology and pilot-release experiments before full-scale releases can begin. The education of the public and policy makers on the need and benefits of a responsible approach is also important. In Florida, pressure is mounting to drop the responsible approach concept involving pilot-scale releases and instead plant millions of red drum fry as a neighboring state has done (Wickstrom 1993). Advocates of the latter approach simply assume that the bigger the numbers planted, the better.

Use Adaptive Management

Adaptive management is a continuing assessment process that allows improvement over time. The key to this improvement lies in having a process for changing both production and management objectives (and strategies) to control the effects of enhancement. Essentially, adaptive management is the continued use of the nine key components above to ensure an efficient and wise use of a natural resource. The use of adaptive management is central to the successful application of the approach outlined above. Some minimum level of ongoing assessment is needed, superimposed over a moderate research framework that provides a constant source of new information. New ideas for refining enhancement are thus constantly considered and integrated into the management process.

Summary

The need for marine stock enhancement has been identified, and we must learn from mistakes made in the past. The necessity and benefit of following a responsible approach in implementing enhancement cannot be overemphasized. Several organizations have started to subscribe to this new approach to marine stock enhancement. The juveniles from their pilot releases are just starting to enter the fisheries, so the results are not known.

The exception is the striped mullet enhancement program in Hawaii at The Oceanic Institute. This program has shown the benefits that can be gained from closely following the approach outlined in this paper. In addition to The Oceanic Institute's decision to develop a proper genetic management plan and to make quantitative assessments of the effects of hatchery releases on wild populations, it performed numerous pilot studies to optimize release strategies.

Without these pilot experiments, Hawaii researchers would not have increased survival rate by over 400% in Kaneohe Bay nor provided a 20% contribution to the catch in the recreational fishery in Hilo Bay. We predict that identifiable fish from each of the other programs referenced will also have a substantial effect on the catch and validate our suggested approach. What is needed now is a concerted effort by the managers of new and existing enhancement programs to use, evaluate, and refine the approach described here.

Given the worldwide decline in fisheries catch rates, bold new initiatives are needed to revitalize fisheries. We need to take care, though, to preserve existing stocks as we work to restore and increase the harvest levels of those stocks using cultured fishes.

Acknowledgments

We wish to thank Devin Bartley, Don Kent, Rich Lincoln, Stan Moberly, and Scott Willis, who have greatly contributed to the development of the ideas expressed herein. We also thank Maala Allen, Paul Bienfang, Churchill Grimes, Gary Sakagawa, Kimberly Lowe, and Dave Sterritt for insightful comments on the manuscript. Order of authorship was determined by the flip of a coin.

This paper is funded in part by a grant from the National Oceanic and Atmospheric Administration (NOAA). The views expressed herein are those of the authors and do not necessarily reflect the views of NOAA or any of its subagencies.

References

Bain, M. B. 1987. Structured decision making in fisheries management: trout fishing regulation on the Au Sable River, Michigan. North American Journal of Fisheries Management 7:475–481.

Bartley, D. M., and D. B. Kent. 1990. Genetic structure of white seabass populations from the Southern California Bight region: applications to hatchery enhancement. California Cooperative Oceanic Fisheries Investigations Report 31:97–105.

Bergman, P. K., F. Haw, H. L. Blankenship, and R. M.

Buckley. 1992. Perspectives on design, use, and misuse of fish tags. Fisheries (Bethesda) 17(4):20–24.

Boggs, C. H., and S. G. Pooley, editors. 1987. Strategic planning for Hawaii's aku industry. NOAA (National Oceanic and Atmospheric Administration) NMFS (National Marine Fisheries Service). Southwest Fisheries Center Administrative Report H-87-1:22, Honolulu, Hawaii.

Buckley, R. M., and H. L. Blankenship. 1990. Internal extrinsic identification systems: overview of implanted wire tags, otolith marks and parasites. American Fisheries Society Symposium 7:173–182.

Crowder, L. B., and W. E. Cooper. 1982. Habitat structural complexity and interaction between bluegills and their pray. Ecology 63:1802–1813.

FAO (Food and Agriculture Organization of the United Nations). 1991. Food and Agriculture Organization of the United Nations yearbook 70(1990): fishery statistics.

Gyllensten, U. 1985. The genetic structure of fish: differences in the intraspecific distribution of biochemical genetic variation between marine, anadromous and freshwater species. Journal of Fisheries Biology 26: 691–699.

Hansen, L. P. 1988. Effects of Carlin tagging and fin clipping on survival of Atlantic salmon (*Salmo salar* L.) released as smolts. Aquaculture (Netherlands) 70: 391–394.

Honma, A. 1993. Aquaculture in Japan. Japan FAO Association. Chiyoda-Ku, Tokyo.

Houde, E. D. 1987. Fish early life dynamics and recruitment variability. American Fisheries Society Symposium 2:17–29.

Isaksson, A., and P. K. Bergman. 1978. An evaluation of two tagging methods and survival rates of different age and treatment groups of hatchery-reared Atlantic salmon smolts. Journal of Agricultural Research in Iceland 10(1):74–99.

Jefferts, K. B., P. K. Bergman, and H. F. Fiscus. 1963. A coded-wire identification system for macro-organisms. Nature (London) 198:460–462.

Jorstad, K. E. 1994. Cod stock enhancement studies in Norway—genetic aspects and the use of genetic tagging. World Aquaculture Society, New Orleans, Louisiana.

Kapuscinski, A. R., and L. D. Jacobson. 1987. Genetic guidelines for fisheries management. University of Minnesota, Minnesota Sea Grant College Program, Sea Grant Research Report 17, Duluth.

Kikuchi, T. 1974. Japanese contributions on consumer ecology in eelgrass (*Zostra marina*) beds, with special reference to tropic relationships and resources in inshore fisheries. Aquaculture 4:145–160.

King, T. L., R. Ward, I. R. Blandon, R. L. Colura, and J. R. Gold. 1995. Using genetics in the design of red drum and spotted seatrout stocking programs in Texas: a review. American Fisheries Society Symposium 15:499–502.

Landsberg, J. H., G. K. Vermeer, S. A. Richards, and N. Perry. 1991. Control of the parasitic copepod *Caligus elongatus* on pond-reared red drum. Journal of Aquatic Animal Health 3:206–209.

Lasker, R. 1987. Use of fish eggs and larvae in probing some major problems in fisheries and aquaculture. American Fisheries Society Symposium 2:1–16.

Leber, K. M. 1994. Prioritization of marine fishes for stock enhancement in Hawaii. The Oceanic Institute, Honolulu, Hawaii.

Leber, K. M. In press. Significance of fish size-at-release on enhancement of striped mullet fisheries in Hawaii. Journal of the World Aquaculture Society.

Leber, K. M., D. A. Sterritt, R. N. Cantrell, and R. T. Nishimoto. In press. Contribution of hatchery-released striped mullet, *mugil cephalus*, to the recreational fishery in Hilo Bay, Hawaii. Hawaii Department of Land and Natural Resources, Division of Aquatic Resources, Technical Report 94-03, Honolulu.

Leber, K. M., N. P. Brennan, and S. M. Arce. 1995. Marine enhancement with striped mullet: are hatchery releases replenishing or displacing wild stocks? American Fisheries Society Symposium 15:376–387.

Mackett, D. J., A. N. Christakis, and M. P. Christakis. 1983. Designing and installing an interactive management system for the southwest fisheries center. Pages 518–527 in O. T. Magoon and H. Converse, editors. Coastal Zone 83. American Society of Civil Engineers, New York.

Main, K. L. 1987. Predator avoidance in seagrass meadows: prey behavior, microhabitat selection and cryptic coloration. Ecology 68(1):170–180.

McEachron, L. W., C. E. McCarty, and R. R. Vega. 1995. Beneficial uses of marine fish hatcheries: enhancement of red drum in Texas coastal waters. American Fisheries Society Symposium 15:161–166.

McFarlane, G. A., and R. J. Beamish. 1990. Effect of an external tag on growth of sablefish and consequences to mortality and age at maturity. Canadian Journal of Fisheries and Aquatic Sciences 47:1551–1557.

McVey, J. P., editor. 1991. Handbook of mariculture, volume 2: finfish aquaculture. CRC Press Inc., Boca Raton, Florida.

Murphy, B. R., and W. E. Kelso. 1986. Strategies for evaluating freshwater stocking programs: past practices and future needs. Pages 303–316 in R. H. Stroud, editor. Fish culture in fisheries management. American Fisheries Society, Fish Culture Section and Fisheries Management Section, Bethesda, Maryland.

NOAA (National Oceanic and Atmospheric Administration). 1991. Our living oceans: first annual report on the status of U.S. living marine resources. NOAA Technical Memorandum NMFS/SPO-1.

NOAA (National Oceanic and Atmospheric Administration). 1992. Our living oceans: report on the status of U.S. living marine resources. NOAA Technical Memorandum NMFS-F/SPO-2.

Olla, B. L., and M. Davis. 1988. To eat or not be eaten. Do hatchery reared salmon need to learn survival skills? Underwater Naturalist 17(3):16–18.

Platt, J. R. 1964. Strong inference. Science 146(3642): 347–353.

Pooley, S. G., editor. 1988. Recommendations for a five-year scientific investigation on the marine resources and environment of the main Hawaiian Islands.

MARINE STOCK ENHANCEMENT

NOAA (National Oceanic and Atmospheric Administration) NMFS (National Marine Fisheries Service). Southwest Fisheries Center Administrative Report H-88-2, Honolulu, Hawaii.

Popper, K. R. 1965. Conjectures and refutations, the growth of scientific knowledge. Harper and Row, New York.

Richards, W. J., and R. E. Edwards. 1986. Stocking to restore or enhance marine fisheries. Pages 75–80 in R. H. Stroud, editor. Fish culture in fisheries management. American Fisheries Society, Fish Culture Section and Fisheries Management Section, Bethesda, Maryland.

Scott, E. L., E. D. Prince, and C. D. Goodyear. 1990. History of the cooperative game fish tagging program in the Atlantic Ocean, Gulf of Mexico, and Caribbean Sea, 1954–1987. American Fisheries Society Symposium 7:841–853.

Shaklee, J. B., C. A. Busack, and C. W. Hopley, Jr. 1993a. Conservation genetics programs for Pacific salmon at the Washington Department of Fisheries: living with and learning from the past, looking to the future. Pages 110–141 in K. L. Main and E. Reynolds, editors. Selective breeding of fisheries in Asia and the United States. The Oceanic Institute, Honolulu, Hawaii.

Shaklee, J. B., J. Salini, and R. N. Garrett. 1993b. Electrophoretic characterization of multiple genetic stocks of barramundi perch in Queensland, Australia. Transactions of the American Fisheries Society 122:685–701.

Soloman, D. J. 1990. Development of stocks: strategies for the rehabilitation of salmon rivers. Pages 35–44 in D. Mills, editor. Strategies for the rehabilitation of salmon rivers. Linnean Society of London, London.

Stoner, A. W. 1982. The influence of benthic macrophytes on foraging behavior of pinfish, *Lagodon rhomboides*. Journal of Experimental Marine Biology and Ecology 104:249–274.

Stoner, A. W. 1994. Significance of habitat and stock pre-testing for enhancement of natural fisheries: experimental analyses with queen conch, *Strombus gigas*. Journal of the World Aquaculture Society 25:155–165.

Svasand, T., K. E. Jorstad, and T. S. Kristiansen. 1990. Enhancement studies of coastal cod in western Norway, part 1. Recruitment of wild and reared cod to a local spawning stock. Journal du Conseil International pour l'Exploration de la Mer 47:5–12.

Svasand, T., and T. S. Kristiansen. 1990. Enhancement studies of coastal cod in western Norway, part 4. Mortality of reared cod after release. Journal du Conseil International pour l'Exploration de la Mer 47:30–39.

Trumble, R. J., I. R. McGregor, G. St-Pierre, D. A. McCaughran, and S. H. Hoag. 1990. Sixty years of tagging Pacific halibut: a case study. American Fisheries Society Symposium 7:831–840.

Tsukamoto, K., and six coauthors. 1989. Size-dependent mortality of red sea bream, *Pagrus major*, juveniles released with fluorescent otolith-tags in News Bay, Japan. Journal of Fish Biology 35(Supplement A):59–69.

Tsukamoto, K. 1993. Marine fisheries enhancement in Japan and the quality of fish for release. European Aquaculture Society Special Publication 19:556.

Waples, R. S. 1987. Multispecies approach to the analysis of gene flow in marine shore fishes. Evolution 41:385–400.

Wickstrom, K. 1993. Biscayne redfish: yes! Florida Sportsman (March):90–91.

Willis, S. A., W. W. Falls, C. W. Dennis, D. E. Roberts, and P. G. Whitchurch. 1995. Assessment of season of release and size at release on recapture rates of hatchery-reared red drum. American Fisheries Society Symposium 15:354–365.

Stocking strategies

I. G. COWX

International Fisheries Institute, University of Hull, Hull, UK

Abstract Stocking, transfer and introduction of fish are commonly used to mitigate loss of stocks, enhance recreational or commercial catches, restore fisheries or to create new fisheries. However, many stocking programmes are carried out without definition of objectives or evaluation of the potential or actual success of the exercise. This paper describes a strategic approach to stocking aimed at maximizing the potential benefits. A protocol is discussed which reviews factors such as source of fish, stocking density, age and size of fish at stocking, timing of stocking and mechanism of stocking. The potential genetic, ecological and environmental impacts of stocking are described.

KEYWORDS: genetic integrity, introductions, stock density, stocking.

Introduction

Stocking, transfer or introduction of fish is frequently used by fisheries owners, managers and scientists in the belief they will improve the quantity or quality of catches and have long-term beneficial effects on fish stocks. Over the past 50 years, large-scale movements of fish have occurred, including a total of 1354 introductions of 237 species into 140 countries (Welcomme 1988). Furthermore, many thousands of stocking events, involving millions of individual fish, take place annually in managed fisheries (Hickley 1993).

Although large sums of money have been invested in stocking activities, relatively few programmes have been properly evaluated and the evidence suggests stocking exercises rarely lead to any long-term tangible benefit. This appears to be the result of indiscriminate stocking, without well-defined objectives or prior appraisal of the likelihood of the success of the exercise. However, if stocking programmes are designed and implemented to satisfy defined goals it should be possible to improve the success rate. For example, catches of Arctic charr, *Salvelinus alpinus* L., in Lake Geneva showed a dramatic improvement following intensive stocking on an annual basis (Champigneulle & Gerdeaux 1994).

More recently, concerns have also been expressed about the potential risks associated with the stocking and introduction of fish, and the subsequent interactions with wild stocks. These include the loss of genetic integrity in indigenous fish stocks and ecological imbalance and consequent shift in community structure.

Thus there is a need for fisheries managers to be more aware of the possible impact of stocking, both in terms of the effects on wild populations and the likelihood of

Correspondence: Dr Ian G. Cowx, University of Hull, International Fisheries Institute, University of Hull, Hull HU6 7RX, UK.

improvements in stocks. Of particular importance is the need to develop a strategic approach to stocking which defines the objectives of the exercise and orients the implementation phase towards meeting these goals.

Objectives of stocking

The reasons for stocking are many and varied (European Inland Fisheries Advisory Commission [EIFAC] 1984a), but generally fall into four main categories which are related to the status of the wild stocks, the impact of anthropogenic activities and the ease with which factors limiting natural production can be removed or ameliorated.

Stocking for mitigation

This encompassess stocking with fish carried out as a voluntary exercise or statutory function for fishery protection schemes, such as reservoir dam construction, land drainage works or similar habitat perturbation. Mitigation stocking has been considered the simplest way to compensate for such activities. However, stocked fish may be released into unaffected parts of the river catchment or lake, and the impact on the wild stocks in these areas must be considered.

Stocking for enhancement

Enhancement stocking is the principal method used to maintain or improve stocks where production is actually or perceived to be less than the water body could potentially sustain, but where reasons for the poor stocks cannot be identified. This type of stocking is used where fishermen express dissatisfaction with the quality of fishing, or to enhance stocks in sections of river where access is restricted by natural barriers, or in the operation of commercial or put-and-take fisheries where the production of exploited species needs enhancing. It also includes activities carried out to strengthen quality and quantity of spawning stock of a given species so as to improve natural reproduction potential.

The majority of stocking in the past probably falls into this category and it is driven by fishermen complaining about the status of the fishing. However, in many cases the fishermen's and fisheries managers' assessments of the state of the stock have probably been unduly pessimistic, resulting from natural fluctuations which can have a profound effect on some fish populations, or merely the estimates of the potential production have been unrealistically high.

If production is already limited or driven by natural population cycles, it is unlikely that stocking will have a beneficial long-term effect.

Stocking for restoration

Stocking for restoration relates to that which is carried out after a limiting factor to stock recovery or improvement has been removed or reduced, e.g. water quality

STOCKING STRATEGIES

improvement, habitat restoration or the easing of passage for migratory fish. Re-establishment of fisheries which have previously been eliminated by poor water quality or habitat degradation are in this category. Stocking in this case is justifiable because the underlying problems limiting production have been tackled and long-term benefits are likely to accrue.

Creation of new fisheries

This category includes attempts to establish a new stock of fish in a river, lake or reservoir which has not previously held that stock because of natural barriers, evolutionary isolation, or where new (exotic) species are introduced into existing fisheries in an attempt to increase species diversity, improve fish yield or fill an apparent vacant niche. This is the most controversial stocking procedure and has led to considerable contention in the past, e.g. the introduction of Nile perch, *Lates niloticus* (L.), into Lake Victoria (Okemwa & Ogari 1994). However, if the new species is able to occupy a vacant niche the impact on the indigenous species may be negligible and the introduced species may become a valuable addition to the fishery. For example, barbel, *Barbus barbus* (L.), introduced into the River Severn, England have become a valuable component of the recreational fishery (Churchward, Hickley & North 1984). Introduction of species which will compete with indigenous species at some stage of their life history, for example mixing chub, *Leuciscus cephalus* (L.), dace, *Leuciscus leuciscus* (L.), roach, *Rutilus rutilus* (L.), or bream, *Abramis brama* L., may lead to the failure of the stocked species to produce a self-sustaining population and/or a reduction in the biomass of one or all the indigenous species. Consequently, stocking of this type needs careful planning to avoid potential catastrophic effects on the natural fish populations.

Protocol for pre-stocking appraisal

Identify need for stock improvement

In view of the many concerns that exist about stocking, a responsible attitude towards the activity is essential. A step-by-step approach to planning, assessing and implementing a stocking programme is shown in Fig. 1. It is based on the 'project approach' to management activities (Gittinger 1972) and has been described in detail by Crean (1994). Essentially, the activities can be broken down into the following phases: identification, preparation, appraisal, implementation and evaluation.

The first step when considering any stock improvement activity must be to ensure proper clarification of the management policy and objectives. It is only then that the project proposal can be properly formulated to achieve the desired effects. Part of this exercise includes establishing whether the stock is below optimum production level or whether the quality of the stock (e.g. in terms of age or size distribution) could be improved. This requires not only an assessment of the status of existing stocks, but an appraisal of the condition of the water body and the natural and artificial factors that may limit production. These assessments must be based on firm evidence from scientific

I.G. COWX

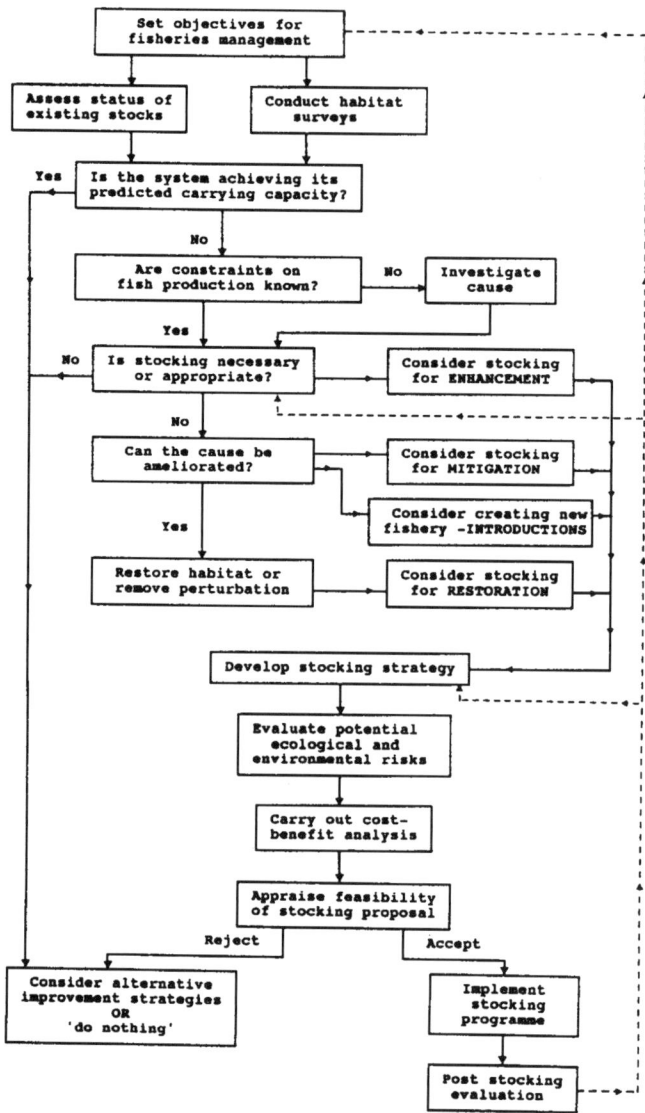

Figure 1. Suggested strategy for planning a stocking exercise to minimize the potential risk, maximize the potential benefit and evaluate the success of the project (modified from Salmon Advisory Committee [SAC] 1991).

studies (preferably of a long-term nature to overcome annual fluctuations) and not on hearsay or unsubstantiated complaints.

To aid the decision-making process, a technique commonly employed in development

STOCKING STRATEGIES

project formulation, the logical framework (Anon. 1982), can be used. This approach is useful in setting out the design of the stocking programme in a clear and logical way so that any weaknesses that exist can be addressed at an early stage, or if these are insuperable, the stocking can be aborted. Details of how the logical framework can be used in development of inland fisheries management projects are provided in Crean (1994) and Cowx (1994).

Establish objectives of stocking

If the fishery is of the desired quality, the need for any stock improvement must be questioned. Stocking generally does not tend to improve the catches in waters where there is adequate natural recruitment. Under these circumstances alternative improvement strategies should be considered or the 'do nothing' approach adopted.

If production is considered to be below the potential of the system it is important to try to identify the constraints and resolve them before stocking is carried out. If no apparent cause can be identified, enhancement stocking could be considered, but there is probably little sense in stocking a water if it is not capable of supporting a sustainable population. In this case it is probably worth while considering alternative improvement strategies or just leave the system alone and concentrate resources on rivers or lakes which can possibly be restored. This does not, however, exclude put-and-take fisheries, which are stocked to provide catchable-sized fish for rapid exploitation by anglers and do not take into account sustainability through natural recruitment.

Where the limiting factor(s) can be isolated, efforts should be made to resolve the problems before resorting to stocking. If remedial action cannot be taken, as in the case of a dam, for example, then mitigation stocking could be considered. This will probably not lead to a sustainable population and stocking may have to be on a continuous basis.

Alternatively, if mitigation stocking is not a viable or statutory option, considerations could be given to creating a new fishery based on species found elsewhere in the catchment or exotic species. If the system has not previously held the species to be stocked, it is important to establish whether it is suitable for the proposed introduction and whether there is likely to be an impact on other indigenous species. In this case the *Protocol for Introduction of Aquatic Species* drawn up by EIFAC (1988) should be strictly followed.

Finally, if it is possible to remove or minimize the cause of the decline in the fishery, this course of action should be taken. The fishery may then recover without stocking. Habitat improvement is the most desirable option because it should lead to long-term sustainable improvement with minimal deleterious ecological impact. It is also an efficient use of resources because it may have greater long-term benefits than enhancement stocking and also have other conservation and ecological benefits, e.g. improved primary or secondary production. In cases where natural recovery may be ineffective, because, for example, the spawning stock has been reduced to an apparently critically low level, restoration stocking may be appropriate to promote stock recruitment.

Development of stocking strategy

If the stocking appears to be worth while then implementation of the proposal must be carefully planned to minimize potential problems and risks (Fig. 2).

This formulation phase should assess the resources available to undertake the project, including availability of stocking material, labour, transport and finance, and identify any constraints which might jeopardize the successful implementation of the stocking programme. The protocol to be adopted to accrue the maximum benefit from the exercise should also be defined. For example, the source of fish, species mix, stocking density, size or age of fish to be stocked, timing and mechanism of release, should all be defined. Details of the factors that need to be considered in devising the stocking strategy are discussed later.

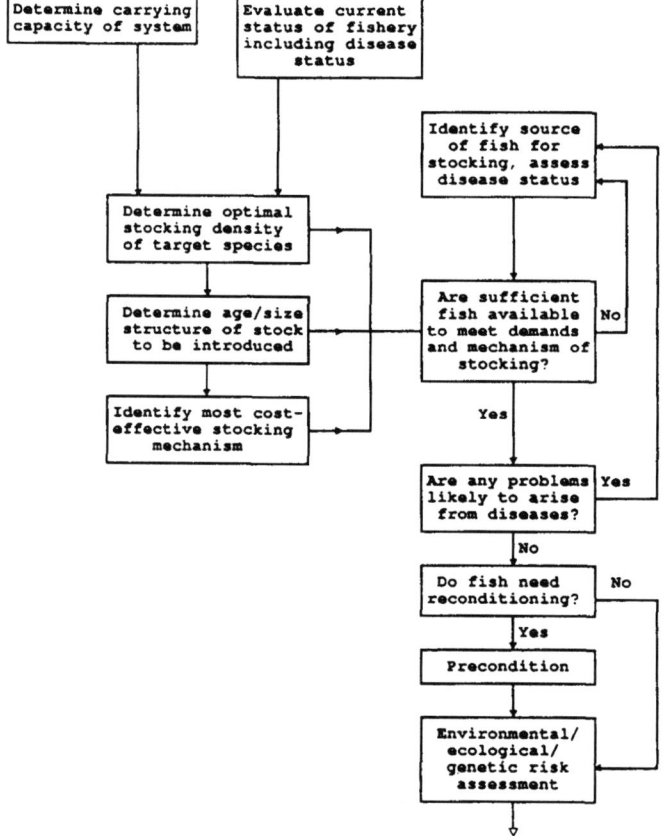

Figure 2. Flow chart illustrating the resource problems that must be considered when planning a stocking exercise.

STOCKING STRATEGIES

Assessment of proposal

All the ecological factors and consequences must be reviewed, with all the known or predicted advantages and disadvantages being accurately identified. The stocking proposal must also be assessed for its possible impacts on the potential yield and stability of both the introduced stock and the resident species. Details of the genetic risks and possible fishery and environmental effects associated with stocking activities must be discussed. In determining whether the stocking or introductions will have any undesirable impact, the decision-making process can be enhanced by compliance with appropriate protocols (e.g. EIFAC 1988).

The socio-economic influence of any change can also be crucial and must, therefore, be evaluated, particularly in terms of justifying costs. Estimation of the costs should include all capital and recurrent expenditure and be evaluated against the predicted benefits of the proposal, for example in terms of improved catches. The assessment of benefits should not include the public relations value of the programme, although in many situations this may provide a major incentive for undertaking the work. However, where benefits are real and substantial, the opportunity for public relations should not be foregone.

Finally, a complete review of all the accumulated information should be carried out by an independent, impartial authority, and only then should a decision be made whether to implement or reject the original proposal.

Post-stocking evaluation

One component of stocking exercises which has been largely neglected is an evaluation of the programme and an audit of the degree of success of the work. Such evaluations should be considered an integral part of the programme, and stocking should not be approved without building in a post-project evaluation. The evaluation should assess the efficiency and long-term benefits of the various stocking practices and regimes, and attempt to identify the factors contributing to their success or failure. It should be noted that demonstrating improved catches alone does not necessarily signal a successful stocking exercise, as it may not lead to a good spawning stock or sustainable recruitment. This information will provide a feedback mechanism (Fig. 1) to improve the formulation of realistic management objectives where stocking is a fundamental activity; to improve stocking strategies and techniques; and to provide a database of experiences against which the risks and the feasibility of new proposals can be appraised.

Potential risks from stocking

Genetic interactions

Recently it has been recognized that fish stocks (particularly salmon) from different catchments, and to a lesser extent from within the same catchment, exhibit genetic variation (Davidson, Birt & Green 1989; Hauser, Beaumont, Marshall & Wyatt 1991;

I.G. COWX

Hindar, Ryman & Utter 1991). These geographic, genetic differences may manifest themselves in, for example, growth potential, age at maturity, fecundity, or, in salmon, season of return to fresh water. They may also have some adaptive significance with respect to the performance of the stocks in a particular environment. Consequently, stock from non-indigenous sources may be less well adapted to the riverine or lacustrine environment into which they are released, and stocking may be less successful than expected. In addition, any fish surviving to reproduce with indigenous individuals may also confer a reduced adaptation upon some or all of their offspring. Several examples of this problem exist in the literature, particularly with respect to returning adult salmon, where pure donor stock have a lower return rate than natural stock, and hybrids exhibit an intermediate rate (e.g. O'Grady 1984). This, and other aspects of genetics in fishery management, have been discussed by Allendorf, Ryman & Uttar (1987).

Consequently, when embarking on any kind of stocking programme, consideration must be given to maintaining the genetic integrity of the indigenous stocks. Unfortunately, the intensity and long-term nature of many stocking programmes will make it difficult to rectify the situation once genetic integrity has been disrupted. However, where possible, stocking should be restricted to those using fish derived from local populations, or failing that habitats which are environmentally similar, or fish that have not been held in captivity for more than one generation.

Ecological interactions

Carrying capacity of the target habitat. Perhaps one of the greatest concerns with stocking programmes is that they rarely take into account the capacity of the recipient system to support the enhanced stocks (Kelly-Quinn & Bracken 1989). If too many fish are present, increased mortality rates, through predation and starvation, reduced growth rates and increased dispersion generally follow. Thus, while stocking may produce large increases in fish numbers at certain times or in localized areas, no more fish will survive than the habitat will allow.

Evidence for such a competitive bottleneck has been provided by Hegge, Hesthagen & Skurdal (1993), where the capacity for enhancing trout stocks in a stream was limited by benthic feeding conditions. Similarly, Timmermans (1967) suggested that the capacity of Belgian canals to support intensive stocking of cyprinids must be limited because it rarely produced an appreciable increase in catch rates. In the worst case scenario, overstocking can lead to a reduction in the performance of the fishery, below that prior to the introduction. For example, when the spawning stock of salmon exceeds an optimal level, the number of smolts produced may decrease (Salmon Advisory Committee [SAC] 1991).

Species interaction. Stocking of one species may have undesirable effects on endemic stocks either through predation or competition. For example, Kennedy (1984) and Kennedy & Strange (1986) found a reduction in the survival of stocked salmon fry due to competition from resident trout fry and salmon parr. Conversely, changes in trout populations have similarly been recorded following salmon parr stocking. In this case the salmon do not affect trout fry survival, but the presence of the salmon parr does cause a reduction in trout stocks.

STOCKING STRATEGIES

The introduction or enhancement of a predator population may have detrimental effects on the recruitment of potential prey species in the receiving system. For example, the introduction of pikeperch, *Stizostedion lucioperca* (L.), into rivers in eastern England allegedly caused a collapse in the cyprinid stocks (Linfield 1984). Similarly, the introduction of Nile perch into Lake Victoria appears to have resulted in the loss of several hundred species of haplochromine species and a decline in the catches of other commercial species (Okemwa & Ogari 1994). There are suggestions that the heavily exploited Nile perch stocks are now declining.

Stocking may also lead to undesirable changes in habitat which may impact on the populations of indigenous species the programme is designed to enhance. For example, the introduction of grass carp may greatly reduce the growth of aquatic macrophytes (Stott 1977), which may be reflected in the productivity of other species which utilize the vegetation either directly or indirectly for food, cover or spawning substrate.

Such habitat and species interactions, which may disturb the ecological balance of the fish community or restrict the potential success of stocking, must, therefore, be considered when planning a stocking programme.

Disease control. With the transfer of stocks between water bodies, there is an obvious risk of disease transmission. In many countries various legislation controls the movement of fish, but this is frequently violated, with consequences that are difficult to reverse. In the ideal situation all fish should be certified disease-free before stocking. This should be possible for fish originating from farms, and must become a statutory function attached to all consent applications.

However, when fish are transferred from one river or lake to another it is improbable that disease-free status can be guaranteed. In this case the disease status of both the donor and recipient stocks should be assessed and stocking only allowed if no new pathogen is being introduced and the stocked fish are healthy and have a low parasite/pathogen loading. If any possibility of disease transfer exists, the fish should be held in quarantine until the risk has been assessed.

Conversely, it is possible that fish introduced from one system to another may not be resistant to an endemic disease or parasite, and the stocking exercise may be unsuccessful. Although this may appear to have little impact on the receiving stock, there is a danger that the introduced fish may act as a reservoir for the proliferation of the disease.

Stocking strategies

When undertaking a stocking programme there are many procedures which should be considered during the implementation phase. It is therefore necessary to plan the stocking exercise to ensure its success. Several of these procedures have been discussed already and will be dealt with only superficially, whilst others, which relate to implementation strategy, will be discussed in more detail. Unfortunately, many of the practical aspects have to be considered in a descriptive way because relatively little information is available about the effect of various procedures on the success of stocking. A schematic approach to the planning exercise, to take on board these points, is suggested in Fig. 2.

I.G. COWX

Source of fish

There is an increasing awareness of the importance of maintaining genetic integrity of fish stocks. Consideration should therefore be given to minimizing the dilution of genetic variation by indiscriminate stocking policies with fish of unknown origin. Before implementing a stocking programme, a number of options relating to the source of fish should be considered.

Options for systems where species is extinct

(1) Donor stock with the same biological characteristics as the recipient system.
(2) Stock chosen from a lake or part of a river with a similar environment (e.g. size of stream, gradient, water temperature, flow regime, altitude, profile).
(3) Artificial propagation based on stock from (1) or (2) (sufficient fish should be used as broodstock to avoid reducing genetic variability of the species).
(4) Presume genetic differences of little adaptive significance and obtain stock from anywhere they are cheaply available, ideally from a number of sources, to maximize range of genetic material.

Options for depleted or relict stocks

(1) Build-up of stock by hatchery production based entirely on local stock and return brood stock to home system.
(2) Redistribution of adults from elsewhere in the catchment (may be unsuited for introduction to other parts with different prevailing conditions).
(3) Choose stock from a system with a similar environment.

Options for rivers where new species to be introduced

(1) Farm-reared fish, certified disease-free.
(2) Stock from a lake or part of a river with a similar environment which have been quarantined and certified clear of parasites and diseases alien to recipient system.
(3) No obvious ecological problems likely to be caused — as with introductions of predators.

Stocked fish should not have been reared in captivity for more than one generation in order to limit the possible effects of selection within the hatchery, thus particular care must be taken when obtaining fish from hatcheries.

Preconditioning and acclimatization

There is a growing body of evidence to suggest that fish should be preconditioned to survive the prevailing conditions in the receiving water body. For example, fish which have been reared or are to be transferred from still water to a river should be exposed to running water conditions for an extended period before their release. This exercising builds up the red muscle tissue in the stocked fish, thus increasing their ability for sustained swimming (Fisher & Broughton 1984).

Acclimatization to temperature is also thought to be important (Philippart & Baras

STOCKING STRATEGIES

1988). Prior to stocking barbel into the River Mehaigas they lowered the temperature of the water in which the fish were held over an extended period (10–11 days) until it approximated to that of the receiving water.

Handling and transportation of stock

Berka (1986) provides an overview of the procedures for transportation of fish.

Handling and transportation inevitably cause stress and possibly damage to fish, which can subsequently affect post-stocking survival. As a result, procedures which minimize handling time or frequency should be adopted from the time of capture of the donor fish to planting into the recipient system.

The techniques employed to capture the fish in the first instance should cause minimal damage; seine netting and 'controlled' electric fishing are the preferred techniques. During collection and transportation handling should be avoided where possible. Fish should be stored at low density and provided with an ample supply of oxygen.

All fish should be starved for at least 24 h prior to transportation to reduce oxygen demand, due to increased respiration rates during digestion, and minimize ammonia production. If the fish are to be transported long distances consideration should be given to reducing the effective toxicity of un-ionized ammonia by lowering the temperature and pH.

The use of suitable anaesthetics should be considered with a view to reducing physical activity and hence both the risk of damage and rate of respiration. Tertiary amyl alcohol ($180–900\,\mathrm{mg\,l^{-1}}$) and benzocaine ($10–40\,\mathrm{mg\,l^{-1}}$) are suggested as being the most suitable agents.

Finally, there is no point in introducing fish that are in poor condition or health, as this will affect the success of the exercise.

Stocking density

As previously discussed, a thorough assessment of the receiving water body should be carried out to determine the optimal stocking density. In lakes a relationship exists between shore line development, depth and predicted fish biomass (Leopold & Bninska 1984). However, no definitive relationship is available for calculating stocking density of different species in rivers; it is generally based on the experience of the managers. If possible, a similar, appropriate relationship should be determined for river fisheries.

Alternatively, a database of experiences should be set up to provide guidance on stocking densities which maximize the benefits in terms of improving stocks (Hickley 1994). This should be based on measured success of stocking at different densities. Such data are more readily available for salmon, e.g. see summary in Table 1, but rarely for other species. Thus, effort should be made to construct tables to indicate the success of stocking of all species at different densities. This can only be achieved if the outcome of stocking programmes is evaluated and reported.

When calculating the stocking density, consideration must be given to the existing stock biomass and allowances should be made for migration/dispersal, predation and

I.G. COWX

Table 1. Performance of stocking salmon into rivers at different densities and life history stages (data compiled from EIFAC 1984b and Kennedy 1988)

Stage of stocking	Density per m^2	% survival to end of growing season	Estimated smolt production per $100\,m^2$
Green ova	6.2–59.0	1.7– 4.0	4.3–10.0
Eyed ova	0.4–11.0	3.5–19.4	8.8–48.5
Unfed fry	0.3–29.3	1.3–38.6	3.3–96.5
Fed fry	0.1– 1.8	6.7–22.7	2.5–56.8

predicted survival of stocked fish. Values of between 10 and 80% annual mortality are given in the literature (EIFAC 1984b), so compensatory densities will be difficult to determine. The most important issue is that overstocking is avoided.

Size or age of stock

There has been much debate over the most appropriate size or age of fish for stocking. Many of these arguments can be removed by drawing up tables which illustrate the success of stocking of different age groups. Table 1 gives a summary example for salmon. This suggests that fed fry is probably the most effective life history stage to stock as the return rates are greatest. However, this information should be evaluated in relation to the advantages and disadvantages of obtaining sufficient fish of the appropriate age or size (Table 2). When this comparison is made, eyed ova or unfed fry are probably a more cost-effective life history stage for stocking.

Similar tables need to be drawn up for all species, although the exercise may be somewhat superfluous because the operation is often dictated by suitability of available donor fish, particularly with cyprinids. In this latter case there is considerable evidence to suggest that fish greater than 10 cm, at least in their second year of life, have a better chance of survival than 0-group fish which have not overwintered (A. Henshaw, personal communication).

Timing of stocking

There is a considerable volume of literature on the most appropriate time for stocking of salmonids. The general conclusion is that stocking in spring is more efficient (4–12 times) than winter (Aass 1984; Cresswell 1981; O'Grady 1984). Very little similar information is available for other species but a common sense approach probably defines the timing reasonably well. Fish should be stocked when the flow rates and water temperature are generally low, to minimize displacement of fish and stress respectively. The stocking should preferably take place when the productivity of the receiving water is high, but not during the spawning period as the stocked fish may interfere with natural reproduction processes. Stocking early in the summer, when natural food availability is good and to allow the fish to adjust to the conditions in the receiving water before overwintering, is preferable.

STOCKING STRATEGIES

Table 2. Comparison of life history stages for stocking (after Kennedy 1988)

Stage	Advantages	Disadvantages
Green ova	(1) No hatchery facilities required, therefore low cost (2) Stocking success can be monitored by follow-up surveys	(1) Artificial redds required therefore distribution is labour intensive and clumped (2) Choice of site is crucial to success, i.e. substrate type (3) Must be stocked out within 48 h of fertilization (4) Large number required, survival rates low (4%) to summer fry
Eyed ova	(1) Three-week period for handling, more robust and easier to transport than fry (2) As (2) above (3) Only limited hatchery facilities required	(1) As (1) above (2) As (2) above (3) Some incubation facilities required (4) Large numbers required, survival rate to summer fry 20%; 11–14% to end of growing season
Unfed fry	(1) Simple to disperse, stocking density can be more easily controlled (2) As (2) above (3) Limited hatchery facilities required (4) Can be stocked into preferential habitat, e.g. riffle where natural spawning is low	(1) Restricted time period for stocking out, therefore river conditions may be unsuitable (2) Facilities required for holding alevins after hatching (3) Survival affected by habitat and interaction with resident stock
0 + parr	(1) Fewer eggs required (2) Relatively easy to disperse (3) Success of stocking can be monitored by tagging and/or follow-up surveys (4) As (4) above	(1) Hatchery and rearing facilities required (2) Higher survival to smolts (15–20%) (3) As (3) above (4) Transport tanks required, more difficult to trickle stock
1 + parr	(1) As 0 + parr (1) above (2) As 0 + parr (2) above (3) As 0 + parr (3) above (4) Interaction with resident population less critical	(1) As 0 + parr (1) above (2) High survival to smolts (25–40%) (3) Good transport required, difficult to trickle stock
Smolt	(1) As 0 + parr (1) above (2) Can be released at critical time and location to maximize migration success (3) Can be tagged to evaluate return rate (4) No requirement for production in river system	(1) Full range of hatchery and rearing facilities required (2) High costs of production (3) Return rates 3–4 times less than for wild smolts (e.g. 2–3%) (4) Site and method of release is critical to survival

Mechanism of release

Three mechanisms for releasing fish are used:

(1) Spot planting — introducing all the fish into the receiving waters at the same site;

(2) Scatter planting – introducing fish into several sites in the same region;
(3) Trickle planting – introducing fish into the same region over a period of time.

Spot planting can lead to competition among the stocked fish, or with natural stocks, and in rivers is often associated with considerable downstream displacement to reduce population interactions (Cresswell 1981). Scatter stocking gives a wider dispersal at the outset and minimizes competitive pressures. Trickle stocking similarly removes competition, but is often constrained by lack of labour, finance and available stock.

Evidence suggests that scatter and trickle stocking (Fjellheim, Raddum & Saegrov 1993; Berg & Jorgensen 1994) are more successful than spot stocking, but the latter is generally carried out because it is easier to undertake.

Resource problems

The previous sections have described many of the issues that must be addressed when designing the stocking strategy. However, they are of little importance if the resources for implementing the programme are not available when required. A recurring problem is that the fish for stocking may not be readily available at the ideal time or in the numbers required. Access to the target zone may also be problematic, making scatter stocking difficult. All these circumstances must be addressed, and a compromise strategy for implementation must be drawn up.

Recommendations

Stocking is an important tool in the management of fisheries, albeit for commercial, recreational or conservation purposes. However, the management rationale and implications of stocking activities have not received the attention desired to support such a commonly used tool. It is recommended that a strategic approach to stocking is adopted. As part of this approach a number of aspects should be addressed (modified from Hickley 1994).

- Whenever, stocking of fish is to be considered, the aims, and specific objectives of the exercise must be clearly defined and adhered to.
- When evaluating stocking as a possible management tool, the relative benefits and cost of all options should be considered. The 'do nothing' option should not be disregarded but should be considered as fully as any of the other options under discussion, despite possible public pressure to stock.
- The strategy for any programme of stocking, transfer or introduction should be carefully tailored to suit the species in question, taking into account its entire suite of ecological prerequisites, so as to maximize the chances of success.
- The potential adverse impacts of stocking in terms of environmental, genetic and ecological interactions should be considered fully and the 'precautionary principle' adopted if any adverse impacts are foreseen.
- All projects should have in place the methodology to enable adequate monitoring of progress, and ultimately, success or failure. This post-stocking appraisal should

STOCKING STRATEGIES

include a mechanism of disseminating the outcome to minimize the risk of any unforeseen adverse effects in future exercises.
- A series of guidelines should be produced for all species which are stocked or introduced, clearly defining the most effective protocol for deciding whether stocking should take place, how it should be implemented and the potential impacts of such activities.

References

Aass P. (1984) Brown trout stocking in Norway. *EIFAC Technical Paper* 42(Suppl. 1), 123–128.
Allendorf F.W., Ryman N. & Utter F.M. (1987) Genetics and fishery management – past, present and future. In: N. Ryman & F.W. Utter (ed.) *Population Genetics and Fishery Management*. Seattle: University of Washington.
Anon. (1982) The logical framework approach (LFA). Norwegian Agency for Development Corporation, Oslo.
Berg S. & Jorgensen J. (1994) Stocking eel (*Anguilla anguilla*) in streams. In: I.G. Cowx (ed.) *The Rehabilitation of Freshwater Fisheries*. Oxford: Fishing News Books, Blackwell Scientific Publications, pp. 314–325.
Berka R. (1986) The transport of live fish. *EIFAC Technical Paper* 48.
Champigneulle A. & Gerdeaux D. (1994) The recent rehabilitation of the Arctic charr (*Salvelinus alpinus* L.) fishery in Lake Geneva. In: I.G. Cowx (ed.) *Rehabilitation of Freshwater Fisheries*. Oxford: Fishing News Books, Blackwell Scientific Publications, pp. 293–301.
Churchward A.S., Hickley P. & North E. (1984) The introduction, spread and influence of barbel (*Barbus barbus*) in the River Severn, Great Britain. *EIFAC Technical Paper* 42 (Suppl. 2), 335–343.
Cowx I.G. (1994) Fish stock assessment – biological basis for sound ecological management. In: D.M. Harper (ed.) *Ecological Basis for River Management*. London: Wiley. In press.
Crean K. (1993) Planning and development of inland fisheries. In: I.G. Cowx (ed.) *Rehabilitation of Freshwater Fisheries*. Oxford: Fishing News Books, Blackwell Scientific Publications, pp. 21–33.
Cresswell R.C. (1981) Post-stocking movements and recapture of hatchery-reared trout released into flowing waters – a review. *Journal of Fish Biology* 18, 429–442.
Davidson W.S., Birt T.P. & Green J.M. (1989) A review of genetic variation in Atlantic salmon, *Salmo salar* L., and its importance for stock identification, enhancement programmes and aquaculture. *Journal of Fish Biology* 34, 547–560.
European Inland Fisheries Advisory Commission (1984a) Report of the EIFAC Working Party on Stock Enhancement. *EIFAC Technical Paper No.* 44.
European Inland Fisheries Advisory Commission (1984b) Documents presented at the symposium on stock enhancement in the management of freshwater fish. Volume 1: Stocking. *EIFAC Technical Paper No.* 42 (Suppl. 1).
European Inland Fisheries Advisory Commission (1988) Code of practice and manual of procedures for consideration of introductions and transfers of marine and freshwater organisms. *EIFAC Occasional Paper No.* 23.
Fisher K.A.M. & Broughton N.M. (1984) The effect of cyprinid introductions on angler success in the River Derwent, Derbyshire. *Fisheries Management* 15, 35–40.
Fjellheim A., Raddum G.G. & Saegrov H. (1994) Stocking experiments with wild brown trout (*Salmo trutta*) in two mountain reservoirs. In: I.G. Cowx (ed.) *The Rehabilitation of Freshwater Fisheries*. Oxford: Fishing News Books, Blackwell Scientific Publications, pp. 268–279.
Gittinger P. (1972) *Economic Analysis of Agriculture Projects*. John Hopkins University Press, Baltimore.
Hauser L., Beaumont A.R., Marshall G.T.H. & Wyatt R.J. (1991) Effects of sea trout stocking on the population genetics of landlocked brown trout, *Salmo trutta* L., in the Conway River system, North Wales, UK. *Journal of Fish Biology* 39 (Suppl. A), 109–116.

Hegge O., Hesthagen T. & Skurdal J. (1993) Juvenile competitive bottleneck in the production of brown trout in hydroelectric reservoirs due to intraspecific habitat segregation. *Regulated Rivers: Research and Management* **8**, 41–48.

Hickley P. (1994) Stocking and introduction of fish – a synthesis. In: I.G. Cowx (ed.) *The Rehabilitation of Freshwater Fisheries*. Oxford: Fishing News Books, Blackwell Scientific Publications. pp. 247–254.

Hindar K., Ryman N. & Utter F. (1991) Genetic effects of cultured fish on natural fish populations. *Canadian Journal of Fisheries and Aquatic Sciences* **38**, 1867–1876.

Kelly-Quinn M. & Bracken J.J. (1989) Survival of stocked hatchery-reared brown trout, *Salmo trutta* L., fry in relation to carrying capacity of a trout nursery stream. *Aquaculture and Fisheries Management* **20**, 211–226.

Kennedy G.J.A. (1984) Factors affecting the survival and distribution of salmon (*Salmo salar* L.) stocked in upland trout (*Salmo trutta* L.) streams in Northern Ireland. *EIFAC Technical Paper* **42** (Suppl. 1), 227–242.

Kennedy G.J.A. (1988) Stock enhancement of Atlantic salmon (*Salmo salar* L.). In: D.H. Mills & D. Piggins (eds). *Atlantic Salmon: Planning for the Future*. London: Croom Helm. pp. 345–372.

Kennedy G.J.A. & Strange C.D. (1986) The effects of intra- and inter-specific competition on the survival and growth of stocked juvenile Atlantic salmon, *Salmo salar* L., and resident trout, *Salmo trutta* L., in an upland stream. *Journal of Fish Biology* **28**, 479–489.

Leopold M. & Bninska M. (1984) The effectiveness of eel stocking in Polish lakes. *EIFAC Technical Paper* **42** (Suppl. 1), 41–45.

Linfield R.S.J. (1984) The impact of zander (*Stizostedion lucioperca* (L.)) in the United Kingdom and the future management of affected fisheries in the Anglian region. *EIFAC Technical Paper* **42** (Suppl. 2), 353–362.

O'Grady M.F. (1984) The importance of genotype, size on stocking and stocking date to the survival of brown trout released into Irish lakes. *EIFAC Technical Paper* **42** (Suppl. 1), 178–191.

Okemwa E. & Ogari J. (1994) Introductions and extinctions of fish in Lake Victoria. In: I.G. Cowx (ed.) *The Rehabilitation of Freshwater Fisheries*. Oxford: Fishing News Books, Blackwell Scientific Publications. pp. 326–337.

Philippart J.C. & Baras E. (1988) The biology and management of the barbel, *Barbus barbus* (L.) in the Belgian River Meuse basin, with special reference to the reconstruction of populations using intensively-reared fish. In: *Proceedings of the 19th (1988) Institute of Fisheries Management Annual Study Course*, 61–82.

Salmon Advisory Committee (1991) *Assessment of Stocking as a Salmon Management Strategy*. MAFF Publications, London.

Stott B. (1977) On the question of the introduction of grass carp (*Ctenopharyngodon idella* Val.) into the United Kingdom. *Fisheries Management* **8**, 63–71.

Timmermans J.A. (1967) Restocking of fishing waters with catchable roach. *Proceedings of the 3rd British Coarse Fish Conference, University of Liverpool*, pp. 30–32.

Welcomme R.L. (1988) International introductions of inland aquatic species. *FAO Fisheries Technical Paper* **294**.

A Review of the Hatchery Programs for Pink Salmon in Prince William Sound and Kodiak Island, Alaska

RAY HILBORN*

*University of Washington, School of Fisheries,
Box 357980, Seattle, Washington 98195-7980, USA*

DOUG EGGERS

*Division of Commercial Fisheries, Alaska Department of Fish and Game,
Post Office Box 25526, Juneau, Alaska 99801-5526, USA*

Abstract.—Five hatcheries in Prince William Sound, Alaska, release more than 500 million juvenile pink salmon *Oncorhynchus gorbuscha* each year, constituting one of the largest salmon hatchery programs in the world. Before the program was initiated in 1974, pink salmon catches were very low, averaging 3 million fish per year between 1951 and 1979. Since 1980 the catch has averaged more than 20 million fish per year. However, catches in three other areas in Alaska with substantial fisheries for pink salmon (southeast Alaska, Kodiak Island, and the southern Alaska Peninsula) also increased equivalently during the same period, and the hatchery production did not become the dominant factor in Prince William Sound until the mid-1980s, long after the wild population had expanded. A hatchery program in the Kodiak area provides useful contrast to the Prince William Sound program because it is smaller and more isolated from the major wild-stock-producing areas of Kodiak Island. The evidence suggests that the hatchery program in Prince William Sound replaced rather than augmented wild production. Two likely causes of the replacement were a decline in wild escapement associated with harvesting hatchery stocks and biological impacts of the hatchery fish on wild fish. Published papers disagree on the impact of the 1989 *Exxon Valdez* oil spill, but none of the estimates would account for more than a 2% reduction in wild-stock abundance, and the decline in wild stocks began well before the oil spill. No evidence in the Kodiak area program suggests any impact on wild stocks. This analysis suggests that agencies considering the use of hatcheries for augmenting salmonids or other marine species should be aware of the high probability that wild stocks may be adversely affected unless the harvesting of the hatchery fish is isolated from the wild stocks and the hatchery and wild fish do not share habitat during their early ocean life.

In response to low salmon abundance in the 1960s and 1970s the state of Alaska began several hatchery programs, including the creation of the Fisheries Rehabilitation, Enhancement and Development division within the Alaska Department of Fish and Game (ADF&G). The state legislature also passed the Hatchery Act (1974) and the Fisheries Enhancement Loan Program, which provided for low-interest loans to regional aquaculture organizations (Hull 1993). Under this legislative framework the Prince William Sound (PWS) Aquaculture Corp. (PWSAC) was formed in December 1974 by a group of commercial fishermen based in Cordova, Alaska. It currently operates three pink salmon hatcheries in PWS, and the Valdez Fisheries Development Association (VFDA) operates a single hatchery (Solomon Gulch) in Valdez Arm (Figure 1A). Approximately 70% of the hatchery production in PWS comes from the three PWSAC hatcheries, but we will use data from the entire hatchery program—that is, both PWSAC and VFDA. Some of the spirit and hope of the early days of salmonid aquaculture in Alaska are captured in Wilson and Buck (1978): "the future potential for significantly increased salmon harvests throughout the state is enormous. Alaska's approach to salmon aquaculture and fisheries enhancement bears watching in the next decade as this multifaceted program attempts to yield larger harvests and bring new stability to a historically cyclical resource."

The PWSAC is a private nonprofit corporation funded both by a 2% tax on landings of fishermen in PWS and by sales of fish captured in cost recovery fisheries. It now operates the largest hatchery program in North America, releasing more than 500 million fry of pink salmon *Oncorhynchus gorbuscha* each year and some juveniles of sockeye salmon *O. nerka*, chum salmon *O. keta*, coho salmon *O. kisutch*, and chinook salmon *O. tshawytscha*. Olsen (1994) and Pinkerton (1994) describe the biological and social history of PWSAC.

---
* Corresponding author: rayh@u.washington.edu

Received February 4, 1999; accepted June 9, 1999

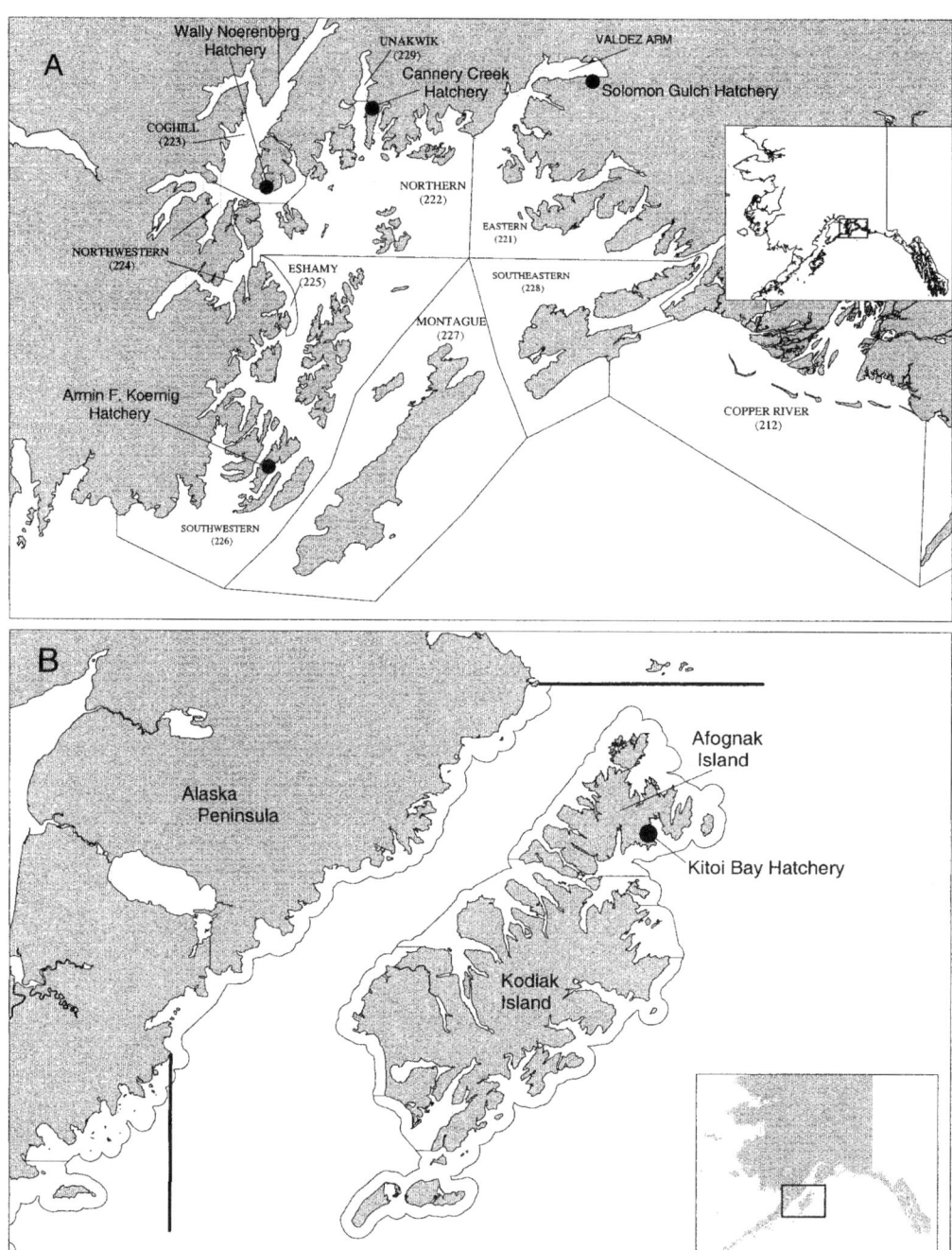

FIGURE 1.—Maps of **(A)** Prince William Sound (PWS) and **(B)** the Kodiak area, Alaska, showing locations of the hatcheries and, in PWS (A), the fishing districts (district numbers in parentheses).

PINK SALMON HATCHERY PROGRAMS

The hatchery run of pink salmon to the Kodiak Island area is entirely supported by the Kitoi Bay Hatchery on Afognak Island (Figure 1B). The ADF&G rebuilt the facility, originally constructed in 1956, after its destruction in the 1964 earthquake. The facility was initially operated as a research facility, but emphasis switched to pink salmon production in 1976; it also produces sockeye salmon, chum salmon, and coho salmon. The ADF&G operated the facility before 1987 and Kodiak Regional Aquaculture Association (KRAA) assumed full operation of the hatchery in 1992. The KRAA is funded by a 2% tax on landings by fishermen in the Kodiak area as well as by earnings on a fund created from the proceeds of a one-time terminal area cost recovery fishery that occurred in 1989. This cost recovery fishery occurred because the *Exxon Valdez* oil spill in 1989 prevented harvest of returning salmon in the traditional fishing areas.

Concern about the biological success and economic viability of hatchery programs is increasing (Hilborn 1992; Meffe 1992; Hilborn and Winton 1993), and the PWS and Kodiak pink salmon programs appear to be excellent subjects for evaluating the biological success of large hatchery programs for four reasons. First, both programs are large and spatially quite discrete. Second, there are four regions of Alaska with significant wild pink salmon production, but only in PWS and the Kodiak area are there large-scale hatcheries. The other two areas provide the opportunity for natural controls that depict changes in wild stocks that occurred while the hatchery program came on line. The ADF&G has maintained a regular program of escapement monitoring throughout the PWS and Kodiak areas so that changes in escapement can be documented. Third, unlike the chinook salmon and coho salmon hatchery programs in Canada and the lower 48 United States, which have been ongoing for more than 100 years, the PWS and Kodiak pink salmon programs began in recent years, and there are reliable data on wild stocks before the program began. Finally, significant physical differences exist between the programs in PWS and the Kodiak area: the location of the Kodiak area hatchery is well isolated from the major wild spawning areas whereas the PWS hatcheries are not.

Previous papers have explored the implications of these hatchery programs. Eggers et al. (1991) compared the pink salmon production in PWS with that in the Kodiak area and with other wild Alaskan pink salmon stocks and noted that PWS production had increased at the same time as the other stocks. They suggested that intense harvest of hatchery fish in PWS had been responsible for the decline of PWS wild stocks, replacing wild production with hatchery production. Tarbox and Bendock (1996) inferred that the hatchery program in PWS was a major contributor to declines in wild stocks. Smoker and Linley (1997) challenged the conclusions of Eggers et al. (1991) and of Tarbox and Bendock (1996) and considered alternatives to replacement of wild stocks by hatchery fish.

The purpose of this paper is to review the biological success of the PWS and Kodiak pink salmon hatchery programs. We now have considerably more years of data than were available to Eggers et al. (1991), and we have examined some additional areas of wild Alaskan pink salmon production. Further we also examined evidence for biological interaction between wild and hatchery fish in PWS and the Kodiak area and changes due to fishing. Finally we consider how our findings from the PWS and Kodiak areas can be applied to other hatchery programs for salmonids and marine species.

Methods

This analysis is strictly retrospective and is based on published data taken primarily from ADF&G reports on wild-stock catches and escapements as well as hatchery runs in southeast Alaska, Prince William Sound, Kodiak Island, and south Alaska Peninsula management areas.

For PWS, total catch numbers and delivery weights of pink salmon for the years 1965–1997 were taken from Morstad et al. (1998). The wild pink salmon peak aerial survey escapement index counts were not reflective of true escapement (Bue et al. 1998b). The escapements in Morstad et al. (1998) were estimated by dividing cumulative spawner-days, based on stream counts from aerial surveys, by the estimated stream residence time of 17.5 d (Helle et al. 1964). Multiyear studies of streams in the PWS aerial survey index program (Bue et al. 1998b) indicate that stream life is similar in streams within districts and between years. These estimates differed from the stream life used in the historical escapement calculations. Stream life estimated for Irish and Hawkins creeks (17.8 d) was used to adjust the index counts for the Eastern and Southeastern fishing districts (Figure 1A), and stream life estimates for the remaining streams were averaged (11.1 d) and applied to the remaining districts.

Runs of pink salmon to PWS hatcheries provide catches in common-property commercial fisheries, cost recovery catches in hatchery terminal harvest areas, and broodstock. Numbers for catch of private nonprofit hatchery fish in mixed-stock commercial and cost recovery fisheries, as well as broodstock and unused fish, were taken from annual hatchery reports provided to ADF&G. Before 1987 the wild and hatchery fish contributions to the mixed-stock commercial fishery were estimated from the relative magnitude of returns to hatchery terminal areas and wild-stock escapement levels. Estimates of hatchery catches from 1987 to 1997 were based on a coded-wire-tagging program (Geiger and Sharr 1990; Peltz and Geiger 1990), and catches of wild stocks were approximated as the total common-property commercial harvest less the estimated hatchery contribution.

For the Kodiak area, total catch numbers of pink salmon for 1965–1996 were taken from Brennan et al. (1998), and those for 1997 were from ADF&G catch records (K. Brennan, ADF&G, personal communication). Catches of hatchery fish were assumed to be the entire commercial catch and cost recovery in the Izhut Bay, Duck Bay, and Kitoi Bay subdistricts. No significant populations of wild pink salmon exist near Kitoi Bay, and the hatchery there is not near traditional fishing areas for wild pink salmon. Catches of wild pink salmon do not occur in the hatchery terminal harvests, and catches of hatchery fish are negligible in fishing areas outside the terminal harvest area. Estimates of the commercial catch, cost recovery, and broodstock for the Kitoi Bay Hatchery, 1972–1997, were compiled from ADF&G catch records and from hatchery annual reports filed with ADF&G (Steve Honnold, Alaska Department of Fish and Game, personal communication). Wild-stock catch was estimated as total catch less hatchery catch.

Wild-stock escapement estimates were determined from cumulated weir counts and expanded peak counts of live fish derived from aerial or foot surveys (Brennan et al. 1998). Peak counts were expanded by a factor of 1.84 based on estimated stream life (Barrett et al. 1990). Escapements for streams not surveyed were interpolated from surveyed streams in the respective year, based on the historical average odd- and even-year escapement distribution among streams.

For the southern Alaska Peninsula area, total catch numbers of pink salmon were obtained from Campbell et al. (1998). Wild-stock escapement estimates were determined from peak counts of live fish derived from aerial or foot surveys (Campbell et al. 1998). Peak counts were expanded by a factor of 1.4 based on estimated stream life (B. A. John-

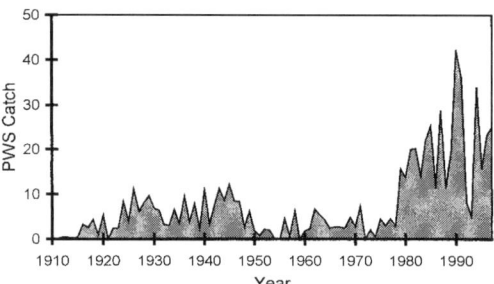
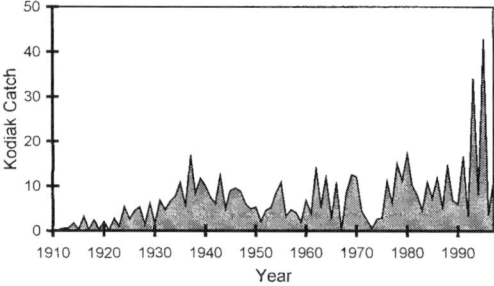

FIGURE 2.—The annual catch (in millions) of pink salmon in Prince William Sound (PWS; top panel) and the Kodiak Island area (bottom panel).

son and B. Barrett, ADF&G, unpublished manuscript).

For southeast Alaska, catches of pink salmon were obtained from ADF&G (1997). Estimates of wild-stock escapement were determined from peak counts of live fish derived from aerial surveys. The index counts were expanded for streams not surveyed in a particular year based on historical estimates of escapement distribution among streams. The index counts were standardized to account for differences in counting bias among individual observers (K. A. Hofmeister, ADF&G, unpublished, 1998). Standardized peak index counts were expanded by 2.5 to account for stream life (Dangel and Jones 1988).

Results

History of Pink Salmon Returns

The long-term history of pink salmon catches in PWS reveals four distinct periods. From 1896 to 1913, annual catch was less 1 million; 1916–1950 catches averaged 5.8 million fish per year; 1951–1979 catches dropped considerably to 3.3 million per year; and since 1980 catch has averaged 20.6 million fish per year (Figure 2). The dramatic rise since 1980 can be taken as evidence for success of the hatchery program. However, the three periods in PWS production since 1916 cor-

PINK SALMON HATCHERY PROGRAMS

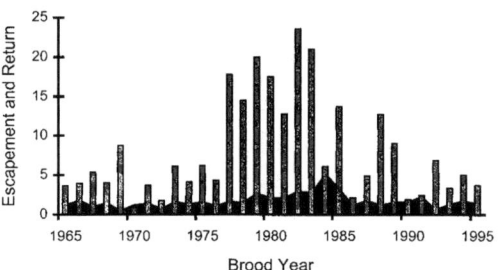

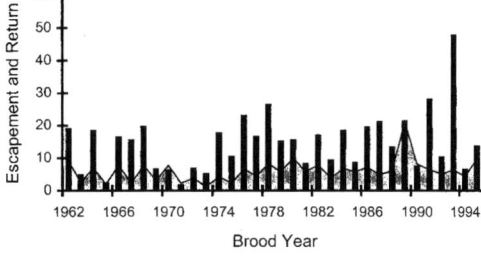

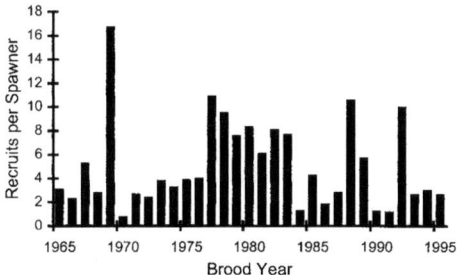

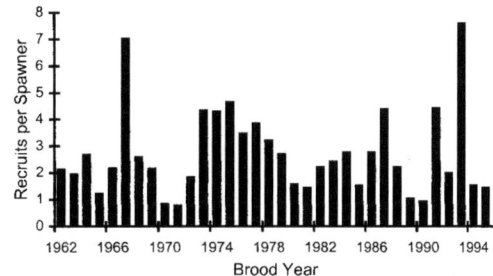

FIGURE 3.—Historical production of wild pink salmon in Prince William Sound, illustrated by **(top)** total return (vertical bars) and escapement (dark shaded area) of wild pink salmon (millions of fish) and **(bottom)** the index of wild recruits per spawner.

FIGURE 4.—Historical production of wild pink salmon in the Kodiak Island area, illustrated by **(top)** total return (vertical bars) and escapement (light shaded area) of wild pink salmon (millions of fish) and **(bottom)** the index of wild recruits per spawner.

respond to general patterns in abundance of pink salmon and sockeye salmon throughout Alaska, and these major changes are generally ascribed to changes in ocean conditions. These three periods are now commonly called "regimes" and fluctuation between regimes is the "interdecadal oscillation" (Francis and Hare 1994; Hare and Francis 1995; Mantua et al. 1997). Interpreting the impact of the hatchery program is closely connected with understanding and interpreting changes in other pink salmon populations in Alaska. Catch from the Kodiak Island area rose less dramatically after 1977 but, on average, was more than double the 1970s levels (Figure 2).

Figure 3 shows a major increase in total run to PWS in the late 1970s followed by an increase in escapement; then in the mid-1980s, wild-stock escapement and total runs declined. The index of wild recruits per spawner was elevated during 1977–1983 then experienced irregular but lower values from 1984 to 1993. In the Kodiak area both escapement and runs began to gradually increase in the mid-1970s (Figure 4).

History of Hatchery Production

The hatchery program in PWS began in the mid-1970s and by the early 1980s produced several hundred million fry per year (Figure 5). The returns from hatchery production kept pace with the releases such that when pink salmon fry production increased to about 500 million in 1987, the subsequent adult returns were 15–35 million. Ocean survival apparently increased early in the program, but survival was poor in 1990 and 1991. In the 1990s, 20–40% of the total return was taken for cost recovery and broodstock.

In the Kodiak area, fry releases rose throughout the late 1970s and 1980s to about 150 million per year (Figure 6, top). The 1991 brood year produced a high of about 10 million fish and the 1987 brood was slightly lower, but only a few million fish were produced annually in other brood years. Although the Kodiak hatchery program is roughly one third the size of the PWS program in releases, survival is much lower, and only the 1991 hatchery brood year (1993 year of capture) produced a significant proportion of Kodiak pink salmon catch. As in PWS, hatchery ocean survival (Figure 6, middle) was more than 6% in the 1987 and 1991 brood years but only 1–2% in other years since 1980. In contrast, survival in PWS hatcheries was at least double the Kodiak average. Only in brood years 1985–1987 (harvest years 1987–1989) was there any cost recovery harvest

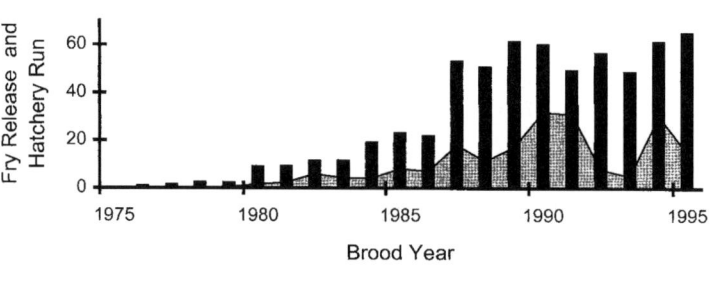
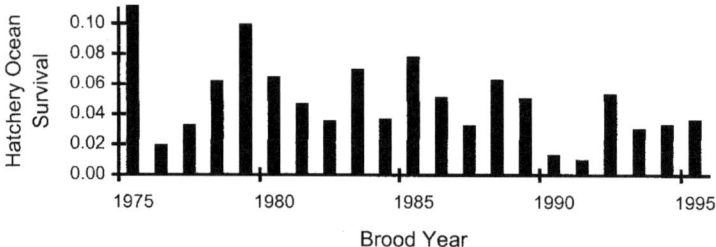
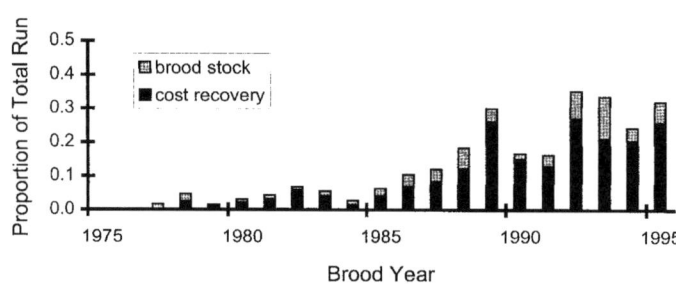

FIGURE 5.—Historical production of pink salmon from hatcheries in Prince William Sound, illustrated by **(top)** fry releases (vertical bars in tens of millions) and hatchery return of adults (shaded area in millions), **(middle)** ocean survival rate for hatchery fish, and **(bottom)** proportion of the total run of pink salmon that has gone to cost recovery fisheries and broodstock.

(Figure 6, bottom), and in 1989 almost all of the run was taken for cost recovery when the ocean salmon fisheries were closed because of the *Exxon Valdez* oil spill.

Pink Salmon Stock Changes Outside PWS

There are two other major pink salmon production areas in Alaska: southeast Alaska (the Alaska panhandle) and the southern Alaska Peninsula. Both of these areas also experienced a major increase in abundance since the 1977 regime shift. Some differences exist in the spawning habitat among areas, PWS having a high proportion of intertidal spawning. Pink salmon in all areas have similar marine life cycles, spending their ocean life in the Gulf of Alaska and northeast Pacific Ocean. Eggers et al. (1991) suggested that other populations of wild Alaskan pink salmon should reflect what would have happened to PWS pink salmon in the absence of a hatchery program.

In southeast Alaska and the southern Alaska Peninsula, high production beginning in 1975–1976 followed low production in the 1960s and early 1970s (Figure 7). The catch in all four pink salmon regions has increased considerably since the mid-1970s. We normalized the data by dividing them by the average for 1976–1985, obtaining a 5-year running average to smooth the data, and then plotted all four pink salmon areas together in Figure 8. The 5-year running averages of total returns (hatchery and wild) to the four areas, divided by the 1976–1985 average for each area, show little clear discrimination among areas; returns increased in all areas with PWS having the lowest relative value in recent years. It is clear that PWS

PINK SALMON HATCHERY PROGRAMS

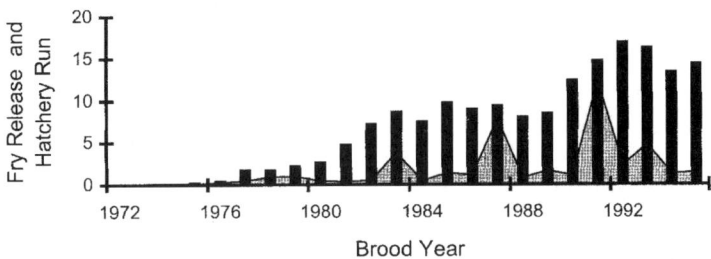

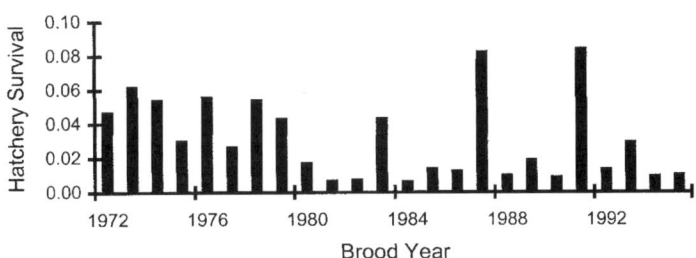

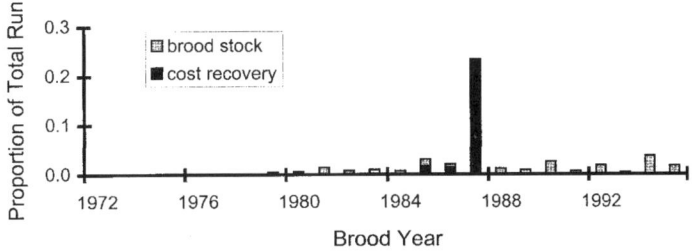

FIGURE 6.—Historical production of pink salmon from hatcheries in the Kodiak Island area, illustrated by **(top)** fry releases (vertical bars in tens of millions) and hatchery return of adults (shaded area in millions), **(middle)** ocean survival rate for hatchery fish, and **(bottom)** proportion of the total run of pink salmon that has gone to cost recovery fisheries and broodstock.

returns increased the most from the period before 1975, but this increase had taken place before 1984 when large-scale hatchery production began. For the 5-year running-average escapement, the general trend indicated increases in all areas except PWS, which has declined dramatically since the mid-1980s. For the 5-year running average of total return and wild return for PWS and Kodiak Island, almost no difference existed between total and wild pink salmon returns in the Kodiak area. In PWS, the wild return declined dramatically beginning in the mid-1980s while the total return stayed roughly constant, indicating that wild stocks were being replaced by hatchery stocks.

When the average return for 1986–1995 was compared with the return for 1965–1975 in each region, south Alaska Peninsula and Prince William Sound both increased roughly sixfold, southeast Alaska increased 3.5-fold, and Kodiak increased about twofold (Table 1). However with the base period of 1976–1985 (after the improvement in ocean conditions and before large-scale hatchery production affected PWS), PWS, southeast Alaska, and south Alaska Peninsula all experienced very similar increases in returns—1.43, 1.55, and 1.37, respectively—while increases in Kodiak returns lagged behind at 1.13. From the pre-regime-shift base period (1965–1975), PWS and south Alaska Peninsula were highest, but this was accomplished by wild stocks in both PWS and south Alaska Peninsula.

Discussion

The purpose of the aquaculture program in Prince William Sound and Kodiak Island was to stabilize natural variability in the pink salmon runs

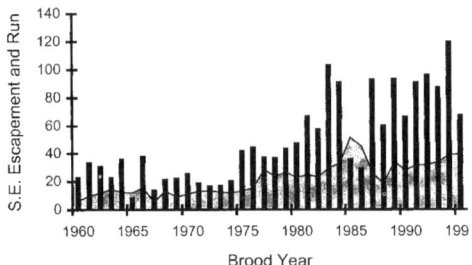

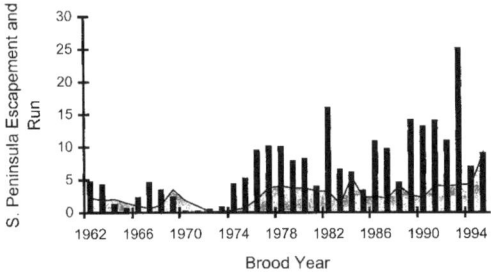

FIGURE 7.—Historical production of pink salmon in **(top)** southeast Alaska (S.E.) and **(bottom)** the southern Alaska Peninsula (S. Peninsula), illustrated by total catch (vertical bars) and escapement (shaded areas); all numbers are in millions.

to the area and to provide for a sustainable and economically viable fishery. The success of any enhancement program depends on meeting a series of biological criteria including (1) the successful production of fish that survive to be captured, (2) adequate survival, sustained for a long period, (3) hatchery production that can be harvested without affecting the production of the wild fish, and (4) production of enhanced fish that does not significantly reduce the survival and production of wild fish (so that there are true net benefits of the enhancement).

The data presented earlier show clearly that criterion 1 has been met: the PWS and Kodiak pink salmon programs produce fish that survive and contribute to the fishery. The survival rates achieved (particularly in PWS) are the envy of hatchery managers for chinook salmon and coho salmon up and down the coast, where a 5% survival rate is considered an incredible success, even for fish reared for a year in the hatchery, fed extensively, and therefore released at a very large size. It is more difficult to determine the long-term success of the fish culture; the middle panels of Figures 5 and 6 provide some indication that survival rates may be declining. However, fish survival rates fluctuate and it is impossible to know whether the lower survivals in 1990 and 1991 broods portend things to come or are part of natural variation. Further, the estimates of survival rates before 1987 were not derived from coded wire tags (as are later survivals), so these periods may not be comparable.

The biological success of the programs is less obvious. If we accept the trends seen in southeast Alaska and southern Alaska Peninsula stocks as indicative of what would have happened in the absence of hatchery programs in PWS and Kodiak, then there appears to be little if any net production. As discussed earlier, pink salmon production in the other areas increased at the same time, and whereas pink salmon increased in PWS more than in two of the three control areas, the greater increase took place before the onset of large hatchery production.

This interpretation is supported by the increase in wild production in PWS that began in the early 1980s, only to have the wild production replaced by the hatchery production in the late 1980s and 1990s. This pattern of replacement in PWS can be interpreted as a classic example of the following concern stated by Brannon and Mathews (1988). "In the first place, rather than supplementing natural populations, hatchery production tended to replace natural production, with the result that naturally spawned fish no longer contributed effectively to the fishery. The net gain from hatchery propagation in this regard may have been very little." There is no evidence of replacement in the Kodiak area.

There are two independent items supporting the replacement theory for PWS. (1) The stocks in other areas without hatcheries increased at the same time, and (2) the wild stocks first increased in PWS, then as hatchery production increased, wild production declined.

These observations do not constitute "proof"; the other areas are not randomized controls, but rather "natural" controls with all of the possibilities of another covariate being responsible. Furthermore, the apparent replacement of wild fish by hatchery fish in the 1980s is based on an effective sample size of 1—that is, we only have one time series of data from hatchery and wild production in PWS.

Alternative Explanations for the Decline in PWS Wild Salmon

Why did the wild stocks decline after the 1985 brood year? There are four possible hypotheses, including harvesting, competition with wild fish,

PINK SALMON HATCHERY PROGRAMS

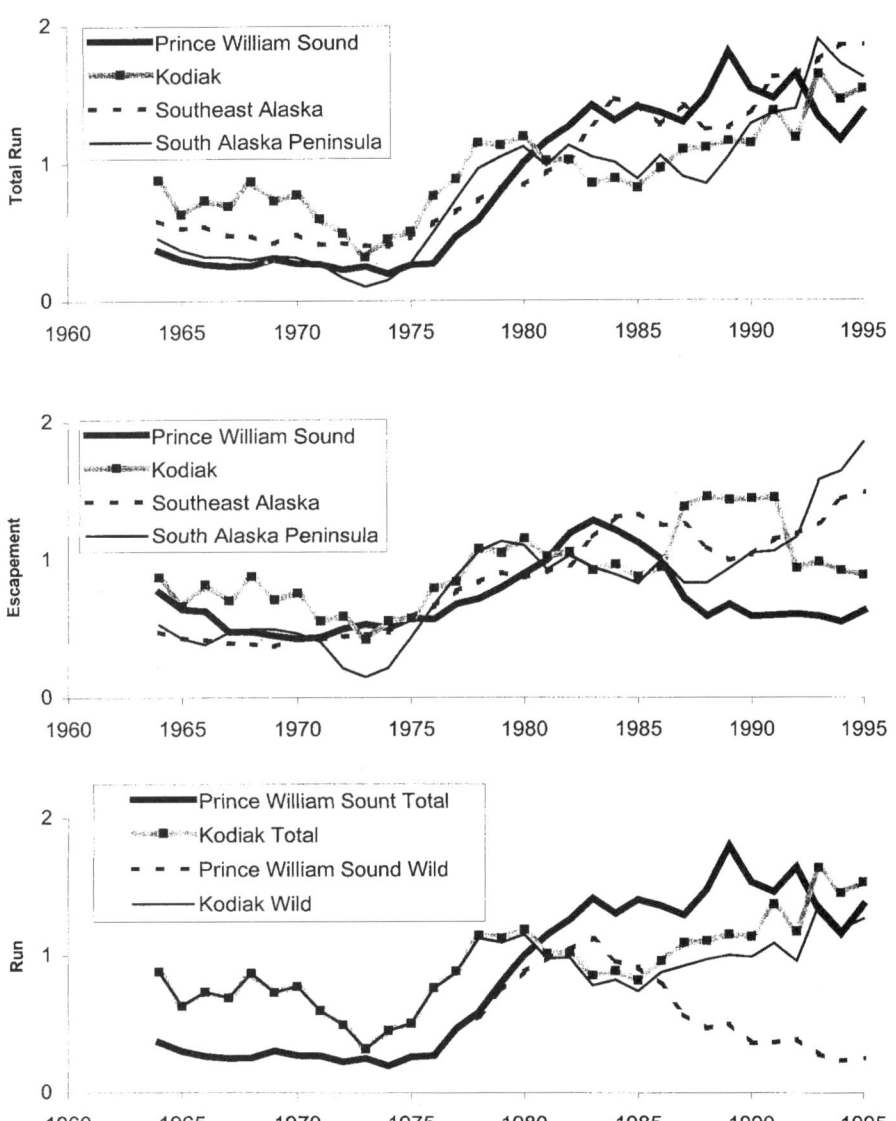

FIGURE 8.—Five-year running averages of run, escapement, and total run for the four pink salmon production areas in Alaska, scaled by the 1976–1985 average values for each area. **(Top)** Running average of total run (hatchery plus wild); **(middle)** running-average escapement; and **(bottom)** running averages of wild pink salmon run and total run including hatchery plus escapements (Prince William Sound and Kodiak areas only) divided by the average total run including hatchery for each area.

natural changes, and straying or genetic impacts of hatchery fish. We will deal with each of these in turn.

Impacts due to changes in escapement.—To examine the decline of the wild stocks in PWS we divided the data into two periods: (1) brood years 1977–1985, characterized by large returns after the rebuilding from the low runs of the 1960s and early 1970s, and (2) brood years 1986–1995, the recent period of low returns of wild fish.

The average wild return to PWS in the later period was 32% of the return in the first period,

TABLE 1.—Ratios of average run for 1986–1997 to averages for 1965–1975 and for 1976–1985, in four Alaskan pink salmon regions.

Base period	Region			
	Kodiak Island	Prince William Sound	Southeast Alaska	South Alaska Peninsula
1965–1975	1.90	5.74	3.54	5.93
1976–1985	1.13	1.43	1.55	1.37

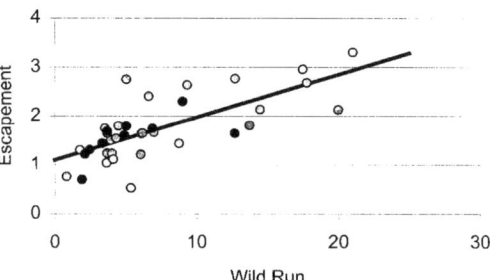

FIGURE 9.—Relationship between wild-stock escapement and total wild-stock return (both in millions of fish) in Prince William Sound for brood years 1960–1985 (gray dots; 1984 data excluded) and 1986–1995 (black dots).

whereas the escapement was 56% and the recruit per spawner was 57% of that during 1977–1985 (Table 2). Thus, we can conclude that part of the decline in wild stocks was due almost equally to a reduction in average escapement and a reduction in recruits per spawner. The escapement goal for PWS during both periods (brood years 1977–1995) was 1.8 million pink salmon; thus the average escapements in the 1977–1985 period were above the goal while the escapements from brood years 1986–1995 were slightly below the goal. Figure 9 shows the pattern, typical of net fisheries management, in which the actual wild-stock escapement during 1960–1985 and 1986–1995 in PWS increased with larger runs rather than the "ideal" of escapement holding constant regardless of run size. A strike by commercial fishing boat operators occurred in 1984, resulting in an escapement of 5.2 million fish, thus the data point for that year was not plotted. Two important conclusions can be drawn from Figure 9. First, the lower escapements in the later period appear to be due to the lower runs. Second, we see no difference in the escapement–return relationship between the two time periods. The analysis, at the PWS-wide scale, does not support a conclusion that the fishery was managed differently after large hatchery returns began.

It has been suggested that the presence of large hatchery runs led to higher exploitation and lower escapements. For instance, Geiger (1994) states "the entire 1992 wild run was needed for spawning escapement. Yet, for a variety of reasons related to the need to harvest the hatchery return, the harvest rate on wild salmon was held to nearly the recent average." The 1992 run is the lowest black dot in Figure 9 and may constitute a single instance of PWS-wide overharvest of wild stocks, but it is clearly not an indication of a systematic pattern of changed harvest policies in recent years.

However, when we look at the spatial pattern of escapements we see more evidence that the presence of hatchery fish led to a changed harvest pattern. The fishing districts in the north and west of PWS were heavily affected by the fishery for the hatchery stocks, whereas districts in eastern PWS were much less affected by these fisheries (Figure 10). The districts where the hatchery stocks passed have shown much stronger declines in escapement than the lightly affected districts.

The passage from Geiger (1994) above suggests that the economic pressure to exploit hatchery stocks in common-property fisheries was a major contributor to the reduced escapements in some parts of PWS, but overall we conclude that the reduced escapements after 1988 would have occurred regardless of the presence of large hatchery returns.

Impacts due to biological competition.—The lower escapement only explains part of the decline in wild stocks. There was also a reduction in the recruits per spawner in PWS to 57% of what it had

TABLE 2.—Data for Prince William Sound wild stocks, fry release, and common-property (CP) harvest rates for a period of high wild-stock runs (brood years 1977–1985) and low wild-stock runs (brood years 1986–1995).

Brood years	Average total wild return (millions)	Average brood year escapement (millions)	Average recruits per spawner	Average fry release (millions)	CP harvest rate
1977–1985	16.3	2.7	6.0	76	0.82
1986–1995	5.2	1.5	3.5	502	0.74

PINK SALMON HATCHERY PROGRAMS

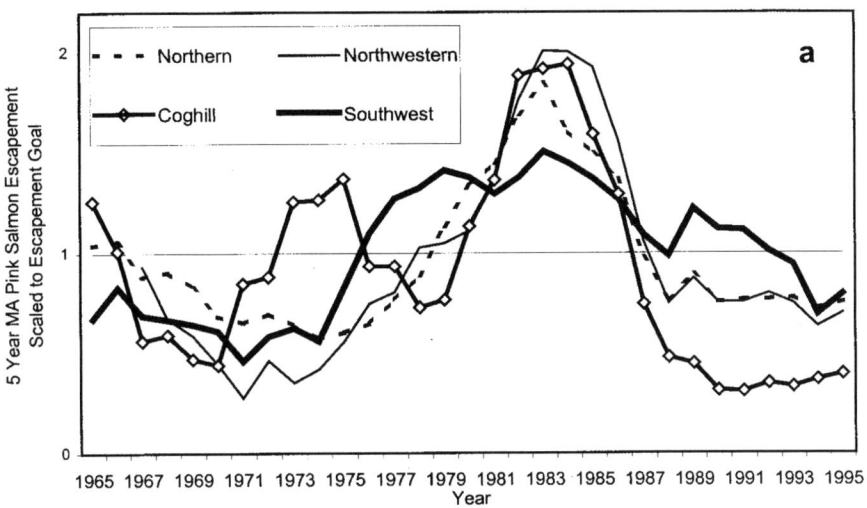

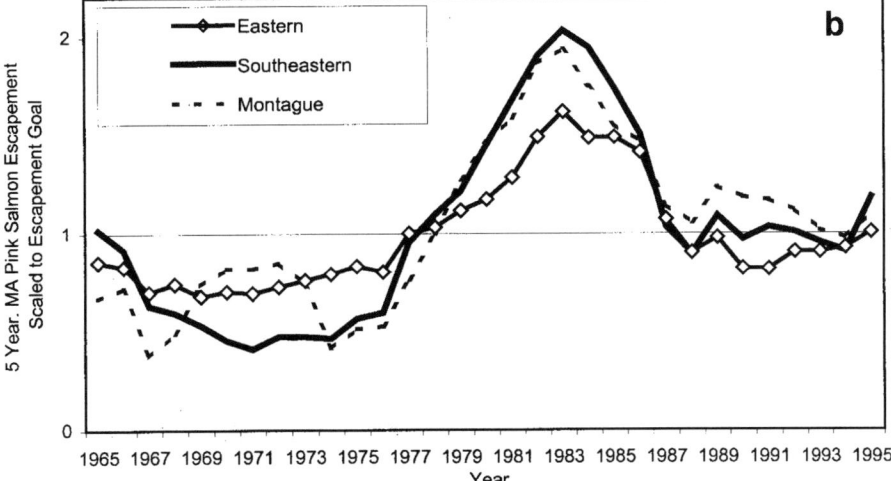

FIGURE 10.—Trends in escapement in different fishing districts within Prince William Sound (PWS). The 5-year moving average (MA) divided by the 1962–1997 average is plotted, which illustrates by contrast (a) four districts in the northwest and southwest of PWS that are strongly affected by hatchery production and (b) three fishing districts in the south and east that are less affected by the hatcheries.

TABLE 3.—Total return divided by escapement index for the four major regions producing pink salmon in Alaska.

Brood years	Region			
	Kodiak Island	Southeast Alaska	South Alaska Peninsula	Prince William Sound
1977–1985	2.32	1.98	2.37	6.03
1986–1995	2.39	2.46	3.03	3.47

been in the earlier period. The escapement numbers are more likely reliable as an index rather than as an unbiased count; therefore it is the change in the ratio of total return to escapement (Table 3), rather than the absolute level, that is of more interest.

In Kodiak Island, southeast Alaska, and the southern Alaska Peninsula, the return per spawner increased after 1985 while it decreased in PWS. A major difference between these regions is the level of hatchery release in PWS and the close proximity of PWS hatcheries to the wild-stock production areas. In the Kodiak area, hatchery and wild stocks are physically separated, thus minimizing interaction and competition. Only PWS saw reduced recruits per spawner and only PWS had a large hatchery program during the more recent period.

Marine competition and freshwater genetic impacts by the hatchery stocks have both been hypothesized as mechanisms for hatchery impacts on survival of wild stocks. Sharp et al. (1994) documented high straying rates of coded-wire-tagged hatchery fish into wild streams in PWS, which suggests that this straying may lead to a decline in wild-stock productivity due to hybridization with hatchery strains. Using thermal marking of hatchery fish, T. Joyce and D. G. Evans (ADF&G, unpublished) confirmed very high rates of straying into streams near the hatcheries. Thus, if the hatchery stocks have poorer fitness when spawning in the wild, the intense straying by these fish is a plausible explanation for the decline in wild recruits per spawner.

In examining the impact of changes in both escapement and hatchery releases, we graphed the relationship between escapement and the natural logarithm of recruits per spawner in PWS (Figure 11, top). This is the traditional graph for fitting the Ricker curve to salmon data. The best-fit linear trend showed a decline in $\log_e$ recruits per spawner as escapement increased, but the data were noisy.

We also graphed the relationship between wild recruits per spawner and the number of hatchery releases in the year the wild fish went to sea and presumably competed with the hatchery releases (Figure 11, bottom). Again we saw a downward trend, but the data were noisy with two outliers representing occurrences of high recruits per spawner in years of large hatchery releases. It happens that both of these outliers correspond to years of low escapement.

We fit a Ricker model treating smolt releases as an auxiliary variable (Hilborn and Walters 1992: equation 7.7.4), which we write as follows:

$$R_{y+2} = S_y \exp\left\{\alpha\left[1 - \frac{S_y}{b} - c(H_{y+1} - \bar{H})\right]\right\},$$

where R is the recruitment, S is the spawning stock, H is the number of smolts released from the hatchery system, $\bar{H}$ is the average smolt release, $\exp(\alpha)$ is the recruits per spawner in the absence of density dependence, b is the value wherein recruits equals

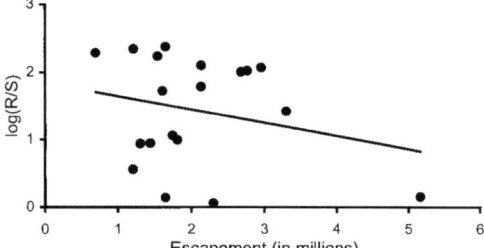

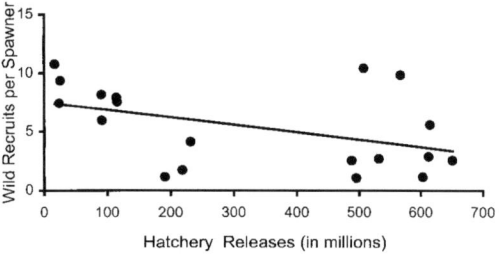

FIGURE 11.—**(Top)** Natural logarithm of recruits per spawner (R/S) for wild pink salmon in Prince William Sound plotted against escapement in the same brood year (for brood years 1977–1995). **(Bottom)** Wild recruits per spawner plotted against the hatchery release in the year the wild fish migrated to sea.

spawners, c is a parameter indicating the magnitude of the decrease in recruits due to smolt releases, and y is the calendar year.

Table 4 shows the results for five recruitment models. Our first model assumes that recruitment is constant with no effect of escapement or smolts. Next we fit the model above assuming no density dependence or hatchery effect; that is, b was set equal to a very large number and c was assumed to be 0. This second model assumes recruitment is proportional to escapement. The improvement in fit is highly significant ($P = 0.0087$), indicating that more spawners do produce more recruits (Figure 12, upper left). Values for P were calculated using a likelihood ratio test (Hilborn and Mangel 1997). Next we fit the normal Ricker model, which assumed $c = 0$ (Table 4, third model; Figure 12, upper right). The improvement in fit was indicated by $P = 0.16$ when compared with the proportional recruitment model. Then we fit a model with proportional recruitment and a hatchery effect; b was set equal to 10^{12} so there was no density dependence, and $P = 0.06$ (again compared with the proportional recruitment model; Table 4, fourth model; Figure 12, lower left). Finally we fit the full model with both density dependence and smolt effect. When compared with the proportional re-

PINK SALMON HATCHERY PROGRAMS

TABLE 4.—Negative log likelihood and P-values for five models predicting pink salmon recruitment for the 1977–1995 brood years.

Model	df	Negative log likelihood	Model compared to	P
Constant recruitment	18	27.18		
Recruitment proportional to escapement	18	22.28	Constant recruitment	0.0018
Regular Ricker model	17	21.69	Proportional recruitment	0.28
Smolt impact only, no density dependence	17	20.31	Proportional recruitment	0.047
Both density dependence and smolt impact	16	17.18	Proportional recruitment	0.006

cruitment, $P = 0.006$ for this model (Figure 12, lower right). These statistics show that the best explanation for what happened to PWS wild pink salmon is a combination of changes in escapement and increasing hatchery releases. The P-level for the model with both effects is impressive, however hatchery releases were highly correlated with year, and the result could be due to any factor that changed with time in a similar fashion. Implications of these model fits are summarized in Figure 13: in the presence of larger smolt releases, expected recruitments are lower. The optimum escapement to maximize harvest of wild stock in the absence of smolt releases is 2.1 million.

We can now use this model to predict what would have happened if no smolts had been released. Table 5 shows the wild escapement, wild recruits, and predicted recruits from the model just presented; "log residual" is the logarithm of observed recruitment divided by the predicted recruitment and is an estimate of the environmentally induced deviation in that year. Brood years 1990 and 1991 had very negative residuals whereas brood years 1989 and 1992 had very positive residuals. Scenario 1 (Table 5, column 6) shows what the run would have been using this model if the escapement had been 2.1 million each year and no smolts were released. Scenario 1 is unrealistic in that we have seen that managers do not control escapement to a fixed target. Scenario 2 (Table 5,

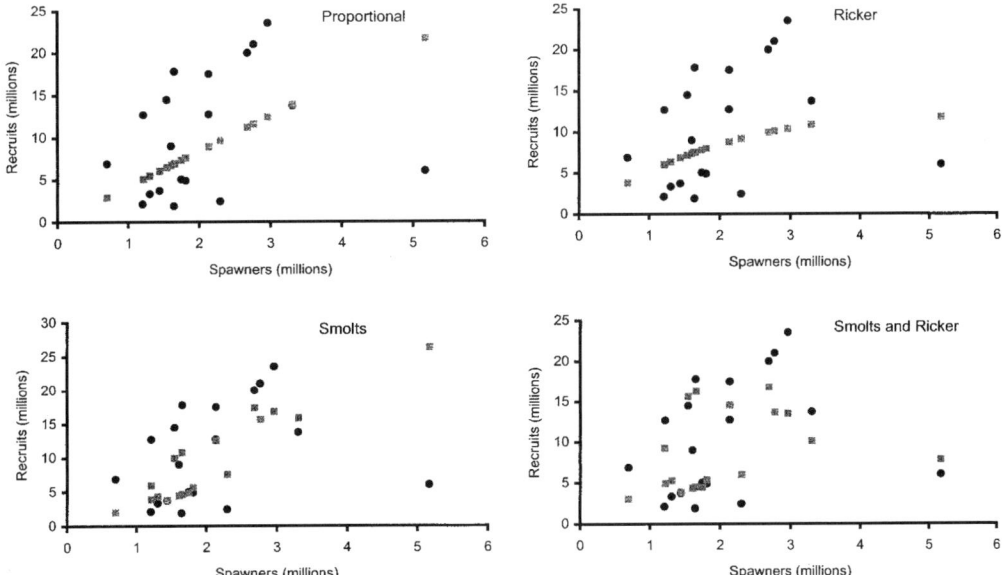

FIGURE 12.—Observed recruitment (circles) and predicted recruitment (squares) for four models (see Discussion, Table 4) of wild pink salmon recruitment in Prince William Sound from brood years 1977–1995.

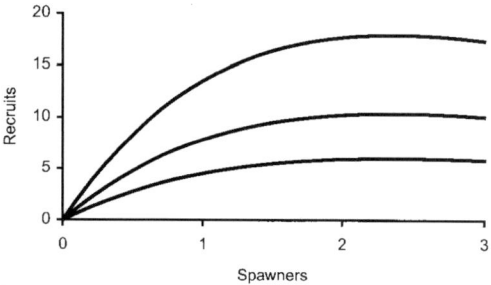

FIGURE 13.—Average expected wild-stock recruitment plotted against wild-stock escapement with releases of 0 (upper line), 250 million (middle line), and 500 million hatchery smolts (lower line).

column 7) shows what the run would have been using the actual escapements in the model and if no smolts were released. Scenario 2 is unrealistic also, because the anticipated higher wild returns without smolts would have led to higher realized escapements. The "predicted escapement" (Table 5, column 8) is what the escapement would have been if the solid line in Figure 9 had been used to predict the escapement based on the predicted total wild run under Scenario 3—what the total return would have been if the simulated escapements had been used and no smolts were released (Table 5, last column). We believe this scenario is most realistic for the actual escapement.

With the averages for brood years 1986–1995, we would have expected 20.57, 17.52, and 19.05 million pink salmon returning under the three scenarios we just discussed. These expectations compare with an actual total return of 24.5 million during those years. Using our Scenario 3 we would thus estimate that the net increase due to hatchery production during this period was 5.5 million fish per year.

However, the other pink-salmon-producing areas all showed increased recruits per spawner in the later period, indicating better ocean conditions than during the earlier period. The average ratio of recruits per spawner in the later period to recruits per spawner in the former period for the other three areas is 1.18, indicating those areas saw an 18% average increase in recruits per spawner during brood years 1986–1995. The bottom row of Table 5 shows the predicted total returns allowing for an 18% increase in the later period. Thus our best estimate of the net production due to the

TABLE 5.—Predicted total returns in selected scenarios, all if no smolts were released. All numbers are millions.

					Predicted run with:			Predicted run with simulated escapement (scenario 3)
Brood year	Wild escapement	Observed recruits	Predicted recruits	Log residual	2.1 M escapement (scenario 1)	Actual escapement (scenario 2)	Simulated escapement using Figure 9	
1975								6.16
1976								4.32
1977	1.65	17.80	16.29	0.09	19.44	18.47	1.63	18.40
1978	1.54	14.48	15.64	−0.08	16.47	15.31	1.47	15.04
1979	2.68	19.99	16.85	0.17	21.10	21.08	2.70	21.06
1980	2.14	17.51	14.60	0.18	21.35	21.38	2.41	21.49
1981	2.13	12.74	14.57	−0.13	15.56	15.58	2.93	15.28
1982	2.96	23.54	13.55	0.55	30.91	30.29	2.97	30.26
1983	2.77	21.00	13.71	0.43	27.27	27.10	2.43	27.45
1984	5.17	6.05	7.88	−0.26	13.66	9.19	5.17	9.19
1985	3.30	13.74	10.13	0.30	24.13	22.82	3.49	22.28
1896	1.21	2.12	9.25	−1.47	4.08	3.42	1.89	4.01
1987	1.81	4.90	5.40	−0.10	16.14	15.72	3.04	15.70
1988	1.22	12.70	4.92	0.95	45.94	38.63	1.44	41.64
1989	1.61	9.00	4.36	0.73	36.78	34.66	2.46	37.01
1990	1.65	1.90	4.50	−0.86	7.51	7.13	4.73	5.56
1991	2.30	2.45	6.04	−0.90	7.21	7.26	4.33	5.80
1992	0.70	6.88	3.08	0.80	39.72	23.86	1.58	37.20
1993	1.31	3.34	5.31	−0.46	11.21	9.76	1.60	10.55
1994	1.75	5.04	4.48	0.12	20.00	19.32	4.34	16.03
1995	1.44	3.71	3.87	−0.04	17.07	15.48	2.01	16.97
1986–1995 average	1.50	5.20	5.12	−0.12	20.57	17.52	2.74	19.05
Average with 18% increase					24.27	20.68		22.48

PINK SALMON HATCHERY PROGRAMS

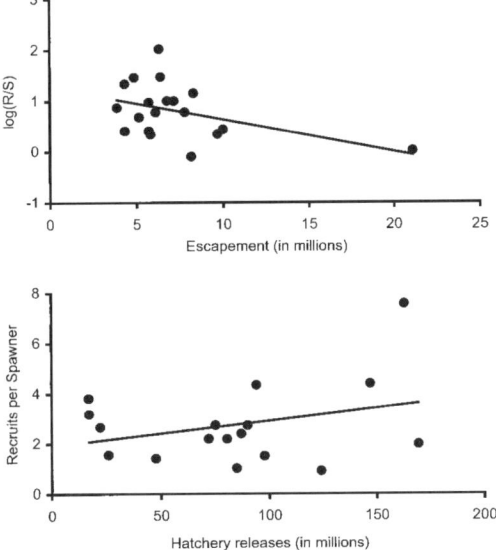

FIGURE 14.—**(Top)** Natural logarithm of recruits per spawner (R/S) for wild pink salmon in the Kodiak Island area plotted against escapement in the same brood year (for brood years 1977–1993). **(Bottom)** Wild recruits per spawner plotted against the hatchery release in the year the wild fish migrated to sea.

hatchery program (using Scenario 3) is 2 million pink salmon per year.

We repeated the same analysis for the Kodiak area, examining the relationship between $\log_e(R/S)$ and escapement and the relationship between recruits per spawner and hatchery releases (Figure 14). There was some evidence for density dependence, but only based on one year (1989) with a very high escapement, and no evidence that higher hatchery releases have led to fewer wild recruits per spawner.

We repeated the range of models for Kodiak that we had used for PWS. The proportional and smolt models (Figure 15, left top and bottom panels, respectively) did not provide an improvement in fit over the hypothesis that returns were constant, and only the Ricker model provided a significant improvement in fit, which was clearly due only to the one data point. We concluded there was no evidence that hatchery production affected wild production in the Kodiak area.

Decline in Wild Stocks in PWS was a Natural Change

This possibility cannot be eliminated. We know of no quantitative way to assess this probability because it depends on the degree to which the other areas serve as effective controls on ocean condi-

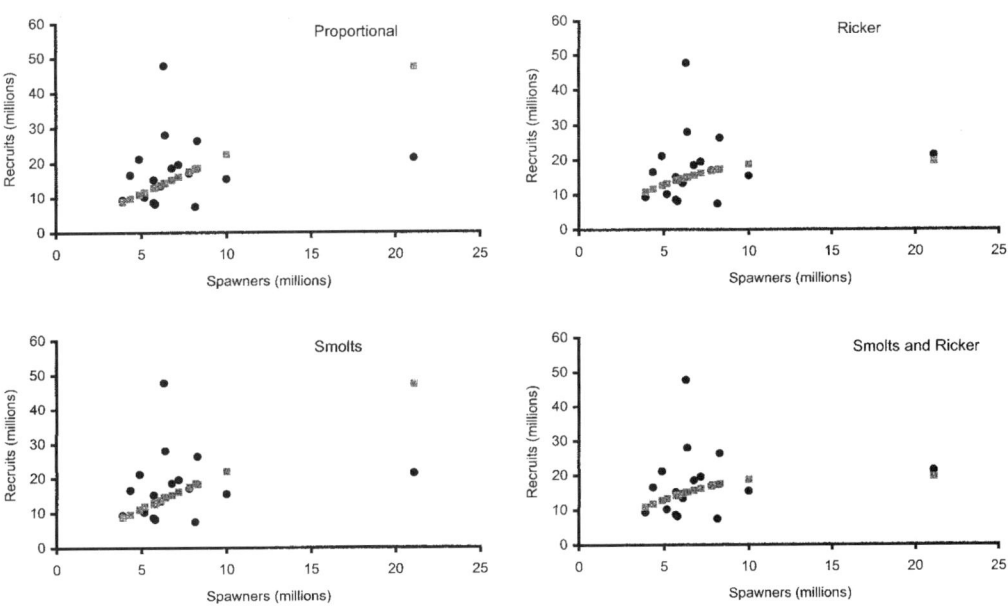

FIGURE 15.—Observed recruitment (circles) and predicted recruitment (squares) for four models of wild pink salmon recruitment in the Kodiak Island area for the 1977–1993 brood years.

tions and we accept that there is an unexplained factor that changed in the mid-1980s in PWS.

Earlier we discussed two plausible mechanisms for the hatchery impacts—genetic degradation due to straying and competition in the early life history. Smoker and Linley (1997) discussed these mechanisms and suggested they are unlikely given the short hatchery rearing period for pink salmon. Similarly, Smoker and Linley discounted the possibility of marine competition. Higher hatchery releases in PWS coincided with lower wild recruits per spawner, but it is possible that something in PWS changed starting in brood year 1986.

It is widely recognized that the *Exxon Valdez* oil spill in 1989 might have affected wild spawning pink salmon. There is disagreement regarding the amount of loss caused by the oil spill. Some investigators (Bue et al. 1996, 1998a; Geiger et al. 1996) estimated pink salmon damage ranged as high as 2% of the total wild return to PWS, whereas others (Brannon and Maki 1996) argued that even this loss was an artifact of the sampling regime and not a real effect. Thus none of the published work has suggested that the loss from the oil spill would even be detectable on a PWS-wide basis. The decline in wild recruits per spawner in PWS, illustrated in Figure 3, began well before the oil spill (beginning in the 1984 brood year); and brood years 1988 and 1989, most affected by the oil spill, had among the highest recruits per spawner in the period after 1986. Thus we found no evidence that the *Exxon Valdez* oil spill could account for the decline in recruits per spawner seen after 1986.

Smoker and Linley provided a defense of the PWS hatchery program, arguing that because escapements declined throughout PWS in the 1990s, it was a phenomenon unrelated to hatchery production. Their argument has a number of problems. First, the escapement clearly declined in the Southeastern District (Smoker and Linley 1997: Figure 1) from a high in the early 1980s, the same pattern as seen in PWS as a whole. We have shown that in areas where the wild stocks pass through the fisheries targeting on hatchery fish (Figure 10), escapement declined more than in areas less affected by the hatchery-oriented fisheries. Given our understanding of the relationship between escapement and total run (Figure 9), we conclude that the decline in escapement was due to the decline in the wild-stock run, which in turn was due to a decline in recruits per spawner, shown to be related to smolt releases.

Conclusions

The Prince William Sound and Kodiak Island pink salmon programs provide what may be the best opportunity to determine if mass production of juvenile fish can increase total fish production. The hatchery systems for chinook salmon, coho salmon, and steelhead *O. mykiss* throughout North America are so ubiquitous that it is difficult, if not impossible, to evaluate the impact of hatchery fish on wild production because there are few areas that can be considered to be controls. Further, escapement of chinook salmon and coho salmon are very difficult to monitor. In the PWS and Kodiak pink salmon fisheries we have the best possible situation: very large programs, which makes impacts more detectable, and areas of pink salmon production without large hatchery programs.

We suggest there was little if any increase in total abundance due to the hatchery program in PWS. Our best estimate is 2 million fish per year. The program was conceived in a period of low abundance of wild fish, but by the time large-scale hatchery production came on-line the wild production had increased. Hatchery production increased and wild production then declined. In contrast, abundance of wild stocks in the three other pink-salmon-producing areas of Alaska increased as much and stayed high while wild production in PWS declined. The Kodiak area appears to have experienced no impact of hatchery fish on wild production for three reasons. (1) The program there was smaller relative to the wild stocks; (2) the hatchery was physically isolated so there was little mixed-stock fishing on hatchery and wild fish, and there was little interaction by these fish during their early life history; and (3) the hatchery survival rates were much lower than in PWS, therefore the ratio of hatchery return to wild return was much lower.

This conclusion has wide consequences—because there are dozens, if not hundreds, of hatchery programs existing or planned—for many marine species around the world. Planners and operators of these programs rarely if ever consider negative impacts on wild production, and no marine hatchery program has any form of experimental design in place that could determine if the hatchery would replace wild production.

To our knowledge no one now argues that existing hatchery programs in the United States and Canada produce fish at a cost comparable with the value of the fish, but it is generally assumed by hatchery operators, politicians, and the public that

PINK SALMON HATCHERY PROGRAMS

hatcheries augment total production. The lesson from PWS, however, is just the opposite: we should expect hatchery production to replace wild production rather than augment it whenever there is biological interaction and mixed-stock fishing. The PWS hatchery program for pink salmon provides by far the most dramatic evidence for this effect.

These conclusions apply to mass hatchery production where wild stocks are present. Obviously, if there are no wild stocks or if they are severely depleted at the onset of the hatchery program, the potential for the loss of wild-stock production is less. Also, these conclusions are not really relevant to various forms of supplementation hatcheries that use hatchery rearing as a short-term measure to rebuild wild production. There are many problems in evaluating supplementation hatcheries (Winton and Hilborn 1994), but we do not believe that the Prince William Sound or Kodiak Island hatchery programs are relevant models.

Acknowledgments

We thank Brian Bue, Brian Bigler, Claribel Coronado, Hal Geiger, Al Maki, Don Rogers, Jim Seeb, Bob Wilbur, John Winton, three anonymous reviewers, and numerous staff members of ADF&G for supplying data, discussion, and comments on the manuscript; R. Hilborn was supported in part by Exxon Corp.

References

ADF&G (Alaska Department of Fish and Game). 1997. Commercial, subsistence, and personal use salmon fisheries southeast Alaska—Yakutat region 1996. ADF&G, Commercial Fisheries Management and Development Division, Regional Information Report 1J96-32, Juneau.

Barrett, B. M., C. Swanton, and P. Roche. 1990. An estimate of the 1989 Kodiak management area salmon catch, escapement, and run numbers had there been a normal fishery without the *Exxon Valdez* oil spill. Alaska Department of Fish and Game, Division of Commercial Fisheries, Regional Information Report 4K90-35, Kodiak.

Brannon, E., and S. Mathews. 1988. Report to the Senate Environment and Natural Resources Committee. Washington State Legislature, Olympia.

Brannon, E. L., and A. W. Maki. 1996. The *Exxon Valdez* oil spill: analysis of impacts on the Prince William Sound pink salmon. Reviews in Fisheries Science 4:289–337.

Brennan, K., D. Prokopowich, and D. Gretsch. 1998. Kodiak Management Area commercial salmon annual management report, 1996. Alaska Department of Fish and Game, Division of Commercial Fisheries, Regional Information Report 4K98-35, Kodiak.

Bue, B. G., S. Sharr, S. D. Moffitt, and A. K. Craig. 1996. Effects of the *Exxon Valdez* oil spill on pink salmon embryos and preemergent fry. Pages 619–627 *in* S. D. Rice, R. B. Spies, D. A. Wolfe, and B. A. Wright, editors. Proceedings of the *Exxon Valdez* oil spill symposium. American Fisheries Society, Symposium 18, Bethesda, Maryland.

Bue, B. G., S. Sharr, and J. E. Seeb. 1998a. Evidence of damage to pink salmon populations inhabiting Prince William Sound, Alaska, two generations after the *Exxon Valdez* oil spill. Transactions of the American Fisheries Society 127:35–43.

Bue, B. G., S. Sharr, D. G. Sharp, J. A. Wilcock, and H. J. Geiger. 1998b. Estimating salmon escapements using area-under-the-curve, aerial observer efficiency, and stream-life estimates: the Prince William Sound pink salmon example. North Pacific Anadromous Fish Commission Bulletin 1:240–250.

Campbell, R. D., A. R. Shaul, M. J. Witteveen, and J. J. Dinnocenzo. 1998. South Peninsula annual salmon management report 1997. Alaska Department of Fish and Game, Division of Commercial Fisheries, Regional Information Report 4K98-29, Kodiak.

Dangle, J. R. and J. D. Jones. 1988. Southeast Alaska pink salmon total escapement and stream life studies, 1987. Alaska Department of Fish and Game, Division of Commercial Fisheries, Regional Information Report 1J88-24, Juneau.

Eggers, D. M., L. R. Peltz, B. G. Bue, and T. M. Willette. 1991. Trends in abundance of hatchery and wild stocks of pink salmon in Cook Inlet, Prince William Sound, and Kodiak Alaska. Alaska Department of Fish and Game, Division of Commercial Fisheries, Professional Paper 35, Juneau.

Francis, R. C., and S. R. Hare. 1994. Decadal-scale regime shifts in the large marine ecosystems of the north-east Pacific: a case for historical science. Fisheries Oceanography 3:279–291.

Geiger, H., and S. Sharr. 1990. The 1988 tag study of pink salmon from the Solomon Gulch hatchery in Prince William Sound, Alaska. Alaska Department of Fish and Game, Division of Commercial Fisheries, Fishery Research Bulletin 90-2: 18–33, Juneau.

Geiger, H. J. 1994. Recent trends in pink salmon harvest patterns in Prince William Sound, Alaska. Proceedings of the 15th Northeast Pacific pink and chum salmon workshop. Alaska Sea Grant College Program, Report 94-02, Vancouver.

Geiger, H. J., B. G. Bue, S. Sharr, A. C. Wertheimer, and T. M. Willette. 1996. A life history approach to estimating damage to Prince William Sound pink salmon caused by the *Exxon Valdez* oil spill. Pages 487–498 *in* S. D. Rice, R. B. Spies, D. A. Wolfe, and B. A. Wright, editors. Proceedings of the *Exxon Valdez* oil spill symposium. American Fisheries Society, Symposium 18, Bethesda, Maryland.

Hare, S. R., and R. C. Francis. 1995. Climate change and salmon production in the northeast Pacific

Ocean. Canadian Special Publication of Fisheries and Aquatic Sciences 131:357–372.

Helle, J. H., R. S. Williamson, and J. E. Bailey. 1964. Intertidal ecology and life history of pink salmon at Olsen Creek, Prince William Sound, Alaska. U.S. Fish and Wildlife Service, Bureau of Commercial Fisheries, Special Scientific Report–Fisheries 483.

Hilborn, R 1992. Hatcheries and the future of salmon in the northwest. Fisheries 17(1):5–8.

Hilborn, R., and M. Mangel. 1997. The ecological detective: confronting models with data. Princeton University Press, Princeton, New Jersey.

Hilborn, R., and C. J. Walters. 1992. Quantitative fisheries stock assessment: choice, dynamics and uncertainty. Chapman and Hall, New York.

Hilborn, R., and J. Winton. 1993. Learning to enhance salmon production: lessons from the salmonid enhancement program. Canadian Journal of Fisheries and Aquatic Sciences 50:2043–2056.

Hull, H. D. 1993. The theory of clubs and the Prince William Sound Aquaculture Corporation: a benefit–cost analysis. Master's thesis. University of Washington, Seattle.

Mantua, N. J., S. R. Hare, Y. Zhang, J. M. Wallace, and R. C. Francis. 1997. A Pacific interdecadal climate oscillation with impacts on salmon production. Bulletin of the American Meteorological Society 78: 1069–1079.

Meffe, G. K. 1992. Techno-arrogance and halfway technologies, salmon hatcheries on the Pacific coast of North America. Conservation Biology 6:350–354.

Morstad, S., D. Sharp, J. Wilcock, and J. Johnson. 1998. Prince William Sound management area 1997 annual finfish management report. Alaska Department of Fish and Game, Commercial Fisheries Management Development Division, Regional Information Report 2A98-05, Juneau.

Olsen, J. 1994. Development of the pink salmon enhancement program in Prince William Sound, Alaska. Proceedings of the 15th northeast Pacific pink and chum salmon workshop. Alaska Sea Grant College Program, Report 94-02, Vancouver.

Peltz, L., and H. Geiger. 1990. A tagging study of the effect of hatcheries on the 1987 pink salmon fishery in Prince William Sound, Alaska. Alaska Department of Fish and Game, Division of Commercial Fisheries, Fishery Research Bulletin 90-2: 1–17, Juneau.

Pinkerton, E. 1994. Economic and management benefits from the coordination of capture and culture fisheries: the case of Prince William Sound pink salmon. North American Journal of Fisheries Management 14:262–277.

Sharp, D., S. Sharr, and C. Peckham. 1994. Homing and straying patterns of coded wire tagged pink salmon in Prince William Sound. Proceedings of the 15th northeast Pacific pink and chum salmon workshop. Alaska Sea Grant College Program, Report 94-02, Vancouver.

Smoker, W. W., and T. J. Linley. 1997. Are Prince William Sound salmon hatcheries a fool's bargain? Alaska Fisheries Research Bulletin 4:75–78.

Tarbox, K. E., and T. Bendock. 1996. Can Alaska balance economic growth with fish habitat protection? A biologist's perspective. Alaska Fisheries Research Bulletin 3:49–53.

Wilson, W. J., and E. H. Buck. 1978. Status report on salmonid culture in Alaska. Fisheries 3(5):10–19.

Winton, J., and R. Hilborn. 1994. Lessons from supplementation of chinook salmon in British Columbia. North American Journal of Fisheries Management 14:1–13.

Testing the Importance of Fish Stocking as a Determinant of the Demand for Fishing Licenses and Fishing Effort in Colorado

John Loomis
Department of Agricultural and Resource Economics
Colorado State University

Peter Fix
Human Dimensions in Natural Resources Unit
Colorado State University

Abstract: A time-series regression analysis of data from 1975 to 1995 showed no statistically significant relationship between number of catchable or subcatchable coldwater fish stocked and number of resident and nonresident season licenses sold. Stocking of warmwater subcatchables was a significant determinant of resident season license sales, although a 100% change in stocking results in only a 7% to 9% increase in license sales. Stocking of warmwater catchable fish did have a significant effect on nonresident season fishing licenses with a 100% change in stocking resulting in a 5% to 7% change in nonresident season license sales. Cross-section regressions across lakes and streams in Colorado found that current season stocking of catchables was a significant determinant of angler use in all regressions, while the previous years stocking of subcatchables was significant in only one regression. A 1% increase in the number of catchable trout stocked resulted in a 0.43% change in lake angler use and 0.23% increase in stream angler use.

Keywords: fishing license demand, trout stocking, angler hours, regression

Introduction

Many fish and game agencies believe they face a "catch-22" situation with regard to stocking trout. The long-run costs of fish stocking are high, and dealing with whirling disease is likely to make them even higher. Yet they believe if they fail to stock or greatly reduce stocking, their license sales will drop precipitously. Furthermore, businesses that depend on fishing have the same belief that angler visitation will drop significantly if stocking is reduced. These businesses and their industry organizations often put pressure on fish and game agencies to maintain stocking levels.

Fish Stocking and Angler Use

Thus, agencies sometimes find themselves on a "stocking treadmill," afraid of the consequences of reduced stocking.

These beliefs are partially consistent with economic theory of consumer demand. The demand for a good is a function of price, attributes of the good, including quality, and substitutes. Quality of fishing is composed of several aspects including social conditions (e.g., solitude, being with friends) and biological conditions (e.g. number and size of fish caught). Believing that license sales and angler use will fall significantly if stocking is decreased is implying that the biological aspects of quality are a major determinant of the demand equation.

However, there appears to be little systematic, empirical research on the effect of stocking on angler license sales or use. In one of few studies, Ward (1991) investigated this relationship for New Mexico reservoirs. Stocking of large trout did have a positive and statistically significant effect on nonresident angler license sales; however, it was not significantly related to resident license sales. Stavins (1993) estimated the demand for fishing licenses in the U. S. by combining data across states. The price of the license and acres of fishable waters were the key variables, while stocking was not. These studies suggest that perhaps the biological aspects of quality are not major determinants of demand.

While to date there is little literature on the relationship between stocking and license sales and angler effort, there are several studies that find a statistically significant link between angler catch rates (of all fish, stocked and natural) and visitation. One such study for fishing in Montana found that a 1% increase in trout catch rates increased visitation by 0.3% (Duffield, Loomis, & Brooks, 1987). Increases in catch rates of warm water species had an even smaller effect, with a 1% change leading to just a 0.11% change in angler visits (Duffield et al., 1987). Loomis and Cooper's (1990) study of trout fishing on the North Fork of the Feather River in California found that a 1% change in trout catch rates resulted in 0.41% to 0.83% increases in number of fishing trips during the years 1981-1985. For steelhead fishing in Idaho and Oregon, a 1% increase in steelhead catch rates resulted in 0.524% and 0.32% increase in angler trips (Loomis, 1992). Johnson and Walsh (1987) found that one additional fish caught per day at Blue Mesa Reservoir in Colorado would increase angler days by 0.46%. Fish catch explained just 10% of the total variation in number of angler days. Size of fish had a larger effect on increasing angler days and slightly higher explanatory power than number of fish caught. Cole et al. (1990) found that while fish catch was statistically significant at the .05 level for reservoir fishing in New Mexico, its coefficient was small. Other important variables in explaining reservoir fishing in New Mexico were the number of boat ramps, accessibility of the shoreline, and surface area of the reservoir.

As can be seen from all of these empirical studies, the relationship between changes in angler catch rates and number of trips is less than proportionate. The relationship between the percent change in angler catch rates and percent change in visitation is defined in economics as an elasticity. A relationship is said to be quite responsive or elastic if a 1% change in a given variable (e.g., price or catch rates) results in more than a 1% change in quantity of use (e.g., angler days, trips). The opposite case of a relatively unresponsive relationship or inelastic relationship occurs when a 1% change in a given variable leads to less than a 1% change in quantity of use. This is the case for trips in response to catch rates.

Perhaps this less than proportionate relationship between fish catch rates and angler use should not be surprising. There are at least two reasons to suspect the relationship would be less than proportionate. First is the economic concept of diminishing marginal returns or diminishing marginal utility. Each additional unit of any non-addictive good results in smaller and smaller increases in enjoyment and satisfaction. For example, the second cup of tea in the morning probably adds less enjoyment than the first cup. The same is true for fishing. While the first fish caught adds a great deal to the enjoyment of a fishing trip, each additional fish caught adds less and less enjoyment than the previous fish caught. Therefore, increases in fish stocking will stimulate smaller and smaller increases in fishing trips as anglers become satiated. This diminishing marginal effect is one reason demand curves for nearly all goods slope downward.

The second reason for less than a proportionate response of angler use to fish catch comes from the literature on the multiple motivations for fishing. As illustrated by the research of Harris and Bergersen (1985), many anglers place greater emphasis on being in the out-of-doors areas of high scenic beauty and opportunities for solitude than they do on catching their limit. The Harris and Bergersen survey of Colorado anglers also found that catching one's limit was ranked only 9th in importance. Other more highly ranked objectives of anglers include catching large fish rather than catching many smaller fish.

Nonetheless, some fish and game agencies such as the Colorado Division of Wildlife (CDOW) continue to make major policy decisions, admittedly *assuming* a proportional response between angler use and stocking (Bennett, Nehring, Krieger, Harris, & Nesler, 1996). These authors recognize "this assumption is an over simplification and may lack validity..." but go on to state "we have not been able to quantify these relationships" (p. 18) between stocking and angler use.

Thus, the specific purposes of this paper are to:

- Determine if there is a statistically significant relationship between stocking of catchables and subcatchables and sales of resident and nonresident angler licenses.

- Determine if there is a statistically significant relationship between stocking of catchables and subcatchables and the angler use at fishing waters in Colorado.

Fish Stocking and Angler Use

- Test whether there is a proportionate relationship between stocking and angler use.

Models for Estimating the Responsiveness of Angler License Sales and Days to Fish Stocking

License Sales Model

Using historic data on past license sales in response to price increases and stocking is one approach to test these hypotheses. As mentioned, this is the approach used by Ward (1991), and it follows a long tradition of time-series statistical analysis. The functional form of the models estimated was double logarithmic so the coefficients could be interpreted as elasticities.

The first license model is:

(1a) $\ln(LICENSE_t) = B_0 - B_1(\ln RPRICE_t) + B_2(\ln COLDCAT_t) + B_3(\ln COLDSUB_{t-1}) + B_4(\ln WWCAT_t) + B_5(\ln WWSUB_{t-1}) + B_6(\ln RPGAS_t) + B_7(YEAR_t) + B_8(\ln COPOP_t)$

where:

$LICENSE_t$ = number of resident or nonresident angler licenses sold in year t.
$RPRICE_t$ = inflation adjusted price of a resident or nonresident fishing license in year t.
$COLDCAT_t$ = # of catchable coldwater fish stocked in year t.
$COLDSUB_{t-1}$ = # of subcatchable coldwater fish stocked in year $t-1$, (the previous year).
$WWCAT_t$ = # of catchable warm water fish stocked in year t.
$WWSUB_{t-1}$ = # of subcatchable warm water fish stocked in year $t-1$.
$RPGAS_t$ = inflation-adjusted price of gasoline in year t.
$YEAR_t$ = a sequential variable to reflect systematic trends in license sales over time.
$COPOP_t$ = Colorado population in year t (a variable used for resident license sales).

To test for the possibility that license sales in the current year is related to stocking levels in previous years, we estimated models that contained variables for the lagged stocking levels. A two-year lag structure was used based on discussions with hatchery managers. Through these discussions, it was determined that the subcatchables should be catchables in one to two years, and thus it was concluded that stocking three or more years ago was not likely to have much effect on current license sales. The second model is:

(1b) $\ln(LICENSE_t) = B_0 - B_1(\ln RPRICE_t) + B_2(\ln COLDCAT_{t-1}) + B_3(\ln COLDSUB_{t-2}) + B_4(\ln WWCAT_{t-1}) + B_5(\ln WWSUB_{t-2}) + B_6(\ln RPGAS_t) + B_7(YEAR_t) + B_8(COPOP_t)$

Due to a 0.97 correlation between YEAR and COPOP, both of them could not be used in the resident season fishing license regression equation. We therefore used COPOP, although the results with YEAR are quite similar. Since this model involves the logarithms of both the dependent variable and the fish catch variables, the coefficients (i.e., B_i's in equation 1) can be directly interpreted as elasticities. As noted below, these inelasticities will be useful for hypothesis testing purposes. To investigate the robustness of our results, we estimated several other regression equations that reflected other combinations of current and lagged stocking and license sales.

Angler Use Model

To test whether angler effort is sensitive to the level of trout stocking, we use a cross-sectional approach. This analyzes whether there is any systematic variation in the level of visitation across sites that is due to different stocking levels at those sites. This statistical analysis is carried out by regressing angler use from creel census information collected at a variety of stocked waters in Colorado against seasonal fish stocking at these same waters. Each site i is a unit of observation in this model. In many cases we have one-two years of previous observations of stocking and angler use so we can test for the significance of lags in this model as well.

The model that pools data across streams and lakes is of the form:

(2) $(AnglerUse_{ij}) = A_0 + A_1(CURRENTCAT_i) + A_2(PREVSUB_i) + A_3(LAKE)$

where:

Angler Use$_{ij}$ = number of anglers or angler hours (both variables used in different models) at site i using fishing mode j, where j is bank or boat.
CURRENTCAT$_i$ = number of catchable coldwater fish stocked at site i in the current year.
PREVSUB$_i$ = number of subcatchable coldwater fish stocked at site i in the previous year.
LAKE = 1 if water is a lake, 0 if stream.

This basic model results in two variations of the dependent variable:

(a) anglers, measured as total number of anglers visiting the particular water over the season (labeled **Total Bank Anglers or Total Boat Anglers**);
(b) angler use, measured as total hours over the season (labeled **Total Angler Hours**).

Given the data, the general specification in equation (2) results in seven distinct regressions: two models that pool data from lakes and streams, two for streams only, and three for lakes. Lakes have more models since there were sufficient number of observations of boaters at lakes to estimate a separate boating model. To save space, all seven models are not presented in this section, but the specifications are presented in the results section.

Due to the presence of frequent zero values for stocking for some years, log transformations are not possible. Therefore, the seven models are estimated in linear form. To calculate the elasticity of angler use with respect to stocking, we relied upon a formula for the elasticity (E) from a linear regression model: E = Slope coefficient* (mean of the independent variable/mean of the dependent variable); see Loomis and Walsh (1997) for more details.

Hypotheses

The first eight hypotheses tested relate to whether there was a statistically significant increase in *license sales* with respect to the level of stocking of: (1) catchable species such as trout, (2) subcatchable coldwater species such as trout, (3) catchable warmwater species, (4) subcatchable warmwater species. For regression equations 1a and 1b on license sales the null and alternative hypotheses are:

(3a) $H_0: B_2(lnCOLDCAT_t) = 0$ vs $H_a: B_2(lnCOLDCAT_t) > 0$
(3b) $H_0: B_2(lnCOLDCAT_{t-1}) = 0$ vs $H_a: B_2(lnCOLDCAT_{t-1}) > 0$
(4a) $H_0: B_3(lnCOLDSUB_{t-1}) = 0$ vs $H_a: B_3(lnCOLDSUB_{t-1}) > 0$
(4b) $H_0: B_3(lnCOLDSUB_{t-2}) = 0$ vs $H_a: B_3(lnCOLDSUB_{t-2}) > 0$
(5a) $H_0: B_4(lnWWCAT_t) = 0$ vs $H_a: B_4(lnWWCAT_t) > 0$
(5b) $H_0: B_4(lnWWCAT_{t-1}) = 0$ vs $H_a: B_4(lnWWCAT_{t-1}) > 0$
(6a) $H_0: B_5(lnWWSUB_{t-1}) = 0$ vs $H_a: B_5(lnWWSUB_{t-1}) > 0$
(6b) $H_0: B_5(lnWWSUB_{t-2}) = 0$ vs $H_a: B_5(lnWWSUB_{t-2}) > 0$

These hypotheses can be tested by whether the coefficient on the particular stocking variable is significantly greater than zero. The statistical test involves a one sided t-test on the coefficient.

The next two hypotheses relate to whether there is a statistically significant increase in *number of anglers* or *angler hours* of effort with respect to the level of stocking of: (a) catchable trout and (b) subcatchable trout planted in the previous year. Referencing the regression equation (2) the null and alternative hypotheses are:

(7) $H_0: A_1(CURRENTCAT_i) = 0$ vs $H_a: A_1(CURRENTCAT_i) > 0$
(8) $H_0: A_2(PREVSUB_i) = 0$ vs $H_a: A_2(PREVSUB_i) > 0$

Testing Elasticities

If the null hypotheses are rejected in favor of the alternative hypotheses of statistically significant positive relationships, then testing the next set of hypotheses regarding the *magnitude* of the relationship is in order. Of course, if the t-tests suggest we should accept the null hypothesis of no significant effect of stocking on license sales or angler days, then the following hypotheses regarding the magnitude of the elasticity are superfluous. That is, if $A_1 = 0$, then clearly the elasticity also equals zero.

The null hypothesis is that there is a proportionate relationship between *license sales* and fish stocking in equations (1a) and (1b). In particular:

(9) H_0: $e_{L,S}=1$ (e.g., $B_2=B_3=B_4=B_5=1$)　　vs　　H_a: $e_{L,S}<1$ (e.g., $B_2<1$, $B_3<1$, $B_4<1$, and $B_5<1$)

where $e_{L,S}$ is the elasticity of license sales with respect to number of fish stocked. This can be statistically tested by calculating the 95% confidence interval around the estimated coefficient in the double logarithmic demand function given in equation (1).

The parallel set of null hypotheses is that there is a proportionate relationship between the number of anglers/angler hours and fish stocking in equation (2). In particular:

(10) H_0: $e_{D,S}=1$　　vs　　H_a: $e_{D,S}<1$
(11) H_0: $e_{H,S}=1$　　vs　　H_e: $e_{H,S}<1$

where $e_{D,S}$ and $e_{H,S}$ is the elasticity of the number of anglers and angler hours, respectively, to fish stocking levels. This is calculated by A_1^* (mean CURRENTCAT/mean of the dependent variable). Unlike the double logarithmic-derived elasticity, the angler use elasticity reflects the ratio of several random variables. In order to calculate the upper bound of the elasticity accurately, a technique for estimating confidence intervals of elasticities from linear equations presented by Miller, Oral, and Wells (1984) was used. This approach estimates the variance of the elasticity (S^2_e) at the mean of the data as:

(12) $S^2_e = e^2 [S^2_{\hat{y}}/\hat{y}^2 + S^2_{B1}/B_1^2]$

where $S^2_{\hat{y}}$ is the variance of the estimate and y is angler use in hours or number of anglers depending on the model. This estimated variance can then be used in the calculation of the 95% confidence interval. If the 95% confidence interval is less than one, we would reject a proportionate relationship between stocking and angler use.

Data Sources

Stocking and License Sales

The first analysis looks at the relationship between fish stocked and the number of licenses sold per year. Therefore, data on the annual

Fish Stocking and Angler Use

number of licenses, both resident and nonresident, sold in Colorado from 1975 to 1995 and Colorado fish stocking data were obtained from the CDOW. The stocking data was aggregated across all water bodies in the state and categorized by warm and cold water and catchables and subcatchables. The license sales figures included resident season, nonresident season, nonresident five-day, and nonresident ten-day. However, throughout the course of the study period there was a switch from nonresident five-day to nonresident ten-day and then back to nonresident five-day. Therefore, there are not continuous data available for the temporary nonresident licenses. Data were obtained on license prices from 1975 to 1995 from the CDOW. The license prices were adjusted for inflation to 1983 dollars using the Consumer Price Index. Data on Colorado population were obtained from the U. S. Bureau of Census.

Stocking and Response in Angler Use

Examining the relationship between stocking and angler use involved matching creel census information for specific waters and stocking information specific to that water. This information was available in two databases, the first database contained the census information collected by the CDOW, and the second database contained the stocking by each water. Both of these databases were sorted by the CDOW's five-digit water code, allowing for the integration of these two databases.

The first step to combining these two databases was to screen the fish stocking to those dates matching the years of the angler-use census. It was decided to use the catchable fish stocked in the same year as the census and the year prior to the census. For subcatchable fish, the year previous to the census and two years prior to the census were used. Further, the angler-use analysis only examined coldwater fishing; thus the bodies of water that were primarily stocked with warmwater fish were not considered. The waters examined were stocked primarily with rainbow trout.

The next step was to combine the number of fish stocked in the current year, the previous year, and two years prior with the corresponding creel census information. Not all waters with census information had stocking for every year. If there were no stocking in one or more years then a zero was entered. This was done because it was not a case of missing data but rather the zero represents valuable data; CDOW did not stock that water body in those years. However, there was large variability in the lakes and streams that were stocked, with ranges of 49,585 and 3,391,984 for current catchables and the previous years subcatchables, respectively. Our dataset included 168 observations for coldwater streams and 45 for coldwater lakes, for a total of 213 observations in the pooled dataset. These data reflect over 100 different water bodies throughout Colorado and a sampling of years from 1980 to 1994. These observations occurred in five regions within the Rocky Mountain region of Colorado. The breakdown of regions for streams was: Central (14%), Northeast (23%),

Northwest (29%), Southeast (24%), and Southwest (10%). The breakdown for lakes was Central (17%) and Northeast (83%).

There are several alternative measures of angler use. **Total bank anglers or total boat anglers** are the seasonal estimates from CDOW for that water body, for that fishing mode, based on the days sampled. The total seasonal estimate represents an expansion based on the number of fishing hours surveyed relative to the total fishing hours for the season. The total fishing hours for the season account for the length of time the water is accessible (e.g., not frozen over, trail not snowed over) as well as the amount of sunlight fishing hours received by the site (e.g., if in a narrow canyon, there are fewer hours of direct sunlight). We estimated seven regressions that reflected combinations of fishing mode (boat versus bank for lakes) and definitions of the dependent variable. We also pooled the lake and stream data while including a intercept shift variable for lake (coded as 1).

Table 1
Resident Season Fishing License Demand

	Dependent Variable is Natural Log of Resident License Sales			
Variable	Model 1A: Current Catchable and Lagged Subcatchables[1]		Model 1B: Two Year Lag Specification[2]	
	Coefficient	Prob.	Coefficient	Prob.
C	7.308	.118	14.330	.0000
LRPRICE	-.318	.079	-.391	.0002
LCOLDCAT	.063	.580		
LCOLDSUB(t-1)	-.014	.887		
LWWCAT	.003	.880		
LWWSUB(t-1)	.021	.762		
LCOPOP	.345	.055		
COPOP			1.67E-07	.018
LRPGAS	-.010	.949		
LCOLDCAT(-1)			-.102	.361
LWWCAT(-1)			.015	.446
LCOLDSUB(-2)			-.009	-.009
LWWSUB(-2)			.027	.487

[1] R^2 = .847 S.E.= .041 F = 7.14 (p = .004) Durbin-Watson= 2.009, 1976–1995; N= 17
[2] R^2 = .816 S.E.= .041 F = 6.66 (p = .0063) Durbin-Watson= 1.858, 1977–1995; N= 17

Stastical Results

Resident Season Fishing Licenses

Table 1 displays the results for equation (1A): the demand for resident fishing licenses from 1976 to 1995 using current plantings of catchables and lagged plantings of subcatchables. As can be seen, the model explained 85% of the variation in license sales over this period. The inflation-adjusted price of the season license was significant at the 10%

Fish Stocking and Angler Use

level. Colorado population was significant at the 5% level. Table 1 also displays the results for equation (1B), which indicates that neither the previous year's plantings of catchables nor two-year lagged planting of subcatchables were significant, while price and population were. Thus, we accept the null hypothesis that stocking has no effect on license sales within the range of our data.

Results of Hypothesis Tests

To investigate the robustness of this result we estimated several other variations on equations 1A and 1B that concentrated only on coldwater fish stocking or emphasized only planting of catchables. Table 2 summarizes the lack of significance for several other regression specifications that are not presented in order to conserve space. As is evident from Table 2, we accept the null hypotheses that within the range of the current level of stocking there appears to be no statistically significant effect from stocking *catchable* coldwater or warmwater species on resident angler license sales over the past 20 years in Colorado. We also accept the null hypothesis that stocking *subcatchable* coldwater species has no statistically significant effect on resident angler license sales over the past 20 years. The only null hypothesis rejected is that stocking of subcatchable warmwater species does appear to have a positive effect on resident angler license sales. In terms of the relative responsiveness or elasticity of license sales with respect to stocking of subcatchable warmwater species, the null hypothesis of proportional response is rejected in favor of the alternative hypothesis of less than proportional response. In particular, a 100% change in stocking of subcatchable warmwater species results in a 7-9% change in resident angler license sales in Colorado. This relationship is not very responsive; rather, it is quite inelastic.

Table 2
Summary of Effects of Stocking on Resident License Sales, 1975-1995

Stocking Variable	#times Significant/ # of Regressions	Significance Level
Coldwater Catchables(t)	0/4	
Coldwater Subcatchables(t)	0/2	
Coldwater Catchables(t-1)	0/3	
Coldwater Subcatchables(t-1)	0/3	
Coldwater Subcatchables(t-2)	0/3	
Warmwater Catchables(t)	0/3	
Warmwater Subcatchables(t)	2/2	.05 and .1
Warmwater Catchables(t-1)	0/3	
Warmwater Subcatchables(t-1)	1/3	.05
Warmwater Subcatchables(t-2)	0/2	

Non-Resident Fishing Licenses

Table 3 presents the full regression model for the demand for nonresident angler season fishing licenses from 1976 to 1995. As with the resident license model, the inflation-adjusted price of a season fishing license is the most consistently significant variable. The number of catchable and subcatchable coldwater fish stocked had no statistically significant effect on sales of nonresident season fishing licenses in Colorado during the period under study. Warmwater subcatchables planted in the previous year had a negative effect. Only current planting of warmwater catchables had a statistically significant positive effect on sale of nonresident season fishing licenses. Stocking of warmwater catchables contributes a substantial amount of the explanatory power of .89 as a separate coldwater catchable and subcatchable model had an R-square of just .39 (with neither coldwater stocking variable being significant).

Table 3
Non Resident Season License Demand

Dependent Variable is Natural Log of Non-Resident Season Licenses

Sample(adjusted): 1976 1995; N = 17

Variable	Coefficient	Std.Error	t-Statistic	Prob
C	18.462	3.766	4.901	.0006
LRPRICE	-1.092	.179	-6.086	.0001
LCOLDCAT	.125	.196	.636	.539
LCOLDSUB(-1)	.019	.100	.189	.853
LWWCAT	.050	.029	1.692	.121
LWWSUB(-1)	-.305	.066	-4.570	.001
LRPGAS	-.418	.110	-3.780	.003

R^2= 0.8897   S.E. = 0.0633 Durbin-Watson = 2.47   F = 13.438 (p = 0.0002)

We also estimated regressions for nonresident short-term licenses such as five- and ten-day licenses. Once again, neither current nor last year's stocking of coldwater fish species had a systematic effect on the sales of licenses. While the lack of significance is consistent, these regressions have small sample sizes as these licenses were only issued for seven to eight years.

The absence of a systematic effect of stocking levels on resident and nonresident license sales may be due to the limited range of stocking numbers in the data for the 20-year time period. That is, insignificance is possible since there may not have been sufficient variability in the number of coldwater fished stocked each year. Our finding of no significant effect of stocking of coldwater fish on license sales is only valid within the range of the stocking data (e.g., for two million to six million catchables and 10

Fish Stocking and Angler Use

million to 22 million subcatchables). One cannot conclude that reducing stocking to zero would have no effect on license sales as zero is well outside of the range of data used in the analysis.

Our general conclusion on the analysis of the responsiveness of license sales to *current levels of* stocking is that there is no relationship for the range of stocking observed in the data over the past 20 years. Thus, stocking additional levels of trout adds to the CDOW costs, but does not generate additional license sales revenue to CDOW. There are, of course, benefits to some anglers from catching the additional stocked fish, but Colorado studies summarized by Johnson, Behnke, Harpman, and Walsh (1995) suggest the benefits are approximately $1 per additional fish.

Thus, we accept the null hypotheses of no effect of planting coldwater catchables in the current year and previous year on license sales in the current year. We also accept the null hypothesis of no effect of planting of coldwater subcatchables in the previous year and two year's on license sales in the current year. We also accept the null hypothesis of no effect on license sales for planting of warmwater catchables in the current and previous year on this year's resident license sales. However, for warmwater subcatchables, we reject the null hypothesis as there does appear to be a statistically significant effect on resident license sales.

Table 4
Summary of T-Statistics on Trout Stocking and Angler Use

Dependent Variable	Sample	Current catchables Coeff.	T-statistic	Previous subcatchables Coeff.	T-statistic	R^2
Total bank anglers	Stream	.357	3.352[c]	-.001	-.567	.065
Total angler hours	Stream	.353	3.253[c]	-.001	-.57	.061
Total bank anglers	Lake	.965	5.277[c]	-.017	-1.075	.404
Total boat anglers	Lake	.12	1.754[a]	.015	2.479[b]	.25
Total angler hours	Lake	1.085	4.915[c]	-.002	-.117	.393
Total bank anglers	Pooled	.535	5.941[c]	-.001	-.641	.24
Total angler hours	Pooled	.592	6.107[c]	-.001	-.371	.305

a, b, c indicate significant at the .1, .05 and .01 levels, respectively.
Total bank (boat) = the total number of people fishing from the bank (boat) during the survey period—this is a CDOW estimated number.
Total hours = the total hours of angling during the survey period—this is a CDOW estimated number.
Stream = a sample with only coldwater streams, N=168; lake = a sample with only coldwater lakes, N=45; pooled = a sample with both coldwater streams and lakes, N=213.

Effect of Stocking on Number of Anglers and Angler Hours

As summarized in Table 4, for the seven regression models of the form given in equation (2), the number of current catchables had a statistically significant effect in all seven, while the number of previous year's subcatchables was only significant in one of the seven. Note that we also tested longer lags in terms of stocking of catchables in the previous year and stocking of subcatchables two years earlier. These two longer lag variables were not significant predictors in combination with the current levels for the total bank anglers and total angler hours in the pooled model regression.

Results of Hypothesis Tests

In terms of our hypotheses, the level of stocking of catchable trout does have a statistically significant effect on the total number of anglers visiting a stream and lake over the season as well as total angler hours. Since several of the catchable trout variables are statistically significant, we can calculate the elasticity of angler days and hours with respect to catchable trout. These elasticities are summarized in Table 5. In general we reject the null hypotheses of a proportional relationship (elasticity =1) and accept the alternate hypothesis that the elasticity is less than one for all but two cases. The two cases where a proportionate relationship is possible are total bank anglers for lakes and total fishing hours for lakes.

Table 5
Summary of Elasticity of Anglers and Hours to Stocking of Trout in Colorado

Dependent Variable	Sample	Mean Elasticity Current Catchables	Upper 95% CI
Total bank anglers	Stream	.236	.891
Total angler hours	Stream	.230	.858
Total bank anglers	Lake	.489	1.26
Total boat anglers	Lake	.213	.706
Total angler hours	Lake	.427	1.06
Total bank anglers	Pooled	.285	.946
Total angler hours	Pooled	.284	.923

Total bank (boat) = the total number of people fishing from the bank (boat) during the survey period.
Total hours = the total hours of angling during the survey period.
Stream = a sample with only coldwater streams, N=168; lake = a sample with only coldwater lakes, N=45; pooled = a sample with both coldwater streams and lakes, N=213.

For the models that pool data across lakes and streams, the mean elasticity for bank angler is 0.285 and the mean elasticity of demand for total hours is 0.284. This elasticity means that a 1% change in the number of catchable trout planted results in a 0.28% change in number of anglers and angler hours. This relationship would be classified by economists as being in the inelastic or relatively insensitive range. However, anglers fishing lakes from the bank appear more responsive to levels of stocking. Number of bank anglers and total angler hours at lakes increases by 0.489% and 0.427%, respectively, for every 1% increase in catchable trout stocked. Lake boat anglers are less responsive, with a 1% increase in stocked trout resulting in only a 0.21% change in number of anglers fishing

Fish Stocking and Angler Use

from a boat. Generally, the responsiveness of lake anglers fishing from the bank to stocking may suggest continued emphasis on stocking small lakes where the predominant fishing mode is from shore and deemphasizing stocking in lakes where the dominant fishing mode is by boat or in streams.

Conclusion

Over the range of stocking levels for coldwater species observed during the last 20 years in Colorado, stocking levels did not have a significant effect on resident and nonresident season license sales. Only plantings of warmwater fish species had any significant effect on fishing license sales. However, stocking of catchable trout did have a significant effect on angler effort. Stocking of catchable trout was a significant variable in seven of seven estimated regressions predicting angler use, while the lag effect of last year's planting of subcatchables was only significant in one of the regressions. Therefore, overall stocking levels are not effecting the numbers of people who fish, but stocking of coldwater catchables is influencing angler effort. This analysis only examined the relationship between stocking and use and did not look at how stocking effects catch rates.

While the level of stocking of catchable trout does have a significant influence on angler use, the relationship is fairly unresponsive, especially for streams. We calculated the responsiveness of season angler use to stocking levels and found that five of the seven elasticities were less than one. The elasticities that were greater than one were for lake anglers fishing from the bank, and total lake angler hours suggest continued targeting of hatchery trout on lakes, particularly lakes with a large proportion of bank anglers.

While reductions in stocking of trout will affect the number of anglers and hence the local economies surrounding heavily stocked lakes, it may not result in elimination of angling at these sites. Therefore, agency policies based on an assumption of proportionate response between angler use and stocking should be reexamined.

The lack of significance of stocking as a determinant of license sales may be due to limited variability in stocking levels in Colorado. Further license sales analysis should be conducted with data sets with more years. One cannot use these models to estimate the effect of zero stocking levels on license sales as zero is well outside the range of the data. Nonetheless, the results are useful for evaluating sizeable changes in stocking that might result from decisions not to replace whirling disease-infected hatcheries. It appears that modest reductions in stocking would have little effect on resident and nonresident license sales.

Due to using available visitation data that aggregate all types of anglers together, we cannot estimate the differential effect that reduced stocking has on different types of anglers. While there is a small net overall effect, for some types/styles of fishing or angler motivations there may be a large drop in use, while other types of anglers may not reduce their fishing at all. However, to determine the differential effect on various types of anglers requires special survey data.

Acknowledgments: This research would not have been possible without the support and assistance of numerous individuals. Without implicating, we would like to thank John Epifanio and David Nickum of Trout Unlimited for their funding and insightful suggestions. William Weiler, Harry Vermillion, Tom Powell, and Mary McAfee of Colorado Division of Wildlife (CDOW) were most generous in providing the raw data used in the statistical analysis. Without their cooperation this study could not have been completed. Three reviewers provided valuable suggestions for improving the clarity of this paper. The opinions expressed are those of the authors and do not represent those of Colorado State University.

References

Bennett, J., Nehring, B., Krieger, D., Harris L., & Nesler, T. (1996). *An assessment of fishery management and fish production alternatives to reduce the impact of whirling disease in Colorado.* Denver, CO: Colorado Division of Wildlife.

Cole, R., Ward, T., Ward, F., Deitner, R., Bolton, S., Fiore, J., & Green-Hammond, K. (1990). *RIOFISH, a fishery management and planning model for New Mexico reservoirs.* Las Cruces, NM: New Mexico State University.

Duffield, J., Loomis, J., & Brooks, R. (1987). *The net economic value of fishing in Montana.* Bozeman, MT: Montana Department of Fish, Wildlife and Parks.

Harris, C. & Bergersen, E. (1985). Survey on demand for sport fisheries. *North American Journal of Fisheries Management, 5,* 400-410.

Johnson, D., Behnke, R., Harpman, D., & Walsh, R. (1995). Economic benefits and costs of stocking catchable rainbow trout: A synthesis of economic analysis in Colorado. *North American Journal of Fisheries Management, 15,* 400-410.

Johnson, D., & Walsh, R. (1987). *Economic benefits and costs of the fish stocking program at Blue Mesa Reservoir, Colorado.* Colorado Water Resources Research Institute Technical Report 49. Fort Collins, CO: Colorado State University.

Loomis, J. (1992). The evolution of a more rigorous approach to benefit transfer: Benefit function transfer. *Water Resources Research, 28,* 701-705.

Loomis, J., & Cooper, J. (1990). Comparison of environmental quality-induced demand shifts using time series and cross section data. *Western Journal of Agricultural Economics, 15,* 83-90.

Loomis, J., & Walsh, R. (1997). *Recreation economic decisions* (2nd Ed.). State College, PA: Venture Press,

Miller, S., Capps, O. Jr., & Wells, G. (1984). Confidence intervals for elasticities and flexibilities from linear equations. *American Journal of Agricultural Economics, 66*(3), 392-396.

Fish Stocking and Angler Use

Stavins, R. (1993, November 3). *Private options to use public goods: The demand for fishing licenses.* Faculty Research Workshop, John F. Kennedy School of Government. Cambridge MA: Harvard University.

Ward, F. (1991, September 17). [Report to Steve Henry, Chief of Fisheries]. Santa Fe, NM: New Mexico Department of Game and Fish.

Population dynamics and potential of fisheries stock enhancement: practical theory for assessment and policy analysis

Kai Lorenzen

Imperial College London, Silwood Park, Ascot SL5 7PY, UK (k.lorenzen@imperial.ac.uk)

The population dynamics of fisheries stock enhancement, and its potential for generating benefits over and above those obtainable from optimal exploitation of wild stocks alone are poorly understood and highly controversial. I review pertinent knowledge of fish population biology, and extend the dynamic pool theory of fishing to stock enhancement by unpacking recruitment, incorporating regulation in the recruited stock, and accounting for biological differences between wild and hatchery fish. I then analyse the dynamics of stock enhancement and its potential role in fisheries management, using the candidate stock of North Sea sole as an example and considering economic as well as biological criteria. Enhancement through release of recruits or advanced juveniles is predicted to increase total yield and stock abundance, but reduce abundance of the naturally recruited stock component through compensatory responses or overfishing. Economic feasibility of enhancement is subject to strong constraints, including trade-offs between the costs of fishing and hatchery releases. Costs of hatchery fish strongly influence optimal policy, which may range from no enhancement at high cost to high levels of stocking and fishing effort at low cost. Release of genetically maladapted fish reduces the effectiveness of enhancement, and is most detrimental overall if fitness of hatchery fish is only moderately compromised. As a temporary measure for the rebuilding of depleted stocks, enhancement cannot substitute for effort limitation, and is advantageous as an auxiliary measure only if the population has been reduced to a very low proportion of its unexploited biomass. Quantitative analysis of population dynamics is central to the responsible use of stock enhancement in fisheries management, and the necessary tools are available.

Keywords: stocking; supplementation; density dependence; mortality; growth; evolution

1. INTRODUCTION

(a) Overview

Stock enhancement is a fisheries management approach involving the release of cultured organisms to increase abundance and yield of natural fish or invertebrate stocks. Releases may be carried out on a long-term basis to raise yields above the level supported by natural recruitment, or temporarily to rebuild depleted populations. Stock enhancement describes a continuum of hatchery release and associated harvest regimes, the extremes of which are culture-based fisheries and supplementation. In culture-based fisheries or ranching systems, recruitment is largely or entirely based on hatchery releases, and release and harvesting regimes may be designed to maximize production. By contrast, in supplementation, hatchery fish are released to bolster the natural spawning stock, and release and harvesting regimes may be designed to maximize natural recruitment. In the current analysis, I deal with enhancement in its full breadth but exclude considerations specific to the supportive breeding of small populations such as depensation, demographic stochasticity and the genetics of low effective population size. Stock enhancement may be implemented under a variety of different institutional settings such as private or communal enterprises, or for public benefit under open access.

Stock enhancement is one of the oldest, yet most controversial and least well-understood approaches to fisheries management. Stocking of hatchery fish has been practised on a large scale since the mid-nineteenth century, and systematic transfers of wild juveniles probably have a much longer history. Current global production by stock enhancement and culture-based fisheries has been estimated at ca. 2 Mt yr^{-1} (Lorenzen et al. 2001). This includes some enhancement programmes conducted on a very large scale by government agencies (notably for Pacific salmon) and many small, often resource-user-led initiatives.

Stock enhancement as a management approach is more common in freshwater than in marine systems, reflecting differences in scale, institutional arrangements and state of hatchery technology (Welcomme & Bartley 1998). For well over a hundred years, stock enhancement has been the subject of fierce controversy regarding its effectiveness and possible adverse impacts on wild stocks (reviewed in Hilborn 1999; Taylor 1999; Smith et al. 2002). Generally, this 'hatchery controversy' has divided stakeholders along disciplinary lines, with aquaculture practitioners, scientists and some fisheries managers broadly in favour, but fisheries ecologists vigorously against the use of stock enhancement. The result has been a plethora of poorly conceived and managed enhancements, and a very uneven development of relevant science. Advances in the science and

One contribution of 15 to a Theme Issue 'Fisheries: a future?'.

practice of hatchery management have allowed increasingly effective production of fish for release, but the crucial, broader issues of using hatchery fish in population management and conservation have received little systematic attention (Hilborn & Winton 1993). The few studies to address the population dynamics of enhancement (reviewed in § 1d) have been largely ignored by both management practitioners and scientists. Poor appreciation of the dynamics of enhancements limits their potential for achieving management objectives (Botsford & Hobbs 1984; Lorenzen 1995), and allows their misuse as an apparent 'quick fix' for management problems they cannot effectively address (MacCall 1989; Hilborn 1999). Without quantitative assessment, it is difficult to gauge the true potential of enhancement and refute unrealistic proposals and claims. The need for a critical and realistic assessment, a 'common version of reality' (Waples 1999) of stock enhancement is now widely recognized (Blankenship & Leber 1995; Hilborn 1999; Leber 2001; Lorenzen et al. 2001). The aim of this paper is to contribute to a common reality by developing and analysing a general model for the dynamics of stock enhancement.

(b) *Rationale for stock enhancement*
In theory, successful stock enhancement can yield significant production, social and ecological benefits. First, it can increase the use of natural aquatic productivity beyond the level achievable by harvesting alone, providing high quality food at relatively low external inputs of energy and protein and with limited effects on aquatic habitats and their competing uses (Lorenzen et al. 2001). Second, enhancement can create new economic opportunities for fisheries-related livelihoods, and provide incentives for active management of fisheries resources (Pinkerton 1994; Lorenzen & Garaway 1998). Third, enhancement can maintain the abundance of exploited stocks above the level supported by natural recruitment alone. This may provide partial mitigation against the ecosystem effects associated with depletion of key species by fishing (Pauly et al. 1998; Jackson 2001; Mehner et al. 2002; Pauly et al. 2002). Fourth, and rather more speculatively, genetic resource management of enhanced stocks could be employed to mitigate against the evolutionary effects of fishing (Stokes et al. 1993; Conover & Munch 2002) by replenishing stocks with offspring from the genotypes most susceptible to harvesting which are otherwise selected against.

The biological rationale for stock enhancement has three key components: recruitment limitation, hatchery advantage, and manipulation of population structure. Fish populations in general are believed to be recruitment limited in the sense that under most conditions, additional recruits will increase the abundance of the recruited stock (Munro & Bell 1997; Walters & Korman 1999; Hixon et al. 2002). This view is also implicit in dynamic pool fisheries models (Beverton & Holt 1957). Recruitment limitation may be exacerbated by anthropogenic factors such as fishing or degradation of juvenile habitat (Blankenship & Leber 1995; Blaxter 2000). If adult abundance is recruitment limited, increasing the level of recruitment through hatchery releases can be expected to increase abundance and yield of the recruited stock. For this to be beneficial overall, hatcheries must be able to produce a higher number of recruits per spawner than are produced in natural stocks.

This 'hatchery advantage' is substantial and well documented. On average, juvenile survival in aquaculture facilities is several orders of magnitude higher than in the wild, and even though this is partially offset by increased mortality upon release an overall advantage is likely to remain (Lorenzen 1996b, 2000). A significant hatchery advantage is of course possible only in organisms of very high fecundity. The hatchery advantage not only allows increased recruitment above natural levels, it paves the way for structural manipulations of fish populations. It enables, for example, the construction of populations of fast growing juveniles harvested at the optimal size for production, replenished with offspring from a relatively small hatchery broodstock. Such stock management strategies are used to raise productivity in extensive aquaculture systems or culture-based fisheries on scales from ponds to large reservoirs (Walter 1934; Lorenzen 1995; Lorenzen et al. 1997). Whether such manipulations are ecologically acceptable and economically viable in natural populations will depend on specific circumstances, but the biological potential is a crucial, and much underexplored aspect of stock enhancement.

(c) *Reality check: problems and progress in addressing them*
Despite clear rationale and potential benefits, the actual performance of stock enhancements has been mixed and, more often than not, disappointing. Many enhancements have failed to deliver significant increases in yield or economic benefits, and/or have had deleterious effects on the naturally recruited components of the target stocks (Hilborn 1998; Levin et al. 2001; Arnason 2001). For enhancement to produce net benefits and avoid unacceptable deleterious effects on the wild-stock component, several conditions must be met. First, only certain stocks offer the potential for biologically effective and economically viable enhancement, even with the best stock management and aquaculture technology (Blankenship & Leber 1995; Travis et al. 1998). Second, where potential exists in principle, appropriate release and harvesting regimes must be developed with respect to both the wild and stocked components of the target stock (Botsford & Hobbs 1984; Lorenzen 1995). Third, hatchery production and release strategies must provide fish that perform well in the wild, at a low cost. Inadvertent developmental and genetic adaptations of the hatchery environment, which are deleterious in the wild, make this a major challenge (Olla et al. 1998; Lorenzen 2000; Fleming & Petersson 2001). Fourth, hatchery and fisheries management strategies must be developed that minimize genetic hazards to the wild stock (Utter 1998). Many stock enhancement programmes have paid little attention to some or all of these conditions, and their success or otherwise has been a hit or miss affair. The need for a more informed and responsible approach to the development of stock enhancements has been widely recognized, however, and various conceptual frameworks proposed to guide the process (Cowx 1994; Blankenship & Leber 1995; Lorenzen & Garaway 1998). At the same time, there has been substantial progress in hatchery production and genetic management of enhancements. Hatchery management and release techniques such as nutrition optimization, behavioural enrichment and conditioning, and soft release can greatly reduce developmental

adaptation to the hatchery environment and improve post-release performance in the wild (Olla et al. 1998; Brown & Dey 2002). Genetic resource management can effectively address, but not entirely eliminate, problems arising from limited effective population size in the hatchery, disruption of the genetic structure of the wild population, and genetic adaptation to the hatchery environment (Utter 1998; Price 2002; Miller & Kapuscinski 2003). Deliberate manipulations of hatchery organisms including hybridization, triploidization and artificial selection provide means of minimizing genetic interactions with wild conspecifics, or improving performance traits of stocked fish (Jonasson et al. 1997; Bartley et al. 2001). Obviously, even a rigorous and responsible development approach using the best available science does not guarantee the emergence of effective and sustainable enhancements. This is well illustrated, for example, by the Alaskan pink salmon and Norwegian cod enhancement programmes, both of which have a history of systematic investigation and enlightened management but have proved uneconomic under current conditions (Boyce et al. 1993; Hilborn 1998; Svasand et al. 2000). By contrast, the equally well-developed Japanese chum salmon enhancement programme, as well as various smaller initiatives in freshwaters, is believed to be effective as well as economically viable (Hilborn 1998; Arnason 2001). Indeed, some enhancements provide very high physical and economic returns to limited investment (Ahmad et al. 1998; Lorenzen et al. 1998). Overall, this suggests a potential for certain, well-conceived and managed enhancements to be technically effective and economically beneficial. In such systems, moderate quantitative differences in biological or economic parameters can make all the difference between success and failure. For example, decline in salmon prices owing to the large supply from aquaculture may have turned many salmon enhancement projects from economic successes into failures (Boyce et al. 1993; Arnason 2001). A good, quantitative understanding of the dynamics of an enhanced fishery is therefore crucial to its sustainable development.

(d) *Understanding the dynamics of stock enhancement*

At the heart of the enhancement system are the enhanced stock and its dynamics in response to harvesting, hatchery releases and environmental factors. These dynamics remain poorly understood beyond the most basic information gleaned from empirical recapture rates for, at best, a small set of management options. A handful of studies, however, have covered significant ground towards a more comprehensive and theory-based assessment. Botsford & Hobbs (1984) conducted the first general, quantitative analysis of stock enhancement as a fisheries management policy. Recognizing that density-dependent processes at different life stages are fundamental to enhancement dynamics but poorly understood, they used a set of alternative and very general assumptions to derive robust insights and decision rules. Cuenco (1994) took a similarly general approach to the problem of supplementing declining salmon populations, providing simple decision rules for populations of semelparous organisms

with non-overlapping generations. An alternative to such general but abstract analyses has been the use of conventional fisheries models incorporating empirically based representations of certain population processes (Polovina 1990). However, conventional fisheries models disregard size and density-dependent processes that are central to the dynamics of enhancements. Simple and empirically robust models for two such processes, density-dependent growth and size-dependent mortality, form the basis of an assessment methodology for culture-based fisheries developed by Lorenzen (1995, 2000) and Lorenzen et al. (1997). The dynamic implications of genetically-based performance differences between wild and hatchery components of enhanced stocks were first analysed by Byrne et al. (1992), and more recently by Ford (2002).

In this paper I build on the earlier work reviewed here to develop a general and practical theory of fisheries enhancement, an integrated framework for the evaluation of release and harvest regimes with respect to yield and abundance of different population components. As a case study, I explore the potential of enhancing the North Sea sole stock, using stock assessment data and integrating basic economic considerations. I close by discussing general implications for the development and management of enhancements, and their future role in fisheries management.

2. POPULATION DYNAMCIS THEORY FOR ENHANCED FISHERIES

A practical theory of stock enhancement must allow analysis of the impacts of management variables such as stocking size and density, post-release performance, and harvest regulations on fisheries yield, as well as the status of the wild and hatchery stock components. It must be based on biologically meaningful process models that are simple, robust and general with parameters that can be estimated from widely available data or inferred from comparative analyses. The dynamic pool theory of fishing (Beverton & Holt 1957) provides a practical and widely used methodology for the assessment of capture fisheries, which can be extended to the analysis of enhancements. Three extensions are necessary to achieve this. First, the stock–recruitment relationship must be 'unpacked' in order to analyse the effect of releasing pre-recruit juveniles. Second, population regulation in the recruited stage must be accounted for because it determines to what extend additional recruits can increase stocks and yields: the potential of enhancement. Third, biological differences between hatchery and wild fish have important implications for the dynamics of enhancements and must be accounted for. The following sections set out how this may be done.

(a) *Unpacking recruitment*

Conventional dynamic pool theory divides the life history of exploited fish and invertebrates into a density-dependent and possibly stochastic pre-recruit phase, and a density-independent and deterministic recruited phase. Recruitment, the transition between these phases, may be associated with identifiable biological processes but is often assumed to occur at a somewhat arbitrary age. I define recruitment as the transition from a juvenile stage subject to density-dependent mortality, to a recruited stage subject to density-dependence in growth and reproductive parameters.

Most stock enhancement efforts are likely to involve releasing fish in the pre-recruit stage; hence unpacking recruitment is a necessary step to analysing the effects of different release sizes and densities.

Precise size or stage-specific data on population dynamics of pre-recruits are available for only a handful of populations (e.g. Elliott 1994). In general, an overall stock–recruitment relationship is the most an analyst can hope for, and this must be unpacked without recourse to more detailed data. Three pieces of information provide the basis for doing this: the general allometry of natural mortality, empirical and theoretical information on density-dependent processes at different life stages, and a mathematical way for breaking stock–recruitment relationships into successive stages.

Natural mortality rates within natural fish populations are strongly size-dependent with an allometric weight exponent of $ca.$ -0.29 to -0.37 (McGurk 1986; Lorenzen 1996b). In other words, natural mortality is approximately inversely proportional to length:

$$M(L) = M_1 \frac{1}{L} \quad (2.1)$$

where $M(L)$ is the natural mortality rate at length L, and M_1 is the natural mortality rate at unit length. Lorenzen (2000) gives survival equations based on this mortality–length relationship for different growth models; these are used in equations (2.10) and (3.1). The average M_1 in wild fish is 15 yr^{-1} at unit length of 1 cm, while that of stocked hatchery fish may be in a similar range or substantially higher (Lorenzen 1996b, 2000). Because models based on an inverse relationship between mortality and length provide good predictions of survival in relation to release size in fish stocking experiments (Lorenzen 2000), average mortality rates in consecutive phases of the recruitment process may be expected to follow this relationship. This is a first major step in unpacking recruitment.

The next question is where and how stochastic and/or density-dependent processes generate variation around the 'average' allometry and thus give rise to variable and often density-dependent stock–recruitment relationships. The general pattern that has emerged in this respect may be summarized as follows. Vital rates of early life stages (eggs and larvae) tend to be highly variable, but density independent (Myers & Cadigan 1993a; Leggett & DeBlois 1994). Small changes in the very high rates of mortality suffered by these stages cause major variation in cohort survival, and are believed to account for a large part of variability in recruitment (Beyer 1989; Rothschild 2000). By contrast, vital rates in juveniles are often density dependent and thereby tend to dampen the variability created at early life stages (Myers & Cadigan 1993b; Elliott 1994). Density-dependent survival at this stage may arise directly from density effects on the mortality rate (Elliott 1994), or indirectly from the interaction of size-dependent mortality with density-dependent growth (Shepherd & Cushing 1980; Post et al. 1999). Either mechanism or a combination may arise from trade-offs between foraging and predation risk-taking in juveniles, and result in density-dependent survival to recruitment (Walters & Korman 1999). Density-dependent growth replaces density-dependent mortality as the dominant regulatory mechanism in larger fish (Walters & Post 1993; Post et al.

1999; Lorenzen & Enberg 2002). Most probably, this transition is gradual and related to declining effects of growth variation on mortality (as overall mortality rates are declining), and increasing effects on biomass (as body mass is increasing). Broadly in parallel with ontogenic changes in regulatory mechanisms, there is a transition from intra-cohort to inter-cohort density dependence. The appropriate metric of density therefore changes from stage-specific numerical abundance to whole population biomass or similar measures that reflect aggregated effects on resources (Walters & Post 1993; Lorenzen 1996a). Even though the transition in mechanisms and metrics of density-dependence is likely to be gradual, it is practical to assume distinct phases of intra-cohort density-dependent mortality before, and inter-cohort density-dependent growth after recruitment. This is unlikely to misrepresent dynamics provided recruitment is assumed to occur at a size most probably within the growth-dominated phase of regulation, and dynamics in the recruited phase takes account of size-dependent mortality as well as density-dependent growth. I now focus on pre-recruit processes and return to regulation in the recruited population in § 2b.

Having established that within the pre-recruit stage, density dependence is most likely to act on juvenile mortality in a manner dependent on stage-specific numerical abundance, it is possible to partition the stock–recruitment relationship into a density-independent larval phase, and a density-dependent juvenile phase. The latter may again be subdivided into a pre- and post-release phase according to the stage or size at which juveniles will be released. Mathematically it is straightforward to partition an overall Beverton–Holt stock–recruitment relationship into consecutive relationships of the same functional form (Beverton & Holt 1957; Walters & Korman 1999). The overall relationship is given by

$$N_r = \frac{a^* S}{1 + b^* S} \quad (2.2)$$

where N_r is the number of recruits, S is spawner biomass, a^* is the maximum number of recruits produced per unit spawner biomass (the product of larval production and subsequent survival) and b^* describes the degree of density dependence in recruitment. This may be partitioned into a three-stage model with density-independent larval production

$$N_0 = fS \quad (2.3)$$

and two consecutive phases of potentially density-dependent survival according to a Beverton–Holt relationship, e.g. for the first stage:

$$s_1 = \frac{N_1}{N_0} = \frac{a_1}{1 + b_1 N_0} \quad (2.4)$$

The parameters f, a_1, b_1, a_2, and b_2 of the three-stage model are related to a^* and b^* by

$$a^* = f a_1 a_2 \quad (2.5)$$

and

$$b^* = f b_1 + f a_1 b_2. \quad (2.6)$$

The three-stage model thus has three free parameters, which are, however, constrained within certain ranges given that a_1 and a_2 are survival rates and thus must be between zero and unity.

The key to unpacking recruitment in a meaningful way, of course, is in relating the abstract phases of the model to actual life stages or sizes, and this requires good biological knowledge of the target organism. In demersal fish with pelagic larvae for example, settlement represents a clear transition to the juvenile stage and often coincides with density-dependent mortality (Van der Veer 1986). It may thus be assumed that the period from settlement to recruitment corresponds to the density-dependent juvenile phase of the model. If this is indeed the case, field measurements or comparative data on survival from settlement to recruitment should be broadly consistent with predictions from the stock–recruitment model at observed levels of spawner biomass. Subdivision of the juvenile phase before and after release may be informed by the allometry of mortality, or further empirical data. If stage-specific survival s_1 is known for some level of initial density N_0^* entering the stage (e.g. the estimated abundance when field measurements were taken), the stage-specific density-dependent parameter b_1 is given by rearranging the Beverton–Holt survival model as:

$$b_1 = \left(\frac{a_1}{s_1} - 1\right)\frac{1}{N_0^*}. \qquad (2.7)$$

Note that b_1 is constrained by $s_1 \leqslant a_1 \leqslant 1$; hence stage-specific survival at N_0^* puts an upper limit on the potential degree of density dependence within the stage. If survival in consecutive stages reflects the general allometry of mortality, this translates into declining potential for density-dependent mortality with increasing size.

The unpacking approach is illustrated with an example in § 3a. It is possible, of course, that survival rates implied by the unpacked stock–recruitment relationship and specific biological data are inconsistent. Where this happens, reviewing fundamental assumptions will probably prove productive, not only as a basis for assessment but in terms of basic biology.

Recruitment variation is a pervasive feature of fish population dynamics. A large share of variability in recruitment appears to be generated in the egg and larval stages, prior to the action of density-dependent processes (Myers & Cadigan 1993a; Leggett & DeBlois 1994; Secor & Houde 1998). However, environmental variability may also affect the intensity of density-dependent processes in juvenile stages (Giske & Salvanes 1999; Levin et al. 2001). Episodes of low larval survival or weak juvenile density dependence may create temporary opportunities to increase recruitment through juvenile releases, but regulation in the recruited stock may limit the overall benefits of such strategies. I do not explore the implications of recruitment variability further, but note that this can easily be done by defining parameters f, a or b in the unpacked model as stochastic variables.

(b) Regulation in the recruited population and recruitment limitation

Regulation in the recruited phase determines the ultimate biological limits of enhancement, particularly (but not only) when hatchery fish are released as recruits or late pre-recruits. Density dependence in the recruited population may act on growth, reproductive traits such as age or size at maturity, and mortality (Rose et al. 2001). Density-dependent growth appears to play a key role in regulating abundance, and is well described by a von Bertalanffy growth function with asymptotic length $L_\infty(B)$ defined as a linear function of population biomass B (Lorenzen 1996a; Lorenzen & Enberg 2002):

$$L_\infty(B) = L_{\infty L} - gB \qquad (2.8)$$

where $L_{\infty L}$ is the asymptotic length in the absence of competition $(B \to 0)$, and g measures the strength of density dependence. Interactions between density-dependent growth and size-dependent mortality only have a weak regulating effect in the recruited stock because overall mortality is low. By contrast, strong density-dependent effects on reproductive traits may arise from interactions of density-dependent growth and size-dependent maturation and fecundity schedules. Rochet (1998) and Beverton (2002) show that many populations respond to increases in fishing effort and concomitant reduction in density with reduced age, but little or no change in size at maturity. Overall reproductive allocation at a given size appears to be largely independent of density, but a tendency to produce more and smaller eggs at low density has been noted (Rijnsdorp et al. 1991; Rochet et al. 2000). Some populations, however, have undergone substantial changes in both age and size at maturity in response to exploitation. These changes defy simple generalizations, and may well reflect a combination of phenotypic plasticity and natural selection by fishing. Life-history theory holds the key to unravelling the proximate dynamics of these responses (Thorpe et al. 1998), but to date satisfactory predictive models remain elusive. I note this as a key area of research interest, and confine my analysis here to populations that show essentially constant size at maturity.

The concept of recruitment limitation is an important element of the biological rationale for enhancement. Recruitment limitation is defined here as a state in which natural recruitment is limited to a level at which the addition of further recruits increases the abundance of the recruited stock (i.e. elicits a less than complete compensatory response). The notion that the abundance of recruited stocks can be increased by additional recruits is borne out by the observation that in many stocks, very large year classes raise biomass and fisheries yield far above the long-term average (Myers et al. 1990; Munro & Bell 1997). That does not mean that direct density-dependent processes are absent in the recruited phase: episodes of strong recruitment can depress growth significantly. The ratio of asymptotic length at current B to asymptotic length at very low biomass $(B \to 0)$, $L_\infty(B)/L_{\infty L}$ is typically above 0.9 at the long-term average biomass $\bar{B}$ of exploited populations, but may decline to less than 0.7 during periods of high abundance (Lorenzen & Enberg 2002). Direct density dependence thus has a significant compensatory effect on biomass, but is not sufficient to effect complete compensation. In extensive aquaculture systems, stocking can maintain high biomass densities that depress $L_\infty(B)/L_{\infty L}$ well below 0.9 on a permanent basis (Lorenzen 1996a; Lorenzen et al. 1997). Why, then, is the long-term average abundance (i.e. carrying capacity) of wild populations reached at a relatively low biomass so that $L_\infty(\bar{B})/L_{\infty L}$ remains above 0.9? The answer must lie in compensatory processes that act on future recruitment, and are stronger than effects on current biomass. The action of such processes is borne out, for example, by the observation that in a highly variable fish population, strong year classes are followed by weak recruitment and vice versa (Marshall & Frank 1999). Compensatory effects on future recruitment may act on reproductive output of the parent generation, or on survival of their offspring. Density-dependent growth combined with constant size at maturity alone implies strong regulation of reproductive output, and there may be further effects on size-related fecundity or egg quality. Density-dependent survival in the juvenile phase appears to be ubiquitous (§ 2a) and probably contributes significantly to the degree of recruitment limitation observed in fish populations. However, recruitment limitation as defined here is likely to arise even without juvenile density dependence, as a consequence of the nature of compensatory processes in the recruited stock. This implies a general potential for enhancing abundance of the recruited stock, and an equally general expectation of significant compensatory decline in natural recruitment.

(c) *Ecological differences between wild and hatchery fish*

Ecological differences between wild fish and hatchery fish derived from a local founder stock arise from plastic developmental responses to, and natural or artificial selection in, the hatchery environment (Price 2002). Experimental evidence for the success of conditioning and soft release in improving performance on the one hand (Olla et al. 1998; Jonsson et al. 1999), and heritability of poor performance on the other (Reisenbichler & Rubin 1999) shows that both developmental and genetic factors can be important. Their relative contribution is likely to vary and must be assessed experimentally for specific fisheries. Differences due to developmental plasticity diminish over the lifetime of a cohort due to increasing adaptation of individuals and the action of natural selection, and are not passed on to offspring produced in the wild (bar possible maternal effects). Differences induced by selection in the hatchery are passed on to the following generation, subject to natural selection that will act in the direction of the wild phenotype and reduce differences over successive generations. The rate at which this phenotypic change occurs is given by the heritability h^2 of the traits in which the wild and hatchery phenotypes differ. Heritability is the change in a quantitative trait due to selection within one generation, relative to the selection differential between the current and the optimal trait value. Heritability of morphological traits is generally ca. 0.2; that of fitness traits tends to be lower at between 0.01 and 0.1 (Mousseau & Roff 1987; Burt 1995).

Which ecological traits are most likely to differ between wild and hatchery fish, and by how much? In general, natural mortality rates of released hatchery fish are higher than those of wild conspecifics of similar size, often by a substantial margin (Lorenzen 2000; Fleming & Petersson 2001). Reproductive success of hatchery fish in the wild also tends to be substantially below that of their wild conspecifics, at least in salmonids (Fleming & Petersson 2001). By contrast, no strong or consistent differences have been reported for growth (Svasand et al. 2000; Fleming & Petersson 2001). Most life-history differences between wild and hatchery fish are expressed even when the two groups do

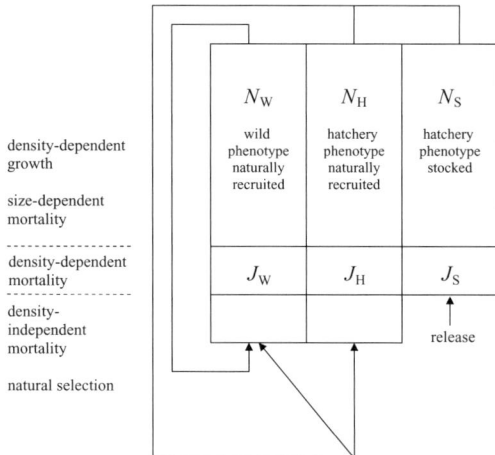

Figure 1. Structure of the fisheries enhancement model, showing the population components, flows and key processes.

not interact ecologically (e.g. Lorenzen 2000), but asymmetric interactions may further modify relative performance. Evidence for the effects of asymmetric interactions is complex and mixed (Weber & Fausch 2003). I will not consider such interactions in this paper, but again note this as an area of further research and population-level analysis.

A simple and straightforward way of accounting for ecological differences between wild and hatchery fish in population dynamics modelling is to disaggregate the population into components with different life-history parameters. Normally the hatchery component will be less well adapted than the wild component, and this will be reflected in poorer values of performance traits. Natural selection will act to move the average performance of the combined population towards that of the wild component, and this process may be modelled as transition of offspring from the hatchery component into the wild component at a rate equal to the heritability h^2. The result is a simple model of phenotypic evolution in the enhanced fishery that can be used to assess the implications of a wide range of possible ecological differences and assumptions about their genetic and/or developmental basis.

(d) *Population model for stock enhancement*

To explore the dynamics of stock enhancement, I use a population model incorporating the key aspects identified above: an unpacked stock–recruitment relationship, regulation in the recruited phase, and a population differentiated into components according to phenotype and origin (figure 1). The three components considered are wild (wild phenotype, naturally recruited), hatchery (hatchery phenotype, naturally recruited) and stocked (hatchery phenotype, stocked). This differentiation allows us to address a range of different questions, including the contributions of stocking and natural recruitment to yield, and the implications of releasing genetically maladapted fish.

Growth is described by the density-dependent von Bertalanffy model defined in equation (2.8), starting with a constant length at recruitment $L(1, t)$. All population components are assumed to share the same growth pattern. A discrete time model to predict mean length $L(a, t)$ of age group a at time t from mean length of the cohort in the previous year $L(a − 1, t − 1)$ is given by

$$L(a,t) = L_\infty(B) − (L_\infty(B) − L(a − 1, t − 1)) \exp(−K) \quad (2.9)$$

where $L_\infty(B)$ is the asymptotic length at biomass density B (equation (2.8)).

I assume that fishing occurs in discrete events once a year, and that natural mortality is size dependent and acts continuously between the fishing events. Population numbers N_I of the different components ($I = W, H, S$) are given by

$$N_I(a,t) = N_I(a − 1, t − 1) \exp(−F(a − 1, t − 1))$$
$$\times \left(\frac{L(a − 1, t − 1)}{L(a − 1, t − 1) + L_\infty(B)(e^k − 1)} \right)^{\frac{M_{1,I}}{L_\infty K}} \quad (2.10)$$

where F is the fishing mortality rate, and $M_{1,I}$ is the natural mortality rate at unit length (Lorenzen 2000). Catch at age $C_I(a, t)$ is given by

$$C_I(a,t) = N_I(a,t)(1 − \exp(−F(a,t))). \quad (2.11)$$

Gear selectivity and proportional maturity are described by length-dependent logistic functions. Fishing mortality is given by

$$F(a,t) = \frac{F_\infty}{(1 + \exp(q(L(a,t) − L_c)))} \quad (2.12)$$

where F_∞ is the fishing mortality at fully selected length, L_c is the length at 50% gear selection and q describes the steepness of the selectivity curve. The proportion mature $Q(a, t)$ is given by

$$Q(a,t) = \frac{1}{(1 + \exp(p(L(a,t) − L_m)))} \quad (2.13)$$

where L_m is the length at 50% maturity and p describes the steepness of the maturity curve.

Total biomass B, spawner biomass S and yield Y of the population components are given by

$$B_I(t) = \sum_a \alpha L(a, t)^\beta N_I(a, t) \quad (2.14)$$

$$S_I(t) = \sum_a Q(a, t) \alpha L(a, t)^\beta N_I(a, t) \quad (2.15)$$

$$Y_I(t) = \sum_a \alpha L(a, t)^\beta C_I(a, t) \quad (2.16)$$

where α and β are parameters of the length–weight relationship.

Natural juvenile production $\mathcal{J}$ up to the stage at which hatchery fish are released is described as follows. Survival of naturally spawned juveniles to the stage at which hatchery fish are released is given by a Beverton–Holt type survival function s_1 dependent on total larval production:

$$s_1 = \frac{a_1}{1 + b_1 f(S_W + r(S_H + S_S))} \quad (2.17)$$

where f is the larval production per unit of spawner biomass, and r is the reproductive performance of the hatchery and stocked components relative to the wild phenotype

($0 \leqslant r \leqslant 1$). Natural selection is assumed to act during the first juvenile stage, described by transition of a proportion h^2 (heritability) of larvae produced by the hatchery and stocked components to juveniles of the wild component. The numbers of wild and hatchery juveniles $\mathcal{J}_W$ and $\mathcal{J}_H$ are thus given by:

$$\mathcal{J}_W = f s_1(S_W + rh^2(S_H + S_S)) \quad (2.18)$$

$$\mathcal{J}_H = f s_1 r(1 − h^2)(S_H + S_S). \quad (2.19)$$

Survival from release to recruitment is subject to the second Beverton–Holt survival function s_2, dependent on the combined abundance of naturally produced juveniles ($\mathcal{J}_W + \mathcal{J}_H$) and stocked fish R.

$$s_2 = \frac{a_2}{1 + b_2(\mathcal{J}_W + \mathcal{J}_H + R)}. \quad (2.20)$$

Recruitment into the different population components at age 1 is then given by:

$$N_W(1, t) = s_2 \mathcal{J}_W \quad (2.21)$$

$$N_H(1, t) = s_2 \mathcal{J}_H \quad (2.22)$$

$$N_S(1, t) = s_2 R \quad (2.23)$$

where R is the number of hatchery fish released. This formulation allows release at any juvenile size or stage to be represented by a particular combination of before- and after-release survival functions, within the extremes of either function being density independent ($b_1 = 0$ or $b_2 = 0$) and the other accounting for the full extent of compensation.

Although the focus of my analysis is on population dynamics, management decision-making almost inevitably involves making trade-offs between inputs and outcomes measured and valued in different ways, such as fishing effort and release numbers or yield and abundance. Valuing inputs and outcomes in monetary terms and combining them in economic performance indicators allows trade-offs to be considered directly, even though valuation may be difficult in practice. I use two simple indicators of economic performance of different management regimes: the overall resource rent generated at equilibrium, and the net present value of stock rebuilding strategies. In both cases I value hatchery releases, fishing effort, and yield in monetary terms. Assuming that the costs and value are proportional to the number of hatchery fish released, fishing effort and yield, respectively, net benefit (or utilty) U^* at equilibrium is given by

$$U^* = \pi Y^* − \gamma_1 R − \gamma_2 F \quad (2.24)$$

where Y^* is the equilibrium yield at release numbers R and fishing mortality F, π is the ex-vessel price of fish, γ_1 is the unit cost of hatchery fish released, and γ_2 is the cost of generating a unit of fishing mortality. The net present value (NPV) of a management strategy implemented from time $t = 0$ is given by

$$\text{NPV} = \sum_{t=0}^\infty \frac{\pi Y_t − \gamma_1 R_t − \gamma_2 F_t}{(1 + \delta)^t} \quad (2.25)$$

where δ is the discount rate.

3. DYNAMICS AND POTENTIAL OF STOCK ENHANCEMENT

I use the above model to explore key issues in the management of stock enhancements: interactions between fishing and release regimes in long-term enhancement programmes and stock rebuilding, and the implications of releasing hatchery fish that are maladapted due to developmental or genetic factors. Throughout I use biological (yield and abundance of stock components) and economic (net benefit and NPV) criteria, as both sets of criteria are required to understand the potential and implications of enhancement. As a case study, I use North Sea sole (*Solea solea*), a candidate stock for enhancement, with good stock assessment data but as yet no experimental releases.

(a) Case study: North Sea sole

Sole (*Solea solea*) is among the most valuable flatfish in Europe and has long been considered as a candidate species for stock enhancement. Culture technology posed some initial difficulties but is now well developed (Howell 1997), and laboratory experiments have been carried out to assess behavioural attributes of hatchery fish relevant to post-release survival (Ellis *et al.* 1997). No experimental releases of sole have been documented, but experiments with other flatfish such as age-1 turbot (*Psetta maxima*) have demonstrated survival in the wild and numerical recapture rates of 1–11% in the commercial fishery (Stottrup *et al.* 2002).

The North Sea sole stock supports a valuable beam trawl fishery, yielding *ca.* 20 000 t yr^{-1}. The fishery has been routinely monitored and assessed for over 40 years. The stock is considered overfished, with yield marginally below maximum sustainable yield (MSY) but spawner biomass (S) estimated at only 20% of unexploited S (ICES 2003). Recruitment is highly variable but virtually independent of S, implying strong density dependence in pre-recruit mortality. In the recruited stock, growth is strongly density dependent, giving rise to density-dependent age at maturity while length at maturity is approximately constant (Rochet 1998; Lorenzen & Enberg 2002).

While many population parameters can be estimated with a high degree of precision from survey data, there is considerable uncertainty about the true natural mortality rate. Because natural mortality is difficult to estimate, it is common practice to use a reasonable 'guesstimate' in stock assessments. The North Sea sole assessment, upon which most of the parameter values used here are based, assumes a constant $M = 0.1$ yr^{-1} in all recruited age groups (ICES 2003). Estimates of most derived quantities such as stock biomass, recruitment and fishing mortality are conditional on the natural mortality rate assumed. To construct a baseline scenario close to the reported assessment, I use a size-dependent natural mortality of $M_1 = 3$ yr^{-1} cm so that $M = 0.1$ yr^{-1} at $L = 30$ cm. It should be noted, however, that the assumed natural mortality rate is very low compared with the wild population average (Lorenzen 1996b) and direct field measurements of juvenile mortality in sole (Jager *et al.* 1995). Underestimating true natural mortality in assessments leads to conservative exploitation regimes for the capture fishery (Punt 1997), but overestimates

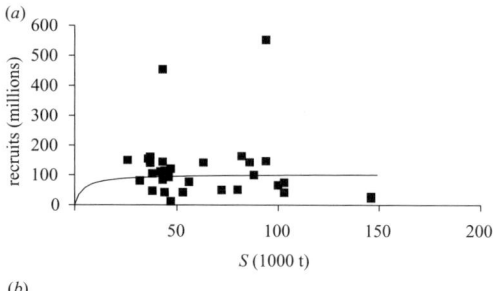

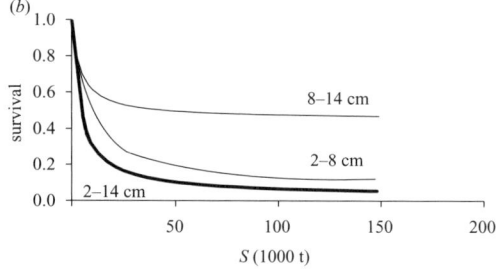

Figure 2. Stock–recruitment relationship and juvenile survival in North Sea sole (1957–1991). (*a*) Observed spawner biomass (S) and subsequent recruits (age 1), and Beverton–Holt stock–recruitment relationship with $a^* = 25\ 500$ t^{-1} and $b^* = 0.000243$ t^{-1}. (*b*) Survival in relation to spawner biomass, predicted by the unpacked stock–recruitment relationship for the full juvenile stage (2–14 cm length) and the sub-stages of 2–8 cm and 8–14 cm (assuming $a_1 = a_2 = 1$). Data from ICES (2003).

potential benefits from enhancement. I briefly explore the implications of different values of M_1 in § 3e.

The relationship between spawner biomass (S) and recruitment (at age 1) in North Sea sole is shown in figure 2*a*. Note that recruitment is virtually independent of S over the observed range, indicating strong density dependence in recruitment, but the relationship is poorly defined for low S. For further analysis, I use a Beverton–Holt type relationship with parameters $a^* = 25\ 500$ t^{-1} and $b^* = 0.000243$ t^{-1} (figure 2*a*). This relationship implies a more gradual increase in recruitment over the range of low S for which no data are available other than the best fitting curve, from which it is not significantly different. In keeping with the stock assessment I assume that recruitment occurs at age 1; hence no density-dependent mortality other than that mediated by growth affects fish aged 1 and older. This may be an overly positive assumption, given that Myers & Cadigan (1993*b*) detected density-dependent mortality in sole up to age 1.5. To unpack the relationship I assume $a_1 = a_2 = 1$; therefore $f = a^*$, and $b^* = a^*(b_1 + b_2)$. The resulting survival rate over the full juvenile period (figure 2*b*) is $s = 0.093$ at spawner biomass $S = 40\ 000$ t, broadly consistent with a field estimate of $s = 0.14$ for juvenile sole from settlement to age 1 (Jager *et al.* 1995). Because growth over the six months from settlement at 2 cm to age 1 at 14 cm is approximately linear, survival s between any lengths L_0 and L_t within this period can be predicted from

$$s = \left(\frac{L_0}{L_t}\right)^{\frac{M_1}{v}} \quad (3.1)$$

where v is the linear length growth rate (Lorenzen 2000). Applying this relationship to the full juvenile period and solving for M_1 gives $M_1 = 29.3$ yr^{-1} cm. Note that this is far higher than the baseline value assumed for the recruited stock, and see § 3f for further discussion. To evaluate release of hatchery fish at an intermediate size of 8 cm, applying equation (3.1) with $M_1 = 29.3$ yr^{-1} cm to the stages from 2 to 8 cm, and from 8 to 14 cm length gives $s_1 = 0.18$ and $s_2 = 0.51$. Together with $a_1 = a_2 = 1$, this implies $b_1 = 4.35 \times 10^{-9}$ and $b_2 = 5.22 \times 10^{-9}$ (equation (2.7)). The resulting stage-specific survival rates are also shown in figure 2b. Note that the constancy of second-stage survival at high S reflects near-constant entry into the second stage due to prior density dependence, rather than absence of density dependence in the second stage.

In the economic assessment I assume an ex-vessel price $\pi = 10$ US\$ kg^{-1} for whole sole, and a cost of $\gamma_1 = 1$ US\$ per piece for 1 year old hatchery fish (Moksness & Stole 1997). The cost of fishing mortality (effort) γ_2 is difficult to estimate, but its precise value is not essential here because my aim is merely to illustrate general trade-offs. For simplicity, I assume that the fishery is currently at its open access equilibrium, i.e. the cost of fishing equals the value of the catch and the resource generates zero rent (Clark 1976). This assumption is arbitrary but not unrealistic: although the North Sea sole fishery is regulated through quotas, the latter assume the character of open-access resources and lead to rent dissipation even if they succeed in conserving the stock. An overview of all parameter values is given in table 1.

(b) Enhancement as a long-term strategy

Key issues in the biological dynamics of enhancement as a long-term strategy for increasing yield concern the effects of releasing fish at different life stages, and trade-offs between harvesting and release regimes. I explore the effects of releasing larvae of 2 cm before, juveniles of 8 cm during, or recruits of 14 cm after juvenile density-dependent mortality, over a wide range of fishing mortality rates. For each life stage released, the numbers of hatchery fish are set to equal the equilibrium numbers of wild fish produced at the same stage given a fishing mortality of $F = 0.6$ yr^{-1}. Equilibrium effects of continuous enhancement on total and naturally recruited yield and spawner biomass are shown in figure 3. The effectiveness of enhancement in terms of raising total yield (figure 3a) increases as more advanced life stages are released. At current levels of fishing mortality ($F = 0.6$ yr^{-1}), increasing the abundance of larvae, juveniles and recruits by 100% raises total yield by 4%, 29% and 81%, respectively. Underlying the differential effects of the same proportional enhancement at different life stages are compensatory responses that differ in their strength and dynamics. Releasing larvae elicits the strongest compensatory response in naturally recruited yield (figure 3b) and spawner biomass (figure 3d), except at low fishing mortality when responses to juvenile and recruit stocking are stronger. Larval releases elicit compensatory responses

mainly through juvenile density-dependent mortality, while releases of recruits elicit growth responses in the recruited stock. Intermediate juvenile stages may elicit strong responses in both juvenile mortality and post-recruit growth, and the combined effect may be stronger than from either larval or recruit releases when exploitation levels are low. Effects on total spawner biomass (figure 3c) mirror those on yield in terms of the relative effectiveness of different life stages. It is striking, however, that the effect of enhancement on spawner biomass is small compared to that of fishing mortality. At the current $F = 0.6$ yr^{-1}, increasing recruits by 100% through enhancement would raise spawner biomass from 44 000 t to 78 000 t, still far below the unexploited spawner biomass of 205 000 t.

Direct and effective (net of compensatory responses in the naturally recruited stock) yield per stocked fish increase with release size (figure 4). Importantly, the two measures also converge as the magnitude of compensatory responses declines with increasing release size. Direct yield per stocked fish as estimated from tag returns can be much higher than effective yield where compensatory processes are strong. Optimizing release size requires assessment of compensatory responses and cannot be based on returns from tagged hatchery fish alone. Density dependence in juvenile mortality is quite ubiquitous and precludes effective enhancement with larval releases except when natural larval production is very low (Secor & Houde 1998). Indeed, rather large juveniles may be required to bypass density-dependent mortality, which is detectable up to age 2.5 in some demersal stocks (Myers & Cadigan 2003b). Even releases of advanced juveniles such as cod yearlings or Pacific salmon smolts have been shown to elicit density-dependent mortality to the extent of complete compensation (Hilborn 1998; Svasand et al. 2000).

It has previously been pointed out that releases of hatchery juveniles will only be effective if regulation in the juvenile phase is either weak (Travis et al. 1998), or can be bypassed by releasing larger juveniles (Hilborn 1999). The current study corroborates this point, but also shows that when enhancement bypasses juvenile density dependence it may face stronger compensatory responses in the recruited stock. It is impossible to evade compensatory responses completely, but it may be possible to develop release and harvesting regimes that provide sufficient net gain in the face of such responses. Quantitative analysis of population dynamics, integrating over the full life cycle and several generations, holds the key to doing this. Field studies testing for displacement of wild by stocked juveniles (e.g. Leber et al. 1995) provide important information, but are not sufficient to establish the full extent of compensation.

Enhancement increases total spawner biomass, but very high levels of enhancement are required to compensate for the reductions in S associated with even moderate levels of fishing mortality. Fishing drastically reduces the proportion of wild and hatchery recruits reaching large size and providing significant reproductive output (see also § 3c). Heppell & Crowder (1998) and Salonen et al. (1998) allude to this problem in the contexts of sea turtle bycatch mortality and biomanipulation. Enhancement is fundamentally an approach to exploitation, allowing increased production while maintaining a high biomass of mostly small and immature fish, but relatively ineffective as

Table 1. Model parameters and their baseline values.
(Population parameter values approximately reflect those of the North Sea sole stock (Lorenzen & Enberg 2002; ICES 2003).)

parameter	baseline value (range)	definition
growth		
$L_{\infty L}$	46 m	asymptotic length at $B \to 0$
K	0.3 yr^{-1}	growth rate
g	4.6×10^{-5} cm t^{-1}	competition coefficient
$L(1)$	14 cm	length at recruitment (age 1)
α	1.0×10^{-8}	coefficient of length–weight relationship
β	3	exponent of length–weight relationship
natural mortality		
$M_{1,W}$	3 yr^{-1} cm	mortality of wild phenotype at $L = 1$ cm
$M_{1,H}$	3 (3–13) yr^{-1} cm	mortality of hatchery phenotype at $L = 1$ cm
b	0.2 (0.1–0.5)	density dependence of juvenile mortality
reproduction		
L_m	26 cm	length at maturity
F	25 500 t^{-1}	juvenile production per unit spawner biomass
p	1	steepness of maturity function
r	1, 0	relative reproductive performance of stocked fish
recruitment		
a_1	1.0 yr^{-1}	survival over first juvenile period at $J \to 0$
b_1	$0, 9.53, 4.35 \times 10^{-9}$	density-dependent parameter
a_2	1.0 yr^{-1}	survival over second juvenile period at $J \to 0$
b_2	$9.53, 0, 5.22 \times 10^{-9}$	density-dependent parameter
fishing		
F_∞	0.6 (0–2) yr^{-1}	fishing effort asymptote
L_c	26 cm	gear selection length
q	1	steepness of selectivity curve
evolution		
h^2	0.2 (0.0–1.0)	heritability of life-history traits
economics		
γ_1	1 (0–2) US$	cost of hatchery fish at age 1 ($L = 14$ cm)
γ_2	330 million US$ yr^{-1}	unit cost of fishing mortality
π	10 000 US$ t^{-1}	ex-vessel price of fish
δ	10%	discount rate

an approach to conserving stocks subject to high mortality on large and mature fish.

(c) **Bio-economics**
Population dynamics theory suggests that release of additional recruits may well allow significant production increases in many stocks, and this is supported by predictions for the North Sea sole case study. The crucial question is under what conditions this would be economically beneficial (Peterman 1991), considering costs of enhancement itself and trade-offs between enhancement and effort regulation. Figure 5 sets out the key considerations and reference points of a basic bio-economic analysis of enhancement, using the North Sea sole example. Point A is the bio-economic open access equilibrium of the non-enhanced fishery, where revenue equals the opportunity costs of fishing. Enhancement as a welfare programme without cost recovery or effective effort restrictions would allow the fishery to expand to a new open access equilibrium B. By contrast, if costs of enhancement were recovered from the fishing sector, for example through a tax, the enhanced open access equilibrium would be at point C. All three open access equilibria are suboptimal in that they imply rent dissipation, albeit to a different degree. The greatest resource rents would be achieved at point D for the non-enhanced, and at point E for the enhanced fishery. Of the two options considered here, enhancement with optimal effort management would generate only marginally higher resource rent than optimal effort management without enhancement. Which of these or other outcomes are considered optimal depends on the economic and social objectives of management. What this simple analysis shows, however, is the importance of considering trade-offs between fisheries regulation and hatchery releases in the overall assessment of enhancement as a management strategy.

Botsford & Hobbs (1984) have shown that optimal fishery policy with enhancement is strongly dependent on costs, prices and biological returns. In figure 6, I analyse optimal policy with respect to resource rent as a function of the price of hatchery fish. If hatchery fish are free or very cheap, the optimal policy would be stocking at over five times the current level of recruitment combined with a fishing mortality of $F = 0.8$ yr^{-1}. This would generate a rent of over 400 million US$, three times the maximum rent obtainable from a pure capture fishery (figure 6b). At the other extreme, no enhancement is feasible at costs of hatchery fish above 1.2 US$ per piece when optimal (with respect to rent) management of the capture fishery would generate a rent of 130 million US$ at a low fishing mortality of $F = 0.2$ yr^{-1}. Total spawner biomass is fairly

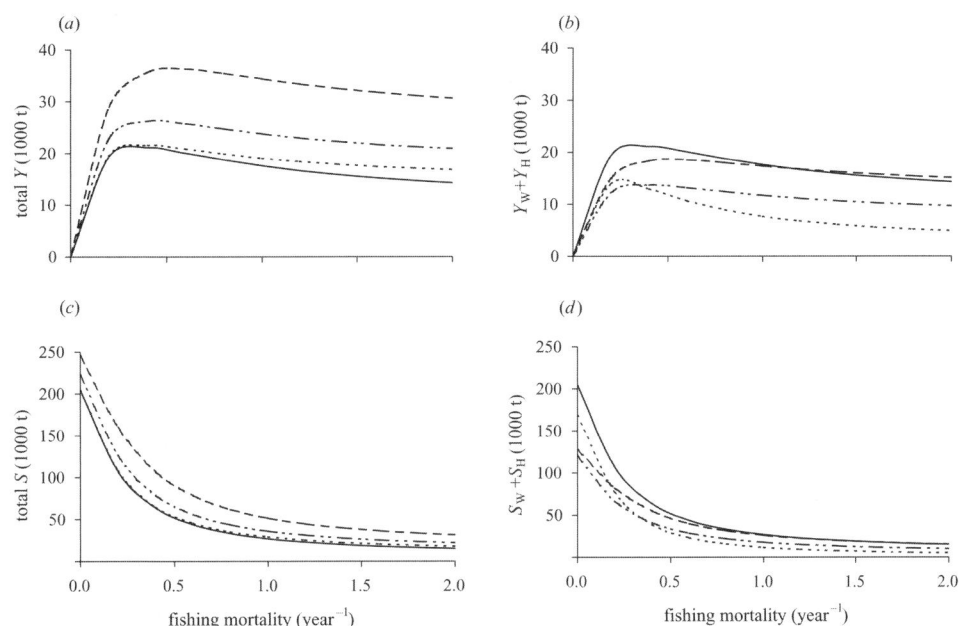

Figure 3. Effect of stock enhancement on total yield (*a*), yield of naturally recruited fish (*b*), total spawner biomass (*c*) and naturally recruited spawner biomass (*d*). Non-enhanced baseline (solid line) and releases of recruits (dashed line), juveniles (dashed-dotted line) and larvae (dotted line). Hatchery releases increase abundance at the stage of release by 100% of the natural level at $F = 0.6$ year^{-1}.

insensitive to hatchery costs and consequent levels of stocking and fishing mortality, but its naturally recruited component (S_W+S_H) is increasingly depressed as costs decline and levels of stocking and fishing mortality increase (figure 6*d*). This illustrates how enhancement can support intensive fisheries while maintaining high population abundance and, thus, key aspects of ecosystem structure and functioning. The trade-off however is that naturally recruited spawners are increasingly replaced with stocked fish. Enhancement can help to reconcile intensive exploitation with certain ecosystem management objectives, but this will be at the expense of the natural component of the target stock.

There are clear trade-offs between production and the conservation of wild stocks in enhancement. When enhancement is biologically effective and stocking costs are low, optimum economic policy may depress the abundance of the naturally recruited stock component even when the concomitant loss of natural production is taken into account. Hence, where wild stock abundance has a value in addition to that of the associated fishery productivity, this must be included explicitly in the economic analysis, and/or direct conservation safeguards need to be introduced in order to maintain an abundant wild stock (at the expense of some production benefit). Hatchery-enhanced stocks can supply many but not all of the production and ecosystem services provided by wild stocks (Holmlund & Hammer 1999). Ecological services provided by juveniles prior to the stage at which hatchery fish are released, the value of fish stocks as indicators of ecological integrity, and the existence value of wild populations are among the attributes at which enhanced stocks will fall short of the value of wild populations. The values attached to different aspects of the enhanced stock will differ between systems (pristine versus highly modified environments, developing versus developed countries), and between stakeholders within systems (fishers versus conservationists). Bio-economic analysis cannot resolve such differences, but it can help greatly to make informed choices.

The analysis presented here remains economically simplistic, but still provides key insights for fisheries policy. It can be extended by integrating the biological models developed here into more sophisticated economic models which, so far, have relied on abstract biological models (Arnason 1991, 2001). It must also be realized that in practice, many of the reference points used in the static bio-economic analysis can only be reached via complex temporal patterns of investment, cost recovery and deliberate effort control.

(d) *Enhancement for stock rebuilding*

What is the potential for enhancement to contribute to rebuilding of spawner biomass in depleted stocks? In the North Sea sole stock, spawner biomass at the present level of exploitation is *ca.* 20% of its unexploited level. Rebuilding spawner biomass to *ca.* 40% is called for by the precautionary approach (ICES 2003) and may also have beneficial ecosystem effects. I therefore explore rebuilding trajectories and evaluate the time needed to rebuild to target biomass and net present value of alternative recovery scenarios with or without enhancement (table 2; figure 7).

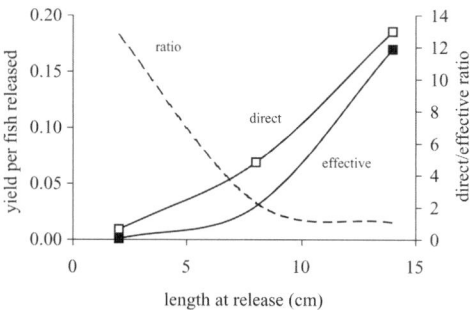

Figure 4. Direct and effective yield per stocked fish as a function of release size. Also shown is the ratio of direct to effective yield, i.e. the factor by which direct returns of tagged fish overestimate their effective contribution to yield. Fishing mortality $F = 0.6 \text{ yr}^{-1}$ and hatchery releases increasing abundance at the stage of release by 100%.

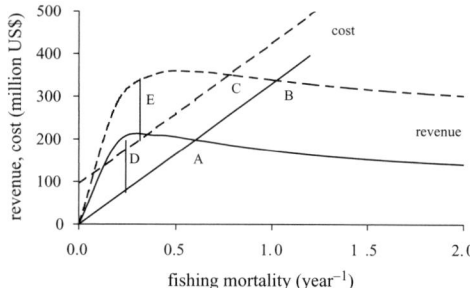

Figure 5. Bio-economic reference points for sole stock enhancement. Revenue and cost curves without enhancement (continuous lines), and for releases increasing natural recruitment by 100% at $F = 0.6 \text{ yr}^{-1}$ (dashed line). Open access equilibria for the baseline case without enhancement (A), enhancement as a welfare programme without cost recovery (B) and as a commercial operation with cost recovery (C). Also shown are the maximum resource rents achievable without (D) and with (E) enhancement.

Simply closing the fishery until target spawner biomass is reached and subsequent harvesting at $F = 0.3 \text{ yr}^{-1}$ (strategy B) achieves rebuilding after only 2 years of closure, and has the highest NPV of all options. The same scenario with temporary enhancement (strategy C) is the second best option, closely followed by reducing exploitation to $F = 0.3 \text{ yr}^{-1}$ without and with enhancement (strategies D and E, not shown in figure 7). The option of enhancing to rebuild spawner biomass before reducing effort (strategy F) avoids temporary yield loss, but foregoes the economic benefits of immediate effort reduction. It delays recovery and has a much lower NPV than the options that involve immediate effort reductions. Reducing effort immediately is far more advantageous in NPV terms than attempting stock enhancement and postponing effort reductions. However, simply closing the fishery will result in effort being redirected elsewhere and/or hardship to fishers, so that combining gradual effort adjustment with compensation and decommissioning programmes may be ecologically and socially advantageous. The contribution of enhancement to rebuilding is likely to be limited in either case, but continuous enhancement could be considered as an alternative to rebuilding the natural spawning stock. The predicted, rapid rebuilding of spawner biomass after effort reduction is based on gains in biomass due to growth and increased survival of already recruited fish. It therefore takes almost immediate effect, while enhancement and increased natural recruitment will become effective only after at least one generation. However, there are situations where enhancement can help to rebuild stocks more quickly than closure of the fishery alone (figure 7c). This is the case principally where stocks have been reduced to such low levels that natural rates of biomass growth are insufficient to achieve rebuilding within one or two generations, or in semelparous species. Enhancement may be particularly beneficial in populations that show depensatory density dependence at low abundance (Liermann & Hilborn 1997; Walters & Kitchell 2001). To be effective in rebuilding stocks from very low abundance, the level of enhancement must be high relative to the natural recruitment capacity of the depleted stock. A high level of enhancement also implies a high level of genetic risks to the target stock, and necessitates careful genetic resource management (see Utter 1998; Miller & Kapuscinski 2003).

Experience with stock rebuilding efforts involving enhancement broadly corroborates the theoretical results obtained here. A retrospective analysis of the successful striped bass (*Morone saxatilis*) stock rebuilding programme in Chesapeake Bay points to a predominant role of effort reduction, and at best a marginal contribution of enhancement (Richards & Rago 1999). Where hatchery releases have played a major role in fisheries restoration, this is typically in the context of bolstering very small, or re-establishing locally extinct populations (Philippart 1995). Overall, this suggests that enhancement is of limited use for rebuilding of overexploited stocks. Any proposals for enhancement as a rebuilding strategy must be carefully evaluated against alternative or additional measures, and the methodology developed here provides the basis for doing this even where data are very limited. Where enhancement may be effective in principle, it must also be considered that developing hatchery production and release protocols and scaling up production to meet requirements for rebuilding large stocks is likely to take years if not decades.

(e) *Maladapted hatchery fish*
Maladaptation of hatchery fish to the natural environment may be reflected in a variety of life-history traits, and be based on developmental and/or genetic factors. I consider increased natural mortality of hatchery fish in the post-recruit phase as an example, using different assumptions on the biological basis of maladaptation and the reproductive competence of hatchery fish. The latter assumptions include release of sterile fish, and releases of reproductively competent fish that either produce wild phenotye offspring (implying that parental maladaptation results from developmental plasticity), or produce maladapted offspring subject to different levels of selection pressure towards the wild (optimum) phenotype. Total yield declines with increasing mortality of hatchery fish under all assumptions, gradually approaching the non-enhanced level (figure 8a). Reproduction of released hatchery fish makes a very slight contri-

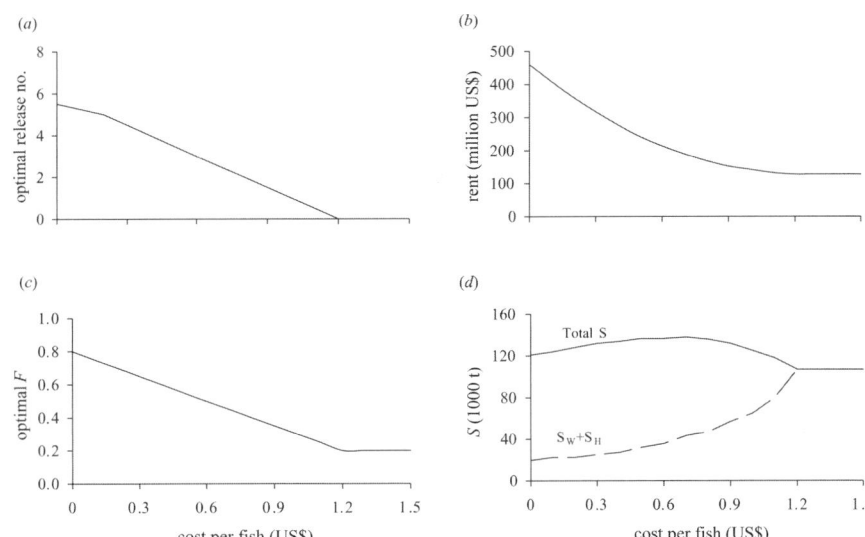

Figure 6. Influence of the unit cost of hatchery fish on optimal fishery policy with respect to resource rent. (a) Optimal release numbers as multiples of natural recruitment at $F = 0.6$ yr^{-1}. (b) Resource rent at optimal release and effort regime. (c) Optimal fishing mortality F. (d) Abundance of total and naturally recruited (wild and hatchery phenotype) spawners.

bution to yield if maladaptation is moderate and arises from developmental plasticity only. If maladaptation has a genetic basis and is perpetuated through reproduction, reproduction depresses equilibrium yield below the level achieved by release of non-reproducing fish. The demographic effect of maladaptation is greatest if the genetically-determined reduction in performance is only moderate (here, a doubling of the base rate of natural mortality). Hatchery releases always depress wild phenotype spawner biomass (figure 8b), and this effect is greatest if maladaptation is weak, genetically based and subject to low heritability. Continuous release of genetically (and phenotypically) fit hatchery fish does not depress productivity because the fish perform as well as their wild conspecifics, but carries a great risk of displacing the wild genotype for precisely the same reason. This is of conservation concern where the hatchery and wild genotypes are not identical. Releasing genetically maladapted individuals reduces yield (figure 8a) but causes less displacement of the wild genotype. Effective yield per released hatchery fish (figure 8c) indicates that, in the case of North Sea sole, no enhancement will be economically viable (produce a yield per hatchery fish above the break-even level of 0.1 kg) if maladaptation causes M_1 to rise above 7 yr^{-1}. If maladaptation is genetically based, the threshold is even lower at approximately $M_1 = 5$ yr^{-1}.

This analysis provides important insights into the genetic risks of enhancement. Continuous release of well-adapted hatchery genotypes is likely to cause introgression to the extent of virtual replacement of the wild genotype, but have no effect on productivity. Moderately maladapted hatchery genotypes pose the greatest combined risk of introgression and loss of productivity. The demographic and genetic impact of poorly adapted genotypes is predicted to be effectively self-limiting, but several caveats are in order. Even poorly adapted genotypes can have significant ecological and genetic effects on wild conspecifics if released in very large numbers, or when maladaptation is manifested only under extreme environmental conditions (Philipp & Whitt 1991). Results are broadly consistent with those obtained by Byrne et al. (1992) and Ford (2002), and illustrate the importance of considering interactions between demographic and genetic processes in the analysis of fisheries enhancement. Due to such interactions, outcomes of enhancement in terms of yield, abundance and the level of introgression are more sensitive to small differences in the performance of released organisms than expected from either demographic or genetic considerations alone.

Releasing sterile fish has the potential of minimizing the risks of both ineffective enhancement if hatchery fish are maladaptated, and displacing the wild genotype if they are not. Moreover, potential benefits from successful reproduction of hatchery fish in the wild are predicted to be small. Release of sterile fish is thus indicated as a management strategy provided they do not compromise the reproductive performance of wild conspecifics, e.g. through behavioural interactions.

(f) *Feasibility of North Sea sole enhancement: conclusions*

While the primary aim of my analysis has been to derive general insights into the dynamics and potential of stock enhancement, it has also provided a preliminary assessment of the potential for enhancing the North Sea sole stock in particular. Overall, results are not encouraging: if the assumptions and parameter values used here are correct, enhancement could be technically effective but would generate only marginal economic benefits. A natural mortality rate more in line with comparative information for other wild stocks (let alone released hatchery fish) would

Table 2. Performance of different options for rebuilding the North Sea sole stock to a spawner biomass of 80 000 t. Economic assumptions as before, i.e. fishery is assumed to be at bio-economic open access equilibrium and cost of seed fish is 1 US$ per juvenile, and discount rate 10%.

strategy	description	time to ($S = 80\ 000$ t)	NPV (million US$)
A	no change ($F = 0.6\ \text{yr}^{-1}$, no enhancement)	∞	0
B	close fishery until target S is reached	2	761
C	close fishery and enhance until target S is reached	2	733
D	reduce exploitation to $F = 0.3\ \text{yr}^{-1}$, no enhancement	10	656
E	reduce exploitation to $F = 0.3\ \text{yr}^{-1}$ and enhance until target S is reached	4	607
F	maintain $F = 0.6\ \text{yr}^{-1}$ and enhance until target S is reached, then set $F = 0.3\ \text{yr}^{-1}$ and discontinue enhancement	10	142

imply far lower returns and all but preclude the prospect of economic feasibility. Any further assessment of enhancement as a management option for North Sea sole should involve a release experiment to assess mortality rates, and estimation of the true costs of hatchery production and fishing. Such new information is easily integrated into the model developed here.

4. IMPLICATIONS AND OUTLOOK

Stock enhancement holds significant potential for raising yields of target stocks where effective hatchery production, release and harvest regimes can be developed. However, both economic and conservation considerations pose strong constraints on the sustainability of enhancements, and only a small subset of technically feasible enhancements will be beneficial overall compared to alternative fisheries management options. Understanding the dynamics of stock enhancements is crucial to identifying such beneficial applications, and the current study provides both general insights in this respect and a methodology for the evaluation of specific systems.

(a) Dynamics and potential of stock enhancement

There appears to be good biological potential for increasing yields through releases of hatchery fish that bypass juvenile density-dependent processes at least partially. Effective enhancement will increase total abundance, but reduce abundance of the naturally recruited component of the stock below its non-enhanced optimum either through compensatory density dependence or through overfishing. The key challenge is thus to design release and harvesting regimes that provide sufficient net returns in the face of compensatory processes acting at all life stages. Whether this is possible at all will depend on specific biological and economic conditions. While enhancement will generally involve negative impacts on the naturally recruited component of the target stock, raising total stock abundance under heavy exploitation may contribute to maintaining structure and functioning of heavily exploited ecosystems.

Despite biological potential, economic benefits of stock enhancement in commercial fisheries will often be marginal or negative given current market prices and post-release performance of hatchery fish. Strong trade-offs exist between the costs of fishing and hatchery releases. Cost and post-release survival of hatchery fish strongly influence optimal policy, which may range from no enhancement at high cost (low survival) to high levels of stocking, to fishing effort and yield at low cost (high survival).

Release of genetically maladapted hatchery fish reduces the effectiveness of enhancement, and is most detrimental overall if fitness is only moderately lower than in the natural population. Releasing sterile fish minimizes risks from maladaptation to both the enhancement programme and the wild stock, provided sterile fish do not interfere with the reproductive performance of wild fish.

As a temporary measure for stock rebuilding, enhancement is beneficial only if the population has been reduced to a very low proportion of its unexploited biomass. Effort restrictions are the most effective short-term measure, and delaying such restrictions in favour of enhancement may incur large economic loss as well as ecological damage. Enhancement may contribute to the rebuilding of overexploited stocks under certain conditions, but cannot substitute for effort restrictions.

These general insights into enhancement dynamics and potential should not substitute for a careful and objective analysis of specific enhancement proposals or programmes. There is no general answer to the question whether stock enhancement is effective or sustainable—it depends on specific circumstances, technology and management, and not least the values that stakeholders attach to outcomes.

(b) Development and management of enhancements

Quantitative assessment of biological and economic outcomes is crucial to the rational evaluation of enhancement and alternative or additional management measures, and should be central to any responsible enhancement programme. The theoretical framework and model developed here provide a powerful and general tool for the evaluation of enhancement programmes, from early planning to full-scale operation. Preliminary assessments (such as the one conducted here for North Sea sole) can and should be carried out before significant investment in experimental research or production facilities, and before any alternative management options are dismissed or delayed in favour of enhancement. Combining population dynamics and bio-economic modelling with participatory planning will promote a broad-based assessment of alternatives, and reduce the influence of unrealistic expectations and partisan views on decisions. Often, preliminary assessments will rule out enhancement as an effective and economically beneficial option. Where this is not the case, further research and development may be justified.

Where available, stock assessments provide information on the values of model parameters pertaining to the wild stock, while release experiments allow the estimation of

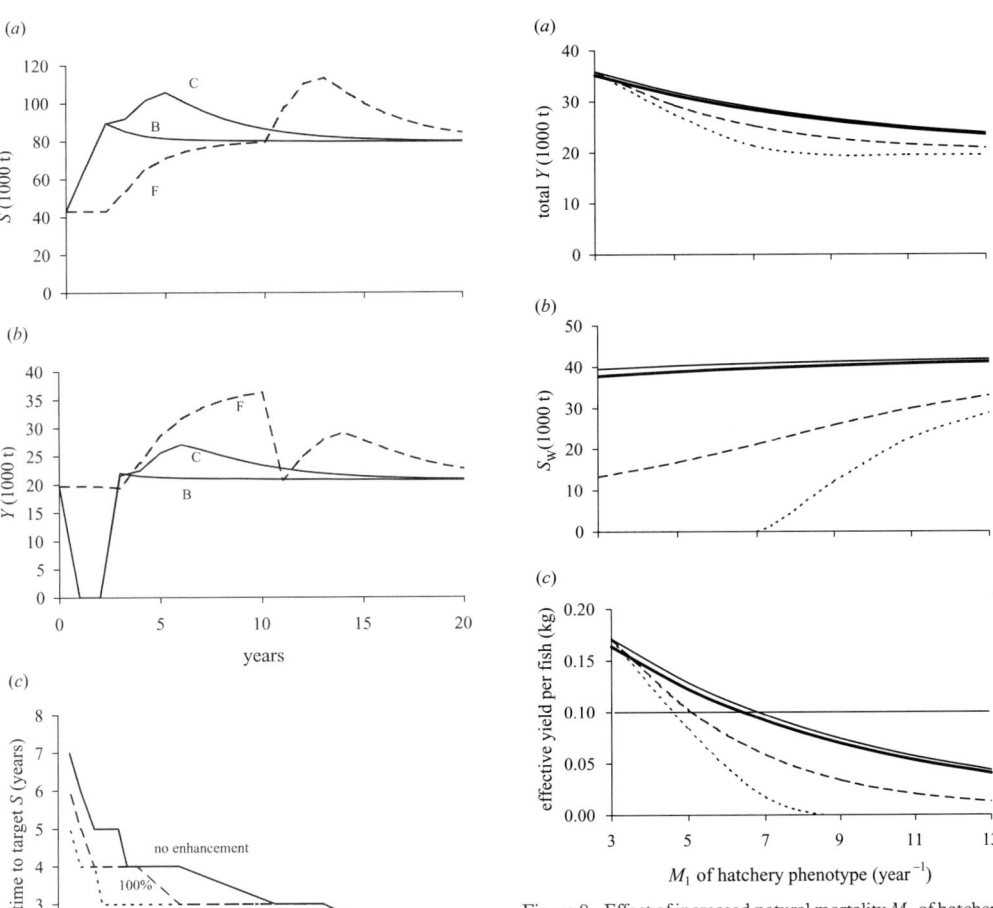

Figure 7. Rebuilding the sole stock to a target spawner biomass S of 80 000 t with and without enhancement. Development of spawner biomass (a) and yield (b) over 20 years following adoption of rebuilding policies B (close fishery until target B is reached), C (close fishery and enhance until target B is reached) and F (enhance until target B is reached, then reduce fishing effort to $F = 0.3 \text{ yr}^{-1}$). See table 2 for further details of rebuilding policies. Also shown (c) is the time needed to rebuild to target S from different levels of initial S (relative to unexploited S) without enhancement and with enhancement at 100% and 500% of natural recruitment at initial S.

Figure 8. Effect of increased natural mortality M_1 of hatchery fish due to developmental or genetic maladaptation on yield (a), abundance of wild phenotype spawners (b), and effective yield per released fish (c). Assumptions: stocked fish do not reproduce (bold solid line), stocked fish reproduce and all offspring are of the wild phenotype (i.e. maladaptation is developmental only, $h^2 = 1.0$; solid line), and stocked fish reproduce and maladaptation is genetically based with heritabilities of $h^2 = 0.2$ (dashed line) and $h^2 = 0.0$ (dotted line). The horizontal line in (c) denotes the economic break-even level of yield per released fish.

others including the mortality rate of stocked fish. Comparative studies can provide invaluable *a priori* information on parameter values including those of the stock–recruitment relationship (Myers 2001), size-dependent mortality in wild and released hatchery fish (Lorenzen 1996a, 2000), density-dependent growth in the recruited phase (Lorenzen & Enberg 2002), and comparative performance of wild and hatchery fish (Fleming & Petersson 2001). Comparative information allows prognostic evaluations to be conducted even in very data-poor situations, exploring alternative management options for a range of scenarios that capture the uncertainty imminent in biological and economic assumptions.

Should a proposed enhancement programme pass the prognostic evaluation and continue to pilot or operational stage, experimental studies will be required to resolve key uncertainties (Leber 1999; Hilborn 2004). Experimental studies must encompass monitoring of the wild, as well as stocked, components of the enhanced population, and be carried out on a large scale to capture the compensatory effects that ultimately determine biological enhancement success (see also Peterman 1991; Hilborn 2004). Population analysis can help to identify the most pertinent uncertainties, and aid the design of experiments to resolve them. Given the dynamic complexity of enhancements and the time-scales involved in responses, experimental approaches alone are unlikely to be efficient in evaluating potential and optimizing release and harvesting regimes. Close integration of population modelling and experimental management is likely to be the most efficient approach to assessing and developing enhancements, and should be a prominent element of planning frameworks (e.g. Cowx 1994; Blankenship & Leber 1995).

(c) *Research*

Further development of the theory presented here is required, in particular with respect to five areas: interaction of size and density-dependent processes throughout the life cycle, proximate basis of life-history plasticity, combined effects of natural selection by hatchery production and fishing, competitive asymmetries between wild and hatchery fish, and community-level interactions. The approach used here is an extension of conventional fisheries stock assessment models, and treats these problems in a separate and largely phenomenological manner (describing measurable responses in population parameters rather than underlying biological processes). However, the emerging evolutionary ecology of fisheries suggests that these aspects are closely connected. Recruitment limitation may arise from natural selection for use of restricted feeding habitats due to predation risk, a multi-species interaction (Walters & Korman 1999). This provides a theoretical framework for linking recruitment to the dynamics of prey and predator species, which in turn may be subject to 'cultivation' effects by the very population whose recruitment is being studied (Walters & Kitchell 2001). An evolutionary perspective will provide a deeper understanding of how the ecological interactions underlying enhancement dynamics arise, and most probably reveal new relationships between key parameters and processes. Fisheries enhancements may provide the most effective, if not the only, way of testing such theories on relevant ecological scales. Hence, enhancement research is likely to make significant contributions to fundamental fisheries ecology.

(d) *Future role of enhancements*

What, if any, role does the future hold for stock enhancement in fisheries management? Conditions for the development of sustainable stock enhancements have never been better than at present. Emerging theory and assessment methodology for stock enhancement will facilitate realistic and quantitative policy analysis, weeding out ineffective or damaging enhancements, identifying new opportunities and optimizing operational systems (this study). Aquaculture technology is increasingly capable of cost-effectively producing fish that perform well in the wild, a crucial precondition for economically viable enhancement (Olla *et al.* 1998; Brown & Dey 2002). Genetic resource management can mitigate, if not fully eliminate, genetic risks to wild populations (Waples 1999; Miller & Kapuscinski 2003). The tendency in many regions of the world to replace open access to fisheries with common or private use rights regimes (Hilborn *et al.* 2003) establishes institutions conducive to investment into fisheries resources, including enhancement approaches. The potential for enhancement to increase productivity and thus reward active stewardship may in itself provide incentives for resource users to cooperate in management, provided they contribute to costs and external institutional arrangements support collective action (Pinkerton 1994; Lorenzen & Garaway 1998). Real prices of fisheries products are high and increasing, as demand will continue to outgrow supply despite a further expansion of aquaculture (Delgado *et al.* 2003).

Even though the general conditions are thus conducive, stock enhancement will remain subject to strong biological, economic and institutional limitations. These arise from natural processes and conditions beyond management control, and inherent difficulties of establishing compatible institutional regimes in larger systems where stakeholders are diverse and often have conflicting interests. Strong public support for conservation of natural aquatic resources makes large-scale manipulations for production ends all but unacceptable. Stock enhancement is therefore likely to remain a niche form of aquatic resource use, dominated in output by both capture fisheries and aquaculture. However, enhancement can make significant contributions to fisheries-related livelihoods where basic biological and economic conditions are met, and help to reconcile intensive exploitation with certain (but not all) ecosystem management objectives. Effective conservation of aquatic resources on a scale beyond individual protected areas and conservation schemes can be achieved only if the burgeoning demand for fisheries products, and the needs of the many people relying on fisheries for all or part of their livelihoods can be satisfied. Where stock enhancement is biologically effective and economically feasible, its environmental and socio-economic impacts may well compare favourably to realistic production and livelihoods alternatives. Research on stock enhancement issues will remain a dynamic and exciting area of fisheries science, and continue to make major contributions to the advancement of fisheries ecology as well as aquaculture science.

I thank the editors for inviting this paper; Ray Hilborn, Marc Mangel, Steve Ralston and Andi Stephens for constructive comments on the manuscript; and Ken Leber for a useful discussion in the course of the work. I carried out this research while on sabbatical leave at the Santa Cruz Center for Stock Assessment Research (CSTAR), and I thank the University of California and the National Marine Fisheries Service for their hospitality and support. The UK Department for International Development, Fisheries Management Science Programme, provided partial financial support.

REFERENCES

Ahmad, I., Bland, S. J. R., Price, C. P. & Kershaw, R. 1998 *Open water stocking in Bangladesh: experiences from the Third Fisheries Project*, pp. 351–370. FAO Fisheries Technical Paper 374.

Arnason, R. 1991 On the external economies of ocean ranching. *ICES Mar. Sci. Symp.* **92**, 218–225.

Arnason, R. 2001 *The economics of ocean ranching: experiences, outlook and theory*. FAO Fisheries Technical Paper 413.

Bartley, D. M., Rana, K. & Immink, A. J. 2001 The use of inter-specific hybrids in aquaculture and fisheries. *Rev. Fish Biol. Fish.* **10**, 325–337.

Beverton, R. J. H. 2002 Fish population biology and fisheries research. In *The Raymond J.H. Beverton lectures at Woods Hole, Massachusetts* (ed. E. D. Anderson), pp. 61–106. US Department of Commerce. NOAA Technical Memo NMFS-F/SPO-54.

Beverton, R. J. H. & Holt, S. J. 1957 *On the dynamics of exploited fish populations*. London: HMSO.

Beyer, J. E. 1989 Recruitment stability and survival—simple size-specific theory with examples from the early life dynamics of marine fish. *Dana* 7, 45–147.

Blankenship, H. L. & Leber, K. M. 1995 A responsible approach to marine stock enhancement. *Am. Fish. Soc. Symp.* 15, 67–175.

Blaxter, J. H. S. 2000 The enhancement of marine fish stocks. *Adv. Mar. Biol.* 38, 1–54.

Botsford, L. W. & Hobbs, R. C. 1984 Optimal fishery policy with artificial enhancement through stocking: California's white sturgeon as an example. *Ecol. Mod.* 23, 293–312.

Boyce, J., Herrmann, M., Bischak, D. & Greenberg, J. 1993 The Alaska salmon enhancement program: a cost/benefit analysis. *Mar. Res. Econ.* 8, 293–312.

Brown, C. & Dey, R. L. 2002 The future of enhancements: lessons for hatchery practice from conservation biology. *Fish Fish.* 3, 79–94.

Burt, A. 1995 Perspective: the evolution of fitness. *Evolution* 49, 1–8.

Byrne, A., Bjorn, T. C. & McIntyre, J. D. 1992 Modeling the response of native steelhead to hatchery supplementation programs in an Idaho river. *N. Am. J. Fish. Mngmt* 12, 62–78.

Clark, C. W. 1976 *Mathematical bioeconomics: the optimal management of renewable resources*. New York: Wiley.

Conover, D. O. & Munch, S. 2002 Sustaining fisheries on evolutionary time scales. *Science* 297, 94–96.

Cowx, I. G. 1994 Stocking strategies. *Fish. Mngmt Ecol.* 1, 15–31.

Cuenco, M. L. 1994 A model of an internally supplemented population. *Trans. Am. Fish. Soc.* 123, 277–288.

Delgado, C. L., Wada, N., Rosegrant, M. W., Meijer, S. & Ahmed, M. 2003 *Fish to 2020: supply and demand in changing global markets*. Washington, DC: International Food Policy Research Institute.

Elliott, J. M. 1994 *Quantitative ecology and the brown trout*. Oxford University Press.

Ellis, T., Howell, B. R. & Hughes, R. N. 1997 The cryptic responses of hatchery-reared sole to a natural sand substratum. *J. Fish Biol.* 51, 389–401.

Fleming, I. A. & Petersson, E. 2001 The ability of released, hatchery salmonids to breed and contribute to the natural productivity of wild populations. *Nord. J. Freshwat. Res.* 75, 71–98.

Ford, M. J. 2002 Selection in captivity during supportive breeding may reduce fitness in the wild. *Conserv. Biol.* 16, 815–825.

Giske, J. & Salvanes, A. G. V. 1999 A model of enhancement potentials in open ecosystems. In *Stock enhancement and sea ranching* (ed. B. R. Howell, E. Moksness & T. Svasand), pp. 22–36. Oxford: Fishing News Books.

Heppell, S. S. & Crowder, L. B. 1998 Prognostic evaluation of enhancement programs using population models and life history analysis. *Bull. Mar. Sci.* 62, 495–507.

Hilborn, R. 1998 The economic performance of marine stock enhancement projects. *Bull. Mar. Sci.* 62, 661–674.

Hilborn, R. 1999 Confessions of a reformed hatchery basher. *Fisheries* 24, 30–31.

Hilborn, R. 2004 Population management in stock enhancement and sea ranching. In *Stock Enhancement and sea ranching: developments, pitfalls and opportunities* (ed. K. M. Leber, S. Kitada, H. L. Blankenship & T. Svasand), pp. 201–209. Oxford: Blackwell Publishing.

Hilborn, R. & Winton, J. 1993 Learning to enhance salmon production: lessons from the Salmonid Enhancement Program. *Can. J. Fish. Aquat. Sci.* 50, 2043–2056.

Hilborn, R., Branch, T. A., Ernst, B., Magnusson, A., Minte-Vera, C. V., Scheuerell, M. D. & Valero, J. L. 2003 State of the world's fisheries. *A. Rev. Environ. Resources* 28, 359–399.

Hixon, M. A., Pacala, S. W. & Sandin, S. A. 2002 Population regulation: historical context and contemporary challenges of open vs. closed systems. *Ecology* 83, 1490–1508.

Holmlund, C. M. & Hammer, M. 1999 Ecosystem services generated by fish populations. *Ecol. Econ.* 29, 253–268.

Howell, B. R. 1997 A re-appraisal of the potential of the sole, *Solea solea* (L.) for commercial cultivation. *Aquaculture* 155, 355–365.

ICES 2003 Report of the working group on the assessment of demersal stocks in the North Sea and Skagerrak. ICES CM 2003/ACFM:02.

Jackson, J. B. C. 2001 What is natural in the coastal oceans? *Proc. Natl Acad. Sci. USA* 98, 5411–5418.

Jager, Z., Kleef, H. L. & Tydeman, P. 1995 Mortality and growth of 0-group flatfish in the brackish Dollart (Ems estuary, Wadden Sea). *Neth. J. Sea Res.* 34, 119–129.

Jonasson, J., Gjedre, B. & Gjedrem, T. 1997 Genetic parameters for return rate and body weight in sea-ranched Atlantic salmon. *Aquaculture* 154, 219–231.

Jonsson, S., Brannas, E. & Lundqvist, H. 1999 Stocking of brown trout, *Salmo trutta* L.: effects of acclimatization. *Fish. Mngmt Ecol.* 6, 459–473.

Leber, K. M. 1999 Rationale for an experimental approach to stock enhancement. In *Stock enhancement and sea ranching* (ed. B. R. Howell, E. Moksness & T. Svasand), pp. 63–75. Oxford: Fishing News Books.

Leber, K. M. 2001 Advances in marine stock enhancement: shifting emphasis to theory and accountability. In *Responsible marine aquaculture* (ed. R. R. Stickney & P. J. McVey), pp. 79–90. Wallingford: CABI Publishing.

Leber, K. M., Brennan, N. P. & Arce, S. M. 1995 Marine enhancement with striped mullet: are hatchery releases replenishing or displacing wild stocks? *Am. Fish. Soc. Symp.* 15, 376–387.

Leggett, W. C. & DeBlois, E. 1994 Recruitment in marine fish: is it regulated by starvation and predation in egg and larval stages. *Neth. J. Sea Res.* 32, 119–134.

Levin, P. S., Zabel, R. W. & Williams, J. G. 2001 The road to extinction is paved with good intentions: negative association of fish hatcheries with threatened salmon. *Proc. R. Soc. B* 268, 1153–1158. (doi:10.1098/rspb.2001.1634)

Liermann, M. & Hilborn, R. 1997 Depensation in fish stocks: a hierarchic Bayesian meta-analysis. *Can. J. Fish. Aquat. Sci.* 54, 1976–1985.

Lorenzen, K. 1995 Population dynamics and management of culture-based fisheries. *Fish. Mngmt Ecol.* 2, 61–73.

Lorenzen, K. 1996a A simple von Bertalanffy model for density-dependent growth in extensive aquaculture, with an application to common carp (*Cyprinus carpio*). *Aquaculture* 142, 191–205.

Lorenzen, K. 1996b The relationship between body weight and natural mortality in fish: a comparison of natural ecosystems and aquaculture. *J. Fish Biol.* 49, 627–647.

Lorenzen, K. 2000 Allometry of natural mortality as a basis for assessing optimal release size in fish stocking programmes. *Can. J. Fish. Aquat. Sci.* 57, 2374–2381.

Lorenzen, K. & Enberg, K. 2002 Density-dependent growth as a key mechanism in the regulation of fish populations: evidence from among-population comparisons. *Proc. R. Soc. B* **269**, 49–54. (doi:10.1098/rspb.2001.1853)

Lorenzen, K. & Garaway C. J. 1998 How predictable is the outcome of stocking? pp. 133–152. FAO Fisheries Technical Paper 374.

Lorenzen, K., Xu, G., Cao, F., Ye, J. & Hu, T. 1997 Analysing extensive fish culture systems by transparent population modelling: bighead carp, *Aristichthys nobilis* (Richardson 1845), culture in a Chinese reservoir. *Aquacult. Res.* **28**, 867–880.

Lorenzen, K., Juntana, J., Bundit, J. & Tourongruang, D. 1998 Assessing culture fisheries practices in small water bodies: a study of village fisheries in northeast Thailand. *Aquacult. Res.* **29**, 211–224.

Lorenzen, K. (and 12 others) 2001 Strategic review of enhancements and culture-based fisheries. In *Aquaculture in the third millennium* (ed. R. P. Subasinghe, P. Bueno, M. J. Phillips, C. Hugh & S. E. McGladdery), pp. 221–237. Rome: FAO.

MacCall, A. D. 1989 Against marine fish hatcheries: ironies of fisheries politics in the technological era. *CalCOFI Rep.* **30**, 46–48.

McGurk, M. D. 1986 Natural mortality of marine pelagic fish eggs and larvae: the role of spatial patchiness. *Mar. Ecol. Progr. Ser.* **37**, 227–242.

Marshall, C. T. & Frank, K. T. 1999 Implications of density-dependent juvenile growth for compensatory recruitment regulation of haddock. *Can. J. Fish. Aquat. Sci.* **56**, 356–363.

Mehner, T., Benndorf, J., Kasprzak, P. & Koschel, R. 2002 Biomanipulation of lake ecosystems: successful applications and expanding complexity in the underlying science. *Freshwat. Biol.* **47**, 2453–2456.

Miller, L. M. & Kapuscinski, A. R. 2003 Genetic guidelines for hatchery supplementation programs. In *Population genetics: principles and applications for fisheries scientists* (ed. E. M. Hallerman), pp. 329–355. Bethesda, MD: American Fisheries Society.

Moksness, E. & Stole, R. 1997 Larviculture of marine fish for sea ranching purposes: is it profitable? *Aquaculture* **155**, 341–353.

Mousseau, T. A. & Roff, D. A. 1987 Natural selection and the heritability of fitness components. *Heredity* **59**, 181–197.

Munro, J. L. & Bell, J. D. 1997 Enhancement of marine fisheries resources. *Rev. Fish. Sci.* **5**, 185–222.

Myers, R. A. 2001 Stock and recruitment: generalizations about maximum reproductive rate, density-dependence and variability using meta-analytic approaches. *ICES J. Mar. Sci.* **58**, 937–951.

Myers, R. A. & Cadigan, N. G. 1993a Is juvenile natural mortality in marine demersal fish variable? *Can. J. Fish. Aquat. Sci.* **50**, 1591–1598.

Myers, R. A. & Cadigan, N. G. 1993b Density-dependent juvenile mortality in marine demersal fish. *Can. J. Fish. Aquat. Sci.* **50**, 1576–1590.

Myers, R. A., Blanchard, W. & Thompson, K. T. 1990 Summary of North Atlantic fish recruitment 1942–1987. *Can. Tech. Rep. Fish. Aquat. Sci.* **1743**.

Olla, B. L., Davis, M. W. & Ryer, C. H. 1998 Understanding how the hatchery environment represses or promotes the development of behavioral survival skills. *Bull. Mar. Sci.* **62**, 531–550.

Pauly, D., Christensen, V., Dalsgaard, J., Froese, R. & Torres, F. 1998 Fishing down marine food webs. *Science* **279**, 860–863.

Pauly, D., Christensen, V., Guenette, S., Pitcher, T. J., Sumaila, U. R., Walters, C. J., Watson, R. & Zeller, D. 2002 Towards sustainability in world fisheries. *Nature* **418**, 689–695.

Peterman, R. M. 1991 Density-dependent marine processes in North Pacific salmonids: lessons for experimental design of large-scale manipulations of fish stocks. *ICES Mar. Sci. Symp.* **92**, 69–77.

Philipp, D. P. & Whitt, G. S. 1991 Survival and growth of northern, Florida and reciprocal F_1 hybrid largemouth bass in central Illinois. *Trans. Am. Fish. Soc.* **120**, 58–64.

Philippart, J. C. 1995 Is captive breeding an effective solution for the preservation of endemic species? *Biol. Conserv.* **72**, 281–295.

Pinkerton, E. 1994 Economic and management benefits from the coordination of capture and culture fisheries: the case of Prince William Sound pink salmon. *N. Am. J. Fish. Mngmt* **14**, 262–277.

Polovina, J. J. 1990 Evaluation of hatchery releases of juveniles to enhance rockfish stocks, with an application to Pacific Ocean perch *Sebastes alutus*. *Fish. B.-NOAA* **89**, 129–136.

Post, J. R., Parkinson, E. A. & Johnston, N. T. 1999 Density-dependent processes in structured fish populations: interaction strengths in whole-lake experiments. *Ecol. Monogr.* **69**, 155–175.

Price, E. O. 2002 *Animal domestication and behavior*. Wallingford: CABI Publishing.

Punt, A. E. 1997 The performance of VPA-based management. *Fish. Res.* **29**, 217–249.

Reisenbichler, R. R. & Rubin, S. P. 1999 Genetic changes from artificial propagation of Pacific salmon affect the productivity and viability of supplemented populations. *ICES J. Mar. Sci.* **56**, 459–466.

Richards, R. A. & Rago, P. J. 1999 A case history of effective fishery management: Chesapeake Bay striped bass. *N. Am. J. Fish. Mngmt* **19**, 356–375.

Rijnsdorp, A. D., Daan, N., van Beek, F. A. & Heessen, H. J. L. 1991 Reproductive variability in North Sea plaice, sole and cod. *J. Cons. Int. Explor. Mer* **47**, 352–357.

Rochet, M. J. 1998 Short-term effects of fishing on life history traits of fish. *ICES J. Mar. Sci.* **55**, 371–391.

Rochet, M. J., Cornillon, P. A., Sabatier, R. & Pontier, D. 2000 Comparative analysis of phylogenetic and fishing effects in life history patterns of teleost fish. *Oikos* **91**, 255–270.

Rose, K. A., Cowan, J. H., Winemiller, K. O., Myers, R. A. & Hilborn, R. 2001 Compensatory density-dependence in fish populations: importance, controversy, understanding and prognosis. *Fish Fish.* **2**, 293–327.

Rothschild, B. J. 2000 'Fish stocks and recruitment': the past thirty years. *ICES J. Mar. Sci.* **57**, 191–201.

Salonen, S., Helminen, H. & Sarvala, J. 1998 Compatibility of recreational fisheries and ecological lake restoration in pikeperch (*Stizostedion lucioperca* L.) management in Lake Koyliojaervi, SW Finland. In *Recreational fisheries: social, economic and management aspects* (ed. P. Hickley & H. Thompkins), pp. 80–87. Oxford: Fishing News Books.

Secor, D. H. & Houde, E. D. 1998 Use of larval stocking in restoration of Chesapeake Bay striped bass. *ICES J. Mar. Sci.* **55**, 228–239.

Shepherd, J. G. & Cushing, D. H. 1980 A mechanisms for density-dependent survival of larval fish as the basis of a stock–recruitment relationship. *J. Cons. Int. Explor. Mer* **39**, 160–167.

Smith, T. D., Gjørsæter, J., Stenseth, N. C., Kittilsen, M. O., Danielssen, D. S., Solemdal, P. & Tveite, S. 2002 A century of manipulating recruitment in coastal cod populations: the Flødevigen experience. *ICES Mar. Sci. Symp.* **215**, 402–415.

Stokes, T.K. & McGlade, J.M., Law, R. 1993 In *The exploitation of evolving resources.* New York: Springer.

Stottrup, J. G., Sparrevohn, C. R., Modin, J. & Lehmann, K. 2002 The use of releases of reared fish to enhance natural populations of turbot *Psetta maxima* (Linne, 1758). *Fish. Res.* **59**, 161–180.

Svasand, T., Kristiansen, T. S., Pedersen, T., Salvanes, A. G. V., Engelsen, R., Naevdal, G. & Nodtvedt, M. 2000 The enhancement of cod stocks. *Fish Fish.* **1**, 173–205.

Taylor, J. E. 1999 *Making salmon.* Seattle, WA: University of Washington Press.

Thorpe, J. E., Mangel, M., Metcalfe, N. B. & Huntingford, F. A. 1998 Modelling the proximate basis of salmonid life-history variation, with application to Atlantic salmon, *Salmo salar* L. *Evol. Ecol.* **12**, 581–599.

Travis, J., Coleman, F. C., Grimes, C. B., Conover, D., Bert, T. M. & Tringali, M. 1998 Critically assessing stock enhancement: an introduction to the Mote Symposium. *Bull. Mar. Sci.* **62**, 305–311.

Utter, F. 1998 Genetic problems of hatchery-reared progeny released into the wild, and how to deal with them. *Bull. Mar. Sci.* **62**, 623–640.

Van der Veer, H. W. 1986 Immigration, settlement and density-dependent mortality of a larval and early post-larval 0-group (*Pleuronectes platessa*) population in the western Wadden Sea. *Mar. Ecol. Progr. Ser.* **29**, 223–236.

Walter, E. 1934 Grundlagen der allgemeinen fischereilichen Produktionslehre, einschliesslich ihrer Anwendung auf die Fuetterung. *Handb. Binnenfisch. Mitteleuropas* **4**, 481–662.

Walters, C. & Kitchell, J. F. 2001 Cultivation/depensation effects on juvenile survival and recruitment: implications for the theory of fishing. *Can. J. Fish. Aquat. Sci.* **57**, 39–50.

Walters, C. & Korman, J. 1999 Linking recruitment to trophic factors: revisiting the Beverton–Holt recruitment model from a life history and multispecies perspective. *Rev. Fish Biol. Fish.* **9**, 187–202.

Walters, C. J. & Post, J. R. 1993 Density-dependent growth and competitive asymmetries in size-structured fish populations: a theoretical model and recommendations for field experiments *Trans. Am. Fish. Soc.* **122**, 34–45.

Waples, R. S. 1999 Dispelling some myths about hatcheries. *Fisheries* **24**, 12–21.

Weber, E. D. & Fausch, K. D. 2003 Interactions between hatchery and wild salmonids in streams: differences in biology and evidence for competition. *Can. J. Fish. Aquat. Sci.* **60**, 1018–1036.

Welcomme, R. L. & Bartley, D. M. 1998 Current approaches to the enhancement of fisheries. *Fish. Mngmt Ecol.* **5**, 351–382.

GLOSSARY

MSY: maximum sustainable yield
NPV: net present value

UNDERSTANDING HOW THE HATCHERY ENVIRONMENT REPRESSES OR PROMOTES THE DEVELOPMENT OF BEHAVIORAL SURVIVAL SKILLS

Bori L. Olla, Michael W. Davis and Clifford H. Ryer

ABSTRACT

Although great strides have been made in the development of techniques for rearing marine fish, methods for improving survival capabilities following release of these fish into the natural environment have not kept pace. Most responsible for impeding progress has been the lack of adequate knowledge of the complex interactions between a species and the myriad physical and biological factors a hatchery fish faces upon release. Being reared in the psychosensorily deprived environment of a hatchery may lessen the innate capability of fish to carry out the basic survival strategy of all fish: to eat and not be eaten. In this synthesis we focus on some of the key behaviors that play a role in predator avoidance and food acquisition and on how the rearing environment may affect the expression of these behaviors. One of the major causes of mortality in hatchery-reared fish is predation, much of which occurs shortly after release. Available evidence seems to indicate that antipredator behavior in hatchery-reared fish is not fully developed. Possible deficits may also occur in feeding after release, when fish are faced with the shift from hatchery-supplied foods to the capture of live prey. Added to possible hatchery-induced deficits in behavior are the alterations in behavior that may result from the stress of handling and transportation. Behavioral capabilities in hatchery-reared fish can be improved in a number of ways, including exposure to predators or predatory stimuli, alteration of spatial and temporal distribution of food, mitigation of rearing and transport stress, and control of the social environment.

Habitat degradation, environmental pollution, and overexploitation have in one way or another contributed to the reduction of fishery populations worldwide. The global marine catch is presently at more than 80 million mt (FAO, 1993), and future increases are improbable, especially of critically overexploited inshore stocks (NOAA, 1995). One possible way to rejuvenate or enhance depleted stocks is to rear large numbers of fish under artificial conditions and then release them into the natural environment. This is not a new approach, having been attempted in Norway and the USA during the late 19th century and in Japan during the early 20th century (for reviews, see Solemdal et al., 1984; Kitada et al., 1992; Edwards and Nickum, 1995), but today as then, population enhancement of marine fish species has had minimal success. Although great strides have been made in the development of techniques for successfully rearing marine fish, methods for successful outplanting have not kept pace. Impeding progress has been inadequate knowledge of the complex interactions between a species and the physical and biological environment into which it is to be released. Without an understanding of the dynamics of those interactions, attempts at enhancement have, in many cases, been more like playing roulette than practicing good science.

An effective enhancement strategy for a given species must be based on a detailed understanding of the natural sequence of recruitment events and habitat associations that characterize its life history (Fig. 1). Most marine fishes undergo a succession of recruitment events as they grow, develop, and change their relationship with their environment.

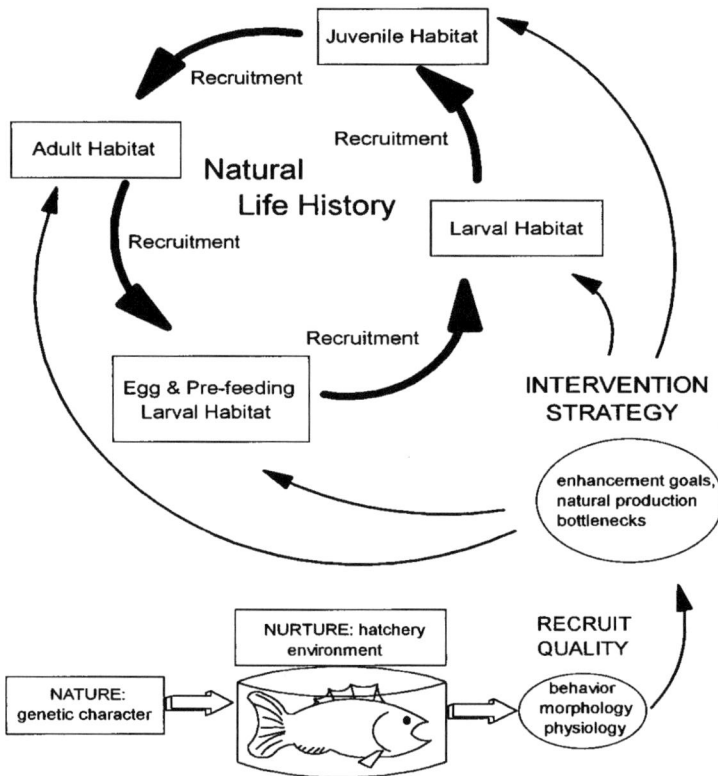

Figure 1. Enhancement of natural stocks with hatchery-reared fish. Fish are reared in a hatchery, subject to influences from nature (genetic character) and nurture (hatchery environment). The quality of the resulting fish is measured against behavioral, morphological, and physiological criteria. Fish are introduced into the wild according to a rational intervention strategy that considers enhancement goals and natural production bottlenecks of wild stocks. Wild stocks undergo a series of recruitment stages moving from egg and prefeeding larval habitat to feeding larval habitat to juvenile habitat to adult habitat.

At any point in this progression, a population may be limited by recruitment or by habitat availability. When the factors that have caused the "bottleneck" in population growth are identified, then an appropriate intervention strategy, e.g., enhancement, can be formulated. For example, it would be useless to enhance a population with larvae when juvenile habitat limits adult population size. Equally important to the success of outplanting is to ensure that released fish are of a quality able to meet the challenges of the natural environment and do not cause degradation of wild stocks. By quality we refer to such characteristics as genetics and behavioral capabilities.

Efforts at outplanting should ensure that long-term cultivation of fish does not lead to a genetic divergence from environmentally adapted wild phenotypic norms (Nævdal, 1994; Blankenship and Leber, 1995; Busack and Currens, 1995). For example, changes in morphological traits of hatchery fish that could influence environmental adaption and sur-

Table 1. Possible behavioral deficits in fish that are naive to predators and predation and possible remedial actions for these deficits.

Possible deficits	Possible remedial actions
—Predator recognition, predator evasion strategies, vigilance	—Exposure to predators and predatory stimuli
—Hyperaggression	—Alteration of fish density and spatial and temporal distribution of food
—Behavioral decrements induced by stress	—Mitigation of rearing and transport stress
—Decrease or absence of schooling in facultatively schooling species	—Alteration of fish density and spatial and temporal distribution of food
—Appropriate use of habitat refugia	—Availability of habitat refugia depending on habitat requirements; provision of habitat refugia within hatchery where appropriate
—Activity rhythms	—Provision of daily and seasonal cycles of conditions that mimic natural conditions to be faced upon release

vival, such as body size and shape, can occur relatively quickly, e.g., within 23 yrs (Petersson et al., 1996). Genetic selection under hatchery conditions can also influence, in addition to morphological traits, the expression of behavioral traits, leading to differences between wild and domestic stocks in, e.g., timidness (Vincent, 1960), surface response (Vincent, 1960; Moyle, 1969), and aggressiveness (Swain and Riddell, 1990). Traits that are most adaptive for life in a hatchery environment and readily selected for may be much less adaptive for the challenges that fish must face when released into the natural environment (Blaxter, 1976).

Fish that have been reared in the psychosensorily deprived environment of the hatchery may be less able to carry out the basic survival strategy of all fish: to eat and not be eaten (Olla et al., 1995). What may be most adaptive in a hatchery tank or pond may have little utility in the natural environment. Our intent in this work is to focus on some of the key behaviors that play a role in postrelease survival of hatchery-reared fish and to examine how the rearing environment affects the expression of these behaviors. We will draw examples from the published literature, strongly emphasizing our own work.

Predation

RECOGNIZING AND RESPONDING TO PREDATORS.—One of the major causes of mortality in hatchery-reared salmonids is predation, much of which occurs shortly after release (see for example Fresh et al., 1982; Healey, 1982; Mace, 1983; Fisher and Pearcy, 1988). Although predator recognition and evasion have a strong innate component, there remains the question of whether these behaviors can be fully expressed in a rearing environment that is free from the threat of predation. Available evidence seems to suggest that antipredator behavior in hatchery-reared fish is not fully developed (Table 1). For example, the ability of hatchery-reared coho salmon, *Oncorhynchus kisutch,* to evade predators can be markedly improved by exposure to predators before release (Olla and Davis, 1989; see also Ginetz and Larkin, 1976, for sockeye salmon, *Oncorhynchus nerka*). Even in the absence of live predators, exposing coho smolts to visual, chemical, and tactile stimuli associated with predation resulted in a rapid decrease in predator-induced mortal-

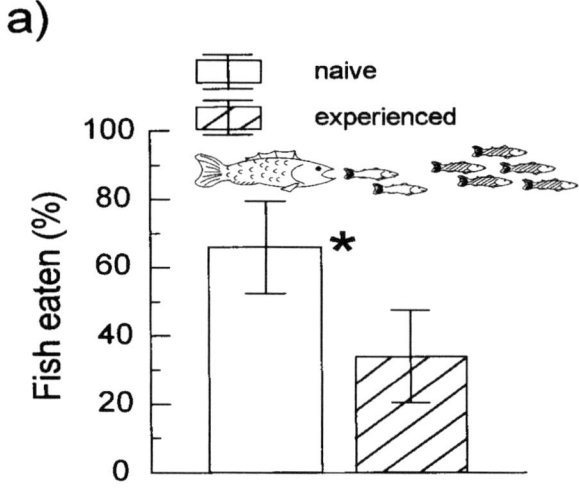

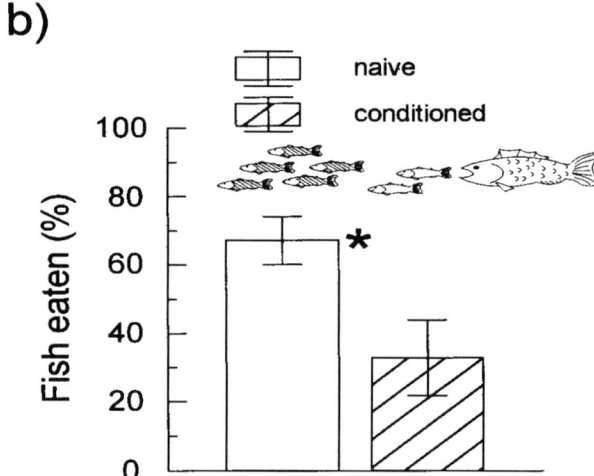

Figure 2. Predation (percent eaten) for juvenile coho salmon that were (a) experienced as survivors of a previous predation bout or (b) conditioned by exposure to predator stimuli. Values are mean ± SE for naive and conditioned fish; *, $P < 0.05$. Modified from Olla and Davis (1989).

ity. The rapidity with which fish acquired antipredator skills (Olla and Davis, 1989; Fig. 2A,B) would speak to the potential success of applying antipredator "training" methods to hatchery fish.

Although it seems clear that fish reared under artificial conditions may not develop their full potential for antipredator behavior, the precise nature of the behavioral deficits is not well understood. Wild fish are able to recognize predators and in some cases are able to discern predator motivation as well (Helfman, 1989). Artificially propagated fish naive to predators may lack this capability or may be unable to distinguish between predatory and nonpredatory species, with the result that they misdirect their vigilance. Furthermore, the disorientation associated with transfer to a natural but novel habitat may further erode vigilance, predator recognition, and the ability to respond behaviorally to threat.

EFFECTS OF SOCIAL INTERACTION ON PREDATOR RECOGNITION AND AVOIDANCE.—Fish engage in many forms of social behavior that directly influence their feeding success. For gregarious species that move about in groups, a principal adaptive advantage of schooling behavior is to reduce the probability of mortality by predation (Pitcher and Parrish, 1993). Among such species, studies have demonstrated that solitary fish are more vulnerable to predators than are fish in schools (Fig. 3A; Neill and Cullen, 1974; Landeau and Terborgh, 1986). Schooling fish have been shown to modify their behavior in the presence of predators, forming more cohesive groups (Fig. 3B; Morgan, 1988; Ryer and Olla, 1996a, unpub. data). For species in which schooling is part of the behavioral repertoire, spatial distribution relative to one another following release into the natural environment may have an impact on vulnerability to predation, especially important at and shortly after release, when fish are probably most vulnerable to predators.

In obligate schoolers, we would assume that innate attraction would override responses to most other factors and that fish would tend to school upon release, but for species that school facultatively, although predisposition to school may be innate, the hatchery environment could play an important role in determining whether schooling behavior is preeminent upon release. Recent evidence offers support for this contention. Manipulation of the spatial and temporal distribution of food can make juvenile walleye pollock, *Theragra chalcogramma,* either act socially and function as a school or act more as individuals and perform activities spatially separated from one another (Fig. 3C; Ryer and Olla, 1992, 1995a). When conditioned to feed on food that was widely dispersed, juvenile walleye pollock separated spatially and foraged as individuals, unresponsive to the behavior of species mates, but when conditioned to food distributed in clumps and presented at random time intervals, fish responded by adopting a group foraging strategy, observing and reacting to the food-discovery behavior of others and showing cohesive, socially interactive schooling behavior. This behavior persisted for up to 7 d even after food was shifted from a clumped to a dispersed distribution. These observations lead us to suggest that, depending on a species' innate behavioral patterns, conditioning to school may offer another avenue for improving the persistence of antipredator schooling behavior and increase the potential survival after release.

Simply being in the presence of conspecifics that have survived predation (Patten, 1977) can also improve antipredator behavior in naive fish (Fig. 4). Because many species of fish are social, they have evolved abilities to interpret conspecific behavior. Through either passive or active information transfer (Pitcher, 1986), individuals in a group are able to share information about their environment (Ryer and Olla, 1991a; Magurran and Higham, 1988), facilitating learning by naive fish of appropriate responses to predatory threat. When naive coho fry were placed with fry that had previously experienced predation, naive fish were better able to avoid predation (Patten, 1977). Similarly, zebra danio, *Brachydanio rerio,* learned to associate antipredator behavior with a novel stimulus through association with trained fish (Suboski et al., 1990).

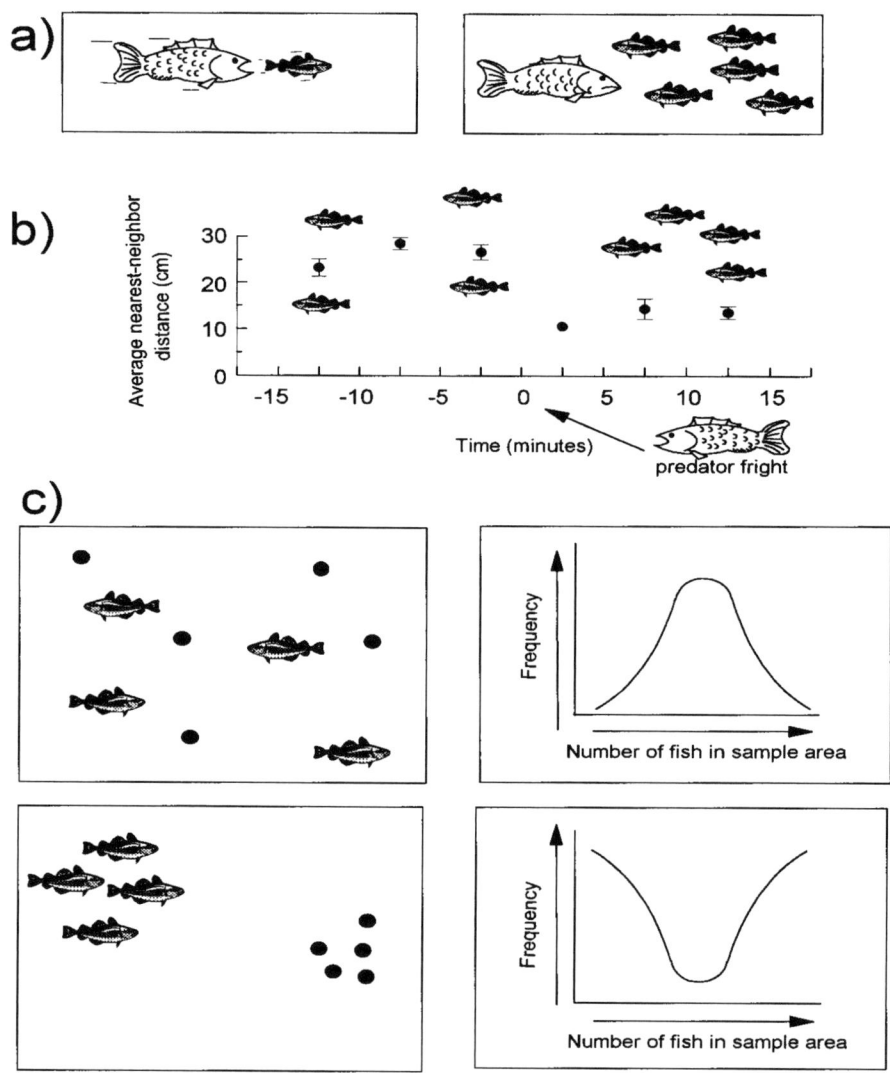

Figure 3. The benefits of schooling behavior for predator avoidance in pelagic marine fish. (a) solitary fish are more vulnerable to predation. (b) fish modify interfish distance (cm, mean ± SE) in response to predator fright stimulus. (c) manipulation of spatial and temporal distribution of food alters schooling behavior. (b) is from Ryer and Olla (unpublished data); (c) is modified from Ryer and Olla (1992, 1995b).

In contrast to the typically nonaggressive behavior of schooling fish, there are species in which aggressive interactions predominate, playing an important role in food acquisition. Even facultative schoolers may, under certain environmental conditions, resort to intense aggressive behavior when competing for food (Ryer and Olla, 1995b, 1996b).

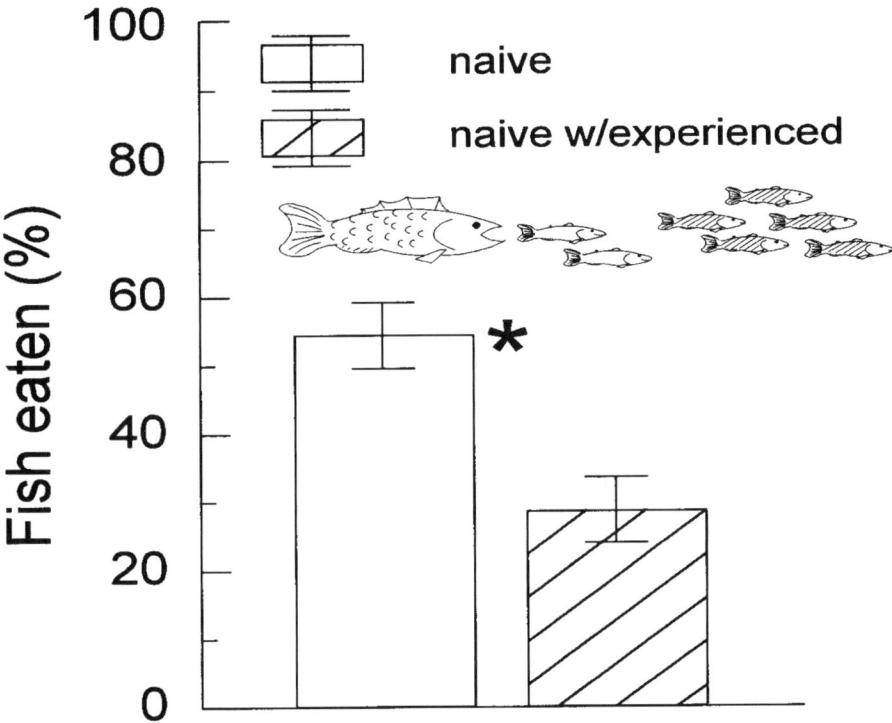

Figure 4. Predation (percent eaten) on juvenile coho salmon with or without previous experience with predators. Values are mean ± SE, *, $P < 0.05$. Modified from Patten (1977).

One concern for species that normally exhibit aggressive behavior is that artificial propagation may lead to an increase in aggressiveness (Swain and Riddell, 1990), which in turn could result in compromised predator-avoidance capabilities. Most animals strike a dynamic balance between vigilance for predators and other activities such as foraging and aggressive interactions (Caraco, 1979; Dill and Fraser, 1984; Lendrem, 1984; Magurran et al., 1985; Fraser and Huntingford, 1986; Metcalfe et al., 1987; Huntingford et al., 1988a,b; Ryer and Olla, 1991a, 1996a), but hyperaggressive fish may devote too little time or attention to predator vigilance and therefore be more vulnerable to predation. In addition, predators often concentrate attacks upon conspicuous individuals (Landeau and Terborgh, 1986; Theodorakis, 1989) or individuals that have separated themselves from other fish (Parrish, 1989; Neill and Cullen, 1974; Morgan and Godin, 1985), both characteristics associated with aggressive behavior.

Mass release of hyperaggressive fish may also have the unintended consequence of displacing wild fish from available habitats. Fenderson et al. (1968) found that hatchery-reared Atlantic salmon, *Salmo salar*, were more aggressive than wild Atlantic salmon at high population densities. Because hatchery fish are accustomed to high densities, were they to linger in the release area rather than dispersing, wild fish might be aggressively excluded from appropriate habitats (Flagg et al., 1995) and possibly made more vulnerable to predation.

Reduced predator-avoidance capabilities may also arise if a hatchery selects for the greater body size or higher metabolic rates that result from aggressive or competitive interactions (Ryer and Olla, 1995b, 1996b). In order to maintain higher foraging rates, these fish may need to forage at times or in places where they are at greater risk from predators (Johnsson, 1993). Laboratory-reared hybrids of wild steelhead trout, Oncorhynchus mykiss, and domesticated rainbow trout, O. mykiss, were more willing to feed on the side of a tank containing a predator than were laboratory-reared wild steelhead, even though standardized encounters with predators demonstrated that the two were equally susceptible to predation (Johnsson and Abrahams, 1991). These results suggest a negative genetic impact of domestication upon a stock, pointing to the unintended damage to wild populations that may occur through interbreeding.

RELEASE TO APPROPRIATE HABITATS.—For many nonschooling species, physical characteristics of the environment provide refuge from predation. For these fish, rapid dispersal into habitats appropriate for reducing predation risk would seem to be the most effective way of decreasing predation vulnerability. Proximity and availability of those habitats as well as the ability of hatchery-reared fish to recognize and use them would determine potential for survival. For example, many juvenile flatfish bury themselves in the substrate to avoid predation (Gibson and Robb, 1992; Keefe and Able, 1994). Release sites for these species should be in areas where sufficient appropriate sediment is available for the number of fish being released (Keefe and Able, 1994; Fujii and Noguchi, 1993). Structurally complex habitats appear to be necessary for settlement and survival of Atlantic cod, *Gadus morhua*, providing refuge from predation (Gotceitas et al., 1995; Tupper and Boutilier, 1995a,b). In the case of coral-reef species, appropriate refugia in the form of occupiable physical niches may be required, especially for newly recruited juveniles (Sale, 1978; Smith, 1978), and should include a complex of factors that are innately recognizable by the species, e.g., appropriate numbers and sizes of niches in which shelter-seeking fish can avoid predation (Keats et al., 1987; Gotceitas et al., 1995). Even if the density of physical niches matches the density of the fish released, competition for niches from wild conspecifics or from other species could in fact prevent the new recruits from utilizing a critical part of the habitat (Sale, 1978; Bachman 1984).

EFFECTS OF STRESS ON PREDATOR-PREY INTERACTIONS.—Contributing to predator-induced mortality at or shortly after release is the stress induced by handling, transportation, and release processes. When stress is not severe enough to cause death directly, its indirect effects may still lead to mortality if behavior such as predator evasion is compromised (Sylvester, 1972; Coutant et al., 1979; Olla et al., 1992a,b, 1995, 1997). Coho smolts, subjected to handling stress and then exposed to predation with an equal number of unstressed conspecifics, experienced twice the predation suffered by unstressed coho (Fig. 5A; Olla and Davis, 1989; Olla et al., 1992a, 1995). The extent of stress-induced predator mortality and the time taken to recover from stress, i.e., to regain ability to evade as well as unstressed fish, depended upon both the intensity of the stress and the particular hatchery stock (Olla et al., 1992a, 1995). There were also differences in vulnerability to predation from stress between species of Pacific salmon; chinook smolts, *Oncorhynchus tshawytscha*, showed a greater sensitivity to handling than did coho smolts (Olla et al., 1995).

The stress induced by handling, transport, and release procedures can also induce departures from physiological and biochemical homeostasis that, if severe enough, can lead directly to mortality (Schreck, 1981; Pickering et al., 1982; Schreck et al., 1997). El-

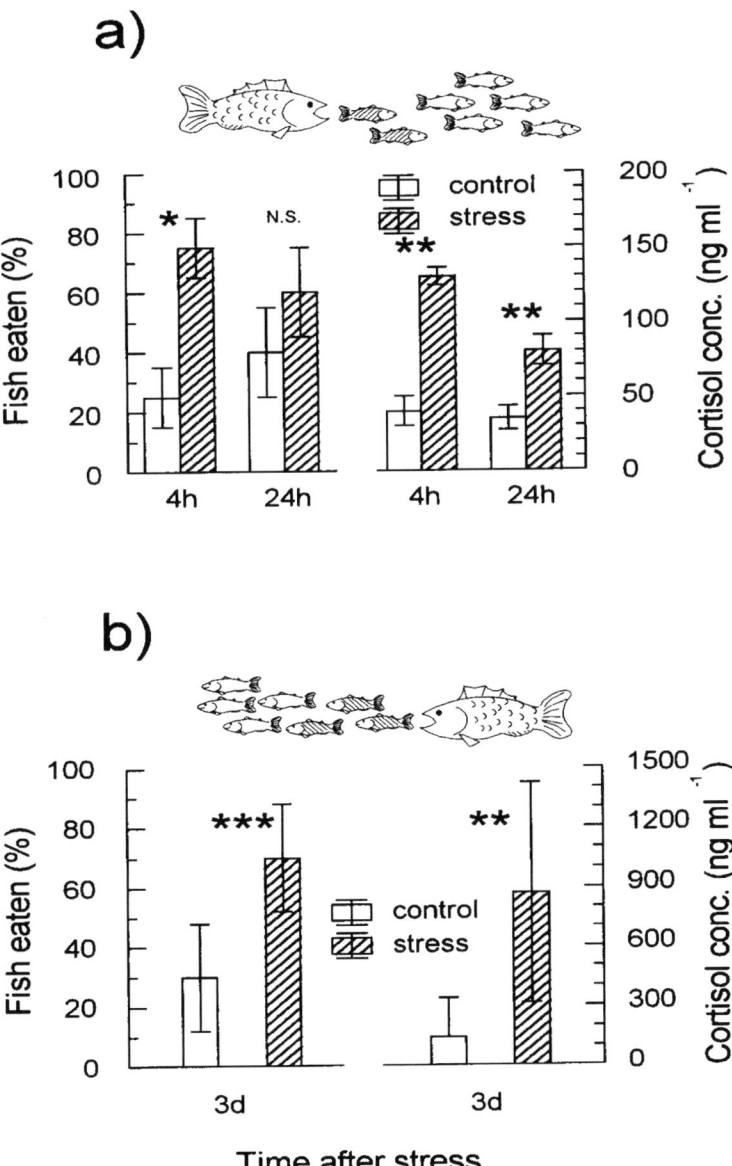

Figure 5. Predation (percent eaten) and cortisol concentrations (ng ml^{-1} plasma) for fish stressed by handling and allowed to recover (a) for 4 and 24 h in juvenile coho salmon or (b) for 3 d in juvenile walleye pollock. Values are mean ± SE for control and stressed fish; *, $P < 0.05$; **, $P < 0.01$; ***, $P < 0.001$; N.S., not significant. (a) is modified from Olla et al. (1995); (b) from Olla et al. (1997).

evated serum cortisol, commonly used for biochemical assessment of stress in fish, has been shown to be a good predictor of mortality, physiological performance deficits, and decreased disease resistance (Schreck, 1981; Schreck et al., 1997). Recent studies have also shown at times a linkage between cortisol concentrations and behavioral alterations (Olla et al., 1995, 1997). In coho smolts, increased vulnerability to predation was correlated with elevated cortisol concentrations 4 h after stress was imposed; after 24 h, although predator vulnerability had returned to control levels, cortisol had not (Fig. 5A; Olla et al., 1995). When juvenile walleye pollock, a pelagic marine species, were subjected to the stress induced by being towed in a net for 15 min, vulnerability to predation increased, as did cortisol concentrations (Fig. 5B; Olla et al., 1997). Cortisol concentrations remained elevated, the fish never recovered from stress, and after 6 d mortality increased sharply, reaching 100% after 14 d.

It is apparent that evaluation of the potential for stress-induced direct or indirect mortality, and methods to mitigate such effects, should be an integral part of any enhancement program. Even from the meager information available, it is clear that sensitivity to handling stress varies markedly among stocks and species, necessitating evaluation on a species-by-species basis (Olla et al., 1994).

CHOOSING APPROPRIATE MANAGEMENT STRATEGIES TO MINIMIZE PREDATION.—The above discussion makes clear the need to understand what factors limit the size of a natural population before enhancement strategies can be evaluated. Where population size is limited by available habitat, enhancement may not be a viable option until the habitat can be expanded. Even when a population is limited by recruitment, and empty habitat is available, the natural-history characteristics of fish species must still be matched with the proposed release strategies. Such a life-history approach was used with Atlantic cod to predict the optimal times for release of hatchery-reared fish into the wild (Salvanes et al., 1994). The model presented in Figure 1 is based on the assumptions that fish should inhabit the habitats providing the lowest mortality rate coupled with the highest growth rate and that shifts in habitat will occur as the risk- benefit ratio changes. In this particular example, release of Atlantic cod was optimal after they had reached the size at which they settle to the benthic habitat (Salvanes et al., 1994). The release time varied seasonally, depending upon availability of forage and presence of predators in the wild.

Fish exhibit daily rhythms of movement, foraging, and shelter seeking in response to light intensity, temperature, presence of predators, and food, and these rhythms can change with seasonal patterns of environmental factors and ontogeny (Helfman, 1993; Keefe and Able, 1994). Release of fish during a period of normal inactivity may lead to their confusion, inability to seek appropriate habitats, conspicuousness, and subsequent loss to predation. Predators in aquatic systems may be particularly active during dawn and dusk (Helfman, 1993), making releases during these periods particularly hazardous. In response to lowered temperatures, which decreased swimming performance, Atlantic salmon shifted activity cycles from diurnal to nocturnal, which may be an adaptation for avoiding avian predators during conditions of low evasion capability (Fraser et al., 1995). Benthic flatfish (plaice, *Pleuronectes platessa,* and dab, *Limanda limanda*) possess a tidally synchronized diurnal activity cycle, thereby avoiding crepuscular and nocturnal predators, including gadoid fishes and crabs (Burrows et al., 1994).

Table 2. Possible behavioral deficits in fish that are naive to live prey and wild foods and possible remedial actions for these deficits.

Possible deficits	Possible remedial actions
—Prey recognition, prey capture and handling	—Exposure to live prey and prey stimuli
—Appropriate use of habitat	—Provision of simulated habitat features appropriate for the species
—Feeding decrements induced by stress	—Mitigation of rearing and transport stress
—Hyperaggression (prolonged subordination affecting resource competition, displacement of wild fish)	—Alteration of fish density, food distribution, provision of simulated habitat features
—Decrease or absence of schooling in facultatively schooling species (cooperative social interactions)	—Alteration of spatial and temporal distribution of food

Feeding

RECOGNIZING AND ADAPTING TO NATURAL PREY.—Deficits in feeding upon release of hatchery-reared fish may result from their having been fed hatchery diets under conditions that require behavior markedly different from what is adaptive under natural conditions (Table 2). One of the greatest challenges facing hatchery-reared fish involves the switch from nonliving food or unnatural live prey, e.g., *Artemia*, to the capture of live natural prey following release. Laboratory studies suggest that fish can adapt to novel or live prey with learning (Ware, 1971; Godin, 1978; Ringler, 1979; Paszkowski and Olla, 1985a). For example, when hatchery-reared coho salmon smolts that had been reared on food pellets were given a choice of live prey and food pellets, they selected live sand shrimp, *Crangon*, over the pelleted food (Fig. 6; Paszkowski and Olla, 1985a). The ability of smolts to capture live prey improved so rapidly that peak capture efficiency was achieved <120 min after first exposure. This rapid transition from inanimate food to live prey was also observed in Atlantic salmon, in which the transition took place within 180 min (Stradmeyer and Thorpe, 1987). The rapidity of this adaptation indicates that an innate attraction to the sensory stimuli characteristic of live prey persists in hatchery stock. This response pattern to live prey generally agreed with those of other species, when naive fish were presented with novel food items under laboratory conditions. In these cases, fish were seen to attack any food type and then to modify foraging tactics to improve success rates (Ware, 1971; Webb and Skadsen, 1980; Gillen et al., 1981; Werner et al., 1981). For example, 43% of pellet-reared tiger muskellunge (*Esox masquinongy* × *E. lucius*) successfully captured live fathead minnows on their first attempt, and by the seventh attempt, the success rate was 100% (Webb and Skadsen, 1980).

Although results for a number of species suggest that hatchery fish can, under artificial conditions, switch with some facility from artificial food to live prey, there are also indications that not all individuals from a given population will successfully make this transition. For example, 31% of coho smolts would not feed on any food after being transferred to the laboratory from the hatchery (Paszkowski and Olla, 1985a). Similarly, 40% of fall chinook salmon were unable to feed on live prey after transfer to the laboratory from the hatchery (Maynard et al., 1996). In hatchery-reared tiger muskellunge, 27% of fish tested were unsuccessful in making the switch from pelletized food to live prey even after 14 d

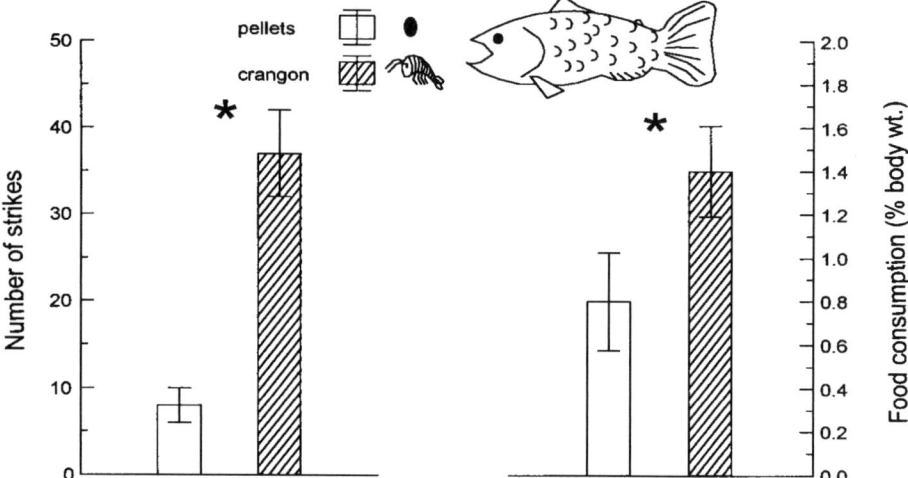

Figure 6. Number of strikes at food pellets or live prey (*Crangon*) and food consumption (percent body weight) for juvenile coho salmon (mean ± SE; *, $P < 0.05$). Figure modified from Paszkowski and Olla (1985a).

(Gillen et al., 1981). These results suggest that some portion of hatchery-reared fish may never adapt to foraging for natural prey items.

Even ability to adapt readily to novel food items under laboratory conditions by no means assures success after release into the wild (Fig. 7). Studies on postrelease feeding success, principally on salmonids, have shown that soon after release hatchery fish often consume less than wild fish (Hochachka, 1961; Sosiak et al., 1979; Ersbak and Haase, 1983; Bachman, 1984), consume fewer prey types (Sosiak et al., 1979), lag behind wild fish in switching to new prey (Ersbak and Haase, 1983), and experience depressed growth and survival (Miller, 1954; Ersbak and Haase, 1983). In Japanese flounder, *Paralichthys olivaceus*, when the dominant prey item for wild fish (mysids) was at a low density, recently released hatchery-reared fish consumed different prey items than did wild fish (Fujii and Noguchi, 1993). Feeding and the growth rate of wild flounder were greater than those of hatchery-reared fish, even in years when mysids were abundant. Similarly, soon after release in the wild, hatchery-reared cod fed on slower-moving prey such as snails and bivalves, whereas wild fish ate gobies and crustaceans, although this difference was not evident after a month (Nordeide and Salvanes, 1991).

Some of these deficiencies may result in part from the microhabitats that hatchery fish occupy after release. In some cases, hatchery-reared salmonids tend to position themselves higher above the substrate than do wild fish, thereby lowering the probability of encountering benthic food types (Sosiak et al., 1979; Maynard et al., 1996). These fish have also been observed to underutilize eddies and pools and other parts of a stream where metabolic demand would be lessened and growth increased, but the varied nature of the studies makes it difficult to assign any one cause to the deficits in feeding after release, which are probably due to a combination of factors. Results do, however, point to the importance of being able to identify specific causes and their interactions, so as to

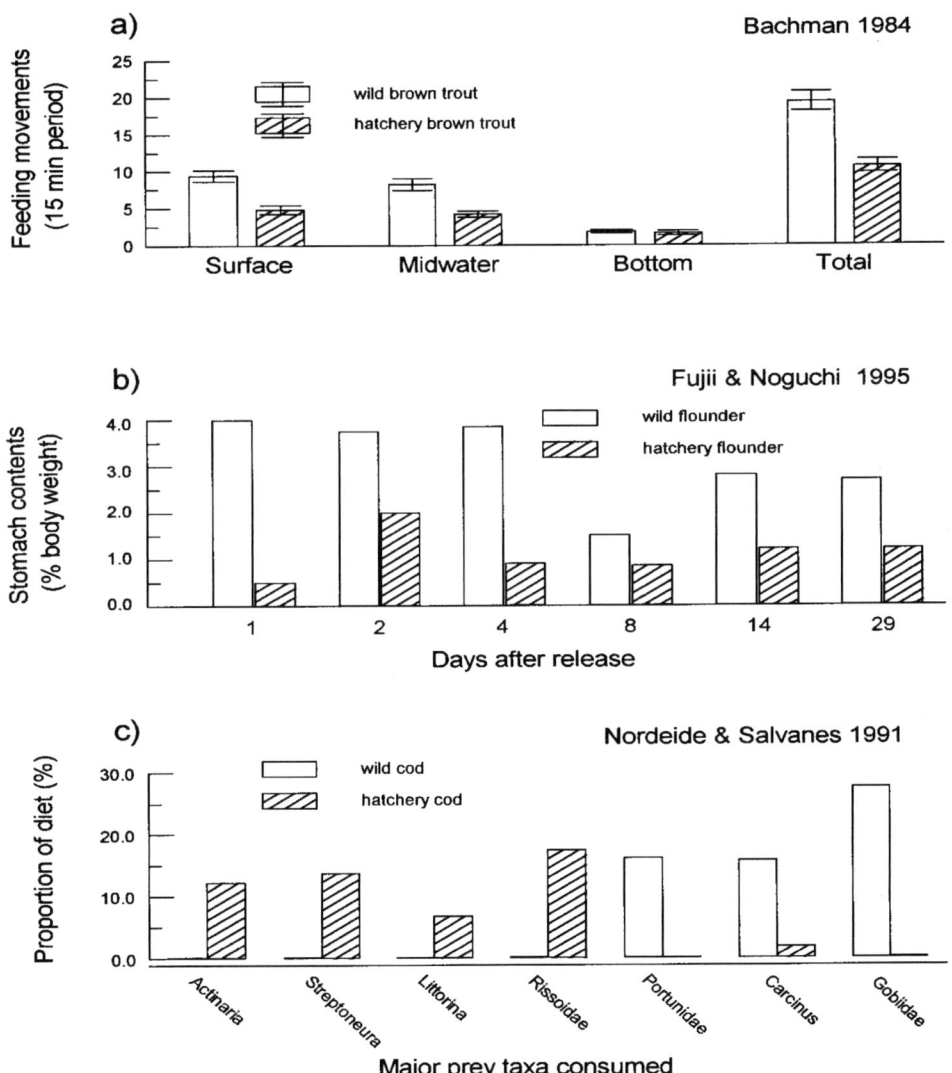

Figure 7. Ability of hatchery-reared fish to adapt to natural food after release into the wild. (a) Number of feeding movements in a 15-min period (mean ± SE) for wild and hatchery-reared brown trout at the surface, midwater, and bottom of a stream. (b) Stomach contents (percent body weight) for wild and hatchery-reared flounder. (c) Prey taxa consumed (percent of diet) for wild and hatchery-reared cod. (a) is modified from Bachman (1984); (b) from Fujii and Noguchi (1993); (c) from Nordeide and Salvanes (1991).

develop ways of improving the transition from hatchery feeding to feeding in the natural environment when those causes derive primarily from nurture.

SOCIAL INFLUENCES ON FEEDING SUCCESS.—Hatchery fish can be more aggressive than wild fish (Swain and Riddell, 1990), particularly at high population densities (Fenderson

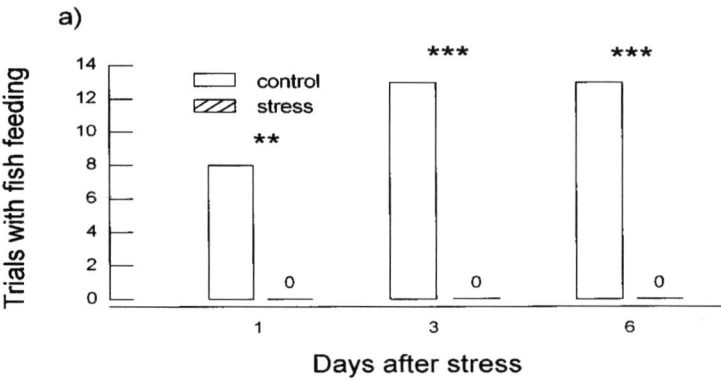

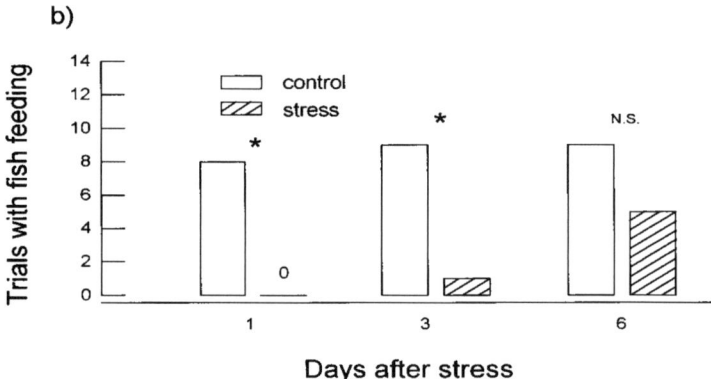

Figure 8. Effect after 1, 3 and 6 d of net-entrainment stress on feeding for (a) juvenile walleye pollock and (b) juvenile sablefish. *, $P < 0.05$; **, $P < 0.01$; ***, $P < 0.001$; N.S., not significant. Figure modified from Olla et al. (1997).

et al., 1968), and as already discussed, this trait can have negative implications for both hatchery and wild fish stocks. Game-theory modeling has been used to evaluate factors that could lead to hatchery selection for aggressive behavior (Doyle and Talbot, 1986). It was concluded that rearing fish on limited rations, where dominant individuals can monopolize food, has the potential to select for aggressiveness. In this scenario, innately aggressive individuals (Huntingford et al., 1990) are able to acquire more food and accelerate their growth at the expense of subordinates. Because large individuals show greater survival and greater reproductive contribution, aggressiveness, a trait known to be heritable in salmonids (Taylor, 1990), will increase over multiple generations. Although experiments to demonstrate genetic changes in aggressiveness have yet to be conducted, it has been demonstrated that food distribution in the rearing environment determines the ability of dominant fish to monopolize food and grow faster than subordinates. Groups of juvenile chum salmon, *Oncorhynchus keta,* reared on food dropped into the tank at a single location, one or two pellets at a time (localized food), experienced greater growth depensation, i.e., greater variation in individual growth, than did groups reared on food that was periodically spread over the entire surface of the tank (dispersed food; Ryer and Olla, 1995b). Behavioral observations documented that dominants were able partially to monopolize the localized food and thereby increase their growth rates but were unable to monopolize dispersed food. Similar experiments with juvenile coho salmon gave the same results (Ryer and Olla, 1991b) but also demonstrated that this effect could be mitigated by increased rations (Ryer and Olla, 1996b). These studies demonstrate the importance of aquacultural practices designed to lessen behavioral problems arising from the rearing of large numbers of fish in confined areas.

Social interactions can induce stress and potentially affect food acquisition long after the stress-producing interactions have ceased. For example, for members of a social hierarchy, prolonged subordination may produce physiological and behavioral deficits that persist after release. In coho smolts, fish that occupied a low-ranking position in a dominance hierarchy for 6–10 wks expressed low levels of aggressive behavior even after dominant fish were removed (Paszkowski and Olla, 1985b). Under the constant stress of subordination, coho may undergo physiological changes associated with the continued production of high levels of corticosteroids, which are irreversible over the short term (Ejike and Schreck, 1980) and whose effects may be manifested in low levels of aggression. Similar results were observed for green sunfish, *Lepomis cyanellus,* where fish that had suffered long-term subordination in the past tended to be subordinates in future encounters (McDonald et al., 1968). If this type of behavior persists after release, subordinate fish would be less likely to compete for resources and therefore have less access to food, resulting in lowered growth rates and a decreased probability of survival. The percentage of released fish vulnerable to this effect will depend on the species, stock, and hatchery environment.

EFFECTS OF STRESS ON FEEDING.—Stress associated with handling, transportation, and introduction into an unfamiliar environment can affect feeding after release. Transfer from the hatchery to the laboratory caused 31% of coho smolts to cease feeding at least over the short term (Paszkowski and Olla, 1985a). Transplanting wild and hatchery cutthroat trout, *Salmo clarki,* to unfamiliar streams caused even wild fish to decrease feeding and loose weight for several weeks (Miller, 1954). Although the wild stock eventually recovered, the hatchery stock never did.

Marked differences between species make it necessary to examine the effects of stress on feeding on a species-by-species basis. For example, towing walleye pollock in a net for 15 min, at speeds exceeding their swimming ability induced stress that resulted in decreased feeding 1, 3, and 6 d afterward (Fig. 8A). The fish never recovered; after 6 d mortality increased sharply, reaching 100% after 14 d (Olla et al., 1997). In contrast, juvenile sablefish, *Anoplopoma fimbria,* proved to be much more resistant to the same stressor; towing for 15 min and even 2 h resulted in no alteration in feeding behavior. Decreased feeding was seen after 4 h of towing, but feeding recovered to pretest levels after 6 d (Fig. 8B).

Conclusions

Predation after release into the wild is a major cause of mortality in hatchery-reared fish. The ability to avoid predation in hatchery-reared fish can be improved by "training," examples of which include exposure to predators and predatory stimuli, manipulation of fish density and social interactions, and spatial and temporal distribution of food and refuges, as well as mitigation of rearing and transport stressors.

Hatchery-reared fish face the challenge of switching from artificial food to live food in the natural environment. Hatchery practices can be altered to "train" fish in more appropriate feeding behaviors. Food can be distributed in varying ways, both spatially and temporally, to enhance social interactions and foraging efficiency. Fish can be exposed to live food and food stimuli. Handling and transport stressors can be mitigated to relieve the stress of outplanting and adaptation to the wild.

It is essential that we understand the factors limiting the size of natural populations before enhancement strategies can be evaluated and implemented. In addition, released fish must be of sufficient quality to meet the challenges of the natural environment and not cause degradation or displacement of wild stocks.

Literature Cited

Bachman, R. A. 1984. Foraging behavior of free-ranging wild and hatchery brown trout in a stream. Trans. Am. Fish. Soc. 113: 1–32.

Blankenship, H. L. and K. M. Leber. 1995. A responsible approach to marine stock enhancement. Am. Fish. Soc. Symp. 15: 167–175.

Blaxter, J. H. S. 1976. Reared and wild fish — how do they compare? Pages 11–26 in G. Persoone and E. Jasper, eds. Proc. 10th European Mar. Biol. Symp. Universa Press, Wetteren.

Burrows, M. T., R. N. Gibson and A. Maclean. 1994. Effects of endogenous rhythms and light conditions on foraging and predator-avoidance in juvenile plaice. J. Fish Biol. 45 (Suppl. A): 171–180.

Busack, C. A. and K. P. Currens. 1995. Genetic risks and hazards in hatchery operations: fundamental concepts and issues. Am. Fish. Soc. Symp. 15: 71–80.

Caraco, T. 1979. Time budgeting and group size: a theory. Ecology 60: 611–617.

Coutant, C. C., R. B. McLean and D. L. DeAngelis. 1979. Influences of physical and chemical alterations on predator-prey interactions. Pages 57–68 in R. H. Stroud and H. E. Clepper, eds. Predator-prey systems in fisheries management: proceedings of the International Symposium on Predator-prey Systems in Fish Communities and Their Role in Fisheries Management. Sport Fishing Institute, Washington, D.C.

Dill, L. M. and A. H. G. Fraser. 1984. Risk of predation and the feeding behavior of juvenile coho salmon (*Oncorhynchus kisutch*). Behav. Ecol. Sociobiol. 16: 65–71.

Doyle, R. W. and A. J. Talbot. 1986. Artificial selection on growth and correlated selection on competitive behaviour in fish. Can. J. Fish. Aquat. Sci. 43: 1059–1064.

Edwards, G. B. and J. G. Nickum. 1995. Use of propagated fishes in Fish and Wildlife Service programs. Pages 41–44 in M. R. Collie and J. P. McVey, eds. Interactions between cultured species and naturally occurring species in the environment. Proc. 22nd U.S.- Japan Aquaculture Panel Symp. Alaska Sea Grant Report AK-SG-95- 03. Homer, Alaska.

Ejike, C. and C. B. Schreck. 1980. Stress and social hierarchy rank in coho salmon. Trans. Am. Fish. Soc. 109: 423–426.

Ersbak, K. and B. L. Haase. 1983. Nutritional deprivation after stocking as a possible mechanism leading to mortality in stream-stocked brook trout. N. Am. J. Fish. Manage. 3: 142–151

FAO (Food and Agriculture Organization of the United Nations), Marine Resource Service. 1993. Review of the state of marine fishery resources. FAO Fish. Tech. Pap. 335. 136 p.

Fenderson, O. C., W. H. Everhart and K. M. Muth. 1968. Comparative agonistic and feeding behavior of hatchery-reared wild salmon in aquaria. J. Fish. Res. Bd. Can. 25: 1–14.

Fisher, J. P. and W. G. Pearcy. 1988. Growth of juvenile coho salmon (*Oncorhynchus kisutch*) off Oregon and Washington, USA, in years of differing coastal upwelling. Can. J. Fish. Aquat. Sci. 45: 1036–1044.

Flagg, T. A., F. W. Waknitz, D. J. Maynard, G. B. Milner and C. V. W. Mahnken. 1995. The effect of hatcheries on native coho salmon populations in the lower Columbia River. Am. Fish. Soc. Symp. 15: 366–375.

Fraser, D. F. and F. A. Huntingford. 1986. Feeding and avoiding predation hazard: the behavioral response of the prey. Ethology 73: 56–68.

Fraser, N. H. C., J. Heggenes, N. B. Metcalfe and J. E. Thorpe. 1995. Low summer temperatures cause juvenile Atlantic salmon to become nocturnal. Can. J. Zool. 73: 446–451.

Fresh, K. L., R. D. Cardwell, B. P. Snyder and E. O. Salo. 1982. Some hatchery strategies for reducing predation upon juvenile chum salmon (*Oncorhynchus keta*) in freshwater. Pages 79–89 *in* R. R. Melteff and R. A. Neve, eds. Proc. North Pacific Aquacult. Symp. Alaska Sea Grant, University of Alaska.

Fujii, T. and M. Noguchi. 1993. Interactions between released and wild Japanese flounder (*Paralichthys olivaceus*) on a nursery ground. Pages 57–65 *in* M. R. Collie and J. P. McVey, eds. Interactions between cultured species and naturally occurring species in the environment. Proc. 22nd U.S.-Japan Aquacult. Panel Symp. Alaska Sea Grant Report AK-SG-95-03. Homer, Alaska.

Gibson, R. N. and L. Robb. 1992. The relationship between body size, sediment grain size and the burying ability of juvenile plaice, *Pleuronectes platessa* L. J. Fish Biol. 40: 771–778.

Gillen, A. L., R. A. Stein and R. F. Carline. 1981. Predation by pellet-reared tiger muskellunge on minnows and bluegills in experimental systems. Trans. Am. Fish. Soc. 110: 197–209.

Ginetz, R. M. and P. A. Larkin. 1976. Factors affecting rainbow trout (*Salmo gairdneri*) predation on migrant fry of sockeye salmon (*Oncorhynchus nerka*). J. Fish. Res. Bd. Can. 33: 19–24.

Godin, J.-G. J. 1978. Behavior of juvenile pink salmon (*Oncorhynchus gorbuscha* Walbaum) toward novel prey: influence of ontogeny and experience. Environ. Biol. Fish. 3: 261–266.

Gotceitas, V., S. Fraser and J. A. Brown. 1995. Habitat use by juvenile Atlantic cod (*Gadus morhua*) in the presence of an actively foraging and non-foraging predator. Mar. Biol. 123: 421–430.

Healey, M. C. 1982. Timing and relative intensity of size-selective mortality of juvenile chum salmon (*Oncorhynchus keta*) during early sea life. Can. J. Fish. Aquat. Sci. 39: 952–957.

Helfman, G. S. 1989. Threat-sensitive predator avoidance in damselfish-trumpetfish interactions. Behav. Ecol. Sociobiol. 24: 47–58.

_____. 1993. Fish behaviour by day, night and twilight. Pages 479–512 *in* T. J. Pitcher, ed. Behaviour of teleost fishes, 2nd ed. Chapman & Hall, New York.

Hochachka, P. W. 1961. Liver glycogen reserves of interacting resident and introduced trout populations. J. Fish. Res. Bd. Can. 18: 124–135.

Huntingford, F. A., N. B. Metcalfe and J. E. Thorpe. 1988a. Choice of feeding station in Atlantic salmon, *Salmo salar*, parr: effects of predation risk, season and life history strategy. J. Fish Biol. 33: 917–924.

_____, _____ and _____. 1988b. Feeding motivation and response to predation risk in Atlantic salmon parr adopting different life history strategies. J. Fish Biol. 32: 777-782.

_____, _____, _____, W. D. Graham and C. E. Adams. 1990. Social dominance and body size in Atlantic salmon parr, *Salmo salar* L. J. Fish Biol. 36: 877–881.

Johnsson, J. I. 1993. Big and brave: size selection affects foraging under risk of predation in juvenile rainbow trout, *Oncorhynchus mykiss*. Anim. Behav. 45: 1219–1225.

_____ and M. V. Abrahams. 1991. Interbreeding with domestic strain increases foraging under threat of predation in juvenile steelhead trout (*Oncorhynchus mykiss*): an experimental study. Can. J. Fish. Aquat. Sci. 48: 243–247.

Keats, D. W., D. H. Steele and G. R. South. 1987. The role of fleshy macroalgae in the ecology of juvenile cod (*Gadus morhua* L.) in inshore waters off eastern Newfoundland. Can. J. Zool. 65: 49–53.

Keefe, M. L. and K. W. Able. 1994. Contributions of abiotic and biotic factors to settlement in summer flounder, *Paralichthys dentatus*. Copeia 1994: 458–465.

Kitada, S., Y. Taga and H. Kishino. 1992. Effectiveness of a stock enhancement program evaluated by a two-stage sampling survey of commercial landings. Can. J. Fish. Aquat. Sci. 49: 1573–1582.

Landeau, L. and J. Terborgh. 1986. Oddity and the "confusion effect" in predation. Anim. Behav. 34: 1372–1380.

Lendrem, D. W. 1984. Flocking, feeding and predation risk: absolute and instantaneous feeding rates. Anim. Behav. 32: 298–299.

Mace, P. M. 1983. Bird predation on juvenile salmonids in the Big Qualicum Estuary, Vancouver Island. Can. Tech. Rep. Fish. Aquat. Sci. 1176: 1–79.

Magurran, A. E. and A. Higham. 1988. Information transfer across fish shoals under predator threat. Ethology 78: 153–158.

_____, W. J. Oulton and T. J. Pitcher. 1985. Vigilant behaviour and shoal size in minnows. Z. Tierpsychol. 67: 167–178.

Maynard, D. J., G. C. McDowell, E. P. Tezak and T. A. Flagg. 1996. Effects of diets supplemented with live food on the foraging behavior of cultured fall chinook salmon. Prog. Fish-Cult. 58: 187–191.

McDonald, A. L., N. W. Heimstra and D. K. Damkot. 1968. Social modification of antagonistic behaviour in fish. Anim. Behav. 16: 437–441.

Metcalfe, N. B., F. A. Huntingford and J. E. Thorpe. 1987. The influence of predation risk on the feeding motivation and foraging strategy of juvenile Atlantic salmon. Anim. Behav. 35: 901–911.

Miller, R. B. 1954. Comparative survival of wild and hatchery-reared cutthroat trout in a stream. Trans. Am. Fish. Soc. 83: 120–130.

Morgan, M. J. 1988. The effect of hunger, shoal size and presence of a predator on shoal cohesiveness in bluntnose minnows, *Pimephales notatus* Rafinesque. J. Fish. Biol. 32: 963–971.

_____ and J.-G. J. Godin. 1985. Antipredator benefits of schooling behaviour in a cyprinodontid fish, the banded killifish (*Fundulus diaphanus*). Z. Tierpsychol. 70: 236–246.

Moyle, P. B. 1969. Comparative behavior of young brook trout of domestic and wild origin. Prog. Fish-Cult. 31: 51–56.

Nævdal, G. 1994. Genetic aspects in connection with sea ranching of marine fish species. Aquacult. Fish. Manage. 25 (Suppl. 1): 93–100.

Neill, S. R. StJ. and J. M. Cullen. 1974. Experiments on whether schooling by their prey affects the hunting behaviour of cephalopods and fish predators. J. Zool. (Lond.) 172: 5499.

NOAA (National Oceanic and Atmospheric Administration). 1995. Our living oceans. U.S. Dept. Commer., NOAA Tech. Memo. NMFS-F/SPO-19. 160 p.

Nordeide, J. T. and A. G. V. Salvanes. 1991. Observations on reared newly released and wild cod (*Gadus morhua* L.) and their potential predators. ICES Mar. Sci. Symp. 192: 139–146.

Olla, B. L. and M. W. Davis. 1989. The role of learning and stress in predator avoidance of hatchery-reared coho salmon (*Oncorhynchus kisutch*) juveniles. Aquaculture 76: 209–214.

_____, _____ and C. B. Schreck. 1992a. Comparison of predator avoidance capabilities with corticosteroid levels induced by stress in juvenile coho salmon. Trans. Am. Fish. Soc. 121: 544–547.

_____, _____ and C. H. Ryer. 1992b. Foraging and predator avoidance in hatchery-reared Pacific salmon: achievement of behavioural potential. Pages 5–12 *in* J. E. Thorpe and F.

A. Huntingford, eds. World Aquaculture Workshops, Number 2: The importance of feeding behavior for the efficient culture of salmonid fishes. Papers Presented at World Aquaculture 90, Halifax, Nova Scotia, 12 June 1990, World Aquaculture Society, Baton Rouge, Louisiana.

_____, _____ and _____. 1994. Behavioural deficits in hatchery-reared fish: potential effects on survival following release. Aquacult. Fish. Manage. 25 (Suppl. 1): 19–34.

_____, _____ and C. B. Schreck. 1995. Stress-induced impairment of predator evasion and non-predator mortality in Pacific salmon. Aquacult. Res. 26: 393–398.

_____, _____ and _____. 1997. Effects of simulated trawling on sablefish and walleye pollock: the role of light intensity, net velocity and towing duration. J. Fish Biol. 50: 1181–1194.

Parrish, J. K. 1989. Re-examining the selfish herd: are central fish safer? Anim. Behav. 38: 1048–1053.

Paszkowski, C. A. and B. L. Olla. 1985a. Foraging behavior of hatchery-produced coho salmon (*Oncorhynchus kisutch*) smolts on live prey. Can. J. Fish. Aquat. Sci. 42: 1915–1921.

_____ and _____. 1985b. Social interactions of coho salmon (*Oncorhynchus kisutch*) smolts in seawater. Can. J. Zool. 63: 240–2407.

Patten, B. G. 1977. Body size and learned avoidance as factors affecting predation on coho salmon, *Oncorhynchus kisutch*, fry by torrent sculpin, *Cottus rhotheus*. Fish. Bull., U.S. 75: 457–459.

Petersson, E., T. Järvi, N. G. Steffner and B. Ragnarsson. 1996. The effect of domestication on some life history traits of sea trout and Atlantic salmon. J. Fish Biol. 48: 776–791.

Pickering, A. D., T. G. Pottinger and P. Christie. 1982. Recovery of the brown trout, *Salmo trutta* L., from acute handling stress: a time-course study. J. Fish Biol. 20: 229–244.

Pitcher, T. J. 1986. Functions of shoaling behaviour in teleosts. Pages 294–337 *in* T. J. Pitcher, ed. The behavior of teleost fishes. The Johns Hopkins University Press, Baltimore.

_____ and J. K. Parrish. 1993. Functions of shoaling behaviour in teleosts. Pages 363–439 *in* T. J. Pitcher, ed. Behaviour of teleost fishes, 2nd ed. Chapman & Hall, New York.

Ringler, N. H. 1979. Prey selection by drift-feeding brown trout (*Salmo trutta*). J. Fish. Res. Bd. Can. 36: 392-403.

Ryer, C. H. and B. L. Olla. 1991a. Information transfer and the facilitation and inhibition of feeding in a schooling fish. Environ. Biol. Fish. 30: 317–323.

_____ and _____. 1991b. Agonistic behavior in a schooling fish: form, function, and ontogeny. Environ. Biol. Fish. 31: 355–363.

_____ and _____. 1992. Social mechanisms facilitating exploitation of spatially variable ephemeral food patches in a pelagic marine fish. Anim. Behav. 44: 69–74.

_____ and _____. 1995a. The influence of food distribution upon the development of aggressive and competitive behaviour in juvenile chum salmon, *Oncorhynchus keta*. J. Fish Biol. 46: 264–272.

_____ and _____. 1995b. Influences of food distribution on fish foraging behaviour. Anim. Behav. 49: 411–418.

_____ and _____. 1996a. Growth depensation and aggression in laboratory reared coho salmon: the effect of food distribution and ration size. J. Fish Biol. 48: 686–694.

_____ and _____. 1996b. Social behavior of juvenile chum salmon, *Oncorhynchus keta*, under risk of predation: the influence of food distribution. Environ. Biol. Fish. 45: 75–83.

Sale, P. F. 1978. Coexistence of coral reef fishes — a lottery for living space. Environ. Biol. Fish. 3: 85-102.

Salvanes, A. G. V., J. Giske and J. T. Nordeide. 1994. Life-history approach to habitat shifts for coastal cod, *Gadus morhua* L. Aquacult. Fish. Manage. 25 (Suppl. 1): 215–228.

Schreck, C. B. 1981. Stress and compensation in teleostean fishes: response to social and physical factors. Pages 295–321 *in* A. D. Pickering, ed. Stress and fish. Academic Press, London.

_____, B. L. Olla and M. W. Davis. 1997. Behavioral responses to stress. Pages 145–170 *in* G. Iwama, A. Pickering, C. B. Schreck and J. Sumpter, eds. Fish stress and health in aquaculture. Cambridge University Press, New York.

Smith, C. L. 1978. Coral reef fish communities: a compromise view. Environ. Biol. Fish. 3: 109–128.

Solemdal, P., E. Dahl, D. S. Danielssen and E. Moksness. 1984. The cod hatchery in Flødevigen — background and realities. Flødevigen rapportser. 1: 17–45.

Sosiak, A. J., R. G. Randall and J. A. McKenzie. 1979. Feeding by hatchery-reared and wild Atlantic salmon (*Salmo salar*) parr in streams. J. Fish. Res. Bd. Can. 36: 1408–1412.

Stradmeyer, L. and J. E. Thorpe. 1987. The responses of hatchery-reared Atlantic salmon, *Salmo salar* L., parr to pelleted and wild prey. Aquacult. Fish. Manage. 18: 51–61.

Suboski, M. D., S. Bain, A. E. Carty, L. M. McQuoid, M. I. Seelen and M. Seifert. 1990. Alarm reaction in acquisition and social transmission of simulated-predator recognition by zebra danio fish (*Brachydanio rerio*). J. Comp. Psychol. 104: 101–112.

Swain, D. P. and B. E. Riddell. 1990. Variation in agonistic behavior between newly emerged juveniles from hatchery and wild populations of coho salmon, *Oncorhynchus kisutch*. Can. J. Fish. Aquat. Sci. 47: 566–571.

Sylvester, J. R. 1972. Effect of thermal stress on predator avoidance in sockeye salmon. J. Fish. Res. Bd. Can. 29: 601–603.

Taylor, E. B. 1990. Variability in agonistic behaviour and salinity tolerance between and within two populations of juvenile chinook salmon, *Oncorhynchus tshawytscha*, with contrasting life histories. Can. J. Fish. Aquat. Sci. 47: 2172–2180.

Theodorakis, C. W. 1989. Size segregation and the effects of oddity on predation risk in minnow schools. Anim. Behav. 38: 496–502.

Tupper, M. and R. G. Boutilier. 1995a. Effects of habitat on settlement, growth, and post-settlement survival of Atlantic cod (*Gadus morhua*). Can. J. Fish. Aquat. Sci. 52: 1834–1841.

_____ and _____. 1995b. Size and priority at settlement determine growth and competitive success of newly settled Atlantic cod. Mar. Ecol. Prog. Ser. 118: 295–300.

Vincent, R. E. 1960. Some influences of domestication upon three stocks of brook trout (*Salvelinus fontinalis* Mitchill). Trans. Am. Fish. Soc. 89: 35–52.

Ware, D. M. 1971. Predation by rainbow trout (*Salmo gairdneri*): the effect of experience. J. Fish. Res. Bd. Can. 28: 1847–1852.

Webb, P. W. and J. M. Skadsen. 1980. Strike tactics of *Esox*. Can. J. Zool. 58: 1462–1469.

Werner, E. E., G. G. Mittlebach and D. J. Hall. 1981. The role of foraging profitability and experience in habitat use by the bluegill sunfish. Ecology 62: 116–125.

DATE ACCEPTED: May 20, 1997.

ADDRESS: *Fisheries Behavioral Ecology Program, Alaska Fisheries Science Center, National Marine Fisheries Service, Hatfield Marine Science Center, Newport, Oregon 97365.*

Genetic Differences in Growth and Survival of Juvenile Hatchery and Wild Steelhead Trout, *Salmo gairdneri*

R. R. REISENBICHLER[1] AND J. D. MCINTYRE

Oregon Cooperative Fishery Research Unit,[2] Oregon State University, Corvallis, Oreg. 97331, USA

REISENBICHLER, R. R., AND J. D. MCINTYRE. 1977. Genetic differences in growth and survival of juvenile hatchery and wild steelhead trout, *Salmo gairdneri*. J. Fish. Res. Board Can. 34: 123–128.

Relative growth and survival of offspring from matings of hatchery and wild Deschutes River (Oregon) summer steelhead trout, *Salmo gairdneri*, were measured to determine if hatchery fish differ genetically from wild fish in traits that can affect the stock–recruitment relationship of wild populations. Sections of four natural streams and a hatchery pond were each stocked with genetically marked (lactate dehydrogenase genotypes) eyed eggs or unfed swim-up fry from each of three matings: hatchery × hatchery (HH), hatchery × wild (HW), and wild × wild (WW). In streams, WW fish had the highest survival and HW fish the highest growth rates when significant differences were found; in the hatchery pond, HH fish had the highest survival and growth rates. The hatchery fish were genetically different from wild fish and when they interbreed with wild fish may reduce the number of smolts produced. Hatchery procedures can be modified to reduce the genetic differences between hatchery and wild fish.

REISENBICHLER, R. R., AND J. D. MCINTYRE. 1977. Genetic differences in growth and survival of juvenile hatchery and wild steelhead trout, *Salmo gairdneri*. J. Fish. Res. Board Can. 34: 123–128.

Nous avons mesuré la croissance et la survie relatives de descendants issus de croisements entre des truites steelhead de pisciculture et des steelheads d'été sauvages, *Salmo gairdneri*, de la rivière Deschutes (Orégon), afin de déterminer si les poissons de pisciculture diffèrent génétiquement des poissons sauvages par des caractères susceptibles d'affecter la relation stock–recrutement des populations sauvages. Des sections de quatre cours d'eau naturels et un étang de pisciculture ont été ensemencés chacun avec des œufs embryonnés génétiquement marqués (génotypes lactate déshydrogénase) ou avec des alevins non nourris au stade de nage vers le haut résultant de chacun de trois croisements : pisciculture × pisciculture (HH), pisciculture × sauvage (HW) et sauvage × sauvage (WW). Dans les cours d'eau, les poissons WW ont le plus haut taux de survie, et les poissons HW le plus haut taux de croissance là où il y a des différences significatives; dans l'étang de pisciculture, les poissons HH ont les taux de survie et de croissance les plus élevés. Les poissons de pisciculture diffèrent génétiquement des poissons sauvages et, quand ils se croisent avec des poissons sauvages, peuvent causer une diminution du nombre de smolts produits. Les méthodes de pisciculture peuvent être modifiées de façon à réduire les différences génétiques entre les poissons de pisciculture et les poissons sauvages.

Received July 6, 1976
Accepted October 5, 1976

Reçu le 6 juillet 1976
Accepté le 5 octobre 1976

INTRODUCTION of artificially propagated salmonids into natural stream systems may influence resident wild populations (Hochachka 1961). The hatchery-reared fish may affect the wild population through competition for food and space resources (Needham and Slater 1944; Reimers 1957; Vincent 1972), or the genetic structure of the wild population may be affected by interbreeding of hatchery fish and wild fish on natural spawning grounds. A genetic effect is contingent upon there being a genetic difference between the wild fish and the hatchery fish spawning together in the wild. Results of studies with brook trout, *Salvelinus fontinalis* (Greene 1952; Flick and Webster 1964), and with Atlantic salmon, *Salmo salar* (Fenderson et al. 1968), indicated the existence of genetic differences between hatchery and wild fish.

For anadromous species on the Pacific coast, it

---

[1]Present address: Oregon Department of Fish and Wildlife, Research Division, Corvallis, Oreg. 97330, USA.

[2]Cooperators are the Oregon Department of Fish and Wildlife, Oregon State University, and the U.S. Fish and Wildlife Service.

Printed in Canada (J4446)
Imprimé au Canada (J4446)

is not known whether adults resulting from release of artificially propagated juveniles differ genetically from wild adults, especially in traits that may affect the stock–recruitment relationship of a wild population (Ricker 1975). This study was designed to test the hypothesis that there are no genetic differences in growth rate or survival between offspring from matings of hatchery × hatchery (HH), hatchery × wild (HW), and wild × wild (WW) summer steelhead trout, *S. gairdneri*.

Materials and Methods

Summer steelhead trout were captured during their upstream migration in a trap at Pelton reregulating dam on the Deschutes River, a tributary of the Columbia River, during fall and winter 1974–75. They were held at the Oregon Department of Fish and Wildlife Round Butte Hatchery until sexually mature. Hatchery fish could be identified because they are finclipped before release into the Deschutes River. As the terms are used in this paper, hatchery fish are reared in a hatchery for about 1 yr before their release, and wild fish result from natural reproduction.

Eyed eggs or unfed swim-up fry from matings of these adults subsequently were placed in four small streams in the upper Trout Creek drainage which flows into the Deschutes River. Average weekly flows in these streams exceeded 28 ℓ/s only during periods of snowmelt or unusually wet weather. Water temperatures ranged from 0°C during winter and spring to 26°C during summer. Steelhead and resident rainbow trout and longnose dace, *Rhinichthys cataractae*, are known to occur in these streams. In January 1975 screens were installed in these streams to delineate four study sections (one each, 1.6 km long, in Opal, Dutchman, and Potlid creeks; and one, 0.8 km long, in Trout Creek) and to prevent steelhead and resident rainbow trout from entering the sections for spawning. Trout longer than 6.5 cm were removed from the sections by repeated electrofishing. Downstream traps were installed at each screen to capture out-migrating fish.

On February 24, 1975, each adult was killed, its eggs or sperm placed in a numbered container and stored at 4°C and a sample of liver tissue analyzed electrophoretically to determine the lactate dehydrogenase-4 (LDH-4) genotype of each fish (Utter et al. 1973). Preliminary analyses indicated the occurrence of three LDH genotypes, BB, $B'B$, and $B'B'$, which are known to exhibit simple Mendelian inheritance (Morrison and Wright 1966). After the genotypes were determined, the stored gametes were combined into three sets of fertilized eggs. Each set consisted of groups of eggs of each LDH genotype and each mating as follows:

	Matings		
Sets	HH	HW	WW
I	$B'B'$	BB	$B'B$
II	BB	$B'B$	$B'B'$
III	$B'B$	$B'B'$	BB

The numbers of adults used in these matings were:

Hatchery				Wild			
Males		Females		Males		Females	
BB	$B'B'$	BB	$B'B'$	BB	$B'B'$	BB	$B'B'$
13	14	13	16	4	4	4	4

Non-LDH genotypic variability was maximized in the offspring by dividing the eggs from each female into a number of subsamples equal to the number of males to be used in the mating. Each subsample of eggs was then fertilized by sperm from one male. Thus, a mating representing m males and f females consisted of $m \times f$ individual matings.

It was necessary to repeat the spawning procedure on a second date (March 18) because of low survival in one mating of set III. Matings for sets II and III were repeated at this time. The numbers of adults used in these matings were:

Hatchery				Wild			
Males		Females		Males		Females	
BB	$B'B'$	BB	$B'B'$	BB	$B'B'$	BB	$B'B'$
13	14	13	20	5	12	4	6

Fertilization was delayed for < 12 h for all matings on both dates. The fertilized eggs were placed in Heath® incubator trays with a 23–27 ℓ/min flow of 10°C water. The eggs were left undisturbed until they reached the eyed stage; dead eggs were then removed.

Eggs from both spawning dates were used to stock the study sections (Table 1). Vibert® boxes, each containing 600 eggs, were marked to identify the eggs they contained, and three boxes of eggs (from each mating: HH, HW, WW) were placed in artificial redds excavated at regular intervals in the substrate of the study sections in Dutchman (15 redds), Potlid (15), and Trout (10) creeks. Each depression was then filled with gravel 1.5–7 cm in diameter. Potlid and Dutchman creeks were stocked with 9,000 eggs from each mating and Trout Creek with 6,000 from each. The Vibert® boxes were removed from the redds after fry emergence from the gravel was complete, and dead embryos were counted.

Opal Creek was stocked with 22,500 swim-up fry and a hatchery pond at Round Butte Hatchery with 18,000 (Table 1). The fry stocked in the hatchery pond were reared according to standard practices by personnel at Round Butte Hatchery.

Samples of juvenile fish were captured from the sections by electrofishing several times during the period from July 15, 1975 to April 8, 1976. These samples and the fish removed from the traps were frozen and taken to the laboratory where they were thawed and measured (fork length). The LDH genotype of each fish was determined from eye-tissue homogenates.

Results

Among eyed eggs from the three streams, survival of WW embryos was significantly greater ($P < 0.05$) than that of HH and HW embryos

TABLE 1. Lactate dehydrogenase genotypes of eyed eggs (E) of steelhead trout (*Salmo gairdneri*) placed in artificial redds and unfed swim-up fry (F) stocked in experimental streams and a hatchery pond. HH = hatchery × hatchery; HW = hatchery × wild; WW = wild × wild.

Steelhead stocked		Matings		
Location	Life stage	HH	HW	WW
Potlid Creek[a]	E	$B'B'$	BB	$B'B$
Hatchery Pond[a]	F	BB	$B'B$	$B'B'$
Dutchman Creek[b]	E	BB	$B'B$	$B'B'$
Trout Creek[b,c]	E	$B'B$	$B'B'$	BB
Opal Creek[b,c]	F	$B'B$	$B'B'$	BB

[a]Eggs fertilized February 24, 1975.
[b]Eggs fertilized March 18, 1975.
[c]Opal and Trout creeks stocked with fish from the same matings.

(Table 2). (These data did not include one Vibert® box of WW embryos from Potlid Creek and five boxes — two HH, two HW, and one WW — from Trout Creek that had been washed out of their redds.) The variance of the arc sine square root of percentage mortality for groups of eyed eggs was analyzed following the technique of least squares analysis of a linear model (Draper and Smith 1966) to overcome the problem of unequal sample sizes.

The traps at the lower ends of each study section were functional and assumed to have captured all downstream migrants during the period from their installation in April to November 8, 1975, except in Trout Creek where high water caused the trap to be inoperative during portions of May and June. The traps were not functional for most of the period after November 8, 1975. Less than 500 experimental fish from each stream were captured in the traps (Table 3).

In all streams where the numbers of juveniles from the experimental matings differed significantly, progeny of WW matings were most abundant (Table 4). In the hatchery pond, HH fish were the most abundant. Overwinter (October–

TABLE 2. Percent survival for different groups of eyed eggs stocked in three experimental streams (HH = hatchery × hatchery; HW = hatchery × wild; WW = wild × wild). *9,000/mating; **6,000/mating.

Creek	Total no. stocked	Matings		
		HH	HW	WW
Potlid	27,000*	89	83	93
Dutchman	27,000*	70	72	78
Trout	18,000**	75	84	86
Combined creeks	72,000	78.4	79.5	86.1

TABLE 3. Frequency of fish from three experimental matings that were recovered from traps before October 1975.

Creek	Matings			χ^{2a}
	HH	HW	WW	
Opal	135	121	120	1.1
Potlid	52	228	105	141.1[b]
Dutchman	122	99	190	23.4[b]
Trout	75	88	81	2.4

[a]Expected values for the chi-square calculations were derived by using data from Table 3 and assuming that eggs from dislocated Vibert® boxes made no contribution to the experimental populations.
[b]Statistically significant at the 0.005 level.

March) shifts in relative abundance occurred in Opal Creek, where survival of WW fish was significantly greater than that of HH fish in the fall but not in the spring. In Potlid Creek, frequencies in the fall sample indicated no significant difference in relative abundance, and frequencies in the spring sample indicated significantly greater survival of WW than of HH fish.

Out-migration rates of 0% for HH, 31% for HW and 69% WW would have produced the observed frequency change between fall and spring in Opal Creek. The only trap sample obtained after September, which was from Opal Creek on December 1, 1975, consisted of only eight fish — two HW and six WW. Deschutes River summer steelhead trout fry from tributary streams (Fessler 1974) and summer steelhead and resident rainbow trout from several Idaho streams (Bjornn 1971) are known to migrate downstream during the fall and winter. A differential out-migration in Opal Creek but not in Trout Creek, both of which contained fish from the same matings, may have been due to the more extreme winter conditions in Opal Creek. Trout and Potlid creeks were covered with ice during most of the winter, but Opal Creek often was not. The recurrent absence of a protective covering of ice resulted in greater fluctuations in temperature in Opal Creek and ice frequently formed on the stream bottom.

In Opal and Potlid creeks, HW fish were larger ($P < 0.005$) than HH and WW fish (Table 4). Mean lengths of fish from Dutchman and from Trout creeks were not significantly different. In the hatchery pond, the HH fish were significantly larger than the HW and WW fish ($P < 0.01$).

Differences in the environmental conditions from stream to stream and sampling error resulting from the relatively small numbers of wild adults used in the matings were possible causes for the lack of consistency observed in the relative

TABLE 4. Frequencies (percent) and mean lengths (millimeters) in parentheses, of fish from experimental matings in samples from streams and a hatchery pond. χ^2 compares numbers of fish from each mating.

Location and collection date	No. fish	Matings			χ^{2a}
		HH	HW	WW	
Opal Creek					
July 14, 1975	222	34(46)	29(46)	37(45)	2.0
September 9, 1975	159[b]	26(59)	32(62)	42(59)	5.6[c]
October 25, 1975	464	26(60)	30(63)	44(61)	27.5[c]
March 3, 1976	528	33(63)	32(65)	35(63)	0.8
Potlid Creek					
September 9, 1975	130	31(54)	37(57)	32(55)	0.7
November 4, 1975	316	39(59)	31(60)	30(60)	2.6
March 3, 1976	296	28(62)	33(65)	39(63)	7.2[c]
Dutchman Creek					
August 17, 1975	453	33(55)	33(54)	34(54)	0.1
November 7, 1975	422	32(62)	36(63)	32(61)	2.7
February 29, 1976	234	29(70)	37(68)	34(70)	2.6
Trout Creek					
November 4, 1975	435	23(51)	29(52)	48(51)	27.4[c]
April 8, 1976	115	22(54)	36(52)	42(52)	7.8[c]
Hatchery Pond					
August 16, 1975	497	40(60)	31(56)	29(56)	10.8[c]

[a]Expected values were calculated by assuming that embryos from dislocated boxes made no contribution to the experimental populations.
[b]Measured lengths of 158 fish.
[c]Statistically significant at the 0.05 level.

lengths and abundances of fish in the different study sections. Analysis of the variance in length for fish obtained in the fall and spring from all streams and inspection of the relative frequency data suggested that there was no selective advantage associated with individual LDH genotypes of fish that were stocked in the streams.

Discussion

There were genetic differences in growth rate and survival between the offspring of hatchery and wild steelhead. Differences in the numbers of fish recovered from the traps, although not consistent, suggested that there were behavioral differences between offspring of hatchery and wild fish; however, the effect of migration differences on the relative growth and survival of fish from these matings was not clear.

The observed differences in survival suggested that the short-term effect of hatchery adults spawning in the wild is the production of fewer smolts and ultimately, fewer returning adults than are produced from the same number of only wild spawners. This effect depends on the particular limiting factors and environmental conditions in a given stream system and on the total number of spawners. In some stream systems, the same total number of smolts may be obtained simply by allowing more hatchery adults to spawn (Fig. 1A), whereas in other systems, numbers of smolts may be reduced regardless of how many hatchery adults spawn (Fig. 1B).

The length data for fish from Potlid and Opal creeks indicated that there were genetic differences between offspring from the different matings surviving after 1 full yr of exposure to natural selection in the stream environment. If these and other genetic differences persist until the offspring of hatchery and wild fish return as adults, there will be an additional effect on the wild population. The expected result is a reduction in the overall reproductive success of the wild population (long-term effect in Fig. 2). The relative magnitude of short-term and long-term effects would be expected to vary through time. In years in which the environment is unusually favorable for survival of fish, the short-term effect would be expected to be reduced, but this reduction in turn would increase the long-term effect in later years when the "mild year offspring" returned as adults and spawned in the stream. Conversely, in years in which the environment is unusually harsh, the short-term effect would be expected to be high, and the long-term effect low.

The results may provide only a conservative

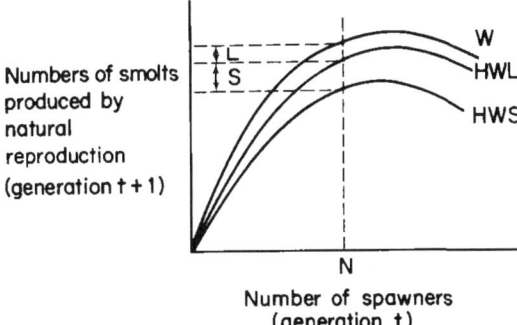

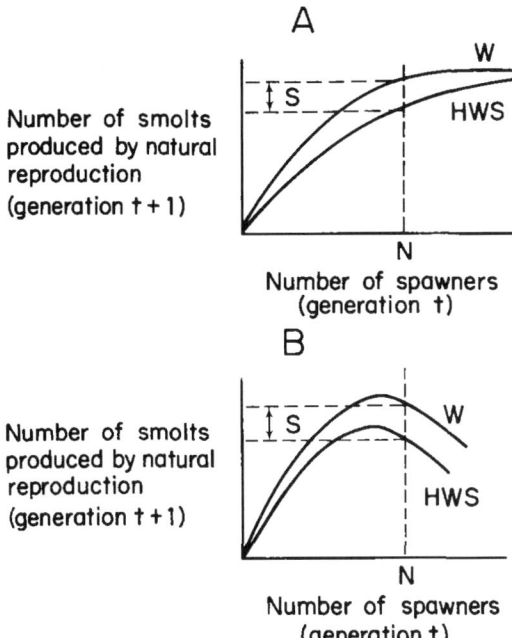

FIG. 1. Hypothetical stock–recruitment relationships. Numbers of smolts produced at high spawner densities are equivalent (A); never equivalent (B). See text for explanation. Short-term effect at an escapement of N spawners is S. W = the relationship for all wild spawners with no previous hatchery introductions. HWS = the relationship due to hatchery adults interbreeding with wild adults. HWS varies with the proportion of the total spawners that are hatchery fish; this proportion increases the distance between HWS and W. Available spawning area not limited (A); limited (B).

FIG. 2. Suggested effects of interbreeding between hatchery and wild fish and a hypothetical stock–recruitment relationship for naturally reproducing fish. Total effect at an escapement of N spawners is L + S. W = the relationship for all wild spawners with no previous hatchery introductions; HWL = the relationship resulting from genetic differences between wild adults that are offspring of hatchery and wild fish; HWS = the relationship resulting from interbreeding of hatchery wild adults. S = short-term effect and L = long-term effect (see text for explanation).

estimate of the effects of interbreeding between hatchery and wild fish. Most of the adult hatchery fish captured in the 1974–75 winter were no more than two generations removed from wild steelhead trout parents. As a stock of fish is subject to the selective forces of the hatchery environment over more generations, the genetic difference between the hatchery fish and the original wild stock will probably increase. In addition, a large number of adult hatchery fish spawn in the river immediately below the Pelton reregulating dam; consequently, some of the wild fish used for this experiment are likely to have had a hatchery fish for a parent. Finally, making all matings for each stream in the hatchery on the same day eliminated any potential effects due to differences in adult behavior. Differences in spawning time and other adult behavior may have a considerable effect on the relative performance of offspring of hatchery and wild adults. Hatchery steelhead from the Deschutes River tend to spawn earlier than the wild fish. If natural selection in the stream system is regulating spawning time for optimal survival, then in most years poorer survival would be expected from fish spawning earlier than wild fish.

The smaller the genetic difference between the hatchery fish and wild fish the smaller the expected effect on the wild population. The genetic difference can be reduced by using native wild fish which are adapted to that particular stream system for brood stock to initiate the hatchery program and in subsequent generations. Hatchery practices could be modified to reduce artificial selective pressures such as selection for early time-of-return and rapid growth in the hatchery. These modifications extend the period of egg collection and prevent destruction of slow-growing individuals (grade-outs). This decrease in efficiency represents the cost of reducing the genetic differences between hatchery and wild fish.

BJORNN, T. C. 1971. Trout and salmon movements in two Idaho streams as related to temperature, food, stream flow, cover, and population density. Trans. Am. Fish. Soc. 100: 423–438.

DRAPER, N. R., AND H. SMITH. 1966. Applied regression analysis. John Wiley and Sons, Inc., New York, N.Y. 407 p.

FENDERSON, O. C., W. H. EVERHART, AND K. M. MUTH. 1968. Comparative agonistic and feeding behavior of hatchery-reared and wild Atlantic salmon in aquaria. J. Fish. Res. Board Can. 25: 1–14.

FESSLER, J. 1974. Determination of factors limiting efficiency of hatchery production and the population characteristics and life history of Deschutes River summer steelhead. Oreg. State Wildl. Comm., Proj. F-88-R-6: 27 p.

FLICK, W. A., AND D. A. WEBSTER. 1964. Comparative first year survival and production in wild and domestic strains of brook trout (*Salvelinus fontinalis*). Trans. Am. Fish. Soc. 93: 58–69.

GREENE, C. W. 1952. Results from stocking brook trout of wild and hatchery strains at Stillwater Pond. Trans. Am. Fish. Soc. 81: 43–52.

HOCHACHKA, P. W. 1961. Liver glycogen reserves of interacting resident and introduced trout populations. J. Fish. Res. Board Can. 18: 125–135.

MORRISON, W. J., AND J. E. WRIGHT. 1966. Genetic analysis of three lactate dehydrogenase isozyme systems in trout: evidence for linkage of genes coding subunits A and B. J. Exp. Zool. 163: 259–270.

NEEDHAM, P. R., AND D. W. SLATER. 1944. Survival of hatchery-reared brown and rainbow trout as affected by wild trout populations. J. Wildl. Manage. 8: 22–36.

REIMERS, N. 1957. Some aspects of the relation between stream foods and trout survival. Calif. Fish Game 43: 43–69.

RICKER, W. E. 1975. Computation and interpretation of biological statistics of fish populations. Bull. Fish. Res. Board Can. 191: 382 p.

UTTER, F. M., F. W. ALLENDORF, AND H. O. HODGINS. 1973. Genetic variability and relationsips in Pacific salmon and related trout based on protein variations. Syst. Zool. 22: 257–270.

VINCENT, E. R. 1972. Effect of stocking catchable trout on wild trout populations. Proc. Annu. Conf. West. Assoc. State Game Fish Comm. 52: 602–608.

Factors Influencing Survival and Growth of Stocked Walleye (*Stizostedion vitreum*) in a Centrarchid-Dominated Impoundment

Victor J. Santucci, Jr.[1]

Ridge Lake Station, Illinois Natural History Survey, R.R. 1, Box 233, Charleston, IL 61920, USA

and David H. Wahl

Center for Aquatic Ecology, Illinois Natural History Survey and Department of Ecology, Ethology, and Evolution, University of Illinois, 607 E. Peabody Drive, Champaign, IL 61820, USA

Santucci, V.J. Jr., and D.H. Wahl. 1993. Factors influencing survival and growth of stocked walleye (*Stizostedion vitreum*) in a centrarchid-dominated impoundment. Can. J. Fish. Aquat. Sci. 50: 1548–1558.

We compared survival and growth of fry and small (mean total length = 48–61 mm), medium (132–145 mm), and large (186–216 mm) fingerling walleye (*Stizostedion vitreum*) stocked for 4 yr in a centrarchid-dominated impoundment. Mean survival based on fall population estimates 1 and 2 yr after stocking indicated highest survival for large fingerlings (mean survival = 31 and 10%, respectively), followed by medium ones (7 and 4%). Few individuals from the fry and small fingerling size groups were recovered in extensive field sampling. Creel census data reinforced these findings. Thermal stress at stocking and predation by largemouth bass (*Micropterus salmoides*) were more important than either hooking mortality or spillway escapement in influencing survival. Walleye diets were dominated by bluegill (*Lepomis macrochirus*) in volume (87%) and frequency of occurrence (84%). Growth rates were slower with bluegill as predominant prey compared with walleye growth in waters containing clupeids and cyprinids and may have been influenced by the abundance and size distribution of bluegill. Based on benefit/cost analysis (survival or catch/cost of rearing), stocking walleye >200 mm provided the highest return on investment.

Nous avons comparé la survie et la croissance d'alevins et d'alevins de moins d'un an, de petite (longueur totale moyenne = 48–61 mm), de moyenne (132–145 mm) et de grande (186–216 mm) taille, chez le doré (*Stizostedion vitreum*), ensemencés pendant 4 ans dans un réservoir dominé par les centrarchidés. La survie moyenne, basée sur des estimations démographiques faites à l'automne 1 et 2 ans après l'ensemencement, indique que le taux de survie est le plus élevé chez les alevins d'un an de grande taille (survie moyenne = 31 et 10 % respectivement); ils sont suivis en cela par ceux de taille moyenne (7 et 4 %). Peu de sujets appartenant au groupe des alevins et des alevins d'un an de petite taille ont été capturés lors d'échantillonnages extensifs. Les données provenant des relevés des prises confirment ces observations. Le stress thermique au moment de l'ensemencement et la prédation par l'achigan à grande bouche (*Micropterus salmoides*) ont plus d'importance que la mortalité attribuable à des blessures causées par les hameçons ou aux échappées par le déversoir, en termes de survie. Le régime alimentaire du doré est dominé par le crapet arlequin (*Lepomis macrohirus*) en volume (87 %) et en fréquence (84 %). Lorsque le doré se nourrit principalement de crapets arlequins, le taux de croissance est inférieur à ce qu'il est lorsqu'il grandit dans des eaux qui abritent des clupéidés et des cyprinidés; le taux de croissance peut être influencé par l'abondance et la distribution selon la taille du crapet arlequin. Une analyse des bénéfices par rapport au coût (survie ou capture/coût de l'élevage), nous apprend que l'ensemencement avec des dorés d'au moins 200 mm est la formule la plus rentable.

Received June 10, 1992
Accepted January 28, 1993
(JB518)

Reçu le 10 juin 1992
Accepté le 28 janvier 1993

Stocking of walleye (*Stizostedion vitreum*) is often required to maintain populations because of overexploitation (Anthony and Jorgensen 1977; Schupp and Macins 1977) or because successful reproduction is precluded by factors such as inadequate spawning substrate (Johnson 1961; Ney 1978; Prentice and Clark 1978), inappropriate water temperatures in winter (Hokanson 1977) or at spawning time (Prentice and Clark 1978), and egg predation (Wolfert et al. 1975). The potential for walleye to contribute to a sport fishery depends on the survival and growth of stocked fish. Survival and growth of various sizes of stocked fry and fingerling walleye have been estimated frequently (Laarman 1978), but not in small centrarchid-dominated impoundments. Because small impoundments are an important and often heavily utilized resource (Anderson 1976), managers are interested in either providing additional sport species or maintaining existing populations in these systems.

To be successful in waters dominated by centrarchids, stocked walleye must avoid predators such as largemouth bass (*Micropterus salmoides*) and consume prey such as bluegill (*Lepomis macrochirus*). Whereas largemouth bass predation is an important cause of mortality of other stocked species (Krummrich and Heidinger 1973; Stein et al. 1981; Wahl and Stein 1989), predation has not been evaluated as a source of mortality of stocked walleye. Walleye will consume centrarchids in lakes lacking other suitable forage (Schneider 1975;

[1]Present address: Max McGraw Wildlife Foundation, P.O. Box 9, Dundee, IL 60118, USA.

TABLE 1. Number, density, and mean total length ($N \geq 50$) of walleye fry and three size groups of fingerlings stocked in Ridge Lake, Illinois, 1987–90.

Year	Date	Size group	Number	Density (fish·ha^{-1})	Mean length ±95% CI (mm)
1987	18 June	Small	675	120	61 ± 1.4
	3 Nov.	Medium	364	65	142 ± 2.0
	3 Nov.	Large	120	21	208 ± 4.5
1988	20 June	Small	765	137	61 ± 1.6
	7 Nov.	Medium	337	60	145 ± 2.3
	7 Nov.	Large	145	26	216 ± 3.7
1989	27 Apr.	Fry	14000	2500	9 ± 0.1
	31 May	Small	670	120	48 ± 1.2
	20 Nov.	Medium	385	69	132 ± 1.5
	21 Nov.	Large	81	14	186 ± 3.1
1990	25 Apr.	Fry	30000	5357	9 ± 0.2
	25 June	Small	702	125	56 ± 1.1
	20 Oct.	Medium	237	42	140 ± 2.4
	20 Oct.	Large	260	46	211 ± 2.2

Beyerle 1978; Paxton and Stevenson 1978), but little is known about the influence of centrarchid prey densities and size distributions on walleye feeding and growth. In predator-free ponds and small lakes, walleye had lower survival and grew more slowly with bluegill forage than with cyprinid forage (Schneider 1975; Beyerle 1978). Because centrarchid predators were lacking in these studies, additional information regarding the stocking success of walleye in small impoundments containing established centrarchid predator and prey populations is needed.

We evaluated survival and growth of walleye fry and fingerlings stocked in an impoundment containing an established centrarchid community. We examined potential sources of mortality of stocked walleye of various sizes, including stocking stress, predation from largemouth bass, hooking mortality from angling, spillway escapement, and use of forage. In addition, we used relationships among production costs for various-sized walleye and estimates of survival to determine the most economical size for stocking. With an understanding of the mechanisms affecting survival of stocked fish and benefit/cost analyses, we make recommendations regarding walleye stocking in centrarchid-dominated systems.

Study Area

Ridge Lake, Illinois (39°27'N, 80°09'W), is a 5.6-ha experimental fishing lake with a maximum depth of 6.5 m and mean depth of 2.8 m. Typically, the lake is thermally stratified at a depth of 1–3 m during late May through early September; temperature in the epilimnion ranges from 19 to 33°C and the hypolimnion is anoxic. Mean summer Secchi disk depths are less than 1 m and moderate standing crops of macrophytes exist in the shallow regions. The primary overflow structure, a tower spillway, discharges water from the lake bottom and can also be used to drain the lake. When the capacity of the tower spillway (0.71 m^3·s^{-1}) is exceeded, water is discharged over an auxiliary surface spillway. Both spillways are equipped with downstream weirs (13-mm-mesh wire screen) designed to hold emigrating fish alive in water-retaining catch baskets.

Methods

Ridge Lake was drained in October 1985 and restocked during 1986 with juvenile and adult largemouth bass (27 kg·ha^{-1}), bluegill (15 kg·ha^{-1}), black crappie (*Pomoxis nigromaculatus*, 5 kg·ha^{-1}), and channel catfish (*Ictaluris punctatus*, 13 kg·ha^{-1}) obtained from area lakes and fish hatcheries. Walleye fry (total length = 9 mm) were stocked in April 1989 and 1990; fingerlings were stocked in May or June (60 mm) and October or November (145 and 215 mm) of each year from 1987 to 1990 (Table 1). Except for individuals of the smallest size group, all fingerlings were reared in ponds with natural forage. The smallest groups of fingerlings were transferred from ponds to raceways at 50 mm and reared to stocking size (about 60 mm) on artificial feed. Before stocking, a subsample of walleye was measured (total length, nearest millimetre) and weighed (nearest gram) and each fingerling group was marked in each year with unique fin clips (left or right pectoral or pelvic) detectable throughout the study; fry were unmarked. Fish were allowed to adjust to lake temperatures for a minimum of 30 min before stocking. Mean lengths and densities of stocked fry and fingerlings were similar among years, except in 1990 when stocking densities were higher (Table 1).

Stocking mortality of walleye was estimated by holding subsamples of fry ($N = 100$) and fingerlings ($N = 30$) in suspended cylindrical cages ($N = 3$ per stocking). Mortality in cages was used to estimate losses associated with hauling, handling, fin clipping, and temperature stress. Fry cages were plastic containers (114 L) whereas fingerling cages (942 L) were constructed of 6.4-mm-mesh plastic screening. Fry and fingerling cages extended below the lake surface to depths of 0.6 and 1.0 m, respectively. Numbers of dead and live fish were counted after 24 h; we observed no additional mortality after that period.

We assessed losses of stocked fingerlings to predation by examining the stomach contents of largemouth bass, walleye (from previous stockings), and black crappie on days 1 and 2 poststocking and for an additional 2 d during the week after walleye were stocked; fry were excluded from this assessment because they were too small to identify accurately. Largemouth bass were collected by anglers and electrofishing (3000-W AC, 230 V, three-phase) whereas walleye and black crappie were collected by electrofishing and trapnetting (1.8 × 0.9 m rectangular frame nets, 13-mm bar mesh netting, single 15-m lead). Stomach contents of largemouth bass and walleye were removed with clear acrylic tubes (Van Den Avyle and Roussel 1980); black crappie were killed for diet analysis.

We estimated population sizes of largemouth bass, walleye and black crappie in September and October of each year. Fish were captured for the marking census by electrofishing and trapnetting and were recaptured within 1 mo by electrofishing, angling, trapnetting, and gillnetting (monofilament nets, 46 m long × 1.8 m deep consisting of six panels with meshes of 19-, 25-, 32-, 38-, 45-, and 51-mm bar mesh). All fish were marked with a fin clip (upper caudal) and the population size of each species or size group was estimated with the Chapman modification of the Petersen formula (Ricker 1975).

Losses of stocked walleye to predators were estimated for each sample day when samples contained more than 10 individuals of a predator taxon. To estimate the number of walleye eaten on each day, we divided the number eaten by the number of predators examined and then multiplied the proportion of predators with walleye by the estimated number of predators in the population. The minimum length of predator included in each population estimate was determined from the maximum prey to predator length ratio (0.57) found for walleye eaten by predators in Ridge Lake. Daily estimates were summed to obtain a total estimate of the numbers of stocked walleye lost to predation (Wahl and Stein 1989).

A creel census measuring total angler effort, catch, and harvest was conducted while the lake was open to public fishing, late April through mid-October. In addition to providing angling statistics, this census allowed us to assess short-term hooking mortality and angling vulnerability of walleye and to supplement the population estimate and diet samples (Santucci and Wahl 1991). Fishing was by permit only and the lake was open 5 d per week (closed Mondays and Tuesdays). A single entry point provided access and only boat fishing (maximum of eight boats, three persons each) was allowed. The minimum legal length limit for largemouth bass and walleye was 357 mm. Before fishing, anglers were questioned as to their species preference and they were instructed to keep all boated fish in live wells. Except for walleye, fish were retrieved at a lakeside laboratory where they were measured (total length, nearest millimetre) and weighed (nearest gram) after which we returned sublegal-size and unwanted fish to the lake. Anglers were given flags to indicate that a walleye had been caught. Walleye were retrieved immediately and sublegal-size and unwanted fish were placed in a floating creel (see previous description of fingerling cages) to determine hooking mortality. After holding overnight (12–15 h), walleye were measured, weighed, and checked for fin clips and survivors were returned to the lake.

We monitored spillway escapement of walleye daily when water discharged over either the tower or the surface spillway. Weir catch baskets were checked frequently to avoid losses of retained fish to mammalian or avian predators; we saw no signs that predators were feeding at either spillway weir. Walleye found in weirs were discarded after they were measured and checked for fin clips.

Walleye were collected for diet analysis each month during April through November 1988–89 and April through mid-July 1990 by electrofishing and gillnetting. Sampling was discontinued in summer 1990 because gizzard shad (*Dorosoma cepedianum*) were accidentally introduced into the lake. Length (nearest millimetre) and weight (nearest gram) were recorded for each walleye for determination of growth. Stomach contents were identified to species for fish and to family for invertebrates; volumes were determined by water displacement. To examine size relationships between walleye and their prey, we measured total lengths (TL) of intact fish and standard lengths (SL) of partially digested fish from stomachs of walleye. For bluegill measured in standard length, total length (10–90 mm) was estimated as

$$TL = 1.28(SL) - 0.698, r^2 = 0.99, N = 180.$$

For the comparison of predator and prey length relationships, walleye were grouped into 25-mm length intervals and lengths of ingested bluegill were averaged for each interval.

The size structure and relative abundance of young bluegill, the most abundant forage fish, were estimated in September 1988–89 from three shoreline rotenone samples spaced equally around the lake. A 4-mm-mesh block net (30.5 m long × 1.8 m deep) was deployed to enclose a rectangular area (53–96 m^2) and rotenone was applied in the enclosed area at a 2–3 ppm concentration (Timmons et al. 1979). All recovered fish were counted and a subsample ($N > 200$) was measured.

The abundance of larval bluegill, a potential source of food for young walleye, was estimated at weekly intervals at night from April through August 1987–90. Two Miller high-speed samplers (1.6 m long × 14 cm in diameter, 0.5-mm mesh) were towed from the bow of a boat at about 2 m·s^{-1}; a calibrated flowmeter was used to estimate the volume of water filtered. Oblique tows sampled from the surface to a depth of 4 m along a transect on the central axis of the lake. All larvae were preserved in 95% ethanol, identified to species, and counted. In addition to providing estimates of larval bluegill abundance, ichthyoplankton samples provided evidence that no walleye reproduced during the study.

We obtained estimates of costs of walleye fry and fingerlings from several sources including commercial producers (American Fisheries Society 1982) and public extensive and intensive culture facilities. Walleye produced extensively are those raised on natural foods in ponds; intensive culture comprises fish raised on artificial feeds in raceways or tanks. All estimates included both direct and indirect costs of rearing. To assess which walleye size group was the most economical to stock, we used the estimated costs of rearing to determine the total cost for each stocking. Dividing the total cost by the numbers of walleye surviving after 1 and 2 yr and averaging these values across years provided mean costs per surviving walleye. Walleye survival estimates were based on the total number of walleye stocked, unadjusted for initial mortality. To determine the cost per walleye caught by anglers, we followed this same procedure substituting numbers caught by anglers for numbers surviving. We used the number of walleye caught by anglers for this assessment because numbers harvested during the study were low.

Except where indicated, statistical analyses were one-way analysis of variance (ANOVA) for randomized complete block designs (blocked by year) and Tukey's multiple comparisons, which allowed us to identify differences among size groups. An arcsine transformation was used on percentage data to stabilize the variance before statistical tests were completed (Steel and Torrie 1980).

Results

Survival

Survival of walleye fry and the smallest group of fingerlings was extremely low; only one individual from each of these size groups was recovered in 4 yr of extensive electrofishing, shoreline rotenone, trapnet, gillnet, seining, and angling

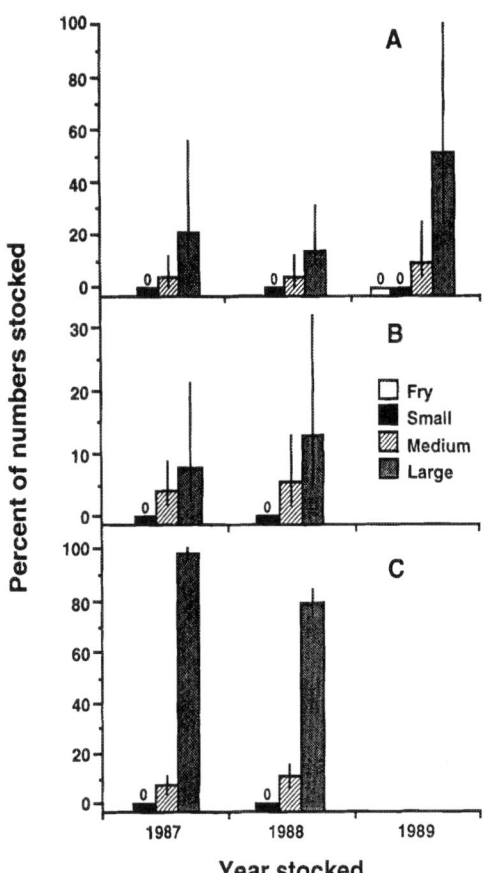

FIG. 1. Survival of walleye in fall during the (A) first and (B) second years after stocking and (C) percentages of stocked walleye caught by anglers in Ridge Lake, Illinois. Walleye were stocked as small (48–61 mm), medium (132–145 mm), and large (186–216 mm) fingerlings in 1987, 1988, and 1989; fry (mean length = 9 mm) were stocked in 1989. Values are percent survival of initial numbers stocked based on Petersen mark–recapture population estimates; angler catch data were obtained from a creel census during 1988–90. Vertical lines represent 95% CI. Note the different scales on each panel.

collections. For fingerlings, survival 1 yr after stocking differed among size groups (ANOVA, $F = 20.82$, df = 2,4, $P = 0.008$; Fig. 1A). Across years, large fingerlings (mean survival = 31%) had higher survival than small fingerlings (0%; Tukey's multiple comparisons, $T = 0.31, P = 0.007$). Survival was intermediate for medium fingerlings (7%) but was not different from either the large ($T = 0.04, P = 0.06$) or small size groups ($T = 0.07, P = 0.09$).

Compared with estimates after 1 yr, survival of medium and large fingerlings was lower 2 yr after stocking (Fig. 1B). Mean survival of walleye stocked in 1987 and 1988 declined from 6 to 4.5% for medium fingerlings and from 20 to 10% for large fingerlings. Despite the decline in survival for these larger size groups, patterns of survival observed 1 yr after stocking were also apparent the second year after stocking. After 2 yr, mean survival was highest for walleye stocked as large fingerlings

TABLE 2. Mean losses (± 95% CI) of walleye fry and three size groups of fingerlings due to hauling, handling, and temperature stress from four stockings in Ridge Lake, Illinois, 1987–90 (fry were not stocked in 1987 or 1988). Mortality was estimated for each stocking by holding subsamples of walleye ($N = 30$ fingerlings or $N = 100$ fry per cage) in cylindrical cages ($N = 3$) suspended in Ridge Lake for 24 h; fish were tempered for 30 min before stocking. Water temperature was recorded at 10 cm at stocking.

Group (mean length, mm)	Mortality (%)	Lake temperature at stocking (°C)	Temperature change from hatchery (range, °C)
Fry (9)	20 ± 1	22 ± 6	+10 to +11
Small (48–61)	22 ± 9	28 ± 24	+3 to +8
Medium (132–145)	1 ± 2	12 ± 6	−1 to +2
Large (186–216)	1 ± 1	12 ± 6	−3 to +2

(10%), followed by those stocked as medium fingerlings (4%) and then small fingerlings (0%; $T > 0.01, P < 0.05$).

Creel census data reinforced the patterns of survival observed after stocking. Anglers caught a higher percentage of walleye from the large size group (mean percentage of number stocked = 91%) than from either the medium (9%) or small size groups (0%; $T > 0.60, P < 0.03$; Fig. 1C). Differences between the means of the small and medium groups did not differ ($T < 0.17, P > 0.1$). To assess the influence of multiple recaptures on the angler catch, we marked walleye after they were caught by clipping dorsal spines and then counted the number of times they were recaptured. From this mark–recapture technique, we estimated that 33% of the walleye caught by anglers were caught more than once. These high initial catches and recapture rates occurred when fishing pressure was high at Ridge Lake (914–1003 angler·h·ha^{-1}). However, angler effort directed toward walleye was low (<10 angler·h·ha^{-1}).

Factors Influencing Survival

Mortality from stocking stress influenced first-year survival of stocked walleye in Ridge Lake. Stocking mortality was higher for fry (mean stocking mortality = 20%) and small fingerlings (22%) than for either the medium (1%) or large fingerlings (1%; $T > 0.04, P < 0.005$; Table 2). Fry and small fingerlings stocked in spring and early summer were exposed both to higher water temperatures at stocking and greater temperature changes between hatchery and lake than the larger fingerlings stocked in the fall (Table 2). However, thermal stress did not explain all of the differences in stocking mortality because we observed a large difference in mortality of small fingerlings in 1989 (7%) and 1990 (55%) when the lake temperatures were similar (27°C).

Neither walleye nor black crappie appeared to prey heavily on stocked walleye fingerlings. Only five walleye from the small size groups were found in other walleye stomachs ($N = 72$ examined), and none were found in black crappie stomachs ($N = 121$). Furthermore, estimated densities of walleye (<13·ha^{-1}) and black crappie (<1·trap net·d^{-1}, effort >45 net·d·yr^{-1}) were low during the study.

In contrast with other predators, substantial numbers of walleye fingerlings were recovered from largemouth bass stomachs. Losses to largemouth bass predation were higher for the small and medium walleye (mean percentage of number stocked = 6 and 17%, respectively) than for the large fingerlings (0%;

TABLE 3. Estimated numbers of stocked walleye fingerlings eaten by largemouth bass in Ridge Lake, Illinois, 1987–90. Estimates of predatory mortality were obtained on each sampling date by multiplying the number of walleye per largemouth bass stomach collected by night electrofishing by the number of largemouth bass in the population (ranges were determined from 95% CI). Summing these daily values provided a minimum estimate of the total number of walleye eaten. Minimum lengths of largemouth bass included in the population estimates were based on the maximum prey to predator length ratio (0.57) found for walleye and largemouth bass in Ridge Lake. Population estimates and 95% CI are based on Petersen estimates (Ricker 1975), except as noted.

Year	Walleye mean length (mm)	Minimum largemouth bass length (mm)	Largemouth bass population estimate (95% CI)	Number of largemouth bass examined	Estimated number of walleyes eaten (range)	Percentage of stocked walleyes eaten (range)
1987	61	107	1261 (903–1826)	117	77 (55–111)	11 (8–16)
	142	249	545 (327–965)	21	42 (25–74)	12 (7–20)
	208	366	57[a]	6	0	0
1988	61	107	1130 (832–1575)	256	73 (46–88)	10 (6–12)
	145	254	380 (248–608)	104	23 (15–36)	7 (4–11)
	216	376	16 (6–39)	5	0	0
1989	48	84	925 (688–1271)	303	0	0
	132	231	712 (501–1047)	46	89 (63–131)	23 (16–34)
	185	325	74 (44–131)	9	0	0
1990	56	98	549 (421–717)	125	79 (60–103)	11 (8–15)
	140	246	274 (199–388)	38	67 (49–95)	28 (21–40)
	211	370	9[a]	2	0	0

[a]Population estimates are based on the proportion of largemouth bass of all sizes greater than the minimum length.

$T > 0.06$, $P < 0.005$; Table 3). The low estimated predation on the small size group in 1989 occurred despite the fact that the number of largemouth bass capable of eating these fish was high ($N = 165 \cdot ha^{-1}$) and that we examined more largemouth bass ($N = 303$, 33% of the estimated population) than in other years ($N = 117$–256, 9–22%). For the largest size group, few largemouth bass ($N < 15 \cdot ha^{-1}$) in any year were large enough to eat these walleye. Predation by largemouth bass on stocked walleye did not appear to be related to the density and size structure of the largemouth bass population. We did not find a relationship between the percentage of stocked fish eaten and the density of largemouth bass of effective predatory size (linear regression, $r^2 = 0.10$, $P = 0.33$).

Vulnerability of walleye to largemouth bass was highest immediately after stocking. Of all the walleye recovered from largemouth bass stomachs in the week after stocking, 76% were eaten within 48 h of stocking. However, we also found evidence of longer term predation by largemouth bass on walleye in 1989 and 1990. Small numbers of walleye from the medium size groups stocked in fall 1988 and 1989 ($N = 3$ and 2, respectively) were found in largemouth bass stomachs up to 7 mo after they were stocked. However, because of the small number of walleye recovered, we were not able to quantify the impact of long-term predation on walleye survival.

We observed moderate losses of walleye (14%) that were caught by anglers and released into holding cages. Nearly all fish that died in the cages did so within 1 h of being caught. For all years combined, hooking mortality did not differ among sizes (Fig. 2; chi-square, $P = 0.54$), but losses were higher for walleye caught with live bait (18%, 95% CI = 13–23%) than for those caught with artificial lures (5%, 1–9%; chi-square, $P = 0.002$). As a source of losses of stocked fish, hooking mortality varied between size groups. Both within and among years, hooking mortality as a percentage of walleye stocked was higher for large fingerlings (4–12%) than for medium (0.3–2%) size groups (chi-square, $P < 0.03$). These differences between size groups were not related to size at cap-

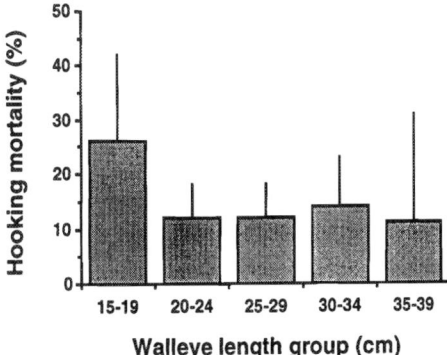

FIG. 2. Angler-induced mortality for 5-cm size groups of walleye (15–39 cm) as a percentage of the total catch in Ridge Lake, Illinois, 1988–90. Vertical lines represent Clopper–Pearson 95% CI.

ture, but resulted from lower stocking densities and higher catches of individuals from the large size groups.

Escapement losses were a small fraction of the number of walleye stocked (<2%); only 10 individuals were collected in the surface spillway weir and 11 in the tower spillway weir. All emigrating fish were from the medium and large fingerling stockings.

Diets and Growth

Forty-two percent of the walleye stomachs ($N = 102$) examined from 1988 through mid-1990 were empty. Diet analysis of young-of-year was precluded by the low survival of fry and small fingerlings and late fall stockings of the larger size groups. However, ichthyoplankton tows indicated that larval bluegill (<15 mm) were available as prey from May through mid-August of each year, mean densities (±95% CI) ranged from

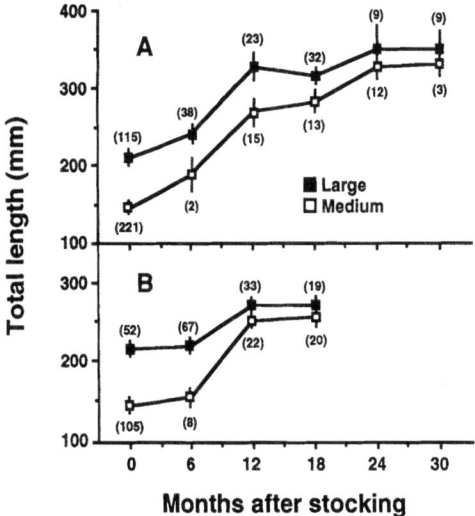

FIG. 3. Mean total lengths of medium (132–145 mm) and large (186–216 mm) walleye fingerlings following stocking in November (A) 1987 and (B) 1988 in Ridge Lake, Illinois. Walleye growth was assessed at 6-mo intervals corresponding to May and October each year after stocking. Vertical lines represent 95% CI; sample sizes are in parentheses.

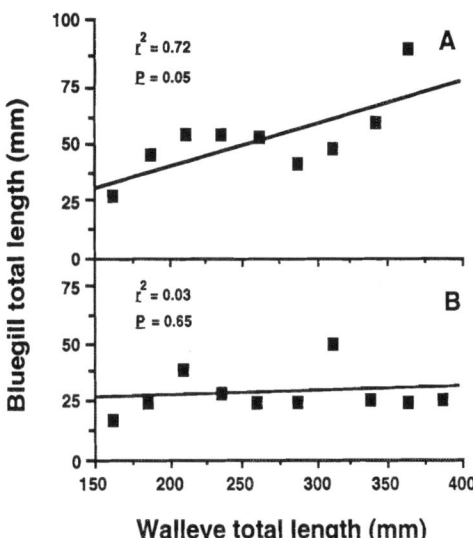

FIG. 4. Mean total lengths for bluegill eaten by walleye during (A) 1988 and (B) 1989 in Ridge Lake, Illinois. Data for walleye were combined within 25-mm length intervals. Walleye were collected monthly from April through November by electrofishing.

12 ± 6 to 64 ± 45 larvae·m^{-3}. For age-1 and older walleye, bluegill made up a higher proportion of the diet (87% of the volume; chi-square, partitioned df, $P = 0.0005$) and occurred in more of the stomachs containing food (84%) than all other prey. Primarily juvenile bluegill (16–90 mm) were eaten. Adult bluegill (>90 mm) were not eaten and larval bluegill were found in only a small proportion of walleye (<2% of the volume; frequency <10%). We did not find differences in diets among years (chi-square, $P = 0.20$), but diets of walleye in the fall differed from those earlier in the year (chi-square, partitioned df, $P = 0.004$) because only young bluegill were eaten in the fall. In spring and summer, other fish (including largemouth bass, black crappie, and unidentifiable fish remains) and invertebates combined to make up 13% of the volume of the walleye diets.

We observed differences in growth between size groups of stocked walleye. Individuals from the medium and large size groups grew from spring through fall, but growth slowed in winter (Fig. 3). Because growth rate was higher for the medium size group, mean lengths converged in 24–30 mo for walleye stocked in 1987 and in 12–18 mo for those stocked in 1988. In addition to the growth rate difference between size groups, growth rates differed between years. Walleye from the large size groups were larger after the 1988 growing season (mean length = 327 mm) than after the 1989 season (269 mm; t-test, $P = 0.001$). Likewise, first-year growth of the medium size groups was faster in 1988 (mean length = 266 mm) than in 1989 (253 mm; t-test, $P = 0.01$).

Differences in abundance and size distribution of young bluegill between years may explain the patterns of growth observed for various size groups of walleye within and between years. There was a positive relationship between mean lengths of ingested bluegill and lengths of walleye (160–380 mm) in 1988 (Pearson correlation, $r^2 = 0.72$, $P = 0.05$; Fig. 4A) but not in 1989 (Fig. 4B). The size of available prey may have limited growth rates of larger walleye during 1989. Our estimates of bluegill abundance in the lake also varied among years. Whereas densities of 10- to 34-mm bluegill in shoreline rotenone samples did not differ between 1988 and 1989 (mean densities = 31.2 and 9.3 bluegill·m^{-2}, respectively; t-test, $P = 0.15$), densities of larger bluegill in these samples 35–70 mm) were higher in 1988 (8.9 bluegill·m^{-2}) than in 1989 (1.3 bluegill·m^{-2}; t-test, $P = 0.0002$).

To further examine the relationship between forage base and growth, we compared literature values for growth rates of walleye from several lakes. For these comparisons, we chose lakes that (1) were located in the north-central region of the United States, (2) had the principal forage species identified, and (3) had growth data available for walleye ages 0–3. Growth rates in Ridge Lake (1987 stocking) were similar to those of walleye from other waters having centrarchid and other spiny-rayed prey species (two-way ANOVA, $F = 0.67$, df = 1,11, $P = 0.40$; Table 4). Length increments in lakes with gizzard shad or cyprinids were higher than those of walleye from lakes with spiny-rayed forage ($F = 20.04$, df = 1,31, $P = 0.001$). On average, walleye from gizzard shad and cyprinid waters were larger at age 0 and had faster growth rates than those from centrarchid lakes during each subsequent year.

Benefit–Cost Analysis

We found substantial differences in the estimated initial cost of walleye among sources; costs were lowest for extensively reared fish and were highest for those purchased commercially (Table 5). For each source except extensive culture, production costs increased approximately twofold for walleye reared to each successively larger size. Costs per survivor after 1 and 2 yr were higher for medium fingerlings than for large fingerlings from all sources (Table 5). Costs per angler-caught wall-

TABLE 4. Annual length increments (mm) for walleye ages 0–3 from Ridge Lake, Illinois, and other selected lakes and reservoirs. The two sets of values for Ridge Lake are for medium and large walleye stocked in 1987. Lakes chosen for comparison had the most abundant forage species identified and were located in the north-central region of the United States.

Water body (source)	Surface area (ha)	Age 0	Age 1	Age 2	Age 3	Available forage
Ridge Lake, Illinois (present study)	6	142	125	60	—	Centrarchids
		208	119	24	50	
Jewet Lake, Michigan (Schneider 1983)	5	139	155	45	9	Yellow perch (*Perca flavescens*)
Kildeer Reservoir, Ohio (Paxton and Stevenson 1978)	115	173	92	23	49	Centrarchids, yellow perch
Ferguson Reservoir, Ohio (Paxton and Stevenson 1978)	123	181	53	59	37	Centrarchids, yellow perch
Clinton Lake, Illinois (IDOC, 1986–88)[a]	2024	199	140	63	109	Gizzard shad
Stockton Lake, Missouri (Goddard and Redmond 1978)	10072	310	142	33	114	Gizzard shad
Pleasant Hill Reservoir, Ohio (Johnson et al. 1988)	344	240	132	84	63	Gizzard shad
Lake Erie, Ohio (Van Vooren and Davies 1974)	2.57×10^6	205	166	72	51	Gizzard shad, cyprinids
McConaughy Res., Nebraska (McCarraher et al. 1971)	14560	185	167	113	57	Gizzard shad

[a]G. Lutterbie, Illinois Department of Conservation, personal communication.

TABLE 5. Initial costs per individual (1989 US$), costs per survivor 1 and 2 yr after stocking (mean of 3 yr), and cost per individual caught by anglers for three sources of walleye fry and three sizes of fingerlings stocked in Ridge Lake, Illinois, 1987–89. Numbers of walleye surviving and numbers in the angler catch were determined from mark–recapture population estimates and a creel census, respectively. NC indicates no walleye were collected.

Size group (mean length, mm)	Intensive culture[a]				Extensive culture[b]				Commercial cost[c]			
	Initial cost per walleye	Cost per survivor at 12 mo	Cost per survivor at 24 mo	Cost per walleye caught by anglers	Initial cost per walleye	Cost per survivor at 12 mo	Cost per survivor at 24 mo	Cost per walleye caught by anglers	Initial cost per walleye	Cost per survivor at 12 mo	Cost per survivor at 24 mo	Cost per walleye caught by anglers
Fry (9)	0.008	NC	NC	NC	0.001	NC	NC	NC	—	NC	NC	NC
Small (48–61)	0.016	NC	NC	33.76	—	NC	NC	—	0.52	NC	NC	1097.20
Medium (132–145)	0.36	4.89	10.34	5.67	0.05	0.68	1.44	0.79	1.18	16.03	33.90	18.57
Large (186–216)	0.80	3.18	6.4	1.13	0.05	0.20	0.40	0.07	2.06	8.19	16.54	2.91

[a]S. Stuewe, Illinois Department of Conservation, personal communication.
[b]J. Daly and B. Parsons, Minnesota Department of Natural Resources, personal communication.
[c]American Fisheries Society (1982).

eye were highest for small fingerlings followed in order by medium and large fingerlings. Costs per survivor and cost per fish caught were substantially higher for intensively as compared with extensively reared walleye, but as expected, both were much lower than for fish purchased from commercial facilities. For all sources and all estimates of survival, large fingerlings (186–216 mm) were the most economical walleye to stock.

Discussion

Mechanisms Influencing Survival

Based on survival rates, stocking success of walleye in small centrarchid-dominated lakes will be low for fry and small fingerlings and moderate for larger fingerlings. In predator-free ponds and small lakes containing bluegill, survival of walleye fingerlings (100 mm) was also moderate (Schneider 1975; Beyerle 1978). Low survival in centrarchid waters may be related to prey availability. Jennings and Philipp (1992) observed a positive correlation between the success of walleye fry stockings and the density of small cladocerans. Because fish may become the major food of walleye as small as 30–40 mm (Maloney and Johnson 1957; Johnson et al. 1988), the abundance of larval and juvenile fish may influence survival of walleye fingerlings more than zooplankton densities. Prey density was not a likely factor influencing survival of stocked walleye fingerlings in Ridge Lake because larval and juvenile bluegill were present in high numbers throughout the summer and fall of each year. However, previous work has shown that prey type can influence survival; Schneider (1975) and Beyerle (1978) observed lower survival of walleye with bluegill than with minnows even though the density of edible-sized bluegill was higher than that of minnows. While low to moderate survival of walleye may be typical with bluegill prey, our results also indicate that other factors contributed to mortality of stocked walleye.

The size and time of year that walleye are available for stocking may influence initial mortality. Although we found mortality associated with thermal stress at stocking to be higher for smaller than for larger walleye, the fry and small fingerlings were always stocked at higher lake temperatures than the medium and large fingerlings. However, thermal stress does not explain all of the differences in stocking mortality because we found within-size-group differences in mortality unrelated to temperature. Factors potentially influencing survival of other stocked species, such as condition and health at stocking (Belusz 1978) and hauling or handling stress (Johnson and Metcalf 1982; Carmichael et al. 1984; Mather and Wahl 1989), might have caused higher losses of the smaller size groups of walleye.

Predation by largemouth bass was an important source of mortality for walleye. Small and medium size groups consistently experienced higher losses to predation and lower survival than large fingerlings. The absence of small fingerlings from stomachs in 1989 may have occurred because these walleye were smaller, digested more quickly, or in better condition than those stocked in other years. Temperature may have also biased our estimates of predation losses. We assumed that predator stomach contents represented 1 d of feeding. Shorter evacuation rates (<24 h at $\geq 27°C$; Hunt 1960; Beamish 1972) would result in underestimates of the numbers of small fingerlings eaten whereas longer evacuation rates (>24 h at $\leq 15°C$) would result in overestimates of the numbers of medium fingerlings consumed. As a result, our estimates of differences in losses to predation among size groups are conservative. Regardless, higher losses for the smaller size groups indicate that vulnerability to largemouth bass was related to size at stocking. Because relationships between size and vulnerability to predation have also been observed for a variety of other stocked species (Krummrich and Heidinger 1973; Shireman et al. 1978; Stein et al. 1981; Wahl and Stein 1989), evaluations of predator populations before stocking appear warranted.

We anticipated that losses would also be related to predator density because the density of largemouth bass capable of consuming each successively larger walleye size group declines. However, we did not observe a relationship between these variables in Ridge Lake. In contrast, Carline et al. (1986) observed a strong positive correlation between losses of stocked tiger muskellunge (*Esox masquinongy* $\times$ *E. lucius*) and largemouth bass densities. Relationships were developed from 14 stockings in six different lakes. With additional lakes and differing predator populations, a stronger link between walleye losses and predator density and size structure may be found.

The effect of angler exploitation on walleye survival was minimal during our study because most walleye had not attained legal size. However, we did observe moderate losses of walleye that were caught and released by anglers. These losses are consistent with estimates of hooking mortality reported for walleye in Minnesota ponds (Payer et al. 1989) but are above those reported in other lakes (Schaefer 1989). As a percentage of walleye stocked, hooking mortality does not explain the pattern of higher survival for larger fish; losses were higher for the large size groups than for the medium size groups. However, hooking mortality does account for a portion of total mortality and may partially explain the decline in walleye survival between the first and second years after stocking.

We observed high angler catches and recaptures of walleye despite the low effort directed toward this species. Walleye were not only vulnerable to angling, but they were vulnerable to anglers fishing for other species. Because angler exploitation rates increase with decreasing walleye age and length (Serns and Kempinger 1981), the small size (most were <350 mm) and young ages (age 3 and less) of walleye available during our study may partially explain the high catch rates. However, high angling vulnerability may be typical for walleye in small lakes because of the relative ease with which fish can be located (Beyerle 1978). Also, angler exploitation rates can be high when prey availability and walleye growth rates are low (Forney 1967). If slow growth is typical for walleye in impoundments containing centrarchid forage, then walleye catchability may also be high in these lakes.

Losses of walleye through reservoir discharges are well documented (Walburg 1971; Smith and Andersen 1984; Jernejcic 1986) and may result in substantial population declines in a lake (Groen and Schroeder 1978). Escapement losses probably depend on spillway design and flow rates. Survival of walleye fry has been shown to be influenced by lake discharges (Willis and Stephen 1987). The high watershed to lake surface area ratio (66:1) in Ridge Lake can result in substantial discharges. We observed high discharges within 2 wk of each fry stocking but were not able to assess losses due to the small size of fry at stocking. Although spillway escapement was not a major factor affecting the survival of walleye fingerlings, it may have been important for walleye fry.

Prey Selection and Growth

In small impoundments where soft-rayed prey is often lacking, abundant species such as bluegill will be an important food of walleye. Young bluegill were more abundant (11–40 bluegill·m^{-2}) than other potential fish prey (<0.2 fish·m^{-2}) and were the principal food of walleye in Ridge Lake. Bluegill were also found to be an important food item in other lakes where the abundance of alternative prey was limited (Dendy 1946; Paxton and Stevenson 1978) or where bluegill were the only piscine prey available (Schneider 1975; Beyerle 1978, Forsythe and Wren 1979). In contrast, diets of walleye from lakes with soft-rayed and spiny-rayed forage fishes contained higher percentages of minnows or clupeids than centrarchids (Range 1973; Boaze and Lackey 1974; Goddard and Redmond 1978; Johnson et al. 1988). Relative abundance and availability of these prey may have influenced walleye prey selection; however, evidence from at least some lakes suggests that walleye will select for soft-rayed taxa even when spiny-rayed forage is abundant (Parsons 1971; Wagner 1972; Knight et al. 1984).

The species available as prey may influence not only diet but growth of walleye. Comparing across several lakes, we found that walleye growth rates were slower in lakes with bluegill or other spiny-rayed prey species than in lakes predominated by gizzard shad or cyprinids. Because lakes with clupeids or cyprinids were larger than those with centrarchids, differences in available habitat among lakes may also have influenced walleye growth. However, previous pond and small lake studies have demonstrated slower growth of walleye with bluegill than with minnows as prey (Schneider 1975; Beyerle 1978). Extensive work with esocids has shown that forage species have inherent differences that can affect growth of predators. Esocids exhibit slower growth in centrarchid impoundments compared with gizzard shad (Weithman and Anderson 1977; Newman and Storck 1986; Wahl and Stein 1988) or fathead minnow impoundments (Gillen et al. 1981). Wahl and Stein (1988) suggested that esocid growth was slower because predators benefit less from bluegill prey; the caloric content of bluegill was lower and the costs of capture were higher than those of gizzard shad and minnows. Capture costs were influenced by prey morphology (body depth and the presence or absence of spines) and antipredatory behavior (Gillen et al. 1981; Moody et al. 1983; Wahl and Stein 1988). These inherent differences among prey species may partially explain the observed growth patterns of walleye among lakes with centrarchid and soft-rayed forage. However, further studies are needed to determine the specific effects of prey morphology, behavior, and energy content on walleye growth.

In lakes lacking soft-rayed forage species, bluegill abundance and size structure may influence walleye diet and growth. Diet analyses from lakes with a diverse size range of abundant prey indicate that walleye are size selective and that prey size typically increases with walleye size (Parsons 1971; Knight et al. 1984; Johnson et al. 1988). We found a positive predator to prey length relationship for walleye and bluegill during one year but not in another. Differences between years appeared to be related to the availability of preferred sizes of bluegill. The availability of appropriately sized bluegill also appeared important in determining growth of walleye. Walleye from medium and large size groups grew at similar rates when a range of bluegill sizes were available, but growth of larger walleye was reduced when primarily small bluegill were present. Although bluegill populations with a high relative abundance of small individuals may be typical in some small impoundments (Coble 1988), bluegill abundance and size structure will likely vary among lakes and years. Predicting walleye growth in these lakes will depend on our understanding of the factors influencing bluegill populations, such as environmental conditions (Stevenson et al. 1969), angling (Coble 1988), competition (Gerking 1966; Werner and Hall 1977), or predation (Mittelbach 1984).

Management Implications

The success of walleye stocking programs will depend largely on the survival and growth of stocked fish. Considerable variation has been observed in survival of stocked walleye among lakes (Laarman 1978; Ellison and Franzin 1992), across years within a lake (Schneider 1983; Jennings and Philipp 1992), and with size at stocking (Laarman 1981; Heidinger et al. 1985; Koppelman et al. 1992). Due to this variability, the success of any walleye stocking practice is largely unpredictable. Further complicating our understanding of walleye stocking success is the fact that variables affecting survival typically have not been identified. Our ability to improve the success and consistency of walleye stocking practices depends on increased understanding of the variables governing survival of stocked walleye. We found large fingerlings to have higher survival than smaller fingerlings or fry and that thermal stress at stocking and predation by largemouth bass were more important than hooking mortality or spillway escapement in determining walleye survival. By stocking walleye at least as large as 200 mm in the fall when lake temperatures have declined, we were able to reduce losses to largemouth bass predation and to thermal stress. Although initial costs are substantially higher for these large fingerlings compared with smaller fingerlings or fry, return on investment increased with walleye size and 200-mm fingerlings were the most economical walleye to stock. Unfortunately, growth of stocked walleye in small impoundments with centrarchid forage will be slower than that of walleye in lakes with other prey populations.

Survival of walleye stocked in some lakes has declined with successive stockings due to predation by survivors from prior stockings (Beyerle 1978; Schneider 1983) and has resulted in recommendations that fingerling walleye not be stocked in consecutive years. We did not observe reduced survival for successive walleye stockings because predation by largemouth bass was consistent across years and predation by walleye was low. By stocking walleye >200 mm, losses to the majority of resident predators may be reduced and stocking in consecutive years may be warranted. However, the effects of predator size structure and abundance on walleye survival rates may modify these recommendations.

The high vulnerability of walleye to angling in Ridge Lake and other small lakes (Beyerle 1978; Schneider 1979) indicates that exploitation could be high in these waters. Moderate hooking mortality for walleye suggests that protective size limits may lower fishing mortality. Minimum length limits are not recommended for populations having slow growth because densities of sublegal-size walleye may increase and further reduce growth rates (Serns 1978; Brousseau and Armstrong 1987). Slow growth in centrarchid impoundments may make protective slot limits more effective for managing walleye in these waters. Increased stocking densities may be warranted when protective slot regulations are enforced because anglers are allowed to harvest the smaller, more easily caught walleye in the population. Lower losses of walleye caught with artificial lures in our work and by others (Payer et al. 1989) suggest that restrictions on the use of live bait may be useful in reducing losses of walleye where size limits are enforced.

Acknowledgments

We thank T. Storck for initiating this study and for offering helpful suggestions throughout. Special thanks to S. Stuewe, L. Willis, and K. Cottrell of the Jake Wolf Memorial Fish Hatchery for providing walleye fry and fingerlings; A. Brandenburg of the Little Grassy Fish Hatchery for providing channel catfish; and J. Mick, D. Burkett, and G. Tickachek for coordinating activities with the Illinois Department of Conservation. K. Brewer, D. Coates, J. Corcoran, J. Finck, R. Mauk, M. Mounce, T. Patterson, and S. Shasteen assisted with field collections and laboratory analysis. R. Hunt provided computer assistance. P. Bayley, D. Clapp, E. Lewis, C. Mayer, J. Schneider, T. Stahl, and R. Stein provided helpful suggestions on earlier drafts of this manuscript. This study was supported by the Illinois Department of Conservation through Federal Aid in Sport Fish Restoration, project F-51-R. Additional support was provided by the Max McGraw Wildlife Foundation.

References

AMERICAN FISHERIES SOCIETY. 1982. Monetary values of freshwater fish and fish-kill counting guidelines. Am. Fish. Soc. Spec. Publ. 13.

ANDERSON, R.O. 1976. Management of small warmwater impoundments. Fisheries (Bethesda) 1(6): 5–7, 26–28.

ANTHONY, D.D., AND C.R. JORGENSEN. 1977. Factors in the declining contributions of walleye (*Stizostedion vitreum vitreum*) to the fishery of Lake Nipissing, Ontario, 1960–76. J. Fish. Res. Board Can. 34: 1703–1709.

BEAMISH, F.W.H. 1972. Ration size and digestion of largemouth bass, *Micropterus salmoides* Lacepede. Can. J. Zool. 50: 153–164.

BELUSZ, L.C. 1978. An evaluation of the muskellunge fishery of Lake Pomme de Terre and efforts to improve stocking success. Am. Fish. Soc. Spec. Publ. 11: 292–297.

BEYERLE, G.B. 1978. Survival, growth, and vulnerability to angling of northern pike and walleyes stocked as fingerlings in small lakes with bluegills or minnows. Am. Fish. Soc. Spec. Publ. 11: 135–139.

BOAZE, J.L., AND R.T. LACKEY. 1974. Age, growth, and utilization of the land locked alewives in Claytor Lake, Virginia. Prog. Fish-Cult. 36: 163–164.

BROUSSEAU, C.S., AND E.R. ARMSTRONG. 1987. The role of size limits in walleye management. Fisheries (Bethesda) 12(1): 2–5.

CARLINE, R.F., R.A. STEIN, AND L.M. RILEY. 1986. Effects of size at stocking, season, largemouth bass predation, and forage abundance on survival of tiger muskellunge. Am. Fish. Soc. Spec. Publ. 15: 151–167.

CARMICHAEL, G.J., J.R. TOMASSO, B.A. SIMCO, AND K.B. DAVIS. 1984. Characterization and alleviation of stress associated with hauling largemouth bass. Trans. Am. Fish. Soc. 13: 778–785.

COBLE, D.W. 1988. Effects of angling on bluegill populations: management implications. N. Am. J. Fish. Manage. 8: 277–283.

DENDY, J.S. 1946. Food of several species of fish, Norris Reservoir, Tennessee. J. Tenn. Acad. Sci. 21: 105–127.

ELLISON, D.G., AND W.G. FRANZIN. 1992. Overview of the symposium on walleye stocks and stocking. N. Am. J. Fish. Manage. 12: 271–275.

FORNEY, J.L. 1967. Estimates of biomass and mortality rates in a walleye population. N.Y. Fish Game J. 14: 176–192.

FORSYTHE, T.D., AND W.B. WREN. 1979. Predator–prey relationships among walleye and bluegill, p. 475–482. *In* R.H. Stroud and H. Clepper [ed.] Predator–prey systems in fisheries management. Sport Fishing Institute, Washington, D.C.

GERKING, S.D. 1966. Annual growth cycle, growth potential, and growth compensation in the bluegill sunfish in northern Indiana lakes. J. Fish. Res. Board Can. 23: 1923–1956.

GILLEN, A.L., R.A. STEIN, AND R.F. CARLINE. 1981. Predation by pellet-reared tiger muskellunge on minnows and bluegills in experimental systems. Trans. Am. Fish. Soc. 110: 197–209.

GODDARD, J.A., AND L.C. REDMOND. 1978. Northern pike, tiger muskellunge, and walleye populations in Stockton Lake, Missouri: a management evaluation. Am. Fish. Soc. Spec. Publ. 11: 313–319.

GROEN, L.G., AND T.A. SCHROEDER. 1978. Effects of water level management on walleye and other coolwater fishes in Kansas reservoirs. Am. Fish. Spec. Publ. 11: 278–283.

HEIDINGER, R.C., J.H. WADDELL, AND B.L. TETZLAFF. 1985. Relative survival of walleye fry versus fingerlings in two Illinois reservoirs. Proc. Annu. Conf. Southeast. Assoc. Fish Wildl. Agencies 39: 306–311.

HOKANSON, K.E.F. 1977. Temperature requirements of some percids and adaptations to the seasonal temperature cycle. J. Fish. Res. Board Can. 34: 1524–1550.

HUNT, B.P. 1960. Digestion rate and food consumption of Florida gar, warmouth, and largemouth bass. Trans. Am. Fish. Soc. 89: 206–211.

JENNINGS, M.J., AND D.P. PHILIPP. 1992. Use of allozyme markers to evaluate walleye stocking success. N. Am. J. Fish. Manage. 12: 285–290.

JERNEJCIC, F. 1986. Walleye migration through Tygart dam and angler utilization of the resulting tailwater and lake fisheries, p. 294–300. *In* G.E. Hall and M.J. Van Den Avyle [ed.] Reservoir fisheries management: strategies for the 80's. Reservoir Committee, Southern Division of the American Fisheries Society, Bethesda, Md.

JOHNSON, B.L., D.L. SMITH, AND R.F. CARLINE. 1988. Habitat preferences, survival, growth, foods, and harvests of walleyes and walleye × sauger hybrids. N. Am. J. Fish. Manage. 8: 292–304.

JOHNSON, D.L., AND M.T. METCALF. 1982. Causes and control of freshwater drum mortality during transportation. Trans. Am. Fish. Soc. 111: 58–62.

JOHNSON, F.H. 1961. Walleye egg survival during incubation on several types of bottom in Lake Winnibigosh, Minnesota, and connecting waters. Trans. Am. Fish. Soc. 90: 312–322.

KNIGHT, R.L., F.J. MARGRAF, AND R.F. CARLINE. 1984. Piscivory by walleyes and yellow perch in western Lake Erie. Trans. Am. Fish. Soc. 113: 677–693.

KOPPELMAN, J.B., K.P. SULLIVAN, AND P.J. JEFFRIES, JR. 1992. Survival of three sizes of genetically marked walleyes stocked into two Missouri impoundments. N. Am. J. Fish. Manage. 12: 291–299.

KRUMMRICH, J.T., AND R.C. HEIDINGER. 1973. Vulnerability of channel catfish to largemouth bass predation. Prog. Fish-Cult. 35: 173–175.

LAARMAN, P.W. 1978. Case histories of stocking walleyes in inland lakes, impoundments, and the Great Lakes — 100 years with walleyes. Am. Fish. Soc. Spec. Publ. 11: 254–260.

LAARMAN, P.W. 1981. Vital statistics of a Michigan fish population, with special emphasis on the effectiveness of stocking 15-cm walleye fingerlings. N. Am. J. Fish. Manage. 1: 177–185.

MALONEY, J.E., AND F.H. JOHNSON. 1957. Life histories and interrelationships of walleyes and yellow perch, especially during their first summer, in two Minnesota lakes. Trans. Am. Fish. Soc. 85: 191–202.

MATHER, M.E., AND D.H. WAHL. 1989. Comparative mortality of three esocids due to stocking stressors. Can. J. Fish. Aquat. Sci. 46: 214–217.

MCCARRAHER, D.B., M.L. MADSEN, AND R.E. THOMAS. 1971. Ecology and fishery management in McConaughy Reservoir, Nebraska. Am. Fish. Soc. Spec. Publ. 8: 299–311.

MITTELBACH, G.G. 1984. Predation and resource partitioning in two sunfishes (Centrarchidae). Ecology 62: 499–513.

MOODY, R.C., J.M. HELLAND, AND R.A. STEIN. 1983. Escape tactics used by bluegills and fathead minnows to avoid predation by tiger muskellunge. Environ. Biol. Fishes 8: 61–65.

NEWMAN, D.L., AND T.W. STORCK. 1986. Angler catch, growth, and hooking mortality of tiger muskellunge in small centrarchid-dominated impoundments. Am. Fish. Soc. Spec. Publ. 15: 346–351.

NEY, J.J. 1978. A synoptic review of yellow perch and walleye biology. Am. Fish. Soc. Spec. Publ. 11: 1–12.

PARSONS, J.W. 1971. Selective food preferences of walleyes of the 1959 year class in western Lake Erie. Trans. Am. Fish. Soc. 100: 474–485.

PAXTON, K.O., AND F. STEVENSON. 1978. Food, growth, and exploitation of percids in Ohio's upground reservoirs. Am. Fish. Soc. Spec. Publ. 11: 270–277.

PAYER, R.D., R.B. PIERCE, AND D.L. PEREIRA. 1989. Hooking mortality of walleyes caught on live and artificial baits. N. Am. J. Fish. Manage. 9: 188–192.

PRENTICE, J.A., AND R.D. CLARK, JR. 1978. Walleye fishery management program in Texas — a systems approach. Am. Fish. Soc. Spec. Publ. 11: 408–416.

RANGE, J.D. 1973. Growth of five species of game fishes before and after introduction of the threadfin shad into Dale Hollow Reservoir. Proc. Annu. Conf. Southeast. Assoc. Game Fish Comm. 26: 510–518.

RICKER, W.E. 1975. Computation and interpretation of biological statistics of fish populations. Bull. Fish. Res. Board Can. 191.

SANTUCCI, V.J. JR., AND D.H. WAHL. 1991. Use of a creel census and electrofishing to assess centrarchid populations. Am. Fish. Soc. Symp. 12: 481–491.

SCHAEFER, W.F. 1989. Hooking mortality of walleyes in a northwestern Ontario lake. N. Am. J. Fish. Manage. 9: 193–194.

SCHNEIDER, J.C. 1975. Survival, growth and food of 4-inch walleyes in ponds with invertebrates, sunfish or minnows. Mich. Dep. Nat. Resour. Fish. Res. Rep. 1833: 18 p.

SCHNEIDER, J.C. 1979. Survival, growth, and vulnerability to angling of walleyes stocked as fingerlings in a small lake with yellow perch and minnows. Mich. Dep. Nat. Resour. Fish. Res. Rep. 1875: 20 p.

SCHNEIDER, J.C. 1983. Experimental walleye-perch management in a small lake. Mich. Dep. Nat. Resour. Fish. Res. Rep. 1905: 30 p.

SCHUPP, D.H., AND V. MACINS. 1977. Trends in percid yields from Lake in the Woods, 1888–1973. J. Fish. Res. Board Can. 34: 1784–1791.

SERNS, S.L. 1978. Effects of a minimum size limit on the walleye population of a northern Wisconsin lake. Am. Fish. Soc. Spec. Publ. 11: 390–397.

SERNS, S.L., AND J.J. KEMPINGER. 1981. Relationship of angler exploitation to size, age and sex of walleyes in Escanaba Lake, Wisconsin. Trans. Am. Fish. Soc. 110: 216–220.

SHIREMAN, J.V., D.E. COLLE, AND R.W. ROTTMANN. 1978. Size limits to predation on grass carp by largemouth bass. Trans. Am. Fish. Soc. 107: 213–215.

SMITH, E.J., ADN J.K. ANDERSEN. 1984. Attempts to alleviate fish losses from Allegheny Reservoir, Pennsylvania and New York, using acoustics. N. Am. J. Fish. Manage. 4: 300–307.

STEEL, R.G.D., AND J.H. TORRIE. 1980. Principles and procedures of statistics: a biometrical approach. McGraw Hill, New York, N.Y.

STEIN, R.A., R.F. CARLINE, AND R.S. HAYWARD. 1981. Largemouth bass predation on stocked tiger muskellunge. Trans. Am. Fish. Soc. 110: 604–612.

STEVENSON, F., W.T. MOMOT, AND F.J. SVOBODA. 1969. Nesting success of the bluegill, *Lepomis macrochirus* Rafenesque, in a small Ohio farm pond. Ohio J. Sci. 69: 347–352.

TIMMONS, T.J., W.L. SHELTON, AND W.D. DAVIES. 1979. Sampling reservoir fish populations in littoral areas with rotenone. Proc. Annu. Conf. Southeast. Assoc. Fish Wildl. Agencies 32: 474–484.

VAN DEN AVYLE, M.J., AND J.E. ROUSSEL. 1980. Evaluation of a simple method for removing food items from live black bass. Prog. Fish-Cult. 42: 222–223.

VAN VOOREN, A.R., AND D.H. DAVIES. 1974. Lake Erie fisheries investigations. Ohio Department of Natural Resources, Federal Aid in Sport Fish Restoration Project F-35-R-12, Columbus, Ohio.

WAGNER, W.C. 1972. Utilization of alewife by inshore piscivorous fishes in Lake Michigan. Trans. Am. Fish. Soc. 101: 55–63.

WAHL, D.H., AND R.A. STEIN. 1988. Selective predation by three esocids: the role of prey behavior and morphology. Trans. Am. Fish. Soc. 117: 142–151.

WAHL, D.H., AND R.A. STEIN. 1989. Comparative vulnerability of three esocids to largemouth bass (*Micropterus salmoides*) predation. Can. J. Fish. Aquat. Sci. 46: 2095–2103.

WALBURG, C.H. 1971. Loss of young fish in reservoir discharge and year-class survival, Lewis and Clarke Lake, Missouri River. Am. Fish. Soc. Spec. Publ. 8: 441–448.

WEITHMAN, A.S., AND R.O. ANDERSON. 1977. Survival, growth, and prey of Esocidae in experimental systems. Trans. Am. Fish. Soc. 106: 424–430.

WERNER, E.E., AND D.J. HALL. 1977. Competition and niche shifts in two sunfishes (Centrarchidae). Ecology 58: 869–876.

WILLIS, D.W., AND J.L. STEPHEN. 1987. Relationship between storage ratio and population density, natural recruitment, and stocking success of walleye in Kansas reservoirs. N. Am. J. Fish. Manage. 7: 279–282.

WOLFERT, D.R., W.D.N. BUSCH, AND C.T. BAKER. 1975. Predation by fish on walleye eggs on a spawning reef in western Lake Erie, 1969–1971. Ohio J. Sci. 75: 118–125.

EXPERIMENTS WITH VARIOUS RATES OF STOCKING BLUEGILLS, *LEPOMIS MACROCHIRUS* RAFINESQUE, AND LARGEMOUTH BASS, *MICROPTERUS SALMOIDES* (LACÉPÈDE), IN PONDS

H. S. SWINGLE
Alabama Agricultural Experiment Station
Auburn, Alabama

ABSTRACT

Results from stocking adult bluegills and largemouth bass, fingerling bass and adult bluegills, and fingerling bluegills and fingerling largemouth bass are compared. Increase in rates of stocking bluegills from 8 to 1,500 per acre in bluegill-bass combinations gave corresponding increases in pounds of harvestable fish per acre and corresponding decreases in the cost per pound. The calculated A_T value (percentage harvestable fish) could be used to predict the success of various stocking rates. Unbalanced populations resulted from stocking with insufficient numbers of either largemouth bass or bluegills. When normally adequate numbers of each species were stocked, unbalanced populations occasionally resulted from extremely high mortality among the stocked fish. Where no fish were removed by fishing, an average of 25.6 percent of the stocked bass died during the first 6 months and an additional 20.4 percent during the following year; similarly an average of 15.4 percent of the stocked bluegills died during the first year and an additional 19.1 percent during the second year.

INTRODUCTION

The objective of various methods of stocking ponds with a combination of bluegills, *Lepomis macrochirus* Rafinesque, and largemouth black bass, *Micropterus salmoides* (Lacépède), is to establish fish populations from which it will be possible subsequently to harvest satisfactory annual crops of fishes. Such a population has been defined as a "balanced population" and is characterized by (Swingle, 1950): (1) a definite range in ratio (F/C) of weight of all forage fishes to the weight of all piscivorous fishes. The range in balanced populations is F/C = 1.4 to 10.0, and the most desirable range is F/C = 3.0 to 6.0; (2) a narrow range in the ratio (Y/C) of the weight of small forage fishes to the total weight of piscivorous fishes. The range in balanced populations is Y/C = 0.02 to 5.0, and the desirable range is Y/C = 1.0 to 3.0; (3) more than 40 percent (A_T value) of the total weight of the population in the form of fish of harvestable size. The desirable range is A_T = 60 to 85; (4) at least 18 and preferably more than 35 percent (A_F value) of the total weight of bluegills in the form of harvestable bluegills. These requirements for balanced populations obviously limit the stocking methods that may be used to obtain such results. Three principal methods of stocking appear possible.

Stocking Bluegills and Bass in Ponds

One method is to stock small numbers of adult or fingerling largemouth bass and bluegills, and to depend upon subsequent reproduction and predation to yield the correct numbers of each species. This method usually requires 2 or more years to establish a balanced population and would appear somewhat uncertain, as its success depends upon two variables, reproduction and predation, in two species of fishes.

A second method is to stock approximately 100 largemouth bass fry or fingerlings per acre and a small number of adult or fingerling bluegills. In this case, reproduction of and subsequent predation upon the latter species must furnish the correct number of bluegills to satisfy the requirements for balance.

The third method requires that the correct number of small bluegills and small bass be stocked so that 1 year later the weights of those which are still living will satisfy the A_T and A_F requirements and at the same time establish a desirable F/C ratio. Methods of calculating the numbers required for stocking when the carrying capacity of the pond is known are described elsewhere (Swingle, 1950). For example, the numbers recommended for stocking ponds in the Southeast (Swingle and Smith, 1940, 1942; Smith and Swingle, 1943) were 100 largemouth bass fry or fingerlings and 1,000 to 1,500 bluegill fingerlings per acre for fertilized ponds where the carrying capacity was determined to be between 300 and 500 pounds per acre. These recommendations have been checked in numerous experiments during the last 12 years and have been reasonably satisfactory in general usage. Balanced populations usually result in 1 year but do not, however, result invariably for reasons to be discussed later.

Experiments at the Alabama Agricultural Experiment Station dealing with these various methods of stocking will be reported and discussed in this paper.

STOCKING WITH ADULT FISH

Several experiments in which a pond was stocked with adult bluegills, adult largemouth bass, and other species were reported (Swingle and Smith, 1940, 1943a, b) to have failed because in such a combination there was no food for large adult bass. Consequently these fish failed to produce eggs the first year. In subsequent years they also failed because the pond was overcrowded with bluegills, crappie, *Pomoxis annularis* Rafinesque, and catfish, *Ameiurus natalis* (LeSueur).

In other experiments, it was found that young 6- to 8-ounce largemouth bass obtained enough food from insects and tadpoles to spawn successfully in every test whereas large adults lost weight on such a diet and failed to spawn. Consequently, it was decided to try adult stocking again, using largemouth bass averaging slightly less than 0.5 pound.

A 1.6-acre unfertilized pond was stocked with 10 largemouth bass (6 to 8 ounces each) in the winter. After they spawned successfully during the following April, 12 adult bluegills (0.3 pound each) were

American Fisheries Society

added. These bluegills spawned several months later and the young bass reduced the numbers of young bluegills to a level which was apparently correct for adequate stocking of the pond. During the following summer 122 bluegills weighing 33.9 pounds and 34 largemouth bass weighing 15.3 pounds, a total of 30.7 pounds of fish per acre, were caught. At the end of that year, the pond was drained and the following fishes recovered:

42 large largemouth bass	40.5 pounds
115 small largemouth bass	21.0
392 large bluegills	86.0
562 intermediate bluegills	29.0
22,451 small bluegills	78.5
27 large yellow bullheads	12.0
Total	267.0

Population values: $F/C = 3.3$; $Y/C = 1.3$; $A_T = 52.0$

This population showed a desirable balance. The only drawback was that the bass caught during the second summer had an average weight of 0.45 pound and would have been too small to harvest in states having legal lengths of 11 or more inches. At the same time they were too large to have been thinned by predation and their removal was necessary to allow those remaining to grow to a more desirable size.

Because adult stocking has given general unfavorable results no additional experiments have been conducted with this method. The stocking rate in this experiment was approximately 5 largemouth bass (6 to 8 ounces each) plus 7 adult bluegills (4 ounces each) per acre. If additional experiments should prove that such a method of stocking could produce as desirable results as the present method of stocking with fingerling bluegill and advanced fry or 1-inch fingerling largemouth, the inability to sex largemouth bass accurately would bar its use in ponds of 1 acre or less. Unless lower rates of stocking could be used in large ponds, the problem of producing sufficient 6-ounce largemouth bass and 4-ounce bluegills in hatcheries for stocking any considerable acreage of ponds would be a much more difficult one than production of sufficient fingerlings to stock the same acreage at rates of 100 bass and 1,000 to 1,500 bluegills per acre.

Stocking with Largemouth Fingerlings and Adult Bluegills

To determine whether hatchery production and distribution of bluegills could be avoided, 100 largemouth bass fingerlings (1-inch) and 3 adult bluegills per acre were stocked in fertilized ponds. A 3.5-acre pond containing no cover and a 2.1-acre pond containing logs and stumps for cover were stocked with this combination. Assuming a carrying capacity of 350 pounds per acre, the calculated A_T value at the end of 1 year should have been approximately 20.3, indicating an unbalanced population. Seining during the second summer showed that bass had spawned in the pond without cover but had failed to spawn in the one with cover. The bluegills in both ponds had failed to reproduce, indi-

Stocking Bluegills and Bass in Ponds

cating unbalance and an overcrowded bluegill population as would have been expected from the calculated A_T value. Consequently, fishing was prohibited in both ponds for an additional summer to continue heavy predation pressure on the bluegills and to bring the A_T value of the populations into the balanced range. A heavy hatch of both bass and bluegills at the beginning of the third summer in the pond without cover indicated that a balanced population had been obtained. Failure of both bass and bluegills to reproduce in the pond containing cover continued the unbalanced condition in that pond. Both ponds were opened to fishing during the third summer. The catch per acre on each was as follows:

Species	Pond with cover Pounds	Pond without cover Pounds
Bluegills	8.9	56.0
Largemouth bass	28.3	70.0
Total catch per acre	37.2	126.0

After this initial catch of large fish in the pond with cover, no more fish of desirable size could be caught and the pond was drained. The population values were found to be $A_T = 18.6$, $F/C = 6.6$, and $Y/C = 5.8$, all indicative of an unbalanced condition. The pond without cover continued to give good fishing the following year with a catch consisting of 182.0 pounds of bluegills and 74.3 pounds of bass, or a total of 256.3 pounds per acre. The catch in these ponds substantiated the conclusion that with this method of stocking, largemouth bass could not reduce the young bluegills sufficiently to achieve balance except in the absence of cover even when no largemouth bass were removed until the third summer after stocking.

It appeared probable, however, that the problem of overcrowding by bluegills could be solved by originally stocking with more bass to insure adequate predation. Consequently, a 6-acre pond containing heavy brush cover in approximately one-half the pond area was stocked in June with 5 adult bluegills (average weight 3.5 ounces), 1 adult shellcracker (or redear) *Lepomis microlophus* Gunther, and 200 largemouth fingerlings (1-inch) per acre. The following November this pond was drained and the following fish were recovered:

730 large bluegills	144.8 pounds
3 shellcrackers	1.8
561 largemouth bass	65.5
Total	212.1

The bass had controlled adequately bluegill reproduction so that the survivors had grown to an average size of 0.20 pound. However, the bass were so overcrowded that their average size was only 0.12 pound, and no small bluegills were left for forage. Furthermore, overcrowding permitted a build-up of only 35.3 pounds per acre in a pond with a carrying capacity of 300 pounds per acre. For these reasons this method of stocking was considered to be a failure.

TABLE 1.—*Production of fish in 2-year experiments on rates of stocking bluegills in bluegill-largemouth bass combinations where 100 bass fingerlings were stocked per acre.*

Stocking rate per acre	Pounds recovered per acre							At Values	Balance	Average pounds per acre
	Largemouth bass		Bluegills				Total			
	Large	Small	Large	Inter-mediate	Small					
8 Bluegill adults	84.0	0.0	4.2	28.8	308.8		425.8	20.7	No	
8 Bluegill adults	45.0	26.2	5.0	8.8	216.8		301.8	16.6	No	
8 Bluegill adults	39.0	0.0	12.2	8.2	285.5		344.9	14.8	No	357.5
100 Bluegill fingerlings	54.5	0.0	8.0	55.0	354.2		471.7	13.2	No	
100 Bluegill fingerlings	52.0	0.2	9.0	7.2	350.0		418.4	14.6	No	
100 Bluegill fingerlings	65.0	23.0	179.0	30.0	53.5		350.5	69.6	Yes	413.5
248 Bluegill fingerlings	91.0	3.8	55.0	18.0	312.4		480.2	30.4	No	
248 Bluegill fingerlings	62.0	0.0	69.5	16.2	301.8		449.5	29.3	No	
248 Bluegill fingerlings	70.5	18.5	127.0	24.8	170.2		411.0	48.1	Yes	446.9
500 Bluegill fingerlings	95.2	0.0	34.5	29.0	276.0		439.3[1]	29.5	No	
500 Bluegill fingerlings	75.5	4.5	63.3	16.0	228.2		387.5	35.8	Doubtful	
500 Bluegill fingerlings	76.0	11.2	197.2	22.0	87.2		393.6	69.4	Yes	406.8
748 Bluegill fingerlings	124.0	7.0	60.5	3.0	269.0		463.5	39.8	Yes	
748 Bluegill fingerlings	50.0	10.2	138.3	48.2	126.8		373.5	50.4	Yes	
748 Bluegill fingerlings	98.5	3.8	74.5	38.5	186.3		401.6	43.1	Yes	412.9
1,000 Bluegill fingerlings	45.0	4.3	119.2	46.2	299.3		514.0	31.9	No	
1,000 Bluegill fingerlings	123.5	31.2	125.5	8.0	91.2		379.4	65.6	Yes	
1,000 Bluegill fingerlings	120.0	8.5	76.0	6.5	115.8		326.8	60.0	Yes	406.7
1,500 Bluegill fingerlings	28.0	7.0	294.0	4.0	53.0		386.0	83.4	Yes	
1,500 Bluegill fingerlings	47.0	7.0	193.0	0.0	103.0		350.0	68.0	Yes	368.0

[1] Includes 4.5 pounds small warmouth, *Chaenobryttus coronarius* (Bartram).

Stocking Bluegills and Bass in Ponds

TABLE 2.—*Average production of harvestable fish and cost of fertilizer per pound of fish with different stocking rates of bluegills.*

Stocking per acre		Harvestable fish	
Bluegills	Largemouth Bass	Pounds per acre	Cost per pound, Cents
1,000	0	57.8	93.7
0	100	81.0	65.9
8	100	63.1	84.6
100	100	122.5	43.7
248	100	158.3	33.8
500	100	180.6	29.7
748	100	181.9	29.4
1,000	100	203.1	26.2
1,500	100	280.6	19.1

Numbers of bass intermediate between 100 and 200 per acre may have given better results. However, since amount of cover influenced the success of this method, and the relative percentages of deep and shallow water in ponds would probably affect predation, stocking requirements would be expected to vary for each individual pond. This does not appear to be a practical method for general use.

STOCKING WITH FINGERLING FISH

Stocking with the correct numbers of small bluegills and largemouth bass to achieve balanced populations is widely practiced throughout the United States. The correct numbers are dependent upon the carrying capacity of the water to be stocked (Swingle, 1950) and are calculated for the average mortality expected within 1 year. In order to obtain additional information on the effect of various rates of stocking, a series of 18 ponds, each having a surface area of 0.25 acre and an average depth of 4 feet, were uniformly fertilized to give an average carrying capacity of approximately 400 pounds of fish per acre.

Ponds in triplicate were stocked on December 1, 1947, with 8, 100, 248, 500, 748, and 1,000 bluegill fingerlings per acre. On May 3, 1948, 100 largemouth bass fingerlings (1-inch) per acre were added to each pond. Because it had been found previously that lower rates of stocking with bluegills usually required approximately two summers to produce balanced populations, the experiment was continued until October 6, 1949, when all ponds were drained and the populations counted and weighed. The results of these experiments, together with two similar 2-year experiments conducted in 1943-44 with 1,500 bluegills plus 100 bass per acre, are shown in Table 1.

It is of interest to compute the average cost of fertilization per pound of the harvestable (large) fish produced by the various rates of stocking and recovered upon draining the ponds. For purposes of comparison, the results from a 2-year experiment with two ponds stocked with largemouth bass only and with three ponds stocked with bluegills only are included (Table 2).

TABLE 3.—*Differences between calculated A_T values and the mean A_T values of 3 tests.*

Stocking rate	Calculated A_T	Mean A_T	Standard deviation of means	Mean-Calculated A_T difference
8 bluegills + 100 largemouth bass	18.1	17.4	± 1.7	− 0.7
100 bluegills + 100 largemouth bass	22.5	32.5	±18.5	+10.0
248 bluegills + 100 largemouth bass	30.0	35.9	± 6.1	+ 5.9
500 bluegills + 100 largemouth bass	42.5	44.9	±12.4	+ 2.4
748 bluegills + 100 largemouth bass	55.0	44.4	± 3.1	−10.6
1,000 bluegills + 100 largemouth bass	67.5	53.5	± 8.1	−14.0
1,500 bluegills + 100 largemouth bass	90.2	76.0	± 7.4	−14.2

Stocking Bluegills and Bass in Ponds

The average production of harvestable fish per acre increased and the cost per pound decreased with increase in the rates of stocking bluegills up to 1,500 per acre in a bluegill-bass combination. Stocking with 100 bass only produced more usable fish than stocking with bluegills only, but both methods were extremely inefficient in the production of usable fish. It must be pointed out that since the experiment ran 2 years, the cost per pound was abnormally high as harvest is usually begun at the end of the first year. During the second year a considerable number of large fishes could normally be expected to die unharvested.

While production of large numbers of harvestable fish is one criterion of a successful method of stocking, a balanced population also must be established if good fishing is to continue after the original crop of large fish is removed.

The necessity for stocking adequate numbers of largemouth bass to prevent overcrowded bluegill populations and to establish balance is generally recognized (Bennett, 1943; Toole, 1947). The limits of the A_T and the A_F values in balanced populations previously discussed point out an additional cause of unbalance. Stocking too few bluegills can cause overcrowded bluegills in bass-bluegill populations during the second summer, and may cause permanent unbalance unless corrective measures are employed.

For the combinations used in these experiments, it is possible to compute the A_T value that might be expected normally to result by the beginning of the second summer and to predict the combinations that would be successful if average mortality occurred among the stocked fishes. The calculated A_T values were derived by assuming that 80 percent of the bluegills that were stocked would survive to reach an average size of 4 ounces at the beginning of the second summer, while 70 percent of the bass would survive to reach an average weight of 1 pound. The combined weight of the estimated surviving fishes was divided by 400 (the carrying capacity of the ponds in pounds per acre) and multiplied by 100 to give the calculated A_T values (Table 3). For comparison, the mean A_T values from three tests with each stocking rate are included.

The calculated A_T values represented conditions expected at the beginning of the second summer after stocking and the mean A_T values were the average of those found in three ponds the following October. It is evident that the calculated A_T value did foretell approximately the average results to be expected. The calculated A_T values indicated that at the beginning of the second summer ponds stocked with 100 bass plus 8, 100, or 248 bluegills would contain unbalanced populations. Seining samples during June and July of the second summer showed that eight of the nine ponds were overcrowded with bluegills. When the ponds were drained that fall, one additional pond had come into balance, and the other seven remained unbalanced in spite of continued predation during the second summer. One pond of the three stocked with 100 bluegills and 100 bass had a very light hatch of bluegills in July of the first summer. Bass apparently thinned this first ha' ' ' the correct number

American Fisheries Society

for stocking, and it was in balance throughout the second year. When drained, it was found to contain 1,176 large bluegills per acre.

The stocking rate of 500 bluegills plus 100 bass per acre had a calculated $A_T = 42.5$, just above the minimum satisfactory value of 40. One pond produced a balanced population, one a population intermediate between balance and unbalance ($A_T = 35.8$), and one an unbalanced population.

Higher rates of stocking, from 748 to 1,500, bluegills plus 100 largemouth bass per acre would be expected from the calculated A_T values to yield balanced populations. Only one of the eight, with a stocking rate of 1,000 bluegills plus 100 bass per acre, failed to do so. In this pond the survival of the stocked bass was only 32 percent and unbalance apparently resulted from this cause.

In most of these experiments excessive natural mortality of bass occurred in unbalanced populations. A regression coefficient of $b = .957$ pound harvestable fish per each 1 percent survival of stocked bass was calculated, but the significance of this value could not be demonstrated (Snedecor, 1946). There was, however, significance at the 1 percent point for the effect of rates of stocking bluegills on production of harvestable fish.

There appeared to be a superficial relationship in these results between survival of bass and rates of stocking bluegills. A regression coefficient of $b = +0.16$ was calculated, but was not significant at the 5 percent level.

Problems Involved in Variable Mortality

Results in the experiments previously discussed showed that balanced populations usually resulted from stocking the calculated numbers of bluegills and bass necessary to produce A_T values and F/C ratios in the balanced range. While the average results appear to be encouraging, the variations in production within treatments are not so satisfactory (Table 1). They appear, however, to be the normal variations which are encountered in experiments on fish production. An analysis of all experiments conducted at this Station leads to the conclusion that the principal cause of differences in ponds receiving the same treatment was variation in the survival rates of the stocked fish.

The percentage survival of largemouth bass in comparable experiments at the end of the first 6 months after stocking with various rates of bass in bass-bluegill combinations is summarized in Table 4. The average survival for the standard stocking rate of 100 bass per acre was 74.4 percent (standard deviation of ± 13.6 percent) and for all rates was 66.5 percent, but the range was from 18.9 to 100 percent. In all cases the bass were handled carefully, accurately counted, transported only a short distance (usually less than 1 mile), reexamined, and all weakened individuals replaced before the ponds were stocked. Therefore, it must be assumed that the observed variations in survival are caused by differences in natural mortality.

Stocking Bluegills and Bass in Ponds

TABLE 4.—*Percentage survival of largemouth bass in ponds with bass-bluegill combinations 6 months after stocking as 1-inch fingerlings in May.*

Year	Number experiments	Number stocked per acre	Percentage survival	
			Average	Range
1943	13	150	75.6	53.0-100.0
1944	4	100	83.0	72.0- 96.0
1944	16	148	45.9	18.9- 73.0
1946	5	100	64.8	52.0- 92.0
	1	120	50.0	
	1	240	61.6	
1947	9	100	76.0	60.0- 92.0
	2	48	95.8	91.8-100.0
1945	16	120	68.3	36.7- 96.7
Total or average	67		66.5	

Further light is shed on this problem by consideration of the survival of largemouth bass in 29 two-year experiments (Table 5). In these experiments the bluegills were stocked in November-December and largemouth bass fingerlings (1-inch) were stocked the following May. All ponds were drained in October-November of the following year, or approximately 18 months after the bass were stocked. While the average survival with the standard stocking rate of 100 bass per acre was 54 percent (standard deviation of ± 18.5 percent) and from all methods of stocking was 54.3 percent, the range was extremely variable, extending from 24.0 to 92.0 percent.

The mortality rate for the standard stocking rate of 100 bass per acre may be summarized as follows:

	Percentage mortality		
Period after stocking bass	maximum	minimum	average
6 months	48.0	4.0	25.6
18 months	76.0	8.0	46.0
Difference (second year)			20.4

It appears that, when there is no fishing, on the average 25.6 percent of the stocked bass may be expected to die within the first 6 months and an additional 20.4 percent during the following year.

TABLE 5.—*Percentage survival of largemouth bass in ponds with bass-bluegill combinations 18 months after stocking as 1-inch fingerlings in May.*

Year	Number experiments	Number stocked per acre	Percentage survival	
			Average	Range
1942-43	3	100	62.0	40.0-92.0
	1	200	44.0	
1943-44	1	100	44.0	
	1	150	61.3	
	1	200	56.0	
	1	400	64.0	
1948-49	21	100	53.3	24.0-84.0
Total or average	29		54.3	

American Fisheries Society

Similarly bluegill mortalities for all rates of stocking were as follows:

Period after stocking bluegills	Percentage mortality		
	maximum	minimum	average
11 months	77.1	0.0	15.4
23 months	76.0	0.0	34.5
Difference (second year)			19.1

Where no bluegills are removed by fishing, approximately 15.4 percent of the stocked fish on the average may be expected to die during the first year and an additional 19.1 percent during the second year.

It should be pointed out that the mortality during the second year produces variable results in 2-year experiments, but is otherwise relatively unimportant unless it is excessive. It merely replaces or supplements fishing as a method of removal of adult fishes.

Mortality during the first year after stocking, however, is extremely important as it must be estimated in calculation of the correct numbers for stocking. Extreme variations below the average survival figure are responsible for the failure of many ponds to achieve balance. Variations above the average cause differences in total production and average size but do not prevent the achievement of a balanced condition within the population. The most serious problem is that of variations in mortality among the stocked bass prior to their reproduction. It is during this period that excessive mortality may prevent adequate predation and cause unbalanced populations.

The most probable cause of excessive variation from the mean mortality of bass would appear to be either cannibalism among the stocked bass or predation by the bluegills. Predation among young bass may be largely prevented as long as adequate food for rapid growth is available. However, in the methods used for stocking ponds in the Southeast, bluegill fingerlings are added in the fall or winter and largemouth bass advanced fry or fingerlings (1-inch) are added the following spring. The small bass must compete with the stocked bluegills for food until the latter reproduce. This period of competition usually lasts from 1 to 2 months.

After the first hatch of bluegills the bass feed largely upon small fish and compete to a minor extent with large bluegills.

Various experiments were run in an attempt to determine some of the causes of mortality among stocked bass (Table 6). Stocking with large bluegills (average 1.5 ounces) 5 months prior to stocking bass resulted in an average bass survival of 62 percent as compared to a survival of 86 percent where smaller bluegills (average 0.14 ounce) were used. Since large bluegills would be expected to have a higher demand for food than small ones, these results might suggest that competition for food between these species was the major cause of high mortality of bass. However, where no bluegills were present, the average survival of bass was 76 percent, an intermediate value. It appears that factors

Stocking Bluegills and Bass in Ponds

TABLE 6.—*Experiments with factors affecting largemouth bass survival in bluegill-bass combinations.*

Experiment	Stocking rate per acre[1]		Percentage survival of bass	Total pounds fish per acre
	Bluegills	Largemouth bass		
1	1,500 (large)	100	60.0	259.7
2	1,500 (large)	100	64.0	305.5
3	1,500 (small)	100	84.0	282.7
4	1,500 (small)	100	88.0	282.2
5	0	100	76.0	40.0
6	0	100	76.0	26.8
7	1,500	100	92.0[2]	671.2
8	1,500	100[3]	68.0	301.9
9	1,500	100[3]	76.0	343.2

[1]Bluegills were stocked November 29, 1946, and averaged 1.5 ounces in experiments 1 and 2. The average size in the remaining experiments was 0.14 ounce. Bass (1-inch) were added May 7, 1947. Ponds were drained October 17-20, 1947.
[2]In this experiment, bluegills were fed poultry laying mash during the spring and summer in addition to pond fertilization. All other ponds received only normal fertilization.
[3]Bass widely scattered around pond in stocking. Added at one place in other ponds.

in addition to competition from bluegills such as lack of young fish for food and consequent increase of cannibalism among bass may have been important. Experiment 9 in which bluegills were fed, gave the highest bass survival (92 percent). This result seemed to emphasize the important effect of sufficient food for rapid growth of both species upon bass survival.

Careful scattering of bass around the small ponds at the time of stocking did not appear better than their addition at one point. The comparative average survivals with the two methods were 74 and 86 percent respectively.

It must be emphasized that the above results merely indicate some of the problems involved in pond management and suggest the necessity for further detailed research.

LITERATURE CITED

BENNETT, GEORGE W.
    1943. Management of small artificial lakes. A summary of fisheries investigations, 1938-1942. Bull. Ill. Nat. Hist. Surv., Vol. 22, Art. 3, pp. 357-376.

SMITH, E. V., and H. S. SWINGLE
    1943. Results of further experiments on the stocking of fish ponds. Trans. 8th N. Am. Wildlife Conf., pp. 168-179.

SNEDECOR, GEORGE W.
    1946. Statistical methods. Iowa State Coll. Press, Ames, Iowa. 485 pp.

SWINGLE, H. S.
    1950. Relationships and dynamics of balanced and unbalanced fish populations. Ala. Agric. Exp. Sta. Bull. 274, 74 pp.

SWINGLE, H. S., and E. V. SMITH
    1940. Experiments on the stocking of fish ponds. Trans. 5th N. Am. Wildlife Conf., pp. 267-276.

American Fisheries Society

1942. Management of farm fish ponds. Ala. Agric. Exp. Sta. Bull. 254, 23 pp. (Revised January 1947).
1943a. Factors affecting the reproduction of bluegill bream and largemouth black bass in ponds. Ala. Agric. Exp. Sta. Cir. 87, 8 pp.
1943b. Effect of management practices on the catch in a 12-acre pond during a 10-year period. Trans. 8th N. Am. Wildlife Conf., pp. 141-155.

TOOLE, MARION
1947. Utilizing stock tanks and farm ponds for fish. Texas Game, Fish and Oyster Commission Bull. 24, 45 pp.

5.3 HONORABLE MENTION FULL CITATIONS AND ABSTRACTS

Anderson, J. L. 2002. Aquaculture and the future: why fisheries economists should care. Marine Resource Economics 17:133–151.

> This paper explores the relationship between traditional fisheries, fisheries enhancement (ranching), and aquaculture. It evaluates why they are different and why fisheries economists have largely neglected aquaculture issues, despite the fact that most of the growth in fish supply over the past two decades has been the result of aquaculture development. It is argued that the core difference between aquaculture and traditional fisheries is the degree of control; control of the environment, production, and marketing systems. It is further argued that the degree of control is closely related to the strength of property rights. Three examples are presented to provide empirical support for the propositions. They focus on the salmon, lobster, and shrimp industries.

Ford, M. J. 2002. Selection in captivity during supportive breeding may reduce fitness in the wild. Conservation Biology 16(3):815–825.

> I used a quantitative genetic model to explore the effects of selection on the fitness of a wild population subject to supportive breeding. Supportive breeding is the boosting of a wild population's size by breeding part of the population in captivity and releasing the captive progeny back into the wild. The model assumes that a single trait is under selection with different optimum trait values in the captive and wild environments. The model shows that when the captive population is closed to gene flow from the wild population, even low levels of gene flow from the captive population to the wild population will shift the wild population's mean phenotype so that it approaches the optimal phenotype in captivity. If the captive population receives gene flow from the wild, the shift in the wild population's mean phenotype becomes less pronounced but can still be substantial. The approach to the new mean phenotype can occur in less than 50 generations. The fitness consequences of the phenotypic shift depend on the details of the model, but a 30% decline in fitness can occur over a broad range of parameter values. The rate of gene flow between the two environments, and hence the outcome of the model, is sensitive to the wild environment's carrying capacity and the population growth rate it can support. The results have two important implications for conservation efforts. First, they show that selection in captivity may significantly reduce a wild population's fitness during supportive breeding and that even continually introducing wild individuals into the captive population will not eliminate this effect entirely. Second, the sensitivity of the model's outcome to the wild environment's quality suggests that conserving or restoring a population's habitat is important for preventing fitness loss during supportive breeding.

Pinkerton, E. 1994. Economic and management benefits from the coordination of capture and culture fisheries: the case of Prince William Sound pink salmon. North American Journal of Fisheries Management 14:262–277.

> Aquaculture developments often create policy conflicts with established fisheries when the two are not coordinated through a common planning framework. The state of Alaska and community-based, fisher-led salmon aquaculture associations have been unusually successful at coordinating, through cooperative management, the traditional salmon capture fisheries and new culture fisheries for pink salmon *Oncorhynchus gorbuscha*, despite predictable problems. The Prince William Sound Aquaculture Corporation in particular has moved from its original involvement in resource enhancement into partnership with the state in harvest planning, allocation, and comprehensive regional planning. Some of the specific economic benefits and the general management benefits of this institutional arrangement are explored. One economic benefit was an 8-year period of price advantage for the association's cost-recovery fish because of large and consistent volume and quality. The ecological, political, and institutional conditions that made these developments possible are analyzed.

Ryman, N., and L. Laikre. 1991. Effects of supportive breeding on the genetically effective population size. Conservation Biology 5(3):325–329.

> The genetically effective population size (N_e) is a key parameter in conservation biology, because the rate of inbreeding (ΔF), and thereby the rate of loss of genetic heterozygosity, is proportional to the inverse of the effective number ($\Delta F = 1/2N_e$; e.g., Crow & Kimura 1970). The importance of maintaining large effective sizes of natural and captive populations is reflected by the considerable fraction of the literature on biological conservation that focuses on that issue (e.g., Flesness 1977; Ryman & Ståhl 1980; Ryman et al. 1981; Frankel & Soulé 1981; Soulé et al. 1986; Allendorf & Ryman 1987; Lande & Barrowclough 1987). This note addresses a problem in conservation genetics that, to our knowledge, has not been previously recognized: the reduction of the genetically effective population size that may result from breeding-release programs aimed at supporting natural populations. Typically, in such programs a fraction of the wild parents (or their offspring) are brought into captivity for reproduction or preferential survival, and the offspring are released into the natural habitat where they mix with wild conspecifics. We refer to this practice as supportive breeding, stressing the fact that no exogenous genes are introduced into the overall population.
>
> The logic of supportive breeding is generally to increase survival through breeding in a protected captive environment. However, this process also implies that the reproductive rate of one segment of the overall population is favored, which results in an increase in the total variance of family size, a parameter of critical importance to the genetically effective size of the population. The amount of change of effective population number that may result from manipulating the variance of family size can be derived from the distributions of family size among the wild and captive breeders (Crow & Kimura 1970), but in many situations either or both of those distributions are unknown. Rather, the manager may have some rough idea of the approximate effective sizes of the wild and captive populations and of their relative contributions to the combined offspring population. Such a situation is frequently encountered in, for example, the field of fishery management; hatchery records provide a basis for estimating the effective number

of parents of the released fish, the effective size of the wild population may be at least crudely approximated from field counts of the number of spawners, and tagging data contribute information on the proportion of hatchery fish in the offspring generation. To facilitate the theoretical treatment of such a situation where the total population is subdivided into segments with different repro- ductive rates, we present a method for estimating the overall effective size from the effective sizes of the con- tributing segment.

The implications of supportive breeding resulting in a reduction of the effective size are important in light of current tendencies in conservation biology. First, the practice of releasing captive-bred animals into the wild is coming to be used increasingly to support weak and endangered populations (e.g., IUCN 1987; Foose & Ballou 1988; Griffith et al. 1989). Currently, such releases appear to be most extensively practiced in the field of fishery management as a means of enhancing wild stocks (e.g., Ryman & Utter 1987), and the issue we address is particularly pertinent to fishes and other species that are characterized by potentially large variations in offspring production. Second, our findings are relevant in the con- text of current contentions regarding genetic and de- mographic concerns in biological conservation and pop- ulation management (Lande & Barrowclough 1987; Lande 1988).

Section 6

Summary

GREG G. SASS AND MICHEAL S. ALLEN

6.1 What Have We Learned?

Undoubtedly, and as evidenced by the breadth of journals represented by the selected articles, fisheries science has been an interdisciplinary endeavor and will likely continue to be. Because no single researcher or manager is likely to be proficient in all of these necessary components of fisheries science, scientifically-defensible fisheries management will be the result of collaboration among fisheries specialists and scientists from other disciplines. Sound fisheries science will also require the understanding of fish communities in an ecosystem context. The need to foster collaboration in fisheries science is evident in the time series of articles selected in this book and is likely representative of the diverse research required to address current and future management needs.

The selected articles in this book were represented by 21 different journals ranging in discipline from fisheries, ecology, human dimensions, and others. This breadth of journals reiterates the interdisciplinary nature of fisheries science through the various outlets chosen for publication of seminal research. Interestingly, American Fisheries Society publications (*Transactions of the American Fisheries Society*; *Fisheries*) only represented about 14% of the articles selected. Although these statistics may appear under-representative of the longest-running and largest fisheries society in the world, we reasoned that this was likely the result of the interdisciplinary nature of fisheries science. The number of authors per selected article increased over time in all of the sections, which is similar to other scientific disciplines (Leimu and Koricheva 2005). Collaboration, clearly evidenced by the increasing number of authors, has been critical for conducting influential fisheries science capable of addressing challenging questions. Regardless of the mechanism(s) explaining this pattern, collaboration has increased over time and will likely be essential in the future to address emerging issues in fisheries science.

Our assessment of the articles selected in this book indicate a continued need for adaptive fisheries management, with increased use of experiments to evaluate the response of anglers and fish populations to changes in management activities (Walters 1998). Examples of such experiments are rare among the articles selected for this book, and through such experiments, we could reduce uncertainty in fishery management outcomes. Continual discussion among fisheries researchers, along with needed allocations of available funding to research and longer-term commitments to this ideal, may facilitate such "experiments" to improve our understanding of fisheries management actions.

The articles selected for reprinting in this book, along with our section editor summaries, suggest that the interdisciplinary nature of fisheries science is most related to the pillars of ecosystem science, as described by Carpenter (1998) with additional challenges outlined by Walters (1998). Per Carpenter (1998), the pillars of ecosystem science may be defined in four, interacting aspects: 1) long-term studies; 2) comparative studies; 3) manipulative experiments; and 4) models and theory. All four aspects of ecosystem science are well represented within the articles selected for this book. Our summary suggests that ecosystem science, which is not possible without interdisciplinary research and interactions, has been incredibly important to the advances made in this discipline and should be used moving forward in fisheries science. We discuss specific challenges related to our five sections below.

Challenges in Managing Fish Stocks

The seminal research focusing on the management of fish stocks has been largely related to preventing growth and recruitment overfishing through the use of predictive models. These models led to the development of regulatory tools (e.g., size limits, bag limits, gear restrictions, closed seasons) that prevent overexploitation in fisheries on the assumption that all fisheries are harvest-oriented. Emerging challenges to managing fish stocks may result from the increasing prevalence of catch-and-release angling. Such practices, dictated by human desires and sometimes by fisheries policy, may lead to imbalances in fish communities being managed for multiple, desirable species. For example, Largemouth Bass *Micropterus salmoides* and Muskellunge *Esox masquinongy* (two apex predator fisheries) have become predominately catch-and-release species throughout North America. Similar situations may arise in marine environments with species that are ecologically important and anglers prefer to catch-and-release. We must also consider non-harvest effects on fish stocks, particularly when fishing effort is very high and in cases when barotrauma is a concern. Thus, we recommend that additional research is needed to explore alternative methods for managing fish communities where catch-and-release dominates the fishery or causes substantial mortality.

Models will always have a role in fisheries management. As such, challenges for future modeling efforts may include the continuation of collecting appropriate data to inform models, developing novel ways to better estimate key parameters to reduce model uncertainty, and to better incorporate and acknowledge compensation and depensation in fish stocks. Relevant models will also need to incorporate single- and multiple-species considerations, which begs for interdisciplinary collaboration. Continued and long-term empirical data collection is essential for models to move them from theoretical to realistic. Thus, it is critical for researchers to stress the importance of long-term commitments to data collection and also to develop novel ways of measuring vital rates (growth, natural mortality) in fish stocks. Appropriate data collection further serves to reduce uncertainty in key parameters used in fisheries stock assessments. Depensation (i.e., lower than expected recruitment at low population levels) has long been recognized as a potential outcome of exploited fish stocks (Ricker 1963, Liermann and Hilborn 1997). However, exploitation is not the only mechanism that may lead to depensatory dynamics. There is a critical need for precautionary approaches to fisheries management that consider a range of perturbations (e.g., invasive species, habitat loss, eutrophication), as numerous anthropogenic stressors (including fishing) can decrease the resiliency of certain fish populations and lead to depensation (Ricker 1963, Liermann and Hilborn 1997, Walters and Kitchell 2001). We must also accept that commercial fisheries are not the only fish stocks that can be overexploited; recreational and subsistence fisheries are also vulnerable to numerous stressors leading to collapse. Lastly, a better empirical understanding of angler effort and behavior dynamics is critically needed in future modeling efforts in order to sustain marine and freshwater fisheries.

Challenges in Managing People

A fair criticism of this book is that we combined economics and human dimensions into a single section. Without question, this section could have been divided into two or more sections. Future syntheses should consider reviewing seminal research more completely in each of these disciplines. Nevertheless, our summary shows some interesting trends and future challenges.

The articles summarized in this section became increasingly more complex through the chronology, beginning with simplistic assumptions about resource use and later expanding to consider many catch and non-catch related metrics to understand fisher behavior. Such increases in complexity make sense, as authors considered more complex management systems and fisher behaviors. This trend suggests that seminal research in this arena of fisheries science will consider even more complex behaviors of fishers and management systems in the future.

Many of the articles in this section are based upon models of angler effort that assume responses to changes in resource conditions. It is apparent that we need more empirical data to better understand angler effort dynamics in recreational fisheries. How anglers respond to catch and non-catch associated utilities will have strong implications on the potential for overexploitation, and this section shows clearly how most research to date uses models with assumed responses about angler effort dynamics. Empirical estimates of fishing effort dynamics would improve our understanding and management for open-access recreational fisheries. We feel that this is an extensive area of future research; moving social and economic studies forward in fisheries science.

Urbanization of human societies is causing angler numbers to decline in many regions. Little of the seminal research presented here addresses issues about shrinking angler numbers and the implications for fisheries in the future. Are fisheries resources going to be more sustainable in developed countries due to fewer recreational anglers and higher rates of catch-and-release? How will our management needs change to address alterations in angler numbers and demographics? In our minds, questions of this type will assuredly be addressed by future seminal research in social and economic aspects of fisheries science.

Challenges in Managing Fish Habitat

Without fishing and human perturbations, habitat has been naturally associated with fish evolution and speciation, and has dictated the emergent species composition and abundances of aquatic and marine fish communities. Abiotic habitat (e.g., water temperature, dissolved oxygen, flow, substrate) and biotic, structural habitats (e.g., aquatic macrophytes, woody debris, periphyton) serve many roles for fishes, which allow them to complete their life histories. For example, water temperature dictates fish metabolism and growth, dissolved oxygen requirements differ among fishes, and structural habitats can serve as spawning substrate, refuge from predation, and allow habitat partitioning among fishes. Habitat complexity also promotes species diversity in aquatic and marine ecosystems. The availability of suitable habitat may be the most critical component for sustainable fisheries. One of the greatest challenges facing fish populations is habitat loss due to anthropogenic perturbations. Therefore, the critical questions are what degree of habitat quality and quantity are required to sustain fisheries, and are there ways to mitigate habitat loss through anthropogenic habitat conservation or restoration efforts?

Future assessments of fish habitat requirements should continue to consider fish behavior and use deliberate experiments that modify habitat and test for population responses. A large body of fish-habitat related research has focused on the behavior of fishes in structured environments and how these relationships influence growth, predation risk, and predator-prey interactions. Earlier generalizations of fish behavior in a food web context were often related to optimal foraging theory. Thus, it was assumed that fishes behave optimally in regards to growth without consideration of risk taking behavior. More recent research has refined those behaviors to address predation risk as a major factor leading to non-optimal foraging (if you do not survive, you have no fitness), as well as prey not always being vulnerable to predation (reaction vat versus foraging arena hypotheses) (Walters and Juanes 1993). Articles selected in this section undoubtedly improved our understanding of how fishes respond to changes in habitat quality and quantity, and how behavior influences predation risk and survival. Future studies should quantify how fish vital rates (growth, recruitment) respond to changes in habitat quality and quantity.

Challenges in Managing Fish Communities and Ecosystems

Besides the great need for a better understanding of human behavior as related to fisheries, this section may be the most complex to summarize because it places fishes in the context of aquatic communities and also expands beyond water. Those realizations call for interdisciplinary fisheries science to not only understand individual fish species life history and ecology, but also to connect those aspects to: 1) human interventions (fishing); 2) human behaviors as they relate to fisheries; 3) the acknowledgement of critical habitat as a key component in determining fish community outcomes and resiliency to fisheries; and 4) challenges related to humans having already induced major effects on fishes and their environments. Arguably, the seminal findings providing evidence for the importance of fish communities and ecosystems to fisheries sustainability have originated directly from ecosystem-scale studies and experiments. Ecosystem manipulations that can isolate key variables at a scale relevant to management (i.e., whole-lake or ecosystem vs. microcosm or mesocosm experiments) have been highly influential to fisheries science. This is further supported and highlighted above by the fact that the major breakthroughs in fisheries science seem to be linked with the pillars of ecosystem science, as described by Carpenter (1998). Those include the interactions among long-term studies, comparative studies (which are unable to determine mechanism, but are important for identifying pattern), modeling and theoretical studies, and experimental studies at an appropriate scale. This section of the book clearly demonstrates that future research should consider large-scale experiments to evaluate how fish communities respond to perturbations.

The articles in this section showed that manipulation of aquatic communities has improved management outcomes for all trophic levels. Biomanipulation has become an accepted method for influencing phytoplankton concentrations in some lake systems, and only through ecosystem experiments have such management options become illuminated. Effects of fishing on multiple trophic levels are now widely recognized in marine and freshwater systems. Challenges that lie ahead include a greater recognition of the importance of forage fishes to overall food web dynamics and sustainable fisheries, a better understanding of "undesirable" fish species and their role in native and invaded food webs, ecosystem consequences of single- versus multiple-species management in a food web context, and increased stock assessment research on lower trophic level fishes that represent very high biomass.

Challenges in Managing Fisheries Enhancements

Articles selected in this section show a clear trend toward more cautious and responsible use of hatchery fish as a fisheries management tool. With fishes being a primary source of protein around the world, use of fisheries enhancements will certainly continue to be an important management method. Expansion of hatcheries as a tool in marine environments is likely, given the focus of the various articles in this section.

Challenges for the future include illegal and/or accidental introductions of fishes into new environments and their genotypes into native stocks. Indeed, the widespread use of hatcheries by management agencies has fostered this method as somewhat justified in the public eye. Education and enforcement of existing laws will be crucial for minimizing the effects of non-native organisms that may be introduced by well-intentioned members of the public. We anticipate that future seminal research regarding fisheries enhancements will include studies to minimize influences from illegal introductions, as well as minimizing the genetic impacts to wild stocks.

Finally, the development of genetic markers will allow for more robust evaluations of stock enhancement programs in the future. It is now feasible to "mark" large numbers of fish, thus allowing more rigorous evaluations of the efficacy of stock enhancement efforts. We expect future research in this area to refine the methods that can maximize benefits to fisheries and minimize negative impacts to wild stocks.

Conclusions

After compiling many of the seminal studies in this discipline, have these and other articles successfully addressed some of the most fundamental questions in fisheries science compared to 50 or 100 years ago? Are we better at managing fisheries? We believe the answer is a definitive yes. Science has moved forward and management has become more effective. The articles highlighted here and others serve as the basis for stock assessment, habitat management, and recognition of the importance of aquatic communities. As just a few examples, effective management has recovered overfished stocks (e.g., Red Drum *Sciaenops occellatus*, Striped Bass *Morone saxatilis*), and research highlighted here has resulted in more responsible use of stock enhancement by agencies around the world.

However, in many ways, we still seek answers to basic questions in our discipline. Researchers today still need better methods to estimate vital rates, like natural mortality and growth, which are critically important to stock assessment and management. We continue to seek a better understanding of how changes in habitat influence fish abundance and growth. For every article highlighted in this book, we can find many more recent studies that seek to address similar questions. Some of this may be due to older studies being overlooked, which was a primary motivation for compiling this book. However, another plausible reason is, that like all studies, the seminal articles presented here still leave many unanswered questions. That is how science works.

If we still ask questions that are similar to many of those addressed in these seminal works today, why were these articles selected? We believe that the articles selected here were aimed to evaluate the most important questions directly. Their approaches included using large data sets that had extensive spatial and/or temporal coverage (e.g., Forney 1974; Tonn and Magnuson 1982), large-scale experiments (e.g., Carpenter et al. 1985), comprehensive meta analyses (e.g., Myers et al. 1999; Winemiller and Rose 1992; Rose et al. 2001), and modeling efforts that were innovative and informed with new data sources (e.g., Kitchell et al. 1977). By attempting to answer the most pressing questions directly, these articles became some of our seminal studies in fisheries science. Now, and in the future, fisheries researchers would be best served by addressing the most important and needed questions directly with well-designed studies at a range of spatial and temporal scales. We encourage resource managers to use those studies that do so to improve fishery resources without disciplinary boundary constraints.

Acknowledgments

We thank the American Fisheries Society (AFS) for providing us with the opportunity to develop this book. Aaron Lerner and Kurt West of AFS publications were critical in the completion of this project through their continual support and in working directly with our librarian interns in seeking and funding permissions for the articles reprinted herein. Special thanks go to the AFS Education and Fisheries Management Sections for funding. This funding enabled us to meet in person to work on the book and to pay salary for librarian interns seeking permissions for reprinted articles. We are greatly indebted to our team of librarian interns, particularly Melissa Allen and Elliott Shuppy, and Liz Krznarich and Lori Steckervetz. We thank the University of Wisconsin-Madison, Center for Limnology for allowing us to collaborate with their librarian interns. Paul Hanson (supervisor of the Center for Limnology library) and Steve Carpenter (Director of the Center for Limnology) also deserve thanks for supporting this project. We also thank Marilyn Larsen (Assistant Director of the Center for Limnology) for managing and administering our funding. Molly Spacapan (M.S., Human Dimensions of Natural Resources, Department of Natural Resources and Environmental Sciences, University of Illinois at Urbana-Champaign) was critical in the development of the survey used to select articles reprinted in this book. Without her participation, this book would not have been possible. Greg Sass thanks the Illinois Natural History Survey and the University of Illinois at Urbana-Champaign for supporting earlier efforts on this project, as well as the Wisconsin Department of Natural Resources (Jen Hauxwell and Jack Sullivan) for supporting later efforts on this project. Lastly, we thank Dr.'s Olaf Jensen (Rutgers University), Andrew Rypel (Wisconsin Department of Natural Resources), Jeffrey Stein (Illinois Natural History Survey), and Cory Suski (University of Illinois at Urbana-Champaign) for reviewing earlier drafts and providing constructive feedback to greatly improve this book.

References

Allan, J. D., R. Abell, Z. Hogan, C. Revenga, B. W. Taylor, R. L. Welcomme, and K. Winemiller. 2005. Overfishing of inland waters. BioScience 55:1041–1051.

Carpenter, S. R. 1998. The need for large-scale experiments to assess and predict the response of ecosystems to perturbation. Pages 287–312 *in* M. L. Pace and P. M. Groffman, editors. Successes, limitations, and frontiers in ecosystem science. Springer, New York.

Costello, C., D. Ovando, R. Hilborn, S. D. Gaines, O. Deschenes, and S. E. Lester. 2012. Status and solutions for the world's unassessed fisheries. Science 338(6106):517–520.

FAO (Food and Agriculture Organization of the United Nations). 2012. The state of world fisheries and aquaculture 2012. Rome, Italy.

Leimu, R., and J. Koricheva. 2005. Does scientific collaboration increase the impact of ecological articles. BioScience 55(5):438–443.

Liermann, R., and R. Hilborn. 1997. Depensation in fish stocks: a hierarchic Bayesian meta-analysis. Canadian Journal of Fisheries and Aquatic Sciences 54:1976–1984.

Nielsen, L. A. 1993. History of inland fisheries management in North America. Pages 33–54 *in* C. C. Kohler and W. A. Hubert, editors. Inland fisheries management in North America. American Fisheries Society, Bethesda, Maryland.

Real, L. A., and J. H. Brown. 1991. Foundations of ecology. University of Chicago Press, Chicago.

Walters, C. J., and F. Juanes. 1993. Recruitment limitation as a consequence of natural selection for use of restricted feeding habitats and predation risk taking by juvenile fishes. Canadian Journal of Fisheries and Aquatic Sciences 50:2058–2070.

Walters, C. J. 1998. Improving links between ecosystem scientists and managers. Pages 272–286 *in* M. L. Pace, and P. M. Groffman, editors. Successes, limitations, and frontiers in ecosystem science. Springer, New York.